SEVERE AND HAZARDOUS WEATHER

An Introduction to High Impact Meteorology

ROBERT M. **RAUBER**
JOHN E. **WALSH**
DONNA J. **CHARLEVOIX**
Department of Atmospheric Sciences
University of Illinois at Urbana-Champaign

KENDALL/HUNT PUBLISHING COMPANY
4050 Westmark Drive Dubuque, Iowa 52002

Book Team

Chairman and Chief Executive Officer Mark C. Falb
President and Chief Operating Officer Chad M. Chandlee
Vice President, Higher Education David L. Tart
Director of National Book Program Paul B. Carty
Editorial Development Manager Georgia Botsford
Developmental Editor Lynne Rogers
Vice President, Operations Timothy J. Beitzel
Assistant Vice President, Production Services Christine E. O'Brien
Senior Production Editor Charmayne McMurray
Permissions Editor Renae Horstman
Cover Designer Jenifer Chapman

Cover images:
Lightning, tornado, and mountain images © istockphotos, 2008.
Housetop and bridge images provided by authors.

ISBN Student Edition 978-0-7575-5043-0
ISBN Special Edition 978-0-7575-5339-4

Printed in the United States of America
10 9 8 7 6 5

DEDICATION

To Ruta, my wife, and Carolyn and Stacy, my daughters—the sunshine in my life (no matter how big the storms!).

Bob Rauber

To Laura, my wife, and Rachel and Emily, my daughters—the hurricanes in my life (no matter how calm the eyes).

John Walsh

To Glen Romine, my husband, and Evelyn and Madeline, my daughters—the stability in my life, no matter how high the pressure or how severe the conditions.

Donna Charlevoix

BRIEF CONTENTS

CONTENTS

PREFACE

In the three years that have passed since the publication of the second edition of this book, severe and hazardous weather has generated unprecedented levels of attention among the public and policy makers. Hurricane Katrina's devastation provided a striking reminder of our vulnerability to severe weather. A persistent drought has affected much of the West, producing water shortages that have affected agriculture and residential users in a dozen states. Heat and wildfires have accompanied the dry weather in many parts of the country. Severe thunderstorms have swept through major urban areas such as New York City, St. Louis, Chicago, and Dallas, leaving tens of thousands without power for days and even weeks. The Pacific Northwest was affected by a severe windstorm in October, 2006. Flooding has caused extensive damage in Texas, Oklahoma, New England, Ohio, and Minnesota during the last few years. Ice storms in the South, citrus-damaging freezes in the West, and snowstorms in the Midwest and East also made headlines during the winters of 2005–06 and 2006–07.

The 2007 report of the Intergovernmental Panel on Climate Change also made headlines with its message that the climate is warming and that humans have contributed significantly to the warming of the past several decades. The record-breaking hurricane season of 2005, the landfall of two category 5 hurricanes in a two-week period of 2007, and the intense heat and drought of recent summers have raised questions about the relationship between climate change and severe weather. While the weather-climate linkages are still being actively researched, some connections have been established and greater impacts of climate change on weather are expected as global warming continues. Accordingly, we have added to this edition of *Severe and Hazardous Weather* a new focus on climate and its associations with severe weather. An entirely new Chapter on Climate, Climate Change, and Global Warming (Chapter 5) introduces the fundamental drivers of climate, summarizes climate variations of the past, and provides an assessment of ongoing global warming. In addition, sections on the effects of global warming have been added to each chapter that covers a particular type of hazardous weather. We believe that these introductory-level discussions of the linkages between global warming and hazardous weather are unique in introductory meteorological textbooks, and they provide students with insights into the state of the art of climate science.

While hazardous weather seems especially prominent during the past few years, its occurrence has been a fact of life throughout history in all parts of the world. Hazardous weather has left lasting imprints, impacting the full range of the world's cultures and economies and leading to significant changes in the affected areas. Several popular nonfiction books in recent years (Flannery's *The Weather-Makers,* Diamond's *Collapse,* Egan's *The Worst Hard Times*) have addressed the effects of weather and climate on societal and cultural change. The drought of the 1930s, for example, led to large westward migrations and to changes in agricultural practices in the central United States. Utilities in major urban centers of the eastern states were placed underground after a devastating blizzard in 1888. The great wars of the twentieth century were profoundly influenced by weather events such as the severe winter of 1941–42, when cold weather helped turn the tide against Germany during the siege of Leningrad.

Severe weather has also been popularized by the entertainment industry. Fictional tornadoes in Kansas and Oklahoma have appeared in popular movies of the 1930s and 1990s, and an actual storm in the North Atlantic was the subject of a popular novel and film in 2000. A fictional outbreak of bitter cold air over North America was the subject of a major film *(The Day after Tomorrow),* and Al Gore's *An Inconvenient Truth* has been especially popular among the same college and university students who are the target audience of this textbook.

Our goal in *Severe and Hazardous Weather* remains to provide a current, relevant, and scientifically accurate discussion of all types of hazardous weather. The new edition of *Severe and Hazardous Weather* contains a number of improvements to achieve this goal. In addition to incorporating climate and climate change into the text, we have significantly revised the chapters on Thunderstorms, Tornadoes, and Tropical Cyclones (Hurricanes). The chapter on Thunderstorms (Chapter 18), for example, has been restructured with an introductory section on non-severe thunderstorms to serve as a foundation for the discussion of the physical and dynamical processes in various types of severe thunderstorms. This chapter also includes new material on the spectacular supercell event of March 2007. Detailed information on the new EF (Enhanced Fujita) system for tornado classification, introduced by the National Weather Service in 2007, has been included in Chapter 19. The coverage of Tropical Cyclones in Chapter 24 includes new text and figures highlighting Hurricane Katrina as the most devastating severe weather event of the present decade. In nearly every chapter, the "Focus" and "Online" boxes have been updated to include recent events that many students are likely to remember. Our experience in teaching Severe and Hazardous Weather is that these recent events are excellent vehicles for stimulating student interest in the underlying physics and dynamics of weather systems.

We have also revised the questions and end-of-chapter problems in many of the chapters to ensure consistency among chapters. In particular, the "Check Your Understanding" questions provided in boxes within the chapter text are now more consistently limited to factual information contained in the preceding sections of the text, thereby serving as better monitors of student comprehension during their reading of the chapter. The "Test Your Understanding" sections at the end of each chapter often entail the synthesis of the chapter material and, in many cases, require students to go beyond the factual information, apply their reasoning, and build upon the basic concepts. Finally, we have updated and redrafted many of the figures in the book to enhance their clarity and unify their appearance.

The popularity of *Severe and Hazardous Weather* has grown tremendously in the past few years, and we will strive to keep it current and relevant to students' experiences with weather while maintaining a solid emphasis on the underlying science. We hope you enjoy this third edition!

Bob Rauber
John Walsh
Donna Charlevoix

PREFACE

to the First Edition of Severe and Hazardous Weather

Few topics capture the public's attention and fascination like severe and hazardous weather. The awesome power of severe weather, and the devastation and destruction it causes, have made lasting impressions throughout recorded history. Virtually everyone on Earth is affected by hazardous weather during their lifetime and almost everyone is curious about how and why hazardous weather develops. In the last fifty years, we have learned more about severe and hazardous weather processes than in all of human history. With this text, we hope to open the doors to an understanding of severe and hazardous weather and to allow you, the student, to develop an appreciation for the complexities and power of severe weather. In a more practical vein, we hope to leave you better prepared for severe weather, more aware of what is happening when severe weather threatens, and better able to react when severe weather strikes.

The text is written for college students taking an introductory course in severe and hazardous weather. The book provides a current, relevant, and scientifically accurate discussion of all types of hazardous weather. The material is presented in a manner that students with a wide variety of backgrounds can understand. It conveys meteorological concepts in a descriptive manner without resorting to mathematics. Although this book is targeted for introductory severe weather courses, it can serve as an excellent alternative to introductory texts in general meteorology courses in which the instructor wishes to emphasize hazardous weather. The book is designed to be used at the university level with students that range in background from nonscience majors to meteorology majors. The book can also be used as a resource for students in advanced high school courses. It is especially suited to university courses that satisfy elective or general education requirements in the physical sciences.

This book evolved out of a class note packet used to teach a course on Severe and Hazardous Weather at the University of Illinois over the past nineteen years. At present, this course is taken by about a thousand students each year at the university. We have tried to be as inclusive as possible with respect to types of severe and hazard weather events so that the text can be adapted for use in any part of the country. Each topic is addressed at a fundamental level to ensure understanding by the reader. The most recent research on each type of hazardous weather has been incorporated, making the text up to date as of the time of this printing. The newest ideas and theories are presented in such a way that even the most science-challenged student will gain some insight into the current state of understanding of severe and hazardous weather. The tone of the text is conversational in an attempt to put the most apprehensive reader at ease and to allow basic explanations of more complex topics.

Because we believe that learning about weather is enhanced by experiencing weather, we have included examples of significant historical and recent severe weather events throughout the text. Actual weather maps, radar and satellite imagery, and upper-air charts are used to explain and describe various events. This allows the reader to examine the weather events from the perspective of a "severe weather meteorologist." Case studies are described in basic meteorological terms so that students become familiar with terminology and weather products and can use these skills to assess and evaluate future weather events long after studying this text.

PEDAGOGY

Students of severe and hazardous weather in this new century are in a unique position to apply what they have learned immediately to weather events occurring near home, across the country, or around the world. Our goal is not only to assist the student in their comprehension of material, but to provide them the means to access data and weather information worldwide and to apply their knowledge to real-world situations. To this end, we have integrated the Internet throughout the text in a way we believe is unique in the earth sciences, at least today, through the *Severe and Hazardous Weather Website.*

Throughout the text we have inserted "Check your Understanding" questions after major topic sections. These fundamental questions let the reader stop and assess their knowledge of the most recently presented material. The questions address the most fundamental concepts within each chapter. Each chapter ends with a series of activities focusing on different aspects of learning. "Test your Understanding" questions are similar to questions in the chapter but are slightly more advanced and test the comprehension of material by requiring synthesis or application of the concepts presented. "Test Your Problem-Solving Skills" questions are designed for more advanced students or for courses in which a more quantitative component is desired. "Use the Severe and Hazardous Weather Website" problems are designed to allow the students to immerse themselves in real-world weather events. This section allows students to apply knowledge learned, analyze the current weather situation, examine weather forecasts, synthesize material from websites worldwide, or examine weather products and evaluate the current weather situation in terms of its severe or hazardous potential.

Special topics that focus on particular case studies, advanced topics, or interesting offshoots of the subject matter are included in "Focus" boxes, set aside from the main text, but found close to related discussions. The "supplemental sources" list at the end of the text provides other sources of information on many of these subjects for instructors who wish to elaborate on a topic.

Every chapter includes "Online" sections that explain animations or simulations available on both the CD-ROM and accompanying website. These animations allow students to develop an appreciation for the dynamic nature of the atmosphere that cannot be conveyed with a static, two-dimensional figure, diagram, or map. The animations provide yet another dimension to the study of severe weather.

The accompanying website, the *Severe and Hazardous Weather Website,* is designed with a non-meteorologist in mind. With a single click, students can access current weather products from various Internet sources. They can also access forecasts from a variety of sources and for a number of models, both short- and long-range. Links to archived weather products are provided for students and instructors to investigate recent or significant events. Finally, all animations and images provided on the CD-ROM of the textbook are also available online. Additional links and information for the various types of severe weather events are also provided. The authors maintain and update all links to ensure that current, relevant material is available at all times.

TO THE STUDENT

You are embarking on an exciting journey into the world of severe and hazardous weather. Weather is something we all experience but often think little about. This text prompts you to be aware of the weather surrounding you and to examine it with more scrutiny. While principles of physics govern the atmosphere, we attempt to provide you with a discussion of weather processes in a way that everyone from music majors to electrical engineers can understand.

The text is a science textbook. Science texts cannot be read like humanities texts, novels, or the newspaper. The best strategy for using this text is to approach your studies in a slow and systematic way. You want to read and retain the new information, processing it in such a way that you will be able to recall it later and apply it to the weather events you experience throughout your life. The following is a suggested strategy for reading this and other science texts:

1. Preview the chapter to see what it contains. Page through the chapter looking for words in quotes or italics; examine tables and figures—read captions; scan the Focus and Online boxes. Read the "Check your Understanding" and the "Test Your Understanding" questions *before* you read the chapter. That way you will know what important points to watch for as you read the text.
2. Read the text. Take the time to carefully read and re-read difficult or unfamiliar material and to answer the "Check your Understanding" questions at the end of each section. If you can't answer a question, stop and find the answer before proceeding to the next section.
3. When you have finished a chapter, answer the "Test Your Understanding" questions at the end of the chapter. If you can answer these questions with confidence, you can be assured you have a fundamental understanding of the key concepts from the chapter.

4. Use the *Severe and Hazardous Weather Website.* Familiarize yourself with a few websites that are relevant to the material you read in the chapter. See if you can find examples of the concepts or weather discussed in the chapter.

You will find that the most exciting aspect of studying severe weather is monitoring a severe weather event as it happens and understanding how the weather system is evolving! You are provided tools to help you do this in the "Use the Severe and Hazardous Weather Website" questions at the end of the chapter and by simply using the *Severe and Hazardous Weather Website* as a launching point for exploration. The website is designed to guide you through the many thousands of weather websites available on the Internet. With only a few clicks, you now have any type of weather product, forecast, or severe weather product at your fingertips. You will also find animations of recent significant weather events on the website and on the CD-ROM accompanying the text. Explore these animations to gain a dynamic perspective of the weather that cannot be gained from reading a static textbook.

Above all, explore, question, and appreciate the power of severe and hazardous weather!

TO THE INSTRUCTOR

Since this text will be used primarily in North American universities and colleges, we have concentrated the text on the hazardous weather of North America, but not to the complete exclusion of other parts of the globe. The text is designed to support "modular teaching." The first eight chapters cover the fundamentals of the atmosphere. Subsequent chapters can be presented in any order the instructor desires. The chapters are organized by type of weather: extratropical cyclones, winter weather phenomena, convective phenomena, tropical weather systems, and large-scale, longer-term weather events. Materials can also be presented by geographic region. For example, instructors teaching in the Rocky Mountain region may wish to focus their attention on Chapter 10 (Extratropical Cyclones Forming East of the Rocky Mountains), Chapter 16 (Mountain Snowstorms), Chapter 17 (Mountain Windstorms), Chapter 20 (Hailstorms), Chapter 22 (Downbursts) and Chapter 25 (Floods). Instructors in the Gulf Coast region, New England, along the West Coast, or in the Midwest can tailor their classes to focus on weather their students will most likely experience. Based on our experience at the University of Illinois, the material in each chapter can be covered in two or three fifty-minute class periods, depending on the depth in which the material is presented.

In the United States, the meteorological community uses both English and Metric units to describe weather conditions, resulting in a systematic inconsistency that (unfortunately) is unlikely to change in the near future. Students of severe and hazardous weather must therefore be able to work with English and Metric units. In writing this book, we recognized this dilemma and made the conscious decision to use a variety of units in our descriptions of weather phenomena. We normally report measurements in the most common unit used in meteorology and then report the corresponding metric or English unit. For example, rainfall is most commonly reported in inches, so we normally report inches, adding the metric equivalent, centimeters, in parentheses. The public and virtually all students with no background in meteorology are unfamiliar with wind speeds in meters per second, but are very familiar with miles per hour. However, all weather maps report winds in knots. We therefore report winds in knots, with miles per hour in parentheses, so that students can have a good feeling for the winds reported on weather maps. Until the use of units in the United States becomes consistent, we as educators will have to deal with this issue. Our choices in this textbook were based on communicating most effectively with a broad range of students who, in general, have no background in meteorology and may or may not have sufficient science background to be comfortable with metric units.

All online materials are provided on the included CD-ROM so that students and instructors can examine an actual archived event while the topic is being discussed. This becomes especially important when the weather event being covered is "out of season." The online material can also be found on the *Severe and Hazardous Weather Website* so it can be accessed remotely even if the CD-ROM is somewhere else. Guidance is provided in the instructor's manual on how best to incorporate animations into lecture presentations or as laboratory exercises.

A variety of ancillary materials are available to help present material effectively, assess students fairly, and provide relevant examples that students can use to help apply knowledge to current and future weather events.

Instructors manual: Each chapter includes tips on presenting concepts, sample lesson plans, a list of key words, answers to all questions in the book, and a test bank (true/false, matching, multiple choice, and short answer questions).

CD-ROM of test bank: All questions in the test bank from the instructor's manual are included in electronic form to assist instructors in student assessment.

CD-ROM of text figures: All figures within the text are provided on a CD-ROM as individual images. We have also included on the CD-ROM Powerpoint presentations containing all figures and all tables, organized by chapter.

Bob Rauber
rauber@atmos.uiuc.edu

John Walsh
walsh@atmos.uiuc.edu

Donna Charlevoix
charlevo@atmos.uiuc.edu

THE SEVERE AND HAZARDOUS WEATHER WEBSITE

http://severewx.atmos.uiuc.edu

The *Severe and Hazardous Weather Website* is designed for both students and weather buffs. To help navigate the sea of weather websites now available we have provided a condensed listing of links to the most popular and common products. These links will allow you to monitor the weather and investigate severe weather events with confidence and ease.

Chapter-by-chapter links to animations and photo galleries discussed in the text as well as chapter-related websites are displayed in an easy to navigate menu. Every animation and photo found on the CD-ROM that accompanies this text can be found online. Many of the animations were developed specifically for students of severe weather to help illustrate key weather processes and phenomena. The Online boxes throughout the text provide a summary of the animation or photo gallery as well as guidance on key features to look for when you are viewing the animations. We have chosen not to include actual URLs of web sites of interest due to the dynamic nature and unpredictability of the majority of web site addresses. A full list of Online topics (and credits) is provided here.

A page of the website is devoted to current weather links. The links are organized to simplify the process of finding weather maps, satellite and radar imagery, and severe storm information. A complete list of forecast products is provided on another page of the website, with additional suggestions for those of you who are amateur forecasters. Finally, links to archived weather products and storm data are provided. The links will be helpful to both students and instructors who wish to analyze past weather events.

CHAPTER 1

Online 1.1 Clouds *(Courtesy of Lourdes Aviles and Glen Romine, University of Illinois)*

Online 1.2 Scales of Atmospheric Motions *(Courtesy of Dept. Atmospheric Sciences, University of Illinois)*

CHAPTER 2

Online 2.1 ASOS Instruments *(Courtesy of National Weather Service)*

Online 2.2 Six Views of Weather from Space *(Courtesy of Dept. Atmospheric Sciences, University of Illinois)*

CHAPTER 3

Online 3.1 Using Surface Maps to Find Significant Meteorological Features *(Courtesy of Dept. Atmospheric Sciences, University of Illinois)*

Online 3.2 Upper Air Weather Maps *(Courtesy of Dept. Atmospheric Sciences, University of Illinois)*

CHAPTER 4

Online 4.1 Uncertainties in Numerical Weather Prediction Models *(Courtesy of Dept. of Atmospheric & Oceanic Sciences, University of Wisconsin)*

Online 4.2 An Illustration of Ensemble Forecasting *(Data courtesy of NCEP/NOAA)*

Online 4.3 Model Simulations Showing the Effect of Wind Shear on Thunderstorms *(Courtesy of Robert Wilhelmson and Brian Jewett, University of Illinois)*

Online 4.4 View a Simulation of a Hurricane, a Thunderstorm, a Tornado, and a Landspout *(Courtesy of Glen Romine and Robert Wilhelmson, University of Illinois; Bruce Lee, Windlogics Inc.; David Porter, University of Minnesota)*

CHAPTER 5

Online 5.1 Evolution of Twenty-first-Century Temperatures (Courtesy of *William Chapman, University of Illinois; data from Program for Model Data and Intercomparison, Lawrence Livermore National Laboratory*)

CHAPTER 6

Online 6.1 Infrared Satellite Imagery Captures the Diurnal Cycle of Atmospheric Stability *(Courtesy of Dept. Atmospheric Sciences, University of Illinois)*

Online 6.2 Clear Daytime Skies Favor Severe Thunderstorm Development *(Courtesy of Cooperative Institute for Meteorology, University of Wisconsin-Madison)*

CHAPTER 7

Online 7.1 Experience the Coriolis Force *(Courtesy of Apple Computers Inc.)*

Online 7.2 The Geostrophic Wind *(Courtesy of Dept. Atmospheric Sciences, University of Illinois)*

CHAPTER 8

Online 8.1 Effects of Friction on the Flow in High- and Low-Pressure Systems *(Courtesy of Dept. Atmospheric Sciences, University of Illinois)*

Online 8.2 Jetstreaks, Flow Curvature, and Low-Pressure Systems *(Courtesy of Dept. of Atmospheric & Oceanic Sciences, University of Wisconsin, and Dept. Atmospheric Sciences, University of Illinois)*

CHAPTER 9

Online 9.1 Finding Cold and Warm Fronts in Surface Data *(Courtesy of Dept. Atmospheric Sciences, University of Illinois)*

CHAPTER 10

Online 10.1 Alberta Clippers *(Courtesy of Dept. of Atmospheric & Oceanic Sciences, University of Wisconsin, and Dept. Atmospheric Sciences, University of Illinois)*

Online 10.2 Colorado Lows *(Courtesy of Dept. Atmospheric Sciences, University of Illinois, and RAP/UCAR)*

Online 10.3 An Upper-Level Cyclone Crosses the Rockies *(Courtesy of Dept. of Atmospheric & Oceanic Sciences, University of Wisconsin, and Dept. Atmospheric Sciences, University of Illinois)*

Online 10.4 Cutoff Lows *(Courtesy of Dept. Atmospheric Sciences, University of Illinois)*

CHAPTER 11

Online 11.1 The Subtropical Jetstream and Water Vapor Satellite Imagery *(Courtesy of Dept. Atmospheric Sciences, University of Illinois)*

Online 11.2 Rare April Nor'easter Strikes the East Coast in 2007 *(Courtesy of Dept. of Atmospheric & Oceanic Sciences, University of Wisconsin, and UNISYS)*

Online 11.3 The Superstorm of 12 to 14 March 1993 *(Courtesy of Dept. Atmospheric Sciences, University of Illinois)*

CHAPTER 12

Online 12.1 The January 2007 North American Ice Storm *(Courtesy of Department of Atmospheric Sciences, University of Wyoming)*

Online 12.2 The 1998 Northeast Ice Storm *(Courtesy of Dept. of Atmospheric & Oceanic Sciences, University of Wisconsin, and RAP/UCAR)*

CHAPTER 13

Online 13.1 Persistent Winds Cause a Major Lake-Effect Snow Accumulation in 2007 (Courtesy of *Department of Atmospheric Sciences, University of Illinois*)

Online 13.2 Lake-Effect Circulations *(Courtesy of Neil Laird, Hobart and William Smith Colleges)*

CHAPTER 14

Online 14.1 A Pacific Storm Builds a Ridge over North America *(Data courtesy of Department of Atmospheric and Oceanic Sciences, University of Wisconsin)*

Online 14.2 The Cold Outbreaks of 2004 in the Northeastern United States *(Data courtesy of NCEP)*

CHAPTER 15

Online 15.1 The Great Plains Blizzard of November 2005 (Courtesy of *National Weather Service and UNISYS*)

Online 15.2 A Ground Blizzard from Space *(Courtesy of SSEC, University of Wisconsin-Madison)*

CHAPTER 16

Online 16.1 The Drought-Busting Blizzard of 2003 *(Courtesy of NOAA)*

Online 16.2 Twin Upslope Snowstorms Wreck 2006 Holiday Season in Denver *(Courtesy of Dept. of Atmospheric & Oceanic Sciences, University of Wisconsin, and WSI)*

CHAPTER 17

Online 17.1 Wave Cloud over the Western Plains *(Courtesy of CIRA)*

Online 17.2 Weather Conditions during the Santa Ana *(Courtesy of Dept. Atmospheric Sciences, University of Illinois and UCAR)*

CHAPTER 18

Online 18.1 An MCS Squall Line *(Data courtesy of NCEP)*

Online 18.2 Parallel Squall Lines *(Data courtesy of NCEP)*

Online 18.3 Vertical Wind Shear and Storm Structure *(Courtesy of Robert Wilhelmson, University of Illinois/NCSA)*

Online 18.4 Views of Supercells *(Courtesy of Brian Jewett, University of Illinois)*

CHAPTER 19

Online 19.1 A Tornado Gallery *(Courtesy of Glen Romine, University of Illinois)*

Online 19.2 Non-Supercell Tornadoes *(Courtesy of Bruce Lee, Windlogics, Inc. and Robert Wilhelmson, University of Illinois)*

Online 19.3 Seasonal Variation of Tornado Frequency in the United States *(Data courtesy of NCEP)*

Online 19.4 The May 3, 1999 Oklahoma City Tornado *(Data courtesy of NCEP)*

Online 19.5 Record May 2003 Tornadoes *(Courtesy of Dept. Atmospheric Sciences, University of Illinois)*

CHAPTER 20

Online 20.1 The Aurora Supercell that Produced World Record Hailstones *(Courtesy of Dept. Atmospheric Sciences, University of Illinois)*

Online 20.2 Seasonal Variation of Hail in the United States *(Courtesy of NSSL)*

CHAPTER 21

Online 21.1 Lightning from Space *(Courtesy of NASA)*

CHAPTER 22

Online 22.1 Animation of a Downburst Simulated by a Numerical Model *(Courtesy of Glen Romine, University of Illinois)*

Online 22.2 Doppler Radar Detects a Microburst over Utah *(Courtesy of NOAA Warning Decision Training Branch)*

CHAPTER 23

Online 23.1 An animation of the Sea-Surface Temperatures during an El Niño Event *(Courtesy of NOAA/CDC)*

Online 23.2 A Three-Dimensional View of the El Niño/La Niña Cycle in the Ocean *(Courtesy of NOAA/CDC)*

CHAPTER 24

Online 24.1 The Eye of Isabel (2003) *(Courtesy of CIMSS-University of Wisconsin-Madison)*

Online 24.2 Hurricane Katrina (2005) *(Data courtesy of the Department of Atmospheric Sciences, University of Illinois, UNISYS, the NOAA Tropical Prediction Center, and the NRL Monterey Marine Meteorology Division.)*

Online 24.3 Global Sea-Surface Temperature *(Courtesy of US Navy FNMOC)*

Online 24.4 Wind Shear *(Courtesy of CIMMS-University of Wisconsin)*

CHAPTER 25

Online 25.1 Tropical Storm Allison *(Courtesy of Southwest Regional Climate Center)*

Online 25.2 Frequent Convective Rain Events in 2007 Produce Flooding in Texas (Courtesy of National Weather Service, Hydrometeorological Prediction Center)

Online 25.3 Flooding in the Central Valley of California *(Data courtesy of NCEP)*

CHAPTER 26

Online 26.1 Evolution of the 1999 Drought in the Eastern United States *(Data courtesy of NCEP)*

Online 26.2 The West Coast Ridge Prevails, then Yields, during the 1991 Drought *(Data courtesy of NCEP)*

Online 26.3 A Satellite Captures the Drying of Lake Mead *(Courtesy of NASA)*

CHAPTER 27

Online 27.1 The Heat Index Soars in July 2005 *(Data courtesy of William Chapman, University of Illinois; data from NCEP/NCAR reanalysis)*

Online 27.2 The Heat Returns in August 2000 *(Courtesy of Dept. Atmospheric Sciences, University of Illinois)*

ACKNOWLEDGMENTS

Although only three authors appear on the cover, this book is really the work of thousands of scientists worldwide who spend their lives out in the weather, and then back in their laboratories and offices struggling to understand what happened. Their work continually advances our science and has saved countless lives. We could not have developed this book without the generosity of our many colleagues at universities, government laboratories and agencies, research laboratories and in private industry who gave us permission to use their photographs, images, and figures in this book.

We are particularly indebted to scientists at the National Oceanic and Atmospheric Administration, the National Aeronautics and Space Administration, and in the U.S. Navy, whose excellent websites have provided us with an inexhaustible source of images. We owe a special thanks to the University of Wisconsin for their permission to use so many images and animations from their websites—a wonderful resource for all the world! Special thanks are also due our colleagues Bruce Lee and Harold Brooks, who provided unique material for various chapters of the book.

The Department of Atmospheric Sciences at the University of Illinois at Urbana Champaign has supported this effort through many years of development. We collectively thank all our current and former faculty and staff for their help in interpreting severe weather events over our years together. A special thanks to Mohan Ramamurthy, Bob Wilhemson, Ken Beard, Larry Di Girolamo, Greg McFarquhar, Mankin Mak, and Harry Ochs, who have endured the brunt of our questions over the years. Our former Department Head, Donald Wuebbles, who is now directing the University of Illinois School of Earth, Society and Environment, has been behind this project from day one and has committed Department support to help us throughout the book's development. Without the continuous help and cooperation of David Wojtowicz, chief developer of Weather World 2010 and computer administrator of the Department, we would still be writing on paper and drawing maps by hand.

Bill Chapman and Diane Portis developed the graphics appearing in several Chapters. Bill Chapman also helped with the development of animations for the CD-ROM and the *Severe and Hazardous Weather Website.* Support for his effort was provided by the National Science Foundation's Climate Dynamics Program. Brian Jewett, Neil Laird, and Glen Romine generously provided many images and animations, and Brian provided the inspiration for the design on many of the pages on the *Severe and Hazardous Weather Website.* Adam Houston provided various photographs appearing in the book. We also thank our department's present and past secretarial staff, Karen Garrelts, Jennifer Drennan, Peggy Cook, Annetta Ashbrook, Norene McGhiey, Nena Richards, and Shirley Palmisano.

Our publisher, Kendall/Hunt, has been a superb partner. It has been a pleasure to work with you and all your staff.

Finally, we want to thank our families. As anyone who has written a book knows, family members must sacrifice so much time, provide so much support, and have so much patience or the project will never get done. So Ruta Rauber, Carolyn Rauber, Stacy Rauber, Glen Romine, Evelyn Romine, Madeline Romine, Laura Walsh, Rachel Walsh, and Emily Walsh, THANKS!!! This book is as much yours as it is ours.

ABOUT THE AUTHORS

Bob Rauber Bob Rauber has been a professor at the University of Illinois since 1987. He maintains active research programs in the fields of mesoscale meteorology and precipitation physics and has been an investigator in over twenty major field research campaigns. His research has appeared in *Nature,* the *Journal of Geophysical Research, Geophysical Research Letters,* the *Journal of the Atmospheric Sciences, Monthly Weather Review,* the *Journal of Applied Meteorology,* the *Journal of Atmospheric and Oceanic Technology,* and *Weather and Forecasting.* He has also authored articles for a number of encyclopedias. Bob has received multiple teaching awards from the University of Illinois, is a Fellow of the American Meteorological Society, and is currently the chief editor of the *Journal of Applied Meteorology and Climatology.*

John Walsh John Walsh taught atmospheric science courses at the University of Illinois for over thirty years, primarily at the introductory undergraduate level. His research emphasizes climate and climate change, particularly in the polar regions. He has been a lead author for the Intergovernmental Panel on Climate Change and for the Arctic Climate Impact Statement. He is a member of the Polar Research Board and has served on various other national panels and committees. He is an associate editor of the *Journal of Climate* and a Fellow of the American Meteorological Society.

Donna Charlevoix Donna Charlevoix joined the faculty at the University of Illinois in 1997. She provides instruction in undergraduate courses in Atmospheric Sciences and graduate courses related to teaching in higher education within the Departments of Atmospheric Sciences and Educational Organization and Leadership. She is also Director of Introductory Courses for the Department of Atmospheric Sciences and conducts research in science education. She is a recipient of multiple teaching awards from the University of Illinois and is a member of the Chancellor's Honors Program Faculty. She serves on numerous academic and teaching and technology-related committees for the University, as well as on the American Meteorological Society Board on Higher Education.

PROPERTIES OF THE ATMOSPHERE

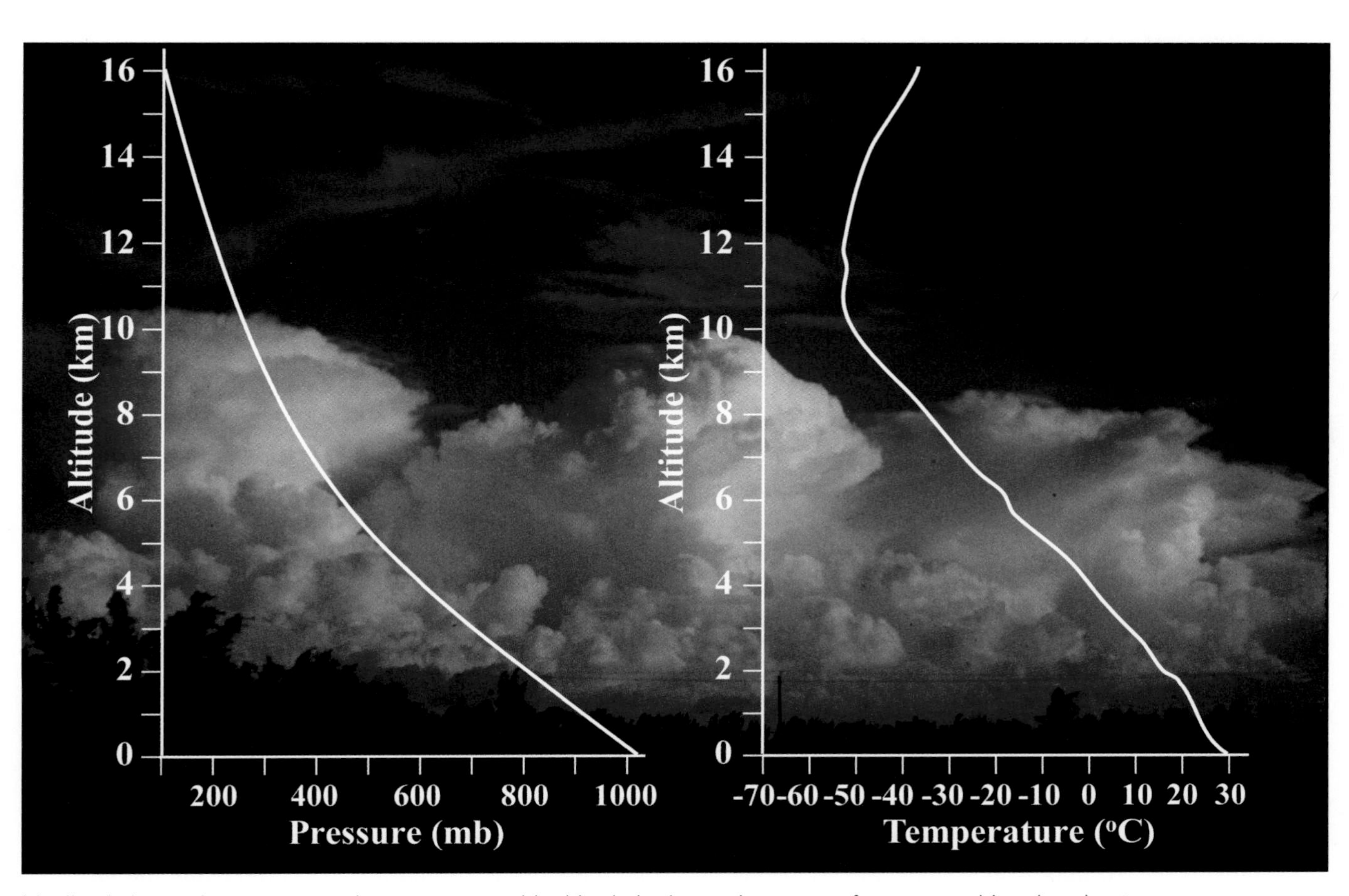

Idealized change in pressure and temperature with altitude in the environment of an approaching thunderstorm.

KEY TERMS

absolute zero
aerosol
altocumulus
altocumulus lenticularis
altostratus
anemometer
anemometer height
atmospheric pressure
barometer
Celsius scale
centigrade
cirrocumulus
cirrostratus
cirrus
cloud droplets
condensation
cooling degree-day
cumulonimbus
cumulus
density
deposition
dewpoint temperature
evaporation
extratropical cyclone
Fahrenheit scale
freezing
general circulation
graupel
hail
haze droplets
heating degree-day
hurricane
ice nuclei
Kelvin scale
latent heat
mean sea level
mean sea-level pressure
melting
mesosphere
millibar

nimbostratus
nimbus
ozone
phase change
polar jetstream
pressure
raindrops
relative humidity
saturated
saturation
saturation vapor pressure
sea-level pressure
snowflake
stratosphere
stratus
sublimation
subtropical jetstream
supercooled
temperature
thermistor
thermometer
thermosphere
thunderstorm
tornado
tropopause
tropopause fold
troposphere
vapor pressure
water vapor
wind
wind barb

The atmosphere is a thin layer of air extending from the Earth's surface to a height of a few hundred kilometers. Ninety percent of the atmosphere's mass, and all of its weather, is confined in a layer between the ground and a height of about 12 km (7.5 miles). The layer of the atmosphere where weather occurs is indeed a thin layer, as we can see by comparing its thickness to the 2.5 km (1.5 miles) length of San Francisco's Golden Gate Bridge—the weather layer's thickness is only about five times the length of the bridge. At highway speeds, it would take only seven minutes to drive through this layer (if roads went straight up!).

The atmosphere is composed of different types of molecules. If we exclude water vapor, the atmosphere is made up primarily of two gases: nitrogen (78 percent) and oxygen (21 percent). Other gases, which include carbon dioxide, ozone, hydrocarbons, and argon, make up the remaining 1 percent. The component of the atmosphere most important to weather is water, which can be present as a gas (water vapor), a liquid (***haze droplets, cloud droplets,*** and ***raindrops***), or a solid (snow, ice pellets, and hail). The atmosphere also contains large numbers of invisible tiny particles called ***aerosol.***

We describe the atmosphere in terms of its properties. These include basic properties such as temperature, pressure, moisture content, wind speed and direction, and other observable properties such as visibility, cloud cover, and precipitation rate. These properties are measured continuously worldwide by a network of instrumentation that has become increasingly complex due to advances in technology. Meteorologists depict the measurements on weather maps to condense and more easily interpret the enormous quantity of weather data. In this chapter, we examine the basic properties of the atmosphere.

TEMPERATURE

We all have an intuitive feeling for temperature because our bodies feel hot and cold sensations. From a scientific viewpoint, ***temperature*** is a measure of the average speed that molecules move in a substance. In a solid, molecules vibrate in place—faster for higher temperatures, slower for lower temperatures. In a liquid, molecules remain in contact with one another, but move freely about—faster for higher temperatures, slower for lower temperatures. Individual air molecules move around rapidly (hundreds of kilometers per hour!) in random directions, continuously bouncing off each other. The faster the average speed of the molecules, the higher the temperature. When substances with different temperatures come in contact, energy associated with the vibration or movement of the molecules will be transferred from the substance with a higher temperature to the substance with a lower temperature. The average speed of the molecules will increase in the colder substance and decrease in the warmer substance. When energy transfer no longer occurs, the substances are at the same temperature.

Thermometers or electronic devices called ***thermistors*** measure temperature. In the United States, temperature is commonly measured using the ***Fahrenheit scale,*** devised by Gabriel Daniel Fahrenheit (1686–1736). The range 0° F to 100° F spans the common temperatures experienced outdoors: 0° F being a very cold day in winter and 100° F being a very hot day in summer. Of course, air temperatures can be much colder; in the continental United States, wintertime extremes often fall below –20° F, and temperatures in the Antarctic fall below –100° F.

A second scale, called the ***centigrade,*** or ***Celsius scale*** after its developer, Anders Celsius (1701–1744), is used in all countries except the United States. The Celsius scale is based on the boiling point of water and the melting point of ice. Celsius devised a scale such that the boiling point of water (at sea level pressure) was 100° C (212° F) and the melting point of ice was 0° C (32° F).

Recall that temperature is related to the energy of motion of molecules. Imagine a situation where we continually remove energy from a substance. Eventually, the molecules will slow down to a point where they are essentially not moving and no more energy ... be extracted. William Thomson (1824–1907), ... ed the title Lord Kelvin, first devised ... le where "zero" was set at this mini... te. ***Absolute zero***, or zero on the ... urs at –459.67° F (–273.15° C). Fig... the Kelvin, Celsius, and Fahrenheit ... s; shows some common and extreme ... at different locations on the Earth; ... ble and simple formulas to convert ... e scales. In the United States, the ... is used for surface temperatures, the ... upper atmospheric temperatures, and ... for scientific applications. In other ... d, the Celsius scale is used for surface ... well as upper atmospheric tempera... ok, we will use both Fahrenheit and ... ures.

... varies every day as the sun rises and ... season to season as the sun moves ... r in the sky. We will see later that these ... iations drive the weather.

FOCUS 1.1

Degree-Days—A Measure of Energy Consumption

Outdoor temperatures determine to a large extent the energy required to heat buildings in winter and cool them in summer. To estimate energy consumption, power companies use a concept called heating and cooling degree-days. A degree-day is a unit and should not be confused with a calendar day. There can be more than one degree-day in a twenty-four hour period. *Heating degree-days* are defined as 65 minus the average of a day's high and low Fahrenheit temperature. For example, if a city's high was 40° F and the low was 20° F, the number of heating degree-days for that day would be 65 – (40 + 20)/2 = 35. If the average of the day's high and low exceeds 65, it is assumed that heating is not required and the city has no heating degree-days for that day. *Cooling degree-days* are the average of a day's high and low Fahrenheit temperature minus 65. For example, if the city's high was 80° F and the low was 60° F, the number of cooling degree-days would be (80 + 60)/2 – 65 = 5. If the average

(continued)

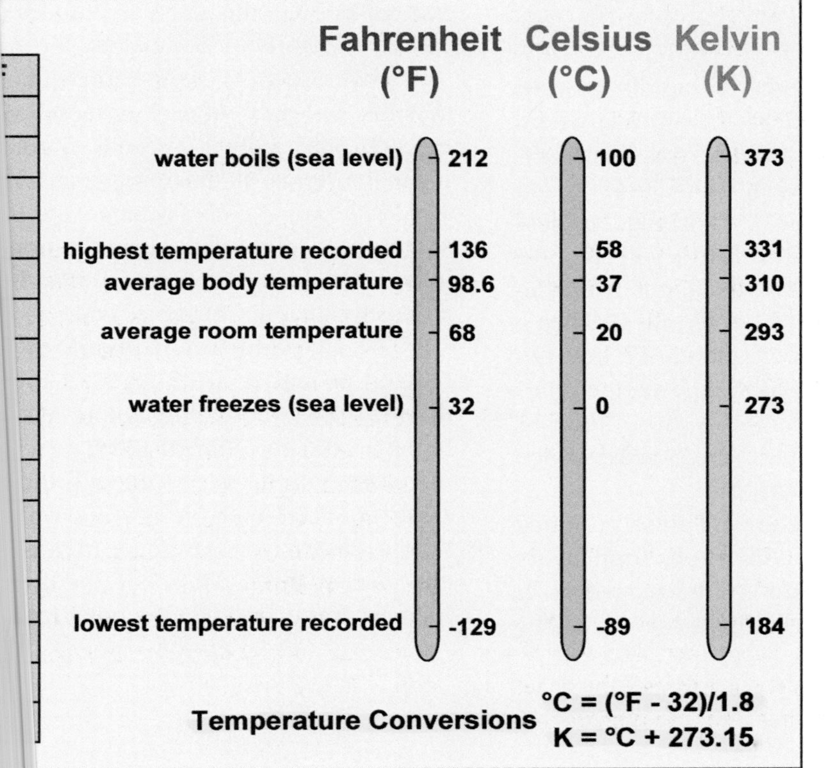

FIGURE 1.1: Examples of common and extreme temperatures observed on Earth as expressed on the Fahrenheit, Celsius, and Kelvin temperature scales. Formulas to convert between scales and a table of common Celsius temperatures and their Fahrenheit counterparts is also shown in the figure.

(continued)

temperature is below 65, the cooling degree-days for that day is zero. The number of degree-days is summed up over all days in a heating season (typically October–April) or a cooling season (typically May–September) to obtain the season's total energy requirement.

The oceans have a moderating influence on temperature. The seasonal change of temperature over the oceans is much smaller than over the interior continent. As a result, the energy required to heat and cool homes is much greater in the continental interior compared to the coasts. This is particularly true on the West Coast, where air arriving at the coast has more moderate temperatures than that over land. The difference between coasts and the interior continent is evident in the seasonal totals of heating and cooling degree-days. For example, in Omaha, Nebraska, the average annual number of heating degree-days (6300) is more than double that of San Francisco, California (3016). The cooling degree-days in Omaha (1072) is over seven times that of San Francisco (145). The East Coast does not differ as much from the interior because air arriving at the East Coast often flows off the continent. Nevertheless, the values of heating and cooling degree-days at Atlantic City, New Jersey (4728 and 824, respectively) are still well below those of Omaha.

Temperature normally decreases rapidly upward away from the Earth's surface. At the altitude where commercial jets fly—about 10 to 12 km (6 to 7 miles)—temperatures are typically between –40° C (–40° F and –50° C (–58° F). Figure 1.2 shows an average vertical profile of temperature from the Earth's surface to space. Temperature decreases with height from the surface to an altitude of about 12 km (7 miles). Then a remarkable thing happens: above about 12 km, the temperature increases with height. The temperature begins to decrease again only above an altitude of about 50 km (31 miles). Temperature increases once again near the fringes of outer space.

Examine the diagram in Figure 1.2 closely. Notice that the atmosphere can be divided into four layers based on the changing temperature. The layer of air in the lower atmosphere where temperature decreases with height is called the ***troposphere***. For practical purposes, all weather that develops on Earth occurs in the troposphere. The layer of air above the troposphere, where temperature increases with height, is called the ***stratosphere***. The boundary between these two layers is called the ***tropopause***. Temperature increases with height in the stratosphere because of the absorption of ultraviolet radiation by ***ozone*** in that layer. The stratosphere has a very important effect on the Earth's weather. While storms can grow vertically in the troposphere, they cannot penetrate far, if at all, into the stratosphere. We will examine the reasons why storms are limited to the troposphere when we study atmospheric stability in Chapter 6.

The tropopause can be thought of as a "lid" on the Earth's weather. In essence, the tropopause prevents vertical air currents (typically associated with clouds and storms) from moving into the stratosphere. As shown in Figure 1.3, the tropopause slopes downward from the tropics to the poles. In tropical regions, the tropopause occurs at an altitude of about 16 to 18 km (~10 to 11 miles). In the middle latitudes, the tropopause is typically at an altitude of 11–13 km (~7–8 miles), while at polar latitudes, the tropopause is found at altitudes as low as 8 km (~5 miles). Because the tropopause is higher in the tropics, tropical storms grow to greater heights than storms that form in middle latitudes and polar regions.

Two jetstreams, the ***subtropical jetstream*** and ***polar jetstream***, circle the globe at latitudes of about 25° and 50°. These rivers of fast moving air in the upper troposphere are critical to the development of Earth's weather systems. Just north of each of these jetstreams, air from the stratosphere often descends in a narrow zone, leading to the "folds" in the tropopause as depicted in Figure 1.3. ***Tropopause folds*** are one way in which air in the stratosphere mixes with air in the troposphere. We will learn more about jetstreams and their importance to hazardous weather in Chapters 7 and 8.

The two additional layers of the atmosphere located at higher altitudes (see Figure 1.2) are the ***mesosphere*** (where temperature again decreases with height), and the ***thermosphere*** (where temperature increases with height). Although these layers have no significant impact on Earth's weather, molecules within these layers intercept high energy solar particles and radiation. These are the layers in which brilliant auroras occur over the polar latitudes.

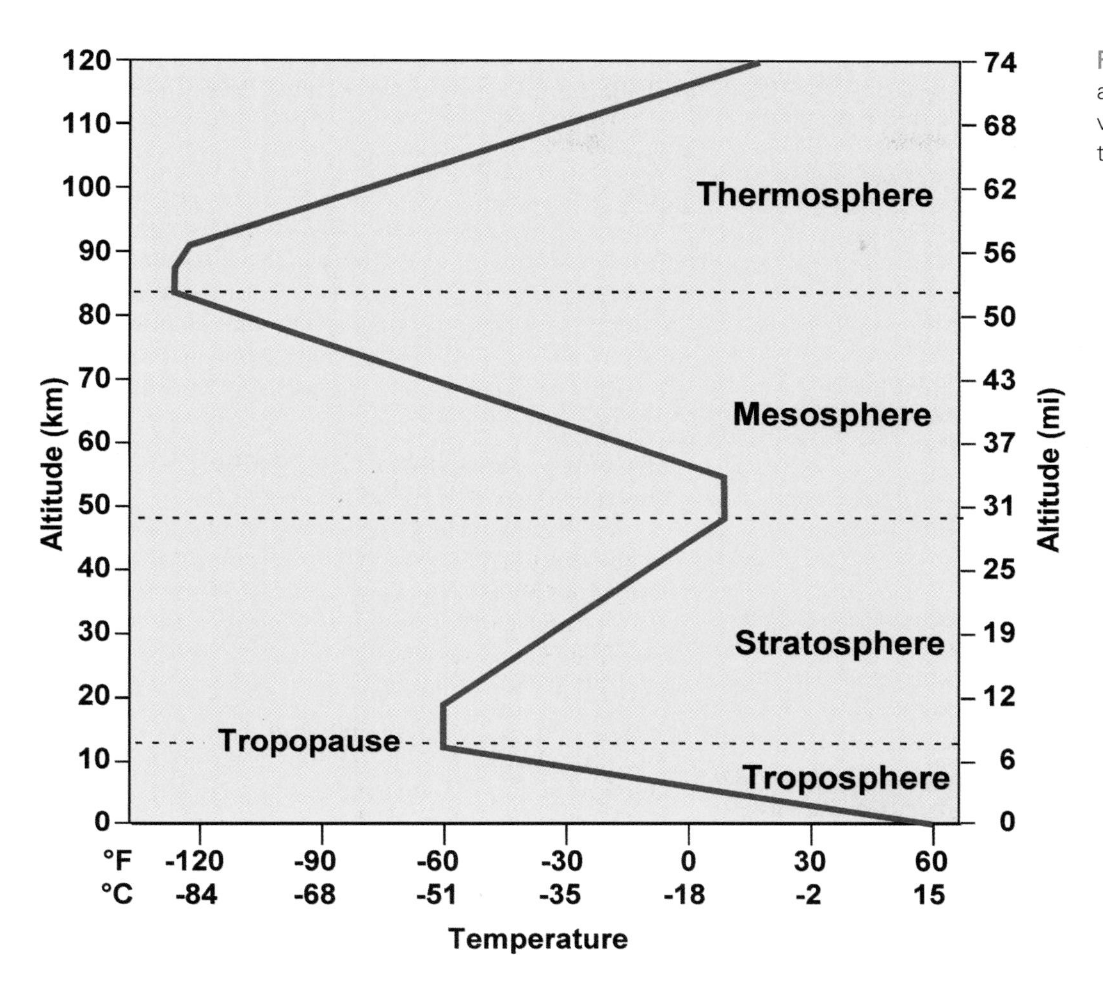

FIGURE 1.2: The global, annual average vertical variation of atmospheric temperature with altitude.

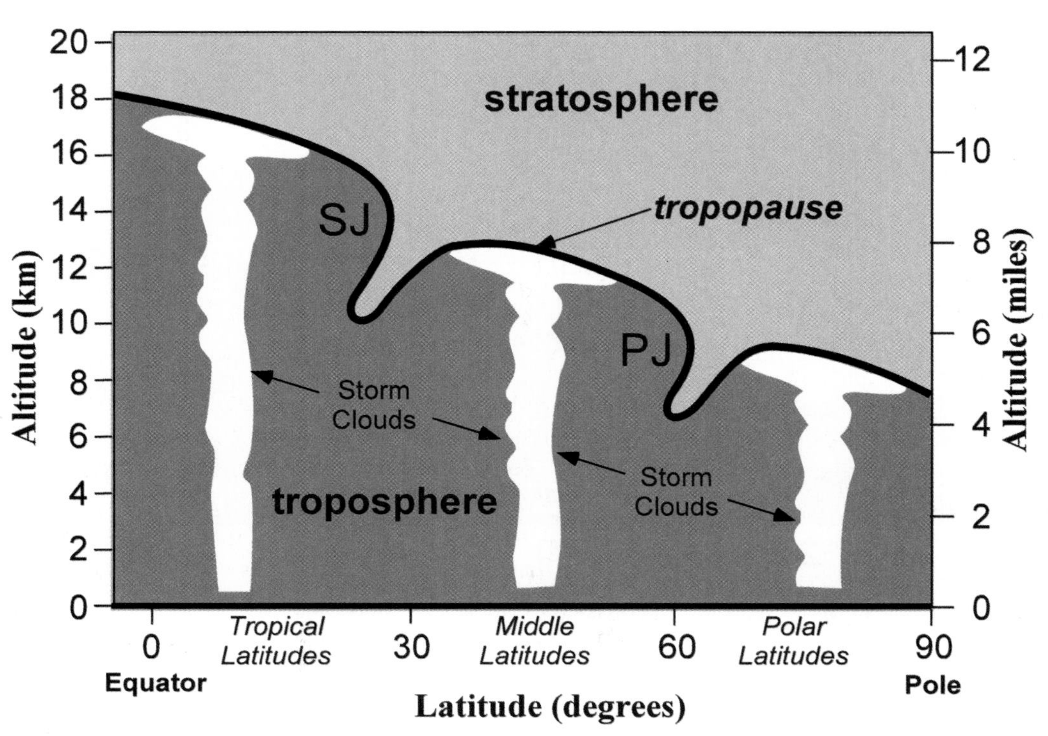

FIGURE 1.3: The annual average depth of the troposphere between the equator and the pole. The depth of storm clouds is limited to the height of the tropopause. The tropopause is located at higher altitudes in the tropics and lower altitudes in the polar regions. The labels SJ and PJ denote the positions of the subtropical jetstream and the polar jetstream, respectively. Stratospheric air sometimes descends on the poleward side of these jetstreams, leading to folds in the tropopause.

CHECK YOUR UNDERSTANDING 1.1

1. How is temperature defined? What units for temperature are used in meteorology in the United States?
2. How does air temperature change with height in the troposphere and stratosphere?
3. What is the tropopause?
4. How does the altitude of the tropopause affect the height of clouds in the troposphere?

PRESSURE

Atmospheric pressure is the force applied by air on a unit area of surface. To understand ***pressure***, think of a wall with air on one side. Each air molecule that strikes the wall will impart a force on the wall. If we divide the wall into unit areas (e.g., square centimeters, square inches), then the force applied per unit area will be the force applied by all of the molecules that strike the wall over a unit area.

One way of thinking about pressure that we will use extensively in this book is to consider the "weight" of a column of the atmosphere above a unit area of surface. Let's suppose we choose a square inch of the Earth's surface at sea level. How much does a column of air weigh above the square inch if we extend our imaginary column from sea level all the way to the top of the atmosphere, as shown at New York City in Figure 1.4? The answer will depend on when we choose to make the measurement, but on average, we will find that at sea level the air in the column weighs 14.7 pounds. The pressure applied on that square inch of surface is 14.7 lbs/in^2. This value is called "average *sea-level pressure.*" This leads us to our first important concept about pressure: *Pressure is equivalent to the weight of a column of air above a unit area.* Let's look at some examples.

Returning to Figure 1.4, we see a profile of the topography across the United States with a few locations identified along with their elevation. The higher the elevation of a particular location, the lower the atmospheric pressure that location will experience. As we noted, above one square inch in New York City, which is at sea level, a column of atmosphere weighs 14.7 lbs and the pressure in New York is 14.7 lbs/in^2. Chicago is about 600 ft above sea level. Chicago is a little closer to the top of the atmosphere than New York, and therefore has fewer air molecules in its column. The air column, on average, weighs 14.4 lbs, and the pressure in Chicago is 14.4 lbs/in^2. Denver

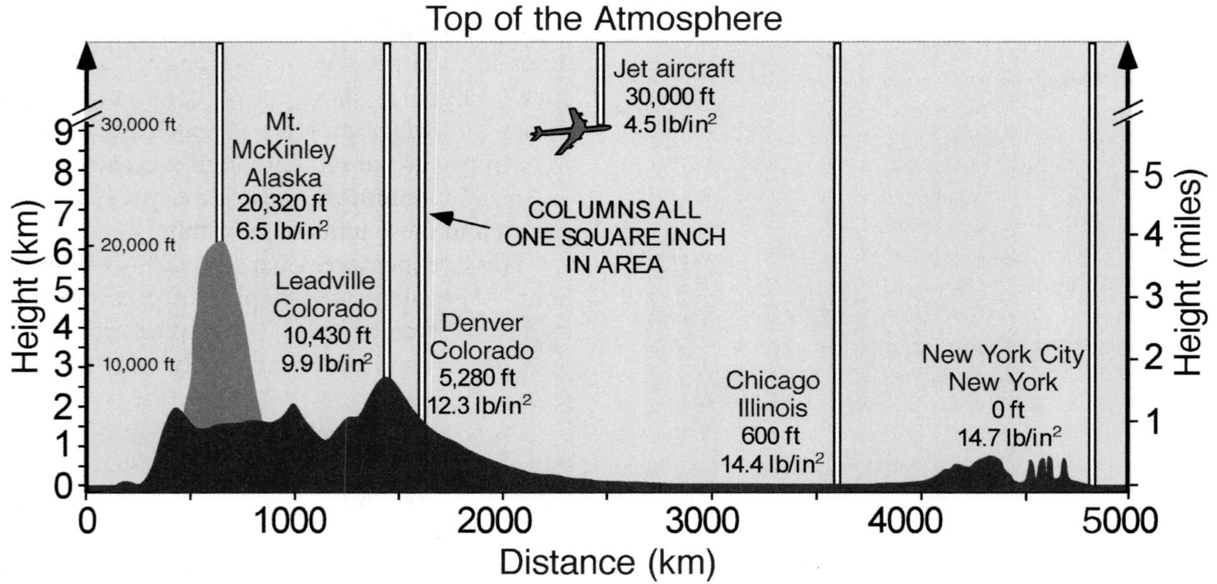

FIGURE 1.4: Columns of air of horizontal dimensions 1 inch^2 extending from the surface to the top of the atmosphere over locations across the United States that are at different elevations. The amount of air and average weight of the air in each column differs, with locations at the highest elevation having the least air in the columns, and consequently, the lowest pressure.

is 5280 ft above sea level and the pressure there is 12.3 lbs/in^2. At Leadville, Colorado, which is 10,430 ft above sea level, the pressure is 9.9 lbs/in^2. In Alaska, at the top of Mt. McKinley (20,320 ft), the pressure is 6.5 lbs/in^2. At 30,000 ft, the altitude that commercial aircraft fly, the air pressure is only 4.5 lbs/in^2. At this altitude, you would have to take three breaths to receive the same amount of oxygen that one breath would supply at sea level. The English units of lbs/in^2 are rarely used in meteorology. Equivalent metric units are $dynes/centimeter^2$ and $newtons/meter^2$. Meteorologists use a special unit called a ***millibar***, which we will abbreviate *mb*. A millibar is equivalent to 1000 $dynes/centimeter^2$.

Average atmospheric pressure at sea level is 1013.25 mb. Pressure decreases rapidly with height, as shown in Figure 1.5. Pressure decreases to about half its surface value (500 mb) approximately 5.5 km (18,000 ft) above sea level because about half of the air molecules are above this level and half are below. The pressure at the tropopause, approximately 10 km (6 miles) above sea level, is about 250 mb—a quarter of its surface value.

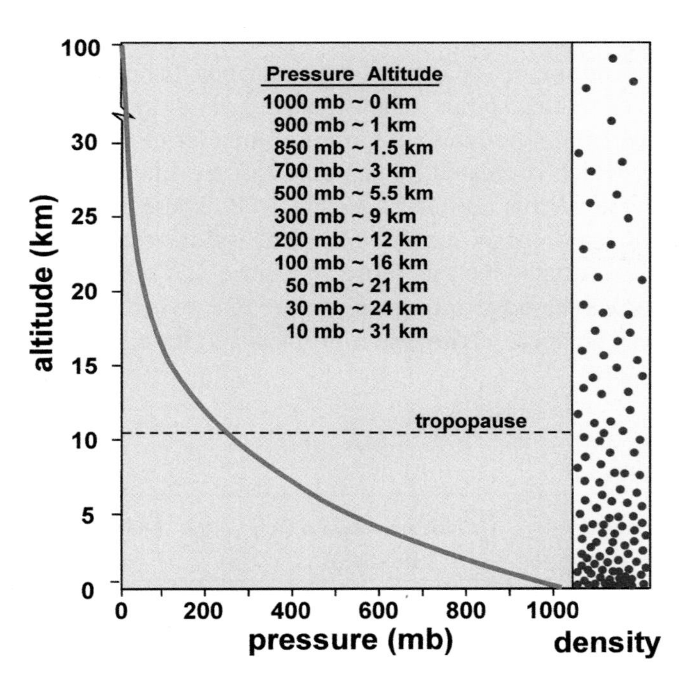

FIGURE 1.5: Average variation of pressure with altitude in the atmosphere. The dot pattern on the right side of the figure represents the air's density, which decreases with height.

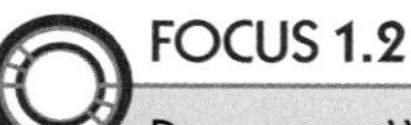

FOCUS 1.2

Dangerous Weather Is Always Nearby

The weather forecast for today is for a temperature of –46° C (–50° F) and 100 knot winds. There will be so little oxygen in the air that suffocation is likely in two to three minutes. Does this sound like a miserable day? On many days, this is an accurate forecast for a location only 10 kilometers (6 miles) from your home. Getting there won't be easy—you have to go straight up. The next time you take a commercial airline flight, think of these conditions—they are right outside the window!

Pressure is measured with instruments called ***barometers***. One of the simplest barometers is a mercury barometer, named because it uses the liquid metal mercury. Figure 1.6 shows a mercury barometer. The dish at the bottom is open so that air can push down on the surface of the mercury. A vacuum exists in the upright tube. Air applying pressure on the surface of the mercury pushes the mercury up into the tube. The higher the pressure, the stronger the push and the higher the column of mercury rises. As atmospheric pressure fluctuates from day to day, the height of the column changes. When the pressure equals the mean sea-level pressure (1013.25 mb), the column height is exactly 29.92 inches. Meteorologists, particularly those reporting in the media, sometimes use the pressure measurement "inches of mercury." Table 1.1 shows some common and extreme values of surface pressure observed on Earth in different pressure units. Although mercury barometers are not in common use today, pressure is often reported in these units to the public.

Pressure measurements are made worldwide every hour. A problem with depicting pressure measurements on a weather chart is that pressures vary slowly in the horizontal but very rapidly with altitude. As a result, a map of station pressures (the actual measured pressures) will look much like a map of the topography. For example, the pressure in Denver, Colorado would always be lower than the pressure in San Francisco, California because Denver is much higher in elevation. To observe the horizontal distribution of

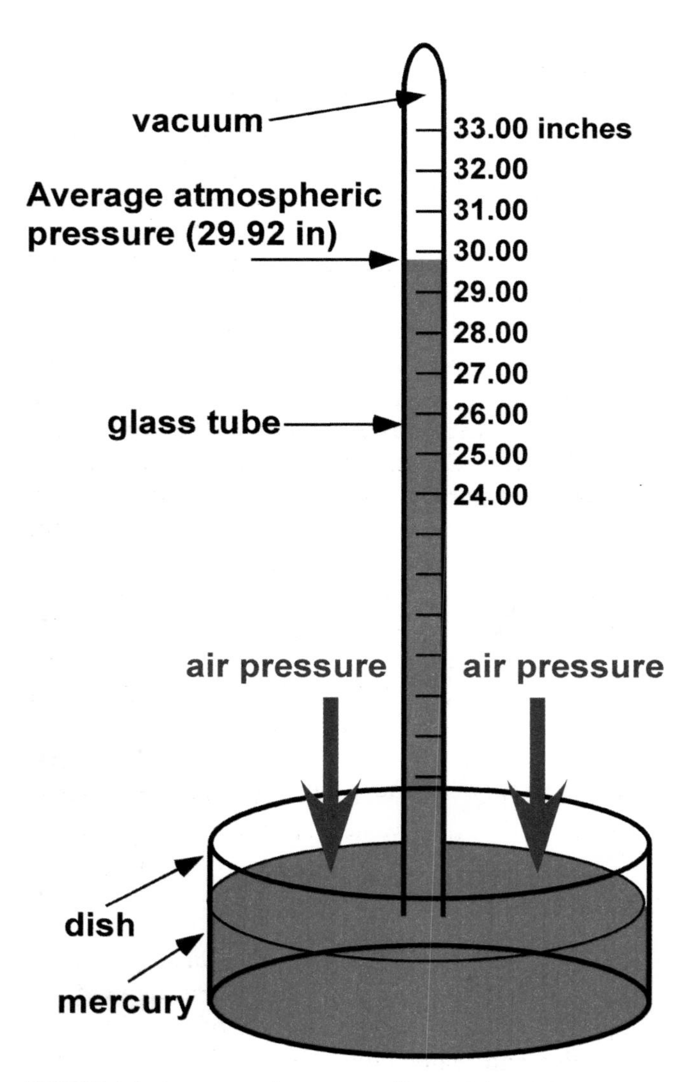

FIGURE 1.6: A mercury barometer. The height of the mercury column is proportional to atmospheric pressure. Units of pressure are given in inches of mercury.

pressure, meteorologists have to convert station pressure to a common altitude. Normally, this altitude is chosen as average sea level, or ***mean sea level.*** Once this conversion is made, meteorologists can then see how pressure varies over a region. Surface maps depicting pressure distributions use ***mean sea-level pressure.***

CHECK YOUR UNDERSTANDING 1.2

1. How is atmospheric pressure defined?
2. What units of pressure do meteorologists use?
3. What is the value of average sea-level pressure?
4. How does a mercury barometer measure atmospheric pressure?

MOISTURE

When we view a photograph of the Earth, such as the Apollo 17 photograph in Figure 1.7A, the most striking feature is Earth's clouds. Earth's storms appear as elegant swirls of clouds that curl upon themselves in ever tighter spirals. The organized cloud systems we see from space are composed of smaller clouds, each of which is created by an updraft of air, like the thunderstorm that appears in Figure 1.7B. These clouds, in turn, are composed of individual water droplets and ice crystals, such as those in Figure 1.7C. When we look at clouds from space, we are really seeing countless numbers of tiny water droplets and ice crystals.

TABLE 1.1 Range of Sea Level Pressures Observed on Earth

	Inches of Mercury (in. HG)	Pounds per Square Inch (lbs/in^2)	Millibars (mb)
Highest recorded sea-level pressure (2001*)	32.06	15.8	1086
Strong high-pressure system	30.86	15.2	1045
Average sea-level pressure	29.92	14.7	1013
Deep low-pressure system	28.94	14.2	980
Lowest recorded sea-level pressure (1979#)	25.69	12.6	870

*Tosontsengel, Khövsgöl Province, Mongolia
#Typhoon Tip, Western Pacific

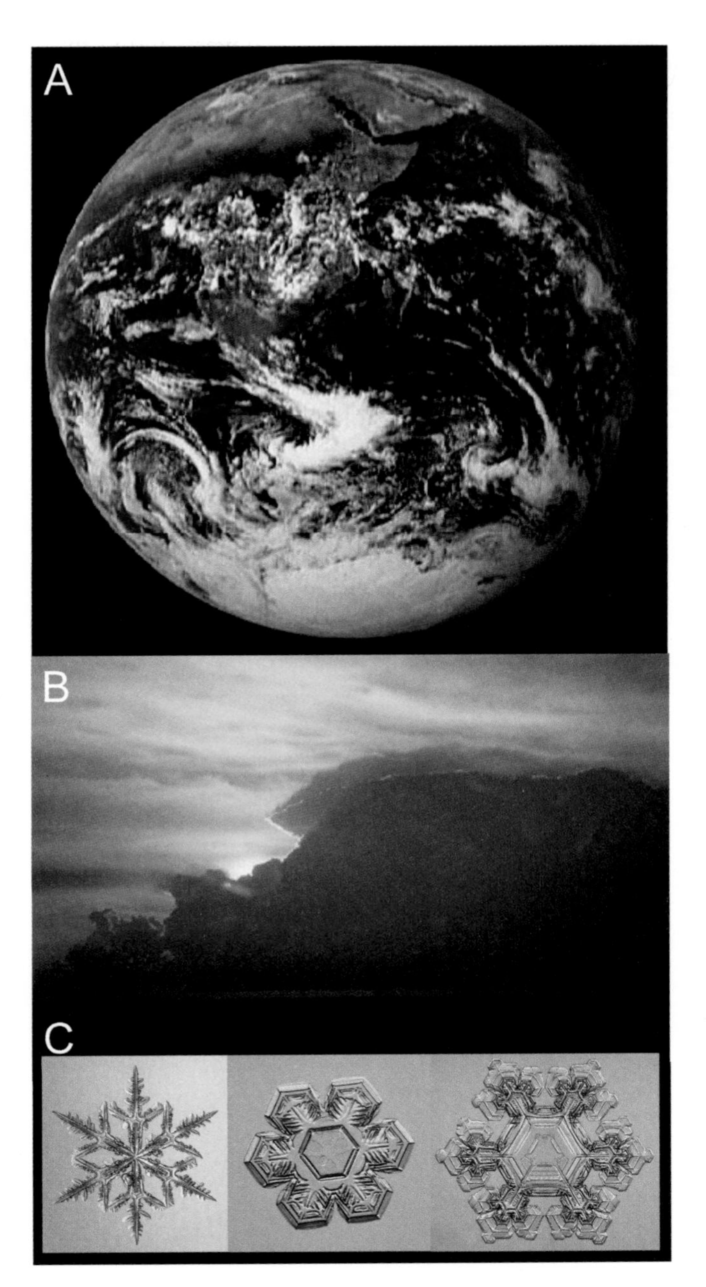

FIGURE 1.7: The storms visible on the photograph of Earth occur across the globe in organized circulations (A). These storms are composed of clouds like the thunderstorm in panel B, which are, in turn, made up of water droplets and ice crystals such as those in panel C.

Moisture Variables

Clouds form from ***water vapor***. Water vapor is an invisible gas composed of individual water molecules. In the atmosphere, water vapor is mixed with other gases, such as nitrogen and oxygen. Recall that pressure is the force applied by air molecules striking a unit area. Some of the molecules in the air are water vapor molecules. The force per unit area applied by only the water vapor molecules is the ***vapor pressure***. The vapor pressure is a measure of the absolute amount of moisture in the air. The vapor pressure at normal atmospheric temperatures ranges from near zero in a dry cold atmosphere to about 60 mb in a humid tropical atmosphere. Figures 1.8A and 1.8B shows the average variation of vapor pressure across the United States in January and July. Note the low value of vapor pressure in the summer in the western desert areas and the high values along the Gulf Coast. Note also that in winter the atmosphere in north-central areas of the United States contains only about a quarter of the moisture that the deserts do in summer. How can this be?

Imagine that we continually add water vapor into a small volume of the atmosphere (you can do this experiment by taking a hot shower in the bathroom with the door closed). What will happen? After some time, a cloud or fog will form. The atmosphere becomes ***saturated*** as the invisible vapor condenses into visible cloud droplets. When the atmosphere cannot contain any more water vapor without condensing into cloud droplets, we say that the atmosphere is saturated. The vapor pressure at which the atmosphere becomes saturated is called the ***saturation vapor pressure***. The atmosphere's capacity for water vapor, and therefore its saturation vapor pressure, depends on temperature. Think of the motion of individual air molecules and water molecules. At higher temperatures, the molecules move faster. It is harder for individual water molecules to coagulate into droplets when they are moving faster and enduring more and stronger collisions with their neighboring molecules.

The relationship between the saturation vapor pressure and temperature is shown in Figure 1.9. We can see from the table and the graph that the atmosphere has little capacity for moisture when the temperatures are very cold. The atmosphere can contain eighty-four times as much moisture at 30° C as it can at –30° C.

Humans are very sensitive to the amount of moisture in the air because our bodies use ***evaporation*** of perspiration to cool. When air is near ***saturation***, our bodies cool inefficiently because water has difficulty evaporating. This is why summer temperatures near the humid Gulf of Mexico Coast can seem suffocating, while the same temperatures in the dry western United States feel pleasantly warm.

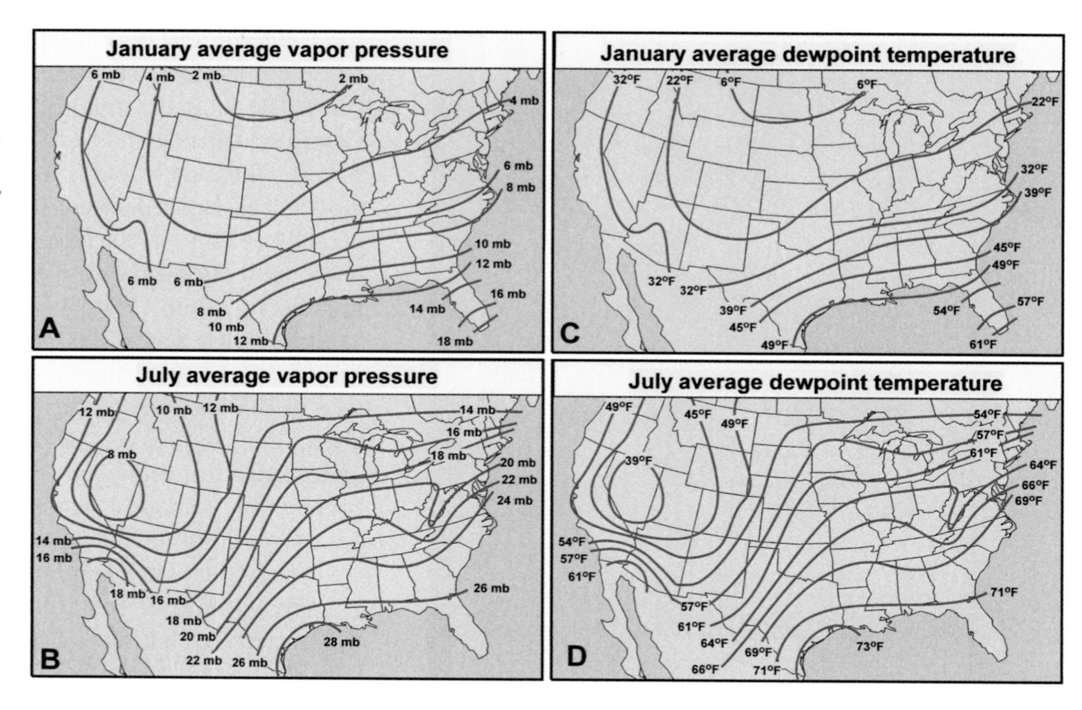

FIGURE 1.8: Average vapor pressure (mb) across the United States in (A) January and (B) July. Average dewpoint temperature (°F) in (C) January and (D) July. Note the correspondence between Panels A and C, and Panels B and D, illustrating that vapor pressure and dewpoint temperature are both measures of the absolute amount of moisture in the air.

Since humans are sensitive to how close air is to saturation, meteorologists frequently use a quantity called the ***relative humidity*** to describe the atmospheric moisture. The relative humidity—the amount of water vapor in the atmosphere relative to the atmosphere's capacity for moisture at a given temperature—is defined as:

Relative humidity = (vapor pressure/saturation vapor pressure) × 100%.

The relative humidity depends on two quantities: the absolute amount of moisture in the air (the vapor pressure) and the amount of moisture the air can contain when it is saturated (the saturation vapor pressure). The saturation vapor pressure is determined by the temperature, as we saw in Figure 1.9. When air is saturated, the relative humidity equals 100 percent.

Let us imagine a location where the atmosphere's vapor pressure was constant for an entire day. What would happen as the atmosphere heats during the day and cools during the night? Figure 1.10 shows the evolution of the temperature and relative humidity over a 24-hour period. As the temperature goes down, the relative humidity goes up and vice-versa. In this case, the change in the relative humidity is due solely to the change in the saturation vapor pressure, which varies with the temperature. The amount of water vapor in the air remains essentially constant and therefore the vapor pressure is constant as well.

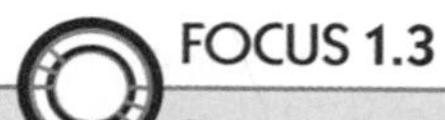

FOCUS 1.3

Extreme Indoor Relative Humidity in Winter

In winter, cold air from outside enters our homes and is heated to a comfortable temperature. Although the amount of water vapor in the air (the vapor pressure) remains the same, the relative humidity of the air changes dramatically as it is heated. This is because the capacity of the indoor atmosphere for water vapor increases as the air is warmed. For example, consider a bitter-cold, foggy winter day when the outside temperature is –20° C (–4° F) and the relative humidity is 100%. From the chart in Figure 1.9, the saturation vapor pressure would be 1.3 mb. Because the air is saturated, the vapor pressure would also be 1.3 mb. Let's assume that the air enters a house and is heated to 20° C (68° F). Now the saturation vapor pressure is 23.4 mb. The amount of moisture in the air did not change, so the vapor pressure remains 1.3 mb. The relative humidity of the indoor air becomes 1.3 mb/23.4 mb × 100 = 6%. Values of relative humidity this low cause physical discomfort for many people. Humidifiers must be used to combat respiratory problems and cracked skin caused by the dry indoor air of winter.

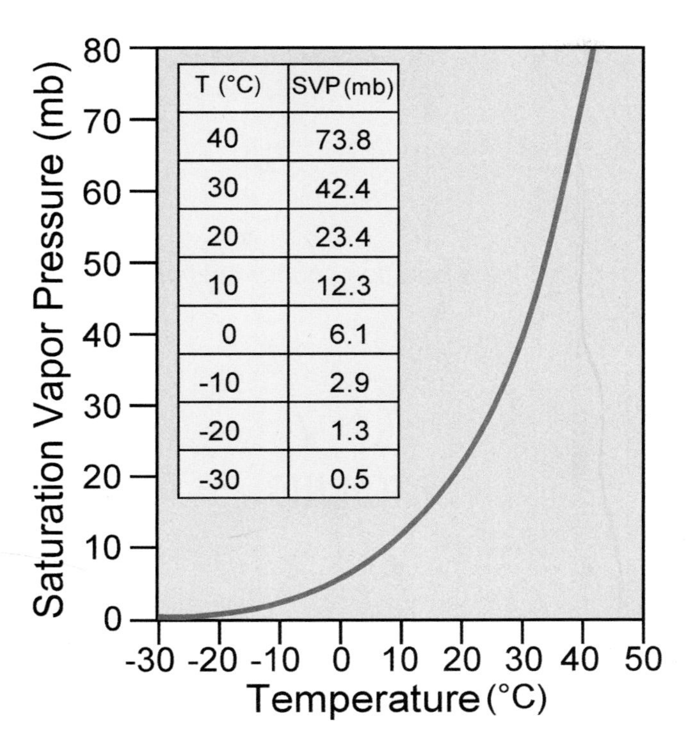

T (°C)	SVP (mb)
40	73.8
30	42.4
20	23.4
10	12.3
0	6.1
-10	2.9
-20	1.3
-30	0.5

FIGURE 1.9: Relationship between saturation vapor pressure (mb) and temperature (°C).

To understand how relative humidity can vary with temperature, let's return to the question concerning why the summer atmosphere in the deserts of the southwestern United States contains more moisture than the winter atmosphere in the north-central plains of the United States, yet the atmosphere in the desert feels dry and the winter atmosphere can feel moist. Suppose that the vapor pressure (the actual moisture content) is 8 mb in the desert and 2 mb in the north-central plains. Let's further suppose that the summertime temperature in the desert is 95° F (35° C) and 32° F (0° C) in the north-central plains in winter. Based on Figure 1.9, the saturation vapor pressure in the desert would be about 60 mb and in the north-central plains only about 6 mb. The relative humidity would be 8 mb/60 mb × 100 = 13 percent in the desert and 2 mb/6 mb × 100 = 33 percent in the north-central plains. Hence, the air over the north-central plains has a higher relative humidity and is closer to saturation. To our bodies, the atmosphere over the north-central plains would feel moist compared to the desert atmosphere, even though the desert atmosphere contains four times as much water vapor. The relative humidity is a measure of the moisture content of the atmosphere *relative to its capacity for water vapor,* which is determined by the temperature.

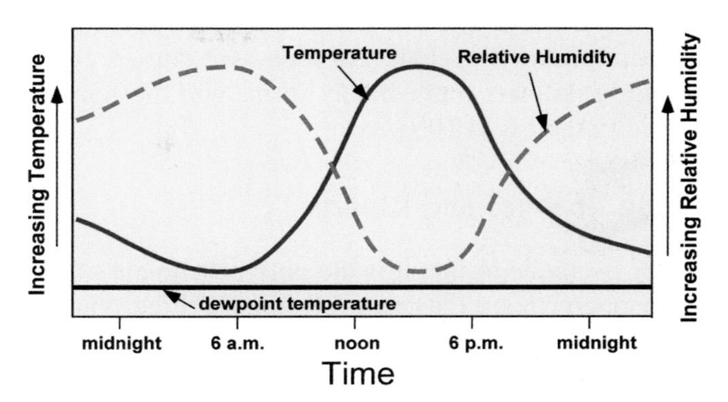

FIGURE 1.10: Daily variation of temperature, dewpoint temperature, and relative humidity on a day when the moisture content of the air (the vapor pressure) does not vary.

While vapor pressure is a good way to describe the actual amount of moisture in the air, it is very difficult to measure. Instead, meteorologists rely on other variables to determine the absolute (as opposed to relative) amount of water vapor in the air. The most common variable used, and the one reported on surface weather maps, is the ***dewpoint temperature.*** The dewpoint temperature is the lowest temperature to which air can be cooled at constant pressure before saturation occurs. To determine the dewpoint temperature, air is cooled until saturation occurs (and dew forms, hence the name) without changing the pressure. Although the dewpoint temperature is a temperature, *it is a measure of moisture content.* This is illustrated in Figures 1.8C and 1.8D, which show the average variation of dewpoint temperature across the United States in January and July. Note the direct correspondence to the vapor pressure in Figures 1.8A and 1.8B.

The relative humidity can be roughly estimated by comparing the dewpoint temperature and the actual temperature. For example, suppose that the dewpoint temperature is 69° F and the temperature is 70° F. This means that air only has to cool 1° F for saturation to occur. Thus, the relative humidity must be high (the air is close to saturation). In contrast, suppose the dewpoint is 40° F. Now the air at 70° F would have to cool 30° F to reach saturation. This means that the air is quite dry and its relative humidity is low. Returning to Figure 1.10, we see that as air cools and heats, the dewpoint temperature is constant although the relative humidity changes. By comparing the dewpoint temperature and the temperature, we can qualitatively estimate the relative humidity. When the dewpoint and the temperature are nearly the same, the relative humidity is high. When they are far apart, the relative humidity is low. The dewpoint can never exceed the temperature for very long. As soon as it does, a cloud of liquid droplets immediately forms and the relative humidity returns to 100 percent.

Phase Changes and Clouds

Water is unique in that it is the only constituent of the atmosphere that exists naturally in all three phases. Air contains water vapor, and clouds in the air are composed of liquid and ice particles. ***Phase changes***, from vapor to water, water to ice, vapor to ice, ice to vapor, etc., occur all the time in the atmosphere as water and ice clouds form and dissipate. Specific terminology is used to describe these phase changes. Figure 1.11 shows all the phase changes that can occur in the atmosphere.

Phase changes of water are critical to the development of storms. The primary reason is that energy is either required or liberated during phase changes. To understand this, consider a simple thought experiment. Take a full pot of water out of the refrigerator (let the temperature of the water be 0° C) and put it on the stove on a cherry red burner. How long does it take for the pot of water to begin to boil? Typically, it will boil in about ten minutes. How long does it take the full pot of water to undergo the phase change and turn from liquid water in the pot to water vapor in the air? Depending on the pot size, it might take hours. Think of all the energy that went into converting the water in the pot into vapor in the air. Where did all that energy go? The energy went to accelerating the molecules to high speeds characteristic of vapor, and to break the strong bonds that link individual water molecules to each other. The energy is called ***latent heat*** because it is the "hidden heat" required for a phase change.

Let's continue our thought experiment. Imagine now that, by some miracle, all of the vapor molecules fly back into the pot and become liquid again. What happens to all the energy they had? The energy actually would reappear as heat (and probably explode the kitchen off the house!). This experiment may seem silly, but it occurs in the atmosphere all the time. Water is evaporated off the oceans, lakes, and other water sources as solar energy heats the water. When clouds form, the water vapor condenses to liquid. At that time all the latent heat required to evaporate the water reappears as heat and warms the air in which the cloud is forming. Heat is also required for other phase changes, such as ***melting*** of ice, and released during the reverse process, the ***freezing*** of water.

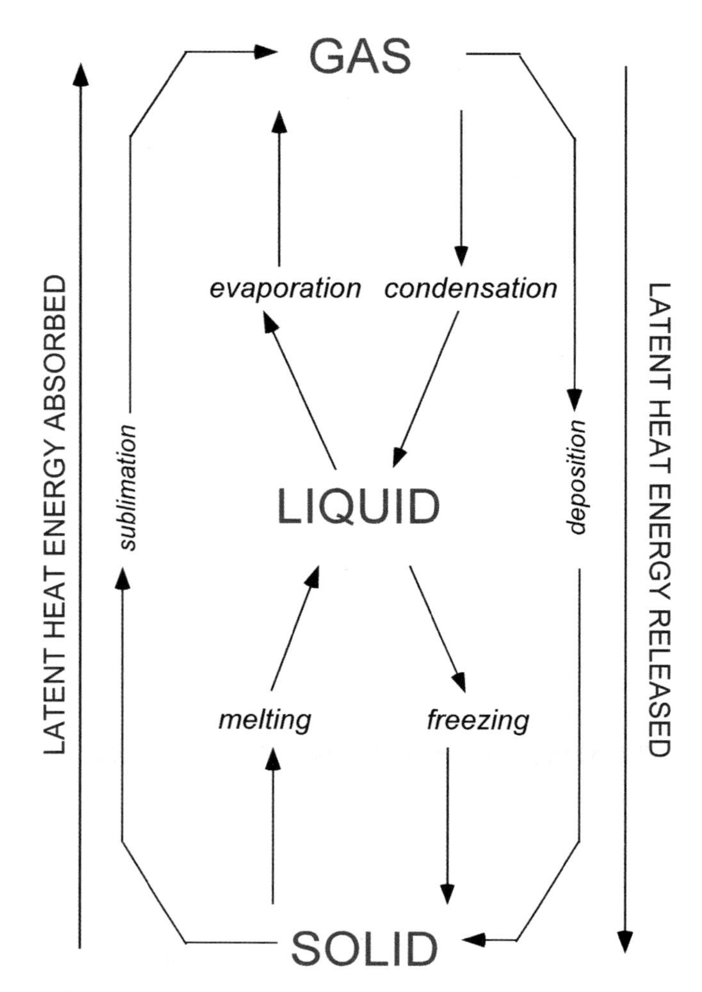

FIGURE 1.11: Phase changes of water. Energy is released into the atmosphere when water vapor becomes a liquid through *condensation* or solid through *deposition*, or water becomes ice by *freezing*. Energy is absorbed from the atmosphere when ice undergoes *melting* or becomes vapor directly through *sublimation*, or when water undergoes *evaporation*.

Latent heat release is vital to the creation of thunderstorms. We will examine the role of latent heat in storm formation in Chapter 6.

Clouds are composed of ice particles, water droplets, or a mixture of both. Some water droplets in clouds can be liquid, but have temperatures well below 0° C. How can this happen? For water to freeze, microscopic impurities that have molecular crystaline structure similar to ice must be present in the water. These impurities, called ***ice nuclei***, are always present in large water bodies (such as the ice cubes in a refrigerator), but are not necessarily present or active in all the tiny droplets that compose clouds. Drops that do not freeze when they are colder than 0° C are ***supercooled***. In many clouds, ice nuclei only become active when the temperature cools below 5° F to 14° F (–15° C to –10° C), so supercooled clouds are most common when the cloud top temperatures are warmer than 5° F (–15° C). Supercooled water is important in the formation of aircraft icing and freezing rain. When ice particles falling through clouds encounter supercooled droplets, the droplets freeze instantly to the ice upon contact. If enough droplets are collected this way, an ice particle can grow to appreciable size and becomes a small (< 3 to 4 mm), soft ball of ice, called ***graupel***. Continued collection of supercooled water will lead to further growth and the formation of ***hail***. Ice crystals can also collect other ice crystals and become ***snowflakes***. At temperatures above freezing, the snow, graupel, and hail melt to form raindrops.

ONLINE 1.1
Clouds

There are many types of clouds. Clouds are classified by their altitude and shape and whether or not they produce precipitation. Clouds are called *cumulus* if they are associated with strong updrafts and are towering, with cauliflower-like lobes. In general, the horizontal and vertical dimensions of these clouds are comparable. If they grow to thunderstorms and produce precipitation, they are called *cumulonimbus* ("nimbus" is the Latin word for rain). Clouds that are layered and widespread are *stratus* clouds. The width of these clouds is much greater than the depth. If precipitation falls from these clouds, they are called *nimbostratus*. High wispy clouds are *cirrus* clouds. This terminology originated with a pharmacist named Luke Howard in 1803. He identified four broad categories: *cumulus* (clouds with vertical development), *stratus* (layer clouds), *cirrus* (high, fibrous clouds), and *nimbus* (precipitating clouds). High (> 6 km (4 miles)) clouds are called *cirrostratus* if they are layered and *cirrocumulus* if they are puffy. Middle clouds (~2 to 6 km (1 to 4 miles)) with similar shapes are called *altostratus* and *altocumulus*. Numerous secondary categories are also used. For example, clouds forming at the crest of waves generated by airflow over mountains (see Chapter 17) are called *altocumulus lenticularis*, because they are often lens-shaped. Photographs of clouds and their classifications appear in Chapter 1 Online.

CHECK YOUR UNDERSTANDING 1.3

1. What does it mean when we say air is saturated?
2. What is the relative humidity?
3. What is the dewpoint temperature?
4. What is the value of the relative humidity if the dewpoint temperature equals the air temperature?
5. What is latent heat?

WIND

Wind is simply the movement of air. To specify the wind, we must refer to two quantities: the wind direction and its speed. Official measurements of wind at surface stations are made at an elevation of 10 meters (~33 feet), which is referred to as the ***anemometer height***. An ***anemometer*** is a device to measure wind speed. Wind direction is measured using north-east-south-west coordinates and is expressed in degrees (Figure 1.12A). *The wind direction, by convention, is the direction from which the wind is blowing.* The easy way to remember this is to think of a north wind. When "a north wind is blowing" the day is likely to be cold, since a north wind is blowing *from the north,* bringing cold air southward. A south wind is blowing *from the south,* bringing warm air northward. The wind direction can also be expressed in degrees. A north wind is from 0 degrees, an east wind from 90 degrees, a south wind from 180 degrees and a west wind from 270 degrees. Meteorologists typically measure wind speed in knots. The exact conversion between knots and miles/hour is 1 knot = 1.15 mph. In metric units, winds are normally measured in meters/second (1 knot = 0.51 m/s).

Meteorologists depict wind speed and direction using ***wind barbs*** on a staff (see Figure 1.12B). The coordinate system used for wind direction, shown in Figure 1.12A, conventionally has north as up, south as down, east to the right, and west to the left. Using this coordinate system, if you move from the tail to the head of the staff, you are moving in the direction the wind is blowing. The barbs on the end of the staff indicate the wind speed. A half barb is 5 knots, a whole barb is 10 knots and a pennant is 50 knots. Combinations of these three are used for other wind

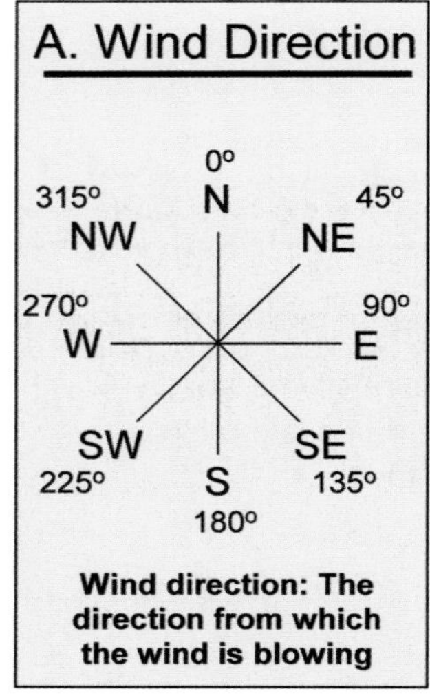

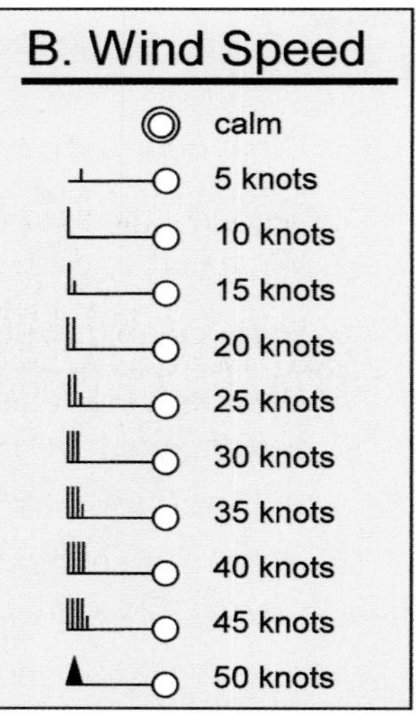

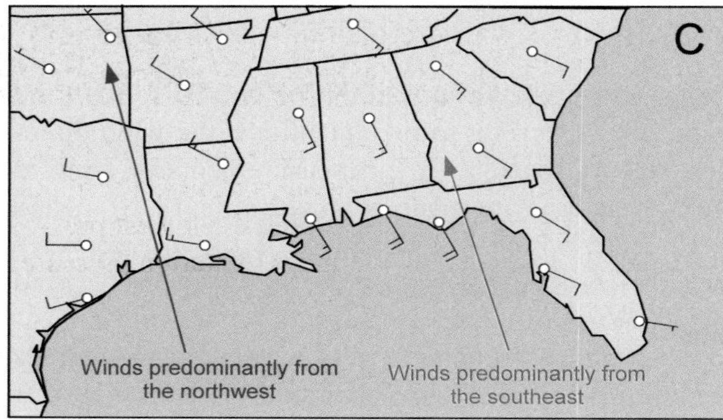

FIGURE 1.12: Conventions for plotting wind direction and wind speed. Wind direction is measured in degrees from north using the coordinate system in Panel A. In Panel B, wind direction is indicated by a staff that points toward the station in the direction the wind is blowing (Panel B). Barbs on the end of the staff indicate the wind speed, with a half-barb representing 5 knots, a whole barb 10 knots, and a pennant 50 knots. Panel C shows examples of winds plotted at several cities in the southeastern United States.

speeds. The map of the southeastern United States in Figure 1.12C depicts the wind direction and speed at several cities. Cities on the eastern side of the map generally have winds from the south or southeast at 10–20 knots, while cities on the western side have winds from the west or northwest at 10 to 15 knots.

Wind circulations in the atmosphere occur on many scales. Winds continually orbit around the Earth in the Earth's ***general circulation***. Embedded within the general circulation in the mid-latitudes are ***extratropical cyclones***. These large storms, with diameters up to 3000 km (~1800 miles), can cover an area as large as about a third of the contiguous United States, and are responsible for many types of severe weather. In the tropics, ***hurricanes*** are much smaller (~300 km/~180 miles diameter), but capable of terrible destruction as they move onto shore. ***Thunderstorms*** occur on an even smaller scale (~30 km/18 miles or less), and ***tornadoes*** spawned by thunderstorms rarely are larger than 0.5 km (0.3 miles) in diameter. We will study all of these wind circulations as we examine specific weather phenomena in later chapters.

ONLINE 1.2

Scales of Atmospheric Motions

Atmospheric circulations are hard to visualize with still photographs. In Online 1.2, animated satellite images allow us to see the motions of storms on a variety of scales. The animations are composite images taken between 0000 UTC 28 May 2007 and 1200 UTC 31 May 2007 from geostationary satellites. The satellites are located at fixed points directly above the Earth's equator: one is over the Pacific Ocean and the other is over South America. The view is the North American continent and adjacent areas of the Pacific and Atlantic Oceans. On these images, high clouds such as cirrus clouds and the tops of deep thunderstorms appear white. Middle clouds such as altostratus clouds appear light gray. Low clouds such as the stratocumulus over the tropical oceans appear dark gray. The ground and ocean appear very dark.

As you run the animation, notice first Earth's general circulation. The high clouds of the middle and high latitudes all have an eastward drift. They are within the belt of westerly winds that continually encircles the planet. Now look at the clouds over the tropical oceans. These clouds are moving westward in a belt of low-level winds called the trade winds. These winds consistently blow westward in the Northern Hemisphere and are also part of Earth's general circulation. Returning to the middle and higher latitudes, we can see large swirling storms, called extratropical cyclones, develop and decay within the westerlies. Extratropical cyclones are the parent storms of many types of hazardous weather, ranging from tornadic thunderstorms to blizzards. Tropical Depression Alvin and Tropical Storm Barbara are located in the eastern

(continued)

(continued)

Pacific just south of Mexico. Tropical cyclones such as Alvin and Barbara are normally much smaller in area than extratropical cyclones, but can have much stronger winds and lower central pressures. Thunderstorms appear in many locations on the animation, particularly over Mexico. Thunderstorms are much smaller than cyclones. Nevertheless, these storms can produce extremely destructive weather, including tornadoes, strong straight-line winds, lightning, hail, and flash floods.

CHECK YOUR UNDERSTANDING 1.4

1. What is wind?
2. What is the convention for reporting wind direction?
3. In what units are wind speed reported?
4. What do wind barbs represent on a weather map?

TEST YOUR UNDERSTANDING

1. Over what altitude range does most of Earth's weather occur?
2. In molecular terms, what is the difference between a solid, a liquid, and a gas?
3. What are the basic properties of the atmosphere used in describing weather?
4. Describe how temperature can be used to define layers of the atmosphere.
5. Sketch the vertical temperature profile of the entire atmosphere. The x-axis should be temperature and y-axis altitude. Label the four layers of the atmosphere and the tropopause.
6. Why does temperature typically increase with height in the stratosphere?
7. How does the height of the tropopause change with latitude? What implication does this have for the depth of storms?
8. What is the pressure in a strong high-pressure system? What about a deep low-pressure system? (Report your answer in mb.)
9. As you move upward away from the Earth's surface, does the pressure increase, decrease, or remain constant? Explain your answer.
10. If a map shows station pressures (the actual pressure measured by an observer) across the contiguous United States, where would the lowest pressures be found? Why are maps of station pressure not very useful for meteorologists?
11. What is meant by "sea-level pressure"?
12. Why do your ears pop when you ride up an elevator in a skyscraper?
13. Clouds are composed of what types of particles?
14. Define vapor pressure, saturation vapor pressure, and saturation.
15. What atmospheric variable determines the maximum amount of moisture that can be present in air?
16. What two variables describe the absolute amount of moisture in the air?
17. How does relative humidity typically change over the course of a day?
18. Describe two ways that the relative humidity of air can be increased.

19. Qualitatively, what is the relative humidity if the dewpoint temperature and the air temperature differ substantially? What if the temperature and the dewpoint temperature are close in value?

20. Latent heat is released to the air during which phase changes? During which phase changes is latent heat removed from the air?

21. Identify several storm types that represent different scales of wind circulations.

22. Name the four primary cloud categories.

TEST YOUR PROBLEM-SOLVING SKILLS

1. On a very hot day you decide to relax with a friend and enjoy a fruity cup of mango juice. You have ice in the cup and also a thermometer-swizzle stick. As the mango juice cup cools (because of the ice), moisture suddenly starts condensing on the outside of the cup. The mango juice cup temperature is 20° C when this happens. The air temperature is 30° C.
   (a) What is the dewpoint temperature?
   (b) Estimate the vapor pressure in the atmosphere. (Hint: use Figure 1.9.)
   (c) Estimate the relative humidity in percent.
   (d) Why did moisture condense on the outside of the cup?

2. Four cities report the following temperatures and dewpoint temperatures:

City	Temperature	Dewpoint Temperature
Newark, New Jersey	59	59
Tucson, Arizona	86	50
Miami, Florida	77	68
St. Louis, Missouri	77	59

   (a) Which city has the highest vapor pressure?
   (b) Which city has the lowest vapor pressure?
   (c) Which city has the highest saturation vapor pressure?
   (d) Estimate the relative humidity of each of the four cities. (Hint: use Figure 1.9.)

3. You and your friend decide to go on a weekend road trip to the nearby mountains. In your car is a barometer that reads 1003 mb as you leave home. Later, you arrive in the mountains and make the long, winding drive up to Summit Lodge, a motel known for its great view of the valley below. You notice that the barometer reads 960 mb. Your friend says, "That's an extremely low pressure, which means a strong storm must be coming. We better head back home right now." How would you reason with your friend in order to save the weekend?

4. Occasionally in the summer in the agricultural regions of the Midwest, the dewpoint temperature climbs to the low 80s. On one particular July day the dewpoint temperature in Des Moines, Iowa, was 85° F.
   (a) Estimate the vapor pressure (mb) at the time the dewpoint was recorded.
   (b) How does this amount of moisture compare to the average amount in the air in Iowa during July? What about during January?
   (c) If the actual air temperature was 90° F, what was the relative humidity?
   (d) Automobile air conditioners running in these conditions were dripping liquid water. Those same air conditioners running in September produced no liquid water. How do you explain the difference?

USE THE SEVERE AND HAZARDOUS WEATHER WEBSITE

http://severewx.atmos.uiuc.edu

1. Use the *Severe and Hazardous Weather Website* to obtain the current weather conditions in the following cities: Chicago, Illinois; Little Rock, Arkansas; San Francisco, California; Bangor, Maine; Los Angeles, California; and New Orleans, Louisiana.
   (a) What is the surface temperature in Chicago, Illinois, and Little Rock, Arkansas?
   (b) What is the surface dewpoint temperature in San Francisco, California, and in Bangor, Maine?
   (c) Does Los Angeles, California, or New Orleans, Louisiana, have a higher relative humidity? Why? Which city has a higher saturation vapor pressure? Why?

2. Use the *Severe and Hazardous Weather Website* to examine clouds via satellite imagery. Find a satellite animation that shows clouds present in the United States.
   (a) Note the date and time of the satellite image. Describe the clouds present across the United States in terms of their spatial scale and any particular shapes or structures you can identify.
   (b) Animate the images. Do the clouds display any rotation? If so, in what direction? Is the mass of clouds moving? If so, in what direction?
   (c) What other notable observations can you make about the present cloud structures?

CHAPTER TWO

METEOROLOGICAL MEASUREMENTS

The National Weather Service Doppler radar at Lincoln, Illinois. Rawinsondes are launched from the building on the left.

KEY TERMS

Aircraft Meteorological Data Reporting (AMDAR)
Automated Surface Observing System (ASOS)
Automated Weather Observing System (AWOS)
Canadian Lightning Detection Network (CLDN)
clear air mode
Doppler radar
Doppler shift
geostationary orbit
Greenwich Mean Time (GMT)
hodograph
infrared channel
inversion
low Earth orbit
meteogram
Meteorological Aviation Report (METAR)
multiple Doppler radar
National Lightning Detection Network (NLDN)
phased array antenna
precipitation mode
radar
radar echo
radar reflectivity (or radar reflectivity factor)
radial velocity
Radio Acoustic Sounding System (RASS)
radio occulation
rawinsonde
satellite
Skew-T/Log P diagram
sounding
storm relative radial velocity
Stuve diagram
sun-synchronous

synoptic
synoptic meteorology
total accumulated rainfall
Universal Coordinated Time (UTC)
vertical wind shear
visible channel
water vapor channel
wind profiler
WSR-88D (Weather Service Radar 1988-Doppler)
Zulu Time (Z)

Meteorologists understand and predict severe and hazardous weather using two scientific approaches: analysis of measurements and numerical modeling of storm behavior. Numerical modeling is described in Chapter 4. This chapter focuses on basic measurements of weather properties and techniques and instrumentation to gather data. Modern instrumentation, such as Doppler radar and satellites, are emphasized because of their importance in detecting and tracking hazardous weather systems.

Atmospheric properties, such as pressure, temperature, dewpoint temperature, wind direction, and wind speed, are measured routinely around the globe. To examine large weather systems and construct weather maps that represent the state of the atmosphere at a given instant in time, these measurements must be made simultaneously. Meteorologists call these measurements ***synoptic*** and studies using these measurements ***synoptic meteorology***. Everyone in the world must use a common time to make measurements simultaneously. The international standard adopted for measurements is ***Universal Coordinated Time*** (UTC). Universal Coordinated Time is also called ***Greenwich Mean Time*** (GMT), because it is the local standard time at the Greenwich Laboratory in London, England, and ***Zulu Time*** (Z), because Zulu, the radio communication code word for the letter Z, is always used following radio time transmissions.

Figure 2.1 shows the relationship between UTC and local time zones in the contiguous United States, southern Canada, and northern Mexico. Standard time in the Eastern Time Zone is UTC minus 5 hours, and in the Pacific Time Zone is UTC minus 8 hours. Most of Alaska is in the Alaskan Time Zone (UTC –9 hours), although St. Lawrence Island and a portion of the Aleutian Islands are in the Hawaii Standard Time Zone (UTC –10 hours). For example, at 7 A.M. in New York City, it is 4 A.M. in San Francisco, and 1200 UTC in both cities. At 7 P.M. on 14 February in New York City it is 4 P.M. 14 February in San Francisco and 0000 UTC 15 February in both cities. Time in UTC is expressed on a 24-hour clock, so 9 A.M. in February in New York City would be 1400 UTC. During daylight saving time, the Eastern Time Zone is UTC –4 hours and the Pacific Time Zone UTC –7 hours. Hawaii and Arizona do not observe daylight saving time.

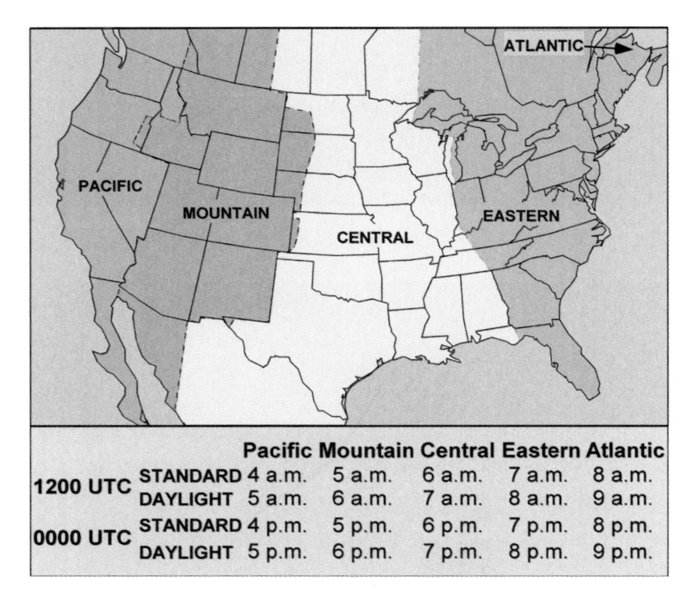

		Pacific	Mountain	Central	Eastern	Atlantic
1200 UTC	STANDARD	4 a.m.	5 a.m.	6 a.m.	7 a.m.	8 a.m.
	DAYLIGHT	5 a.m.	6 a.m.	7 a.m.	8 a.m.	9 a.m.
0000 UTC	STANDARD	4 p.m.	5 p.m.	6 p.m.	7 p.m.	8 p.m.
	DAYLIGHT	5 p.m.	6 p.m.	7 p.m.	8 p.m.	9 p.m.

FIGURE 2.1: Conversion between Universal Coordinated Time (UTC) and local time for time zones in the contiguous United States, northern Mexico, and southern Canada.

FOCUS 2.1

Historical Perspective on Technology in Meteorology

Our ability to predict and to understand hazardous weather has been intertwined with advances in technology. Although the Greeks and other ancient peoples put forth ideas about the atmosphere, the absence of instruments prevented the quantitative methods that are at the core of modern science. Until the invention of the first thermometer around 1600, it was not even possible to document day-to-day changes or to compare levels of warmth at different locations. The invention of the barometer by

(continued)

(continued)

Torricelli in the 1640s permitted the detection of changes in atmospheric pressure, a first step toward an awareness that stormy weather is generally associated with lower pressure.

With the development of the anemometer for wind measurements in the 1660s, and the invention of the hygrometer for the measurement of humidity in the late 1700s, the primary surface weather variables could be measured. However, comparisons of current weather information at different locations required that information be transmitted faster than was possible in the days of horseback travel. Ben Franklin's famous deduction that a low-pressure center traveled from Philadelphia to Boston was made only after the several days it took for the needed information to reach him from several hundred miles away. The invention of the telegraph in 1854 made possible the instantaneous transmission of weather information and the construction of the first surface weather maps.

Because the atmosphere is three-dimensional, meteorologists require far more than surface maps. Upper-air information did not become routinely available until the 1930s, when rawinsonde launches were first made at regular time intervals. After radar was declassified in the mid-1940s following World War II, it came into use for mapping precipitation.

The invention and rapid development of electronic computers in the 1950s spurred the development of numerical models for weather prediction. By the early 1960s, computer-based models were providing useful information to weather forecasters. Also during the 1960s, satellites began to transmit images of clouds and other weather information, providing the first truly worldwide coverage of even the data-sparse oceans and polar regions.

The technological revolution of meteorology continues to accelerate. New satellites with advanced measurement capabilities, the widespread use of Doppler radars, wind profilers, lightning detectors, and other sensors have all become part of the operational suite of observing systems. The proliferation of personal computers and Internet access in the 1990s brought these data to the world. Weather data, maps, and forecasts for all types of severe and hazardous weather are now a click away on your very own laptop.

SURFACE MEASUREMENTS

Automated weather stations now have the capability to make standard measurements of atmospheric properties continuously at most locations in North America, although data are normally reported hourly. Worldwide, nonautomated measurements are made by observers every three hours, at 0000, 0300, 0600, 0900, 1200, 1500, 1800, and 2100 UTC.

The U.S. National Weather Service, Department of Defense, and the Federal Aviation Administration have automated surface measurements of weather conditions in the United States. The U.S. National Weather Service automated stations are called ***Automated Surface Observing Systems***, or ***ASOS***, while Federal Aviation Administration and Department of Defense stations are called ***Automated Weather Observing Systems***, or ***AWOS***. ASOS and AWOS stations work nonstop, updating observations twenty-four hours a day, every day of the year.

The ASOS suite of instruments reports cloud height and amount, visibility, precipitation type, intensity, and accumulation; obstructions to vision such as fog or haze; sea-level pressure, altimeter setting, temperature; dewpoint temperature; and wind direction, speed, and character (gusts, squalls). The ASOS stations also provide selected significant information such as variable cloud height, visibility, precipitation beginning/ending times, rapid pressure changes, pressure tendency, a wind shift, or a peak wind speed.

These data are communicated to the control tower, aircraft, and to the National Weather Service for dissemination to the public. The data are available via the Internet for U.S. stations within a few minutes of collection, and can be easily accessed through this text's *Severe and Hazardous Weather Website*.

FOCUS 2.2

The U.S. National Weather Service

The U.S. National Weather Service (NWS) provides weather, hydrologic, and climate forecasts and warnings for the United States, its territories, and adjacent ocean areas. The NWS is part of the National Oceanic and Atmospheric Administration (NOAA) under the Department of Commerce of the federal government. NWS data and products are an information database and infrastructure used by the government, researchers, private companies, the public, and the global community.

The NWS is headquartered in Silver Spring, MD, and has six major operating centers. Among these six centers is the *National Centers for Environmental Prediction (NCEP),* itself comprised of nine specialized centers that analyze and forecast the global atmosphere and maintain the overall national warning system for severe and hazardous weather (Figure 2A).

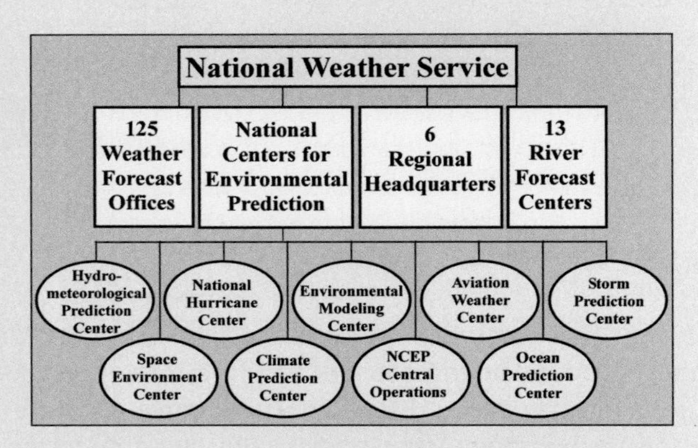

FIGURE 2A: The organization of the National Weather Service, including the nine centers that make up that National Centers for Environmental Prediction (NCEP).

The NCEP includes the *National Hurricane Center,* which is responsible for forecasting the behavior of tropical storms and hurricanes and issuing watches and warnings for these storms; the *Storm Prediction Center,* which monitors and forecasts severe and non-severe thunderstorms, tornadoes, and other hazardous weather phenomena across the Unites States; the *Hydrometeorological Prediction Center,* which issues precipitation forecasts focused particularly on heavy rain, heavy snow, and areas of potential flash flooding; and the *Ocean Prediction Center,* which issues forecasts and warnings for marine interests. The NCEP *Environmental Modeling Center* develops, tests, and operates numerical prediction models that are the basis for weather forecasting (see Chapter 4). The *Aviation Weather Center* identifies existing or imminent weather hazards to the aviation community and creates warnings for transmissions to pilots. In addition to the national centers, the NWS maintains six regional headquarters, six regional climate centers, 125 forecast offices, and thirteen river forecast offices around the country. The forecast offices, in addition to forecast responsibilities, issue severe thunderstorm and tornado warnings. River forecast offices monitor the nation's rivers and issue flood warnings, and climate centers maintain climate databases and monitor and report current climate conditions.

ONLINE 2.1

ASOS Instruments

The ASOS suite of instruments includes a rain sensor, temperature, dewpoint temperature, and pressure sensors; a device to determine precipitation type and amount; a wind vane to measure wind direction; an anemometer to measure wind speed; and devices to measure sky conditions (cloud height and amount), freezing rain, and visibility (Figure 2B). Photographs of each of the instruments in the ASOS suite appear online, along with a link to a National Weather Service site where the operating principles of each instrument are explained.

FIGURE 2B: The Automated Surface Observing System instrument suite.

Figure 2.2 shows the location of surface weather stations in North America. Heavily populated regions of North America, such as the northeast United States, have a high density of stations, while few stations exist in sparsely populated regions such as the western deserts and the Canadian Arctic. Over the ocean, surface data are obtained from ships and weather buoys. Satellites also infer information about surface winds and sea surface temperature.

Surface data from ASOS and AWOS stations are available in a number of different formats. The original data are coded using the ***Meteorological Aviation Report*** format called a ***METAR***. These codes are difficult to read without some training, but are commonly translated on many weather websites to make them accessible to the public. The data are also plotted on ***meteograms;*** graphs that show how several atmospheric properties change with time. Figure 2.3 shows a meteogram from Buffalo, New York. This twenty-four-hour period was part of a several day lake effect storm that left Buffalo buried under four to five feet of snow (see Chapter 13). The meteogram for Buffalo shows two passages of a band of heavy snow over the station, the first time between 0100 and 0400 UTC, and the second between 1200 and 1800 UTC. The number of stars appearing in the "weather" category indicates the intensity of the snow. Each band was accompanied by a shift in the wind direction, a reduction in visibility, a lowering of cloud base, and a decrease in the spread between the temperature and dewpoint temperature. Over 17 inches of snow accumulated during the 24-hour period. The sea level pressure fell during the entire period. Meteorologists use meteograms to analyze the structure of hazardous weather events and track their progress as they move between stations.

RAWINSONDES

The atmosphere extends many kilometers above the Earth's surface. To understand weather systems and predict their future behavior, measurements are required through the depth of the troposphere and well into the stratosphere. ***Rawinsondes*** are the primary instrument packages used to make these measurements. A rawinsonde is a balloon-borne instrumentation system that measures pressure, temperature, dewpoint temperature, wind direction, and speed (Figure 2.4).

Rawinsondes are launched worldwide twice a day at 0000 UTC and 1200 UTC. Figure 2.5 shows the location of rawinsonde launches in North America. The density of the data is sparse compared to the surface measurements, primarily because of cost considerations. The average spacing of rawinsonde launch locations is about 500 km (~200 miles). Rawinsondes are normally launched 50 minutes prior to the standard time (1200, 0000 UTC) so that they

FIGURE 2.2: Locations of surface meteorological observation stations in North America.

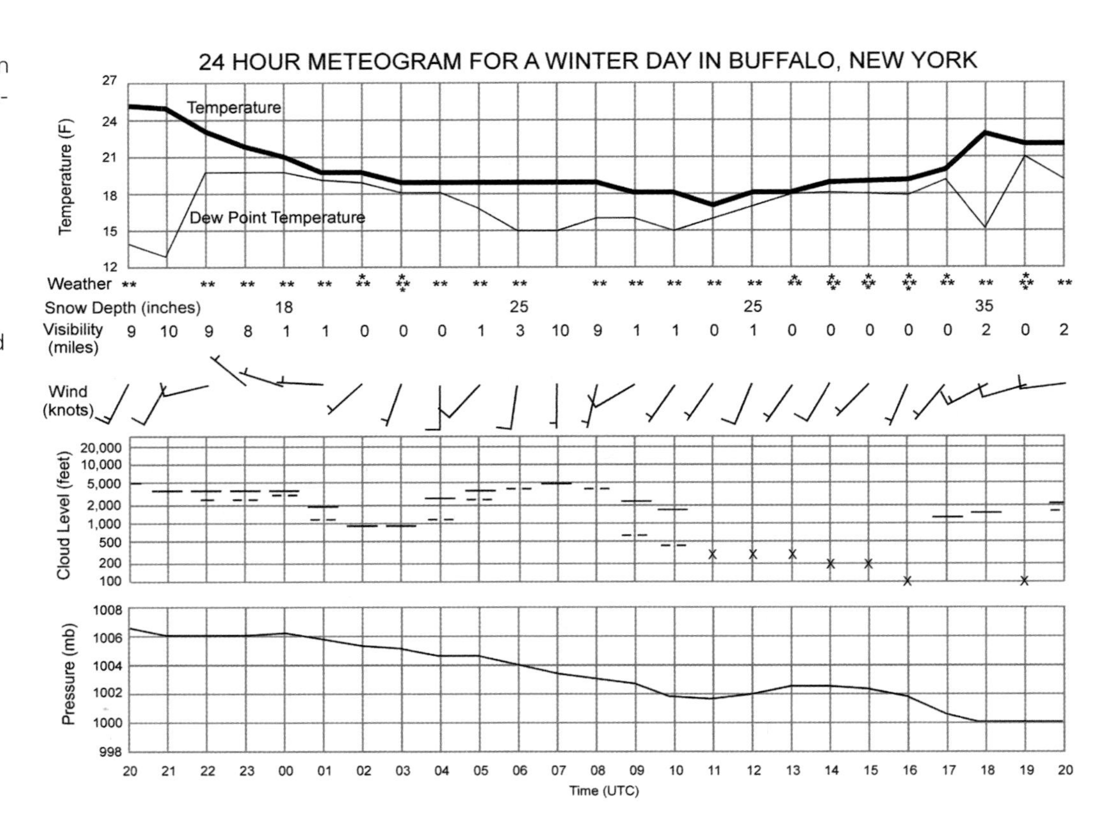

FIGURE 2.3: A meteogram showing the weather conditions in Buffalo, New York for a 24-hour period in winter. In the row labeled "weather," the number of stars denotes the snowfall intensity, with two stars indicating light, three moderate, and four indicating heavy snow. In the cloud level panel, two short lines indicate broken cloud cover, a long line, complete overcast, and an "X", obscured (by snow) so that clouds cannot be seen.

FIGURE 2.4: A rawinsonde launched from a National Weather Service office. Inset shows the instrumentation package.

sample the jetstream level around 250 mb close to 1200 and 0000 UTC. When launched at a site under the jetstream, balloons occasionally are carried over the horizon by the wind. When this occurs, the launch site loses line-of-sight contact with the transmitter attached to the rawinsonde, which can result in loss of wind data, or in some cases, all data. The balloons typically rise about 20 km (about 65,000 ft, to about the 60 mb pressure level) before they burst. Since the rate of the balloon's ascent is about 300 meters/minute (~1000 ft/minute), a balloon typically rises about an hour before it bursts. When a balloon breaks, the instrument parachutes back to the earth.

A ***sounding*** is a depiction of the vertical structure of the atmosphere above a location on the Earth, as measured by a rawinsonde. Data collected from the rawinsonde's instruments are plotted on a diagram that depicts pressure on the vertical axis and temperature (°C) on the horizontal axis. Figure 2.6 shows a ***Stuve diagram,*** one example of a diagram commonly used to plot soundings. Note that pressure is on the vertical axis of the diagram. Since pressure decreases with height, this axis also represents altitude. The pressure does not decrease at a

FIGURE 2.5: Location of rawinsonde launch sites in North America. Rawinsondes are launched twice a day at 1200 UTC and 0000 UTC.

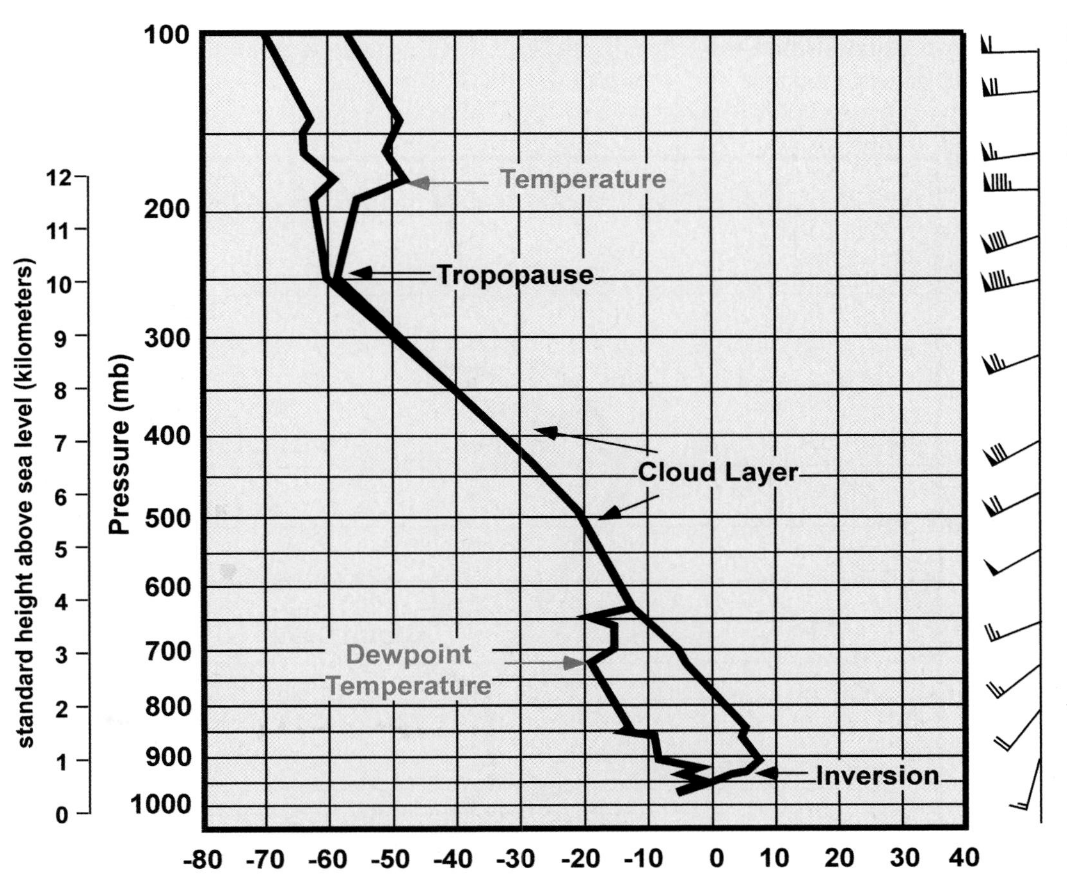

FIGURE 2.6: A sounding taken at Minneapolis, Minnesota and plotted on a Stuve diagram. The thick right line is the temperature (°C) and the thick left line is the dewpoint temperature (°C) measured by the rawinsonde. (Note that the values of temperature and dewpoint temperature are equal between 640 and 250 mb). A scale that converts pressure to average height above sea level appears on the left side of the diagram. Horizontal winds measured by the rawinsonde are plotted as wind barbs on the right side of the diagram, with the same convention used for surface reports.

steady rate, but instead decreases logarithmically (note, for example, that the distance between 200 and 100 mb is much greater than the distance between 1000 and 900 mb). By plotting pressure decreasing logarithmically upward on the vertical axis of the Stuve diagram, distance on the vertical axis is directly proportional to altitude above the Earth's surface. Figure 2.6 illustrates this point—the altitudes marked on the left side of the Stuve diagram increase steadily with height.

Soundings on the Stuve diagram are depicted by plotting two lines that correspond to the temperature and dewpoint temperature measured by a rawinsonde as it ascends in the atmosphere. Winds are normally plotted along a vertical line located on the side of the diagram using standard wind barb conventions (see Chapter 1). Figure 2.6 shows an example of a sounding from Minneapolis, Minnesota. Several features appear on this sounding. An ***inversion*** layer, where the temperature increases with height, is located between the surface (located at 975 mb) and 905 mb. A cloud layer is present between 640 and 250 mb. In this layer, the temperature and dewpoint temperature are equal so that the relative humidity is 100 percent. The tropopause is located at 250 mb. Winds are out of the south-southwest in the inversion layer in the lower atmosphere, but switch to southwesterly above the inversion and westerly in the stratosphere. The jetstream, represented by the 95 knot winds between 250 and 180 mb, is located near the tropopause.

A second diagram in common use in meteorology is the ***Skew-T/Log P diagram***. The diagram gets its name from the temperature and pressure coordinates of the diagram. Lines of constant temperature, which are vertical on the Stuve diagram, are slanted, or skewed, on the Skew-T diagram. As on the Stuve diagram, pressure is logarithmic. Figure 2.7 shows the same Minneapolis, Minnesota sounding plotted on a Skew-T diagram. The Skew-T is the most common diagram used in meteorology to plot soundings, and many sites on the World Wide Web provide soundings only in the Skew-T format.[1]

[1]Both the Stuve and Skew-T diagrams have many uses that are beyond the scope of the material covered in this book. These applications require an understanding of other lines that normally appear on the diagram. Students interested in advanced applications of the Stuve and Skew-T diagrams should consult an advanced meteorological text for details on their use.

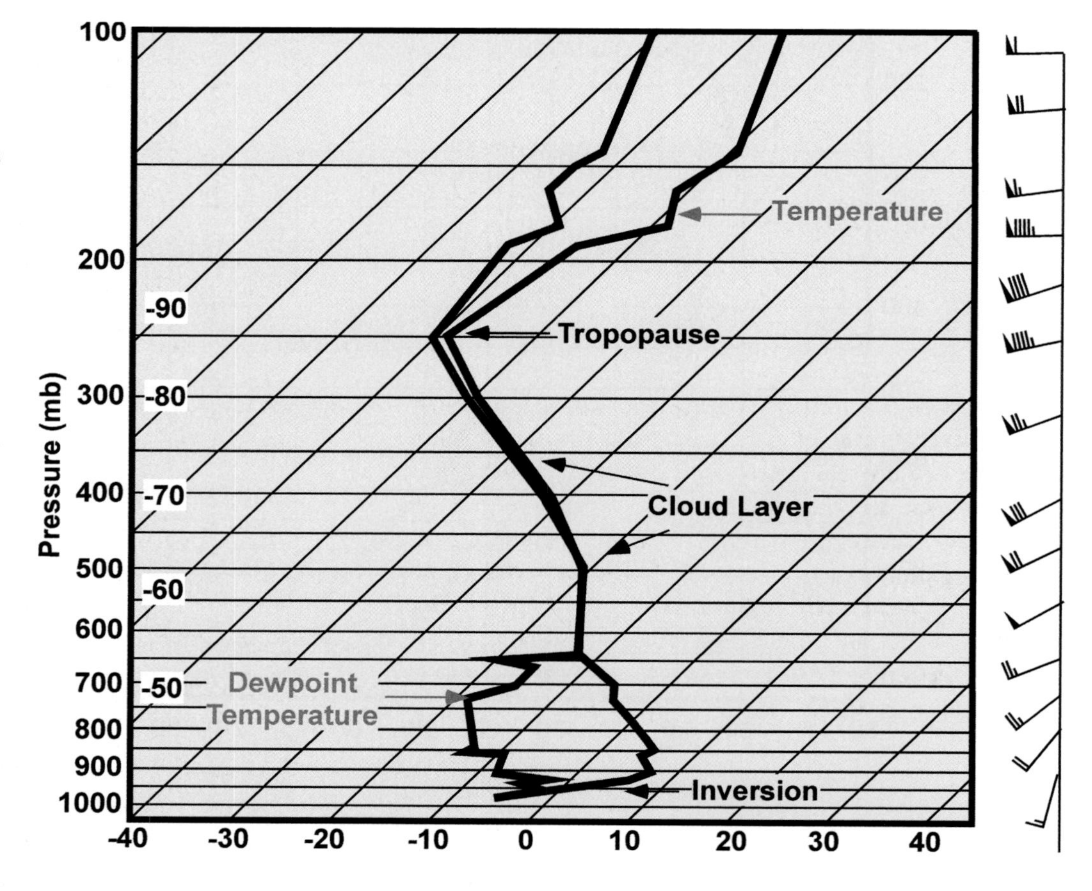

FIGURE 2.7: The same sounding as in Figure 2.6, except that the data are plotted on a Skew-T/Log-P diagram. Note that lines of constant temperature are skewed to the right rather than vertical.

FOCUS 2.3

Hodographs

The change of wind direction and speed between two altitudes is called *vertical wind shear*. To quickly assess vertical wind shear from rawinsonde data, meteorologists often use a diagram called a *hodograph* (Figure 2C). A hodograph is a polar diagram conventionally designed with north (0°) at the bottom, south (180°) at the top, east (90°) to the left, and west (270°) to the right. This coordinate system is flipped from the one used for wind barbs (Figure 1.12) because arrows, rather than wind barbs, are used to denote winds on hodographs. Distance from the center of the hodograph denotes wind speed, with rings about the center of the hodograph marking specific speeds. In Figure 2C, the rings denote 20, 40, and 60 knots. Wind data from soundings are normally plotted on a hodograph at evenly spaced altitudes—for example, at 0.5, 1.0, 1.5, 2.0 kilometers, etc. A point is placed on the hodograph denoting the wind speed and direction at each altitude, as shown in Figure 2C. For example, the point in Figure 2C with the label "2.5" near it would be a wind from 240° (a west-southwesterly wind) at 50 knots. The altitude (kilometers) of the observation (the 2.5 in the example) is usually noted as a number near each point. The points are then connected with a line. The length of a line between two points denotes wind speed shear. The angle between the line and a radial on the hodograph denotes directional wind shear. Severe thunderstorms form most often in an environment where winds rapidly change direction and speed with height. In the classic severe thunderstorm environment, winds increase rapidly above the surface and change direction from southeast near the surface to southwest aloft (e.g., Figure 2C). The hodograph is an important tool forecasters use to determine whether thunderstorms erupting later in the day will be severe. On many Internet sites, hodographs are shown together with soundings or are available as a separate option.

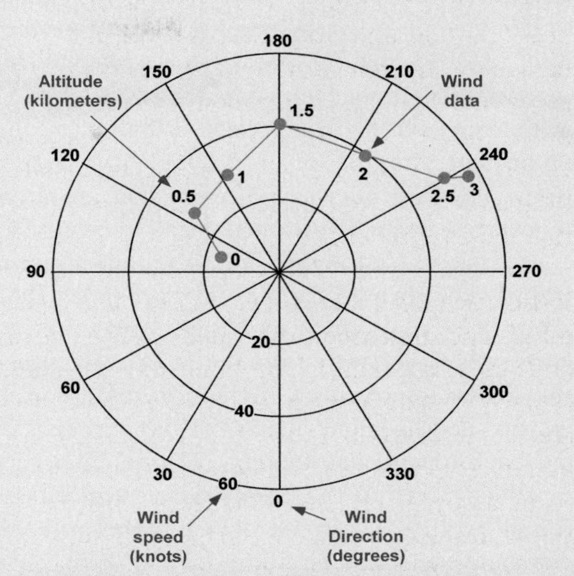

FIGURE 2C: Example of a hodograph characteristic of a severe thunderstorm environment. Each dot can be regarded as the head of an arrow pointing from the diagram center in the direction the air is moving.

The surface and rawinsonde measurements described above are common measurements that have been made for many decades. Although instruments today are more sophisticated and automated, they still obtain the same type of information that meteorologists recorded routinely for well over half a century. We turn our attention now to new instrumentation that has revolutionized meteorology in the last few decades, making it possible to issue accurate and timely warnings about hazardous weather.

CHECK YOUR UNDERSTANDING 2.1

1. What are synoptic measurements? How are such measurements coordinated for simultaneous measurement?
2. What types of observations are available from automated surface observing systems (ASOS)?
3. What is a rawinsonde? How frequently does it provide information?
4. What atmospheric variables does a rawinsonde measure?

RADAR

Precipitation is often a key ingredient in hazardous weather. One way that meteorologists monitor precipitation is by examining data from weather radars. A ***radar*** is a device that transmits pulses of microwave energy. An electronic component of the radar called a transmitter creates microwaves (just like the microwaves in a microwave oven) that are focused into a narrow beam by an antenna and transmitted into the atmosphere (see Figure 2.8). The transmitter only transmits microwaves for a very short time (about 1 millionth of a second) and then waits while the microwaves travel away from the radar at the speed of light. When microwaves encounter objects such as raindrops and hailstones, some of the energy is scattered back toward the antenna. The antenna gathers this energy, called the ***radar echo,*** and passes it through another electronic device called a receiver. From there the signal is processed and displayed. All this takes about one thousandth of one second.

The radar then sends out another pulse of microwaves and does it all over again. The radar electronics can measure very accurately how long it took the pulse of microwaves to travel out to the raindrops and back. Knowing the speed of light, it is easy to calculate the distance to the rain. This information, and the pointing angles of the antenna, indicate exactly where the rain is located in space.

The receiver gathers information about the intensity of the returned signal. We know from theory that the amount of energy returned to the radar depends on three parameters: the size of the precipitation particles, the type of particles (ice crystals, hail, rain), and the number of particles in the beam. The greater the size and number of particles the beam intercepts, the larger the returned signal will be. For this reason, large values of radar reflectivity are associated with heavy rain or hail and small values with non-precipitating clouds.

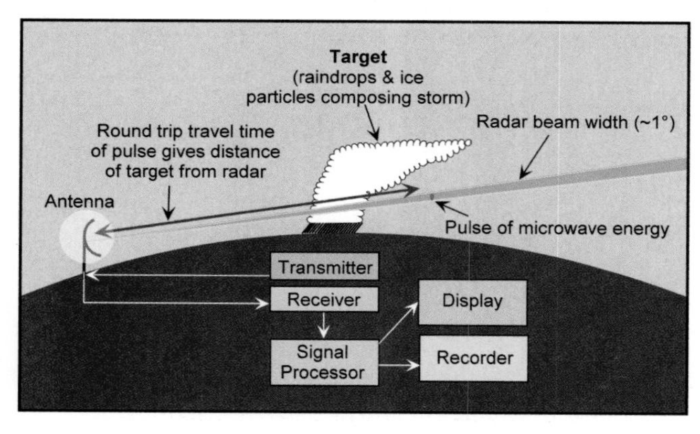

FIGURE 2.8: The principal components of a radar, and the path of a microwave pulse as it travels outward to a storm and is scattered back toward the antenna.

Studies show that there is a close relationship between the returned signal and the rate at which rain is falling. This is because larger drops are associated with heavier rainfall rates. The intensity of the returned energy, called the ***radar reflectivity factor,*** or more simply the ***radar reflectivity,*** allows meteorologists to estimate the rain rate. Adding up (integrating) the reflectivity measurements over time allows meteorologists to estimate the total amount of rain that fell during the period of observation. This capability has revolutionized flash flood forecasting.

The power of the reflected signal depends on the sixth power of the diameters of the raindrops viewed by the radar, so the amount of energy reflected back to the radar varies enormously as a radar beam moves from a heavy rain shaft to cloud-free regions. Radar meteorologists use the symbol "Z" to denote the radar reflectivity. To display radar reflectivity, meteorologists use a logarithmic, or decibel scale, which is denoted *dBZ.* The color scale the National Weather Service radars use depends on whether or not precipitation is present. When significant precipitation is within view of the radar, the radar is placed in ***precipitation mode.*** The color scale used for precipitation mode appears in Figure 2.9. Reds, pinks, purple, and white denote intense precipitation. Yellows denote moderate rain, while greens and blues indicate light precipitation. The table in Figure 2.9 shows the approximate relationship between the rainfall rate and dBZ values the National Weather Service uses.

When no significant precipitation systems are within the range of the radar, the radar is switched to ***clear air mode.*** In this mode, the radar only receives energy scattered back from insects, birds, turbulence, and ground objects. Sometimes in winter, the clear air mode is also used during light snow events to better distinguish the structure of the snowfall. The reflectivity scale used for the clear air mode, which covers a lower dBZ range, also appears in Figure 2.9.

Radars transmit electromagnetic energy at specific frequencies. When energy is scattered back to the radar, the frequency of the returned signal is gen-

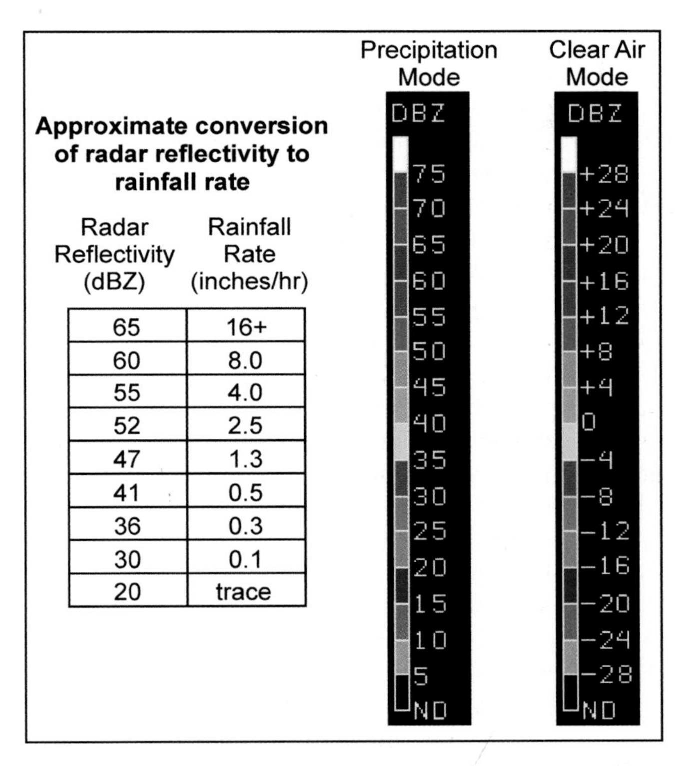

FIGURE 2.9: Radar reflectivity color values used by National Weather Service Doppler radars when a radar is in "precipitation mode" and "clear air mode."

erally shifted slightly. The frequency shift is related to the movement of the raindrops and ice particles along the direction of the radar beam. Radars that measure the frequency shift are called ***Doppler radars*** (see Focus 2.4). Since raindrops and ice particles move with the wind, the frequency shift is a measure of the wind component along the direction of the beam. Unfortunately, Doppler radars cannot measure any air motion across the beam. Nevertheless, by mapping out the wind along the beam (called the ***radial velocity***), a Doppler radar can identify strong straight-line winds and also detect rotation of the wind, which can sometimes be used to infer the location of tornadoes. In severe storm situations, the radial velocity can be adjusted so that the storm motion, independently determined from animation of the reflectivity field, can be subtracted out. The new field, called the ***storm relative radial velocity***, often more clearly will show rotation from which it is easier to identify tornadoes. Doppler radars have revolutionized our ability to provide warnings of tornadoes and other severe weather events. Table 2.1 summarizes the measurement capabilities of a Doppler radar.

TABLE 2.1 **Doppler Radar Measurements**

Measurement	Derived Quantity
Time it takes for the microwave energy to travel from transmitter to the target (precipitation) and back to the receiver	*Distance to the precipitation*
Pointing angles of the antenna	*Altitude of the precipitation and its geographic location*
Fraction of transmitted microwave energy scattered back to the antenna by the target	*The intensity of the precipitation, and when added over time, the total precipitation*
Frequency of transmitted signal and the signal received from the target	*Speed of the wind toward or away from the radar*

FOCUS 2.4

The Doppler Shift and Doppler Radars

A common experience we all have had is the change in the sound of a train blowing a whistle as it approaches and then moves away. The whistle has a high pitch as the train approaches and a lower pitch as it departs. The train whistle did not make a different sound, but the frequency of the sound heard by the stationary observer changed. What happened? When the train approached, the sound waves were "compacted" because both they and the train were moving in the same direction (i.e., successive waves had to travel shorter distances so their arrival times were compacted).

(continued)

(continued)

Their frequency was increased, which the observer's ear perceived as a high-pitched sound. When the train was moving away, the sound waves were still moving toward the observer, but they were "stretched" because the train and the waves were moving in opposite directions (i.e., successive waves had to travel longer distances so that the arrival times were stretched out). The frequency was decreased, lowering the pitch. The faster the train is moving, the greater the frequency shift the observer will hear. This shift in frequency is called the *Doppler shift,* named after the Austrian physicist Hans Christian Doppler, who first explained the phenomenon.

All electromagnetic energy (the scientist's term for visible light, x-rays, ultraviolet energy, microwaves, radio waves, etc.) travels through space as waves with oscillating electric and magnetic fields. Just like waves on water or sound waves in the air, microwaves and other forms of electromagnetic energy have a frequency (number of wave crests passing a point in a second) and a wavelength (distance between wave crests). Microwaves emitted by a radar undergo a Doppler shift in frequency when they are scattered by raindrops moving toward or away from the radar. The frequency of the returned microwaves changes slightly from the transmitted frequency. The faster the drops are moving, the greater the shift. The shift will be in one direction (higher frequency) if the drops are moving toward the radar and the opposite (lower frequency) if they are moving away. The drops blow along with the wind, so the change in frequency is a measure of the wind speed along the direction of the beam. Doppler radars can measure the frequency of the microwaves comprising the returned signal, and from this information, estimate the wind speed in the direction of the radar beam, toward or away from the radar.

Radars detect other targets beside precipitation. Energy transmitted by radars is scattered by flying insects, birds, bats, and any other airborne object. The dominant source of clear air signals in summertime is from insects. We can think of the insect's motion in two parts: the motion of an insect due to its own energy, and the motion of an insect due to the wind. Since radars view many insects simultaneously, and insects fly in random directions, their motion due to their own energy tends to average out to zero velocity. The radar measures their common motion, which is the wind blowing the insects along. Radars also detect energy scattered by small variations in atmospheric density associated with regions of the atmosphere that are turbulent. These turbulent blobs of air also move with the wind. Energy scattered from insects and from turbulent regions of the atmosphere constitutes a "clear air signal" that, together with the precipitation signal, is used by Doppler radars to determine information about the wind.

The U.S. National Weather Service Doppler radars are called ***WSR-88D*** (***Weather Service Radar 1988-Doppler***) radars. The Doppler radar network was installed in the early and mid 1990s. Figure 2.10 shows the location of the radars in the United States and their coverage at the 10,000 ft level. Radars were placed so that their coverage areas overlap whenever possible so that when technical problems develop, one of the other nearby radars can provide coverage. Each radar can monitor the atmosphere a distance of approximately 250 km (155 miles) from the radar location.

Researchers have developed methods for combining data from more than one Doppler radar viewing the same storm at different angles. Generally, these types of measurements are confined to short research programs. Much of our understanding of severe storm structure and dynamics comes from these ***multiple Doppler radar*** studies. With more than one Doppler radar, the entire wind field can be measured throughout storms at very high resolution, typically a few hundred meters (about 1000 ft). The data obtained can be spectacular. For example, Figure 2.11A shows a satellite image of a storm with radar echoes superimposed on the image. There is a thunderstorm over northeast Kansas that passed by two Doppler radars during a research program. The small box in Figure 2.11A is expanded in Figure 2.11B to show the radar reflectivity (colors), which is related to the precipitation rate, and the winds (arrows) 2 km (1.2 miles) above the ground in this storm. It is easy to see that the showers occurred along a line where the wind shifted from southerly to southwesterly. Researchers use Doppler radars to construct wind

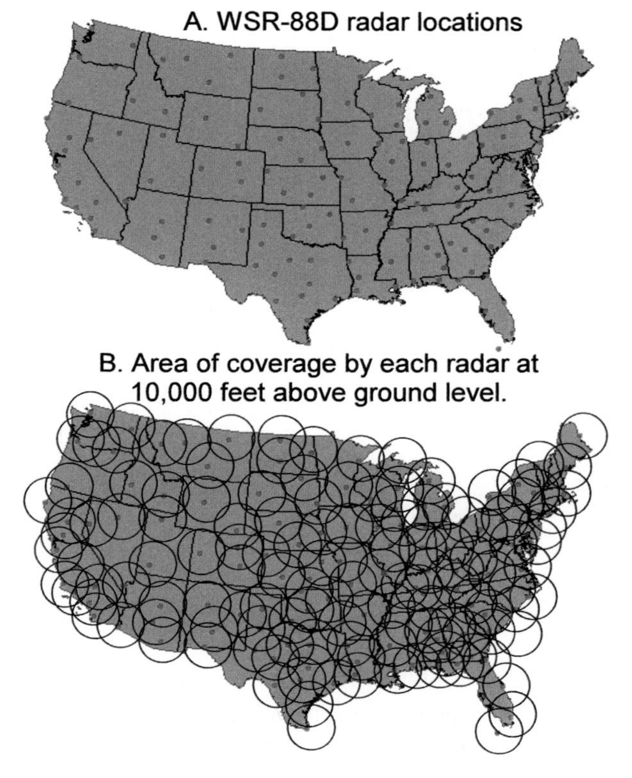

FIGURE 2.10: Location of the National Weather Service radars in the contiguous United States and their coverage at the 10,000 ft level.

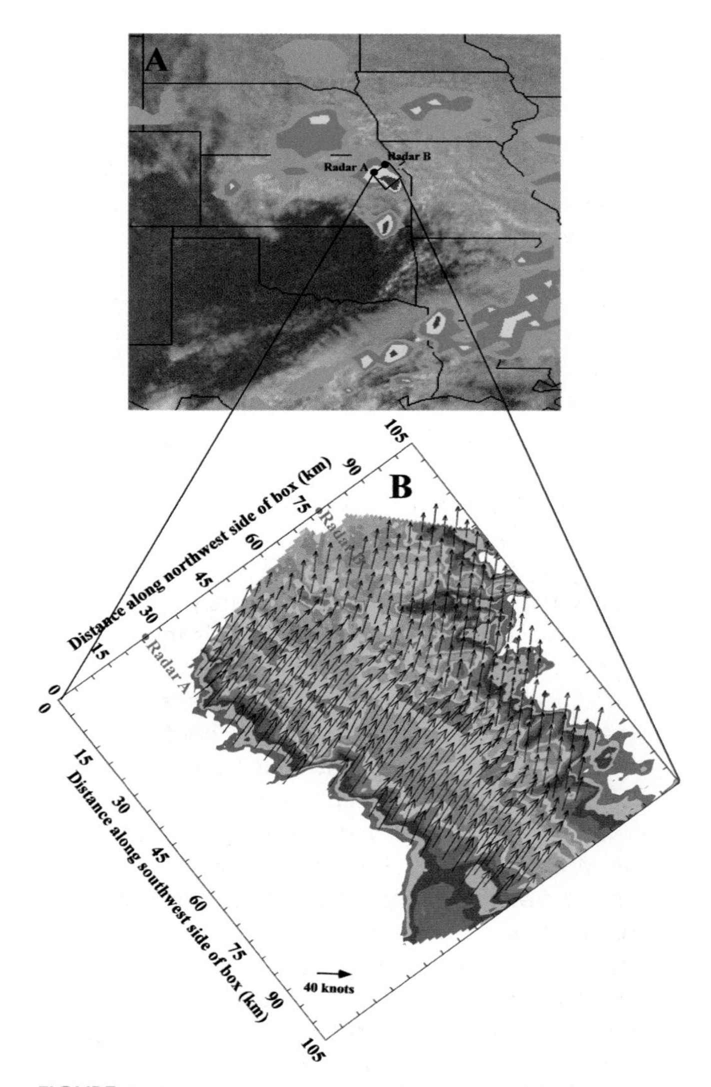

FIGURE 2.11: (A) Infrared satellite image overlaid with radar data showing a weather system over the central plains of the United States. Note the area of storms over northeastern Kansas. In (B), the details of the wind field (arrows, with length of arrow proportional to wind speed—see 40 knot reference arrow), and the radar reflectivity 2 km (1.2 miles) above the ground within this line of storms appear in detail (colors). The radial velocity data from two nearby Doppler radars were used to develop the wind analysis.

fields such as this at all levels, leading to full three-dimensional depictions of the wind within storms.

WIND PROFILING

A ***wind profiler*** is another type of Doppler radar that operates in very high frequency (VHF) and ultra high frequency (UHF) radio bands. Unlike the typical Doppler radar, which has a dish type antenna, the antenna of a wind profiler is an array of cables, as shown in the photograph in Figure 2.12. The antenna is called a ***phased array antenna,*** because the array transmits electromagnetic radiation with a slight time delay from one side to the other across the array. The delay has the effect of creating a beam of radiation that points in a specific direction. With traditional Doppler radar, energy is scattered primarily by raindrops and ice particles; with profilers, energy is scattered primarily by small variations in atmospheric density associated with regions of the atmosphere that are turbulent. The wind profiler senses the motion of the air along the beam by determining the Doppler shift that occurs in the returned signal as these turbulent regions of air move toward or away from the profiler.

By using three beams and some trigonometry, a vertical profile of the wind (wind speed and direction at altitudes above the surface) can be obtained, similar to the winds measured by a rawinsonde. Wind profilers work best in clear air. Profiler winds are measured between the altitudes of 0.5 and 16.5 km (0.3 to 10.3 miles). Below 9.5 km (5.9 miles), data are available every 250 meters (820 ft), while above

FOCUS 2.5

Doppler Radar and Tornadoes

The central United States experienced a record-breaking week of tornadoes from 4 May through 10 May 2003, when nearly 400 tornadoes occurred in 19 states and caused 42 deaths during the seven days. On 8 May, a large outbreak of tornadoes occurred in eastern Kansas. In Figure 2D, high-resolution images of the reflectivity (Panel A) and storm-relative radial velocity (Panel B) fields from a tornadic thunderstorm during this outbreak illustrate typical radar characteristics of tornadic thunderstorms. Severe thunderstorms often develop a hook-like appendage in the radar reflectivity field. Tornadoes typically develop near the center of the hook. Rotation of the storm appears as a tight couplet of adjacent strong inbound (blue) and outbound (yellow/red) radial motions on the radial wind velocity image, with the rotation corresponding in location to the center of the reflectivity hook. The tornado is most likely located where the red and light blue pixels touch. At this location, the radar measures a 70 kt change in radial velocity between the two pixels. In Chapter 19, we explore why the hook shape develops and why the tornado occurs within the hook. Animations of the reflectivity and radial velocity images are used to track hook positions and radial velocity couplets, allowing meteorologists to estimate the time of arrival of dangerous conditions and issue specific warnings.

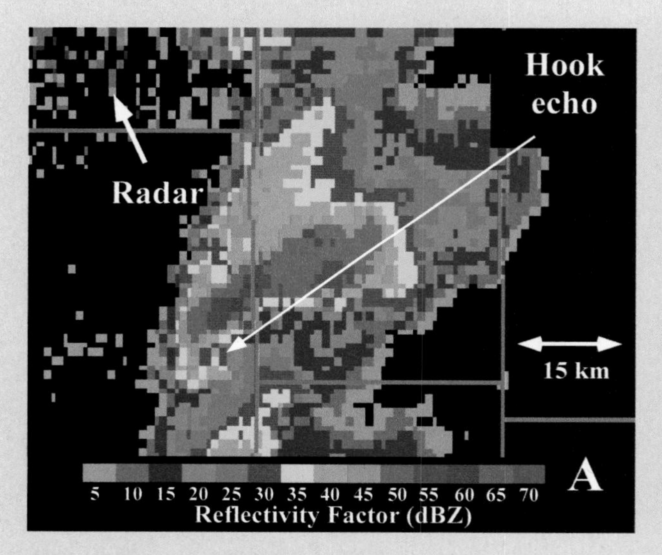

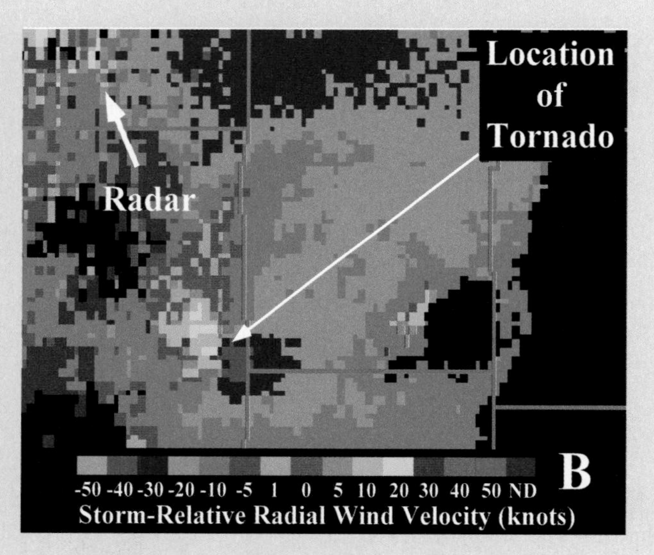

FIGURE 2D: Radar images of (A) the radar reflectivity factor (dBZ) and (B) the storm-relative radial wind velocity (knots) for a supercell thunderstorm that produced a tornado near Lawrenceville, Kansas on 8 May 2003. On the radial velocity display, yellow and red (blue and green) colors represent air motion away from (toward) the radar. Orange lines are county boundaries.

9.5 km, data are available every 500 meters (1640 ft). Both hourly and six-minute wind profiles are now publicly available over the Internet.

Some wind profiler instruments also are supplemented with a temperature profiler called a ***Radio Acoustic Sounding System***, or ***RASS*** (see the corners of the wind profiler in Figure 2.12). The RASS transmits an acoustic (sound) wave in the vertical direction. The wind profiler then sends out its own pulse that backscatters off of the RASS acoustic wave. The wind profiler measures the Doppler shifted frequency off the acoustic wave, determining the speed of sound. The speed of sound is related to the temperature of the air through which the acoustic wave is passing. The network of wind profilers operated by NOAA primarily covers the Central Plains of the United States as shown in blue in Figure 2.13. Nearly thirty other agencies also operate wind profilers, some of which

FOCUS 2.6

Radar Estimated Precipitation and Flood Forecasting

Radar reflectivity is a general indicator of precipitation intensity. Although an exact relationship between radar reflectivity and precipitation rate does not exist, research has shown that the two quantities are related sufficiently that the precipitation rate can be estimated from the radar reflectivity. In general, radar estimates of the short-term precipitation rate over an area can deviate by more than a factor of two; however, radar estimates of *total accumulated rainfall*—determined by adding up, or integrating, the values of precipitation rate over time—tend to be more accurate. Figure 2E, the radar-estimated precipitation during the landfall of Hurricane Georges along the Gulf Coast of Louisiana, Alabama, and Florida in 1998, illustrates the high resolution with which rainfall patterns can be determined from radar imagery. The eye of the hurricane made landfall just west of Mobile, Alabama. The precipitation in Georges was asymmetric about the eye of the storm, with over 20 inches (50 cm) of rain falling at locations on the east side of the eye and much less on the west side. These types of precipitation estimates allow hydrologists to estimate the total precipitation over a watershed, which can be used to determine stream runoff. For this reason, radar is an important tool for issuing flash flood warnings.

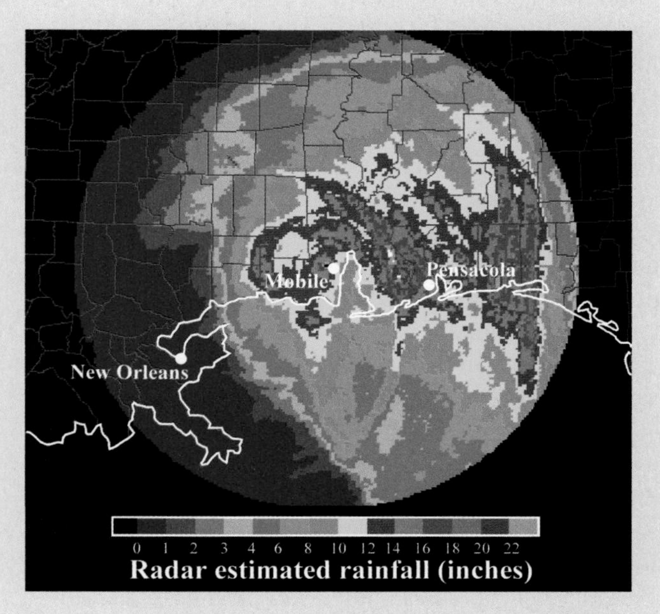

FIGURE 2E: The total rainfall estimated by a radar located near Mobile, Alabama during the landfall of Hurricane Georges in 1998.

FIGURE 2.12: Photograph of a wind profiler site. The wire mesh is the profiler. The large devices in the corners of the array are parts of a radio acoustic sounding system (RASS). The RASS emits sound chirps that are used in conjunction with the profiler to measure the speed at which the sound moves upward. From the speed of sound, vertical profiles of temperature can be derived.

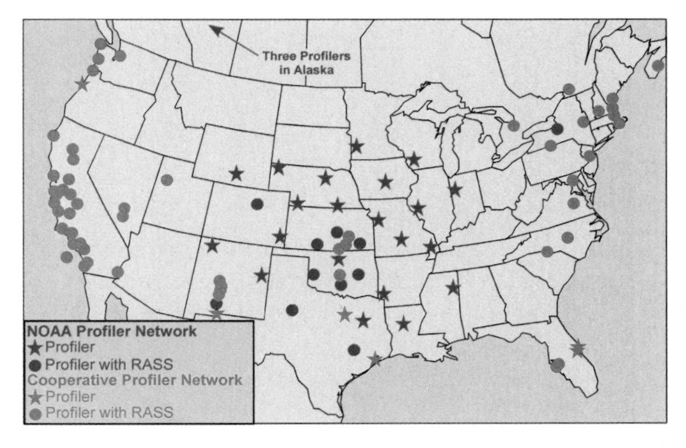

FIGURE 2.13: Locations of profilers in the NOAA profiler network (blue) and the Cooperative Profiler network (red). Stars denote profilers and circles denote profilers equipped with a radio acoustic sounding system.

are shown in red on Figure 2.13. These wind profilers are operated in many different ways, are optimized for different applications, are sometimes moved, and may not always be available. NOAA also ingests data from these Cooperative Agency Profilers to continually monitor wind and temperature aloft.

Similar to a wind profiler instrument, WSR-88D Doppler radars also have the capability of measuring a vertical wind profile. The wind direction and speed above a WSR-88D radar are obtained through mathematical manipulation of the radial velocity measurements all around the radar. These vertical wind profiles are also obtained every six minutes. To obtain a wind profile, raindrops and ice particles have to be present to scatter radar energy back to the antenna. For this reason, WSR-88D wind profiles are best obtained in precipitation. However, because WSR-88D radars are very sensitive, they can also detect motions within about 20 km (~12 miles) of the radar site in clear air, provided that the air is turbulent or contains insects. These "clear air" capabilities allow low level vertical wind profiles to be obtained even when there is no threatening weather nearby. The WSR-88D wind profiles complement vertical profiles from the wind profiler network since WSR-88D operate best in precipitation and wind profilers operate best in clear air.

Figure 2.14 shows an example of thirteen wind profiles lined up alongside one another. These were measured sixty minutes apart, one-tenth of the actual time resolution available from WSR-88Ds and wind profilers. Conventionally, the earliest (oldest) profile is placed on the right and the most recent on the left, which has the effect of making changes in time look the way they appear in space. For example, notice the wind shift that occurs within the shaded region in Figure 2.14. It occurs first 1 km above the ground at 0300 UTC, and progressively moves upward as time passes. This is a front passing the radar site, with winds switching from southerly ahead of the front to northwesterly behind the front. Note that a rawinsonde network, which provides data at 0000 and 1200 UTC,

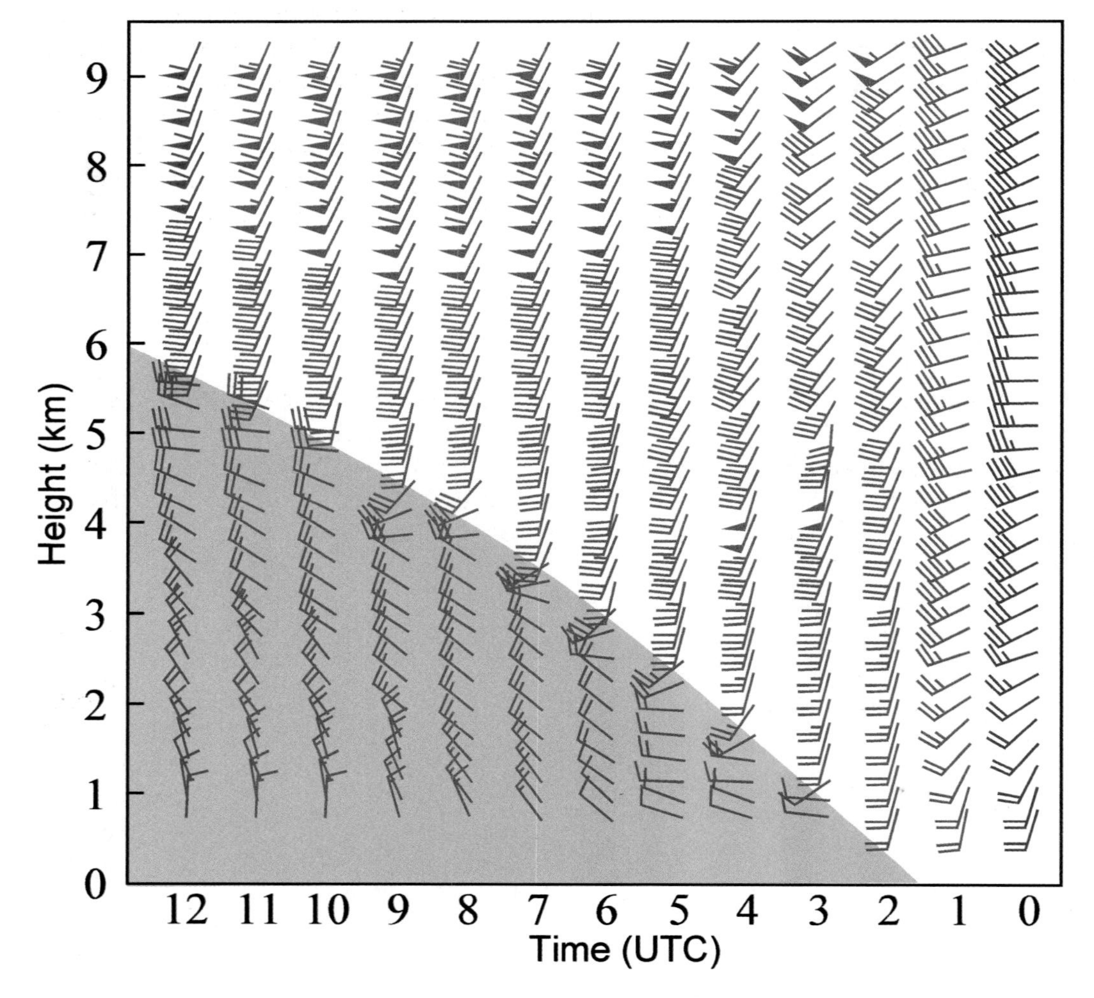

FIGURE 2.14: Wind profiles derived from measurements from a Doppler radar. The vertical wind profiles here are shown every hour for a 12-hour period. The gray area denotes a cold airmass that moved over the radar. Winds within this airmass are from the northwest, while ahead of the airmass, the winds are from the south. The blue columns are located at the times when equivalent wind data would be available from rawinsondes. The wind barbs represent horizontal winds and use the same convention for direction and speed as described in Chapter 1 (see Figure 1.12).

only would have provided the blue profiles on the far left and right. Wind profile displays such as those shown in Figure 2.14 allow meteorologists to rapidly detect changes occurring in the atmosphere. The extra wind soundings also improve forecasts.

CHECK YOUR UNDERSTANDING 2.2

1. What atmospheric variables does a radar monitor?
2. How does a radar determine rainfall rates?
3. In addition to rainfall rates, what else does a Doppler radar measure?
4. What is a wind profiler? What information does it provide?

SATELLITES AND SATELLITE IMAGERY

The first successful meteorological ***satellite*** to acquire an image of Earth was the Television and Infrared Observational Satellite 1 launched into orbit on 1 April 1960. Today, many satellites are used in daily operations of weather services around the world. They are found in two types of orbits: ***geostationary orbits*** and ***low Earth orbits***. A geostationary orbit is a circular orbit lying in Earth's equatorial plane in which a satellite has the same rotational velocity as Earth. In this orbit, a satellite remains essentially motionless relative to a point on the Earth's equator. A geostationary satellite must be 35,800 km (22,300 miles) above the Earth's surface in order for it to maintain the same rotational velocity as the Earth. From this altitude, a satellite has a good view of the entire Earth's disk except for the polar regions. Because the view of the Earth is fixed from a geostationary satellite, images of the Earth can be animated to watch weather systems evolve. Images of the Earth are normally available every fifteen minutes, although the satellites can be programmed from the Earth to "rapid scan" an area of particularly important weather, such as a landfalling hurricane.

A satellite in a low Earth orbit is normally several hundred to several thousand kilometers above the Earth's surface. These meteorological satellites often are placed in a near-polar orbit that is ***sun-synchronous***, meaning that the orbit crosses the equator at the same local time every day. Low Earth orbiting satellites only view a small part of the Earth at any one time, and pass over any point on the Earth only twice a day as the Earth rotates underneath the orbital plane of the satellite. These satellites, which are used primarily for research and specialized applications, are the only satellites that give a clear view of the poles.

FOCUS 2.7

Soundings from GPS and COSMIC Satellites

On 14 April 2006, a new era in atmospheric sounding technology began when a constellation of six satellites was launched from Vandenberg Air Force Base in California. Although the satellites have the official formidable name "Constellation Observing System for Meteorology, Ionosphere and Climate and Taiwan's Formosa Satellite Mission #3", scientists in the United States use their acronym "COSMIC." The satellites are a joint Taiwan-United States project to improve forecasting by dramatically increasing the number of soundings available throughout the world. The satellites use signals from the Global Positioning System satellites and a technique called *radio occultation* that was developed by the NASA Jet Propulsion Laboratory and Stanford University in the late 1960s to study planetary atmospheres.

Specialized GPS receivers aboard COSMIC's six satellites track signals from 24 U.S. Global Positioning System satellites. COSMIC intercepts GPS radio signals as the line of sight between the satellites passes through the atmosphere when one satellite passes below the horizon of the other. As radio signals from the GPS satellites pass through the atmosphere, the signals' paths are bent (like light passing into a prism) and their progress is slowed. The degree of bending and change in speed of the radio signal depends on the atmosphere's density along the signal's path. By measuring these changes, scientists can deduce the temperature and moisture content of the atmosphere along the path of the radio signal, effectively producing a sounding similar to one obtained with a rawinsonde. Because there are

(continued)

(continued)

twenty-four GPS satellites and six cosmic satellites, soundings are obtained continually at positions all over the globe as the satellites move along their orbits. Figure 2F shows the number and locations of COSMIC soundings for one day (green triangles) compared to the standard rawinsonde network (red circles). The COSMIC soundings have already improved weather forecasts, particularly because of the excellent coverage they now afford over remote areas like the world's oceans, deserts, and unpopulated areas such as the Arctic and Antarctic.

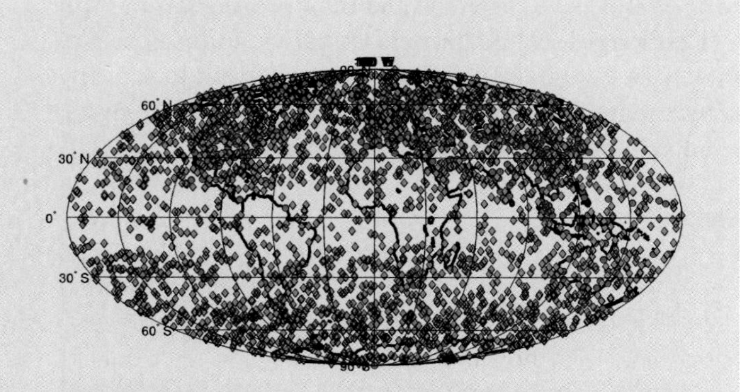

FIGURE 2F: Typical 24-hour coverage of COSMIC soundings (green diamonds) compared to existing rawinsonde launch sites (red circles).

Geostationary satellites sense electromagnetic radiation coming from the Earth in several frequency bands or channels. Three primary channels used in weather monitoring are called the ***visible, infrared,*** and ***water vapor channels.*** The visible channel measures solar radiation reflected from the Earth or atmosphere at a frequency visible to the human eye. Images from this channel appear similar to what a human would see while sitting on the satellite and viewing the world in black and white. The major disadvantage of the visible channel is that the Earth appears dark at night. Figure 2.15A shows a visible image of central North America taken at 2100 UTC on 2 January 2007. The western and central part of North America is in daylight, while the northeastern part is not. Since the visible channel detects radiation reflected from cloud tops or the Earth's surface, the brightest regions on visible images are surfaces that reflect a great deal of solar energy, such as clouds or snow. Dark gray regions on a visible image are surfaces that absorb most incoming solar radiation, such as oceans and forests. Black indicates no reflected radiation, as occurs at night.

All objects emit radiation, with the amount and type of radiation determined by the object's temperature. At temperatures corresponding to the Earth's surface and cloud tops, radiation emitted by objects is virtually all infrared energy. Our eyes are unable to detect infrared radiation. The infrared channel is tuned to be sensitive to infrared electromagnetic radiation emitted by the Earth. Since the whole globe emits radiation, the whole Earth can be imaged both day and night. *The warmer an object is, the more radiation it emits.* Clouds do not transmit infrared radiation well; rather, they absorb infrared radiation and then reemit radiation at a rate that corresponds to their temperature.

Infrared radiation radiating to space from clouds originates in a narrow layer near the cloud top. Since temperature decreases with height in the troposphere, the temperature at the top of clouds will depend, in general, on the altitude of the cloud tops. High altitude clouds such as cirrus and deep thunderstorms with high tops radiate little infrared radiation from the cloud tops because the tops are very cold. Shallow clouds and the Earth's surface emit more infrared radiation because they are much warmer. By gray shading (or coloring) an image based on the intensity of the received signal, high and low clouds become visible on the image.

Figure 2.15B shows an infrared image of the same scene as Figure 2.15A. All of central North America appears, including the part that is experiencing darkness. The darkest shading on this image indicates the warmest surfaces (lower clouds, ground) and brighter regions indicate colder surfaces (higher clouds). For example, the clouds over the Pacific in the lower left corner of the images are bright in the visible, but dark gray in the infrared, implying that they are warm-topped (low altitude) clouds. In winter, when the Earth's surface is cold, there can sometimes be ambiguity about whether the energy emitting surface is a cloud or simply cold ground or snow, since both appear the same gray shade in an infrared image.

The third channel, the water vapor channel, measures radiation at a specific infrared frequency that is sensitive to radiation emitted from water vapor molecules in the atmosphere. This channel is most sensitive to water vapor at altitudes between 2 and 18 km (1.2 and 11 miles). Since water vapor is ubiquitous but variable in concentration in the Earth's atmosphere, animation of water vapor imagery allows meteorologists to monitor atmospheric circulations in both clear and cloudy situations. This channel also provides useful imagery during both day and night. Figure 2.15C shows a water vapor image of the same scene as Figure 2.15A. A sharp change in water vapor concentration often will occur along the axis of a jetstream, a river of fast moving air in the upper troposphere. The jetstream location often can be identified on water vapor images by this sharp contrast. For example, the subtropical jetstream axis in Figure 2.15C is located along the bright/dark boundary extending from just south of the Texas border in Mexico, across the Gulf of Mexico to the western tip of Florida, and northeastward into South Carolina. Plumes of water vapor also appear emerging from the tops of thunderstorms in water vapor imagery. The brightest regions on a water vapor image mark regions of moist air and clouds, while the darkest regions correspond to very dry air.

Table 2.2 summarizes the measurements from each of the primary channels on geostationary satellites. Together, these images provide exceptional views of storms as they evolve and allow meteorologists to continually monitor storm evolution. At the beginning of 2007, the United States had four functional satellites in geostationary orbit (GOES-10, GOES-11, GOES-12, and GOES-13), with two in operation (11, 12) and two in storage (10, 13). GOES-12 (called GOES-East) was located over the Amazon providing a view of eastern North America (near 75° W) and the western Atlantic. GOES-11 (called GOES-West) was located over the eastern Pacific Ocean (near 135° W), providing a view of western North America and the eastern Pacific.

The Japanese had one geostationary satellite in operation: MTSAT-1R over the mid Pacific at 140° E, providing a view of the western Pacific, eastern Asia, and Australia. The Europeans had Meteosat-6, Meteosat-8, and Meteosat-9 over the Atlantic Ocean all near the prime meridian at 0°, with a view of Europe, Africa, and the eastern Atlantic. They also had Meteosat-5 over the Indian Ocean. India and China also operate geostationary satellites that carry meteorological as well as other instruments for meteorological purposes. The most recent, launched in October 2004, operates at 105° E. Together these satellites provide a complete view of the world's weather, excluding the polar regions. Figure 2.16 shows the view of each of five geostationary satellites, giving a complete view of weather around the world. Image processing techniques exist to re-map satellite

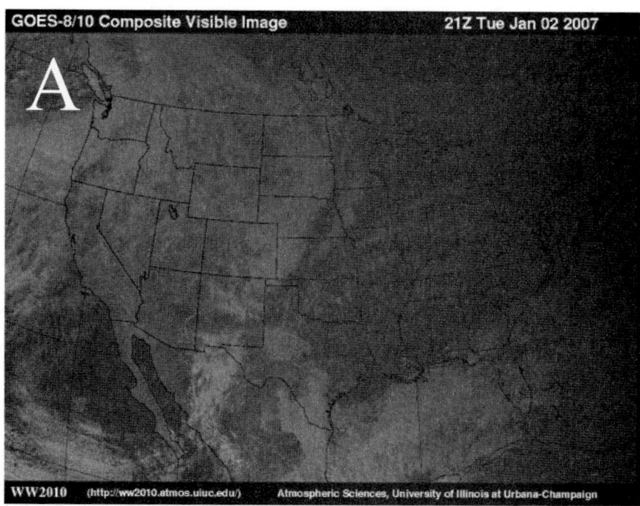

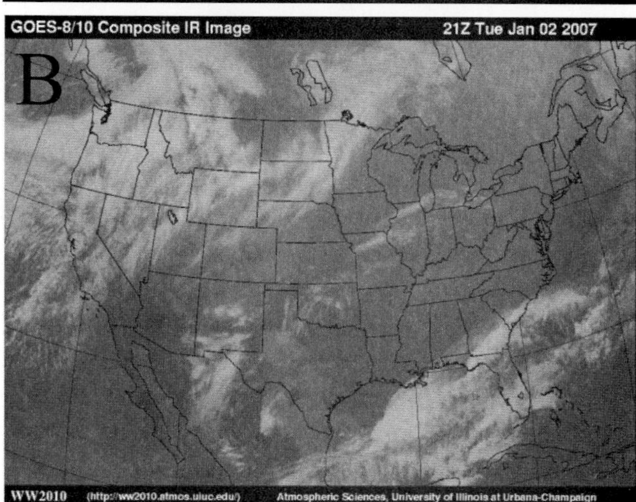

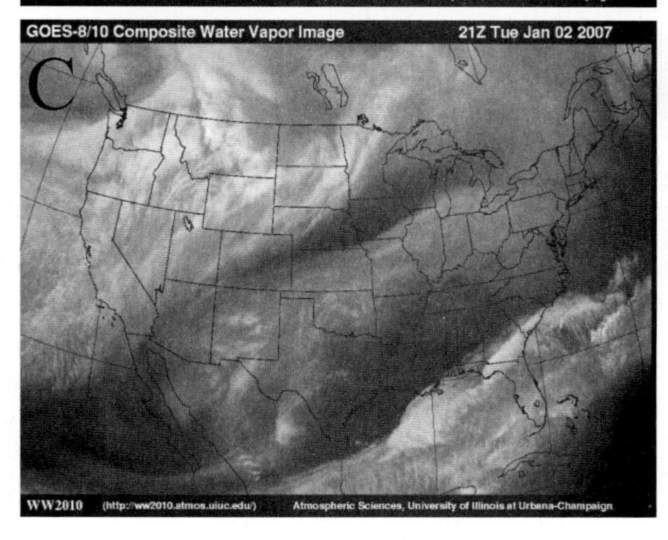

FIGURE 2.15: (A) Visible, (B) infrared, and (C) water vapor images of central North America on 2100 UTC 2 January 2007.

TABLE 2.2 **Interpreting Satellite Imagery**

	Visible	Infrared	Water Vapor
Satellite Measures	Reflected solar radiation	Emitted infrared radiation (temperature)	Infrared radiation emitted by water vapor only
Brightest Regions	Thick clouds, snow	Cold cloud tops (high clouds)	Moist air
Darkest Regions	Oceans, forests, unfrozen rivers in winter	Warm cloud tops (low clouds) and warm regions of the earth's surface	Dry air

ONLINE 2.2

Six Views of Weather from Space

Two geostationary satellites are normally available operationally to view weather over North America and adjacent oceans. Satellite animations of the weather for the 48-hour period beginning 0000 UTC 30 December 2006 from each of these satellites appear online. During this period, a large cyclone develops over the central United States and moves northeast. A second cyclone develops just off the East Coast and moves over the Atlantic. At the same time, a large cyclone over the Pacific develops and moves toward the West Coast of North America. These animations illustrate the different views of the GOES satellites and highlight the differences among the visible, infrared, and water vapor channels. As you watch these animations, the most prominent feature is the progression of day and night in the visible channel. In fact, 9 hours of complete darkness has been omitted to speed the visible channel animations, so the sunlit portion of the Earth appears sooner. Cloud features are sharpest on the visible channel where the contrast between the cloud and ground is greatest. Brightness on the infrared channel allows us to separate high from low clouds. In the animation, you can often see nearby high and low clouds streaming in different directions. The long boundaries between bright and dark areas on the water vapor animations align with the axes of jetstreams. Watch how fast water vapor streams along at the boundary between bright and dark on these images. Note how the jetstream interacts with each of the cyclones. What other features can you identify on these animations?

images into different projections and combine data from several satellites. The views of central North America in Figure 2.15, for example, are a different projection from the original global view similar to that in Figure 2.16.

COMMERCIAL AIRCRAFT MEASUREMENTS

In recent years it has become evident that significant valuable meteorological data can be obtained from large areas of the world by collecting the data using commercial aircraft fitted with appropriate instrumentation and software packages. Many of the world's airlines and cargo transport company aircraft now transmit meteorological information from air to ground or satellite. The various systems used are collectively named ***Aircraft Meteorological Data Reporting (AMDAR)*** systems. The meteorological data is reported to the National Weather Service and is used to develop weather forecasts. Every time a commercial aircraft carrying an AMDAR system equipped with meteorological sensors takes off or lands, a sounding is obtained in the vicinity of the airport. A wealth of data is also collected in the upper troposphere and lower stratosphere at aircraft cruising altitudes. These data are important because the jetstreams—bands of high winds that encircle the earth and strongly influence storm development—are found at these altitudes. Figure 2.17 shows the typical coverage of AMDAR data for a six-hour period over the United States between the altitudes of 300 and 150 mb. These data improve our ability to forecast weather.

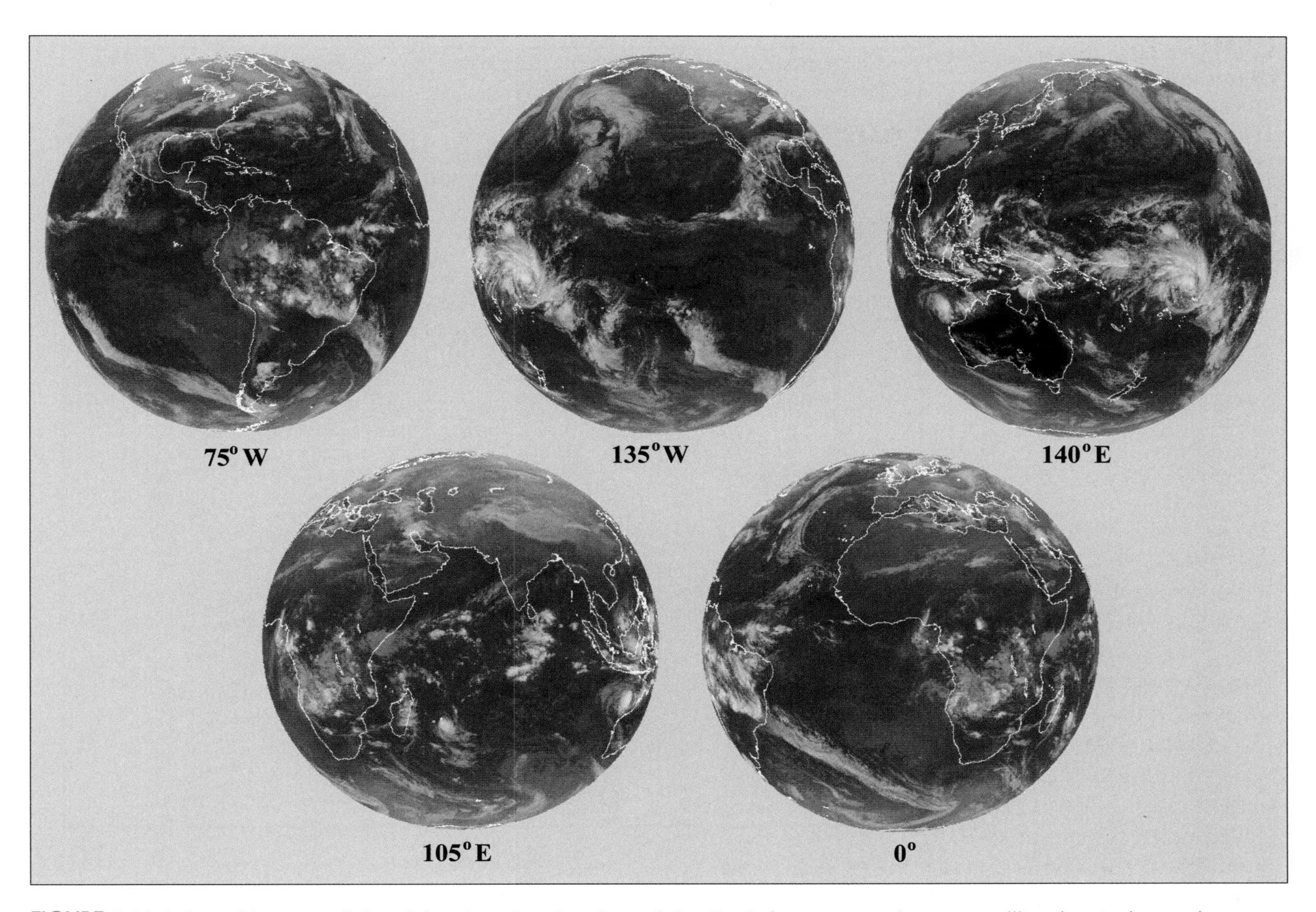

FIGURE 2.16: Infrared images of the globe showing the view of the Earth from geostationary satellites located over the Amazon at 75° W longitude, the eastern Pacific at 135° W longitude, the western Pacific at 140° E longitude, the Indian Ocean at 105° E longitude, and the eastern Atlantic at 0° longitude.

LIGHTNING DETECTION

The ***National Lightning Detection Network*** **(NLDN)** and ***Canadian Lightning Detection Network*** **(CLDN)** map out the location of all cloud-to-ground lightning strikes in the United States and Canada (Figure 2.18). The NLDN and CLDN consist of remote, ground-based sensing stations that monitor cloud-to-ground lightning activity. From the direction, time of arrival, and other characteristics of electromagnetic waves created by the lightning stroke, the system triangulates using three or more receivers to determine the position, time, strength, and polarity of the lightning stroke.

A large database on lightning strokes in the United States and Canada has been assembled. An example of lightning strike mapping in the Northwest United States is shown in Figure 2.19. The NLDN is used to track electrically active regions of thunderstorms. Lightning mapping is particularly important in forested areas such as the Northwest United States and Canada because it speeds the detection and control of fires.

CHECK YOUR UNDERSTANDING 2.3

1. What are the two types of orbits in which weather satellites monitor the Earth?
2. What are the three primary channels used to create weather satellite images?
3. How do commercial aircraft contribute to weather data collection?
4. What important information does the National Lightning Detection Network provide?

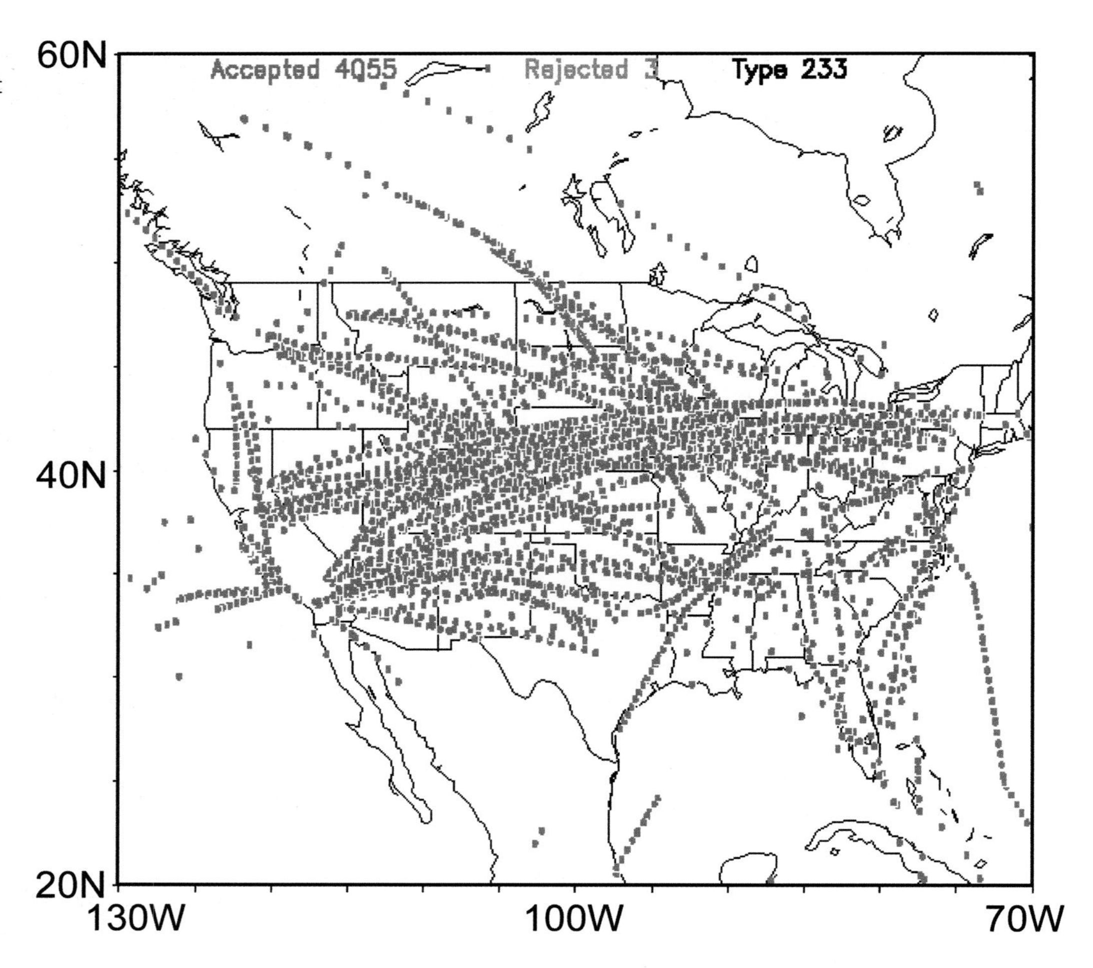

FIGURE 2.17: Locations where commercial aircraft collected AMDAR data during a 6-hour period ending at 0600 UTC 2 January 2007.

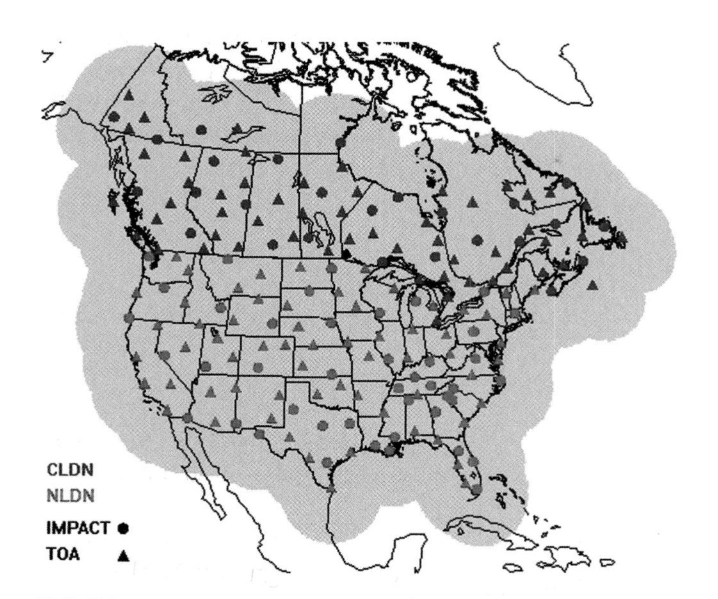

FIGURE 2.18: The United States and Canadian National Lightning Detection Networks. The symbols denote two different types of sensors that are used for lightning detection.

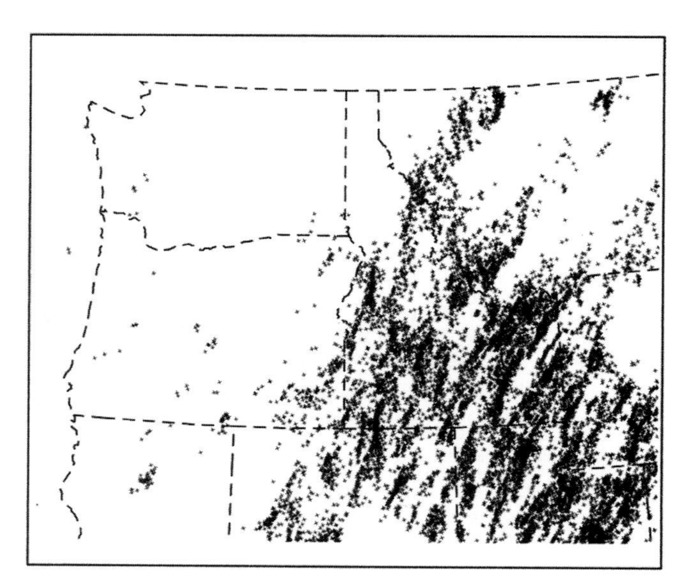

FIGURE 2.19: Position of cloud-to-ground lightning strikes over the northwest United States on 1 September 2000.

TEST YOUR UNDERSTANDING

1. What is the advantage of making measurements of atmospheric properties simultaneously?
2. What do the abbreviations UTC, GMT, and Z mean?
3. How do you convert from universal coordinated time to your local time?
4. What does the National Weather Service call their automated weather stations?
5. What is a meteogram?
6. How does the density of rawinsonde launch locations compare to the density of automated surface observing systems?
7. What layers of the atmosphere does a rawinsonde sample during its ascent?
8. What is a sounding?
9. How does a weather radar detect precipitation?
10. What information does a radar reflectivity image provide?
11. What do the hot colors (reds, yellows) typically represent on a radar reflectivity image when the radar is in precipitation mode?
12. What information does a radar radial velocity image provide?
13. What do the hot colors (reds) typically represent on a radar radial velocity image?
14. What four characteristics of precipitation can Doppler radars detect?
15. What is the typical horizontal range of a radar?
16. Give a non-meteorological example of the "Doppler shift" that you may encounter in your everyday life.
17. What is a WSR-88D?
18. Identify at least one advantage and one disadvantage of radar estimated precipitation compared to traditional rain gauge measurements.
19. What is a wind profiler?
20. How have Doppler radars and wind profilers revolutionized our ability to observe winds aloft?
21. What is a vertical wind profile?
22. How many years have weather satellites been providing information about the atmosphere?
23. What is a geostationary orbit?
24. What is a weather satellite measuring when it creates a visible image? An infrared image? A water vapor image?
25. If a cloud is dark on an infrared image but bright on a visible image, is this a high cloud or a low cloud? Why?
26. How would you use radar imagery and/or satellite imagery (visible, infrared, and water vapor) to identify the following atmospheric phenomenon?
    (a) Fog
    (b) Thunderstorms
    (c) Cirrus clouds
    (d) Jetstream
    (e) Dust storm
    (f) Widespread steady precipitation (Nimbostratus clouds)
    (g) Cloud vs. snow cover
27. Identify at least two advantages and two disadvantages of AMDAR. (Hint: refer to Figure 2.17.)
28. Why is more than one lightning detector needed to pinpoint the location of a lightning stroke?

TEST YOUR PROBLEM-SOLVING SKILLS

1. You are the coordinator for a program in which elementary school children collect weather data to share with other schools. The schools are located in the states listed below. For each state, identify the local time the schools would need to collect data during Daylight Saving Time so that all the data are taken at the same instant. Assume that data will be collected twice daily to correspond to standard rawinsonde times.

   Participating states are: Alaska, Utah, California, Hawaii, Illinois, Florida, Missouri, New Mexico, Ohio, Oklahoma, and Texas.

2. The table on the next page shows data collected from a rawinsonde. Plot the pressure, temperature, dewpoint temperature, and wind data versus altitude on a Stuve diagram (make a copy of the diagram from Appendix B). Plot the wind direction and speed using standard wind barbs.
   (a) Between what pressure levels are clouds likely to be present?
   (b) Suppose you were standing at the site where the sounding was launched. What might the cloud cover look like?
   (c) Estimate the relative humidity at the surface.
   (d) Where are inversions located on the sounding? (Identify the pressure levels that indicate the top and bottom of each inversion.)
   (e) At what pressure level is the tropopause located? How did you determine this?
   (f) If you were to examine an infrared satellite image, how might the gray shading appear at the time and location where this sounding was taken?

3. Assume that a rawinsonde ascends 65,000 feet from sea level and experiences an average wind of 40 knots during the ascent. Assume also that the ascent and descent rates of the rawinsonde were both 5 meters/second. How far downwind (in miles) did the rawinsonde travel by the time it landed back on the surface? (Use Appendix A for unit conversions.)

4. Make a copy of the blank hodograph in Appendix B. Using the sounding data from question 2, plot a hodograph using the data from the surface up through and including 300 mb. According to the hodograph you developed:
   (a) As you ascend from the surface during this time, how does the wind direction change?
   (b) During severe thunderstorms, wind speeds typically increase with height and wind direction typically changes from southeast near the surface to southwest in the lower troposphere and to west in the upper atmosphere. Does the hodograph you plotted show wind conditions conducive to a severe thunderstorm? Why or why not?

5. You are a meteorological intern at the National Weather Service's Phoenix, Arizona office. It is raining during your shift and the Phoenix Doppler radar data suddenly becomes unavailable. How would you obtain information regarding the amount of precipitation that fell in the Phoenix-metro area?

6. Examine Panel A of Figure 2D. Assume that you are driving from east to west through the center (bright red) echo area of the thunderstorm (against the advice of your passengers and your better judgment).
   (a) Qualitatively describe how the precipitation intensity (rainfall rate) will change during your journey.
   (b) If this storm contained hail, where would it most likely be falling?

7. During the winter in California, a variety of weather conditions are often influenced by the local topography. (See Appendix C for a map of physical geography.) Assume there is fog in the Central Valley and snow covering the Sierra Nevada, and that skies are clear in the mountains and above the fog.
   (a) What would these two areas look like on a visible satellite image? How did you determine your answer?

Rawinsonde Data

Level	Pressure (mb)	Altitude (m)	Temperature (°C)	Dewpoint (°C)	Wind Direction (deg)	Wind Speed (kts)
0	1000	60				
SFC	986	178	3.4	1.6	340	10
2	984	194	3.4	3.4	341	11
3	925	693	2.6	2.6	5	16
4	872	1171	2.2	1.6	330	10
5	857	1312	4.2	–0.7	310	10
6	850	1379	3.8	–1.2	300	10
7	786	2010	0	–10	236	10
8	727	2629	–4.9	–13.9	225	21
9	700	2925	–7.3	–12.3	220	22
10	625	3795	–14.3	–14.4	212	34
11	583	4320	–16.7	–22.7	220	45
12	525	5101	–20.5	–22.8	213	51
13	500	5460	–23.3	–25.6	210	56
14	474	5848	–26.9	–30.7	213	60
15	464	6002	–27.9	–41.9	214	61
16	456	6127	–28.7	–46.7	215	62
17	410	6885	–31.1	–59.1	223	70
18	400	7060	–32.7	–58.7	225	72
19	375	7512	–35.9	–59.9	225	72
20	332	8351	–40.1	–60.1	229	86
21	317	8666	–41.1	–61.1	230	90
22	300	9040	–42.3	–62.3	230	85
23	253	10190	–43.9	–68.9	239	80
24	250	10270	–42.7	–67.7	240	80
25	243	10462	–40.9	–67.9	242	80
26	230	10835	–41.9	–69.9	246	79
27	207	11548	–42.3	–70.3	250	77
28	200	11780	–43.5	–71.5	250	76
29	150	13680	–56.9	–76.9	260	65
30	131	14548	–56.7	–80.7	252	59
31	122	15001	–55.9	–79.9	247	57

(b) What would these two areas look like on an infrared satellite image? How did you determine your answer?

(c) During summer, the Central Valley typically has clear skies. Assuming that skies are clear over the Sierra, what would the region look like on an infrared satellite image? Explain.

8. You are examining a visible satellite image during January. You notice a large bright region over the Northern and Central Plains of the United States. The bright region has what appears to be dark, crooked lines across it. What are you observing? How did you determine your answer?

USE THE SEVERE AND HAZARDOUS WEATHER WEBSITE

http://severewx.atmos.uiuc.edu

1. On the *Severe and Hazardous Weather Website*, navigate to the page of "Meteorological Measurements." The bottom of the page provides links to additional resources related to instrumentation described in this chapter. Hourly surface data around the world is reported using a standard coded format called METAR. Each station that reports data has a four letter identifier code. For example, Chicago's O'Hare Airport is KORD.
   (a) Find the surface station closest to your hometown. What is the four letter identifier?
   (b) Locate the latest METAR coded data for the location you selected in (a). Copy the coded data and write out the decoded equivalent, using units when appropriate. Include the date and time of the data.
   (c) Using the additional resources on the website, find a meteogram for the same location that includes time and date of the METAR data. Print a copy of the meteogram. On the meteogram, identify the variables and properties that were provided in the METAR data. Do any of the data not match? What data is included in the meteogram but not the METAR? What data is provided in the METAR but not the meteogram?

2. Using the "Current Weather" page of the *Severe and Hazardous Weather Website,* find the last two soundings taken closest to your location. (This question will be easiest to answer on a day that thunderstorms or other precipitating weather systems pass the location between soundings.)
   (a) Print or save each of the soundings.
   (b) What is the date and time of the soundings? What was the location from which the rawinsondes were launched?
   (c) What is the format of the sounding (Stuve or Skew-T)?
   (d) Describe how each of the following has changed during the past twelve hours: vertical wind profile, temperature profile, dewpoint profile, and relative humidity.
   (e) Identify any inversions present on the latest sounding. Were they present on the sounding twelve hours prior? If not, hypothesize as to why they formed. If there were inversions present twelve hours prior but not on the current sounding, explain these changes as well.
   (f) Examine the soundings for cloud layers and exceptionally dry layers. If they are present, identify their location by listing the pressure levels defining them.
   (g) At what pressure level does the tropopause occur on each sounding?
   (h) Describe how the winds change with height on each sounding.
   (i) What is the fastest wind over the sounding site at each time? At what pressure level was the fastest wind recorded on each sounding?

3. Using the "Current Weather" page of the *Severe and Hazardous Weather Website,* find a hodograph and a sounding from the same time from the nearest rawinsonde launch site to your location.
   (a) Print or save the hodograph and sounding.
   (b) Compare the wind direction at each point on the hodograph with the sounding. Are there any discrepancies between the two?
   (c) Repeat (b) using wind speed.
   (d) In your opinion, is one diagram easier to read than the other? Why or why not?
   (e) Using the information provided in Focus 2.3, would you expect severe thunderstorms based on this hodograph?

4. Using the "Current Weather" page of the *Severe and Hazardous Weather Website,* find the National Weather Service radar summary and zoom in on a region where precipitation is being detected.
   (a) Explain how the intensity of the precipitation varies across the area.
   (b) What would you estimate the rainfall rate to be? Explain how you determined this.
   (c) Locate a surface map for the same region that has station model data. Is precipitation being reported? If not, what do you think is happening to cause this discrepancy?
   (d) Are there regions of differing intensity? If so, what might be the reason for this?

5. Using the "Current Weather" page of the *Severe and Hazardous Weather Website,* find a National Weather Service radar image where heavy rain is falling. Compare the radar reflectivity field to the hourly total or storm total precipitation.
   (a) In comparing the images, what can you determine about the general direction of movement of the rainfall?
   (b) Based on these images, do you think a meteorologist would issue a watch or warning? Why or why not?

6. Use the "Current Weather" page of the *Severe and Hazardous Weather Website* to find a visible and an infrared satellite image of the same area for the same date and time. The satellite imagery should have some clouds present.
   (a) Is there any fog present? If so, how did you determine this?
   (b) Identify regions of high clouds and low clouds. Explain how you determined each.
   (c) How could you use a series of these images, such as an animation, to distinguish clouds from underlying snow cover in winter?

7. Use the "Current Weather" page of the *Severe and Hazardous Weather Website* to find both rawinsonde and profiler data. Choose locations that are close to each other for each of the instruments and examine the vertical wind profile measured by each. Identify similarities and differences in the profiles.

WEATHER MAPS

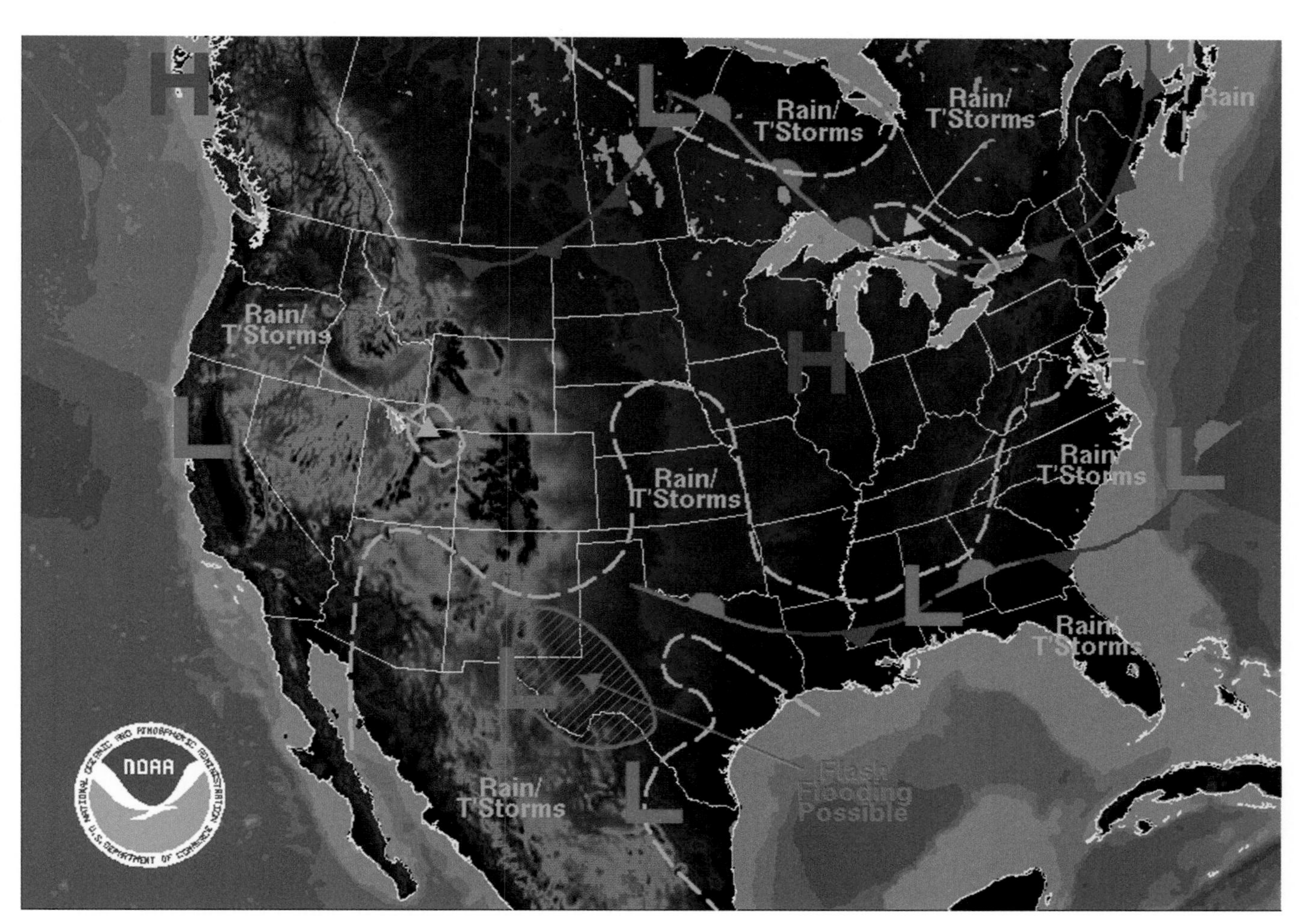

Forecast for sea-level pressure analysis and fronts.

KEY TERMS

absolute vorticity
altimeter
constant pressure map
contour data
cross section
cyclone
dewpoint depression
extratropical cyclone
front
isobar
isodrosotherm
isotach
isotherm
jetstreak
jetstream
long wave
low-level jet
pressure gradient
pressure surface
relative vorticity
ridge
short wave
station model
temperature gradient
trough
vorticity

We live on the Earth's surface and are naturally interested in weather conditions where we reside. Maps depicting sea-level pressure, surface temperature, dewpoint temperature, wind, and precipitation can inform us about conditions that directly affect our lives and alert us to the potential arrival of hazardous weather. Although surface maps depict what weather is happening, they do not generally help us understand *why* specific weather conditions are occurring.

To understand the "why" of hazardous weather—the processes that make weather happen—meteorologists require maps and diagrams that depict atmospheric structure at levels above the Earth's surface. This chapter introduces basic weather maps and diagrams that we will routinely use in this text to examine surface meteorological conditions and the structure of the "upper atmosphere." It is essential to be able to use these maps and diagrams to interpret and understand the causes of hazardous weather.

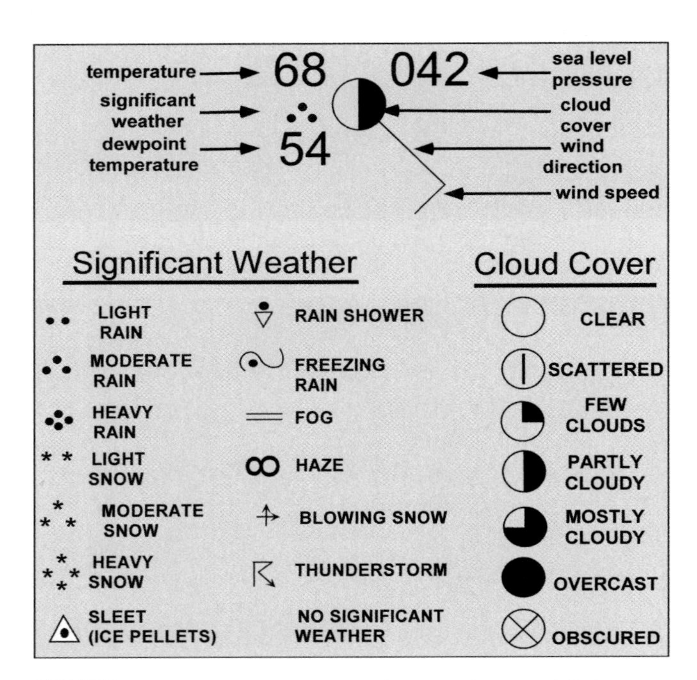

FIGURE 3.1: An example of a surface station model showing temperature, dewpoint temperature, pressure, wind speed, wind direction, significant weather, and cloud cover. Common symbols for significant weather and the convention for plotting cloud cover are also shown.

THE SURFACE STATION MODEL AND SURFACE WEATHER MAPS

Temperature, pressure, moisture, and wind measurements are reported at the surface every hour. Although we say "at the surface," most measurements are actually made two meters above the ground. On weather maps, these data and other parameters are plotted using a ***station model***. An example of a station model is shown in Figure 3.1. The circle denotes the location of the station. The number to the upper left of the station symbol is the temperature. On maps used in the United States, the temperature on the station model is reported in Fahrenheit. The dewpoint temperature appears on the lower left of the station symbol and is also reported in Fahrenheit.

The number to the upper right of the station model is a coded number denoting the sea level pressure. If the number is > 500, a "9" must be added in front and a decimal point placed before the rightmost number to obtain the value of sea-level pressure in millibars. If the number is < 500, a "10" must be added in front and a decimal point placed before the rightmost number. For example, a 998 on a station model would be a pressure of 999.8 mb, while an 042 would be a pressure of 1004.2 mb. Wind on the station model is depicted by staffs and wind barbs, as described in Chapter 1 (see Figure 1.12). Looking at the example in Figure 3.1, the wind is blowing from the southeast toward the northwest. This is a southeast wind at 10 knots. If winds are calm, no staff is plotted. Instead, a second circle is drawn around the station.

Significant weather and cloud cover are also routinely reported and coded on station model reports. There are over a hundred symbols for various weather conditions. Common symbols appear in Figure 3.1. The weather symbol is plotted to the left of the station between the temperature and dewpoint temperature. If no symbol is plotted, no significant weather is occurring at the station. The cloud cover is reported by darkening the station symbol. Figure 3.1 shows how to interpret the cloud coverage. An obscured sky is one in which the reporter or automated sensors can't determine the cloud cover. It might be foggy or smoky, or there might be blowing snow obscuring the view.

Figure 3.2A shows a map of central North America with the station reports from many stations. It is tedious to look at this map because so much data is presented. To examine the distribution of temperature across the area depicted on the map, a meteorologist would have to carefully look at each station. To simplify interpretation of data, meteorologists ***contour***

data, that is, they draw lines connecting points on the map with the same value of temperature, pressure, dewpoint temperature, or other quantities.

Figure 3.2B shows the sea-level pressure contoured in 4 mb intervals. The lines of constant pressure on this map are called ***isobars***. Contouring the data using isobars reveals a strong low-pressure system located over northern Iowa. Look at the winds in Figure 3.2A. There is a close relationship between the pressure pattern in Figure 3.2B and the wind. The wind flows counterclockwise around the low-pressure center. A ***pressure gradient*** exists where the pressure changes over distance. Where the pressure gradient is strong (see locations where the pressure changes rapidly over a short distance, such as in South Dakota and Nebraska), the winds are faster. We will explore this relationship between pressure and wind when we study forces in the atmosphere in Chapter 7.

Figure 3.2C shows the temperature contoured at 5° F intervals. With this map, one can easily see the cold air over the western United States, warm air extending northward from the Gulf Coast, and a strong ***temperature gradient***—a rapid change in temperature with distance—across the states of Texas, Arkansas, and Missouri. Temperatures along the Gulf Coast are in the 70s and in the 20s over the Rocky Mountains. The lines of constant temperature on this map are called ***isotherms***.

The dewpoint temperature is contoured in Figure 3.2D. Note that the warm air is also moist (high dewpoint temperatures), and the cold air over the mountains is dry (low dewpoint temperatures). Lines of constant dewpoint temperature are called ***isodrosotherms***.

ONLINE 3.1

Using Surface Maps to Find Significant Meteorological Features

Animating maps showing atmospheric properties allows meteorologists to track important atmospheric features as they move across a region and trigger hazardous weather. Online are animations of temperature and dewpoint temperature during a forty-eight hour period in June 2007. Note at the beginning of these animations the rapid drop in temperature and dewpoint temperature across Colorado. The temperature gradients in these regions extend northward and southward. Run the animations. How do the temperature gradients evolve during the forty-eight hours? Does the cold air move, and in which direction? Where is the boundary between the warm and cold air after twenty-four hours? How about after forty-eight hours? Now look at the animation of pressure and wind. Note the presence of a low-pressure center and the winds flowing counterclockwise around it. Are these winds consistent with the movement of the cold, dry airmass that was initially over the western United States? As we will learn in Chapter 9, the boundary between the cold and warm air, marked by the strong temperature gradient, is a cold front. Look at the satellite and radar animation of this data. How would you characterize the clouds and precipitation associated with the front as it moves across the Great Plains?

CHECK YOUR UNDERSTANDING 3.1

1. How do you read temperature, pressure, dewpoint temperature, wind speed, and wind direction on a station model?
2. What are the primary symbols for significant weather and cloud cover used on a station model?
3. What is contouring? How does contouring simplify the reading of weather maps?
4. What are lines of constant pressure, temperature, and dewpoint temperature called?

PRESSURE AS A VERTICAL COORDINATE

In Chapter 1 we learned that pressure decreases with altitude. Since each altitude above a point on the Earth's surface has a unique value of pressure, pressure can be easily substituted for altitude as a coordinate to specify locations in the vertical. The use of pressure as a vertical coordinate in meteorology has roots with aviation, rawinsonde measurements and meteorological theory. Most aircraft measure altitude with ***altimeters***, devices that use pressure to estimate altitude above sea level. Aircraft using altimeters actually fly at a constant pressure level, which is interpreted (incorrectly) as a constant altitude.

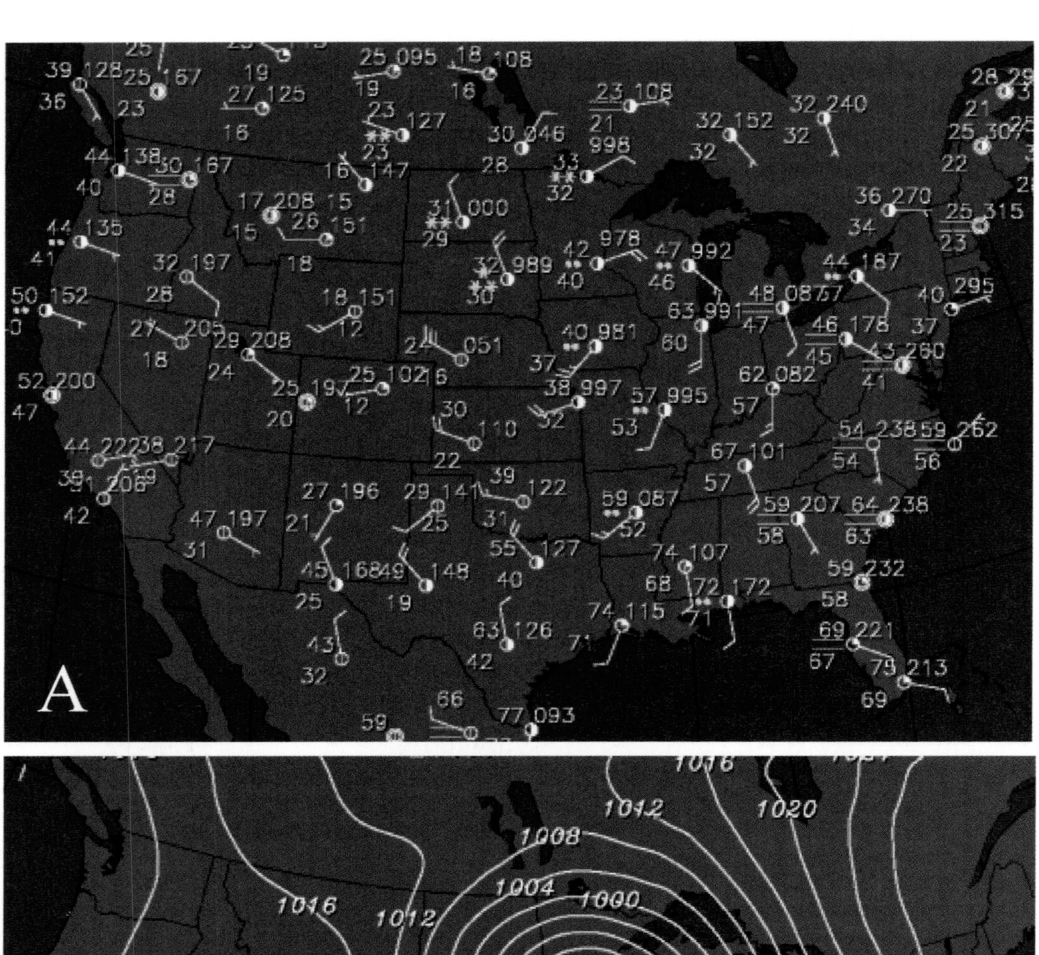

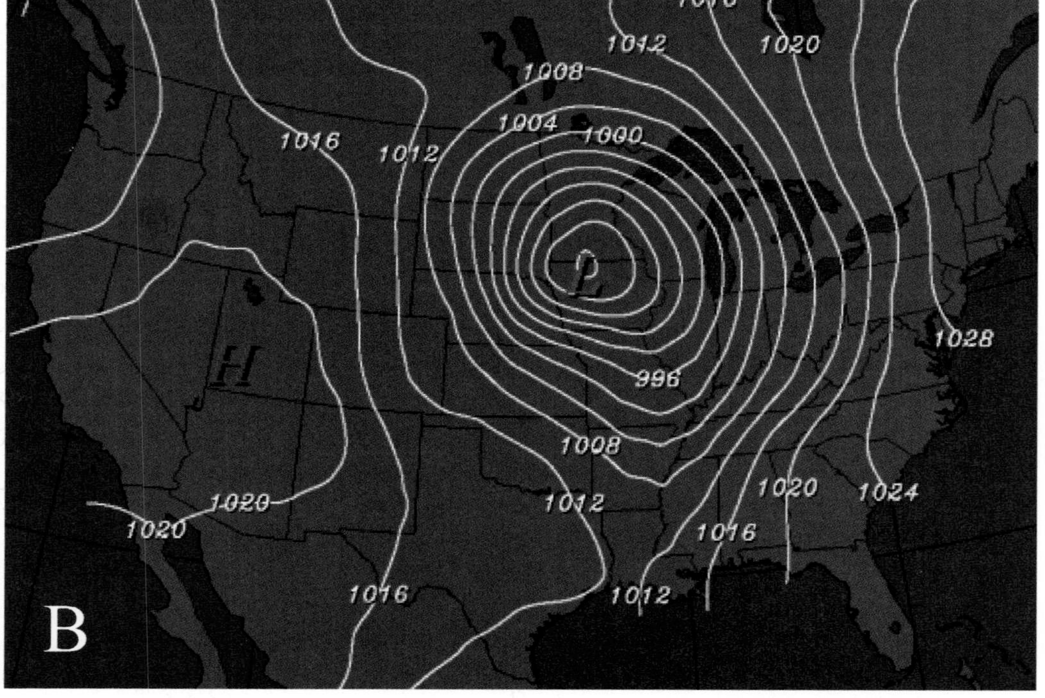

FIGURE 3.2: Maps of central North America showing (A) surface observations from stations within the region; (B) sea level pressure contoured every 4 mb.

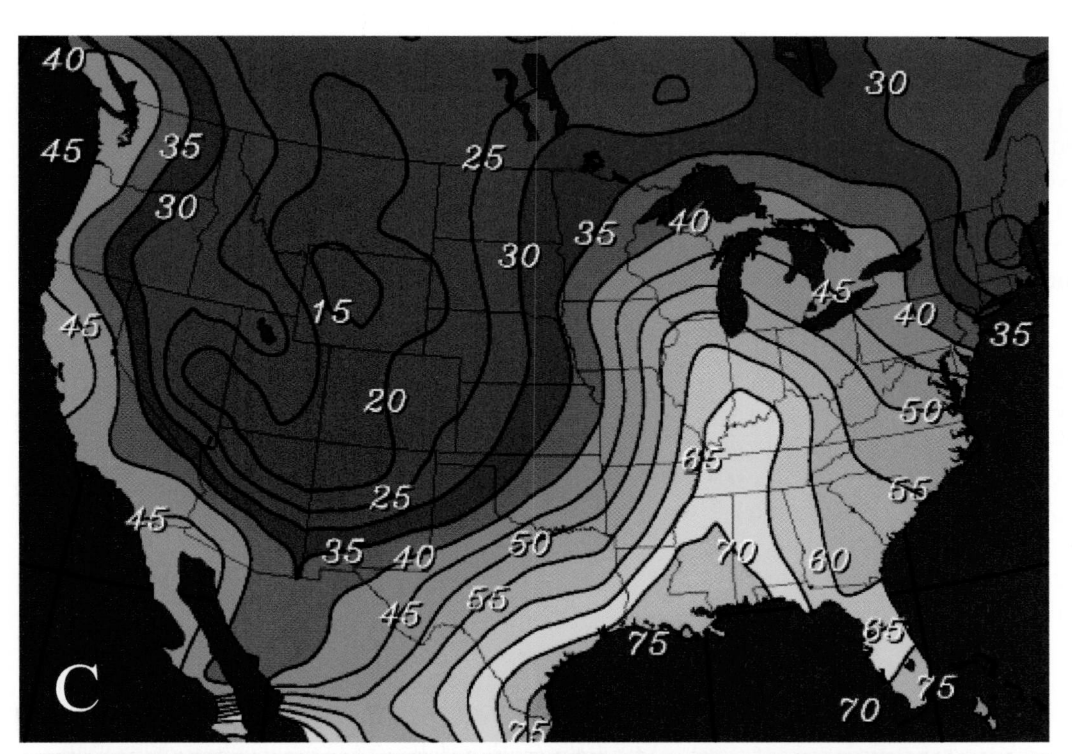

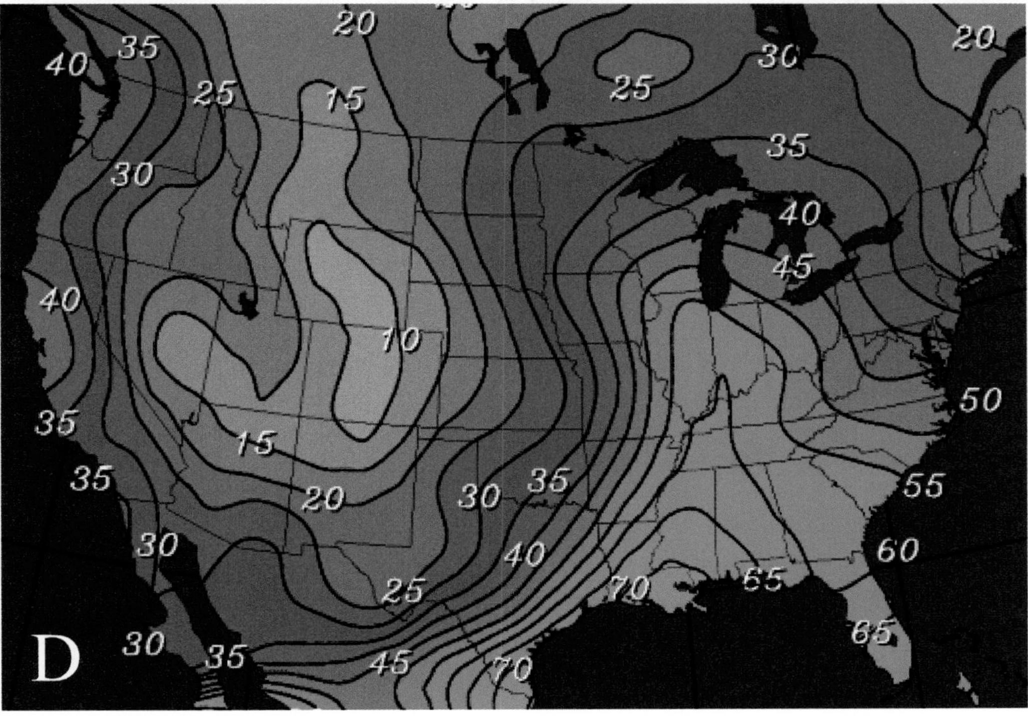

FIGURE 3.2 (continued): (C) temperature contoured every 5° F; and (D) dewpoint temperature contoured every 5° F.

Fortunately, the errors in altitude estimation have no impact on aviation safety. Nevertheless, because aircraft fly on constant pressure surfaces, upper air weather maps, first used extensively during World War II, traditionally have been plotted on constant pressure surfaces. In fact, rawinsondes, the balloon-borne instruments that measure upper atmospheric properties, determine the height of the instrument above the Earth's surface by measuring pressure, not altitude. The altitude of a rawinsonde is calculated from the pressure and temperature data. Finally, in fluid dynamics theories that explain atmospheric motion, equations are often cast in a framework where pressure is used as the vertical coordinate. For these reasons and others, meteorologists use pressure interchangeably with altitude as a vertical coordinate. As students of severe and hazardous weather, we will frequently use pressure to represent altitude above the Earth's surface.

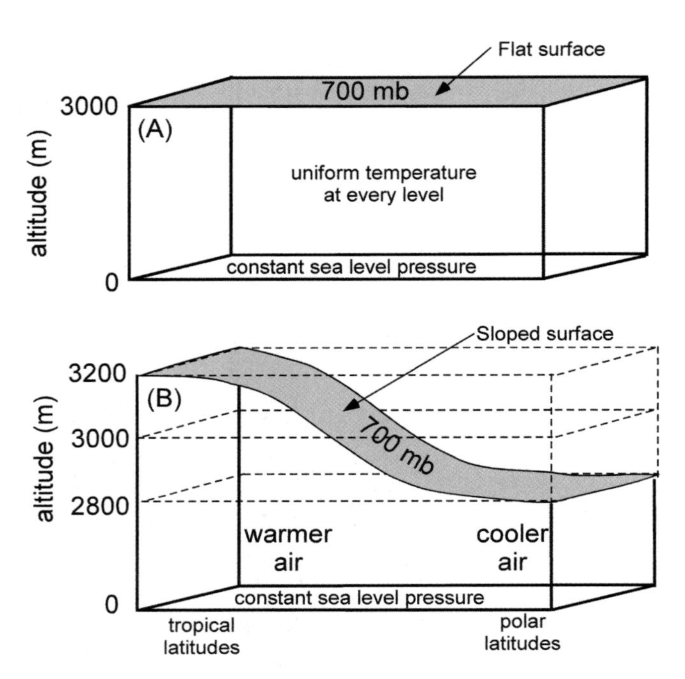

FIGURE 3.3: (A) A pressure surface aloft will be flat (parallel to sea level) if the temperature below that surface does not vary in the horizontal. (B) When temperature varies from one region to another, pressure surfaces will slope downward from warm to cold air. Since the tropics are warmer than the polar regions, pressure surfaces slope downward from the tropics to the poles.

CONSTANT PRESSURE MAPS

Imagine an experiment where we ascend through the atmosphere over a point on the Earth's surface until we reach a particular pressure level, say 700 mb. We leave a floating marker there. Then we make a second ascent at another point and leave another marker. Imagine we do this all over the world, leaving markers everywhere we encounter a pressure of 700 mb. Now imagine a surface connecting all the markers. When we speak of a ***pressure surface,*** we mean a surface above the ground where the pressure has a specific value, in the case of this example, 700 mb. What does such a surface look like?

Consider a small section of the atmosphere within the box shown in Figure 3.3A. Assume for simplicity that the bottom of the box is at sea level, temperature at any level in the box is uniform, and the sea-level pressure is constant across the box. Under these conditions, the 700 mb pressure surface will be "flat," that is, parallel to the ground. Let us assume that the 700 mb surface is located 3000 meters above sea level as shown in Figure 3.3A.

Now, as an experiment, let's heat the left side of the box and cool the right side of the box, below the 700 mb level. Furthermore, let's artificially constrain air from moving sideways. As a result of the heating and cooling, the air on the left side will expand while the air on the right will contract. The result is shown in Figure 3.3B, an example of a ***constant pressure map.***

Recall that pressure is the weight of the air above a unit area. Since we constrained the air to not move sideways, the weight of air in any column above sea level remained the same as it was before the experiment began. In other words, the surface pressure did not change anywhere in the box. Let us compare the pressure on the warm and cold sides of the box at the 3000-meter level in Figure 3.3B. Since air is now compacted on the cold side, but expanded on the warm side, the 700 mb surface will lower in altitude on the cold side and rise in altitude on the warm side. At 3000 meters, the pressure is now higher (> 700 mb) on the warm side and lower (< 700 mb) on the cold side. If we view the entire 700 mb surface in the box, the surface slopes downward from the warm to the cold side. Coming out of the box and back to the Earth, we know that the atmosphere in the polar regions is cold and the tropical atmosphere is warm. Figure 3.3B implies that the 700 mb surface, and indeed *all other pressure surfaces in the troposphere,* slope downward from the tropics to the polar regions.

You may have used topographic maps that show a mountain's topography by displaying contours of elevation. Meteorologists use similar maps that show the

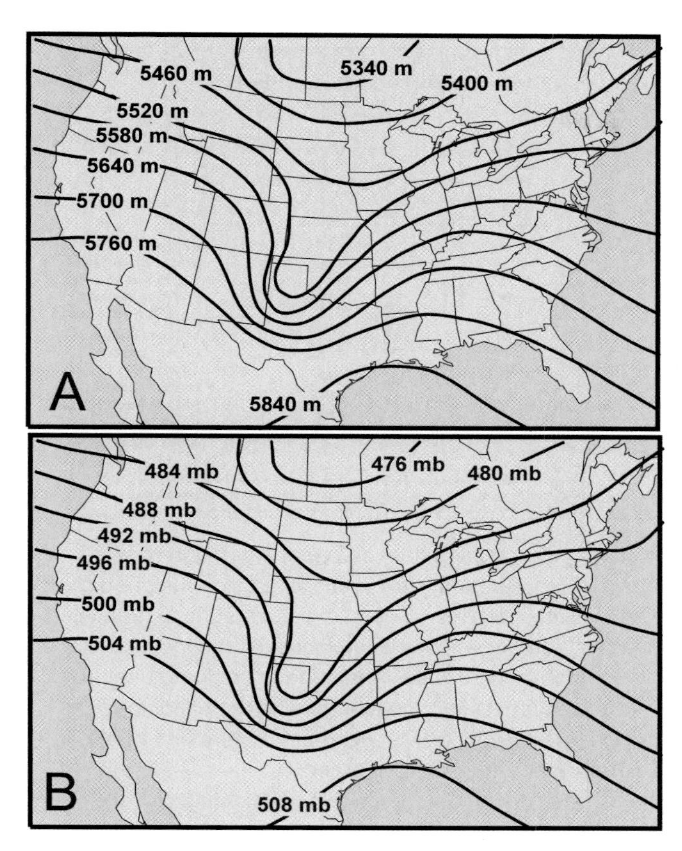

FIGURE 3.4: (A) An analysis of the height of the 500 mb surface. (B) An analysis of the pressure at an altitude of 5700 m above sea level. Note the similarities between the two maps.

height of a pressure surface above sea level. For example, Figure 3.4A shows the topography of the 500 mb surface over the United States on a particular day in December. The altitude of the pressure surface ranges from more than 5840 meters over the Gulf of Mexico to under 5340 meters over the central provinces of Canada. A valley, or ***trough***, in the pressure surface runs from western Texas northward to Minnesota, and ***ridges*** in the surface's topography lie on either side of the trough. We can see from Figure 3.4A that a trough is an elongated region where the height of a pressure surface reaches minimum values and a ridge is an elongated region where the height of a pressure surface reaches maximum values. What is the significance of such a map in weather analysis?

In the first section of this chapter we examined maps in which the pressure pattern at sea level was illustrated by plotting lines of constant pressure called isobars (see Figure 3.2B). The isobars used to display the pressure pattern allowed us to visualize where surface high- and low-pressure centers are located, the intensity of the pressure gradients, and by inference, the strength of the winds. *The height contours of a pressure surface convey the same information.* In fact, height contours on a surface of constant pressure appear almost exactly like pressure contours would appear if plotted on a nearby surface of constant altitude above sea level. Figure 3.4B, a map of the pressure field at an altitude of 5700 meters, illustrates this point.

As Figure 3.4B shows, the pressure gradient (change in pressure over distance) at a constant altitude is strong where the height gradient on a constant pressure surface is strong. Constant height lines have their highest pressure where constant pressure lines have their highest height and vice-versa. The maps look essentially the same, except one depicts the pressure field at a constant altitude (in the example, 5700 m), while the other depicts the height contours of a constant pressure surface (in the example, 500 mb). High pressure occurs at the same location as the ridges in the height field, low pressure at the same locations as the troughs, and strong and weak gradients in pressure at identical locations to strong and weak gradients in the height field.

Meteorologists use maps at constant pressure levels to depict conditions in the upper atmosphere. Because of the direct relationship between the pressure field at a constant altitude and the height contours on a constant pressure surface, we can infer that a strong pressure gradient exists where a strong height gradient exists, and that low- and high-pressure systems correspond in position on both maps. Furthermore, we can identify the positions of the axis of a trough (a line of minimum heights) and a ridge (a line of maximum heights). These features are dynamically related to ***cyclones***, the parent storms of many types of hazardous weather, as we shall see in future chapters.

Table 3.1 shows the most commonly available upper atmosphere constant-pressure maps and a typical altitude that each pressure surface can be found over the middle latitudes. These maps are created primarily from data collected by rawinsondes and are available twice each day corresponding to the worldwide time of rawinsonde launches, 0000 UTC and 1200 UTC. The National Weather Service chose the six levels listed on Table 3.1 as standard analysis levels because they enable meteorologists to resolve the most important features of the atmosphere.

The station model used for upper level maps appears in Figure 3.5. Temperature on upper level maps is always given in Celsius, and is placed in the upper left corner. In the lower left corner, meteorologists use a quantity called the ***dewpoint depression***, which is the

TABLE 3.1 Commonly Available Constant-Pressure Maps

Pressure Level	Approximate Altitude (feet)	Approximate Altitude (km)
850 mb	5,000 ft	1.5 km
700 mb	10,000 ft	3.0 km
500 mb	18,000 ft	5.5 km
300 mb	30,000 ft	9.0 km
250 mb	35,000 ft	10.5 km
200 mb	40,000 ft	12.0 km

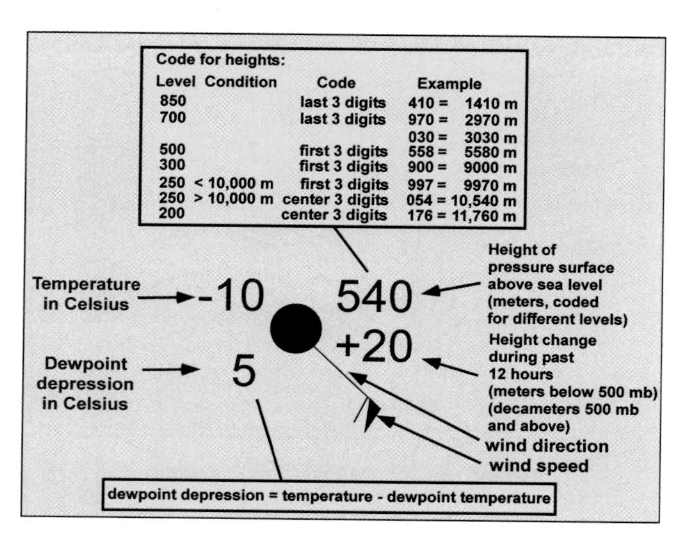

FIGURE 3.5: An example of an upper air station model showing temperature, dewpoint depression, height of pressure surface, height change of pressure surface, and wind direction and speed. The circle indicates the location of the rawinsonde launch. Unlike the surface station model, there is little standardization today regarding the shading of the circle at the station. Some weather charts found on the Internet omit the circle altogether.

difference between the temperature and the dewpoint temperature (°C). A value of zero for the dewpoint depression implies that the atmosphere is saturated (relative humidity of 100 percent), while a large value implies the atmosphere is dry. When the dewpoint depression exceeds 30° C, it is marked as an "X" on some maps. When the dewpoint depression is less than about 4° C, the air is close enough to saturation that clouds are likely to be close to the observation point.

The upper right corner of the upper atmosphere station model gives the height of the pressure surface in meters above sea level. On standard maps, this value is coded using three digits according to the table in Figure 3.5. The missing one or two digits must be inferred based on the pressure level of the map and the three digit value. In the lower right corner, the change in the height of the pressure surface during the last twelve hours is given in meters for levels below 500 mb and decameters (meters divided by 10) for 500 mb and above. Winds are presented using conventional wind barbs (see Chapter 1), with the value of the winds in knots. On modern maps appearing on the Internet, there is no convention for shading of the station circle. In fact, some maps no longer use a circle at all (e.g. Figure 3.6).

The 850 mb map is particularly useful to identify the location of ***fronts***, the boundaries between large airmasses. Fronts appear as strong temperature gradients. The 850 mb map in Figure 3.6 shows a front extending from New Mexico northeastward to Wisconsin. A less distinct front also extends from Wisconsin to New Jersey, while a stronger front (inferred by the sharpness of the temperature gradient) extends across eastern Canada near the top of the map.

The ***low-level jet***, a band of strong winds that flows parallel to a front on its warm side, is a second feature of interest sometimes found on an 850 mb map. The low-level jet acts to transport heat and moisture northward in the lower atmosphere, and contributes to thunderstorm rotation if thunderstorms erupt along the front (see Chapters 18 and 19). In Figure 3.6, a low-level jet flows from Texas northeastward to Iowa and Illinois with winds in the core of the jet reaching 60 knots.

Note the close association between the height contours and the winds on the 850 mb map. Except in mountainous areas, winds flow nearly parallel to the height contours, and their strength is closely related to the height gradient. We will explore the reason behind this relationship between the winds and the height contours in Chapter 7. Values of dewpoint depression on an 850 mb map can be used to determine the distribution of moisture in the lower atmosphere. In Figure 3.6, a band of nearly saturated air (low dewpoint depression) appears within the low-level jet ahead of the front.

Note in Figure 3.6 that several stations in the western United States have no data other than height. Much of the western United States is at such a high altitude that the reporting station's pressure is less than 850 mb. The 850 mb surface does not exist at these locations. The height of a "fictitious" 850 mb surface located below the ground is estimated for these stations from the station's surface pressure and temperature.

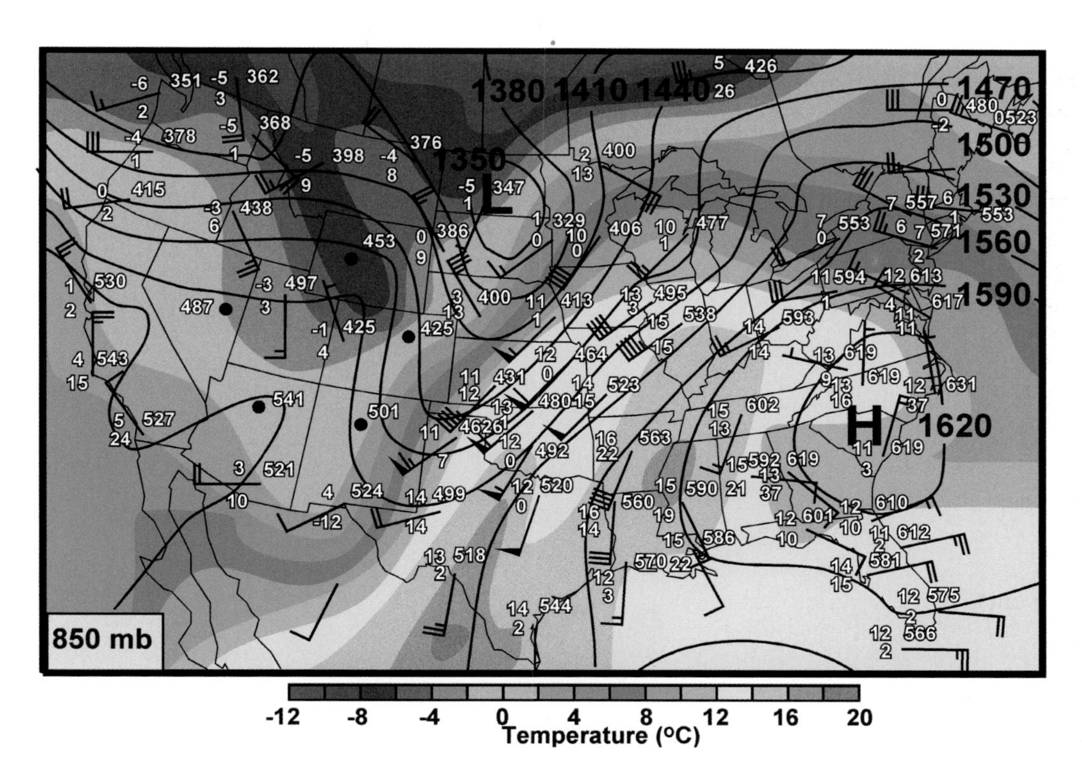

FIGURE 3.6: An 850 mb analysis. The solid lines denote height of the 850 mb surface above sea level, and the colors denote temperature (°C). The 12-hour height change has been omitted from the station model for clarity.

Height contours and isotherms are traditionally analyzed on 700 and 500 mb maps using similar conventions as the 850 mb map. Fronts become less distinct in the temperature field at 700 and 500 mb, but can sometimes still be identified. The 700 mb surface is located at an altitude that intersects many clouds, so the moisture distribution is an important feature on this map. The 500 mb map is used to determine the location of ***short waves*** and ***long waves*** associated with the ridges and troughs in the air-flow pattern. These waves are important because they directly influence the development of cyclones.

FOCUS 3.1

500 mb Maps and Vorticity

In the upper atmosphere, waves in the flow, marked by the troughs and ridges, are important because they can trigger the development of cyclones and smaller weather features such as thunderstorms. The location and intensity of waves in the flow can be measured using a quantity called *vorticity,* which is a measure of the local rotation in a flow. We can understand vorticity simply by placing a paddlewheel horizontally in a flow. Figure 3A uses the paddlewheel technique to show three sources of rotation of air in the atmosphere.

The first source of flow rotation is horizontal shear. In panel (A), the wind in the middle of a channel of flowing air is fast while the wind on the sides of the channel are slower (similar to the winds across a jetstream). A paddlewheel placed on the north side of the flow would spin counterclockwise, and on the south side clockwise.

The second source of rotation is flow curvature. Panel (B) shows a paddlewheel floating through a wave in the flow that consists of a ridge, a trough, and a second ridge. The paddlewheel rotates clockwise as it passes through the ridge and counterclockwise as it passes through the trough.

The final source of rotation in flow is the Earth itself. Panel (C) looks down on the Earth from the North Pole. As shown by the paddlewheel, even air

(continued)

(continued)

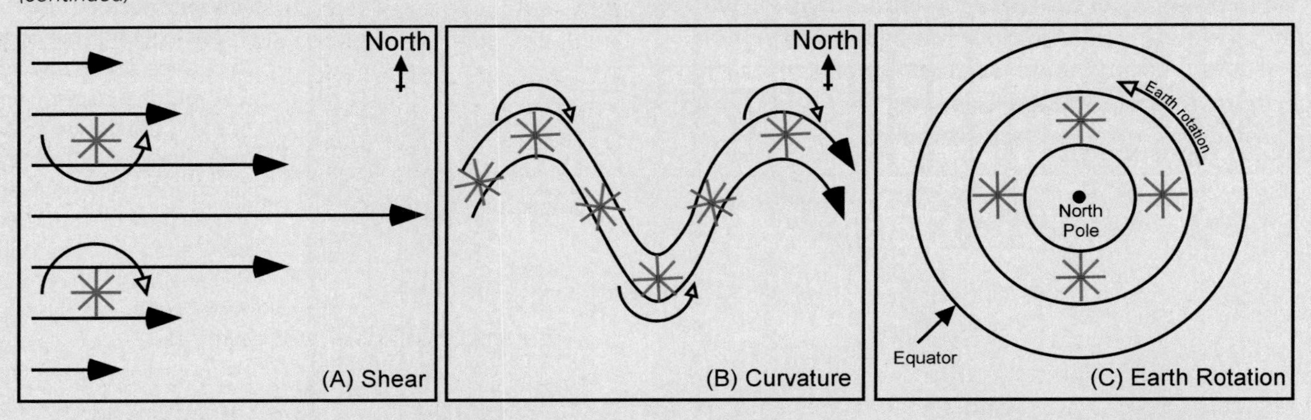

FIGURE 3A: Mechanisms that induce local rotation in a flow and create vorticity: (A) horizontal wind shear; (B) flow curvature; (C) the rotation of the Earth.

that is not moving relative to a point on the Earth's surface has a sense of rotation because the Earth is rotating.

The rate at which a paddlewheel turns in a flow due to the first two effects is a measure of the flow's *relative vorticity* (relative to the Earth), and due to all three effects is its *absolute vorticity.* Vorticity is measured as the change in wind speed (meters/second) divided by distance (meters). Since meters cancel, the unit becomes "inverse seconds" or per second. For large weather systems, the value of vorticity is a small number, of the order of 0.000010 per second, or in scientific notation 1×10^{-5} per second. A large value might be 5×10^{-5} per second. For example, the wind speed northwest of the jetstream axis in the Dakotas in Figure 3.7 decreases from 125 knots (63 meters/second) to 80 knots (40 meters/second) between the rawinsonde sites at Sioux Falls, South Dakota, and Bismarck, North Dakota—a distance of 480 km. The vorticity associated with the horizontal wind shear in this region would be

$$\text{Vorticity} = \frac{63 \text{ meters/second} - 40 \text{ meters/second}}{480{,}000 \text{ meters}}$$

$$\approx 0.00005 \text{ per second} = 5 \times 10^{-5} \text{ per second.}$$

Counterclockwise turning of the paddlewheel is conventionally considered positive and clockwise turning is negative. Absolute vorticity is frequently contoured on 500 mb maps to help meteorologists identify waves in flow and zones of horizontal shear. Vorticity patterns are particularly useful to identify very short waves in the flow that may be difficult to see in the height field. Figure 3B, for example, shows a vorticity analysis at 500 mb. The maxima in the vorticity pattern pinpoint several waves in the flow and "ribbons" of vorticity show horizontal shear zones. The 500 mb surface, which is in the center of the troposphere, is typically the altitude meteorologists use to examine vorticity patterns. On the Internet, 500 mb analyses, particularly those developed from numerical weather prediction models (see Chapter 4), will often display the contoured vorticity field rather than the temperature field.

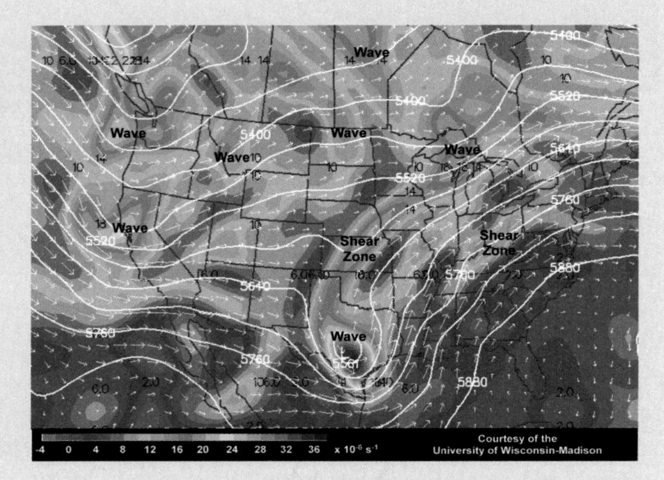

FIGURE 3B: An analysis of height (white contours), winds (arrows), and absolute vorticity (colors) on the 500 mb surface.

The 300, 250, and 200 mb maps are located near the top of the troposphere, and in the lower stratosphere. ***Jetstreams,*** bands of exceptionally strong winds that encircle the earth in the middle latitudes, are found at these pressure levels. In future chapters, we will examine how storm systems in the middle latitudes, whether they are large cyclones that cover half a continent or supercell thunderstorms that cover a small county, depend on processes that occur at jetstream level. Maps that depict the jetstream are very important in the study of hazardous weather.

Figure 3.7 shows a 300 mb map that corresponds in time to the 850 mb map in Figure 3.6. Instead of temperature, contours of wind speed, called ***isotachs,*** are presented on this map. Figure 3.7 clearly shows the jetstream, which flows southeastward along the west coast of the United States, eastward over Southern California and Arizona, northeastward into the Great Lakes region, and then eastward to the Atlantic Coast of Canada. Embedded within the jetstream are regions of exceptionally strong winds, one just off the Pacific Coast near Washington, a second over northeast Canada, and a third weaker region over Nebraska and South Dakota. These regions of exceptionally strong winds, called ***jetstreaks,*** directly influence the development of cyclones. Their location, intensity, and relationship to the position of the waves in the height field are all key features meteorologists examine using maps at 300, 250, and 200 mb. Three levels are examined because the core of the jetstream can be located at different altitudes during different storms. We will see, for example, that East Coast and Gulf Coast cyclones (Chapter 11) are often influenced by two or even three jetstreaks that occur simultaneously at different altitudes. We will learn how meteorologists interpret information on these maps in Chapter 8.

FOCUS 3.2

Upper Air Maps and the Internet

With the dawn of the Internet and the widespread availability of data from numerical forecast models, a profusion of formats and designs for constant-pressure maps has appeared on the World Wide Web. The maps appearing on the Internet differ substantially in both the quality of analyses and readability. Constant-pressure maps appearing on the Internet are either based directly on analysis of the rawinsonde data, or are developed from data grids used to initialize numerical weather prediction models. These data grids are created using mathematical techniques that quality control rawinsonde data and interpolate the data to grids where they can be

(continued)

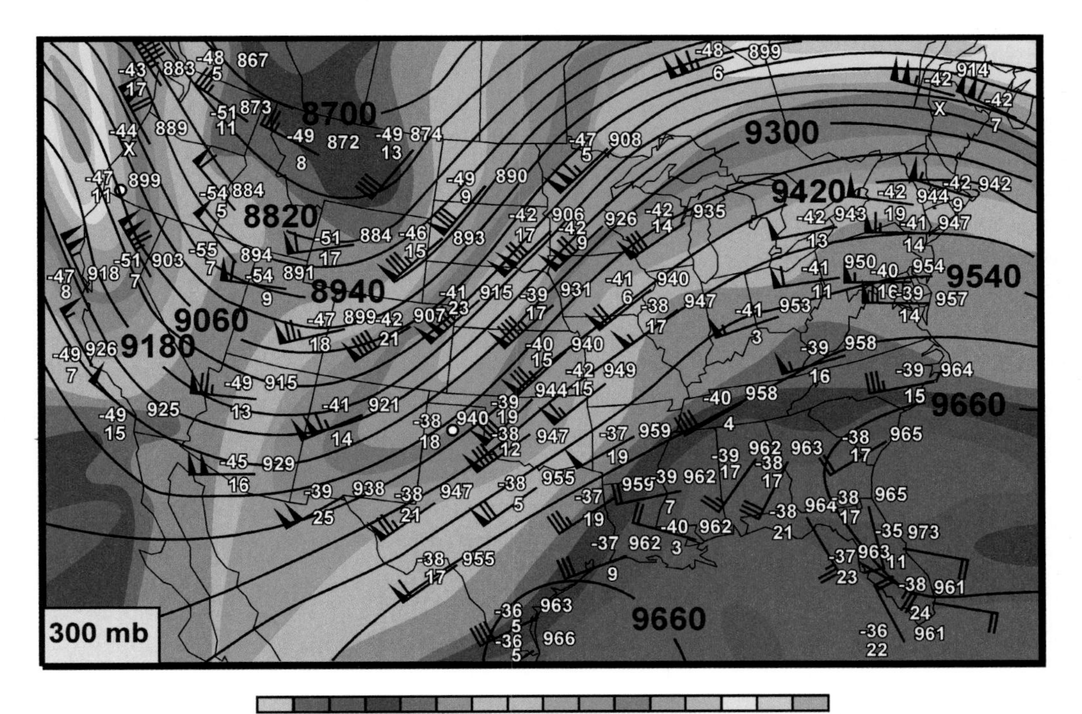

FIGURE 3.7: A 300 mb analysis. Solid lines denote height of the 300 mb surface above sea level, and the colors denote wind speed (knots). The 12-hour height change has been omitted from the station model for clarity.

(continued)

used in the numerical prediction model (see Chapter 4). Many maps on the Internet display partial station models, while others do not display station models at all. Most maps conform to the conventions described in this chapter; however, users must be careful and check the conventions for each map. For example, some maps now plot the dewpoint temperature instead of the dewpoint depression. Some format winds as arrows, with the length of the arrow proportional to wind speed, while others use standard barb conventions described in Chapter 1. Be sure to take the time to examine units and other plotting conventions on maps the first time you access them through the *Severe and Hazardous Weather Website.*

Now examine the 850 mb animation. Note the strong temperature gradient, indicated by the packing of the green lines near Colorado, which denotes the position of the cyclone's cold front at 850 mb. The precipitation that develops later in the cyclone's life develops along this front. We will investigate lifting and triggering of instability in Chapter 6 and the process of thunderstorm and tornado formation in Chapters 18 and 19.

ONLINE 3.2

Upper Air Weather Maps

In the animations in Online 3.1, we saw a low-pressure system move northeastward toward Canada, and a cold front, marked by the sharp temperature gradient, move across the Plains in response to the winds flowing counterclockwise around the low. As we will learn in future chapters, this low-pressure center is part of the circulation of an *extratropical cyclone,* a common swirling storm system found in the middle latitudes of the Earth. Why did this cyclone form? Why did the thunderstorms later erupt where they did? Upper level maps provide clues. Examine the animations of the 300 mb height and wind field which extends back two days before the animations begin in Online 3.1. Note the trough moving eastward from the Pacific and the jetstreak, the region of exceptionally strong winds, embedded in the trough. You can also see this trough moving eastward on the 500 mb animation. The low-pressure center is located east of the trough axis ahead and to the left side of the jetstreak. In Chapter 8, we will learn how troughs and jetstreaks at jetstream level are directly tied to the formation and evolution of cyclones.

CROSS SECTIONS

Consider slicing a birthday cake—after a cake is sliced from top to bottom and a piece removed, we can examine the cake's interior layering. In a similar way, we can slice the atmosphere from top to bottom, revealing its interior structure. A diagram that allows meteorologists to examine the atmosphere's vertical structure is called a ***cross section***. Cross sections permit meteorologists to visualize the vertical structure of fronts, clouds, jetstreams, and other features. In this text, we will use cross sections extensively to illustrate the structure of storms.

Figure 3.8 shows a horizontal map of the radar reflectivity measured in the infamous Oklahoma City supercell thunderstorm that produced a devastating tornado on May 3, 1999. How did this storm appear to the radar in the vertical? Figure 3.9 shows a cross section extending through the storm's hook echo, which marks the location of the tornado (see Focus 2.5), and its main precipitation area. (The line AB on Figure 3.8 marks the exact position of the cross section location). Note the weak echo region (green area) between the hook echo and the main body of the storm. This is the location of the storm's updraft. The intense radar echo in the hook itself is caused by debris from the tornado. The strong (red) echo in the main body of the storm is indicative of hail. In future chapters, we will use cross sections like those in Figure 3.9, as well as constant pressure maps and soundings, to illustrate and understand the storms that create severe and hazardous weather.

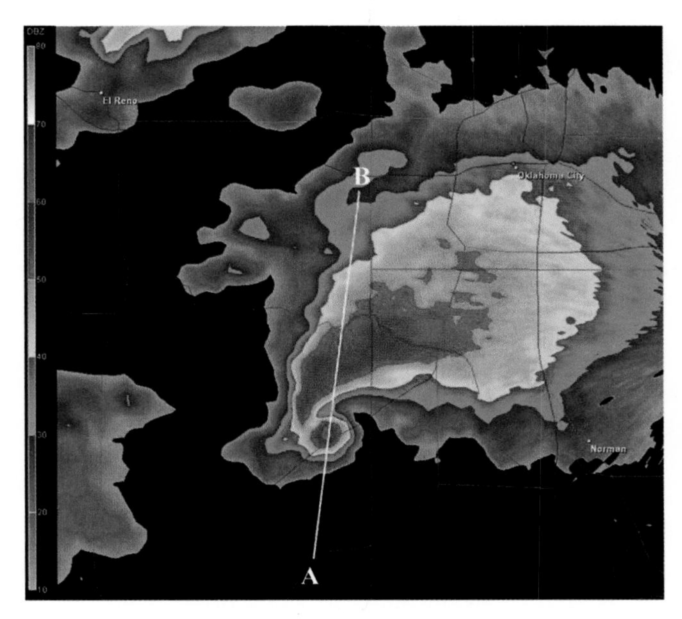

FIGURE 3.8: Horizontal projection of radar reflectivity through the supercell thunderstorm that produced a devastating tornado on May 3, 1999 in Oklahoma City, Oklahoma. The cross section in Figure 3.9 is along line AB.

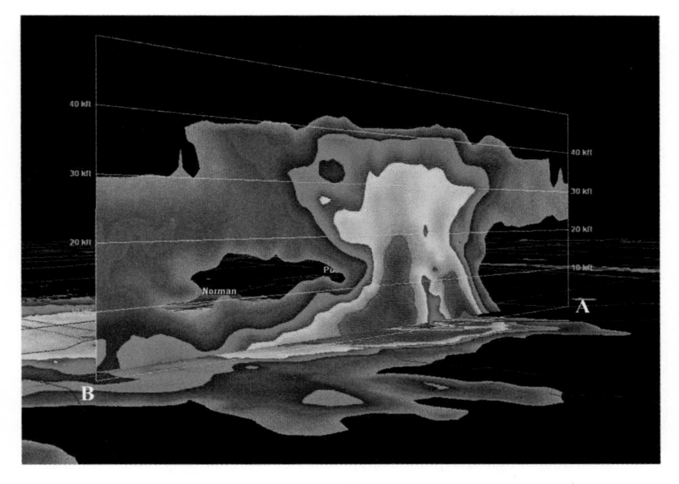

FIGURE 3.9: Vertical cross section of radar reflectivity through the supercell thunderstorm that produced a devastating tornado on May 3, 1999 in Oklahoma City, Oklahoma. The location of the cross section is shown on Figure 3.8.

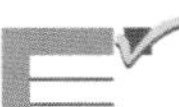

CHECK YOUR UNDERSTANDING 3.2

1. How are topographic maps of the Earth's surface similar to constant pressure maps of the upper atmosphere?
2. How does the upper air station model differ from the surface station model?
3. What constant-pressure maps might you examine to locate the jetstream? How about fronts?
4. What information does a cross section provide that is different from an upper level map or a surface map? – Shows Vertical Structure

TEST YOUR UNDERSTANDING

1. Decode the station model for Huron, South Dakota in Figure 3Q1.

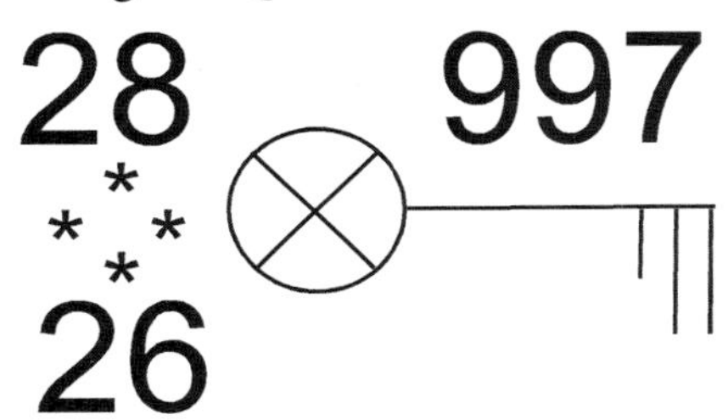

2. How would you identify a strong temperature gradient on a surface map? How about a strong pressure gradient?
3. What is the convention for wind barbs on a station model?
4. Why can pressure be used as a vertical coordinate in meteorology?

5. Altimeters measure pressure, not altitude. Why doesn't this pose a problem for aircraft safety, given that there are over 500 commercial aircraft in the air over the United States at any time?

6. During what decade were upper air maps first used extensively?

7. At what levels (pressure and approximate altitude) are constant-pressure maps typically available?

8. How do upper-level troughs and ridges appear on a constant height map compared to a constant-pressure map?

9. Is the slope of the 500 mb surface steeper during the Northern Hemisphere winter or summer? Why?

10. In Figure 3.6, only height data are available at Reno, Nevada. In Figure 3.7, no wind data is available at Amarillo, Texas. Give a possible reason why these stations have only partial data.

11. Decode the upper air station models for Green Bay, Wisconsin in Figures 3.6 and 3.7.

12. What is the dewpoint depression? What information does the dewpoint depression provide about the atmosphere?

13. What are the key features of interest on an 850 mb map? A 300 mb map?

14. What is a cross section?

15. How do constant-pressure maps differ from cross sections in the information they convey?

TEST YOUR PROBLEM-SOLVING SKILLS

1. Aviation instructors use the phrase "High to low, look out below!" as a cautionary warning for student pilots when their departure airport reports higher sea-level pressure than their destination airport. Explain this simple warning.

2. Use Figure 3.7 to answer the following questions.
   (a) Name the state in the continental United States where the 300 mb surface is at its lowest altitude.
   (b) Name the state in the continental United States where the 300 mb surface is at its highest altitude.
   (c) If you were to draw a map of the pressure pattern at 9500 meters above the surface of the Earth, would it differ substantially from what you see on this map? Why or why not?
   (d) What is the direction and speed of the wind at Denver, Colorado?
   (e) What is the temperature and dewpoint temperature at Springfield, Missouri?
   (f) Suppose you flew an aircraft from International Falls, Minnesota, to Miami, Florida, and used your altimeter to stay at a constant pressure of 300 mb. How much would your altitude change?
   (g) What is the pressure on this map at Phoenix, Arizona?

3. Use the data on Table 3.1 and your knowledge of average sea-level pressure to estimate how far a person living in the coastal city of Jacksonville, Florida, would have to climb a ladder to experience a pressure drop of one millibar. Now, use the station model data for Jacksonville and Miami, Florida, in Figure 3.2A to estimate the distance a person would have had to walk southward along the beach on the day this map represents to experience a pressure drop of one millibar. How do these two distances compare? (Note: The distance between Jacksonville and Miami is 549 km [341 miles].)

4. (a) Using data from Figure 3.7, estimate the vorticity associated with horizontal wind shear between:
      (i) Boise, Idaho and Reno, Nevada
      (ii) Caribou, Maine and Boston, Massachusetts
   (b) What is the relationship between the sign of the relative vorticity and the curvature of the flow?
   (c) How might a meteorologist use the values of vorticity to forecast changes in the weather?

USE THE SEVERE AND HAZARDOUS WEATHER WEBSITE

http://severewx.atmos.uiuc.edu

1. Use the *Severe and Hazardous Weather Website* to obtain a map of surface station observations of the U.S.
   (a) Print or save the map.
   (b) Record the date and time of the map.
   (c) Locate the station closest to your town. Decode the station model including units when appropriate.
   (d) Examine the entire map. List all the types of significant weather occurring across the United States.

2. The term "gradient" is used to indicate a change in some quantity (like temperature) over distance. For example, if the temperature in Philadelphia, Pennsylvania, is 60° F and the temperature in New York, New York, 90 miles (175 km) away, is 20° F, we would say that there is a strong gradient, or change, in temperature between the two cities. A strong gradient in temperature or dewpoint temperature marks the boundary between different airmasses. These boundaries are called fronts.
   (a) Obtain a contoured map of current temperature across the United States. Save or print the map.
   (b) Record the date and time of the map.
   (c) Examine the temperature map and identify regions where a front might be present. Draw a line(s) on the map to mark the position of the front(s). Your line should parallel the isotherms and be located along the "warm" edge of the temperature gradient.
   (d) Obtain a contoured map of current dewpoint temperature across the United States for the same date and time as the temperature map. Save or print the map.
   (e) Draw a line(s) on the dewpoint map in the same location as the line(s) in (c). Do you see a gradient in dewpoint temperature in the same region? Provide a reason why you think there is or is not a gradient.
   (f) Are there any examples where a strong dewpoint gradient is present for a location but not a strong temperature gradient?

3. Use the *Severe and Hazardous Weather Website* to find a current 850 mb map and a 300 mb map for the same date and time that has data at individual stations.
   (a) Print or save the maps.
   (b) What is the date and time of the maps?
   (c) Find the upper air station closest to your town. Decode the conditions at both 850 mb and 300 mb. Be sure to include units when appropriate.
   (d) What are the coldest and warmest temperatures reported on each map?
   (e) What is the fastest wind speed reported on each map?
   (f) Estimate the average slope of each pressure surface between International Falls, Minnesota, and Corpus Christi, Texas. How do the slopes of the two maps compare? Is this what you would have expected? Why or why not?
   (g) Using the 850 mb map, estimate where low-level clouds might be present. List these locations. What variable did you use to determine this?
   (h) Examine the 850 mb map for fronts. Identify the location of any fronts by drawing a line indicating their position.
   (i) Examine the 300 mb map for strong winds. Identify the location of the polar jetstream by shading the region where it occurs. If there are any jetstreaks present, draw a line outlining their location.
   (j) Is there a relationship between the fronts you may have found in (h) and the location of the polar jetstream as identified in (i)? If so, describe what you observe.

4. Use the *Severe and Hazardous Weather Website* to find a 500 mb analysis with the vorticity field contoured.
   (a) Print or save the map.
   (b) What is the date and time of the map?
   (c) Identify waves in the flow from the vorticity analysis. Label each wave "T" in the trough and an "R" in the ridge.
   (d) Find zones of strong horizontal shear. Place a circle around each area of strong shear and explain why you chose each selection.

5. Use the *Severe and Hazardous Weather Website* to find a current 300 mb map with winds analyzed.
   (a) Print or save the map.
   (b) What is the date and time of the map?
   (c) Locate the polar jetstream by shading in or circling the location.
   (d) Using the information in (c) above, find four soundings that together form a line that crosses the jetstream.
   (e) Print or save each sounding.
   (f) List the time, date, and location of each sounding.
   (g) Create a vertical cross section using the sounding data. Make a graph that has height as the vertical axis and distance as the horizontal axis. Let the distance on the horizontal axis be the distance between the most distant rawinsonde sites on your line. On your graph, write the value of each wind speed reported on each sounding at the altitude and location where the observation was made.
   (h) Highlight the vertical structure of the jetstream by drawing isotachs (lines of constant wind speed) on your cross section.

FORECASTING AND SIMULATING SEVERE WEATHER

Numerical model forecast for 850 mb six days in advance of a major cold outbreak in the United States.

KEY TERMS

analysis
analysis map
computer model
consensus forecast
ensemble forecasting
ensemble member
initialization
long-range outlook
medium-range model
mesoscale
mesoscale model
model domain
model grid
Model Output Statistics
model resolution
model timestep
numerical model
Numerical Weather Prediction (NWP)
parameterize
prog
research model
short-range model

Severe weather occurs in various forms in nearly every part of the world. However, accurate weather forecasts can minimize its impact. In many cases, accurate forecasts can reduce the loss of life and, if preparation is feasible, can reduce some types of property damage. While the occurrence of severe weather has always been a factor in human activities, there have been dramatic changes in its impacts in recent decades. On the one hand, our vulnerability to severe weather has increased because population and associated infrastructure have increased. On the other hand, our vulnerability has been reduced by the improvement of weather forecasts and warnings. Because of this improvement, we are now much more able to anticipate and prepare for severe weather than were previous generations. Indeed, Dorothy may never have ended up in the land of Oz if today's weather forecasts had been available to her!

Today's weather forecasts made four days in advance are about as accurate as two-day forecasts were in the 1970s. This substantial improvement in the accuracy of weather forecasts has occurred in about one human generation. Why have forecasts improved so much? The answer lies in the fact that today's forecasting relies heavily upon predictions generated by ***computer models,*** which are often referred to simply as ***numerical models***. These models are actually systems of mathematical equations that describe the behavior of the atmosphere. Forecasters examine and compare the models' predictions, adjust the predictions based on knowledge of local weather and other information they have available, and finally issue forecasts. While new observational tools such as satellites and wind profilers have helped forecasts, most of the enhancement of forecast accuracy in recent decades is attributable to improvements in computer models. The same types of models are used to simulate climate, and they provide the basis for projections of future climate change (Chapter 5). In this chapter, we describe numerical models and their vital role in weather prediction.

Numerical models are also used as research tools. Research simulations of phenomena such as hurricanes, severe thunderstorms, and tornadoes are invaluable in illustrating scientific principles and clarifying the processes responsible for severe weather. In fact, the U.S. National Weather Service has recently (late 2006) implemented a model known as the Weather Research and Forecasting model (WRF), which is used for research as well as for forecasting. As an example of a model's use in research, Figure 4.1 illustrates a line of thunderstorms simulated by a numerical model. By capturing the development and the major structural features of weather systems, numerical models can increase our understanding of hazardous weather. Improved understanding, in turn, leads to improved prediction and warnings. This chapter includes examples of numerical model simulations of various types of hazardous weather phenomena that will be discussed more thoroughly later in the book.

WEATHER FORECASTING BY COMPUTER MODELS

The science of predicting future weather using computer models is called ***Numerical Weather Prediction (NWP)***. The use of NWP models is really only part—albeit a crucial part—of a sophisticated process by which weather forecasts are produced. Figure 4.2 summarizes this process schematically. The collection, transmission, and synthesis of observations are essential to the production of current weather maps, or ***analysis maps***, and also for the start-up of NWP models. While NWP models are run in an automated mode, humans do have some input to the forecasts that are distributed to users, as shown in the green box on the left side of Figure 4.2. For example, human forecasters decide how much weight to give NWP model forecasts in a particular situation. Such decisions are often based on the forecasters' experience with particular NWP models.

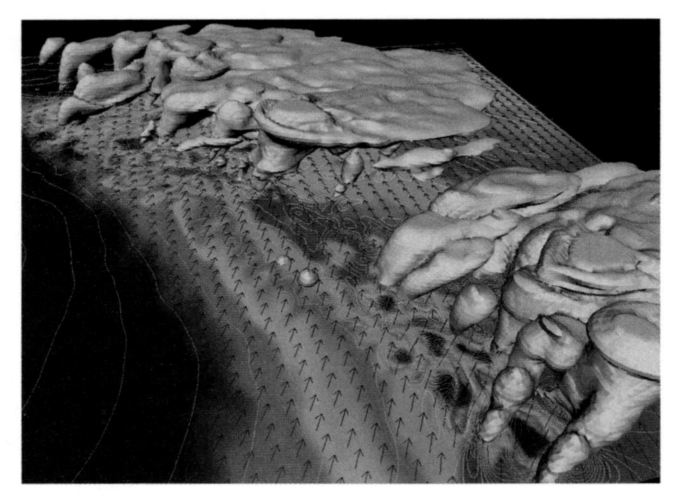

FIGURE 4.1: Numerical model simulation of a line of severe thunderstorms.

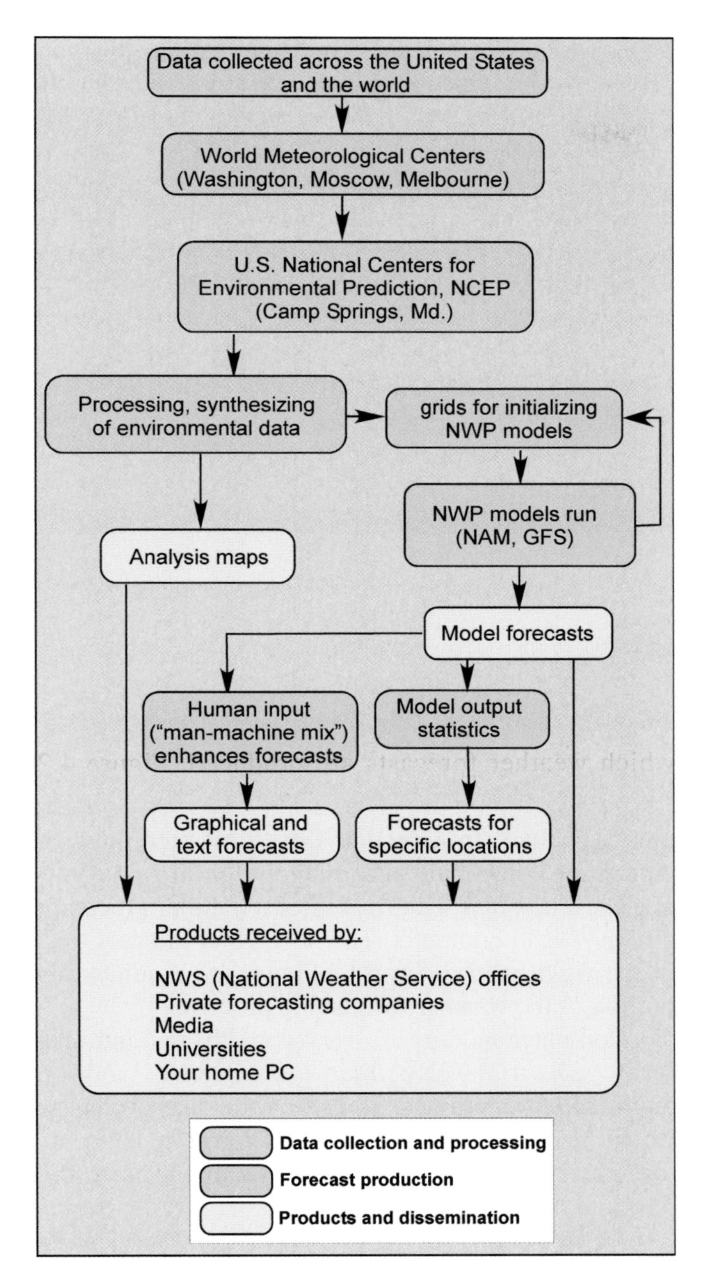

FIGURE 4.2: Schematic representation of the process by which weather forecasts are produced.

A typical computer model consists of an interconnected set of mathematical equations describing the atmosphere's behavior. Three of these equations describe how forces, such as gravity and friction, cause air to accelerate. These equations represent Newton's second law of motion (F = ma: Force = mass × acceleration) applied to the atmosphere. Other equations represent mathematically the facts that mass and energy can neither be created nor destroyed in the earth-atmosphere system; in the case of energy, the equation is a statement of the first law of thermodynamics. Some equations describe how heat and moisture are transferred from the Earth's surface to the atmosphere, others describe the transfer of water from its vapor form to liquid and ice, and still others describe how the sun's radiation heats the Earth's surface and the atmosphere. Some equations have a special property: they predict the future state of the atmosphere, given knowledge of the present state. These equations involve rates of change and spatial gradients, so they are generally written in terms of "derivatives" used in calculus.[1] The equations, when packaged together in a computer code, are called a computer model. For the model to be used, information about the present state of the atmosphere must be available on a regular array of points called a ***model grid***. Figure 4.3 shows two examples of such grids, one covering the Northern Hemisphere and the other covering the contiguous United States. The distance between grid points is approximately 380 km (235 miles) in Figure 4.3a, and approximately 20 km (12 miles) on the smaller grid in Figure 4.3b.

Initializing and Running a Model

The atmosphere is three-dimensional, so numerical models require data from different altitudes. Figure 4.4 shows the sixty vertical levels used to represent altitude in a typical weather prediction model. This particular model represents the atmosphere at each of its sixty levels by a grid of points covering the domain of Figure 4.5 with a horizontal spacing ***(model resolution)*** of 12 km. We know that data are not collected on such a three-dimensional grid. They are collected at the surface in cities and at airports, and aloft using irregularly spaced rawinsondes, profilers, radars, aircraft, and satellites (Chapter 2).

Complicated methods have been developed to take all of this data and interpolate it on to the grids used in numerical models (the fourth tier from the top in Figure 4.2). These methods include a series of checks to catch any gross errors in the measurements or in the transmissions of the observations. The initialization of a model grid begins with a "first guess" at the initial state of the atmosphere. This guess is normally the twelve-hour forecast made by the same model during its previous forecast cycle. The observational data are used to adjust

[1]Numerical weather prediction is one of the most practical applications of calculus, although—to the dismay of students of calculus—the solution of the differential equations is generally achieved by rewriting the equations in terms of finite differences.

A

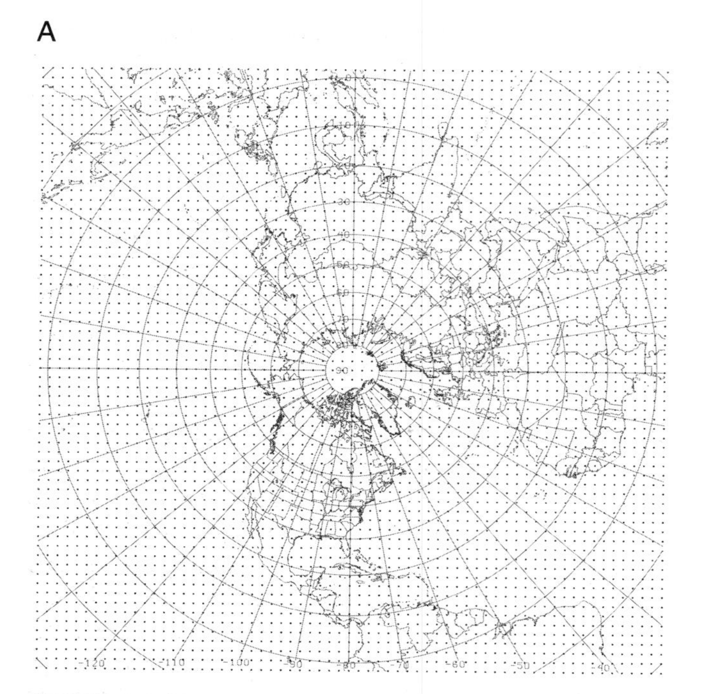

B

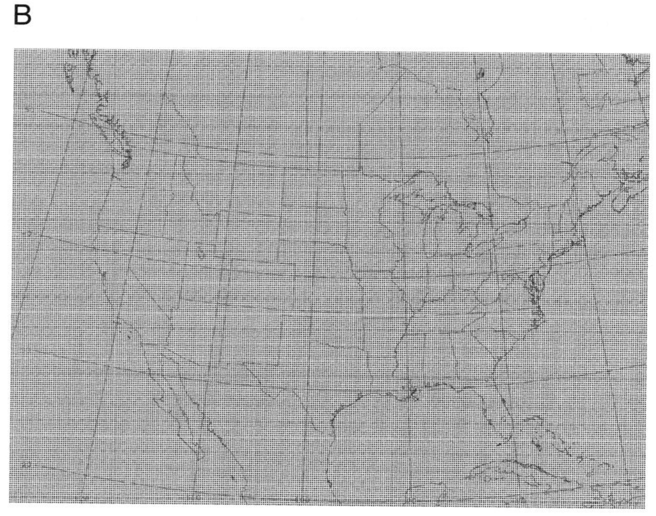

FIGURE 4.3: Examples of grids used in numerical weather prediction models. Weather variables are evaluated at each timestep at all points on grids such as those shown here.

this guess, bringing the model grid as close as possible to the true atmospheric state. The adjusted values are also smoothed to ensure that they do not contain any unrealistic "jumps." Once these steps are performed, a final adjustment is made to insure that the data on the model grids satisfy the equations that govern the atmosphere—otherwise the model forecast will quickly go awry. In Figure 4.2, the fourth and fifth tiers illustrate how this blend of information produces the initial state from which an NWP forecast is obtained. While this ***initialization*** process may seem cumbersome, NWP forecasts would fare much worse if the initialization process were not carried out with care. Sample output from the initialization of a weather prediction model is shown in Figure 4.6.

From the initial values at each grid point, the rates of change of variables like wind speed, temperature, and pressure are evaluated from the equations and are calculated over a period of time called the ***model timestep,*** which is typically about five minutes. The change of a quantity (e.g., temperature) accumulated over this five-minute timestep is then added to the initial or "old" value of that variable, creating a "new" value of the variable. The new value then becomes the old value, and the procedure is repeated again and again, advancing the forecast into the future, five minutes at a time. In actuality, this "time-stepping" or "marching" procedure is carried out for all grid points at all levels, often with the aid of a parallel processor on a powerful computer. The box "NWP models run" in the right-hand column of Figure 4.2 denotes this portion of the weather forecasting process.

Computer models are very complicated and take many years to develop. Teams of scientists work on these models continually, incorporating ways of better using the data to initialize the model grid, to calculate the terms in the equations, and to incorporate the physics of the atmosphere.

As an example of the difficulty faced by scientists developing these models, consider two problems: snow cover and topography. We know from experience that when the ground is snow-covered and the sky is clear, the temperatures at night will drop very rapidly. When the ground is snow-free, the temperatures drop more slowly. For a computer model to correctly predict a low temperature, the physics of the snow's ability to radiate energy must be incorporated. Even if the physics are included in the model, the model has to use the correct distribution of snow cover. Accurate data on the extent of snow cover must therefore be available for the model's initial state, and the model must be able to predict changes of snow cover as the forecast evolves. In practice, the snow depths used for the model's initial

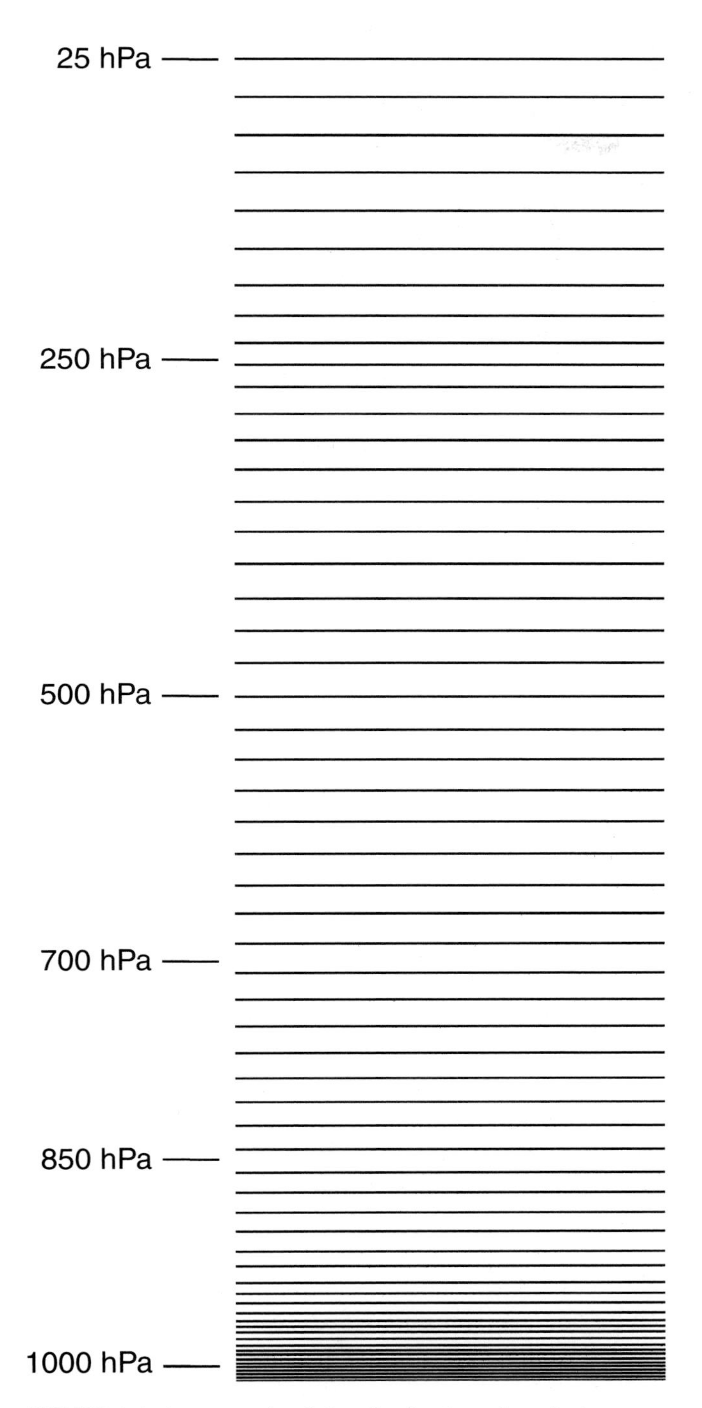

FIGURE 4.4: An example of the distribution of vertical pressure layers used in a numerical prediction model. The layers are spaced closer together near the ground, and farther apart in the upper atmosphere.

state are estimated, since it is very difficult to measure directly all the variations of snow depth and coverage, especially in remote or rural regions.

As our second problem, consider topography. We know that mountains have a tremendous effect on weather patterns. It is important in a model to represent topography as accurately as possible. Figure 4.7 shows the eastern United States topography used in the National Weather Service's North American Model (NAM), which runs at 12 km resolution, in comparison with the topography over the same area when the resolution is 22 km. The representation of the Appalachians and other topography is clearly sharper and more realistic at 12 km model resolution than at 22 km resolution. Although the 12 km pattern appears very complicated, it still misses many small-scale features that are below the resolution of the grid. In actuality, the representation of topography in a computer model is akin to using huge Lego building blocks to represent the landscape—each grid cell, ranging from about 12 km on a side in an NAM-like model to about 100 km in a global model, has a single elevation that differs from the elevation of each surrounding grid cell (except over the ocean, where many adjacent grid cells have the same elevation, as in Figure 4.7). Since individual mountains and valleys are not represented in the model, the model cannot "predict" localized weather effects created by these individual features.

A computer must make a forecast quickly, otherwise the weather events will occur before the forecast of the events is available. To provide timely forecasts, supercomputers (or, more recently, clusters of powerful computers) are currently used to run most models. Consider the problem: the present North American Model has about 700 × 800 grid points per level × 60 levels × 10 variables = 336 million pieces of information. That's just for one time. The model then calculates the state of the atmosphere a few minutes into the future, generating another 336 million pieces of data. And a few minutes later, another 336 million, each of which required perhaps a hundred arithmetic operations on the computer. As forty-eight hours of prediction occur, trillions of calculations are done, creating new grids of numbers at successive times. It is no wonder that weather prediction centers were among the first regular customers of the original supercomputer companies back in the 1960s and 1970s! Even with the fastest computers now available, computer power is still a limitation in numerical weather prediction, requiring the coarser-than-desirable resolution of 50 to 100 km in models covering the entire globe. Fortunately, computer models can be run for any size region ***(model domain)*** of interest to the forecaster. While some models are run for the entire globe, others are run for a small area,

FOCUS 4.1

An Example of the Effect of the Initialization of a Model

The top panel in Figure 4A shows the hourly surface reports for a particular time at a number of stations in the United States. Recalling the decoding of surface station reports in Chapter 3, note the reported sea-level pressures in the upper right portion of each report. As these and other reports were processed into an initial field of sea-level pressure for the initialization of a numerical weather prediction model, the smoothing and balancing resulted in the pressure field shown in the bottom panel of Figure 4A. This field, used directly by the model to start its forecast run, differs slightly from the actual observations. What specific examples of differences between the two panels in Figures 4A can you find?

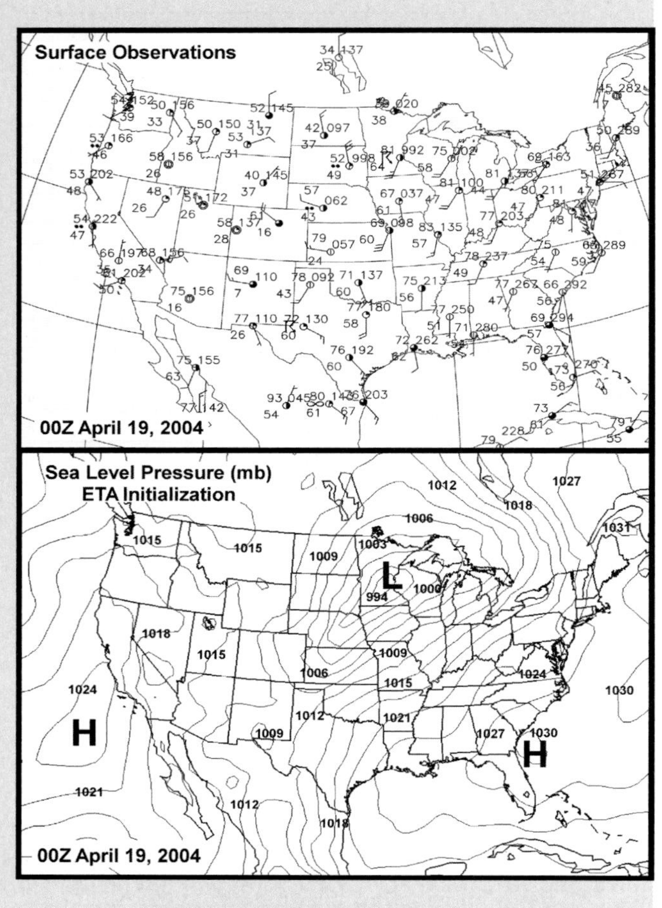

FIGURE 4A: Top: Map of sea-level pressures reported by weather stations. Bottom: Numerical model initialization of sea-level pressure (solid contours) for the same time as the station reports in the top figure.

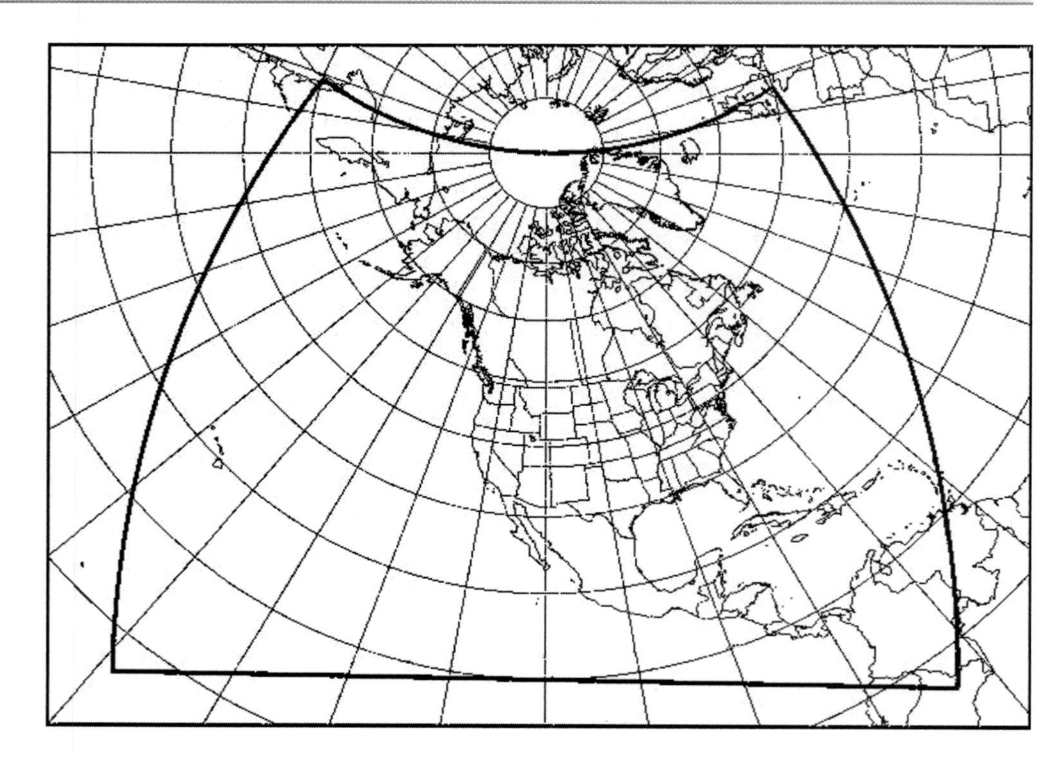

FIGURE 4.5: An example of a regional domain for a model predicting weather over North America.

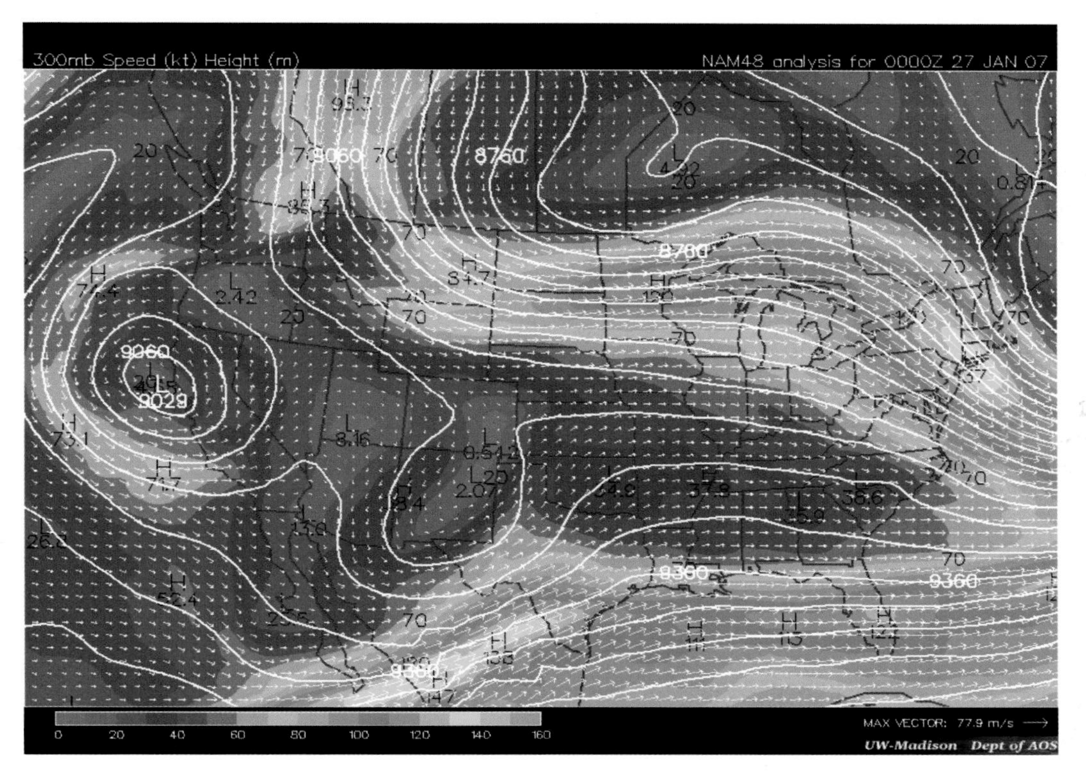

FIGURE 4.6: An example of the North American Model's initialization for 300 mb. This analysis of 300 mb conditions is based on interpolated observational data and a short-range forecast from the model's previous run. The analysis is used to initialize the computer model, i.e., to provide it with consistent information about the current state of the atmosphere so that it can predict the future.

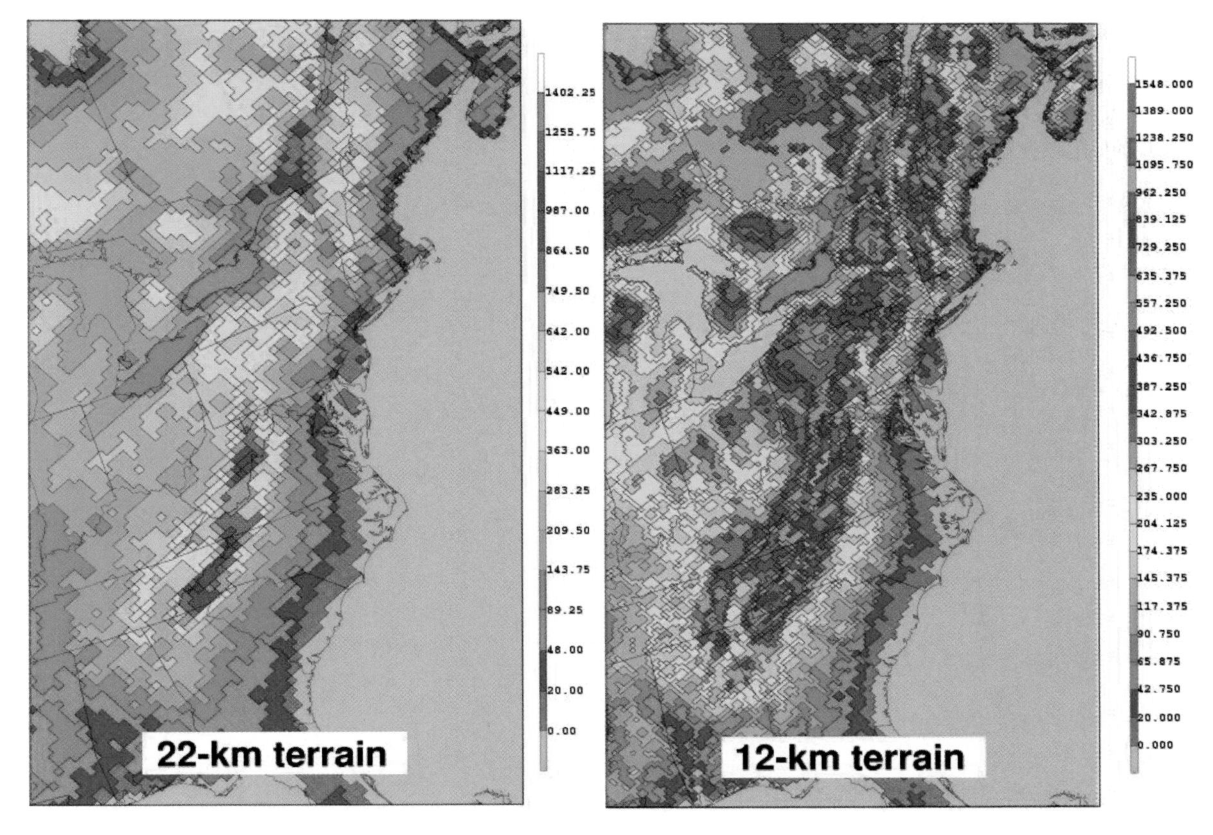

FIGURE 4.7: Representation of surface elevation (meters, color-coded) over the eastern United States as used in weather prediction models with resolutions of 22 km (left) and 12 km (right). 12 km is the resolution of the North American Model in early 2008.

such as a portion of the United States, in order to provide more detailed forecasts of specific conditions within that geographic region. Other models (the NAM model, for example) are run for larger areas such as all of North America and portions of the Pacific and Atlantic Ocean in order to forecast large-scale weather features at relatively fine resolutions of 10 to 20 km. Figures 4.3 and 4.5 illustrate different domains over which a model might be run.

FOCUS 4.2

The Origin of Numerical Weather Prediction

The concept of weather prediction by solving the dynamical and physical equations governing the atmosphere can be traced back to the early 1900s, when V. Bjerknes of Norway proposed the application of prognostic equations of fluid mechanics and thermodynamics to the atmosphere. The first attempt to implement this approach was made in 1922 by a British meteorologist, L. F. Richardson. Computers had not been invented in 1922 (nor had electronic calculators), so Richardson envisioned a huge room full of people working feverishly with pencil and paper—and perhaps an abacus or slide rule—to perform the calculations that are now done by computers. In view of the huge number of calculations, it is not surprising that Richardson saw a need for hundreds of "human calculators" just to compute the forecast as fast as the weather was actually happening, and only over a portion of the globe. Unfortunately, a prototype forecast by Richardson was a dismal failure because inaccuracies in the initial state created imbalances that produced completely unrealistic forecasts. As a result, the development of Numerical Weather Prediction had to wait until the 1950s, when meteorologists led by John von Neumann at Princeton University capitalized on the availability of one of the first electronic computers to show that numerical models could actually produce plausible forecasts. An NWP model first came into routine use in 1960 at the National Meteorological Center of the U.S. National Weather Service, which led the way in implementing numerical models for short-range weather forecasts. In the 1970s, NWP models were used to obtain useful forecasts out to ranges of five to seven days at the new European Center for Medium-Range Weather Forecasts (ECMWF), which remains one of the major producers of numerical weather forecasts. Most of the developed nations now have NWP centers.

CHECK YOUR UNDERSTANDING 4.1

1. What is numerical weather prediction?
2. What does a computer model consist of?
3. Describe how a computer model is initialized and run.
4. Provide two examples of the problems scientists face in developing computer models.

Output from Models

The grids of numbers that a computer model generates are used to create weather maps of the future state of the atmosphere, typically 6, 12, 18, 24, 30, 36, 42, 48, 54, 60, 72, and 84 hours after the initial time. These maps must be generated in a timely fashion and distributed to forecasters quickly. Forecasts by most NWP models are initialized twice daily, at 0000 UTC and 1200 UTC, corresponding to the time that data are available from the rawinsondes.[2] Hence the six-hour forecast maps from the model "run" started at 0000 UTC are typically produced for 0600 UTC, 1200 UTC, 1800 UTC and 0000 UTC of each twenty-four-hour period of the forecast. The maps from the run started at 1200 UTC are typically produced for 1800 UTC, 0000 UTC, 0600 UTC, etc. The first forecast model results usually become available about three hours after the initial time, i.e., at about 0300 UTC or 1500 UTC. Figure 4.8 shows a forecast map, for thirty-six hours into the future, of surface pressure and precipitation from the North American Model. The NAM model is run four times per day (starting at 0000, 0600, 1200, and 1800 UTC). Figure 4.9 shows a six day forecast of 300 mb height and winds from the Global Forecasting

[2]For some models, initializations are also performed at 0600 UTC and 1800 UTC. These initializations receive little, if any, information from rawinsondes, which are launched only at 0000 UTC and 1200 UTC at most locations. The 0600 UTC and 1800 UTC initializations are based primarily on the six-hour forecasts from 0000 UTC and 1200 UTC, modified by surface observations and aircraft, profiler and satellite information on the upper atmosphere. During significant weather events, additional rawinsondes may be launched at 0600 and 1800 UTC.

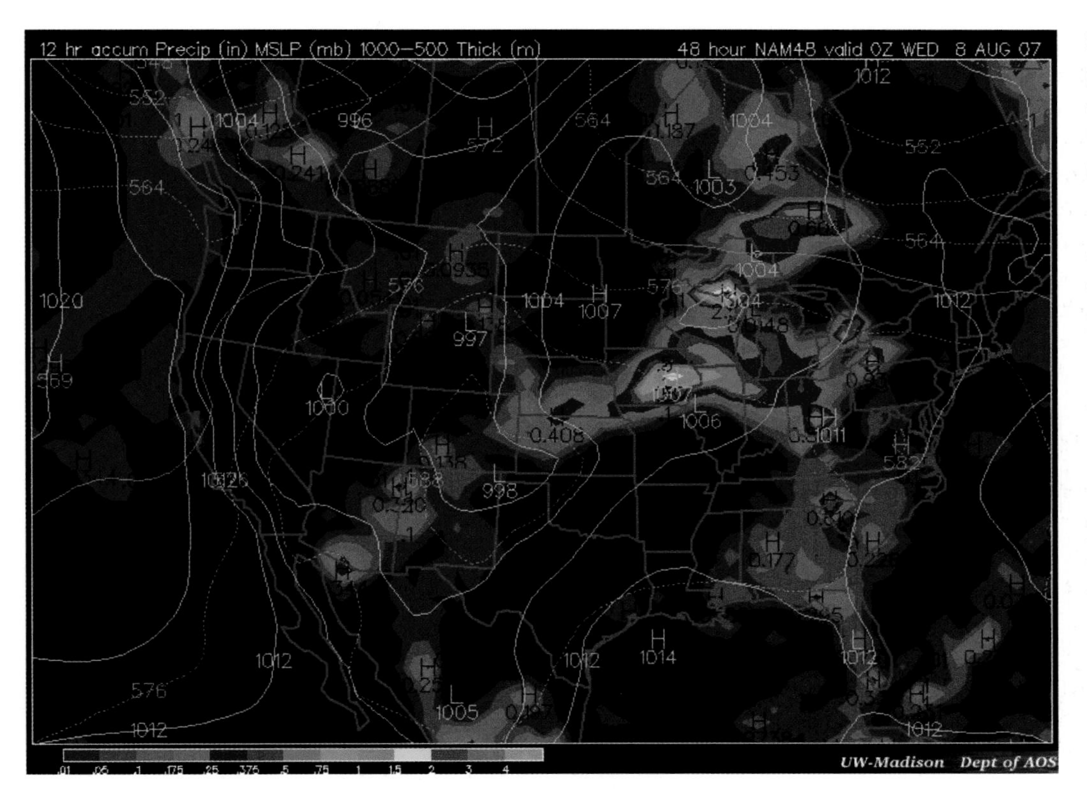

FIGURE 4.8: Sample output from the North American Model showing a 48-hour forecast of sea-level pressure (blue lines), 1000–500 mb thickness (dashed yellow lines), and the precipitation forecast to accumulate during the previous 12 hours (color bar, inches). The 1000-500 mb thickness is simply the depth (in tens of meters) of the layer between 1000 mb and 500 mb. The thickness of a layer is directly proportional to the layer's mean temperature. The forecast is displayed in a format similar to that of weather maps showing the current state of the atmosphere.

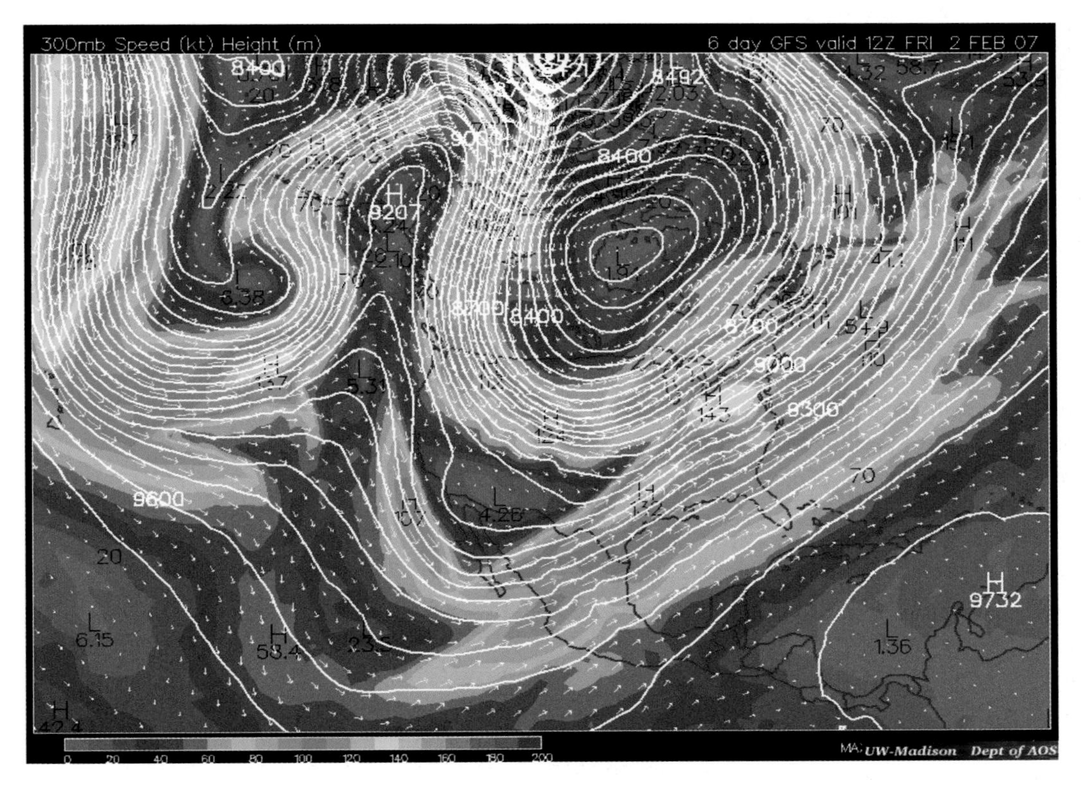

FIGURE 4.9: Sample output from the GFS model showing a 6-day forecast of 300 mb height (meters, white lines), wind speed (knots, colors), and wind vectors.

System (GFS) model. (This forecast is for the same time as the 850 mb forecast shown on the chapter title page). Today, forecasts from these and other models are available at various Internet sites, which can be accessed through links provided at the *Severe and Hazardous Weather Website.* Displays of NWP model output can be a valuable resource for the public, whose access to the products is essentially as easy and rapid as it is for official forecasters. Such access by the public was virtually unheard of before the late-1990s.

Certain models are designed for short-range forecasts and others are for long-range forecasts. The National Weather Service model forecasts available on the Internet are from ***short-range models*** that forecast in the 0 to 84 hour range (e.g., the North American Model), and from ***medium-range models*** (e.g., the GFS model)[3] that forecast for 0 to 16 days ahead. The short-range models cover regional domains, while the medium-range models are global in coverage. As you might expect, the accuracy of forecasts declines rapidly the further in time the forecast proceeds. For ranges beyond about five days, forecasts are used primarily as guidelines rather than to predict specific weather events.

The maps issued by the National Weather Service and by private forecasters are often referred to as either analyses or progs. An ***analysis*** is a map containing contours (e.g., isobars), high and low centers, and sometimes fronts or other information for the time from which a forecast is made. If the analysis is a fully automated depiction of an NWP model's initial state, it is the same as the model's initialization discussed earlier. A ***prog*** (shorthand for "prognostic map") shows similar information, but for a time in the future. Hence forecasters use terms such as "twenty-four-hour surface prog" or "60-hour 500 mb prog." Note that analyses and progs can both include human forecasters' adjustments of model output, resulting in the "human-machine mix" that produces today's best available short- and medium-range forecasts (e.g., Figure 4.10).

The National Weather Service also issues outlooks for 30-day and 90-day periods. Unlike the specific forecasts of day-to-day changes for the short and medium ranges, these ***long-range outlooks*** merely attempt to characterize the coming months or seasons as "above-normal" or "below-normal" in temperature and precipitation. These outlooks are accompanied by estimated confidence levels based on the skill shown

[3]The GFS and many other global models actually represent some of the atmospheric fields as superpositions of various waves, for which the amplitudes increase and decrease over time, consistent with the wave-like character of upper-air maps; hence these models are known as "spectral models."

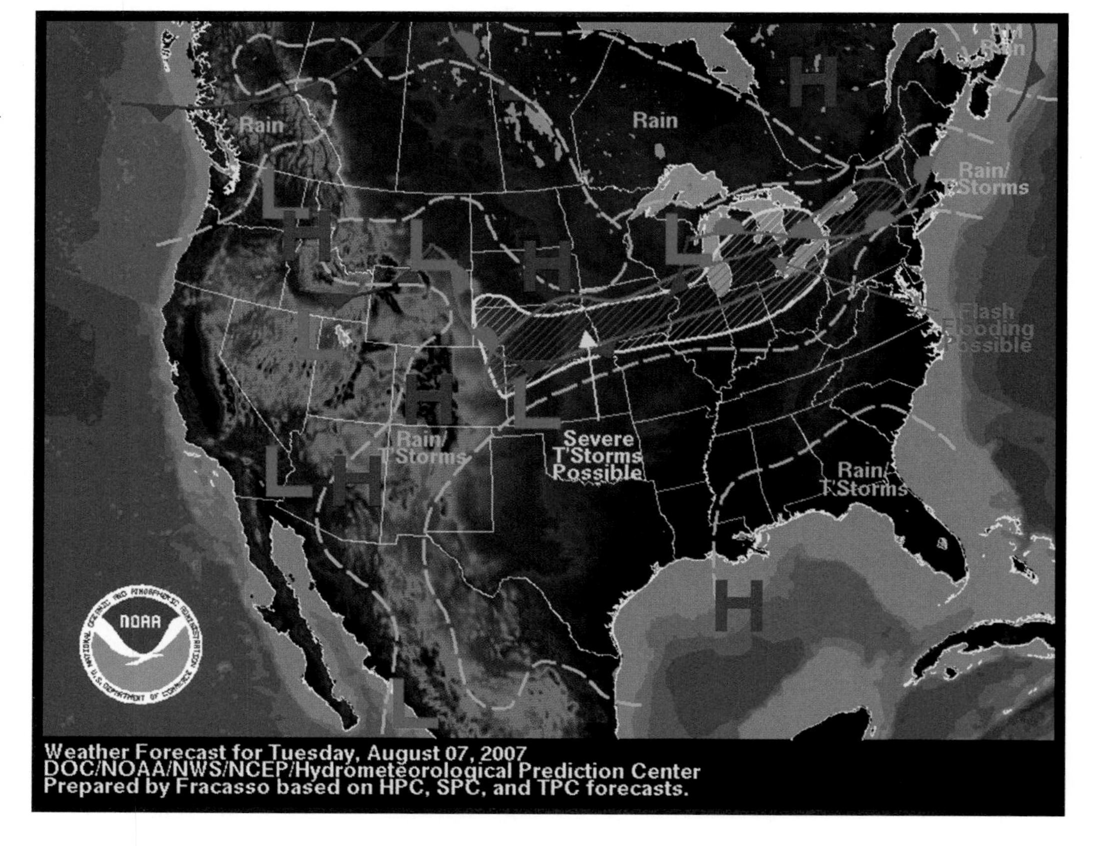

FIGURE 4.10: A forecast for significant weather prepared by human forecasters at the Hydrometeorological Prediction Center based on their interpretation of the output of numerical weather prediction models.

FOCUS 4.3

Synopsis of Operational NWP Models

Table 4A summarizes the characteristics of some of the primary NWP models that are run every day, generally at least twice per day. Most of these models are run by national (or international) weather services, although some models are run at universities and other institutions. Data from most of these models are available through the *Severe and Hazardous Weather Website.* You will notice that the global models generally have coarser resolution than the models that cover limited areas, i.e., a continent or particular country. The differences in resolution reflect the trade-off between areal coverage and resolution: computing limitations place constraints on the total number of grid points if the forecasts are to be available in a timely manner.

Since today's forecasters have access to output from essentially all the models in Table 4A, and even additional models, "information overload" can be a very real problem. Models will disagree among themselves, and different models will produce the "best" forecasts in different situations. In many respects, forecasters today earn their

TABLE 4A Models Used in Numerical Weather Prediction

Model Name	Responsible Agency	Approximate Resolution	Approximate Domain	Range of Forecasts	Comments
North American Model (NAM)	National Centers for Environmental Prediction (NCEP)	12 km	North America	84 hr	Four runs per day: 00, 06, 12, 18 UTC
Global Forecasting System (GFS)	NCEP	80 km	Global	16 days	23 Ensemble members also run at different resolutions
Rapid Update Cycle (RUC)	NOAA/NCEP	20 km	United States	12 hours (operational)	Version at NOAA/FSL runs to 24 hr
Navy Operational Global Atmospheric Prediction System (NOGAPS)	U.S. Navy	55 km	Global	8 days	
Coupled Ocean Atmosphere Mesoscale Prediction System	U.S. Navy	6/18/54 km	Regional	24 hours	Interactive ocean-atmosphere model
European Center for Medium Range Forecasting (ECMWF)	European Center in Reading, U.K.	40 km	Global	10 days	51 ensemble members run to 10 days at 80 km resolution
Global Environmental Mesoscale (GEM)	Environment Canada	24 km	Canada	48 hrs	
Weather Research and Forecasting Model (WRF)	Various U.S. Agencies	Adaptable	Regional	Adaptable	Used in research; also the basis of the North American Model
Mesoscale Model Version 5 (MM5)	National Center for Atmospheric Research/ Penn State	Adaptable	Regional	Adaptable	Used in research; regular forecasts; also posted on web by various universities
Advanced Research Prediction System (ARPS)	University of Oklahoma	Adaptable	Regional	Adaptable	Used in research

(continued)

(continued)

salaries by sifting through the tremendous amount of available NWP guidance and determining which combination of this information will optimize their forecast in a particular situation. Indeed, today's weather forecasts can be described as a "human-machine mix," and much of the "human" component lies in the judicious selection and utilization of the most relevant machine-based guidance.

ONLINE 4.1

Uncertainties in Numerical Weather Prediction Models

An excellent example of the uncertainties inherent in Numerical Weather Prediction products was provided by the model forecasts for 4 February 2007, a day near the start of one of the coldest stretches of weather of the past decade in the northern United States. Online you will find forecasts of the sea-level pressure (and precipitation), 850 mb and 300 mb fields for this time made on each of the preceding seven days. The sequences, beginning with the seven-day forecast, show how the forecasts changed as the range of the forecast decreased, i.e., as the time between the model run and 0000 UTC 4 February became progressively smaller.

The outstanding feature of the actual weather situation on 4 February was a pool of extremely cold air in southern Canada and the Northern Plains. This feature is especially apparent in the 850 mb fields. However, the earlier forecasts (7, 6, 5 days in advance) show an even more intense cold pool, centered farther east in southern Canada. At the 300 mb level, the key features in the verification field (0000 UTC 4 February) are a trough over central North America and a ridge off the western coast of the United States and Canada. While these general features were consistently forecast as far back as seven days in advance, details such as the amplitude and location varied slightly on a day-to-day basis. In particular, the earlier (7-, 6-day) forecasts called for a stronger trough than was observed. Nevertheless, the GFS and other models were generally consistent in their depiction of this cold-air pattern for central and eastern North America. Forecasters who issue outlooks for the medium range (one to two weeks) capitalized on this information and predicted below-normal temperatures with relatively high levels of confidence well in advance of the cold air outbreak.

by the forecasts for particular regions and seasons over past years. While these forecasts do not capture individual weather systems or hazardous events occurring over periods of hours to days, the forecasts can be useful in anticipating longer-duration phenomena such as El Niños (Chapter 23) and droughts (Chapter 26) that can be associated with hazardous weather impacts over months or seasons.

LIMITATIONS OF WEATHER FORECAST MODELS

Forecasts can contain errors even at short ranges, as forecasters are occasionally reminded when an unpredicted snowstorm strikes a region or when a predicted snowstorm fails to live up to its predicted severity, much to the consternation of the public and the press. Why do NWP model forecasts contain errors? Several factors contribute to the degradation of the forecasts as their range increases.

1. Inexact Equations

The equations used in the models represent approximations to the actual physics that occurs, primarily because the mathematics must capture the effect of physics averaged over an entire grid cell, which can be tens to hundreds of kilometers on a side and typically several hundred meters deep. The underlying physics is, in many cases, molecular in scale. Radiative heating, for example, occurs when a photon of energy splits apart a molecule or excites the molecule to a more energetic state. Friction occurs when faster-moving air molecules lose momentum in collisions with slower-moving molecules or with solid objects like the ground. Because countless molecules are present in a grid cell, computer models can never hope to capture each molecular interaction contributing to these processes. Nor can a computer model keep track of the billions of growing or evaporating cloud droplets or raindrops that may be present in a cloudy grid cell. Rather, meteorologists often ***parameterize*** such processes by including estimates of the aggregate effects of the processes rather than detailed formulations of the precise physics. For example, in an NWP model, air in a grid cell near the ground might be made to lose a certain fraction of its speed over a certain period of time; in this way, the effect of surface friction is parameterized—the key parameter is the fractional reduction in the wind speed during each timestep of the model.

2. Errors in the Initial Weather Conditions

The initial weather conditions on which a forecast is based will inevitably contain errors because measurements of the atmosphere are not made at the locations of all grid points. The air over the oceans and the polar regions, for example, is poorly sampled by surface stations and rawinsondes. Satellite techniques permit the estimation of vertical profiles of temperature and moisture around the world, but the derived temperatures typically have errors of one to several degrees; humidities estimated from satellite data can have very large errors. Many studies have shown that small errors in the initial weather conditions will grow over time, so that they contaminate and eventually destroy the accuracy of a forecast. In fact, theoretical studies indicate that even the tiniest of errors will grow and degrade a forecast so that the predictions for specific days become useless after several weeks.

3. Inadequate Resolution

Certain weather phenomena cannot be represented by models with resolution of 10 to 100 km. Models are unable to treat mountains and coastal boundaries with the precision required to reproduce their real-world effects. For example, the surface elevation in a model's grid cell is the average elevation of all points within the grid area, 10 to 100 km on a side; a narrow higher-elevation mountain range within the grid cell will not be "seen" by the modeled air. In Figure 4.7, the highest elevations are approximately 1050 m (3465 ft) at 12-km resolution and 1000 m (3300 ft) at 22-km resolution, both on the Tennessee-North Carolina border. Clingman's Dome in the Smoky Mountains has an actual peak elevation of 2025 m (6643 ft). As a result, the model's air will ascend a smaller vertical distance to cross the model's "mountains" than the air would in the real world. The effect of resolution is apparent in NWP model forecasts of precipitation, especially in mountainous areas, where the amounts are "sharpened" and enhanced by higher resolution. For example, compare the precipitation forecasts for California in Figure 4.11. As a second example, coastlines can be represented only to an accuracy of the grid, confounding predictions of weather systems in which large gradients develop near the coastline, as is often the case during snowstorms in the northeastern United States. Similarly, small weather systems such as thunderstorms (Chapter 18) and lake-effect circulations (Chapter 13) occur on too small a scale to be represented or "resolved" by the typical NWP model.

Because every NWP model forecast contains errors arising from the limitations described above, forecasters use various techniques to extract as much skill as possible from the model forecasts. One such technique is the statistical "post-processing" of the model output in order to adjust for common errors of a particular model or to capture the effects of processes not explicitly included in the model (e.g., the effects of a city's location in a valley). This approach, called ***Model Output Statistics***, is used by the National Centers for Environmental Prediction to obtain specific forecasts for cities and site-specific probabilities of thunderstorms at short-range forecast intervals. Another powerful technique is ***ensemble forecasting***, in which a model is run repeatedly for the same case but with slight changes to the initial state and/or slight changes to the model's formulation. Each of the individual forecasts in this collection of model runs is called an ***ensemble member***. The forecasts for the same time from the various ensemble members can then be averaged into a ***consensus forecast***, which is generally more accurate than the individual forecasts. Moreover, the spread or "scatter" among the various ensemble members is often a good measure of the uncertainty inherent in the model's forecast of a particular situation. The forecasts that show the least scatter tend to be the ones in which forecasters can place the greatest confidence, although there are no guarantees that this strategy will work in a specific case.

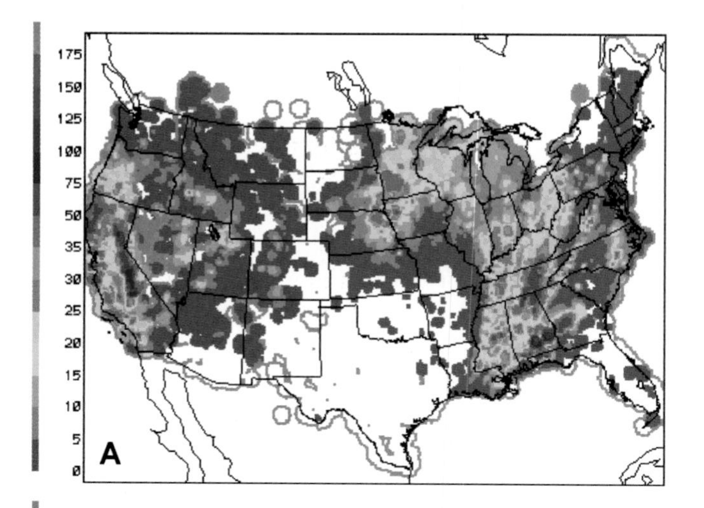

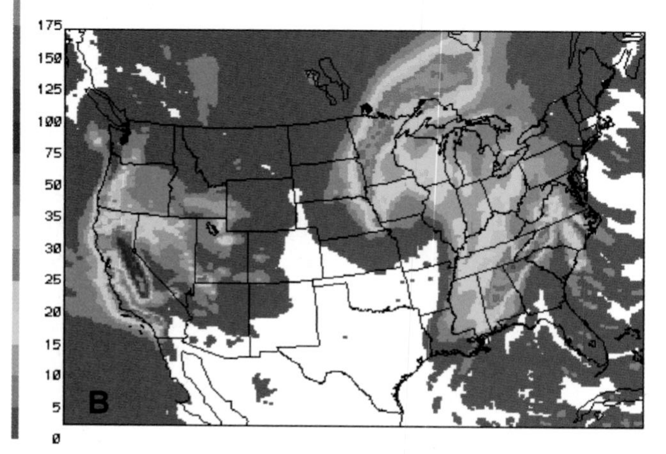

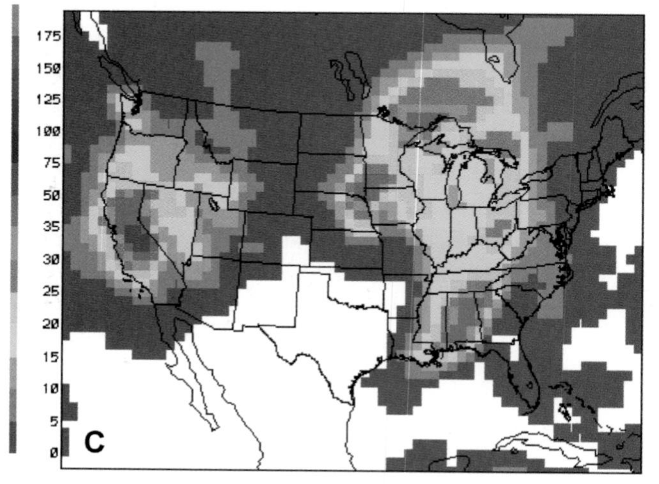

FIGURE 4.11: Twenty-four hour accumulations of precipitation (in millimeters), ending at 1200 UTC 25 November 2001, from (A) an analysis of rain-gauge data, (B) the forecast of a numerical model run at 22-km resolution, and (C) the forecast of the GFS model run at 81-km resolution.

CHECK YOUR UNDERSTANDING 4.2

1. The North American Model is used for weather forecasting. How frequently is it run?
2. For how many days ahead are short-range forecast models typically used? How about medium- and long-range models?
3. What are three main factors that cause NWP model forecasts to degrade as the forecast range lengthens?
4. What is meant by "ensemble forecasting"?

RESEARCH MODELS

There are numerical models that are used only for research. These ***research models*** are designed for specific purposes, such as simulating thunderstorms, tornadoes, and hurricanes. Figures 4.12 and 4.13 illustrate the capabilities of today's models in simulating hurricanes and tornadoes. Such simulations can be directed at a specific severe weather event in

FIGURE 4.12: Numerical model simulation of Hurricane Opal as it approached the Gulf Coast in October 1995. The area of strongest convection is denoted by orange.

ONLINE 4.2

An Illustration of Ensemble Forecasting

The National Centers for Environmental Prediction routinely present the results of ensembles of forecasts in order to convey to users the uncertainty in forecasts for various ranges. For example, an ensemble of forecasts of 500 mb height is displayed in a so-called "spaghetti plot" (e.g., Figure 4B). Two different 500 mb height contours (one in red, one in blue) from all members of an ensemble of 23 forecasts are plotted in the diagrams. Online, you can view an animation of these fields for a 15-day period from an ensemble of forecasts made by the National Weather Service's GFS model. On the first day of the forecast, the contours from the different runs are close together in most locations, essentially overlapping each other. However, as the forecast extends beyond a few days, the contours spread out, giving the plot its "spaghetti" appearance. The rate of spreading varies geographically, implying that the uncertainty increases at different rates in different regions. For example, at days 7 and 8 (168 and 192 hours), the forecasts for the central North Pacific and eastern North America are much closer together than over Europe and the Mediterranean. Even in the initial analysis (the "Day 0" forecast), you will notice a slight separation of the 500 mb contours; this separation represents the initial differences that are prescribed in order to create the differences in the forecasts comprising the ensemble.

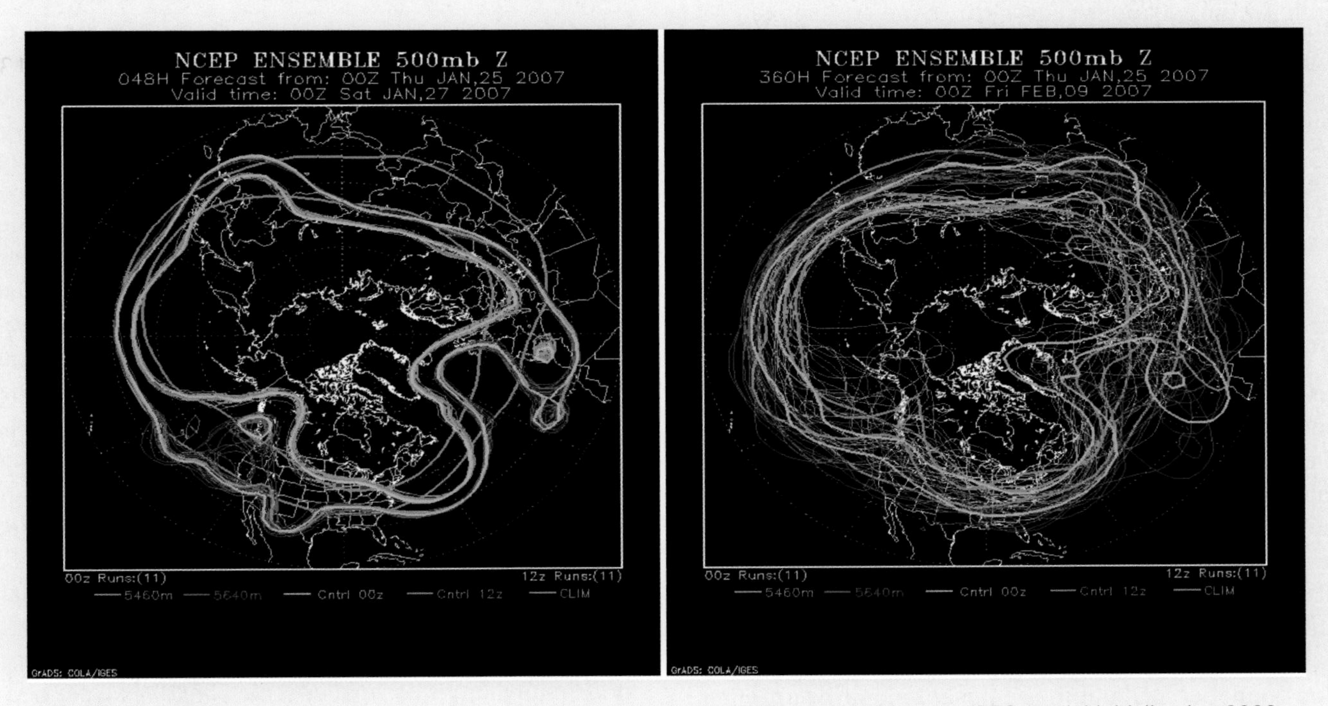

FIGURE 4B: 500 mb "spaghetti plots" for 2-day (left) and 15-day (right) forecasts from the GFS model initialized at 0000 UTC 25 January 2007.

FIGURE 4.13: Numerical model simulation of a tornado. The transparent shading denotes the cloud outline. Red lines are paths followed by rising air parcels, while blue lines are paths followed by descending parcels. Orange and blue arrows at surface denote wind direction in warm and cool air ahead of and behind the numerically modeled storm's gust fronts (see Chapter 18). The tornado is located near the center, where air parcel paths wrap tightly in a vortex.

ONLINE 4.3

Model Simulations Showing the Effect of Wind Shear on Thunderstorms

One of the key factors that enhances the likelihood of severe thunderstorms is a rapid increase of the horizontal wind speed as one goes higher in the atmosphere. The wind is said to exhibit strong "vertical shear" in this situation (Chapter 18). Numerical models can provide striking illustrations of the importance of wind shear. In the animation, the same mesoscale model is used to simulate cloud formation in two cases that are otherwise identical, except that wind shear is present in one case but not in the other. In the animations, you can view the evolution of the two clouds over a two-hour period. How does wind shear affect the size and structure of this simulated storm?

an attempt to explore why the event evolved the way it did. Research models can also be used in a more generic way to perform experiments that shed light on the essential physics of a particular phenomenon. By running the same model with and without the inclusion of a certain physical process, a researcher can study the effects of that particular process on the simulated phenomenon (see Online 4.3). For example, we know that tornadoes can occur with or without visible condensation funnels (Chapter 19). Does the latent heat release associated with the condensation funnel contribute to the development or ultimate intensity of a tornado? If researchers have available a model capable of realistic simulations of a tornado, they can use the model to perform two simulations: one that includes the release of latent heat in the tornado vortex, and the other that excludes the release of latent heat in the vortex. The differences between the two simulations will indicate whether the model's tornado was affected by the release of latent heat in the funnel cloud.

The models that simulate phenomena such as tornadoes, thunderstorms, downslope winds, and related types of severe weather are known as ***mesoscale models***, since the term ***mesoscale*** refers to scales ranging from about 1 kilometer to several hundred kilometers. There are important differences between mesoscale models and the large-scale models used for weather forecasting and climate simulations. For example, the effects of cumulus clouds are parameterized in large-scale models, but these clouds must be fully resolved in a simulation in which the clouds themselves are the features of interest. Other types of parameterizations that "work" in large-scale models may be seriously inadequate in smaller-scale models.

In order to resolve a phenomenon such as a tornado, the spacing of a model's grid points must be extremely small, perhaps several tens of meters, as in the simulation that produced the results illustrated in Figure 4.13. Hence, in comparison with large-scale NWP forecasts, model simulations of smaller-scale phenomena are necessarily limited by computer resource availability to grids that cover relatively small areas. Even when the domain of a simulation is quite small, the computing time required for a sophisticated research model to simulate a particular feature may be longer than the feature actually lasts in reality—for example it might take twelve hours of computer time to simulate the lifetime of a typical tornado (five to ten minutes) or even a supercell thunderstorm (one to two hours). Such simulations would certainly not be useful for forecasting!

Another factor that constrains the application of these models to actual forecasting is the availability of observational data to resolve the initial weather conditions that give rise to a severe event in a particular area. Rawinsonde data, for example, can indicate that instability is present at a single location, but it is unlikely that the rawinsonde's location coincides with the most unstable point in the surrounding area, perhaps a hundred kilometers on a side, in which there were no other rawinsonde launches. A mesoscale model, initialized everywhere with the information from a single sounding, can indicate that severe thunderstorms are likely somewhere within the area, but the model cannot pinpoint the precise location where the most severe weather will develop. Additional fine-scale information about the surrounding environment is required if the location of severe events is to be predicted more precisely.

Research models can contribute to an improved scientific understanding of a phenomenon even without a direct improvement of forecasts of that phenomenon. Modern visualization techniques, such as those used for Figures 4.12 and 4.13 and for the animations in Onlines 4.3 and 4.4, are especially helpful in enabling scientists to study and make sense of the billions of output numbers. Together, models and visualization tools are adding to our understanding of the workings of severe weather phenomena. Improved understanding will ultimately lead to better forecasts of severe weather. In fact, more advanced research models are already being used by operational forecasters, including some that forecast consequences of severe weather systems. Storm surge models, for example, provide input to hurricane forecasters who need quantitative guidance about the likely height of storm surge and coastal flooding at hurricane landfall (Chapter 24). Lake-effect snow models are also providing guidance to forecasters who must pinpoint the location and timing of lake-effect snowstorms in the vicinity of Lakes Erie and Ontario (Chapter 13).

ONLINE 4.4

View a Simulation of a Hurricane, a Thunderstorm, a Tornado, and a Landspout

Research models have been used to simulate a wide variety of severe weather events. Online, you can view results of simulations by several research models, including two different simulations of a particularly severe type of thunderstorm known as a "supercell" (Chapter 18). Supercells produce the majority of strong tornadoes. One of the two online simulations of supercells produces a tornado-like feature that reaches the ground. As you view the simulation of the tornado-producing supercell, keep in mind that the simulated features are generally smoother and/or more symmetric than their real-world counterparts. The smoother appearance is partly a consequence of the finite distance between the grid points of the model.

Other simulations available for your viewing include a series of "landspouts" (a type of tornado, as explained in Chapter 19) and a hurricane (Chapter 24).

FOCUS 4.4

A Perspective on Weather Prediction

Despite the improvements in weather prediction by numerical models, everyone has been inconvenienced by inaccurate forecasts. Occasionally, the errors in the forecasts are sufficiently serious that weather forecasters are the subject of jokes and criticism (or even worse!), particularly if hazardous weather develops when it was not forecast. Are weather forecasters unique in this respect? To answer this question, we need only consider the fields of economics and medicine.

In any developed nation, forecasts of economic trends not only abound, but they are topics of popular conversation. There is little doubt, however, that forecasts of economic trends have a mixed record of success. For example, despite all the economic expertise and prognostications now available, do we know whether the national economy will be stronger or weaker in six months compared to its present state? On a shorter timescale, consider the stock market.

(continued)

(continued)

Although many predictions of stock market trends are available, it is readily apparent that there is always a wide range in the forecasts—some of which are directly contradictory—of the short-term (as well as long-term) trends of the stock market.

The field of medicine also has a prognostic element. For example, if a patient is diagnosed with a certain fatal disease, a physician may state that the patient has six months to live. In all likelihood, that estimate will be higher or lower than the actual time until death, even though the general trend of a patient's welfare is generally captured by a physician's diagnosis and "prediction."

What do weather, economics, and medicine have in common? All involve exceedingly complex systems consisting of huge numbers of interacting parts or "pieces." Air molecules, like individual purchasing decisions, and cells of the human body are far too large in number to be captured by a model. Even if every "piece" could be included in a model, the pieces and their interactions are too complex and poorly understood to permit perfect models or perfect predictions of the evolution of the overall system—the economy, the stock market, the human body, or the atmosphere.

CHECK YOUR UNDERSTANDING 4.3

1. How do weather prediction models differ from research models in terms of the physical processes included in the model's formulation?
2. What types of phenomena do "mesoscale" models simulate?
3. What is the major constraint that prevents mesoscale models from being used for actual forecasting?

TEST YOUR UNDERSTANDING

1. What is the main reason for improved accuracy of weather forecasts in recent decades?
2. If you are examining an "analysis" map, what are you looking at?
3. What are some of the physical processes that numerical models must account for in order to make accurate forecasts?
4. What role do grids have in relation to observational data used to initialize a computer model?
5. What is meant by the "time step" of a computer model? How large is a typical time step?
6. Explain how computer models can approximate complex terrain.
7. How are rawinsonde measurements used when a computer model makes a forecast?
8. Why did numerical weather prediction not come into operational use until the 1960s?
9. Why do the grids of high-resolution models generally cover only a limited area, rather than the entire Earth?
10. How can the public obtain the output (forecast maps) from numerical prediction models?
11. What is meant by "parameterization" in numerical models? Give some examples of processes that are parameterized.
12. For which part of the United States would you expect a twenty-four-hour forecast from an NWP model to be most affected by errors in the model's initial state? Why?

13. Does a model such as the North American Model simulate cumulus cloud formation and dissipation? Explain your answer.

14. Describe two ways in which ensembles of model forecasts can be useful to forecasters.

15. Explain why the effect of inadequate model resolution is likely to be greater for New York City than for Columbus, Ohio.

16. How can a researcher use a numerical model to determine whether a particular physical process is important for the evolution of a certain weather phenomenon?

17. Why do operational weather prediction models generally have coarser resolution than research models?

18. Can mesoscale models predict when and where a tornado will occur? Explain.

19. How would a weather forecaster's job be different if numerical models did not exist?

20. How is the job of a physician similar to the job of a weather forecaster?

21. If you wanted to become a specialist in numerical weather prediction, what college-level classes would you need to take?

22. Lego building blocks (a popular toy for children) are sometimes used in explaining the topography of numerical weather prediction models. Why?

23. Suppose that a weather prediction model is initialized on 1 December 2009 and run for ninety days in order to obtain a long-range outlook for the winter. What can that simulation tell you about the weather to be expected on 14 February 2010?

24. If computing capabilities were unlimited, would NWP model forecasts be perfect? Give at least one reason for your answer.

TEST YOUR PROBLEM-SOLVING SKILLS

1. The initial states (analyses) and forecasts produced by numerical prediction models are often displayed on maps for different levels of the atmosphere. Three such charts are shown in Figures 4.6 and 4.8 and the chapter title page. The variables displayed in the charts are summarized in the table below.

Chart Type	Contour Lines	Colors (Background)
sea-level pressure (Figure 4.8)	isobars (mb) in blue	precipitation on forecasts (none on analysis)
850 mb (chapter title page)	850 mb heights (m)	temperature (°C)
300 mb (Figure 4.6)	300 mb heights (m)	wind speed (knots)

(a) What is the date and time (local for your area) of each map?
(b) Is each map an analysis or a forecast?
(c) Where is the highest sea-level pressure in Figure 4.8?
(d) Where is the largest amount of precipitation forecast to occur? What is the largest amount in this area?
(e) Approximately what altitude (in meters) is represented by the 850 mb chart? What is this altitude in miles?
(f) Approximately what altitude (in meters) is represented by the 300 mb chart? What is this altitude in miles?
(g) Where on the 850 mb map is the hottest air? What is the approximate temperature of this air?
(h) Where on the 850 mb map is the coldest air? What is the approximate temperature of this air?
(i) Where on the 300 mb map are the strongest winds? What are the approximate speed and direction of the strongest winds?

2. Precipitation forecasts from NWP models are generally displayed as accumulated amounts (liquid equivalent) over 6-, 12-, or 24-hour periods. Shown in Figure 4.11 are 24-hour accumulations

of precipitation from (A) an analysis of rain-gauge data, (B) a forecast from a model run at 22-km resolution, and (C) a forecast from a model run at 81-km resolution.

(a) Summarize the general performance of the forecasts in terms of the areal coverage of the precipitation and the locations of the largest amounts. State, together with your reasons, which of the two model forecasts was more accurate in an overall sense.
(b) Where in the United States was the greatest "over-prediction" of precipitation (i.e., forecast was too high) by each model? By approximately how much was the precipitation over-predicted at that location?
(c) Where in the United States was the greatest "under-prediction" of precipitation (i.e., forecast was too low) by each model? By approximately how much was the precipitation under-predicted at that location?
(d) Are the differences between the two forecasts consistent with the differences in their resolution? Explain.

3. A grid cell covering southeastern California in a numerical model has an average elevation of 7000 ft. The area covered by this grid cell includes Mt. Whitney (14494 ft) and Death Valley (282 ft below sea level). The model's forecast of a daily high temperature for this grid cell is 68° F. If the troposphere's temperature on that day decreases with elevation at the typical rate shown in Figure 1.2, what would you forecast for the day's high temperature at the summit of Mt. Whitney? At the bottom of Death Valley?

4. During a recent project in Great Britain, thousands of ordinary citizens contributed to atmospheric research by running a global model as a "background job" on their personal computers (PCs) when their computers would otherwise have been idle. In this way, a huge ensemble of model simulations was created. Use the information below to estimate how time-consuming a global model simulation is on a PC, which you can assume (for purposes of this problem) to have a speed of one million arithmetic operations (+, –, ×, ÷) per second. The model's resolution is quite coarse: grid points are spaced at intervals of 2° latitude and 4° longitude; there are 30 levels. The model contains 7 equations, each of which requires about 100 arithmetic operations per timestep. The timestep is 10 minutes. How long will it take to simulate a month of weather on this PC?

5. In Figure 4B, an ensemble of forecasts made at 0000 UTC 2 February 2007 is summarized in "spaghetti plots" of two height contours, 5460 m and 5640 m, from the 2-day (48 Hour) forecasts (for 27 January 2007) and the 15-day (360 Hour) forecasts (for 9 February 2007).

(a) Where, in the 2-day forecasts, are the major ridges and troughs over North America and the nearby ocean areas?
(b) Where do the models show the most disagreement in the 2-day forecasts? (Consider the entire hemisphere).
(c) Where, in the 15-day forecasts, do most of the simulations indicate ridges and troughs over North America and the nearby ocean areas?
(d) Can you associate any of the ridge-trough features in the 15-day forecast with corresponding features in the 2-day forecast?
(e) Where do the models show the most agreement and the least agreement in the 15-day forecasts? (Consider the entire hemisphere.)

USE THE SEVERE AND HAZARDOUS WEATHER WEBSITE

http://severewx.atmos.uiuc.edu

1. The "Forecast Links" of the *Severe and Hazardous Weather Website* contains links to many NWP model output products. In this exercise you will compare forecasts of sea-level pressure from different models for forecasts made at the same time.

(a) Identify two models to examine for the same time period (e.g., NAM and GFS). Find the 12-hour U.S. forecast map for each model. Identify the date and time the forecast was made and the time for which it is valid. Print or save each map.

(b) Identify any differences in the 12-hour forecast pressure, including the pressure pattern and magnitude of pressure systems.

(c) Find the 60-hour U.S. forecast for the same two models. Identify the date and time the forecast was made and the time for which it is valid. Print or save each map.

(d) Identify any differences in the 60-hour forecast pressure, including the pressure pattern and magnitude of pressure systems.

(e) Go back to the *Severe and Hazardous Weather Website* at the time for which the 12-hour forecast was valid. Find the sea-level pressure analysis from the "Current Weather" page. Compare the actual sea-level pressure pattern and magnitude of pressure systems with the 12-hour forecasts. Which model was more accurate? In which respects was it more accurate?

(f) Go back to the *Severe and Hazardous Weather Website* at the time for which the 60-hour forecast was valid. Find the sea-level pressure analysis from the "Current Weather" page. Compare the actual sea-level pressure pattern and magnitude of pressure systems with the 60-hour forecasts.

  (i) Which model was more accurate and why?

  (ii) How did the accuracy of each model compare for the 12-hour forecasts?

  (iii) Which model did a better overall job of forecasting sea-level pressure during the time period you examined? Support your answer with a quantitative comparison for two locations.

2. The National Weather Service (NWS) provides MOS (Model Output Statistics) forecast products. MOS is a technique used to objectively interpret numerical model output and produce site-specific forecasts. On the NWS website, they provide MOS forecast products and discussions for periods called "short-range" (up to 72 hours) and "extended-range" (up to 192 hours). Use the "Additional Resources" links for Forecasting and Simulating Severe Weather on the *Severe and Hazardous Weather Website* to navigate to the NWS Maps and Models page to view the short- and extended-range MOS products.

   (a) Examine the forecast graphics for short- and extended-range forecasts. Compare and contrast the information available for each type of forecast. Identify any similarities and differences. Provide a brief explanation of any differences you observe.

   (b) MOS forecast products also include text forecasts. Navigate to a short-range text forecast. Using the guidance provided, decode the forecast. Include the location for which the forecast was made, the date and time of the forecast, and units of the forecast variables as appropriate.

3. The U.S. Navy has a strong interest in weather forecasts in marine areas of the world. One model that focuses on marine forecasts is the Naval Operational Global Analysis and Prediction System (NOGAPS). Use the *Severe and Hazardous Weather Website* to navigate to Global Models from the Navy (via the Additional Resources on the Forecasting the Simulating Severe Weather page). At this site you will find output of an ocean wave model that computes waves heights on the basis of wind field forecasts by a numerical weather prediction model.

   (a) Choose one ocean region of the world to examine; record your choice.

   (b) Examine the forecast of wave height for the next few days. Save or print the forecast maps.

   (c) What are the highest waves forecast and where are they expected to be located?

   (d) Examine a forecast of sea-level pressure for the same region. How does the pressure pattern correspond to the wave height forecast? Explain how the two are related.

4. The prediction centers that are a part of NCEP (National Centers for Environmental Prediction) provide a text discussion to accompany their forecasts. The discussion provides insight into the key features the meteorologists considered from the various models when they created their forecasts. Navigate to the extended forecast discussion from the HPC (Hydrometeorological Prediction Center) using the Additional Resources for Forecasting and Simulating Severe Weather on the *Severe and Hazardous Weather Website.*

   (a) Save or print the forecast discussion.

   (b) For what date and time is the forecast valid?

   (c) Discussions are often written in "code" to save on space and time. "Decode" the discussion by translating it to full text.

(d) What models did the forecaster discuss? Was one model favored over another? If so, provide a brief summary why.

5. Precipitation is one of the most difficult quantities to forecast, especially because models use grid cells, but precipitation amounts often vary greatly over scales smaller than the grids. Use the "Forecast Links" of the *Severe and Hazardous Weather Website* to locate forecasts of precipitation totals made for the same time period by a coarse-resolution NWP model (e.g., GFS) and by a fine-resolution model (e.g., NAM, MM5).
   (a) Save or print the maps.
   (b) What is the date and time of the forecast maps you are comparing?
   (c) Typically, precipitation totals are forecast for either the 6-hour or 12-hour period prior to the time the forecast map is valid. What is the time frame for the maps you are examining?
   (d) Focus on one specific region where both models are forecasting precipitation. Record this location.
   (e) Compare the precipitation totals for your region of focus for each of the models. How do the forecasts compare in terms of structural detail? Provide a list of similarities and differences.
   (f) Examine the entire forecast map for each of the models. How does the broad, overall pattern compare? Provide a list of similarities and differences.

6. The ridges and troughs introduced in Chapter 3 are key features of upper-air forecasts from NWP models. Use the "Forecast Links" of the *Severe and Hazardous Weather Website* to locate the 500 mb maps covering a 10-day forecast period from the GFS model.
   (a) Examine the 500 mb forecasts at 12-hour increments through the 10-day period. Record the time period of the forecast.
   (b) Find at least one example of the phenomena listed below. For each example identify the geographic region where the trough or ridge is located and the time period over which it strengthens or weakens.
      (i) a trough that strengthens (amplifies)
      (ii) a ridge that strengthens
      (iii) a trough that weakens
      (iv) a ridge that weakens
   (c) Do the ridges and troughs generally move eastward or westward during the 10-day period?
   (d) Large ridges and troughs make up what are called "long waves" in the flow pattern. Typically there are also "short waves" embedded in the flow. Can you tell whether the shorter or longer waves move more rapidly? If so, which one moves faster?

7. As discussed in this chapter, ensemble forecasting is an excellent method to compare forecasts over the forecast period. The Climate Diagnostics Center (CDC) website contains 500 mb "spaghetti plots" from the GFS model. These plots provide information about the extent to which the different forecasts agree about weather patterns over the next fifteen days. Use the Additional Resources for Forecasting and Simulating Severe Weather on the *Severe and Hazardous Weather Website* to navigate to the Ensemble Forecast Products of the CDC.
   (a) Locate the 500 mb spaghetti plots. When were the plots last updated?
   (b) Select the 00 hr forecast (the analysis). Two plots appear, one contoured and the other spaghetti. What does each plot show?

   Examine the spaghetti plots through the 360-hr (fifteen day) forecast. (The animation link may be helpful.)

   (c) What models are represented in this animation?
   (d) How does each model compare to climatology during the fifteen-day forecast period?
   (e) Approximately how far out into the forecast period does the spaghetti plot truly look like spaghetti? What implication does this have for weather forecasts?

CLIMATE, CLIMATE CHANGE, AND GLOBAL WARMING

Global warming is likely to increase sea surface temperatures, which affect the intensity of hurricanes.

KEY TERMS

absorbed
aerosol
Aleutian low
anthropogenic
climate
climate change
climate model
cold wave
conduction
continental climate
continental drying
convection
crustal plates
eccentricity
energy budget
glacier
greenhouse effect
Hadley Cell
heat capacity
ice age

Icelandic low
infrared energy
internal variability
Intertropical Convergence Zone (ITCZ)
land-sea distribution
latent heat
marine climate
Medieval Warm Period
Milankovich cycles
natural variability
orography
Paleoclimates
Pleistocene
polar amplification
precession
projection
proxy data
radiation
scattered
solar energy
specific heat
subpolar low
subtropical high
trace gases
trade winds
varves

Climate is commonly associated with averages of weather variables such as temperature, precipitation, and wind. These averages are tied to the Earth's ***energy budget*** and the atmospheric circulation. In fact, an area's climate encompasses more than simple averages of weather variables—it also includes ranges and extremes that characterize an area's weather. Characteristics of storms and other types of hazardous weather are all part of a location's climate. A region's climate provides the background or setting for its weather, including its hazardous weather. It follows that changes of climate, whether ***anthropogenic*** (human induced) or naturally occurring, will be associated with changes in the frequency and intensity of hazardous weather, including storms, extreme temperatures, floods, and drought. In this chapter, we survey the fundamentals of climate and climate change, and the general associations between changes of climate and hazardous weather. In later chapters, we draw upon these associations by discussing how specific types of weather systems have varied or are likely to vary with climate change.

CONTROLS OF CLIMATE

The uneven distribution of heating from the sun drives the geographical variation of Earth's climate and weather. To understand Earth's climates, we must begin with an understanding of how the sun heats our planet, how the absorbed ***solar energy*** is exchanged between the Earth's surface and the atmosphere, and how energy is redistributed by the winds and ocean currents. Other factors, such as the land-sea distribution and mountains, play important roles in shaping climate, but it is helpful to begin with the energy budget of the Earth as a whole.

The Earth receives energy in the form of ***radiation*** from the sun at a relatively constant rate of 1366 Watts (joules per second) for every square meter perpendicular to the sun's rays. The total solar energy intercepted by the Earth is huge: about 13 quadrillion Watts (a quadrillion is 1000 trillion), as the incoming solar rays strike an area that is 3.14 times the square of the Earth's radius (6371 km). As shown in Figure 5.1, over half of this incoming solar radiation (51 of an arbitrary 100 units) is absorbed at the surface when averaged over the entire Earth. Approximately 30 units are reflected to space (mainly by clouds) and 19 units are absorbed by the air and clouds. The percentage absorbed at the surface and used to heat the ground or ocean is even greater than 51 percent when the sky is cloud-free, when the atmosphere is dry and clean, and when the sun is more directly overhead; and much less than 51 percent when the sky is polluted or cloudy or when the sun is low on the horizon.

What happens to all the heat absorbed by the surface? Some of it is re-radiated upward as ***infrared energy*** (the same radiation detected by the infrared satellite sensors described in Chapter 2). Much of the infrared radiation emitted by the Earth's surface is absorbed by the atmosphere, which then re-emits the radiation both upward and downward, as shown in Figure 5.1 (middle). This trapping of infrared radiation by the atmosphere is known as the atmosphere's ***greenhouse effect***. The main gaseous contributors to the greenhouse effect are water vapor, carbon dioxide (CO_2), and methane ($CH4_2$); clouds are also very effective at trapping the Earth's infrared radiation and re-emitting it back down to the surface or to space. It is important to note that the atmosphere *absorbs and re-radiates infrared energy;* it does not reflect it back to the surface.

In the right-hand portion of Figure 5.1, the important contributions of non-radiative heat transfer processes are shown. The Earth's surface transfers heat to the overlying air by ***conduction***. Rising air carries this heat upward by ***convection*** (7 units). Even more important, condensation of water into the atmosphere's clouds releases ***latent heat*** energy that the surface supplied when the water evaporated. By providing 23 units of energy to the atmosphere as latent heat, clouds supply the global atmosphere with even more energy than does the direct absorption of solar radiation (19 units).

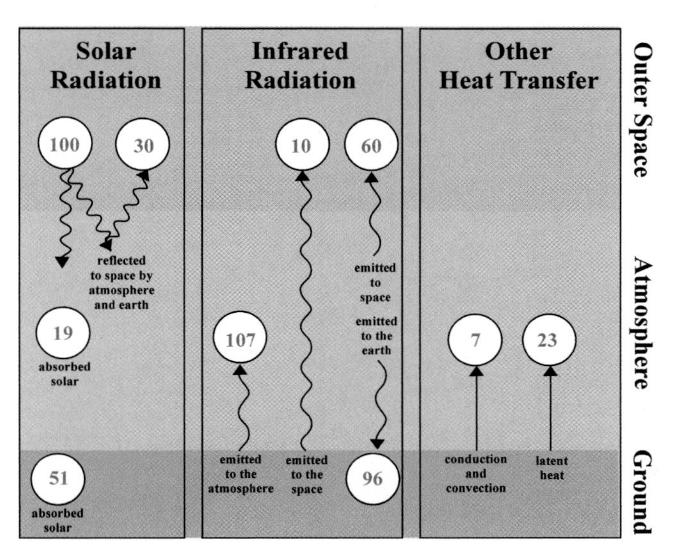

FIGURE 5.1: Annual average distribution of 100 units of incoming solar radiation over day and night over the entire globe. Each number can be regarded as a percentage of the 100 units of incoming solar radiation. The transfer of heat by rising air is depicted in the right-hand portion of the diagram.

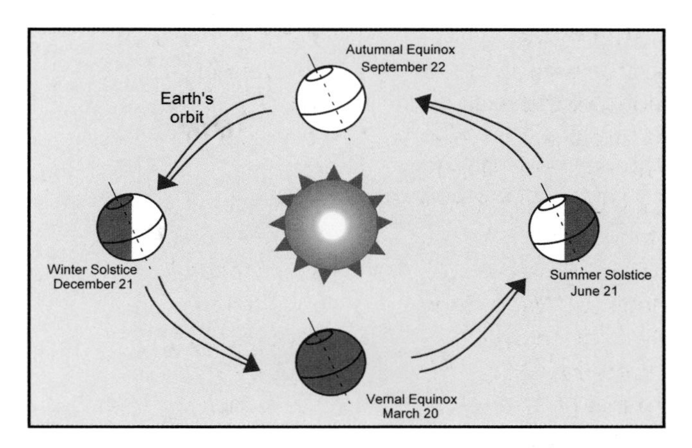

FIGURE 5.2: Earth's position as it revolves around the sun. Because the rotational axis of the Earth tilts 23.5° with respect to the perpendicular to the Earth's orbital plane, the North Pole is in total darkness on the winter solstice, total sunlight on the summer solstice, and has twelve hours of day and night on the vernal and autumnal equinoxes.

SEASONAL CYCLE

Figure 5.2 shows the orientation of the Earth's axis of rotation as it moves around the sun in its yearly orbit. The axis of rotation of the Earth is tilted at an angle of 23.5° to an axis normal to the plane of the Earth's orbit. The tilt is fixed as the Earth orbits around the sun. This tilt affects how much energy reaches the Earth's surface at any particular location. First, during the Northern Hemisphere summer, the Earth's axis tilts so that the polar region of the Northern Hemisphere faces the sun. As a result, direct sunlight falls on the northern polar region all day long. At middle latitudes, the days are long and the nights are short. In the Northern Hemisphere winter, the north pole faces away from the sun. The northern polar region receives no solar energy at all, and at middle latitudes the days are short and the nights are long. The seasonal change in the number of hours of daylight is the primary effect that determines when summer and winter occurs. Note that the opposite scenario occurs in the Southern Hemisphere. The seasons of the two hemispheres are reversed.

The second reason for the seasonal change in solar heating is related to the amount of atmosphere through which the sun's rays pass. Think about how strong the sun feels on a hot summer day when the sun is directly overhead, and how much weaker it feels later in the day, near sunset. As the sun's energy passes through the atmosphere, part of the energy gets absorbed or scattered back to space by air molecules. When the sun's rays pass through the atmosphere at a steep angle from the horizon—as in summer—the rays encounter fewer air molecules before reaching the ground and more solar energy reaches the Earth. In contrast, when the sun's rays pass through the atmosphere at a shallow angle from the horizon—as in winter—the rays pass through more atmosphere before reaching the Earth. More of the sun's energy is ***absorbed*** or ***scattered*** back to space by air molecules and less energy arrives at the Earth's surface.

Because the Earth is a sphere and is rotating about an axis that is tilted with respect to the incoming solar radiation, the absorbed solar energy is not distributed evenly over the Earth. Take a flashlight and shine it directly downward. The light forms a small circle on the ground. Now tilt the flashlight so that the circle widens into a larger oval. The same amount of energy from the flashlight is spread over a larger area because the light is approaching the surface obliquely rather than vertically, so less energy falls per unit area (e.g., on one square centimeter) on the surface. The same thing happens with the sun's energy. When the sun is lower in the sky, as it is in the Earth's higher latitudes, its energy is spread across a larger area, so any unit area receives less energy.

Figure 5.3 shows that the annual average amount of solar radiation absorbed by the Earth is much larger in the tropics than in the polar regions. Infrared energy returned to space also decreases

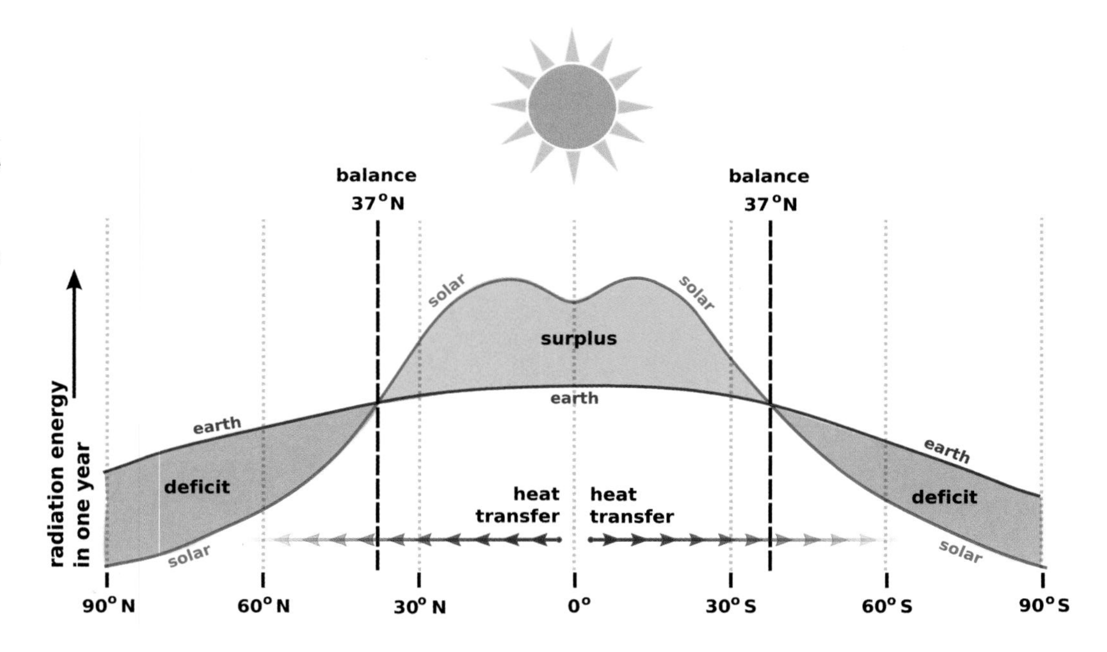

FIGURE 5.3: The average solar radiation absorbed (blue line) and infrared (earth) radiation (red line) lost at each latitude. There is a net gain of radiation equatorward of about 37° in each hemisphere, and a net loss of radiation poleward of about 37°.

poleward, but not nearly as rapidly as the absorbed solar radiation, as shown in Figure 5.3. Since more radiation is gained than lost equatorward of about 37° N and S, while more radiation is lost than gained poleward of 37° in each hemisphere, why don't the tropics become increasingly hotter and the polar regions increasingly colder? The atmosphere's winds and the ocean's currents transport warmer air and water poleward (and colder air and water equatorward), removing the radiative imbalance shown on Figure 5.3. In this respect, the large extratropical cyclones of middle latitudes (Chapters 10, 11) and tropical cyclones (Chapter 24), as well as the atmosphere's high-pressure systems, are essential to maintain the Earth's energy balance.

PRESSURE PATTERNS

The tilt of the Earth's axis enhances the contrast in temperature between the tropics and the poles in winter and reduces the contrast in summer. As a result, the large-scale storms that transport energy poleward are strongest and most frequent during winter. This seasonality is apparent in Figure 5.4, which shows the average patterns of sea-level pressure for January (Northern Hemisphere winter) and July (Northern Hemisphere summer). On the January map, the low-pressure centers over the Atlantic near Iceland and over the Pacific near the Aleutian Islands of Alaska are known as the ***subpolar lows***. While these regions are active storm tracks, they are also the destinations or "graveyards" of storms that develop over eastern coastal regions of North America and Asia, respectively. The ***Icelandic low*** and ***Aleutian low*** are much weaker in the northern summer, consistent with the relative absence of large-scale storms during the warm months. The Southern Hemisphere shows a similar pattern, with a low-pressure belt around Antarctica that is stronger in southern winter (July) than in southern summer (January).

The Northern Hemisphere's major high pressure features in Figure 5.4 are found over Siberia in January and over the eastern Atlantic and Pacific Oceans in July. These features arise in part from the relative coolness of the underlying surface, as described in Chapter 8. The Southern Hemisphere shows similar features over the eastern oceans during its summer (January). These oceanic high pressure features are often referred to as the ***subtropical highs*** because they are located at the northern fringes of the tropics. They represent the northern limits of features known as the ***Hadley Cells*** (Figure 5.5). These cells consist of rising motion along a belt near the equator, poleward flow of air in the upper troposphere, descending motion in the subtropical highs, and return flow near the surface toward the equator.

The return flow toward the equator has an east-to-west component, producing the ***trade winds*** that blow from the northeast in the Northern Hemisphere and from the southeast in the Southern Hemisphere. These two bands of trade winds meet in the ***intertropical convergence zone***, where the ascending branches of the two Hadley cells are located. The maximum of solar heating

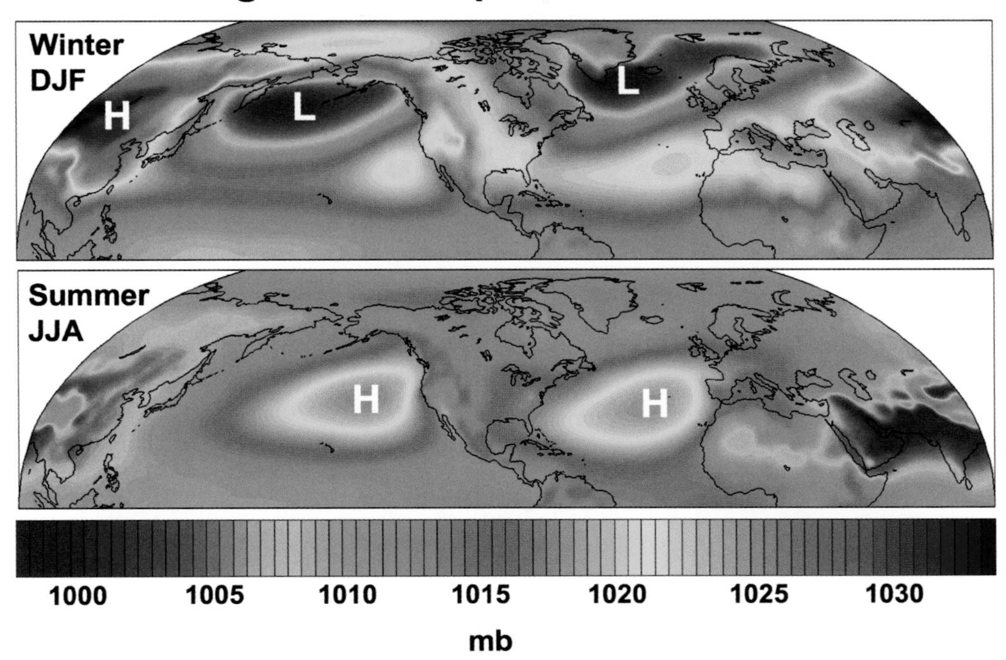

FIGURE 5.4: Average patterns of sea-level pressure for the Northern Hemisphere winter (DJF—December, January, February) and Northern Hemisphere Summer (JJA—June, July, August). Note locations of semi-permanent low- and high-pressure systems.

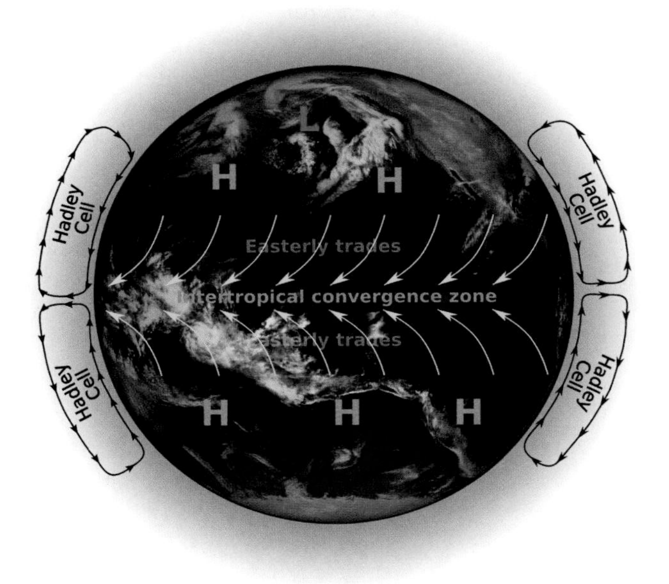

FIGURE 5.5: Average locations of the Hadley cells, including the trade winds and the intertropical convergence zone: Also shown are typical locations of the subtropical highs and subpolar lows. The blue shading denotes the troposphere, exaggerated vertically for illustrative purposes.

in this region creates the buoyancy and upward motion that drive the Hadley cell circulations. The Hadley cells migrate northward from January to July and southward from July to January in response to the seasonal variation of the latitude of maximum solar heating.

LAND AND ITS EFFECTS ON CLIMATE

A major factor in the geographical pattern of climate is the ***land-sea distribution***. Figure 5.6 shows that the average temperatures in both winter and summer are clearly shaped by the distribution of the continents. Land surfaces experience much larger seasonal ranges of temperature than oceans. The smaller seasonal temperature changes over the oceans are consequences of the oceans' ability to store heat. This large *heat capacity* arises from (1) the oceans' ability to mix heat downward so that heat can be stored at greater depths, (2) the low reflectivity of ocean surfaces, and (3) the larger ***specific heat*** of water (i.e., the amount of energy it takes to raise the temperature of one gram of water by 1° C). Significantly more energy is required to raise the temperature of water compared to land material.

Figure 5.7 shows the average daily high and low temperatures through the year at the coastal cities of Miami and San Francisco, as well as the inland cities of Washington DC, St. Louis, Denver, and Fairbanks. Miami and San Francisco are said to have a ***marine climate***, while the other cities display the large seasonal amplitude of a ***continental climate***. Washington, St. Louis, Denver, and San Francisco are all at approximately the same latitude but have some striking differences of climate because of the effect of continentality. For example, the seasonal (January to July)

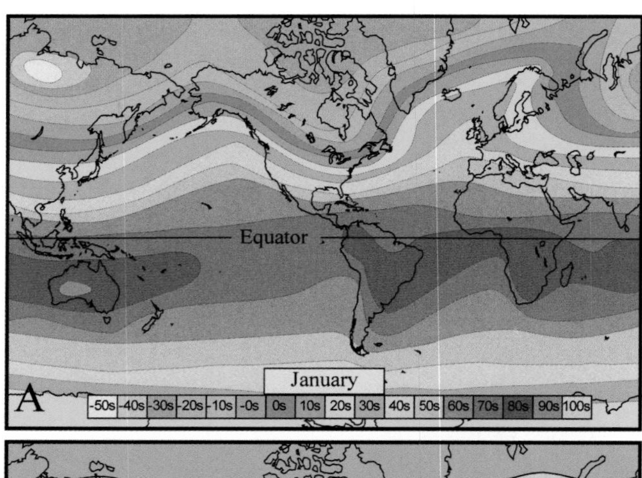

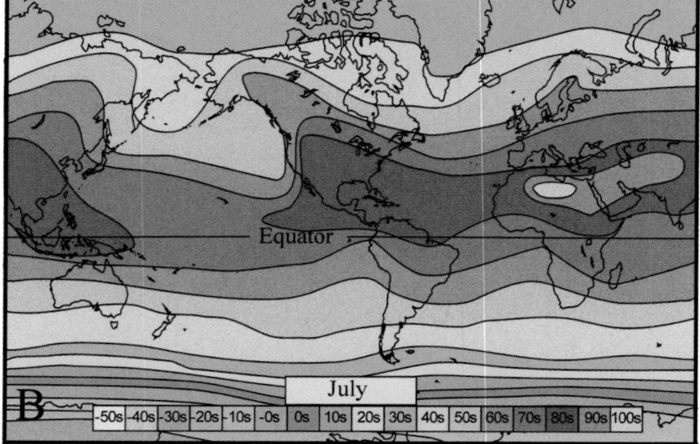

FIGURE 5.6: The average global temperature (°F) in (A) January and (B) July. Large seasonal temperature variations occur in polar regions and over continents in the middle latitudes. In contrast, minimal seasonal temperature variations occur over the tropics and middle latitude oceanic regions.

ranges of average high temperatures are only 13° F (7° C) at San Francisco and 15° F (8° C) at Miami, while they are 45° F (25° C) at Denver and Washington, 51° F (28° C) at St. Louis, and 73° F (41° C) at Fairbanks.

The average daily range—the difference between the average daily high and low temperatures—is also greater at inland cities, increasing from 14° F (8° C) at San Francisco to 29° F (16° C) at Denver. The larger daily range at Denver is due not only to the absence of the moderating influence of onshore flow of maritime air, but also to the dryness of the air. This is a remarkable example of the atmosphere's natural greenhouse effect. In the absence of water vapor and its trapping of infrared energy, nighttime cooling is greater as indicated by Denver's lower minimum temperatures in comparison with more humid locations, such as St. Louis and Washington DC, which are at essentially the same latitude. The wider range of temperatures over the continental interior also means that the extremes of heat and cold tend to increase with distance from the coastline, as we will see in later discussions of cold waves (Chapter 14) and heat waves (Chapter 27).

Orography, the distribution of mountains and hills, is a primary determinant of the climatological distribution of temperature and precipitation. High-altitude locations are generally colder than low-altitude locations because temperature decreases with elevation in the troposphere (Figure 1.2). Leadville, Colorado, for example, at 10,000 feet elevation, is colder than nearby Denver at 5,200 feet by approximately 15° F (8° C) on the average. Precipitation patterns are dramatically affected by mountain ranges. Regions upwind of mountains typically have much greater precipitation, while downwind the climate is dry.

Snow cover also affects the climate of a region. Snow tends to reduce an area's temperature, reflecting incoming solar energy and enhancing the rate at which infrared energy is radiated to space. Areas that experience episodic snow cover are typically colder by 5 to 15° F (3 to 8° C) when the ground is snow-covered. At locations such as Fairbanks, Alaska snow cover is continuous for approximately six months from October to April, contributing to extremely low temperatures during winter (Figure 5.7). Areas with continuous snow cover are indeed the source regions of the airmasses that occasionally affect middle latitudes during cold waves (Chapter 14).

Precipitation is affected not only by the distribution of mountains, but also by continentality, altitude of the terrain, and the atmospheric circulation. The latter includes low-pressure centers (extratropical and tropical cyclones) and high-pressure centers. There is indeed a strong correspondence between the patterns of average precipitation shown in Figure 5.8 and the major atmospheric circulation features in Figure 5.4, with bands of maximum precipitation over the storm tracks leading to the Aleutian and Icelandic lows, and in the intertropical convergence zone that fuels the Hadley cells. Areas of minimum precipitation correspond to the subtropical highs, including the extension of the North Atlantic subtropical high into the Sahara region of Africa. The polar regions, with their cold air and low saturation vapor pressures (Chapter 1), are also areas of little precipitation.

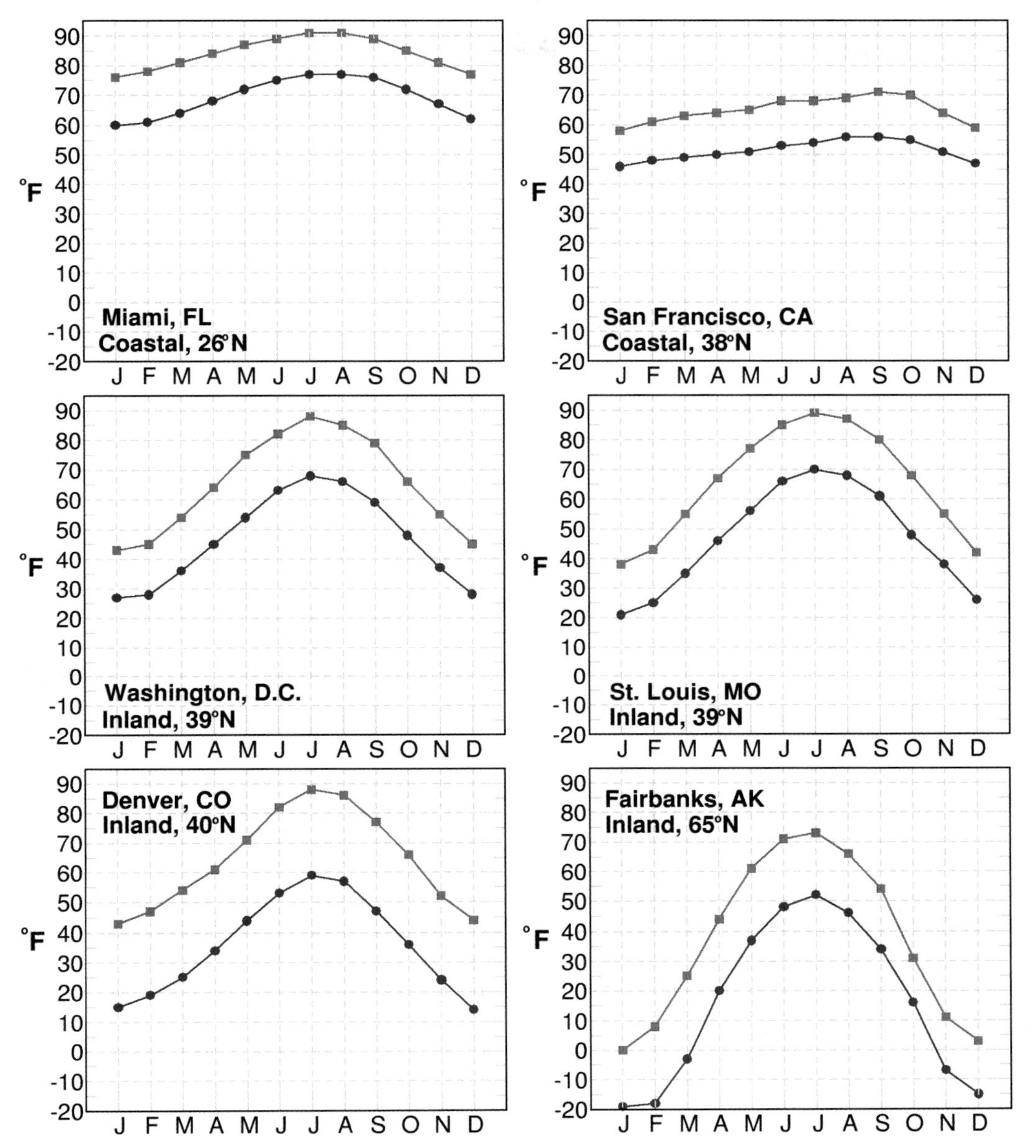

FIGURE 5.7: The average daily high (red) and low (blue) temperatures (°F) by month at the coastal cities of Miami (26° N) and San Francisco (38° N), and at the inland cities of Washington DC (39° N), St. Louis (39° N), Denver (40° N), and Fairbanks (65° N).

The effect of terrain altitude can be seen along the west coast of North America from the Pacific Northwest to southeast Alaska. At these locations, the steep mountains along the coast give onshore flow an upslope component, resulting in tremendous amounts of condensation and precipitation (Chapter 16). Orographically induced maxima of precipitation also occur in other regions, particularly in parts of southeast Asia, where the onshore flow toward the Himalayas during summer provides a large influx of moist air and extensive rainfall.

CHECK YOUR UNDERSTANDING 5.1

1. What is the primary driver of Earth's climate?
2. Give two reasons why the solar heating of Earth's surface varies seasonally.
3. Describe air flow in a Hadley Cell.
4. What are three reasons why oceans have a larger heat capacity than land?

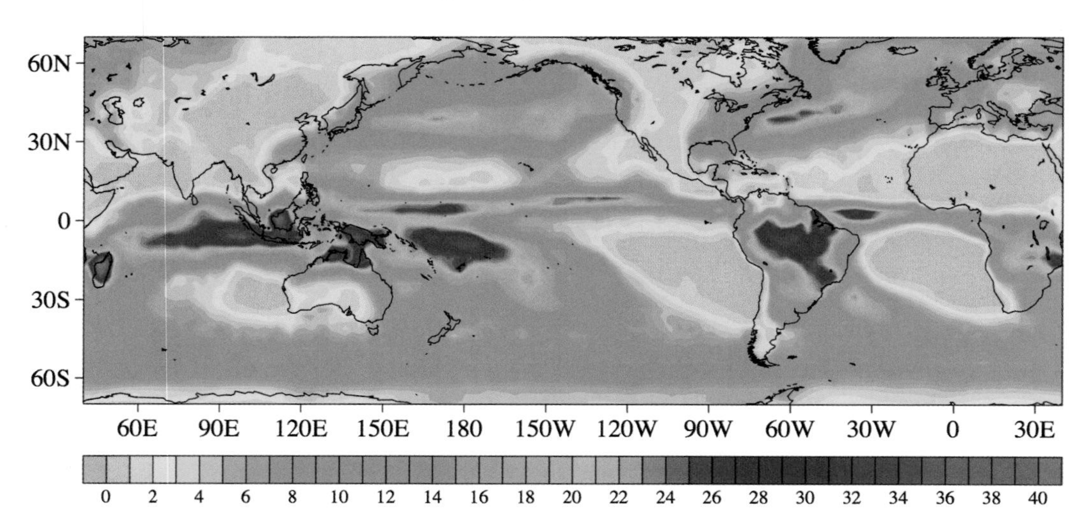

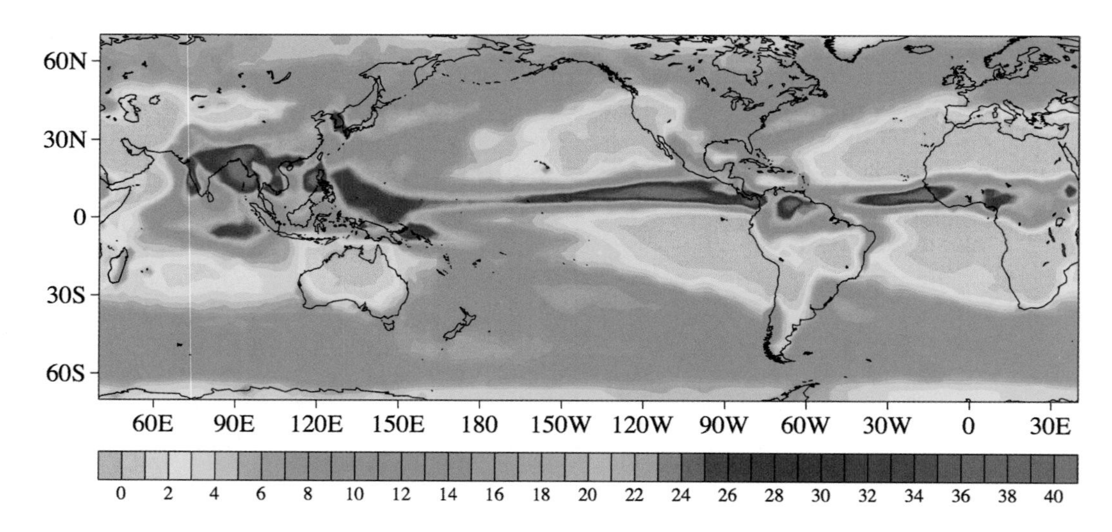

FIGURE 5.8: Average precipitation (cm) for January (Northern Hemisphere winter) and July (Northern Hemisphere summer).

CAUSES OF CLIMATE CHANGE

Climate and its associated weather have changed significantly in the past, and these changes have had profound effects on humans and ecosystems. What causes these changes? The precise causes of changes during specific periods are not always known, but climate scientists have identified a number of factors, listed in Table 5.1, that most likely serve as triggers, or drivers, of ***climate change***. We will examine these factors, starting with the longer timescales (tens to hundreds of millions of years) and work down to the shorter timescales of the past several decades. It is the changes over the timeframe of 10 to 100 years that will provide the background for our discussions in future chapters of potential changes in hazardous weather.

TABLE 5.1 **Timescales of Drivers of Climate Change**

Continental drift	1 million to 1 billion years
Solar variability	10 years (sunspot cycle) to 1⁺ billion years
Atmospheric composition	10 years (greenhouse gases) to 1⁺ billion years
Earth's orbital variations	10 thousand to 100 thousand years
Volcanic eruptions	1 year to 10 years
Natural or internal variability	1 year to 100s of years

FOCUS 5.1

Determining Past Climates

The past few million years have seen about thirty glacial-interglacial cycles with major cooling and warming, and shorter climate variations superimposed on these 100,000+ year cycles. How do we know about past climates? As noted in Chapter 2, instruments for measuring temperature, pressure, and humidity have been in use for only the past few hundred years. Fortunately, the study of climates of the past, or *paleoclimates,* is able to draw upon a variety of indirect methods to deduce climate variations going back hundreds of thousands of years.

Tree rings, concentric layers of annual growth of wood, provide records of the yearly variations of growing conditions in the tree's vicinity. Annual tree growth depends primarily on precipitation and temperature during the summer season, and paleoclimatologists have used information about growth dependencies to develop methods to retrieve estimates of yearly values of temperature and, in some cases, precipitation. While individual trees generally provide useful information for up to a century or so, the use of overlapping growth records of young and old trees can extend the records back several centuries. Paleoclimatologists use a similar approach with ocean corals, which add annual growth layers that depend on the water temperature, nutrient availability, and other factors. Because coral lifetimes are a century or so, overlapping records from different corals can be extended to several centuries with annual resolution.

For longer periods, the pollen in layers of sediments found in lakes or bogs can be analyzed to deduce the types of vegetation growing in the area, and hence the climate characteristics of the region. In some cases, lake sediments have identifiable annual layers *(varves).* Ocean sediments can provide information for even longer time periods—up to tens of millions of years. The organic component of the ocean sediments provides information on the marine life (plankton and higher organisms) with implicit information on the ocean temperature, currents, and the broader climate. The sediment's inorganic component, which originated primarily from land, can provide information about the temperature and precipitation over the nearest continent. Because sedimentation rates in most ocean regions are quite small, the temporal resolution of the climatic information is generally quite coarse.

The ice sheets and glaciers of both hemispheres contain tiny trapped air bubbles that can be analyzed for concentrations of carbon dioxide, methane, and other gases. The ice samples can also be analyzed for their dust content. Temperatures at the time the snow was deposited can be inferred from ratios of isotopes (forms of a molecule containing different numbers of neutrons). Ice cores obtained from the two-mile deep Greenland ice sheet provide information spanning the past 110,000 years, while a core obtained from Vostok, Antarctica contains information from the past 400,000 years (Figure 5.9).

Finally, scientists can make some inferences about past climates by using computer models (Chapter 4) to simulate climate under conditions that differ from the present in significant ways. For example, a global climate model can be run with the seasonal and geographical patterns of solar radiation adjusted for variations in the Earth's orbital parameters *(Milankovitch cycles).* The models can also be run with a reconfiguration of the Earth's surface, e.g., with ice sheets and a lowered sea level, or with an altered distribution of the continents. Climate model simulations are likely to provide the best estimates of climates of the distant past (tens to hundreds of millions of years ago) when the configuration of the continents was different and/or the atmospheric composition was quite different from the present.

Whether information on past climates is deduced from proxy data or from computer model simulations, there are uncertainties in the information. These uncertainties can be as large as the changes in which climate scientists are interested. For example, as we will see later in this chapter, the Northern Hemisphere temperatures reconstructed from proxy data for the period from 1000 A.D. to the beginning of the instrumental record in the 1800s have uncertainties of about 1° C, which is larger than the range of the estimated hemispheric temperatures over the same period. Such uncertainties confound efforts to place recent climate changes into a longer-term context.

Long Term (Millions of Years)

As shown in Table 1, the drivers of climate change over the longer timescales, millions of years, include continental drift, changes of solar output, and atmospheric composition. The continents, or *crustal plates* of the Earth, are in slow motion in response to fluid motions of the Earth's liquid core generated deep below the Earth's surface. The North American and Eurasian plates, for example, are drifting apart at a rate of an inch or two per year, slowly increasing the distance between the two landmasses. This spreading is part of a global pattern of plate motion that began 225 million years ago with the break-up of a single "super-continent" called Pangaea (Greek for "all lands"). Given the importance of the land-sea distribution for present climate, as discussed in the preceding section, this super-continent must have experienced extremely large ranges of temperature.

The output of the sun itself varies slightly. During the 11-year sunspot cycle, the solar output can change by 0.1 percent, or about 1.2 W per m^2. The amplitude of the sunspot cycles undergo a slower cycle lasting about 80 years. Many attempts have been made to link these solar cycles to weather and climate, but the associations are tenuous at best and are much weaker than the associations between climate and the other drivers discussed above. However, on the very long timescales of the sun's evolution as a star, solar variability is undoubtedly important to Earth's climate. For example, the sun's output has increased by about 25 percent over the last 3.5 billion years of the Earth's 4.5 billion year history.

Because the absorption, transmission, and emission of radiation vary widely among different gases and particulates (aerosol particles), the atmosphere's climate depends strongly on its composition. The surface of Venus, for example, is extremely hot (between 800° and 900° F) because the atmosphere of Venus consists primarily of carbon dioxide, producing a runaway greenhouse effect. The composition of the Earth's atmosphere has evolved significantly over the past several billion years, and only in the past several million years has reached its present composition of 78 percent nitrogen, 21 percent oxygen, 1 percent argon, and small percentages of water vapor, carbon dioxide, and other so-called ***trace gases***. The Earth's climate was undoubtedly very different billions of years ago when the atmosphere consisted primarily of hydrogen and helium. Changes in atmospheric composition are also likely contributors to recent and ongoing climate changes, as discussed later in this chapter.

Medium Term (Thousands to Hundreds of Thousands of Years)

When examining past climates, we are particularly interested in explaining ***ice ages***, which are periods of expanded continental ice sheets, polar ice sheets, and mountain ***glaciers***. There may have been several ice ages early in Earth's history, but the first well-documented ice age occurred from 800 to 600 million years ago. Another minor ice age occurred from 460 to 430 million years ago. There were extensive polar ice caps between 350 to 260 million years ago. In the intervals between these times, scientists believe that there were no permanent ice sheets present on Earth.

Our present ice age, which continues today, began 40 million years ago with the growth of an ice sheet in Antarctica, and intensified during the period geologists call the ***Pleistocene*** (starting around 3 million years ago) with the spread of ice sheets in the Northern Hemisphere. Cycles of extensive glaciation and interglacial ice retreats have characterized the Earth's climate during the Pleistocene. Glacial advances are strikingly apparent in the temperatures deduced from ice cores obtained in Greenland and Antarctica (Figure 5.9), and from studies of ocean temperatures based on marine sediment.

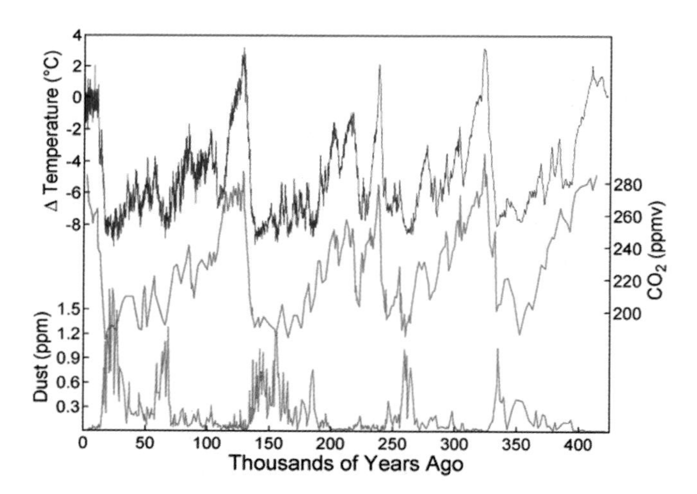

FIGURE 5.9: Variations in temperature (blue), carbon dioxide (green), and dust (red) over the past 400,000 years based on information in the Vostok (Antarctica) ice core. Time is shown on the horizontal axis in thousands of years before the present, which is at the left.

Glaciers have carved many of the geological features present on Earth today, such as the Great Lakes. During the last glacial period, which peaked about 18,000 years ago, an ice sheet several kilometers thick covered much of North America, extending southward to the Ohio River and the central Plains (Figure 5.10). The lowering of sea level by more than 300 feet during such a glaciation exposes many areas that are now shallow ocean (continental shelves) offshore of the present continents. For example, the Gulf of Mexico and Atlantic shorelines of the United States were generally 50 to 200 miles seaward of their present positions during the last glacial maximum. Glacial maxima are characterized by dry climates and high atmospheric dust content, which is likely a result of the large areas of bare ground. To understand climate change, and human impacts on climate change such as global warming associated with greenhouse gas emissions, we must first understand what causes natural cycles of glaciation and interglacials.

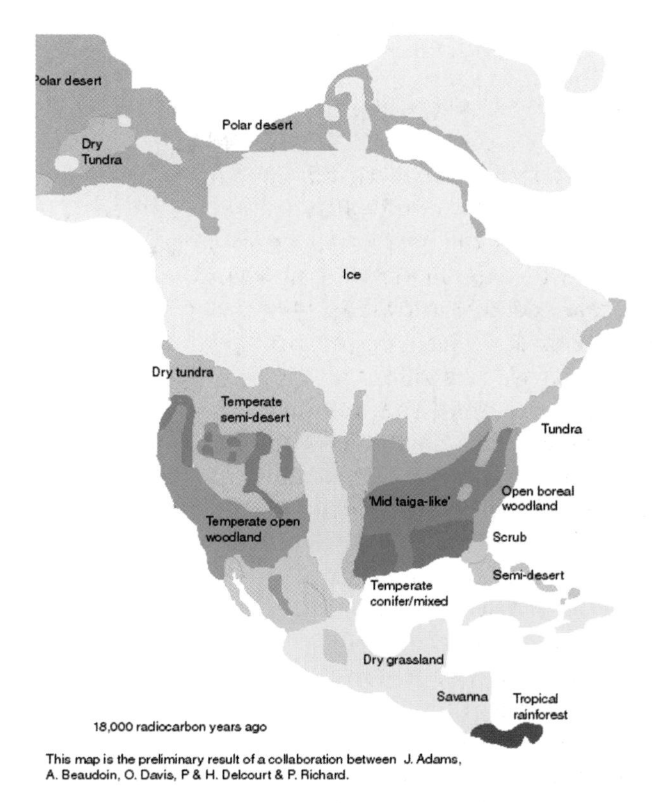

FIGURE 5.10: Reconstruction of primary vegetation types and the configuration of the North American ice sheet and coastline at 18,000 years before the present.

FOCUS 5.2

Ice-Age Weather

The North American ice sheets of the past million years have extended to about 40 to 45° N over Canada and the northern United States, extending from Long Island westward to today's Ohio River, through Missouri and the Central Plains, and into the Pacific Northwest (Figure 5.10). With sea level lower than today by more than 300 feet, the continental margin extended well offshore of today's coasts. One such area, the Bering Sea "land bridge", connected Asia and North America. What was weather like for the animals and perhaps the humans who lived beyond the ice sheet in what is now the southern and western United States? It was almost certainly colder, drier, and dustier, based on the information retrieved from ice cores, although pollen records indicate that today's Desert Southwest was temperate open woodland and hence wetter than today. Storms from the Pacific most likely tracked further south than today, into California and the southwestern states. While the Sierra Nevada mountain range would shield Nevada and Utah from much of the moisture (hence their semi-desert vegetation in Figure 5.10), wintertime precipitation and the snowpack accumulation in Arizona and New Mexico was evidently sufficient to support temperate open forests—a far cry from today's landscape in this area. Water was undoubtedly less scarce than today for the residents of the Southwest, although springtime floods may have been a frequent occurrence as the snow melted. Summers were most likely pleasant in comparison with today's excessively hot and dry conditions. Even Death Valley may have been habitable. Farther north, however, the unglaciated areas of the Pacific Northwest were dry tundra or semi desert; their proximity to the ice sheet probably led to frequent occurrences of cold, dry katabatic winds of

(continued)

(continued)

the type that now occur at the Antarctic periphery (see Chapter 17). Mixed conifer and deciduous forests covered the area of today's Gulf Coast states, while the area near and south of the Ohio River was boreal forest, suggesting the drier cooler climate of today's southern Siberia.

Computer models have been used to simulate the climate at the time of the glacial maximum, about 18,000 years ago, and they indicate an eastward shift of winter storms away from the southeastern United States, perhaps because the Gulf of Mexico was substantially smaller and cooler, and less able to provide the heat and moisture that feed today's Gulf Coast storms (Chapter 11). On the other hand, models show an increase of storminess along the eastern margins of North America and into the northern North Atlantic. The enhanced temperature contrasts between the ice sheet and the western Atlantic ocean waters likely favored the frontal formation associated with extratropical cyclones. Model simulations also suggest that frontal cyclones during summer provided rain events along the southern margin of the ice sheet, whereas nearly all the summer precipitation in this region presently occurs as showers and thunderstorms.

As the ice sheets retreated, the newly exposed land covered by glacial till was prone to blowing dust events. While this blowing dust formed the beds of loess that characterize today's Corn Belt and Wheat Belt, the blowing dust events were most likely hindrances to animals and humans who tried to venture northward into the newly exposed land areas of the central United States.

For an ice age to occur, substantial areas of continents must be present at high latitudes. The land would be far too warm for snow to accumulate, if it fell to the ground at all, when continents are located near the equator. Glaciation can also take place only when the continents lie well above sea level. This is controlled both by the depth and breath of the oceans and the distributions of mountains and high terrain, both affected by processes associated with continental drift. Finally, ice ages can't develop when ocean currents carry too much heat to high latitudes. Again, currents are controlled in part by the distribution of continents and oceans.

The second major long-term control on ice ages may be atmospheric composition. The concentration of carbon dioxide in the atmosphere, a major contributor to the atmospheric greenhouse effect, has varied substantially due to natural processes over millions of years. Increased concentrations of carbon dioxide will warm the atmosphere sufficiently that they may inhibit glaciation. Natural variations in carbon dioxide can occur, for example, if there are changes in the number of marine organisms that extract it to make shells, changes in the amount of chemical weathering on land (which reduces carbon dioxide), or changes in the abundance of plant life.

Variations of the Earth's orbit about the sun have been implicated in the cycling of glacial and interglacial periods within the Pleistocene. The key orbital parameters are ***eccentricity*** (ellipticity), which is associated with a seasonal variation of Earth-sun distance and has a cycle length of about 110,000 years; the ***tilt of the Earth's axis***, which varies between 21.8° and 24.4° with a cycle length of 41,000 years; and the ***precession*** of the Earth's axis (with a cycle of about 22,000 years), which determines whether the Earth is closest to the sun during the northern hemisphere or southern hemisphere winter. (The Earth is closest to the sun now in early January, the northern hemisphere winter). These orbital cycles, originally linked to ice ages by the Serbian mathematician Milutin Milankovich, can change the solar radiation reaching 60° N in a particular season by as much as 25 percent, and they show correlation with the ice ages of the past million years. However, it is likely that feedbacks, perhaps involving carbon dioxide changes (Figure 5.9), the reflection of sunlight to space by enhanced snow cover, and changes in the oceanic circulations because of sea ice amplify the effect of the ***Milankovich cycles*** to produce the full swings between glacial and interglacial periods.

Short Term (Several Years to Thousands of Years)

Aerosols are small particulates in the solid or liquid phase. The atmospheric aerosol includes wind-blown dust, sea salt particles from ocean spray, organic particulates derived from vegetation, combustion products, and other anthropogenic particulates. Many of these particulates serve as nuclei for haze and cloud droplets, which reflect sunlight in much the same way as clouds. The effects of aerosols are complicated by their ability to absorb as well as reflect radiation, especially when black carbon (soot) is a significant component of the aerosol.

Volcanic eruptions provide dramatic examples of the input of aerosols to the atmosphere. Large eruptions provide sufficient amounts of aerosols that they have been implicated in climate changes over timescales of several weeks to several years. The volcanoes with the longer impacts on climate are those that inject large amounts of mass into the stratosphere, where the stable stratification and lack of clouds (see Chapter 6) prevents the downward mixing of the particulates and gasses or their incorporation into precipitation.

Notable eruptions of the past few centuries include Mt. Tambora (1815) and Krakatoa (1883) in Indonesia, as well as Mt. Pinatubo (1991) in the Philippines. Mt. Tambora has been associated with the "Year without a summer" across parts of North America and Europe in 1816 (see Focus 5.3), and the cooling effects of Krakatoa and Mt. Pinatubo have been well documented in the meteorological records. These and other eruptions are prominent in Figure 5.11, which is a reconstruction of the post-1850 history of volcanic impacts on radiation reaching the surface. In addition to highlighting the eruptions of Krakatoa and Pinatubo, Figure 5.11 shows how volcanic signals persist for several months to several years. The figure also shows that much of the twentieth century was volcanically inactive, at least through the 1960s. As of late 2007, there have been no atmospherically significant volcanic eruptions since the 1991 Pinatubo eruption, which impacted surface radiation globally for several years (Figure 5.11).

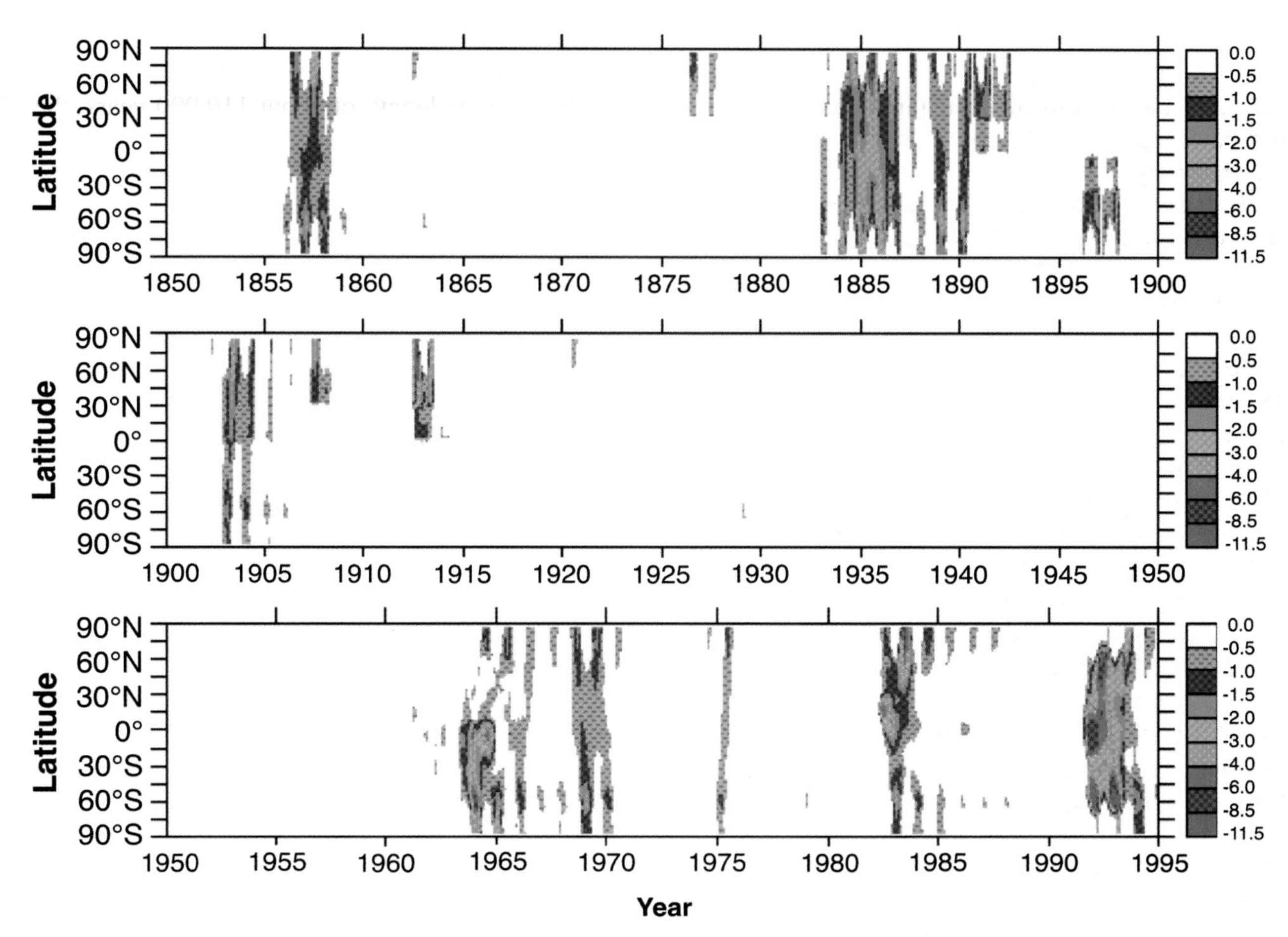

FIGURE 5.11: Reconstruction of volcano effects on the atmosphere's radiation balance. Horizontal axis is time, dating back to 1850; vertical axis is latitude. Colors show impacts on surface radiation in Watts per m^2; negative values denote reductions in the radiation reaching the surface. Major volcanic eruptions occurred in 1883 (Krakatoa), 1963 (Agung), 1982 (El Chichon), and 1991 (Pinatubo).

FOCUS 5.3

The Year Without a Summer

One of the most famous climatic episodes of the early nineteenth century occurred in 1816, known as the "Year without a summer". The regions most severely affected were New England, southeastern Canada, and northern Europe. While the event is commonly associated with the 1815 eruption of Mt. Tambora in Indonesia, other large volcanic eruptions had occurred in 1812 and 1814. Moreover, 1816 was one of a series of exceptionally cold summers during 1811–1818, leading many New England farmers to abandon their land and move westward. As a result, today's reforested landscape in New England is partially attributable to the extreme cold of that decade.

The "Year without a summer" was not without its warm days, with temperatures above 90° F during several days in late June. The outstanding feature, however, was a series of summer *cold waves* that brought frost and snow to much of the region, damaging or destroying crops that were already behind schedule because of a cold spring. The cold spring was compounded by dry conditions that put the region on the verge of a drought. The tone for the summer was set with two extreme cold outbreaks in May. During mid-May, frosts struck all of New England, including its southern coast, extending into Pennsylvania where the average date of the last frost is in April. At the end of the month, frosts again struck New England and areas to the west, freezing ponds with ice up to a quarter inch thick in Erie, Pennsylvania. A subsequent cold outbreak on June 6 to 9 brought a killing freeze and as much as five to six inches of snow to New Hampshire and Vermont, with snow reported even in Connecticut and Massachusetts. Birds, sheep, and crops fell victim to the cold weather in New England and southern Canada, especially in the Quebec city area.

A cold, dry July with temperatures in the 40s during many days ended any hopes of salvaging crops. Northern New England even experienced crop-destroying frosts on July 8 and 9, which is normally close to the time of the year's hottest temperatures. Frosts during the second and third weeks of August completed the series of devastating summer weather events.

The summer of 1816 was the coldest in 200 years of record at New Haven, Connecticut. It would likely have had a similar distinction elsewhere if weather records of this duration were available from other locations. Grain prices rose substantially, and cattle prices fell as farmers sold their cattle in the absence of grain for feed. Northern Europe was also severely impacted by cold in the summer of 1816, which marked the beginning of a famine and typhus epidemic that killed millions in France during the next several years.

While there is no rigorous proof concerning the cause of the extreme cold of the summer of 1816, it is highly likely that a contributing factor was the series of volcanic eruptions during 1811–1815, culminating with Mt. Tambora in April of 1815. The fact that the 1811–1818 period as a whole was exceptionally cold makes it difficult to dissociate the cold from the volcanic activity. Nevertheless, one cannot rule out contributions from other factors, including decadal-scale natural variability.

Primary concerns in the study of recent and modern climate are the proportions of the primary greenhouse gases (water vapor, carbon dioxide, methane, ozone), as well as aerosols. There is no doubt that greenhouse gas concentrations have increased in recent centuries as the world has industrialized. Histories based on combinations of direct measurements from recent decades and ice core samples from earlier centuries, show that greenhouse gas concentrations are increasing rapidly and are higher now than at any time in the past 2000 years (Figure 5.12) and, in the case of carbon dioxide, higher than at any time in the past 400,000 years. Carbon dioxide concentrations have increased by more than 35 percent from their pre-industrial values of about 280 ppm (parts per million by volume) to the present value of about 380 ppm. Radiative calculations show that a doubling of carbon dioxide increases the downward infrared radiation at the surface by 4 to 5 Watts per m^2, an amount that is much greater than century-scale solar variability (about 1 Watt per m^2). On a molecule-per-molecule basis, methane is even more effective than carbon dioxide in the trapping of infrared radiation. However, the concentration of methane is still so much lower than that of carbon dioxide that methane's contribution to the greenhouse effect is smaller.

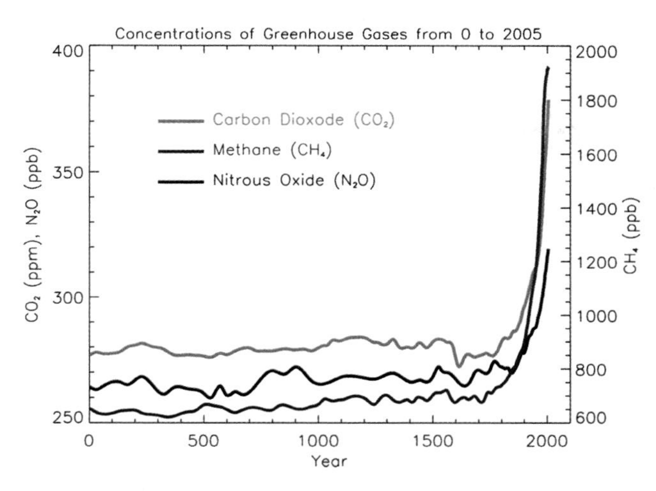

FIGURE 5.12: Concentrations of carbon dioxide, methane, and nitrous oxide over the past 1000 years.

The strongest gaseous contributor to the greenhouse effect by far is water vapor. In this respect, any scenario of global warming is highly dependent on the change of water vapor, which has the potential to amplify greenhouse changes if its concentrations increase. The role of water vapor is complicated by clouds. If increased water vapor also leads to an increase in clouds, the increase of cloud cover could conceivably act to reduce global warming by reflecting larger fractions of incoming solar radiation, or increase global warming by trapping infrared radiation. The quantitative effects of changes in water vapor and clouds represent major uncertainties in understanding future climate change.

A final important consideration in the study of natural climate change is the so-called ***natural variability*** of the climate system. Even without the variable forcing from volcanoes, greenhouse gases, solar variability, and the Earth-sun orbital modulations, climate would vary from year to year, decade to decade, and century to century. These variations arise from air-sea interactions and other poorly understood processes internal to Earth's climate system and are referred to as "natural variability" or ***internal variability.*** Common examples of natural variability include the El Niño/La Niña cycle (Chapter 23); the multi-decadal warming and cooling that occurred in the early and middle twentieth century, especially over high latitudes; and certain time periods during which the Northern Hemisphere cooled while recovering from the most recent ice age. These cooling episodes have been hypothesized to be in response to changes in the North Atlantic Ocean associated with pulses of freshwater from melting ice sheets. The importance of natural variability arises from its potential to amplify or offset other changes over timescales of several decades. The "masking" of climate changes by natural variability is especially important for the interpretation of human induced climate changes associated with enhanced warming from anthropogenic emissions of greenhouse gases.

CHECK YOUR UNDERSTANDING 5.2

1. What are the long-term drivers of climate change?
2. What is an ice age? How long have the Earth's ice ages lasted?
3. What are aerosols? How do they affect the atmosphere's energy budget?
4. By approximately what percent have the atmosphere's carbon dioxide concentrations increased since the pre-industrial period?

RECENT AND ONGOING CLIMATE CHANGE

As implied by the preceding discussion, the Earth's climate varies over many timescales in response to different causes. Scientists use various types of paleoclimatic information to determine how climate has changed in the past (Focus 5.1). For example, Figure 5.13 shows Northern Hemisphere temperatures reconstructed for the past 1000 years. This reconstruction is based on instrumental (thermometer) measurements for the past 100 years, and on tree rings, corals, and ice core data for the preceding centuries. The uncertainties in the estimates, shown by the gray shading, are considerably larger in the early centuries. The reconstruction indicates that the Northern Hemisphere is now warmer than at any time in the past 1000 years, largely because of the twentieth-century warming. Prior to 1900, the record shows a slow cooling from the so-called ***Medieval Warm Period.*** It should be noted that this reconstruction has been the subject of debate because of its reliance on ***proxy data*** such as tree rings, for which the derived temperatures are limited primarily to the growing season. Hence the winter season may be under-represented in the temperatures.

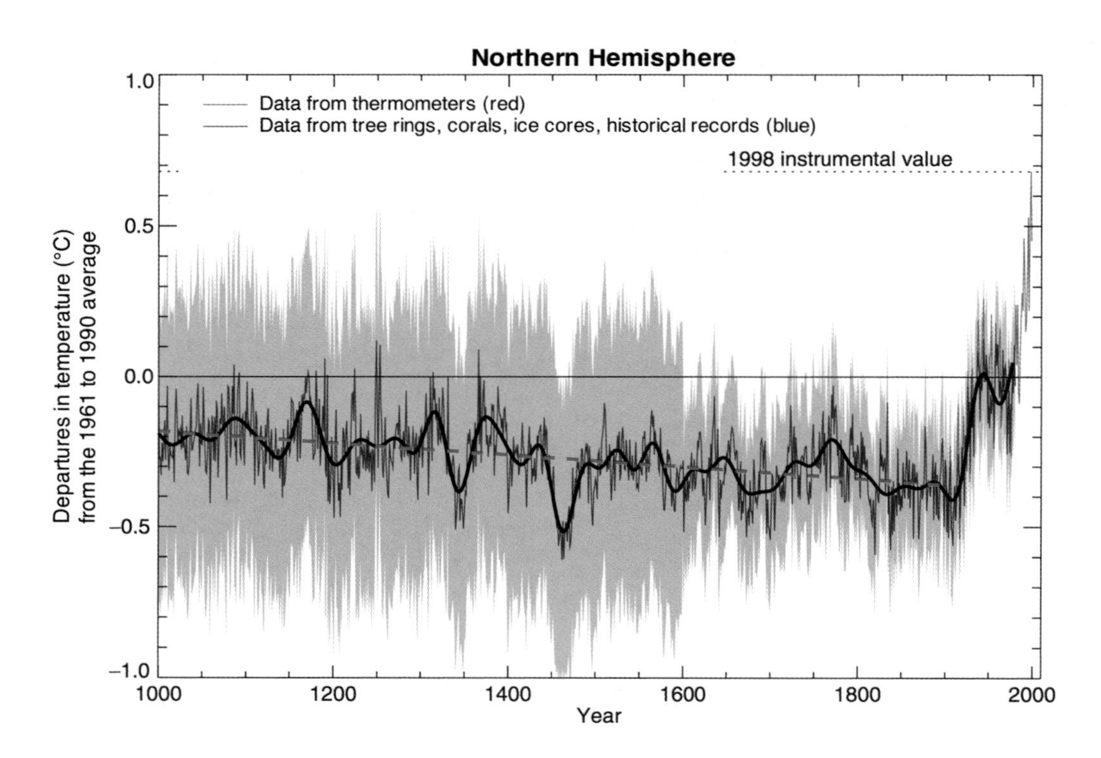

FIGURE 5.13: Reconstruction of Northern Hemisphere temperature, relative to the average for 1961 to 1990, for the past 1000 years. Yearly values in red are from instrumental data (thermometers), values in blue are from tree rings, corals, and other proxy information. Smooth black line is a running average; gray shading denotes range of uncertainty.

The more recent (1880–2006) temperatures, based on actual measurements, are shown for both hemispheres in Figure 5.14. Both hemispheres have warmed since 1880, the Northern Hemisphere by about 1.4° F (0.8° C), the Southern Hemisphere by about 1.0° F (0.6° C). In each hemisphere, the six or seven warmest years have all occurred since 1990. Figure 5.14 also shows that the Northern Hemisphere experienced a cooling from the 1940s to the 1970s. While the explanation of this cooling is still the subject of research, many scientists regard it as an example of the natural variability that occurs on multidecadal timescales. This variability is superimposed on a longer warming trend.

The recent warming is far from geographically uniform. Figure 5.15 shows that the warming has been strongest in northern North America and northern Asia, as well as over the Antarctic Peninsula. In these areas, the warming has exceeded 3° F (2° C) and has been even stronger in winter. By contrast, the oceans show much less warming, and even pockets of cooling. This pattern is consistent with the ability of the oceans to mix heat downward, much the same as mixing reduces the seasonal cycle of temperature over the ocean. In addition, the atmospheric circulation is known to have contributed to the strong warming over the northern land areas, i.e., there has been a greater frequency of southerly winds into northwestern North America and westerly winds into northern Asia. The warming of the past several decades is most apparent near the surface and in the lower troposphere. Evidence of this warming, in addition to the direct measurements of temperature, is the remarkable retreat of most of the world's mountain glaciers. Trends in the upper troposphere are less conclusive, so it is difficult to determine whether there has been a change in the vertical gradients of temperature or the troposphere's stability (Chapter 6). There are indications that the stratosphere has cooled over the past several decades The decrease of the stratosphere's ozone, which heats the stratosphere by absorbing ultraviolet radiation, may also have contributed to the cooling of the stratosphere. The cooling of the stratosphere is consistent with satellite measurements of an apparent increase since 1979 of about 200 meters in the height of the tropopause.

Precipitation, which is the key variable for floods and droughts (Chapters 25 and 26), has increased since 1900 by about 2 percent over the land areas of the world. Figure 5.16 shows the geographical distribution of these trends, which are most coherent over the northern land areas, especially the northern United States and Canada. Drying has occurred in central Africa, western South America, and parts of the southwestern United States. Trends of precipitation over the ocean during the same time period cannot be determined because the rain gauge network used to measure precipitation is primarily land-based. These precipitation trends over land are generally consistent with changes in river flows and lake levels. However,

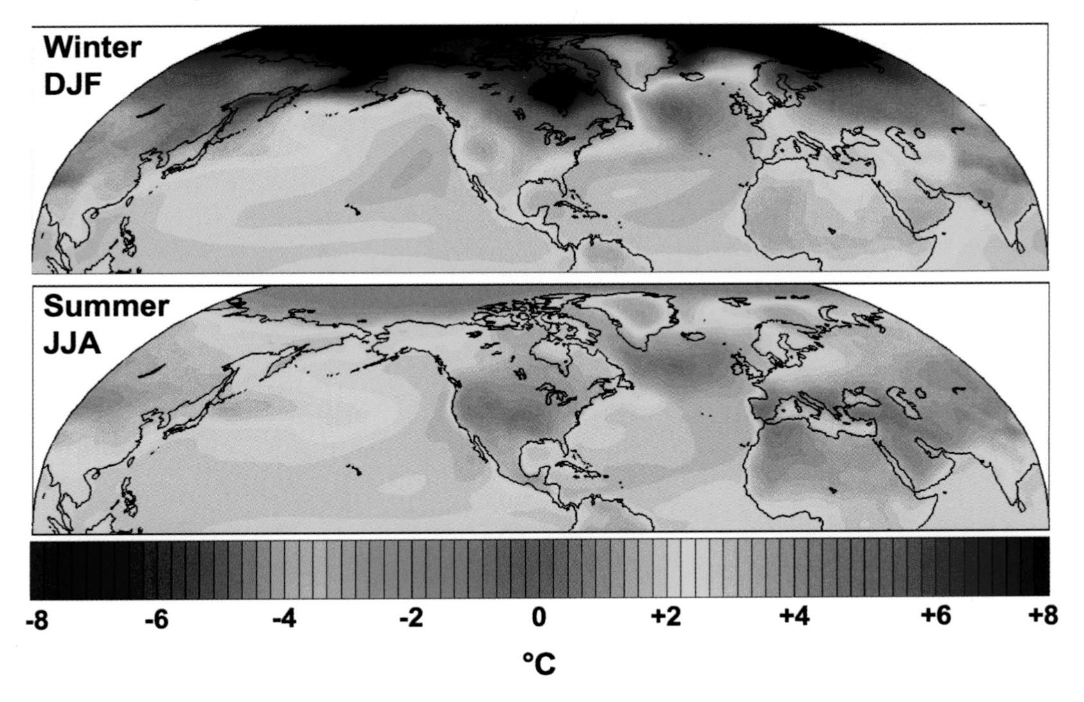

FIGURE 5.17: Projected changes of surface air temperature (°C) by 2070–2090 (relative to 1980–2000) for Northern Hemisphere winter (December–February) and summer (June–August) in the A1B scenario. Results are averaged over fifteen climate models. Yellow, orange, and red denote progressively greater warming.

ONLINE 5.1

Evolution of Twenty-First Century Temperatures

Climate warming associated with increasing greenhouse gases will not be a steady progression toward higher temperatures. As indicated by the discussion of natural variability, there will be years and even decades when cooling occurs. Online, you can see the year-by-year progression through 2100 of temperatures around the Northern Hemisphere. Notice that the areas of greatest warming change from year to year and decade to decade, in ways that lead to an occasional cooling. Nevertheless, warming becomes progressively dominant in the middle and late portions of the century, leading to the broad pattern of warming shown in Figure 5.17. Because the animation is based on monthly temperatures averaged over about 15 models, and the different models' "hot spots" are not in the same areas at any particular time, the pattern in the animation is actually smoother in space and time than in any individual model (and, by implication, smoother than the evolution will be in the real world).

toward more precipitation, especially in western coastal areas where storms encounter mountain ranges (e.g., western North America). The summer pattern (Figure 5.18) shows an increase of precipitation in higher latitudes and over the ocean, but a decrease over large parts of the United States and western Eurasia. This so-called ***continental drying*** during summer has long been a characteristic of greenhouse climate simulations by models, and it has major implications for summer growing conditions in some of the world's most productive agricultural areas. A trend toward drier surface conditions during summer would favor drought (Chapter 26) and, because evaporation would be reduced, increased heating of the surface and potential for heat waves (Chapter 27).

Other changes projected by the climate models include poleward shifts of the major circulation features, including the storm tracks of the North Atlantic and North Pacific Oceans. These shifts are consistent with a northward movement of the thermal contrast between retreating snow-and-ice cover and the warmer ice-free ocean. In addition to this shift of the winter storm tracks, there is a tendency in all seasons for more frequent heavy-precipitation events. This

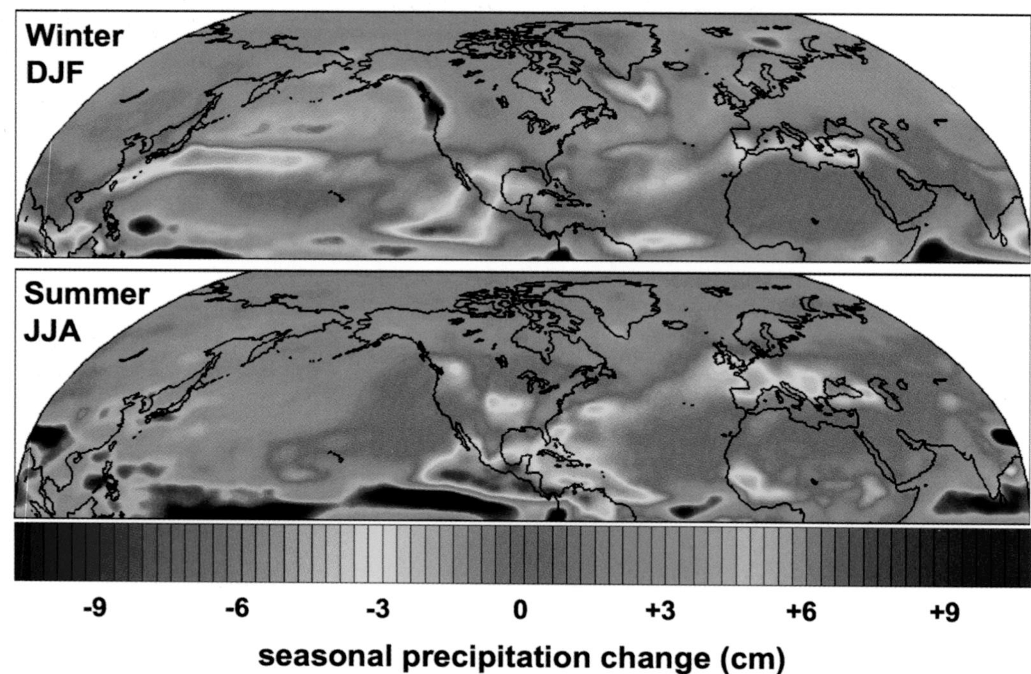

FIGURE 5.18: Projected changes of precipitation (cm) by 2070–2090 relative to 1980–2000 for Northern Hemisphere winter (December–February) and summer (June–August) in the A1B scenario. Results are averaged over fifteen climate models. Blue and green denote increases of precipitation (wetter climate); yellow and orange denote less precipitation (drying).

tendency is consistent with the increase of the atmosphere's moisture content as temperature increases, and the decrease in atmospheric stability. Analyses of observational data from recent decades indicate that there is already an ongoing trend for a greater portion of precipitation to occur in extreme (heavy-precipitation) events. This trend, together with the projected drying of the mid-latitude continents during summer, raises the possibility that threats from both floods (Chapter 25) and droughts (Chapter 26) could increase in the coming decades.

The additional warming projected for the next century and beyond, together with the warming that has already occurred in the past century, presents significant risks as well as opportunities for society. The risks are receiving increasing attention in the policy and planning arenas. Threats to agriculture and water supplies are obvious consequences of the changes summarized above. However, climate change will also bring new opportunities to some areas as, for example, growing conditions become more favorable in regions where agriculture is now limited by low temperatures. There is also the intriguing consideration that the present interglacial has lasted as long as, or even longer than, most interglacials of the past million years. It is not inconceivable that greenhouse warming will counter a return to a colder climate over the coming centuries. Such speculation must be tempered by the realization that climate has undergone rapid but poorly understood shifts in the past, especially in the North Atlantic sector where a delicate balance of air-sea interactions and ocean currents combine to drive much of the world's deep-ocean circulation.

Because of the connection between climate and hazardous weather, changes of climate have important implications for the frequency and intensity of almost all types of severe weather. In the remainder of the book, we will highlight the changes in hazardous weather due to climate changes that are anticipated in the coming decades as a result of increasing greenhouse gas concentrations.

CHECK YOUR UNDERSTANDING 5.3

1. How has the Northern Hemisphere's temperature varied during the past century?
2. How has global precipitation changed since 1900?
3. Is greenhouse warming expected to be greater over the continents or the oceans? Why?
4. How are the Northern Hemisphere's major storm tracks projected to shift in scenarios of global warming?

TEST YOUR UNDERSTANDING

1. How are climate and weather related?
2. What happens to the solar radiation that is absorbed by Earth's surface?
3. What is the greenhouse effect?
4. What is the role of latent heat in the atmosphere's energy budget?
5. How would climate be different if the Earth's axis were not tilted relative to a line perpendicular to the plane of Earth's orbit around the sun?
6. Why don't the tropics become progressively hotter even though more radiation is gained than is lost equatorward of about 40° N and S?
7. How do the intensities of the subpolar lows and the subtropical highs vary from winter to summer?
8. How are the trade winds related to the Hadley Cells?
9. How does the climate of a coastal city such as San Francisco compare with the climates of inland cities at similar latitude (e.g., Denver, St. Louis)? Why?
10. What factors affect the geographical distribution of precipitation?
11. What are some of the tools that scientists use to determine how climate varied in the past?
12. Why does the movement of the continents over long-time scales affect climate?
13. How has the sun's output of energy varied over the Earth's history? By how much does it typically vary during an eleven-year sunspot cycle?
14. How does the climate of Venus compare to the climate of Earth? Why?
15. During the peak of the last glaciation, about 18,000 years ago, how were sea level and the coastlines of the United States different from today?
16. What factors are thought to have played a role in the cycling of glacial and interglacial periods of the past two or three million years?
17. At the time of the last glacial maximum, about 18,000 years ago, how did the climate of the Desert Southwest of the United States most likely compare to today's climate in the same area?
18. What can be said about the cause of New England's "Year Without a Summer" in 1816?
19. Which gas is the strongest contributor to the greenhouse gas effect in our atmosphere?
20. Why is it difficult to anticipate the role of clouds in future climate change?
21. How does a "projection" of climate change differ from a "prediction"?
22. What are the characteristics of the polar amplification of climate change? Why do climate models show a polar amplification?
23. If global warming continues, how are the temperatures of the stratosphere expected to change? Given that the lower troposphere is projected to warm, what does this imply for the vertical variation of temperature?

TEST YOUR PROBLEM-SOLVING SKILLS

1. The Earth's orbit around the sun is not a perfect circle, but an ellipse. The Earth is closest to the sun in early January and farthest from the sun in early July. The Earth-sun distance is presently about 3 percent larger in July than in January. (Note that this orbital parameter is different from the tilt of the Earth's axis, which causes much greater seasonal variations than does the variable Earth-sun distance). Suppose that the Earth's orbit were instead circular, with a radius that is the average of the present-day Earth-sun distance. How would temperatures with a circular orbit compare to present-day temperatures during
   (a) winter in the Northern Hemisphere?
   (b) summer in the Northern hemisphere?
   (c) winter in the Southern Hemisphere
   (d) summer in the Southern Hemisphere?

2. Large amounts of fresh water are presently locked up in the Greenland and Antarctic ice sheets. Suppose, for simplicity, that Greenland's ice sheet covers an area equal to 1 percent of the Earth's ocean area and that its average thickness is 0.75 km. Suppose further that the Antarctic ice sheet covers an area equal to 4 percent of the Earth's ocean area and that its average thickness is 1.75 km.
   (a) By how much would sea level rise if the Greenland ice sheet melted completely?
   (b) By how much would sea level rise if the Antarctic ice sheet melted completely?
   (c) By consulting a topographic map, determine whether the following cities would remain above sea level if both ice sheets melted: Orlando, FL; Washington, DC; Philadelphia, PA; and New York, NY.

3. Figure 5.7 shows the average high and low temperatures for each calendar month at six cities in the United States.
   (a) How does the timing of the warmest and coldest months compare to the timing of the maximum and minimum solar radiation at these cities? Provide an explanation for any lags.
   (b) Do the coastal and inland cities show any differences in the timing of the warmest months? of the coldest months?

4. Figure 5.8 shows the average present-day precipitation amounts for January and July, while Figure 5.18 shows the projected changes of precipitation by 2070–2090 based on a moderate scenario of increasing greenhouse gas concentrations. (The winter and summer maps in Figure 5.18 can be assumed to be representative of January and July, respectively.)
   (a) Construct a table showing the actual and percentage changes (including the signs) of the precipitation changes in January and July at Miami, Florida, New York City, New York, Dallas, Texas, Denver, Colorado, Los Angeles, California, Seattle, Washington, and Fairbanks, Alaska. For which city is the *actual* change projected to be largest in January? in July?
   (b) For which city is the *percentage* change projected to be largest in January? in July?
   (c) Do your answers to (b) suggest a latitudinal pattern in the percentage changes?

5. Suppose global warming continues and the global average surface temperature increases from 15° C (59° F) to 19° C (66° F) by the year 2110. Suppose further that the average relative humidity of the air near the surface is 60 percent and remains so as the world warms.
   (a) What is the vapor pressure that corresponds to the present-day global average temperature? (Hint: use Figure 1.09.)
   (b) What would be the corresponding vapor pressure in 2110?
   (c) What is the percentage increase of the vapor pressure by 2110?
   (d) How would such a change alter the atmosphere's greenhouse effect?

USE THE SEVERE AND HAZARDOUS WEATHER WEBSITE

http://severewx.atmos.uiuc.edu

1. An excellent source of information on recent global temperatures is maintained by NASA's Goddard Institute for Space Studies.
   (a) Use this site to examine the worldwide pattern of surface temperature anomalies (departures from normal, relative to average for 1971–2000) for the most recent calendar year. Which areas showed the greatest abnormal warmth? Were any areas cooler than their 1971–2000 average temperatures? Repeat your examination by displaying the anomalies averaged over the past decade.
   (b) This site also lets you plot a time series of the yearly global average temperatures back to 1880. By viewing such a plot, determine how last year's global mean temperature compares to the warmest years on record.

2. Figure 5.2 shows the Earth's position in different seasons. The winter pole is always in darkness, while the summer pole experiences continuous sunlight. At the equinoxes, both poles receive twelve hours of daylight and darkness. The progression of night and day in the polar regions is easy to observe from satellites. Use the *Severe and Hazardous Weather Website* to navigate to the Colorado State University (CSU) Cooperative Institute for Research in the Atmosphere (CIRA). Access their satellite imagery and examine an image of the GOES 11 or GOES-12 "full disk" view of the Earth in the "visible" channel.
   (a) What is the date of the image?
   (b) In what geographic region is the day-night boundary located?
   (c) Do your responses to (a) and (b) conform to Figure 5.2? Explain why or why not.

3. Where do the most extreme temperatures and precipitation occur on Earth? Use the *Severe and Hazardous Weather Website* to find the National Climatic Data Center's summary of Global Extremes.
   (a) Which continent has the highest number of extreme reports?
   (b) Note the dates given for each extreme report. Based on these dates, could someone argue that these reports are, in fact, not the most extreme values ever experienced on Earth? Why or why not?
   (c) What correlations, if any, can you infer between elevation and the type of extreme report?

4. A variable that provides an integral measure of recent climate change is the area of sea ice in the northern polar regions. *The Cryosphere Today* website of the University of Illinois allows you to see today's coverage of sea ice and snow cover based on the latest satellite information. It also provides an animation of the daily ice and snow coverage of the most recent thirty-day period. Based on the information provided by this site, how does the present coverage of sea ice compare to coverage one year ago? How does it compare to the average for the past several decades. What trends are shown by sea ice coverage over the past thirty years? Have the recent changes been largest in summer or winter?

5. The National Climate Data Center maintains a Climate Monitoring website that includes a "State of the Climate" assessment for the current year and for each year of the past decade. Which events are highlighted in the past month's "Selected Global Significant Events?" Which events are highlighted for the United States? By accessing the "Special Reports" link, you can see which climatic events of the past several have had particularly large impacts on the United States. What is the most recent such event?

6. The "Global Warming" site of the National Oceanic and Atmospheric Administration contains answers to frequently asked questions about global warming. From the information at this website, provide brief answers to the following questions that you may be asked by friends or relatives:
   (a) Is climate becoming more variable or extreme?
   (b) Can recently-observed climate changes be explained by natural variability, including changes in solar output?
   (c) Is the hydrologic cycle changing?

ATMOSPHERIC STABILITY

A thunderstorm over central Illinois.

KEY TERMS

adiabatic process
air parcel
capping inversion
Convective Available Potential Energy (CAPE)
Convective Inhibition (CINH) Index
cold front
condensation level
conditionally unstable
convection
convergence
dry adiabatic lapse rate
environment
environmental lapse rate
inversion
K Index
lapse rate
level of free convection
Lifted Index (LI)
lifting condensation level
lifting mechanism
moist adiabatic lapse rate
neutral stability
sea breeze
Showalter Index
Severe Weather Threat Index (SWEAT)
stable
Total Totals Index
unstable
upslope

Thunderstorms can produce violent weather, yet we know from experience that severe thunderstorms occur only occasionally, not every day in every location. Evidently certain atmospheric conditions are required for the development of these storms. What are these conditions, and how do we explain why they are associated with thunderstorms? Our first clue is that these storms contain air that is rapidly rising in a process known as ***convection***. If we wish to understand the storms, we must understand why air rises so rapidly. Our understanding begins with the concept of stability. Because stability determines the location and intensity of convective storms in the atmosphere, it is one of the most fundamental and important concepts for understanding severe weather.

THE CONCEPT OF STABILITY

Let's begin with a simple example. Figure 6.1 shows three balls, one located in a valley, one on a flat surface, and one on the top of a hill. All three balls are initially stationary. Suppose we bump each ball ever so slightly, giving it a tiny "displacement" from its original position. The ball in the valley will roll back to its original position after we bump it. The second ball will stop at its new position after the bump. The ball on the hill will roll away, speeding down the hill.

We say that the ball is stable if, for a small displacement, it returns to its original position.

We say that the ball is neutral if, for a small displacement, it stays in its new position.

We say that the ball is unstable if, for a small displacement, it accelerates away from its original position.

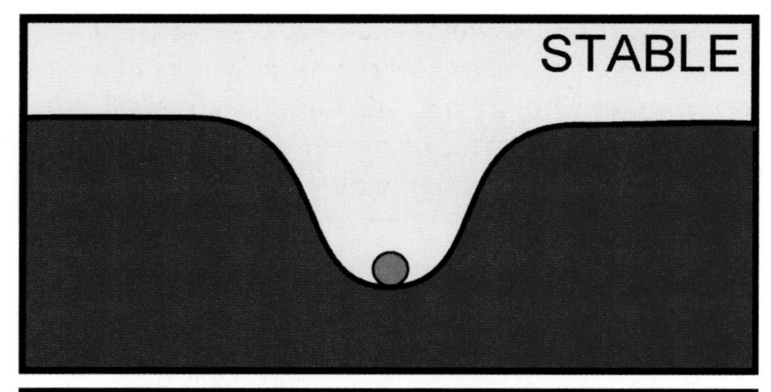

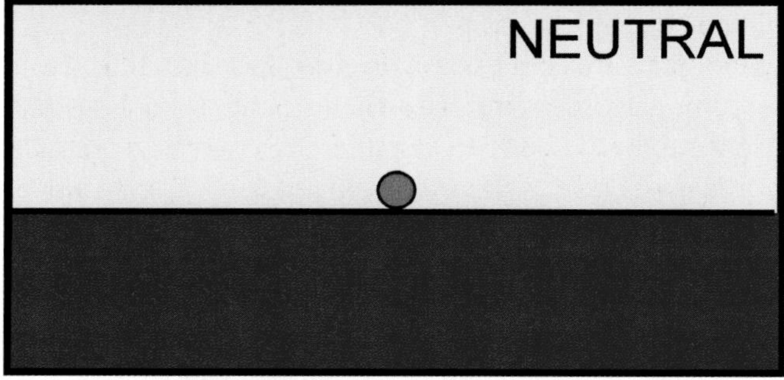

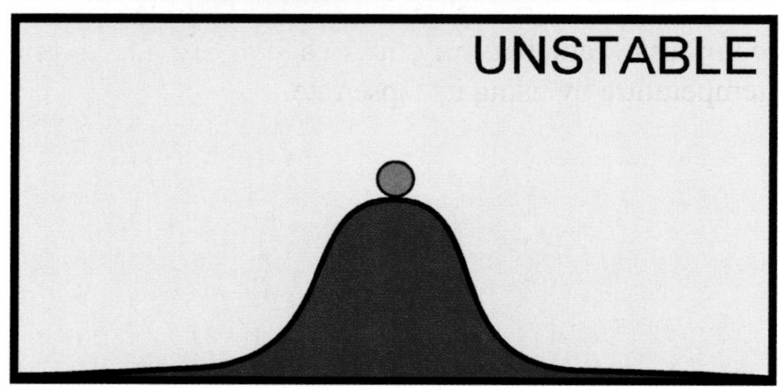

FIGURE 6.1: Stable, neutral, and unstable positions of a ball on a smooth surface.

Now apply this same concept to a parcel of air in the atmosphere. An ***air parcel*** is a distinct blob of air that we will imagine we can identify as it moves through the atmosphere. You can think of an air parcel as the air contained within an imaginary flexible balloon-like membrane that enables the air within it to expand or contract freely. We distinguish three possibilities for the stability of an air parcel:

1. Stable: If the parcel is displaced vertically, it will return to its original position.
2. Neutral: If the parcel is displaced vertically, it will remain in its new position.
3. Unstable: If the parcel is displaced vertically, it will accelerate away (upward or downward) from its original position.

Now consider what actually happens to a parcel of air moving vertically (either rising or sinking) through the atmosphere. To do so, we need to first understand the concept of an adiabatic process. An ***adiabatic process*** is one in which a parcel of air does not mix with its environment or exchange heat energy with its environment. To imagine an adiabatic process, we can imagine a parcel in a balloon whose skin will not permit the exchange of mass or energy. What will happen if this parcel moves upward or downward?

If you ever have pumped up a bicycle tire, you know that as you hold the pump, it gets hot. Compression heats the air, which then heats the pump. On the other hand, if you have ever used spray deodorant, hair spray, or some other spray that comes out of an aerosol can, you know that the spray always feels cool, even if the can is at room temperature. The decompressing air cools as it expands coming out of the can. The same effect occurs when the air is let out of a tire: the escaping air expands and cools. Hence we have two processes with opposite effects: compression heats air; expansion cools air. Air parcels rising through the atmosphere will expand as they rise because they encounter lower pressure (Figure 6.2), just like the air escaping from a tire. The rising parcels cool because they must do work to expand; this work consists of "pushing" away the surrounding air. If air parcels descend in the atmosphere, they compress and heat. In this case, the surrounding air does work on the air parcel by compressing it, and the work done during compression shows up as an increase of energy (temperature) of the air parcel. If a parcel of air expands or compresses adiabatically, it is easy to predict its final temperature by using its lapse rate.

LAPSE RATES

A ***lapse rate*** is the rate at which temperature decreases (lapses) with increasing altitude. Air parcels that do not contain cloud (are not saturated) cool at a very specific rate as they rise through the atmosphere. This rate of cooling with increasing altitude, called the ***dry adiabatic lapse rate,*** is about 10° C/km. This rate is the same no matter if the air starts at the surface and rises 1 km, or starts in the upper troposphere and rises 1 km. The 10° C/km dry adiabatic lapse rate is one of the most important numbers in atmospheric science (not only is it a constant, but it is a nice round number and hence easy to remember!). A rising air parcel will cool at a rate of 10° C per km of ascent, and a sinking air parcel will warm at the same rate, 10° C per km of descent.

What if air parcels are saturated (contain clouds)? In Chapter 1, we discussed the concept of latent heat. Recall that when water undergoes a phase change, heat is either required (water to vapor) or liberated (vapor to water). Recall the boiling water experiment in Chapter 1. When we (conceptually) boiled a pot of water, we found that it took a long time for our pot of water to undergo the phase change and turn to water vapor in the air. The energy from the burner went into accelerating the water molecules to high speeds characteristic of vapor, and into breaking the strong bonds that bind individual water molecules to each other. We called the energy "latent heat" because it was the "hidden heat" required for a phase change. We continued the conceptual experiment, allowing all of the

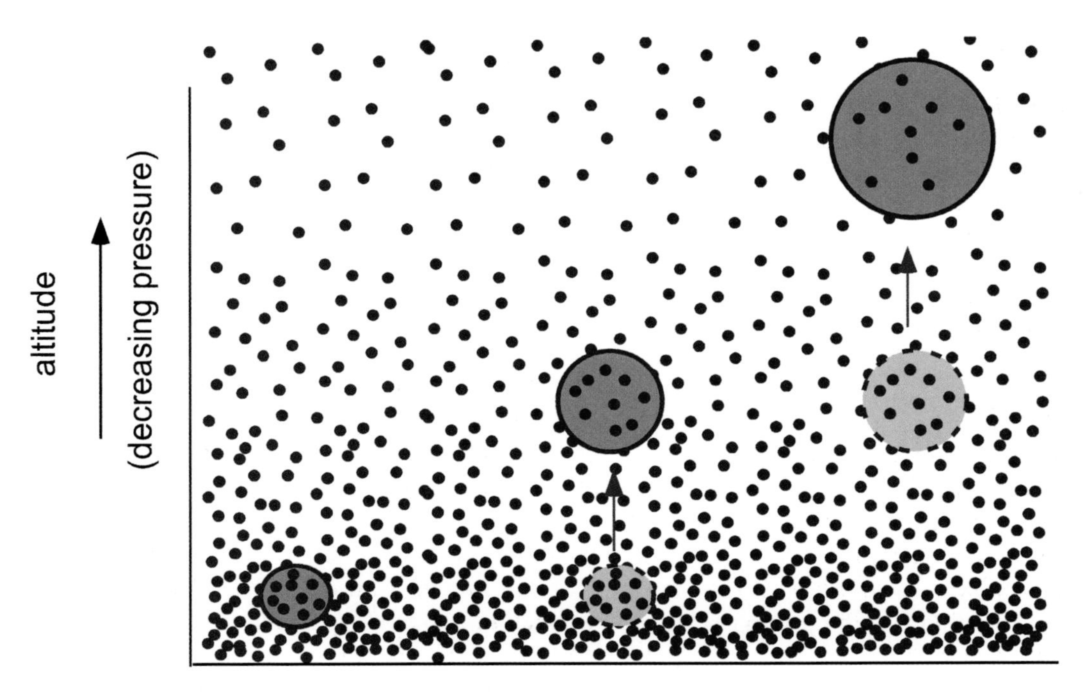

FIGURE 6.2: Illustration of an air parcel rising. As the air parcel rises, over time it encounters air of lower and lower pressure, causing the parcel to expand and its temperature to decrease. A sinking air parcel would contract and warm. Note that the number of air molecules in the air parcel remains constant.

vapor molecules to fly back into the pot and become liquid again. We learned that all the energy that went into vaporizing the water would be liberated as heat. When clouds form, water vapor condenses to liquid, and latent heat warms the air in which the cloud is forming.

Now consider an air parcel rising adiabatically, but this time let's have the air be saturated (i.e., relative humidity = 100%) so that further cooling will lead to condensation. As the air rises, the cooling of the parcel due to the expansion is 10° C/km. However, since a cloud forms in the parcel as the air cools, latent heat is released into the parcel, warming the parcel. By how much? It turns out that latent heating warms the air at a variable rate. In the lower atmosphere, the heating rate is about 4° C/km. In the middle troposphere, latent heat warms the air at a rate of about 2° C/km, and near the tropopause, where there is so little moisture, the latent heating rate is negligible. The ***moist adiabatic lapse rate*** is determined by considering the combined effects of expansional cooling and latent heating. In the lower troposphere it is about 10° C/km – 4° C/km = 6° C/km. In the middle troposphere it is about 10° C/km – 2° C/km = 8° C/km. Near the tropopause the moist and dry adiabatic lapse rates are essentially the same—about 10° C/km. In the examples that follow, we will assume that the moist adiabatic lapse rate is 6° C/km for illustration. You should keep in mind that this rate does vary with altitude. The moist adiabatic lapse rate, 6° C/km, is the second key number to emerge from this discussion. While it is not a constant like the dry adiabatic lapse rate, its approximate value (6° C/km) should be remembered.

In addition to the two adiabatic lapse rates, one more concept is important to an understanding of atmospheric stability: the ***environment***. The environment is simply the atmosphere outside the parcel. Using a balloon analogy, everything in the balloon is the parcel, everything outside the balloon is the environment. A rawinsonde's thermometer measures the environment's temperature as a function of height. From many such measurements, we know that temperature generally decreases with height in the environment, at least up to the tropopause. How fast does temperature decrease with height in the environment? Well, that depends on where you are on the Earth, the time of day, the season of the year, etc. In other words, the environment differs from place to place and time to time—no two rawinsondes give exactly the same vertical profiles of temperature. However, each rawinsonde profile of temperature permits the calculation of the rate at which the environment's temperature decreases with increasing altitude. This rate is known as the ***environmental lapse rate***. In general, the environmental lapse rate varies from about 4 to about 9° C/km. It is important to remember that the environmental lapse rate is variable—it varies from location to location, from day to day at a fixed location, and even from one layer of the atmosphere to another at a particular location and time.

Figure 6.3 shows an example of an idealized sounding in which there is a well-defined troposphere below 10 km and a well-defined stratosphere above 10 km. The temperature in the troposphere decreases from 20° C at the surface (elevation = 0 m) to –50° C at 10 km above the surface. The environmental lapse rate, which is the decrease of temperature divided by the increase of elevation, is 70° C/10 km, or 7° C per km. This value lies well within the range given in the preceding paragraph. Note that the environmental lapse rate in the stratosphere in Figure 6.3 is negative, since

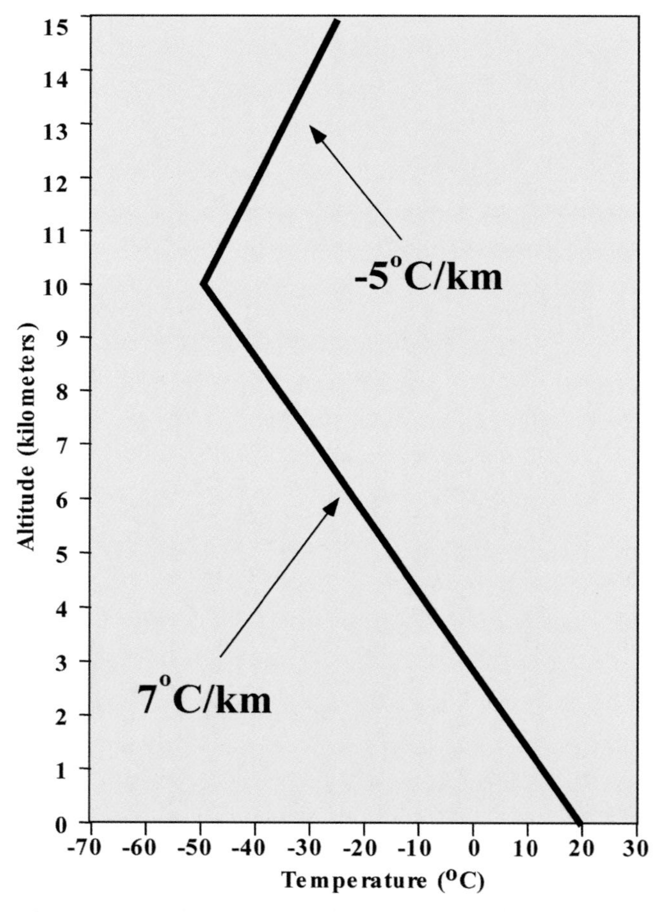

FIGURE 6.3: Idealized example of a vertical temperature profile having a constant environmental lapse rate of 7° C/km in the troposphere and a constant environmental lapse rate of –5° C/km in the stratosphere.

the environment's temperature increases with increasing elevation in the stratosphere (as it does whenever an inversion exists). The actual value of the stratosphere's environmental lapse rate in Figure 6.3 is –5° C/km. Recall that when the temperature of a layer of air increases with height, an ***inversion*** is said to be present. An inversion will always have a negative lapse rate.

CHECK YOUR UNDERSTANDING 6.1

1. What is the significance of stability with respect to severe thunderstorms?
2. What is an adiabatic process?
3. What is the approximate value of the dry and moist adiabatic lapse rates in the lower atmosphere?
4. What is the name of the lapse rate obtained from rawinsonde measurements?

DETERMINING STABILITY

Our goal in this chapter is to be able to determine the conditions under which convective clouds and storms may form—where the atmosphere is unstable. Before we put our ideas about vertical motions and lapse rates together to help determine instability, consider one more experiment. Suppose a drop of oil is put at the bottom of a glass of water. Is the oil drop stable or unstable? It is unstable because it will accelerate up to the top of the glass. Why did it do that? The oil drop is less dense than the water, making it buoyant.

In the atmosphere, a parcel of air is less dense than its environment if the parcel is warmer than the environment, i.e., if the parcel's temperature exceeds the environment's temperature. Think of a balloon. (In this discussion we will assume that the balloon is made of material light enough to be considered weightless.) If the air inside the balloon is heated, as when the flame is turned on under a hot-air balloon, the balloon becomes unstable, accelerating upward away from its position on the ground. If the air inside the balloon is permitted to cool, the balloon will eventually stop rising. If the air inside the balloon is made cooler and hence denser than the air in the surrounding environment, the balloon will sink. Thus an air parcel will rise or sink if its temperature is warmer or cooler than the environment's temperature. The stability of the atmosphere is determined precisely by such comparisons.

The following examples illustrate the use of temperature lapse rates in determinations of stability.

Figure 6.4 shows two examples of environmental temperature as a function of height. These temperatures might be measured by a rawinsonde ascending through

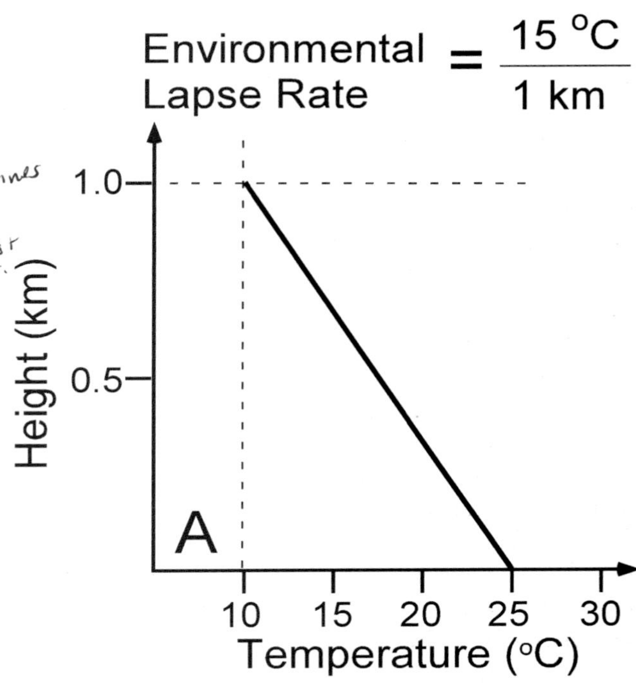

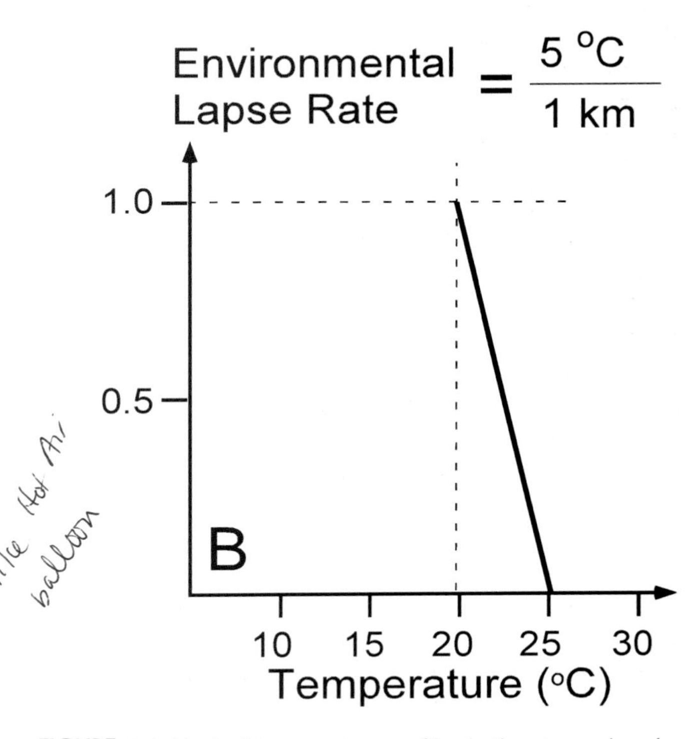

FIGURE 6.4: Vertical temperature profiles in the atmosphere's lowest kilometer with constant lapse rates of (A) 15° C/km and (B) 5° C/km.

the atmosphere. The environmental lapse rates in these examples differ from typical soundings in two ways. First, the atmosphere rarely (if ever) has a constant lapse rate for such large depths (the same is true of Figure 6.3). Lapse rates normally vary quite a bit with height. The second difference is the magnitude of the lapse rate in (A). The 15° C/km environmental lapse rate is way too large and is never observed in the atmosphere. We use it here because it makes it easy to see the point we are trying to illustrate about stability.

When we displace a parcel of air upward adiabatically, it will cool at the rate of 10° C/km or 6° C/km, depending on whether it is unsaturated or saturated, respectively. Let's examine the unsaturated case first. Recall that unsaturated air parcels are said to undergo "dry" adiabatic ascent (the parcel is "dry" only in a relative sense—it is drier than if it were saturated at the same temperature).

Look first at Figure 6.5A, where the surface temperature is 25° C and the environmental lapse rate is 15° C/km. Let's lift a parcel of air adiabatically from the surface (we just pick any blob of air out of the environment and lift it). At the surface, the air parcel will have the same temperature as its surrounding environment. During lifting, the air parcel will cool at the dry adiabatic lapse rate of 10° C/km, so its temperature at 1 km will be 25° – 10° = 15° C. According to the chart, the temperature of the undisturbed environment at 1 km is 10° C. We determine the stability of the parcel by comparing the parcel's temperature after lifting to the temperature of the surrounding environment at the same altitude. If the air parcel is warmer than the surrounding air, the parcel will continue to rise. If the air parcel is cooler than the surrounding air, the parcel will sink back toward the surface. At 1 km, the temperature of the air parcel is 15° C and the temperature of the surrounding air is 10° C. Since the air parcel is warmer than the environment, it will continue to rise. This example illustrates instability, and we say that the environment is ***unstable***.

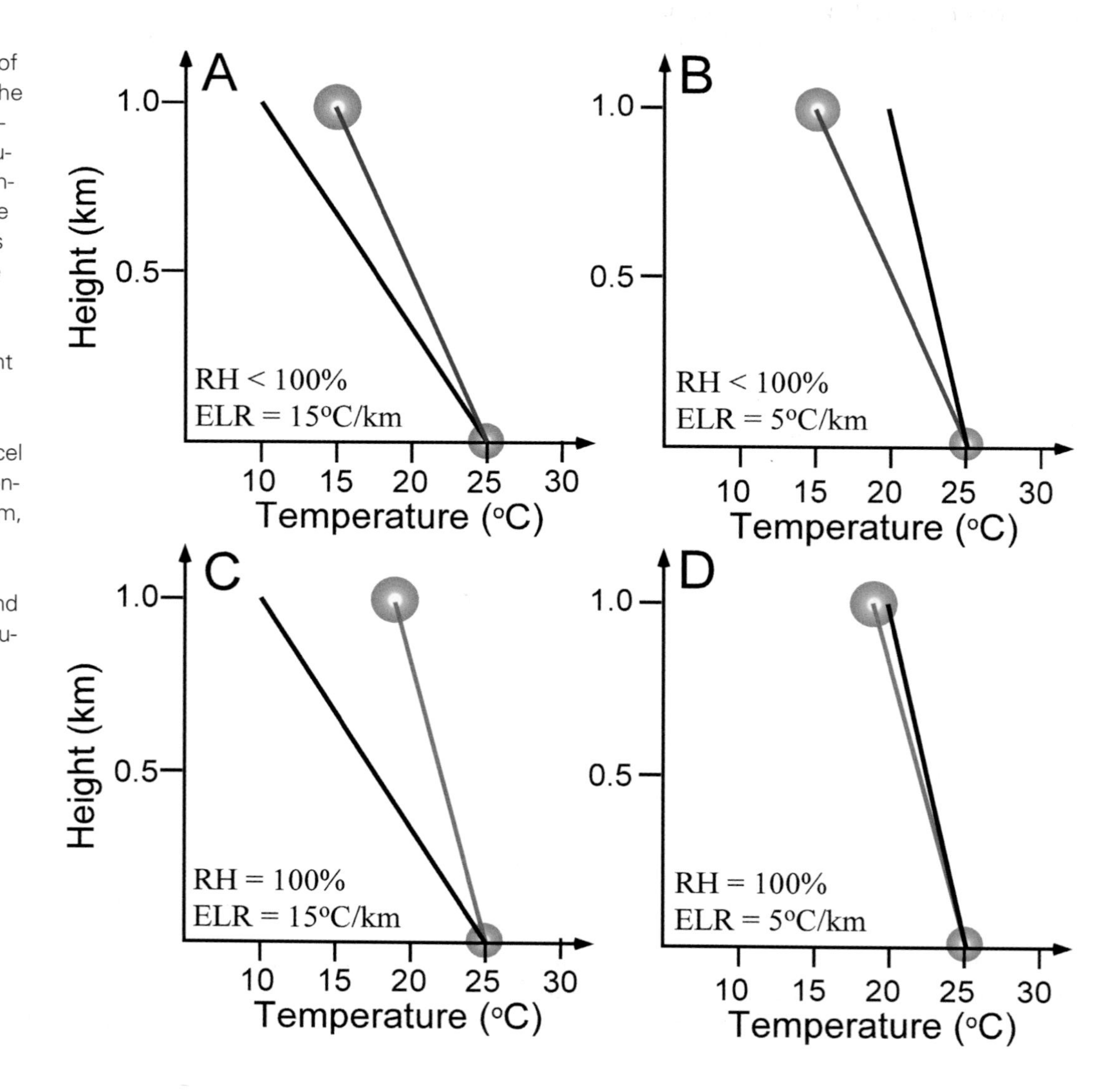

FIGURE 6.5: Examples of atmospheric stability. The environmental temperature is black, the unsaturated rising parcel's temperature is blue, and the saturated rising parcel's temperature is red. The four examples include: (A) unstable parcel in an unsaturated environment with an environmental lapse rate (ELR) = 15° C/km, (B) stable parcel in an unsaturated environment with ELR = 5° C/km, (C) unstable parcel in a saturated environment with ELR = 15° C/km, and (D) stable parcel in a saturated environment with ELR = 5° C/km.

What about the example in Figure 6.5B? Here the surface temperature is also 25° C, but the environmental lapse rate is 5° C/km. If we displace a dry (unsaturated) air parcel upward from the surface to 1 km, the air parcel will cool at 10° C/km and the temperature of the air parcel again will be 15° C. However, the environment is 20° C at 1 km. The air parcel is colder than its environment, so the parcel will sink back to its original position. The example illustrates stability, and we say that the environment is ***stable.***

It should be apparent from the two examples above that a rising air parcel will have the same temperature as its environment at all levels only if the environmental lapse rate is exactly 10° C/km. This situation is referred to as ***neutral stability,*** and it corresponds to the ball on the flat surface in Figure 6.1.

Now consider examples for saturated parcels (Figures 6.5C and 6.5D). A saturated parcel cools at 6° C/km, the moist adiabatic lapse rate. Let's assume that we displace a saturated parcel of air upward 1 km as shown in Figure 6.5C. The environmental lapse rate in this case is the same as in Figure 6.4A. Since the parcel is now saturated, water vapor will condense to form clouds. After ascending 1 km, the temperature of the air parcel will be 25° – 6° = 19° C. Since the environmental lapse rate is 15° C/km, the temperature of the environment at 1 km is 25° – 15° = 10° C. The temperature of the air parcel is much warmer than its environment and it will continue to rise. This parcel is unstable. Alternatively, we say that the environment in this case is unstable for rising air parcels.

What about Figure 6.5D? Again, we start with a parcel of saturated air with a surface temperature of 25° C. We displace it 1 km. It will cool at the moist adiabatic lapse rate. As before, the temperature of the air parcel at 1 km will be 19° C. Since the environmental lapse rate is 5° C/km, the temperature of the surrounding environment at 1 km will be 25° – 5° = 20° C. Here, the parcel will be cooler than the environment and the parcel will sink back to its original position. This parcel is stable. Alternatively, we say that the environment in this case is stable for rising air parcels.

The cases we looked at so far are generally extreme or even unrealistic. Let's look at a more realistic case. In this case, the environmental lapse rate falls between the dry and moist adiabatic lapse rates. Figure 6.6 shows an example. Here, the environmental lapse rate is 8° C/km. Let's displace an unsaturated air parcel upward 1 km. Again, at the surface the air parcel and the environment have the same temperature, 25° C. At 1 km, the unsaturated air parcel, cooling at the dry adiabatic lapse rate, will have a temperature of 15° C. The air in the surrounding environment will be 25° – 8° = 17° C. The air parcel is cooler than the environment and will sink back to its original position. The unsaturated parcel is stable. What if we displace a saturated parcel? In this case the rising air parcel will cool at the moist adiabatic lapse rate and have a final temperature of 19° C. The environment that surrounds the parcel at 1 km will be 17° C. The saturated parcel is warmer than its surrounding environment and will continue to rise. The saturated parcel is unstable. We describe the environment in this case as ***conditionally unstable.*** In a conditionally unstable atmosphere, the condition required for the instability is that the displaced parcel of air be saturated.

It is important to note that you can determine the stability of any atmospheric layer by displacing parcels within that layer, either upward or downward. It is not necessary to start with parcels at the surface. In the examples above, we have chosen the surface as a starting point for the sake of simplicity.

Table 6.1 summarizes the various possibilities for the stability of an atmospheric layer in the lower troposphere in which the environmental lapse rate is constant. However, the situations described so far are oversimplified in the sense that the real atmosphere rarely has the same lapse rate throughout its troposphere. Rather, the lapse rate typically changes with altitude, although there may be distinct layers in

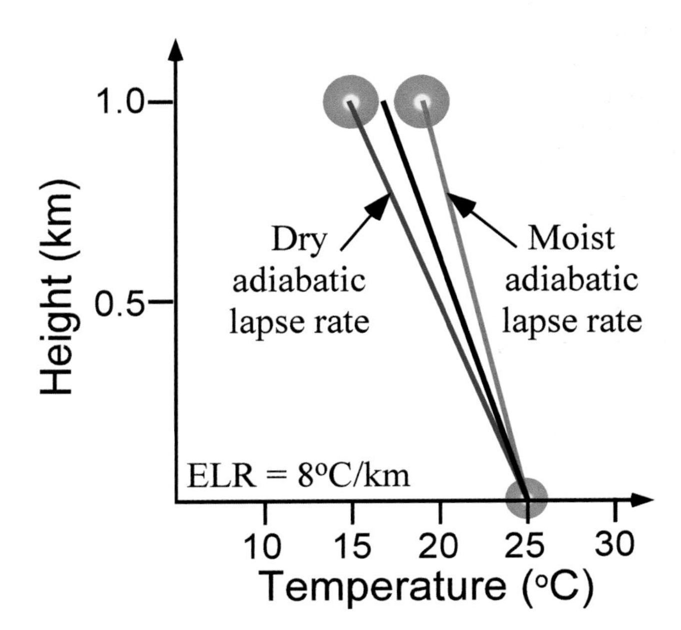

FIGURE 6.6: Example of a conditionally unstable atmosphere in which the environmental lapse rate is 8° C/km. Note that the parcel is stable if unsaturated and unstable if saturated. Parcel lapse rates are blue (unsaturated) and red (saturated); environmental lapse rate is black.

TABLE 6.1 **Summary of Categories of Atmospheric Layer Stability**

Environmental Lapse Rate (ELR)	Stability
ELR > 10° C/km	Unstable
6° C/km < ELR < 10° C/km	Conditionally unstable (Unstable if saturated, stable if unsaturated)
ELR < 6° C/km	Stable
ELR = 10° C/km	Neutral if unsaturated, unstable if saturated
ELR = 6° C/km	Neutral if saturated, stable if unsaturated

which the lapse rate is nearly constant over several hundred meters. Also, recall that the moist adiabatic lapse rate is not always 6° C/km, but changes with altitude to nearly 10° C/km near the tropopause. Figure 6.7 shows two actual soundings from Topeka, Kansas. While the May sounding (Figure 6.7A) shows a change of environmental lapse rates near 800 mb, the November sounding (Figure 6.7B) dramatically illustrates the tendency for environmental lapse rates to change abruptly, especially near the surface. The inclusion of temperature and dewpoint soundings on the same plot enables a meteorologist to determine the vertical extent of saturated layers, i.e., clouds; Figure 6.7B indicates extensive cloudiness in the middle troposphere (800–400 mb) where the dewpoint depression is close to zero.

Figure 6.7 also shows (in orange) the paths that would be followed by air parcels lifted from 50 mb above the surface. The parcel in Figure 6.7A would be warmer than its surroundings above approximately 650 mb, from which point it could rise by its own buoyancy (parcel warmer than its environment) to 155 mb. By contrast, the parcel in Figure 6.7B would be considerably colder than its environment at all levels above 900 mb, indicating a stable environment. Hence the reasoning about parcel-environment temperature differences still provides a basis for stability assessments, even when the environmental lapse rate is not constant with altitude.

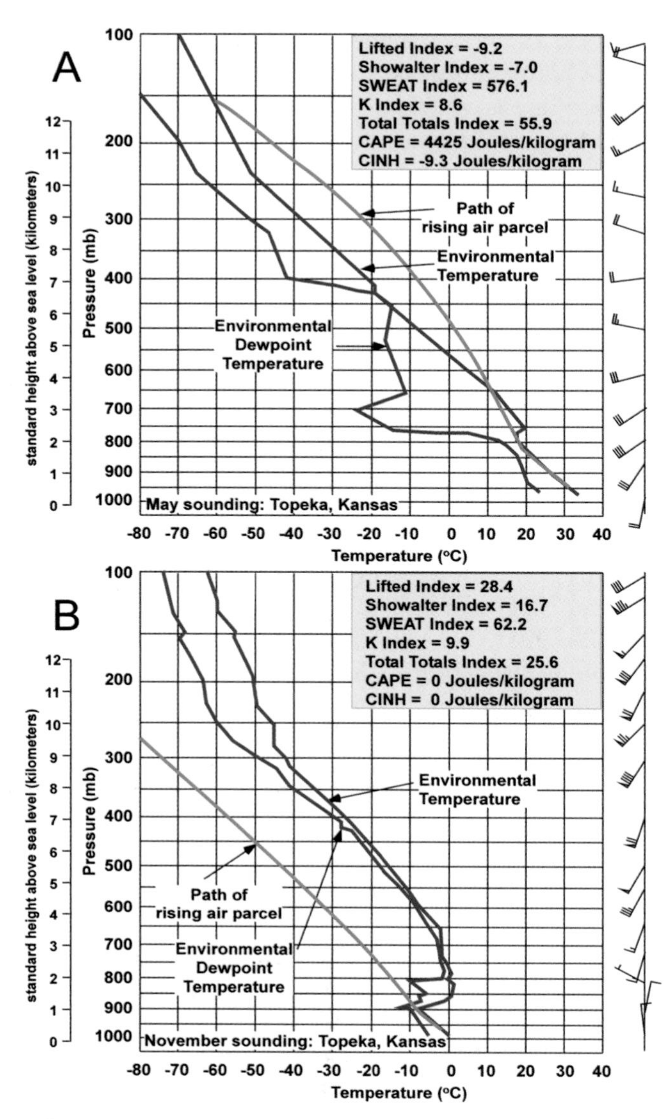

FIGURE 6.7: A sounding from Topeka, Kansas taken in May (A) and November (B). Solid blue lines in each panel are profiles of temperature (right) and dewpoint (left); solid orange lines show temperatures of air parcels rising from 50 mb above surface. Sounding (A) is conditionally unstable, sounding (B) is very stable. The shaded boxes list different indices or measures used by meteorologists to estimate stability; the indices are explained later in the final section of this chapter.

How can the atmosphere's stability at a particular location be changed? According to the preceding discussion, changes of the environmental lapse rate will change the stability. A change of the lapse rate of an atmospheric layer can be achieved by warming or cooling either the top or the bottom of the layer. As long as the temperature does not change uniformly throughout the layer, the layer's lapse rate and stability will change. The most common change in stability is the diurnal variation that occurs as air near the surface warms by day and cools by night, resulting in decreased stability in the afternoon and enhanced stability at night. You can view this effect in satellite imagery almost any day in the spring, summer, and autumn. During these seasons, cumulus clouds, showers, and thunderstorms are more common in the afternoon, when the ground is warmest, and less common around sunrise, when the ground is coolest. This diurnal variation of stability is most common over land, where the day-night temperature variation is greatest (see Online 6.1).

ONLINE 6.1

Infrared Satellite Imagery Captures the Diurnal Cycle of Atmospheric Stability

Time-lapse animations of infrared satellite images can provide insights into the diurnal changes of atmospheric stability in two ways. First, and most obvious, is their depiction of the enhanced development of cumulus and cumulonimbus clouds during the afternoon, illustrating the consequences of daytime destabilization of the lower atmosphere. Second, infrared images track the surface temperatures (in areas that are not cloud-covered), showing directly the changes in heating that are responsible for the destabilization. Online, you can view a twenty-four-hour cycle of infrared images of the contiguous United States on a warm day in mid July. In the animation, the darkening of the surface indicates the warming, which is especially apparent over the western United States and northern Mexico. This warming is followed almost immediately by widespread outbreaks of cumulus clouds, showers, and thunderstorms over large parts of the Mountain West, especially the Four Corners region. As the ground cools during the evening and early nighttime hours, the cumulus activity dies off dramatically, although the residual high clouds from a few tall thunderstorms persist. This daily cycle is repeated over the major landmasses during the warm season, when most clouds are convective rather than stratiform, layered clouds.

What does all this have to do with thunderstorms? Thunderstorms consist of unstable air parcels rising through an environment (the air around the storm) that is cooler than the rising parcels. The atmosphere is normally conditionally unstable in the thunderstorm environment. Since air parcels at the surface are normally unsaturated, they are stable. They must be lifted high enough to cool sufficiently so that their relative

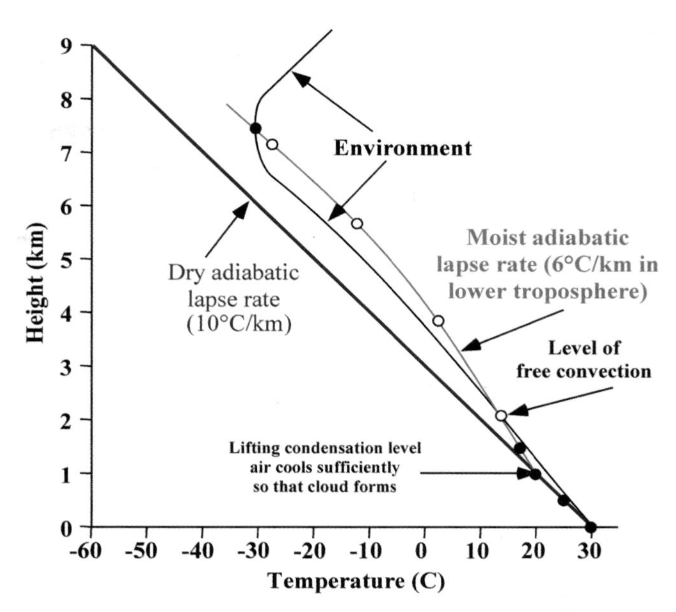

FIGURE 6.8: Schematic representation of the temperature of an air parcel rising from the surface. The parcel cools at the dry adiabatic lapse rate (heavy blue line) until reaching saturation at the lifting condensation level. During remainder of ascent, the parcel cools at the moist adiabatic lapse rate (thin red line). The thin black line shows the temperature profile of the environment. The parcel becomes warmer than the environment at the level of free convection. Filled circles indicate that parcel is negatively buoyant (colder than environment) and must be lifted; open circles indicate that parcel is buoyant (warmer than environment) and can rise freely.

humidity reaches 100 percent and they become saturated. The level where saturation first occurs is called the ***condensation level***, which is seen as the cloud base. If the air is lifted to cloud base, that level is also called the ***lifting condensation level***. Air parcels normally do not become unstable until they rise some distance beyond the base of the cloud. The level at which an air parcel first becomes buoyant (its temperature first exceeds the surrounding environment's temperature) is called the ***level of free convection***. We use the term convection to describe buoyantly rising air. The lifting condensation level and the level of free convection are sometimes close to one another, while in other environments these levels are separated by a large vertical distance. Figure 6.8 illustrates the temperature of an air parcel that is lifted through its lifting condensation level to the level of free convection, and then subsequently rises buoyantly to the tropopause, forming a thunderstorm. The parcel of air is unsaturated at the surface, so it cools at the dry adiabatic lapse rate during its initial ascent. When it cools to its dewpoint temperature, the parcel becomes saturated and a cloud forms (at the lifting condensation level). If the parcel continues to be lifted, it will cool at the moist adiabatic lapse rate because it is now saturated. (Note the change in the slope of the parcel's temperature at the lifting condensation level in Figure 6.8.) Shortly thereafter, the parcel becomes warmer than the environment (at the level of free convection), from which point it can rise on its own. We can witness this process by watching cumulus clouds over a period of time; the process is usually more dramatic in time-lapse video.

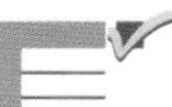

CHECK YOUR UNDERSTANDING 6.2

1. How is the density of an air parcel related to its temperature and to atmospheric stability?
2. As the environmental lapse rate becomes larger, does the atmosphere become more stable or less stable? Why?
3. What is meant by "conditional instability"?
4. What is the difference between the lifting condensation level and the level of free convection?

MECHANISMS THAT CAUSE AIR TO RISE

The previous section refers frequently to "lifting." What "lifts" air parcels or causes them to rise in the atmosphere? Various processes responsible for lifting, termed ***lifting mechanisms***, are illustrated in Figure 6.9. One process involves the horizontal movement of relatively cold dense air into a region occupied by warmer, less dense air. Air parcels are lifted when a dome of cold air (a cold "airmass," in the terminology of Chapter 9) moves forward, displacing the air ahead of it upward. The cold air can be a large airmass moving south from Canada—in this case, the leading edge of the cold air would be a ***cold front*** (see Figure 6.9A). The cold air may simply move onshore from the ocean displacing the hot air over land upward—meteorologists call this a ***sea breeze*** front (Figure 6.9B), a common trigger for thunderstorms in Florida and along the East Coast in summer. There are other examples, but the principle is the same: cool air, moving forward, lifts warmer air ahead of it. If the warmer air is conditionally unstable, and the lifting sufficient to produce clouds, instability may occur and thunderstorms may develop.

FOCUS 6.1
Thunderstorms over Mountains

If one examines statistics on the frequency of thunderstorms, a relative maximum is apparent over the Rocky Mountains. In this area, some thunderstorm activity occurs nearly every day from late spring through summer, especially over the central Rockies. Residents of Colorado, for example, are accustomed to summer days that begin with sunny mornings, only to end with clouds and gusty thunderstorms in the late afternoon and evening. Figure 6A shows a typical midday buildup of thunderstorms viewed from the Continental Divide northwest of Denver. Why are thunderstorms so common near the mountains? One important factor is the enhanced solar heating of the east-facing slopes during the morning. Relative to flat ground, the eastern slopes are more nearly perpendicular

FIGURE 6A: Developing mountain thunderstorms, as seen from the Continental Divide in Colorado around noon on a July day.

(continued)

(continued)

to the rays of the morning sun. Hence they are heated more quickly than the surrounding terrain. The buoyant air near the surface of the east-facing slopes then moves upslope, giving an early start and a "boost" to the lifting process illustrated in Figure 6.9C. When the rising thermals reach saturation and their level of free convection over the mountains, conditional instability can be released, resulting in the development of thunderstorms that still have several hours of solar heating on which to feed. These storms occasionally become severe, producing hail, strong gusty winds, and small tornadoes. Since the prevailing winds are generally from the west,—albeit weak during the summer,—the thunderstorms that form over the Rocky Mountains generally move eastward over the Plains during the evening.

Lifting also occurs when air flows over a mountain (Figure 6.9C). If the winds carry conditionally unstable air upslope, clouds, showers, and thunderstorms can develop. All that is required for this type of lifting is that the prevailing wind have a component perpendicular to the mountain, so that the air is forced upward on the windward or ***upslope*** side of the mountain.

Lifting occurs whenever low-level winds blow toward each other ***(convergence)***, even in the absence of fronts, airmass contrasts, and mountains. As we will see in Chapter 8, surface low-pressure systems are characterized by an inflow of low-level air arising from frictional deflection of the winds (Figure 6.9D). This behavior occurs in extratropical cyclones as well as in tropical storms and hurricanes. While the resulting clouds are generally widespread, they often contain convective regions if the inflowing air at low levels is conditionally unstable.

There is yet another, even simpler, way to make air parcels unstable—the air near the surface can be heated during the day by solar radiation to the point where the air becomes very warm and rises spontaneously through the cooler environment aloft. In this case, no lifting is required at all. The parcels become absolutely unstable (the environment's lapse rate actually becomes greater than the dry adiabatic lapse rate in this case). We often see thunderstorms erupting randomly in midafternoon. These storms begin as parcels of air that "overheat" near the surface, and rise buoyantly into the upper troposphere.

ONLINE 6.2

Clear Daytime Skies Favor Severe Thunderstorm Development

On many summer days when the sky is clear in the morning, cumulus clouds develop during midday and can cover much of the sky by midafternoon, when the day's maximum temperature is reached. If the atmosphere is conditionally unstable, these cumulus clouds can develop further into towering cumulus and cumulonimbus clouds, bringing showers and thunderstorms to an area where the morning had been clear and sunny. Wherever a preexisting cloud cover is present during the morning, the solar heating of the surface is reduced, so the development of showers and thunderstorms is less likely. A spectacular example occurred over the Great Plains on 9 October 2001.

In the online animation, you can see a clear area over Kansas and Nebraska, with a preexisting cloud cover to the east and north. A line of strong thunderstorms develops in the clear area in midafternoon (around 2000 UTC), at the time of maximum heating of the ground. The explosive development occurred in a conditionally unstable atmosphere, with the lifting triggered by solar heating and focused by a "dry-line" boundary (Chapter 9). Several of these storms actually produced tornadoes. You will notice that thunderstorms did *not* develop in the area with preexisting cloud cover, where the solar heating of the ground was weaker.

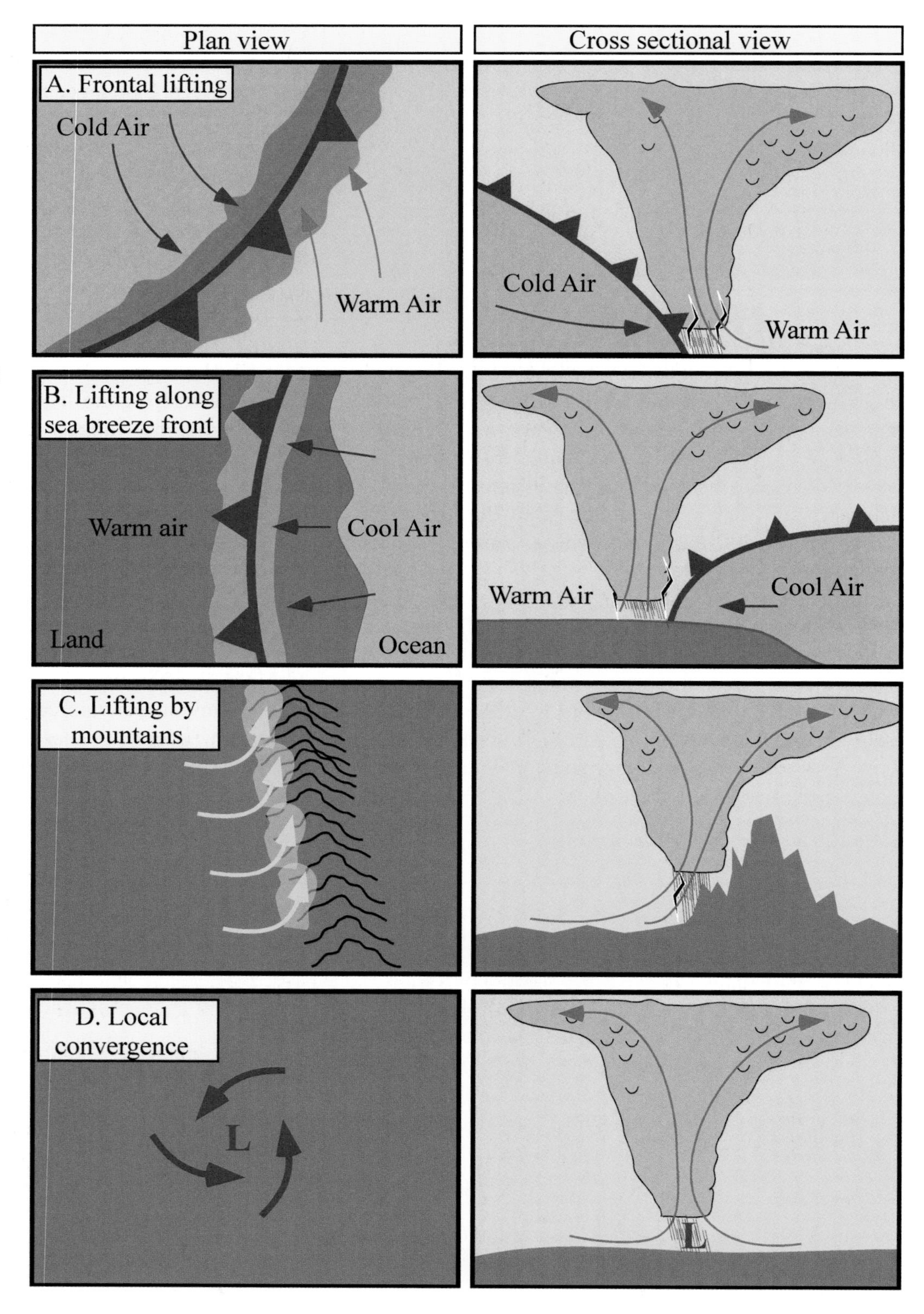

FIGURE 6.9: Examples of mechanisms by which air can be lifted to the level of free convection. Left panels show view from above, right panels show vertical cross sections (e.g., looking north). (A) lifting at a frontal boundary between warm and cold airmasses, (B) lifting by a sea breeze front, (C) lifting by flow over mountains, (D) lifting as a result of convergent horizontal winds at low levels.

STABILITY INDICES

One way meteorologists can determine the stability of the atmosphere is by calculating the lapse rate as discussed in the earlier part of this chapter. Often, we want to be able to quickly evaluate just how stable or unstable a large region of the atmosphere is, and calculating the lapse rates for many different areas manually is not practical. Meteorologists have several "tools" or indices that they use to describe the stability of the atmosphere. These indices are computed using data from rawinsondes or output from computer models. Each index provides slightly different information about the potential for severe weather—they do not forecast precisely where severe weather will occur.

One such tool is called the ***Lifted Index***. The Lifted Index (LI) is a measure of the instability of the atmosphere. The LI is calculated by subtracting the temperature of a parcel lifted to 500 mb from the temperature of the environment at 500 mb:

LI = T (environment at 500 mb) – T (parcel lifted to 500 mb)[1]

The lower the value of the Lifted Index, the warmer the parcel is relative to the environment. A negative LI indicates instability. Note that the calculation of the lifted parcel's temperature must allow for the transition from dry adiabatic ascent to moist adiabatic ascent when the parcel reaches saturation. Table 6.2 shows the ranges of the LI that are often associated with different types of convective weather.

While the Lifted Index is a common and useful measure of stability, other indices are often used in an attempt to capture the likelihood of various types of convective storms. The ***Showalter Index*** is similar to the Lifted Index, except that a parcel is lifted from the 850 mb level in the evaluation of the Showalter Index. Severe weather enthusiasts often monitor the ***Convective Available Potential Energy (CAPE)***, which is derived from soundings such as those shown in Figure 6.7; the CAPE is proportional to the chart area over which the parcel's temperature exceeds the environmental temperature, so it is a measure of the parcel's *positive* buoyancy accumulated over its trajectory above its level of free convection on a chart such as Figure 6.7 or 6.8. The CAPE is also proportional to the maximum updraft speed a parcel can achieve during unstable buoyant ascent. The "flip side" of the CAPE is the ***Convective INHibition (CINH) Index***, which is a measure of the near-surface negative buoyancy that must be overcome by heating before parcels can reach the Level of Free Convection. The popular ***K Index*** is often used in assessing the likelihood of non-severe convective precipitation. The K Index and the somewhat similar ***Total Totals Index*** are based on different combinations of the temperatures and dewpoints at standard upper-air levels (850 mb, 700 mb and 500 mb). Finally, a favorite of some meteorologists is the ***SWEAT Index*** (shorthand for ***"Severe Weather Threat,"*** not for perspiration). The SWEAT Index is a weighted sum of the 850 mb temperature, the Total Totals Index, the wind speeds at 850 mb and 500 mb, and the change in wind direction between 850 mb and 500 mb. The inclusion of wind speeds and direction in the SWEAT Index is an indication of the importance of horizontal winds, especially their

[1]When the Lifted Index is calculated, the lifted parcel is sometimes assumed to have the average temperature and humidity of the lowest several hundred meters (50 mb) of the atmosphere, thereby allowing for the fact that not all "thermals" originate precisely at the surface. In our discussion of the Lifted Index, we will not distinguish between parcels originating at the surface and those originating slightly above it.

TABLE 6.2 Stability Categories and Likelihood of Severe Convective Storms for Various Ranges of the Lifted Index (LI), Showalter Index (SI), Convective Available Potential Energy (CAPE), Total Totals (TT) Index and SWEAT (SW) Index

Stability	LI	SI	CAPE	TT	SW
Very stable (no significant activity)	> +3				
Stable (Showers possible; T'showers unlikely)	0 to +3	> +2	<0		
Marginally unstable (T'showers possible)	–2 to 0	0 to 2	0 to 1000	45 to 50	
Moderately unstable (Thunderstorms possible)	–4 to –2	–3 to 0	1000 to 2500	50 to 55	250 to 300
Very unstable (Severe T'storms possible)	–6 to –4	–6 to –3	2500 to 3500	55 to 60	300 to 400
Extremely unstable (Severe T'storms probable; tornadoes possible)	< –6	< –6	> 3500	60	> 400

vertical variations, in the development of severe thunderstorms. The reasons for the importance of horizontal winds, and their changes with elevation, will be covered in Chapters 18 and 19.

Table 6.2 shows typical ranges of the stability indices used for anticipating severe thunderstorms. The ranges are listed for various categories of atmospheric stability. The K Index is not included in Table 6.2 because it is used primarily in assessing the potential for non-severe convective precipitation; the K Index typically ranges from values below +15 (little or no chance of thunderstorms) to values greater than +40 (thunderstorm probabilities of 90 to 100 percent).

FOCUS 6.2

Stability, Capping Inversions, and Severe Storms

On 3 May 1999, a historic tornado outbreak occurred in central Oklahoma and southern Kansas. Tornadoes developed from severe thunderstorms that formed in an environment that was conditionally unstable. The soundings in Figure 6B were taken from Norman, Oklahoma, on (a) 1200 UTC 3 May 1999 and
(b) 0000 UTC 4 May 1999. On the morning of 3 May, the atmosphere was already conditionally unstable as evidenced by the sounding. Note that if a parcel were lifted to 675 mb, it would become buoyant and continue to rise on its own. The Lifted Index at this time was –2.5. At the same time there was an inversion present from 850 mb to 825 mb. This type of inversion is often referred to as a *capping inversion*. An inversion is a very stable layer: a parcel lifted through this layer will cool while the temperature in the environment around it becomes warmer. Therefore, the parcel can never become warmer than the surrounding environment. If a parcel were to be lifted at this time, it would encounter this stable layer and its rising motion would be inhibited.

Twelve hours later, during the early evening of 3 May (Figure 6B(b)), we can see that the temperature inversion has disappeared. How did this happen? During the day, the heating of the surface by the sun increased the temperature throughout the lower troposphere, eliminating the inversion that capped any vertical motion. (Note that the surface temperature on the 0000 UTC sounding is considerably warmer than on the 1200 UTC sounding.) Now, a lifted parcel of air can more easily rise throughout the atmosphere, eventually leading to tall, vertically developing cumulonimbus clouds, which often lead to severe weather. The Lifted Index at this time had dropped to –8.0, an extremely negative value indicating a highly unstable atmosphere. At approximately the time this sounding was taken, multiple severe thunderstorms producing strong tornadoes were in the vicinity. The tornadoes resulted in widespread destruction and damage, and nearly fifty deaths.

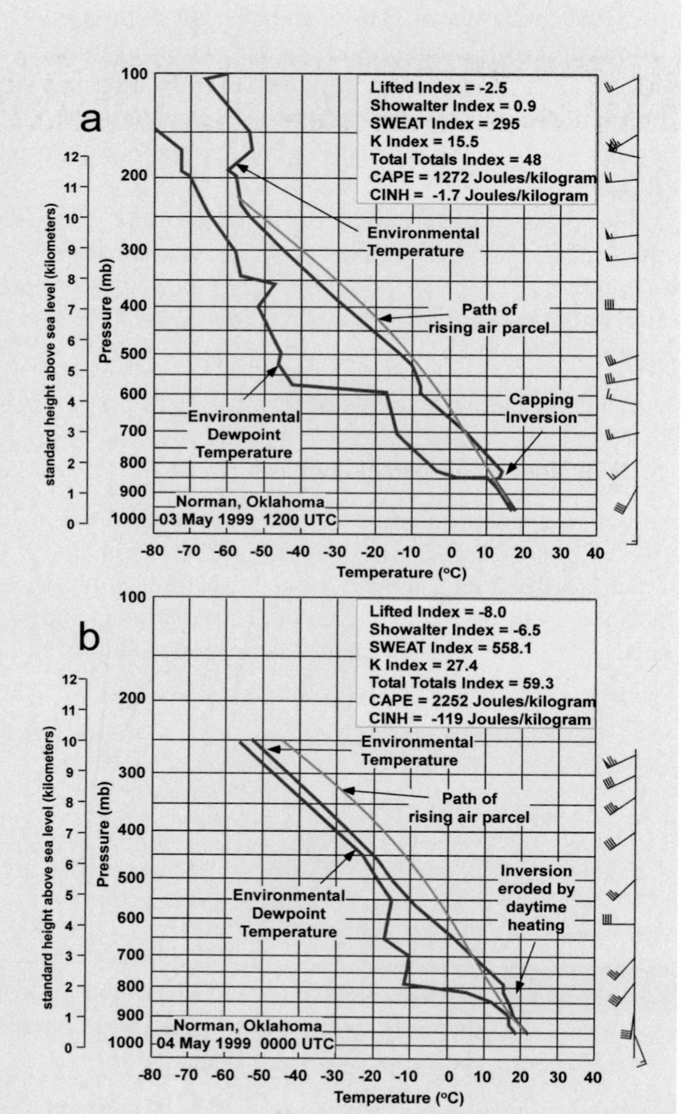

FIGURE 6B: Soundings from Norman, Oklahoma for (a) 1200 UTC 3 May 1999 and (b) 0000 UTC 4 May 1999.

CHECK YOUR UNDERSTANDING 6.3

1. What are four ways that air parcels can be lifted?
2. How does heating of the ground by sunlight affect the stability of the lower atmosphere?
3. Why do meteorologists assess atmospheric stability by using stability indices rather than explicit calculations using lapse rates from soundings?

TEST YOUR UNDERSTANDING

1. What does it mean if air is "stable" or "unstable"?
2. What is convection?
3. What is meant by the term "air parcel"?
4. What do the inflation of a tire and the descent of an air parcel have in common?
5. Why does an air parcel cool as it rises?
6. Which of the following is (are) a constant: dry adiabatic lapse rate, moist adiabatic lapse rate, environmental lapse rate?
7. Describe the motion of a vertically displaced air parcel in a stable atmosphere, an unstable atmosphere, and a neutral atmosphere.
8. What is the reason why the moist adiabatic lapse rate is different from the dry adiabatic lapse rate?
9. An inversion is indicative of a stable layer in the atmosphere. Briefly explain why this is so.
10. If the environmental lapse rate in the lower troposphere is approximately 5° C/km, would you expect strong convection to develop? Why or why not?
11. Would strong convection be expected to develop if the environmental lapse rate was 8° C/km? Explain.
12. Explain why atmospheric stability at a particular location can change.
13. Is the atmosphere likely to be more unstable in mid-afternoon or in early morning? Why?
14. What is the lifting condensation level?
15. What is the level of free convection?
16. List five ways that an air parcel can be brought to its level of free convection.
17. Cumulus clouds are present in Denver, Colorado. What lifting mechanism discussed in the text is least likely to have caused the clouds to form? Why?
18. Would you expect air parcels to be lifted in areas where surface winds are convergent or divergent? Why?
19. Why are cumulonimbus clouds less common in the winter than in summer in the United States?
20. What is the Lifted Index? How is the value of the Lifted Index related to the severity of storms?
21. Determine if the data for each sounding indicates that thunderstorms will develop and if so, if the storms will be severe.
    (a) The Davenport, Iowa July sounding shows: K-Index = 38, LI = –1.5, SI = –1.6, CAPE = 150, TT = 48, SW = 225
    (b) The Nashville, Tennessee July sounding shows: K-Index = 34, LI = –4.2, SI = –1.3, CAPE = 2027, TT = 46, SW = 244
    (c) The International Falls, Minnesota July sounding shows: K-Index = –11, LI = +13.6, SI = 15.6, CAPE = 9, TT = 25, SW = 108

TEST YOUR PROBLEM-SOLVING SKILLS

1. You are taking a hot-air balloon ride on a clear, calm November evening. The environmental lapse rate is 5° C/km. As the balloon rises, you compare the temperature inside the balloon (via digital readout) with the reading on a thermometer you hold in the passenger basket. After a few bursts of flame at an altitude of 1 km (3280 ft), the temperature inside the balloon is 68° F (20° C), and the temperature in the basket is 32° F (0° C). All of the balloon's controls (including the ballast control) then freeze and become inoperative, causing the balloon to enter a state of free drift. Assuming that the balloon's expandable skin does not conduct any heat, what is the maximum altitude you could reach before the balloon stops rising? (Hint: The maximum altitude would be reached if the balloon's skin, basket, and passengers had negligible weight.)

2. Consider air flowing onshore at the surface from the Pacific Ocean with a temperature of 50° F (10° C) and a dewpoint of 41° F (5° C). The onshore airflow is forced to ascend the mountains of the western United States, where its trajectory reaches a maximum elevation of 3 km (9540 ft). The air then descends to the plains of Colorado, where the elevation is 1.5 km (4920 ft). Assume that the air's dewpoint decreases at 2° C/km of ascent prior to saturation, increases at the same rate during descent, and decreases at the moist adiabatic lapse rate of 6° C/km during saturated ascent.
   (a) Calculate the relative humidity of the air as it comes onshore. Hint: Use Fig. 1.9.
   (b) Determine the temperature, dewpoint, and relative humidity of the air at the top of the mountains (3 km).
   (c) Determine the temperature, dewpoint, and relative humidity of the air when it reaches the Plains of Colorado (1.5 km).

3. The following data are from a hypothetical sounding.

Pressure (mb)	Height (m)	Temp. (° C)	Dewpoint (° C)
1000	0	14	10
980	200	16	12
950	500	14	12
900	1000	11	11
850	1500	5	5
800	2000	2	−3
700	3000	−9	−16
600	4000	−16	−18
500	5500	−23	−25
400	7000	−28	−30
300	8500	−30	−34
200	10,000	−30	−35
100	12,000	−27	−36

   (a) Plot the sounding (temperature and dewpoint) on a copy of the Stuve diagram given in Appendix B.
   (b) Are any inversions present in this sounding? If so, in which layers? (Identify "layers" according to the pressures at their top and bottom, e.g., the "400 to 500 mb" layer.)
   (c) Are any layers absolutely unstable? If so, which one(s)?
   (d) Are any layers neutral? If so, which one(s)?
   (e) Where is the level of free convection for a parcel that is lifted from the surface? (You may assume that the dewpoint of an *unsaturated* air parcel decreases during lifting at a rate of 2° C/km; a *saturated* parcel must have temperature = dewpoint.)
   (f) If the parcel in (e) reaches the level of free convection, to what altitude will it rise? (Assume the moist adiabatic lapse rate stays constant at 6° C/km.)
   (g) What is the Lifted Index for this sounding? (Assume the parcel is lifted from the surface.)

USE THE SEVERE AND HAZARDOUS WEATHER WEBSITE

http://severewx.atmos.uiuc.edu

1. The output products of the NAM model (presented in Chapter 4) include an analysis and forecasts of the Lifted Index (LI) and Relative Humidity (RH) at six-hour intervals through the sixty-hour period of the forecast. Use the *Severe and Hazardous Weather Website* "Forecast Links" to locate this product. Examine the forecast for LI and RH for your location.
   (a) Do the values of LI tend to be higher in the evening (0000 UTC) or in the early morning (1200 UTC)?
   (b) When and where during the next sixty hours are the most negative LI values forecast to occur?
   (c) When and where during the next sixty hours are the highest RH values forecast to occur?
   (d) What, if any, relationship exists between the forecast values of LI and the values of RH?
   (e) Save or print one of the forecast maps. Note the time for which the forecast is valid. On that date and time, examine the radar (from the *Severe and Hazardous Weather Website*) to locate any showers and thunderstorms present. Record the location of any precipitation. Did precipitation develop in the regions where the LI and RH forecasts were favorable?

2. At the *Severe and Hazardous Weather Website,* you may view the latest soundings for the United States (and much of the rest of the world) within about two hours of the rawinsonde launch times. If you choose the "Stuve" diagram as your display option, the soundings will be shown in the same format as the examples presented in this book. In addition to displaying the temperature and dewpoint profiles obtained from the rawinsonde ascent, the diagram contains secondary sets of lines representing the paths of air parcels rising dry adiabatically or moist adiabatically. The particular path followed by a parcel is also shown on some soundings.
   (a) Locate the sounding launch location closest to your location. Record the date and local time of the sounding.
   (b) Examine the sounding and identify the lines that indicate the temperature, dewpoint, and parcel temperature.
   (c) Are there any layers where the parcel temperature exceeds the environmental temperature? If so, identify these layers using the pressure at the top and bottom of the each layer.
   (d) Is there a capping inversion present? If so, identify the layer by listing the pressures that define the layer.
   (e) You will be more likely to find a capping inversion on a sounding from 1200 UTC. Explain why.
   (f) Often there is a list of abbreviations and numbers to the right of the sounding. This list contains many of the variables discussed in this and previous chapters. If the sounding you choose has this list, identify the following:
      (i) Lifted Index
      (ii) Showalter Index
      (iii) Total Totals Index
      (iv) K-Index
      (v) CAPE
      (vi) CINH
      (vii) Lifting Condensation Level of a parcel
      (viii) Level of Free Convection of a parcel
   (g) For each of the variables you identified in (f), compare their value to Table 6.2. Based on your findings, what kind of weather might you expect to develop in this area?

3. Satellite measurements of infrared radiation are used to construct atmospheric soundings that in turn can be used to estimate atmospheric stability by the Lifted Index and other indices mentioned in this chapter. The advantage of satellite-derived soundings is that they can fill in gaps between rawinsonde stations and between the twice-daily rawinsonde launch times. A disadvantage is that clouds prevent the retrieval of infrared information from below the tops of clouds; also, errors of a few degrees Celsius are typical in satellite-derived upper-air temperatures.

   Navigate to the *Severe and Hazardous Weather Website* "Additional Links" under the "Atmospheric Stability" link. Access the "NESDIS

GOES Temperature and Moisture Sounding Fields." This site is especially useful on spring and summer afternoons when atmospheric instability is conducive to thunderstorms.

(a) Access the latest U.S. display of CAPE Index, Lifted Index (LI), Convective Inhibition (CINH), and Total Precipitable Water (TPW), a measure of the total water vapor in a column of air of unit area. Save or print each plot. Note the time of the plot.
  - (i) Over what states or region is the highest CAPE found?
  - (ii) Over what states or region is the lowest LI found?
  - (iii) Over what states or region is the lowest CINH found?
  - (iv) Over what states or region is the highest TPW found?
  - (v) Based on your answers to (i) through (iv), in what region or states, if any, would you expect to find severe thunderstorms?

(b) Access a radar or satellite image for approximately the same time. Is your answer to (v) in Part (a) supported by the radar or satellite image?

FORCES AND FORCE BALANCES

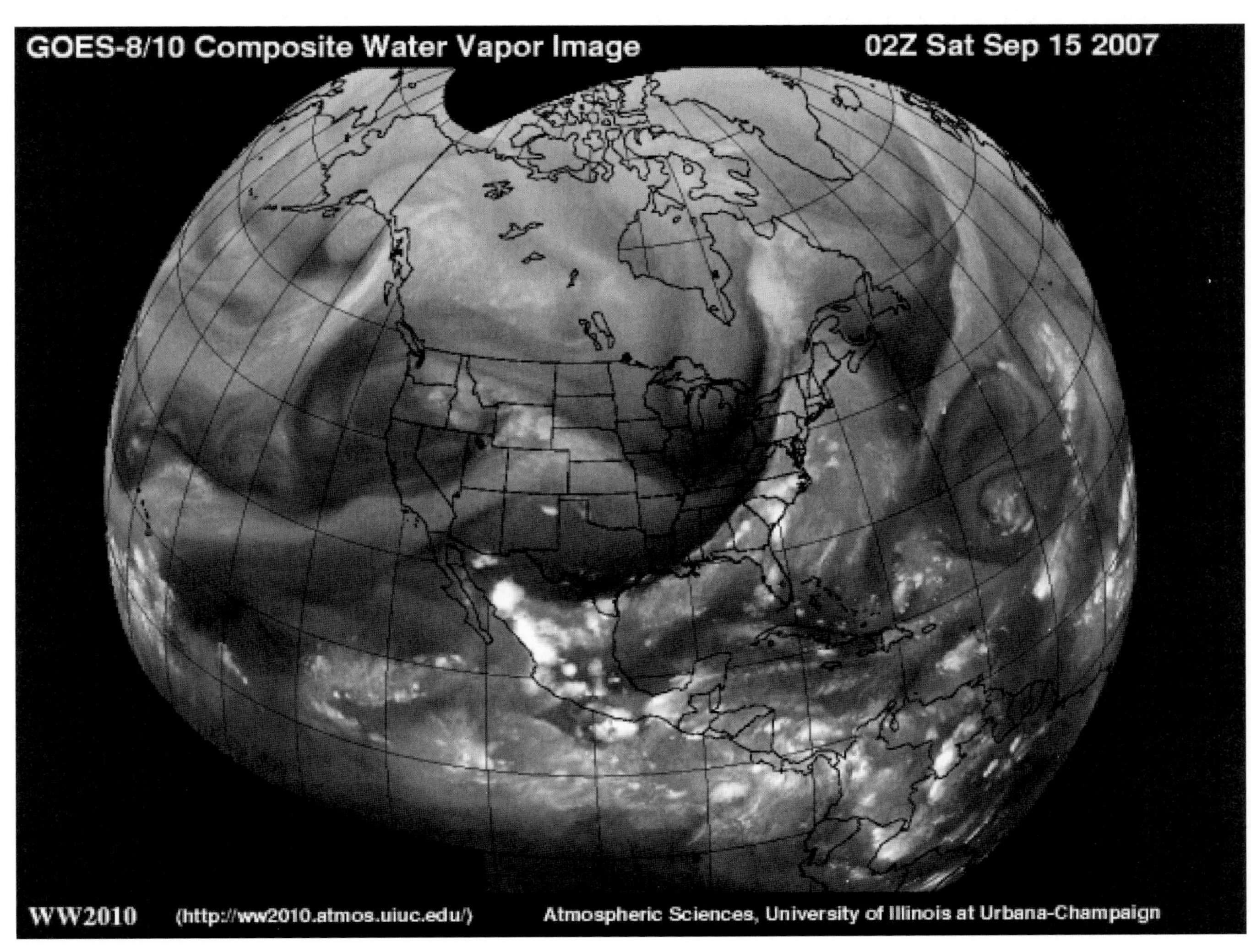

Forces control the Earth's winds, and govern the behavior of the magnificent storms we see from satellite views high above the Earth.

KEY TERMS

angular momentum
boundary layer
conservation of angular momentum
Coriolis force
frictional force
friction layer
geostrophic balance
geostrophic wind
gravitational force
hydrostatic balance
jetstreak
jetstream
mechanical turbulence
pressure gradient
pressure gradient force
shear-induced turbulence
thermal turbulence
turbulent eddies
unit area
unit mass
wind shear

Why does air move and what controls the speed and direction of its movement? These fundamental questions lie at the core of understanding hazardous weather. From the strongest hurricane to the smallest swirl of leaves, four forces control the wind: the ***pressure gradient force,*** the ***gravitational force,*** the ***frictional force,*** and the ***Coriolis force.*** The first three forces are "fundamental" because they occur in the atmosphere whether or not the Earth is rotating. The Coriolis force results from the Earth's rotation. In this chapter we will study forces and force balances, and how these force balances relate to the presence of ***jetstreams*** in the upper troposphere. In following chapters, we will learn how imbalances between these forces create the world's storm systems and most of its severe weather.

FORCES THAT AFFECT ATMOSPHERIC MOTION

The Pressure Gradient Force

Consider an imaginary thin wall, such as in Figure 7.1, with air molecules located on the left and right sides. The molecules, in random thermal motion, strike the wall on both sides, imparting momentum to the wall with each collision, much like two balls striking each other on a billiard table. Under what conditions will the air molecules apply a net force, so that the wall will begin to move from left to right?

Let's consider two situations. In the first, there are more air molecules on the left side of the wall than the right, with each molecule, on average, imparting the same energy per collision to the wall. Since more molecules strike the left side of the wall than the right, the wall will accelerate to the right. The force per unit area, or the pressure, is greater on the left side of the wall.

By ***unit area*** we mean an area with a dimension of one unit, such as one square meter, or if we were to measure pressure in English units, one square inch. Suppose instead that the same number of molecules are present on both sides of the wall, but the molecules are moving faster, on average, on the left (i.e., the temperature is higher on the left). Each impact on the left side will impart more energy than an impact on the right. The force per unit area is again stronger on the left and the wall will accelerate from left to right. *The wall moves in response to the difference in pressure between its left and right sides.* This difference in air pressure can result from a difference in air density (more molecules), air temperature (more energy per collision), or both.

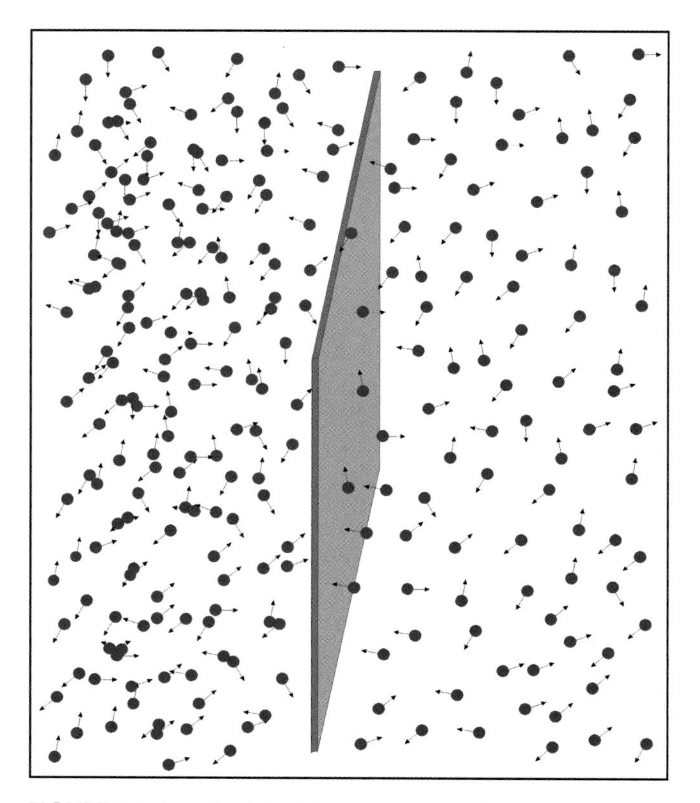

FIGURE 7.1: A wall with higher pressure on its left side (indicated by greater molecular density on the left side). The arrows indicate the direction of motion of the molecules.

Now consider a random molecule in the atmosphere, as shown in Figure 7.2. If there are more molecules on the left side of this molecule than on its right (the air density is greater on the left), and/or the molecules on the left, on average, have more thermal energy (the air temperature is higher on the left), then collisions, on average, will drive the random molecule from left to right. On a larger scale, if the pressure is higher on the left than on the right each of the molecules in the air will be accelerated to the right, and the wind, which is the average motion of all the molecules, will be from left to right. The rate at which the air will accelerate depends on the rate at which pressure changes with distance, or the ***pressure gradient.*** The force applied to a small parcel of air due to pressure differences is called the pressure gradient force.

The most extreme horizontal pressure gradients on the Earth occur in tornadoes and the eyewalls of hurricanes. For example, Figure 7.3 shows the pres-

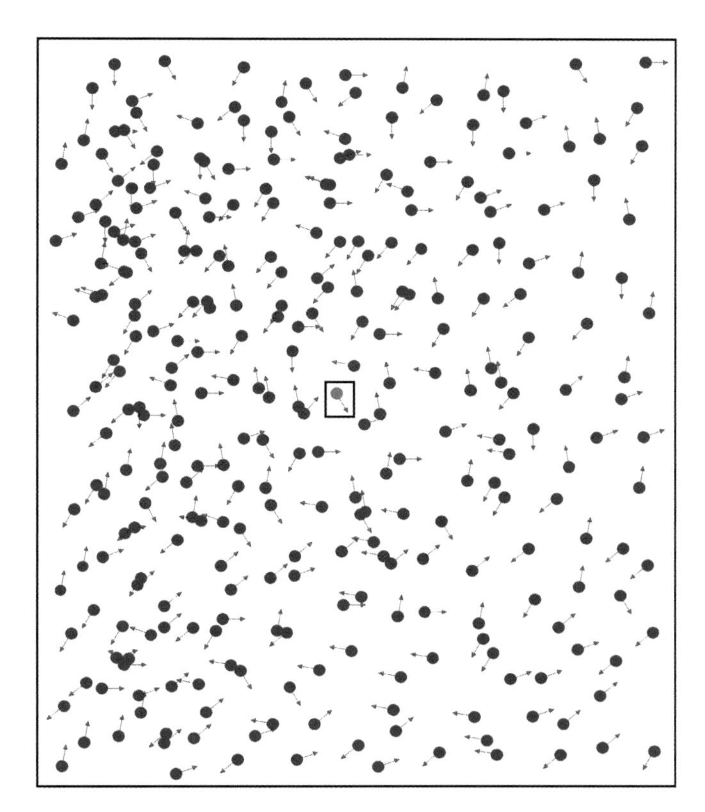

FIGURE 7.2: An air molecule with higher air pressure on its left (indicated by greater molecular density on the left side).

sure distribution measured by a network of barometers in southern Florida as Hurricane Andrew moved over Homestead, Florida in 1992. The pressure difference across the eyewall between Homestead and Miami Beach, Florida, a distance of 42 km, was 78 mb. This pressure gradient of 78 mb/42 km = 1.9 mb/km led to sustained winds of 85 to 115 knots (98 to 132 mph) with gusts as high as 150 knots (173 mph). Strong pressure gradients can also exist in extratropical cyclones. Figure 7.4 shows the pressure distribution during an exceptionally strong cyclone over the east coast of North America where the minimum central pressure was 960 mb. The pressure gradient along line AB between New York City and Boston was 0.1 mb/km. This seemingly small change in atmospheric pressure produced sustained winds of 25 to 40 knots (29 to 46 mph), with gusts as high as 65 knots (75 mph) along the coast of New England. The direction of the pressure gradient force is always perpendicular to the isobars, from high to low pressure. For example, the arrows in Figure 7.4 indicate the direction of the pressure gradient force at several locations. Note that the longest arrows, which indicate the strongest pressure gradient force, occur where the isobars are closest together.

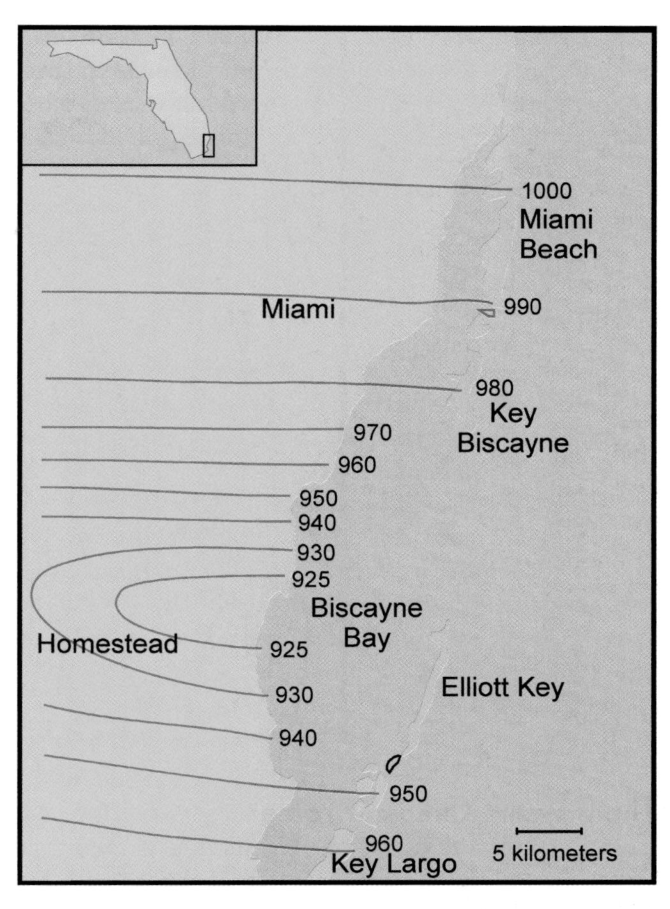

FIGURE 7.3: Observations and analysis of the minimum pressure during Hurricane Andrew's landfall in Florida in 1992.

FOCUS 7.1

Experience the Change of Pressure Across a Hurricane Eyewall

The pressure gradients across hurricane eyewalls are some of the strongest horizontal pressure gradients found on Earth. A typical strong hurricane might have a central pressure of 950 mb and a pressure outside the eyewall of 1000 mb, a difference of 50 mb. How big is this pressure change? The easy way to experience it directly is to visit Chicago and take an elevator to the top of the Sears Tower, one of the world's tallest buildings. The pressure change from the bottom to the top of the building, a distance of 412 meters (1353 feet), is about 50 mb. Expect your ears to pop!

In contrast to horizontal variations, pressure varies rapidly in the vertical. Pressure decreases from its mean sea-level value (1013.25 mb) to 100 mb at 16 km

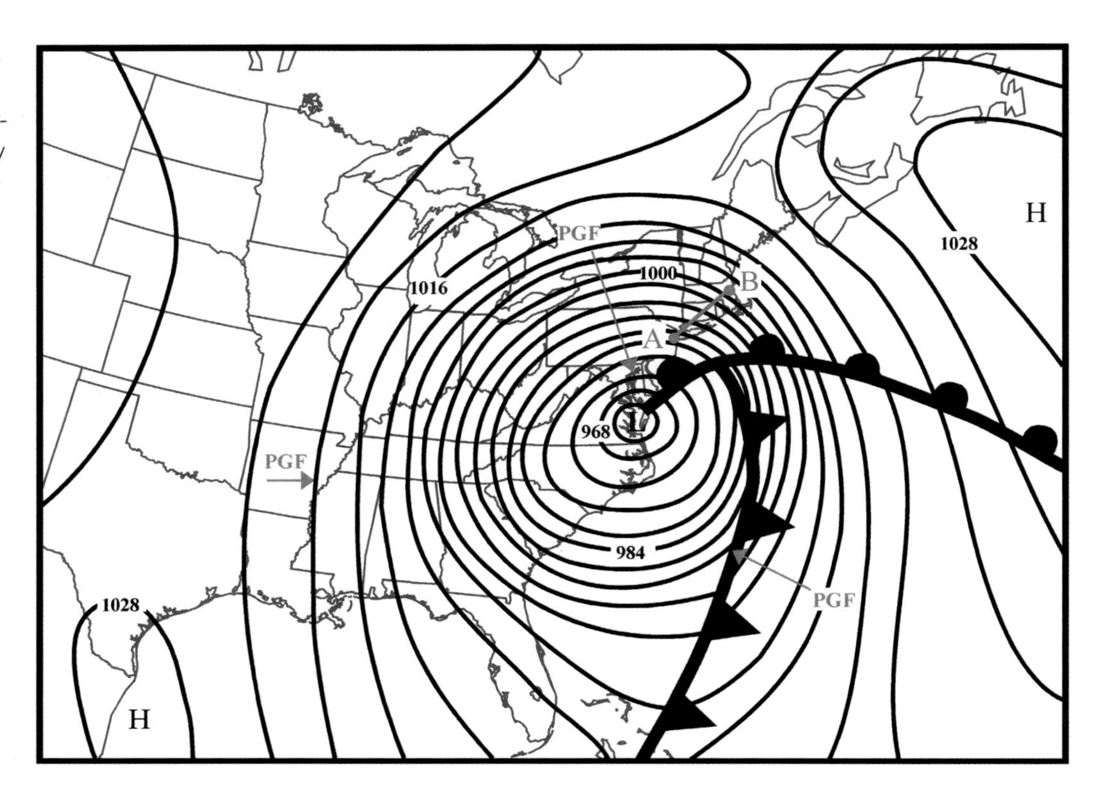

FIGURE 7.4: Surface pressure distribution over the East Coast of North America during an exceptionally strong cyclone. The direction and strength of the pressure gradient force (PGF) is indicated at three locations by the orange arrows. Longer arrows imply a stronger force. The pressure gradient between "A" and "B" is 0.1 mb/km, leading to a 25- to 40-knot sustained wind in that region.

and 1 mb at 31 km above the Earth's surface. The vertical pressure gradient, about 120 mb/km in the lower atmosphere, is obviously very large. Were it not for gravity acting in the opposite direction, the Earth's atmosphere would quickly disperse into space.

The Gravitational Force

Any two elements of mass in the universe are attracted to each other by gravity. The magnitude of the gravitational force is proportional to the mass of each object, and inversely proportional to the square of the distance between the objects. The Earth is much more massive than any parcel of air within the atmosphere. Furthermore, the difference in distance between the center of the Earth and a parcel of air at the top of the troposphere versus the center of the Earth and a parcel of air at the bottom of the troposphere is so small that for weather applications, the magnitude of the gravitational force acting on any air parcel can be considered constant throughout the troposphere. The direction of the gravitational force is toward the center of the Earth.

The Frictional Force

The force of friction acts in a direction opposite the motion of air, and therefore always acts to reduce the speed of the flow. In the atmosphere, the force of friction is ultimately realized on the molecular scale as individual faster-moving air molecules collide with slower-moving air molecules or with the Earth. Except very close to the surface of the Earth, friction acts primarily through the mixing of parcels of air moving at different speeds. The turbulent motions that lead to the mixing of air are called ***turbulent eddies,*** and arise primarily from three sources, as shown in Figure 7.5. ***Mechanical turbulence*** develops when air encounters obstructions associated with ground roughness. Trees, buildings, and terrain features all deflect air in different directions, mixing air down from aloft and up from the surface. ***Thermal turbulence*** develops when the air near the Earth's surface is heated sufficiently during the day that it becomes unstable and rises to higher altitudes. When convection carries slower-moving surface air to altitudes where the winds are stronger, the mixing that occurs reduces the air motion at the higher altitude. Thermal turbulence may also transport fast moving air downward from higher altitudes, producing gusty surface winds. ***Wind shear*** exists when winds change (increase or decrease) over some distance. ***Shear-induced turbulence*** occurs when wind speed changes rapidly with distance, typically with height. When the vertical shear of the wind becomes large, tumbling motions begin to develop in the flow that mix the lay-

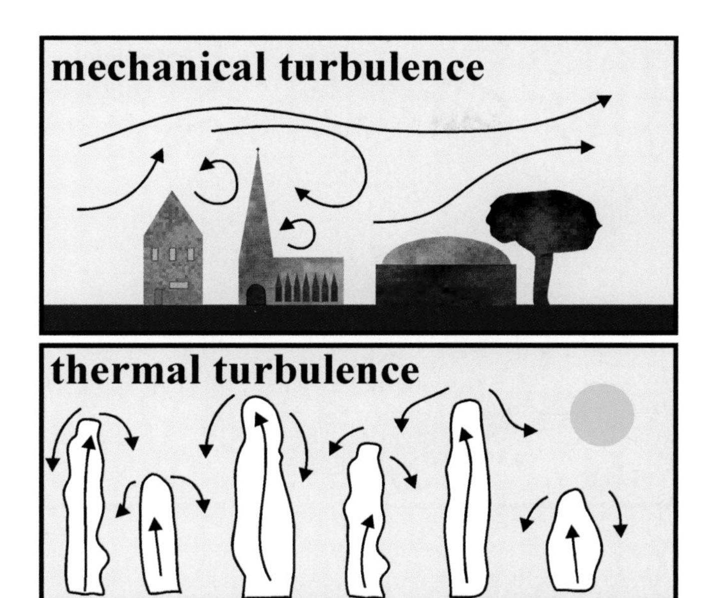

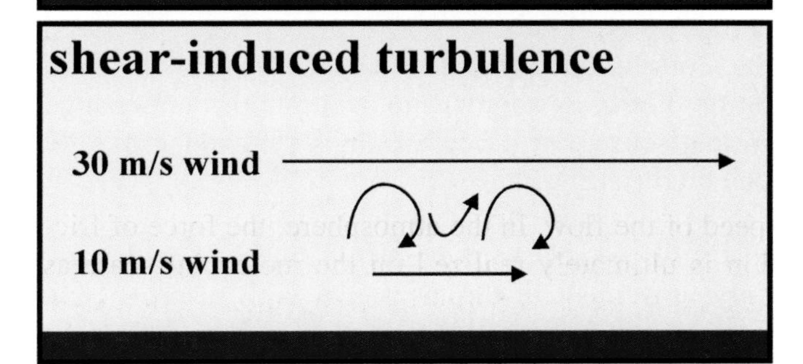

FIGURE 7.5: Principal mechanisms that lead to turbulent mixing in the lower atmosphere.

ers of faster and slower moving air, smoothing the vertical wind profile. These tumbling motions are what commercial aircraft often experience when the captain turns on the "fasten seat belt" sign in mid-flight. Although turbulent motions can also be generated by horizontally sheared flow, such motions rarely occur because horizontal shear is typically weak.

Frictional drag is strongest near the Earth's surface and decreases rapidly with height. The atmospheric layer in which friction is an important force is called the ***boundary layer***, or sometimes the ***friction layer***. The depth of the boundary layer depends on the underlying surface roughness (hills, buildings, trees, etc.), surface heating, atmospheric stability, and the wind speed. On a night with weak winds, the boundary layer over a large lake may only extend upward a few hundred meters, while the boundary layer over a city on a hot, windy afternoon may extend upward a few thousand meters.

The Coriolis Force

The Coriolis force, an apparent force associated with the rotation of the Earth, was first described mathematically in 1835 by the French scientist Gustav Gaspard de Coriolis (1792–1843). To understand the effect of the Coriolis force on atmospheric flows, it is helpful to first examine a simple rotating system like a playground merry-go-round. Suppose two people are seated across from one another on a stationary merry-go-round playing catch with a ball (Figure 7.6A). When the first person throws the ball to his partner, the ball takes a straight line path across the merry-go-round. What happens if the merry-go-round is rotating counterclockwise (as the Earth does when looking from space at the north pole)? When the ball is thrown, it will still follow a straight line path, but as it crosses the merry-go-round, the partner will rotate out of the path of the ball (Figure 7.6B). In the reference frame of the person throwing the ball, the ball appears to veer to the right. In a playground, the riders of the merry-go-round can tell they are rotating because they can see the background. However, suppose the same experiment were done in the dark with glowing balls. The riders would no longer sense visually that they were rotating and the behavior of the balls would be strange indeed! We live on a rotating Earth, but aside from the daily progress of the sun and stars across the sky, we have no sense at all that we are rotating. Objects, such as air parcels, moving across the rotating Earth will move in a strange manner, much like the glowing balls on the darkened merry-go-round ride.

ONLINE 7.1

Experience the Coriolis Force

To help understand the Coriolis force, experiment with releasing balls on a rotating disc online. Watch what happens to balls crossing the disk at varying speeds and at varying disk rotation rates. What is really happening? Viewed from above the rotating disk—the balls take a straight-line path. From the perspective of being on the disk, the balls appear to curve and, in some cases, actually return to their original position. In this case, the ball did go straight—the point of release rotated to the other side and intercepted the ball!

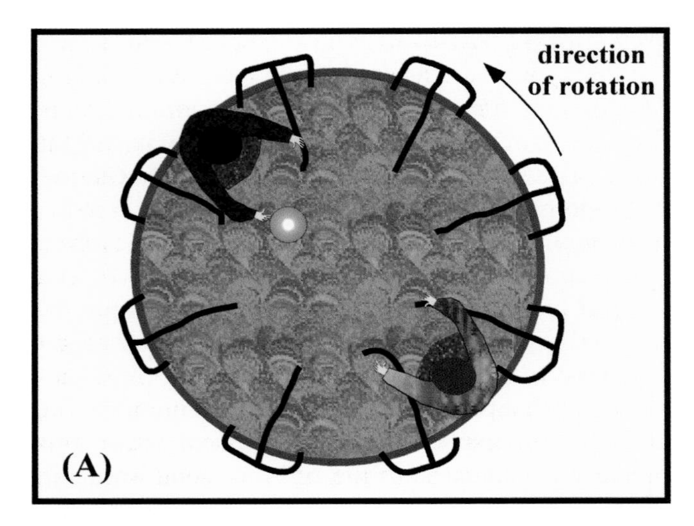

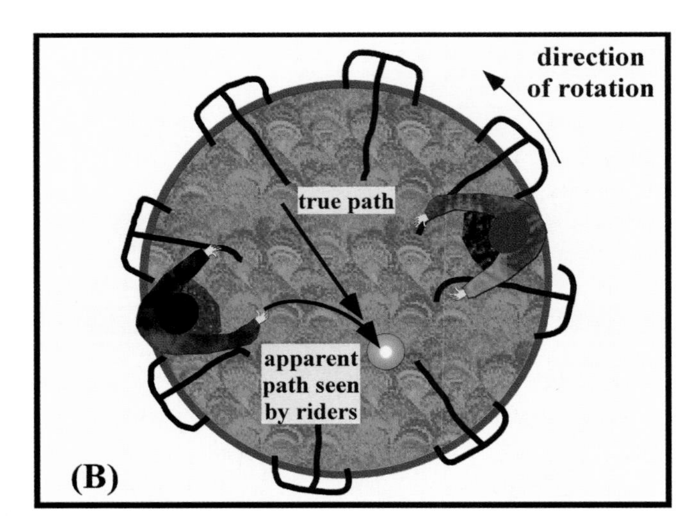

FIGURE 7.6: The Coriolis force illustrated on a rotating platform. Viewed from above, the ball takes a straight path. From the perspective of the person throwing the ball, the ball appears to veer off to the right of his partner when the platform is rotated counterclockwise.

To understand the Coriolis force on the Earth, it is helpful to first consider our simple experience of swinging a ball in a circle on a string. We all know that if the string is shortened while the ball is rotating, the ball will turn faster. Skaters use this same principle to go from a slow turn to a rapid spin—in this case, a skater brings her arms and legs in toward the axis of rotation to speed up. The ball rotating on a string and the skater in a spin are both examples of an important principle of physics called the ***conservation of angular momentum***. The ***angular momentum*** of an object is defined as the product of its mass (M), its rotational velocity (V), and its radius from the axis about which it is rotating (R), or

$$\text{Angular momentum} = M \times V \times R.$$

Except for air located directly over the poles, all air on the Earth has angular momentum, even if the air is still with respect to a point on the Earth, because the air is rotating with the Earth. On Earth, the angular momentum of air is defined as the product of the air's velocity with respect to the Earth's axis of rotation (V_{air}), the air's distance from the Earth's axis of rotation (R_{air}, see Figure 7.7), and the air's mass (M_{air}), or

$$\text{Angular momentum} = M_{air} \times V_{air} \times R_{air}.$$

In the absence of a twisting force called a torque, air's angular momentum does not change as it moves about the Earth. Although torques exist in the atmosphere (caused, for example, when air encounters mountains), we can assume for our investigation of the Coriolis force that as a parcel of air moves about on the Earth, its angular momentum does not change. In that sense, our parcel of air is very much like the ball on the string.

Let's assume first that an air parcel is initially at rest with respect to the Earth's surface, so that the only movement of the air parcel with respect to the Earth's axis is due to the rotation of the Earth itself (see Figure 7.7). How fast is the air parcel rotating about the Earth's axis? That depends on where on the Earth's surface it is located. The radius of the Earth is 6380 km, and the Earth rotates once every twenty-four hours, so at the equator, the speed of rotation of a point on the Earth's surface is 1670 km/hour (1018 miles/hour!). At 30° North Latitude, the distance from the surface of the Earth to the Earth's axis is 5525 km. At this latitude, the speed of rotation of a point on the Earth is 1446 km/hour (882 miles/hour). At 60° North, the speed of rotation is half that of a point at the equator, 835 km/hour (508 miles/hour). At the pole, there is no rotational speed.

Suppose that an air parcel initially still with respect to a point on the equator (but moving 1670 km/hour with respect to the Earth's axis) impulsively is pushed northward at 20 km/hour toward a latitude of 30° North. As it does so, it moves ever closer to the Earth's axis of rotation (see Figure 7.7). Like the ball on a shortened string, its rotational velocity *with respect to the Earth's axis* will increase, but how fast? Since its mass doesn't change, the conservation of angular momentum tells us that its new rotational velocity with respect to the Earth's axis will be its old

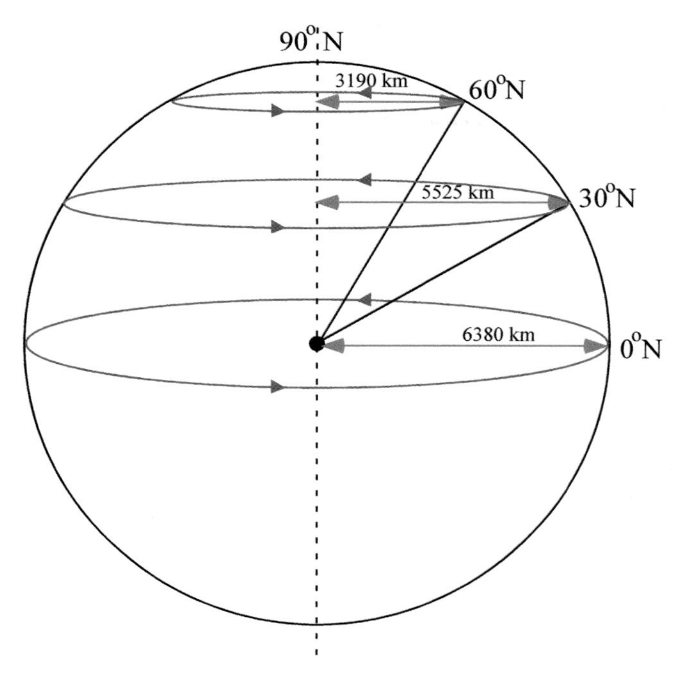

FIGURE 7.7: The distance of points on the Earth's surface from the Earth's axis of rotation at three latitudes: 0° North, 30° North and 60° North. A point at the pole at 90° North is on the axis of rotation.

rotational velocity times the ratio of its distance from the Earth's axis at the equator, to its distance from the Earth's axis at 30° North, or

New Rotational Velocity =

$$1670\ km/hour \times \frac{6380\ km}{5525\ km} = 1928\ km/hour$$

Recall from above that the rotational velocity of a point on the Earth with respect to the Earth's axis at 30° North was 1446 km/hour. That means that, *relative to the Earth's surface,* the air parcel is now moving eastward at 1928 – 1446 = 482 km/hour (294 miles/hour)! To a person standing at the equator watching the air parcel depart northward, it would appear that some force caused the air to turn to the right—that "force" is the Coriolis force. To a person in space observing the air, the change in speed is no more complicated that the experiment with the ball on the string—the air moved in toward the axis of rotation and sped up. To the observer on the rotating Earth, the increase in speed of the air would appear bewildering.

We know that air never attains such speeds as the 482 km/hour quoted in the last paragraph. In fact, *the Coriolis force alone will not change the speed of the air relative to the Earth's surface.* We can understand why this is so if we consider what happens to air as it begins moving eastward with respect to the Earth. Consider your experience sitting on a rapidly rotating playground merry-go-round. If you do not hold on to a rail, an outward centrifugal force throws you off the ride. Air at rest on the Earth (a rather large spherical merry-go-round) also experiences a centrifugal force, but it is held on to the surface by gravity. However, when air moves eastward across the Earth, its total rotational speed (Earth + air speed) is faster than the Earth's speed below it. The eastward moving air experiences a centrifugal force that "throws" it outward like the ball on the merry-go-round. In the Northern Hemisphere, on our spherical Earth, this outward movement is to the right, or southward. An observer on the Earth watching an air parcel moving east would see it deviate to the right (south), again driven by some force—the Coriolis force. The reason that our air parcel that left the equator for 30° North would never attain an eastward velocity of 482 km/hour is that it would never make it to that latitude. As soon as it departed the equator, it would begin deviating east and before long would be going directly east. It would then turn back south! As it moved, it would trace out a path that kept its speed with respect to the Earth's surface constant, in this example, 20 km/hour. The air would continually change direction, initially northward, then eastward, then southward. Indeed, it would return to the equator and cross into the southern hemisphere where it would deviate to the left (the opposite of the Northern Hemisphere, but still eastward) and start another round-trip.

To the inhabitants of our rotating planet, the Coriolis force is very real, and indeed one of the most important forces affecting atmospheric motion. There are several important facts about the Coriolis force that will be important to our studies of severe and hazardous weather. The Coriolis force (1) causes objects to deviate to the right of their direction of motion in the Northern Hemisphere (and to the left in the Southern Hemisphere); (2) affects the direction an object will move across the Earth's surface, but has no effect on its speed; (3) is strongest for fast-moving objects and zero for stationary objects; and (4) is zero at the equator and maximum at the poles.

The Coriolis force deflects all objects moving across the Earth's surface, including air. Consideration of the Coriolis force is essential to understand large weather systems such as cyclones, hurricanes, and jetstreams. The Coriolis force is less important, but cannot be neglected, in smaller scale weather systems such as thunderstorms, since the distances and time scales involved do not allow the deflection to become noticeable. In very small circulations, such as a tornado, the Coriolis force becomes insignificant compared to other forces.

CHECK YOUR UNDERSTANDING 7.1

1. Which four forces control the movement of air?
2. What causes the pressure gradient force? In which direction does it act?
3. What causes friction in moving air? In which direction does the force of friction act?
4. What is the Coriolis force? In what direction does it act in the Northern Hemisphere?

NEWTON'S LAWS OF MOTION

The physical laws governing motion were first stated by Sir Isaac Newton (1642–1727), the renowned English mathematician, astronomer, and physicist who also discovered, among other things, calculus and the universal law of gravitation. Newton's first law of motion states that *an object at rest will remain at rest and an object in motion will remain in motion traveling at a constant speed in a straight line as long as no force is exerted on the object.* Imagine for a moment a world in which there is no gravity or friction. A ball thrown horizontally by a child living on such a world would continue into space, traveling endlessly in a straight line at the speed at which it left the child's hand. In the real world, forces act all the time to change the motion of objects. In the case of the ball, friction with the air would slow the ball's horizontal motion and gravity would cause it to fall to the ground. The effect of forces on the motion of objects is the subject of Newton's second law, which states that *the force exerted on an object equals its mass times its acceleration,* or F = ma. An object experiences an acceleration any time it changes speed or direction.

When we study motions in the atmosphere, we will always consider a ***unit mass*** of air, such as 1 kilogram. In this case, the force/unit mass equals the acceleration. We will from this point forward in the text assume the term "force" to mean "force/unit mass." In reality, more than one force is always acting to influence air motions, so the net acceleration of a parcel of air will be the sum of the forces acting on it. In the atmosphere, we can consider horizontal motions and vertical motions independently when considering the effects of forces. Furthermore, we will divide horizontal motions into two classes: those above the boundary layer and those within the boundary layer. We will ignore friction when considering vertical motions and horizontal motions above the boundary layer. With these considerations, Newton's second law provides three fundamental relationships that will guide our study of atmospheric motions and the development of storms. These relationships are summarized in Table 7.1.

FORCE BALANCES

Air motions are driven by pressure differences that develop due to the uneven solar heating that occurs across the Earth's surface. This difference in heating, as well as other factors such as the disruption of airflow as air crosses the world's mountain ranges, continually induce pressure gradients that keep the atmosphere from ever achieving an exact state of balance. As we shall see, the imbalances between the forces affecting atmospheric motion cause the storm systems of the world to develop. We begin our investigation of atmospheric flows; however, by first considering the types of flows that can exist in the atmosphere when

TABLE 7.1 **Newton's Second Law for Horizontal and Vertical Motions**

(PGF denotes the pressure gradient force)

Type of Motion	Newton's Second Law
Horizontal motions above the boundary layer	Acceleration = Horizontal PGF + Coriolis Force
Horizontal motions within the boundary layer	Acceleration = Horizontal PGF + Coriolis Force + Friction
Vertical motions	Acceleration = Vertical PGF + Gravity

the forces are exactly in balance so that there are no accelerations.

Hydrostatic Balance

The Earth's average pressure decreases rapidly with altitude from its mean sea-level value of 1013.25 mb to about 1 mb at 31 km. Despite the very strong upward pressure gradient force, air does not rush off into space because the upward force is, to a very good approximation, balanced by the downward force of gravity, as shown in Figure 7.8. Whenever the upward pressure gradient force is *exactly* balanced by gravity, we say that the atmosphere is in ***hydrostatic balance***. Except in thunderstorms, the atmosphere is essentially in hydrostatic balance everywhere on the Earth.

Geostrophic Balance

Imagine a stationary air parcel located in the Northern Hemisphere in a region of the atmosphere where the pressure decreases uniformly from south to north (Point A in Figure 7.9). The parcel is initially stationary, so the only force acting on it is the pressure gradient force (PGF). The PGF, which acts from high pressure (south) toward low pressure (north), perpendicular to the isobars, accelerates the air parcel northward. However, as soon as the parcel begins to move across the Earth, it immediately is deflected by the Coriolis force, which acts to the right of the air parcel's motion. The Coriolis force, which is proportional to the air parcel's speed, will be weak, since the air accelerates to point B from rest at point A. Since the air parcel experiences a strong PGF pulling it northward and a weaker Coriolis force pulling it eastward, it will move north-northeast, deflecting away from B to C. At C, the forces are still not in balance, so the air parcel will continue to accelerate, now in the direction of the *net* combined force, which is toward the north-northeast. As the air parcel speed increases and direction of motion changes from northward to northeastward, the Coriolis force increases and progressively changes direction from eastward to southeastward (D), always to the right of the parcel's motion. The air parcel will continue to accelerate in this way, turning further to the right, until the PGF and Coriolis force are equal and opposite (E). At E, the PGF and Coriolis force exactly balance, so that there is no net force acting on the parcel. The air parcel will no longer accelerate and, following Newton's first law of motion, *will remain in motion traveling at a constant speed.*

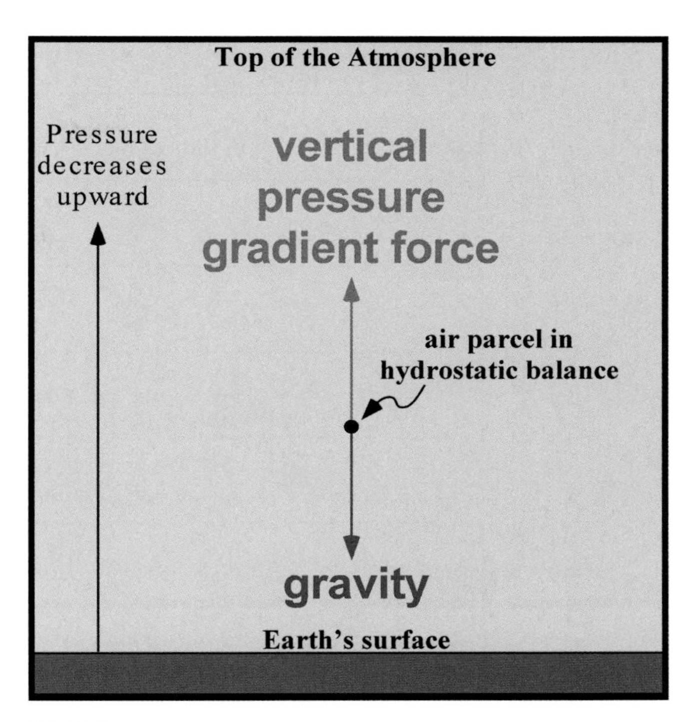

FIGURE 7.8: Hydrostatic balance exists nearly everywhere in the atmosphere because the upward-directed vertical pressure gradient force balances the downward force of gravity.

ONLINE 7.2

The Geostrophic Wind

The process of air accelerating from rest to a state of geostrophic balance can be understood by viewing an animation of the process online. Watch as an air parcel initially at rest in the presence of a pressure gradient in the Northern Hemisphere accelerates, deflecting to its right under the influence of the Coriolis force, until it flows parallel to the isobars and achieves a state of geostrophic balance.

The balance between the pressure gradient force and Coriolis force is called ***geostrophic balance***. The wind that would exist if the atmosphere was in geostrophic balance is called the ***geostrophic wind***. As illustrated in Figure 7.9, the geostrophic wind flows parallel to *straight isobars* (or height contours on an upper level map), and its strength is related to the pressure gradient (the distance between the isobars). In the northern hemisphere, higher pressures (heights) are to the right of the geostrophic wind. The atmosphere is nearly, but rarely, in *exact* geostrophic balance. We can see this by considering the flow field

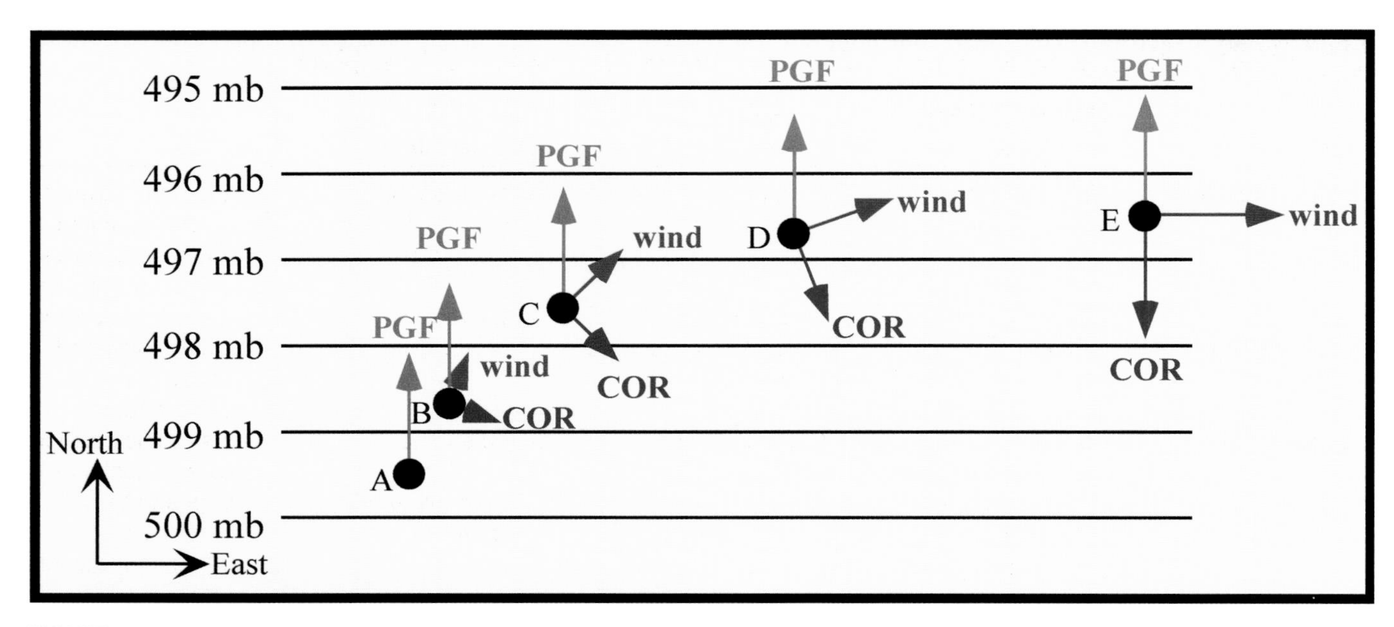

FIGURE 7.9: Pressure gradients create a net force directed from higher to lower pressure. In this example, the arrow points in the direction the force is acting. The pressure gradient force is always directed from high to low pressure, perpendicular to the isobars. The Coriolis force acts to deflect the air to the right relative to its direction of movement. The changes that occur in the forces as air accelerates from point A to point E are described in the text.

on the 500 mb chart in Figure 7.10. The arrows on this figure indicate the direction and relative speed of the airflow. The flow near point A is *curved,* so that air turns counterclockwise, or "cyclonically," about the trough. We see from the *orientation* of the contours that the PGF an air parcel would experience moving through this flow is continually changing, since the contours are curved. If an air parcel was in geostrophic balance at point A, it would no longer be in balance just downstream since the orientation of the PGF would have changed. At point B, although the orientation of the isobars is relatively constant, the distance between the isobars is continually changing along the direction of flow, implying that the *magnitude,* or strength, of the pressure gradient force is continually changing. An air parcel moving through either flow field will continually have to accelerate to respond to the changing PGF. The atmosphere is in exact geostrophic balance only when the isobars are straight and parallel, and the air flows parallel to the isobars. These conditions rarely exist over large regions, or for long times. These conditions also do not apply in the boundary layer, where friction is also an important force and acts to keep air from ever achieving a state of geostrophic balance. Nevertheless, for air above the boundary layer, we can see from the wind arrows superimposed on the map in Figure 7.10 that the atmosphere is nearly in geostrophic balance, since the speed and direction of the wind are closely related to the distance and orientation of the contours. The air flows nearly parallel to

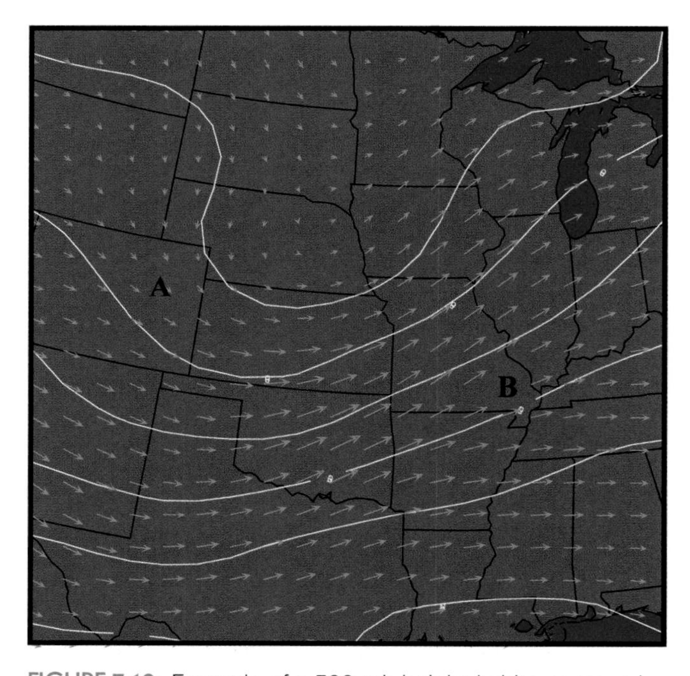

FIGURE 7.10: Example of a 500 mb height (white contours) and wind field (arrows) over the central United States. Note that the airflow is nearly in geostrophic balance. Longer arrows indicate faster wind speeds.

TABLE 7.2 **Summary of the Properties of Forces Acting on Air in the Earth's Atmosphere**

Force	Direction in Which Force Acts	Strength Depends on	Effect on Air	Balances
Vertical pressure Gradient force	Upward, from higher to lower pressure	Magnitude of the vertical pressure gradient	Accelerates air vertically toward lower pressure	Hydrostatic balance when equal and opposite to gravitational force
Horizontal pressure Gradient force	Horizontally, from higher to lower pressure	Magnitude of the horizontal pressure gradient	Accelerates air horizontally	Geostrophic balance when equal and opposite to Coriolis force
Coriolis force	To the right (left) of the wind direction in the Northern (Southern) Hemisphere	Wind speed and latitude	Affects wind direction, but no effect on wind speed	Geostrophic balance when equal and opposite to horizontal pressure gradient force
Frictional force	Opposite the direction of the flow	The roughness of the underlying surface	Reduces air velocity, important primarily in boundary layer	———
Gravitational force	Toward the center of the Earth	Essentially constant in the troposphere	Accelerates air downward	Hydrostatic balance when equal and opposite to vertical pressure gradient force

the contours and increases in speed when the contours are closer together. We will see shortly that the large cyclones of the middle latitudes develop rapidly below regions of the jetstream where the PGF and Coriolis forces are not balanced.

SUMMARY OF FORCES AND FORCE BALANCES

Table 7.2 summarizes the properties of the forces acting on parcels of air as they move about in the atmosphere. In Table 7.2, the pressure gradient force is divided into its vertical and horizontal components so that they can be compared more easily with the other forces that lead to geostrophic and hydrostatic balance. Although the magnitude and direction of each force depends on different factors, the forces, when considered together, always determine the speed and direction of the wind as air moves within Earth's atmosphere. By applying Newton's second law, we can learn about the effects that force balances have on the winds, and even more importantly, the accelerations that occur when forces are out of balance.

GEOSTROPHIC BALANCE AND THE JETSTREAM

The *jetstream* is a narrow band of strong winds that encircles the Earth in the mid-latitudes. The band of strongest winds is typically 300 to 500 km (200 to 300 miles) wide and can extend from near the tropopause at 250 mb down to about 500 mb. The jetstream typically follows a wavelike pattern, as depicted on the 300 mb map in Figure 7.11. Its maximum speed occurs just below the tropopause. There can be as many as three, or as few as one jetstream present at any longitude at one time between the equator and the pole. When they occur, the three jets are called the subtropical, polar, and arctic jetstreams. They are found at different latitudes and altitudes but often merge and interact, dramatically altering the storm systems associated with them. Jetstreams vary in intensity. Regions of exceptionally strong winds within the jetstream, such as the region over Arizona and New Mexico in Figure 7.11, are called ***jetstreaks***.

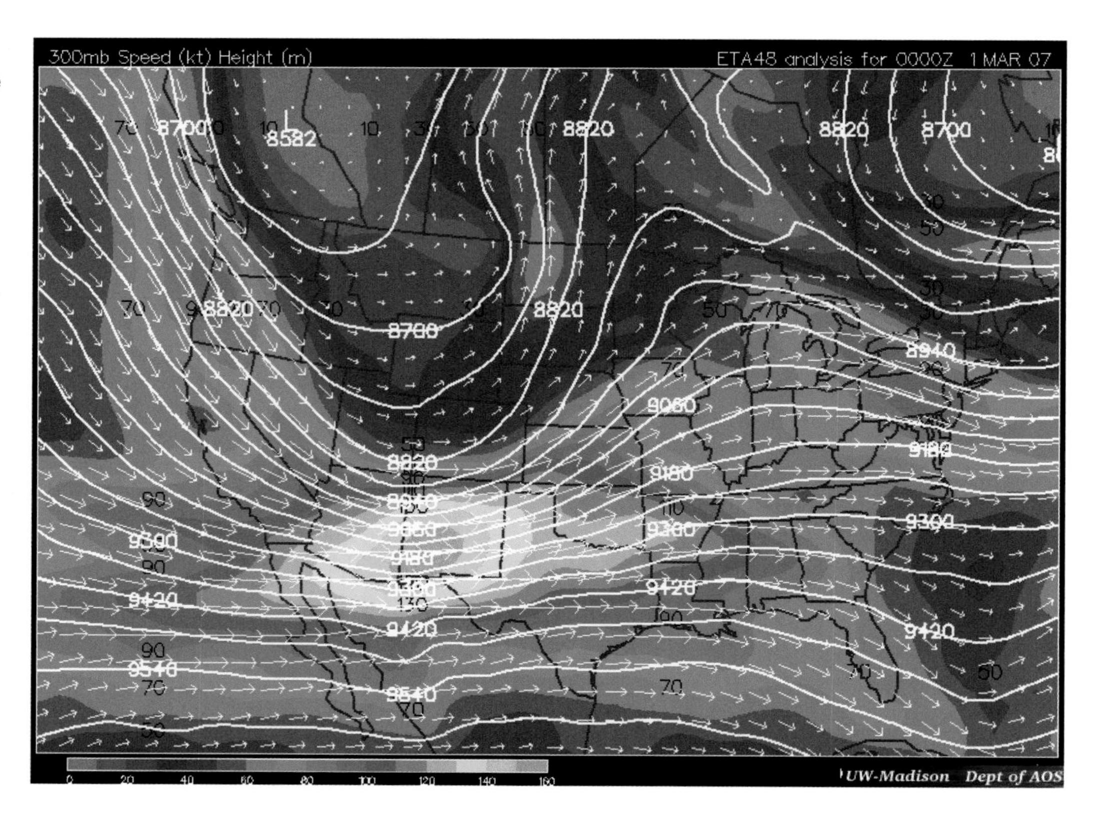

FIGURE 7.11: 300 mb height field (meters, white lines), wind speeds (knots, colors), and winds (arrows). The southern branch of the jetstream, indicated by light blue, green, and yellow colors, flows southeastward over the west coast of North America, northeastward over the center of the continent, and eastward offshore of the East Coast. A jetstreak is embedded in the jetstream over the southwestern United States with the core of the jetstreak near southwestern New Mexico. Note that the winds are strongest where the contours are closest together.

FOCUS 7.2

Flying and the Jetstream

For most of the cool season, the jetstream lies over the United States and southern Canada. The fastest winds in the jetstream are found very close to the altitudes where commercial jets fly (about 9 kilometers [30,000 ft] above sea level). The jetstream, on average, flows from west to east. Aircraft at cruising altitude can fly at about 500 miles/hour (800 km/hour). The jetstream flows at roughly 100 miles/hour (160 km/hour). Aircraft flying eastbound pick up extra speed (500 + 100 miles/hr = 600 miles/hr) as they fly with the wind, while those flying westward lose speed (500 – 100 miles/hr = 400 miles/hr) flying into the wind. If an aircraft flew directly within a jetstream with 100-mph winds, an eastbound coast-to-coast trip of 3000 miles would take 5 hours and a westbound trip 7.5 hours! To reduce this difference, flight-planning computers take into account jetstream winds, placing westbound aircraft at altitudes above or below the jetstream to avoid the strongest winds, while placing eastbound aircraft in the jetstream to pick up the additional speed.

We can understand why jetstreams exist in the atmosphere by considering what happens when the atmosphere is in both hydrostatic and geostrophic balance in the presence of a temperature gradient, such as the gradient that always exists between the equator and the poles. Figure 7.12 shows a situation where a temperature gradient is present, with cold air to the north and warm air to the south. Let us assume that the pressure at the surface is the same everywhere in Figure 7.12. If we consider some other altitude above the surface (for example, 1 km), will the pressure be higher in the warm or the cold air? We know from Chapter 1 that the cold air is denser than the warm air. Recall that pressure is the weight of air *above* a unit area. Since the surface pressure is the same in both the cold and warm air, the same number of molecules must, on average, reside in an entire air column above a unit area in the cold and warm air. However, since the cold air is more dense, more molecules must reside between the surface and 1 km in the cold air than in the warm air. This means that fewer molecules must be present between 1 km and the top of the atmosphere in the cold air, implying that the pressure is lower at 1 km in the cold air. In fact, surfaces of constant pressure will slope downward from the warm air to the cold air, as we learned in Chapter 3.

Figure 7.12 shows a sloping pressure surface. Pressure surfaces at higher altitudes slope more

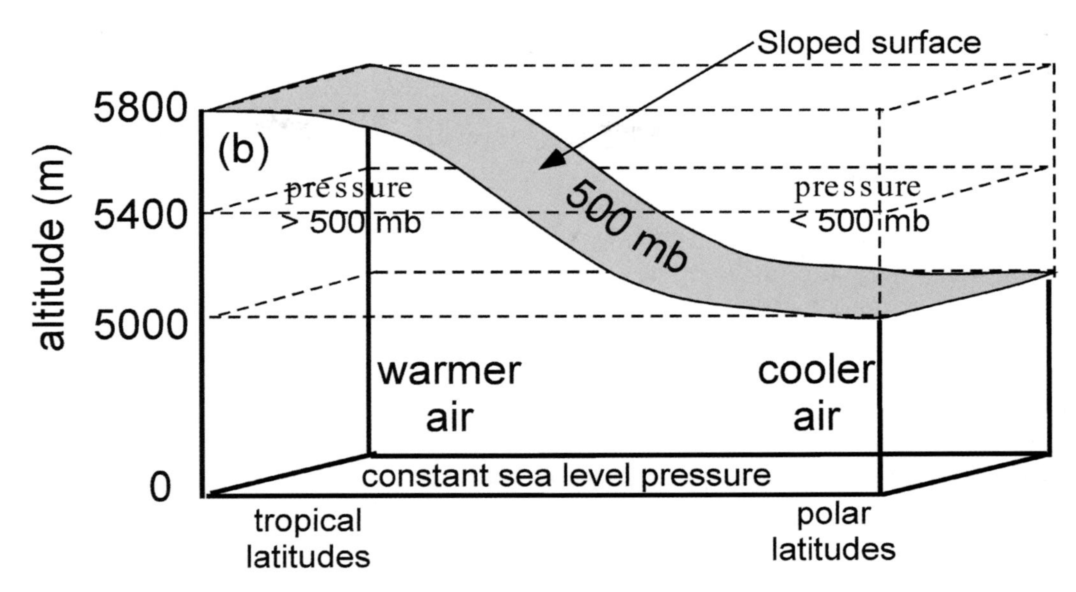

FIGURE 7.12: The relationship between the slope of a constant pressure surface aloft (in this example, 500 mb), the pressure gradient aloft, and the average temperature gradient in the air below the pressure surface. In this example, the pressure found at an altitude of 5400 m is not the same at all latitudes. The pressure exceeds 500 mb at 5400 m in the tropics, but is less than 500 mb at 5400 m in the polar regions. As a result, a pressure gradient must exist at the 5400 m level directed from high pressure (the tropics) to low pressure (the pole).

steeply than those at lower altitudes. For example, the altitude of the 850 mb surface in the wintertime typically slopes from about 1600 meters (1 mile) near the equator to 1200 meters (0.75 miles) near the poles, while the 300 mb surface slopes from about 9600 to 8200 meters (6 to 5 miles) over the same distance. The slopes are steepest across the middle latitudes (30 to 60°), where strong north-south temperature gradients typically are found. Figure 7.12 shows that sloping pressure surfaces are associated with horizontal pressure gradients. The steeper the slope, the stronger the horizontal pressure gradient, the stronger the pressure gradient force and the stronger the geostrophic wind. It is for this reason that winds increase with height in the troposphere. Figure 7.12 shows that the pole-equator temperature gradients imply that the pressure gradient force will have a component directed poleward toward the cold air in both hemispheres. Considering geostrophic balance, this implies that upper tropospheric winds will generally have a west to east component in both hemispheres.

Temperature gradients in the atmosphere are typically not uniform. Sharp temperature gradients typically occur within narrow regions called fronts, which are boundaries between cold and warm airmasses. We will learn more about fronts in Chapter 9. The top of the cold airmass normally slopes away from the warm air in a dome-like shape (Figure 7.13), with the concentrated temperature gradient along the leading edge of the cold air. Figure 7.13 shows schematically what happens to pressure surfaces at various altitudes across a front. The sharp temperature

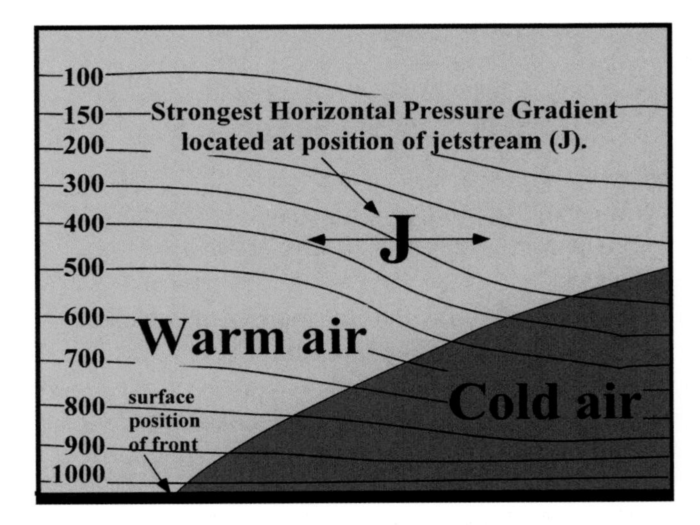

FIGURE 7.13: Pressure surfaces slope downward toward cold air in regions of strong temperature gradients (the slope on this diagram is exaggerated for clarity). Note that the strongest slope occurs over the front. The horizontal pressure gradient force, and strongest winds (e.g., the jetstream), occur where the slope of the pressure surfaces is steepest. Winds are directed out of the page if the diagram is for the Northern Hemisphere.

difference across the front induces a locally strong pressure gradient (implied by the strongly sloped pressure surfaces) over the front. Assuming geostrophic balance, this implies that a band of strong winds, the jetstream, must reside over the outer portion of the cold air dome. For this reason, fronts and jetstreams are closely related. Jetstreams in the upper troposphere are found above fronts in the lower troposphere.

CHECK YOUR UNDERSTANDING 7.2

1. What is geostrophic balance?
2. What is hydrostatic balance?
3. What is the jetstream?
4. Between what altitudes is the jetstream typically found?

TEST YOUR UNDERSTANDING

1. How is the pressure gradient force related to the wind speed?
2. What does a strong pressure gradient look like on a sea-level pressure map?
3. At the surface, in what weather phenomena are the most extreme horizontal pressure gradients found?
4. Where in the atmosphere would you find a pressure gradient of 120 mb/km?
5. Given that the vertical pressure gradient force acts upward, why don't air molecules fly off into space?
6. Why can the gravitational force be considered constant for weather applications?
7. Name three types of turbulence that occur in the atmosphere. Briefly describe how each develops.
8. On a hot summer day, air flows off the Southeast Coast of the United States and over the Atlantic Ocean. What should happen to the depth of the friction layer once the air moves over the ocean?
9. What is a typical depth of the friction layer in the atmosphere?
10. What is the principle of "conservation of angular momentum"? What role does this principle play in the movement of air?
11. Does the Coriolis force affect the direction that air moves, the speed air moves, or both?
12. Summarize the four key properties of the Coriolis force that are important to understand severe and hazardous weather.
13. What forces must be included in Newton's second law of motion for horizontal air motion in the boundary layer? At 7 km above the ground?
14. Identify one meteorological phenomenon where the atmosphere is unlikely to be in hydrostatic balance.
15. When air is in geostrophic balance, what is the relationship between the direction of flow and the height gradient on a constant pressure map? What about the direction of the flow and isobars on a constant height map?
16. If the same pressure gradient exists near the 10 km level at 20° N and 60° N, where will the geostrophic wind be stronger? (Hint: consider the strength of the Coriolis force.)
17. If air in geostrophic balance moves into a stronger pressure gradient, what will happen to the wind speed?
18. Can moving air be in geostrophic balance on the equator?
19. What is a jetstreak? At what atmospheric level would a jetstreak most likely be found?
20. What is the relationship between the low-level temperature gradient and the pressure gradient aloft?
21. Why are jetstreams found above regions of strong temperature gradients in the atmosphere?

TEST YOUR PROBLEM-SOLVING SKILLS

1. Use the *surface map* below to answer the following questions.
   (a) Estimate the lowest and highest sea-level pressure on the map. Be sure to include units.
   (b) Determine the direction of the pressure gradient force at: (i) Boise, Idaho; (ii) Sacramento, California; (iii) Phoenix, Arizona; (iv) Laramie, Wyoming; (v) Omaha, Nebraska; (vi) Little Rock, Arkansas; (vii) Chicago, Illinois; and (viii) Atlanta, Georgia.
   (c) Identify the city experiencing the strongest wind speeds.
   (d) Compute the pressure gradient over a 500 km distance in (i) Idaho–Utah, (ii) Oklahoma, and (iii) Indiana–Ohio. What is the relationship between the spacing of the isobars and the value of the pressure gradient?

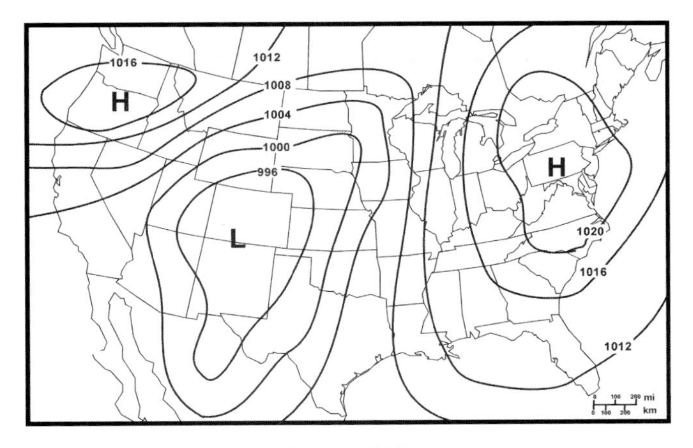

Chapter 7 Test Problem-Solving Skills 1

2. Use the 300 mb map below to answer the following questions.
   (a) In what region would the coldest air be found in the layer of air below 300 mb?
   (b) Assuming that the air is nearly in geostrophic balance, what direction would the wind be blowing at (i) Sacramento, California; (ii) Phoenix, Arizona; (iii) Laramie, Wyoming; (iv) Omaha, Nebraska; (v) Little Rock, Arkansas; (vi) Chicago, Illinois; and (vii) Atlanta, Georgia?
   (c) Locate Denver, Colorado. Assume that the air over Denver at 300 mb is in geostrophic balance. In what direction is each of the following pointing? (i) pressure gradient force, (ii) Coriolis force, (iii) geostrophic wind.
   (d) Where would you find the strongest winds on the map? How about the weakest winds?

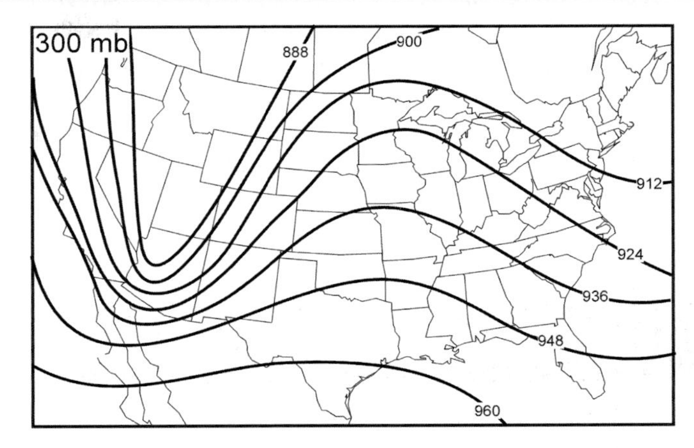

Chapter 7 Test Problem-Solving Skills 2

3. You and your friend are returning from a summer road trip across the United States. At a rest area, your friend comes out of the restroom with a very puzzled look on his face and says to you, "That was strange. After I washed my hands, the water had a clockwise swirl as it drained out of the sink. The Coriolis force is supposed to make water drain clockwise in the Southern Hemisphere and counterclockwise in the Northern Hemisphere. Where are we?"
   (a) Is your friend correct in asserting that the Coriolis deflection in the Southern Hemisphere is opposite in direction to the Coriolis deflection in the Northern Hemisphere?
   (b) Should the Coriolis force play a role in the way sinks drain? Why or why not?
   (c) Fill a sink with water. Let the water settle (no movement), and then open the drain. Repeat the procedure a total of six times. How many times did the water drain counterclockwise? How does this reinforce your answer to (b)?

4. Examine Figure 7.11 and answer the following:
   (a) In Figure 7.11, what is the ratio of the height gradients at 300 mb over Oklahoma City and Salt Lake City? If the wind speeds are proportional to the lengths of the arrows, what is the ratio of the 300 mb wind speeds over the two cities?

(b) In Figure 7.11, what is the ratio between the height gradients at 300 mb over Great Falls, Montana and Los Angeles, California? What is the ratio of the 300 mb wind speeds over these two cities?

(c) By generalizing from your answers to (a) and (b), determine how an upper-air wind speed should change when an upper-air pressure gradient strengthens by a factor of two.

5. The Earth's rotation speed is about 700 mph at 45° N and about 500 mph at 60° N. Suppose a missile is fired from 45° N, aimed at a target due north at 60° N. Unfortunately, the launch crew has carelessly entered "zero" in the setting for the Earth's rotation rate. If the missile's speed is 2070 mph, and its accuracy is perfect on a non-rotating Earth, where will it land relative to its target on a rotating Earth? (Useful information: 1° latitude = 69 miles.)

USE THE SEVERE AND HAZARDOUS WEATHER WEBSITE

http://severewx.atmos.uiuc.edu

1. Use the *Severe and Hazardous Weather Website* "Current Weather" to navigate to a surface weather map where pressure is analyzed.
   (a) Save or print the map. Record the date and time of the map.
   (b) Examine the isobar pattern of the map. Are the strongest pressure gradients found in the vicinity of high-pressure systems or low-pressure systems?
   (c) If wind data is not available on the map, navigate the website to a map of the same date and time that displays wind data. Compare the pressure gradients on the surface map with the wind speeds. Describe the relationship between pressure gradients and wind by noting the magnitude of the gradients and wind speeds for several locations across the United States.
   (d) Locate your hometown on the map.
      (i) What direction is the pressure gradient directed at your location?
      (ii) What is the direction of the wind at your location?
      (iii) Is the pressure gradient directed perpendicular to the wind? Provide reasoning for what you observe.
   (e) Compare the wind direction across the United States with the pressure pattern. Are the two variables related in the same way that they are presented in this chapter? Discuss why or why not. What other force may be influencing the wind direction?

2. Locate a 300 mb map that shows both the height field and isotachs (contours of wind speed) using the *Severe and Hazardous Weather Website* "Current Weather" link.
   (a) Save or print the map. Record the date and time of the map.
   (b) Compare the height gradient to the wind speed. What relationship exists?
   (c) Compare the height gradient to the wind direction. Does the air appear to be in geostrophic balance? Why or why not?
   (d) In question 1, a surface pressure map was examined. How does the wind direction relative to the height gradient on the 300 mb map compare to the wind direction and pressure pattern on the surface map? Comment on any similarities and/or differences.

3. Use the *Severe and Hazardous Weather Website* to navigate to Environment Canada's website (choose "Forces and Force Balances" from the menu, then "Additional Resources"). At this website you can view satellite data including full-disk animations. Navigate to the North and South America IR satellite view for either GOES-East or GOES-West.
   (a) Record the image date and times and view (GOES-East or GOES-West).
   (b) Play the animation several times and examine the cloud movement carefully. What is the prevailing wind direction in the tropical regions? What is the prevailing wind direction in middle latitudes?

(c) Identify any storm systems that are present over several frames of the animation. For each storm system:
   (i) Record the hemisphere and approximate latitude at which they are located.
   (ii) What direction are the storm systems moving?
   (iii) What direction are the storm systems rotating?
   (iv) What forces might play a role in the movement and rotation of the storms you identified?

4. In this exercise you will be comparing wind flow patterns at differing heights in the atmosphere and at different latitudes. Navigate the *Severe and Hazardous Weather Website* via the "Forecast Links" to a website that provides North American forecast output (The GFS model is a good choice.)
   (a) Locate maps of the upper troposphere (300, 250, or 200 mb) and the lower troposphere (850 mb) that are valid for the same date and time. Save or print the maps and record the date and time of the maps.
   (b) On the upper tropospheric map identify an area of extremely strong winds and an area of relatively weak winds. Record both locations. Examine these same locations on the lower tropospheric map. The lower tropospheric map likely provides temperature. How would you characterize the lower tropospheric temperature gradients below the strong and weak upper tropospheric winds? Is the relationship consistent with the text's explanation for the development of the jetstream?
   (c) Now find two widely separated latitudes (e.g., 30° N and 70° N) where the pressure gradients are approximately the same. How do the wind speeds differ at the two locations? Why might these wind speeds be different?
   (d) On the upper tropospheric map, at a single latitude find areas where the height gradient curves in a clockwise and counterclockwise manner (this will be referred to as "curvature" in Chapter 8) but in which the pressure gradients are the same. Can you detect consistent differences in wind speed between the regions of opposite curvature? If so, you have identified the "curvature effect," a key topic in the next chapter and one that leads to the formation of surface pressure systems.

5. Navigate the *Severe and Hazardous Weather Website* to the "Current Weather" page and locate a map of the upper troposphere (300, 250 or 200 mb).
   (a) Save or print the map. Record the date and time of the map.
   (b) Locate any jetstreams that are present by identifying the regions where they are present. What is the fastest wind speed in each of the jetstreams?
   (c) Are there any jetstreaks present in any of the jetstreams? If so, what is the strongest wind speed in the core of each jetstreak?

THE DEVELOPMENT OF HIGH- AND LOW-PRESSURE SYSTEMS

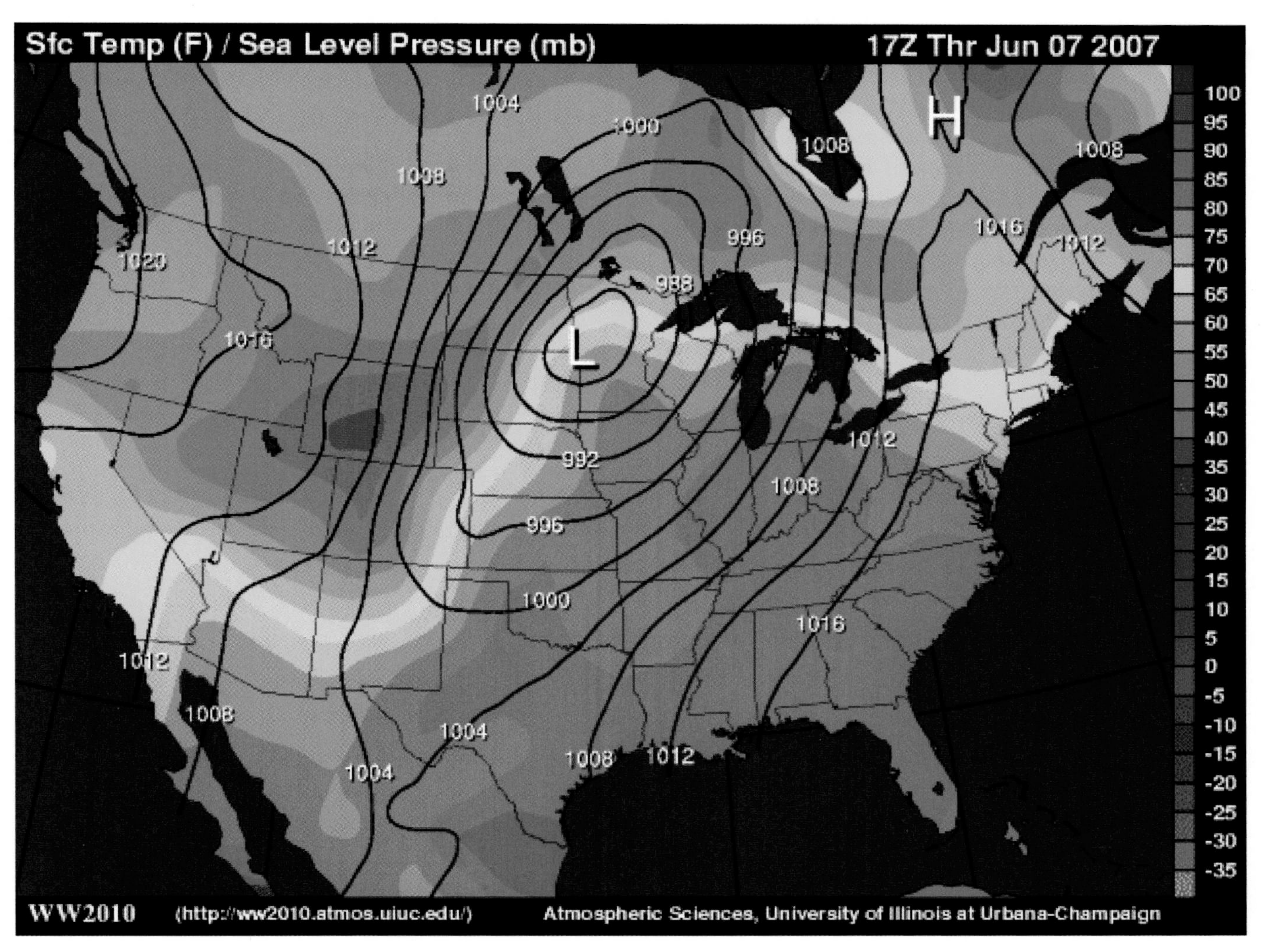

Surface map showing a low-pressure system over western Minnesota.

KEY TERMS

anticyclone
anticyclonic curvature
centripetal acceleration
convergence
curvature effect
cyclone
cyclonic curvature
diabatic process
divergence
dynamic process
entrance region
exit region
extratropical cyclone
friction layer
gradient wind balance
high-pressure center
jetstreak
low-pressure center
polar front jetstream
ridge

semi-permanent high
semi-permanent low
subgeostrophic
subtropical jetstream
supergeostrophic
thermodynamic process
trough

In Chapter 7, we examined the concept of the geostrophic wind—a wind that occurs when the pressure gradient force and Coriolis force exactly balance. In nature, the Coriolis and pressure gradient forces are generally not in perfect balance, and often are far from balanced. ***Extratropical cyclones***, the parent storms of many types of hazardous weather ranging from severe thunderstorms to snowstorms, develop as a direct result of accelerations created by imbalances between the pressure gradient force and the Coriolis force, primarily at the level of the jetstream. Friction, primarily acting within the friction layer near the Earth's surface, ultimately destroys extratropical cyclones. Because extratropical cyclones are the parent storms for many types of hazardous weather, it is essential that we understand the processes that create these storms and lead to their demise. High-pressure systems, which are generally associated with benign weather, also evolve in response to force imbalances, although cooling and heating play particularly important roles in the evolution of high-pressure centers.

FORCE IMBALANCES

As a first step, we will examine how ***convergence*** of air in the atmosphere changes surface pressure and affects vertical air motion. Air is said to "converge" into an air column whenever the flow of air is such that the mass of air in the air column increases with time. A simple convergent flow field is shown in Figure 8.1A. In the atmosphere, pure flows such as this

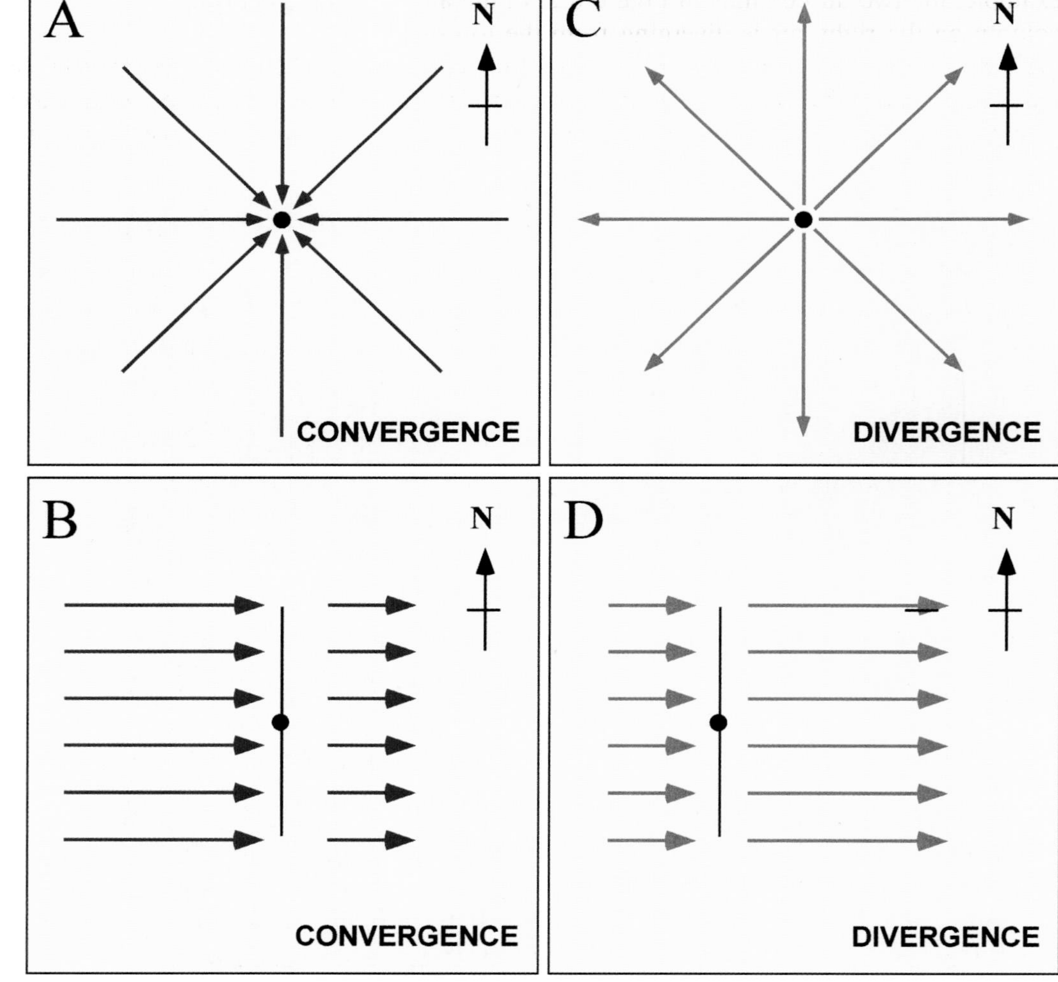

FIGURE 8.1: Simple flow fields associated with convergence and divergence. Convergence occurs as air flows toward a central point (A), or decelerates (B). Divergence occurs as air flows away from a central point (C), or accelerates (D). In each panel, the arrows depict air movement.

are rarely observed. Convergence is normally part of a flow that may also include translation (straight, uniform flow) and rotation. Figure 8.1B shows another example of a convergent flow field. In this case, air is decelerating as it flows from left to right, leading to convergence over the area of deceleration. *Convergence within an air column is always associated with increasing surface pressure, since the mass/unit area, or weight of the air column, will increase with time.*

Divergence, the opposite of convergence, occurs when the mass of air in an air column decreases with time. Figures 8.1C and 8.1D show corresponding examples of divergent flow fields. *Atmospheric divergence is always associated with decreasing surface pressure. This occurs because mass is taken out of the column of air, decreasing the weight of the air above a unit area, and hence the surface pressure.* Understanding the mechanisms by which divergence occurs in the atmosphere is central to our understanding of storms such as extratropical cyclones and hurricanes.

Vertical motions in the atmosphere are also related to the convergence and divergence fields. Consider, for example, the two air columns in Figure 8.2. In the air column on the right, air is diverging from the top of the column, and converging into the bottom of the column. To maintain hydrostatic balance, air within the column must rise in response to the compression of air into the bottom and the evacuation of air from the top. We will see shortly that this is exactly what happens in low-pressure centers. In the air column on the left, where air is converging into the top and diverging from the bottom, hydrostatic balance is maintained as air descends. The right column is typical of a ***low-pressure center,*** or ***cyclone,*** and the left of a ***high-pressure center,*** or ***anticyclone.*** Recall that rising motion is associated with clouds and precipitation, while descending motion is associated with clear skies. Therefore, we would expect to find clouds and precipitation associated with low-pressure centers, and relatively clear skies with high-pressure centers.

Divergence and convergence above the friction layer result from imbalances between the pressure gradient and Coriolis forces. These imbalances force air to accelerate into some regions of the atmosphere and out of others. To understand how these processes occur, we will consider what happens to the balance between these two forces as we follow air parcels moving through different flow regimes.

Curved Flow

Consider a simple pressure pattern consisting of a circular, symmetric low-pressure center located in the upper troposphere in the Northern Hemisphere, as shown in

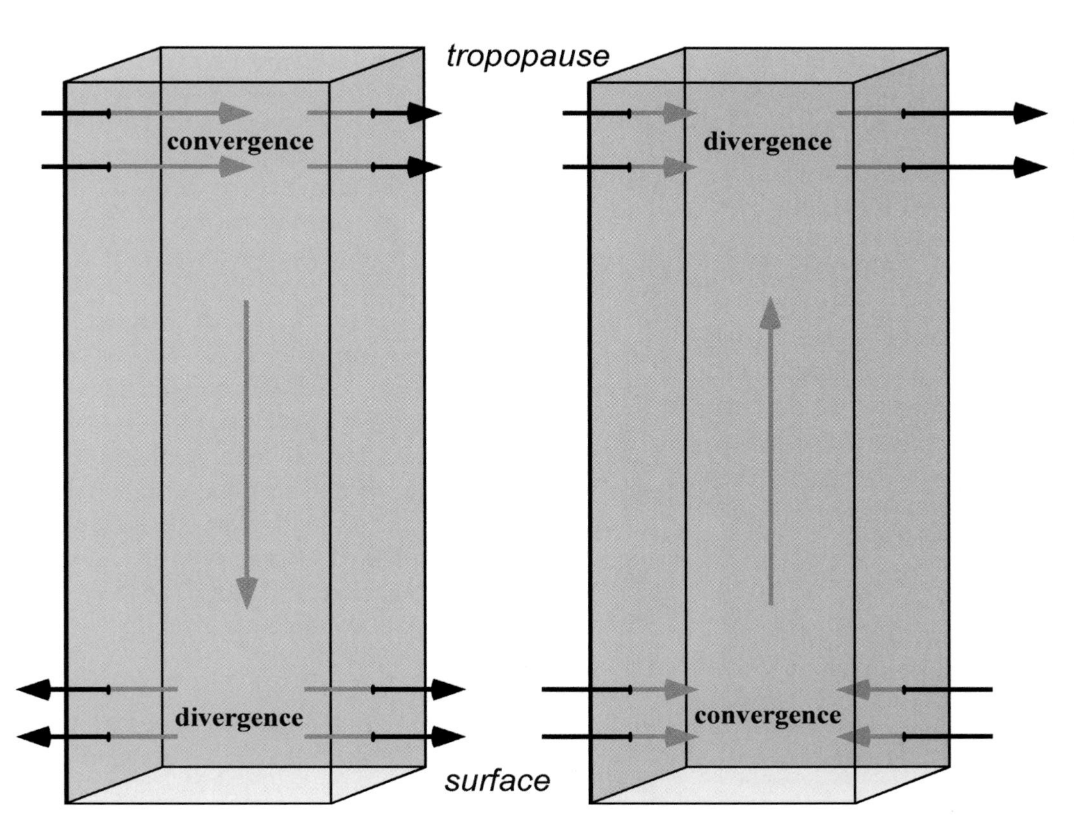

FIGURE 8.2: The relationship between divergence, convergence, and vertical air motions in air columns. Black portions of arrows are outside of columns, while gray portions are inside.

Figure 8.3. Let's further assume for simplicity that a northward-moving air parcel located at point A is in geostrophic balance, so that the inward-directed pressure gradient force exactly balances the outward-directed Coriolis force. Since the forces exactly balance at point A, there is no net force on the air parcel and we can invoke Newton's first law: the air parcel will remain in motion traveling at a constant speed in a straight line. Let us move the air parcel forward a very small distance and consider the change that occurs in the force balance. For clarity, we will exaggerate this distance in Figure 8.3 and move the parcel to point B, keeping in mind that the slightest advance forward by the air parcel will be sufficient to alter the force balance. Since the air parcel moves forward traveling at a constant speed in a straight line, the Coriolis force will not change magnitude or direction. However, the orientation of the pressure gradient force will change so that it is southwestward, still directed toward the center of the low (from higher pressure to lower pressure). We can think of this southwestward force as having two components, one directed southward and one directed westward. The southward component, directed opposite the direction of the flow, will slow the flow relative to its geostrophic value, reducing the magnitude of the Coriolis force. Recall that the Coriolis force is proportional to the speed of the flow. This slowing of the flow also causes the inward-directed pressure gradient force to always exceed the outward directed Coriolis force at any point in the flow field. The flow is no longer in geostrophic balance. Indeed, our assumption of geostrophic balance in the beginning was incorrect; there will always be a net inward-directed force in our simple circularly symmetric flow field. This net inward-directed force is a ***centripetal acceleration***, similar to the acceleration experienced by a ball whirled on a string or by a satellite orbiting the Earth under the influence of gravity. Our air parcel will "orbit" the low-pressure center. When the air parcel orbits the low-pressure center in a manner such that the flow remains parallel to the isobars, we say that it is in ***gradient wind balance***.[1] Most flow fields in the atmosphere possess some curvature, and, above the friction layer, are nearly in gradient wind balance. Reiterating a key point: *in cyclonically curved flow (counterclockwise in the Northern Hemisphere), the true flow will be slower than its geostrophic value.*

Now let's consider what happens if we have a simple flow field consisting of a circular, symmetric high-

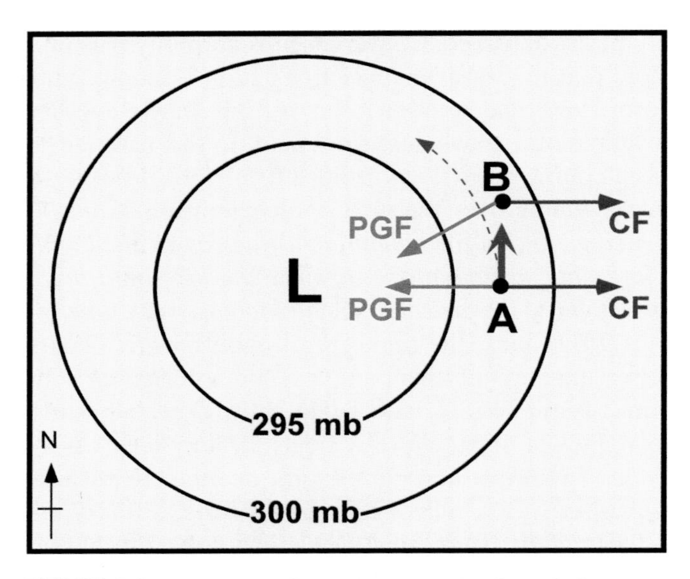

FIGURE 8.3: Illustration of the change in the force balance that occurs when air moves through a curved flow pattern associated with a low-pressure center located above the friction layer. Air accelerates around the low in a counterclockwise manner, flowing parallel to the isobars (or height contours) at a speed less than the geostrophic value.

pressure center located in the upper troposphere in the Northern Hemisphere, as shown in Figure 8.4. Let's again assume for simplicity that a southward-moving air parcel located at point A in Figure 8.4 is in geostrophic balance, so that the outward-directed pressure gradient force exactly balances the inward-directed Coriolis force. We again allow our air parcel to move forward a very small distance, exaggerated again for clarity. Invoking Newton's first law, the Coriolis force remains constant since the direction and speed of the air parcel will not change. However, we see that the pressure gradient force will now be directed southeastward. It will have a component eastward, and a component southward, *in the direction of the flow.* This will accelerate the flow relative to its geostrophic value, increasing the magnitude of the Coriolis force. This acceleration of the flow causes the inward-directed Coriolis force to always exceed the outward directed pressure gradient force at any point in the flow field. The flow is no longer in geostrophic balance, and, as in the flow around the low-pressure center, there will always be a net inward-directed centripetal acceleration. Note, however, that with curved flow around the high, *the true flow will be faster than its geostrophic value.*

Look again at Figure 7.11 in Chapter 7. We see from the direction of the wind and the contours that as air flows around the troughs and ridges, air parcels progress from regions of ***cyclonic curvature*** (counterclockwise in

[1]The term "gradient wind *balance*" is a slight misnomer, since the curved flow results from the imbalance between the pressure gradient force and the Coriolis force.

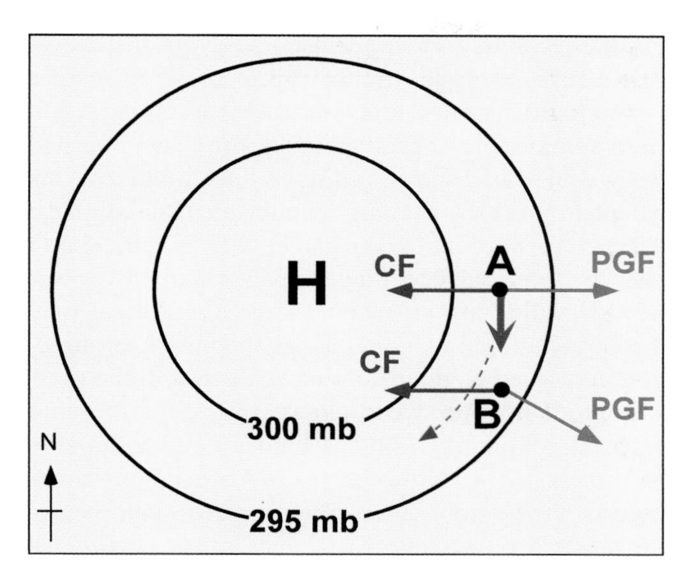

FIGURE 8.4: Illustration of the change in the force balance that occurs when air moves through a curved flow pattern associated with a high-pressure center located above the friction layer. Air accelerates around the high in a clockwise manner flowing parallel to the isobars (or height contours) at a speed greater than the geostrophic value.

Northern Hemisphere) to ***anticyclonic curvature*** (clockwise in Northern Hemisphere). Let us consider a part of this flow field, consisting of a trough, a ridge, and another trough, as shown in Figure 8.5. Recall that a trough is an elongated area of low pressure (or heights on a constant pressure chart), and that a ***ridge*** is an elongated area of high pressure. On upper air maps of middle latitudes such as Figure 8.5, troughs generally appear as southward dips and ridges as northward bulges in the jetstream. Figure 8.5 was carefully drawn so that the distance between the isobars is the same everywhere, implying that the *magnitude* of the pressure gradient force and the geostrophic wind is constant. (We will ignore the effects of changing latitude on the Coriolis force for simplicity.) At the base of the trough, the cyclonically curved flow will be slower than its geostrophic value ***(subgeostrophic)*** while at the crest of the ridge, the anticyclonically curved flow will be greater than the same geostrophic value ***(supergeostrophic).*** To satisfy this condition, air must accelerate from the base of the trough to the crest of the ridge, and decelerate from the crest of the ridge to the base of the next trough.

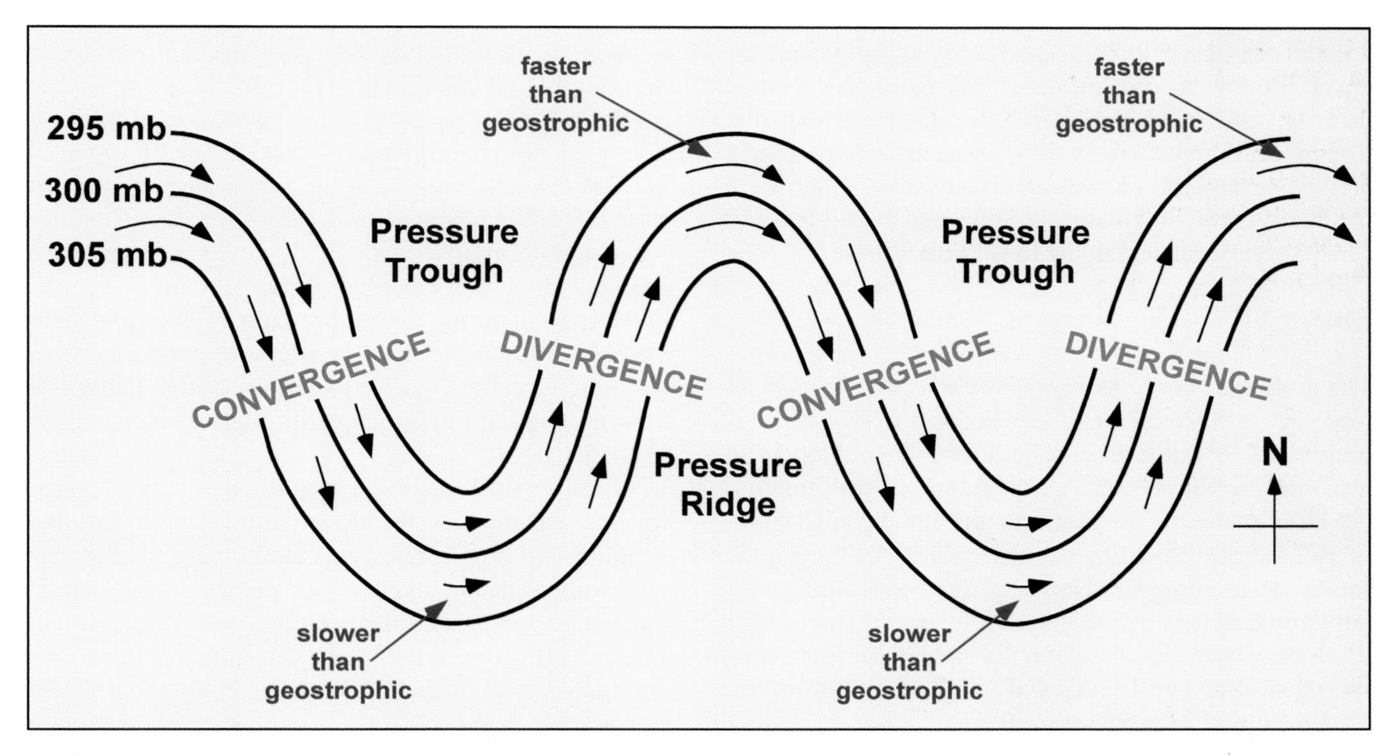

FIGURE 8.5: An upper tropospheric flow pattern consisting of two troughs and a ridge. Arrows indicate the direction of the airflow. The length of the arrow is proportional to the wind speed with short arrows representing slower winds and long arrows representing faster winds. As air flows from the base of a trough to the crest of a ridge, air accelerates due to a change in the curvature, leading to divergence. Convergence occurs as air decelerates from the crest of a ridge to the base of a trough. A surface low-pressure center will develop directly under the region of maximum divergence, and a surface high-pressure center will develop under the region of maximum convergence.

Instead of air, lets imagine for a minute that the contour pattern in Figure 8.5 is actually a highway with bumper-to-bumper traffic, with cars crossing through the "ridge" moving faster than cars in the troughs. We can easily imagine that the distance separating cars between the first trough and the ridge will increase with time—the cars "diverge" away from each other. Between the ridge and the second trough, cars will pile up as the faster-moving cars in the rear crash, or "converge" into the slower-moving cars in the lead. Individual air molecules moving from the first trough to the ridge, on average, also diverge, or separate from one another. Convergence occurs on the opposite side of the ridge. This effect is most pronounced at the jetstream level where the flow is strongest.

Divergence removes mass from air columns, while convergence adds mass to air columns. Since the surface pressure is the weight of a column of air above a unit area, the surface pressure must decrease under the regions of divergence and increase under the regions of convergence. When surface pressure decreases, a surface low-pressure center will strengthen or a surface high will weaken; conversely, when surface pressure increases, a surface low-pressure center will weaken or a surface high will strengthen. The most rapid changes associated with curvature of the flow will occur directly underneath the inflection points between the troughs and ridges, where divergence and convergence are strongest. The ***curvature effect*** is one important mechanism by which surface low- and high-pressure centers can be created in the lower atmosphere.

Jetstreaks

The jetstream contains regions where locally strong pressure gradients produce exceptionally strong winds called ***jetstreaks***. A jetstreak is located in a region of strong pressure gradient, and is indicated by the large values of the isotachs and the close spacing of the height contours. Figure 7.11 in Chapter 7 shows an example of a jetstreak over the southwestern United States.

Figure 8.6A is an idealized schematic of a jetstreak similar to the jetstreak in Figure 7.11. As a starting point, let us assume that at point A, at the ***entrance region*** of the jetstreak, a parcel of air is in geostrophic balance with the pressure gradient force acting northward and the Coriolis force southward. Since the forces exactly balance at point A, there is no net force on the air parcel and we can invoke Newton's first law: the air parcel will remain in motion traveling at a constant speed in a straight line. Applying the technique we used when studying the curvature effect, let us let the air parcel move forward a very small distance and consider the change that occurs in the force balance. Again, for clarity, we will exaggerate this distance in Figure 8.6A and move the parcel to point B, keeping in mind that the slightest advance forward by the air parcel will be sufficient to alter the force balance. Since the air parcel moves forward traveling at a constant speed in a straight line, the Coriolis force will not change magnitude or direction. However, as the parcel moves forward, the pressure gradient force will increase, since the pressure gradient continually tightens until we reach the core of the jetstreak. Now the pressure gradient force exceeds the Coriolis force. The air parcel, while moving eastward, will begin to accelerate northward across the jet in response to the force imbalance. As the air parcel accelerates and moves east-northeast toward the core of the jetstreak, the Coriolis force and the pressure gradient force both increase, but in such a way that the pressure gradient force always "leads" the Coriolis force.

At the core of the jetstreak the balance of forces shifts. Consider an air parcel in the core of the jet. Let us assume that the air has returned to a state of geostrophic balance in the jet core at point C. We invoke Newton's first law, allowing the parcel to move a very small distance forward (exaggerated to point D in Figure 8.6A). Since the air parcel moves forward with a constant speed and direction, the Coriolis force will remain constant. However, the pressure gradient force now decreases, since the pressure gradient is weakening in the direction of the flow. The Coriolis force now leads the pressure gradient force, and the air parcel will accelerate southward while moving eastward as it passes through the ***exit region*** of the jetstreak.

Figure 8.6B shows schematically the effect of force imbalances on the flows within a jetstreak and resulting areas of convergence and divergence. As air moves through the jetstreak, air parcels are displaced northward in the entrance region, and southward in the exit region. Divergence occurs in the right entrance region of the jetstreak (looking in the direction of flow) and convergence in the left entrance region as air is displaced from the right (south) to left (north) side of the jet. The opposite pattern occurs in the exit region of the jetstreak. Divergence aloft will result in lower surface pressure, while convergence will result in higher surface pressure. As we shall see, jetstreaks strongly influence the formation and evolution of extratropical cyclones.

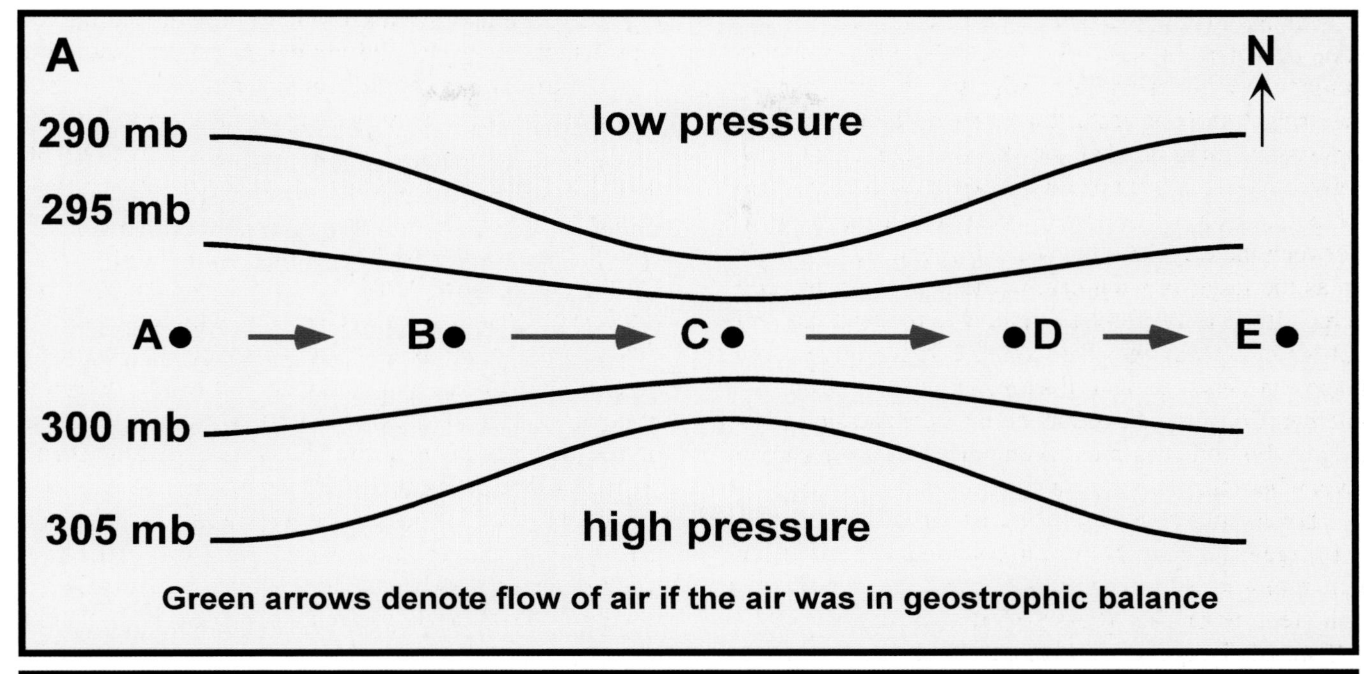

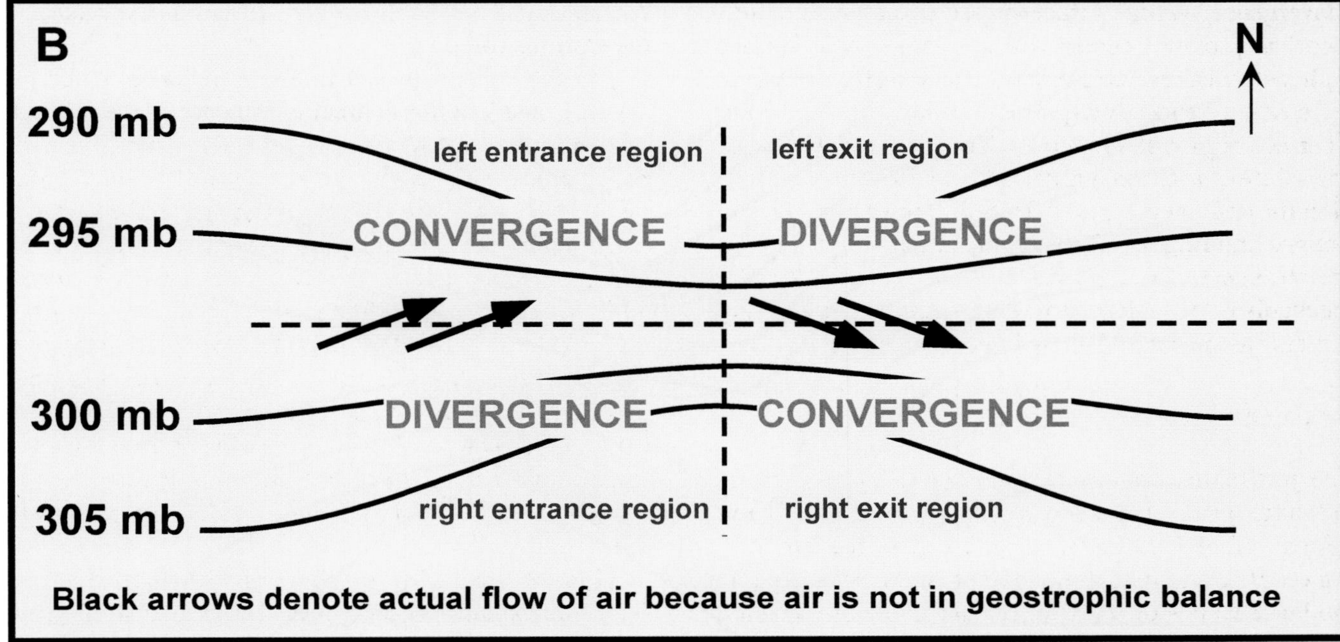

FIGURE 8.6: Schematic of the pressure pattern in the vicinity of a jetstreak embedded in a jetstream. The green arrows in Panel A show the direction and speed of the geostrophic wind at the center of the airstream. In Panel B, the dashed lines bisect the jetstreak in both directions, defining the right and left entrance and exit regions, and the black arrows denote the actual airflow. As air accelerates from Point A toward Point B in the entrance region of the jetstreak (Panel A), the pressure gradient force increases, and air accelerates across the isobars from the right entrance to the left entrance quadrants (Panel B). As a result, divergence occurs in the right entrance quadrant and convergence in the left entrance quadrant. The opposite pattern develops in the exit region of the jetstreak as the pressure gradient force decreases along the direction of the airflow, leading to divergence in the left exit quadrant and convergence in the right exit quadrant.

FOCUS 8.1

Multiple Jetstreaks

The jetstream is not always a single ribbon of fast-moving air encircling the pole. In nature, a single jetstream can split into two branches, and merge again at a location downstream. In fact, in winter two distinct jetstreams often develop: one called the *polar front jetstream* because it is closely associated with the position of the frontal boundary between the cold polar and warm semi-tropical air-masses, and the other called the *subtropical jetstream* because it originates in the tropics and flows northeastward into the middle latitudes. Jetstreaks, cores of exceptionally high winds within the jetstream flow, are typically present within individual jetstreams, and within branches of a split jetstream. The force imbalances associated with each jetstreak create divergence and convergence patterns as illustrated in Figure 8.6B. In situations where branches of a jetstream merge, such as the flow pattern illustrated in Figure 8A, two jetstreaks can align in a manner that the divergent quadrants of each jetstreak overlap. For example, in the flow shown in Figure 8A, the divergent right entrance region of jetstreak J1 overlaps the divergent left exit region of jetstreak J2. When such alignment occurs, exceptionally strong divergence occurs aloft (in the vicinity of point D in the example), causing air aloft to rapidly evacuate the air column and the surface pressure to rapidly deepen (lower). The most extreme low pressures associated with cyclones in the middle latitudes usually occur when two (or even three) jetstreaks, each embedded in different branches in the jetstream, interact with one another as their parent jetstreams merge. An example of jetstreak interaction in the formation of a low-pressure system appears in Chapter 11.

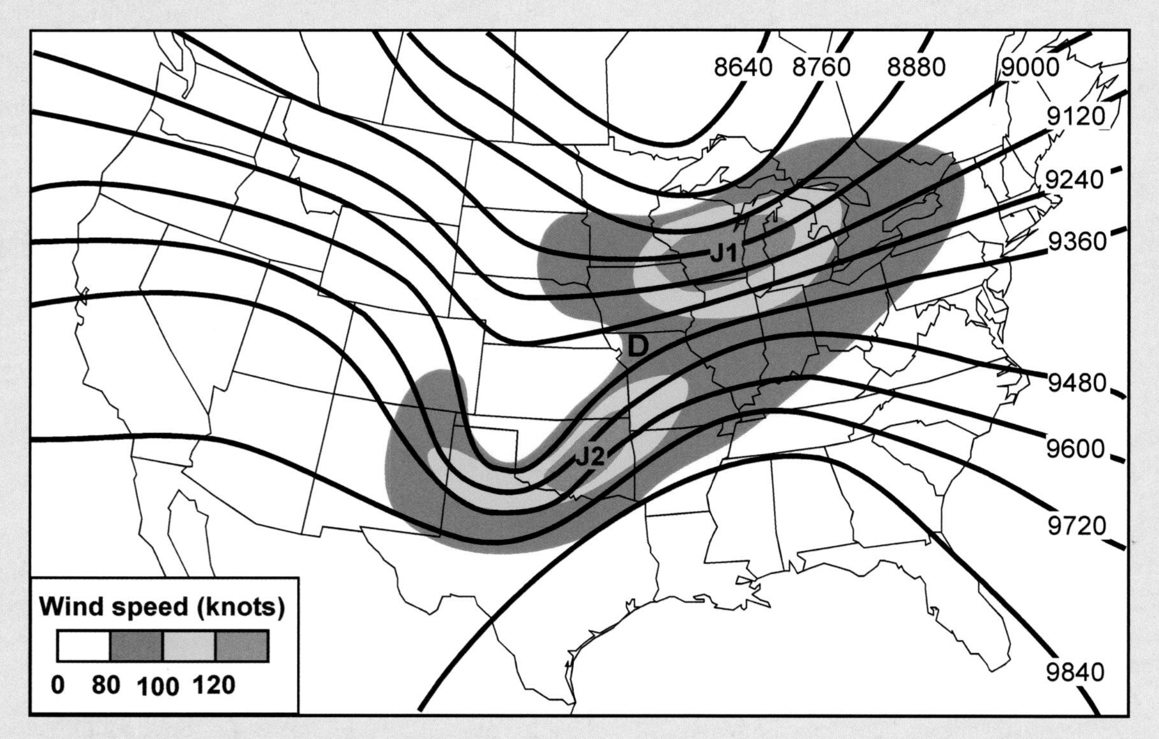

FIGURE 8A: An example of two jetstreaks aligned so that the divergent quadrants of each jetstreak superimpose. In this example, the divergent right entrance region of jetstreak J1 overlaps the left exit region of jetstreak J2. This would lead to very strong divergence centered on point D. A strong low-pressure center would develop at the surface under point D.

Combined Effect of Curvature and Jetstreaks

The jetstream normally contains jetstreaks that migrate through curved flow patterns. The resulting divergence and convergence patterns that develop depend on both the dynamics of curved flow and the dynamics of jetstreaks. A simple way to understand these patterns is to superimpose the divergence and convergence patterns associated with a jetstreak (Figure 8.6) with the divergence and convergence patterns associated with curved flow (Figure 8.5). Figure 8.7 shows a 300 mb chart with height contours. A ***trough*** is present east of the Rocky Mountain states, and a jetstreak, indicated by the strong height gradient, is located in the base of the trough. The trough axis and the jetstream axis are noted in the figure. The red letters denote convergence (C) and divergence (D) due to changes in flow curvature, while the corresponding green letters denote convergence and divergence due to the effect of the jetstreak. The superimposition of the divergence and convergence patterns are such that south of the jetstream axis, the divergence and convergence effects counteract each other while north of the jetstream axis, the effects reinforce one another. The strongest divergence aloft occurs on the northeast side of the trough, north of the jetstream axis in the left exit region of the jetstreak, while the strongest convergence occurs on the northwest side of the trough, in the left entrance region. Strong surface cyclones develop below the region of strongest divergence (see Figure 8.8). Although high pressure will develop at the surface below the region of strongest convergence, other processes more directly influence the position of high-pressure centers. These processes

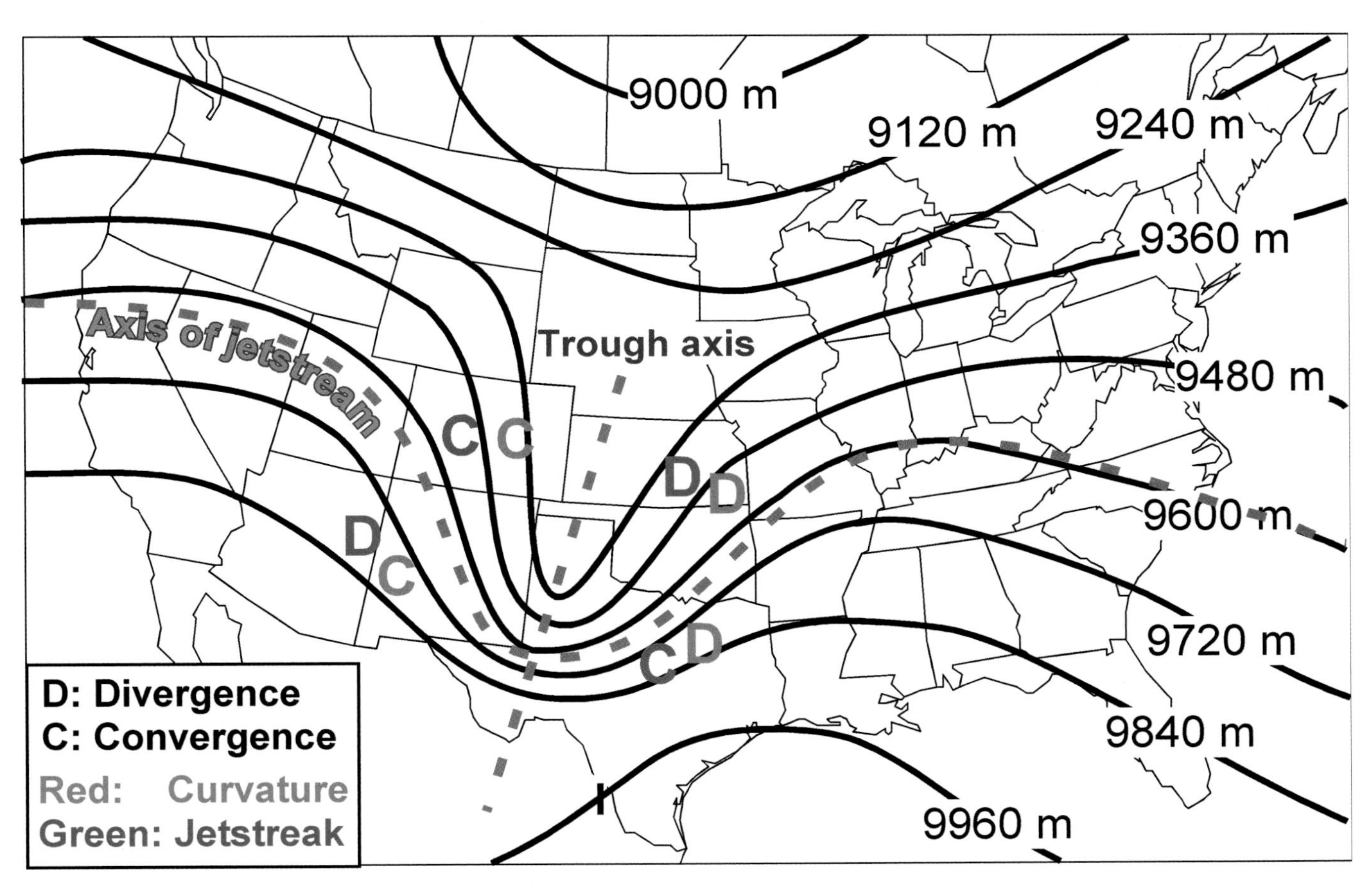

FIGURE 8.7: Illustration of the combined effects of a jetstreak and flow curvature. The jetstreak is located in the base of the trough. The strongest divergence occurs north of the jetstream axis east of the trough where divergence due to the jetstreak effect (D, green) and curvature (D, red) superimpose. A strong surface low-pressure center will develop under the region of strong divergence. The strongest convergence occurs north of the jetstream axis on the west side of the trough, where convergence due to the jetstreak (C, green), and curvature (C, red) superimpose. South of the jetstream axis the divergence/convergence patterns associated with the jetstreak and curvature oppose one another. In this example, a surface low-pressure center would develop over the eastern Kansas-Oklahoma border.

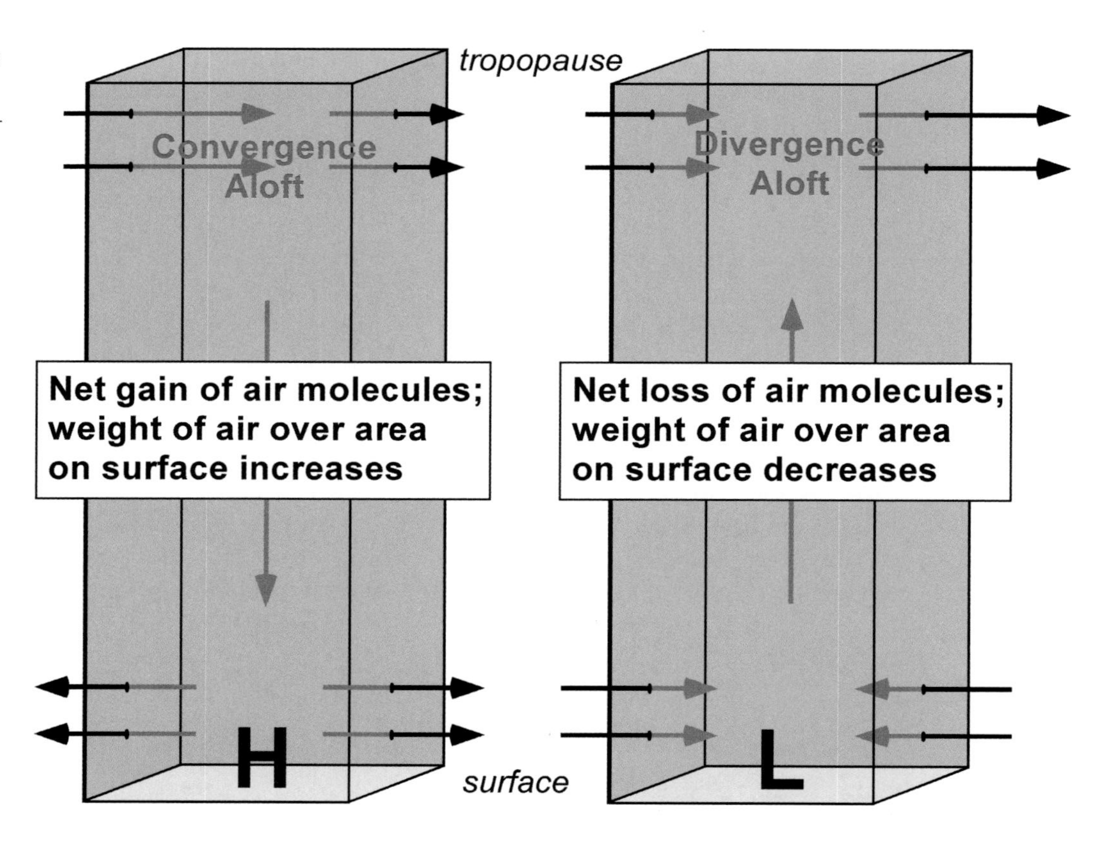

FIGURE 8.8: Convergence aloft increases the mass of air in a column, increasing the column weight and creating high-surface pressure. Divergence aloft creates low-surface pressure by decreasing the weight of air in a column. Black portions of arrows are outside of columns, while gray portions are inside.

will be discussed in the section on "effects of heating and cooling" below.

CHECK YOUR UNDERSTANDING 8.1

1. How does a change in the curvature of airflow affect the speed of the flow as air flows from a ridge to a trough? A trough to a ridge?
2. What is a jetstreak?
3. Where will a low-pressure center form at the surface relative to an upper air trough and ridge? (Assume a jetstreak is in the base of the trough.)

THE FRICTION LAYER

In the discussion above, we considered force imbalances above the friction layer. As we move now to the friction layer, we must consider the effects of friction. Let's consider what happens to the balance of forces in the friction layer by introducing friction and examining its effect on the direction of the wind. Panel A of Figure 8.9 shows air in geostrophic balance. Panel B shows how the force of friction disrupts this balance. Friction always reduces the wind speed, and therefore reduces the Coriolis force, which is proportional to wind speed. Since the pressure gradient force is unchanged, it becomes the dominant force.

Hence friction leads to a deflection to the left of the geostrophic wind direction, toward the low-pressure side of the isobars. The Coriolis force always remains to the right of the wind, and the frictional force is always opposite the wind. When a new balance is established between the three forces, the wind crosses the isobars toward lower pressure. Summarizing, *friction always acts to turn the wind such that the flow has a component from higher pressure toward lower pressure.*

The left side of Figure 8.10A shows the flow pattern associated with a surface low-pressure system. With a low-pressure center, air will turn inward, spiraling into the center of low pressure. The stronger the force of friction, the greater the turning will be. Over very rough surfaces, the effect of friction can be so strong that the surface winds nearly flow down the pressure gradient, perpendicular to the isobars. In most cases over land, the turning is of the order of 20 to 40°. Over water, the turning of the wind is typically 10 to 20°. The right side of Figure 8.10A shows that with

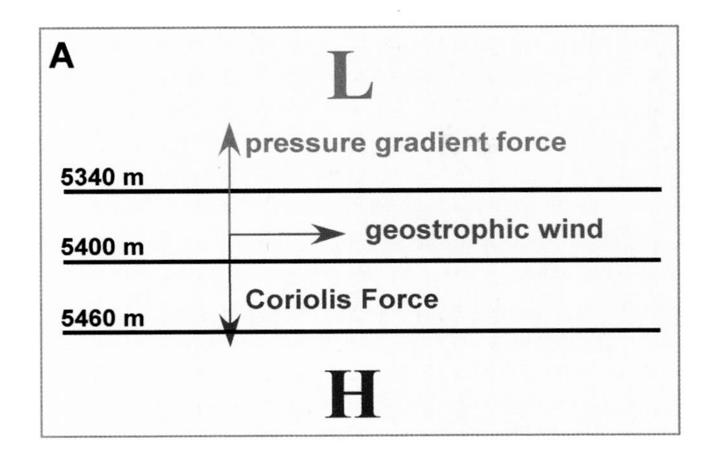

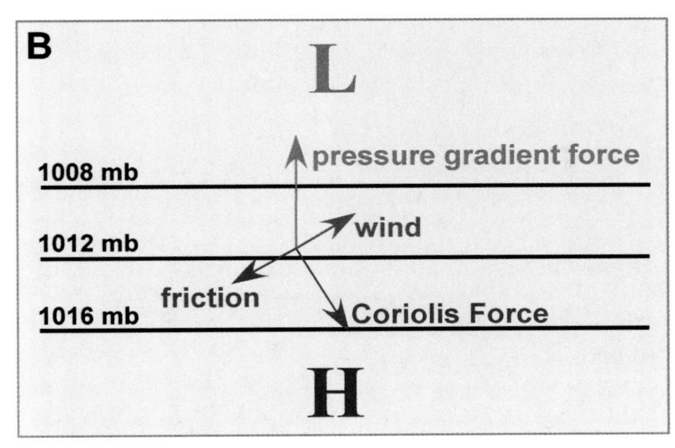

FIGURE 8.9: (A) The orientation of the pressure gradient force and the Coriolis force when air is in geostrophic balance in the northern hemisphere. (B) When the frictional force acts to slow the air velocity, the balance of forces is altered so that air flows across the isobars from high pressure to low pressure.

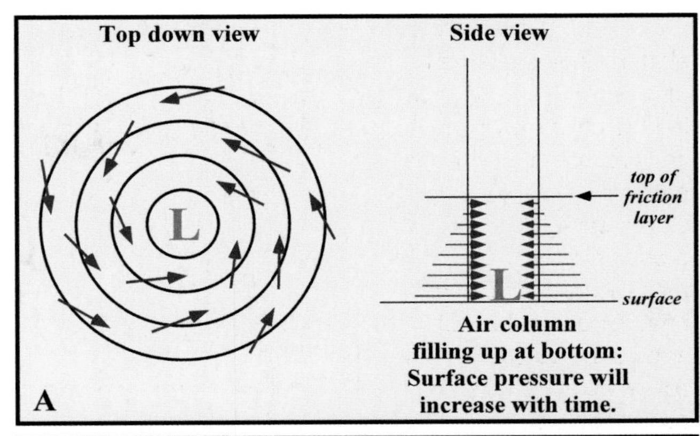

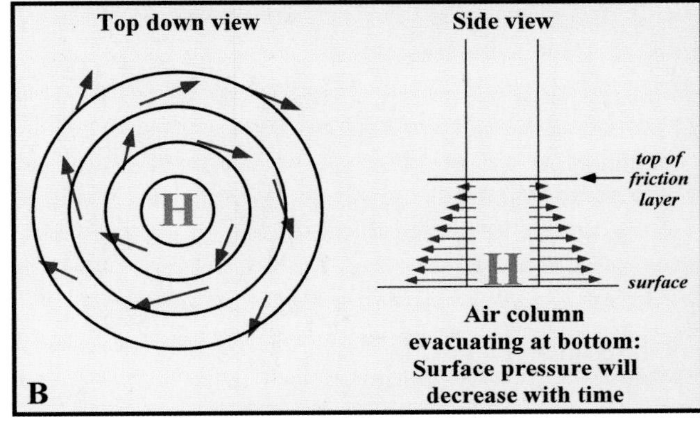

FIGURE 8.10: A) The left side shows a surface low-pressure center and wind flow. The right side shows a vertical column of air with the wind converging into the low at different altitudes. B) Same as A, but for a surface high-pressure center. Friction slows wind speed, causing air to spiral into lows (A) and out of highs (B). The effect is to "fill" the low and "evacuate" the high. As a result, the central pressure of a low will rise, eventually leading to the dissipation of the low. Similarly, the central pressure of a high will lower, eventually leading to the dissipation of the high. As shown by the air columns in each panel, the effect of friction is strongest near the surface and diminishes with altitude to the top of the friction layer.

altitude, the effect of friction decreases until, at the top of the friction layer, the winds approach their geostrophic value and air no longer converges into the low. The top of the friction layer is generally 1 to 3 km above the surface, depending on surface roughness (water vs. land, forest vs. plains etc.), and the amount of surface heating due to solar radiation.

The same arguments hold for high-pressure centers (Figure 8.10B). In the case of a high, friction slows the wind, again reducing the strength of the Coriolis force. Again, the pressure gradient force dominates but now the imbalance causes the air to spiral outward from the high. Friction is strongest near the surface, so the divergence out of the high is strongest near the ground. Higher in altitude, the effect of friction diminishes until, at the top of the friction layer, air no longer diverges from the high-pressure center.

Figure 8.11 shows a wind field associated with a low-pressure system over eastern Minnesota and a high-pressure system over Utah. This example clearly shows air converging into the low-pressure center and diverging out of the high—a direct result of surface friction. Recall that convergence into an air column adds mass to the column, *increasing the surface pressure.* Divergence out of an air column subtracts mass from the column, *decreasing the surface pressure.* As a result, *friction always contributes to weakening of both surface high-pressure centers and low-pressure centers.* If no divergence or convergence would occur aloft, surface low- and high-pressure systems would quickly dissipate due to the effect of friction in the friction layer.

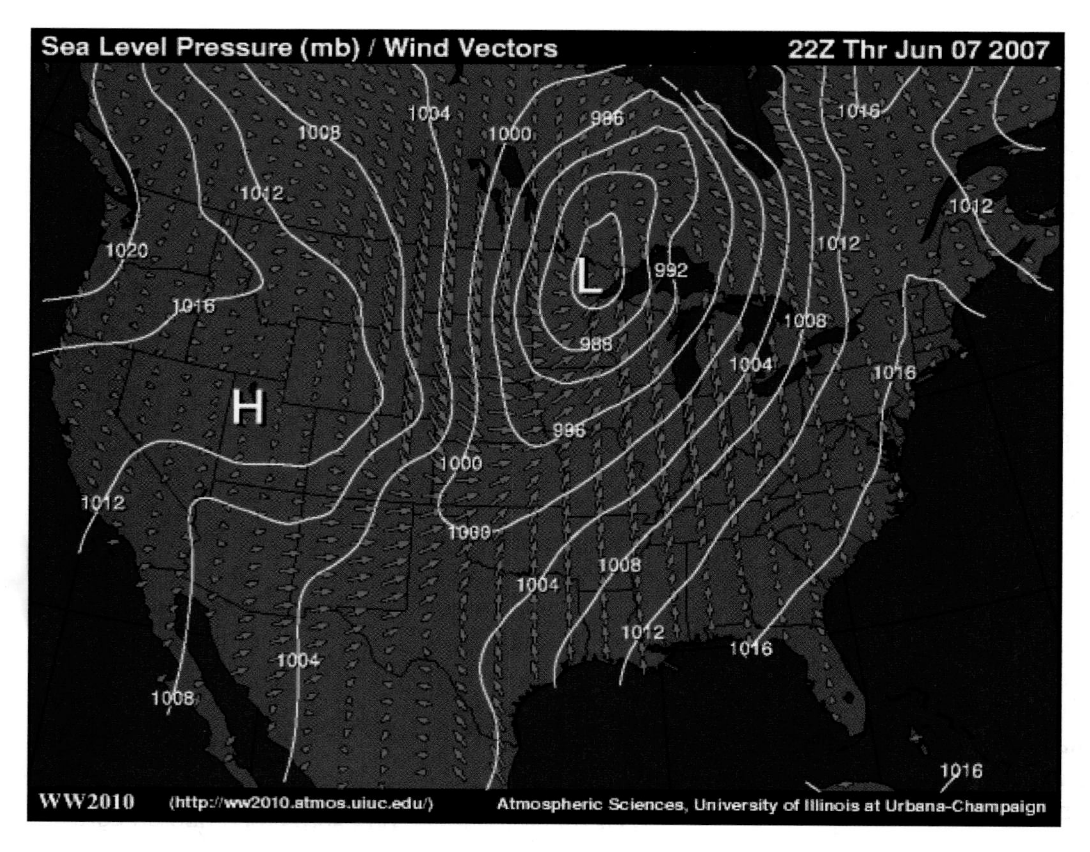

FIGURE 8.11: Surface winds (green arrows) and pressure pattern during the development of a strong low-pressure system over eastern Minnesota on 7 June 2007. Note that winds flow across the isobars out of the high-pressure area in Utah and into the low-pressure system over Minnesota in a manner similar to the schematic in Figure 8.9.

ONLINE 8.1

Effects of Friction on the Flow in High- and Low-Pressure Systems

The force of friction within the *friction layer* causes air to spiral into low-pressure systems and out of high-pressure systems. The effect of friction decreases with height and changes depending on the surface over which the air is moving. Online you can see how changing the force of friction affects both the strength of the wind and the direction the wind flows relative to high- and low-pressure centers. On the animation, the force of friction is progressively increased from zero until it equals the magnitude of the pressure gradient force. Arrows at the bottom show the relative strength of the pressure gradient force, Coriolis force and friction, and the direction these forces act to maintain balance for a point located directly between the high-and low-pressure centers (the red arrow). Compare the flows on this animation with winds in a real surface low-pressure system. The angle that the winds cross the isobars in the real weather system and in the animation can be used to estimate the strength of the force of friction compared to the pressure gradient force.

EFFECTS OF HEATING AND COOLING

There is a third mechanism by which surface low- and high-pressure systems can be created. Heating and cooling of the atmosphere create circulations that redistribute air and can lead to the formation of high- and low-pressure centers. Processes that involve the transfer of heat energy are called ***diabatic processes***.[2] Diabatic processes are extremely important to the formation of hurricanes and cold waves and significantly influence the intensity of extratropical cyclones and high-pressure systems.

[2] "Diabatic" can be loosely regarded as the opposite of "adiabatic." Recall from Chapter 6 that an air parcel rising adiabatically does *not* exchange heat with its surrounding environment.

The sun continually heats our atmosphere, depositing most of its energy in the tropics and less in polar areas. The sun's energy, for the most part, does not heat the atmosphere directly, since the atmosphere is nearly transparent. Most solar energy is absorbed by the Earth's surface. This energy is ultimately distributed within the atmosphere through direct transfer of heat from the Earth's surface (conduction), buoyant motion of air parcels (convection), radiation, and through latent heat release associated with condensation (see Figure 5.1).

We can understand how heating affects surface pressure by considering Figure 8.12A, which shows an atmosphere for which there are no horizontal pressure gradients. Let us heat the atmosphere in the central part of the figure, but not on the edges, and, for a moment, constrain the atmosphere so that no horizontal motions occur (Figure 8.12B). The heated air column will expand, causing individual pressure surfaces to "bow" upward, creating an outward-directed pressure gradient force above the surface (note that because we have artificially constrained our atmosphere to have no horizontal motions, the surface pressure will remain unchanged, since the total mass of air in any vertical air column is still the same). Now let's free the atmosphere to move horizontally. The outward-directed pressure gradient force will cause air to diverge away from the middle of the air column, reducing the surface pressure. The redistribution of mass will lead to a new pressure distribution, as shown in Figure 8.12C.

For illustration, we artificially divided the process into two steps. In nature, the adjustment to heating occurs continually. As long as heating continues, the outward circulation in the upper troposphere continually carries mass out of the heated air column, intensifying the surface low, while the inward-directed circulation created by the low pressure in the lower atmosphere carries mass into the air column, working to dissipate the low. The intensity of the two branches of the circulation depends on the intensity of the heating.

Cooling of the atmosphere produces the opposite effect. In Figure 8.12D, the air column in the center of the figure has now been cooled, causing the air column to contract, and forcing individual pressure surfaces to dip downward. If we again constrain the horizontal motions, we see that there is an inward directed pressure gradient force everywhere above the surface. When we let the atmosphere adjust (Figure 8.12E), the increase in mass in the center caused by the acceleration of air toward lower pressure aloft will lead to the development of a high-pressure center at the surface. In low levels, a compensating outward circulation will develop in response to the horizontal pressure gradient. As with heating, the intensity of the two branches of the circulation will depend on the intensity of the cooling.

FOCUS 8.2

The World's Semi-Permanent High-Pressure Centers

The most significant atmospheric cooling occurs over the polar landmasses of Asia, North America, and Antarctica in winter, and over the relatively cool North and South Atlantic and Pacific Oceans in summer. In winter, the polar landmasses receive very little solar radiation and cool rapidly as they radiate infrared energy to space, especially when the sky is cloud-free. Heat from the atmosphere radiates to the land surfaces and space, continually cooling the atmosphere. Very widespread and strong high-pressure systems can develop as cooling occurs (see Figure 5.4). The highest surface pressures recorded on Earth occur in Siberia in winter. The same process of high-pressure development occurs over cool oceans in summer. Together, these regions are the source areas for the world's major high-pressure centers. High-pressure centers over these regions are often called *semi-permanent.* Although the high-pressure systems move in response to storms, they quickly reform as radiative processes continually cool air and cause high pressure to redevelop.

THE DEVELOPMENT OF HIGH- AND LOW-PRESSURE CENTERS

In this chapter, we have considered how ***dynamic processes*** (the jetstreak and curvature effects associated with force imbalances) and ***thermodynamic processes*** (heating and cooling) lead to the redistribution of mass in the atmosphere and create the world's low- and high-pressure systems and its jetstreams. In nature, dynamic and thermodynamic processes work simultaneously. Let us now consider the effect of both types of processes.

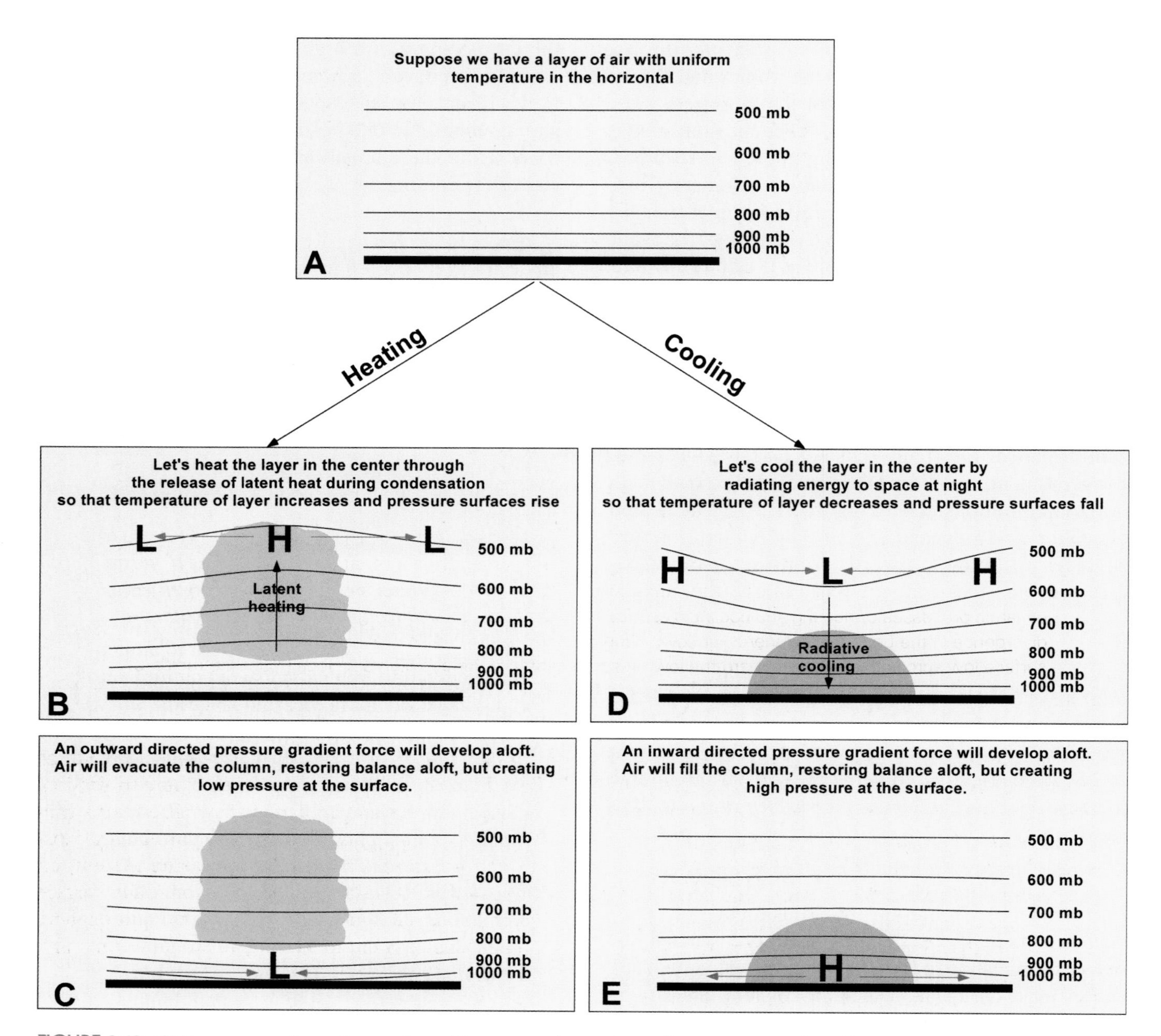

FIGURE 8.12: (A) Pressure surfaces in a region of the atmosphere where the temperature is uniform everywhere. In (B), the atmosphere in the center of the diagram is heated. The region between the individual pressure surfaces expands, leading to an outward pressure gradient aloft. As air accelerates out of the heated column to balance the pressure aloft, the weight of the column, and thus the surface pressure, decreases (C). In (D), the atmosphere in the center of the diagram is cooled. The region between the individual pressure surfaces contracts, leading to an inward pressure gradient aloft. As air accelerates into the cooled column to balance the pressure aloft, the weight of the column, and thus the surface pressure, increases (E).

When divergence occurs within a column of air, the surface pressure will decrease with time, and a low-pressure center will develop (Figure 8.13). *Divergence is associated with changes in the curvature in the flow, jetstreaks, and heating of the atmospheric column.* The effects of curvature and jetstreaks favor upper air divergence on the east side of a trough to the north of the jetstream axis. In storms, heat is provided through latent heat released during cloud formation. Once a low-pressure system forms at the surface, air will circulate around the low, but the force of friction will cause the air to turn toward the low, converging into the low-pressure center. This friction-induced convergence will work to destroy the low, since it

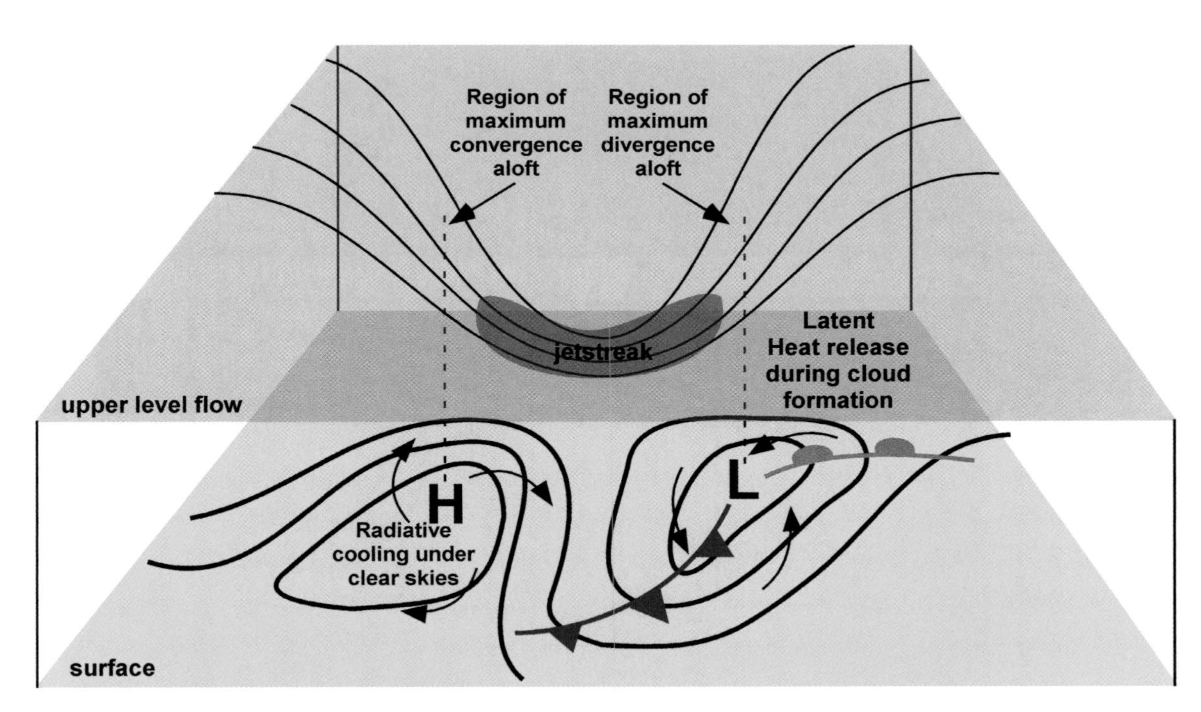

FIGURE 8.13: The combined effects of all processes related to the dynamics of the jetstream and the thermodynamic processes of heating and cooling. A surface low develops under the region of maximum upper-level divergence in the upper atmosphere. Air flows counterclockwise (in the Northern Hemisphere) around the surface low across the isobars toward the low-pressure center. A surface high develops under the region of maximum convergence in the upper atmosphere.

"fills" the atmospheric column. The low will intensify (pressure will drop) if the divergence aloft exceeds the convergence near the surface. The low will weaken (pressure will rise) if the opposite occurs. With convergence into the low, and divergence aloft, there must be rising motion (Figure 8.14). Said simply, air flows in and spirals up and out in the vicinity of low pressure. Rising motion is associated with clouds and precipitation. Low-pressure systems are therefore typically associated with stormy weather.

Surface high-pressure centers are also created by dynamic and thermodynamic processes. Air converges into an air column aloft to form a high-pressure system. In the case of highs, convergence associated with the cooling of air is generally more important than convergence associated with curvature and jetstreak processes. Cooling occurs over broad areas, such as the cold North Atlantic and Pacific Oceans in summer, or the Canadian and Asian Arctic in winter. These areas are where strong high-pressure systems develop (see Figure 5.4). High-pressure systems are enhanced by convergence associated with the jetstream as these cool airmasses move out of their source regions (Figure 8.13).

Once a high forms, air flows clockwise around the high, and outward away from the high due to friction. The divergence near the surface associated with friction acts to destroy the high-pressure system (by causing pressure to lower). In a high-pressure system, air flows in aloft, downward, and outward at the surface (Figure 8.14). Downward motion in the atmosphere produces clear skies. Highs, therefore, are generally associated with fair weather.

ONLINE 8.2

Jetstreaks, Flow Curvature, and Low-Pressure Systems

Low-pressure systems develop in the middle latitudes as a result of divergence created by force imbalances in the jetstream. These imbalances develop due to jetstreaks and changes in flow curvature. Online, a series of 300 mb maps show the evolution of the jetstream during the development and evolution of a surface low-pressure system. The evolution of the surface

(continued)

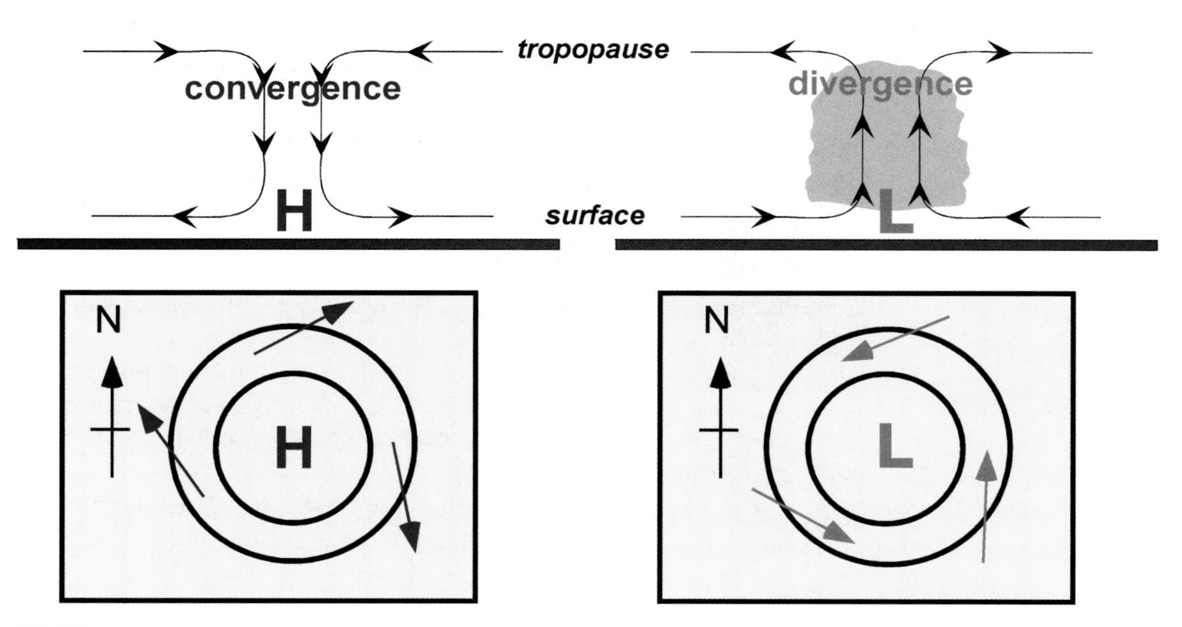

FIGURE 8.14: Upper-level divergence leads to the creation of a surface low-pressure system. Clouds develop as air converges into the low near the surface, rises, and diverges aloft, as indicated by the arrows on the figure. Convergence aloft leads to the creation of a surface high-pressure system. Clear skies develop as air converges aloft, sinks and diverges out of the high near the surface.

(continued)

low-pressure system appears in a companion animation. The low-pressure system created blizzard conditions in Iowa, Nebraska, and northern Illinois. Compare the position of the low-pressure system with the curvature and jetstreaks as they evolve. Does the position of the low-pressure center correspond to that expected from the discussion in the text? Does the timing of the low pressure's intensification correspond to the timing of the jetstreak's progression through the trough? What are some of the complexities you see in the real atmosphere (for example, multiple jetstreaks) compared to the idealized patterns discussed in the text?

CHECK YOUR UNDERSTANDING 8.2

1. How does surface friction modify the pressure at the center of low- and high-pressure systems?
2. If an air column is heated through latent heat release, how will the surface pressure change?
3. If an air column is cooled through radiative cooling, how will the surface pressure change?

TEST YOUR UNDERSTANDING

1. Explain how a change in wind speed can create divergence or convergence.
2. Air flowing around an upper-level cyclone moves slower than the geostrophic wind. Explain why.
3. If the jetstream is flowing in a curved pattern, where would regions of convergence and divergence be found?
4. If a jetstreak is present in the upper-level flow, where would regions of convergence and divergence be found?
5. How do flow curvature and jetstreaks combine to produce divergence and convergence at jetstream level?
6. What is the relationship between a jetstreak and the jetstream?
7. Where is the entrance region of a jetstreak? The exit region?
8. Which force, the pressure gradient force or the Coriolis force, is stronger in the entrance region of a jetstreak? In the exit region?
9. Assume a ridge/trough/ridge configuration is present in the jetstream and a jetstreak is located in the base of the trough. Where in this flow pattern will the maximum divergence aloft be found? Convergence?
10. How does friction disrupt geostrophic balance near the surface?
11. How does surface roughness influence air flow in the friction layer?
12. If surface isobars are oriented north-south with low pressure to the west, approximately what direction will the surface wind be blowing?
13. Where and in what seasons do semi-permanent high-pressure centers form at the surface?
14. Identify one dynamic and one thermodynamic mechanism responsible for the intensification of surface high-pressure systems.
15. Identify one dynamic and one thermodynamic mechanism responsible for the intensification of surface low-pressure systems.
16. Low-pressure centers over the ocean tend to be stronger than low-pressure centers over continents. Why might this be so?
17. Does air rise or descend in low-pressure systems? In high-pressure systems? Why?
18. Consider a situation in which upper-level divergence exceeds low-level convergence. Is the surface system a high or a low, and is the system becoming stronger or weaker?
19. Consider a situation in which upper-level convergence exceeds low-level divergence. Is the surface system a high or a low, and is the system becoming stronger or weaker?
20. Why is cloudiness typically associated with surface low-pressure systems?
21. Why are clear skies typically associated with high-pressure systems?

TEST YOUR PROBLEM-SOLVING SKILLS

1. Suppose that on a particular day, the pressure distribution in the lower atmosphere looked like the figure below.

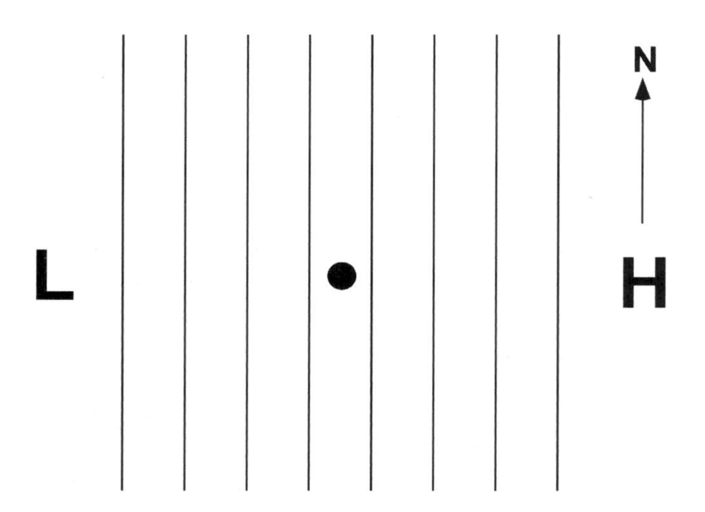

A rawinsonde is launched at the dot. Assume that the friction layer is 2 kilometers deep, the surface wind speed is 10 knots, and that friction has reduced the wind speed by 50 percent from its geostrophic value. Describe qualitatively how the wind speed and direction measured by the rawinsonde would change as the balloon ascended from the surface to the top of the friction layer.

2. Use the 300 mb map below to answer the following questions. For each question below, identify whether *divergence, convergence,* or *neither convergence nor divergence* will occur due to the curvature or jetstreak effects.
   (a) The curvature effect will result in ______ at point A.
   (b) The curvature effect will result in ______ at point E.
   (c) The curvature effect will result in ______ at point F.

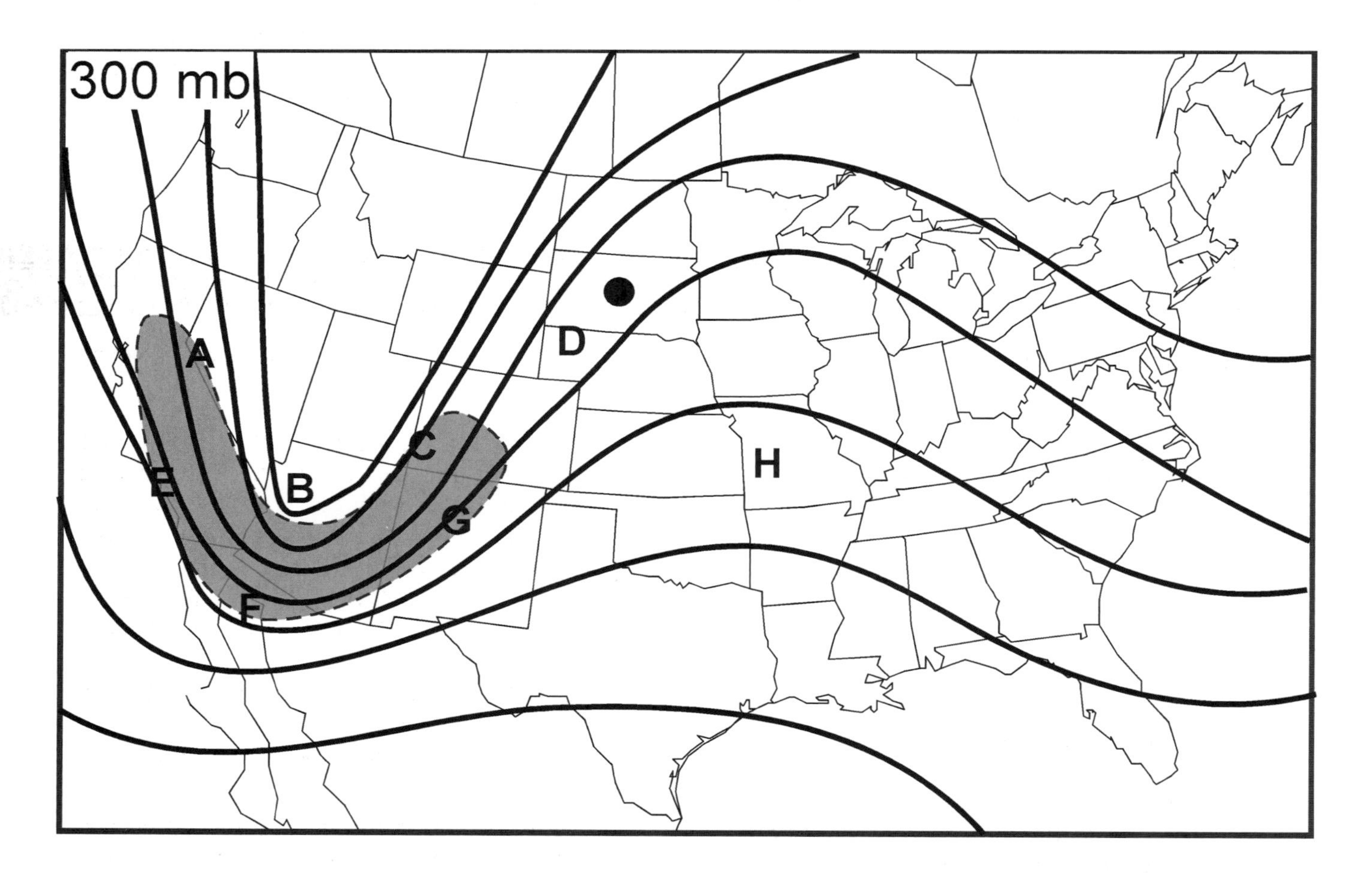

(d) The curvature effect will result in ______ at point C.
(e) The jetstreak effect will result in ______ at point A.
(f) The jetstreak effect will result in ______ at point B.
(g) The jetstreak effect will result in ______ at point C.
(h) The jetstreak effect will result in ______ at point G.
(i) At which location would you expect to find the maximum upper-level divergence?
(j) At which location would you expect to find the maximum upper-level convergence?
(k) At which location would you expect a surface low to develop?
(l) At which location would you expect a surface high to develop?

3. Figure 8.7 shows the height contours on the 300 mb surface at an instant in time, for example at 0000 UTC. Imagine that you're riding in a balloon that exactly follows an air parcel as it moves west to east across the continent. Your balloon starts its journey on the 300 mb surface over San Francisco, California, with the flow pattern shown in Figure 8.7. Assuming that this flow pattern and the height gradients do not change over time or over the vertical range of your journey, qualitatively describe your position *in three dimensions* as the balloon traverses the country. (To do this problem, you can copy Figure 8.7 and draw the trajectory of the balloon. On a separate sheet of paper, graph its position in the vertical as it traverses the country.)

4. Three common locations for surface low-pressure systems to develop are eastern Colorado, along the Gulf Coast and along the East Coast near Cape Hatteras.
   (a) Make three copies of the blank U.S. map found in Appendix B. On each map, sketch in a 300 mb height pattern that has a trough in a position that would lead to the development of a surface low at one of the three locations listed above.
   (b) Jetstreaks also contribute to the development of a surface low. On each map, sketch in isotachs showing a jetstreak. Place the jetstreak so that it would be in a position to contribute to the development of each low-pressure center.
   (c) Often the storms that develop along the Gulf Coast and East Coast are more intense than the storms that develop in Colorado. Why might this be?

USE THE SEVERE AND HAZARDOUS WEATHER WEBSITE

http://severewx.atmos.uiuc.edu

1. Navigate the *Severe and Hazardous Weather Website* to a current 300 mb chart for North America and adjacent oceans.
   (a) Save or print the map. Record the date and time of the map.
   (b) Identify all troughs, ridges, and jetstreaks on this map.
   (c) Based on your understanding of the relationship between jetstreaks, changes in flow curvature and surface low-pressure systems, estimate the locations where surface low-pressure systems should exist.
   (d) Navigate the *Severe and Hazardous Weather Website* to a surface map for the same date and time. Were your predictions of the location of surface low-pressure systems correct? Why or why not?

2. Use the *Severe and Hazardous Weather Website* to locate a map of sea-level pressure with wind data (wind barbs or wind arrows). Examine the angle between the isobars and the wind data. Based on your analysis, identify regions where surface roughness is high and where it is low. Are your results what you would expect? Why or why not?

3. Monitor the weather across the United States. On a day when a strong low-pressure center and strong high-pressure center are both present, complete the following:
   (a) Navigate the *Severe and Hazardous Weather Website* to find a plot of sea-level pressure,

radar imagery, and visible and infrared satellite imagery for the same date and time. Save or print the images. Record the date and time.

(b) Contrast the intensity of the pressure gradient in the region surrounding the low-pressure and high-pressure centers. Which is associated with stronger pressure gradients?

(c) Describe the pattern of cloud and precipitation associated with the high-pressure center and the low-pressure center.

4. Some of the strongest low-pressure systems in the middle latitudes develop when the divergent regions of two jetstreaks overlap.

   (a) Copy the blank U.S. map found in Appendix B.

   (b) Sketch a plausible contour pattern at the 300 mb level that has two jetstreaks positioned such that there are overlapping regions of divergence.

   (c) Navigate the *Severe and Hazardous Weather Website* to a map of the jetstream level (try 300 mb or 250 mb) and see if you can identify a similar pattern either in the current flow or in forecasts. If you find an example that matches, compare and contrast it with what you have sketched. Are there indications of a developing surface low beneath the overlapping regions of divergence aloft?

CHAPTER NINE

AIRMASSES AND FRONTS

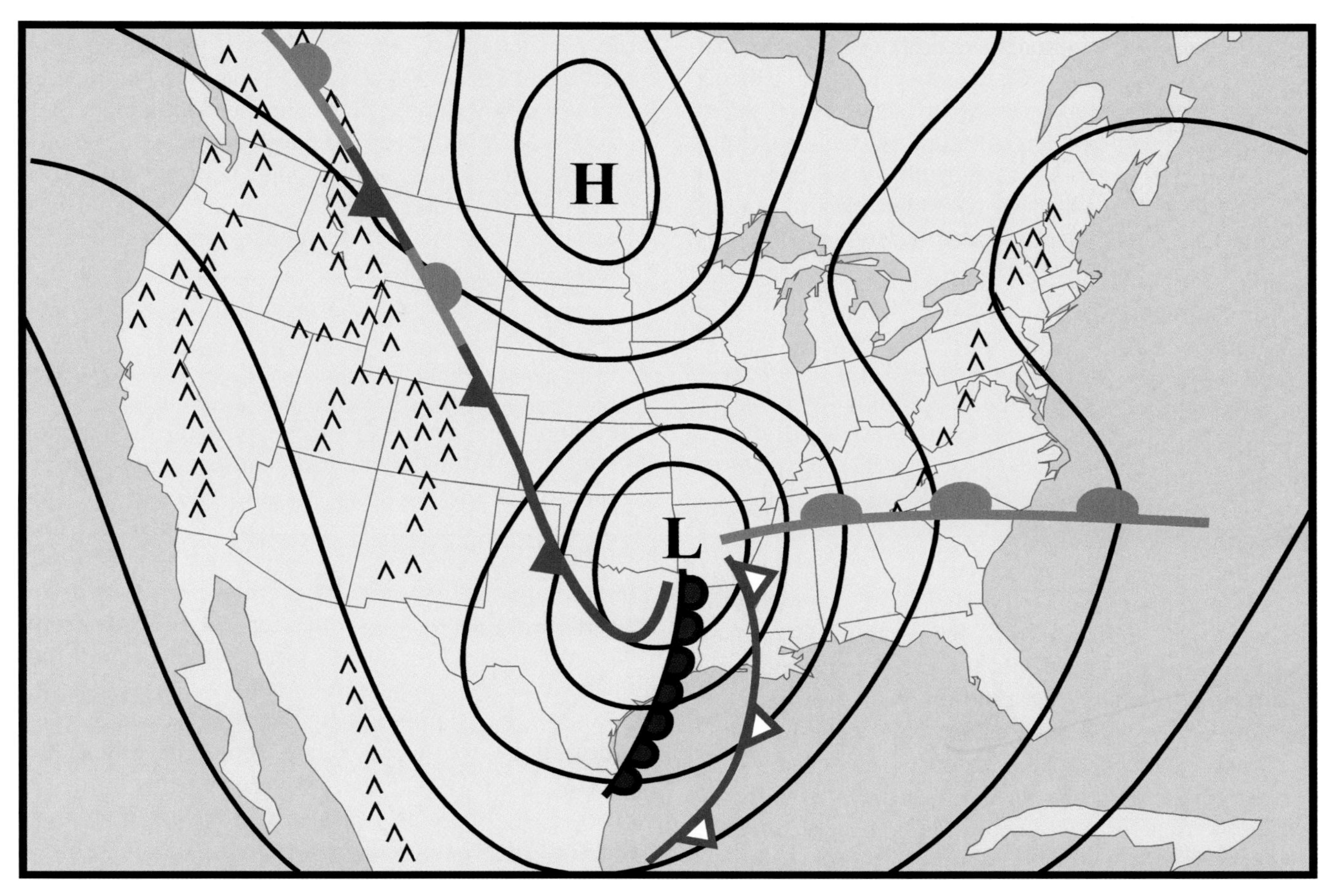

The sea-level pressure pattern and fronts as a cyclone moves across the Great Plains.

KEY TERMS

airmass
Aleutian low
arctic airmass
arctic front
Bermuda high
cold front
cold front aloft
cold occlusion
continental airmass
dry line
front
Icelandic low
maritime airmass
occluded front
Pacific high
polar airmass
stationary front
tropical airmass
upper-level front
warm front
warm occlusion

AIRMASSES

An ***airmass*** is a large body of air with relatively uniform thermal and moisture characteristics. Airmasses cover large regions of the Earth, typically several hundred thousand square kilometers, and can be as deep as the depth of the troposphere or as shallow as 1 to 2 km. Airmasses form when air remains over a relatively flat region of the Earth with homogeneous surface characteristics for an extended period of time. Areas with relatively uniform surfaces include continental arctic regions such as Canada and Siberia, cool oceanic regions such as the North Atlantic and Pacific, deserts such as the Sahara and the American southwest, and tropical oceanic regions including the equatorial Atlantic and Pacific, as well as smaller water bodies such as the Caribbean Sea and the Gulf of Mexico. The atmosphere "acquires" the thermal and moisture properties of the underlying surface as heat is transferred between the surface and the atmosphere and moisture evaporates into the air from the surface.

Meteorologists classify airmasses according to their thermal and moisture properties. Airmasses characterized by bitter cold temperatures are classified as ***arctic airmasses***, cold temperatures (or cool in summer) as ***polar airmasses***, and warm-hot temperatures as ***tropical airmasses***. Airmasses that form over oceans are called ***maritime airmasses*** and those that form over continents are ***continental airmasses***. Continental airmasses are normally drier than maritime airmasses since there is relatively little surface water evaporating into air over the continents.

Figure 9.1 identifies the typical airmass source regions of North America. Arctic airmasses develop over Canada and the frozen Arctic Ocean in wintertime, generally poleward of about 60° N. Extremely cold airmasses can form as air continually cools in the near-perpetual darkness of winter. Continental polar airmasses continually develop over the northern parts of the North American continent in all seasons of the year. Maritime polar airmasses develop over the North Atlantic and North Pacific Oceans. The oceans moderate wintertime temperatures within these airmasses. Maritime tropical airmasses originate over the tropical Atlantic and Pacific, the Gulf of Mexico, and the Caribbean Sea. Continental tropical airmasses develop over the desert regions of Mexico and the southwestern United States.

The centers of cold airmasses are associated with high pressure on surface weather maps. High pressure develops in response to cooling, as discussed in Chapter 8. In winter, high-pressure centers form and are the dominant feature over the northern parts of the continents of Asia and North America. In summer, when the oceans are cooler than the landmasses, large high-pressure centers are the dominant feature of the atmosphere over the North Atlantic and Pacific Oceans. The high-pressure center over the Atlantic is called the ***Bermuda high*** because it is centered near Bermuda, while its Pacific counterpart is called the ***Pacific high***.

The centers of very warm airmasses appear as semi-permanent regions of low surface pressure. These low-pressure areas appear over desert areas such as the American Southwest in summer, and in Southeast Asia, central Africa, and near the equator. In winter, semi-permanent low-pressure centers appear over the northern oceans. The semi-permanent low over the Pacific is called the ***Aleutian low***, while its Atlantic counterpart is called the ***Icelandic low***. Semi-permanent low-pressure centers differ substantially from migrating tropical and extratropical cyclones, although the latter often migrate into the areas of the semi-permanent lows.

Airmasses do not remain over their source regions indefinitely. As storms move about the globe, airmasses move out of their source regions and over other areas. Arctic airmasses can move as far south as the Gulf of Mexico during a cold air outbreak. Maritime tropical air can extend well into Canada in midsummer. The arrows in Figure 9.1 show the typical paths North American airmasses take as they move out of their source regions. Once out of their source regions, the temperature and moisture characteristics of airmasses change as they begin to acquire the thermal and moisture characteristics of the new surfaces over which they reside. For example, in summer warm humid air over the Gulf of Mexico typically moves northward over the central United States. Over land, this air may heat several degrees and acquire additional moisture from crops through a process called evapotranspiration. By the time the air moves far inland, it can be considerably more uncomfortable than along the coast.

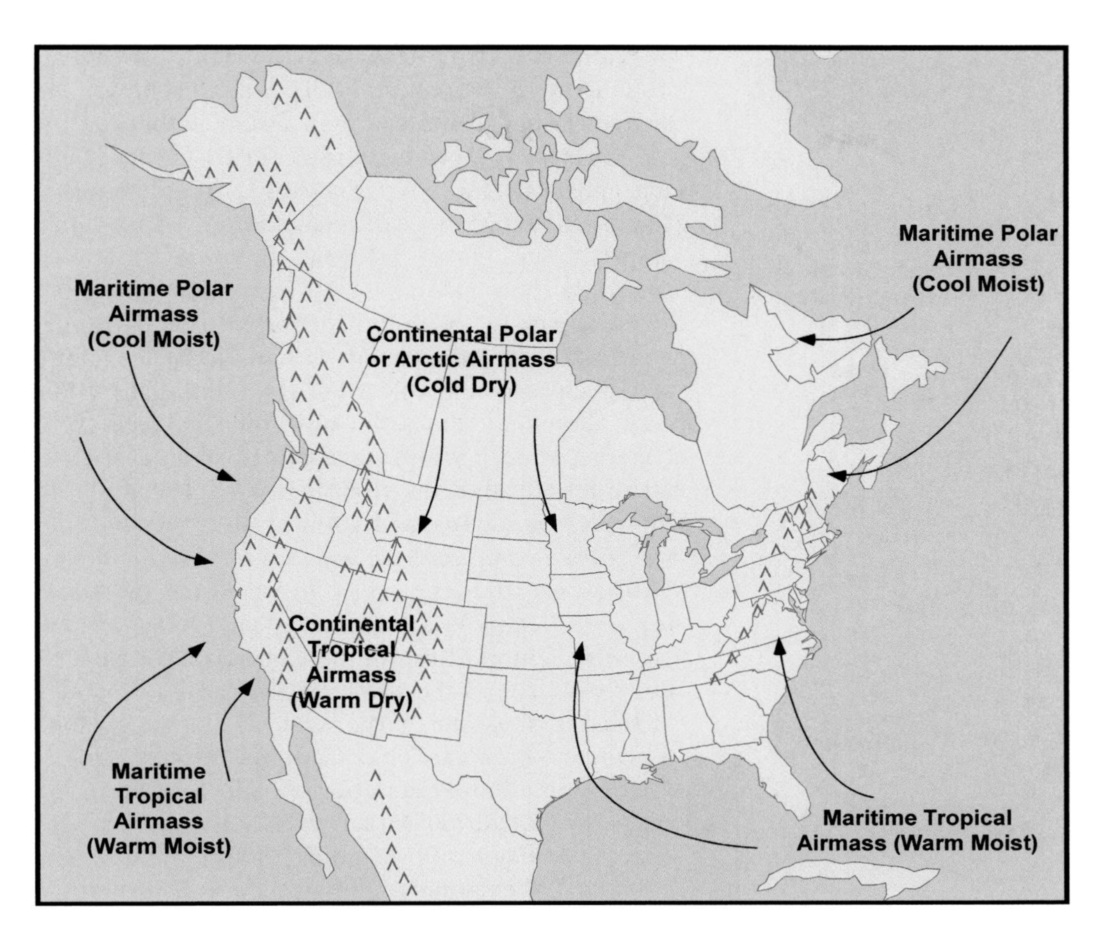

FIGURE 9.1: Airmass source regions for North America and the typical tracks airmasses take out of their source regions.

Not all airmasses are in contact with the ground. An example is the stratosphere, a large airmass that covers the entire globe. Airmasses exist aloft in the troposphere, residing on top of other airmasses in contact with the surface. Airmasses are three-dimensional, and the boundaries between airmasses are often quite sharp and distinct. These boundaries, called ***fronts***, are meteorologically important because much of the precipitation in the middle latitudes, and most severe weather, occurs in their vicinity.

CHECK YOUR UNDERSTANDING 9.1

1. How large and how deep is a typical airmass?
2. What characteristics make a region a good source for airmasses?
3. What is the relationship between airmasses and fronts?

FRONTS

Meteorologists classify fronts based on the thermal and moisture characteristics of the airmasses, the direction of movement of the airmasses, and whether the boundary between the airmasses is in contact with the ground (a surface front), or can only be found aloft (an upper level front). In most cases, two airmasses in contact with one another will have different thermal properties, one cold and the other warm.

Cold Fronts

We call the boundary between two airmasses a ***cold front*** if the cold air is advancing forward, lifting the warm air. The leading edge of the cold airmass typically has a shape like a dome, as shown in Figure 9.2A. Note that, because of the tilt of the airmass boundary in Figure 9.2, the front's horizontal

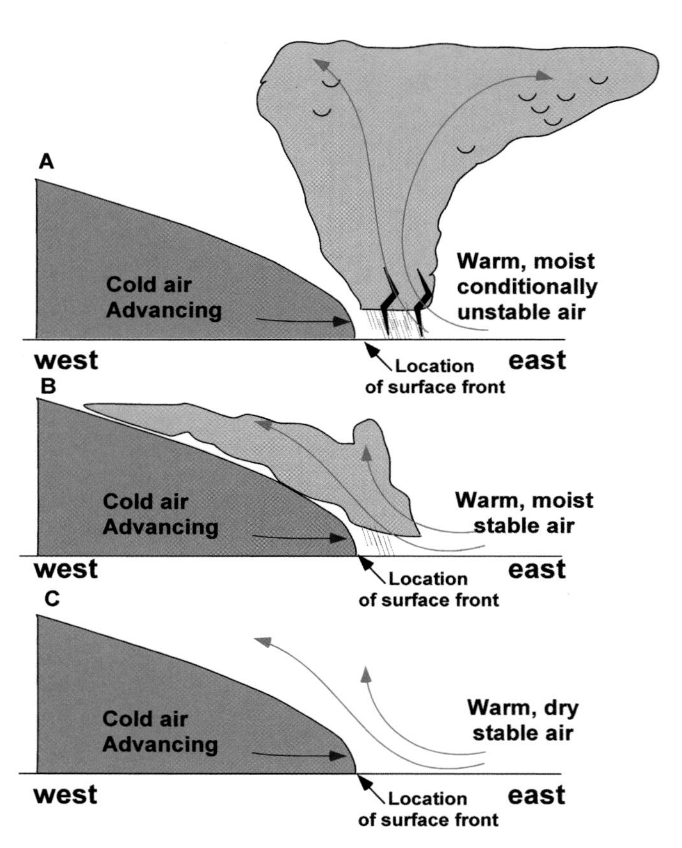

FIGURE 9.2: Cross section through three cold fronts. In A, the air ahead of the front is conditionally unstable and forms thunderstorms when lifted. In B, stable moist air is lifted along the frontal surface. In C, dry air is lifted, so clouds do not form. In all cases, the cold air is advancing into warmer air.

position varies with elevation. The type of precipitation that will occur along a cold front depends on the characteristics of the warm air ahead of the front. If the warm air is moist and conditionally unstable, lifting ahead of the front can trigger thunderstorms. These thunderstorms will often form in a line, called a squall line, along the front. In some situations, supercell thunderstorms may form along the front (see Chapter 18), with tornadoes, hail, and damaging winds. On the other hand, the atmosphere may be stable ahead of the cold front (Figure 9.2B). In this case, the clouds that form by lifting may only produce light rain, or no rain at all. In some cases, the warm air ahead of the front is dry (Figure 9.2C). In this case no clouds will form at all. *Regardless of the cloud formations, we say that the front is a cold front if the cold air is advancing, lifting the warm air ahead of it.* When extratropical cyclones (low-pressure centers) form in the Northern Hemisphere, air flows around the low-pressure system counterclockwise. Cold air on the west side of the cyclone advances southward and then southeastward. For this reason, cold fronts are typically located in the southwest quadrant of a cyclone early in its lifetime (see Figure 9.3). Cold fronts that mark the boundary of extremely cold airmasses in wintertime are called ***arctic fronts***. Arctic airmasses are normally much more shallow than airmasses associated with spring and fall season cold fronts.

FOCUS 9.1
Frontal Symbols

Meteorologists indicate the location and type of front on surface weather charts using a heavy line with either barbs or half circles along the line. Solid lines are used when a front is obvious, while broken lines are used to indicate that a front is either forming or dissipating. Although a frontal symbol exists for an upper-level front, it is rarely used. Recognition of the importance of upper-level fronts in severe weather is relatively recent in meteorology. The symbols for all of the fronts discussed in this chapter appear in Figure 9A.

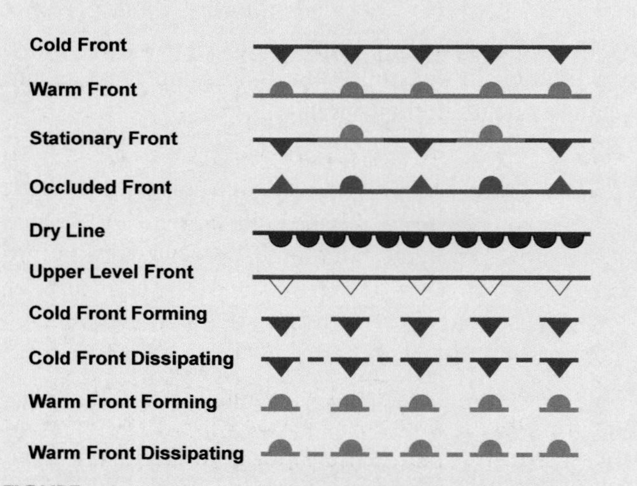

FIGURE 9A: Frontal symbols used on surface weather maps.

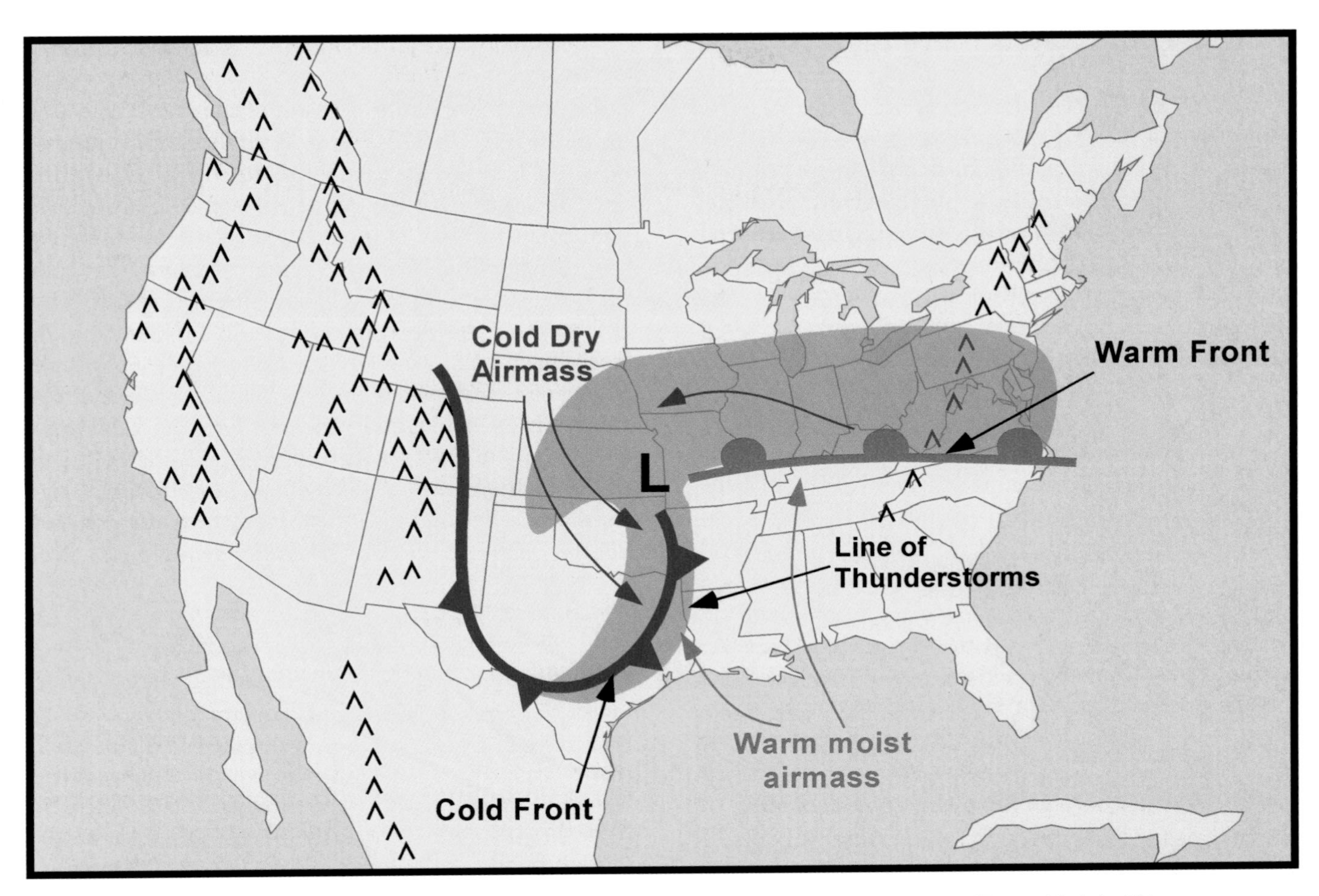

FIGURE 9.3: Example of a cold front moving southeastward across the central United States. The cold air is lifting warm, moist conditionally-unstable air, creating a line of thunderstorms along the front. This cold front would have structure similar to Figure 9.2A. The figure also shows a warm front. Note that the flow in the cold air north of the warm front has a component toward the north, so that the cold air is retreating northward with time. Gray shading denotes cloud cover.

Warm Fronts

A boundary where the cold air is retreating and the warm air is advancing is called a ***warm front***. Since warm air flows northward on the east side of a cyclone, warm fronts are normally found in this location (see Figure 9.3). The cold air, in this case, still has a dome-like structure, although the cold air dome typically slopes more gradually ahead of a warm front than behind a cold front (see Figure 9.4). The warm air flows toward the cold air, rising over the dome as it progresses northward. The type of precipitation that occurs along a warm front again depends on the moisture and thermal characteristics of the warm air. In eastern North America, warm air gliding upward along a warm front typically originates over the Gulf of Mexico or the Atlantic and is normally moist. As the warm air glides upward over the cold air dome, widespread layered clouds will develop. These clouds are deepest just north of the frontal boundary and progressively become thinner and higher toward the north. Precipitation is heaviest closer to the frontal boundary where the clouds are deep, and lighter to the north where the clouds are shallow. The intensity of the precipitation depends again on the stability of the warm air. If the warm air is conditionally unstable, thunderstorms may develop over the warm front (Figure 9.4A). These are often embedded within, or emerge from, the widespread layered clouds. If the warm air is stable, the clouds will be layered (Figure 9.4B). If the temperature in the cold air is below freezing, snow, ice pellets, or freezing rain may occur (see Chapter 12). *Regardless of the cloud formations or precipitation, we say that a front is a warm front if the cold air is retreating and the warm air is advancing.*

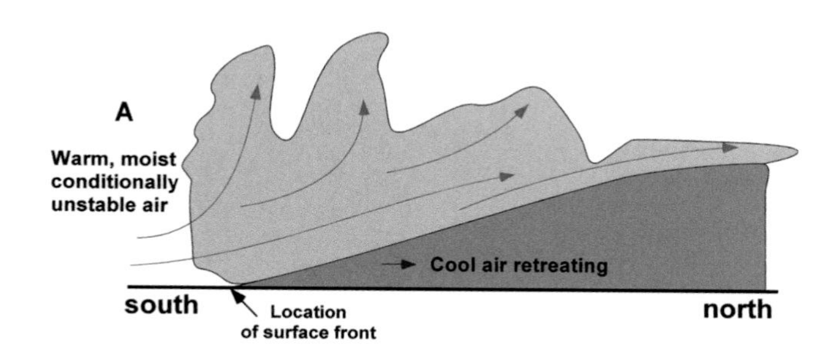

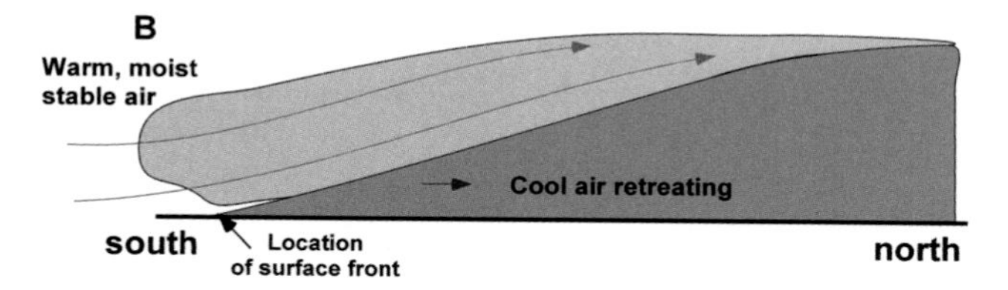

FIGURE 9.4: Cross section through two warm fronts. In A, the air south of the front is conditionally unstable and forms showers and thunderstorms as it rises over the cool air. In B, the moist air south of the front is stable. In this case, the clouds form a wide stratus layer that is deepest near the surface position of the front and thin farther north.

ONLINE 9.1

Finding Cold and Warm Fronts in Surface Data

Online, four-day animations of analyses of sea-level pressure overlaid with (1) temperature, (2) dewpoint temperature, and (3) wind speed and infrared satellite data are presented for the weather event depicted in Figure 9.10. An animation of radar data is also provided. During the event, an arctic front swept across the United States, a cyclone developed along the front, and blizzard conditions developed between Iowa and New England with almost fifteen inches (40 centimeters) of snow falling in Chicago. A second storm developed in Texas, spreading heavy freezing rain across Arkansas and Tennessee. Missouri through the Ohio Valley received 8–15 inches (20–40 centimeters) of snow! Examine these animations. Identify and track the position of the frontal boundaries. Determine what types of airmasses were on either side of the fronts you identify. Use Figure 9.10 as a guide to get started.

Stationary Fronts

Airmass boundaries are sometimes stationary. Although the boundary is stationary, air on both sides of the boundary can be moving. With a ***stationary front***, air on the cold side of the front will always be flowing parallel to the front (see Figures 9.5 and 9.6). If this is the case, the cold air is neither advancing nor retreating. The warm air normally flows toward the front. As the warm air encounters the cold air, it may be lifted along the boundary. If the air is conditionally unstable, a line of showers and thunderstorms may develop in the warm air over the front (Figures 9.5A and 9.6A), a situation that can lead to flash flooding (see Chapter 25) when the front's stationarity allows heavy rain to persist over a particular location. If the warm air is stable, widespread layered clouds may form over the front, with rain falling on the cold side of the front. *Regardless of the cloud formations or precipitation, we say that the front is a stationary front if the cold air is neither advancing nor retreating.*

Occluded Fronts

As cyclones develop and go through their life cycle, the cold air to the west of the cyclone advances rapidly southward around the center of low pressure, while the air to the north of the warm front, which is also cold, retreats northward slowly. Because the cold front typically moves rapidly, it will progress around the south side of the low-pressure center. The cold air will then progress northeastward, approaching the warm front. Eventually, the cold front can actually catch up to the warm front. When this happens, the cold air comes in direct contact with the cold air north of the warm front, creating a new airmass boundary. This boundary, between the cold air behind the cold front and the cold air north of the warm front, is called an ***occluded front***. The process described above is depicted conceptually in Figure 9.7. We see the cold front progress southward (Figure 9.7A) where it eventually encounters the warm front (Figure 9.7B). The boundary between the two cold air regions becomes longer as the cold air behind the cold front continues to advance (Figure 9.7C). As the fronts meet, warm air is forced aloft over both the cold airmasses.

The three dimensional structure of an occluded front depends on the temperatures within the cold air behind the cold front and north of the warm front. Figure 9.7D shows two sets of cross sections through the fronts appearing in Figure 9.7C. The set on the left represents a ***cold occlusion***. In this case, the air behind the cold front is colder than air north of the warm front. The lower cross section (u to v) is at a point where the cold front has not yet caught up to the warm front (see Figure 9.7C). The middle cross

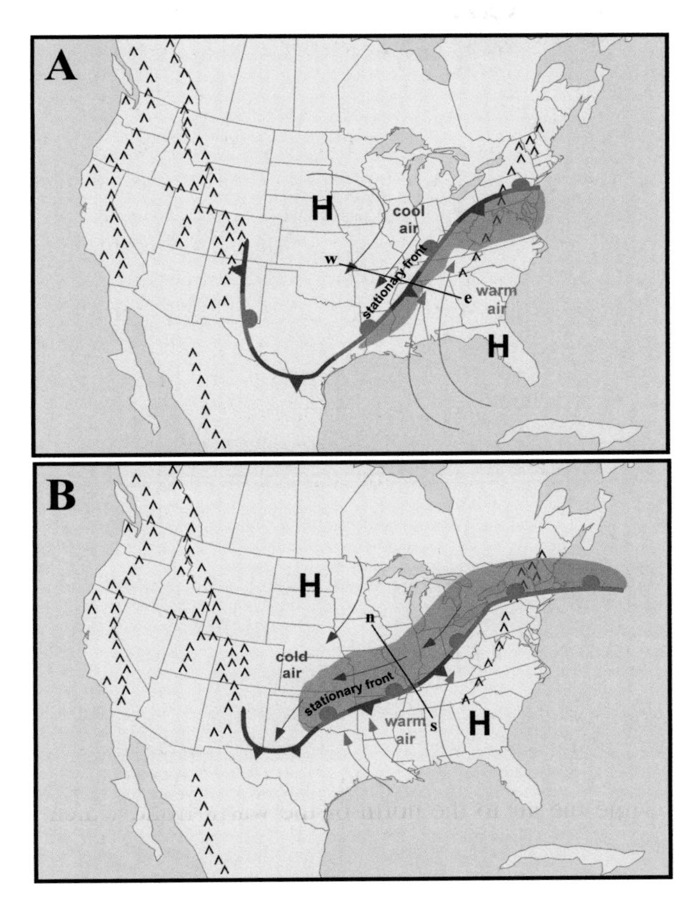

FIGURE 9.5: Maps showing two examples of stationary fronts and associated clouds (gray shading). Note that in both examples the cold air on the north side of the front is moving parallel to the front. In (A) clouds form on the warm side of the front and precipitation would fall in this region. In (B) clouds form on the cold side of the front and any precipitation would fall out on the cold side.

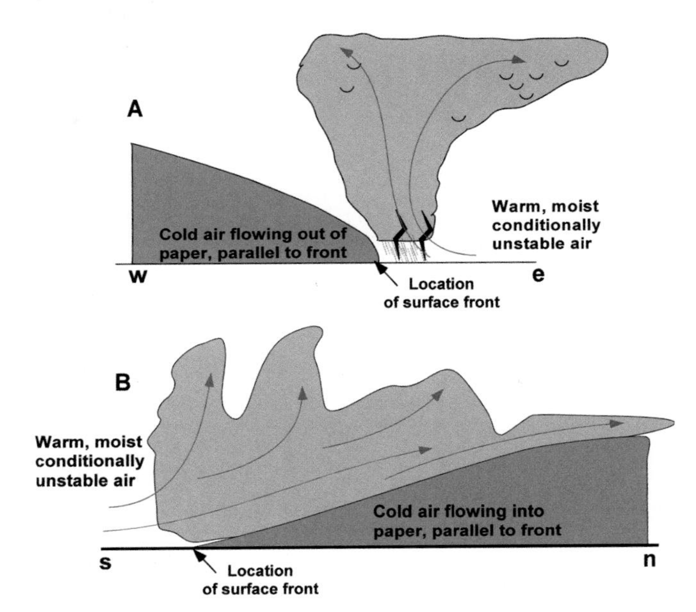

FIGURE 9.6: Two examples of cross sections through stationary fronts. In each case the cold air is flowing parallel to the front, so the front is neither retreating nor advancing. In both cases shown, the warm air is conditionally unstable. In panel A, which corresponds to cross section w-e in Figure 9.5A, the warm air is lifted at the leading edge of the cold air dome and showers erupt ahead of the front with rain falling in the warm air. In panel B, which corresponds to cross section s-n in Figure 9.5B, warm air flows up over the cold air dome so that the showers occur over the front, with the rain falling out on the cold-air side of the front.

section (w to x) is located right at the point where the cold, warm, and occluded fronts meet. The top cross section (y to z) shows the occluded front. In the occluded region, cold air behind the cold front lifts the (less cold) air north of the warm front and the warm air riding on top of the cold airmasses. The right set of cross sections represent a ***warm occlusion***. Warm occlusions develop when air north of the warm front is colder than the air behind the cold front. In this case, air behind the cold front ascends over the cold air north of the warm front. The warmest air again is found above the cold airmasses. Warm occlusions appear to be more common than cold occlusions. Occluded fronts develop during the mature and dissipating stages of cyclones. They are typically characterized by widespread cloudiness and rain or snowfall.

Dry Lines

The fronts discussed so far all separate airmasses with different thermal properties, one airmass cold and the other warm. In the south-central United States (the Texas, Oklahoma, Kansas region) and northern Mexico another type of airmass boundary develops that is primarily marked by a sharp moisture, rather than temperature difference between the two airmasses. A front characterized by a sharp moisture difference, but little temperature change, is called a ***dry line***. Dry lines develop when air flowing eastward from the high desert plateau regions of Arizona, Colorado, New Mexico, and Mexico descends the Rockies into the southern plains and encounters moist air flowing northward from the Gulf of Mexico (Figure 9.8). The desert air typically has low moisture content because of its source region. When this air descends the east slope of the Rockies on its way to the Plains, the air compresses, warms, and dries even more, so by the time it reaches the High Plains, its relative humidity can be quite low (~20 to 30%). Air from the Gulf of

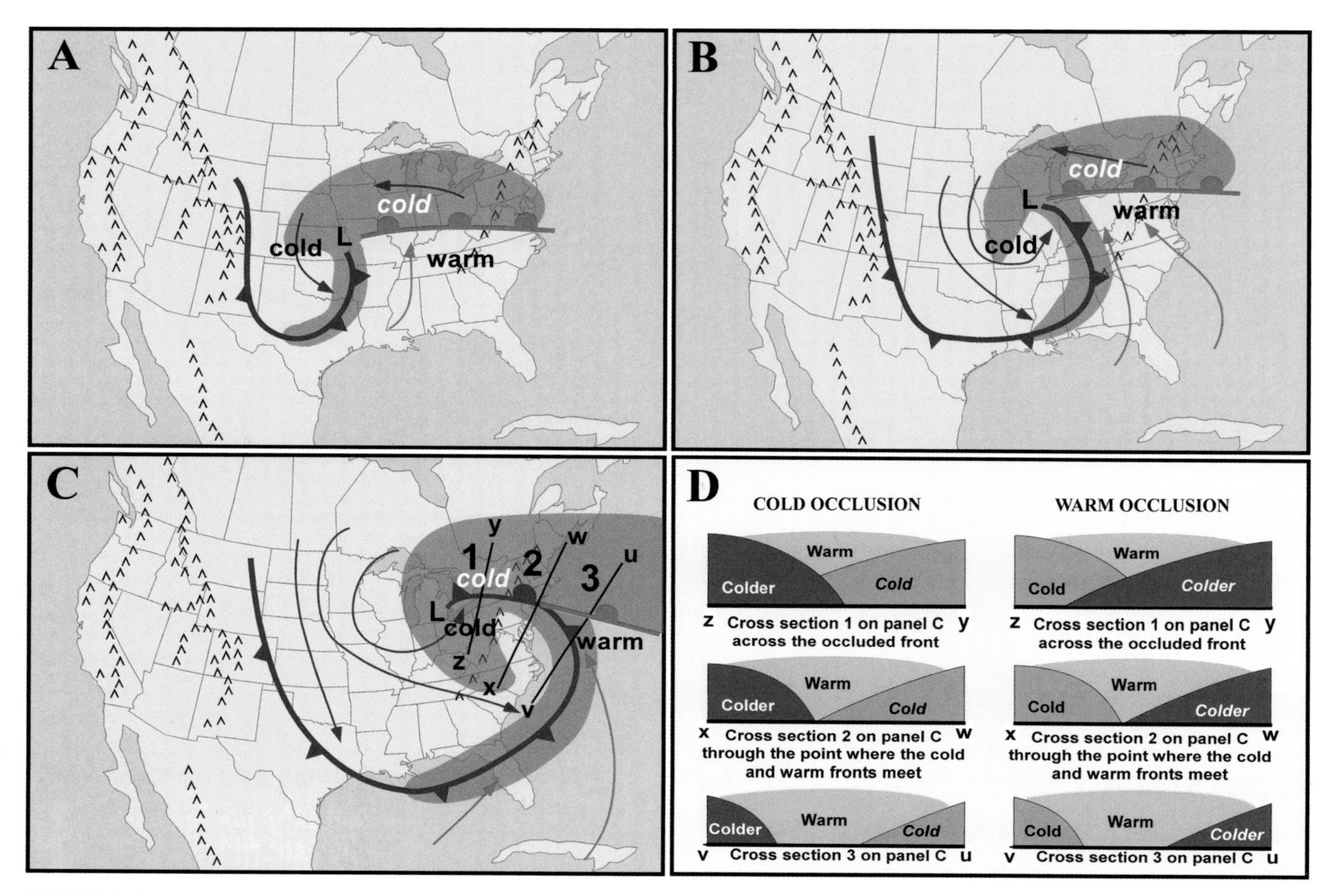

FIGURE 9.7: The development of an occluded front. (A) The cold front moves faster than the warm front because the winds in the cold air behind the cold front are moving rapidly toward the front, while the cold air north of the warm front is primarily moving westward, with only a small component of motion away from the front toward the north. (B) As a result, the cold front "catches up" to the warm front. (C) Cold air completely surrounds the low-pressure center, with the occluded front marking the boundary between air behind the cold front and ahead of the warm front. (D) The vertical structure of an occlusion. Dark blue denotes the coldest air; light blue, moderately cold air; and pink, the warm airmass. Each set of three cross sections correspond to the cross sections shown in (C). The set to the left show the alignment of airmasses when the coldest air is behind the cold front (a cold occlusion). The set to the right show the alignment of airmasses when the coldest air is behind the warm front (a warm occlusion). In all cases, the warm air in the "warm sector" of the storm is forced aloft over both fronts.

Mexico, on the other hand, can be extremely humid. Moist air is less dense than dry air at the same temperature and pressure. This is true because the water molecule (H_2O, molecular weight 18) has less mass than an oxygen molecule (O_2, molecular weight 32), or a nitrogen molecule (N_2, molecular weight 28). As water molecules displace nitrogen and oxygen molecules in a volume of air, air becomes less dense. When the dry air descending the Rockies meets the moist air on the plains, the moist air will rise over the dry air. If the moist air is conditionally unstable, lifting over the dry air may trigger instability and produce thunderstorms. Often a line of thunderstorms will develop along a dry line. These thunderstorms can be severe if the moist air is sufficiently unstable. Thunderstorms tend to develop along dry lines in the mid-afternoon, when the moist air is warmest.

Upper-Level Fronts

Airmasses do not necessarily extend to the surface of the Earth. Often airmasses can be "stacked" in the vertical, so that an airmass boundary will be present aloft, but not at the surface. Meteorologists call this type of airmass boundary a ***cold front aloft*** or ***upper-level front*** (Figure 9.9). In this text, we will adopt the term "upper-level front" for this type of front to avoid confusion with surface cold fronts. Upper-level fronts

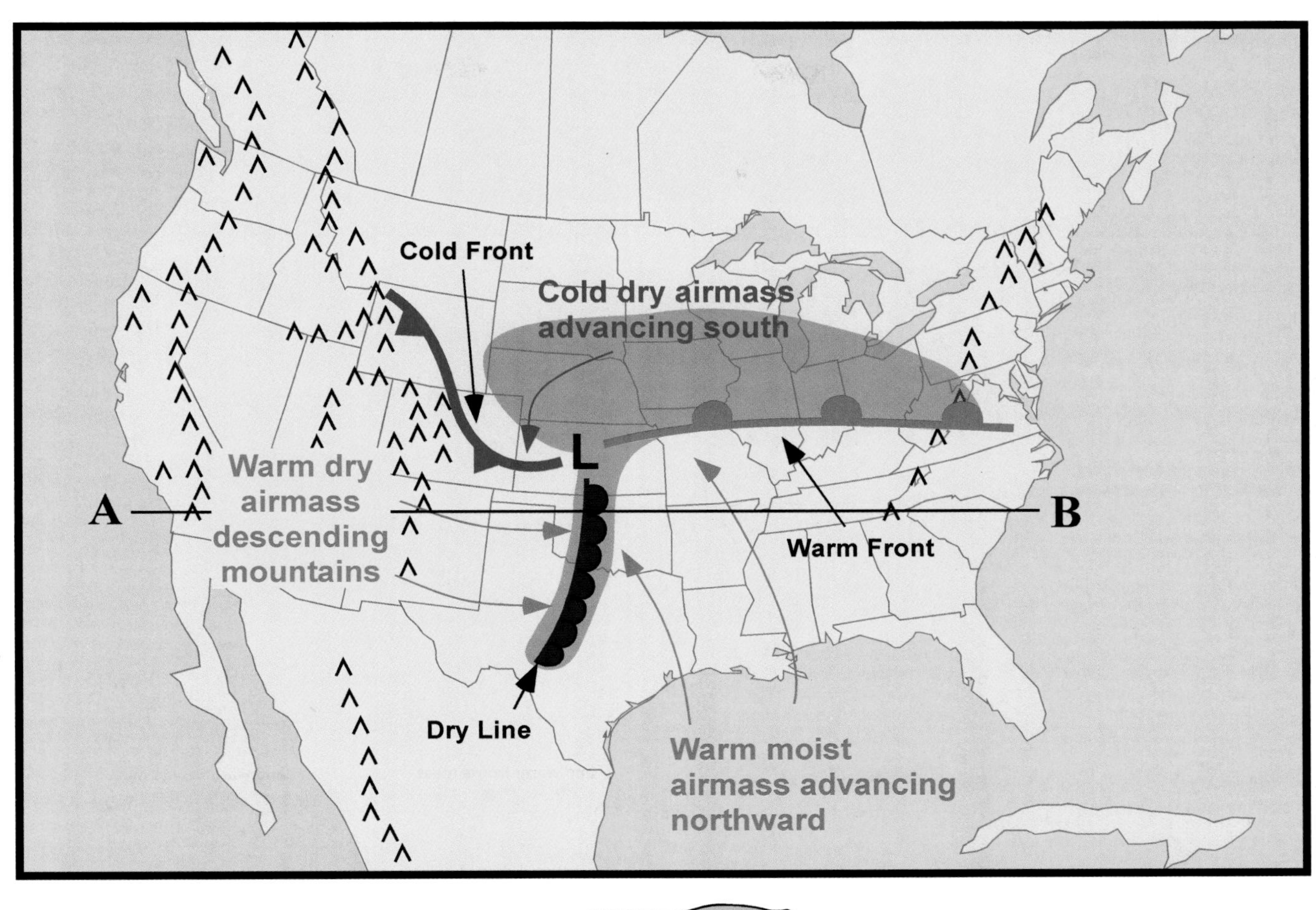

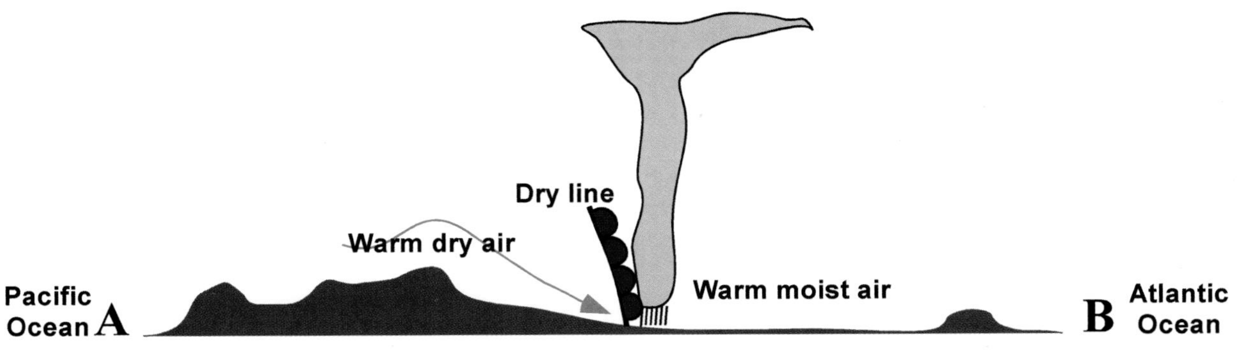

FIGURE 9.8: A dry line develops as warm dry air descends the east side of the Rocky Mountains. Ahead of the dry line, warm moist air, which originated over the Gulf of Mexico, is lifted. In this example, the warm moist air ahead of the front is conditionally unstable so that lifting at the dry line triggers a line of thunderstorms as illustrated in the cross section below the map.

are quite common features within cyclones that form east of the Rockies.

Upper-level fronts develop when two airmasses aloft meet. The first airmass originates in the convergent (west) side of a trough in the upper troposphere. This air, located near the top of the troposphere or even in the lower stratosphere, is very cold. However, as the air descends in response to the convergence aloft, it warms at the dry adiabatic lapse rate, dries, and is carried eastward by the jetstream winds. The airmass ahead of the upper-level front originates in the lower troposphere, east of the trough axis. This air, which initially is quite warm, rises and becomes saturated, cooling at the moist adiabatic lapse rate. The boundary where these two airstreams meet is characterized by a very sharp change in humidity and modest to little change in temperature. The reason there is often not a sharp temperature contrast across the

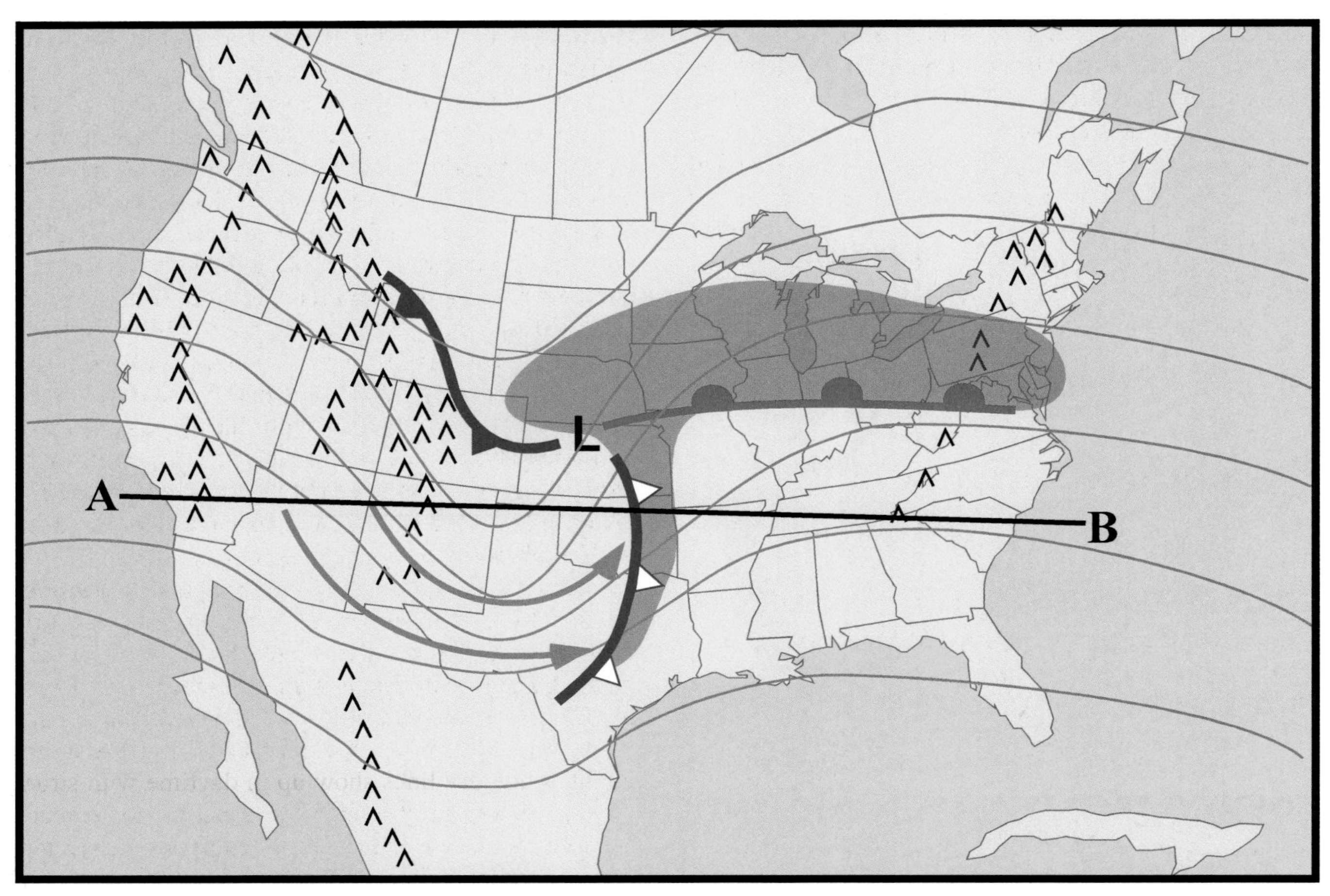

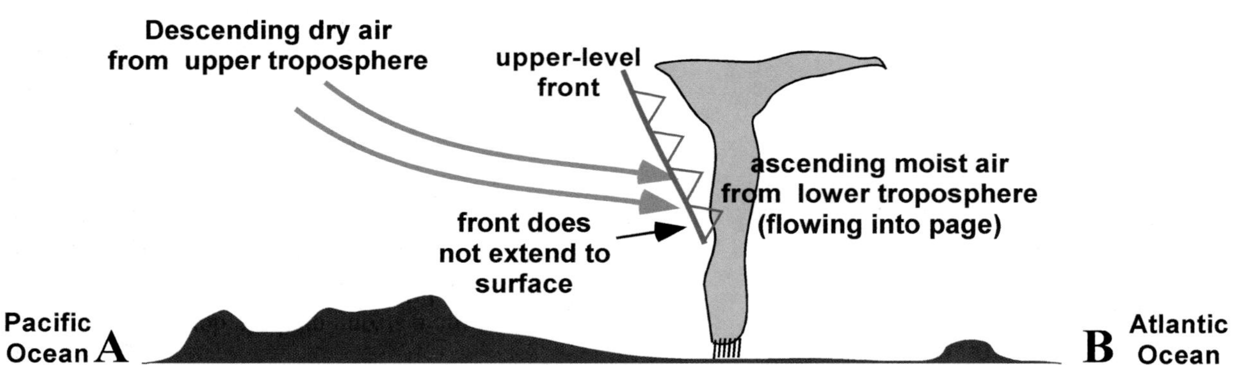

FIGURE 9.9: This figure illustrates both surface conditions and the upper tropospheric height pattern when an upper-level front is present. Cloud cover is depicted as gray shading. The solid orange lines denote height contours on the 500 mb surface. The lower panel of the figure shows the upper-level front in cross section AB.

upper-level front is that the descending airstream has warmed dry adiabatically, and the ascending airstream cooled moist adiabatically, to near the same temperature at the boundary where the airmasses meet. The upper-level front does *not* appear as a sharp boundary in *surface* temperature or moisture fields, since it does not extend to the surface. As the warm moist air east of the upper-level front is lifted by the dry air advancing from the west, strong thunderstorms can be triggered along the upper-level front.

In general, an upper-level front is likely to be present if (1) a sharp line of storms exists with clear air to the west of the line and (2) no significant change in the temperature or dewpoint temperature appears at the surface across the line of storms. Upper-level fronts develop in the same general region as dry lines. The primary distinction between these two fronts is that dry lines are surface based, while upper-level fronts are only found aloft. Air west of the dry line originates at low levels over the high plateau of the desert southwest. Air west of an upper-level front originates in the upper troposphere and lower stratosphere. In Chapters 10 and 11, we will examine how all of the fronts discussed in this chapter form and how they move and interact as extratropical cyclones develop and progress through their life cycles.

CHECK YOUR UNDERSTANDING 9.2

1. What are the three criteria used to classify fronts?
2. What are the six different types of fronts? How are they depicted on weather maps?
3. For each of the six fronts, which airmass is most dense and which airmass is lifted?

FINDING FRONTS ON WEATHER MAPS

Imagine driving on a highway toward a cold front. The temperature outside the car would remain warm until the moment the car crossed into the cold airmass. At that point, the outside temperature would begin to drop, continuing to fall as the car moved farther into the cold airmass. Where would the position of the cold front be? The cold front would be located at the point on the highway where the temperature *began* to fall rapidly. On surface maps such as Figure 9.10, meteorologists often plot contours of temperature. The position of the leading edge of a cold airmass, and therefore the front, can be found on such maps by examining temperature gradients (the change of temperature with distance). *A cold, warm, or stationary front will be located on the warm edge of a sharp temperature gradient.* In many cases, lower dewpoint temperatures will also be found on the cold air side of a cold, warm, or stationary front. For this reason, the leading edge of a sharp gradient in the dewpoint temperature can also be used to identify fronts. Other data can also be used to identify fronts. Fronts often are marked by a sharp shift in wind direction, and often a change in wind speed. Sharp changes in wind direction and speed between nearby stations on weather maps are often the best indicators of frontal positions. Fronts typically align with troughs in the surface pressure field (see cold front in Figure 9.10). Characteristic frontal weather, such as a line of showers or thunderstorms, or a transition from a clear to a cloudy sky, also help mark frontal locations. One can determine whether a front is a cold, warm, or stationary front by carefully examining the wind direction at stations near the front on its cold side. If the wind on the cold side blows toward the front, it is a cold front; away from the front, a warm front; and along the front, a stationary front. Occluded fronts develop late in the life of cyclones. Occluded fronts typically coincide with a sharp wind shift, with cold air on both sides of the boundary marked by the wind shift. Widespread clouds and light to moderate precipitation often accompany occluded fronts.

Dry lines appear as sharp gradients in dewpoint temperature, and are most common in the south-central United States east of the Rockies. Be careful not to confuse cold fronts with dry lines when looking at dewpoint temperature maps. Cold fronts have sharp surface temperature and dewpoint gradients day and night, while dry lines show up in daytime with strong surface dewpoint gradients but weak or no temperature gradients. At night, dry lines appear in surface data more like cold fronts. The reason for this diurnal change has to do with the difference between the daily cycle of heating and cooling on the west and east side of a dry line. The surface temperature on the dry side of a dry line undergoes rapid change from day to night because the skies are clear and dry. The moist side of a dry line is often hazy or cloudy and typically experiences small diurnal changes in surface temperature. Dry lines are typically marked by a wind shift, with southerly winds to the east, and westerly winds to the west of the boundary.

Upper-level fronts are only found aloft and are therefore difficult to identify on surface maps. An important signature of an upper-level front is a pressure trough in the surface data. This pressure trough may also have with it a slight wind shift. Upper-level fronts also coincide with lines of showers and precipitation. In general, if a line of showers moves over a region and surface conditions change little from before to after the passage of the line, there is a good chance that an upper-level front has passed. If it appears from surface data that an upper-level front has passed, the existence of the front aloft can be confirmed by examining the humidity and temperature on upper-level charts such as the 700 or 500 mb chart.

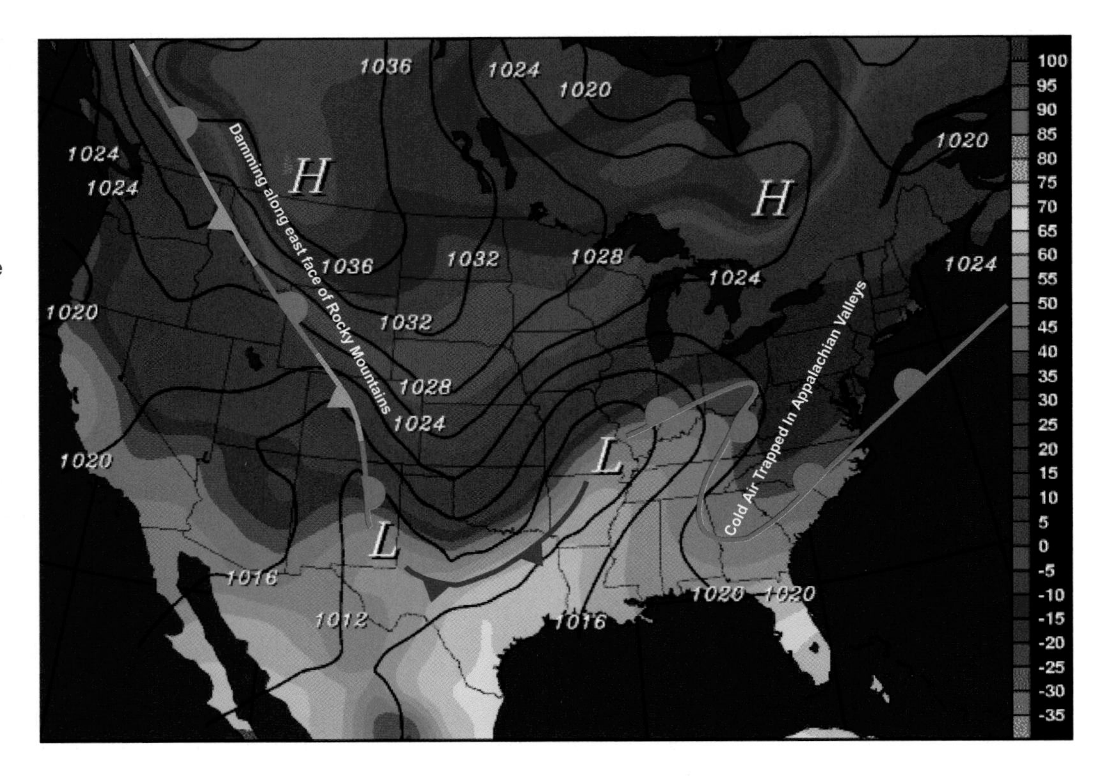

FIGURE 9.10: Analysis of sea-level pressure and surface temperature on a winter day in December. Note the effects of cold air trapping in the Appalachians and cold air damming on the east side of the Rocky Mountains.

The Rocky Mountains in western North America extend from northern Canada into Mexico. Cold airmasses flowing southward from central Canada are often so shallow that they cannot flow westward over the Rockies, which act like a dam, trapping cold air on their east side. On surface weather maps, meteorologists draw stationary fronts to indicate the western edge of cold airmasses dammed along the mountains. These stationary fronts are not boundaries between airmasses, but rather boundaries between air and rock! (see Figure 9.10). Cold air damming also occurs on the east side of the Appalachians. When cold continental air arrives on the east side of the Appalachians, warm moist air moving westward off the Atlantic sometimes forces the cold air into a narrow wedge between the coastline and the Appalachian mountain chain.

Cold air damming and trapping also frequently occurs along and within the Appalachian Mountains. The Appalachians consist of parallel ridges and valleys that extend from Georgia and Alabama to Maine. Often after a cold airmass moves across the eastern United States, cold air will settle into the Appalachian valleys. When a new storm system approaches from the west, warm air will advance northward on both the west and east sides of the Appalachian chain, leaving the denser cold air trapped in the interior valleys. Meteorologists indicate cold air damming and trapping by drawing distorted warm fronts that wrap southward around the west side of the Appalachians and northward again on the east side. The effects of cold air damming by the Rocky Mountains and damming and trapping in the Appalachians both appear on the wintertime surface map shown in Figure 9.10. Figures 9.7 through 9.9 also imply damming of cold air by the Rocky Mountains.

CHECK YOUR UNDERSTANDING 9.3

1. List at least five variables that can be used to identify the position of fronts on weather maps.
2. Where do cold air damming and cold air trapping typically occur?
3. How would you identify a dry line on a map of station reports?

TEST YOUR UNDERSTANDING

1. What are the primary airmass source regions affecting North America?
2. What type of surface weather (temperature, intensity and type of precipitation, wind direction) would a person experience during the passage of a cold front, a warm front, an occluded front, a dry line and an upper level front?
3. What is the difference between a cold front and an arctic front?
4. Draw a vertical cross section through a cold front. Label the cold and warm airmasses, and draw in clouds and precipitation for different stability conditions.
5. Draw a vertical cross section through a warm front. Label the cold and warm airmasses, and draw in clouds and precipitation for different stability conditions.
6. What direction does the cold air move relative to a cold front and a warm front?
7. How does an occluded front develop?
8. What is the difference between a warm occlusion and a cold occlusion?
9. Where is warm air found in the vicinity of an occluded front?
10. What is a dry line?
11. Why are dry lines most commonly found in Texas, Oklahoma, and Kansas?
12. Explain why air can be lifted along a dry line when there is no temperature difference between the airmasses on either side of the boundary.
13. Why is a dry line often more detectable during daytime than at night?
14. Draw a vertical cross section through a dry line. Draw in clouds and precipitation where they would occur.
15. Draw a cross section through an upper-level front. Draw in clouds and precipitation where you would expect them to occur.
16. How do the surface temperature and dewpoint temperature vary across an upper-level front?
17. What type of front is often associated with flash flooding events?
18. Meteorologists often look at maps depicting data in station model format (rather than contoured data of temperature, dewpoint, or pressure) to help pinpoint frontal locations. Explain why.
19. Explain why maps depicting surface temperature contours are not very helpful in identifying dry lines and upper-level fronts.
20. Why does cold air trapping occur in regions such as the Appalachians?

TEST YOUR PROBLEM-SOLVING SKILLS

1. The forecasts listed below describe an overnight frontal passage. Identify the type of front expected to pass through the area.
   (a) Cool today with light, steady rain and winds from the northeast. Warmer tomorrow with clearing skies and precipitation ending. Winds will shift to southerly.
   (b) Scattered showers today with temperatures in the 70s and light southeast winds. A chance of thunderstorms this evening with some becoming severe. Tomorrow will be much colder with temperatures in the 50s, clearing skies, and winds shifting to northwesterly.

(c) Hot and humid today with scattered thunderstorms. Tomorrow expect continued high temperatures but with drier conditions and clearing skies. Overnight winds will shift from southeasterly to southwesterly.
(d) High temperatures today only near 45 with overcast skies and rain showers. Winds will be out of the east. Tomorrow will be cooler with highs only in the high 30s. Continued cloudy skies and rain with winds shifting to northwesterly.
(e) Hot and humid today with temperatures reaching into the high 80s. Winds from the south. Strong thunderstorms tonight. Hot and humid conditions will persist after the thunderstorms pass.

2. The following table contains the 1:00 PM temperatures and dewpoints for a twenty-day period during May at Indianapolis, Indiana. Use the tabulated data to determine the following:
   (a) a day when Indianapolis was influenced by an airmass originating in the North Atlantic.
   (b) a day when Indianapolis was influenced by an airmass originating in central Canada.
   (c) a day when Indianapolis was influenced by an airmass originating in the Gulf of Mexico.
   (d) a day when Indianapolis was influenced by an airmass originating in the desert Southwest of the United States.
   (e) a twenty-four-hour period when a cold front passed through Indianapolis.
   (f) a twenty-four-hour period when a warm front passed through Indianapolis.

Date	Temp. (° F)	Dewpoint (° F)
May 1	58	46
May 2	60	40
May 3	62	41
May 4	77	58
May 5	82	67
May 6	83	72
May 7	68	42
May 8	55	32
May 9	61	41
May 10	63	48
May 11	57	51
May 12	52	51
May 13	59	52
May 14	67	60
May 15	72	52
May 16	81	47
May 17	92	45
May 18	91	53
May 19	87	62
May 20	75	58

3. The warm front shown in Figure 9.3 stretches from Cape Hatteras, North Carolina to St. Joseph, Missouri, passing through southern Illinois (latitude 37° N). Assume a typical warm frontal slope of 1:200 (see Figure 9.4):
   (a) Approximately how far above the surface is the warm front at Peoria, Illinois (40° N), Milwaukee, Wisconsin (43° N), and Marquette, Michigan (46° N)?
   (b) What types of clouds are likely to be present over Peoria, Milwaukee, and Marquette?
   (c) The cold front in Figure 9.3 stretches from the Oklahoma/Arkansas border through Southern Texas. Oklahoma City is approximately 200 miles behind the surface front. Assuming a typical cold frontal slope of 1:75, how deep is the cold air over Oklahoma City?

4. Figure 9.7B shows a frontal system in which the occlusion process is just about to begin. Suppose the cold front is moving at 40 mph and the warm front is moving at 15 mph. Using a ruler to estimate distances, determine the approximate length of the occluded front six hours later. (Hint: The cold front will catch up to the warm front wherever the initial separation distance of the two fronts is less than 6 hours × the difference in frontal speeds. For reference, the cold front's shaded cloud band in Figure 9.7B is approximately 150 miles wide in Indiana and Kentucky.)

5. Consider a dry line moving downslope (west to east) across Texas. In the air behind the dry line, suppose that 0.5 percent of the air molecules are water vapor; while in the moist airmass ahead of the dry line, 3 percent of the molecules are water vapor. (In both cases, suppose the remaining molecules are all O_2 and N_2 in their usual ratio, 3.7 nitrogen molecules per 1 oxygen molecule.) If the temperatures and pressures are the same on either side of the dry line, what is the ratio of the densities of the two airmasses? (Be sure to indicate which airmass is the denser one.)

USE THE SEVERE AND HAZARDOUS WEATHER WEBSITE

http://severewx.atmos.uiuc.edu

1. Use the *Severe and Hazardous Weather Website* to navigate to a U.S. map of surface station observations.
   (a) Print the map.
   (b) Analyze the map by locating any cold and warm fronts. (Hint: It is easiest to identify the center of low pressure first, and then any fronts that may be associated with the low.) Draw in the fronts using the correct symbols.
   (c) Compare and contrast the observations ahead of and behind the cold front with the idealized weather conditions described in Figure 9.3 (temperature, dewpoint, and wind direction).
   (d) Compare and contrast the observations ahead of and behind the warm front you analyzed with the idealized conditions presented in Figure 9.3.

2. Watch the weather patterns across the United States for several days. During the next frontal passage on the Great Plains, examine wind profiler data to identify wind shifts associated with the front.
   (a) Use the wind shift to determine how deep the front extends into the troposphere.
   (b) How did you measure the depth?
   (c) Can you estimate the type of frontal passage based only on wind speed and direction?
   (d) Based on the wind profiler data, what type of front would you expect this to be?

3. Monitor the weather conditions across the United States. The next time a front forms, examine the surface conditions and the U.S. radar.
   (a) What type of front is present?
   (b) Where is the front located?
   (c) Where is the precipitation found relative to the front?
   (d) What is the intensity and type of precipitation that is falling?
   (e) Is this what you might expect for the type of front and time of year?

4. Pay close attention to weather conditions across the United States. When a strong surface front develops you will notice a strong temperature contrast. When this occurs:
   (a) Use the *Severe and Hazardous Weather Website* to find 850 mb, 700 mb, and 500 mb maps for the same time. Save or print the maps. Record the date and time of the maps.
   (b) What evidence can you find for the same front on each of the three upper-level maps?
   (c) If you found any evidence of the front on the upper-level maps, estimate the approximate slope of the front. Does this match the model presented in this chapter?

5. The next time a strong cyclone develops in the central United States, save or print the surface analysis and soundings located ahead of and behind the cold and warm fronts. (You should find at least one sounding ahead of each front and one sounding behind each front; two would be better.)
   (a) Examine the surface temperature and dewpoint as recorded on the soundings. Are these values what you would expect given the time of year and strength of the front?
   (b) Examine the vertical wind profiles in the vicinity of each front. Describe how wind direction changes with height across the cold front and across the warm front.
   (c) Sometimes airmasses can be identified on soundings. The bottom of an airmass aloft is often marked by an inversion and a wind shift across the inversion. Examine the soundings for distinct airmasses. Note the altitude range where each airmass is present and the temperature and moisture characteristics of the airmass.

EXTRATROPICAL CYCLONES FORMING EAST OF THE ROCKY MOUNTAINS

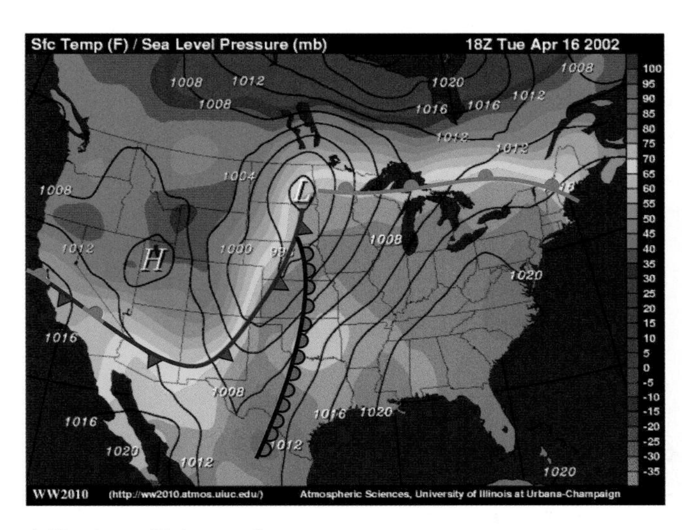

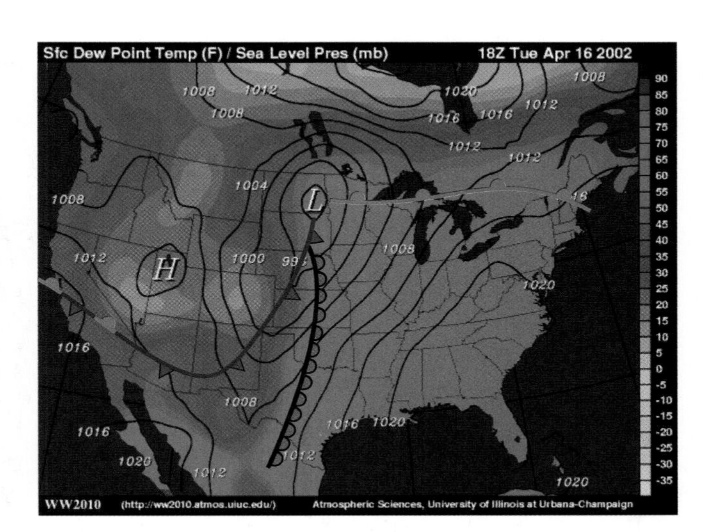

A Northern Plains cyclone.

KEY TERMS

Alberta Clipper
cold front
comma cloud
convergence
cutoff low
cyclone
divergence
dry line
dry slot
extratropical cyclone
feedback process
front
leeward side
low-level jet
occluded front
overrunning
trowal
upper-level front
upslope flow
warm front

Extratropical cyclones are large swirling storm systems that form along the jetstream between about 30° and 70° latitude. ***Cyclones*** are the parent storms from which much of the severe weather of the middle latitudes develops. The entire life cycle of an extratropical cyclone can span several days to well over a week. These storms cover areas ranging from several hundred to a thousand miles across. They are one of nature's mechanisms to balance the temperature differences between the poles and the equator and between the cold upper troposphere and the warm lower troposphere. As the poles continually cool and the tropics continually heat, especially near the surface, cyclones act to move the atmosphere back toward thermal equilibrium by transporting warm air northward and upward, and cold air southward and downward. Pole-to-equator temperature contrasts are most dramatic in late fall, winter, and spring and it is during these seasons that strong cyclones commonly occur. Cyclones are not all death and destruction. Indeed, the central plains of North America are the world's breadbasket because they are blessed with a profusion of these storms, each bringing the rain needed to sustain agriculture.

Figure 10.1 shows a visible satellite image of a cyclone over the central United States. Two days before this image was recorded, there was no storm there at all. The weather was clear and pleasant.

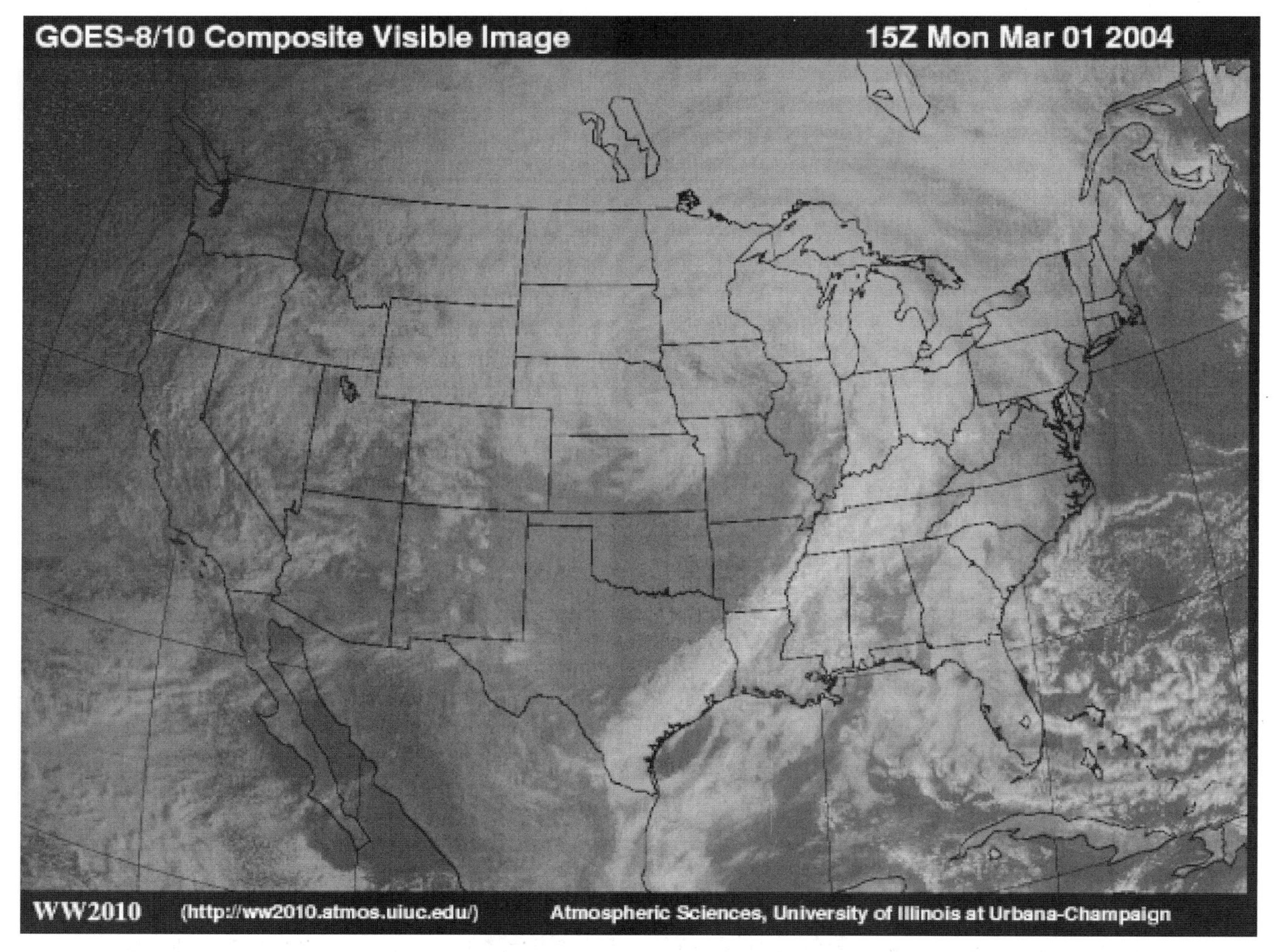

FIGURE 10.1: Visible satellite image of an extratropical cyclone covering the central United States. Note the distinct comma shape to the clouds associated with the cyclone: the "head" of the comma cloud extends northward from Colorado and Kansas to the Canadian border and eastward into Minnesota and Wisconsin, while the "tail" extends in an arc from Michigan through eastern Texas.

Where did the storm come from? How did the storm, with all of its energy and capability to produce destructive weather, appear on the Great Plains? In this chapter, we will use concepts discussed in Chapters 6 to 9 to answer these questions.

The center of a cyclone is coincident with a center of low pressure. Air flows counterclockwise around a cyclone (in the Northern Hemisphere) so that warm air flows northward east of the cyclone center and cold air flows southward west of the cyclone center. You can see in Figure 10.1 that the storm circulation covers an area about one-third the size of the contiguous United States (from the Rocky Mountains to the Appalachian Mountains), and has a shape similar to a "comma." The "tail" of the ***comma cloud*** on the satellite image consists of a line of clouds that typically produce showers and thunderstorms. The head of the comma consists of clouds that can produce rain, ice pellets, freezing rain, and snow. In Figure 10.1, the head of the comma includes clouds from Colorado and Kansas north through the Dakotas and southeastward into Wisconsin and Michigan.

Cyclones undergo distinct life cycles. In general, cyclones form and intensify quickly, typically reaching maximum intensity (lowest central pressure) within thirty-six to forty-eight hours of formation. The storms can sometimes maintain a central low pressure near peak intensity for one or two days. Dissipation of a cyclone to a point where its clouds and circulations are no longer coherent can take several more days to over a week.

Extratropical cyclones produce much of the severe weather across the United States. Tornado outbreaks, lightning, hail, and strong straight-line winds develop from squall lines and supercell thunderstorms that form along the "tail" of the comma. Heavy rain and flooding occur locally when cyclones move slowly, so that thunderstorms form repeatedly over the same area. Heavy rain, heavy snow, ice pellets, and freezing rain all can occur within the "head" of the cyclone. Blizzards occur in winter in the northwest quadrant of cyclones as cold air, strong winds, and heavy snow combine to produce whiteout conditions.

Cyclones preferentially form in five locations in North America. These are (1) just east of the Rocky Mountains in the central United States, particularly in eastern Colorado; (2) just east of the Canadian Rockies, particularly near southern Alberta; (3) just off the Texas-Louisiana coastline along the Gulf Coast of the United States; (4) along the East Coast of the United States, particularly near North Carolina; and (5) over the Bering Sea and Gulf of Alaska in the Pacific Ocean. In this chapter, we focus on cyclones originating just east of the Rockies, particularly in the central United States. In Chapter 11, we will examine cyclones that affect the East and Gulf Coasts of the United States. Pacific cyclones bring snow to the mountains and are responsible for flooding on the west coast of North America. Discussion of these storms will be deferred to Chapters 16 and 25.

CHECK YOUR UNDERSTANDING 10.1

1. What is the lifetime of a typical extratropical cyclone?
2. What is a typical size of an area affected by a mature extratropical cyclone?
3. What does a cyclone look like on a satellite image?
4. Where do cyclones typically form in North America?

THE ENVIRONMENT PRIOR TO THE DEVELOPMENT OF A CYCLONE

Figure 10.2 shows a map depicting the primary airmasses typically present over central North America prior to the development of a cyclone. These are (1) a cold, dry airmass over Canada and the northern United States; (2) a warm, humid airmass over the southern and eastern United States; (3) a warm, dry airmass over the higher elevations of the western United States and Mexico; and (4) a cool moist airmass over the northeast United States and Canada. This "typical" situation varies from season to season and even day to day. For example, the warm humid air might extend to the Canadian border in the summer. In winter, the cold air may extend to the Gulf of Mexico. Nevertheless, we will use this "four airmass" model, recognizing that variations from this model commonly will be found on daily weather maps. A fifth airmass does not appear in Figure 10.2, but is very important in the evolution of cyclones. This airmass originates in the upper troposphere, and is typically only found aloft. The leading edge of this airmass is the "upper-level front" discussed in Chapter 9.

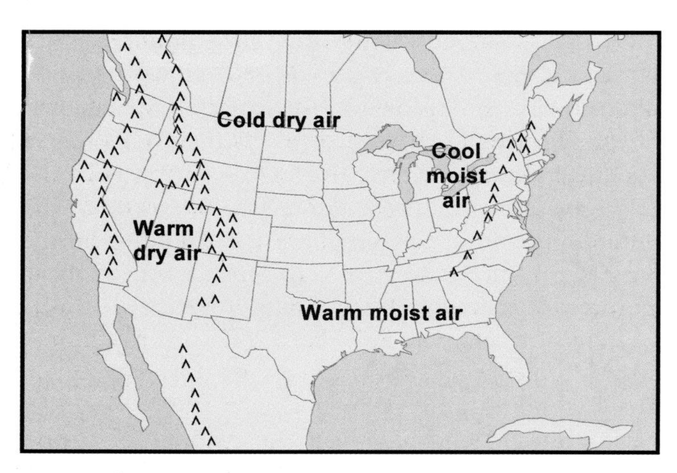

FIGURE 10.2: Typical airmasses found over North America prior to the development of an extratropical cyclone east of the Rocky Mountains.

THE INITIAL DEVELOPMENT OF THE CYCLONE

A cyclone appears as a center of low pressure on surface charts. The cloud patterns associated with a cyclone develop during and after the time that the center of low pressure forms. We know from Chapter 8 that low pressure forms when ***divergence*** occurs aloft associated with embedded jetstreaks and changes in curvature within the jetstream. These features (curvature change, jetstreaks) occur within waves in the jetstream, similar to the wave on the 300 mb analysis depicted in Figure 10.3. Note that in Figure 10.3 the contours are packed together in the base of the wave, indicating a jetstreak, and that a change in curvature occurs as air flows from the west through the crest of

FOCUS 10.1

The Environment on the Great Plains During Cyclone Formation

Cyclones form east of the Rockies when a wave and/or a jetstreak move across the Rockies from the west at jetstream level. The environmental conditions on the Great Plains at the time the wave crosses the Rockies can vary substantially from storm to storm. Conditions on the Great Plains significantly affect both the intensity of a developing cyclone and the type of hazardous weather the cyclone produces.

Consider, for example, the contrast between cyclones forming during daytime and nighttime. If a cyclone forms east of the Rockies during the afternoon, the moist warm air east of a dry line or cold front will be at its maximum daily temperature. Lifting of this air along the front may trigger violent thunderstorms. However, if the same cyclone were to form in the early morning hours, the air ahead of the front would be at its minimum temperature. The buoyancy of this air when lifted would be significantly reduced and only weak thunderstorms, or possibly no storms at all, may form.

The seasonal contrast in the environment on the Plains has dramatic effects on cyclone structure. During the winter months of December through February, strong arctic fronts often develop as bitter cold airmasses form over central Canada. As these fronts become incorporated into a cyclone's circulation, extreme blizzard conditions can develop. In the late spring and early summer months of April through June, the contrasts are most intense between moist tropical airmasses moving northward from the Gulf of Mexico and dry airmasses moving off the Rockies, or cool airmasses moving southward from Canada. Severe thunderstorms are more common along fronts in cyclones forming in this season.

The environment on the Plains just prior to cyclone formation is the product of many other factors. Past storms moving across the Plains cause the airmasses to be situated differently for each cyclone that develops. In some cases, dry arctic air may extend far southward at the time of new cyclone formation, while in others, warm, moist air may extend far northward. Fronts from previous storms may or may not be present in the vicinity where the new cyclone forms. The ground may be wet from previous rains, snow cover may be present, or the earth may be parched dry. These latter factors influence the amount of moisture available to the new storm through evaporation. The variations in the environment of the Great Plains make each cyclone unique and a challenge to forecasters as they attempt to predict the future behavior of the atmosphere as a cyclone is forming.

the wave (the ridge), the base of the wave (the trough), and back to the crest (the next ridge to the east).

Waves in the jetstream are always present in the atmosphere. They are generated in many ways. Flow across topography, such as the Rockies, Alps, and Himalayas, creates waves in the jetstream. Extreme gradients in surface temperatures and associated atmospheric heating or cooling, such as occurs when a warm ocean current flows next to a cold landmass (e.g. Greenland, Gulf of Alaska), can create wave motion in the upper-level flow. In this text, we will not consider the source of these waves. We will assume that these waves exist in the jetstream and that, on occasion (at intervals of about five to fourteen days in fall, winter, and spring, depending on the weather pattern), a wave will move across the Rocky Mountains and over the Great Plains of the United States, triggering the formation of low pressure at the surface.

In Chapter 8, we learned that the divergence aloft necessary to create surface low pressure can arise from changes in flow curvature within waves in the jetstream and/or from jetstreaks. Either effect individually can produce a low-pressure center east of the Rockies as a wave or a jetstreak moves over the plains east of the Rockies. However, the strongest cyclones, particularly those associated with hazardous weather, develop when a jetstreak is located within the base of a trough, as illustrated in Figure 10.3. Recall from Chapter 8 that a trough containing a jetstreak at its base will have a region of strong upper-level divergence in its northeast quadrant, coinciding with the position of the left exit region of the jetstreak. When this quadrant of a wave moves over the plains east of the Rockies, divergence aloft will lead to the formation of low pressure at the surface. As soon as low pressure begins to develop, air in the lower troposphere will start to circulate about the low-pressure center. A cyclone has begun to form.

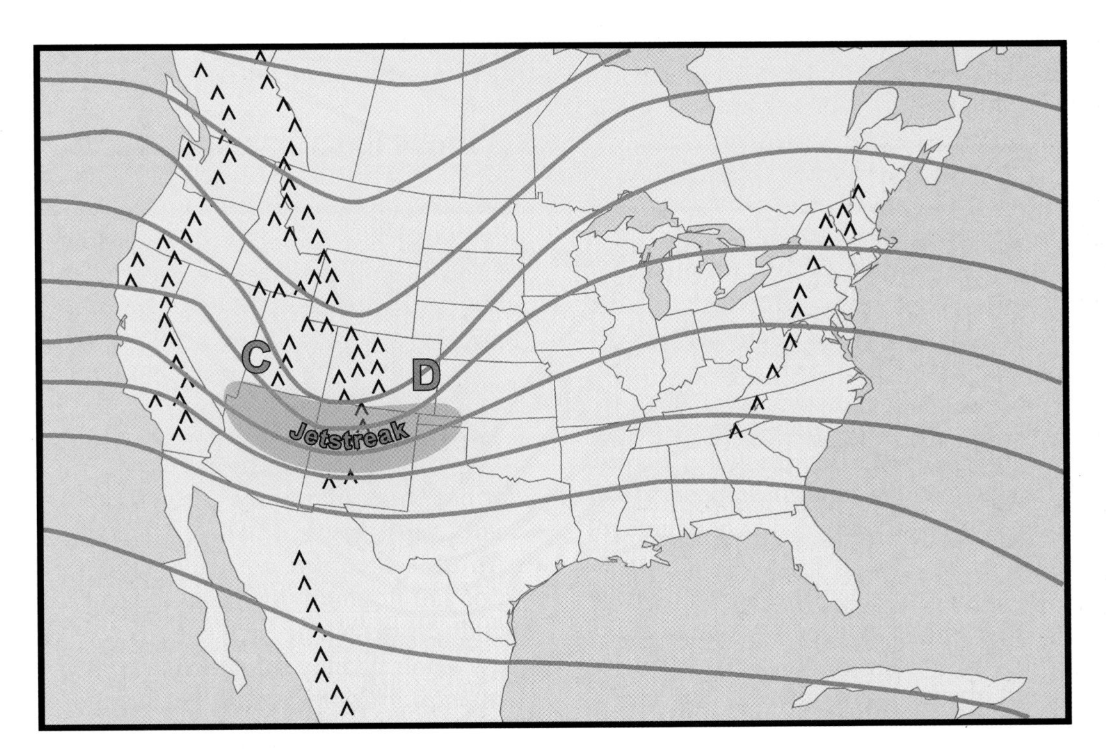

FIGURE 10.3: A typical upper tropospheric 300 mb height pattern just prior to the development of a low-pressure center on the Great Plains east of the Rocky Mountains. The shaded region denotes the position of the jetstreak, and C and D denote the location of the maximum upper-level convergence and divergence, respectively, in the wave. As the divergent region of the wave moves east of the Rocky Mountains, low pressure will begin to form at the surface on the Great Plains, air will begin to circulate around the low, and the cloud patterns and fronts associated with the cyclone will begin to form and move.

INITIAL MOVEMENT OF THE AIRMASSES

Figure 10.4 shows the movement of the airmasses and the development of ***fronts*** as the surface low-pressure center forms on the plains east of the Rockies. On this figure, blue and red arrows denote the movement of surface cold and warm airmasses, respectively, while thicker green and orange arrows denote the movement of moist and dry airmasses in the middle and upper troposphere. East and south of the low-pressure center, warm humid air from the Gulf of Mexico moves northward and northwestward, and encounters the cooler air to the north. The boundary between these two airmasses, the ***warm front***, becomes more distinct as the two airmasses "collide." The warm front extends eastward from the low-pressure center and becomes more prominent as warm air continues its northward movement. The warm air rises upward over the cool air (a process meteorologists call ***overrunning***), producing the characteristic cloud and precipitation pattern discussed for warm fronts in Chapter 9. The cool air remains in contact with the surface and, under the influence of the developing low-pressure center, moves west and around the low.

To the southwest of the low-pressure center, the dry line develops as surface air from the desert southwest flows eastward, descends the Rockies onto the plains of Texas, Oklahoma, and Kansas, and meets warm moist air flowing northward from the Gulf of Mexico. Recall from Chapter 6 that descending air warms at the dry adiabatic lapse rate and its relative humidity decreases. By the time the desert air arrives on the Great Plains, it is normally quite warm and dry, a sharp contrast to the humid air arriving from the Gulf of Mexico.

Cold air typically lies along the east slope of the Rockies northwest of the low-pressure center. As the low-pressure center moves eastward on to the Great Plains, the cold air flows southward to the west of the low-pressure center. The cold air is often not deep enough to cross the Rockies, and therefore must flow between the low-pressure center and the mountains, as shown in Figure 10.4. The leading edge of the cold air is the cold front. It typically first encounters the dry air descending from the desert.

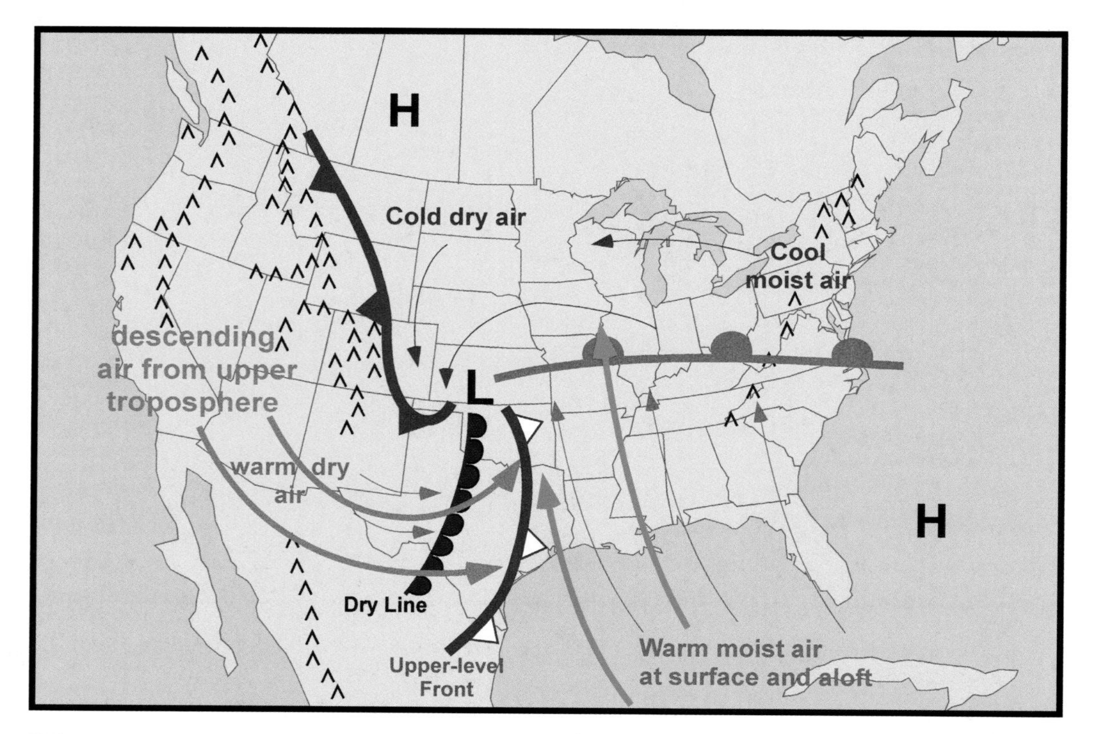

FIGURE 10.4: The airmasses that are part of a developing cyclone. Surface airmass boundaries (the cold front, warm front, and dry line) are shown using standard frontal symbols (see Figure 9A). The thin blue (red) arrows show the direction of the near-surface airflow in the cold (warm) airmass. The blue line with open triangles denotes the upper-level front, the boundary between the dry air descending into the middle troposphere from the upper troposphere (orange arrows), and the moist air rising from the lower troposphere into the middle troposphere (green arrows).

In the upper atmosphere, convergence is occurring on the west side of the wave in the jetstream (Figure 10.3). As we learned in Chapter 8, air in this region is descending. This dry descending upper tropospheric air flows eastward as it moves through the jetstream, wrapping around the base of the wave (orange arrows in Figure 10.4). To the east of this descending airstream, moist air is flowing northward in the lower and middle troposphere. The boundary between these two airmasses, the ***upper-level front***, can normally only be found aloft, typically no lower than the 700 mb level (~3000 m or 10,000 ft). Figure 10.4 shows the typical position of the upper-level front early in the storm's development. The dry air behind the upper-level front appears on visible satellite images (e.g., Figure 10.1) as a region of exceptionally clear air south of the cyclone. Meteorologists often refer to this clear area of a cyclone as the ***dry slot***.

EARLY WEATHER ALONG THE FRONTS

Figure 10.5 shows the type of weather that may develop along each of the fronts during the early stage of cyclone development. We will consider each front individually.

East of the Cyclone Center

A wide region of clouds develops north of the warm front as warm moist air glides upward over the cool air north of the front. The clouds are typically deepest close to the surface position of the front, and become thin and high far north of the front. If one were to drive from south to north, one would first typically encounter deep stratus with precipitation (nimbostratus), then stratus without precipitation, then altostratus, high cirrostratus, and finally wispy high cirrus. During

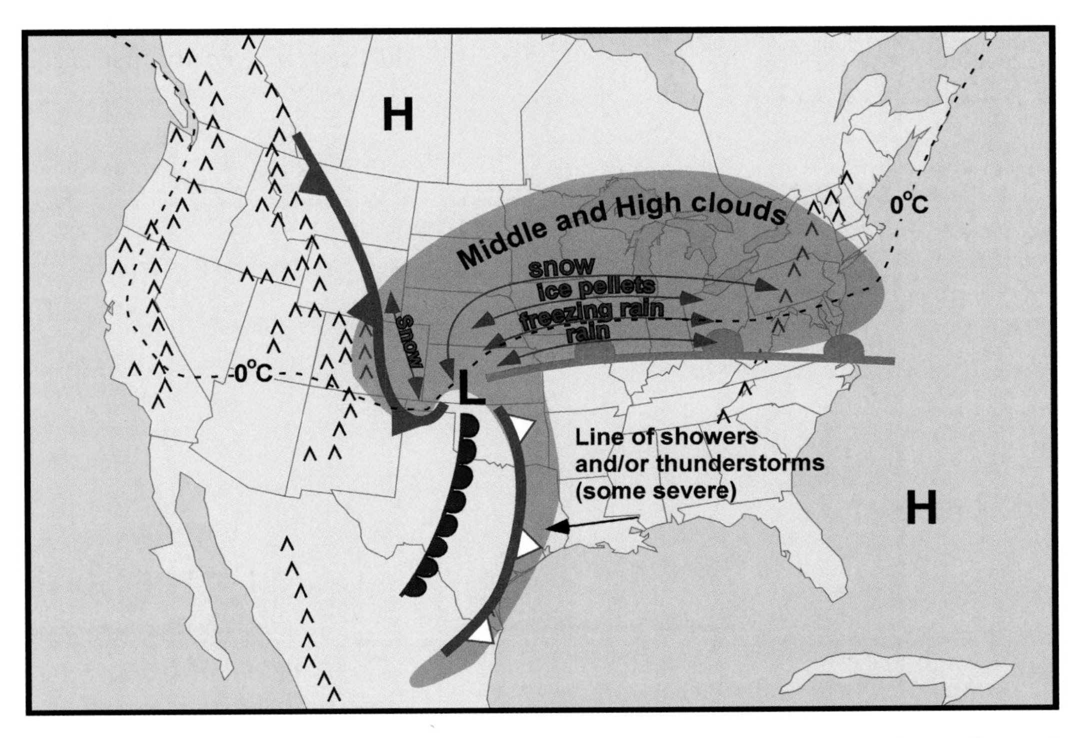

FIGURE 10.5: Precipitation types that develop as a low-pressure center forms and the gradients of temperature and moisture along the fronts intensify. The gray region is cloud cover. Along the upper-level front, and/or the dry line, a line of showers and thunderstorms may form. Some of these storms may be severe. North of the warm front, the precipitation type depends on the temperature. In the case shown, surface air just north of the warm front is warmer than 0° C, but farther north, the temperature is well below freezing. All winter precipitation types occur in this situation. In warmer seasons, the precipitation may be entirely rain, while in mid-winter, it might all be snow. Snow falls along the Rockies as air moves upslope toward the mountains from the Mississippi Valley and then rises more rapidly along the east slope of the Rockies.

the transit, a driver might travel 500 to 800 km (300 to 500 miles). Recall that the warm front's slope is typically about 1:200, implying that the warm air will be located about one kilometer above the surface at a location about 200 km north of the surface warm front. Depending on the vertical temperature structure, precipitation north of the surface warm front may fall as rain, freezing rain, ice pellets, or snow. The precipitation may be steady or showery depending on whether the warm humid air rising along the frontal surface is stable or conditionally unstable.

South of the Cyclone Center

This region is complicated because there are three airmasses at the surface in this region and an additional airmass aloft. The position of the airmass boundaries relative to one another, the time of the day, and the time of the year all affect how showers and thunderstorms will organize along these boundaries. One of three possible scenarios typically occurs. These are shown in Figure 10.6.

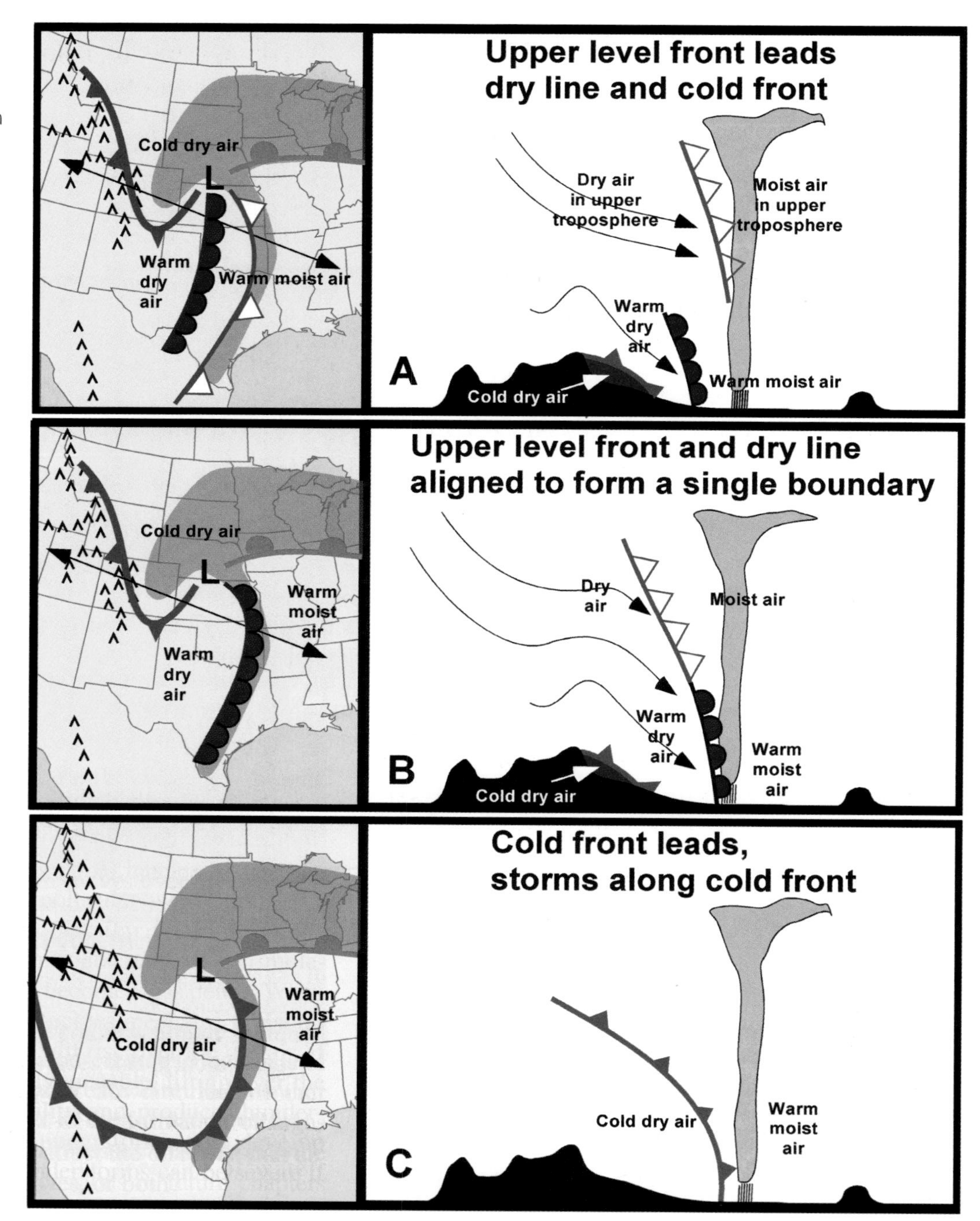

FIGURE 10.6: The frontal boundaries south of the low-pressure center take on different orientations in different cyclones. Panels A through C show three possible orientations of the airmass boundaries. The maps on the left of each panel show the location of the fronts and the location of the cross section that appears on the right side of the panel. (A) The upper-level front is east of both the dry line and the surface cold front; (B) the dry line and upper-level front form a single boundary that leads the cold front; (C) the cold front is the leading boundary.

Upper-Level Front, Dry Line, and Cold Front Present

In this scenario, the warm humid airmass moving northward from the Gulf extends to the dry line at the surface, and to the upper-level front aloft (Figure 10.6A). ***Convergence*** at the leading edge of the upper-level front forces the warm moist air ahead of the upper-level front to rise, triggering a line of thunderstorms and showers. The intensity and severity of these storms depend on the stability of the warm moist air and how the wind direction and speed change with height in the warm air. When the air possesses large conditional instability and the winds increase rapidly with height in the low levels, the thunderstorms can become severe and produce strong straight-line winds, hail, and tornadoes (see Chapters 18 and 19).

Lifting also occurs at the dry line. Aloft, between the dry line and the upper-level front, there is often an inversion at the base of the dry air descending from the upper troposphere (see Figure 10.7). This inversion exists because the air above the inversion has descended from aloft and has warmed during the descent at the dry adiabatic lapse rate. This inversion is a stable layer that acts like a lid, preventing storms from developing along the dry line. If the cyclone forms during the daytime, heating of the humid surface air east of the dry line can sometimes make this air quite unstable so that updrafts can break through the inversion. When this happens, a second line of strong thunderstorms can develop along the dry line (Figure 10.7A). These thunderstorms can be very strong, often producing baseball-sized hail and tornadoes. If the cyclone occurs at night, or the daytime heating is not sufficient (e.g., mid-winter), the dry line may not have active weather (Figure 10.7B). Sometimes, in winter, the air ahead of the upper-level front is stable or very weakly conditionally unstable. In this case, a line of weaker showers will form along the upper-level front instead of the deep thunderstorms that often are along this boundary (Figure 10.7C). The third boundary, the ***cold front***, brings up the rear in this system. The cold front lifts dry air and generally produces no precipitation (Figure 10.6A).

The Upper-Level Front and Dry Line Aligned

This scenario often develops when an old cold front from the Pacific Ocean moves across the mountains from California. The air behind the front cools at the moist adiabatic lapse rate as clouds form and precipitation falls on the windward (west) side of the mountains. The air is then warmed at the dry adiabatic lapse rate during the descent on the ***leeward side*** (east) side of the mountains. By the time the air crosses the mountains and descends on the plains, the air behind the front is quite dry. The net effect is that the Pacific cold front appears on the plains as a deep ***dry line***—with the upper-level front and dry line forming a single boundary (Figure 10.6B). If the air east of the dry line is moist and conditionally unstable, thunderstorms and showers appear ahead of the dry line. The cold front associated with the Canadian air (not the Pacific cold front) again is northwest of the dry line. As it moves forward, it lifts dry air and generally produces no precipitation.

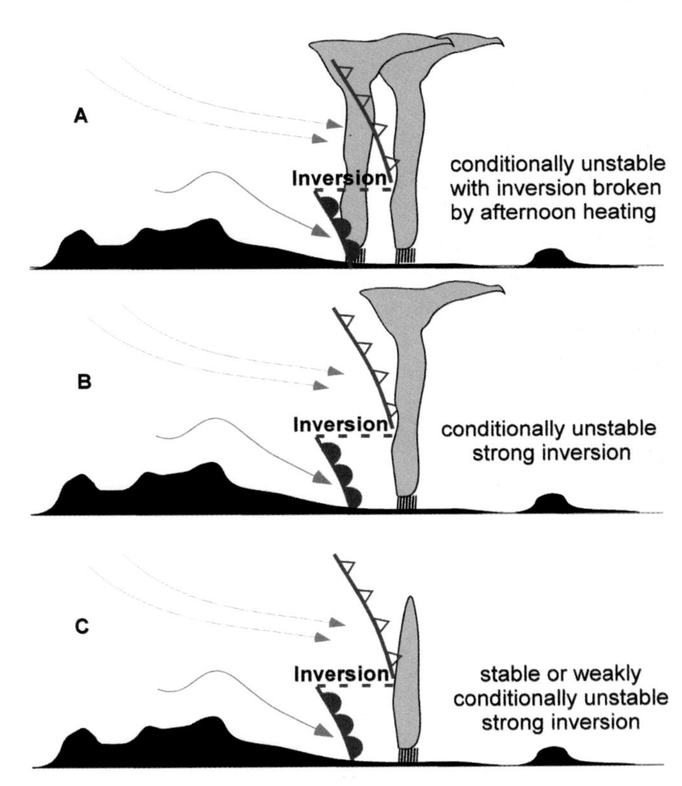

FIGURE 10.7: Three possible scenarios for storm development that can occur when fronts have an orientation as in Figure 10.6A: (A) When the low-level air is moist, warm, and sufficiently conditionally unstable that a rising updraft can break the inversion, thunderstorms may develop along both the dry line and upper-level front; (B) the inversion provides a "lid" that updrafts along the dry line cannot break—under these conditions, storms will not form along the dry line; (C) in mid-winter, the air ahead of the upper-level front may be stable or very weakly conditionally unstable. In this case a weak rainband will develop along the upper-level front.

Cold Front Only

In some cases, Canadian cold fronts move rapidly southward and quickly overtake a dry line, if one is present. This is particularly true when very cold air, associated with very high pressure in Canada, moves rapidly southeastward. In this case, lifting of the warm moist air will occur at the cold front, and the line of showers and thunderstorms will appear ahead of the cold front (Figure 10.6C). Secondary rainbands may develop west of the surface cold front. These form aloft within the warm air rising over the cold front.

Northwest of Cyclone Center

During a cyclone's early development, air north of the low-pressure center is flowing westward (Figure 10.4), a direction that puts this air on a direct collision course with the Rocky Mountains. As air flows westward from the Mississippi Valley to the Rockies, it rises almost 1.6 km (1 mile) due to the slope of the terrain. When the air encounters the Rockies, it has the potential to rise much more over a very short distance. This ***upslope flow*** often produces heavy snow along the east side of the Rockies and eastward onto the Great Plains (Chapter 16). Combined with strong winds and cold temperatures, this region of the storm can quickly develop into blizzard conditions (Chapter 15).

Summary of Early Weather

East of the cyclone center, widespread clouds and precipitation will develop to the north of the cyclone's warm front. This precipitation may take the form of rain, freezing rain, and/or snow depending on the season and the local temperatures. To the south of the cyclone center, there will be a line of showers or thunderstorms along the leading eastern-most boundary, whether it is the upper-level front, the dry line, or the cold front. This line of showers forms the "tail" of the comma cloud. There may be a second line of thunderstorms along the next boundary to the west, if the moist air extends to this region and the air is sufficiently conditionally unstable. In some cases, lines of thunderstorms can also develop ahead of the leading boundary. The mechanism for formation of these lines of thunderstorms will be discussed in Chapter 18. To the northwest of the cyclone center, air flowing westward will rise along the sloping terrain toward and along the Rocky Mountains, creating clouds and precipitation, which, in the cold season, will be in the form of snow.

ONLINE 10.1

Alberta Clippers

An *Alberta Clipper* is a name media forecasters have given to cyclones forming in wintertime east of the Canadian Rockies. These storms are called clippers because they move rapidly across southern Canada and the northern United States to the East Coast. These cyclones develop as divergence occurs at jetstream level in a wave, similar to their southern Rocky Mountain counterparts. Unlike their Rocky Mountain counterparts, Alberta Clippers normally form in an environment where the surface air on the plains east of the Rockies is much colder and drier. Alberta Clippers typically form after cold air is already in place over the central North American continent. These storms typically produce 2–5 inches (5 to 12 cm) of snow, but can produce up to 6 to 8 inches (15 to 20 cm) locally as they pass across the continent. Arctic fronts associated with Alberta Clippers reinforce the cold air with fresh new arctic air, keeping the central part of the continent in the grip of winter. The evolution of the surface pressure and temperature, radar echoes, the 850 mb height and temperature field, and the 300 mb height and wind field during the development and movement of an Alberta Clipper appear in animations online. Note the rapid movement of the low-pressure center from west central Canada through the U.S. Midwest, the relatively weak radar echoes characteristic of snow, and the arctic front that moves across the north central Canada and the United States to the rear of the low-pressure center. Note also that the low-pressure center is located under the left exit region of a jetstreak moving in the polar jetstream.

CHECK YOUR UNDERSTANDING 10.2

1. Where are warm fronts, cold fronts, and dry lines found relative to the center of an extratropical cyclone?
2. What is the role of waves in the jetstream in the development of extratropical cyclones?

(continued)

3. Name all of the boundaries that can trigger thunderstorm development south of a low-pressure system.
4. Where does upslope flow occur relative to the center of a low-pressure system?

ONLINE 10.2

Colorado Lows

On 11 April 2001, a cyclone developed along the New Mexico—Colorado border and tracked northeastward through Kansas, Nebraska, Iowa, and Wisconsin, across Lake Superior, and into Ontario, Canada. This cyclone produced all types of severe and hazardous weather. Over a foot of snow fell, and three to five foot drifts developed in cities along the foothills of eastern Colorado. Thirty-seven tornadoes were reported in Iowa, Missouri, and Michigan. Large hail and high winds were reported in many states. On 29 February 2004, a cyclone developed in a similar location and tracked along a similar path. In contrast to the April cyclone, only one tornado was reported along with a few reports of hail and high winds in the vicinity of Lake Michigan. Online are animations of 1) surface temperature and pressure, 2) dewpoint temperature and pressure, 3) radar reflectivity, 4) infrared satellite, 5) visible satellite, 6) water vapor satellite, 7) 850 mb heights and temperature, and 8) 300 mb heights and isotachs for these cyclones. Use these animations to compare and contrast the structure of these cyclones with each other and with the conceptual model of cyclone structure presented in this chapter. Examine the surface temperature, dewpoint, and wind fields carefully. Can you identify the fronts in these cyclones? Compare the radar evolution with the frontal evolution. What fronts are producing the active weather? The tornadoes during the 2001 cyclone all occurred in the afternoon of 11 April in Iowa. Examine the thunderstorms that produced the tornadoes. Where were they relative to the frontal boundaries? How did this differ from the 2004 cyclone? From the satellite images and the upper-level maps, can you identify dry slots, low-level jets, and the trowal north of each of the cyclone centers? Closely compare the 300 mb maps and the surface pressure maps. Identify the divergent region of the jetstream in each case based on the flow curvature and the position of jetstreaks. Compare the positions of the low-pressure centers at the surface in each case with the likely regions of maximum divergence aloft. Does the surface low-pressure center in each storm follow the forcing associated with the jetstreak and flow curvature?

STORM INTENSIFICATION

A cyclone intensifies as its central pressure lowers and the pressure gradients surrounding the low-pressure center strengthen. Intensification of the low-level pressure gradients causes an increase in wind speed throughout the cyclone, tightening the temperature and moisture gradients in the vicinity of the fronts, and creating heavier precipitation and more possibility for severe weather. The central low pressure within a cyclone will intensify if the divergence aloft due to the effects of jetstreaks and changes in flow curvature exceeds the convergence into the low in the boundary layer due to friction. Intensification occurs most efficiently when the divergence aloft increases with time, a process that occurs through a feedback between the lower and upper troposphere as cold air moves southward west of the cyclone center and warm air moves northward east of the center.

To understand this ***feedback process*** (a process that increases the rate of amplification of a disturbance, such as a cyclone) let us recall principles discussed in Chapter 8 (see Figure 8.12). Consider Figure 10.8A, which shows an atmosphere where there are no horizontal temperature variations. The pressure surfaces are flat. Let us replace the air in the center of the figure with colder air, but leave the air on the edges unchanged. We can cool the air in the center and get the same effect. What happens to the pressure surfaces? Figure 10.8B shows that the cooler air in the middle occupies a smaller volume (e.g., contracts when cooled) and the pressure surfaces move closer together. Individual pressure surfaces drop in altitude as cold air moves into the center of the figure. In other words, *the altitude of an individual pressure surface decreases.*

Now consider what happens when warm air is transported into the region represented in the middle

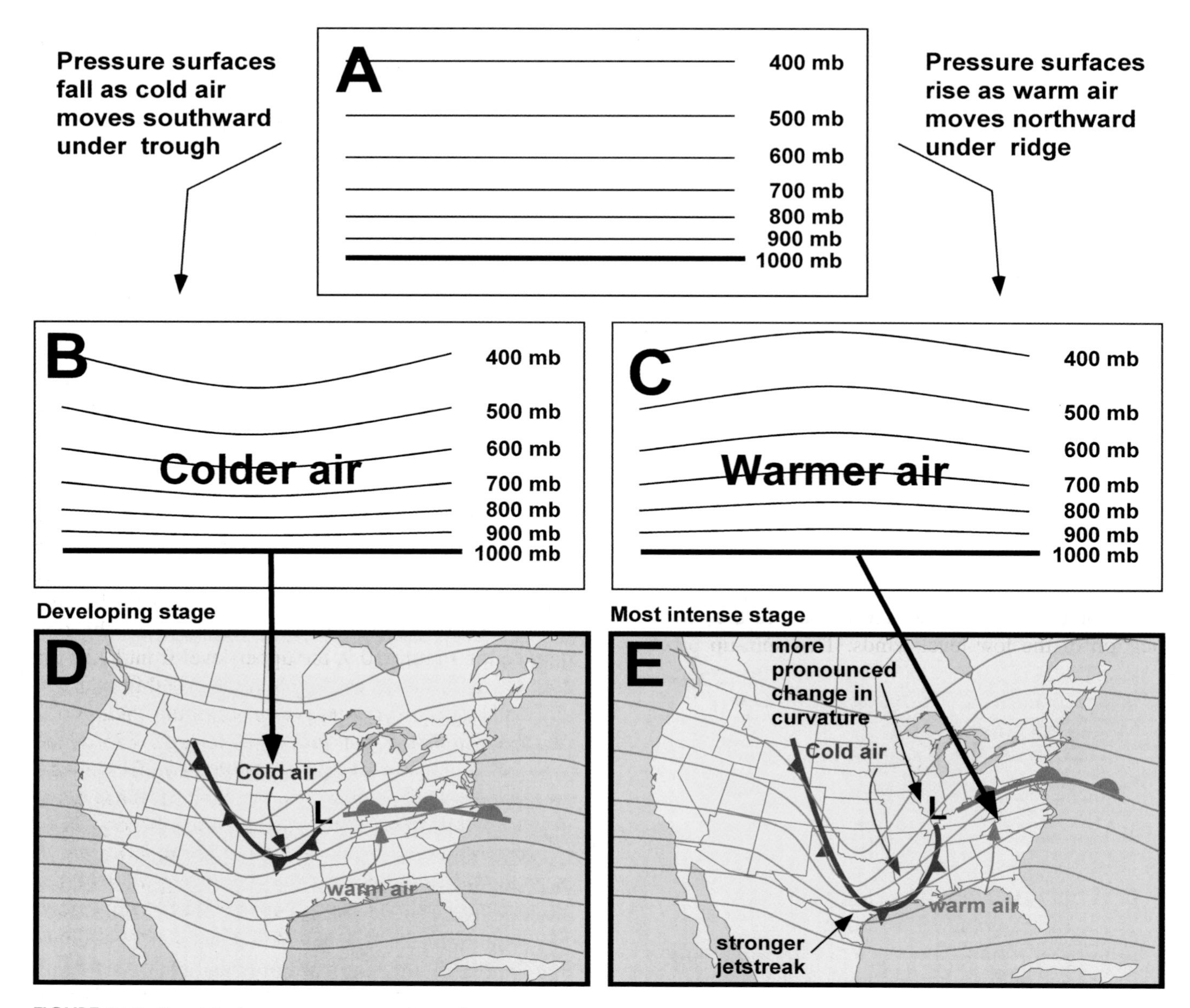

FIGURE 10.8: Panel A shows pressure surfaces for a situation where there are no horizontal gradients of temperature. Panels B and C show the change that occurs in the altitude of pressure surfaces as cold air (B) or warm air (C) moves into region. A drop in the altitude of the pressure surfaces occurs where cold air intrudes while a rise in the altitude of the pressure surfaces occurs where warm air intrudes. Panels D and E show the surface cold and warm fronts and a 300 mb height field (orange) for two times as cold air moves southward west of the surface low-pressure center and warm air moves northward to the east. The altitude of the 300 mb surface drops to the west of the low-pressure center as cold air moves southward. The opposite occurs to the east. These changes intensify the trough and ridge aloft, increasing the intensity of the jetstreak and creating a more pronounced change in curvature. As a result, upper-level divergence increases and the low-pressure center intensifies.

of Figure 10.8A. The air occupies more volume, the distance between the pressure surfaces increases, and individual pressure surfaces rise to higher altitudes (Figure 10.8C). *The altitude of the pressure surfaces increases.*

As warm air moves northward on the east side of a cyclone, the altitude of the pressure surfaces aloft increases with time. The opposite occurs on the west side as cold air moves southward. On an upper-level chart, these two effects appear as a *deepening of the trough west of the low,* and an *intensification of the ridge east of the low* (see the transition in the upper-level flow between Figures 10.8D and 10.8E). Note that this has two important effects:

- the change in curvature between the trough and the ridge is more dramatic, leading to an intensification of the divergence associated with the curvature effect; and
- the pressure gradient intensifies at the base of the trough, leading to stronger winds and an enhanced jetstreak.

Together, these changes intensify the divergence aloft, leading to an intensification of the surface low-pressure center.

This is the first step in the feedback process. The second step follows. As the low-pressure center intensifies, winds become stronger in the lower troposphere, transporting warm air northward faster on the east side of the cyclone, and cold air southward faster west of the cyclone. More rapid transport of cold air south and warm air north in turn causes the pressure surfaces to rise more rapidly east of the low and fall more rapidly west of the low. The net effect: a further intensification of the trough and ridge, a further intensification of the divergence aloft, and a further intensification of the low-level winds. This spin-up process typically occurs over a period of a few hours in a weak cyclone, to as long as twenty-four to thirty-six hours in a very strong cyclone.

THE MATURE CYCLONE

In a mature cyclone, such as the cyclone depicted in Figure 10.1, there are two areas where strong low-level pressure gradients typically lead to a band of strong winds at and just above the surface. These are ahead of either the upper-level front or cold front (ahead of the comma tail), and just northwest of the low-pressure center (under the comma head). These are, respectively, the typical locations of "warm" severe weather (thunderstorms and tornadoes) and "cold" severe weather (blizzards and ice storms). Ahead of the tail of the comma cloud, winds increase rapidly with height in the lower troposphere, forming a feature meteorologists call the ***low-level jet***. In Chapter 19, the low-level jet will be shown to be important for thunderstorm rotation and tornado formation. The line (or lines) of thunderstorms along the tail move eastward with time with the airmass boundaries as the boundaries move eastward across the United States.

North of the cyclone center, as the cyclone spins, a band of moisture and clouds is trapped aloft and is progressively deformed into a narrower and narrower region (e.g., Figure 10.1) as it is wrapped around the cyclone. Air rises throughout the band as the deformation occurs, leading to the production in wintertime of steady, and sometimes very heavy, snow (see Chapter 15). Meteorologists often use the term "trough of warm air aloft," or ***trowal***, to describe the wrap-around band of warm moist air.[1] The term "trowal" is in common use by forecasters and scientists studying the "wrap-around" band of precipitation. We will adopt the term here as well. Let us try to understand the possible structures that the atmosphere can have when a trowal develops.

Figures 10.9 and 10.10 show examples of two cyclones during their mature, most intense stage. First consider the cyclone in Figure 10.9. Figure 10.9A shows the height contours at 300 mb superimposed on the cloud pattern. Figure 10.9B shows the position of the surface fronts, an upper-level front, and the surface pressure pattern. In cyclones such as the one depicted in Figure 10.9, the upper-level front leads the cold front. When a cyclone evolves in this manner, dry air behind the upper-level front in the middle troposphere advances northward on the east side of the cyclone, eventually moving over the warm frontal surface (Figure 10.9B). In this case, a band of warm, moist air becomes trapped between the advancing dry air to the south and the warm front to the north, as shown in cross section in Figure 10.11A (see Figure 10.9 for location of cross section). This band of warm, moist air, the trowal, corresponds to the wrap-around cloud on satellite images.

In other cyclones, the cold front may advance around the south side of the low and begin to wrap northeastward toward the warm front. Eventually it may meet the warm front, at which point an ***occluded***

[1]The term "trowal" is misleading, since troughs in pressure or height fields are associated with cold air aloft, not warm air. In fact, a trowal appears as a ridge in the height field. Meteorologists using the term "trowal" are referring to the potential temperature field rather than the height field. The potential temperature is the temperature a parcel of air would have if it were compressed adiabatically to a pressure of 1000 mb. You can think of potential temperature as the temperature adjusted to account for pressure variations in the atmosphere. The trowal appears as a *thermal* trough, a minimum in the potential temperature field.

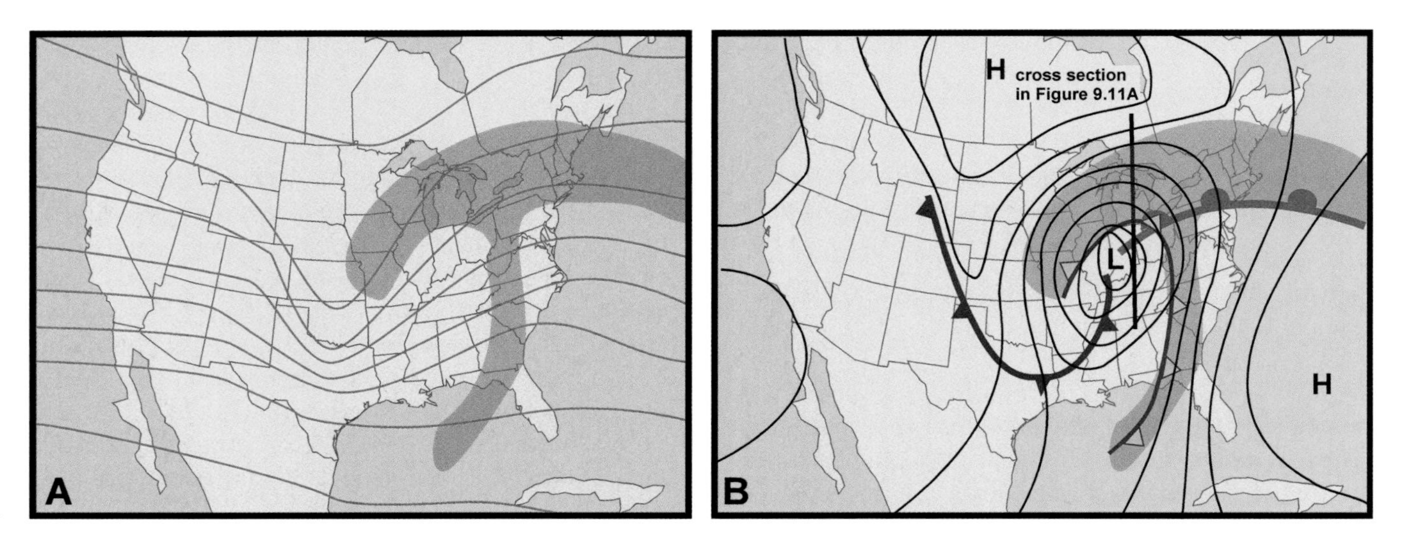

FIGURE 10.9: (A) Jetstream level (e.g., 300 mb) map showing the upper-level trough for a cyclone in its mature stage in a case where an upper-level front is the leading front; (B) surface pressure analysis and fronts. In (B), the upper-level front's position is shown in light blue with open barbs. The gray region on both panels is cloud cover. The blue line denotes the cross section in Figure 10.11A.

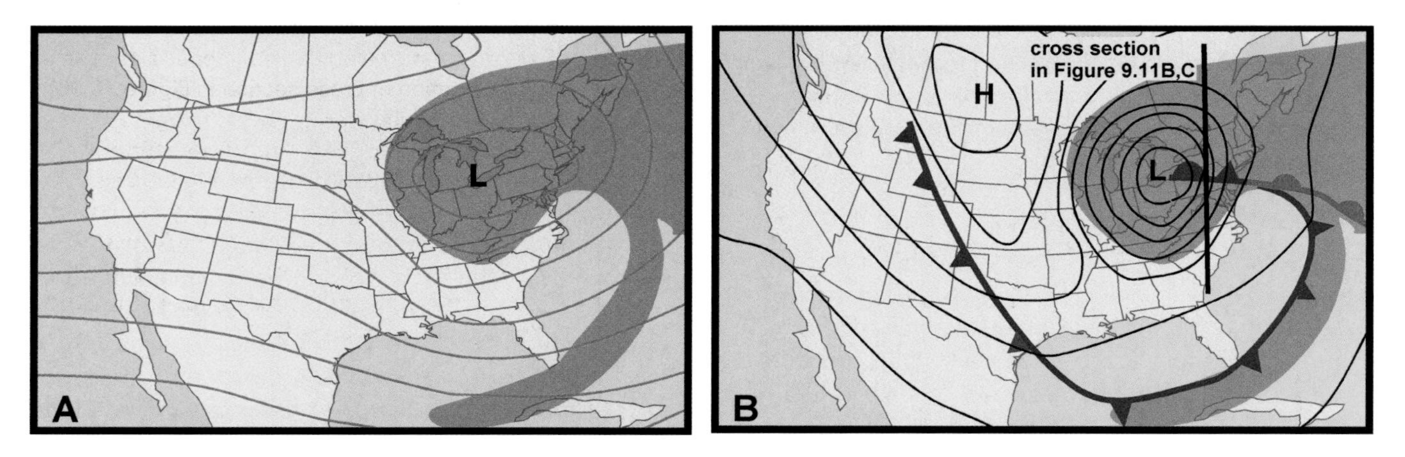

FIGURE 10.10: (A) Jetstream level (e.g., 300 mb) map showing an upper level cutoff low during the mature stage of a cyclone where the cold front leads and an occluded front forms; (B) the corresponding surface pressure analysis and fronts. The gray region on both panels is cloud cover. The black line denotes the location of cross sections corresponding to a cold occlusion (Figure 10.11B) or a warm occlusion (Figure 10.11C).

front will develop (Figure 10.10B). The structure of the occluded front will depend on whether the air behind the cold front is colder or warmer than the air north of the warm front. Figure 10.11B shows the case where air behind the cold front is the coldest air (see Figure 10.10 for cross-section location). In this situation, air behind the cold front will undercut the cool air north of the warm front, creating the type of occluded front referred to as a cold occlusion (see Chapter 9). Figure 10.11C shows an alternative situation where the air north of the warm front is the coldest air. In this case, the air behind the cold front ascends over the air north of the warm front, creating a warm occlusion. In both situations, the air in the warm sector of the storm is displaced upward and is wedged between the two colder airmasses. On satellite images, this warm moist air, the trowal, again appears as the wrap-around band of clouds. Clouds within the trowal in any scenario can produce steady snowfall in winter and rain in other seasons. We will consider the trowal further when we study Great Plains Blizzards in Chapter 15.

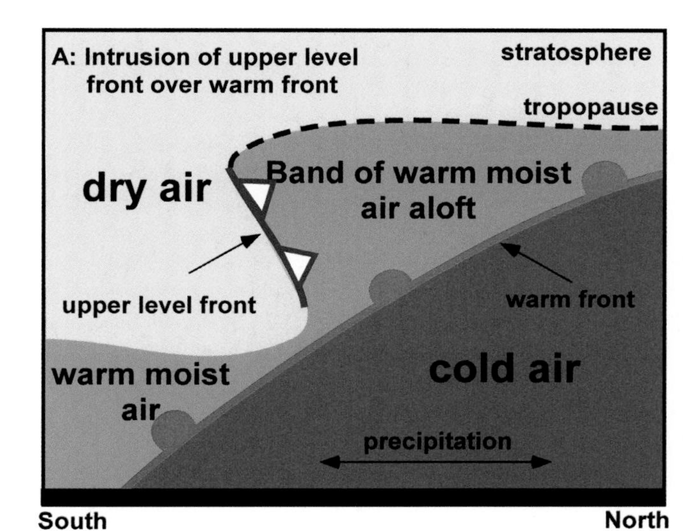

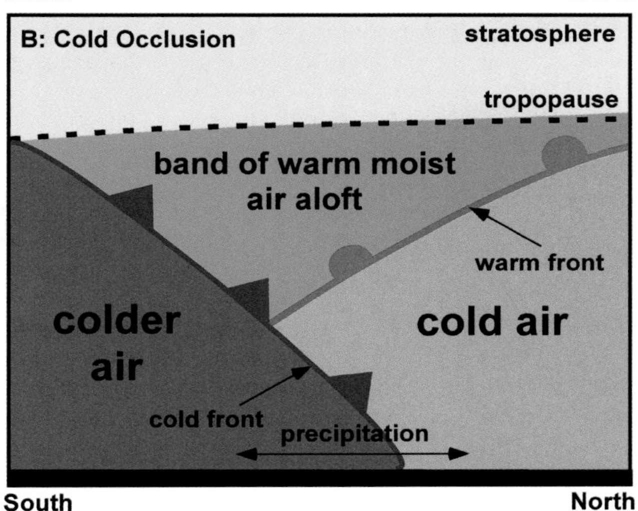

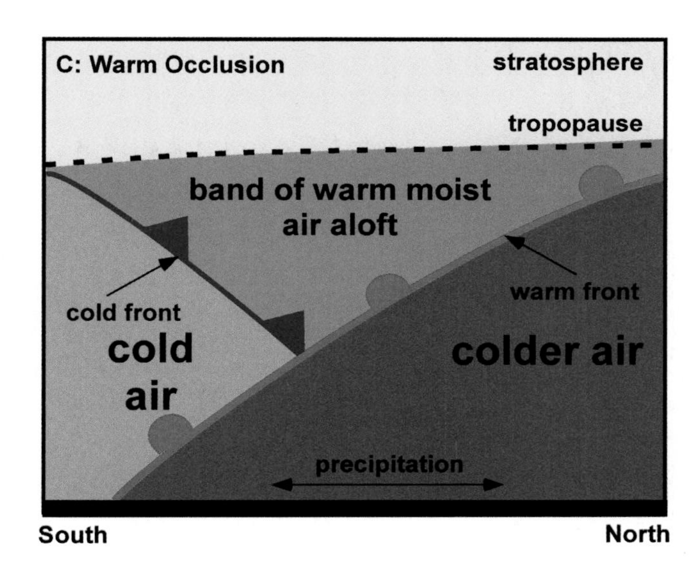

FIGURE 10.11: (A) Schematic cross section looking west through the upper-level front and warm front showing how the band of warm moist air aloft can be trapped between these fronts (see location of cross section in Figure 10.9B); (B) Schematic cross section looking west through an occluded front with "cold occluded" structure showing the positions of the cold and warm fronts and the band of warm moist air trapped aloft (see location of cross section in Figure 10.10B); (C) Cross section similar to (B), but for an occluded front with "warm occluded" structure. In all three cases, meteorologists refer to the wrap-around band of warm, moist air aloft as the trowal.

The "wind down" process of a cyclone occurs as the storm continues eastward. Dissipation of the storm can take several days to over a week depending on the maximum intensity of the storm. During this time, active weather still occurs along the frontal boundaries. As cold air continues to move southeastward, the upper-level trough will continue to deepen. The trough eventually may become so large that a pocket of cold air in the center of the trough becomes completely cut off from the main flow. In this process, illustrated in Figure 10.10A, the jetstream wraps into a completely closed vortex aloft around the cold air pocket, with the vortex "pinched off" (a ***cutoff low***) from the main branch of the jetstream. During this stage, the trowal will typically continue to produce steady precipitation—snow in wintertime—as it is progressively narrowed, stretched, and wound between the colder/drier airmasses to its north and south.

THE DISSIPATING CYCLONE

Recall from Chapter 8 that divergence develops aloft when flow transitions from counterclockwise to clockwise, as happens when air moves from the base of a trough into the crest of a ridge. Once the upper-level trough develops into the cutoff low, the effect of changing curvature becomes insignificant because the flow is always counterclockwise around the cutoff low and there is no longer a *change* in flow curvature. The net effect: *the divergence associated with the curvature becomes insignificant.* Also, as the cutoff low is forming aloft, the jetstreak migrates through the trough to its east side and then weakens dramatically as it moves into the ridge to the east. The net effect: *the divergence associated with the jetstreak becomes insignificant.* At this point, with divergence aloft insignificant, the surface low-pressure center weakens rapidly as surface friction causes air to converge into the low-pressure center at low levels.

FOCUS 10.2

Tracks of Cyclones Forming East of the Rockies

The track of an individual cyclone forming east of the Rocky Mountains depends upon the orientation and configuration of the jetstream during the cyclone's evolution. The most common wintertime cyclone tracks, and the area over which these tracks typically vary, are shown in Figure 10A. Cyclones forming over the southern Rockies in the Colorado/New Mexico region typically follow a track northeastward toward the Great Lakes and into the Canadian provinces of Ontario and Quebec. These cyclones form on the east side of troughs crossing the Rockies, and the jetstream aloft over the cyclone center is normally oriented southwest to northeast. Cyclones forming over the northern Rockies typically develop when the jetstream flows from Canada southeastward into the United States (see Online 10.1: Alberta Clippers). These northern Rocky Mountain cyclones typically track southeastward toward the Great Lakes region, and then turn northeast as they approach the East Coast. The intersection of these two climatological tracks occurs over the Great Lakes. It is also at this location that many cyclones reach their maximum intensity. For these reasons, states of the United States and provinces of Canada bordering the Great Lakes often receive the lion's share of winter snows from continental cyclones. These snowfalls are compounded by lake effect snows, which continue well after the cyclones have departed (see Chapter 13).

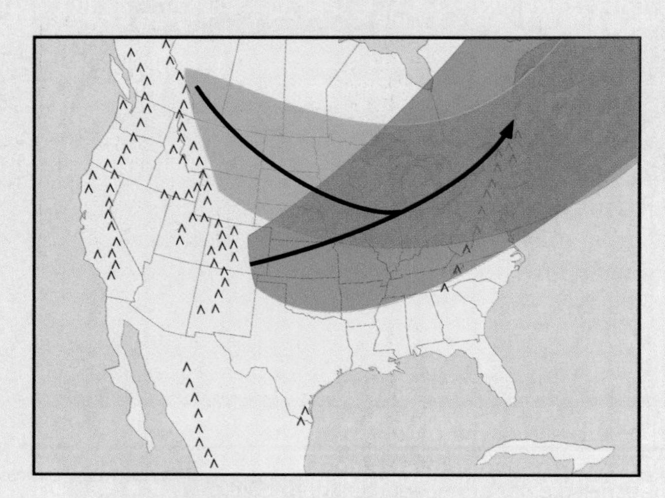

FIGURE 10A: Common tracks of cyclones developing east of the Rocky Mountains. The arrows denote typical tracks and the shaded regions denote the range of the tracks for a large number of cyclones.

ONLINE 10.3

An Upper-Level Cyclone Crosses the Rockies

Most cyclones forming east of the Rockies follow a development process similar to that described in this Chapter—an upper-level wave in the jetstream with an associated jetstreak triggers cyclone formation. Cyclones can form over the plains yet another way. On occasion a deep cutoff low-pressure center will progress across the southern Rockies from the West Coast (Figure 10B). When this fully-developed upper-level low-pressure center emerges on to the plains, bands of precipitation typically develop on its east side and wrap around its northern quadrant. These systems often will develop or will already have well-defined upper-level frontal boundaries along which the bands of precipitation organize. As the upper-level low pressure moves across the plains, the bands organize into the more traditional "comma cloud" shape along the primary upper-level front. Two dramatic

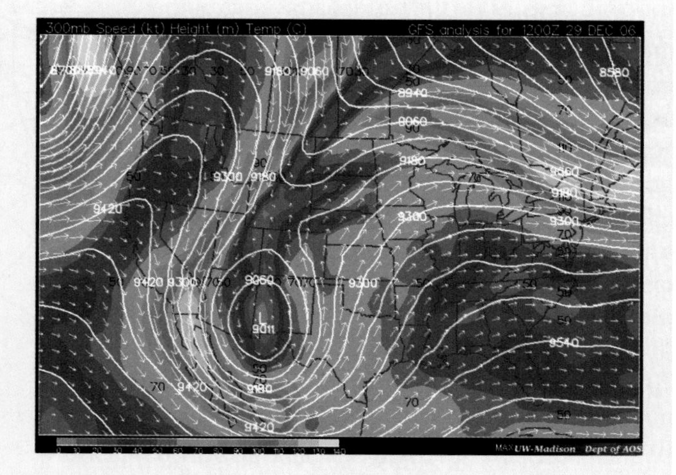

FIGURE 10B: 300 mb chart for 29 December 2006 at 1200 UTC showing a deep cutoff low over the Arizona-New Mexico border prior to the development of the New Year's Eve cyclone over the Great Plains.

(continued)

(continued)

examples of this process occurred just before the Christmas holiday and just before the New Year's holiday in 2006. Both storms produced heavy snow in Denver, snarling the holiday air traffic around the country (see Focus 16.4). Online you can view the complete evolution of the New Year's holiday storm. Satellite animations show the upper-level low approaching the plains from Arizona and New Mexico. Radar animations show the bands of precipitation developing and organizing over the next day into the comma shape along the storm's upper-level front. Upper-level maps show the progress of the cyclone on to, and across, the plains. Compare the position of the surface fronts (look for sharp gradients in temperature and dewpoint) with the precipitation distribution. This storm was remarkable for the lack of coherence between the surface fronts and precipitation, and is an excellent example of the importance of upper-level fronts in cyclones east of the Rockies.

ONLINE 10.4
Cutoff Lows

Dying cyclones that develop a cold-core deep vortex late in their lifetimes can sometimes spin for a week or more in the middle and upper troposphere. Although the low-pressure center at the surface weakens substantially due to frictional convergence of air, the vortex aloft just keeps spinning. This is particularly true when the cutoff low is displaced sufficiently far from the main jetstream that there is no background flow to "push it along" or absorb it. The low heights in a cutoff low are indicative of cold air aloft in the core of the vortex, so the air in a cutoff low is very susceptible to destabilization by surface heating (Chapter 6). As the cutoff low spins, showers and thunderstorms can erupt within the circulation. These storms tend to develop along lines that wrap in a spiral toward the center of the vortex. Sometimes the showers in cutoff lows can even produce brief funnel clouds called "cold air funnels" that rarely reach the ground, but often trigger tornado warnings and warning sirens (see Chapter 19). A cutoff low-pressure center similar to that described above parked over Illinois between 22 to 25 May 2001, bringing a period of showery cool weather. Online satellite and radar animations of clouds and precipitation show the vortex spinning. Watch for the diurnal cycle—storms erupt within the low during the day and dissipate at night, only to return again the following day when solar heating of the surface resumes.

In the dissipation stage, a cyclone becomes a deep vortex with a cold center. The low at the surface is directly underneath the cutoff low aloft, and the whole system slowly spins down as frictional convergence raises the pressure of the surface low. The upper-level vortex may remain, spinning for days as clouds and precipitation continue to develop throughout the cyclone. Eventually, the whole system spins down, often while drifting toward and though the northern United States or Canada and out over the North Atlantic Ocean.

CONTINENTAL CYCLONES AND GLOBAL WARMING

Extratropical cyclones develop as a result of divergence at the level of the jetstream. As shown in Chapter 7, the presence and intensity of the jetstream is directly related to the location and magnitude of temperature gradients in the lower troposphere. The strongest jets occur in the vicinity of sharp fronts. We know from common experience that extratropical cyclones are more frequent, more intense, and occur over a broader latitude range in the winter compared to summer. The annual cycle of cyclone development is directly tied to the temperature contrast between the poles and the tropics, which is greatest in winter and least in summer. Given this basic information about cyclones, can we deduce what might happen to Rocky Mountain cyclone frequency, intensity, and tracks in a world influenced by global warming?

The key data to understand potential effects of global warming on Rocky Mountain cyclones appears in Figure 5.17. Data from the past century and future projections from global climate models all indicate that the most significant warming due to an enhanced greenhouse effect will occur in the polar regions. Since polar regions are expected to heat more significantly than the tropics, the *temperature contrast* between the polar and tropical regions should be reduced during the winter season. The reduction in the temperature gradient across the middle latitudes should result, on average, in a reduction in jetstream intensity and a northward shift in the average jetstream position, not unlike what we see in the natural annual cycle of jetstream strength and location discussed in the previous paragraph. In addition, in a world with warmer polar regions, Spring will arrive somewhat earlier, Fall will transition to winter somewhat later, and winters will be somewhat milder across the middle latitudes. As a result, we can hypothesize that, on average, continental cyclones such as the Rocky Mountain cyclones discussed in this chapter should experience a decrease in frequency and intensity, and a northward shift in track in the warmer climate projected for the twenty-first century by global climate models. This hypothesis has been examined now in several studies that simulate global weather patterns using global climate models. These studies support the basic ideas presented above. In general, studies show a decrease of cyclone frequency of 5 to 10 percent in the cool season in the late twenty-first century compared to the twentieth century.

The societal impacts of a decrease in cyclone frequency and intensity over the continent remain poorly understood. Consider, for example, one primary concern; whether precipitation amount and distribution associated with cyclones might decrease in the twenty-first century. To determine this, we must not only consider cyclone frequency and intensity, but other factors such as the speed at which a cyclone propagates. We know from common experience that a weak, slow-moving cyclone can sometimes produce very heavy rainfall over a broad area, while a fast-moving cyclone may sometimes produce considerably less precipitation. If cyclones in a warmer world progress more slowly across the continent (a possibility with a weaker jetstream), the increased time precipitation regions reside over an area may compensate for the reduced number of times precipitation occurs over the area. Cyclone speed is just one of many complexities that make estimation of impacts on precipitation uncertain at this time. There are many other concerns in addition to precipitation, among these the frequency of occurrence of hazardous weather. We will explore these potential impacts of global warming in future chapters as we consider various phenomena that occur in association with cyclones.

SUMMARY

A cyclone brings many changes to the weather over the continent. Warm humid air is transported northward and upward away from the surface. During its ascent, the water vapor in the air is condensed into clouds and then precipitation, producing the rains required for life but also the hazards of convective and winter weather. Replacing the warm air is cold air, which moves southward from the polar latitudes as it descends from aloft. In this manner, each cyclone contributes to the never-ending effort to rebalance an atmosphere that is continually unbalanced by uneven solar heating between the poles and the tropics and between the Earth's surface and the atmosphere aloft.

CHECK YOUR UNDERSTANDING 10.3

1. Explain the feedback process between air flow of the jetstream and the strengthening of the surface low.
2. What are the typical locations and orientations of fronts during the mature stage of a Rocky Mountain cyclone?
3. What is the trowal and where is it located relative to the center of an extratropical cyclone?
4. How is global warming expected to affect Rocky Mountain cyclones?

TEST YOUR UNDERSTANDING

1. What type of weather may be found in the "head" of an extratropical cyclone? What about along the "tail" of an extratropical cyclone?
2. How do cyclones balance temperature differences caused by unequal solar heating of the Earth?
3. Explain how the jetstream and jetstreaks cause the development of a surface cyclone.
4. What is overrunning? Where would you expect to find overrunning in an extratropical cyclone?
5. What is the dry slot? Where is it found relative to the center of an extratropical cyclone?
6. What front is typically found east of the extratropical cyclone center? Describe the weather conditions in this region.
7. What airmasses may be found south of a low-pressure system? How can they be positioned relative to one another?
8. What role do moisture and stability of the air have in determining the likelihood and severity of thunderstorms along an upper-level front? What about along a dry line? What about a cold front?
9. What is upslope flow? What kind of weather does it help to produce?
10. What types of fronts can be responsible for the "tail" of a comma cloud?
11. Often the skies will clear and the temperatures will actually rise after a strong line of thunderstorms passes in midday in the Midwestern United States. Does this fit the conceptual model of the "tail" of a cyclone? Why or why not?
12. Where are the strongest low-level winds found in a mature cyclone? What type of severe weather can be associated with these strong winds?
13. Describe physically how cold air moving southward west of a low and warm air moving northward east of a low lead to an amplification of the wave in the jetstream.
14. Describe how the clouds and precipitation would change if you were to drive southward through a warm front during a mature cyclone.
15. What happens to the position and intensity of cold and warm fronts as a cyclone matures?
16. What type of weather would you experience if the trowal of a cyclone was overhead in winter? How about summer?
17. Figure 10.1 shows a cyclone at its mature stage. Suppose you drove from Atlanta, Georgia to Glasgow, Montana through this cyclone. What kind of weather would you encounter? You then drive from Glasgow to Milwaukee, Wisconsin. What kind of weather would you encounter? (Assume for simplicity that the cyclone does not move during your drive.)
18. What is a "cutoff low"?
19. Would an extratropical cyclone intensify or dissipate if the upper-air trough were located to the east of the surface low? Explain your answer and include a diagram that illustrates your reasoning.
20. Why does a cyclone eventually dissipate? Include in your explanation the role of the cutoff low and the role of friction.

TEST YOUR PROBLEM-SOLVING SKILLS

1. An extratropical cyclone has formed east of the Rocky Mountains and is moving northeastward across the United States.
   (a) The airmass located over the Gulf Coast has temperatures in the 70s, dewpoints in the 60s, and lapse rates on the order of 8° C/km. Would you expect thunderstorms to develop as the cyclone and fronts move through this area? Why or why not?
   (b) If the airmass located along the Gulf Coast had temperatures in the 70s, dewpoints in the 40s, and lapse rates of approximately 6° C/km, would you expect thunderstorms to develop as the front moves through the area? Why or why not?

2. A surface analysis shows a low-pressure center developing in eastern Colorado. A quick look at the 300 mb chart shows that there is near zonal (west to east) flow in the jetstream, although the winds are much stronger over Arizona and New Mexico than farther to the north. What is causing this surface low to develop?

3. An extratropical cyclone is centered over Tennessee with a cold front extending southwestward into eastern Texas and a warm front east through North Carolina. A line of showers and thunderstorms extends south of the surface low-pressure center through eastern Alabama and into the Gulf of Mexico. What upper air charts would you examine to identify the lifting mechanism for these showers? What variables on the chart would you specifically examine?

4. It is mid-February and an extratropical cyclone in its developing stage is positioned over central Kansas as shown in the figure on the next page. Typical weather conditions are present in the vicinity of the cyclone. Where would you expect to find each of the following weather reports?
   (a) Overcast with freezing rain, relative humidity near 100 percent, easterly winds.
   (b) Clear skies, relative humidity = 25 percent, westerly winds.
   (c) Severe thunderstorm nearby, relative humidity = 93 percent, southeasterly winds.
   (d) Clear skies, relative humidity = 90 percent, southerly winds.
   (e) Heavy snow, relative humidity = 100 percent, northeasterly winds.

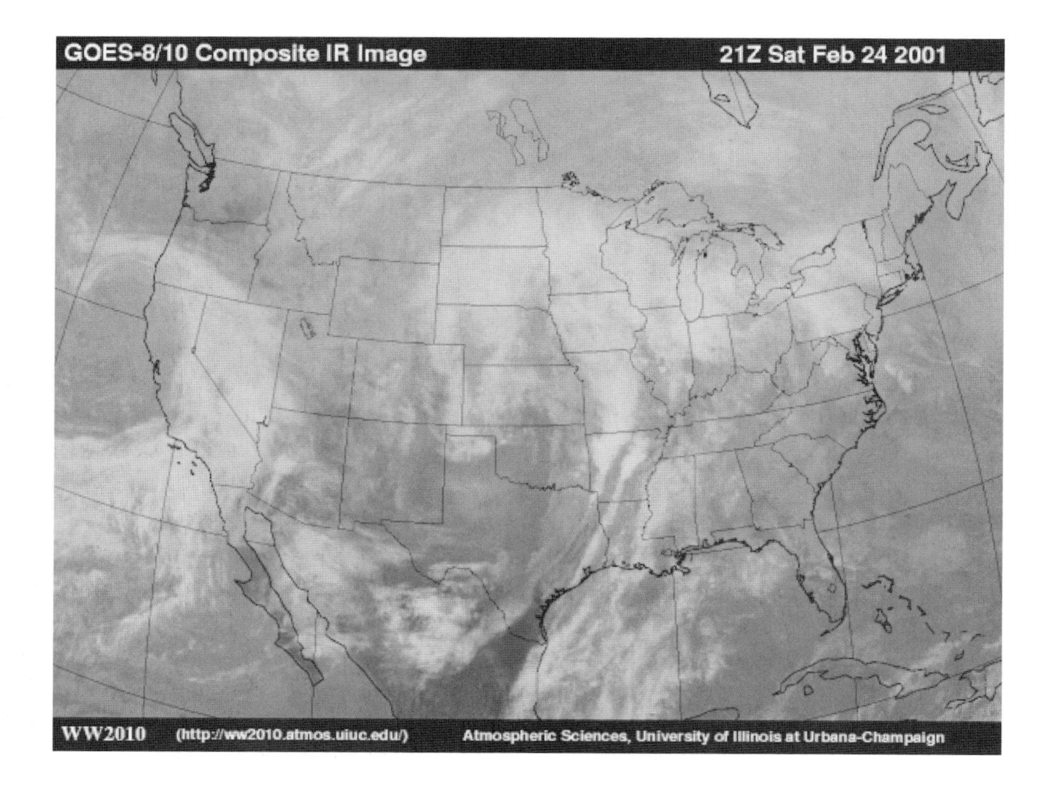

USE THE SEVERE AND HAZARDOUS WEATHER WEBSITE

http://severewx.atmos.uiuc.edu

1. Watch for the next extratropical cyclone that develops east of the Rockies and navigate the *Severe and Hazardous Weather Website* to current maps that depict the cyclone.
   (a) View a map of surface station data for the United States. Locate all frontal boundaries by using the station data. Now examine a surface analysis (with fronts and isobars) for the same time. Were you able to identify all the fronts present? Does your frontal analysis agree with the surface analysis you found on the web? Why or why not?
   (b) Find an infrared satellite image for the same time. Is a comma cloud present? Identify the location of the head and the tail of the comma cloud. Does the tail of the comma correspond to where you located the leading front south of the low-pressure center? If not, why not? Is the front an upper-level front? How did you determine if it was?
   (c) Bring up an upper air map that shows the jetstream. Identify the curvature change in the flow and/or the jetstreak that is supporting the surface low. Estimate and record the location where the maximum divergence should be occurring based on your understanding of Chapter 8.
   (d) Examine a radar image for the same time. Record the locations and types of precipitation associated with the cyclone. Identify the likely lifting mechanisms for all the precipitation you recorded.
   (e) Find soundings for stations located southeast, southwest, northeast, and northwest of the surface low. How does the temperature profile, relative humidity, and average lapse rate in the lower troposphere vary from location to location?

2. During the next severe weather event in a season other than summer, navigate to the *Severe and Hazardous Weather Website* and find the following maps and charts: surface analysis, satellite image, radar image. After the event is over, collect all the severe weather reports (try local National Weather Service offices for storm summaries). Compare the severe weather report locations with the position of the extratropical cyclone that was the parent storm to the severe weather and determine if the severe events developed where you would have expected. Explain any discrepancies you find. (Pay close attention to the time of the maps and the time of the reports.)

3. Shortly after the passage of an extratropical cyclone through the Southern Plains states, navigate the *Severe and Hazardous Weather Website* to wind profiler data.
   (a) Examine a wind profile display from one of the profilers in the Wind Profiler Network. Record the profiler location and time.
   (b) Identify the times of any frontal passages.
   (c) Summarize the changes in the wind pattern associated with the passage of fronts.
   (d) Identify the approximate altitude of the jetstream. Explain how you arrived at your answer.
   (e) Examine the data for evidence of the low-level jet. What evidence do you have for the existence of a low-level jet? (Hint: Look at the winds below 850 mb.)

4. During the mature stage of an extratropical cyclone, navigate the *Severe and Hazardous Weather* Website to locate a satellite image of the cyclone.
   (a) Print or save the satellite image. Record the date, local time, and type of satellite image.
   (b) Identify the trowal associated with the cyclone. List the states or regions where the trowal is located.
   (c) Navigate to soundings on the "Current Weather" page of the *Severe and Hazardous Weather Website.* Identify at least one upper-air station located within the trowal. Examine a sounding collected when the trowal was over the station. Record the date, local time, and location of the sounding.
   (d) Carefully examine the sounding and identify the warmer air aloft. Between what pressure levels is the warmer air located?
   (e) Examine a radar image during this same time. Is the type and intensity of precipitation in the region of the trowal what you would expect, based on the material presented in this chapter? Why or why not?

EXTRATROPICAL CYCLONES FORMING ALONG THE EAST AND GULF COASTS

A late season East Coast cyclone moves over New England on 31 March 1997.

KEY TERMS

angular momentum
beach erosion
blizzard conditions
bomb cyclone
coastal front
cold air damming
dry slot
East Coast cyclone
explosive cyclogenesis
Gulf Coast cyclone
Hadley cell
jetstreak
latent heat
meteorological bomb
Nor'easter
polar jetstream
sensible heat
subtropical jetstream
wrap-around cloud band

Extratropical cyclones form along the East and Gulf Coasts of North America several times each year during late fall, winter, and early spring. Over the centuries, these storms have had tremendous impact on political events and the local economy, influencing everything from the outcome of the American Revolutionary War to the early establishment of underground transportation systems in northeastern cities. Coastal cyclones have the greatest economic impact of any storm type on the northeast United States and Canada, even more than hurricanes. The combined effects of heavy snowfall, high winds, coastal flooding, and cold temperatures in a single storm can create hazardous conditions over a widespread region. The greatest impact of heavy snow is within the urban corridor stretching from Boston, Massachusetts, to Washington, D.C.

Figure 11.1 shows the number of snowfall events exceeding 4 and 10 inches (10 and 25 cm) in the five most populous cities along the East Coast (Boston, New York, Philadelphia, Baltimore, and Washington) between the winters of 1949–50 and 1998–99. During these 50 winters, Boston, at the north end of the urban corridor, experienced 158 storms with snowfall exceeding 4 inches and 35 storms exceeding 10 inches, or about one major and two to three moderate snowfall events each year.[1] At the south end, Washington D.C. averages a major event every four to five years and about one moderate event per year.

Coastal cyclones preferentially form in two regions along the U.S. eastern coastline: the first just off the central East Coast near Cape Hatteras, North Carolina, and the second just off the Gulf Coast centered near the Texas-Louisiana border. Although cyclones can form at other points along the coast, they develop so commonly in these two regions that we will consider separately why each region is favored for cyclone formation. We will also consider separately the impacts of cyclones originating in each of these areas. As shown in Figure 11.2, ***East Coast cyclones*** typically track northeast along the coast and out to sea northeast of Canada's Atlantic Provinces. ***Gulf Coast cyclones*** commonly track either eastward along the Gulf Coast and then northward along the Atlantic coastline, or inland along the Mississippi and Ohio River Valleys.

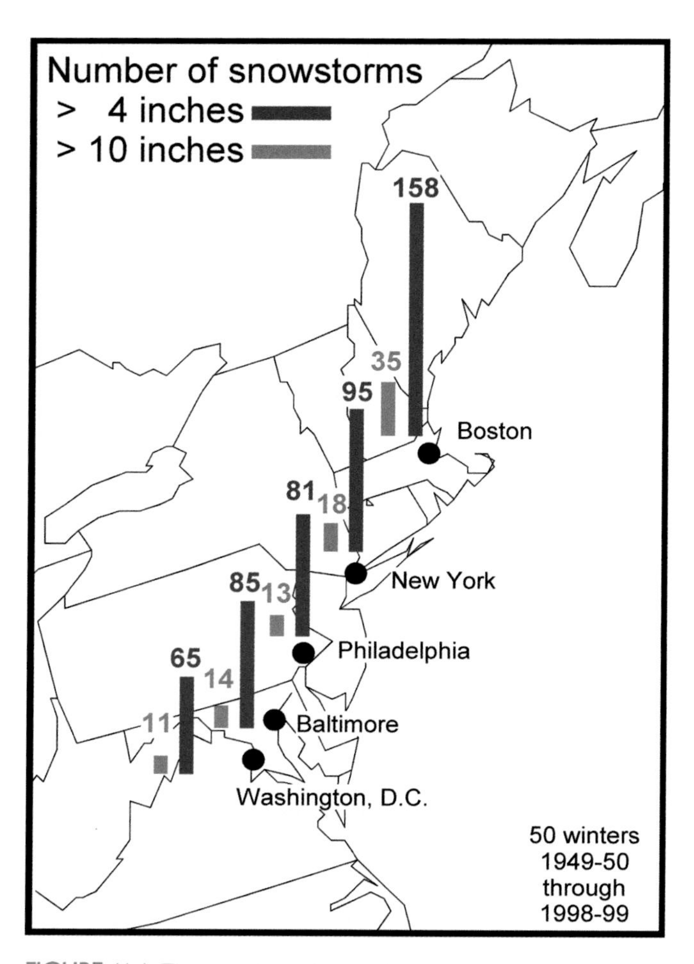

FIGURE 11.1: The number of snowfall events exceeding 4 inches (blue) and 10 inches (red) at 5 cities along the northeast coast during 50 winters beginning in 1949–50 and ending in 1998–99.

Cyclones forming along the East and Gulf Coasts of the United States are often more intense than their Rocky Mountain counterparts. The difference in intensity arises from five factors:

1. *Latent heat released during condensation in the clouds contributes more to storm intensification.*

[1]Note that storms with snowfall > 10 inches are also storms with snowfall > 4 inches. A city having one storm > 10 inches and three > 4 inches would have a total of three storms, not four.

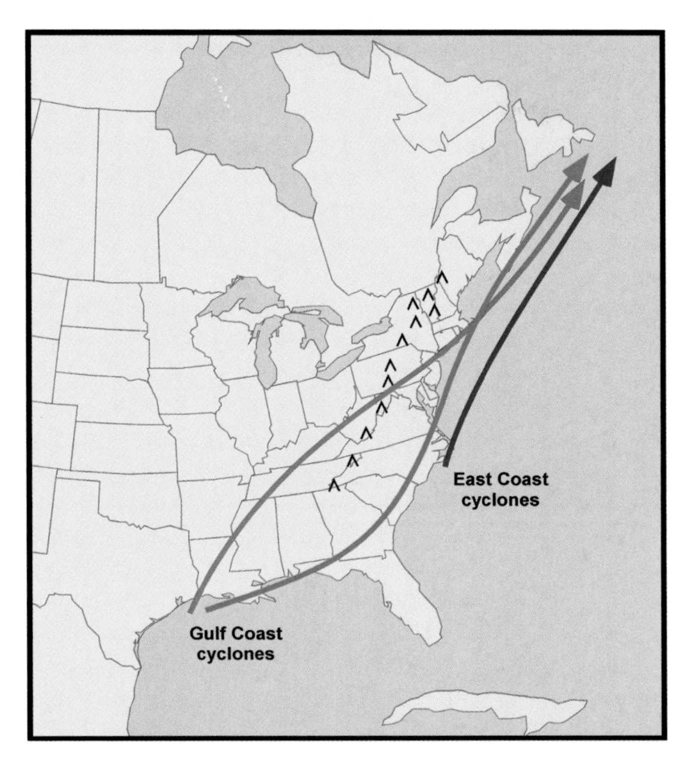

FIGURE 11.2: Typical tracks of cyclones originating along the East and Gulf Coasts of the United States.

Unlike Rocky Mountain cyclones, coastal cyclones form over or near warm water. Warm water is present in the Gulf of Mexico, and is also found in the Atlantic Ocean within the Gulf Stream, a warm ocean current that flows out of the subtropics northward along the Atlantic coast. Warm water provides a strong local source of moisture, so that the clouds within the cyclone can produce heavy precipitation. Latent energy released during condensation heats the air in the core of the cyclones, contributing to further intensification of the low-pressure center (see Chapter 8).

2. *Sensible heat from the ocean surface acts to heat the atmosphere, contributing more to storm intensification.* Energy that is transferred directly to the atmosphere from the surface through conduction is called ***sensible heat***. During the cold season, land and snow surfaces over the North American continent are often much colder than the adjacent ocean surfaces. The air in cyclones forming over water is heated to a greater degree by the ocean surface compared to air in cyclones over land.

3. *Strong thermal contrasts between the ocean and land enhance and maintain a sharp thermal boundary along the coastline.* Very sharp contrasts in temperature can develop in winter along the coast due to differences in surface heating between the cold continent and the warm ocean. These contrasts are most pronounced when a very cold airmass arrives at the coast from the interior of the continent. Cyclones forming along the sharp thermal boundary draw energy from this thermal contrast as the cold and warm air move around the developing low-pressure center.

4. *There is often more than one jetstreak acting to create divergence in the upper level flow.* These jetstreaks are found within two jetstreams: the ***polar jetstream*** and the subtropical jetstream.

5. *The flow in cyclones over water experiences less frictional turning into the low.* Since the frictional force is reduced over water compared to land, the rate at which air converges into the low-pressure center will be reduced, allowing the low to maintain a greater intensity.

In this chapter, we will study how these factors control the intensity of coastal cyclones and focus on the reasons why these storms produce ***blizzard conditions*** and heavy snowfall across such a wide geographic region.

EAST COAST CYCLONES

Figure 11.3 illustrates common weather patterns during the development and evolution of East Coast cyclones. The left panels illustrate the sea-level pressure field, fronts, and cloud patterns at three times: prior to the development of the cyclone (A), during its intensification (B), and at its most intense stage (C). The right panels (D through F) show upper tropospheric height fields and jetstreaks at times corresponding to panels A through C. The core of the jetstreak within the subtropical jetstream is normally centered near the 200 mb level, while the core of the jetstreak within the polar jetstream is typically found near the 300 mb level.

Each cyclone is different from all others. The presence and juxtaposition of the jetstreaks appearing in Figure 11.3 will differ from case to case. Their strength and orientation will strongly affect the intensity of any particular cyclone. In some cyclones, one of these jetstreaks may either not be present or not be positioned to contribute significantly to the divergence

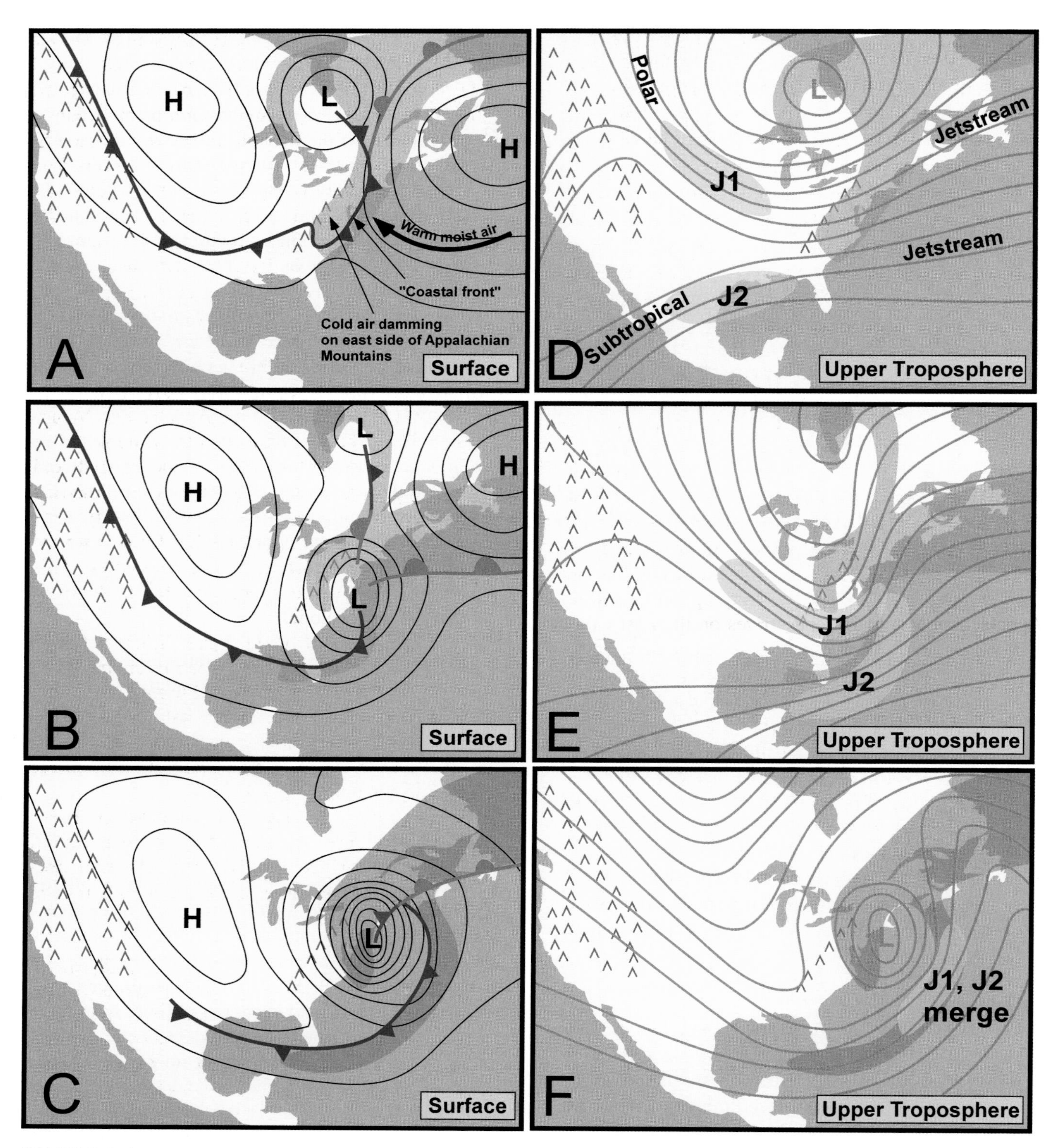

FIGURE 11.3: Figures A through C: A sequence of schematic charts showing the sea-level pressure, fronts, and cloud patterns (gray shading) (A) prior to the development of an East Coast cyclone, (B) during the rapid deepening of the low-pressure center, and (C) at the most intense stage of an East Coast cyclone. Figures D through F: Schematic charts of the upper tropospheric flow patterns that correspond to Figures A through C. The yellow shaded regions (J1, J2) denote jetstreak locations.

occurring over the developing surface low-pressure center. The weather patterns in Figure 11.3 and subsequent figures are meant to be archetypical—they represent a range of possible weather patterns that can lead to the formation of an East Coast cyclone.

The Environment Prior to the Development of an East Coast Cyclone

East Coast cyclones typically develop after an earlier cyclone originates east of the Rockies and progresses across the continent. In some cases, the East Coast storm will begin to develop near the time that the Rocky Mountain cyclone approaches the eastern Great Lakes region. At other times, as shown in Figure 11.3A, the older cyclone will already have progressed into Canada. The older cyclone is normally dissipating as the East Coast cyclone develops. The coupling between these two storms is not coincidental—the first storm provides the necessary environment for the second to form. The Rocky Mountain cyclone moves cold air southeastward across the continent behind its cold front. The cold air eventually spills across the Appalachian Mountains and arrives on the East Coast (Figure 11.3A). The southeastward movement of the cold front is accompanied by a southeastward displacement of the jetstream aloft (recall the relationship between fronts and jetstreams discussed in Chapter 7), so that the polar jetstream now flows across the central Atlantic coast (Figure 11.3D).

Often during the development of an East Coast cyclone, an area of high pressure will be located over the North Atlantic Ocean. This high-pressure center is normally the remains of a cold airmass that had moved out to sea off the North American continent a few days earlier. South of this high-pressure center, air flows westward toward the East Coast of the United States, as depicted in Figure 11.3A. The westward moving air transits over a large region of the Atlantic Ocean, during which it moistens due to evaporation of ocean water and warms due to sensible heating. The temperature of air moving across the Atlantic can approach 60° F (~15° C) by the time it arrives at the East Coast, a sharp contrast to the cold, often subfreezing air spilling over the Appalachians. As cold continental air arrives on the east side of the Appalachians, the warm moist flow moving westward off the Atlantic forces the cold air into a narrow wedge between the coastline and the Appalachian mountain chain, a phenomenon called ***cold air damming*** (see Figure 11.4). The Appalachian Mountains dam the cold air on the east side, while warm air coming off the ocean flows up over the trapped cold air. This whole process leads to the formation of what meteorologists call a ***coastal front***, a stationary boundary between cold air dammed on the east side of the Appalachians and warm air over the ocean. This boundary becomes the focal point for the development of the East Coast cyclone.

In the upper troposphere (Figure 11.3D), a trough typically is present over the east-central United States associated with the older Rocky Mountain cyclone and the cold air that flowed southward out of Canada behind it. Prior to the development of an East Coast cyclone, a strong jetstreak is normally present on the west side of this trough.

When many East Coast cyclones develop, a second jetstream is present across the southern United States and Mexico (Figure 11.3D). This jetstream, centered at about the 200 mb level, is at a higher altitude

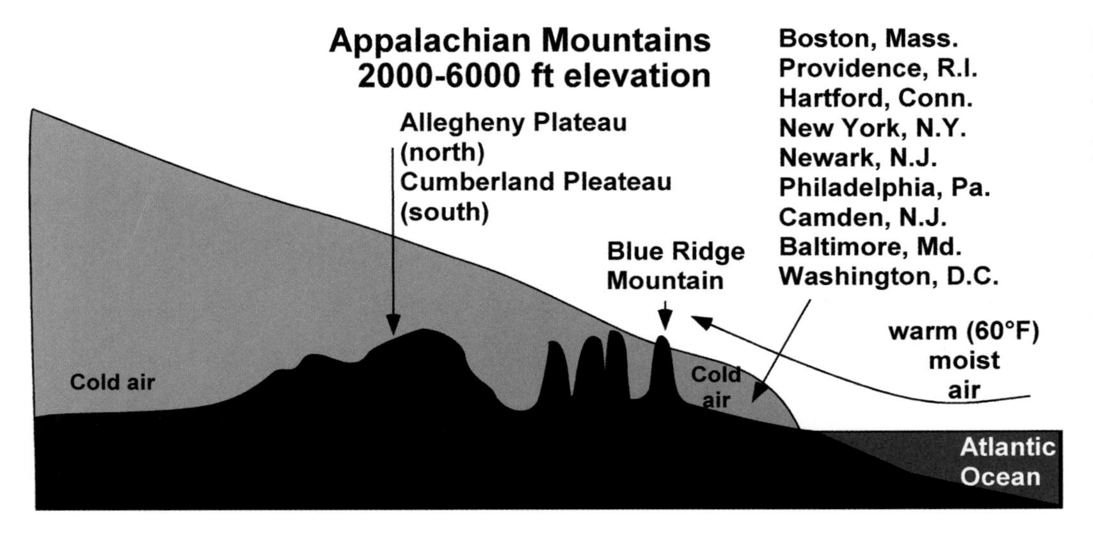

FIGURE 11.4: Illustration of cold air damming on the east side of the Appalachian Mountains. The warm moist air flows in from the Atlantic, trapping the colder air along the coast. The urban areas along the coast are under the cold air, which is typically near or below freezing. Under these conditions, precipitation can fall as heavy wet snow, ice pellets, or freezing rain.

ONLINE 11.1

The Subtropical Jetstream and Water Vapor Satellite Imagery

The subtropical jetstream over the southern United States forms on the poleward side of a circulation called the Northern Hemisphere *"Hadley Cell"*, one of two large cells in Earth's tropical atmosphere (Chapter 5). Figure 11A shows the circulation within the Hadley Cells. The difference between solar heating in the equatorial areas and the middle latitudes generates pressure gradients aloft that drive the flow of the Hadley cell. Under the influence of these pressure gradients aloft, air aloft can flow from the tropics into the middle latitudes. As it does, it accelerates eastward due to conservation of *angular momentum*, creating the subtropical jetstream. Typical maximum values of winds in the subtropical jetstream are about 100 knots. Online, you can "see" the subtropical jetstream by watching high clouds streaming out of the subtropics over the southeastern United States on 2 January 2002.

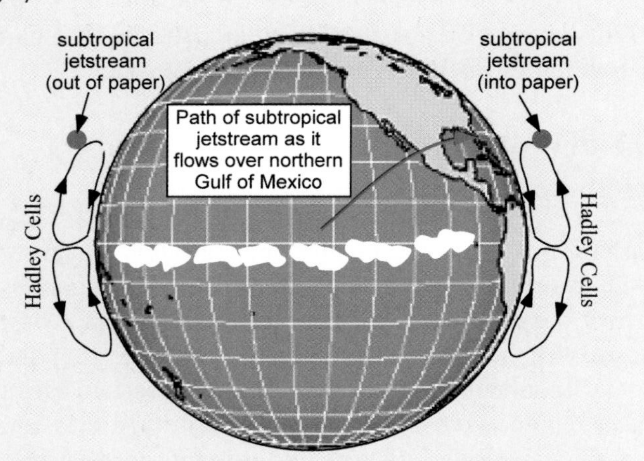

FIGURE 11A: The Hadley Cell in each hemisphere consists of rising motion located near the equator, poleward flow aloft near the tropopause to approximately 30° latitude, descending motion, and return flow back toward the equator at the surface. As air flows northward, it accelerates eastward due to the Coriolis force, creating the subtropical jetstream.

than the polar jetstream, which typically is centered near 300 mb. The southern jetstream is called the ***subtropical jetstream*** because it originates in the tropics and moves northeastward across the subtropics toward the middle latitudes. During the development of a strong East Coast cyclone, the subtropical jetstream often moves over and merges with the polar jetstream, as shown in Figure 11.3D. The subtropical jetstream can have a ***jetstreak*** embedded within it that acts in concert with the jetstreak in the polar jetstream to enhance divergence aloft.

The Initial Development of the Cyclone

At the time of initial development of an East Coast cyclone, the jetstreak on the west side of the polar jetstream migrates toward and into the base of the trough (J1 in Figure 11.3E). As the jetstreak approaches and rounds the base of the trough, its left exit region moves over the coastal front. Curvature also plays a role as divergence associated with the change in curvature east of the trough combines with the jetstreak to maximize divergence on the east side of the trough.

In many storms, the jetstreak in the higher altitude subtropical jetstream (J2 in Figure 11.3E) also can align in such a way that divergence in its left exit region superimposes over the divergence region of the jetstreak in the polar jetstream. When both jetstreaks align in this manner over the coastal front, very strong divergence occurs in the upper atmosphere. The divergence is further enhanced by ***latent heat*** release in the clouds, and the transport of sensible heat upward from the ocean surface by the action of the wind circulations. The low-pressure center will rapidly form over the ocean just east of the coast (Figure 11.3B) directly underneath the region of maximum upper-level divergence. The pressure can fall dramatically in the cyclone center, as much as 30 mb in 24 hours (see Focus 11.2). The strongest cyclones can develop central pressures in the range of 960 to 970 mb, values comparable to the central pressures in moderate hurricanes. The accompanying strong horizontal pressure gradients create strong winds that, north of the cyclone center, can approach hurricane intensity, with gusts as high as 60 to 80 knots in the strongest storms. Unlike a hurricane, these winds are often accompanied by heavy snow, leading to blizzard conditions, heavy surf, and ***beach erosion*** along the coast, and treacherous conditions for shipping and air traffic along the Atlantic seaboard.

In New England, mariners call these storms ***Nor'easters*** because the winds blow from the northeast over New England during the worst of the storm when blizzard conditions rage. The media has popularized this term and the meteorological community and the public have adopted it to describe East Coast cyclones throughout their lifetimes.

FOCUS 11.1

Beach Erosion

The beaches of the Atlantic and Gulf Coasts change constantly, advancing and retreating in response to winds, tides, waves, breakers, and swell. Rapid change can occur during strong storms as high waves and swell advance inland, carry sand seaward, and deposit it offshore. Beach erosion and shoreline retreat would not be a problem in the absence of human structures, since new beaches would reform inland relative to their original position. However, structures such as hotels, highways, boardwalks, houses, and seawalls all create barriers to shoreline retreat. During storms, beach erosion causes beaches to become smaller, sometimes disappearing completely and leaving manmade structures at the water's edge. Beach erosion during storms has become a serious and expensive problem for communities along the East and Gulf Coasts. Loss of beachfront ruins the attractiveness of a community, reduces its potential to draw tourists, and leaves shoreline structures more vulnerable to damage during storms. To fight erosion, jetties are often built along the coasts to reduce ocean currents. These engineering solutions often offer only temporary respite from erosion, since they interfere with the natural flow of sand. Beaches are also replenished through "beach nourishment," a euphemistic term for dredging sand from offshore and dumping it back onto the beach. Beach nourishment is expensive. The cost to New York State alone, adjusted to 2006 dollars, has exceeded $0.6 billion during the last century. Along the Gulf and East Coasts, the cost exceeded $3.0 billion. On the Atlantic Coast, it has been estimated that one-quarter of the replenished beaches have had a useful durability of less than one year, and 62 percent last between one and five years. This is but one of the many high costs of storms along the coastlines of North America.

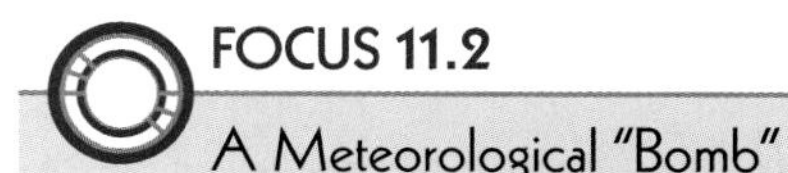

FOCUS 11.2

A Meteorological "Bomb"

Meteorologists frequently use the term *"explosive cyclogenesis"* to describe the rapid deepening of a low-pressure center during cyclone formation. In 1980, two Massachusetts Institute of Technology scientists, Drs. Frederick Sanders and John Gyakum, carried the "explosive" analogy one step further, using the term "bomb" to describe cyclones that have a central pressure fall of 24 millibars in 24 hours[2]. The term *"meteorological bomb"* was further popularized by East Coast meteorologists and later by the national media. *"Bomb cyclones"* form primarily along the U.S. East Coast, over the western Pacific, and over the Mediterranean Sea. Occasionally, when a Rocky Mountain cyclone meets the Sanders and Gyakum criteria, meteorologists refer to the cyclone as a "land bomb." Although the bomb analogy is technically incorrect (bombs explode outward, and therefore are centers of high pressure, not low pressure), the terminology has become very popular and has joined the rich meteorological lexicon used to describe hazardous weather events.

[2]Strictly speaking, the criterion for a bomb cyclone, as defined by Sanders and Gyakum, depends on latitude. A fall of 24 mb in 24 hours is required ta 60° latitude, while only 19 mb in 24 hours is required at 45° latitude. The difference is related to the strength of the Coriolis effect at these latitudes.

CHECK YOUR UNDERSTANDING 11.1

1. Describe the primary surface and upper-air features that are typically present just prior to the development of an East Coast cyclone.
2. What five factors contribute to the intensity of East Coast cyclones?
3. Describe how a coastal front develops prior to the formation of an East Coast cyclone.
4. What is a Nor'easter?

Fronts, Airflow, and Weather Conditions

As the low-pressure center develops, cold air southwest of the low flows out over the ocean, its leading edge marking the surface cold front (Figure 11.3B). The weather along the cold front poses no threat except to ships, since it occurs entirely over the Atlantic Ocean. Of more importance is the weather north and west of the developing low-pressure center. A warm front, the boundary between the cooler air over the North Atlantic and very warm air moving northward from the subtropics, develops east of the cyclone center. Warm air flows upward over the cold air north of the warm front, creating a wide shield of clouds and precipitation that extends from the front northward (Figure 11.3B). Along the coast, easterly flow north of the warm front keeps cold air dammed against the Appalachian Mountains. In the region north of the low-pressure center, warm oceanic air rising upward along the warm front turns westward in the counterclockwise flow about the cyclone. As this happens, the warm moisture-laden air rises over the cold air trapped against the mountains, and over the mountains themselves. Heavy precipitation is produced as the air rises. In winter, precipitation takes the form of heavy snow and/or freezing rain and ice pellets. In some cases the air flowing onshore is conditionally unstable, and can produce thunderstorms that accentuate the snow with an occasional lightning strike. Recall that an East Coast cyclone typically develops after an older cyclone moves across the United States. The cold front associated with this older cyclone can extend from Canada southward along the Appalachians from the older storm to the low-pressure center (Figure 11.3A, B). Additional lifting of the moist airstream aloft can occur along this frontal boundary, further enhancing the overall snowfall from the storm.

FOCUS 11.3

The Blizzard of 1888

The Nor'easter of 1888 was one of the most devastating cyclones to ever affect the northeast coast of the United States (Figure 11B). The heaviest snow on record (40 to 60 inches) occurred in many locations in the Northeast. Winds near the coast averaged 50 to 70 mph with gusts to 80 mph and temperatures were 0 to 10° F. Snowdrifts 30 to 40 feet high were common. Upper tropospheric data were not available in 1888; however, surface reports were available from various sources around the country.

On 11 March 1888 a cold front, with a secondary low-pressure center over Georgia, progressed eastward across the country. The air was quite cold (in the teens) behind the front, while along the coast, air was flowing inland from over the ocean and the temperature was in the 70s. The coastal flow was driven by high pressure northeast of New England. By 3 pm on 11 March (Figure 11C) the cold front moved over the coast and a low-pressure center began to form just offshore of North Carolina. The central pressure dropped 28 mb by 10 pm the following day as the low-pressure center tracked northward along the coast (Figure 11D). For such large pressure drops, strong upper-atmospheric divergence must have been occurring, along with heating of the lower troposphere. The low-pressure center stalled for nearly two days off Long Island, bringing hurricane force winds with below freezing temperatures and blinding snow to the entire New York/New England region. We can

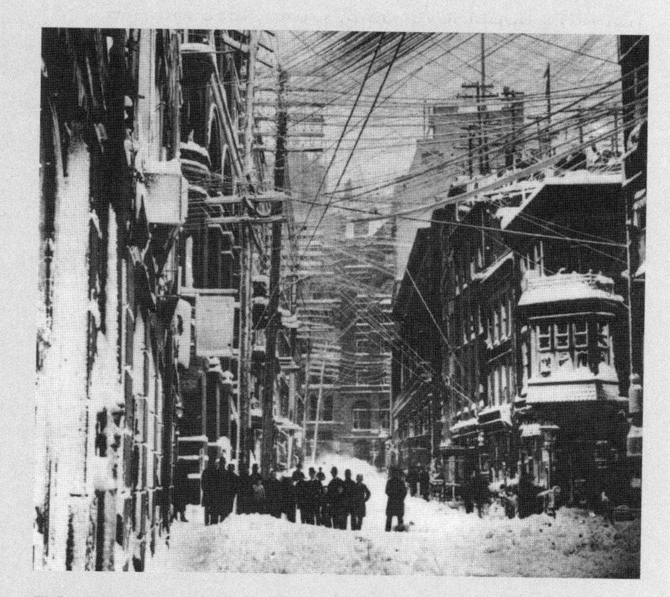

FIGURE 11B: Brokers and financiers survey the damage done by the Blizzard of 1888 to the telegraph and telephone lines connecting Wall Street to the world. The photograph was taken on New Street looking toward Wall Street.

(continued)

(continued)

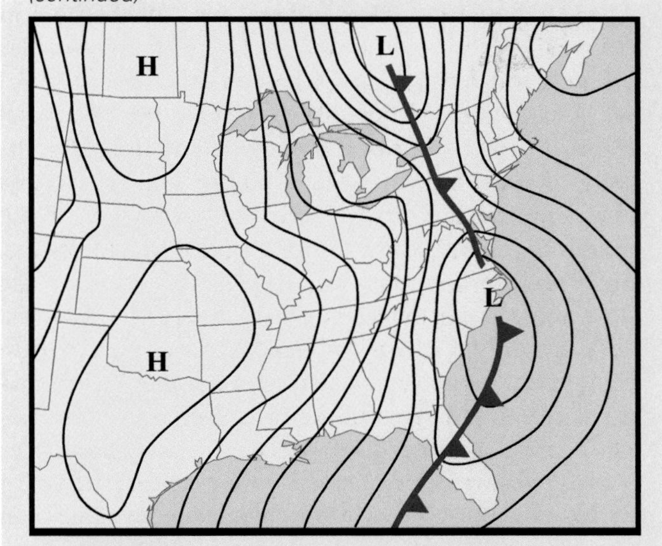

FIGURE 11C: Surface Analysis for 3 PM 11 March 1888. The low-pressure system associated with the Blizzard of 1888 is developing off the central Atlantic Coast. A cold front extends southward over the Atlantic Ocean.

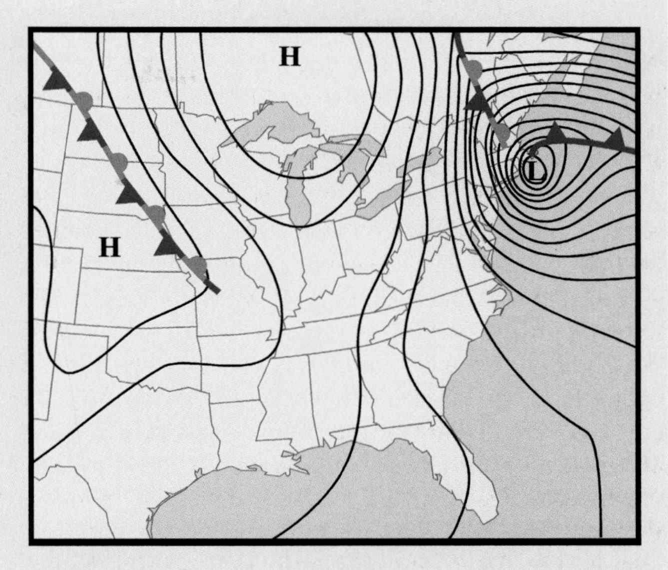

FIGURE 11D: Surface Analysis for 10 PM 12 March 1888 at the height of the 1888 Blizzard.

only surmise why this storm stalled since we do not have upper-level data. Other, less intense storms in the twentieth century have been found to move slowly or not at all when the cutoff low, the center of circulation aloft, aligns vertically with the center of circulation near the surface. This usually occurs late in the lifetime of these large storms when they are out to sea. Such "vertical vortices" can spin for days with no appreciable translation. The longevity of the 1888 blizzard made it the most devastating storm ever to hit New York to date. The Blizzard of 1888 led to substantial changes in urban planning. The utter destruction of the overhead power distribution and communication systems in the big cities of the northeast was the impetus for the installation of underground wiring. The difficulty with transportation started initiatives that eventually led to the creation of the New York subway system.

ONLINE 11.2

Rare April Nor'easter Strikes East Coast in 2007

Nor'easters are most common in the winter months and are generally associated with heavy snowstorms. In April 2007, a late season Nor'easter roared up the East Coast. Instead of snow, the storm brought flooding rains. Nearly 7.6 inches (192 mm) of rain fell in New York City and 9.3 inches in Riverdale, New Jersey on April 15, causing the worst flooding since Hurricane Floyd in 1999. The storm produced several tornadoes (including a fatal EF3 tornado near Mulberry, South Carolina), steady winds of 70–80 mph in some areas, and an amazing gust of 156 mph atop Mount Washington in New Hampshire. Snow fell at higher elevations further inland with Locke, New York receiving 23 inches. The storm developed from the remnants of a Rocky Mountain cyclone that had traversed the United States in the previous days. Visible, infra-red, and water vapor satellite animations and radar animations of this powerful storm can be found online, as well as animations of the surface pressure field, and the 850, 500 and 300 mb analyses. Compare this storm to the conceptual models presented in this chapter. How is it similar, and in which ways does it differ?

Storm Evolution

East Coast cyclones track northeast along the coast, typically reaching their most intense stage twenty-four to forty-eight hours after initial development (Figure 11.3C). The upper atmosphere undergoes a similar evolution to that discussed for Rocky Mountain cyclones in Chapter 10. Upper tropospheric convergence to the west of the trough forces dry air to descend from the upper troposphere. This dry air flows with the jetstream around the base of the trough and then northeastward creating the prominent ***dry slot*** appearing on satellite images (e.g., Figure 11.5). The air north of the dry slot originates at low elevations over the ocean. This moist airstream produces clouds northwest of the low as it cools moist adiabatically on its ascent into the mid-troposphere. Trapped between the advancing dry air in the dry slot and the cold air to the north and west, the warm moist air wraps around the west side of the cyclone, continually ascending and generating the heavy snow that blankets the land below. The wrap-around band of moisture northwest of the cyclone center develops a structure similar to the trowal discussed in the previous chapter for Rocky Mountain cyclones. During this phase of cyclone evolution, the polar and subtropical jetstreaks typically merge and propagate northeastward along the east side of the storm (Figure 11.3F).

The upper-level trough continually deepens as cold air is transported southward west of the low. In many storms, a cold pocket of air eventually pinches off aloft and forms a cutoff low. In this latter stage of storm evolution, the upper-level vortex (the cutoff low) and lower-level vortex (the surface low) align vertically to form a deep vortex (Figure 11.3F). By this time, the "merged" jetstreak has propagated out of the trough and into the ridge to the east and has weakened. Since the vortex consists only of counterclockwise flow, there is no longer any curvature change. In the absence of the effects of jetstreaks and curvature change, the upper-level divergence ceases. The low-pressure center begins to fill as surface friction causes air to converge into the low. The "filling" process is slower over water than over land because the effect of friction is reduced over water. As a result, cyclones can remain very strong for several days while they move up the East Coast. Occasionally, as happened in the Blizzard of 1888 (Focus 11.3), a storm will stall on the coast. If this happens, the duration of snowfall can be extreme to the northwest of the storm center. Moisture from over the ocean is continually fed into the stalled storm, leading to record snows. Nor'easters produce hazardous conditions as far north as Newfoundland before they move out to sea. They normally spin down over the North Atlantic south of Greenland, during which time they remain a major threat to shipping.

FIGURE 11.5: Enhanced infrared satellite image of the 25 January 2000 Nor'easter. Red and yellow colors indicate the coldest (highest) cloud tops. The dry slot is the wedge of dry air descending from the upper troposphere, south of the low-pressure center. At the base of the dry slot (white offshore region), low-level cumulus are forming as cold surface air moves out over the warm Atlantic and is heated. The moist trowal airstream aloft (red and orange region) north of the dry slot produces the heavy snowfall.

CHECK YOUR UNDERSTANDING 11.2

1. As an East Coast cyclone develops, where are the cold and warm fronts relative to the surface low and relative to the Atlantic coastline?
2. Where does air in the dry slot originate?
3. Where does air in the trowal originate?
4. Why do cyclones over water take longer to dissipate than cyclones over land?

GULF COAST CYCLONES

Gulf Coast cyclones develop most frequently during years when the subtropical jetstream is a persistent strong feature in the upper troposphere over northern Mexico and the Gulf of Mexico. The strength of the subtropical jetstream varies substantially from one winter season to the next with an apparent (but not well-understood) relationship to the cycle of sea-surface temperature variations in the tropical east Pacific associated with El Niño and La Niña events (see Chapter 23). The subtropical jetstream is normally strongest during the El Niño phase, when tropical sea-surface temperatures are warm and the flux of moisture into the tropical atmosphere is greatest. Since the genesis of Gulf Coast cyclones is tied to the subtropical jetstream, Gulf Coast cyclones are more common during El Niño years. As shown in Figure 11.2, Gulf cyclones typically follow one of two tracks: the first along the Gulf Coast and then northeast along the Atlantic Seaboard, and the second inland along the Mississippi and Ohio River Valleys. The intensities of Gulf cyclones following the inland and coastal tracks differ both because of different jetstream configurations and because of proximity to the ocean. Nevertheless, their beginnings are tied to the subtropical jetstream.

The East Coast Storm Track

Gulf Coast cyclones that track toward and along the East Coast typically develop after a cold front or arctic front moves across the United States and arrives at the Gulf Coast (Figure 11.6A). The position of the front along the north shore of the Gulf of Mexico marks a zone of sharp contrast between warm air over the Gulf and colder air over the continent. Aloft, a large trough is often present over the entire eastern United States (Figure 11.6E). The trough typically develops in association with an earlier Rocky Mountain cyclone that was responsible for the southern advance of the cold front to the Gulf of Mexico. A primary feature aloft is the subtropical jetstream, which flows from the tropical Pacific Ocean, across Mexico and over the Gulf Coast. A jetstreak in the subtropical jetstream normally triggers storm formation. As illustrated in Figures 11.6B and 11.6F, the development of low pressure at the surface begins as the left exit region of the subtropical jetstreak moves toward the Gulf of Mexico and over the position of the cold front. Divergence within the left exit region of the jetstreak triggers the development of low pressure. As the surface low-pressure center forms and intensifies along the cold front, cold air begins to move out over the warm Gulf waters west of the low-pressure center. Warm air is lifted southeast of this advancing cold air, often triggering a line of thunderstorms over the Gulf of Mexico. The front east of the low-pressure center also migrates northward—the "old" cold front now becomes a warm front.

Like their East Coast counterparts, Gulf Coast cyclones are often influenced by more than one jetstreak. This is particularly true for cyclones that track toward and along the East Coast. As a surface low-pressure center continues to develop along the Gulf Coast, the center of low pressure will move eastward remaining under the divergent left exit region of the jetstreak in the subtropical jetstream (J1 in Figure 11.6G). In some storms, as this evolution proceeds, a jetstreak on the west side of the trough in the polar jetstream may also propagate toward and into the base of the trough (J2 in Figure 11.6G). Should this occur, the divergent left exit region of this jetstreak can move into a position where it "superimposes" on the divergent left exit region of the subtropical jetstreak. Strong upper-level divergence, enhanced further by latent heat release within the clouds, will cause the surface low-pressure center to deepen rapidly under these conditions. Like their east-coast counterparts, when the jetstreaks in both the subtropical and polar jetstreams act in concert, the central low pressure of the storm can fall below that measured in many hurricanes. Unlike hurricanes, the winds in an extratropical cyclone extend over a much broader area. Again, it is important to note that it is not necessary to have two jetstreaks present for a storm to develop; however, the magnitude of the divergence, and intensity of the low-pressure center will increase if two jetstreaks are present and their divergent quadrants are superimposed over one another.

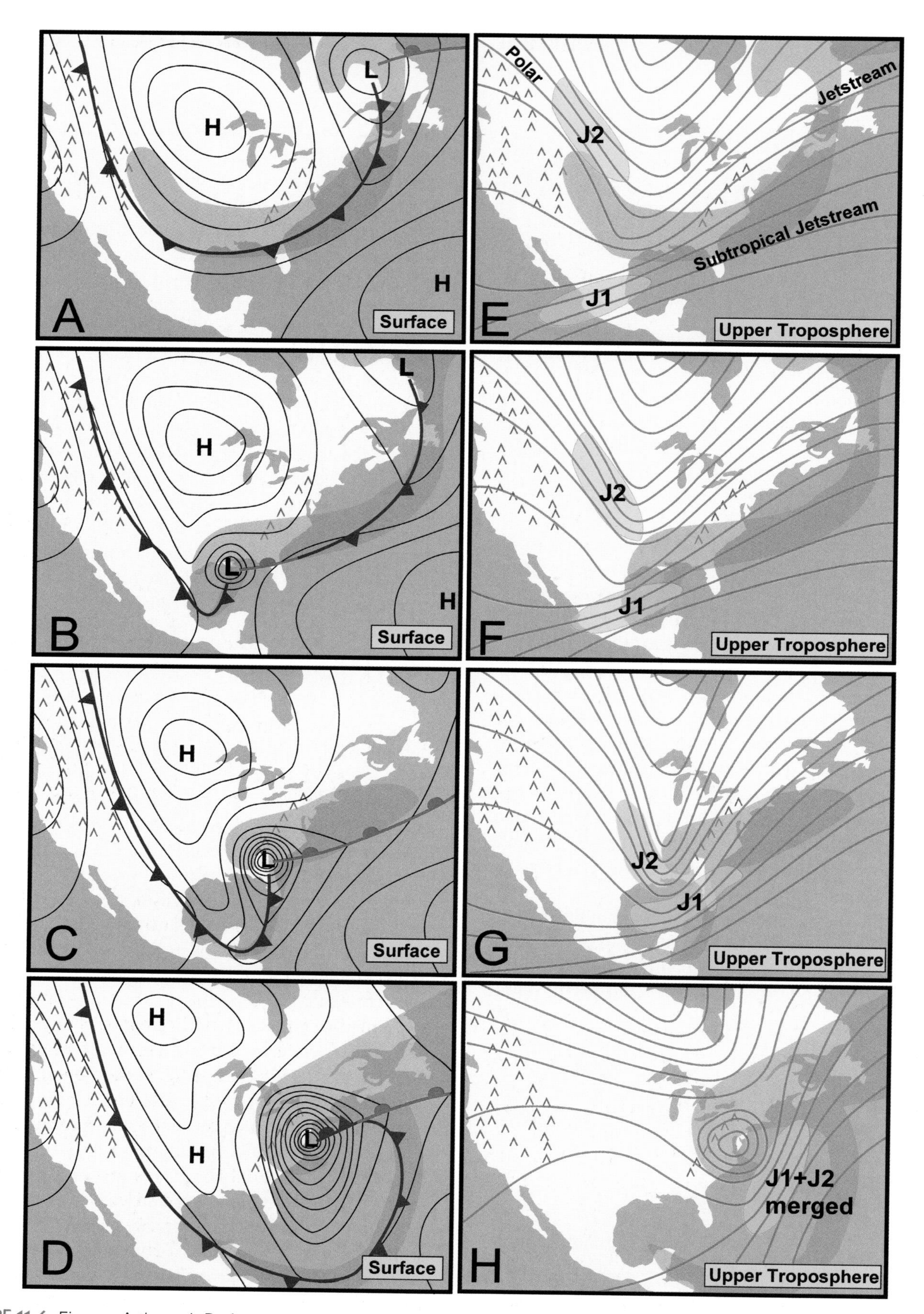

FIGURE 11.6: Figures A through D: A sequence of schematic charts showing the sea-level pressure, fronts, and cloud patterns (gray shading) (A) prior to the development, (B) during the early development, (C) during the rapid deepening, and (D) at the most intense stage of a Gulf Coast cyclone that tracks along the coast. Figures E through H: Schematic charts of the upper tropospheric flow patterns that correspond to Figures A through D: The yellow shaded regions (J1, J2) denote jetstreak locations.

ONLINE 11.3

The Superstorm of 12 to 14 March 1993

The Superstorm of 1993, often referred to as the "Storm of the Century," developed off the coast of Texas, moved east along the Gulf Coast, and then northeastward along the eastern seaboard of the United States. The storm produced heavy snow from the Gulf States northward through New England and into Canada's Atlantic Provinces. It also produced high winds, tornadoes, and even a storm surge: an abnormal rise in sea-level that produces coastal flooding similar to hurricanes (see Chapter 24). Numerous cities set record low pressures. The minimum sea-level pressure, observed near Washington, D.C., was 962 mb. Record snowfall fell in Birmingham, Alabama (17 inches), Chattanooga, Tennessee (20 inches), and at Mt. Mitchell, North Carolina (50 inches). Thousands of people were isolated by record snowfalls in Georgia and the Carolinas. Winds along the Gulf and Atlantic Coasts and in the mountains gusted between 70 to 95 knots as the storm moved up the coast. Between one and three feet of snow fell at locations along the eastern seaboard, with most of the large cities, from Washington to Boston, receiving about one foot. The estimated volume of water that fell during the storm was equal to twenty days of the average flow of the Mississippi at New Orleans. Two hundred forty-three deaths were attributed to the storm—three times the combined total from Hurricanes Hugo and Andrew. Every major airport on the East Coast was closed by the storm. Overall costs are estimated at $7.6 billion in 2006 dollars. Online you can view an enhanced infrared satellite image loop of this disastrous storm from its formation in the Gulf of Mexico to its departure from North America north of the Canadian coastline.

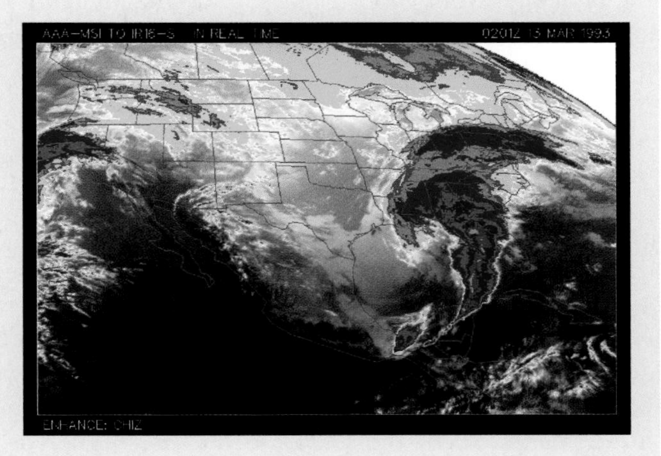

FIGURE 11E: Enhanced infrared image of the March 1993 Superstorm.

Cyclones tracking along the Gulf Coast and then northeastward along the Atlantic Seaboard can produce severe winter conditions across the southern states and across the Appalachian mountains. While the low-pressure center is along the Gulf Coast (Figures 11.6B and 11.6C), warm air flowing northward from the Gulf rises over the warm front producing (from south to north) rain, freezing rain, ice pellets, and heavy wet snowfall across the states of Louisiana, Alabama, Arkansas, Georgia, and Tennessee. In some storms unusual heavy snowfalls or devastating ice storms can develop. As a storm tracks along the East Coast later in its life cycle, heavy snow can fall throughout the Appalachian Mountains and across the urban corridor on the Coastal Plain. Along the storm's cold front, which extends south from the storm's center, severe thunderstorms and tornadoes can develop. Because of the storm's southern track, these thunderstorms pose the most significant threat in southeastern states such as Florida and Georgia. When strong Gulf Coast cyclones track northeastward along the East Coast, the media will sometimes refer to them as Nor'easters because they produce similar weather conditions to East Coast cyclones.

Gulf Coast cyclones intensify as long as the divergence aloft associated with jetstreaks, changes in flow curvature, and latent heat release in the storm clouds exceeds the convergence at the surface due to friction. As a cyclone progresses through its life cycle, the subtropical and polar jetstreams merge east of the cyclone (Figure 11.6H). The evolution from this point forward appears much like the East Coast cyclones described earlier in the chapter. A band of clouds continues to wrap around the low-pressure center to the north and west (Figure 11.6D). The airstream producing the trowal ***wrap-around cloud band*** originates over the ocean and ascends in a wedge between the cold air to the north and west and the advancing cold dry air to the south, in a manner similar to Figure 11.5. Heavy snow falls from the trowal region. Gulf Coast storms dissipate in the same manner as the other cyclones discussed in this and the previous chapter. A cutoff low typically forms aloft (Figure 11.6H) as the cold air flows southward in low levels and wraps around the low. With the development of the closed vortex, the divergence aloft ceases and friction at the surface causes the low to eventually fill.

The Mississippi–Ohio River Valley Storm Track

Gulf Coast storms are more likely to track northward along the Mississippi Valley when the upper-level trough is farther west over the central United States prior to their formation and the airflow across the eastern third of the United States is southerly (Figure 11.7C). With a trough in this flow configuration, the subtropical jetstream merges into the polar jetstream east of the trough axis, with the "merged" jetstream flowing northward over the eastern United States. Cyclones forming in this environment are generally weaker in intensity than their coastal counterparts. They are more commonly influenced by a single jetstreak, typically found within the subtropical jetstream. The cyclones typically first develop along a cold front that has reached the western U.S. Gulf Coast (Figure 11.7A). As these storms track northward between the Mississippi River Valley and the Appalachian Mountains, the most hazardous weather—freezing rain and heavy snow—typically occurs north of the advancing warm front and in the wrap-around precipitation

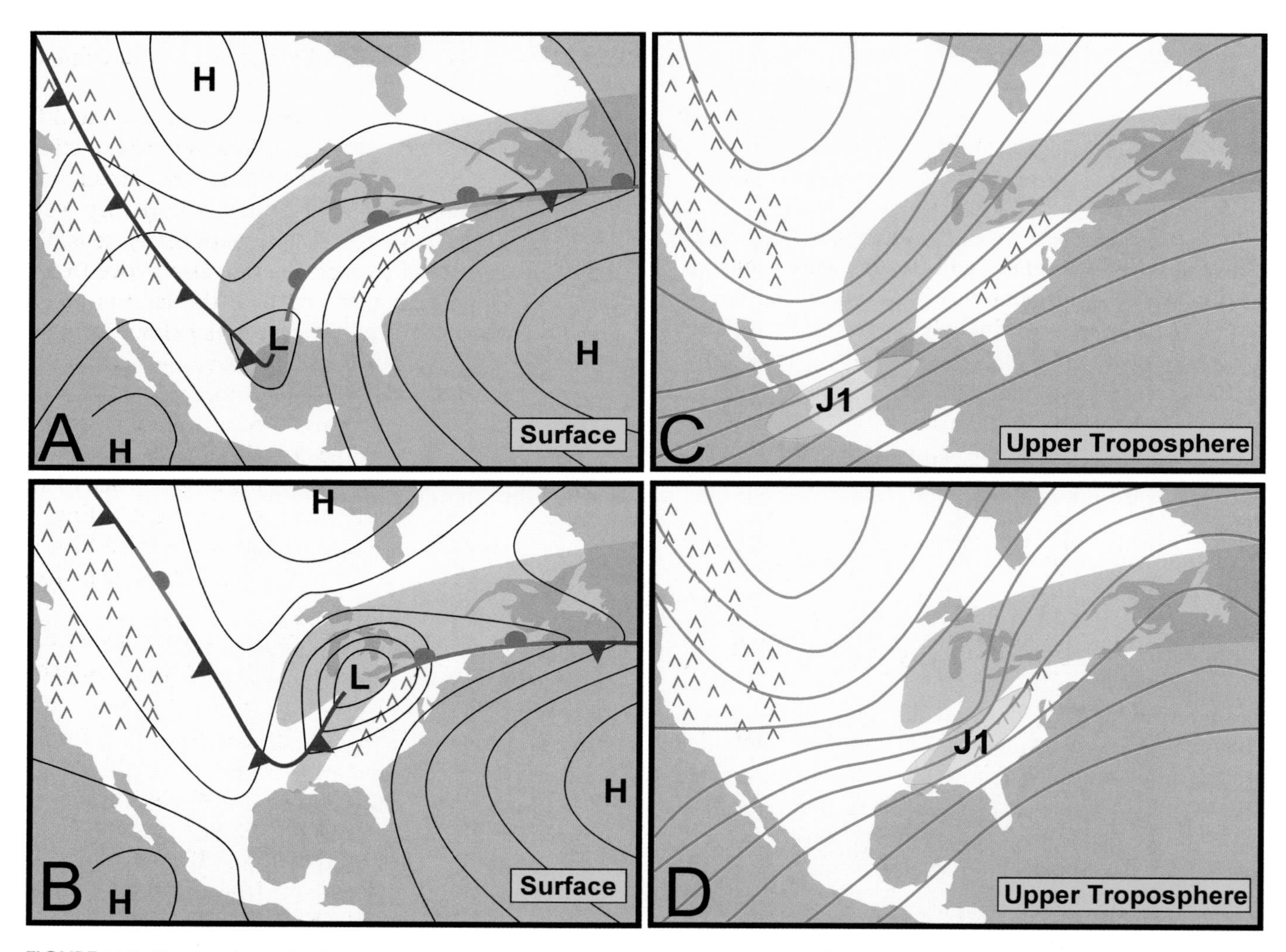

FIGURE 11.7: Figures A and B: Schematic charts showing the sea-level pressure, fronts, and cloud patterns (gray shading) (A) during the early development, and (B) at the most intense stage of a Gulf Coast cyclone that tracks along the Mississippi and Ohio River Valleys. Figures C and D: Schematic charts of the upper tropospheric flow patterns that correspond to Figures A and B. The yellow shaded region (J1) denotes the jetstreak in the subtropical jetstream.

band extending northwest of the surface low-pressure center (Figure 11.7B). For example, the Great Northeast Ice Storm of 1998, to be discussed in the next chapter, developed in association with a Gulf Coast cyclone that took a track inland along the west side of the Appalachian chain. In the southeast sector of these cyclones, strong southerly winds (Figure 11.7B) transport moist air northward from the Gulf of Mexico and over the Appalachians, where persistent rains can lead to local flooding.

FORECASTING AND ASSESSING THE IMPACT OF COASTAL CYCLONES

The distribution of snowfall from coastal cyclones is affected by the track of the cyclone, its intensity, the speed at which the cyclone moves, the topography of the Appalachian mountains, air temperatures, ocean temperatures off the coast, and other factors such as the moisture distribution. The interplay of each of these factors makes each cyclone different from all others. Along the heavily populated northeast corridor between Washington D.C. and Boston, Massachusetts, a forecast of significant snowfall sets in motion an expensive and disrupting chain of public actions designed to protect people and property. Transportation departments must activate armadas of snowplows to clear roads. Public safety departments must call in extra police to deal with accidents and to rescue stranded motorists. School administrators must decide whether and when to cancel school. Businesses have to make decisions whether to shut down or operate, and airlines have to prepare for cancelled flights and closed airports. Millions of dollars are spent or lost—all tied to the accuracy of forecasts that depend crucially on guidance provided by numerical forecast models.

Numerical forecast models generally do an excellent job of forecasting the development, intensity, and track of coastal cyclones. For example, over a decade ago numerical model forecasts for the 1993 Superstorm all accurately predicted the path and intensity of the storm five days in advance of its occurrence. The agreement between different models led forecasters to put out very strongly worded warnings about the Superstorm 24 to 36 hours in advance alerting that this was the "big one" and for municipalities to begin preparations. Damage, fatalities, and injuries due to that storm no doubt would have been substantially higher had accurate forecasts not been made and appropriate warnings issued.

Yet forecasts of heavy snowfall can go incredibly wrong. Such was the case on 4 to 6 March 2001. Numerical forecast models all predicted that a very strong Nor'easter would bear down on the East Coast (Figure 11.8), migrate slowly northward, and bring major heavy snowfall to coastal cities. Public officials, businesses, schools, and airlines reacted quickly to prepare for what some forecasters claimed could be the worst snowstorm along the northeast coast in decades, canceling most activities and urging the public to stay at home. The cyclone was indeed strong. The storm dropped several feet of snow in a band that reached across north-central Pennsylvania, upstate New York, central Massachusetts and southern Vermont, New Hampshire, and Maine. But in the populated areas along the coastal corridor, including New York, Washington, and Philadelphia, most precipitation fell as rain.

What made the forecast for the urban areas so difficult? In simple terms, the problem is the prediction of the position of the rain-snow line. The ocean surface northeast of a cyclone center is normally well above freezing and oceanic air, warmed by sensible heat from the ocean surface, is often also above freezing. In contrast, air near and over the land is much colder and typically has temperatures below freezing. The boundary between ocean-warmed air and cooler air inland is near the position of the rain-snow line. The rain-snow line at the surface in different storms may be located just off the coastline, along the coast, or inland of the coast, sometimes as much as 10 to 50 km (6 to 30 miles). When the rain-snow line is inland, coastal cities receive rain, while inland locations receive heavy snow. This is what happened in the March 2001 cyclone.

The problem of quantitative snow forecasts for the coastal cities is indeed very subtle. Although numerical models may forecast the position, intensity, and track of a cyclone extremely well, an inaccurate placement of the rain-snow line a distance of only 20 km can make what looks like the worst snowstorm in decades become just another rainy day in Manhattan. Twenty kilometers is smaller than the distance between two adjacent grid points in many current models used to forecast weather (see Chapter 4), so uncertainty about the position of the rain-snow line is the rule, not the exception. Significant problems in predicting snowfall amounts in the northeast coastal cities are unlikely to disappear, even as computing

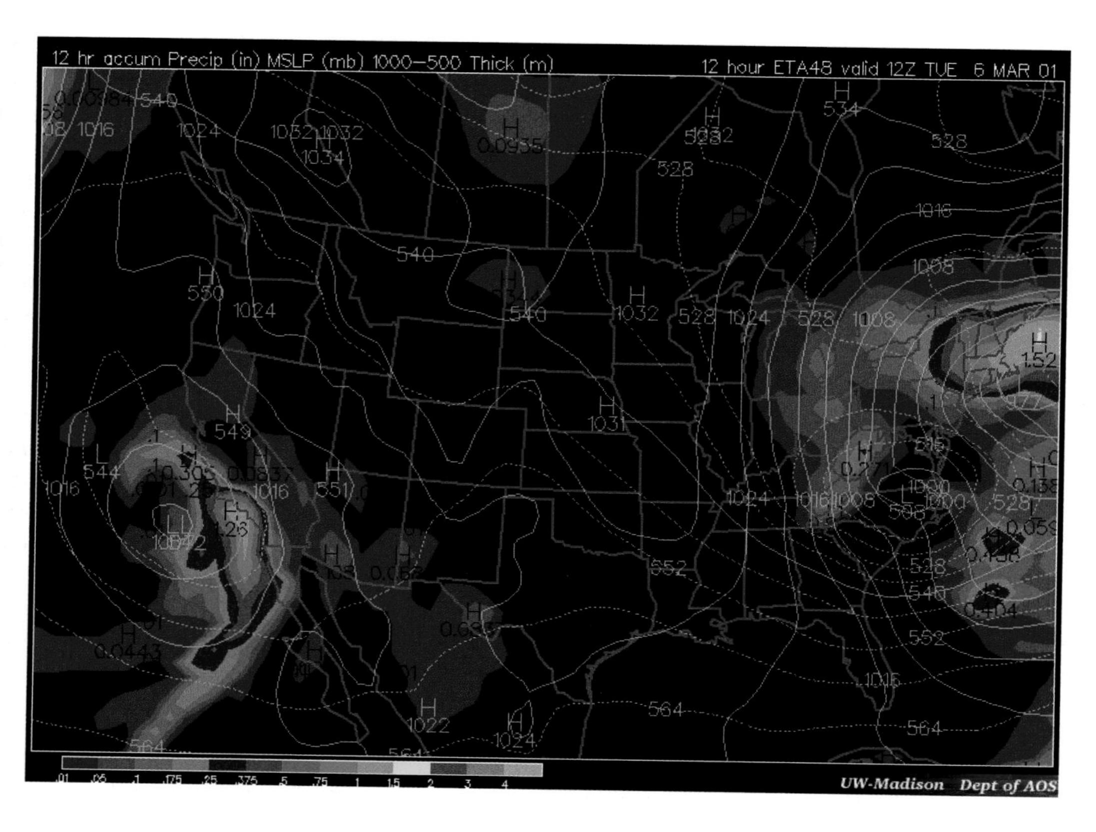

FIGURE 11.8: The twelve-hour forecast of sea-level pressure and precipitation from the Eta Model valid at 1200 UTC 6 March 2001. Blue lines are isobars of sea-level pressure, dashed yellow lines represent 1000-500 mb thickness (proportional to the average temperature of the 1000-500 mb layer), and colored shading represents twelve-hour accumulated precipitation (see color bar at bottom).

resources and models improve. Uncertainty will always be a major part of the forecasting problem. Because of the inherent uncertainty, forecasters are now looking toward ensemble forecasts, based on many models (see Chapter 4), to provide decision makers with an estimate of the *probability* that a snowstorm will occur. This probabilistic approach, rather than a simple yes/no forecast, expresses forecast uncertainty more clearly and properly shifts the responsibility concerning public safety from forecasters to decision makers.

Classifying the strength of a storm is important for communicating a storm's strength to the public and for assessing a storm's impact after it occurs. Meteorologists use the Saffir-Simpson scale for hurricanes, which characterizes storms on a scale of one (weak) to five (devastating) based on their winds (see Chapter 24). The Enhanced Fujita tornado scale, which classifies tornadoes or a scale of zero (weak) to five (devastating), estimates the maximum winds within a tornado during its lifetime based on damage surveys (see Chapter 19). Meteorologists have also been searching for better ways to provide benchmarks to assess the destructive potential of winter cyclones and to communicate this potential to the public.

Two scales have recently been proposed. The first, developed by Professor Gregory Zielinski of the University of Maine, uses a one to five rating similar to the Saffir-Simpson scale for hurricanes. His scale is based on the central low pressure of a cyclone, its twelve hour deepening rate (how fast the central pressure falls), and the maximum pressure gradient between the low pressure center and the nearest high-pressure location. The advantage of this scale is that a rating can be assigned while a storm is in progress and used to communicate a storm's strength to the public. The disadvantage is that it does not communicate information about snowfall, a key element of the forecast for winter cyclones.

An alternative scale, proposed by Drs. Paul J. Kocin of the Weather Channel and Louis W. Uccellini, Direction of the National Weather Service's National Centers for Environmental Prediction, focuses specifically on snowfall. This index, called the Northeast Snowfall Impact Scale (NESIS), takes into account both the snowfall distribution and amounts, and the population distribution and density (based on the 2000 census), and classifies storms based on the impact of the snowfall on population centers. The NESIS scale also uses a one (notable storm) to five (extreme snowfall event) rating to characterize a snowstorm's impact. The advantage of this scale is that it provides a clear way to classify storms based on their impacts on society. The disadvantage is that it cannot be applied while a storm is in progress, since the final snowfall distribution is known only after a storm is over. In the future, it may be possible to use forecast snowfall to calculate this index while a storm is in progress. These scales are both new and it is difficult to tell whether either of these indices will become as commonly used as the Safir-Simpson index for hurricanes and the Enhanced Fujita scale for tornadoes. The efforts of forecasters and scientists to better quantify the nature of these storms provide one example of the many ways that meteorologists are working to convince residents of the urban northeast corridor that adequate preparation is prudent and indeed essential when these great storms approach.

COASTAL CYCLONES AND GLOBAL WARMING

In the previous chapter we noted that continental cyclones were likely to undergo a decrease in both frequency and intensity as a result of global warming. These changes are expected because the temperature contrast between the polar and tropical regions will be reduced during the winter season based on climate model projections of more concentrated warming in polar regions. Can we make the same prediction for cyclones originating over the ocean along the East Coast of North America (or the East Coast of Asia)? The answer is no—there are at least three additional factors that must be taken into account, two that affect cyclone intensity and one that affects cyclone frequency.

One of the consequences of global warming associated with increased greenhouse gas concentrations is that sea-surface temperatures are expected to increase. The Gulf Stream (the ocean current along the East Coast of North America) and the Kuroshio Current (the corresponding current along the East Coast of Asia) transport warm water north from the tropics under the region's rapid cyclogenesis. In a warmer world, these currents will transport yet warmer water. We learned in this chapter and in Chapter 8 that latent heat release in clouds within cyclones contributes to the rapid deepening of the low-pressure center. In a warmer climate, the amount of moisture evaporated from the ocean surface will increase because of higher sea-surface temperatures, leading to more evaporation, latent heat release, and an intensification of the cyclones forming over the warm waters.

A second factor that will accelerate this process is the reduced stability of the atmosphere. One effect of increased greenhouse gases is a warming of the lower troposphere. This effect will be more pronounced locally when air moves over a warm ocean region, as occurs in regions where coastal cyclones develop. Global climate models also predict a cooling of the stratosphere. These effects together—warming below and cooling above—will lead to a reduction in atmospheric stability, and a likelihood that a greater contribution of the precipitation in cyclones will come from convection. More rapid transport of moisture upward in thunderstorms will increase precipitation rates and latent heat release, further intensifying storms.

The third factor we must consider with coastal storms is their environment. In this chapter, we learned that coastal storms typically develop after cold air from a previous continental cyclone arrives along the coastline. The cold front of the previous cyclone provides the necessary environment for the development of the coastal storm. In a warmer climate, particularly with warmer polar regions, the intensity and frequency of cold air outbreaks are predicted to decrease (see Chapter 14). Accordingly, one might expect that the number of strong cold fronts reaching the coastlines would be reduced, particularly along the coast of the Gulf of Mexico. The reduced frequency of cold air outbreaks and cold air advances to the coasts will reduce the number of times that conditions are even appropriate for the development of coastal cyclones.

These three effects together suggest that oceanic cyclones such as those forming along the East Coast will be reduced in frequency, but, somewhat paradoxically, those that do occur will have the potential to grow to a greater intensity. The basic physical ideas presented above are consistent with studies of cyclone frequency and intensity based on analyses of global climate model simulations.

CHECK YOUR UNDERSTANDING 11.3

1. Qualitatively, what is the relationship between El Niño and the frequency of Gulf Coast storms?
2. List all the types of severe and hazardous weather that can affect southern states during Gulf Coast cyclones.
3. Which jetstream is dominant in Gulf Coast cyclones that track west of the Appalachians?
4. What factor complicates forecasts of snowfall in coastal cities during East and Gulf Coast storms?

TEST YOUR UNDERSTANDING

1. Coastal cyclones are typically more intense than Rocky Mountain cyclones. Why?

2. What are the typical tracks of coastal cyclones?

3. What is the importance of a dissipating Rocky Mountain cyclone during the early development of cyclones on the East and Gulf Coasts?

4. What is the importance of the surface high pressure often found over the North Atlantic Ocean during the development of an East Coast cyclone?

5. What is cold air damming? Where does it occur?

6. What is the subtropical jetstream? Where is it typically found? Why does it exist?

7. Sketch the upper tropospheric flow and jetstreak positions for an East Coast cyclone that is affected by two jetstreaks. Identify the divergent quadrants of each jetstreak by marking them with a "D."

8. What is the typical range of sea-level pressures found in intense coastal cyclones?

9. What are all the possible meteorological conditions that contribute to strong divergence aloft during the initial development of East Coast cyclones?

10. New England mariners call some storms "Nor'easters." Where might this terminology come from and what type of weather conditions would you expect to occur over New England during a Nor'easter?

11. Approximately how long does it take for an East Coast cyclone to reach its most intense stage?

12. What types of hazardous weather might residents of the East Coast experience during an East Coast cyclone?

13. Why does severe weather along the cold front of an East Coast storm pose no threat to residents of the East Coast?

14. Describe how jetstreaks within the polar and subtropical jet can strengthen an East Coast cyclone at the same time.

15. Describe at least two ways that air is forced to rise in East Coast cyclones.

16. In what seasons do East Coast cyclones typically develop?

17. Why do snowfall totals of the northeastern United States typically increase as you move north and as you move inland?

18. Compare and contrast the initial surface map during the Blizzard of 1888 with the surface maps in Figure 11.3.

19. For a blizzard to occur in New England, where would the low-pressure center and fronts be located? Why? What about a blizzard in New York City?

20. What are the two reasons why the intensity of inland tracking Gulf Coast cyclones typically differs from that of coastal tracking Gulf Coast cyclones?

21. A front is typically present along the Gulf Coast just prior to cyclone formation. Explain where this front comes from.

22. How does the position of the trough over the United States differ prior to the development of Gulf Coast storms that track through the Mississippi Valley rather than up the East Coast?

23. What jetstream typically triggers the formation of a Gulf Coast storm?

24. Describe weather conditions that residents of Atlanta, Georgia, and Baltimore, Maryland, might experience during (a) an East Coast cyclone; (b) a Gulf Coast cyclone tracking up the Atlantic Seaboard; and (c) a Gulf Coast cyclone tracking northward west of the Appalachians.

25. What causes Gulf Coast cyclones to follow an "inland" track?
26. What aspects of coastal cyclones do numerical prediction models do a good job of forecasting? What phenomena are more difficult to predict?
27. Compare and contrast the two proposed cyclone intensity scales.
28. How might global warming affect cyclones forming along the East and Gulf Coasts?

TEST YOUR PROBLEM-SOLVING SKILLS

1. Photocopy the blank U.S. map found in Appendix B.
   (a) Draw a plausible location for the low-pressure system and associated fronts during a time when Boston, Massachusetts is experiencing a severe Nor'easter.
   (b) Draw isobars to indicate the pressure gradient associated with such a storm. Pay close attention to the number and spacing of the isobars.
   (c) Draw any fronts that would be present and use the appropriate symbols to denote the type of front.
   (d) Draw at least four arrows to indicate the direction of the surface airflow at several locations in the vicinity of the storm.
   (e) Based on your sketch, where would you expect the highest snowfall totals to be? Why?
   (f) Where would you expect the coldest air to be present? Why?
   (g) Where would you expect to find the fastest winds? Why? Did your sketch of isobars reflect this?
2. The average annual snowfall in the northeast United States is shown in the following figure. Examine this figure closely and explain how the snowfall distribution is influenced by coastal cyclones in general versus other factors.

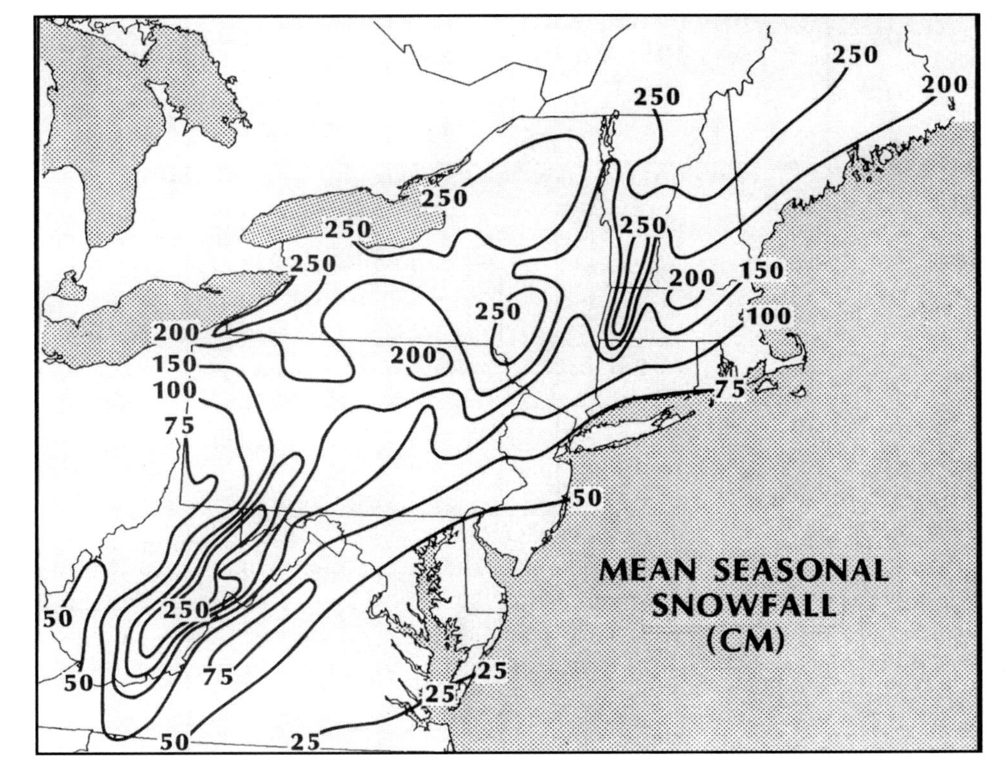

Average snowfall (cm per year) in the northeastern United States for 1955–1985.

3. Figures 11C and 11D in Focus 11.3 show the surface conditions during the Blizzard of 1888. At this time no upper air data was available. Make two copies of the blank U.S. map in Appendix B and sketch a plausible upper tropospheric flow pattern to correspond to each of the surface analyses. Show the positions of the jetstreams and any jetstreaks that may have influenced the storm's development.

4. Explain how ensemble forecasting can provide forecasters with information that can be useful in issuing forecasts for snowfall in coastal cities during coastal cyclones. What aspects of forecasting coastal cyclones are likely to improve in the next decade and why? What aspects of forecasting coastal cyclones are not likely to improve? Why?

5. You are a forecaster in the National Weather Service office responsible for New York City. You examine the three-day forecast and see a strong Nor'easter developing.
   (a) What maps and charts would you examine to determine how intense the cyclone is likely to become? Support your answer.
   (b) What data would you use to estimate the position of the rain-snow line? Support your answer.
   (c) Your best friend is the anchor of the nightly news at a major television station in New York City. She is hoping for a "big story" on this "superstorm". What information would you pass on to her with high confidence? Is this the information your friend and the public want to know?

6. Strong coastal cyclones develop when jetstreaks from both the polar and subtropical jetstreams interact. The strongest coastal cyclones (like the 1993 Superstorm) develop when three jetstreaks are present in the two jetstreams. Make a copy of the blank U.S. map in Appendix B and sketch the upper air pattern and plausible locations of three jetstreaks during the most intense stage of a coastal cyclone. (Hint: The low will be under the right entrance region of the third jetstreak).

USE THE SEVERE AND HAZARDOUS WEATHER WEBSITE

http://severewx.atmos.uiuc.edu

1. Navigate the *Severe and Hazardous Weather Website* to the "Current Weather" page and examine maps of sea-surface temperature for the western Atlantic Ocean (the Gulf Stream) and the Gulf of Mexico.
   (a) If a coastal cyclone were to develop along the eastern seaboard at this time, how would the ocean impact the development of the surface low, clouds, and precipitation?
   (b) If a Gulf Coast cyclone were to develop at this time, how would the Gulf of Mexico impact the storm development?
   (c) Compare and contrast the effect of sea vs. land temperature on the likely intensity of the cyclone if the Gulf Coast cyclone were to take an inland track as opposed to a coastal track.
   (d) Briefly discuss how the observed sea-surface temperatures could influence snowfall totals associated with either a Gulf Coast or East Coast cyclone.

2. Gulf Coast cyclones are more likely to develop during winters when there is a strong El Niño. Navigate the *Severe and Hazardous Weather Website* to the "Current Weather" page to examine the current state of El Niño.
   (a) Is there an El Niño occurring at this time?
   (b) If there is an El Niño occurring, what is its relative intensity? How do you think this El Niño may affect the development of Gulf Coast storms during the winter season?

3. The position of the rain-snow line is a critical factor determining precipitation type in the cities of the northeast United States during Nor'easters:
   (a) Pay attention to the media weather reports the next time a coastal cyclone or nor'easter is forecast, and follow the forecast for the storm from one media source. Record their temperature and precipitation type and amount forecast for one coastal city and one inland city.
   (b) During the storm, use the *Severe and Hazardous Weather Website* to gather information every six hours about the temperature and precipitation type in the coastal and inland cities.
   (c) Compare the forecast to the actual observed conditions. Was the forecast a "hit" or a "bust"? Explain.
   (d) Provide a critique of the media forecast. If you were the forecaster would you have provided the same forecast in light of the information available? Why or why not?

4. Examine water vapor satellite imagery from the *Severe and Hazardous Weather Website* during winter. Look closely at the Gulf of Mexico region. (A wide-scale view will be easiest to assess.)
   (a) Is the subtropical jetstream present? Explain how you determined your answer.
   (b) Animate the water vapor imagery to examine the evolution and track of the jetstream over the past several hours. Describe what you see.
   (c) On the animation of the water vapor imagery, locate the polar jetstream. Discuss how the two jetstreams are interacting (or acting independently).
   (d) On a day when the subtropical jetstream is present, locate a sounding in the vicinity of the subtropical jetstream. Estimate the altitude of the subtropical jet using the sounding data.
   (e) For the same day and time as in (d), examine a sounding in the vicinity of the polar jetstream. Estimate the altitude of the jetstream. Compare this to the altitude of the subtropical jetstream. Are your results what you would expect? Explain.

FREEZING PRECIPITATION AND ICE STORMS

Ice coats trees and fields in the wake of an ice storm in Illinois.

KEY TERMS

aircraft icing
cold air damming
cold air trapping
freezing rain
freezing drizzle
glaze
ice storm
ice nuclei
ice pellets
inversion
melting level
melting process
sleet
supercool
supercooled warm rain process
supercooled water
warm rain

Freezing precipitation is rain or drizzle that freezes on surfaces and leads to the development of an ice glaze. Freezing precipitation produces hazardous winter conditions that significantly impact public safety and the power, insurance, and transportation industries. Severe ice accumulation can lead to extensive power outages, halt air and ground transportation, and cause considerable property damage. During the period since 1950, freezing precipitation has caused more than $16 billion in property losses in the United States. Even minor ***glaze*** accumulation often leads to traffic and pedestrian accidents.

Freezing precipitation is responsible for about 20 percent of all winter weather-related injuries. Freezing precipitation aloft is also responsible for ***aircraft icing***. Freezing rain or drizzle occurs somewhere in about a fourth of all winter weather events in the continental United States. About half of these events qualify as ***ice storms***, based on National Weather Service criteria of either structural damage or ice accumulations of at least 0.25 in (0.64 cm). Ice storms in the United States were especially severe during the winter of 2006–07, affecting Missouri and Illinois in early December, 2006, Nebraska in late December, and a large swath from Texas to New England and southeastern Canada in January, 2007. In all three cases, damages were in the tens of millions of dollars.

SUPERCOOLED WATER

To understand ice storms, we must first understand ***supercooled water***. From elementary school onward, we are customarily taught that water freezes at 0° C (32° F). This is not always true; ice melts at 0° C, but water does not necessarily freeze at this temperature. The molecules in water are in constant motion, moving about rapidly within the fluid. The molecules in ice are locked in a lattice, vibrating, but fixed in place relative to their neighbors. It is very difficult for molecules in water to undergo the transition from rapid random motion in the liquid to the locked-in-place vibrational motion in the ice lattice. In fact, until the temperature reaches about –40° C (–40° F), it is nearly impossible for this transition to occur in pure water. So how does ice form?

In the atmosphere, there are many microscopic particles, including wind-blown clay eroded from soil, organic particles from rotting leaves and car exhaust, and even microscopic creatures such as bacteria. Certain of these particles have a crystalline structure that, from a molecular point of view, is very similar to ice. When these particles are present in liquid water, they provide sites to which water molecules can attach and begin building an ice lattice. Particles that promote the formation of ice are called ***ice nuclei***. Ice nuclei are most effective at promoting the formation of ice crystals when the temperature is colder than –15° C (5° F), marginally effective when the temperature is between –15° C (5° F) and –5° C (23° F), and hardly effective at all when the temperature is between –5° C (23° F) and 0° C (32° F). Ice nuclei are sufficiently rare in the atmosphere that in a small droplet in a cloud, or a small raindrop falling through air, there may not be any nuclei that can activate ice formation between temperatures of about –10° C (14° F) and 0° C (32° F). Without active ice nuclei, the small droplets will not freeze when their temperature drops below 0° C. Instead, the droplets ***supercool*** and remain liquid. When supercooled droplets come in contact with a surface that has a temperature below 0° C, such as a road, a car, a tree, a wire, or an airplane, they immediately freeze onto that surface. Glaze develops as many supercooled drizzle or raindrops fall onto objects and freeze.

VERTICAL STRUCTURE OF THE ATMOSPHERE DURING FREEZING PRECIPITATION

Freezing precipitation can form in two ways. The first way, called the ***melting process***, leads to freezing precipitation when snow falls from high in the clouds into an atmospheric layer where the temperature exceeds 0° C. The snowflakes melt into raindrops within this layer. The raindrops then continue their fall back into a subfreezing (< 0° C) layer of air near the ground. If refreezing occurs only after the drops make contact with surface objects, the result is freezing rain.

The second way freezing precipitation forms is through the ***supercooled warm rain process***. This process occurs when tiny cloud droplets grow to precipitation size by colliding and coalescing with each other. The term ***warm rain*** simply means that snow was not involved in the formation of precipitation. The melting process commonly occurs in deep clouds, while the supercooled warm rain process occurs in shallow clouds. Typically, precipitation formed through the melting process is reported as freezing rain, while precipitation formed through the supercooled warm rain process is reported as freezing drizzle. These processes are discussed separately here under the headings of ***freezing rain*** and ***freezing drizzle*** because clouds that support an active supercooled warm rain process and produce freezing drizzle often pose significant aircraft icing hazards, particularly for small aircraft.

Freezing Rain

As we saw in Chapter 1, temperature normally decreases with height in the troposphere. In freezing rain events, temperature increases with height above the surface, and then decreases with height aloft. The layer in which temperature increases with height is called an ***inversion***. We can see how an inversion forms by examining Figure 12.1, which represents a cross section through a warm front. In the top panel of Figure 12.1, the cold air mass below the front is shaded. The red line in the warm air above the front encloses that part of the atmosphere where the temperature is warmer than freezing. Warm, moist air is flowing upward over the cold air. The lines A, B, C, and D show the locations where the soundings in the middle panels of the figure were obtained. The middle

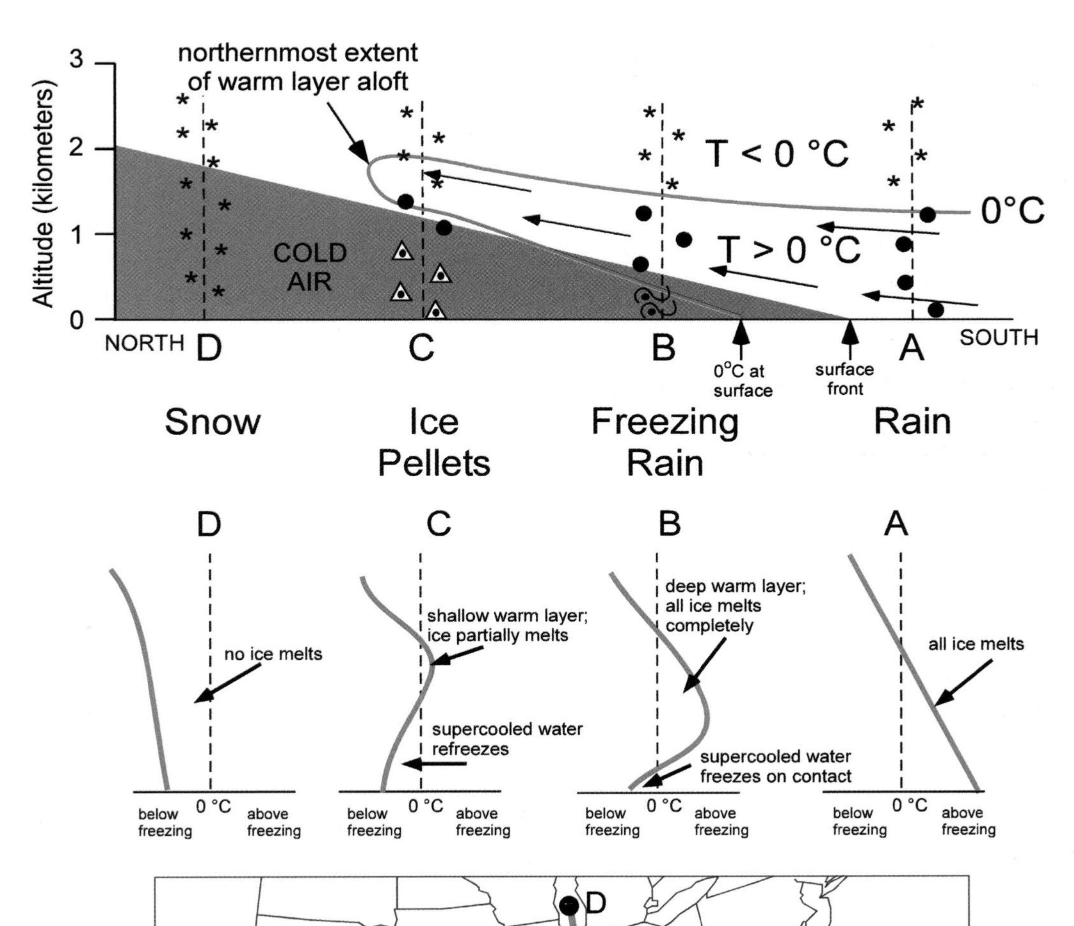

FIGURE 12.1: Top: cross section through a warm front. The shaded region denotes the cold air. The solid line encloses the region of the atmosphere where the temperature exceeds 0° C. The middle diagrams denote temperature soundings taken at points A through D. The type of precipitation at each point is indicated. The lower diagram shows the location of the cross section and points A-D in the case of a warm front. The orange shaded region is the location where freezing rain is falling.

panels show the vertical temperature structure at points A through D, and the lower panel shows a typical location for the cross section, in this case across a warm front.

Precipitation forms as snow in the cold upper troposphere above locations A through D. At point D, the temperature is below freezing at all levels in the atmosphere (see middle diagram), so the precipitation falls to the ground as snow. At point A, south of the front, the temperature in the upper atmosphere is below freezing and the temperature of the lower atmosphere is above freezing. Snow falling through the ***melting level*** (0° C) will melt into raindrops and continue to fall to the ground as rain.

At points B and C, there is a layer in the middle of the atmosphere where the temperature is warmer than 0° C, but a layer at the surface where the temperature is colder than 0° C. The difference between points B and C is that at point B, the layer where the temperature exceeds 0° C is warm and deep, while at point C, the layer is shallow and not as warm. Consider first what happens to a snowflake falling into the warm layer at point B. The snowflake will begin to melt as soon as it crosses the 0° C melting level aloft. Because the layer with temperatures exceeding 0° C is warm and deep, the flake will melt completely before falling back into the subfreezing layer at the surface. The liquid drop will supercool when it falls through the surface cold layer, strike the surface, and freeze on contact.

Temperature profile B supports the formation of freezing rain. Consider now what happens to a snowflake at point C. As the flake falls into the warm layer aloft, it too will begin to melt, but because the layer is shallow and not as warm, it will fall through the layer before melting completely. When it enters the cold air near the ground, it will still contain some ice. The liquid in the mixture immediately will refreeze in the air, creating a frozen raindrop. Frozen raindrops are called ***ice pellets*** or ***sleet***. Ice pellets pose little problem because the particles do not stick to anything—they just accumulate like snow.

Freezing Drizzle

Freezing drizzle refers to light, almost misty precipitation that freezes on contact with surfaces. Drizzle drops are tiny, having diameters between about 0.2 and 0.5 mm (0.01 and 0.02 inches). Although freezing drizzle can form in conditions described in the previous section, it more often forms in cloud layers with depths in the range of 1 to 3 km, such as the cloud layer shown in Figure 12.2. Often the entire cloud will have temperatures colder than freezing. Cloud layers such as the one illustrated in Figure 12.2 often form as a layer of moist air is lifted over a cold air dome associated with an arctic front. For freezing drizzle to form, it is essential that the cloud top temperature be no colder than about –10° C to –15° C. With colder cloud top temperatures, ice crystals (snow) will form

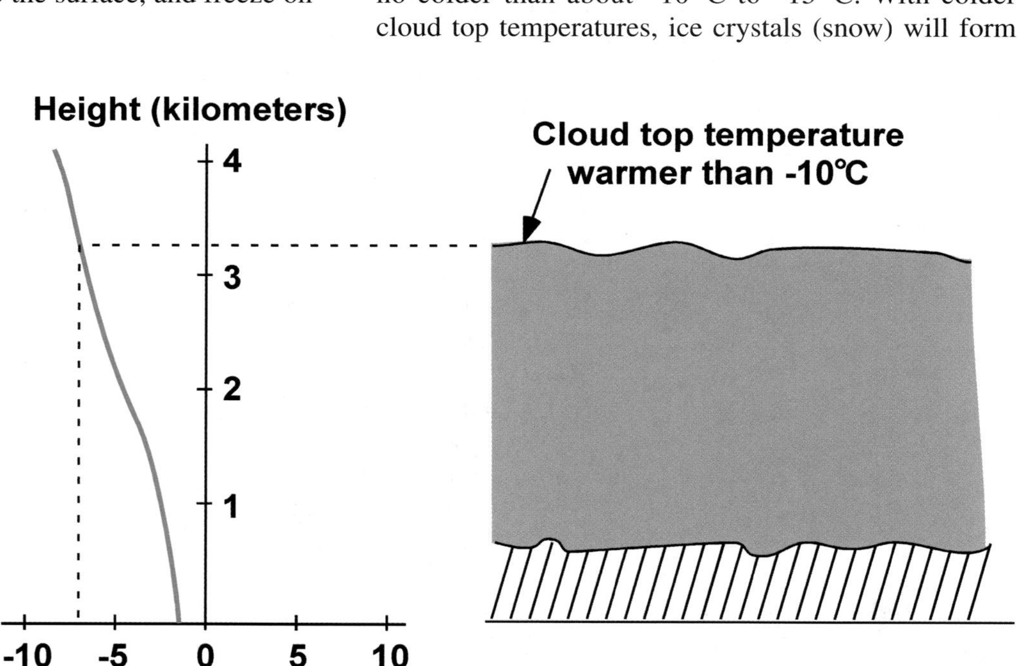

FIGURE 12.2: Example of a temperature sounding through a cloud that might produce freezing drizzle. The cloud produces freezing drizzle drops as tiny supercooled cloud droplets collide and grow to sizes large enough to fall.

in significant numbers in the upper part of the cloud and will collect and remove much of the supercooled water while falling through the lower part of the cloud. In clouds with tops warmer than –10° C to –15° C, ice nuclei are not effective at freezing the cloud droplets. With time, the supercooled droplets can grow to drizzle-size by colliding and coalescing with one another.

Freezing drizzle results in minor glaze accumulation, typically causing traffic and pedestrian accidents. The more significant danger posed by freezing drizzle occurs aloft. Because the entire cloud is often supercooled during freezing drizzle events, aircraft ascending or descending through the cloud often must spend several minutes in icing conditions. The rate at which ice accumulates on the airframe depends on factors related to the aircraft design and on the size of the droplets in the clouds. If droplets in the cloud grow to diameters exceeding about 0.04 mm, they will collect rapidly on the wings and other parts of an aircraft. Aircraft flying through clouds containing supercooled droplets can rapidly accumulate ice on the airframe, degrading flight performance (Figure 12.3). In some cases, aircraft have lost the capability to stay airborne and crashed. Many aircraft counter ice buildup by heating the leading edge of the wings or using mechanical methods to break off ice along the wing. In severe icing, these methods may fail. Because of the danger of icing, aviation forecasters pay particular attention to situations where freezing drizzle may occur.

FIGURE 12.3: Accumulation of ice on an aircraft after flying through freezing drizzle.

FOCUS 12.1

The Roselawn, Indiana ATR72 Aircraft Accident and Freezing Drizzle Aloft

An aircraft flying through a cloud containing supercooled drizzle droplets can rapidly accrue ice on the airframe, including the underside of the wings. The accumulation of ice has two effects: it increases an aircraft's weight, thereby reducing its ability to climb, and it changes the flow characteristics of air moving over the aircraft, increasing drag and stall speed. On 31 October 1994, an ATR72 commuter aircraft was in a holding pattern over northern Indiana on approach to Chicago's O'Hare Airport. Sometime during the holding pattern, the pilot lost control of the aircraft and crashed near the town of Roselawn, Indiana, killing all 68 on board. The meteorological evidence suggests that conditions at the ATR72's flight level on that fateful day supported the formation of supercooled drizzle and severe *aircraft icing.* Studies of this accident have found that the holding pattern was located near the tops of stratiform clouds that had cloud top temperatures of –10° C to –15° C. The layer near the cloud top had substantial vertical wind shear.

Scientists have found that turbulent motions created by strong vertical wind shear promote the growth of tiny cloud droplets to drizzle-size by enhancing the rate at which droplets collide. Radar data showed the region of the flight to have very low reflectivity values that were consistent with the presence of drizzle and the absence of snow. Other aircraft in the general area also reported icing that ranged from very light to severe. As a result of the ATR72 accident, significant efforts are now being made to use numerical weather prediction models to forecast locations where supercooled drizzle drops are likely to develop—regions where the model predicts warm cloud tops (temperatures greater than –10° C to –15° C) and wind shear—and provide this information to pilots for flight planning.

CHECK YOUR UNDERSTANDING 12.1

1. Why might the general public be concerned with freezing precipitation?
2. What is supercooled water?
3. What are the names of the two processes that lead to freezing precipitation?
4. Distinguish between freezing rain, freezing drizzle, and ice pellets.

WEATHER PATTERNS ASSOCIATED WITH FREEZING PRECIPITATION

Several weather patterns are commonly associated with freezing precipitation east of the Rockies in the United States and Canada. Some general weather patterns are not associated with topographic features, while others are associated with the Appalachian Mountains.

A weather pattern commonly associated with freezing precipitation is the arctic front and a high-pressure system. Normally arctic fronts advance southeastward from Canada across the United States in winter ahead of strong high-pressure centers and eventually stall as they approach the coast of the Gulf of Mexico (Figure 12.4A). In many cases, warm air rising over the cold-air dome associated with the high pressure creates a shallow cloud layer that produces freezing drizzle. In some cases, deep clouds form in the air overrunning arctic fronts and producing freezing rain—particularly in southern states where the warm air ahead of the front is more likely to be conditionally unstable. With arctic fronts, freezing precipitation typically develops along a narrow band just on the cold side of the surface position of the 0° C isotherm (Figure 12.4A). The width of the band of freezing precipitation generally does not exceed about 160 km (100 miles), and often can be much narrower. Arctic fronts are associated with about a third of all freezing precipitation events east of the Rocky Mountains in the United States.

Freezing precipitation also commonly occurs north of warm fronts in cyclones. Regions of freezing precipitation associated with warm fronts again can be found just north of the surface location of the 0° C isotherm in a narrow band oriented approximately parallel to the front. The band of freezing precipitation can also extend northwest around the low-pressure center associated with the cyclone. Freezing precipitation falls north of warm fronts in about a third of all freezing precipitation events in the United States. About half of these events have a strong high-pressure system to the north of the cyclone (Figure 12.4C) while half do not (Figure 12.4B). The distinction between these situations is noteworthy

ONLINE 12.1

The January 2007 North American Ice Storm

Most of the United States east of the Rocky Mountains was affected by a series of three winter storms that began during the second week of January 2007 and continued for about ten days. The number of ice-related traffic fatalities during this period was close to 100, and hundreds of airline flights were cancelled in the South and East. About two inches of ice accumulated from Texas to the Midwest, and parts of New York received more than three inches of ice. The precipitation was fed by a strong southwesterly flow of moist subtropical Pacific air in the middle troposphere. This flow is apparent online in the soundings for Norman, Oklahoma during the period January 11 to 16.

Prior to the onset of frozen precipitation, the mild Pacific air reached the surface on January 11. However, a shallow northeasterly flow of cold air from a continental high-pressure system undercut the moist Pacific air on January 12 to 14, creating the lower tropospheric inversion characteristic of freezing rain. The southwesterly winds above the inversion layer approached 100 knots on the 14th and 15th. The shallow inversion persisted until January 16, when the layer of cold air deepened. On January 14, the Norman and Oklahoma City areas experienced thunderstorms while the surface temperatures remained below freezing, creating a heavy accumulation of freezing rain. As the soundings show, the subfreezing layer above the surface was less than 1 km deep at this time.

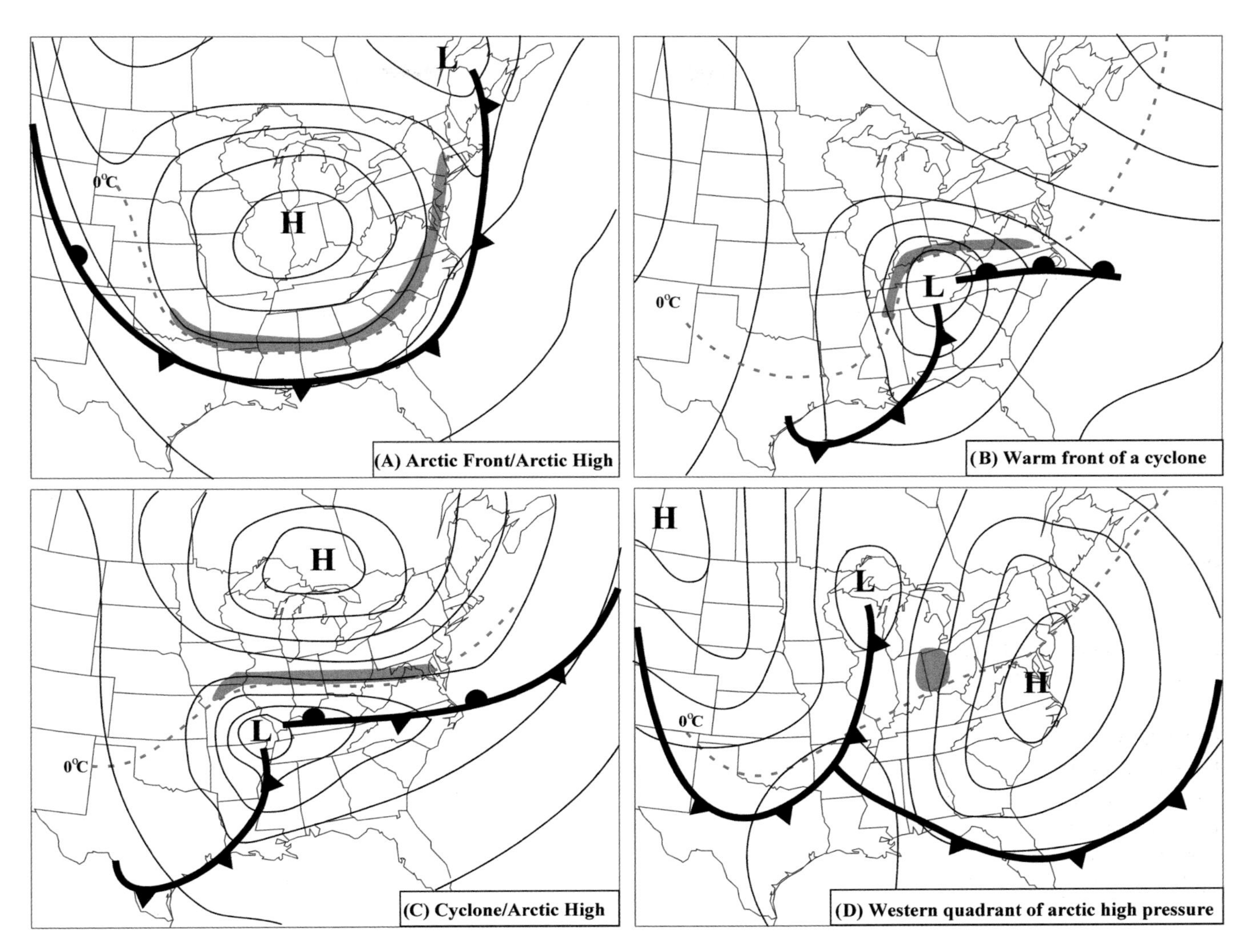

FIGURE 12.4: Common weather patterns associated with freezing precipitation. Dashed red line is the 0° C isotherm and the orange highlighted areas indicate regions where freezing precipitation would develop. Note that in all cases the freezing precipitation is concentrated in a narrow zone on the cold side of the 0° C isotherm.

FOCUS 12.2

Wind and Ice Storms

In a major ice storm, ice accumulation is only part of the story. Wind dramatically increases the stress on trees and structures burdened with ice. Strong winds, coupled with heavy ice accumulation, can quickly destroy trees, electrical transmission towers, and other structures. The strong winds do not even have to occur simultaneously with the freezing precipitation. As long as structures are burdened with the extra weight, winds can continue to cause destruction well after the freezing rain stops falling.

because a stronger pressure gradient will exist across the region of potential glaze accumulation when a high-pressure center is present to the north of the cyclone. Strong surface winds driven by the strong pressure gradient increase the potential for destruction if glaze accumulation becomes significant on trees and power lines.

FOCUS 12.3

Warm Air Aloft!

Arctic air sometimes spreads across the central United States and reaches the shores of the Gulf of Mexico in winter. When cyclones develop along the Gulf Coast, very warm air can flow northward over the arctic airmass. The temperature contrast between the surface arctic air and the warm air aloft can be extreme. An illustration of a remarkable temperature contrast appears in Figure 12A, the 1200 UTC 30 December 1990 sounding from Stephenville, Texas. The surface air temperature at Stephenville was –8.1° C (17.4° F). Only 780 meters above the ground the temperature was a balmy 14.2° C (57.5° F). Southwest winds 780 meters above Stephenville approached 40 knots ahead of the developing storm. When rain from the clouds over Stephenville fell into the subfreezing surface layer, the drops supercooled and arrived at the ground as freezing rain.

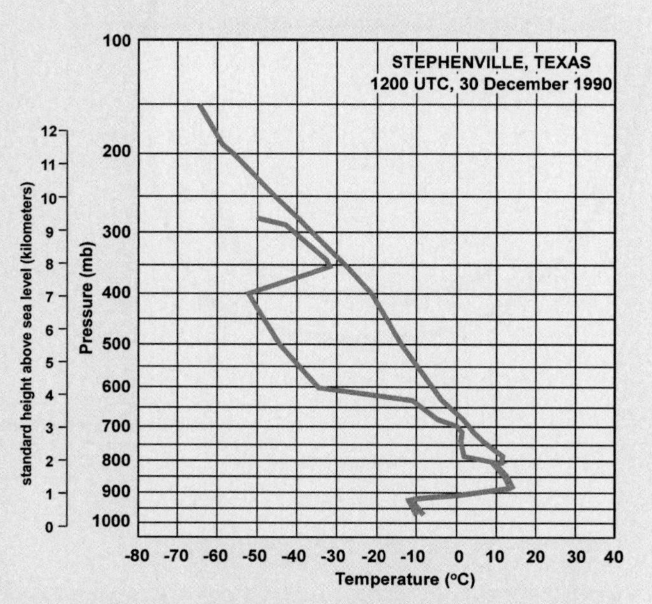

FIGURE 12A: Sounding taken at Stephenville, Texas, on 30 December 1990 at 1200 UTC during a freezing rain event.

Southerly flow normally occurs on the western side of high-pressure centers associated with arctic airmasses as these airmasses move toward the eastern seaboard of the United States and Canada. Surface air west of the high is part of the arctic airmass, and the 0° C isotherm is often located in the central part of the United States when this weather pattern is present. Southerly flow also occurs aloft. As warm moist air is transported northward over the arctic cold-air dome, a wide band of clouds and precipitation often develops. When this weather pattern occurs, freezing precipitation can develop within the region of stronger southerly surface flow. The freezing precipitation typically extends no more than about 200 km (120 miles) north of the 0° C isotherm (Figure 12.4D). Unlike the previous patterns, the freezing precipitation does not occur in an elongated narrow band, but rather over a more circular area. About one-tenth of all freezing precipitation events in the United States are associated with this weather pattern.

Freezing precipitation occurs frequently along the east side of the Appalachian Mountains associated with a process called ***cold air damming***. About 15 percent of all freezing precipitation events affecting the United States east of the Rockies are associated with this process. Cold air damming can develop in two ways.

As arctic airmasses move across the eastern United States, subfreezing air can extend southward along the East Coast of the United States as far as Georgia. As the high-pressure center associated with an arctic airmass moves toward and over the North Atlantic Ocean, easterly flow will develop to its south. The air within the easterly flow arrives at the East Coast after residing over the Atlantic Ocean, where the air acquires both moisture and heat. The warmer Atlantic air rises over the denser cold air remaining on the Coastal Plain east of the Appalachian Mountains, forcing cold air to remain between the mountains to the west and the onshore flow to the east (Figure 12.5A). The cold surface air dams along the east slope of the Appalachians and drains northeastward around the west side of the high-pressure system. In this pattern, the area of potential freezing precipitation is bounded by the Appalachian Mountains to the west, and the 0° C isotherm to the south and east (Figure 12.5A).

Cold air damming also can occur when cyclones develop along the eastern coast of the Gulf of Mexico or the Atlantic Coast and move northeastward along the Atlantic seaboard. Under these conditions, moisture associated with warm air rising over the cyclone's warm front can lead to an enhancement of the freezing precipitation in the region of cold air damming

ONLINE 12.2

The 1998 Northeast Ice Storm

One of the most destructive ice storms in North America occurred over the period of 4 to 9 January 1998 from Lake Ontario eastward along the Canadian-United States border. Several Canadian Provinces, including eastern Ontario, southern Quebec, New Brunswick, and Nova Scotia, as well as U.S. states including New York, Vermont, New Hampshire, and Maine, experienced significant glaze accumulation. In the worst affected areas in Canada, glaze accumulations exceeded three inches (80 mm, see Figure 12B). Two large cities, Montreal and Ottawa, were nearly shut down under the weight of ice. Near the end of the ice storm it was dangerous to even walk in these cities as huge chunks of ice continually fell from buildings. Thirty fatalities were attributed to the storm. In Canada, 1.7 million homes and businesses were without power, many for several weeks in the dead of winter. Figure 12C shows the damage to power transmission lines in southeastern Canada. An additional 0.5 million were without power in the United States. Overall damages approached $3 billion for Canada and were at least $1.4 billion for the United States. The storm was the worst weather disaster in Canadian history. The Northeast ice storm was unusual both in its duration and in the weather pattern that created it. Online you can view several animations of the weather conditions during the ice storm.

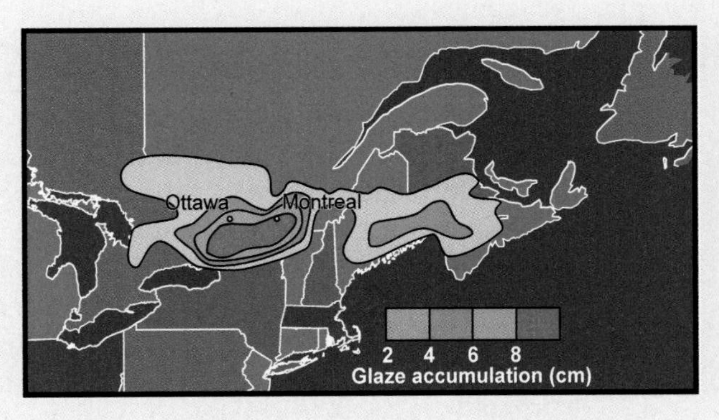

FIGURE 12B: Distribution of glaze during the great Northeast ice storm of 1998.

FIGURE 12C: The aftermath of an ice storm.

(Figure 12.5B). Another effect of a cyclone is to intensify the pressure gradient along the coast, strengthening the easterly flow from the Atlantic. The stronger pressure gradient enhances the wind and the potential for destruction in areas of glaze accumulation.

Finally, freezing precipitation can also occur in the Appalachian Mountain valleys during ***cold air trapping*** events. Cyclones originating east of the Rockies often track toward the Great Lakes following an outbreak of arctic air over the eastern United States. As a cyclone tracks eastward, warm air east of the cyclone will advance northward on either side of the Appalachian Mountains. Often, cold air will remain trapped within the interior Appalachian Mountain valleys. On surface weather charts, analysts often indicate this trapped air by drawing distorted warm fronts that deviate southward along the west slope of the Appalachians and then northward along the east slope (e.g., Figure 12.5C). Freezing rain or drizzle can develop when the trapped air has temperatures below freezing and air aloft is warm.

A similar process occurs in the Columbia River basin in the Pacific Northwest. In that region, cold air remains trapped in valleys as warm air moves northward

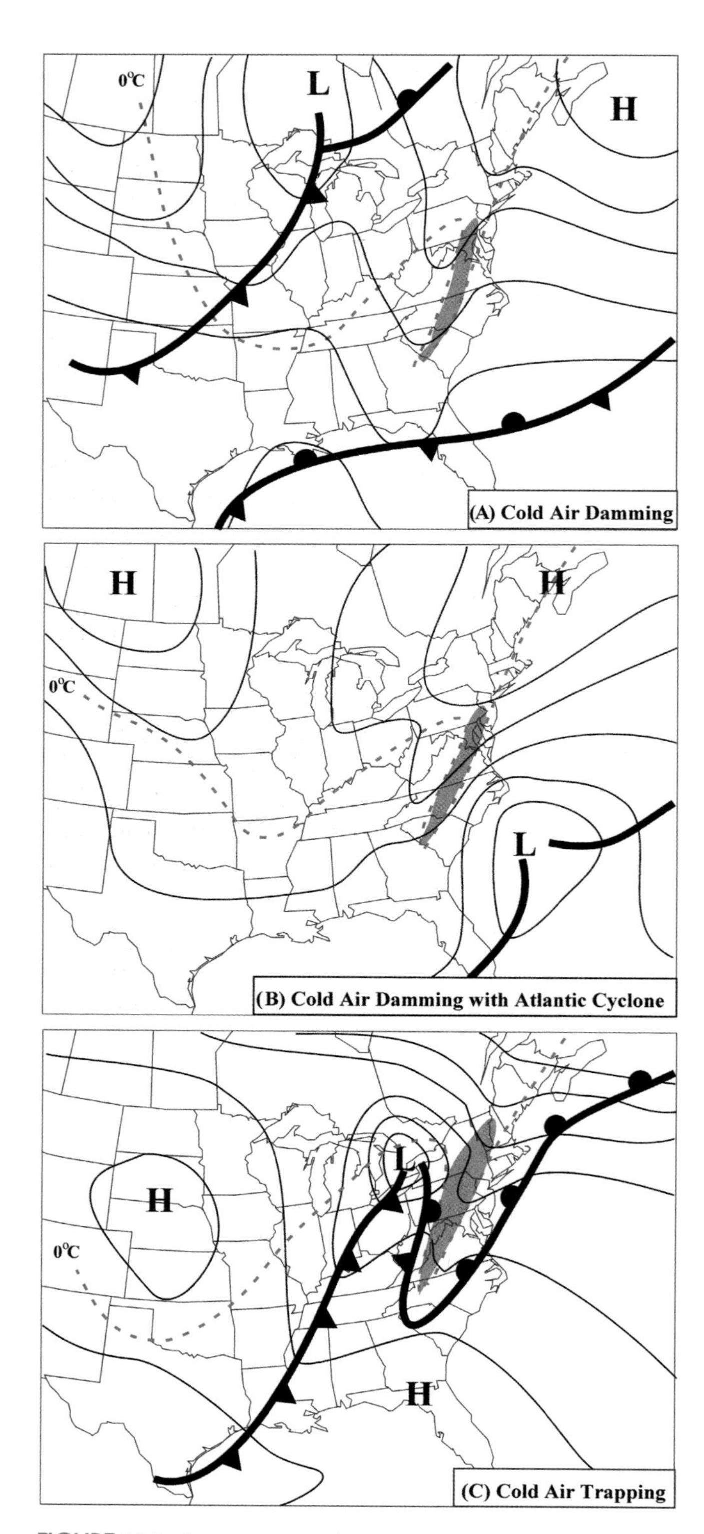

FIGURE 12.5: Common weather patterns associated with freezing precipitation along and within the Appalachian Mountains.

over the coast. Cold air trapping is associated with about 10 percent of all freezing precipitation events affecting the United States.

In all of the weather patterns described above, the zone of freezing precipitation is narrow, generally less than 160 km (100 miles). As fronts and their parent cyclones move across a region, the zone of freezing rain continually moves with the evolving weather system. In most cases, this results in a transition either from freezing precipitation to ice pellets and snow, or freezing precipitation to rain over the course of a few hours. The extreme glaze accumulation characteristic of a major ice storm requires that the zone of freezing precipitation remain over the same region for an extended time period. This only occurs when a weather system propagates very slowly and the fronts associated with the freezing precipitation are nearly stationary. The online sections of this chapter focus on two storms that had such characteristics: the January 2007 ice storm that caused extensive damage from Texas to New England and the 1998 Northeast ice storm that was the worst weather disaster in Canadian history.

FORECASTING FREEZING PRECIPITATION

How do forecasters predict freezing precipitation? As the preceding sections show, the vertical profiles of temperature and humidity are crucial in determining the type of precipitation that will fall. Soundings (Chapter 2) are indispensable in providing information on the vertical temperature structure, particularly the presence of any above-freezing layers in the lowest several kilometers. Soundings are supplemented by the nearly-continuous information from the network of vertical profilers described in Chapter 2. Soundings obtained from satellites are generally less useful because the profiles from satellites cannot be accurately obtained from areas of extensive cloudiness.

During a freezing precipitation event, radar can provide information on the location of layers in which solid precipitation is melting as it falls. These layers often appear as "bright bands" in vertical cross sections obtained from radars. Finally, pilot reports of aircraft icing can provide information on zones of supercooled water in the atmosphere, while the stations in the ASOS and AWOS network have freezing rain sensors to enable forecasters to map the locations of freezing rain at the surface.

Because numerical weather prediction models provide three-dimensional forecasts, their output not only includes forecast weather patterns (including storms), but it permits the construction of "forecast" soundings as well as the presence of clouds and precipitation. Using this information, forecasters can anticipate the formation of freezing precipitation in a general sense. However, the models' vertical resolution (Chapter 4) is sometimes inadequate to capture the important details of the vertical profiles that can spell the difference between freezing precipitation, snow, and rain. Moreover, if models do not include information about the presence or absence of snow on the ground, their forecasts of low-level temperatures can be in error by several degrees C. As a result, forecasts of freezing precipitation demand that meteorologists use modern technology to closely monitor the latest observational information as they are guided by model forecasts of the large-scale weather pattern.

DISTRIBUTION OF FREEZING PRECIPITATION IN NORTH AMERICA

Figure 12.6 shows the median number of hours per year that freezing rain occurs within the United States and Canada. Freezing rain occurs most frequently over eastern Canada and New England with many locations averaging between twenty and forty hours per year. An axis of higher freezing rain frequency extends into the southeastern United States along the Appalachian Mountains, and a second axis eastward from New York and Pennsylvania into Illinois. Freezing rain frequency decreases southward to near zero along the Gulf of Mexico coast and westward to near zero in the western Great Plains. Though infrequent, freezing rain in the southern United States can be devastating. Of all the regions in the United States, the South has the greatest percentage of ice storms that produce insured property losses exceeding $1 million. Freezing rain rarely occurs west of the U.S. Rocky Mountains except in the Pacific Northwest where a local maximum occurs in the Columbia River Basin due to trapping of cold air in the Columbia River Valley.

The distribution of freezing rain corresponds well with the weather patterns described in the previous section. The Appalachian lobe of higher freezing rain frequency is related to cold air damming and trapping events. The maximum in freezing rain extending from Illinois into the Canadian Maritime Provinces

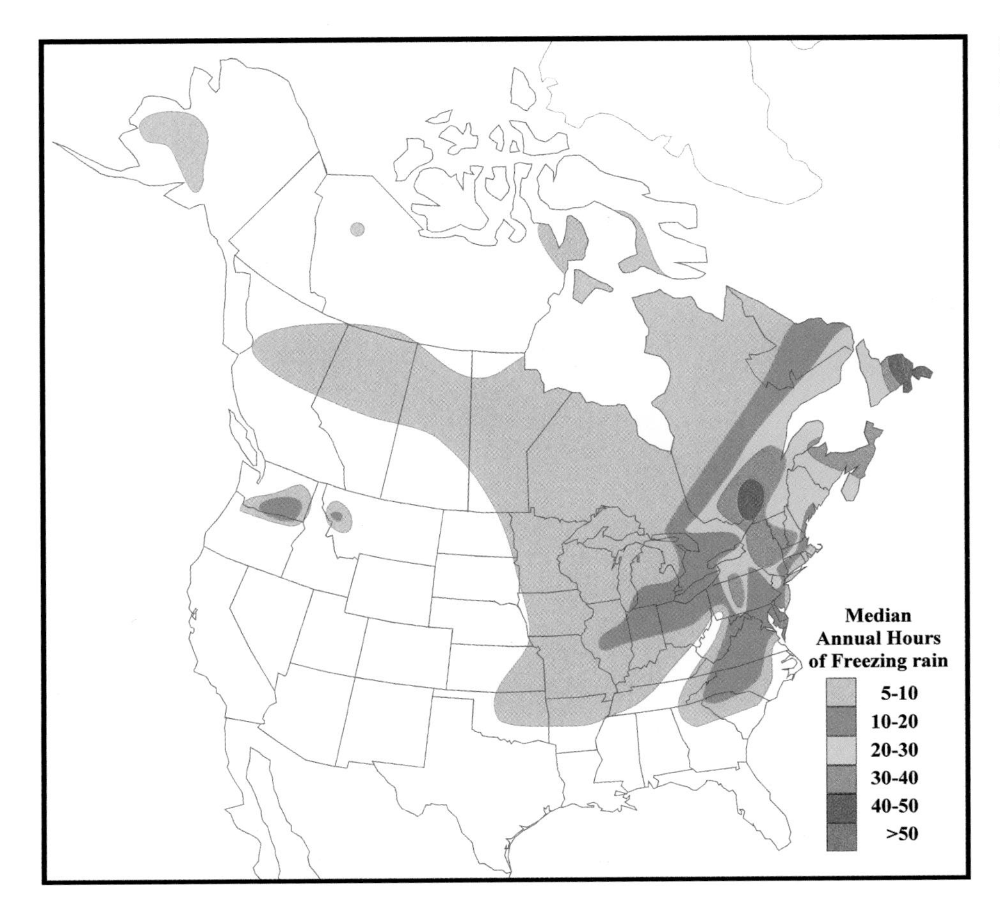

FIGURE 12.6: Median number of hours per year that freezing rain is reported at stations across North America.

corresponds to locations where warm frontal precipitation is common as cyclones track across North America from the Rockies or along the East Coast. The large maximum in New England and eastern Canada is associated with warm fronts of both East Coast and Rocky Mountain cyclones that pass over the area.

Figure 12.7 shows the average number of hours per year that freezing drizzle occurs within the United States and Canada. The pattern for freezing drizzle differs substantially from the pattern for freezing rain in Figure 12.6, although both patterns show the highest frequencies over eastern Newfoundland. A much larger area of relatively high frequencies is found over the central United States and Canada, with a maximum in central Canada south of Hudson Bay. Freezing drizzle typically develops in this area when warm air is lifted aloft over arctic airmasses, creating a wide layer of stratus that produces light, sporadic precipitation. The distribution of freezing drizzle corresponds to the typical location of arctic fronts, which normally originate in central Canada and move southward across the central United States. Freezing drizzle is rare in the western United States, except locally over the river basins from eastern Washington to northwestern Montana. There is also a maximum (thirty to forty days per year) in northwestern Alaska, where freezing drizzle can occur even during the summer months. The minimum in freezing drizzle in the middle of Alaska is due to the isolation of this region from moisture sources by topography and the extreme cold temperatures that occur in the interior during wintertime.

While freezing precipitation can occur at any time of day in the areas shown in Figures 12.5 and 12.6, its frequency of occurrence is about twice as high in the hours before sunrise as in the late afternoon. Because the ground surface is usually coldest and surface-based temperature inversions are most common in the pre-dawn hours, the diurnal cycle of freezing precipitation frequencies is consistent with the soundings in Figure 12.1.

FREEZING PRECIPITATION AND GLOBAL WARMING

Global warming is expected to lead to generally milder winters in middle latitudes. Warmer surface conditions will likely reduce the incidence of freezing

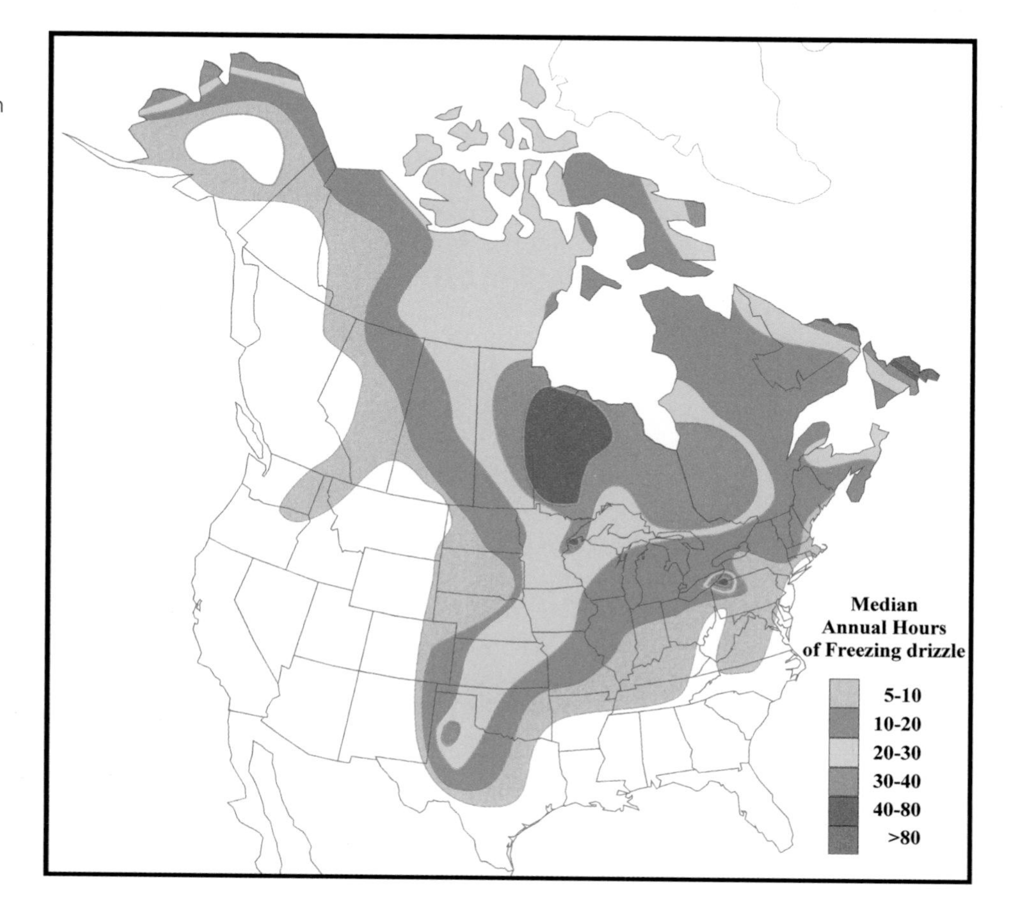

FIGURE 12.7: Median number of hours per year that freezing drizzle is reported at stations across North America.

rain in the southern areas that now experience freezing rain several times in a typical winter. However, one would expect that the patterns in Figures 12.6 and 12.7 will shift northward, resulting in increased frequencies of freezing rain and freezing drizzle over portions of eastern Canada. A generally warmer climate would also increase the altitudes at which freezing drizzle forms by the supercooled warm rain process. A critical question for future changes of freezing precipitation is: Will there be systematic changes in the vertical temperature profiles, particularly in the frequency of shallow temperature inversions near the surface? Projections of future climate change are not yet sufficiently precise to resolve changes in the vertical temperature gradients of the lower troposphere.

An additional consideration with regard to future trends in ice storm occurrences is the change in frequency and intensity of extratropical cyclones affecting central and eastern North America. As noted in Chapters 10 and 11, the most likely scenarios of cyclone change are reduced frequency and intensity because of the weaker temperature contrast between tropical and polar regions. Such changes would decrease the likelihood of major freezing-rain events. However, if cyclones move more slowly, as hypothesized in Chapter 10, the duration of freezing rain at a particular location could increase.

CHECK YOUR UNDERSTANDING 12.2

1. What is a typical width (distance in the direction perpendicular to a front) of the band of freezing rain associated with a frontal cyclone?
2. What is meant by the term "cold air damming"? Where does it typically occur?
3. Where in North America is freezing rain most common? Where is freezing drizzle most common?

TEST YOUR UNDERSTANDING

1. Briefly explain why freezing precipitation can be considered a severe-weather event.
2. What are the National Weather Service's criteria for an "ice storm"?
3. What is the temperature at which a pure liquid water droplet will freeze? Below what temperature do most small liquid water drops freeze in the atmosphere?
4. What is the role of ice nuclei in the formation of freezing precipitation?
5. Briefly explain the melting process for formation of freezing rain. Does this occur in deep or shallow clouds?
6. Describe qualitatively how the temperature profile conducive to ice pellets differs from that conducive to freezing rain.
7. Which process(es) require a temperature inversion for the formation of freezing precipitation?
8. Sketch a vertical temperature profile that would support the formation of freezing rain via the melting process.
9. How are clouds likely to be different if freezing rain, rather than freezing drizzle, forms?
10. What is the sequence of precipitation types that you would typically experience as you move from south to north through an east-west oriented warm front in winter?
11. Briefly explain the supercooled warm rain process for formation of freezing drizzle. Does this occur in deep or shallow clouds?
12. What type of freezing precipitation is most hazardous to aircraft? How do you think the aviation industry might address this hazard?
13. What type of weather pattern is most often associated with ice storms in the Southern Plains and the Gulf Coast States?

14. Where does freezing rain and freezing drizzle fall relative to the 0° C isotherm?

15. Seven weather patterns may result in freezing precipitation during winter. Name each of these patterns and briefly explain where freezing precipitation forms relative to the pressure centers and/or fronts.

16. How does cold air "trapping" differ from cold air "damming"?

17. What is an essential characteristic of any weather pattern in which the glaze accumulation is extremely high (e.g., > 1 inch)?

18. What role does wind play as a destructive force in ice storms?

19. Where is freezing rain and drizzle most common in North America?

20. What type of cloud poses the most serious aircraft icing threat? Why?

21. Why do the most damaging ice storms often occur when a high–pressure system is located to the north of a cyclone?

22. How might global warming affect the occurrence of freezing precipitation?

TEST YOUR PROBLEM-SOLVING SKILLS

1. On a dark and stormy night in early February, you check the weather forecast for the next day. The forecast calls for rain with temperatures just above 0° C overnight, clearing around sunrise, followed by a sunny day with a high temperature near 10° C. When you awaken the next morning, the sky is indeed clearing. The temperature at sunrise is 0° C. However, the overnight storm brought heavy freezing rain, leaving a coating of ice approximately 2 centimeters thick.
   (a) Is the daily high temperature likely to reach 10° C? Explain why or why not.
   (b) Use the following information to estimate the time that the temperature at the surface will start to rise. Assume that all the solar radiation reaching the surface is absorbed by the ice. (Use one square meter to simplify your calculations.)

   Energy required to melt one kilogram of ice = 334,000 joules per kg

   Density of ice = 917 kilograms per cubic meter

   Hourly totals of solar radiation reaching the ground in mid-February:

Time	Solar Radiation Reaching Ground (Joules Per Square Meter)
7–8 A.M.	100,000
8–9 A.M.	400,000
9–10 A.M.	800,000
10–11 A.M.	1,100,000
11–12 noon	1,300,000
12–1 P.M.	1,300,000
1–2 P.M.	1,100,000
2–3 P.M.	800,000
3–4 P.M.	400,000
4–5 P.M.	100,000

   (c) Ice surfaces normally reflect solar radiation. Assume that half of the solar energy is actually reflected to space and repeat (b). Will the temperature rise that day?

2. As frozen precipitation falls through an atmospheric layer that is above freezing, the precipitation will melt completely (favoring freezing rain) if the above-freezing layer is deep enough and warm enough. The warmer the layer, the smaller the layer thickness required for melting the falling precipitation. The precise values of the thickness and the mean temperature of an above–freezing layer can be used to discriminate situations in

which precipitation will reach the surface as ice pellets or freezing rain. Consider the following values to represent the layer depth and temperature required for an ice particle to melt completely to form a raindrop.

Depth of Above-Freezing Layer (Meters)	Mean Temperature of Layer (° C)
4000	0.3
3000	0.4
2000	0.7
1000	1.4
500	3.2
300	9.0

(a) Plot the transition temperature between freezing rain and ice pellets as a function of the layer's depth. Use the data points to make a smooth curve spanning temperatures from 0° C to 10° C, and layer depths from 300 to 5000 meters.
(b) On which side of your line will freezing rain occur? On which side will ice pellets occur? Label your diagram accordingly.
(c) Explain why the transition line is curved the way it is, i.e., interpret the fact that the line becomes parallel to each axis.

3. The following are data from a sounding at an East Coast location on a morning when winter precipitation was forecast.

Pressure (mb)	Temperature (°C)	Dewpoint (°C)
1005	–5	–7
960	6	3
900	3	0
870	3	0
810	1	–1
790	–1	–3
700	–7	–8
647	–10	–12
580	–13	–15
500	–20	–22
410	–28	–30
200	–55	–58

(a) Plot the sounding (temperature and dewpoint) on a Stuve diagram.
(b) What type of precipitation would you predict? State your reasoning.

4. The surface map provided here shows the frontal positions on a winter day, along with the surface temperatures (isotherms labeled in °C) and areas of cloud cover (gray shading). Assume precipitation is falling everywhere there are clouds.

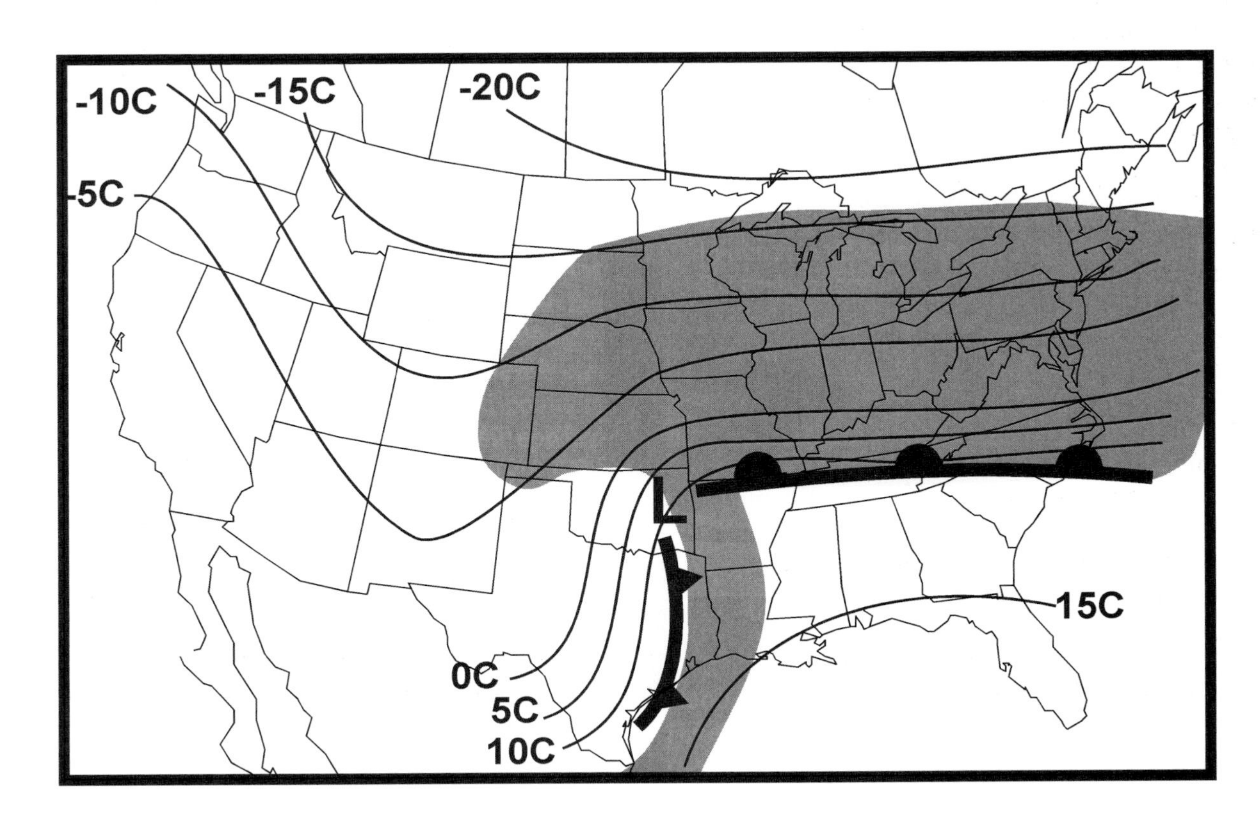

(a) Carefully draw a boundary around the region where freezing rain is most likely to occur. (Assume a typical north-south "width" of the freezing rainband).
(b) If the band of ice pellets accompanying this pattern has the same width as the freezing rainband, draw a boundary around the region where ice pellets are most likely to occur.

5. Freezing precipitation may develop in any of the cities listed below. For each city, identify the large-scale weather patterns that are likely to lead to freezing rain. Make copies of the blank U.S. map in Appendix C. Sketch one surface weather pattern for each city (position of lows, highs, fronts, and 0° C isotherm) that is consistent with freezing precipitation falling at that location.

   Cities:
   (a) Dallas, Texas
   (b) Nashville, Tennessee
   (c) Richmond, Virginia
   (d) Des Moines, Iowa
   (e) Bangor, Maine
   (f) Albany, New York

USE THE SEVERE AND HAZARDOUS WEATHER WEBSITE

http://severewx.atmos.uiuc.edu

1. Figure 3.1 shows the ice pellet and freezing rain symbols used in plots of surface weather reports. During the winter season, navigate to the *Severe and Hazardous Weather Website* "Current Weather" page and find surface maps of station observations and surface analyses.
   (a) Identify locations and times at which ice pellets or freezing rain are indicated as "significant weather." Record the dates, times, locations, and precipitation types at several locations over a twelve-hour period.
   (b) Examine a surface analysis (with isobars, fronts, and isotherms) for the same time period and determine which of the seven weather patterns shown in Figure 12.4 and 12.5 are responsible for the freezing precipitation.
   (c) For locations reporting ice pellets and freezing rain at the surface, obtain the sounding from the nearest rawinsonde site. Examine the sounding to determine the following:
      (i) Is a surface-based inversion present? If so, how deep is it?
      (ii) How deep is the layer of air in which the atmosphere is saturated or near saturation?
      (iii) Does the temperature profile show one or more above-freezing layers that are consistent with the precipitation type? Explain.
   (d) Using all of the above information, determine if the freezing precipitation formed via the melting process or the supercooled warm rain process. Justify your answer with evidence.

2. The detection of freezing rain at automated stations is a challenge with which the ASOS system has had to contend. Use the *Severe and Hazardous Weather Website* to navigate to the National Weather Service's website for freezing rain. Find information about the instrumentation used to detect freezing rain at ASOS sites. Briefly summarize your findings. What percentage of ASOS reports of freezing rain did human observers confirm in side-by-side tests?

3. The ice storm of 5 to 9 January 1998 was one of the most severe on record in southeastern Canada and the northeastern United States. The regions of southern Quebec and northern New England were particularly hard hit, as indicated in Figure 12B. Navigate the *Severe and Hazardous Weather Website* to a site of archived soundings. Examine the soundings for WMW (Maniwaki, Quebec) and GYX (Gray, Maine) over the period 5 to 9 January 1998.
   (a) Are the temperature-dewpoint differences throughout the troposphere consistent with deep clouds and heavy precipitation? Provide several examples to support your answer.

(b) For each location, are the temperature profiles consistent with freezing rain or freezing drizzle?

(c) Find at least one example of an abrupt vertical variation of wind direction and check whether the vertical wind profile helps to explain vertical temperature variations in the lowest several kilometers.

(d) Excellent photographs of the ice accumulation and destruction during the 1998 Northeast ice storm are on many internet sites and can be found with a simple image search. Look at these photographs to get a better impression of the devastation caused by major ice storms.

4. Online 12.1 highlights the ice storm of 2007 that impacted central Oklahoma. Navigate the *Severe and Hazardous Weather Website* to a site of archived soundings. Examine soundings from Oklahoma during the period and find soundings with temperature profiles that correspond to the idealized profiles for freezing rain that forms via the melting process. Save or print the soundings.

LAKE-EFFECT SNOWSTORMS

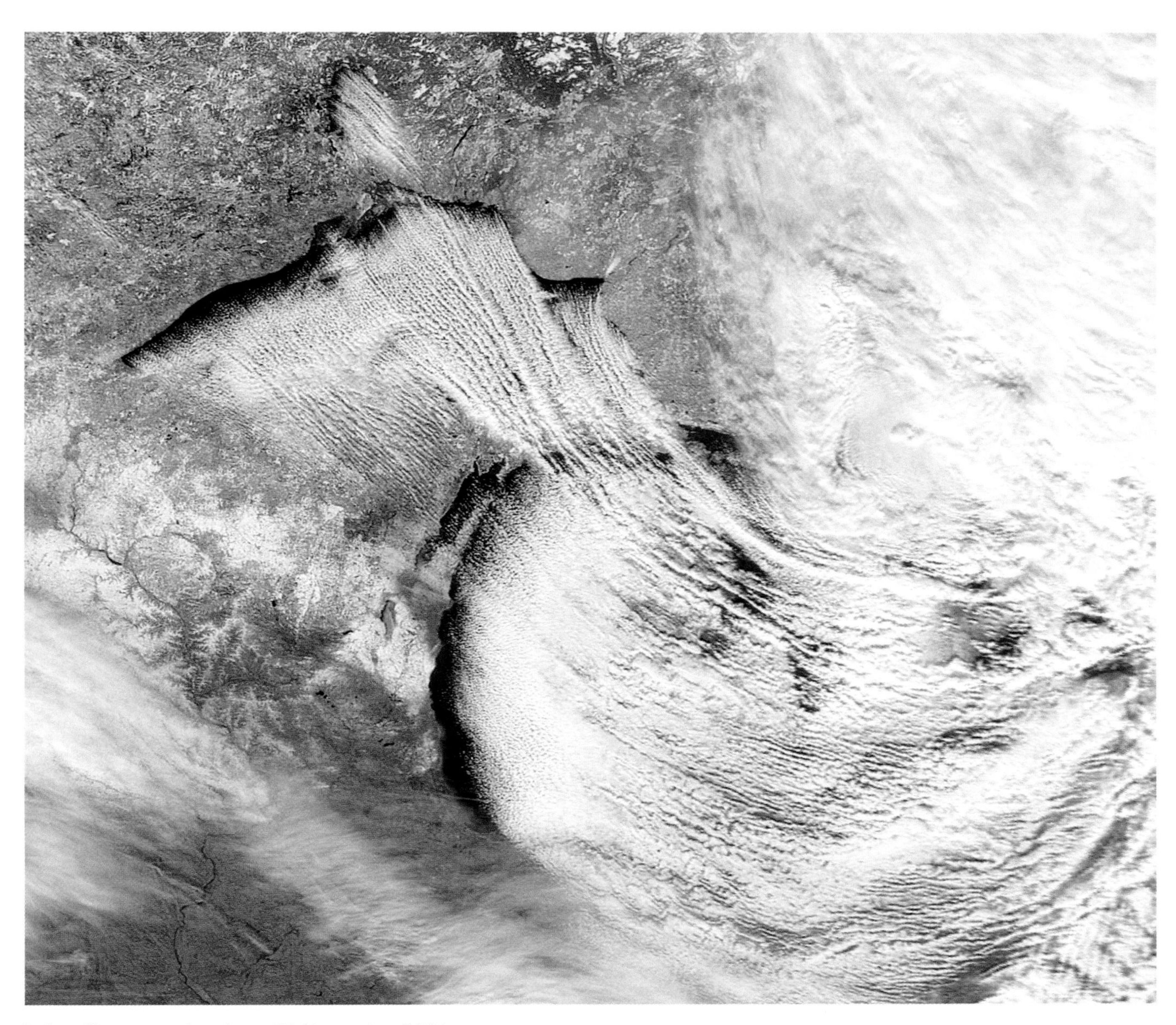

Lake-effect snow bands on 30 November 2004.

KEY TERMS

destabilization
lake-effect snowstorms
lake-enhanced snow
residence time
shore-parallel band
snow belt
snow squall
vortices
wind-parallel roll

Regions just downwind of the Great Lakes receive very heavy snow each winter. The large cities of Syracuse, Rochester, and Buffalo, New York; Cleveland, Ohio; Erie, Pennsylvania; and London, Ontario are all located in lake-effect ***snow belts***. Over 1.5 million residents of the state of Michigan live in regions influenced by lake-effect snows. The snowstorms can produce 0.3 to 1.5 meters (1 to 5 feet) of snow in single extreme events. Snowfall can continue for days in some areas, in some cases at the rate of more than one inch per hour!

Lake-effect snowstorms lead to high expenditures for snow removal, hazardous driving conditions, and cause many lost work and school days. The storms have major economic impacts on the Great Lakes region, crippling industry, recreation, school, and city functions. The Great Lakes snow belts do not extend very far inland, as shown in Figure 13.1. Virtually all of the precipitation falls between the lake shores and about 50 to 80 km (30 to 50 miles) inland. Eighty kilometers is generally the point at which most of the moisture supplied to the air by the lake has been removed as precipitation.

Although the Great Lakes are most widely known for the ***lake-effect snowstorms*** they cause, other inland water bodies, such as the Great Salt Lake in Utah and the Aral Sea in Asia, produce local snowstorms on their downwind shores. Even the smaller Finger Lakes in upstate New York occasionally produce lake-effect snow. In this chapter, we focus on the Great Lakes to examine how lake-effect snowstorms develop and how precipitation organizes within these storms.

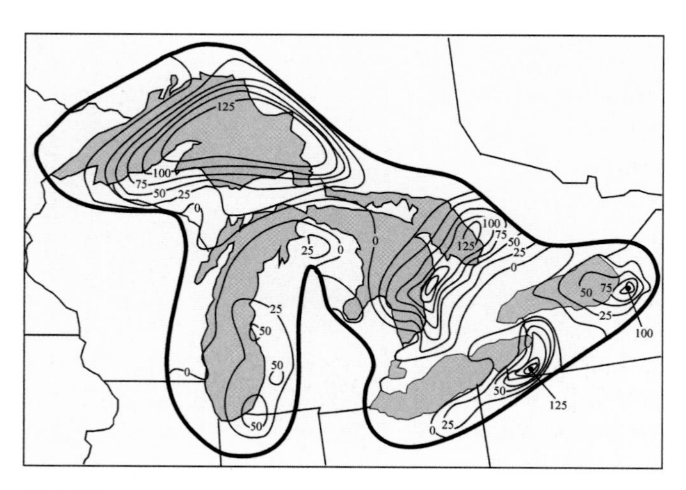

FIGURE 13.1: The additional wintertime precipitation (expressed in millimeters of melted precipitation) attributable to the Great Lakes. The dark line surrounding the lakes represents the 80 km boundary around the shoreline.

THE LARGE-SCALE WEATHER PATTERN FOR LAKE-EFFECT SNOWSTORMS

Very cold air must move across a lake surface for lake-effect snowstorms to develop. The warmer the lake temperature and the colder the air, the more extreme the lake-effect snows will be. The weather pattern that best supports lake-effect snowstorms is shown on Figure 13.2. Typically, lake-effect snow begins after an extratropical cyclone has passed over the region and the cyclone's cold front is well east of the Great Lakes. Cold air behind the front then flows southeastward across the lakes. The strength of this flow is enhanced if an arctic high has moved into the central United States. Under these conditions, a strong pressure gradient develops across the lakes and drives cold air southeastward from Canada. Lake-effect storms occur most often between late November and mid-January when very cold air can move across the lakes, but the lakes remain relatively warm and ice-free.

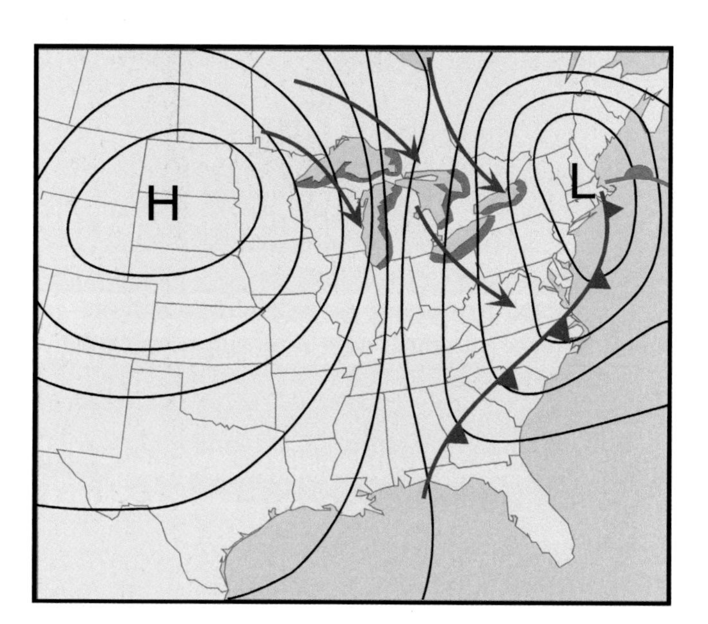

FIGURE 13.2: Typical weather pattern associated with lake-effect snow in the Great Lakes Region. An extratropical cyclone is located east of the Great Lakes, with high-pressure to the west. Major Great Lakes snow belts are shaded red.

LAKE-EFFECT SNOWSTORM DEVELOPMENT

Figure 13.3 shows the key processes that lead to lake-effect snowstorms. During lake-effect storms, the air approaching the lake typically has temperatures between about –5° C (23° F) and –25° C (–13° F). The lake is unfrozen with water temperatures between 0° C (32 °F) and 4° C (39° F). As air moves out over the surface of the lake it accelerates due to a reduction in surface friction caused by the absence of objects such as trees, buildings, and hills. Because air moves faster over the lake compared to over land, divergence occurs near the surface along the upwind shoreline. To compensate for divergence, air descends in a zone along the shoreline. Descending air will remain clear, so the windward side of the lake is often a zone of clear skies. The satellite photos on the chapter cover page and Figure 13.4 show examples of clear conditions on the upwind side of Lake Michigan during lake-effect events. The wind was from the west-northwest when these satellite images were taken.

When cold air first moves over the warm lake, heat is transferred from the lake surface to the air just above the lake surface, raising the air temperature. As we learned in Chapter 1, an increase in air temperature is accompanied by an increase in saturation vapor pressure, which represents air's capacity for water vapor. Water from the lake surface rapidly evaporates into the air as the air temperature increases and the wind blows across the lake surface. In this way, the surface layer of air over the lake rapidly warms and moistens. Aircraft measurements during lake-effect storms over Lake Michigan have shown that air directly above the lake surface can warm as much as 20° C (36° F) while crossing the lake!

Above the warming surface layer, air remains cold, a condition that leads to rapid ***destabilization***. As air near the lake surface heats, it soon becomes unstable and rises to form cumulus clouds. Clouds typically begin to form soon after air moves out over the lake, growing in height and intensity closer to the downwind shoreline, as illustrated in Figures 13.3 and 13.4. Snowfall normally commences well before air reaches the downwind shoreline. Lake-effect clouds typically grow in depth to altitudes of 2 to 3 km (~1 to 2 miles), shallow compared to the thunderstorms of summer, but deep enough to produce heavy ***snow squalls***.

As air crosses the downwind shoreline, friction with the land surface reduces the wind speed, resulting in convergence near the shoreline and forcing air upward (see Figure 13.3). This upward motion triggers additional convection in the now unstable air and strengthens the upward air motion within the cumulus clouds. Snow falls from the clouds, with the heaviest snow falling within and just downwind of this convergence zone. The heaviest snow occurs when air resides over the warm water for a long time, such as might occur when the flow is more along the long axis of individual lakes.

The transfer of moisture and heat from the lake to the air increases with the lake-air temperature

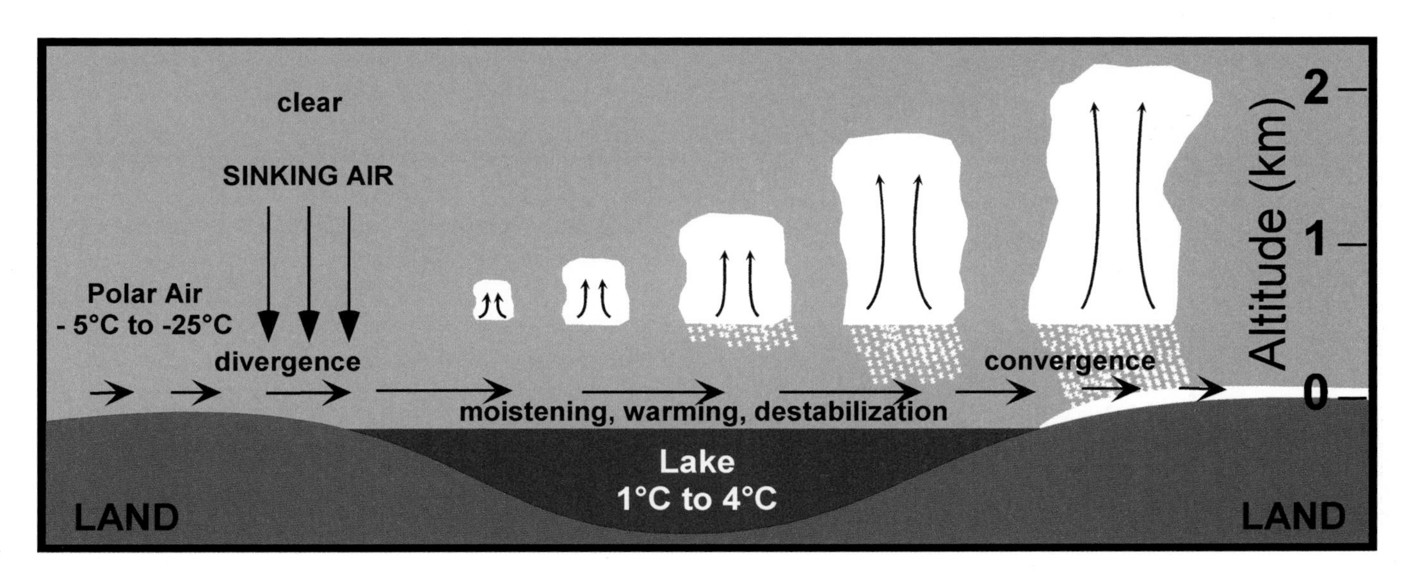

FIGURE 13.3: Physical processes that lead to lake-effect snow: as cold polar air travels over the warm lake surface, heat and moisture are transferred from the water to the air. This warming and moistening of the air near the surface destabilizes the air and leads to rising motions, resulting in the formation of clouds and precipitation. Rising motions are enhanced as the air reaches the downwind shore due to frictional convergence and topographic effects.

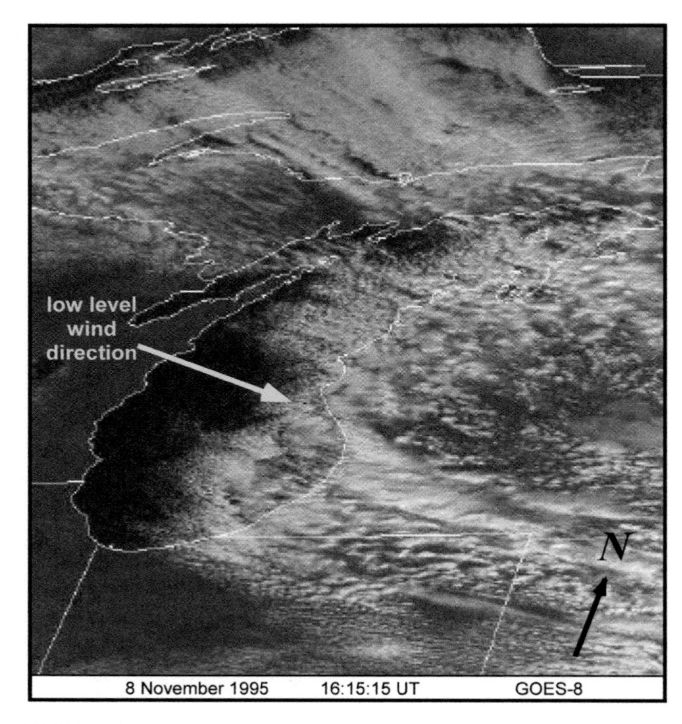

FIGURE 13.4: Visible satellite image of Lake Michigan and southern Lake Superior. The clouds shown here were produced by cold air flowing from the west-northwest over the lakes. The clouds in the upper left were produced by Lake Superior, while those in the lower right were produced by Lake Michigan.

difference. If the temperature difference between the lake and the air flowing over it is less than about 10° C (18° F), there is generally insufficient evaporation and destabilization for lake-effect snow to develop. Because winds from the south and east usually bring relatively mild air that is warmer than, or less than 10° C colder than the lakes, lake-effect snows are rare when the wind is from these directions. Consequently, the northern and western shores of the lakes rarely receive lake-effect snow (see Figure 13.2).

CHECK YOUR UNDERSTANDING 13.1

1. Where along the Great Lakes' shores does lake-effect snow typically occur?
2. Where are high- and low-pressure systems located relative to the Great lakes during a lake-effect snow event?
3. Why is it often clear upwind of a lake during a lake-effect snow event?

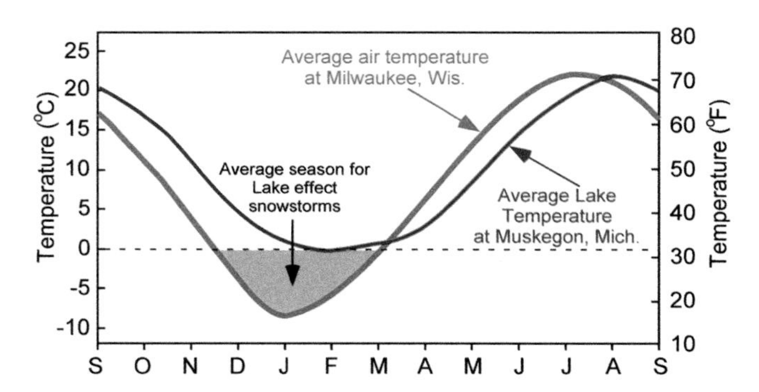

FIGURE 13.5: Mean monthly difference between air temperature at Milwaukee, Wisconsin, on the west shore of Lake Michigan and the lake temperature near Muskegon, Michigan, on the east side of the lake. The time period when lake-effect storms occur is shown in orange.

CLIMATOLOGY OF LAKE-EFFECT SNOWS

The amount of snow that falls during lake-effect storms depends on the temperature of the lake, the temperature of the air about to cross the lake, the wind direction, wind speed, the amount of ice cover on the lake, and the topography downwind of the lakes. As air passes over a lake, it warms very rapidly. The higher the temperature of the lake, the more heat and moisture will be transferred to the air. The colder the air, the more quickly the air will destabilize as heat is added from the lake.

Figure 13.5 shows the average temperature of air at Milwaukee, Wisconsin (just west of Lake Michigan), and the lake water temperature at Muskegon, Michigan (on the east shore of Lake Michigan), for each month of the year. The lake-air temperature difference, on average, is largest in late December and early January, the most favorable time for lake-effect snow. Lake temperatures and ice cover vary from lake to lake and even over a single lake. For example, Figure 13.6 shows the distribution of temperatures and ice cover for the Great Lakes in early December 2003 and late March 2004. Note the cooling that occurs during the period and the more extensive ice cover in March, particularly over Lake Erie. In general, the lakes are coldest in February. Figure 13.7 compares the mean snowfall for December and February for sites downwind of Lake Michigan and southern Lake Superior. The differences in snowfall amounts during the two months are largely due to differences in ice cover and temperature of the lakes. In February, part or all of some lakes can become ice covered. Virtually all heat and moisture transfer stops when the lakes become ice covered, effectively shutting off lake-effect snows.

FOCUS 13.1

Christmas in Buffalo

Record-breaking snows fell on Buffalo, New York, during the Christmas holiday period of 24 to 28 December 2001. Figure 13A shows the daily snow totals. On three of the five days, over twenty inches of snow fell per day! Buffalo lies on the end of Lake Erie. During cold air outbreaks over the Great Lakes, the larger scale flow will sometimes align west to east along the lake. This wind pattern supports the development of a shore-parallel band down the middle of the lake (Figures 13B, C). During the five-day period of intense snowfall, the larger scale flow maintained a southwest-northeast orientation, and the mid-lake band kept dumping snow on Buffalo. The persistent nature of the band can be seen in the storm total precipitation calculated from the Buffalo WSR-88D radar (Figure 13D). The orientation of the high precipitation values is telling—the total precipitation lines up with the band in Figure 13C, with the maximum located over Buffalo. Note the rapid reduction in precipitation in all directions away from Buffalo—areas outside the reach of the mid-lake band received much less snow.

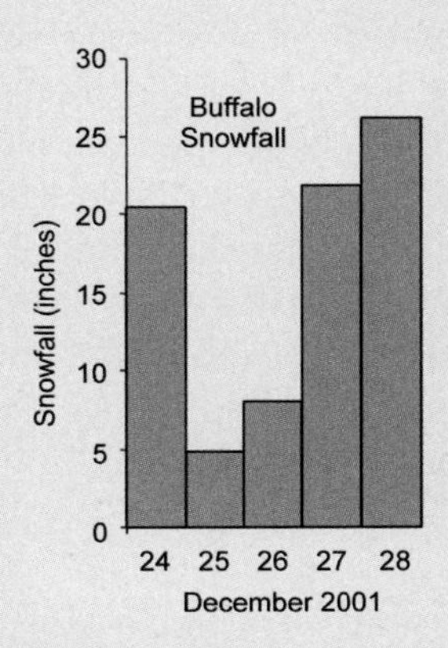

FIGURE 13A: Daily snowfall totals at Buffalo, New York, during the Christmas snowstorm of 2001.

FIGURE 13B: The mid-lake, shore-parallel snow band that buried Buffalo, New York, during the Christmas snowstorm of 2001.

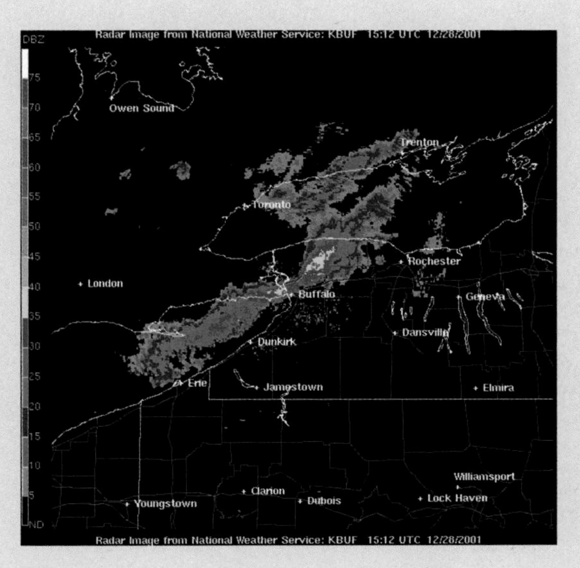

FIGURE 13C: Radar reflectivity from the Buffalo, New York, radar at 15:12 UTC on 28 December 2001.

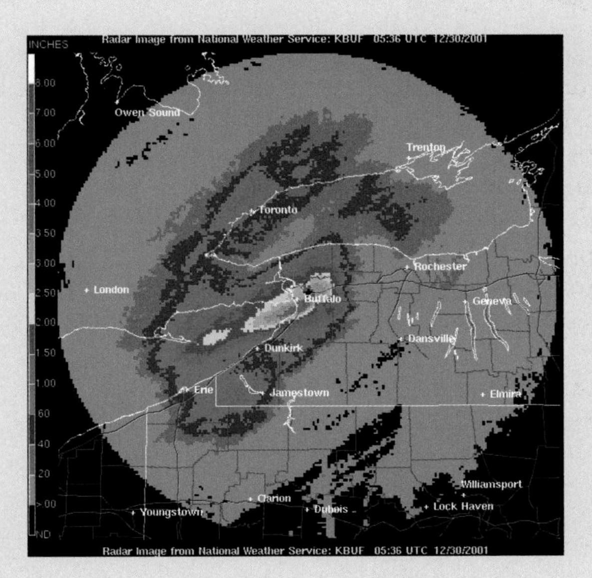

FIGURE 13D: Storm total precipitation (melted equivalent, in inches) estimated by the WSR-88D radar at Buffalo, New York. The precipitation is for the period from 7:10 P.M. 22 December 2001 to 12:38 A.M. 30 December 2001. Although the values are incorrect because the radar is calibrated for rain rather than snow, the distribution of precipitation shows the importance of the mid-lake, shore-parallel band during the storm.

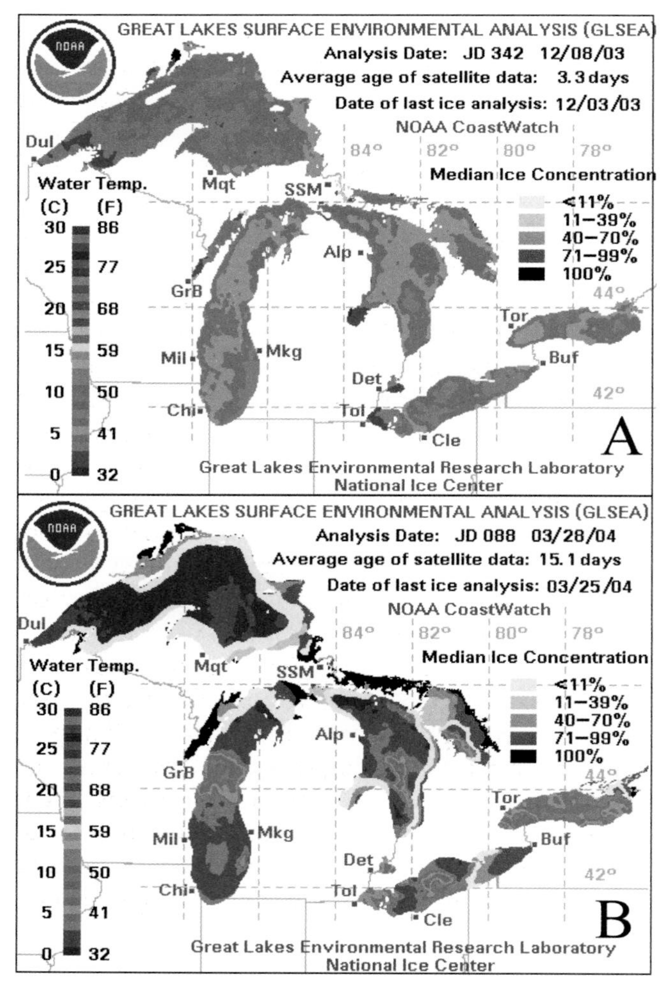

FIGURE 13.6: Lake temperatures and ice concentrations of the Great Lakes estimated from satellite measurements on (A) 8 December 2003, and (B) 28 March 2004.

Topography also influences the distribution of lake-effect snowfall. Figure 13.8 shows the topography in the vicinity of the Great Lakes. Compare this figure to the lake-effect precipitation shown in Figure 13.1. It is clear from these two figures that lake-effect snowfall is locally enhanced by topography, particularly downwind of the lower Great Lakes. The topography plays two roles. The rough terrain enhances surface friction, contributing to increased convergence and lifting on the downwind side of the lake. The hills also force air to rise, contributing to the upward air motion necessary to produce clouds and precipitation.

Air ***residence time*** over a lake is an important factor in lake-effect snows because it affects the total evaporation of water from the lake and the amount of heat transferred from the lake to the air. Both the wind speed and direction determine the residence time of air over a lake. Longer paths to the downwind shore increase the heat and moisture transfer, and therefore the amount of snow falling downwind. Figure 13.9 demonstrates this both schematically (Panel A) and for an actual storm (Panel B). The precipitation is significantly greater downwind of path 2 compared to path 1 because air arriving at the downwind shoreline has a longer residence time over the lake. The effect of wind speed is more complicated. Slower winds allow air to have a longer residence time over the lake, increasing the flux of heat from the lake to the air. However, faster winds create waves and enhance evaporation of moisture. Wind speed and direction also influence how clouds organize over the lake.

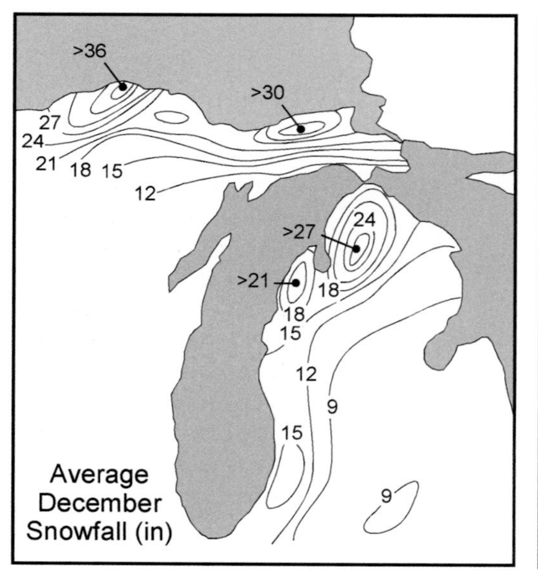

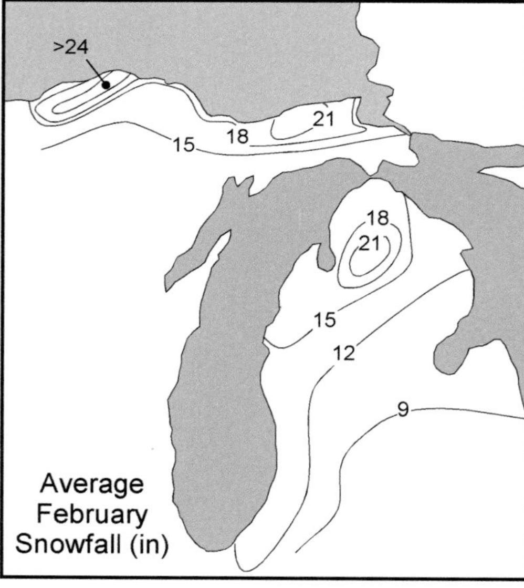

FIGURE 13.7: Average snowfall in inches downwind of Lakes Superior and Michigan for the months of December and February.

ORGANIZATION OF LAKE-EFFECT SNOWFALL

Lake-effect clouds and precipitation organize in three primary ways: in wind-parallel rolls, shore-parallel bands, and vortices. Each depends on the speed and orientation of the wind relative to each lake.

Wind-Parallel Rolls

In cases where winds are strong and blow with a component across the short axis of a lake, the heat transfer between the lake and the air will cause air to rise. However, all air cannot rise at the same time (or a vacuum would be left over the lake!). Cooler air aloft must sink to replace the warmer rising air. These rising and sinking motions often take the form of rolls that align parallel to the wind. Figure 13.10 shows how the circulations in the rolls appear. The clouds and precipitation develop in the upward branches of the rolls, while areas of weak or no precipitation occur in the downward branches. The upward branches of the rolls are typically 1 to 2 km (0.5 to 1.5 miles) wide and are typically spaced about 4 to 6 km (2 to 4 miles) apart, but can be spaced up to 10 km (5 or 6 miles) apart. Figure 13.11 shows a visible satellite image of the Great Lakes during a cold air outbreak when strong winds were blowing from the north across the lakes. Clouds associated with ***wind-parallel rolls*** appear over Lakes Superior, Huron, and northern Lake Michigan. Figure 13.12 shows a radar image of precipitation during a different episode when wind-parallel roll circulations developed on the eastern shore of Lake Michigan. The wind direction during this event was from the northwest. Note that the precipitation forms linear bands separated by relatively precipitation-free areas. The precipitation develops over the lake and increases in intensity along the shoreline and inland.

FIGURE 13.8: Topography of the Great Lakes Region. Higher elevations are shown in red-brown. Hills enhance lifting along the downwind shores leading to increased snowfall in these regions.

Shore-Parallel Bands

The cloud systems over the lakes sometimes do not take on roll structure. When winds are relatively weak, the heat of the lake will force air over the lake to rise,

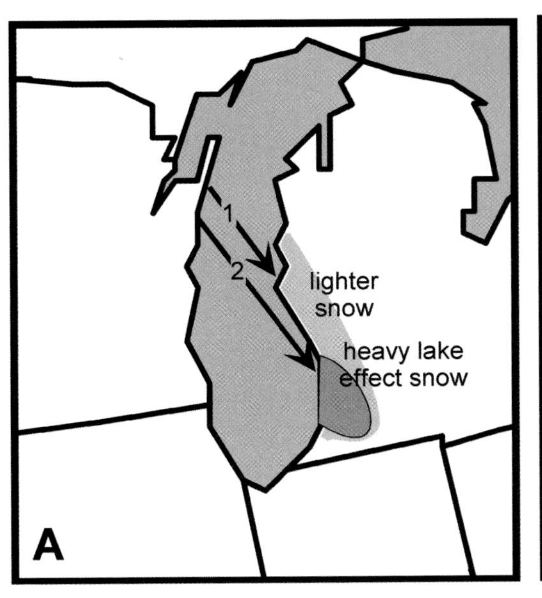

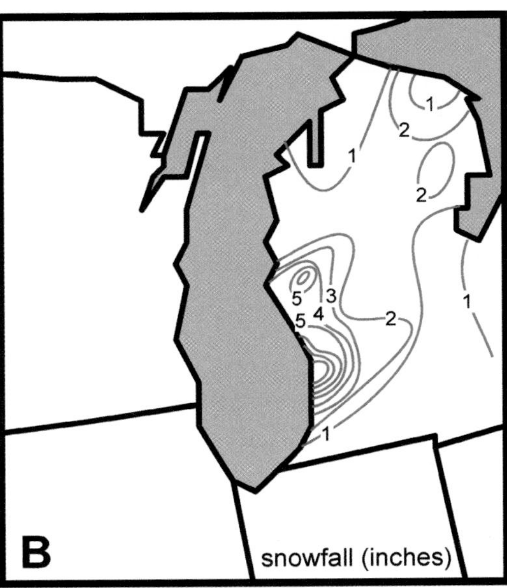

FIGURE 13.9: The residence time of the air over Lake Michigan is much longer along path 2 compared to path 1 for the airflow depicted in Panel A. Air picks up more heat and moisture along path 2 as the air travels over a wider expanse of water. Map B shows total snowfall during a wind regime similar to that shown in Panel A. Snowfall is given in inches.

ONLINE 13.1

Persistent Winds Cause a Major Lake-Effect Snow Accumulation in February 2007

After an unusually mild first-half of the 2006–2007 winter, the atmospheric circulation shifted to a pattern that brought cold weather to much of the eastern half of the United States from late January through February. The Great Lakes region was sandwiched between high pressure over the Northern Plains and a persistent area of deep low pressure in northeastern Canada. As shown online, the northwest-to-southeast orientation of the isobars and the associated southeastward sweep of the cold air from Canada continued for days on end. Because of frictional deflection, the surface flow of cold air was generally west-to-east, parallel to the long axes of Lakes Erie and Ontario. The persistence of the flow pattern is also apparent in the online satellite animation for the same period, which shows bands of precipitation just offshore of the East Coast. The immediate coastal zone generally remained clear for the same reasons (accelerating surface winds, downward motion) that the upwind shores of lakes remain clear during lake-effect snow events. The period 3 to 12 February saw incredible snowfall totals (Figure 13F): On the lee side of Lake Ontario, Parish, New York received 121 inches—more than ten feet!—in the ten-day period, while North Osceola and Mexico, New York received 106 inches. On the five-star rating scale used by the Buffalo office of the National Weather Service, this event was given the full five stars—one of only two such events in the winter of 2006–07.

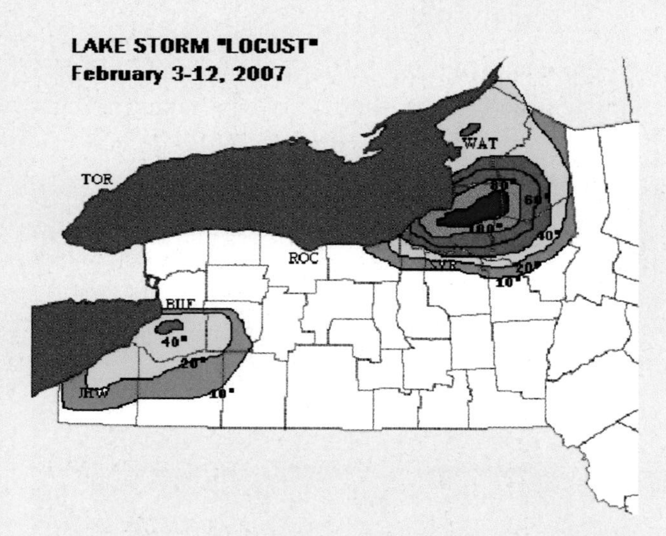

FIGURE 13E: Snowfall (inches) during the 3–12 February 2007 lake-effect snow event, nicknamed "Locust" by the National Weather Service's Buffalo office.

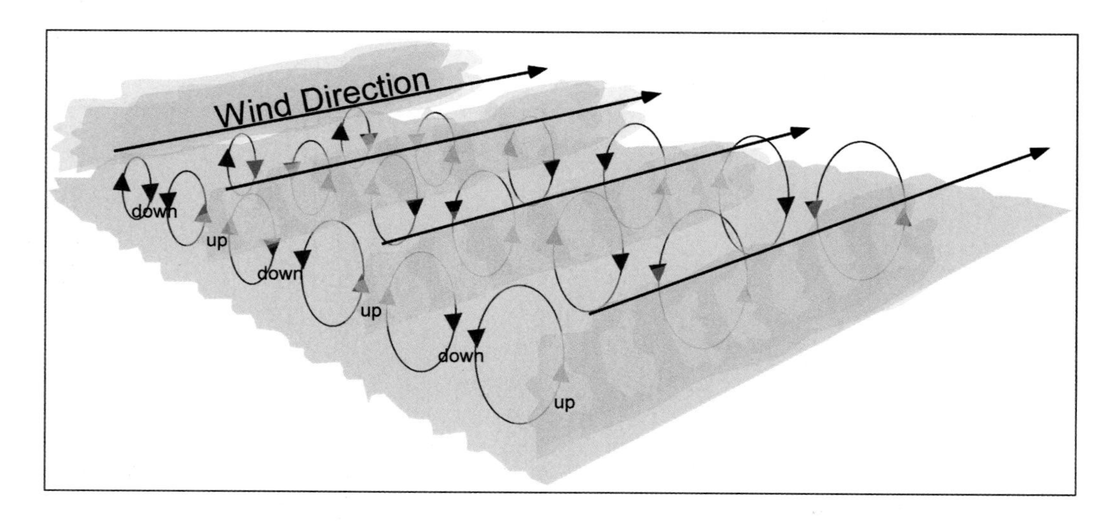

FIGURE 13.10: Circulations and clouds associated with wind-parallel rolls. Note that clouds form in the rising branches of the rolls and clouds dissipate where the air is sinking. Also note that the cloud bands develop parallel to the wind.

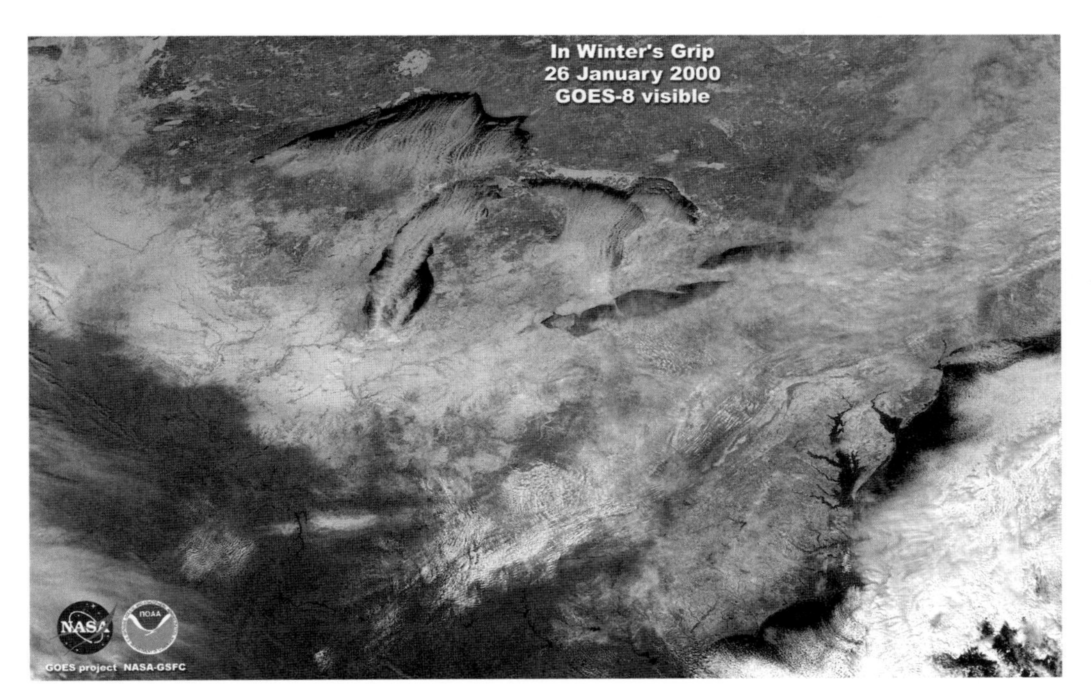

FIGURE 13.11: Visible satellite image of the eastern United States and Canada on 26 January 2000. The white areas south of the Great Lakes and in the southern Appalachian Mountains are snow cover. Wind-parallel rolls are present over Lakes Superior and Huron. The rolls over Lake Superior extend across the northern part of Lake Michigan. A shore-parallel band is present over southern Lake Michigan near the center of the lake.

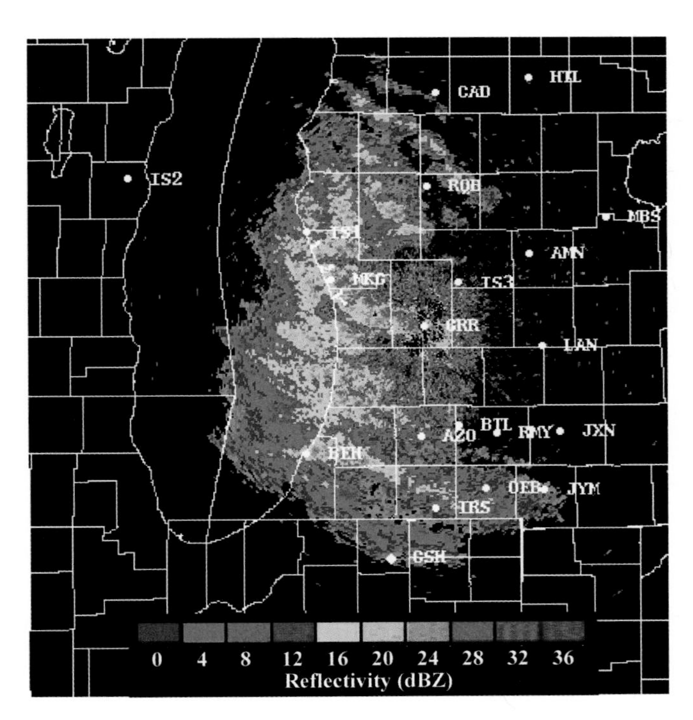

FIGURE 13.12: Radar image of the precipitation associated with wind-parallel rolls over the eastern side of Lake Michigan.

drawing air in from both shores toward the center of the lake. A snow band develops where these airflows meet (see Focus 13.2). This type of band formation can also happen when the winds are strong, but are parallel to the long axis of the lake. On radar, shore-parallel bands appear as a single line of reflectivity oriented along the long axis of the lake (Focus 13.2). ***Shore-parallel bands*** are typically deeper than wind-parallel bands, often extending to a depth of 3 km. Shore-parallel bands are quite common over Lakes Erie and Ontario, where the lake axis is west-east, but also occur over the other lakes. Northerly winds often produce bands parallel to Lake Michigan's north-south axis. Figure 13.13 depicts the circulation associated with shore-parallel bands.

Depending on the strength and orientation of the background wind, a shore-parallel band can be in the middle of the lake or near the shoreline. A shore-parallel band is evident over the southern part of Lake Michigan on Figure 13.11. In this case, the flow over the lake was strong and parallel to the long axis of the lake. Heavy snow falls from shore-parallel bands, and the snow is localized under the narrow cloud feature. Ten inches of snow may fall in one location, while the day may be sunny only a few miles away on either side of the band.

Vortices

Vortices sometimes develop over the Great Lakes. These circulations, which are still being investigated, may be related to many variables including horizontal wind shear (the variation of wind speed with distance), the magnitude of the wind speed, effects of variations in shoreline topography, atmospheric stability, and lake-air temperature differences. They normally

FOCUS 13.2

A Forecaster's Nightmare

Shore-parallel bands are notoriously difficult to forecast. The bands are only a few kilometers wide, may last a long time or dissipate rapidly, and at times are the only precipitation feature over a lake. These bands sometimes migrate during the day, while at other times they remain stationary. When the ends of bands such as the one in Figure 13F happen to position themselves over a large city like Chicago, Cleveland, or Buffalo, parts of the city may receive ten inches of snow between morning and evening rush hours, while other parts just a short drive away receive almost nothing. The skies may even be sunny! Forecasts must sound vague, especially to nonnatives driving through the area, since the possibility of everything from sunny skies to a major winter snowstorm all exist and indeed occur within the range of every local radio and television station.

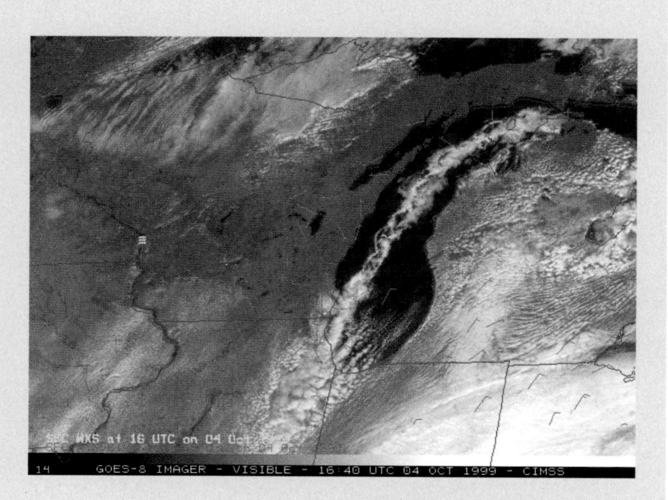

FIGURE 13F: A visible satellite image with surface wind barbs showing a shore-parallel band over Lake Michigan. Note the surface winds are nearly parallel to the lake axis.

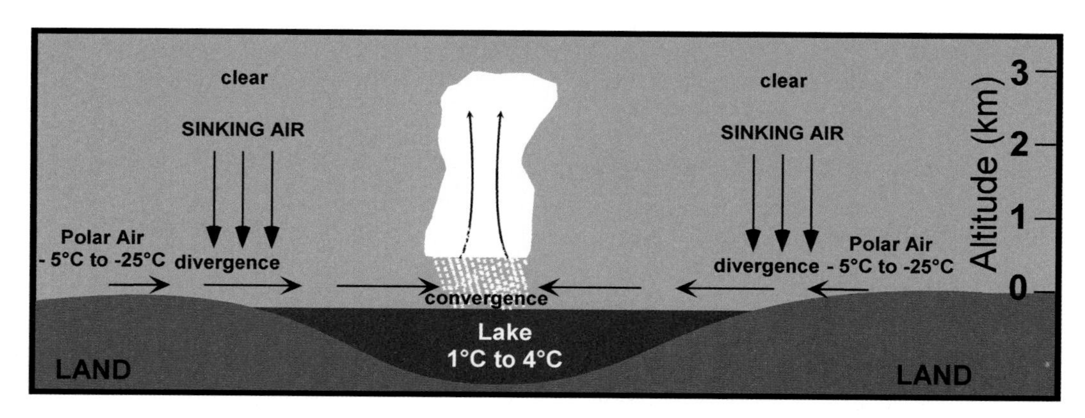

FIGURE 13.13: Physical processes that lead to lake-effect snowfall with a shore-parallel band.

develop over the lake, sometimes near one shoreline, and can maintain closed circulations for several hours. Lake-effect vortices typically drift slowly with the background flow. They seldom maintain coherence once they move inland. The vortices sometimes have radar structures that appear qualitatively similar to hurricanes, including a precipitation-free center called an "eye," a ring of precipitation around the eye similar to the hurricane eyewall, and spiral snow bands. Of course the winds are nothing like a hurricane—normally no more than 5 to 15 knots (6 to 17 mph). The diameters of these vortices range from about 10 to over 100 km (6 to 60 miles), about the order of the width of southern Lake Michigan. The vortex shown on Figure 13.14 has all of these features of the wind field and the reflectivity field. Note the sharp convergence zones along the snow bands that spiral inward toward the center. The vortex even has an "eye." Figure 13.15 shows an example where four vortices were present simultaneously over the Great Lakes. All of these vortices have eyes. When vortices stall off the lake in a position such that one of the snow bands surrounding the center of the vortex is located over shore, the band can often deliver heavy snow.

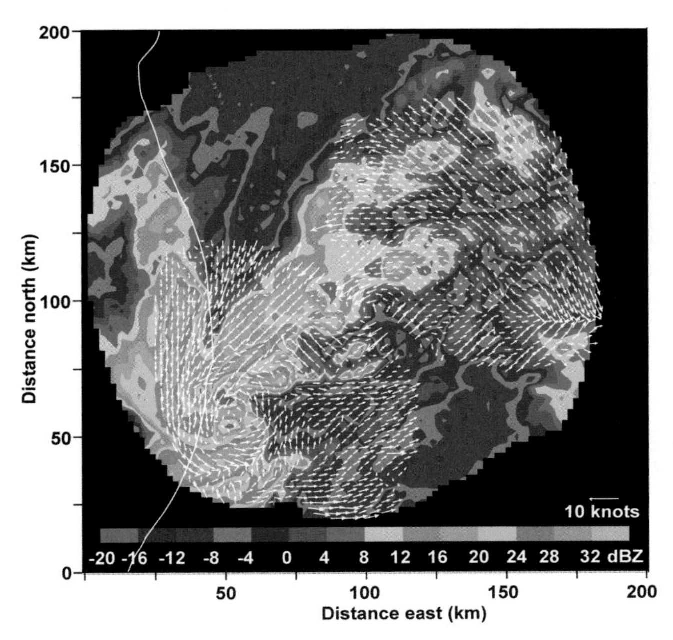

FIGURE 13.14: Radar image of a vortex over eastern shore of Lake Michigan. The wind arrows, which were determined by analyzing the Doppler radial velocity motions, show the circulation around the vortex. Note the snow bands circling around the "eye" of the vortex, which is relatively snow-free.

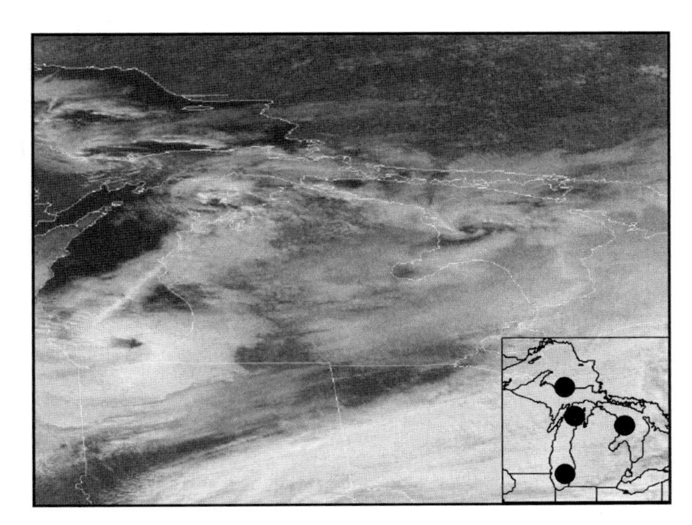

FIGURE 13.15: Satellite image of the upper Great Lakes region showing four separate vortices over Lakes Michigan, Huron, and Superior. The black dots on the insert map indicate the locations of the vortices.

LAKE-ENHANCED SNOWFALL

The lake-effect storms described in the previous sections all occur in otherwise clear cold air that typically follows the passage of a larger scale cyclone. Can the lakes also contribute to snowfall during a cyclone passage? Under the right temperature conditions, the answer is yes. In fact, the worst snowstorms in Chicago, Illinois are often result from heavy snowfall from a combination of cyclone-generated snow and snow generated by Lake Michigan. Figure 13.16 shows an example of a weather pattern that leads to ***lake-enhanced snow*** in Chicago. In Figure 13.16, the trowal region (see Chapter 10) of a large cyclone is passing over the southern Great Lakes, with the heaviest snowfall centered over Chicago. Note that air from eastern Canada is drawn across Lake Michigan by the surface low-pressure center. If this low-level air is sufficiently cold, as it can be when surface winds are from the northeast, evaporation of moisture and heat transfer from the lake will lead to the formation of lake-effect clouds below the clouds generated within the trowal. This lake-generated snow combines with the snow from the larger cyclone circulation, resulting in extremely heavy snowfall for the city. Lake-enhanced snowfall can occur in other locations around

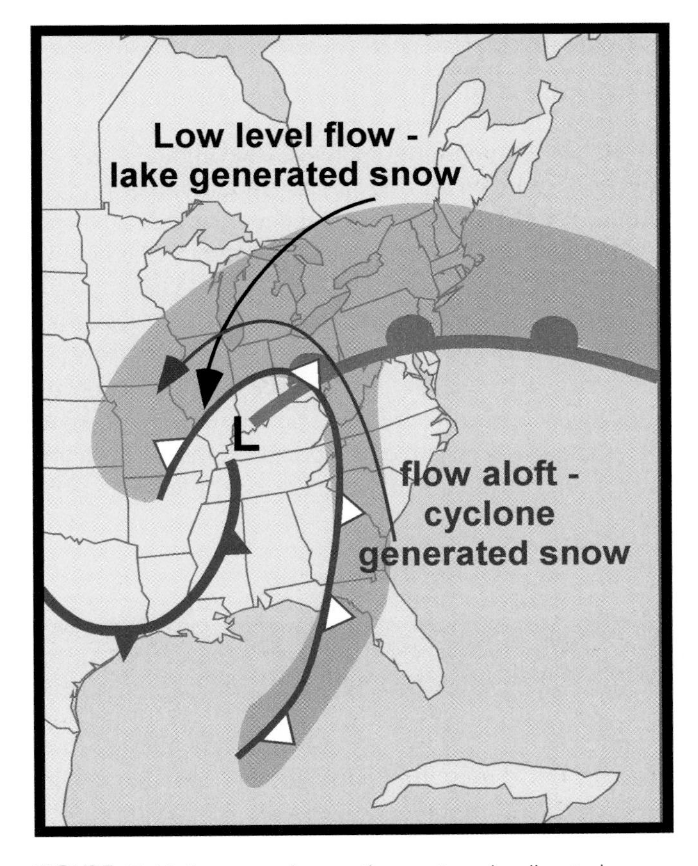

FIGURE 13.16: Large-scale weather pattern leading to heavy lake-enhanced snowfall over Chicago and the southwest shore of Lake Michigan. The dark arrow denotes the direction of the low-level flow. The green arrow denotes the flow aloft ascending into the trowal region of the cyclone. The gray region denotes clouds.

ONLINE 13.2

Lake-Effect Circulations

The lake-effect circulations described in this chapter can readily be observed with radar. Online animations of each of these circulations, taken from different events, show how the snow bands associated with wind-parallel rolls, shore-parallel bands, and vortices evolve with time. The wind-parallel bands on the western shore of Lake Michigan align along rolls that progressively move in to the Michigan shoreline. The shore-parallel band over Lake Michigan drifts eastward, the end of the band over Gary, Indiana, just southeast of Chicago. The vortex on the east side of Lake Michigan formed distinct spiral bands and an eye as it moved onshore. Note that the bands in all of these circulations seem to weaken or disappear farther from the radar site. The bands are not really weakening—the radar beam, which follows a slant path upward through the clouds as it moves away from the radar, simply progresses toward and through the top of the clouds, and then above the clouds farther from the radar.

the Great Lakes as cyclones move across the Midwestern United States and Canada, or along the East Coast.

LAKE-EFFECT SNOW AND GLOBAL WARMING

As global temperatures increase, the higher latitudes are projected to warm by larger amounts than middle latitudes (Figure 5.17). If this indeed happens, the frequency and intensity of cold air outbreaks reaching the Great Lakes is likely to decrease. Assuming that the climatic warming of the lake surface is less than the warming of the polar air, the reduced temperature difference between the lake surface and the overlying air will decrease the frequency and severity of lake-effect snows. A possible complication in this scenario is that the duration of lake ice cover will also be shorter in a warming climate, so the general decrease of lake-effect snows may be partially offset by the greater availability of surface heat and moisture in late winter (February, March). Nevertheless, the most reasonable expectation is that the present-day areas of enhanced snowfall east and south of the Great Lakes (Figure 13.1) will become less prominent if global warming continues.

CHECK YOUR UNDERSTANDING 13.2

1. List six factors that influence the amount of snow that falls during lake-effect storms.
2. What are the three ways that lake-effect snowstorms typically organize?
3. What is the difference between lake-effect snow and lake-enhanced snow?

TEST YOUR UNDERSTANDING

1. Describe the impacts of and damage caused by lake-effect snowstorms.
2. How far inland does lake-effect snow typically extend?
3. How does wind speed change when air flows over a lake? Explain what effect this has on convergence and divergence and how it influences the development of lake-effect snow.
4. How deep are typical clouds that produce lake-effect snow? Do they reach the tropopause?
5. Describe the physical processes that lead to the formation of lake-effect snowstorms.
6. What time of year is lake-effect snow most likely? Why?

7. What effect does the freezing of a lake have on lake-effect snow development?

8. What is the role of topography in the production of lake-effect snow?

9. How does the wind direction affect where lake-effect snow will be most intense?

10. Why do small lakes in the Great Lakes region, such as the Finger Lakes in New York, seldom produce lake-effect snows?

11. What is the preferred wind direction to produce lake-effect snow in each of the following cities?
    (a) Chicago, Illinois
    (b) South Bend, Indiana
    (c) Cleveland, Ohio
    (d) Buffalo, New York
    (e) Marquette, Michigan
    (f) Muskegon, Michigan

12. What are the three ways that lake-effect cloud systems organize over the Great Lakes, and what is the relationship of this organization to the wind speed and orientation of the wind relative to any particular lake?

13. Explain how strong winds across the short axis of a lake can produce wind-parallel rolls.

14. When cold winds blow offshore along the U.S. East Coast, would you expect wind-parallel rolls to develop?

15. Often wind-parallel rolls are observed over Lake Michigan at the same time a shore-parallel band appears on Lake Erie. Why?

16. What are some of the atmospheric variables that may affect the formation of lake-effect snow vortices?

17. How can you distinguish a lake-effect vortex from an extratropical cyclone on a satellite image? How might you distinguish it from a hurricane (aside from its location)?

18. Why do lake-effect snows not develop on the northern side of the Great Lakes when strong southerly winds are blowing during winter?

19. How can satellite imagery be used to determine if a snowfall event in the Great Lakes is due to lake-effect or lake-enhanced snow?

20. How might global warming affect the occurrence of lake-effect snows?

TEST YOUR PROBLEM-SOLVING SKILLS

1. Forecasters often use the difference between a lake's surface water temperature and the 850 mb temperature over the lake to determine the likelihood of lake-effect snowfall and to estimate its intensity. Approximately what would this temperature difference be for the air moving over a lake to eventually become absolutely unstable? (Helpful information: Great Lakes average altitude ≈ 150 m; average winter 850 mb height over Great Lakes ≈ 1450 m.)

2. Like many other severe weather events, lake-effect snow and lake-enhanced snow are influenced by conditions in the upper atmosphere as well as at the surface. Make copies of the blank U.S. map and sounding diagram found in Appendix B.
    (a) On one map sketch the surface conditions that would lead to strong lake-effect snow in Buffalo, New York. Include pressure systems, labeled isobars, fronts, and approximate temperatures.
    (b) On the second map sketch an upper tropospheric flow pattern that would support the surface system you sketched in (a). Identify any troughs, ridges, or jetstreams that would likely be present.
    (c) Estimate air temperatures that would be found approximately 1 to 2 km above the surface in

the vicinity of Buffalo to support lake-effect snow.

(d) On the sounding diagram, sketch a representative temperature and dewpoint temperature profile for Buffalo for the time of lake-effect snowfall. Be sure to include a wind profile to the right of the diagram.

(e) What organization of lake-effect snow would most likely develop in Buffalo based on your analyses?

3. (a) Examine Figure 13.6. Panel A shows conditions on 8 December 2003. Conditions are favorable for lake-effect snow throughout the Great Lakes region. If the air temperature at this time was 20° F and winds were from the northeast, which city or cities listed on the figure would likely experience lake-effect snow? Why?

   (b) Panel B shows conditions on 28 March 2004. Again, if the air temperature was 20° F and winds were from the northeast, which city or cities would experience lake-effect snow? Why does your answer differ from your answer in (a)?

   (c) What is the wind direction and maximum air temperature needed to get lake-effect snow on 8 December 2003 and 28 March 2004 in each of the following cities? (If lake-effect snow is not possible, state why.)

   (i) Chicago, Illinois
   (ii) Marquette Michigan
   (iii) Cleveland, Ohio
   (iv) Buffalo, New York
   (v) Alpena, Michigan

USE THE SEVERE AND HAZARDOUS WEATHER WEBSITE

http://severewx.atmos.uiuc.edu

1. During late fall or early winter, as a cold air outbreak develops across the Midwestern United States, navigate to the *Severe and Hazardous Weather Website* "Current Weather" and locate surface observations for the Great Lakes region and visible satellite imagery for the same time.

   (a) Examine the surface observations. Based on the wind direction and speed, what type of lake-effect snow organization would you expect to develop on each lake, if any?

   (b) Examine the satellite image to verify if lake-effect snow clouds are present. Are the cloud patterns consistent with your determination in (a)? Why or why not?

2. Monitor the National Weather Service radars in the vicinity of the Great Lakes region during late fall and early winter. When you detect a lake-effect snow event occurring, navigate the *Severe and Hazardous Weather Website* to maps of surface conditions. Determine the surface pressure pattern, temperatures, and winds of the region. It may be helpful to draw a composite map using the blank maps found in Appendix B. Compare and contrast your surface analysis to the examples provided in this chapter.

3. Use the *Severe and Hazardous Weather Website* to navigate your browser to the National Weather Service radar in Grand Rapids, Michigan during a lake-effect snow event over Lake Michigan. Animate the radar data.

   (a) Record the date and time of the event.

   (b) What type of organization do you observe in the snowfall?

   (c) How does the observed radar pattern relate to the prevailing surface wind direction? (You will need to examine surface observations for the region.)

4. During the lake-effect snow season, monitor the weather conditions in northern Utah near the Great Salt Lake. When precipitation develops, use the *Severe and Hazardous Weather Website* to examine the precipitation patterns and surface conditions. Determine if the precipitation is lake-effect snow. Provide a coherent argument for your determination using maps and images as justification for your conclusion.

5. Navigate the *Severe and Hazardous Weather Website* to the Additional Resources for Lake-Effect Snowstorms. Access the link to the NOAA Coast Watch Great Lakes Node. Here you will find satellite-derived images of Great Lakes water

temperature and ice cover. Examine a GLSEA (Great Lakes Surface Environmental Analysis) animation for a pervious year.

(a) Which of the Great Lakes changes temperature most rapidly?
(b) Which of the Great Lakes develops the most ice cover during winter?
(c) Which of the Great Lakes remains warmest latest into the winter season?
(d) Which of the Great Lakes would you expect to produce the most lake-effect snow over an entire year? Provide justification for your answer.

CHAPTER **FOURTEEN**

COLD WAVES

Ice fog over Fairbanks.

KEY TERMS

Arctic airmass
channeling effect
cold wave
cold wave warning
departure from the normal temperature
frostbite
hypothermia
leads
polar airmass
steering winds
trajectory
trigger mechanism
subsidence
temperature inversion
wind chill advisory
wind chill factor
wind chill index
wind chill temperature
wind chill warning

The term ***cold wave*** is used to describe an influx of unusually cold air into middle or lower latitudes. Temperatures during extreme cold waves can kill vegetation and fall below the thresholds for which buildings and other infrastructure components were designed, causing structural damage in addition to human suffering. Compared to blizzards, ice storms, and other winter hazards, cold waves generally affect much larger areas. Since 1989, when the National Weather Service began keeping statistics of cold wave fatalities, an average of about thirty deaths per year have been directly attributed to extreme cold. More generally, the Centers for Disease Control and Prevention estimate that approximately 600 deaths per year are attributable to ***hypothermia*** (abnormally low body temperature) in the United States, although the vast majority of such deaths do not occur during cold waves. These numbers do not include deaths caused indirectly by cold, such as fires originating in over-worked furnaces and space heaters. Cold-related deaths in the United States occur disproportionately among the elderly, in the South, and among males. Nearly 75 percent of recent cold-related deaths in the United States have been males.

As with fatalities, the economic losses due to cold waves are also greatest in the South. The greatest direct economic losses from severe cold result from damage in the agricultural sector, especially the citrus industry. During the cold outbreaks of 1983 and 1985, Florida citrus growers suffered losses of $3.6 billion and $2.9 billion, respectively. More recently, a cold wave during January, 2007 caused more than $1 billion of damage to citrus crops in the Central Valley of California, leading to sharp rises in prices of orange juice and other citrus products. Only three months later, unusually cold weather in the Southeast during April 2007 caused widespread damage to peach, apple, and other fruit trees that had blossomed several weeks earlier. The economic losses from cold waves are also high from broken water pipes, commercial slowdowns (e.g., shoppers and moviegoers remain in their homes), and substantially greater heating costs in the residential and commercial sectors. The South is especially vulnerable to cold waves because buildings are not designed for extreme cold, nor are residents generally equipped to deal with cold conditions. Ironically, the actual temperatures during record-setting cold outbreaks are far warmer in the southern states than in the northern states (Figure 14.1).

The "wave" in a cold wave is apparent in the upper-air flow (the jetstream), which is usually amplified into a strong ridge-trough pattern during a major cold outbreak. In the Northern Hemisphere, cold waves occur when very cold, dense air near the surface moves out of its source region in northern Canada or northern Asia. The actual temperature itself is not the most meaningful measure of a cold wave's

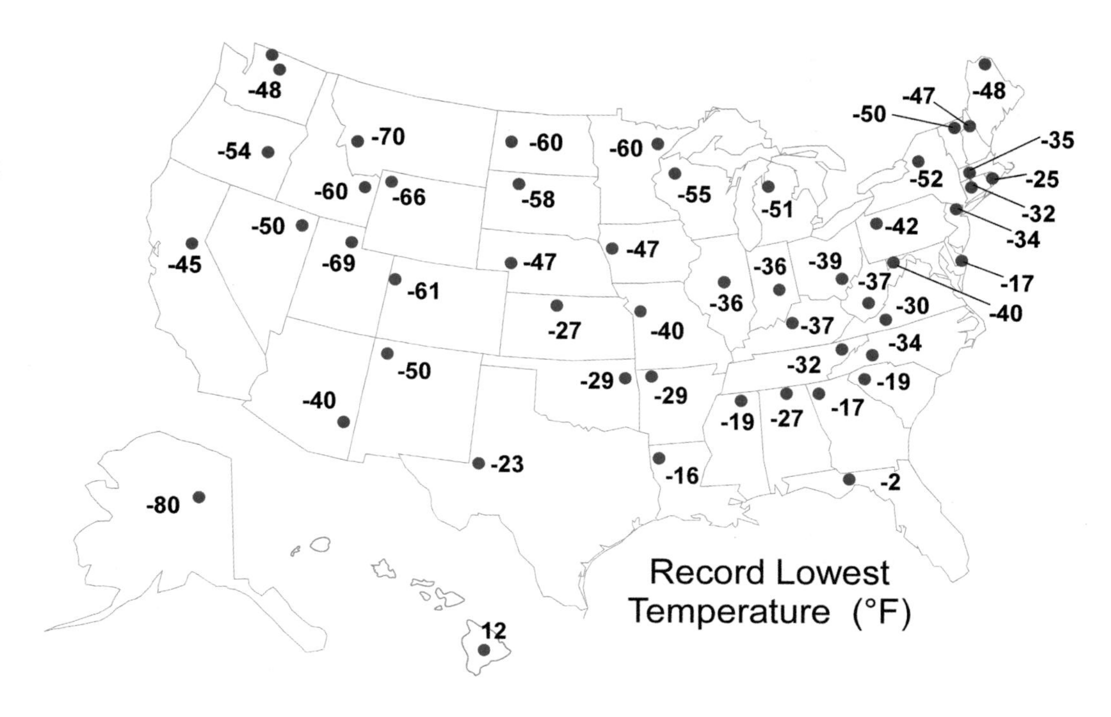

FIGURE 14.1: Lowest temperatures (°F) ever recorded in each state as of the end of the 2006–2007 winter. Black dots denote locations of each state's record low temperature.

intensity and impact. Rather, it is the *departure from the normal temperature* that is the meteorologist's measure of a cold wave. For example, a severe cold wave might bring temperatures of 20° F (−7° C) to central Florida or −10° F (−23° C) to southern New England. However, even a temperature of −10° F would not be unusual for some portions of the northern Great Plains and would actually be warmer than normal for a winter morning in Fairbanks, Alaska, where nobody would consider such temperatures indicative of a "cold wave."

FOCUS 14.1

Extreme Cold in the Arctic

In the high Arctic, the sun provides essentially no heating for several months during winter. In perpetual night, the Earth and atmosphere cool, sometimes to extreme temperatures. The lowest temperatures develop over the snow-covered land of Siberia, Alaska, and northern Canada, especially in low-lying areas. Ironically, these subarctic land areas tend to be colder than the North Pole, where the air gains some heat from the underlying ocean through cracks in the sea ice cover.

While typical January temperatures in Siberia and Alaska are between −10° F and −40° F (−23° C and −40° C), temperatures as low as −70° F (−57° C) sometimes occur. Alcohol thermometers must be used in such situations because mercury freezes at −38° F (−39° C). Extended periods below −40° F severely disrupt the lifestyles of residents of cities like Fairbanks, Alaska. Automobile tires lose their flexibility at such temperatures, and engine belts often snap when it is colder than −50° F (−45° C). Diesel fuel begins to congeal at −40° F, and gaps develop in railroad tracks when the metal contracts. Worst of all, the air is so cold that a thick ice fog forms directly from the moisture in automobile exhaust. The combination of bitterly cold temperatures and thick fog makes for dangerous and depressing conditions at the ground. Fortunately, the extremely cold and foggy layer is usually less than a hundred meters deep, so residents can experience clear skies and warm up by 10° to 20° F (6° to 11° C) by going up a nearby hill.

FORMATION OF COLD AIRMASSES

The core of a cold wave at the surface is a strong high-pressure center that forms during winter in high latitudes. As described in Chapter 8, the coldest surface high pressure centers form by the cooling of air in the lower troposphere. This cooling is favored by the long polar nights, especially when the winds are light (favoring airmass formation) and the sky is cloud-free (favoring the escape of infrared radiation to space). The loss of energy by infrared radiation cools the surface, which in turn cools the lower atmosphere as heat is conducted to the ground from the air immediately above the ground. This cooling of the low-level air increases the air's density and raises the surface pressure. The result of this process is a continental polar or an Arctic airmass (Figure 14.2).

The Northern Hemisphere's coldest airmasses are the ***polar airmasses*** that form over snow-covered northern land areas in Siberia, northern Canada, and Alaska. We will use the term ***Arctic airmass*** to denote an airmass that forms farther to the north, over the Arctic Ocean[1]. While the Arctic Ocean is generally snow- and

[1] In some texts, the distinction between "arctic" and "polar" airmasses is simply a matter of the coldness of the airmass.

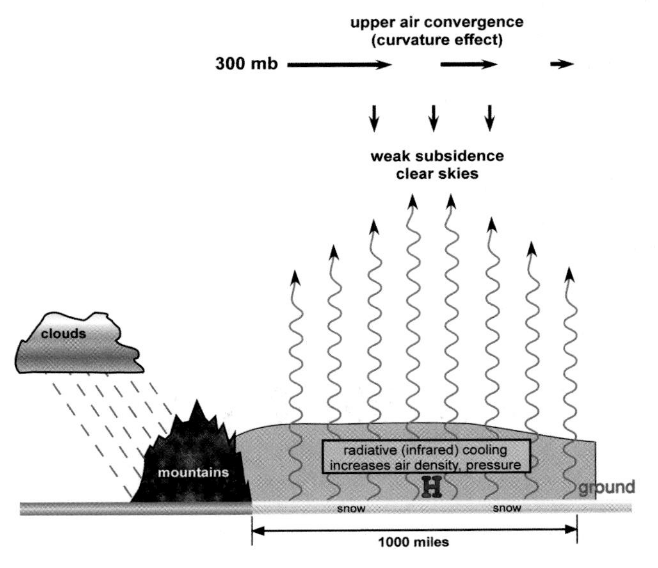

FIGURE 14.2: Summary of key physical factors contributing to intensification of a polar airmass: upper-air convergence, subsidence, clear skies, and radiational cooling over snow-covered surface lead to an increase of density and pressure of air near the surface, resulting in the development of a surface high-pressure system. View is looking northward over northwestern North America; blue shading denotes polar airmass.

ice-covered, its surface is continually fracturing because of the sea ice motion; consequently, numerous cracks, called ***leads***, develop and release significant amounts of heat to the lower atmosphere over the ice. This release of heat prevents the attainment of the lower temperatures that develop over land. Temperatures at the surface in a polar (continental) airmass can reach −50° F to −70° F (−45° C to −56° C), while typical temperatures over the Arctic Ocean in similar meteorological situations are only −40° F to −50° F (−40° C to −45° C).

Cold polar or Arctic airmasses are relatively shallow, often only extending one to several kilometers above the surface. The surface high-pressure systems that form under these dense airmasses weaken with height (see Chapter 8). These airmasses are also characterized by strong ***temperature inversions*** in the lowest several hundred meters. The long polar nights, in combination with light winds and clear skies, can lead to situations in which the surface is colder by 20° F to 30° F (12° C to 18° C) than the air several hundred meters aloft. Such strong inversions are most common when topography serves as an additional factor in "trapping" the cold dense air.

Once the cold high-pressure center has formed, the clear skies and calm winds that characterize a high-pressure system favor additional cooling that further intensifies the surface high. Intensification is also favored by upper-air convergence. Recall from Chapter 8 that this will occur if an upper-air ridge is centered to the west of the surface high. When the surface high is located over northwestern Canada, the ideal location of an upper-level ridge is over Alaska or the West Coast of North America. This flow pattern results in upper-air convergence directly above the center of the surface high pressure. Figure 14.3 shows this ideal configuration of the surface and upper-air systems. The development of a ridge over or immediately offshore of western North America is, in turn, often associated with a storm system in the North Pacific Ocean, as discussed in the following section.

CHECK YOUR UNDERSTANDING 14.1

1. What are the major impacts of cold waves?
2. Why is the phrase "cold wave" used in connection with a cold-air outbreak?
3. What are three factors that favor the formation of the coldest airmasses over high latitudes?
4. What is a typical depth of a cold polar airmass?

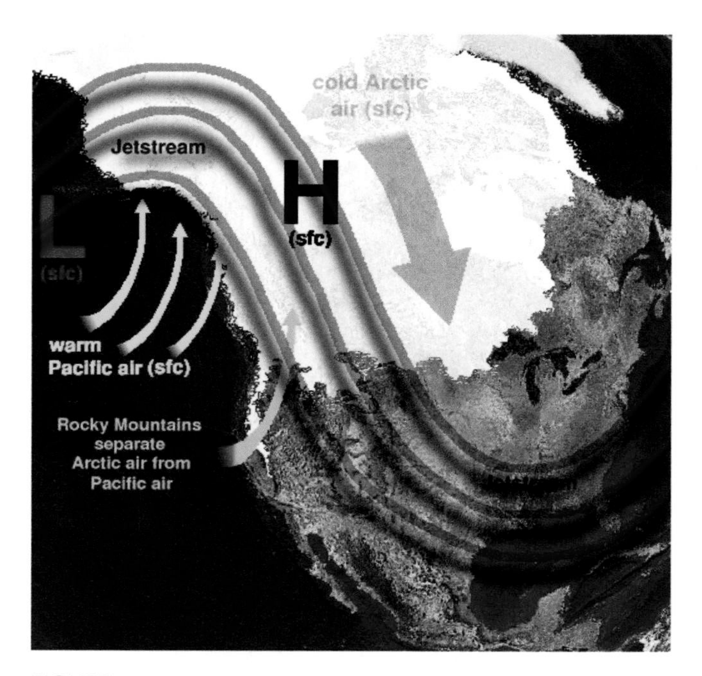

FIGURE 14.3: The typical weather pattern prior to a cold outbreak over central North America. The jetstream develops a large wave pattern, with a surface low and high located downstream of the trough and ridge. Warm air transported northward east of the surface low in the North Pacific contributes to the ridge intensification. Cold air transported southward from Canada east of the high amplifies the trough. The white region in the figure indicates snow cover, while "sfc" denotes surface conditions.

OUTBREAKS OF COLD AIR INTO MIDDLE LATITUDES OF NORTH AMERICA

While the formation of a cold airmass is one requirement for a North American cold wave, another requirement is the southward movement of the airmass into middle latitudes. Two factors contribute to the southward plunge of a cold airmass into the United States. The first is the tendency for a denser fluid to sink relative to a less dense (i.e., warmer) fluid and to spread laterally at low levels, much as a pool of molasses or chocolate syrup will spread horizontally when poured onto a flat surface.

The second factor is the movement of the low-level airmass in response to stronger ***steering winds*** in the middle and upper troposphere. As we will see later in this chapter, the relative importance of these two mechanisms varies among cold outbreaks. In either case, the more rapidly the cold airmass plunges into middle latitudes, the less it will be modified (warmed) by solar radiation or by its passage over warmer surfaces.

Equatorward motion of an airmass in response to jetstream winds will occur when the winds aloft have a southward component, as they do in the area to the east of an upper-air ridge. Hence, the upper air ridge not only enhances convergence that intensifies the surface high, but it also aids the southward plunge of the cold airmass. Several factors contribute to the intensification of a ridge over western North America and the eastern North Pacific. First, air over North Pacific waters is generally much warmer than continental air at the same latitude during winter. Strong cyclones in the Aleutian region of Alaska transport warm air northward on their eastern sides (Figure 14.3). The northward flow of maritime air leads to warming of the lower troposphere and a shift of the jetstream to higher latitudes. If this shift occurs over a longitudinal sector 30° to 60° wide, the resulting northward bulge of the jetstream will appear as a wave-like ridge. For this reason, a precursor of west-coast ridge development is often an unusually strong Pacific cyclone. An example of this process, which led to unusually cold conditions in the central United States during February, 2007, can be viewed in Online 14.1.

The second factor that intensifies the ridge over western North America is the flow associated with strong cyclones originating east of the Rockies or along the East Coast. These cyclones transport cold air southward on their western side, deepening upper-air troughs over the eastern portion of the continent. This trough intensification enhances the southward component of the jetstream flow on the trough's western side and adds to the prominence of the upstream ridge. In this respect, strong cyclones over the central or eastern United States can indirectly augment the polar outbreaks that may occur over subsequent days. Finally, the north-south alignment of the Rocky Mountains favors the enhancement of ridges and downstream troughs in the west-to-east airflow that normally occurs aloft in middle latitudes.[2]

Figure 14.4 shows the surface and 500 mb patterns characteristic of a plunge of cold air from northwestern Canada into the central United States. The surface high that has developed over its source region (Figure 14.4A) begins its southward movement as the ridge intensifies. The movement of the strong surface cyclone from the central to the eastern United States contributes to the intensification of the midwestern

[2]The ridge-trough enhancement by the Rocky Mountains follows from a principle known in fluid mechanics as the *conservation of potential vorticity*. When the prevailing westerlies are forced over a mountain range such as the Rockies, the depth of the air column decreases; the conservation of potential vorticity then requires a decrease of counterclockwise curvature or an increase of clockwise curvature, resulting in southeastward flow downstream of the Rocky Mountains.

ONLINE 14.1

A Pacific Storm Builds a Ridge over Western North America

An example of a Pacific cyclone's role in ridge development over western North America took place during late January 2007. The ensuing cold wave brought early-February cold air to much of the central and eastern United States, with temperatures as low as the −30s (°F) in the northern Plains. The cold weather persisted through much of February, resulting in a month that was much colder than normal and ending what had otherwise been an unusually mild winter in the East. Many of the affected locations in the eastern United States had near-record warmth in January, including several record daily high temperatures. After the arrival of the polar continental airmass, it was spring before 70° F was seen again in many locations.

The maps online are sequences of forecast sea-level pressure, 850 mb temperatures and heights, and 300 mb heights and winds for a period of ten days beginning on 21 January 2007. The initial 300 mb field shows only a weak ridge over northwestern North America. The key ingredient in this sequence is the series of strong North Pacific cyclones that are especially apparent on the 850 mb maps.

As these systems intensify, strong south-to-north airflow (evident in the closely packed isobars over the Gulf of Alaska and the West Coast) carries relatively warm Pacific air into Alaska and the Yukon. This northward transport of warm air is especially apparent on Days 3, 5, 6, and 9. In association with these influxes of warm air, you can see the development of a 300 mb ridge, with its axis generally along the west coast of Canada.

This ridge becomes a primary feature of the 300 mb flow pattern, and it favors the development of a surface high over western Canada (1043 mb by 1200 UTC January 26). The upper-level ridge eventually drives cold air with 850 mb temperatures in the −20s °C (0° to −20° F) into the northern United States. This pattern persisted for the next few weeks, bringing one reinforcing blast of cold air after another to the central and eastern United States.

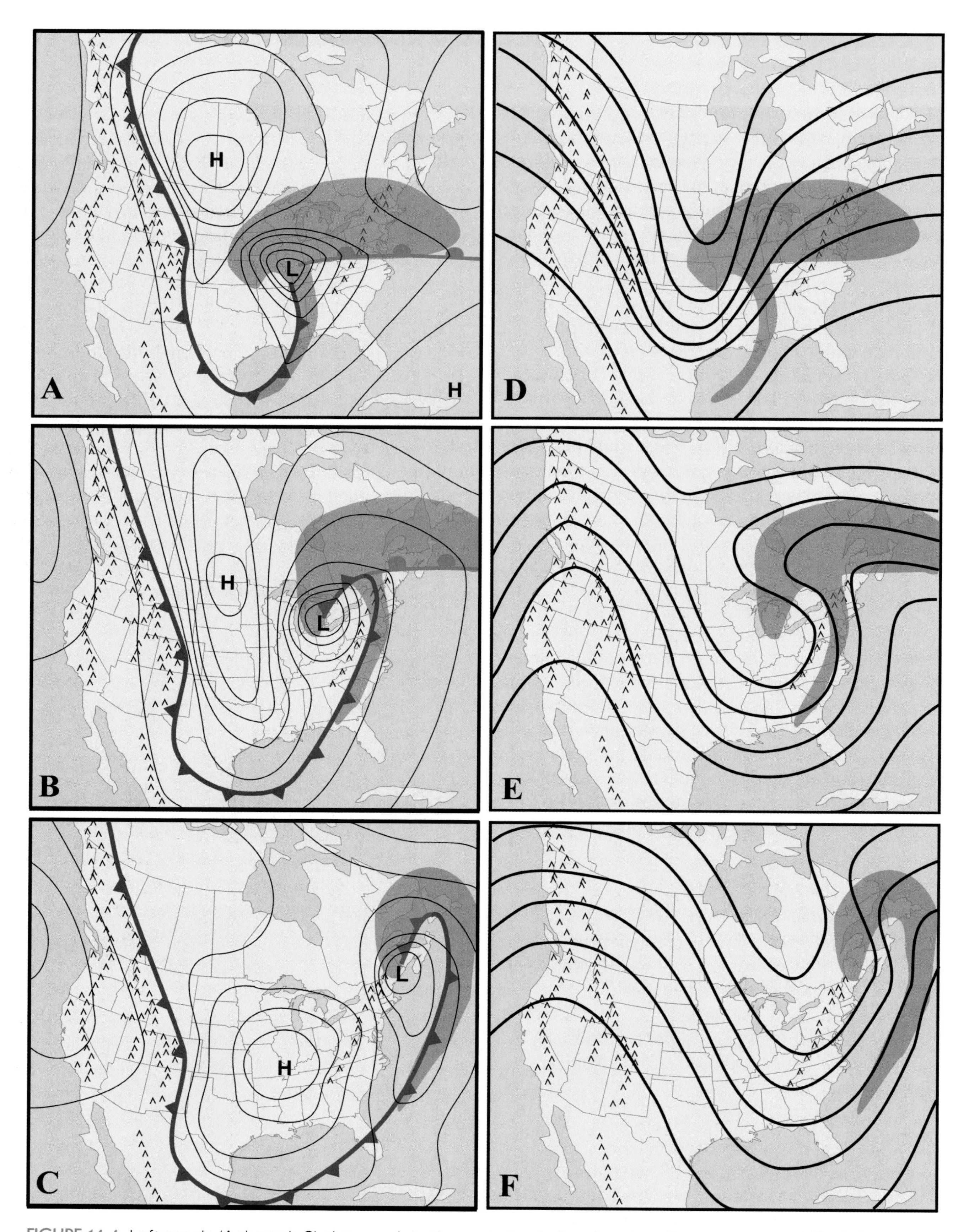

FIGURE 14.4: Left panels (A through C) show a wintertime extratropical cyclone traveling across central North America over several days. Strong northerly winds on the west side of the cyclone transport cold air associated with the polar airmass southward. This southward penetration of cold air also deepens the upper-air trough, as shown in the 500 mb maps in Panels D through F. Shading denotes cloud cover.

trough (Figure 14.4B and 14.4E). As the cyclone moves northeastward and occludes, the surface high moves southeastward (Figure 14.4C). The northerly winds between the occluding low and the southeastward-moving high bring cold air into the central and eastern United States, deepening the trough and moving it eastward (Figures 14.4D through 14.4F). The eastward movement of the upper-air ridge is also favored by the northward flow of milder air on the western side of the surface high. This milder air replaces the bitterly cold air of the polar airmass. The progression in Figure 14.4 occurs over a period of only two to three days, during which time much of eastern North America cools while the western part of the continent, especially the Rocky Mountain region, warms.

Where has the cold air originated on a bitter cold day during a cold wave on the Plains? Figure 14.5 depicts the ***trajectories*** of air samples that reached the Midwestern United States with temperatures in the range of −15° F to −25° F (−25° C to −30° C) in two cold waves, one in 1972 and the other in 1996. The top panels show the locations of the air over the twelve days prior to its arrival in the Midwest; the bottom panels show the air's vertical motion in terms of its changes of pressure during those same twelve days. In each case, the air lingered over northern Canada, where it lost heat during the airmass formation process (Figure 14.2). In both cases, the coldest air was forced by topography to follow a track just to the east of the Rocky Mountains—the ***channeling effect***. Finally, ***subsidence*** (descent) of the air by several hundred millibars is apparent in the trajectories from the polar latitudes to the Midwest. The subsidence was due to the "spreading" of the airmass southward, much like a pool of syrup spreading out, and to radiative cooling of the airmass. However, the subsidence also results in warming as the air sinks and compresses (at the dry adiabatic lapse rate—Chapter 6). Were it not for this warming, the air would have arrived at its final destination with even colder temperatures. We will see additional examples of subsidence in later sections of this chapter.

ONLINE 14.2

The Cold Outbreaks of 2004 in the Northeastern United States

January and February 2004 will long be remembered by residents of the northeastern United States as an exceptionally cold period. After an unusually mild December and first several days of January, bitterly cold air from central Canada drained southeastward into the United States in a series of pulses. As shown in Figure 14A, the temperatures over New England and New York state remained below freezing for almost two entire months. During the three coldest episodes in January, most of the region remained below 0° F (−18° C) even during the afternoon. Boston recorded its coldest January since 1893. In terms of the statewide average temperature, Massachusetts experienced its coldest January on record. The −45° F (−43° C) measured atop Mount Washington on 13 January narrowly missed New Hampshire's state record-low temperature of −47° F (−44° C) set at the same location in 1934.

The cold outbreaks of 2004 occurred despite the absence of a major upper-air ridge over western North America. Rather, a pool of extremely cold air developed over central Canada during December 2004. As shown online, temperatures at 850 mb fell below −40° F (−40° C) on many occasions during January. This shallow mass of cold air was so dense that it was able to drain southward underneath an upper-air flow that was primarily west-to-east, a dramatic illustration of the effect of density on airflow. The southward movement was aided by northerly winds on the west sides of several cyclonic systems that moved up the eastern seaboard. The winds behind these cyclones combined with the low temperatures to produce wind-chill temperatures that made the cold outbreak even harsher.

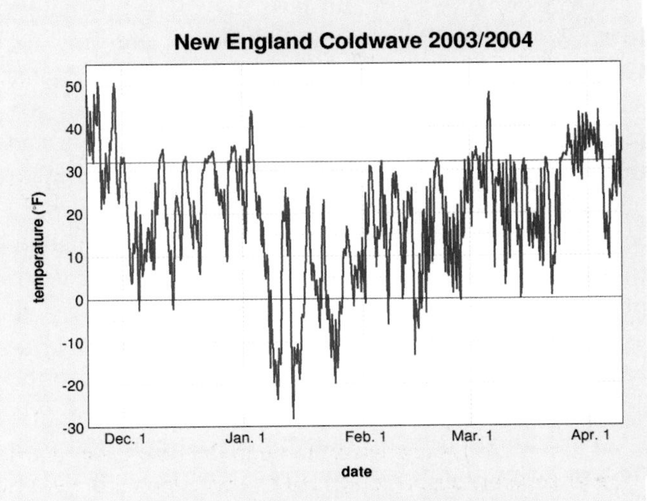

FIGURE 14A: Temperatures (°F) averaged over New England and New York during the winter of 2003–2004.

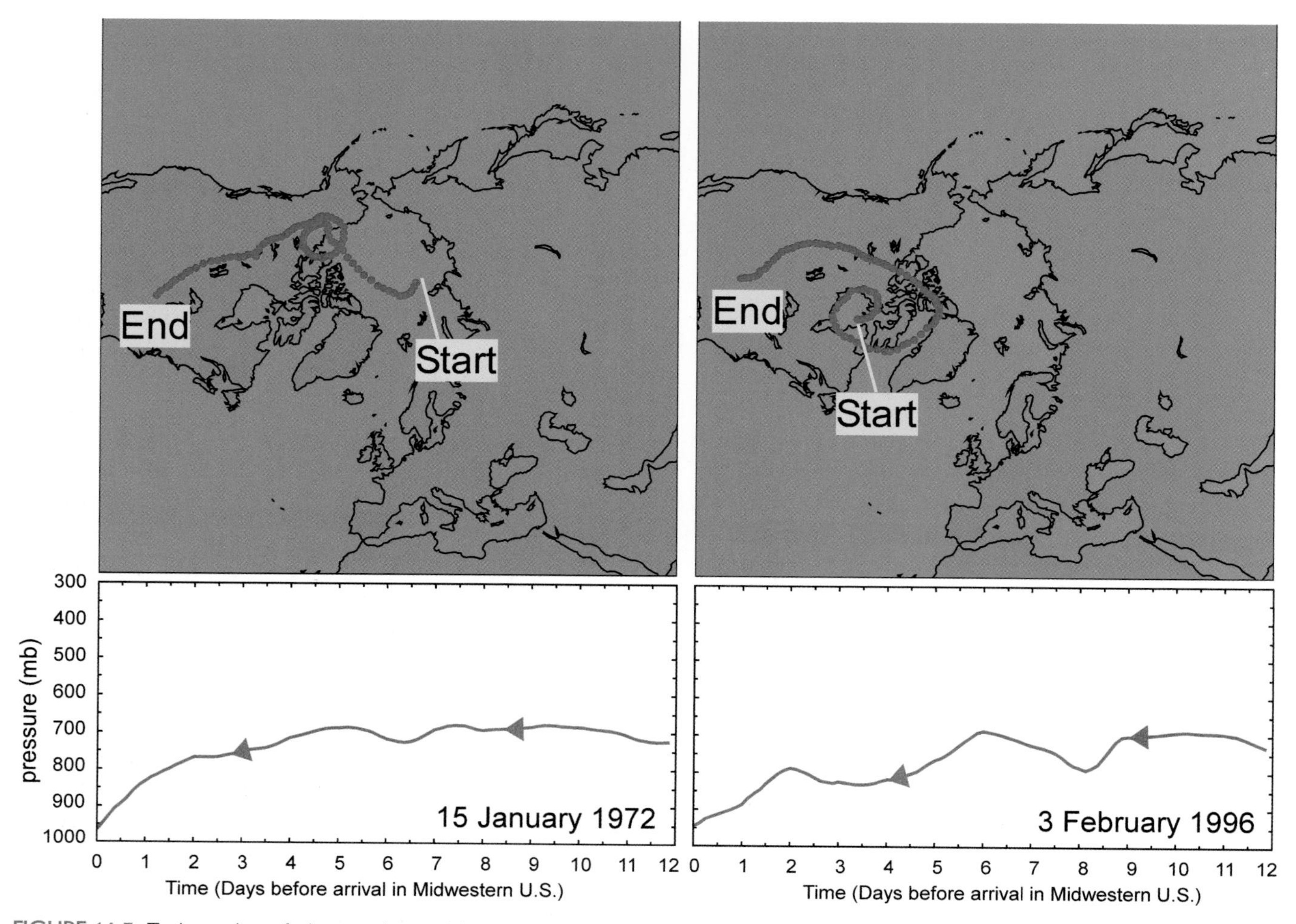

FIGURE 14.5: Trajectories of air parcels reaching the central United States in the core of polar airmasses during two major cold waves. Top panels show the horizontal positions of the air parcels that represent the coldest air over twelve days prior to arrival of the air in the central United States. The bottom panels show the altitudes (as pressure, mb) of the same air parcels over the twelve days prior to arrival at the surface.

Occasionally, a cold airmass may be deep enough to spill over the Rockies, with cold air entering the Great Basin from the northeast and affecting states such as Nevada and Utah. In rare cases, a cold airmass can spill westward over the Sierra Nevada range and into California. In such situations, the sub-freezing air can damage the crops grown in California's Central Valley, as in January 2007. However, the downslope (adiabatic) motion results in sufficient adiabatic compression that the air temperature in coastal cities such as San Francisco, Los Angeles, and San Diego typically will warm considerably from temperatures in the continental interior.

An additional factor favoring an extreme outbreak of cold air in middle latitudes is extensive snow cover. Snow radiates infrared energy very effectively and reflects most incoming solar radiation, rapidly removing heat from the overlying air and lowering the air's temperature. Polar continental airmasses traveling over snow-covered land are kept "refrigerated" by the snow-covered surface, while air passing over snow-free land gains some heat from the underlying ground. Many of the record-breaking cold outbreaks of the United States and Europe have been preceded by the buildup of an extensive snow cover on the land over which the cold air migrated. A strong cyclone passing across the United States (see Figure 14.4) may leave extensive snow cover in its wake. A deep snow cover was a major factor contributing to the record-setting cold of early January 1999, and an extensive snow cover also appears to have contributed to the extreme cold of the first portion of the next winter, especially December 2000 (see Focus 14.2).

FOCUS 14.2

Snow Cover: Nature's Refrigerator

December 2000 will be remembered for its persistent cold and heavy snow. Many cities east of the Rockies set December records for cold, snow, or both. For example, the coldest December on record occurred in Lynchburg, Virginia; Jackson, Mississippi; Louisville and Paducah, Kentucky; South Bend, Indiana; and Chicago (Midway Airport) and Rockford, Illinois. Illinois as a whole experienced its coldest December since the 1800s. Figure 14B shows that December 2000 was 10° F to 15° F (6° C to 9° C) colder than normal over much of the central United States. Except for extreme northeastern Maine, the entire area east of the Rockies was colder than normal.

Why was the central United States so cold? There was no El Niño or La Niña to blame. However, a clue lies in the previous month's weather. November 2000 was also extremely cold, but the cold was centered in the western United States (Figure 14B). In this case, the monthly average temperatures were as much as 10° F colder than normal over a broad area including Montana, Idaho, Wyoming, Colorado, and Utah, and nearly the entire contiguous United States was colder than normal. In fact, for the contiguous United States as a whole, the November–December period was the coldest on record.

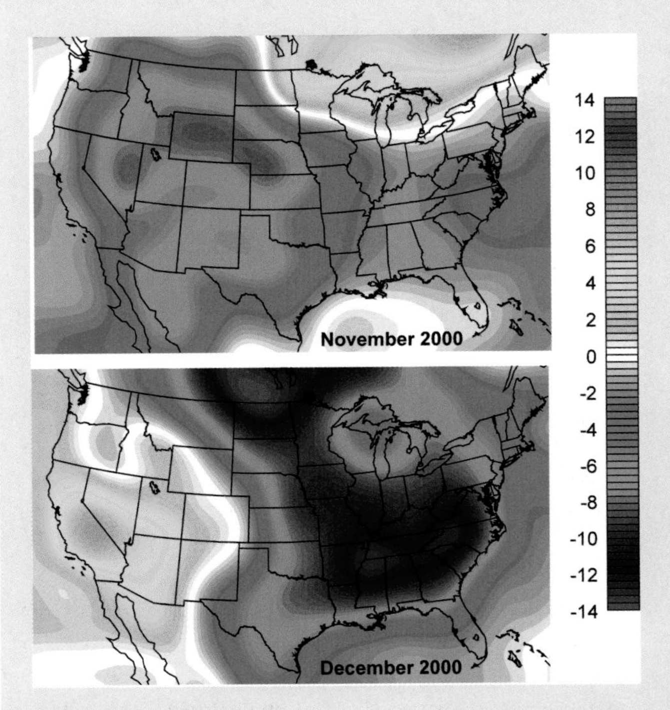

FIGURE 14B: Departures from normal temperatures (°F) for November 2000 (upper) and December 2000 (lower). Purple and blue denote colder-than-normal, orange and yellow denote warmer-than-normal.

The unusually cold conditions in November of 2000 favored an early and extensive buildup of snow in the Great Plains (Figure 14C). In addition, an upper-air trough was centered over the Rockies, coinciding with the axis of coldest air. Because of the trough's position, frontal cyclones were common over the central United States. These cyclones provided the fuel for the unusually heavy snow cover in November: Aberdeen, South Dakota, had its snowiest November on record (30.5 inches), while Williston, North Dakota, tied its November record. Single-storm snowfalls exceeding a foot occurred in Glasgow, Montana, Salt Lake City, Utah, and Flagstaff, Arizona. Farther east, more than forty inches of snow fell downwind of the Great Lakes in cities such as Erie, Pennsylvania, and Buffalo, New York.

The end result was a snowpack that was much deeper and extensive than normal over the northern United States by the end of November (Figure 14C). Even at midday, the December sun is largely ineffective at melting snow over the Dakotas, Minnesota, and the remainder of the northern states and Canada. Thus the stage was set for the continued refrigeration of any airmasses that moved southward or eastward into the remainder of the United States. The December weather pattern supplied such airmasses, and the result was the bitter cold December in the Midwest, East, and South. The impressive temperature statistics were largely the result of repeated intrusions of these refrigerated airmasses, rather than a single extreme but brief cold wave.

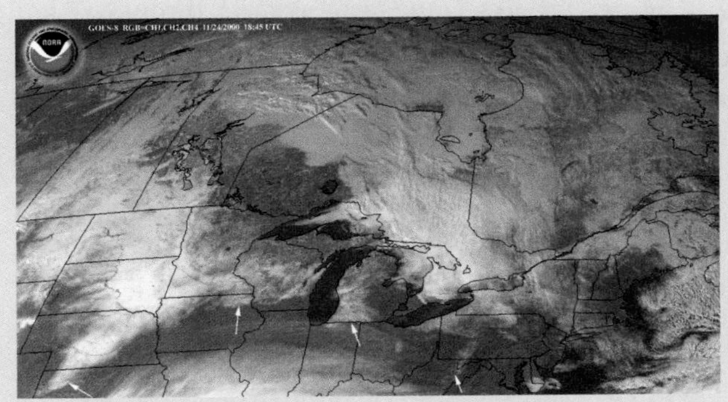

FIGURE 14C: Visible satellite image for 1845 UTC 24 November 2000 showing snow cover (arrows) over the northern United States and southern Canada.

FOCUS 14.3

A Generational Perspective on Cold Waves

Did a grandparent of yours ever make a comment such as "Back in my day, winters were a lot worse than they are now?" Were the coldest days really colder in the days of your grandparents? Figure 14D provides an answer to that question. It shows the years and actual daily mean temperatures for the twenty to twenty-five coldest days in weather history at an Illinois location where weather observations have been made continuously since 1888. The average (midpoint between the day's high and low) temperature for each of these days was colder than −5° F. By contrast, the record-tying low temperature of −25° F, which occurred on 5 January 1999, was accompanied by a daily maximum temperature of 22° F, so that the day's average was merely −1.5° F.

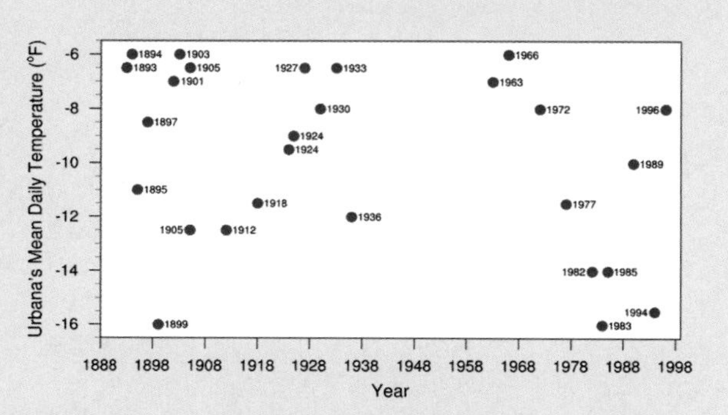

FIGURE 14D: Mean daily temperatures (°F) for the coldest days since 1888 at Champaign–Urbana, Illinois.

Figure 14D indeed has clusters of extremely cold days back in the 1890s, 1900s, and 1920s. However, it also shows clusters in the 1970s, 1980s, and 1990s. So who had the coldest days while they were young? If you are the typical student in your teens or twenties, both you and your great-grandparents. The ones who lucked out in their youth and generally missed the extreme cold days were your grandparents and parents, whoever grew up in the 1930s through the 1960s. Since no days in the years since 1996 have qualified for Figure 14D, today's elementary school students in the Midwest have yet to experience the extreme cold of previous generations.

In summary, major cold outbreaks over the central or eastern United States result from a combination of most or all of the following factors:

- The buildup of a ridge in the jetstream over western North America, often as a result of warm air transported northward in the lower troposphere east of a cyclone in the North Pacific.
- Formation of a surface high-pressure center over northern Canada or Alaska through rapid cooling of the air near the surface and convergence aloft downstream of the ridge.
- Movement of the cold airmass southeastward in response to steering by the upper-level winds and channeling of the cold-air pool by the Rocky Mountains.
- A mechanism to enhance the winds that transport the cold air southeastward, thereby reducing the transit time of the cold air. The ***trigger mechanism*** is often a strong winter cyclone crossing central or eastern North America.
- Extensive snow cover over central North America to keep the polar airmass "refrigerated."

CHECK YOUR UNDERSTANDING 14.2

1. List two factors that favor the intensification of an upper-air ridge over western North America.
2. What is the most common wind direction at the jetstream level above a North American polar airmass?
3. What is the channeling effect?
4. What five factors contribute to the development of major cold waves over the central and eastern United States?

FIGURE 14.6: Major topographic features of Eurasia. Note that Siberia is far from the Pacific Ocean and surrounded to the east and south by mountains.

TABLE 14.1 Climatology for Verkhoyansk, Russia (Elevation 328 ft)

	Jan.	Feb.	Mar.	Apr.	May	June	July	Aug.	Sep.	Oct.	Nov.	Dec.
Temp. (°F)	−58	−48	−26	4	32	54	56	48	36	4	−35	−54
Temp. (°C)	−50	−44	−32	−16	0	12	13	9	2	−16	−37	−48
Precip. (in.)	0.2	0.2	0.2	0.2	0.3	0.9	1.1	1.0	0.5	0.3	0.3	0.2

EXTREME COLD IN EUROPE AND ASIA

Because the Eurasian landmass is the largest in the world, it is not surprising that the coldest airmasses of the Northern Hemisphere develop over this region during winter. The most extreme area of all is northern Asia (Siberia), where the formation of cold surface air is enhanced by (1) the large distance to the nearest unfrozen ocean, isolating the area from warmth and moisture, and (2) the presence of mountains to the east and south, serving as barriers to trap and further isolate the cold surface air once it has formed (Figure 14.6). In the interior lowlands of Siberia, cold surface air can remain entrenched for months during winter. Table 14.1 shows the average monthly temperatures at the Siberian city of Verkhoyansk (67.5° N, 134° E). The harsh conditions in Siberia have hindered the extraction of its vast stores of natural resources (oil, gas, coal, metals) and have made it a dreaded destination of prisoners.

Despite a relatively pleasant summer (average daily highs of 60° F to 70° F (15° C to 20° C), and average daily lows of 40° F to 50° F (5° C to 10° C), the Siberian winter can be likened to a perpetual cold wave: average daily temperatures are about −60° F (about −50° C) for several months. Because of the relatively high latitude, solar radiation is so weak that there is little diurnal cycle of temperature, i.e., the temperature changes little from day to night. Grave-digging is

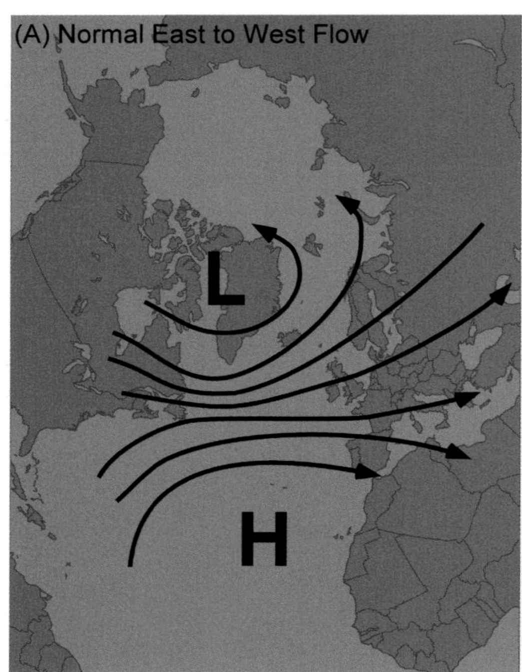

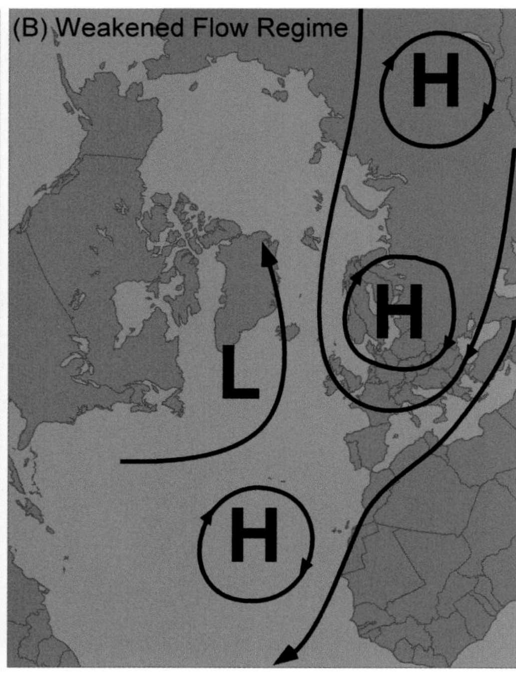

FIGURE 14.7: Regimes of the atmospheric circulation at the surface over the North Atlantic: (A) Normal pattern of west-to-east airflow, (B) weakened flow regime conducive to cold air outbreaks over Europe.

such a challenge that undertakers hope the severely ill can "hang on" until spring!

The heavily populated areas of Europe experience their most extreme cold when the frigid air from Asia spills *westward* into Europe. Ordinarily, Europe is fairly mild for its latitude because the prevailing airflow is west-to-east, bringing mild maritime air from the North Atlantic Ocean over the European land areas. As shown in Figure 14.7A, the west-to-east airflow is a consequence of the pressure gradient between the subpolar low near Iceland and the subtropical high near the Azores. Occasionally, however, these two features weaken simultaneously, slowing or eliminating the eastward flow. Figure 14.7B shows a situation in which the elimination of onshore winds has enabled a pool of cold Siberian air (low temperatures, high-surface pressure) to migrate westward over Europe.

While Figure 14.7 is an idealized schematic diagram, Figure 14.8 shows a map of actual sea-level pressures from a record-setting cold-air outbreak over Europe in January 1987. During 10 to 12 January, the air temperatures averaged over all of western Europe were colder than normal by 20° F to 30° F (11° C to 17° C), paralyzing Europe's transportation systems and creating severe hardship in poorly insulated homes. Some locations were 40° F colder than normal. Figure 14.8 is an example of a broken-down circulation over the North Atlantic Ocean (corresponding to Figure 14.7B) and also shows an unusually strong 1045 mb high-pressure center over northern

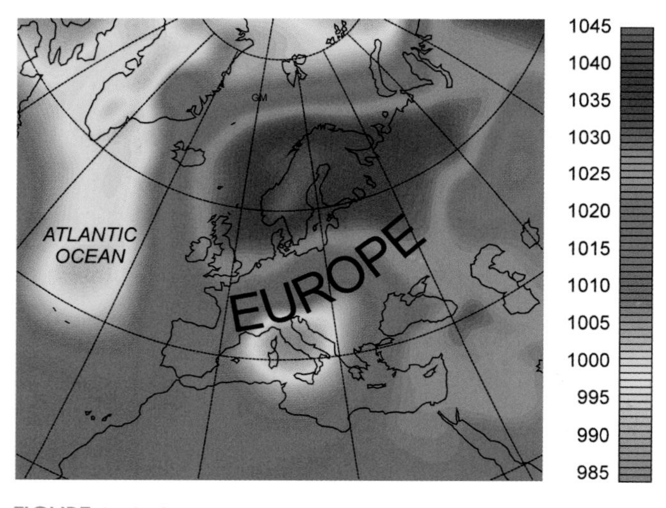

FIGURE 14.8: Sea-level pressure map for the North Atlantic and Europe on 11 January 1987, when northern Europe was experiencing a cold wave. Purple and blue denote high pressure, orange and yellow denote low pressure. Note the strong high-pressure center (1045 mb) over Scandinavia.

Europe. Where did this high-pressure center originate? Figure 14.9, a trajectory plot covering the previous twelve days, shows that the core of the cold air progressed westward toward Europe from the eastern Siberian region where Verkhoyansk is located—where the Northern Hemisphere's coldest air is often found (Table 14.1).

Figure 14.9 also shows the vertical component of the twelve-day trajectory of the air that ended up near

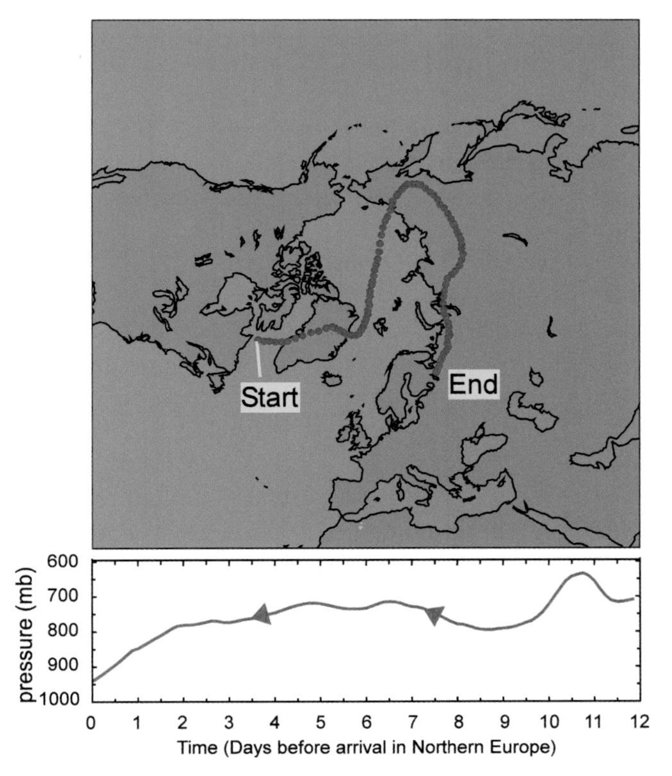

FIGURE 14.9: Trajectory of air reaching northern Europe in the core of the cold outbreak of January 1987. Top panel shows horizontal locations over 12 days prior to arrival in northern Europe; bottom panel shows elevations (as pressure, mb) of the same parcels over the twelve days prior to arrival at the surface.

the surface over northern Europe on 10 January 1987 (Day 0). During the preceding five to six days, the air had slowly descended from a pressure of about 800 mb, implying a total descent of 1 to 2 km. Without that subsidence and its adiabatic (compressional) warming at 18° F (10° C) per km of descent, the air would have been even colder by 20° F to 30° F (11° C to 17° C)! Mid-latitude residents who experience cold outbreaks rarely appreciate the fact that their situation would be significantly worse were it not for subsidence of air in the high-pressure systems that bring them the cold air.

CHECK YOUR UNDERSTANDING 14.3

1. What is the relation between the coldest Siberian air and nearby mountains?
2. From which direction does Europe receive its coldest airmasses?

WIND CHILL

The media, in wintertime, often discuss the ***wind chill factor***. The wind chill factor accounts for the effect of both temperature and wind on the rate at which exposed flesh will cool, and is reported numerically in terms of the ***wind chill temperature***. The wind chill has come into use because solid and liquid surfaces lose heat more rapidly at a given temperature as wind speed increases. This effect arises because an object surrounded by colder air loses heat to the air by conduction. The rate of conductive heat loss is proportional to the differences in temperature between the object's surface and the surrounding air. Since the conducted heat warms the air, the temperature gradient (and hence the rate of conductive heat loss) will decrease if the same air remains around the object and warms. Wind removes the heated air and replaces it with cold air. The stronger the wind, the greater the rate at which heat is carried away by the air. Evaporation of moisture from the skin also increases as the wind's speed increases, resulting in an additional loss of heat. Since exposed skin is subject to both these effects, our skin loses heat at a faster rate as the wind speed increases. Thus, we do indeed "feel colder" as the wind speed increases at a given temperature.

A ***wind chill index*** provides an estimate of perceived temperature based on wind speed and actual temperature. Scientists first pursued the notion of a wind chill index experimentally during the 1940s in Antarctica, where the length of time required for human flesh to freeze is a very practical concern. In order to quantify the rates of heat loss under various wind and temperature conditions, the Antarctic scientists measured the time required for the cooling and freezing of known volumes of water starting at various temperatures. Empirical formulas were developed by equating the rate of heat loss at a given temperature and wind speed to the rate of heat loss with no wind. The equivalent rate of heat loss with no wind would occur at the wind chill temperature.

Since then, several refinements were made to the formula for determining the wind chill "equivalent" (zero wind speed) temperature corresponding to a measured wind and temperature. The most recent of these refinements, made by the United States and Canadian weather services, was implemented in November, 2001. The revision of the index is based on advances in science, technology, and computer modeling of heat loss, and the revised index has even been tested in clinical trials. The new formula is based on a model of the human face, and it uses wind speeds adjusted from their measurement level (typically thirty-three feet) to a height of five feet (about the average height of an adult human face). The new formula does not, however, allow for the effects of sunshine, so it effectively assumes that it is nighttime. For wind speeds in the 20 to 30 mph range, the new formulation (see Table 14.2) produces wind chill values that are considerably higher, often by 10° F to 20° F (6° C to 11° C), than the values produced by the previous formula. So when you are outside on a cold windy day in winter, the wind chill reading won't sound as impressively cold as it would have under the same conditions in years of the past century.

Table 14.2 shows the new wind chill temperatures, together with the actual formula on which they are based. The heavily shaded area of the table (deepest purple) denotes the temperature-wind combinations for which frostbite will occur on exposed skin in five minutes or less. The threshold for this dangerous situation is a wind chill of −18° F (−28° C). As an example, Table 14.2 shows that the combination of an air temperature of 5° F and a wind speed of 30 mph produces a wind chill of −19° F. In this situation, the rate of heat loss from a person's skin is equivalent to that with no wind and a temperature of −19° F, even if the thermometer actually reads 5° F. These conditions would produce ***frostbite*** in thirty minutes or less if the person does not take precautionary measures, e.g., by covering exposed skin, preferably with several layers of clothing.

The issuance of a ***Wind Chill Advisory*** or a ***Wind Chill Warning*** by the National Weather Service is based on criteria that vary with location. In general, wind chill advisories are issued when the resulting cold can be dangerous for long exposure times, while a wind chill warning is issued when the cold temperatures are life threatening for a person not well prepared for cold. For example, in the Rochester, NY area, wind chill warnings are issued when the wind chill temperature is expected to be at or below −25° F, while wind chill advisories are issued when the wind chill temperature is expected to range from −15° F to −24° F.

TABLE 14.2 **Wind Chill Chart**

Wind (mph)	Temperature (°F)																	
Calm	**40**	**35**	**30**	**25**	**20**	**15**	**10**	**5**	**0**	**−5**	**−10**	**−15**	**−20**	**−25**	**−30**	**−35**	**−40**	**−45**
5	36	31	25	19	13	7	1	−5	−11	−16	−22	−28	−34	−40	−46	−52	−57	−63
10	34	27	21	15	9	3	−4	−10	−16	−22	−28	−35	−41	−47	−53	−59	−66	−72
15	32	25	19	13	6	0	−7	−13	−19	−26	−32	−39	−45	−51	−58	−64	−71	−77
20	30	24	17	11	4	−2	−9	−15	−22	−29	−35	−42	−48	−55	−61	−68	−74	−81
25	29	23	16	9	3	−4	−11	−17	−24	−31	−37	−44	−51	−58	−64	−71	−78	−84
30	28	22	15	8	1	−5	−12	−19	−26	−33	−39	−46	−53	−60	−67	−73	−80	−87
35	28	21	14	7	0	−7	−14	−21	−27	−34	−41	−48	−55	−62	−69	−76	−82	−89
40	27	20	13	6	−1	−8	−15	−22	−29	−36	−43	−50	−57	−64	−71	−78	−84	−91
45	26	19	12	5	−2	−9	−16	−23	−30	−37	−44	−51	−58	−65	−72	−79	−86	−93
50	26	19	12	4	−3	−10	−17	−24	−31	−38	−45	−52	−60	−67	−74	−81	−88	−95
55	25	18	11	4	−3	−11	−18	−25	−32	−39	−46	−54	−61	−68	−75	−82	−89	−97
60	25	17	10	3	−4	−11	−19	−26	−33	−40	−48	−55	−62	−69	−76	−84	−91	−98

Frostbite Times 30 minutes 10 minutes 5 minutes

$$\text{Wind Chill (°F)} = 35.74 + 0.6215T - 35.75(V^{0.16}) + 0.4275T(V^{0.16})$$

Where, T=Air Temperature (°F) V=Wind Speed (mph)

Effective 11/01/01

It is important to remember that various regions of the United States are affected differently by cold air outbreaks. As noted at the start of this chapter, southern states suffer more deaths and greater economic losses from cold-air outbreaks, whereas the northern states' infrastructure has generally been designed for bitter cold temperatures and heavy snows. Perhaps the most severe cold to affect the South in its recorded history was the 1899 cold-air outbreak.

While the Great Arctic Outbreak of 1899 was well forecast, other cold waves of the late 1800s and early 1900s were poorly forecast, resulting in large numbers of fatalities (Chapter 15, Focus 15.4). Today, cold waves are generally forecast at least several days in advance because the surface observing network can detect the build-up of extreme cold air-masses in the northern land areas. Numerical weather prediction models are also able to forecast the evolution and movement of the jetstream patterns and surface features that accompany cold waves. For example, the cold outbreak of February 2007 was forecast nearly a week in advance by numerical models (Chapter 4 cover figure and Figure 4.9). In such cases, the National Weather Service is able to issue a *cold wave warning* for areas in which the temperature is expected to fall rapidly to values well below normal.

FOCUS 14.4

The Greatest Arctic Outbreak in the United States: February 1899

Which cold wave was the greatest of them all? While the answer varies by region within the United States, the outbreak that produced the most severe cold conditions over the largest area occurred in the first half of February 1899. This event has been referred to as "The greatest Arctic outbreak in history" and "The mother of all cold waves"[3] (The reference frame for such statements is the period of recorded American history; the last Ice Age, as described in Chapter 5, almost certainly had some colder events.) This Arctic blast affected two-thirds of the nation, including the entire area east of the Rockies, resulting in tremendous losses of livestock and crops. It produced the lowest temperatures ever recorded at many locations. For example, the temperature at downtown Washington, D.C. fell to −15° F (−25° C), the lowest that has ever been recorded in that city.

The cold was especially severe in the band of southern states from Texas to the Carolinas. Figure 14E shows a sample of the low temperatures reached in the South. To make matters worse for Southerners, these temperatures were accompanied by strong winds and snow. Even Miami, Florida recorded subfreezing temperatures with wind chills in the teens (°F). Snow fell as far south as Fort Myers, Florida, and Washington, D.C. reported a snow depth of more than thirty-four inches, a record that still stands today. For only the second time in recorded history, ice flowed into the Gulf of Mexico from the Mississippi River.

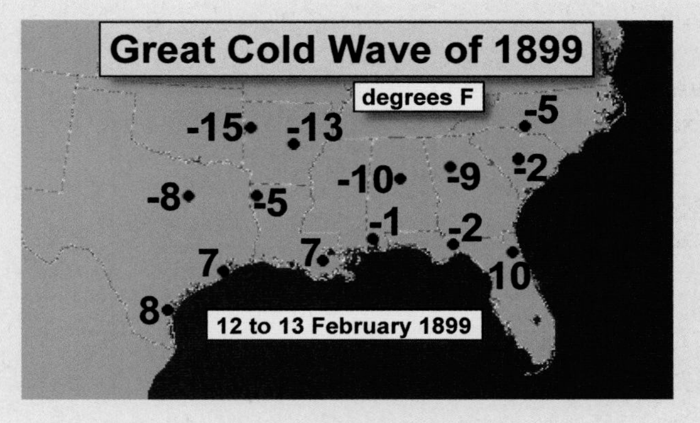

FIGURE 14E: Low temperatures (°F) recorded during 12 to 13 February 1899 during the Great Arctic Outbreak.

Because there were no rawinsondes or other upper-air data in 1899, the meteorological information about the evolution of this event comes from twice-daily surface weather reports. Nevertheless, there are indications that this event conformed to, and was an extreme manifestation of, the sequence of events outlined earlier in this chapter. Several preceding snowstorms during early February had built up an unusually extensive snow cover in the central and eastern states. (Ironically, these earlier snowstorms missed Chicago; the ensuing

[3] D. Ludlam. *Weatherwise* 23 (1970):191.

(continued)

(continued)

bitter cold then froze the ground to a depth of five feet, severely damaging the city's water and gas lines.) A polar airmass of Siberian origin evolved into a huge and intense high-pressure center over northwestern Canada. As this cold airmass spilled southward, sea-level pressure reached a spectacular 1064 mb in Alberta. On 11 February the 1060 mb isobar reached the northern United States (Figure 14F), giving the United States some of its highest pressures ever recorded (cf. Table 1.2). With nothing but a few barbed wire fences between the Dakotas and the Gulf Coast, the stage was set for a record-setting cold outbreak. The low-pressure region in the Gulf of Mexico in Figure 14F then evolved into a major Gulf Coast cyclone (see Chapter 11), producing blizzard conditions in the eastern states and serving as the "trigger" that enabled the core of the cold airmass to be carried rapidly by strong winds to the southern states.

Subsequent analyses have shown that cold waves of the 1980s and 1990s, such as those represented in Figure 14D, produced lower temperatures in various locations of the central, eastern, and southern United States. However, many records set in the 1899 event have yet to be broken. Moreover, in terms of the area affected and the severity of the temperature-wind-snow combination, the 1899 event still stands as the benchmark against which other cold waves are compared.

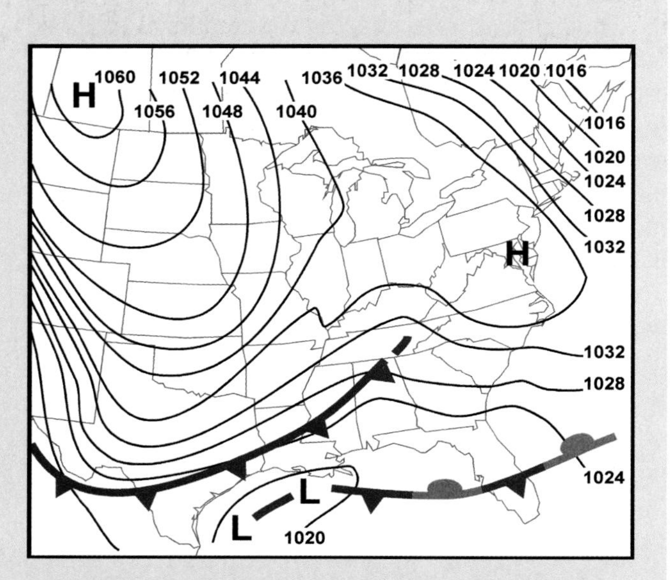

FIGURE 14F: Surface weather map for 1300 UTC 11 February 1899.

COLD WAVES AND GLOBAL WARMING

There are two ways in which global warming can affect the occurrence of cold waves. First, one would expect cold waves to become less severe if greenhouse-driven global warming is the primary climate change of the next century. For the world as a whole, climate models indeed point to a decrease in the intensity of cold waves. However, the role of the atmospheric circulation in shaping cold waves is a second consideration that complicates the future scenario.

We saw earlier in this chapter that the atmospheric circulation is a key factor in the formation and movement of cold airmasses. When the atmospheric circulation drives a polar continental airmass into middle latitudes, the temperature at a particular location can drop by 40 to 50° F. By contrast, the direct radiative impact of a doubling of CO_2 causes a warming of only a few degrees in the average temperature (Chapter 5). Changes of the atmospheric circulation pattern can therefore dominate the effect of increasing greenhouse gasses on a local basis. So, while the world may indeed warm, increased frequencies of cold waves may occur in specific areas. Climate models indeed indicate that this will be the case as the dominant patterns of variability change under greenhouse warming. While the coldest airmasses may be slightly less cold in the future, the areas most frequently affected by cold waves may change. Unfortunately, different climate models show different changes of the atmospheric circulation in greenhouse simulations, so there are no clear indications about precisely which locations may experience more frequent cold waves.

CHECK YOUR UNDERSTANDING 14.4

1. What is the wind chill temperature?
2. How does the newly revised wind chill differ (qualitatively) from the old value under identical weather conditions?
3. Explain the difference between a wind chill advisory and a wind chill warning.
4. Will global warming mean the end to cold waves?

TEST YOUR UNDERSTANDING

1. Which region of the United States suffers the greatest losses from cold waves? Why?
2. Explain why airmasses that develop over Canada are colder than airmasses that form over the Arctic Ocean.
3. What is the ideal trough-ridge pattern of the jetstream for the development of polar airmasses?
4. Explain how a cyclone over the Gulf of Alaska can intensify a ridge over western North America.
5. What causes cold polar airmasses to typically move southeastward over North America?
6. Prior to a severe cold wave, a strong surface cyclone moves across the central and eastern United States. Explain the role of the surface cyclone in the development of the cold wave.
7. How can the temperature of a polar airmass warm as it migrates from Canada into the contiguous United States?
8. What role does the Rocky Mountains play in the occurrence of a cold wave in the central United States?
9. Why is snow cover referred to as "nature's refrigerator"?
10. What two factors help explain why Siberian wintertime temperatures are extremely cold?
11. Why do cold polar airmasses generally move westward over Europe?
12. Where do the Northern Hemisphere's coldest airmasses develop during winter.
13. What happens to the North Atlantic surface pressure pattern prior to a cold outbreak in Europe?
14. Discuss the two ways in which wind enhances loss of heat from the skin.
15. How did scientists in Antarctica obtain estimates of the wind's effect on heat loss?
16. Name at least one important factor that the wind chill index does not take into account.
17. Suppose a thermometer is held outside in a strong wind. After the thermometer has equilibrated, does it show the actual temperature or the wind chill temperature? Why?
18. How would you expect the wind chill to differ from the actual air temperature during the formation of a polar airmass?
19. Discuss how wind chill warnings and advisories are influenced by geographic location.
20. Would you expect Australia to be affected by cold waves during the Southern Hemisphere winter? Why or why not?

TEST YOUR PROBLEM-SOLVING SKILLS

1. Air cools primarily by a net loss of infrared energy, and it warms primarily by a gain of energy from solar radiation. (Much of the solar radiation is absorbed by the ground and then conducted to the air, but the source is still the sun's energy—see Chapter 5.) Suppose for simplicity that the loss of infrared energy, by itself, causes the air near the ground to cool by 5° F per day. Suppose also that the solar radiation, by itself, warms the air by 0.25° F per day for each degree latitude south of 65° N. (North of 65° N, the heating by sunlight is negligible during winter.) Assume that there is no horizontal or vertical wind to change the air's temperature.

(a) As you go south from 65° N, at which latitude do you cross from a zone of net cooling to a zone of net warming?
(b) If an airmass is initially at a temperature of 0° F at a latitude of 50° N, and there is no wind to remove the air from its location, how long would it take that air to cool to −40° F (–40° C)—a typical temperature of polar continental air?
(c) Suppose now that a surface high-pressure center develops, and the air slowly sinks at a rate of 250 meters per day. Recalling the dry adiabatic lapse rate, how would the answers to (a) and (b) change?

2. You have been ice fishing in a warm heated shelter in the exact center of a lake with a 1-mile radius. While you were fishing, the leading edge of a polar airmass arrived, dropping the temperature to −10° F (−22° C). To make matters worse, the wind is now blowing from the north at 20 mph. Unfortunately, your car is parked on the north shore of the lake. Fortunately, the lake is surrounded by trees that reduce the wind speed in the trees by 50 percent from the speed over the frozen lake. You can walk or run at any speed up to 5.4 mph (11 min per mile). If you are to avoid frostbite, what is the safest route to your car? Why? (Hint: When walking or running against the wind, your speed must be added to the wind speed. Use this fact, together with the frostbite times that accompany the wind chill chart).

3. Consider a hypothetical polar airmass having a depth D and a surface air temperature of −30° C (−22° F). In the lowest third of this airmass, there is a strong inversion with the temperature increasing upward at 3° C per 100 m. In the remainder of the depth D, the temperature decreases with elevation at 0.5° C per 100 m. Suppose that this airmass then moves up against a large mountain range in which the lowest point is Gonner Pass (elevation = 1800 m). The temperature at Gonner Pass is −20° C (−4° F) prior to the arrival of the polar airmass.
(a) If D = 2 km, what will be the temperature in the polar airmass at elevations of 1 km, 2 km, and 3 km?
(b) Will the polar airmass spill through Gonner Pass to the other side of the mountains?
(c) How do your answers to (a) and (b) change if the temperature in the inversion layer increases at only 1° C per km?

4. A strong polar airmass 2000 miles in diameter is centered over Yellowknife in northwestern Canada at noon on Sunday. The airmass is moving at 30 mph directly toward St. Louis, Missouri. Create a meteogram (use qualitative axis labels rather than distinct values) to indicate the behavior of the following variables over the next four days at St. Louis:
(a) Temperature
(b) Dewpoint temperature
(c) Surface pressure
(d) Wind direction
(e) Wind speed
(f) Wind chill temperature

USE THE SEVERE AND HAZARDOUS WEATHER WEBSITE

http://severewx.atmos.uiuc.edu

1. Various websites, including those of the National Weather Service and the University of Cologne (Koln), provide synoptic reports from stations around the world. Navigate to these sites using the *Severe and Hazardous Weather Website.* Record the date and time of the data. What is the lowest temperature in the world at this time? Estimate the lowest wind chill temperature from the temperature and wind data.

2. During the winter months, cold airmasses can be tracked by using maps of surface temperature, 850 mb temperature, or sea-level pressure as described in this chapter.
(a) Find at least one example of a cold airmass using the links provided on the "Current Weather" page of the *Severe and Hazardous Weather Website.*
(b) Another tool for locating cold airmasses is infrared satellite imagery, provided that an

area is relatively cloud-free. Examine infrared satellite imagery for the same date, time, and region of the cold airmass you identified. Explain how the infrared image for the region is consistent with the pattern of surface air temperatures.

(c) View two soundings for locations close to the center of the airmass. Is a surface-based temperature inversion present (i.e. the inversion starts at the surface)? If so, compare the soundings with respect to the following:
   (i) What is the inversion strength (temperature difference between the top and bottom of the inversion)?
   (ii) What is the depth in mb of the surface-based inversion?
   (iii) What is the mean lapse rate in the inversion layer?

3. Research the topic of hypothermia using the Internet. Identify the various stages of hypothermia, the symptoms, treatment of hypothermic victims, and ways in which the risk of hypothermia can be minimized. Reference all your sources. Be prudent in choosing the sites that you use to obtain information to ensure that the information is accurate.

4. Use the "Current Weather" page of the *Severe and Hazardous Weather Website* to view the latest winter weather warnings issued by the National Weather Service. Over a period of several weeks during winter, monitor this site for wind chill warnings or advisories. Compare the temperatures or wind chills mentioned in the warnings and advisories to the actual observed values. Discuss how the values vary from one case to another and how the warnings and advisories are geographically dependent.

GREAT PLAINS BLIZZARDS

March blizzard in North Dakota in 1966 nearly buries utility poles.

KEY TERMS

Alberta Clipper
arctic front
black ice
blizzard
blizzard zone
blizzard warning
Colorado cyclone
dry slot
extratropical cyclone
frostbite
ground blizzard
hypothermia
snowrollers
stockmen's warning
traveler's warning
trowal
upslope flow
wind chill temperature index

Severe *blizzards* paralyze the Great Plains of the northern United States and southern Canada nearly every winter. Blinding snowstorms, with 35 to 55 mph (30 to 50 knot) winds and temperatures between 20° F and –30° F (–7° C and –34° C) make travel impossible, strand residents and motorists, kill livestock, and freeze the landscape in a sea of white. Conditions so dangerous that ***hypothermia*** and ***frostbite*** can onset in minutes can keep even seasoned snowplow drivers from attempting their task of snow removal, sometimes for days. Many hundreds of residents of the Great Plains have become stranded and perished in blizzards. Despite the low population density of the states and provinces most affected by severe blizzards, the costs are staggering, especially to agriculture. The blizzards of January 1997, for example, killed nearly $4.7 million worth of livestock, caused $6 million in farm property damage, and cost farmers $21 million for extra feed consumption.

FOCUS 15.1

Hypothermia, Frostbite, and Blizzards

Hypothermia is a decrease in the body's core temperature to a level at which normal muscular and cerebral functions become impaired. The body's core includes the internal organs, especially the brain, heart, and lungs. Mild hypothermia occurs when the body's core temperature drops to between 98.6° F and 96° F (37° C and 35.5° C), a stage that is accompanied by involuntary shivering and minor impairment of motor functions. Moderate hypothermia occurs when the body's core temperature falls between 95° F and 93° F (35° C and 33.8° C), a stage where violent shivering, dazed consciousness, loss of coordination, and irrational behavior are common. Life-threatening severe hypothermia onsets when the body's core temperature drops to 92° F (33.3° C) or below. Shivering will eventually stop, and the victim can no longer stand. The body begins shutting down to preserve its remaining heat. If a person is not warmed soon, death can occur.

Blizzards create conditions where hypothermia can kill quickly. People have frozen to death short distances from their cars and homes. Whiteout conditions are very disorienting, since the sun and even nearby landmarks are obscured. Victims who get lost in blinding snowstorms can become confused and lose control under the influence of the cold in a short time. Even when people are found and can be revived, they often suffer from frostbite.

Frostbite is the term describing the condition where the flesh of the body actually freezes. The human body tries to preserve heat by restricting blood flow to the extremities. As a result, the ears, nose, fingers, and toes are usually the first body parts to freeze. Deep frostbite can cause freezing of muscle tissue or even bone tissue. Frostbite can destroy tissue, and can require amputation if not treated immediately.

The unique geography of the North American continent creates an environment that makes these destructive winter storms possible. The north-south mountain massif that stretches from the Rockies to the West Coast effectively isolates the lower atmosphere on the Great Plains from the warm air over the Pacific. Midwinter's long arctic night permits air over the Canadian Arctic to cool rapidly, creating the bitterly cold air that contributes to both blizzards and cold waves in lower latitudes. Cyclones, the parent storms of blizzards, form just east of the Rockies and reach their greatest intensity on the Plains, creating the strong winds necessary to drive blizzards. The Gulf of Mexico often provides the moisture necessary to fuel snowfall.

The National Weather Service issues a ***blizzard warning*** when winds are expected to exceed 35 mph (30 knots), and falling or blowing snow will reduce visibility to less than a quarter of a mile for at least three hours. Although cold air is a key component of all blizzards, specific temperature criteria are not used to define a ***blizzard***. Since heat loss from the body depends on temperature, wind speed, and wetness, assigning a specific temperature to delineate whether or not a blizzard is occurring has little value when providing warnings to the public. ***Stockmen's warnings*** are issued when severe winter weather threatens cattle in major ranching areas, and ***traveler's warnings*** are issued when winter conditions make travel difficult or impossible.

Figure 15.1 shows the probability of an annual occurrence of a blizzard in each county in the United States. Although blizzards can occur virtually any-

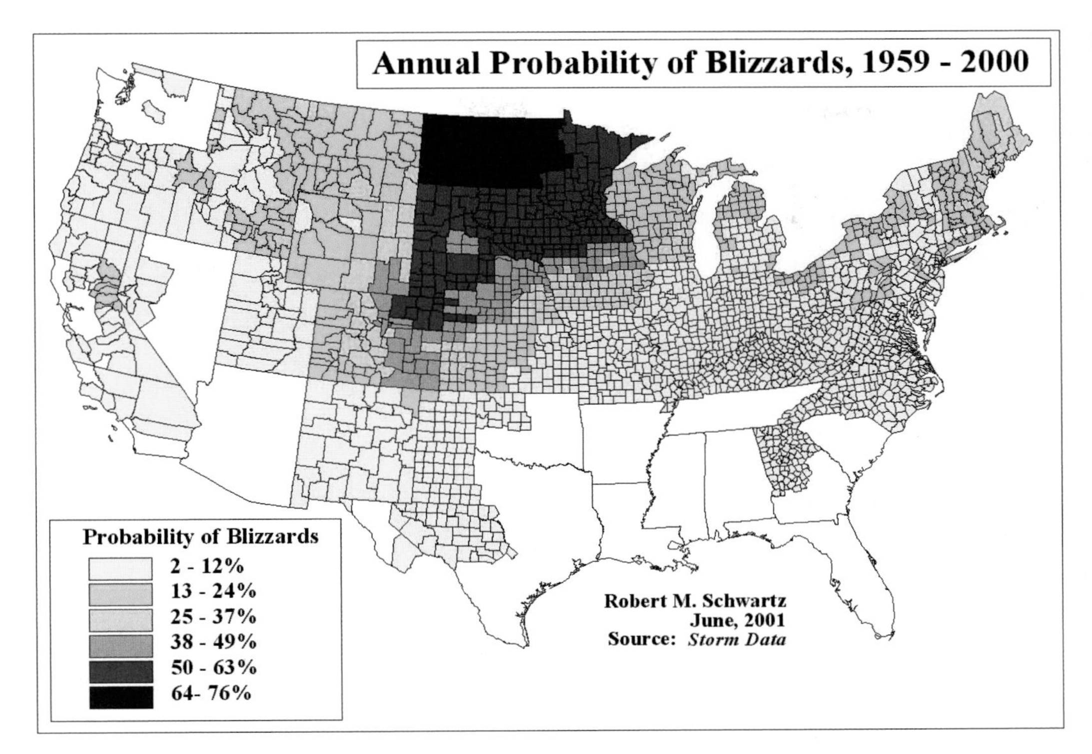

FIGURE 15.1: The probability of the annual occurrence of a blizzard in each county in the United States. The abrupt changes that appear along some state borders are due to differences in blizzard reporting between National Weather Service offices in neighboring states.

where in the northern half of the United States, they occur with highest frequency over North and South Dakota, Minnesota, Nebraska, northern Iowa, and eastern Colorado. Blizzards also have a high frequency on the plains in the neighboring provinces of central Canada including Alberta, Saskatchewan, and Manitoba. This region of North America lies just north of the climatological track of cyclones forming over the southern Rockies and just south of the Canadian Arctic, the primary formation region of bitterly cold air. The abrupt changes in blizzard frequency along some state borders in Figure 15.1 are not meteorological, but rather are due to differences in blizzard reporting between National Weather Service offices in neighboring states.

Figure 15.2 shows the total number of blizzards reported in each county during the forty winters between 1959 and 2000. In counties in the ***blizzard zone*** of North and South Dakota and western Minnesota, between thirty and seventy-four blizzards were reported in this time period, and average of about one to two blizzards every year.

COLD, WIND, AND SNOW

The first ingredient of a severe blizzard, extremely cold air, originates on the plains of west-central Canada in winter. As we learned in Chapter 14, a snow-covered landscape acts to rapidly reduce the air temperature by radiating infrared energy and reflecting solar energy to space. During the long winter nights, especially under clear skies, the temperature of air over the Canadian Plains in the core of a cold airmass can drop to –30° F to –50° F (–34° C to –45° C). Cooling creates high pressure at the surface, a consequence of air converging aloft over the developing cold air dome (see Chapter 8). The strong high-pressure system, which typically has a central pressure exceeding 1030 mb, is a critical component of a blizzard—the presence of high pressure increases the magnitude of pressure gradients that develop when a cyclone emerges onto the plains of the United States.

Blizzards occur within the circulation of ***extra-tropical cyclones***. Cyclones provide wind and snow, the second and third ingredients of a blizzard. In Chapter 10, we learned that cyclones primarily develop in two locations east of the Rockies, the first centered on eastern Colorado and the second centered on the plains of Alberta. Either cyclone type, ***Colorado cyclones*** or ***Alberta Clippers***, can lead to blizzard conditions. The worst blizzards typically occur with Colorado cyclones because of their potential for greater snowfall and stronger winds associated with deeper central low pressure. However, blizzards associated with Alberta Clippers normally have the coldest temperatures.

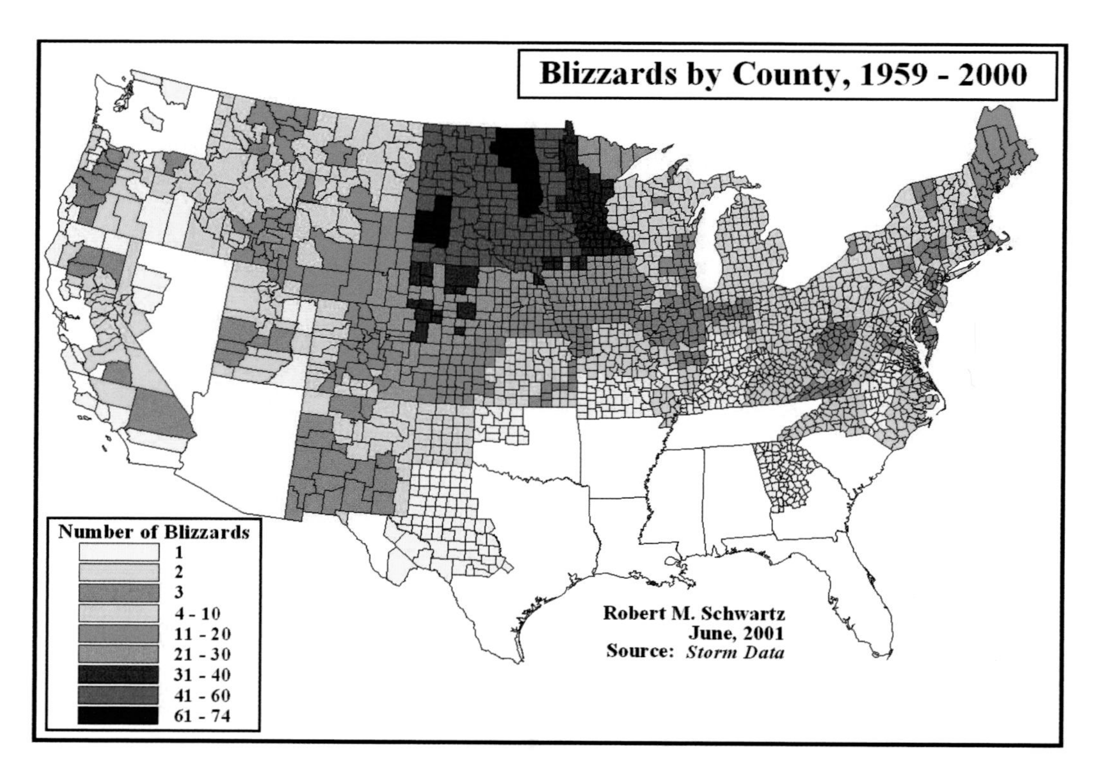

FIGURE 15.2: The number of blizzards reported by each county in the contiguous United States between 1959 and 2000.

CHECK YOUR UNDERSTANDING 15.1

1. What types of warnings does the National Weather Service issue for blizzards and what are the criteria used?
2. Where do blizzards occur most commonly in the United States?
3. What are the three "ingredients" necessary for a blizzard to form?

COLORADO CYCLONES AND BLIZZARDS

As discussed in Chapter 10, a cyclone will develop in the eastern Colorado region when a wave in the polar jetstream moves across the Colorado/New Mexico Rockies. Divergence at jetstream level, associated with curvature change within the wave and/or with jetstreaks (Figure 15.3A), triggers the formation of surface low pressure. For blizzard conditions to develop, bitter cold air and associated high pressure must already be in place over the Canadian plains at the time a cyclone is developing. Such an airmass is shown behind an ***arctic front*** approaching the Canada-U.S. border in Figure 15.3B. Normally, when arctic fronts develop over central Canada, midwinter cool temperatures are already in place across the United States, a result of previous intrusions of cold air from Canada. The southern boundary of the cool airmass over the United States, indicated by the southernmost cold front on Figure 15.3B, will typically lie somewhere across the southern United States.

The arctic front and the forming Colorado cyclone are initially distinct features of the atmosphere over the Plains (Figure 15.3B). However, as the cyclone moves eastward, the arctic airmass moves southward under the influence of the developing pressure gradient. With time, the arctic front approaches the cyclone center, wraps around the west side of the cyclone, and becomes incorporated into the cyclone's structure (Figure 15.4). As the cyclone progresses northeastward and intensifies in the manner discussed in Chapter 10, extremely strong pressure gradients develop on the northwest side of the storm center due to the large difference in pressure between the arctic high to the northwest and the cyclone center (Figure 15.4). These pressure gradients drive the bitterly cold air southward west of the cyclone center, creating the strong, cold winds of the blizzard.

As a cyclone progresses through its life cycle, warm air rising east of the upper-level front and over the warm front becomes wrapped aloft around the northwest side

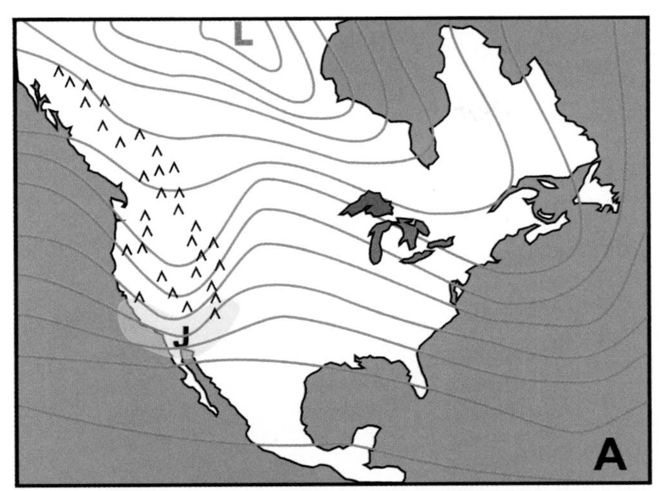

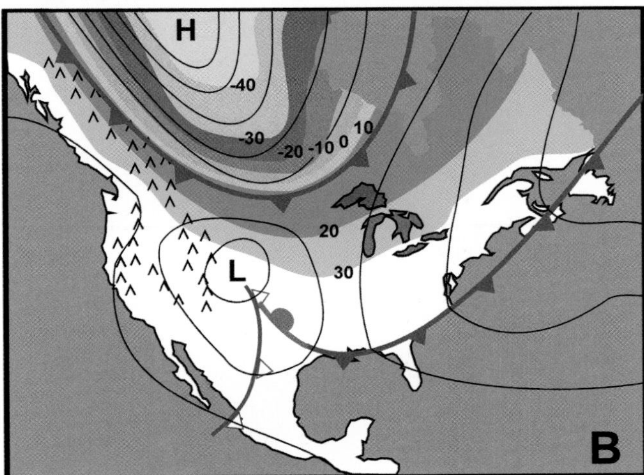

FIGURE 15.3: (A) Example of a 300 mb height pattern during the development of low pressure east of the Rockies associated with a Colorado cyclone. The shaded region denoted "J" represents a jetstreak in the flow. (B) Corresponding surface temperature (colors, °F), and sea-level pressure (black lines) during the development of a Colorado cyclone.

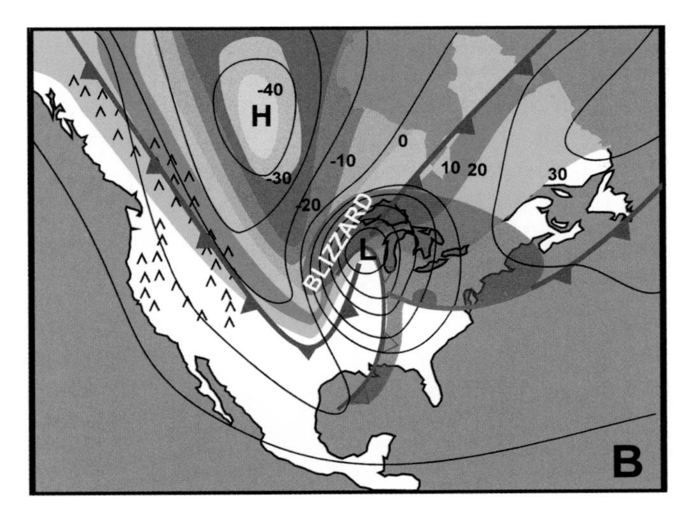

FIGURE 15.4: Surface temperature (colors, °F), and sea-level pressure (black lines) during the mature phase of a Colorado cyclone as the cyclone moves toward the Great Lakes. The gray shading denotes cloud cover.

of the low-pressure center, producing a band of clouds and precipitation around the northwest side of the cyclone (Figure 15.4). In Chapter 10, we learned that meteorologists call the wrap-around band a "trough of warm air aloft," or ***trowal***. Moisture in the trowal is the source of snowfall in blizzards associated with Colorado cyclones. As the trowal deforms into a narrower band northwest of the cyclone center, air within the trowal continually rises, condensing snow crystals that fall into the arctic air below. For example, Figure 15.5 shows two infrared satellite images of the trowal airstream during the 6 to 7 April 1997 blizzard that buried North Dakota. Note on the figure how the clouds deform and narrow in the twelve hours between the images.

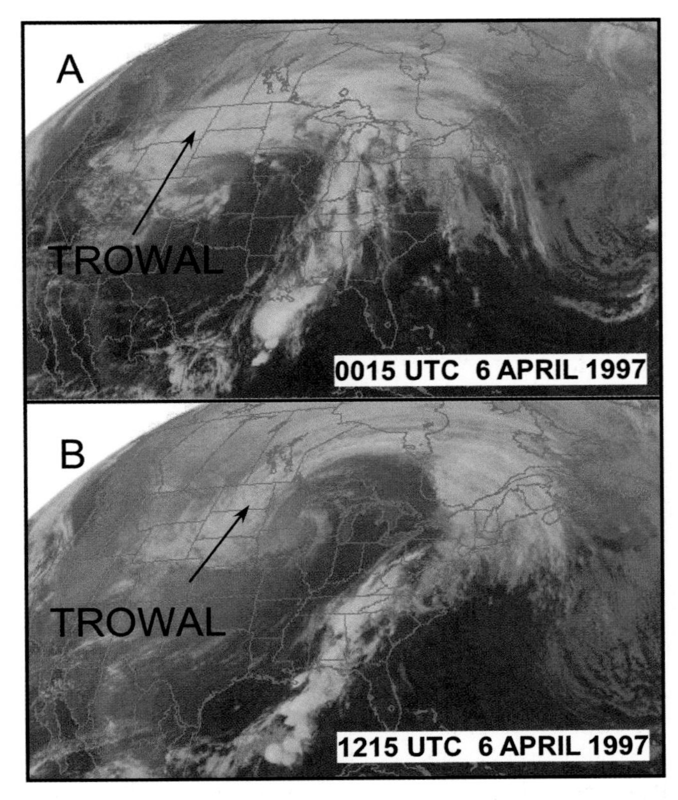

FIGURE 15.5: Infrared satellite images at (A) 0015 UTC and (B) 1215 UTC 6 April 1997 showing the trowal airstream during the 6 to 7 April 1997 blizzard that buried North Dakota. Note how the clouds deform and narrow in the twelve hours between the images.

The snowfall created by the trowal airstream can be extreme, exceeding ten or even twenty inches locally in a single event. The snow falls in a swath that is typically about 300 miles (500 km) wide. Figure 15.6, for example, shows a snow swath created by a storm on 30 January 2002. Because the snow in a blizzard falls from aloft into very cold arctic air, it remains fluffy and dry, characteristics that allow it to blow freely, creating whiteouts and huge snowdrifts at the surface. Even after the trowal airstream passes to the east, strong bitterly cold winds can continue to blow at the surface. The loose snow blows and drifts, making roads impassable. Drifts as high as ten to twenty feet can occur in some storms.

FIGURE 15.6: Visible satellite image on 2 February 2002 showing a large snow swath created by a storm that moved over the region two days earlier.

FOCUS 15.2

The Winter of Blizzards

The worst series of blizzards on record in the Great Plains occurred across North and South Dakota and Minnesota during 1996–1997. North Dakota, which suffered the worst damage, had not one, but *two* presidential disaster declarations for winter weather—the first two ever issued for winter weather in the state's history. Many communities set all-time records for snowfall. Bismarck received 101.7 inches of snow, ten inches above its previous record. Fargo, North Dakota, received 117 inches of snow. The *average* snow depth in Fargo between 1 January and 15 March 1997 was twenty-two inches (Figure 15A). Nine storms met blizzard criteria during the winter, four in January. Table 15A summarizes conditions at Fargo, North Dakota, during each of these storms. The blizzards were produced by both Colorado and Alberta Clipper cyclones.

Six of the blizzards occurred between 15 December and 24 January, an average of one per week. The high temperature was below 0° F (–18° C) on sixteen of these days, and was below freezing on all but one day. After these blizzards, the high temperature barely rose above freez-

(continued)

(continued)

FIGURE 15A: North Dakota buried in snow in 1997.

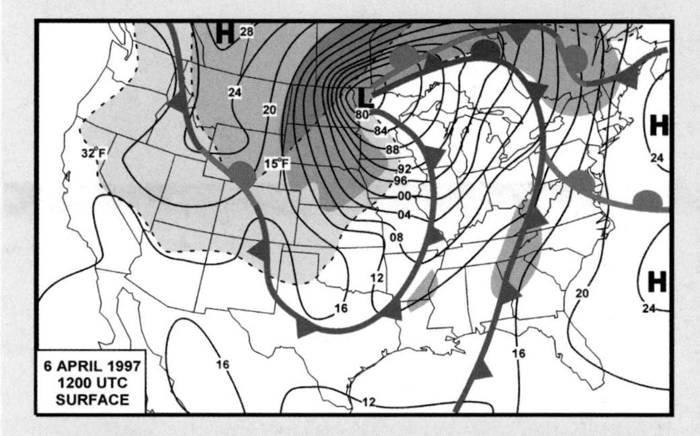

FIGURE 15B: Sea-level pressure field at 1200 UTC 6 April 1997. The region of precipitation is indicated with gray shading. The region with surface temperatures below 32° F (0° C) is shaded light blue, and below 15° F (–9.4° C) dark blue.

ing until the middle of March. Spring finally arrived at the beginning of April, with high temperatures just above 10° C (50° F) near Fargo, North Dakota. It didn't last long. The final blizzard of the season hit on 5 April, when a cyclone developed over eastern Colorado and rapidly deepened as it moved northeastward across South Dakota and into eastern North Dakota (Figure 15B).

A strong high-pressure system over northwest Canada to the northwest of the storm led to an extremely strong pressure gradient over North Dakota. Winds exceeded 58 mph (50 knots) in places and the hardest hit areas received sixteen to twenty-four inches of snow. Snowfall rates were one to two inches per hour during the period of strong winds. When the blizzard ended, 10 percent of the state's cattle herd was lost, $21.5 million in damage was done to farm structures, and 200,000 pounds of milk were dumped because it could not be delivered. Drifts as high as fifteen feet were common and all major roads were closed. Still the snows were not finished. Rapid melting in middle to late April forced the Red River, the border between North Dakota and Minnesota, to record flood levels (see Chapter 25), inundating the cities of Grand Forks and Fargo, North Dakota, a disastrous finale to the "Winter of Blizzards."

TABLE 15A **Summary of Surface Data at Fargo, North Dakota, during Nine Blizzards in the 1996–97 Winter**

Date of Blizzard (1996–97)	Lowest Temperature (°F)	Highest Temperature (°F)	Greatest Hourly Sustained Wind (knots)	Fasted Reported Wind Gusts (knots)	Snowfall (inches)
16–17 November	9	26	34	42	13.5
16–18 December	–8	22	33	43	7.9
23 December	–14	–8	21	—	4.9
4–5 January	0	20	33	40	10.7
9–11 January	–15	15	24	35	4.8
15 January	–8	12	38	49	1.0
22–23 January	–17	24	34	—	4.9
3–4 March	–5	31	25	—	15.5
5–6 April	9	38	24	56	7.0

ONLINE 15.1

The Great Plains Blizzard of November 2005

A powerful Colorado low created widespread blizzard conditions across Nebraska, South Dakota, Iowa, and Minnesota during November 27–29, 2005. Many locations experienced wind gusts in excess of 60 mph, sustained winds of 40 to 45 mph, snowfall totals up to fourteen inches with drifts exceeding ten feet, and near-zero visibility. It had been almost ten years since a storm of this intensity had occurred in parts of Nebraska.

Online, you can see animations of the hourly surface reports from the Central Plains and infrared satellite imagery for November 27 and 28. The center of the surface low, which deepened to 984 mb over Kansas, can be tracked by the center of the cyclonic pattern of winds in the surface reports, which indicate wind speeds of 35 to 40 knots with blowing snow in the storm's northwest quadrant. The infrared satellite animation shows the storm's development in eastern Colorado and its intensification as it tracks across Kansas, Nebraska, Iowa, and Minnesota. A well-defined trowal, beneath which the most severe surface weather occurred, is apparent to the north and northwest of the low center. South of the low center is a cloud-free *dry slot,* preceded by an area of cold cloud tops indicative of strong thunderstorms.

FOCUS 15.3

Snowrollers

Winds blowing over the snow during blizzards transform the landscape into a sea of drifts and curious snow formations. *Snowrollers* are one of the more interesting phenomena that are left behind by blizzards. Snowrollers are log-shaped large snowballs that form when wind scoops out and rolls a chunk of snow, much like you do when you build a snowman (Figure 15C). They can become rather large, like the five-pound roller in Figure 15D.

Snowrollers develop most easily when the ground has an icy, crusty surface on which falling snow cannot stick. When about an inch or so of loose, wet snow accumulates, and the winds are very strong and gusty, chunks of snow can be scooped out and rolled along.

FIGURE 15C: Snow rollers left behind following a blizzard.

FIGURE 15D: A five-pound snow roller.

A second factor creating snow north and west of the cyclone center has to do with the topography over which the storm is moving. The Mississippi River Valley is less than 500 feet (0.15 km) above sea level, while the plains in eastern Colorado just east of the Rockies are about 5000 feet (1.6 km) above sea level. The rise in topography between the Mississippi and Rockies is gentle, hardly noticeable to a traveler driving from St. Louis to Denver. Most of the rise takes place in western Kansas, Nebraska, and the Dakotas. Air flowing westward north of a cyclone must climb the slope of the topography. If an air parcel would traverse the entire distance, it would cool between 16° F (9° C) and 27° F (15° C), depending on whether it was saturated or unsaturated (recall the value of the moist and dry adiabatic lapse rate from Chapter 6). Although it rarely happens that air traverses the entire distance, an ***upslope flow*** occurs when air does flow upslope over a large enough region that significant cooling can occur, particularly across the geographic region where the slope of the plains is significant. The adiabatic cooling causes cloud formation and enhances the snowfall rates.

ALBERTA CLIPPERS AND BLIZZARDS

Alberta Clipper cyclones typically develop after cold air is entrenched over the central United States and the jetstream over Canada and the northern United States is oriented from northwest to southeast as shown in Figure 15.7A. Waves in the jetstream, such as the wave over the Canadian Rockies in Figure 15.7A, trigger the development of the Clipper cyclone (Figure 15.7B). Note that the upper-air wave in Figure 15.7A is considerably farther north than in the case of the developing Colorado cyclone in Figure 15.3A. Although arctic air is normally established over the United States before the Alberta Clipper forms, air over the plains of northern Canada continues to cool rapidly, renewing the bitterly cold temperatures that triggered the earlier cold-air outbreak. When the Clipper cyclone forms and traverses southeastward across Canada and into the Great Lakes region, a new arctic front moves southward to the west of the low-pressure center. Behind this front, temperatures can be bitterly cold, with temperatures in the extreme range of –30° F to –40° F

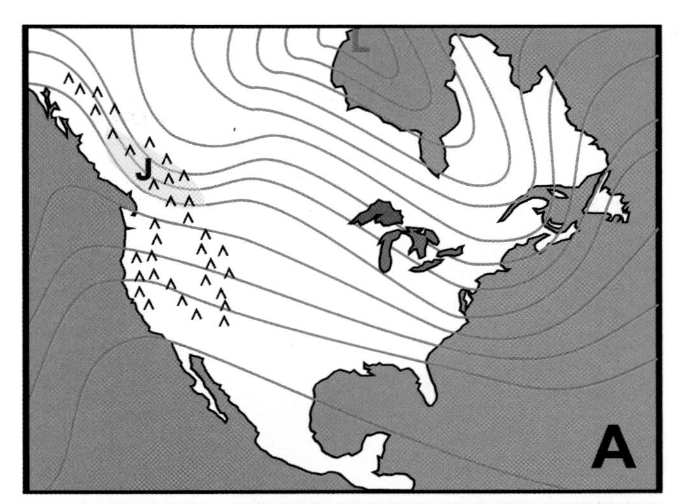

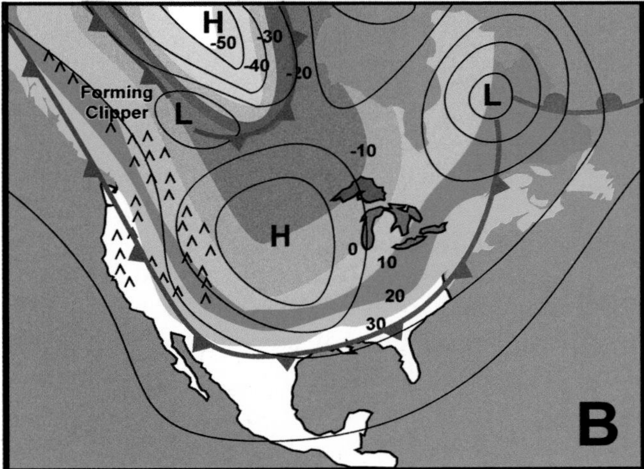

FIGURE 15.7: (A) Example of a 300 mb height pattern during the development of low pressure east of the Rockies associated with an Alberta Clipper cyclone. The shaded region denoted "J" represents a jetstreak in the flow. (B) Corresponding surface temperature (colors, °F), and sea-level pressure (black lines) during the development of an Alberta Clipper cyclone.

(–34° C to –40° C) occasionally reaching southward across the U.S.-Canadian border (Figure 15.8).

Alberta Clippers produce less snow than Colorado cyclones, typically two to five inches. The snow is low density, dry, and easily windblown. Most of the snow forms in air lifted along and over the advancing arctic front. Strong winds to the northwest of the cyclone are again driven by the pressure gradient between the arctic high pressure and the cyclone center (Figure 15.8). Despite the smaller snowfall amounts, drifting and whiteouts can effectively shut down travel. The bitter cold and strong wind can drop

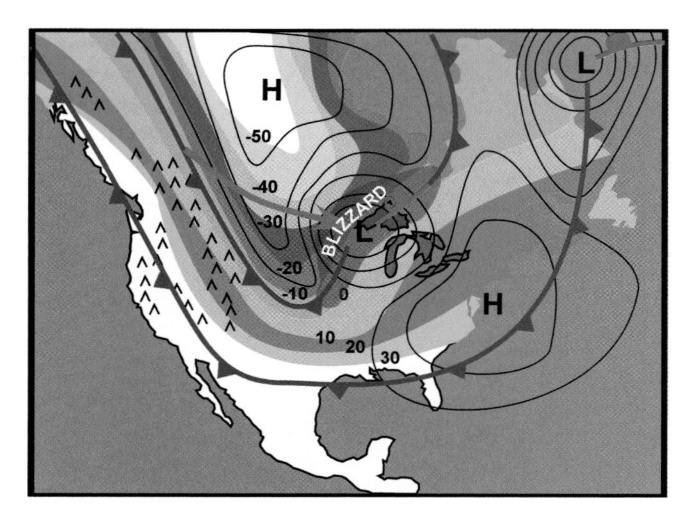

FIGURE 15.8: Surface temperature (colors, °F), and sea-level pressure (black lines) during the mature phase of an Alberta Clipper cyclone as the cyclone moves toward the Great Lakes. The red arrow shows the track of the low-pressure center and the gray shading denotes cloud cover.

wind chill temperatures into the –50° F to –60° F (–45° C to –51° C) range, creating conditions where hypothermia and frostbite become serious risks for anyone caught outside for more than a very short time.

CHECK YOUR UNDERSTANDING 15.2

1. Where do blizzards typically form relative to the center of a Colorado cyclone? What about Alberta Clippers?
2. Summarize the key differences between blizzards associated with Colorado cyclones and Alberta Clippers.
3. Where within the circulation of a cyclone on the Plains does upslope flow occur?

GROUND BLIZZARDS

Blizzard conditions can develop even when the sky is clear. When extremely cold winds blow across snow-covered ground, winds can whip surface snow into the air, creating a shallow layer of blowing snow and whiteout conditions. Blizzards caused solely by blowing snow are called ***ground blizzards***. Ground blizzards are frequent on the Great Plains in the wake of snowstorms. They are generally localized within areas prone to drifting and high winds, but can extend across broad regions (see Online 15.2).

Visibility in a ground blizzard can be as bad as or even worse than in an actual snowstorm. Poor visibility can persist for hours in some cases. Road conditions often deteriorate quickly as drifting snow accumulates on highways, obscuring slick regions of ***black ice*** that are difficult for motorists to see, particularly at night. Ground blizzards typically extend only tens of feet above the surface. They are most common when the surface snow is dry and loose. Light, dry snow falls most commonly when atmospheric temperatures during a snowstorm are well below freezing—precisely the conditions common during severe blizzards. Ground blizzards are especially dangerous because they are unexpected, occurring after the storm has passed.

BLIZZARD SAFETY

Great Plains blizzards transform highways into disaster zones as iced roads cause cars to slide into roadside ditches, and snow and wind reduce the visibility to close to zero (Figure 15.9). Automobiles stranded on the roadside can be buried in snow and isolated for hours or even days before the blizzard abates. Fatalities occur every year as people either die of hypothermia while seeking help, or of carbon monoxide poisoning while running the car's engine for warmth.

The ***wind chill temperature*** index provides a measure of equivalent temperature that the human body experiences in conditions of both cold and wind (see Chapter 14). In blizzards, the wind chill temperature is an *underestimate* of the heat loss experienced on exposed areas of the human body. The reason is the snow, which, when contacting exposed flesh, melts and creates a layer of water on the skin. Melting requires latent heat—energy that is taken directly from the body through the exposed skin. Once the snow melts, the air blowing across the water on the skin causes the water to evaporate. Evaporation also requires latent heat, and the source of the heat again is the body. The calculation of the wind chill temperature does not take into account heat loss through melting and evaporation, so when the snow is blowing, you can expect to feel a lot colder than the wind chill temperature.

Simple rules of winter travel can save your life if you are trapped on the road in a blizzard. Travelers

FIGURE 15.9: Near-zero visibility along a highway during the blizzard of February 12–13, 2007.

ONLINE 15.2

A Ground Blizzard from Space

Ground blizzards are often localized phenomena. On 15 January 1997, however, strong winds created ground blizzards that covered much of the state of North Dakota. A spectacular animation of visible satellite images of North Dakota, obtained from the Space Science and Engineering Center of the University of Wisconsin, shows features of these ground blizzards. On the lower part of the animation, clouds can be seen moving southward. Snow-covered ground is visible north of these clouds. Blowing snow in the "clear" area appears as narrow streaks on the animation. These streaks are aligned with the wind, which was from the north-northwest at 20 to 30 knots. As you view this animation, consider the conditions travelers experienced as they crossed these streams of blowing snow. A whiteout would engulf them suddenly—a surprise snowstorm in an otherwise clear sky.

should: (1) Carry a winter storm survival kit that includes blankets and sleeping bags, a flashlight and batteries, a first aid kit, a knife, high-calorie nonperishable food, extra dry clothing, drinking water, a large can with plastic cover for sanitary purposes, a small can for melting snow for additional drinking water, waterproof matches or a lighter, and several large candles. Burning a single candle inside a car can both provide light and warm the interior several tens of degrees and reduce the need to run the car to keep the interior warm, which will prevent carbon monoxide poisoning from the exhaust. (2) Carry basic winter automobile safety items including a sack of sand or cat litter for traction, a shovel, a windshield scraper and brush, a tool kit, a tow rope, booster cables, and accident flares. (3) Stay with your vehicle. People seeking help often become disoriented in blinding snow only feet from their vehicle and cannot find their way back. A car provides shelter from wind and wet during a snowstorm. Even if you wait for days, the odds of surviving are much higher than they would be if you venture into a blizzard on foot, which is often a fatal decision. Rescuers look for stranded vehicles and are likely to find you alive within one. (4) Carry a cell phone. Obviously, the modern convenience of a cell phone will allow communication with authorities and family and hasten rescue.

FOCUS 15.4

The Children's Blizzard of 1888

While 1888 may be best known for the late-winter blizzard that struck New York and New England in March (Chapter 11, Focus 11.3), a storm that occurred two months earlier (January 12–13, 1888) is the most famous and deadliest blizzard in the recorded history of the Northern Plains. Known as the "Blizzard of 1888" and the "Children's Blizzard," the number of fatalities in this storm is estimated to have been between 250 and 500—comparable to the death toll in the northeastern blizzard of March, 1888. The fatalities included more than 100 school children, who became stranded while trying to return from their rural schoolhouses. The experiences of these children are described in David Laskin's *The Children's Blizzard,* a book that provides a broad perspective on the role of weather in the history of the Great Plains.

The Blizzard of 1888 was exceptional for the suddenness and ferocity with which it struck. The low-pressure center intensified along a cold front that moved rapidly from the central Dakotas to Minneapolis in twelve hours on January 12, 1888. Behind the front, the temperatures dropped by forty to fifty degrees (F) in as little as 4½ hours. Ahead of the front, unusually mild air had led to a mid-winter thaw over much of the Plains, enticing residents to leave their homes ill-prepared for the sudden onslaught of snow, wind, and cold. The combination of sustained gale-force winds (gusting to hurricane force), blowing snow, and plummeting temperatures was accompanied by widespread electrical discharges (St. Elmo's fire, cf. Chapter 21). While the winds and snow trapped many residents (including the schoolchildren trying to return home) out in the open, the ensuing cold sealed their fate. The continental polar airmass that originated in Canada brought temperatures of –40° F (–40° C) to Minneapolis-St. Paul, while North Platte, Nebraska fell to –35° F (–37° C) with a sea-level pressure of 1043 mb. This extremely high pressure, together with the track of the deep low-pressure center through eastern Nebraska and Iowa, indicates that the blizzard conformed closely to the pattern sketched in Figure 15.4. The cold airmass penetrated as far south as Mexico and caused major damage to winter crops along the Gulf Coast and in Florida.

The storm's high death toll was partly a result of the poor weather forecasts, which gave the Great Plains residents essentially no warning of the severity of the impending storm. Two months later, the great Blizzard of 1888 in the Northeast was also badly forecast (Chapter 11). The absence of adequate warnings of these two events led President Benjamin Harrison in 1889 to transfer the weather service from the U.S. Army's Signal Corps to the Department of Agriculture, where it remained for much of the twentieth century.

FORECASTING BLIZZARDS

Some of the most disastrous blizzards in the history of the United States have been poorly forecast (Focus 15.4). The absence of advance warnings has contributed directly to the number of casualties in many of these cases. In recent decades, numerical weather prediction models (Chapter 4) are increasingly able to provide predictions of the low-pressure systems, strong winds, and extreme temperatures associated with blizzards. While the details of the storms' tracks and intensities are not always apparent until a day or two prior to a blizzard, timely warnings can be issued on the basis of these forecasts. Even at lead times of three to five days, numerical models often provide general indications of a developing cyclone associated with extremely cold air. The most difficult element to predict in a blizzard is often the intensity of the blowing snow. While models can successfully forecast the low temperatures and strong winds, the amount of precipitation and especially its dryness are more difficult for models to capture in their forecasts. The accuracy and precision of precipitation forecasts is less than for the temperature forecasts obtained from numerical models, and the wetness of a snowfall is not explicitly forecast by models. Moreover, if a pre-existing snow cover has experienced above-freezing temperatures, it will be less susceptible to blowing and drifting. Such information is generally not included in models. Despite these limitations, the general improvement of weather forecasts during the past several decades, together with the greatly enhanced communication and warning systems, have reduced the frequency with which Great Plains residents are trapped outdoors when blizzards strike.

BLIZZARDS AND GLOBAL WARMING

How might the frequency and intensity of blizzards change as increasing greenhouse gas concentrations lead to a warmer climate? For a particular region such as the Northern Plains, the answer depends on changes in at least two key factors: (1) the effect of temperature on the type of precipitation and (2) the tracks and intensity of major winter storms. A generally warmer climate will favor the occurrence of rain rather than snow in areas of precipitation that are presently climatologically close to the freezing temperature. In addition, areas receiving snow can expect snow that has a greater moisture content, i.e., the snow will have a larger water equivalent. Wetter snow is generally less prone to blowing than is dry snow, so temperature-moisture factor will tend to favor a reduction of blizzard intensity.

The big factor that confounds predictions of changes in blizzards is the response of the major storm tracks to greenhouse warming. As we saw in Chapter 10, there are reasons to expect winter storm tracks to shift northward and cyclones to weaken. A tendency toward weakening storms is consistent with a wintertime greenhouse warming that is greater in polar regions than in middle latitudes (Chapter 5), thereby reducing the contrast between polar and tropical airmasses. Extratropical cyclones draw upon frontal temperature contrasts for their energy. If these contrasts become weaker, there would be a second reason to expect a general reduction in the occurrence of severe blizzard conditions in the northern United States (although southern Canada might experience more frequent cyclones if the tracks of Colorado lows and Alberta Clippers shift northward). However, as noted by the Intergovernmental Panel on Climate Change in its 2007 report, changes of storm tracks and intensities vary considerably among climate models, so there is a large uncertainty concerning future changes in blizzard occurrences.

CHECK YOUR UNDERSTANDING 15.3

1. Why do ground blizzards occur?
2. Why does the wind chill temperature underestimate the body's heat loss during a blizzard?
3. What are two ways in which global warming might affect the frequency and intensity of blizzards in the Great Plains?

TEST YOUR UNDERSTANDING

1. What are the primary impacts associated with blizzards of the Great Plains?
2. What geographic features of North America and surrounding water bodies support the development of blizzards over the Great Plains? What is the role of each of these features in creating conditions for blizzards?
3. A blizzard warning is issued for your area. What type of weather conditions should you expect?
4. What role does high pressure over northwestern Canada play in creating blizzard conditions on the Great Plains?
5. Explain why blizzards associated with Colorado lows have more snow than Alberta Clippers, yet blizzards associated with Alberta Clippers have colder temperatures.
6. What are the roles of the arctic front and the trowal in blizzards?
7. How does the topography of the Great Plains enhance snowfall in Colorado low blizzards?
8. Why do Alberta Clipper blizzards typically produce small snowfall totals?
9. The forecast for Bismarck, North Dakota, is for blizzard conditions with a morning low temperature of –25° F, 25 knot winds from the north-northwest, and three inches of snow accumulation expected by evening, with two to three foot drifts. Without looking at a weather map, what type of

cyclone is likely to have produced this blizzard? Justify your answer.

10. From the point of view of a person caught outside, what is the difference between a ground blizzard and a blizzard?

11. You live in Chicago, Illinois, and are packing your car for a January winter ski vacation in Banff, Alberta. Describe as completely as possible everything you should pack in your car to protect yourself from winter weather during your trip.

12. Why do winter safety specialists recommend that you not run your car for long periods to keep yourself warm when trapped in a blizzard? What is a better approach to keeping warm while inside a trapped car?

TEST YOUR PROBLEM-SOLVING SKILLS

1. Suppose that a person wearing a winter parka will experience hypothermia in still conditions in 120 minutes at 0° F, and one minute sooner for every degree the wind chill factor is lower than 0° F. Use the wind chill chart in Chapter 14 to determine whether this person would experience hypothermia faster at –20° F in calm air or at 0° F in a 20 mph wind. For all cases where the numerical values of the wind and the negative of the temperature are the same (e.g., –15° F and a 15 mph wind), is there any realistic situation where your answer to the first question would change? What can you deduce about the relative role of strong winds and low temperatures in blizzards?

2. You are a forecaster in North Dakota. Heavy snow blanketed the state three days ago. Skies cleared a day later. Forecast maps suggest that strong southerly winds are expected overnight. The overnight low temperature was 20° F, and the afternoon high 35° F. The overnight temperature is expected to drop into the middle 20s across the state. Interstate 94, which crosses the state from west to east, has several stretches of road that are prone to ground blizzards when they occur. Would you include the possibility of ground blizzards in your forecast? Why or why not?

3. The visible satellite image in Figure 15.6 shows a large snow swath extending from eastern Colorado across Kansas, Nebraska, and Iowa. The track of the low-pressure center associated with the cyclone that produced this swath approximately followed the southern boundary of the swath. The width of this snow swath is typical for blizzards over the plains. Suppose a cyclone forms over eastern Colorado. The forty-eight-hour NAM model forecast indicates that the cyclone center will be at the southeast corner of Iowa, and blizzard conditions will prevail everywhere that snow is falling. However, ensemble forecasts for the storm indicate that the uncertainty in the north-south position of the low-pressure center in forty-eight hours is 180 miles (300 km). To be safe, you forecast blizzard conditions for all locations that have the possibility of snow. Use the satellite image and the ensemble forecast information to make a rough estimate of the area (in square kilometers) that will receive a blizzard forecast, but will not receive any snow.

4. You are a forecaster for the National Weather Service Forecast Office in Bismarck, North Dakota. There is a strong cyclone just to the southeast of your county warning area. You are trying to figure out if you should issue a blizzard warning. Since the region is sparsely populated you cannot rely on storm spotter reports and must look at the various maps and images available to you at the forecast office. Explain what maps and images you would examine to determine if blizzard criteria were in fact being met.

USE THE SEVERE AND HAZARDOUS WEATHER WEBSITE

http://severewx.atmos.uiuc.edu

1. Navigate the *Severe and Hazardous Weather Website* to the link for "Great Plains Blizzards." Under "Additional Resources" you will find a link to the Color Landform Atlas provided by Johns Hopkins University.
   - (a) Using the elevation key, calculate the distance air will rise as it travels from east to west across each of the Great Plains states (listed below).
     - (i) North Dakota
     - (ii) South Dakota
     - (iii) Nebraska
     - (iv) Kansas
     - (v) Oklahoma
     - (vi) Texas
   - (b) Determine the amount that air would cool as it traveled across each state. Assume that air would saturate if it ascended 500 meters.
   - (c) Across which state would air temperature cool the most? Explain how this might influence the amount of snowfall found in blizzards.
   - (e) Repeat parts (a), (b) and (c) but for states along the Rocky Mountains except for the distance use the eastern border of the state to the highest point in the mountains.
     - (i) Montana
     - (ii) Wyoming
     - (iii) Colorado
     - (iv) New Mexico
   - (f) Compare and contrast your results for the Rocky Mountain states with the Great Plains states.

2. Over several days during midwinter, examine the conditions found with an extratropical cyclone as it develops and matures (either an Alberta Clipper or a Colorado Low). Use the links under the "Current Weather" page of the *Severe and Hazardous Weather Website* to locate maps that will help you answer the following.
   - (a) Once per day locate the position of an arctic front over Canada at both the surface and the 850 mb level. Either print or save the maps you examined or recreate the maps using the blank diagrams found in Appendix C. Be sure to note the date and time associated with each frontal position.
   - (b) After the cyclone has passed across North America, go back and examine storm reports, snowfall totals, and general weather conditions found in the vicinity of the cyclone. Where did the worst winter conditions occur relative to the arctic front and the cyclone center?
   - (c) Where did the strongest pressure gradients occur at the surface?
   - (d) Were blizzard conditions present at any time during the cyclone's life cycle? Justify your answer.

3. Use the *Severe and Hazardous Weather Website* to examine current or archived surface observations during a winter storm. Look for the blowing snow symbol for significant weather (see Figure 3.1). Based on the wind speed at the stations reporting blowing snow and the radar echoes in the vicinity of the station, would you say that this station is experiencing a ground blizzard? Why or why not?

4. Right after an early winter cyclone progresses across the Great Plains, use the *Severe and Hazardous Weather Website* to navigate to current and archived weather maps of the Great Plains.
   - (a) Record the dates of the storm and the storm type (Alberta Clipper or Colorado Low).
   - (b) Examine a visible satellite image and identify the snow swath produced by the storm. Estimate the length and width of the swath.
   - (c) View animations of radar reflectivity during the storms trek across the country. Does the heaviest precipitation fall in the region where the snow swath is found? Explain.
   - (d) View animations of satellite imagery during the storm's life cycle. Identify the trowal if one exists. Explain how you determined if the trowal was present.
   - (e) Examine storm reports from National Weather Service offices located within the snow swath. Identify all cases of ground blizzards including location, date, and time.
   - (f) View surface air temperature data for the time of the storm. Based on the surface air temperature, would you expect the snow associated with the cyclone to be light and fluffy or heavy and wet? Why?

MOUNTAIN SNOWSTORMS

Snowplow on Donner Pass, California.

KEY TERMS

atmospheric river
avalanche
Chain Law
cloud seeding
Four Corners low
melting level
orographic clouds
orographic lifting
Pineapple Express
powder snow avalanche
slab avalanche
upslope storm
water equivalent (of snow)
weather modification
windward slope

KEY MOUNTAIN RANGES

Bitterroot Mountains
Cascades
Coast Range
Rocky Mountains (Front Range, Wind River Range, Teton Range)
Sierra Nevada
Wasatch Range

Winter snowstorms in the mountains of the western United States are the lifeblood of the economy of the West, and, in many ways, the country itself. Snow fuels the tourist industry, drawing visitors to ski and enjoy the spectacular scenery. More importantly, water from winter snowfall is the primary source of agricultural and urban water supplies. In the western United States, particularly southwestern states that depend on the Colorado River, water from annual snowfall is now barely adequate for water supplies during years of normal snowfall.

Water from melting snows, flowing in great western rivers such as the Colorado and Columbia, also provides hydroelectric power as it makes its way from the high country to the Pacific Ocean. Figure 16.1 shows the percentage of total power generation attributable to hydroelectric plants in each of the western States. Over one-third of the 200 billion kilowatt hours of power generated in the western states is hydroelectric. Hydroelectric power constitutes half of the power generated in Pacific Coast states and over 75 percent of all power generated in Oregon, Washington, and Idaho.

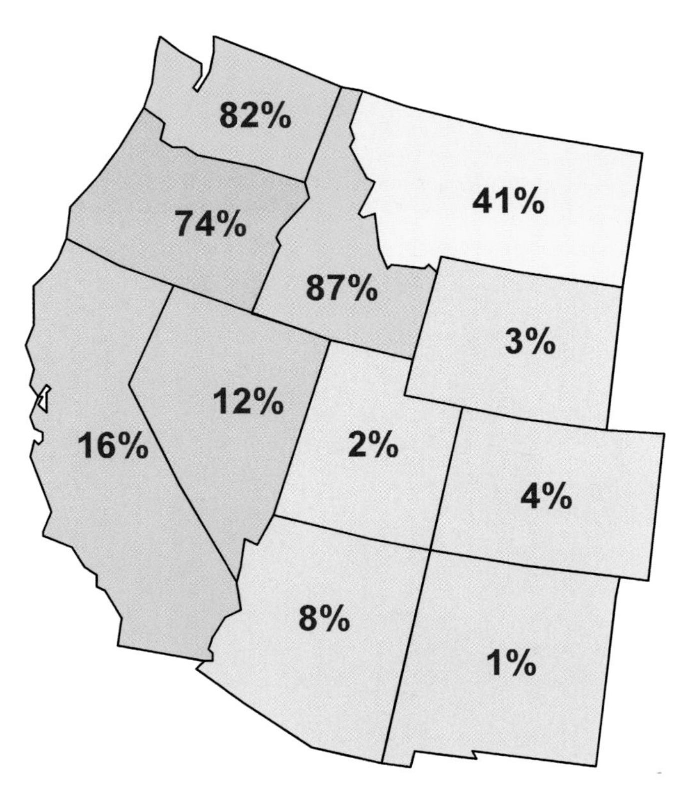

FIGURE 16.1: Percentage of power generated in each of the states in the western United States that was hydroelectric in 2006.

Mountain snowstorms are beneficial, but they can also be destructive when they lead to prolonged highway closures, traffic accidents, and ***avalanches***. Many major west-east U.S. highways cross mountain passes. During heavy snowstorms, Donner Pass in the Sierra Nevada, the conduit for Interstate 80, can close for a week or more, stranding trucks and automobile traffic on both sides of the mountain. Interstate 70, the primary highway leading west from Denver, Colorado, rises to 11,260 ft (3,431 meters) above sea level before crossing the summit of the Rocky Mountain's Continental Divide at the Eisenhower Tunnel.

Heavy snowstorms regularly cause major traffic jams as travelers attempt to drive between Denver and resorts on the west side of the range. Closure of passes due to heavy snow can result in enormous losses for the transportation, tourism, and other industries. Extremely heavy snow can trigger avalanches. Nowhere is this more evident than in the Little Cottonwood Canyon, which connects Salt Lake City, Utah, with the Alta and Snowbird Ski Resorts in the Wasatch Mountains of northern Utah. The road is nine miles long and traverses forty-two avalanche paths, with half the road in avalanche path runout zones. Between 7000 and 10,000 vehicles traverse the canyon daily during the ski season. Even in years of below average snowfall avalanches can trap or bury vehicles on the highway.

In addition to routine meteorological measurements and satellite data, mountain snowstorms are monitored by the National Weather Service, State Departments of Transportation, and the Forest Service using networks of snow gauges and human observers. Although radars are deployed throughout the West, they are not so useful to determining precipitation in mountainous regions for two reasons. First, the relationship between radar reflectivity and snowfall rates are not reliable because ice crystals have a variety of shapes and snow can vary in density depending on its water content. Even if such a relationship existed, mountains block radar beams, limiting coverage in mountainous areas.

FOCUS 16.1

Avalanche!

Heavy mountain snowstorms set the stage for avalanches. Avalanches are large masses of snow that slide down mountainsides. *Powder snow avalanches* are avalanches of loose, freshly-fallen snow. This type of avalanche typically occurs during or shortly after a heavy snowfall. Powder avalanches move rapidly downhill at speeds approaching 100 miles/hr (160 km/hr), generate strong winds at their leading edge, and appear billowy as snow is thrown upward in the wind. *Slab avalanches* usually occur when a temperature rise causes more recent heavy snow, which has not properly bonded to the older snow beneath, to slip away as one large slab. The older snow normally has a crust from a previous melt, allowing the more recent snow to glide over the underlying snowpack. Over 2000 avalanches are reported in an average year in Colorado alone. Figure 16A shows the number of fatalities associated with avalanches in the United States during the period from 1951–2007. The number of fatalities has generally increased as more people visit mountains for winter recreation.

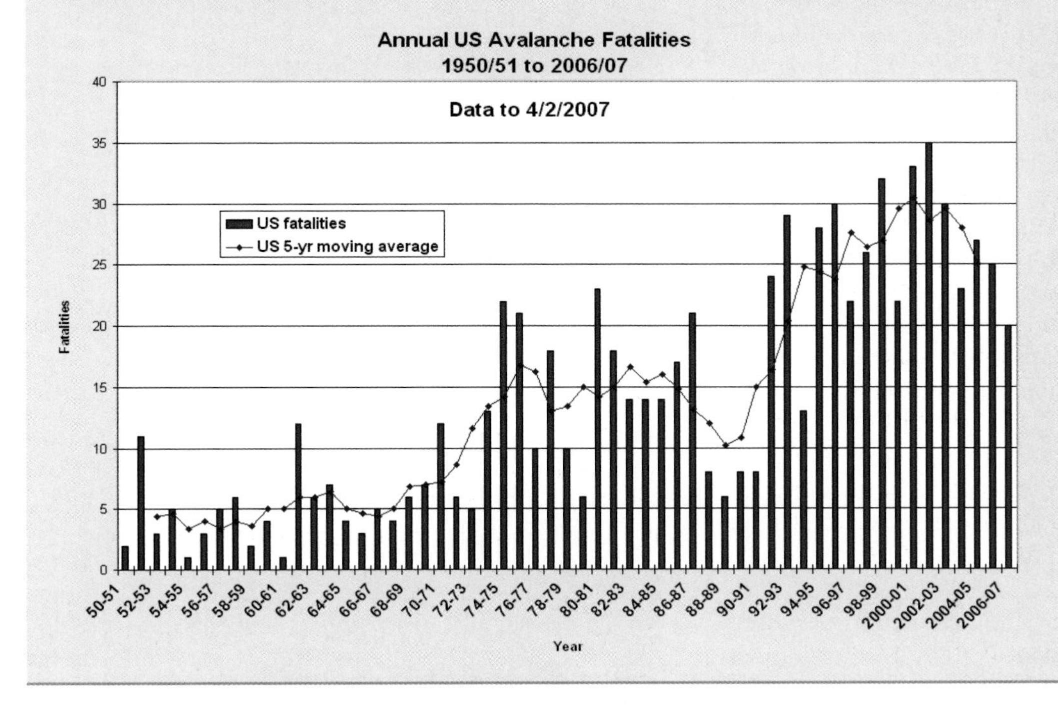

FIGURE 16A: Avalanche fatalities in the United States during the period 1950–1951 through 2006–2007.

Figure 16.2 shows the distribution of mean annual snowfall across the United States and the average number of days that at least one inch of snow cover is present on the ground. East of the Rockies, except locally around the Great Lakes and Appalachian Mountains, snowfall increases from south to north, the distribution controlled primarily by latitude and temperature. From the Rockies westward to the Pacific, the primary controls on snowfall are elevation and mountains—their location, height, and steepness. The role of the mountains in generating snowstorms is evident—all the mountain ranges appear on the figure as regions of heavy snowfall (compare this figure with the physical geography map found in Appendix C). In much of the high country, snow cover is present over a third of the year, and in some locations is present year-round. Figure 16.3, a schematic cross section of the topography across the western United States from the California Pacific Coast to the Mississippi Valley, shows the major mountain ranges. From west to east, these include the ***Coast Range***, the ***Sierra Nevada*** of California, the ***Wasatch Range*** of Utah, and the ***Rocky Mountains*** of Colorado. The characteristics of heavy snowstorms on each of these ranges vary because of their proximity to moisture sources and their elevation. For this reason, we will consider the characteristics of heavy snowstorms along each range separately. First, let's examine the conditions that typically lead to mountain snowstorms.

FOCUS 16.2

Cloud Seeding

Unlike most regions, heavy snowstorms are welcome in the West. When the storms are infrequent, water and energy can be in short supply, particularly in the dry summer months. In fact, clouds over mountains of the West are often "seeded" in an attempt to produce additional snowfall. Ice crystals and water droplets grow at very different rates in the atmosphere under the same temperature and humidity conditions. The difference in growth rates arises because of the difference in the strength of the bonds between molecules on the surfaces of ice crystals and water droplets. Ice crystals grow much more rapidly. For example, in saturated air at a temperature of 14° F (–10° C), ice crystals will grow quickly while water droplets will not grow at all. Sometimes the water droplets will actually evaporate while the ice crystals grow.

Clouds that form over the mountains of the western United States in winter often consist of both ice crystals and water droplets. Because cloud temperatures are below freezing, any droplets in the clouds are supercooled (see Chapter 12). Because the supercooled water droplets in mountain clouds grow very slowly and are tiny, they are swept along with the air currents as air passes over the mountain. In fact, if a cloud forming over a mountain in winter consisted entirely of supercooled water, virtually all of the water would be carried over the mountain and evaporated on the downwind side. The droplets simply would never grow large enough to precipitate. Ice crystals, on the other hand, grow very quickly, collect each other (and some of the supercooled water droplets), and have a good chance of falling to the ground as they are carried along by the wind toward the mountain crest. In short, ice clouds are very efficient at producing precipitation while supercooled water clouds are not.

In the 1940s, scientists discovered that ice crystals could be formed in a cloud of supercooled water droplets by either dropping dry ice pellets (frozen carbon dioxide) through the cloud, or introducing microscopic particles of silver iodide into the cloud. This discovery led quickly to the idea that clouds could be made more efficient precipitation producers by converting their supercooled water to ice. Because of the need for water and hydroelectric energy in the West, mountain cloud systems became a prime target to test the newly developed technology of *weather modification*. Both the government and the water and power industries realized that if even a small percentage of the supercooled water passing over the mountains could be converted to additional precipitation, *cloud seeding* would be cost-effective.

Methods were developed to deliver seeding material to clouds using ground-based silver iodide generators (Figure 16B), silver iodide flares mounted on aircraft, and aircraft-delivered dry ice

FIGURE 16B: Device used to generate microscopic silver iodide aerosol for cloud seeding. The plume is carried from the ground into the clouds as air rises over the mountains in the distance.

(continued)

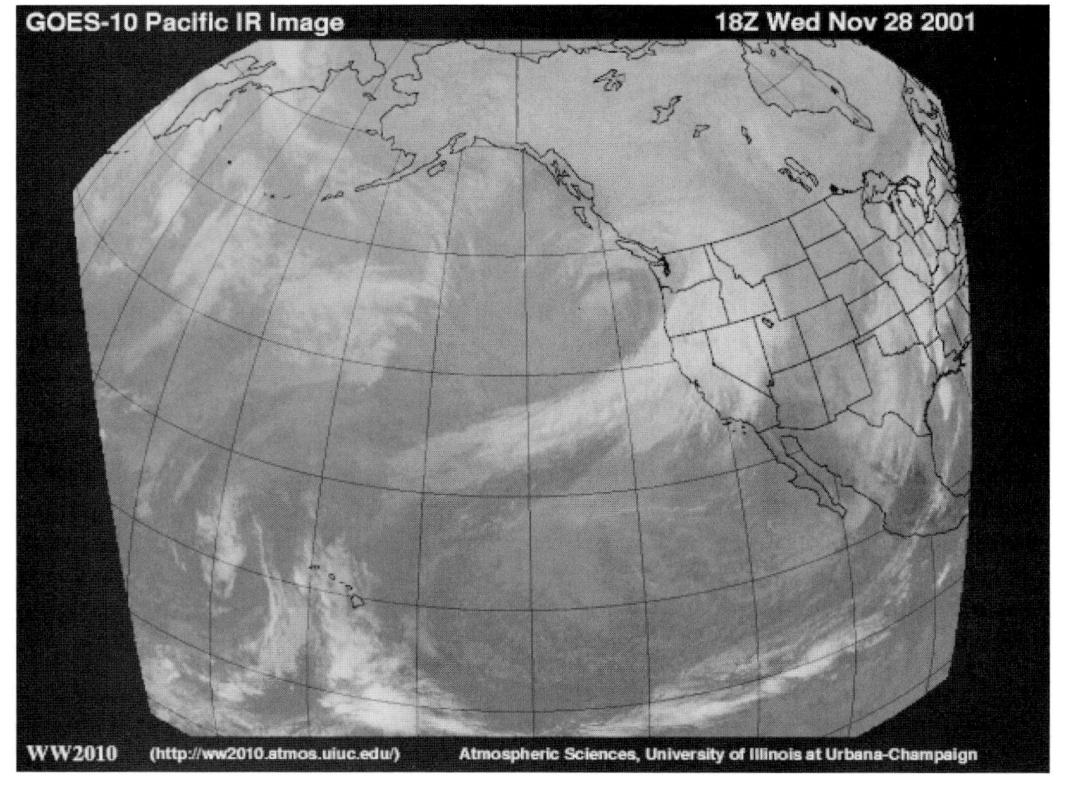

FIGURE 16.4: Infrared satellite photo of a cyclone approaching the mountains of the West Coast at 1800 UTC 28 November 2001. An atmospheric river is evident, extending from near the Hawaiian Islands to the California Coast.

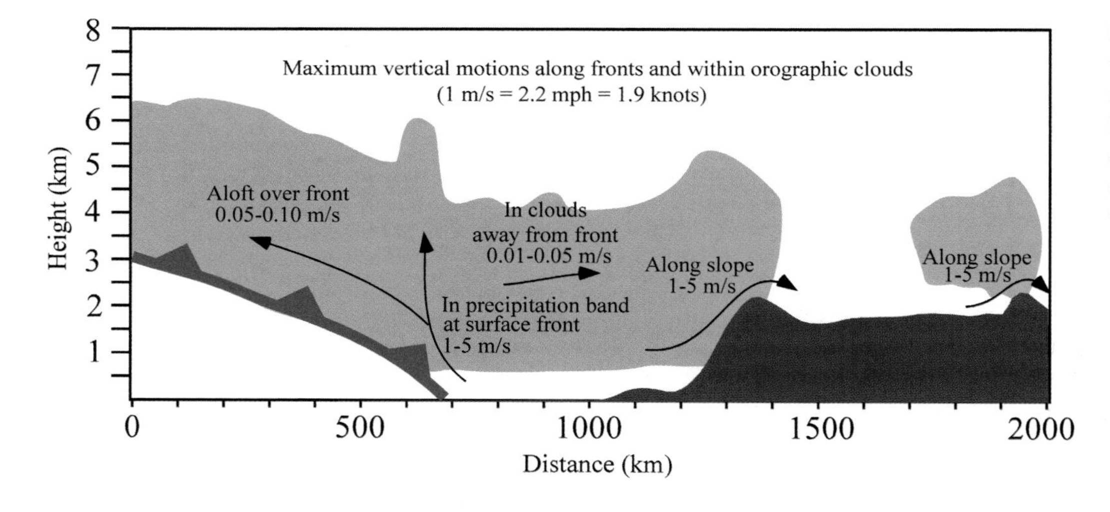

FIGURE 16.5: Maximum vertical motions (in meters/second) occurring in clouds along fronts and in orographic clouds along the Pacific Coast in wintertime.

mountain falls continually in one location. Of course, when the front itself ascends the mountain, enhancement of vertical motions by both frontal and orographic lifting can cause very heavy snowfall locally on the mountain slope.

Precipitation can be generated directly by orographic lifting even in the absence of precipitation from a larger-scale storm system. ***Orographic clouds*** often form well in advance of a large-scale weather system and linger for a day or more after the weather system passes. The effects of mountains on precipitation in the western United States are obvious from Figure 16.6, a map of the average annual precipitation in the United States. In most areas of the western United States, mountains receive five to seven times the precipitation falling in neighboring valleys.

As Pacific storms move over the continent and into the mountains, the mountain barriers and high topography disrupt their flow, modify frontal structures, and weaken their low-level features. The high

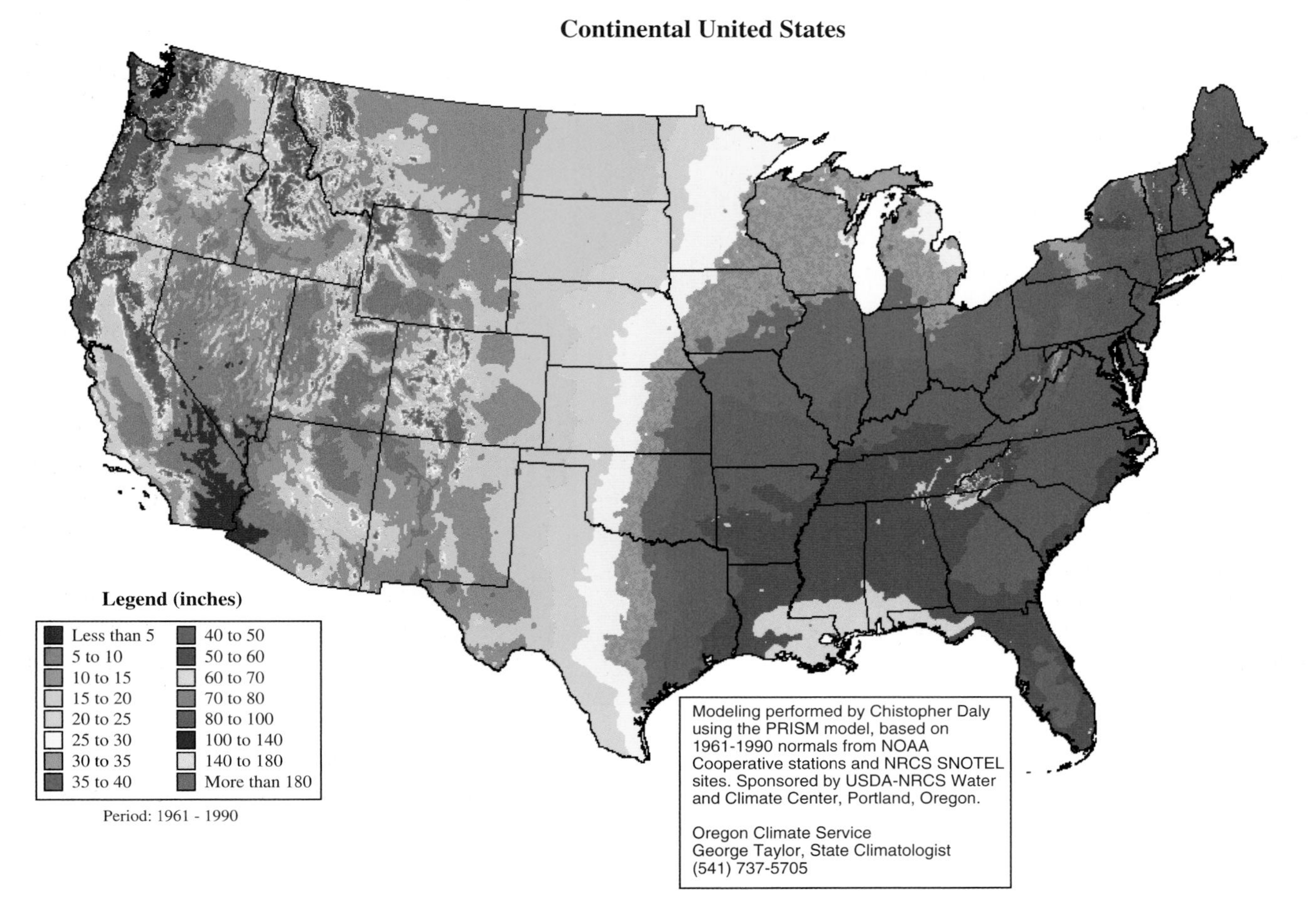

FIGURE 16.6: Average annual precipitation (rain and snow combined) in the United States.

plateaus to the lee of the Sierra Nevada and Cascades are over 2000 m (6500 ft) in elevation (see Figure 16.3), so little of the storm's low-level structure evidenced over the Pacific remains as the weather systems move across this region. Although low-level storm structure is disrupted, upper-level forcing associated with troughs and jetstreams normally continues to trigger cloud formation. Passage of the air over the Sierra and Cascades also significantly reduces the moisture content of air over the high plateaus to the east, the primary reason why these plateaus have a desert climate. Over the interior mountain regions, the primary forcing for precipitation is the orographic lifting provided by mountain ranges.

CHECK YOUR UNDERSTANDING 16.1

1. List three ways in which mountain snowstorms benefit the economy of the western United States.
2. What are the primary controls on snowfall from the Rockies to the Pacific?
3. In what ways do mountains modify storms coming onshore from the Pacific Ocean?

STORMS ALONG THE WESTERN SLOPES OF THE UNITED STATES

The Coast Range is the first mountain range encountered by air approaching from the Pacific. From Figure 16.6, we can see that north of San Francisco, California, the Coast Range receives the heaviest precipitation in the continental United States, with over 180 inches (450 cm) falling on Washington's Olympic Peninsula. The differences between Figures 16.6 and 16.2 allow us to infer that much of this precipitation falls as rain rather than snow. During winter, the ***melting level,*** the altitude in the atmosphere where the temperature equals 0° C, typically ranges between 2000 and 8000 ft (600 and 2500 m) above sea level along the Pacific Coast. The peak altitudes of the mountains along the Coast Range are often at an elevation below the melting level, so that even the mountain peaks receive rain rather than snow. In fact, the locations of the highest mountains along the Coast Range are obvious from the snowfall distributions on Figure 16.2.

After passing over the Coast Range, air next encounters a high, nearly unbroken barrier consisting of the Sierra Nevada of California and the Cascades of Oregon and Washington. The Cascades merge with the Coast Range north of the U.S. border in British Columbia. Air crossing this mountain barrier must rise nearly two miles (3.2 km), and much of the moisture in the atmosphere is condensed out as precipitation during the ascent. Precipitation in the lower elevations falls as rain, while above the melting level at higher elevations, the precipitation falls as snow. Exceptional amounts of snow can fall from a single storm at high elevations. Local forecasters normally predict snowfall in feet, rather than inches.

Figure 16.7 shows the average and extreme values of annual snowfall at several snow gauges located at high elevation sites along the Sierra Nevada and ***Cascades***. In average years, these sites received between 140 and 700 inches (356 and 1775 cm) of snow, in extreme years, 236 and 1122 inches (600 and 2850 cm). Values even larger than this occur in higher terrain and remote areas where no gauge measurements exist.

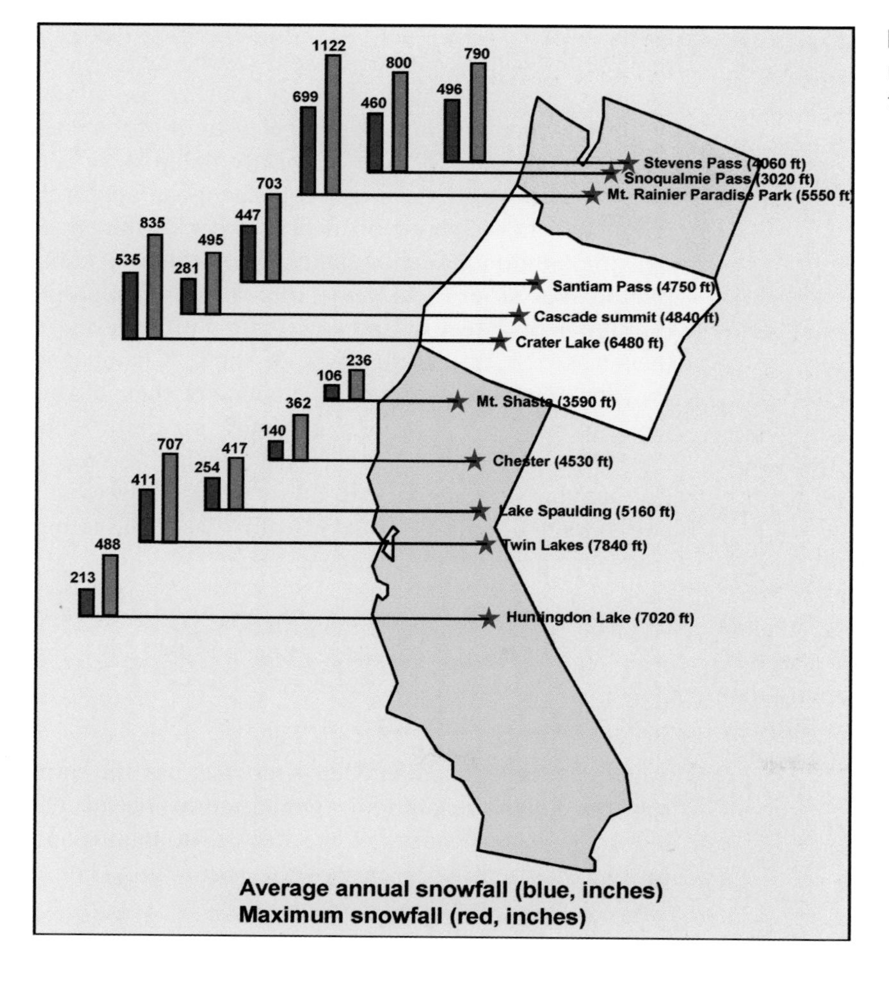

FIGURE 16.7: Average annual (blue) and maximum (red) snowfall in inches at locations in the Sierra Nevada and Cascade Mountains.

Farther to the north in western Canada and southeastern Alaska, so much snow falls on some mountains near the coast that the snow does not melt completely in the summer. In these areas, glaciers slowly transport the unmelted snow to the sea or to lower and warmer elevations.

FOCUS 16.3
The Chain Law

Steady streams of traffic attempt to cross the major mountain ranges of the western United States on highways each winter. During heavy snowstorms, snow piles up so quickly on mountain passes that road crews can't keep the highways clear. To minimize traffic accidents, traffic jams, and stranded vehicles, all western states have a form of the *Chain Law.* The Chain Law requires that individual and commercial vehicles use tire chains or adequate snow tires on mountain passes during heavy snow conditions.

Although each state has slightly different requirements for trucks and automobiles, two levels of the law are generally invoked. The first requires that vehicles use either tire chains or adequate snow tires. A snow tire must have a "mud-and-snow designation" on its sidewall to comply with the law. The second level, invoked during heavier snows, requires that all vehicles use chains. During mountain snowstorms, vehicles are stopped and checked by the highway patrol at the base of the pass. Those that don't comply are sent back down the hill!

The major interior mountain ranges that air encounters on its eastward trek are the Wasatch Range and its northern extension, the ***Bitterroot Mountains,*** and the Rockies, which include the Front, Wind River, and Teton Ranges, and several others. In midwinter, the temperatures along interior ranges are normally colder than those along the Sierra and Cascades. Precipitation amounts are somewhat lower across these ranges than along the coast since moisture has already been precipitated as air passed over the upstream coastal ranges. In Colorado, approximately 200 to 400 inches (~ 500 to 1000 cm) of snow typically falls annually at altitudes between 8000 and 11,000 ft (~2450 and 3350 m), except in the San Juan Mountains in the southwest, which receive somewhat heavier snow, particularly during periods of southwest flow. With southwest flow, air arriving from the Pacific originates west of the Mexican coastline and experiences only minor blocking by upstream topography. Because of the colder temperatures, snow often is composed of crystals that have lower density. The lighter density allows the snow to still be quite deep, even though it has less water mass than snow falling on mountains closer to the Pacific Coast.

The ***water equivalent*** of snowfall is the depth of water that would be obtained if snow is melted. Studies of the water equivalent of snowfall in the mountains of the western United States have found that fresh snowfall typically ranges from 11 to 14 inches of snow per inch of water. The density of freshly fallen snow, the mass per unit volume of water in snow, in fact depends on many environmental conditions. These include the temperature and moisture supply in the clouds, how ice crystals grow and collect each other as they fall through clouds, whether or not the ice crystals encounter supercooled water droplets during their fall, whether or not they partially melt near or on the surface, and the compaction by wind near the surface. All of these factors vary from storm to storm and within snowstorms. As a result, the water equivalent of snow falling in the western mountains can vary widely. At lower elevations, particularly when the surface temperature is near the melting level, the water equivalent can be as low as 4 inches of snow per inch of water. During the winter months, particularly in the higher elevations of the interior mountains, the density of freshly fallen snow can become so low that the water equivalent can approach 25 to 30 inches of snow to 1 inch of water. At these low densities, individual snowflakes are so loosely compacted that skiers can glide rapidly down a slope through 1 or 2 feet of snow creating fluffy waves in their wake. Skiers call this type of snow "Champagne Powder," and Utah ski resorts trademark their champagne powder as "The Greatest Snow on Earth®." In the winter of 2003–04, the National Weather Service in Fairbanks, Alaska, reported an amazing snow water equivalent of 75:1 to 100:1. In this case, shoveling snow is almost like shoveling air!

STORMS ON THE EAST SLOPE OF THE ROCKIES

American and Canadian Meteorologists use the term ***upslope storm*** to describe a winter storm that occurs along the *eastern* slopes of the Rocky Mountains and on the Plains directly east of the mountains. These storms occur with low-level winds that have an easterly component. Although storms on the west slope are also upslope storms, the meteorological community generally does not use the term upslope to describe these storms. The reason probably has to do with the fact that, climatologically, the wind is generally westerly and blows downslope on the east side of the Rockies, so an easterly "upslope" event is less common. Although upslope storms can occur along the east slope of the Rockies from Canada to New Mexico, we will focus on Colorado. In Colorado, the mountains are high and rise abruptly from the plains. In addition, the plains just east of the Rockies support a rapidly growing metropolitan area that extends from Colorado Springs on the south, through Denver, to Fort Collins on the north (see Figure 16.8).

Upslope storms can produce enormous amounts of snow. Nearly 3 ft (~ 1 meter) of snow can fall during a single event in the foothills west of Denver, Colorado; the city itself (which is on the plains east of the Rockies, not in the mountains) can receive 2 ft or more during an event. Severe blizzards along the east slope of the Rockies occur every few years and close down all traffic. Air traffic through the busy Denver International Airport slows or stops, affecting air traffic patterns nationwide.

The Rocky Mountains of Colorado are essentially an unbroken north-south wall that rises about 7000 ft (~2100 meters) above the plains to the east. The plains themselves are about 5000 ft (1500 meters) above sea level, so the mountains reach about 11,500 ft (~3500 m) above sea level, with isolated peaks exceeding 14,000 ft (4267 meters). The plains gently slope downward to the east to the Mississippi River (see Figure 16.3), which is essentially near sea level. Air flowing westward along the surface of the Earth across the plains from the Mississippi to Colorado will rise nearly 5000 ft (1500 meters). When the air reaches the Rockies, it can rapidly ascend this same distance again as it rises up the slopes. In terms of pressure, air rising from the Mississippi to Colorado will decompress from about 1000 to 850 mb. Ascending the slopes of the Rockies, the air may further decompress to 700 mb. All this decompression occurs without the air ever leaving the surface of the Earth! From Chapter 6, we know that decompression of air leads to cooler temperatures, higher humidity, and cloud and precipitation formation. Upslope flow is a very effective cloud and snow generator.

Upslope storms occur when relatively moist, *easterly* winds blow westward across the plains and up the east slope of the Rockies. Two pressure patterns can

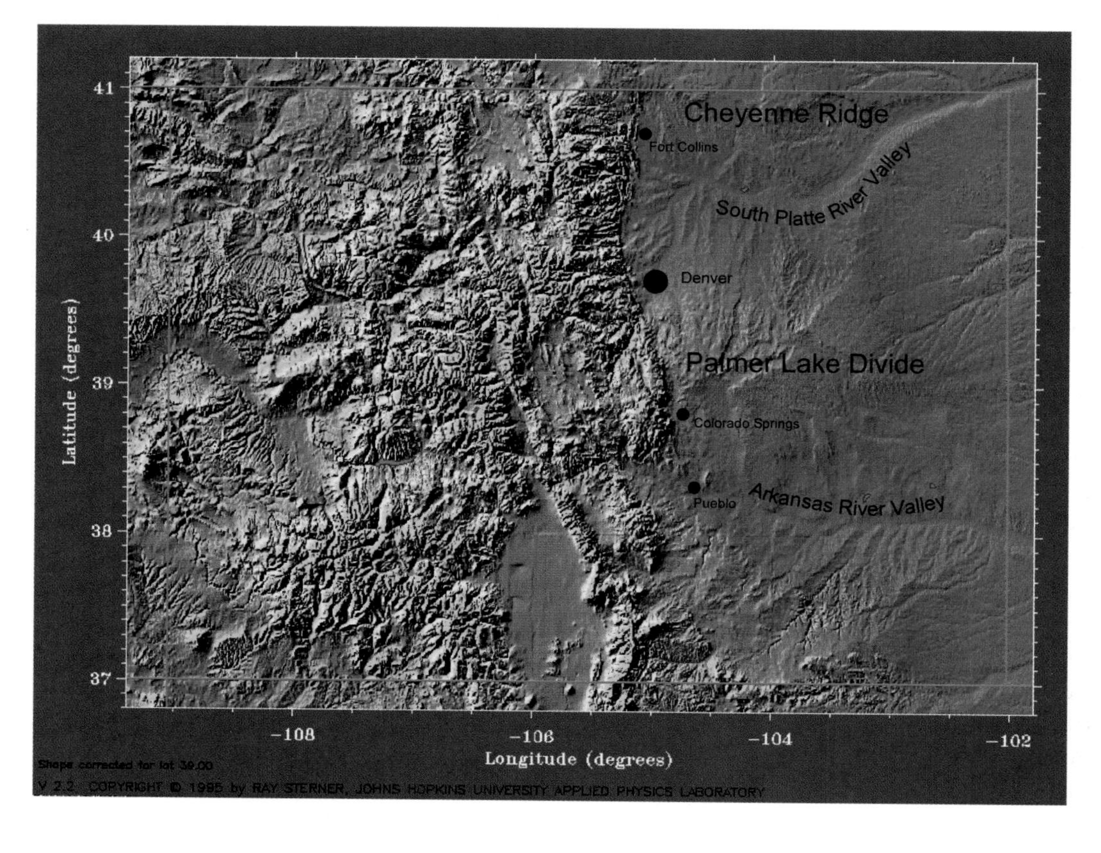

FIGURE 16.8: Geographical features of Colorado.

produce easterly winds, either individually or in tandem. The first pressure pattern that generates upslope winds in Colorado occurs when a high-pressure system is located to the north of the state. Winds circulate clockwise around a high-pressure center, so that south of the high, the winds are easterly. When a high-pressure center is located north of Colorado, air from Canada is brought southwestward into the Colorado mountains. In winter, this air is typically cold and does not contain a significant amount of moisture. Light snow accumulations, typically less than 4 in (~10 cm), occur on the plains and the clouds are typically about 3000 to 6000 ft (~1 to 2 km) deep. From the peaks of the mountains, one can often look down on the cloud deck below.

The second pressure pattern that produces upslope winds occurs when a low-pressure center is located south of Colorado, particularly along the southern Colorado border. Such a low is often called a ***Four Corners low,*** because it passes over the point where Utah, Colorado, New Mexico, and Arizona intersect. When a low occurs in this position, air is drawn northwestward from the Gulf of Mexico into Colorado. This air is warmer than the Canadian air and moisture laden. Often as much as a foot of snow will fall along the plains, and more in the mountains, as a low-pressure center passes south of the state. Because the air from the Gulf is relatively warm and moist, the snow usually has a higher water equivalent than when the upslope motion results from high pressure to the north.

Exceptional blizzards occur when both pressure patterns occur simultaneously (e.g., Figure 16.9), particularly when the low to the south is strong, and the air circulating about the high to the north is very cold. Blizzards are further accentuated when the upslope flow is deep. This occurs when a cutoff low-pressure center is present aloft and moves slowly across the Four-Corners region (see Online 16.1). As illustrated in Figure 16.10, warm air approaching from the Gulf of Mexico must rise, not only over the topography, but also over the cold air circulating southward around the high. Under these circumstances, blizzard conditions on the plains can extend well east of the mountains. Clouds can be as much as 20,000 to 30,000 ft (~6 to 9 km) deep. Snowfall is often reported in feet, the winds are very strong and the temperatures are well below freezing.

An interesting aspect of upslope storms is the variability of the snowfall amounts over short distances. Understanding upslope storms and their variability requires an understanding of the details of the terrain on the plains directly east of the mountains. To the casual observer, the plains east of the Rockies might appear flat, particularly in the presence of the imposing wall of mountains to the west. In reality, the topography of the plains varies substantially (see Figure 16.8). The area between Fort Collins and Denver in the South Platte River Valley is generally at an altitude of 5000 ft (~1500 m). However, north of Fort Collins, along the Colorado Wyoming Border, *perpendicular to the Rocky Mountain chain,* is a line of hills called the Cheyenne Ridge that rise to about 7000 ft (2100 m).

South of Denver, a second line of hills, called the Palmer Lake Divide, also lies perpendicular to the

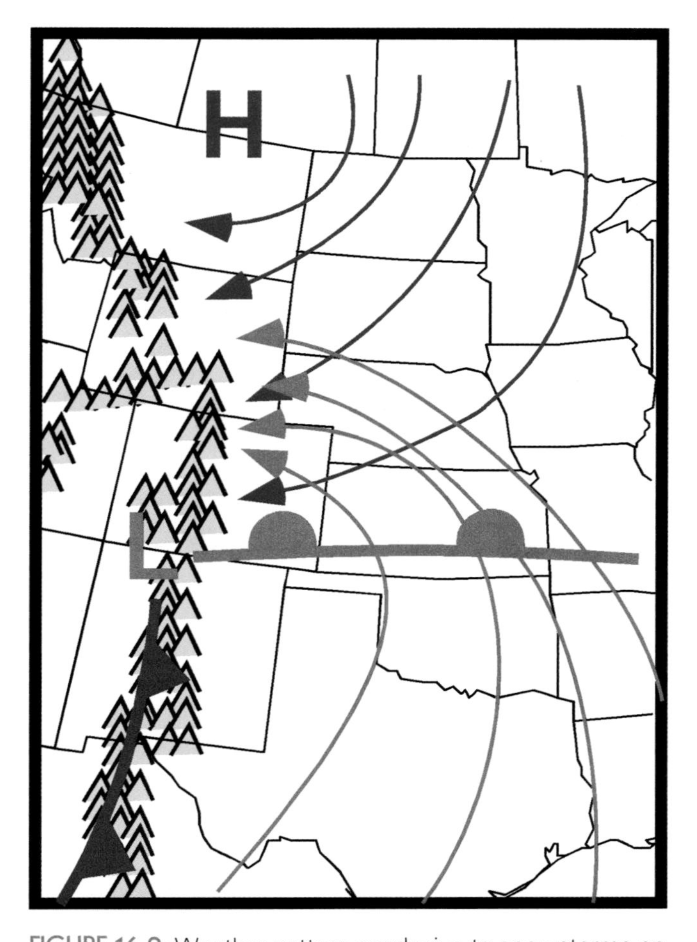

FIGURE 16.9: Weather pattern conducive to snowstorms on the east slope of the Rockies. A low-pressure center in the Four Corners region of Colorado creates a flow pattern that draws moist air northward from the Gulf of Mexico and carries the air upslope across the plains and into the mountains (red arrows). A high-pressure center to the north of Colorado causes dry cold air to flow southward and westward into the mountains (blue arrows). Either pattern alone, high or low pressure, can produce snow along the east slope of the Rockies.

ONLINE 16.1

The Drought-busting Blizzard of 2003

In mid-March of 2003, the snowpack in the Colorado Mountains was well below normal and communities along the Front Range of Colorado were facing an extreme drought with dim prospects for summer water supplies. The period from 17 to 19 March 2003 changed the landscape (Figure 16C). A large cyclone formed over the Colorado Rockies, with a deep cutoff low-pressure center over the Four Corners region of the Southwest. Over the next two days, this cutoff low progressed very slowly eastward (Figure 16Da–d), keeping eastern Colorado under the influence of deep upslope flow for two straight days. During this time, wave after wave of clouds and precipitation moved into eastern Colorado. The net effect: a huge snowfall as illustrated in the snowfall distribution in Figure 16E. Online, you can view the entire storm sequence in infrared satellite imagery.

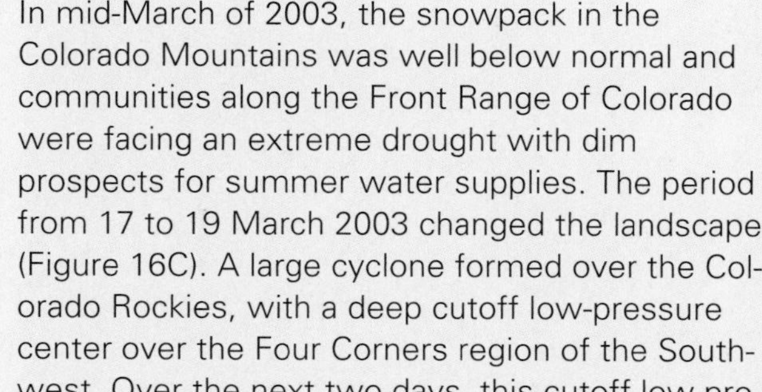

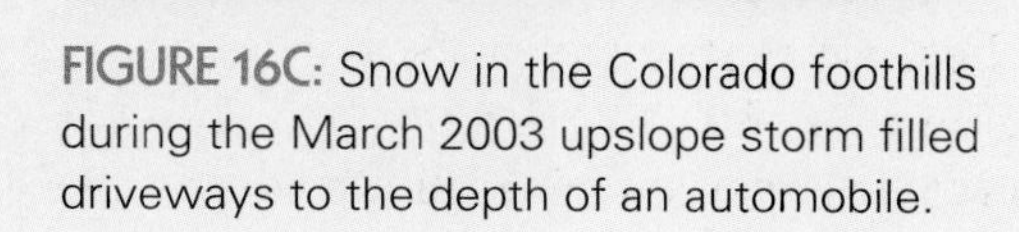

FIGURE 16C: Snow in the Colorado foothills during the March 2003 upslope storm filled driveways to the depth of an automobile.

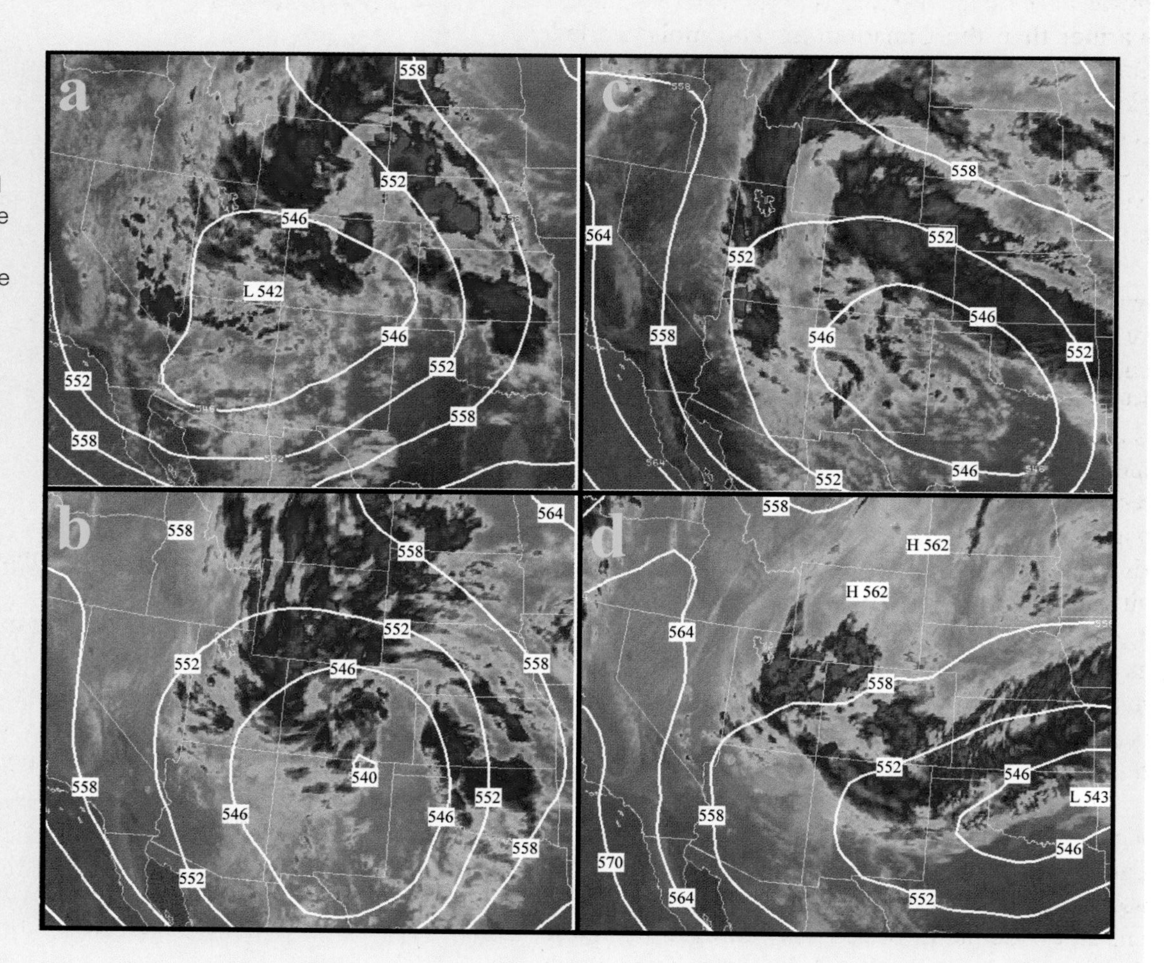

FIGURE 16D: Sequence of 500 mb analyses overlaid on color-enhanced infrared satellite imagery during the Front Range upslope blizzard. (A) 0000 UTC 18 March 2003, (B) 1200 UTC 18 March 2003, (C) 0000 UTC 19 March 2003, (D) 1500 UTC 20 March 2003.

(continued)

(continued)

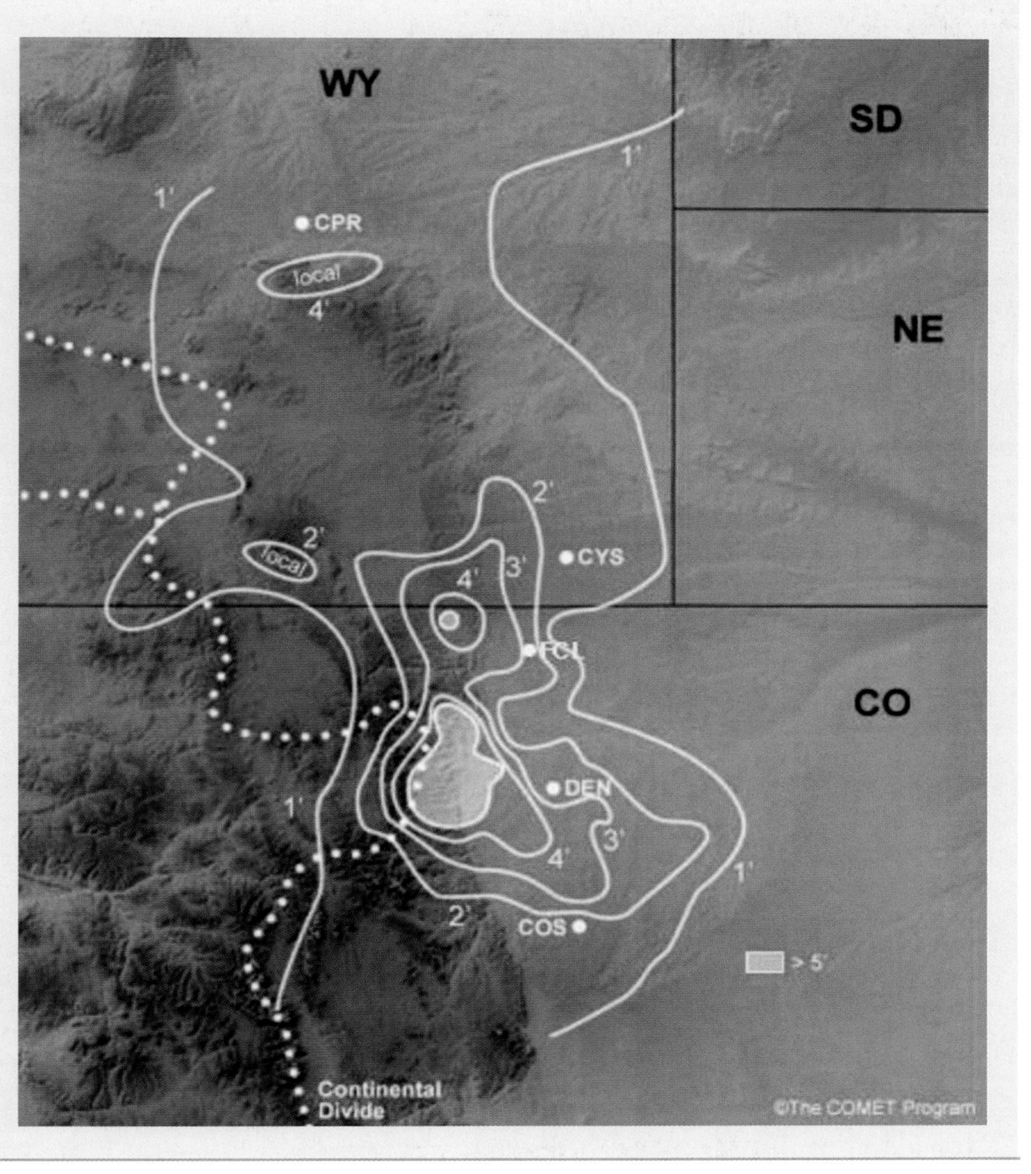

FIGURE 16E: Distribution of snowfall (inches) during the Front Range upslope blizzard of March 2003. Shading denotes area with snowfall in excess of five feet.

Rockies. These two ridges have a tremendous impact on the distribution of snowfall during an upslope event in northern Colorado. The nature of this impact depends on wind direction. From Figure 16.11A, we see that if winds blow from the northeast, air must *descend* the Cheyenne Ridge as it approaches the Rockies. Descending air compresses, warms, and dries. As a result, little or no snow may fall in the northern half of the Denver-Fort Collins corridor. However, as the same air approaches the Palmer Lake Divide, it is forced to rise. In the "wedge" where the Palmer Lake Divide and the Rockies join, air rises very rapidly. Denver (particularly its western and southern suburbs) is right under this rising air. When northeasterly flow continues for 1 to 2 days, 2 to 3 ft of snow can fall locally, even though 50 miles to the north, hardly a flake has fallen. The opposite occurs when the wind direction is from the southeast (Figure 16.11B). Air descends the Palmer Lake Divide, and rises along the Cheyenne Ridge. Under these conditions, northern communities such as Fort Collins will experience heavier snow, while the snowfall in Denver will be reduced.

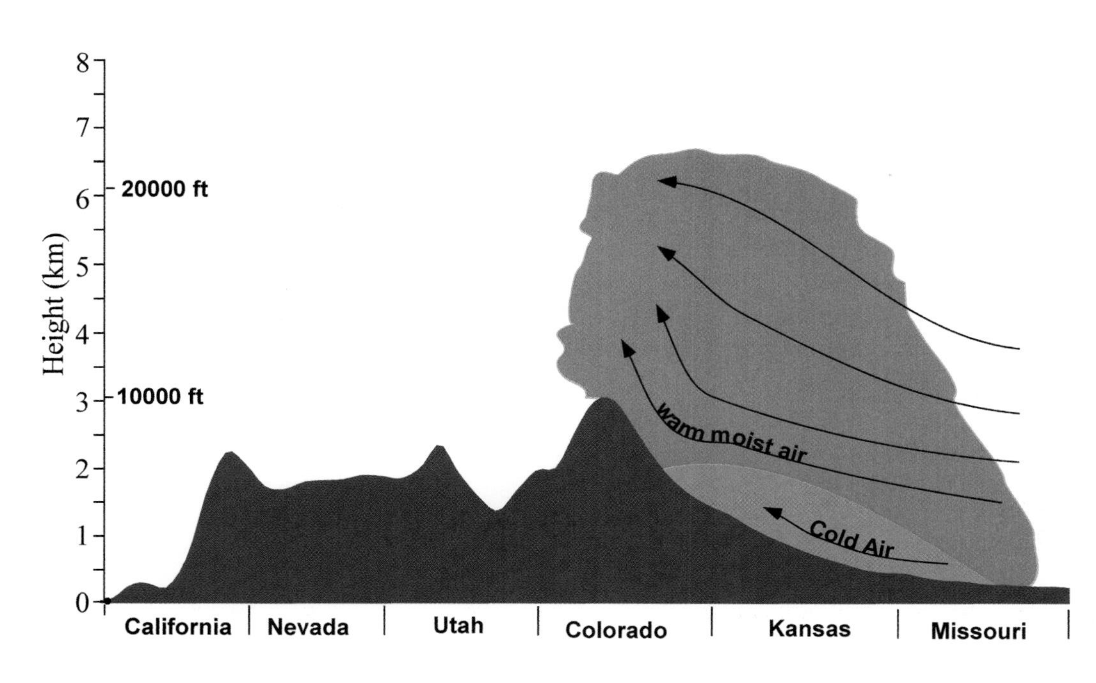

FIGURE 16.10: Profile of an upslope storm. The cold air from the north appears in the low levels. The warm air circulating northward and westward east of the low-pressure system appears in the upper levels. The warm, moist air flows upward over the cold air dome, upward across the High Plains, and up the east side of the Rockies.

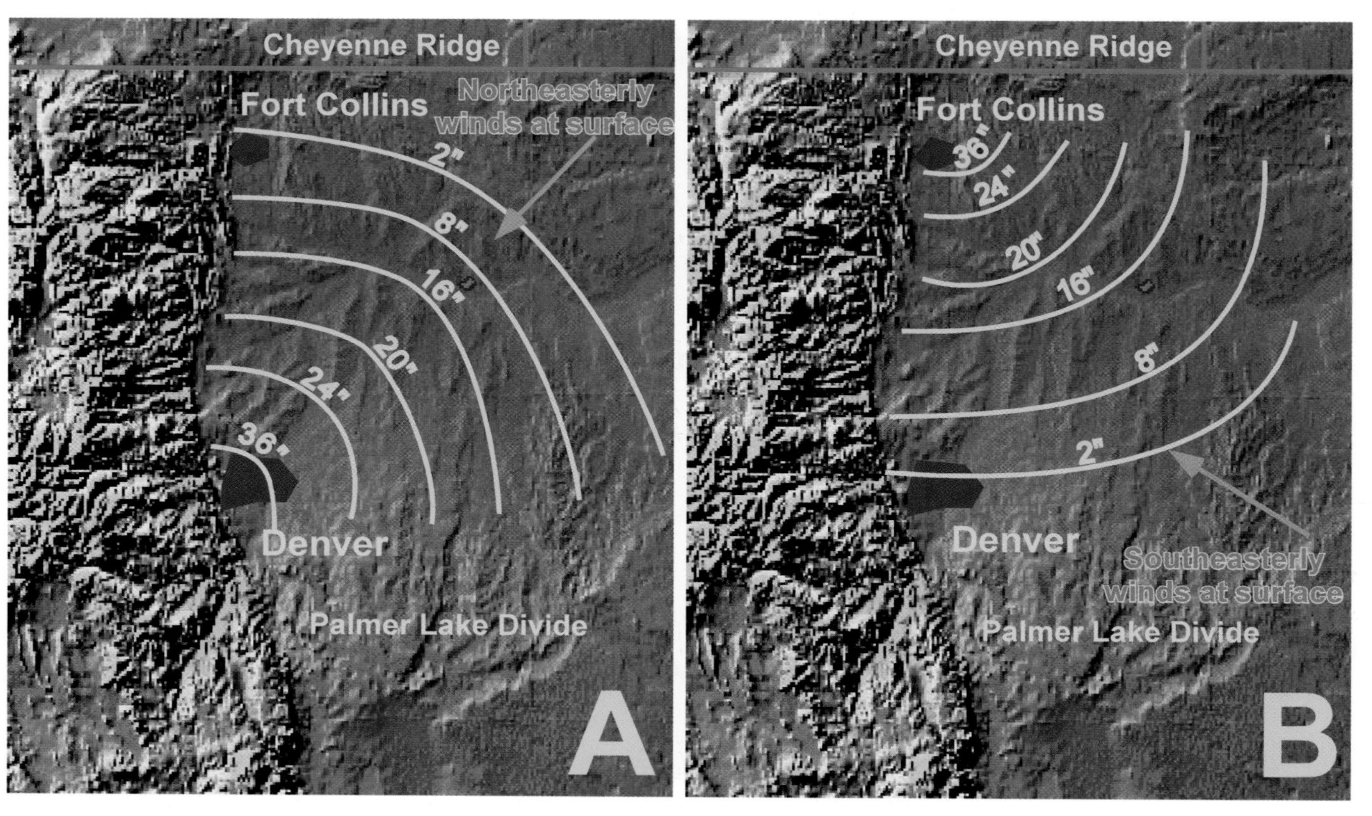

FIGURE 16.11: The snowfall distribution in the Denver-Fort Collins urban corridor changes depending on the wind direction. For northeasterly winds (A), air flows downslope off the Cheyenne Ridge and upslope on the Palmer Lake Divide, favoring heavy snow in the Denver area. With southeasterly winds (B), air flows downslope off the Palmer Lake Divide and upslope at the Cheyenne Ridge, favoring heavier snow in the Fort Collins area.

ONLINE 16.2

Twin Upslope Snowstorms Wreck 2006 Holiday Season in Denver

Denver International Airport, a central hub for holiday season travel, was closed for days in late 2006 as two major blizzards made back to back appearances just before Christmas and again just before New Years. Figures 16F and 16G shows the snow totals from the two storms. The first, on 20–21 December, produced the seventh highest recorded snowfall in Denver. This storm was accompanied by strong surface winds, which led to so much drifting that the airport and the road connecting it to the city were closed for days, canceling over 2000 flights. These storms were the beginning of a near record snow season. With January 2007 as the third snowiest month on record, Denver had sixty-one consecutive days of snow cover—a very unusual event.

Online you can view an animation of the 500 and 850 mb flow patterns for the period of 18 to 31 December 2006. Both December storms had very similar upper-level flow patterns. In each case, a cutoff low developed over the West Coast and moved eastward across the Four Corners region. This pattern led to deep upslope easterly flow over the plains east of the Colorado Rockies, creating the heavy snowfall. Note the cutoff low development over the West Coast and the subsequent movement of each low across the southern Rockies and the Four Corners region. Radar animations for each storm show the evolution of the snowfall as each of the cutoff lows proceeded across the Rockies.

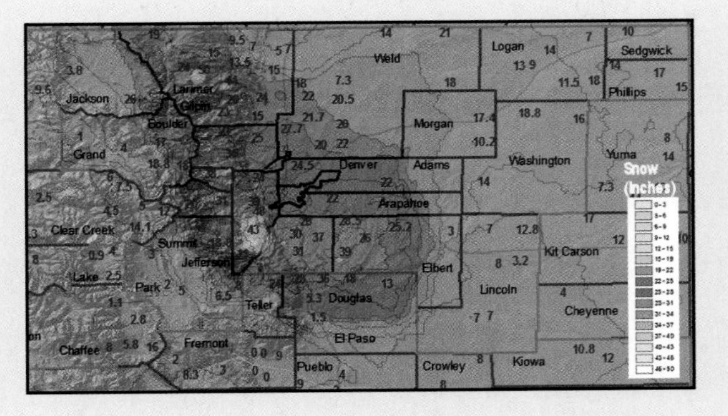

FIGURE 16F: Snowfall amounts (inches) during the blizzard on 20–21 December 2006.

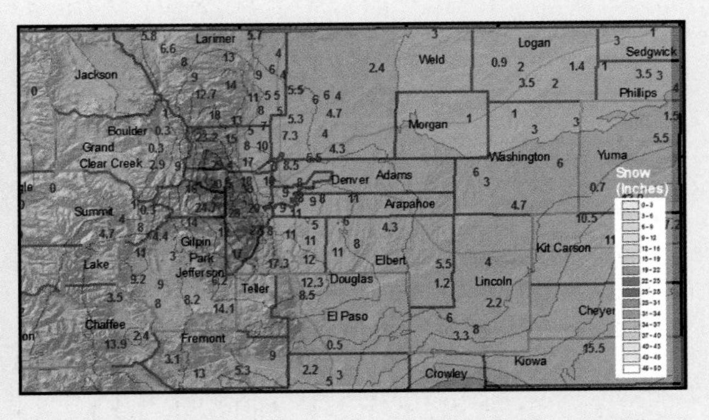

FIGURE 16G: Snowfall amounts (inches) during the blizzard on 28–30 December 2006.

GLOBAL WARMING: MOUNTAIN SNOWS AND DWINDLING WATER SUPPLIES

The snowpack in the mountains of western North America provides water necessary for power generation, agriculture, and urban water supplies. The snow is a natural reservoir, storing water in the winter, and releasing it for use throughout the warm season. Recent assessments of snowpack measurements in the western United States, made over many decades, suggest an alarming trend—the water content of the snowpack, particularly along the coastal mountain ranges, has dropped substantially in response to an 0.8° C (1.4° F) average temperature rise since the 1950s. Of great concern in the West and indeed the country is what may happen to the snowpack in the future if global warming continues at rates predicted by global climate models.

The size and water content of the mountain snowpack is determined by both precipitation and temperature. Global climate models provide mixed information about future precipitation in the mountains—some predict more precipitation in a warmer climate while others predict less, with the average of many models suggesting little change. However, all models consistently predict warmer temperatures over the mountains, with increases ranging from 2 to 7° C (4 to 13° F). These increases in temperature have three direct effects on the snowpack, all bad news for water storage. The first is a change in the time of first snow accumulation in autumn, and first snowmelt in spring. Simply, with warmer temperatures, snows will start to accumulate later in autumn, and begin to melt earlier in spring, reducing the time that the snowpack accumulates.

The second effect of warming temperatures is to decrease the altitude range over which snow falls. Along the coastal ranges, precipitation falls as rain in the lower altitudes and snow at higher altitudes. A warmer climate will be accompanied by an increase in the altitude of the freezing level—in short, more of the mountains, particularly the lower altitudes, will experience rain instead of snow. The third effect of a global temperature increase is on the frequency of the extratropical cyclones that bring the snowstorms to the mountains. As we learned in Chapters 10 and 11, extratropical cyclones develop in response to the temperature differences that develop between the polar and tropical regions of the Earth, particularly in the winter season. Global climate models predict that the temperature contrast between the polar and tropical regions will be reduced during winter, suggesting that the frequency and intensity of extratropical cyclones, particularly in the transition periods in autumn and spring, will be reduced.

The impacts of these future changes in the snowpack should be cause for concern—in the dry summer months, when demand for water is highest, less water will be available. In many years, the demand for water in the West already exceeds supply. The lack of water has the potential to lead to economic disruption. Worse, with the higher temperatures, the drier conditions also have the potential to lead to an increased incidence of catastrophic forest and grass fires (see Chapter 17). Paradoxically, with the precipitation flowing downstream concentrated more in the winter season, the potential for winter flooding is increased (see Chapter 25). For all these reasons, the western United States, with its thirst for water, is one of the most climate sensitive regions of the country.

The problem is not unique to the United States. Worldwide, one-sixth of the Earth's population relies on glaciers and seasonal snowpacks for their water supply. Glaciers and snowpacks have been decreasing in extent for decades worldwide in most mountainous regions. With predicted future global warming, this reduction is expected to continue, with the inevitable decrease in fresh water supplies. Where storage capacities, such as dams and reservoirs, do not exist or are insufficient, summer water supplies may no longer sustain many regions. If global warming predictions are correct, the coming water crisis, tied uniquely to mountain snowstorms, may well be one of the great challenges of society in the coming decades.

CHECK YOUR UNDERSTANDING 16.2

1. Mountain snowfall totals along the Sierra Nevada are typically larger than in the interior mountains. Why?
2. What is an "upslope storm"?
3. Is global warming expected to increase or decrease the snowfall of mountain snowstorms?

TEST YOUR UNDERSTANDING

1. Identify three destructive effects of mountain snowstorms.
2. Why are regions of the West that experience extremely heavy snowfall often areas of significant hydroelectric power generation?
3. What characteristics of mountains influence snowfall?
4. What is an "atmospheric river"?
5. Where do the cyclones that are responsible for mountain snowstorms originate and at what stage in their life cycle do they typically reach the continent?
6. Why are thunderstorms rarely found in cyclones that arrive along the West Coast of North America?
7. What is *orographic lifting?*

8. How would you expect snowfall rates in mountain snowstorms to compare to snowfall rates along a cold front? Support your answer with a brief explanation.

9. Would more snow be expected to fall on the windward or leeward side of a mountain range? Why?

10. How does the melting level impact where snow falls in mountains?

11. What is the "water equivalent value of snow" and why might this value be different for snow falling on interior versus coastal mountains?

12. What factors influence the water equivalent of snowfall in the mountains?

13. The region just downwind of a mountain range is sometimes said to be in a "rain shadow." Based on the precipitation distribution in Figure 16.6, what might this term mean?

14. Describe how the topography of the Great Plains contributes to the amount of snowfall in upslope storms of the Rockies.

15. Where would be the ideal locations for a low-pressure system and a high-pressure system for a heavy snowstorm on the east slope of Colorado's Front Range?

16. What general direction is the surface wind blowing during an upslope storm in eastern Colorado?

17. What surface wind direction favors heavy snowfall in Denver, Colorado?

18. Explain how a small change in wind direction during upslope storms can significantly impact the distribution of snowfall in the urban corridor along the plains bordering the east face of the Colorado Rockies.

19. You decide to earn some extra cash by shoveling snow from your neighbor's sidewalk after every snowfall of the winter. Your neighbor says "I will pay you a dollar for each inch of snowfall that you shovel." What would be a better way to set the pay scale to compensate for the work you actually do?

20. Mountain snowpack water content and depth is determined by what two atmospheric variables?

21. Identify three impacts warmer temperatures may have on snowpack.

TEST YOUR PROBLEM-SOLVING SKILLS

1. The density of water is 1 gram/cubic centimeter and the density of pure ice is 0.917 grams/cubic centimeter. What is the density of snow? To find out, during the next snowfall, take a cardboard tube (e.g., from the inside of a roll of paper towels or bathroom tissue) and calculate the inside volume by measuring its inside diameter (D) and the length (L) of the tube using the equation Volume = $1/4\pi D^2 L$. Take the tube and carefully push it into the snow without compressing the snow entering the tube from the bottom. Fill the tube by pushing it all the way into the snow. Remove the tube, bring it inside, and transfer the snow to a glass container. Melt the snow and then measure its volume (you can use standard kitchen measuring devices). The ratio of the volume of the melt water to the volume of the tube is the snow density. Try this for fluffy dry snow and for wet snow. How do your answers differ?

2. Sacramento, California, is near sea level in the Central Valley upstream of the Sierra Nevada, while Blue Canyon, California, is located in the Sierra Nevada on the windward slopes. A cyclone sweeps through California, with the cyclone's cold front passing by both locations at the same speed. Which location is likely to receive more precipitation and why? What type of precipitation would be most likely at each location? Explain.

3. A strong Pacific storm brings air onshore with a temperature of 41° F (5° C) and a dewpoint of 41° F (5° C). Above what elevation in the coastal mountains would you expect the precipitation to be snow? Explain how you arrived at your answer.

4. What is the ideal location of a surface low-pressure center for "upslope" snow at each of the following locations?
   (a) Cheyenne, Wyoming
   (b) Park City, Utah
   (c) Calgary, Alberta
   (d) Mt. Wilson, California
   (e) Anchorage, Alaska

5. On a particular day, air over Kansas City, Missouri, has a temperature of 59° F (15° C) and a dewpoint of 41° F (15° C). Because of a large high-pressure system to the north, the winds are easterly all the way from Kansas City to Denver, Colorado. By using a topographic map, estimate where you would first encounter snow if you were driving westward from Kansas City. (You may assume that the dewpoint temperature of unsaturated air decreases at about 2° C per kilometer of ascent.)

6. Figure 16.7 shows snowfall amounts in the Sierra Nevada and Cascade Mountains. Examine the correlation between elevation and snowfall amounts:
   (a) Create a line chart that displays annual average snowfall versus elevation.
   (b) On the same chart, plot the maximum snowfall amount versus elevation.
   (c) For each set of data, calculate the linear trend line. Plot each line on the chart and label them appropriately.
   (d) Briefly explain your findings and include a hypothesis for what you observe.

USE THE SEVERE AND HAZARDOUS WEATHER WEBSITE

http://severewx.atmos.uiuc.edu

1. Navigate the *Severe and Hazardous Weather Website* to the Western Regional Climate Center (WRCC) at the Desert Research Institute of the University of Nevada. This site contains snowfall data from snow gauges throughout the western United States. The site also has detailed precipitation maps from each state and climatological maps that illustrate how precipitation amounts depart from normal. Examine data from the WRCC site for the most recent winter. (The Standardized Precipitation Index is a good place to start.)
   (a) For each of the western states, determine if the last year was wetter or drier than normal. Record the anomalies (departures from normal) in a table.
   (b) Examine whether temperature anomalies for the past winter were positive or negative for each state. Record this data in the table.
   (c) Do the precipitation and temperature anomalies tend to have the same sign? Provide a plausible reason why or why not.
   (d) Compare the data for each state. Based on your knowledge of physical and political geography of each state, identify which states are likely to have a surplus of water in the summer and which states are likely to have a deficit of water.

2. During the winter season, use the *Severe and Hazardous Weather Website* to examine infrared satellite imagery when a cyclone is approaching the West Coast of the United States.
   (a) Over a period of several days note the location and structure of clouds associated with the cyclone by outlining the extent of cloud cover on a blank map of the United States (see Appendix B). Record the date next to each outline.
   (b) During the same time period, examine the listing of current winter storm watches and warnings available from the National Weather Service and the Storm Prediction Center. Save the images or text listing, or record the type and location of each watch or warning.

(c) After the storm passes, locate data on actual snowfall totals. Compare the elevation with the snow totals observed and describe your findings.
(d) Compare the snowfall totals with the duration of cloud cover the region experienced. Discuss how the duration and extent of cloud cover may have influenced snowfall amounts.
(e) Find the locations that were under winter storm warnings or watches. How much snow did these areas actually receive? In retrospect, were the watches or warnings justified?

3. Relatively few highways traverse the mountains of the West. Examine highway maps of Colorado to determine the location of the major mountain passes in that state.
   (a) During the passage of the next winter storm, use the *Severe and Hazardous Weather Website* "Current Weather" to identify the regions of the state that are experiencing heavy snowfall. Record the date, time, and location of heavy precipitation.
   (b) During this event, examine the National Weather Service website to determine if watches and/or warnings have been issued for the region. Record the type of watch or warning, the region for which it is valid and any other identifying information.
   (c) Navigate to the Colorado Department of Transportation website to see how the local government is responding to the weather. Has the Chain Law been applied to the affected passes? Are there restrictions on travel?

4. Fronts are modified significantly as they move into the Sierra Nevada from the Pacific and subsequently pass across the plateau between the Sierra Nevada and the Rocky Mountains. During the next winter storm event, navigate to the *Severe and Hazardous Weather Website* to examine surface and low-level troposphere conditions to determine how fronts are modified by topography.
   (a) Identify a cold front using surface data and satellite imagery. Record the surface temperatures on either side of the front at least four times as it passes from the Pacific Coast to the Great Plains.
   (b) For the same times as in (a), examine an 850 mb map and a 700 mb map. Are the wind and temperature data provided on the low-level tropospheric maps helpful in identifying the front? Why or why not?
   (c) Briefly summarize the changes in intensity of the front as it passes west to east. Include in your summary changes in temperature gradient, wind shifts and any other identifying data.

CHAPTER SEVENTEEN

MOUNTAIN WINDSTORMS

Chinook wall spans the Colorado Rockies during a downslope windstorm.

KEY TERMS

altocumulus lenticularis
bora
breaking waves
California Norther
Chinook
Chinook wall
clear air turbulence
downslope windstorm
foehn
ground blizzard
hydraulic jump
inversion
katabatic wind
lee trough
lee wave
leeward side
rotor
Santa Ana
shooting flow
turbulent eddies
upwelling
windward side

OTHER WINDSTORMS:

Afganet
Berg
Elvegust
Germich
Helm
Ibe
Kachchan
Mistral
N'aschi
Nor'Wester
Sno
Puelche
Vardarac
Zonda

Downslope windstorms occur along many of the major mountain ranges of the world. Local names for these windstorms include the ***Chinook*** of the Rockies, the ***Foehn*** of the Alps, the ***Bora*** of the Dinaric Alps of Croatia, the ***Zonda*** and ***Puelche*** of the Andes, the ***Nor'Wester*** of New Zealand, the ***Santa Ana*** of Southern California, the ***Berg*** of South Africa, the ***Kachchan*** of Ceylon, the ***Afganet*** and ***Ibe*** of central Asia, the ***Germich*** of the Caspian Sea, the ***Elvegust*** and ***Sno*** of Norway, the ***Mistral*** of Southern France, the ***N'aschi*** of the Persian Gulf, the ***Vardarac*** of the northern Aegean Sea, and the ***Helm*** of the northern Pennines.

Hurricane force winds at the base of the mountains can persist for a day or more and cause damage to buildings, trees, and other structures. Early documentation of downslope windstorms dates to the Roman Empire. During the Roman incursions into northern Europe, Roman scribes noted that the winds brought warmth to the northern slopes of the Alps. Today warm, dry windstorms of the Alps are called ***foehn*** winds, a name that may have been derived from the Latin ***favonius***, or west wind. Cold downslope windstorms also occur in Europe, particularly along the Adriatic Sea just southeast of the Dinaric Alps of Yugoslavia, where they are called the ***bora*** after the North Wind of Greek mythology, ***Boreas***. Meteorologists now use the terms **foehn** and **bora** to describe warm and cold downslope windstorms in other regions of the world.

In North America, downslope windstorms occur along the east slope of the Rocky Mountains from Alberta to New Mexico, along the east slope of the Sierra Nevada and Cascade Ranges in the western United States, and on the west slope of San Bernardino, Santa Ana, and San Gabriel Mountains and the Sierra Nevada of California (see Figure 17.1). Downslope windstorms also occur on the leeward side of the Alaskan and Chugach mountain ranges in Alaska. In addition, downslope windstorms occur at the base of high-altitude icefields in Alaska, Greenland, and Antarctica, although these winds originate from different processes.

Downslope windstorms along the east slope of the Rocky Mountains are often called Chinooks, a name derived from legends of the Chinook Native American tribe of the Pacific Northwest. The Chinook is normally a warm, dry wind, similar to the Alpine foehn. Chinook winds, which can be violent, are always

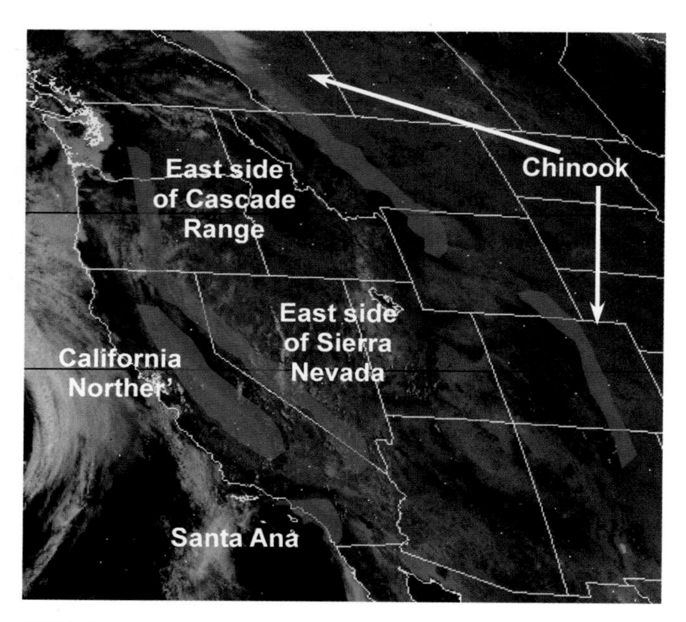

FIGURE 17.1: Locations of downslope windstorms in western North America.

extremely gusty and occur almost every year, particularly in late fall and winter. Wind gusts may exceed 100 knots (115 mph) in the worst storms. The largest metropolitan area affected by Chinooks extends from north to south along the plains of eastern Colorado from Fort Collins to Colorado Springs, and includes the cities of Denver and Boulder. The worst downslope winds are experienced in Boulder, which is located directly at the base of the mountains and only 30 km (20 miles) from the Continental Divide, the crest of the Rockies.

Figure 17.2 shows an example of wind speed measurements taken at the National Center for Atmospheric Research in Boulder during a Chinook that occurred on 16 to 17 January 1982. Downslope winds first reached the surface on 16 January one hour before midnight. Just after midnight, the winds hit with full force. Note the gustiness on this wind trace—at some times the winds varied from over 100 to less than 10 miles per hour and back again in less than a minute. Strong Chinooks such as the January 1982 storm produce widespread roof damage, overturn mobile homes and light aircraft, blow trucks off highways, damage weak structures, and tear down utility poles and trees. Damage in Boulder from a windstorm of this strength can exceed several million dollars. Pressure fluctuations associated with the gusty winds are unpleasant, and produce physiological and psychological reactions in humans such as headaches and stress. Similar damage and human

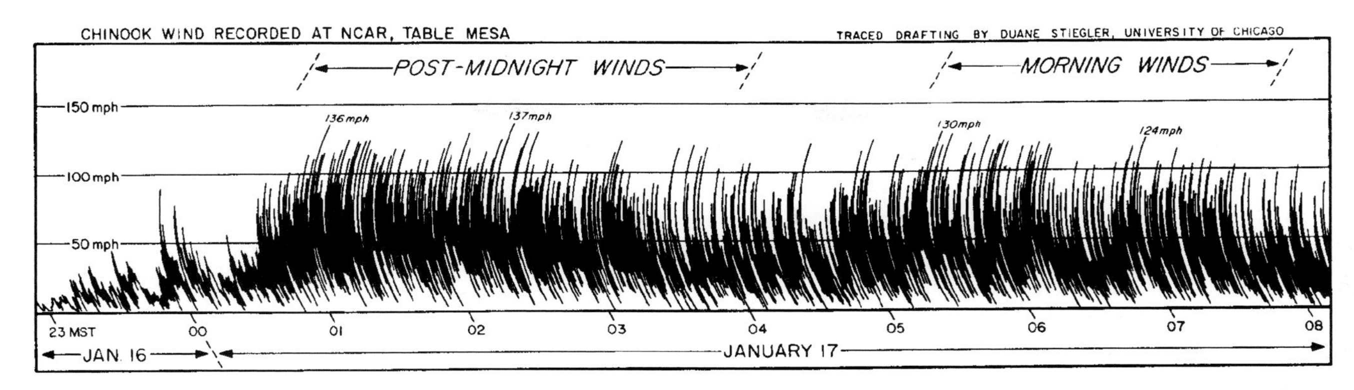

FIGURE 17.2: Wind speed measured at the National Center for Atmospheric Research located in Boulder, Colorado, on 16 to 17 January 1982 during a downslope windstorm.

impacts occur with mountain windstorms in other mountainous regions of the world. Near Zermatt in the Swiss Alps, for example, chalet owners in Foehn regions often place large rocks on their roofs to keep them from blowing off during downslope winds.

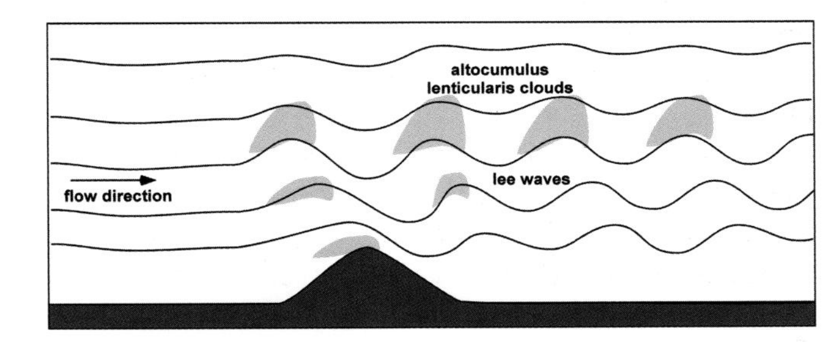

FIGURE 17.3: A series of lee waves downwind of a mountain range. The clouds forming at the crest of each wave are called lenticular clouds because of their characteristic lens shape. The lines are streamlines, which represent the path of the airflow.

THE DYNAMICS OF DOWNSLOPE WINDSTORMS

Mountain windstorms can arise in different ways. We will consider first dynamically driven winds, that is, winds driven by strong pressure gradients that develop across mountain ranges. Mountains act as a barrier to airflow. As air approaches a mountain range under the influence of a strong cross-mountain pressure gradient, it must rise on the ***windward side*** and descend on the ***leeward side***. The mountain creates a wave in the flow, much like waves commonly seen in water. Often a series of waves will form downstream of mountains as the air continues to flow downstream as shown in Figure 17.3. These waves, called ***lee waves***, are commonly observed east of the Rockies. The crests of these waves are often marked by elegant lens-shaped clouds called ***altocumulus lenticularis*** (e.g., Figures 17.3, 17.4 and 17A). The character, intensity, shape, and steadiness of the mountain wave changes depending on the stability of the air and the strength of the winds upstream of the mountains.

When the cross-mountain flow near the mountaintop is strong, and an ***inversion*** is present upstream at a level just above the mountaintop, the mountain wave pattern will take a form called a ***hydraulic jump***. Recall from Chapter 6 that a layer of air that is stable, such as an inversion, will inhibit vertical motions. Any air rising into an inversion will cool more quickly than the surrounding environment and be forced back downward. The effect then is that the inversion acts like a "lid" on vertical motions (as was discussed in Chapter 1).

FIGURE 17.4: Lenticular cloud east of the Colorado Rockies.

Figure 17.5 shows a simplified schematic diagram of a hydraulic jump. The solid lines on the diagram are streamlines, marking the path of the airflow. The

ONLINE 17.1

Wave Cloud over the Western Plains

An extensive wave cloud developed just east of the Colorado Rockies on 8 January 2002. A photograph of this cloud (Figure 17A), taken from the Campus of the University of Northern Colorado at sunset, illustrates many features of wave clouds. The smaller lens-shaped features are called lenticular clouds. In the photograph, these clouds mark the crests of small waves in the flow. The broader cloud over the small lenticulars developed as part of a much larger wave in the flow. The extent of this larger wave can be envisioned by examining an Online animation of visible satellite images.

FIGURE 17A: View of the 8 January 2002 wave cloud looking south-southwest from Greeley, Colorado, near sunset.

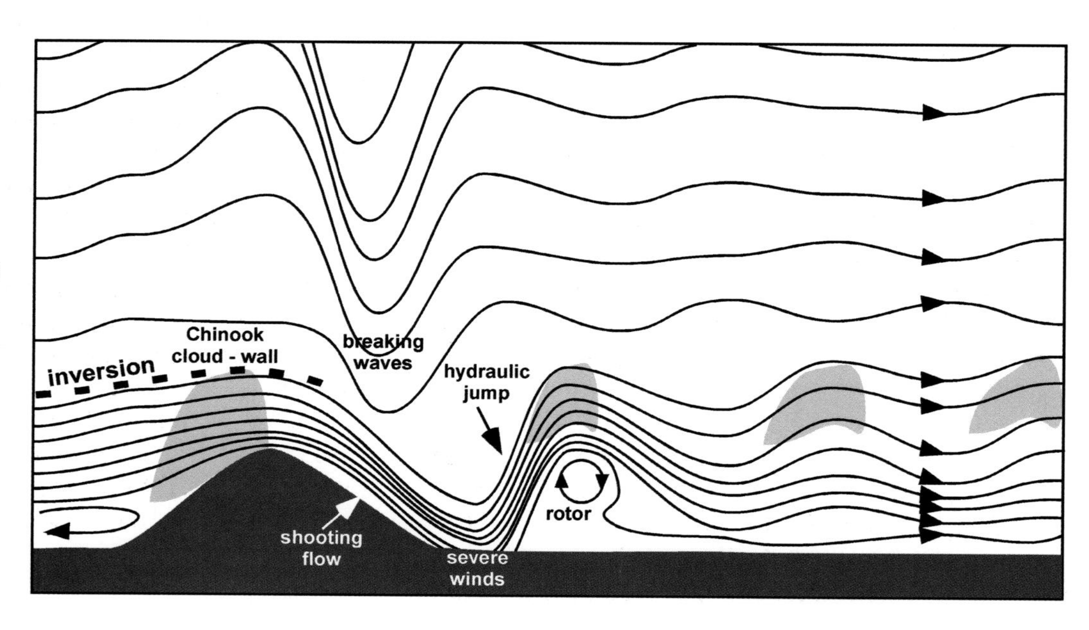

FIGURE 17.5: Schematic of a hydraulic jump located east of a mountain range. The solid lines are streamlines, which mark the path of the airflow. The airflow is fast where the streamlines are packed together, and slow where they are far apart. Large waves, and wave breaking, occur above the shooting flow and cause turbulence. Turbulence, and sometimes a rotor, also appear downstream of the jump.

airflow is fast where the streamlines are close together, and slow where they are far apart. Air passing over the mountain between the inversion level and mountaintop accelerates downslope in a ***shooting flow*** toward the base of the mountain. The strongest winds in downslope wind storms occur as the shooting flow reaches the mountain base.

The shooting flow develops as a result of the large volume of air that must pass between the inversion and mountaintop, the strong pressure gradient across the mountain, and the wave motion in the flow induced by the mountain itself. Near or just beyond the mountain base, the flow abruptly rises (the jump), and becomes turbulent as it adjusts within the deeper layer over the plains downstream of the mountain. Numerical simulations of downslope windstorms suggest that the jump can be quite sharp, with vertical air velocities in the jump region sometimes exceeding 25 meters/second (56 mph). The smoothness of the streamlines in Figure 17.5 masks the extreme turbulence that accompanies these storms.

As air above the inversion spills over the mountain, waves are created in the flow. These waves can become quite large and often break, in a manner similar to ocean waves when they break upon approaching the shoreline. In the atmosphere, ***breaking waves*** create turbulence, much like the smooth waves approaching a shore break and become a jumble of foam and rolling water. These ***turbulent eddies***, small-scale fluctuations and rolls in the flow that are associated with breaking waves, modify the shape and channel of the shooting flow, causing that flow to fluctuate in intensity, creating some of the observed gustiness.

Between the mountain and the jump, or east of the jump, areas of rotation called ***rotors*** can develop. Aircraft passing through these rotors or other regions of turbulent flow encounter ***clear air turbulence***. In downslope flows, such extreme turbulence can be encountered that injuries can occur to passengers on large aircraft and lighter aircraft can experience loss of control. The turbulence can extend high into the troposphere because of wave breaking and strong rising and descending motions associated with the waves.

Changes in the upstream wind speed and the height and strength of the inversion continually influence the character of a downslope windstorm. The position and intensity of the jump is closely tied to the values of these parameters. The jump may at times move over the foothills, so that the winds on the plains will abruptly calm. Later, the jump may move back on the plains, with the winds appearing again with full force.

CHECK YOUR UNDERSTANDING 17.1

1. Where do mountain windstorms occur in North America?
2. What maximum speeds can downslope windstorms attain?
3. Where is the hydraulic jump relative to a mountain during a downslope windstorm?

CHINOOK WINDSTORMS OF THE ROCKY MOUNTAINS

Three primary factors, summarized in Figure 17.6, determine the final temperature of the descending air as it arrives on the plains. The first is the temperature of the air upstream of the mountains. In Europe, air comprising the northern Alpine foehn originates over the Mediterranean Sea, flows northward, and has moderate temperatures prior to ascending the mountains. Air comprising the bora, on the other hand, originates as cold air over east central Europe and flows southward over the Dinaric Alps. In North America, the situation is complicated because the mountains run north-to-south. Air upstream of the Rockies can be warm when it originates over the Pacific, or cold when it originates over western Canada and Alaska. Chinooks east of the Rockies, therefore, can be warm and dry, similar to the foehn, or may be cold and dry, similar to the bora.

The second factor affecting the final temperature of the air on the plains is latent heat release in the clouds upstream. When air approaching the mountains from the west is moist, clouds will form during the ascent on the windward side. These clouds are often sufficiently deep to produce heavy snow on the west side of the mountains. Latent heat is released into the air warming the airstream as condensation occurs on the windward side. Of course, when evaporation occurs on the leeward side, latent heat will be reabsorbed, cooling the airstream. However, if precipitation

FOCUS 17.1

T-REX and the Hunt for Rotors

In the winter of 1951–52, the U.S. Air Force funded the Sierra Wave Project, a major meteorological study in the Owens Valley to the lee of the Sierra Nevada, to investigate turbulence above the mountains that posed hazards to military aircraft. In this project, instrumented gliders were flown through the lee waves generated by the Sierra Nevada. Sierra Wave Project researchers were the first to observe a dangerous atmospheric phenomenon closely related to these waves—the rotor (see Figure 17.5), an intense vertical whirlwind of air on the mountain lee side, located underneath the crest of the mountain wave paralleling the ridge.

In the spring of 2006, an international field campaign involving approximately sixty scientists, called the Terrain-induced Rotor Experiment, or T-REX, went back to the Owens Valley with a sophisticated array of equipment to study the properties of rotors and mountain waves once again. The project was the first deployment of the nation's newest and most advanced meteorological research aircraft (Figure 17B). This aircraft, along with other aircraft, a network of radars, lidars (a radar-like device that uses laser light), weather stations, wind profilers, and weather balloons probed the atmosphere downwind of the mountains in search of rotors and waves to try to understand their properties and better predict their dangerous flows.

To see these flow features yourself, stare at the stereo pictures in Figure 17C. If you cross your eyes just right, you will see three pictures. The "photo" in the middle will be three-dimensional. The long cloud extending into the picture on the left formed as air moved over the crest of the Sierra Nevada. The long cloud on the right is a rotor cloud, which forms in the upper part of the rotor circulation downwind of the mountain wave. Note the raggedness of the cloud, a mark of the extreme turbulence in the airflow. The T-REX campaign goal is better understanding and prediction of aviation hazards that exist in the vicinity of the world's mountain ranges. Scientists today are unraveling the large quantity of data collected during T-REX.

FIGURE 17B: The National Science Foundation/National Center for Atmospheric Research High Performance Instrumented Airborne Platform for Environmental Research (HAIPER) aircraft used in the T-REX campaign.

FIGURE 17C: Stereo images of a cloud formed as air flowed over the Sierra Nevada (cloud on the left) and a rotor cloud (cloud on the right) observed during the T-REX campaign. To see the picture in three dimensions, cross your eyes so that three pictures appear side by side. The middle picture will be three-dimensional.

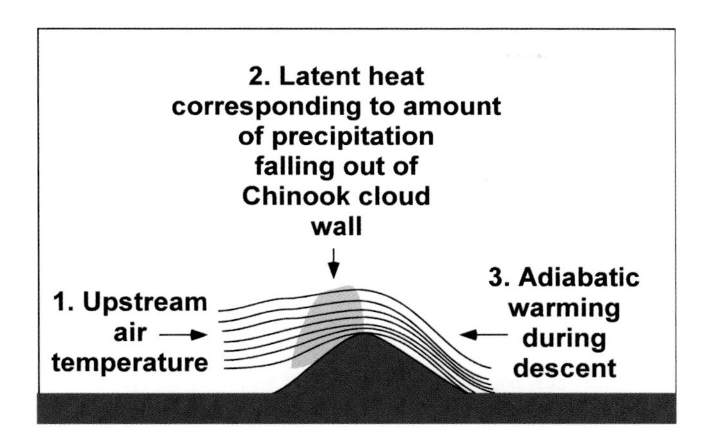

FIGURE 17.6: Three factors determining the temperature of air in downslope windstorms: upstream air temperature, latent heat release, and adiabatic descent.

FIGURE 17.7: The Chinook wall, a wall of clouds that follows the mountain crest, is often present during downslope windstorms.

falls out of the cloud, less water will be available to evaporate on the lee side than was condensed on the windward side.

Under these conditions, air descending the lee side will be warmer than air at the same altitude on the windward side. Clouds normally develop on the windward slopes of the Rockies during Chinooks, and blinding snow often occurs near the mountain crests. During the descent on the east side, the clouds rapidly evaporate so that, viewed from the east side, the clouds along the mountain crest appear to form a wall (see Figure 17.7 and Chapter cover photograph). Meteorologists call this cloud a ***Chinook wall***.

The third factor that determines the temperature of air arriving on the plains is adiabatic warming during descent. Recall that when air descends, it warms at the dry adiabatic lapse rate, approximately 10° C/km. Air originating near mountaintop, as it descends from the crest of the Rockies to the plains, drops about 1.5 to 2.0 km (~4900 to 6600 ft), and warms 27° F to 36° F (15° C to 20° C). With this warming comes substantial drying of the air.

As discussed in Chapter 1, the relative humidity of air decreases as temperature rises. Air reaching the base of the mountains can have relative humidities lower than 10 percent. If snow cover is present on the ground, the snow will quickly sublimate. The rapid decrease in snow cover experienced during some Chinooks calls to mind the Native American legend of Chinook Wind's son, who could melt ice and snow.

Just how the residents of the plains perceive the temperature of Chinook winds depends on conditions prior to their onset. Often in winter, a shallow layer of very cold air associated with an arctic airmass may reside over the plains. When the Chinook begins, this air may be flushed eastward, resulting in a dramatic temperature rise. Studies of windstorms in Boulder, Colorado, have shown that the temperature at the onset of some Chinooks can rise as much as 27° F to 36° F (15° C to 20° C). On the other hand, if warm air resides on the plains at the onset of a Chinook, and the air originating west of the mountains is cold, the temperature at the onset of a Chinook can actually drop. In some cases in Boulder, the temperature has dropped as much as 27° F (15° C). The Chinook therefore can have characteristics similar to either foehn or bora winds.

FORECASTING CHINOOKS

Chinooks occur with several weather patterns, but virtually all of these patterns have three common features. The first is a stable layer upstream of the mountain a kilometer or two above the mountaintop. This stable layer typically appears as an inversion (see Figure 17.5), and acts as a flexible lid on the flow crossing the mountain beneath it. It is important in creating the wave structure above the inversion and in forcing the shooting flow and hydraulic jump on the lee side of the mountain.

The second feature of the atmosphere is a strong surface-pressure gradient across the mountain. Strong pressure gradients occur most often when a surface high-pressure system is located to the west over the Pacific or Great Basin regions, and a low-pressure center or trough of low pressure is present on the plains. Finally, the background air flow at levels near mountaintop (~ 700 mb for the Rockies) is typically

FOCUS 17.2

Rapid Temperature Fluctuations

Chinook winds can produce rapid temperature fluctuations. When warm Chinook winds descend a mountain and encounter a cold air mass on the plains, the cold airmass can be displaced away from the mountain, resulting in a rapid temperature rise. As the mountain flow evolves, the position of the hydraulic jump can sometimes move westward back into the foothills, allowing the cold air to return on the plains. Residents then experience a sharp temperature drop (see Figure 17D).

World-record temperature fluctuations are believed to have been caused in this way in 1943 in Spearfish, South Dakota, just east of the Black Hills, where the temperature rose 49° F (27° C) in two minutes. On the same day in Rapid City, South Dakota, the temperature rose from –4° F to 54° F (–20° C to 12° C) over four hours, dropped to 11° F (–12° C) about an hour later, then rose to 55° F (13° C) fifteen minutes later. Less dramatic temperature fluctuations are observed in cities in Colorado's Front Range communities when a shallow arctic air mass is present over the plains prior to the onset of Chinook winds. Temperature changes of 18° F (10° C) in less than a minute have been observed in other parts of the world, such as the lee side of the Norwegian coastal mountains, resulting in damage to some types of vegetation.

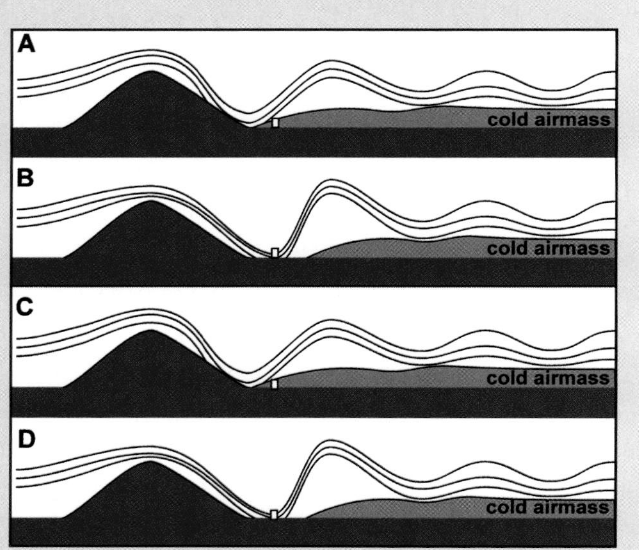

FIGURE 17D: (A) During some downslope windstorms, arctic air is located on the plains prior to the onset of downslope winds. (B) The downslope flow displaces the cold air eastward as the mountain wave and hydraulic jump move out onto the plains. (C) In some cases, the wave may later retreat into the foothills, allowing cold air to return, (D) and then move back to the plains, driving the cold air eastward once again. A city located at the box would experience rapid temperature fluctuations as the warm descending air and cold arctic air moved in and out of the city.

strong and westerly. However, a strong jetstream in the upper troposphere is not required. In fact, a strong jetstream in the upper troposphere may actually impede the development of downslope windstorms because it disrupts the wave structures necessary to support downslope winds.

As a way of illustration, let us examine the weather pattern for the Chinook windstorm illustrated in Figure 17.2. Figure 17.8 shows the 1200 UTC (5 A.M. Mountain Standard Time) 17 January 1982 surface temperature and pressure map. A 130 mph gust was recorded in Boulder, Colorado, about twenty minutes after the time of this map. Note the strong pressure gradient across the Rockies between the high-pressure center over Utah and the low-pressure trough extending from Canada through New Mexico just east of the Rockies. This trough, called a ***lee trough***, exists in part because of the warming of air as it descends the east side of the mountains.

This warming is apparent in the surface temperature field in Figure 17.8. Prior to the onset of the strong downslope winds, an arctic airmass had covered the continent east of the Rockies. Morning temperatures were about 23° F (–5° C) as far south as the Gulf Coast. Frigid temperatures (<0° F), –18° C) were recorded in the interior of the continent. With the onset of the Chinook winds, a narrow band of exceptionally warm air appeared along the east side of the Rockies where the downslope winds reached the surface. Denver's morning temperature was a balmy 52° F (11° C) a sharp contrast to 18° F (–7° C) temperatures just east of the Colorado border in western Kansas.

The 700 mb chart for 1200 UTC 17 January 1982 (Figure 17.9) shows an exceptionally strong gradient in the 700 mb height field between the northern and southern borders of the United States. The 700 mb level corresponds approximately to mountaintop. The

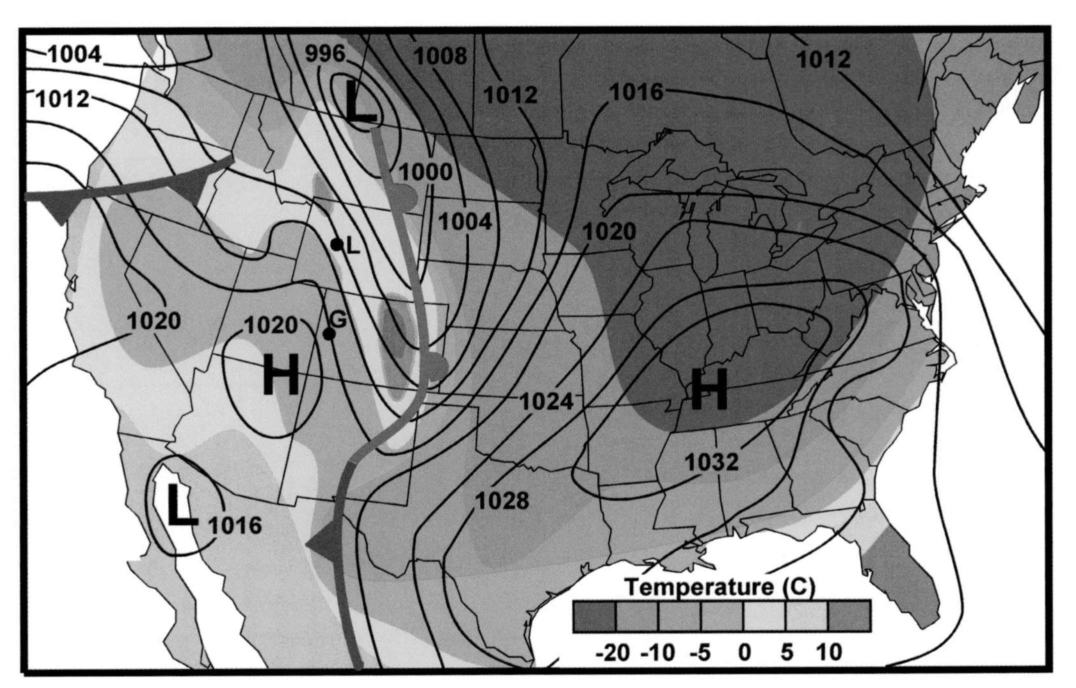

FIGURE 17.8: The surface temperature (°C) and sea-level pressure field (millibars) at 1200 UTC (5 A.M. Local Mountain Time) on 17 January 1982 during the Chinook winds illustrated in Figure 17.2. The points marked G and L are the locations of the Grand Junction, Colorado and Lander, Wyoming, soundings appearing in Figure 17.10.

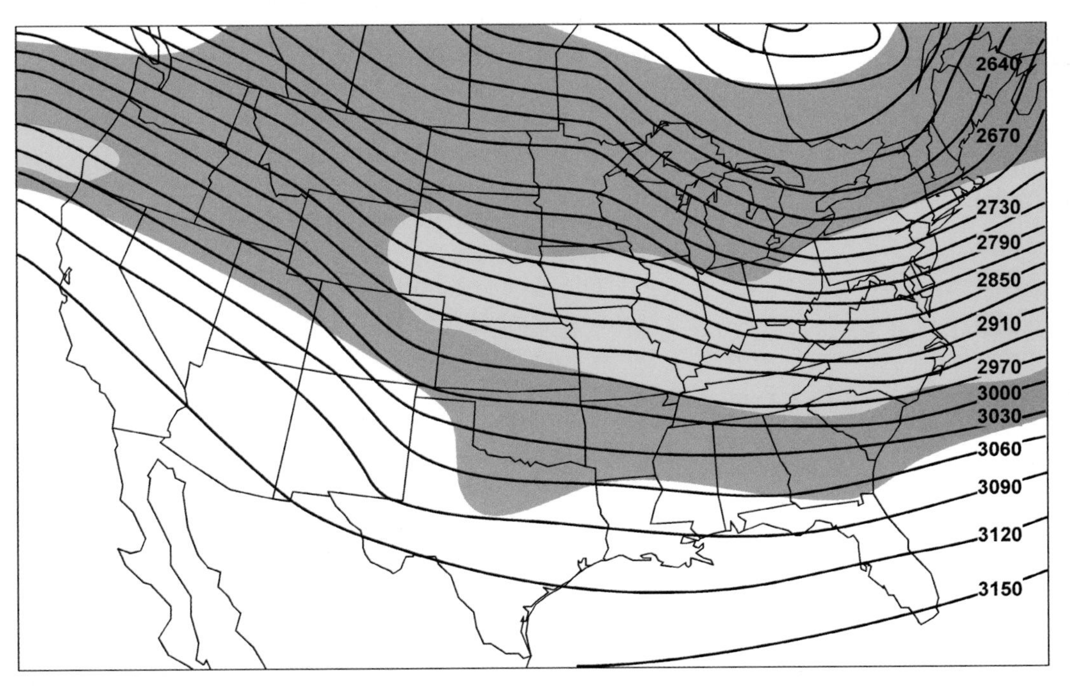

FIGURE 17.9: The 700 mb height field (meters) for 1200 UTC 17 January 1982. The blue shaded area denotes regions where winds exceeded 40 knots (46 mph); the yellow shaded regions denote winds exceeding 50 knots (58 mph).

strong height gradient is indicative of strong background winds at mountaintop, the second of the three common features of weather patterns during downslope windstorms.

The final feature, a stable layer just above mountaintop upstream of the mountain range, is illustrated for the 17 January Chinook by the 1200 UTC soundings from Grand Junction, Colorado, and Lander, Wyoming (Figure 17.10, see locations on Figure 17.8). Both soundings have inversions near 600 mb, with the Lander sounding showing an exceptional inversion of 8° C over a depth of only 100 meters. These inversions were important in creating the destructive winds and hydraulic jump east of the Rockies.

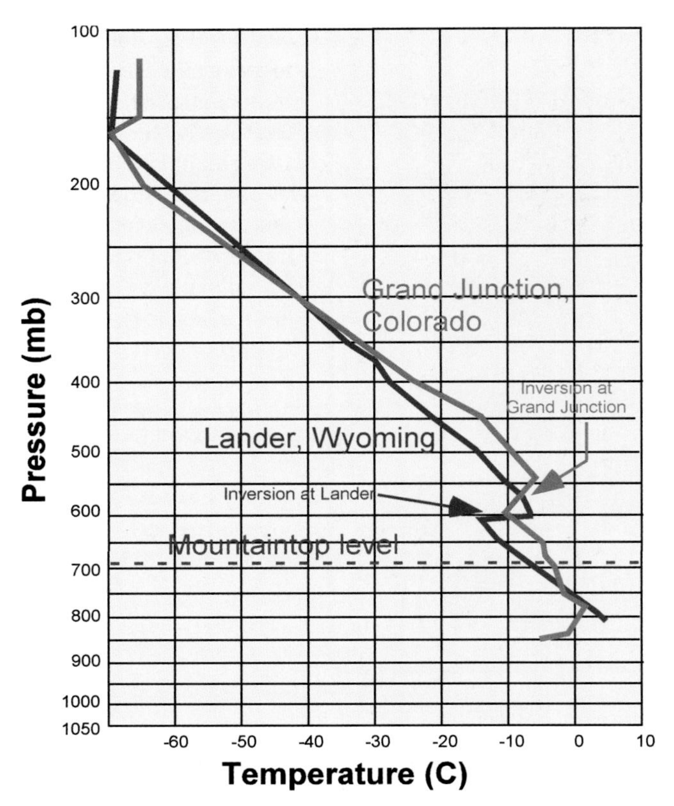

FIGURE 17.10: Soundings taken upstream of the mountains at Grand Junction, Colorado, and Lander, Wyoming, at 1200 UTC 17 January 1982. Note the inversions near 600 mb, which is about 1 km (3200 ft) above the summit of the Rockies.

CHECK YOUR UNDERSTANDING 17.2

1. What three factors determine the temperature of Chinook winds?
2. Why is the dewpoint temperature typically lower on the leeward side of a mountain compared to the windward side during a downslope windstorm?
3. What three factors are common to the weather patterns that produce Chinook winds?

SANTA ANA WINDSTORMS OF CALIFORNIA

Downslope windstorms of the Southern California mountains are called the Santa Ana, a name taken from the Santa Ana River Canyon, which runs from the San Bernardino Mountains southwest through the cities of Riverside, Anaheim, and Santa Ana. The canyons of Southern California experience episodes of Santa Ana winds every year, often with very strong gusts (Figure 17.11). Unlike the westerly downslope winds of the Rockies, the Santa Ana has an easterly component, flowing from the desert westward down the San Bernardino and San Gabriel Mountains and into the Los Angeles basin. Santa Ana winds are generally not as strong as the Chinooks of Colorado, but are dangerous because of their ability to spread wild fires.

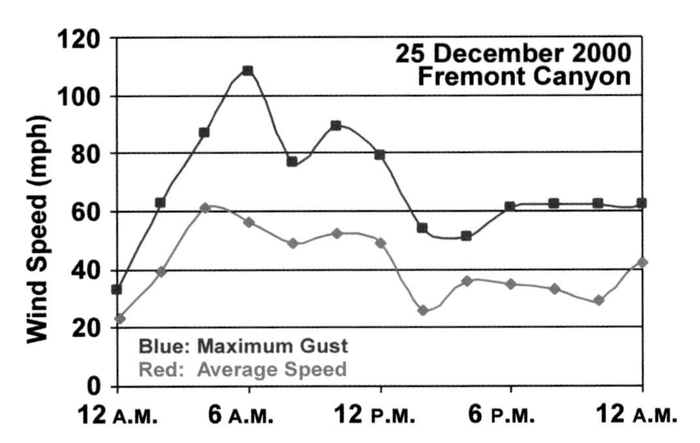

FIGURE 17.11: Wind speed (mph) measured in Fremont Canyon, east of Los Angeles, during the Christmas Day 2000 Santa Ana windstorm. The lower line is the average wind speed and the upper line is the maximum recorded gust.

The wet season in southern California extends from December through March. Vegetation, primarily grasses, scrub, and trees, grows during this short period. The remaining eight months are southern California's dry season, a time period when no rain may fall at all. Conditions for Santa Ana winds are limited to the period September to April, but are most common during late fall and winter. An average of 20 Santa Ana wind events occur each year, lasting on average about 1.5 days. Humidity levels vary during events, with the driest months having the highest frequency of events. During the months of September through November, when natural vegetation is tinder dry, the Santa Ana is most dangerous. A spark from an automobile, campfire, cigarette, or natural phenomenon such as occasional lightning from a rare thunderstorm, can trigger enormous wildfires as the wind blows fire across the landscape.

During a ten-day period in September 1970, Santa Ana winds fanned fires that destroyed nearly a half-million acres. Property damage during fires that burned 80,000 acres in late November 1980 was estimated at $40 million. In 1993, twenty-six major fires

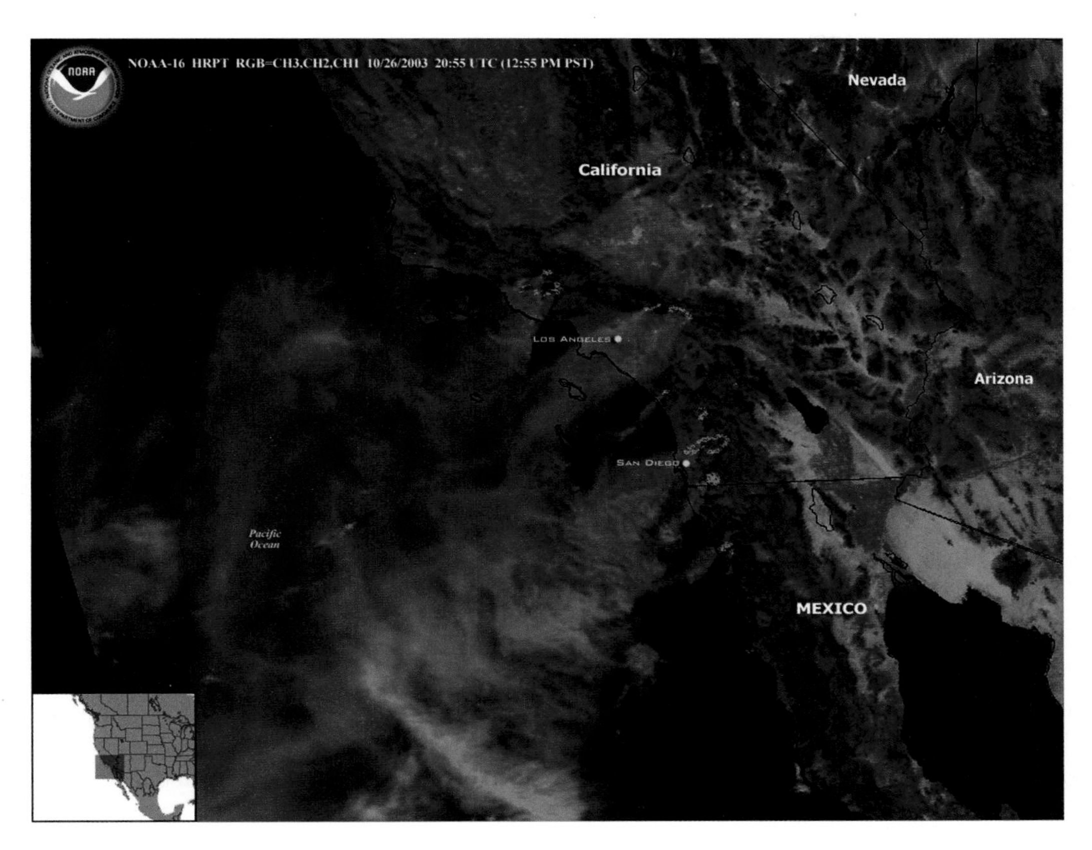

FIGURE 17.12: Color enhanced satellite image of southern California taken by the NOAA-16 Satellite at 2055 UTC on 26 October 2003 showing the smoke plumes associated with the Santa Ana driven fires in the Los Angeles Basin.

fanned by Santa Ana winds killed four people, destroyed or damaged over 1,200 structures, and resulted in almost $1 billion in damage. The damage to residential property has continued to climb as more people in Southern California build expensive homes in the most vulnerable regions of the foothills and mountains. During strong Santa Ana events, air traffic is also at risk as strong winds create clear air turbulence and strong wind shear.

While the Santa Ana wind can cause significant damage with strong winds and spreading of wildfires, scientists have recently learned that they have a beneficial aspect as well. When a strong Santa Ana wind blows out to sea, it pushes the upper-most layers of the ocean away from shore. This allows the deeper, colder water from below to move upward toward the surface (a process called ***upwelling***). This cold water is rich in nutrients and is highly beneficial to local fisheries.

Strong Santa Ana windstorms occur when strong high-pressure systems are located over the Great Basin of the interior western mountains, and a strong north-south pressure gradient is present along the southern California mountains. An inversion above mountaintop upstream (east) of the mountains is probably also necessary for strong Santa Ana windstorms, although studies have yet to conclusively show this. The strong pressure gradient drives an easterly wind across the mountains toward the Los Angeles basin, and the inversion cap aids in the development of downslope flow. As the air descends, it warms at the dry adiabatic lapse rate, arriving in the Los Angeles basin with temperatures that can exceed 77° F to 86° F (25° C to 30° C). The strongest winds occur where air is funneled through canyons. Winds in canyons often gust well above 50 knots (58 mph), and can locally gust to speeds approaching 100 knots (115 mph).

In late October 2003, the Santa Ana struck with full force. Huge wildfires swept through over 200,000 acres of public wildlands, including the Cleveland, Los Padres, and San Bernardino National Forests. Advancing walls of flame swallowed over 1,900 homes, including whole communities, causing at least sixteen fatalities. Figure 17.12, an enhanced satellite image of Southern California, shows the position of the fires and the smoke plumes flowing westward with the Santa Ana winds. A classic Santa Ana, the windstorm was driven by a strong high pressure over the Great Basin of Nevada and Utah (Figure 17.13). During the windstorm, low pressure was also present over Baha, California, enhancing the pressure gradient across the southwestern United States and creating exceptionally strong surface winds.

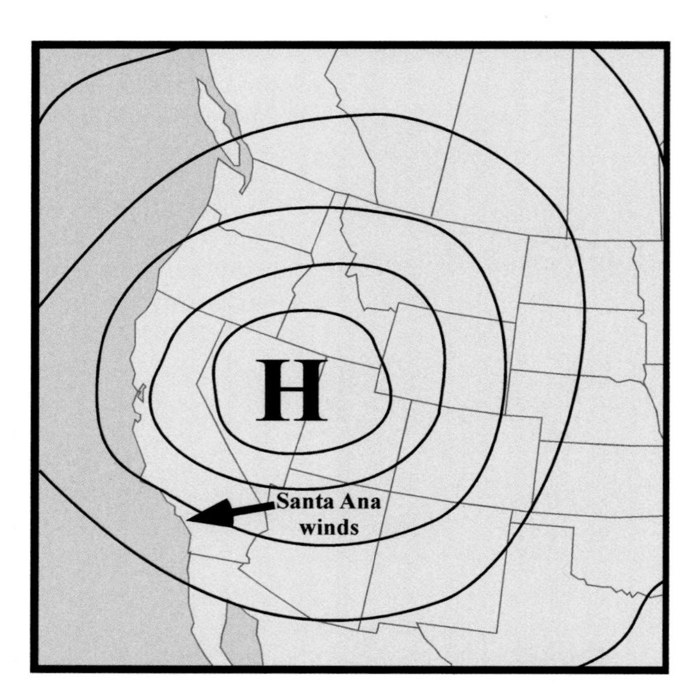

FIGURE 17.13: Typical sea-level pressure pattern over the southwest United States during a Santa Ana windstorm.

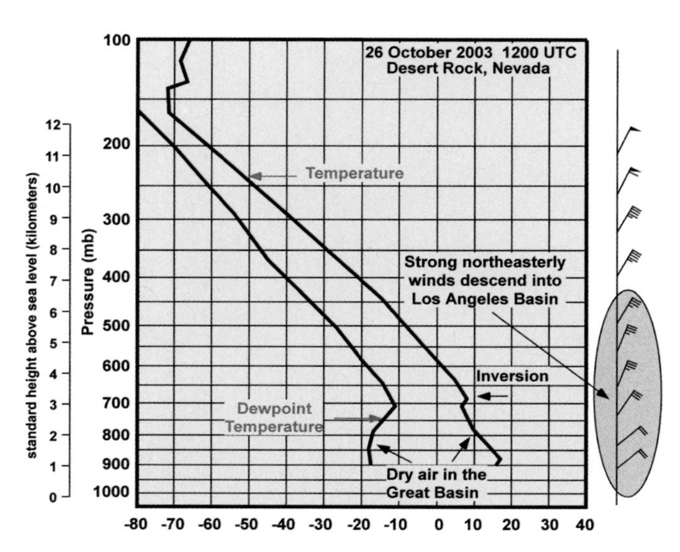

FIGURE 17.14: Sounding taken upstream of the Los Angeles basin at Desert Rock, Nevada at 1200 UTC 26 October 2003.

Figure 17.14, a sounding taken east of the mountains at Desert Rock, Nevada, at 1200 UTC 26 October 2003 shows other key features of the Santa Ana—an inversion layer near 700 mb, exceptionally strong winds with an easterly component, and very dry air in the lower atmosphere. Keep in mind that this sounding sampled the air *before* it descended the mountain canyons. Upon descending the mountain and arriving near sea level, the air would be moving much faster and would be very dry—the worst possible situation for fire. An animation of the weather patterns associated with this disastrous event, and a series of satellite images that graphically illustrate the smoke plumes, can be examined in Online 17.2.

Winds similar to the Santa Ana occur occasionally, but more rarely, along the west slope of the Sierra Nevada. These winds are sometimes referred to as ***California Northers,*** a name that probably originated from the writings of the famous naturalist John Muir who described them in this way in his book *The Mountains of California:*

> This was in the winter of 1873, when the snow-laden summits were swept by a wild 'norther.' I happened at the time to be wintering in Yosemite Valley, that sublime Sierra temple where every day one may see the grandest sights. Yet even here the wild gala-day of the north wind seemed surpassingly glorious. I was awakened in the morning by the rocking of my cabin and the beating of pine-burs on the roof. Detached torrents and avalanches from the main wind-flood overhead were rushing wildly down the narrow side cañons, and over the precipitous walls, with loud resounding roar, rousing the pines to enthusiastic action, and making the whole valley vibrate as though it were an instrument being played.

As these winds descend from the mountaintops into California's Central Valley, which is near sea level, the air often becomes very hot, in extreme cases, exceeding 110° F (43° C). They bring the same fire danger as their southern counterpart, the Santa Ana, often fueling wild fires both in the Sierra Nevada foothills and along the Coastal Range.

ONLINE 17.2

Weather Conditions during the Santa Ana

A loop of twelve hourly surface pressure and temperature weather charts throughout the period of the October 2003 Santa Ana windstorm appears online. These data show the strong pressure gradient associated with the high-and low-pressure system in the northern and southern parts of the western United States respectively. A second animation shows a series of visible satellite images which graphically depict the smoke plumes advancing westward with the Santa Ana.

KATABATIC WINDS

The winds discussed in the previous sections are dynamic in origin, driven by strong pressure gradients across mountain ranges during times when atmospheric stability conditions lead to strong downslope flow. Another type of severe windstorm occurs in cold regions of the world, particularly where vast ice sheets cover high landmasses. The winds are called ***katabatic winds*** because the cold air is dense and flows downslope off the ice sheets. These winds can occur any time of the year, but are most common during winter.

Severe katabatic winds, with gusts exceeding 100 knots (115 mph) normally develop following a period of relative calm over ice sheets. During the calm period, as an ice sheet emits infrared radiation to space, the surface of the ice cools (see Figure 17.15A). Air adjacent to the ice sheets cools by conduction of heat to the ice, and also cools by emitting infrared radiation to space. With time, often a day or more, a dome of extremely cold dense air can build up over the ice sheet. At some point the cold air spills down the slope off the ice sheet. Driven by the force of gravity, these icy winds descend the slope and spill out onto the land or sea below (Figure 17.15B). Although air in katabatic winds warms as it descends from high plateaus, the air is so cold to begin with that, in spite of adiabatic warming, it arrives at the base as a cold wind.

Infamous katabatic winds occur in Antarctica. The continent is a huge elevated ice sheet and the interior of the continent becomes bitterly cold as the ice sheet continually radiates energy to space during the long Antarctic night. Cold air regularly descends the ice sheet, producing winds near the coast at the edge of the sheet that exceed hurricane strength. The onset of these winds is sudden. The winds may be smooth, but more often they are turbulent, carrying loose snow and creating ***ground blizzards*** which drop visibility to zero. Ground blizzards in Antarctica have prevented the landing of aircraft on resupply and rescue missions to scientists stationed on the continent. These dangerous ground blizzards often occur while the skies above are clear. Severe katabatic winds also occur along the coast of Greenland at the base of the world's second largest ice sheet.

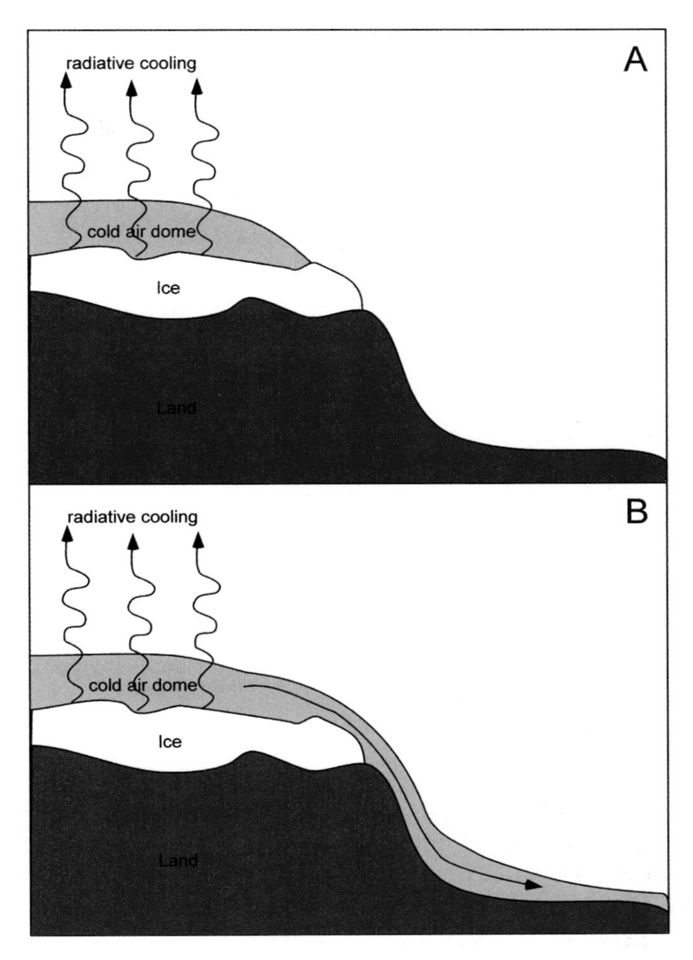

FIGURE 17.15: Schematic representation of severe katabatic winds showing (A) the development of a dome of cold air on an ice sheet and (B) the subsequent drainage of the cold air to low elevations.

MOUNTAIN WINDSTORMS AND GLOBAL CLIMATE CHANGE

Mountain windstorms are phenomena local to the downwind side of mountains and are relatively rare. Not much attention has been given to these storms by scientists studying global climate change. As a result, we know little about how the frequency or intensity of these storms may change in a future warmer world. The primary concern with these events is that they bring warm dry winds, and with that, the potential for forest fires. Nowhere is this more evident than in Southern California, where the Santa Ana winds have been associated with disastrous fires many times over

FOCUS 17.3

Katabatic Winds in North America

In North America, strong katabatic winds occasionally develop at the base of the few remaining elevated ice caps in the mountains of Northwest Canada and Alaska and at the base of glacial valleys. For example, the 300 square mile Harding Ice Field, which blankets mountains of the Kenai Peninsula south of Anchorage, Alaska and feeds over thirty active glaciers, lies nearly 2 km (~6,000 ft) above sea level. Katabatic flows off this ice field descend to the coast to the east. On the west side of the ice field, winds often descend from the ice field and cross Tustumena Lake. The calm lake can become dangerous for small watercraft as strong cold winds appear suddenly and whip waves to heights of over 3 ft (1 m) or more!

the last few decades. As discussed in Chapter 16, changes in precipitation amount and type in the mountains of southern California may lead to a drier landscape, particularly in autumn when the winter rains have yet to fall and the Santa Ana winds begin to blow. The increased threat of forest fires associated with winds such as the Santa Ana is probably the single greatest threat from mountain windstorms in a future, warmer climate.

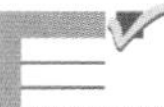

CHECK YOUR UNDERSTANDING 17.3

1. Where do Santa Ana winds occur?
2. What dangers do Santa Ana winds pose?
3. Why are katabatic winds most common in Greenland and Antarctica?

TEST YOUR UNDERSTANDING

1. What is the difference between a foehn and a bora?
2. What kind of damage can downslope winds cause?
3. What is a lee wave?
4. Explain how altocumulus lenticularis clouds form.
5. Why is an inversion important in creating strong downslope winds?
6. What is meant by a hydraulic jump?
7. What causes the observed gustiness in downslope winds?
8. Why should pilots be concerned about downslope windstorms?
9. Explain why some Chinooks are more like the foehn, while others are more like the bora.
10. What role does latent heat release play in the temperature experienced on the plains east of the Rockies during a Chinook?
11. Explain how air at one-kilometer altitude that is descending the leeward side of a mountain can be warmer than the same air at one-kilometer altitude as it ascended the windward side of the mountain earlier in the day.
12. What is a Chinook wall? Where is it located relative to the strongest winds in a mountain windstorm?
13. What type of surface weather pattern is conducive to Chinook winds in Denver, Colorado? What type of temperature profile west of the mountains is most conducive to Chinook winds in Denver?
14. Explain how a Santa Ana may contribute to the intensity of wild fires.
15. What impacts of Santa Ana windstorms are beneficial?
16. What type of surface pressure system is required for Santa Ana windstorms and where is this pressure system typically located?

17. Describe how a katabatic windstorm develops.

18. A katabatic wind is a downslope wind. Explain why it is a "cold" wind in spite of warming due to adiabatic compression during descent.

19. How might global climate change affect mountain windstorms?

20. Compare and contrast Chinook, Santa Ana, and katabatic windstorms in terms of:
    (a) temperature
    (b) geographic location
    (c) typical wind speeds
    (d) meteorological processes that lead to the windstorms
    (e) hazards associated with the windstorms

TEST YOUR PROBLEM-SOLVING SKILLS

1. Assume an air parcel with a temperature of 10° C is located near the ground and is approaching a mountain range that is 2 km high. The air parcel ascends the mountain range and then descends the other side to the same level. During the ascent, at 1 km above the base of the mountain, the air parcel saturates and forms a cloud. Assume for simplicity that all of the condensate falls out of this cloud as precipitation as the air parcel ascends the mountain. What is the final temperature of the air when it reaches the bottom of the mountain on the lee side? Explain how you arrived at your answer.

2. You are a flight controller at the Denver International Airport, which is located on the Great Plains thirty-five miles east of the base of the Rockies and fifty-five miles east of the mountain crest. Westbound aircraft departing Denver normally pass across the crest of the Rockies at 18,000 ft, about 4,000 ft above Mount Evans and Longs Peak, the highest peaks in the area. A severe downslope windstorm has been reported west of the airport near Boulder, Colorado, although surface winds at the airport are only 15 knots from the west, with gusts to 25 knots. Devise a flight plan for an aircraft to take off and fly to San Francisco that would minimize the likelihood of encountering severe turbulence. (Hint: Find a map of the topography of Colorado and adjacent states and look for a path across the mountains where the aircraft would not encounter steep terrain below its path.)

3. The Alps are an east-west oriented mountain range. What type of weather pattern might trigger a downslope windstorm on the south side of the Alps? (Hint: Where would low-and high-pressure centers be located to produce a strong pressure gradient across the Alps with downslope flow on the south side?)

USE THE SEVERE AND HAZARDOUS WEATHER WEBSITE

http://severewx.atmos.uiuc.edu

1. Use the *Severe and Hazardous Weather Website* to monitor the cloud structures in the lee of the Rocky Mountains during winter to determine if lee waves are present. Look for rows of clouds parallel to the mountains. When animated, the clouds appear to be stationary. Record the location of the clouds and the date and times when they are present. Note the extent of the cloud cover and compare the size to extratropical cyclones and to thunderstorm complexes.

2. Downslope windstorms occur several times each winter along the Rockies. Use the *Severe and Hazardous Weather Website* to monitor the map of watches and warnings issued by the National Weather Service each day. On a day when high winds are expected, examine surface observations

and soundings within the forecast region. (Note: When a warning is issued a particularly good web site to monitor the conditions in the foothills of the Rockies is the National Center for Atmospheric Research (NCAR) in Boulder, Colorado. NCAR is located atop a foothill exposed to the full force of the wind.)

(a) Within the forecast region of high winds, record the observed temperature for at least six consecutive hours. Comment on how the temperature fluctuated during the time period and how the temperature changes can be related to the downslope wind event.

(b) Locate the sounding upstream of the mountain from the closest location to the wind event. Record the date and time of the sounding. Does the sounding exhibit the classic characteristics of a wind event? Why or why not?

3. The next time wildfires in Southern California are reported on the news, examine the weather pattern using maps from the "Current Weather" page of the *Severe and Hazardous Weather Website.*

(a) Are Santa Ana winds fanning the flames? How did you determine this?

(b) What evidence can you find from wind reports in the Los Angeles basin that the Santa Ana winds are blowing? What weather products did you examine to determine this?

(c) Identify the stations reporting the strongest winds. Determine what effect orography has on the wind by consulting a topographic map to determine if the high wind reports are coming from canyons or valleys.

(d) Look at a visible satellite image during the wild fire event. Use the smoke to estimate the wind direction. Is there evidence of strong offshore winds indicative of a Santa Ana?

THUNDERSTORMS

Supercell thunderstorm in eastern Colorado.

KEY TERMS

airmass thunderstorm
anvil
Bow Echo and Mesoscale Connective Vortex Equipment (BAMEX)
bookend vortice
boundary
boundary intersection
Bounded Weak Echo Region (BWER)
bow echo
bright band
capping inversion
cloud droplets
cold pool
comma cloud
Convective Available Potential Energy (CAPE)
convective region
cumulonimbus
cumulus stage
derecho
dissipation stage
downdraft
echo-free vault
entrainment
extratropical cyclone
fine line
forward flank
forward flank downdraft (FFO)

forward flank gust front
front-to-rear flow
frontal squall line
gust front
heavy rain
HP supercell
level of free convection
low-level jet
LP supercell
mammatus cloud
mature stage
mesocyclone
mesoscale
mesoscale convective system (MCS)
mesoscale convective vortex (MCV)
mini-bow
ordinary thunderstorm
overshooting top
rain-free base
rear flank
rear flank downdraft (RFD)
rear flank gust front
rear flanking line
rear inflow jet
roll cloud
shelf cloud
squall line
stratiform region
stratosphere
supercell thunderstorm
thunderstorm
training
trigger mechanism
tropopause
updraft
vertical wind shear
virga
wall cloud

Thunderstorms, also called ***cumulonimbus*** clouds, are tall, vertically-developing clouds that produce lightning and thunder. They occur continually in the atmosphere, are important in the redistribution of moisture and heat, and are essential for agriculture and water resources. Although the majority of thunderstorms are not severe, thunderstorms can produce a wide range of severe weather phenomena. Figure 18.1, for example, shows the distribution of tornado, large hail, and wind damage reports for just one year, 2006. The reports are concentrated in the Plains states; however, the maps show that some damage from thunderstorms is likely to be reported in any populated area east of the Rocky Mountains in a given year. Annual property and crop losses from thunderstorm phenomena (lightning, hail, severe winds, and tornadoes) in the United States have averaged $2.4 billion over the twelve years ending in 2006. In the same period, an average of 128 fatalities and 1690 injuries have been reported annually associated with these phenomena.

The U.S. National Weather Service reserves the term "severe" for thunderstorms that have the potential to threaten lives and property from wind or hail. A thunderstorm is considered severe if it contains one or more of the following: hail with diameter of three-quarter inch or larger, wind damage or gusts of 50 knots (58 mph) or greater, and/or a tornado. Because lightning occurs in all thunderstorms, it is not used to differentiate between severe and non-severe conditions.

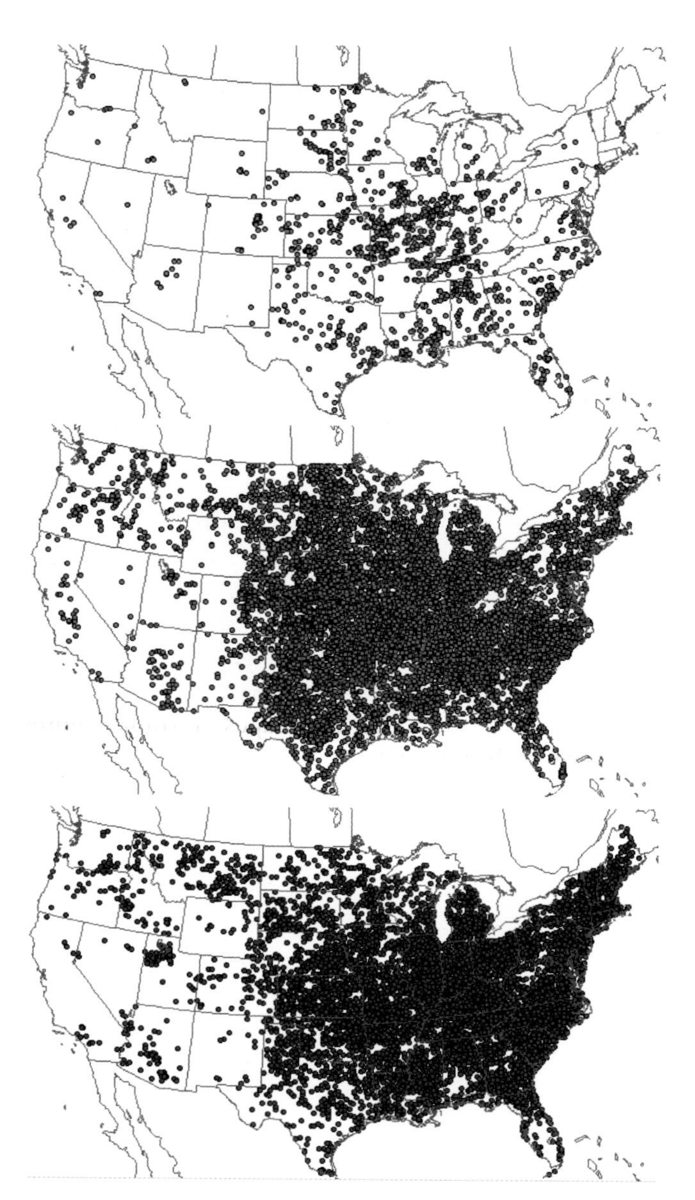

FIGURE 18.1: Locations of severe weather events (top: tornadoes, middle: large hail, bottom: damaging winds) reported to the NOAA Storm Prediction Center in the year 2006. Each circle represents a report.

Severe thunderstorms normally require four elements for formation: (1) a source of moisture, (2) a conditionally unstable atmosphere, (3) a mechanism to trigger the thunderstorm's updraft, either through lifting, or heating of the surface, and (4) ***vertical wind shear***—a rapid change in wind speed and/or wind direction with altitude. Thunderstorms organize differently depending on the lifting mechanism that triggers the storms, the vertical wind shear, and the degree of instability in the atmosphere. Destructive thunderstorms develop most often in an environment characterized by large conditional instability and strong vertical wind shear.

Lifting occurs along ***boundaries*** between airmasses. In past chapters, we have studied fronts—the boundaries between very large airmasses. Fronts are most distinct and easily identifiable on weather maps in the cool season (late fall, winter, and early spring) when the airmasses bounding a front typically have sharply contrasting temperature and moisture characteristics. When strong fronts are present, lifting occurs along the frontal boundary as the front advances and thunderstorms tend to organize along and align with the advancing front.

In recent years, severe thunderstorm experts have paid increasing attention to other less distinct boundaries in the atmosphere. These boundaries can arise from a number of processes. Boundaries are created by differential heating of air over surfaces with different properties, such as water and land, forests and fields, or urban and rural landscapes. Terrain features may create boundaries, such as when cooler air resides over a river valley, or when solar radiation heats one side of a ridge more than another. Often, a boundary may be the leading edge of a cool air outflow—the remnant air of a past thunderstorm. These more subtle boundaries become increasingly important in the warm season (late spring, summer, and early fall) as the contrast between polar and tropical airmasses is reduced and larger scale fronts become less distinct or non-existent. Thunderstorms that develop along subtle boundaries in the warm season often undergo a mode of self organization that eventually leads to the formation of what severe storm meteorologists call a ***mesoscale convective system***, or ***MCS***. The term ***mesoscale*** refers to atmospheric processes that occur on a scale of a few to a few hundred kilometers, or more simply to scales ranging from the size of an individual thunderstorm to the size of the complex of clouds and precipitation composing a group of thunderstorms. In the cool season, when fronts are more distinct and the lifting is focused along fronts, very long lines of thunderstorms can develop along frontal boundaries. We will refer to these storms as ***frontal squall lines***.

If winds increase rapidly with height ahead of a strong front or a less distinct boundary, thunderstorms triggered along that boundary may organize in yet another mode—violent storms called ***supercell thunderstorms***. Supercells can occur along a line or individually, but always have their own individual circulations that are related to the storm's rotation and the vertical shear in the atmosphere. They most often develop when strong winds are present in the upper troposphere and winds in the lowest kilometer or two of the atmosphere increase rapidly and change direction with altitude. Supercell thunderstorms are very strong, rotating cells that produce the most violent weather and the majority of large tornadoes. Supercells can occur in both the warm and cool seasons, provided the conditions exist to support their development.

In this chapter, we will consider each mode of thunderstorm organization—mesoscale convective systems, frontal squall lines, and supercells—examining their structure and evolution as well as the types of severe weather conditions that they create. However, before we examine these more complex thunderstorm systems, we will first consider the life cycle of a single isolated thunderstorm, one that forms in the absence of vertical wind shear.

AIRMASS THUNDERSTORMS

Isolated thunderstorms that form in the absence of vertical wind shear are sometimes called ***airmass thunderstorms*** because they often occur well within an airmass, rather than along a front. They are also sometimes referred to as ***ordinary thunderstorms***. Airmass thunderstorms often form along weak boundaries such as cool air outflows from other thunderstorms, but they can be triggered by surface heating, or lifting along slopes in mountainous regions. They tend to form during hot afternoons—typical conditions found, for example, during the warm, humid summer in the southern and eastern United States. Precipitation from an individual storm may cover an area about 15 to 20 km (9 to 12 miles) in diameter late in the storm's lifetime. Single airmass thunderstorms have a typical lifetime of about an hour. Their short lifetime is directly related to the fact that the environmental winds in the vicinity of the thunderstorm do not change substantially with height.

The primary difference between the environment for airmass thunderstorms and the environment for severe thunderstorms is the strength of the winds aloft. With airmass thunderstorms, the winds aloft are typically weak, and the vertical wind shear minimal. As we know from Chapter 7, winds increase in speed rapidly with altitude in the vicinity of frontal zones, while far from frontal zones winds tend to be weaker and more uniform with height. We therefore would generally expect to find airmass thunderstorms far from frontal boundaries. In the absence of vertical wind shear, airmass thunderstorms grow vertically without any significant tilt (e.g., Figure 18.2).

Airmass thunderstorms undergo three stages of growth as shown in Figure 18.3. In the first stage, called the ***cumulus stage***, the cloud consists of a warm, buoyant plume of rising air—the ***updraft***. The updraft velocity in the cloud increases rapidly with height. The updraft depicted in Figure 18.3 is idealized—the common "cauliflower" appearance of real growing thunderstorms is evidence that true updrafts have more complicated structure. Nevertheless, air within a developing thunderstorm in the cumulus stage is rising vertically through the troposphere. The clouds are composed primarily of small liquid ***cloud droplets***, with little or no raindrops or ice crystals. As the storm rises to altitudes corresponding to temperatures between –10° C and –20° C (14° F and –4° F), small ice crystals begin to form. Eventually these particles grow large enough to precipitate and they begin to fall through the storm.

The air in most thunderstorms is sufficiently buoyant that it will rise to the ***tropopause***. In fact, the updrafts in many storms are strong enough that the clouds will extend a short distance into the stratosphere, producing a bulge at the cloud top called an ***overshooting top*** (Figure 18.3) As the clouds reach the tropopause, the storm enters its second stage of growth, the ***mature stage***.

FIGURE 18.2: A single-cell airmass thunderstorm. The cloud is erect with little tilt and a relatively symmetric anvil.

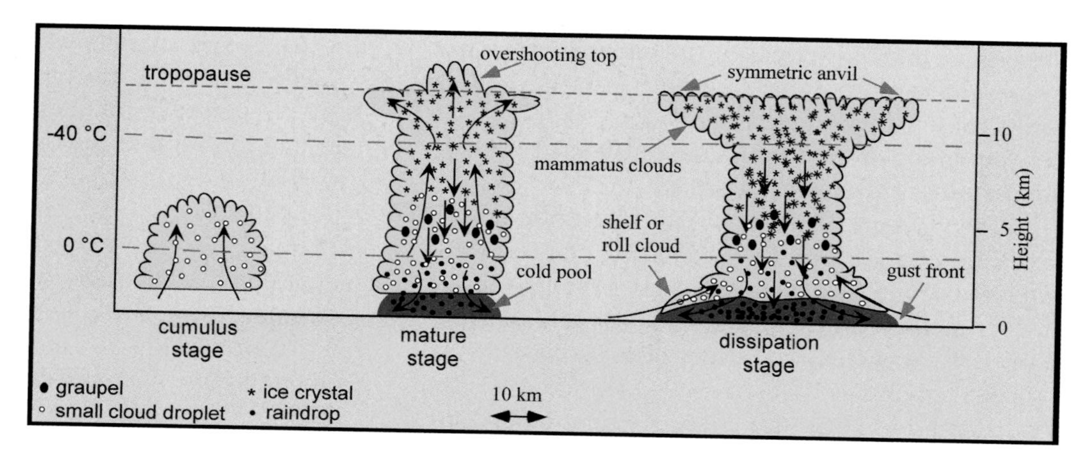

FIGURE 18.3: The life cycle of a non-severe, single-cell summertime airmass thunderstorm. A typical life cycle is on the order of an hour.

Airmass thunderstorms at this stage can be identified visually by examining the upper part of the storm. When a thunderstorm updraft reaches the tropopause, the strong stability of the stratosphere quickly inhibits further ascent. Air from the updraft exhausts horizontally at the tropopause, forming a cloud feature called the ***anvil***. In an environment with little vertical wind shear, the exhaust has no preferential direction and the anvil generally will appear symmetric with a flat top (see Figures 18.2 and 18.3). In a strongly sheared environment, the anvil is typically asymmetric, the cloud debris carried downstream by the strong environmental winds aloft. The anvil cloud will be composed of cirrus farther from the storm, and cirrostratus and altostratus closer to the storm. ***Mammatus clouds*** often form at the base of thunderstorm anvils. These clouds, which appear like rounded "bags" hanging from the anvil, form as particles evaporate in the anvil. Mammatus are composed of pockets of evaporationally-cooled air that descend, transporting cloud particles downward within them.

During the second stage of growth, the precipitation particles cascade downward through the cloud. Each precipitation particle has weight, and displaces air ahead of it as it falls, effectively "dragging" air earthward. The effect of the falling precipitation is to create ***downdraft*** circulations in the middle and higher parts of the cloud. These downdrafts are enhanced by evaporation as precipitation encounters dry air mixing in from the sides and top of the cloud (a process known as ***entrainment***). Evaporation requires latent heat energy, which is supplied by the air containing the precipitation. As cloud and precipitation particles evaporate, air is cooled, becomes denser, and begins to descend faster, enhancing the downdrafts. Near the surface, this rain-cooled air accumulates, forming a ***cold pool***. As the precipitation falls, the downdraft circulations descend deeper into the cloud and counteract the updrafts bringing the warm buoyant air upward from the surface.

An airmass thunderstorm grows vertically since the wind speed does not increase substantially with height in the environment of the storm. Without any tilt downstream in upper levels, *precipitation forming in the upper part of the thunderstorm must fall vertically through the updrafts in the lower levels.* The downdrafts produced by evaporation and the drag force of falling precipitation act to suppress the updrafts, eventually shutting off the source of moisture to the storm. As this occurs, the storm enters its final stage of evolution, the ***dissipation stage***. During this stage, the thunderstorm is composed of downdraft circulations. Heavy rain and rain-cooled air descend from the base of the storm as the clouds composing the storm slowly evaporate.

During its dissipation stage, an airmass thunderstorm will typically produce an outflow of cool air. The cool air, generated by evaporation of rain within downdrafts, spreads outward away from the thunderstorm after reaching the surface, producing a ***gust front***. Clouds will typically form over the gust front as warm air is lifted over the spreading cold pool. These cloud features are called ***shelf clouds***, or ***roll clouds***, depending on their shape. As the cool air rushes outward, new thunderstorms cells may trigger near the cool air's leading edge as the warm air is lifted. When new thunderstorm cells develop, they will undergo similar life cycles, often creating even more thunderstorm cells. When thunderstorm cells perpetuate in this manner, they may eventually organize into a coherent structure—a mesoscale convective system—the topic of our next section.

FOCUS 18.1

How Much Water Is in a Thunderstorm?

We can roughly estimate the "weight" of the water and ice in an airmass thunderstorm by considering that it is about 15 × 15 × 15 kilometers in size and contains an average of about 1 gram of water/cubic meter throughout this volume. A 15 × 15 × 15 kilometer box contains 3.4 trillion cubic meters, so a thunderstorm contains about 3.4 trillion grams of water and ice. In common English units, this converts to about 3.7 million tons.

CHECK YOUR UNDERSTANDING 18.1

1. How does the National Weather Service determine if a thunderstorm is severe?
2. What is the importance of boundaries in thunderstorm development?
3. Why is wind shear important to a severe thunderstorm?
4. Where and when would you expect airmass thunderstorms to form?

MESOSCALE CONVECTIVE SYSTEMS

Mesoscale convective systems (MCSs) produce much of the summer rainfall on the Central Plains of North America. They can progress over a large geographic area during their lifetime, and their cloud shield can often cover an area greater than a large state (Figure 18.4). During their lifetime, MCSs often produce damaging straight-line winds and occasional weak tornadoes.

The life cycle of a typical MCS is illustrated in Figure 18.5. The sequence of radar images in Figure 18.5 were taken from three WSR-88D radars as an MCS formed over eastern Illinois, crossed Indiana, and moved into central Ohio between 4 PM and 11 PM local time on 4 July 2003. The structural features of the MCS appearing in Figure 18.5 are common of many MCSs.

During the warm season, lifting of air along weak airmass boundaries often will trigger thunderstorms, particularly in late afternoon as air on the warm side of the boundary reaches its warmest temperatures of the day. Sometimes the initial thunderstorms may align along the position of the boundary where they were triggered. In other cases, the storms will appear as a disorganized cluster (Figure 18.5A). As time progresses, the storms will begin to organize (Figure 18.5B and 18.5C), often becoming more intense and progressively aligning into an arc-shaped line called a ***squall line*** (Figure 18.5D). A squall line is a long line of thunderstorms in which adjacent thunderstorm cells are so close together that the heavy precipitation from the cells falls in a continuous line. The term "squall," meaning "a violent burst of wind," is fitting, since squall lines can produce exceptional straight-line winds.

One or more segments along the squall line may bow outward, producing a ***bow echo*** (Figure 18.5E) on radar (recall from Chapter 2 that an echo is the radar return as displayed on the radar screen). During this time, a region of widespread less intense precipitation develops to the rear (typically west or northwest) of the squall line (Figure 18.5E and 18.5F). This trailing ***stratiform region*** becomes progressively larger and more widespread as the thunderstorm cells spread along a progressively widening arc (Figure 18.5G). Eventually the thunderstorms decay, leaving in their wake a wide stratiform region (Figure 18.5H) that itself may continue to produce rainfall for several more hours. During this time new thunderstorm cells may also develop ahead of, on the periphery, or even to the rear of the original MCS (Figure 18.5H) and these cells in turn may later organize into a new MCS.

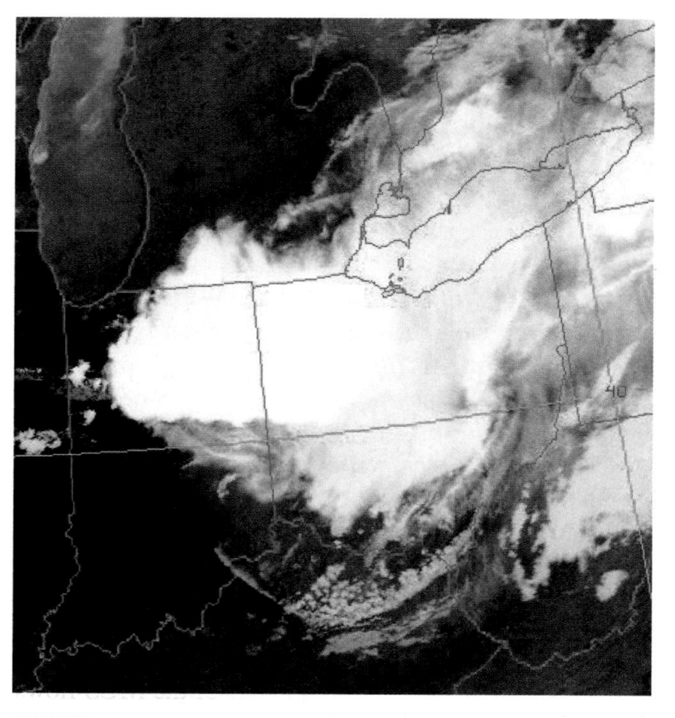

FIGURE 18.4: Visible satellite image of the cloud shield of the mesoscale convective system illustrated in Figure 18.5. The MCS, which occurred on 4 July 2003, was studied during the BAMEX project (see Focus 18.2).

Figure 18.6 shows a series of schematic cross sections that illustrate the evolution of an MCS from the time that initial thunderstorms form through the time when the system has developed a widespread trailing stratiform region. The eight panels in Figure 18.6 correspond roughly in time to the eight panels of Figure 18.5. When the initial thunderstorms form (Figure 18.6A), the updrafts are relatively upright or slightly tilted. Rain from these storms evaporates as it falls into the lower atmosphere, cooling the air through which it falls (Figure 18.6B). This leads to the development of a cold pool (shown in blue in each of the panels of Figure 18.6).

As the cold pool develops, the cooler air spreads outward toward the warm moist air feeding the thunderstorms and also spreads rearward (Figure 18.6C). New updrafts preferentially form along and over the advancing cold pool air, since the cool air, which is more dense, lifts the warm air ahead of it (Figure 18.6D). This reorganization of the updrafts along the leading edge of the developing cold pool leads to the formation of the squall line (note the transition between Figures 18.5B and 18.5D). As the cold pool develops and advances forward into the warmer air, the updrafts forming along and over it progressively tilt rearward. As this tilting occurs, rain falling from the storms falls further to the rear of the convective updrafts and into the relatively drier air in the lower atmosphere to the rear of the line of thunderstorms (often on the cool side of the boundary that initially triggered the thunderstorms). This enhances the rate of evaporation and, as a consequence, dramatically increases both the depth and temperature deficit within the cold pool (Figure 18.6E). At some critical point in the evolution of the MCS, the cold pool becomes deep enough and cold enough that the air begins to rush outward toward the warm air feeding into the thunderstorms (Figure 18.6D-F). The leading edge of this outrushing air is the gust front. It is at this time that severe straight-line winds can occur at the surface. As the cold pool strengthens and the outrush of air begins, air in the evaporation region at middle altitudes of the MCS flows forward toward the line of storms creating a feature called the ***rear inflow jet*** (long blue arrows in Figures 18.6E-H). The rear inflow jet evolves and strengthens as the MCS develops.

The surface winds behind the gust front can be severe, sometimes reaching 80 to 100 knots (92 to 115 mph), but in most cases, the winds range from 20 to 50 knots (23 to 58 mph). A shelf cloud (also called a roll cloud if it has a tube-like appearance) will often develop over the gust front (see Figure 18.7). On radar, the shelf cloud sometimes appears as a ***fine line*** of lower radar reflectivity (Figure 18.5G). Although the shelf cloud does not produce much precipitation, the droplets composing the cloud are generally large enough to be just detectable with radar, creating the fine line return. As the cold pool air rushes out, it lifts the warm air flowing into the line of storms, triggering new thunderstorm updrafts. These thunderstorms align along and over the out-rushing air. When this occurs, the radar return from these storms creates the bow echo visible on radar (Figure 18.5E). On a radar image, a bow echo provides a distinct signature of strong straight-line winds. Figure 18.8 illustrates the relationship between the bow echo, the rear inflow jet, and the strong straight-line winds at the surface.

The character of the outward rush of air depends on the low-level wind shear in the environment ahead of the squall line. Sometimes the outrush of air will initially occur in local regions (about 10 to 20 km (6 to 12 miles) long) creating smaller bows called ***mini-bows***. In other cases, the outrush will occur along a large segment of the line creating a large bow echo 150 to 200 km (90 to 125 miles) long (Figure 18.5E-G). In some MCSs, severe winds can extend over a large area. When widespread severe thunderstorm generated windstorms occur, they are called ***derechos***. The bow often develops rotating eddies on either end that meteorologists refer to as ***bookend vortices*** (Figure 18.8). Small tornadoes sometimes have been observed within the northern bookend vortex. Small tornadoes also sometimes can form along ‘cusps’ that develop along the gust front. Typically tornadoes that develop in MCSs are short-lived, small, and weak.

Aloft above the rear inflow jet, air flows upward and rearward over the cold pool, a feature called ***front-to-rear flow*** (long red arrows on Figure 18.6E–H). As the cold pool deepens and the outrush of air proceeds, the updrafts within the front-to-rear flow of the MCS attain a progressively greater and greater tilt, and the clouds extend further and further to the rear of the original line of thunderstorms. These clouds become the trailing stratiform region. Although rainfall is less intense in this region, the widespread nature of a mature trailing stratiform region can lead to significant rainfall. In many MCSs, the cold pool eventually rushes out beyond the line of storms (Figures 18.5G and 18.6G). Sometimes a completely new line of storms will then develop along the leading edge of the cold pool (e.g. Figure 18.5H). Although we have focused on the cold pool’s rush toward the warm air, it also spreads rearward and toward the sides of the MCS. New thunderstorm updrafts may trigger along these boundaries (for example, note the storms developing rearward and to the north of the MCS in Figures 18.5D–H).

Figure 18.9 summarizes the key structural features of a thunderstorm in a mature MCS. The ***heavy rain***, the location where the rain falls out from the ***convective region*** of the storm, is located just to the rear (west) of the updraft region. Lighter rain falls farther to the west from the *stratiform region* of the squall line where air aloft is rising slowly, and to the east of the convection from the anvil. A radar cross section illustrating the convective and stratiform regions of an MCS appears in Figure 18.10A. On radar cross sections, the stratiform region is characterized by a ***bright band*** of radar reflectivity (BB in Figure 18.10A) at the level where snowflakes falling from aloft melt into raindrops (note that the bright band is responsible for the higher reflectivity appearing in the trailing stratiform region of Figures 18.5F–H). The rear inflow jet is

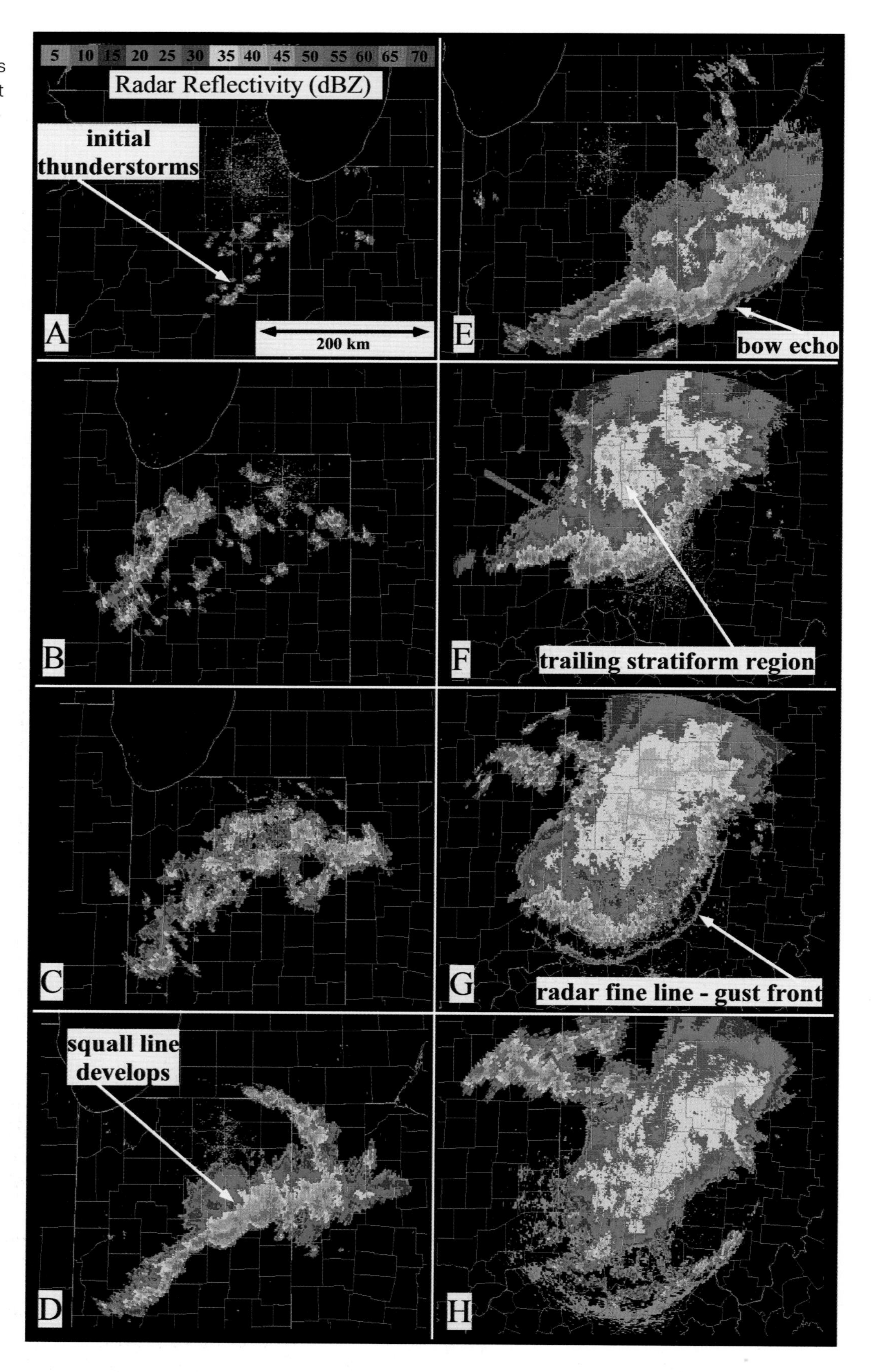

FIGURE 18.5: A sequence of radar reflectivity images showing the development of an MCS. Early thunderstorms evolve to form a squall line and eventually a widespread trailing stratiform region. The sequence spans the late afternoon and evening of 4 July 2003 beginning at 4 pm local time and ending at 11 pm. Each image is one hour later than the previous image.

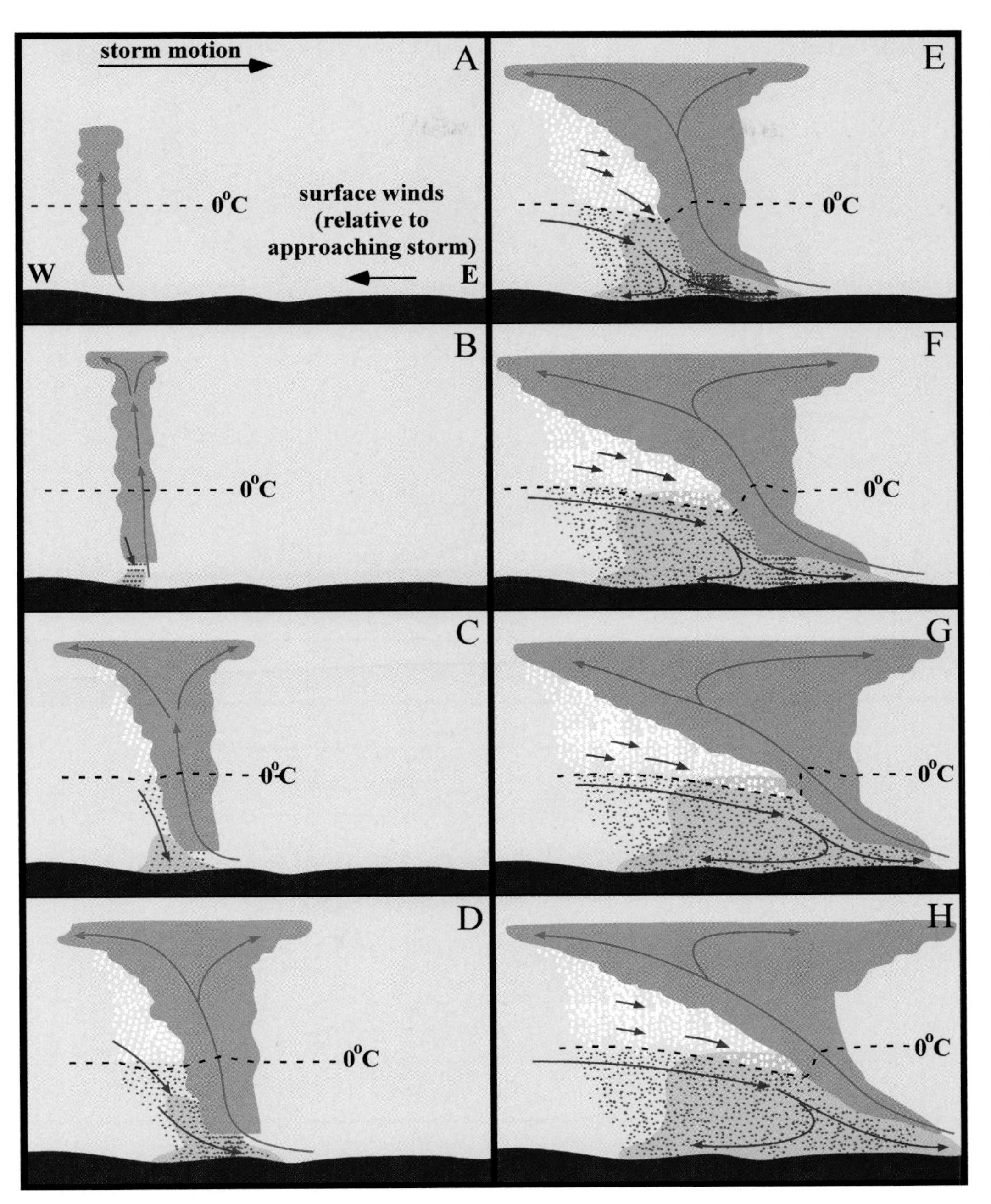

FIGURE 18.6: A sequence of cross sections depicting the evolution of a mesoscale convective system. The cross sections correspond approximately to each stage of evolution depicted in the radar images in Figure 18.5. The gray areas denote clouds. The small white symbols denote precipitation in the form of snow and ice, and the circular blue symbols rain. The melting level is shown as the dashed line. Note that the melting level lowers with time in the storm's center as evaporation and sublimation of ice cools the air (Panels E through H). Red arrows denote rising motion and blue arrows descending motion. The blue shading denotes the cold pool.

FIGURE 18.7: A shelf cloud produced as air is lifted along a gust front ahead of an MCS.

dramatically evident in the radial velocity field in the radar cross section in Figure 18.10B. Air within the rear inflow jet reaches the ground, and spreads out behind the advancing gust front. The width of the heavy precipitation region of an MCS squall line typically ranges between 5 and 15 km (3 to 9 miles). The width of the entire precipitation region of a mature MCS typically ranges between 100 and 200 km (60 and 120 miles).

Near the end of the MCS lifecycle, the trailing stratiform region decays, leaving a large area of clouds. During the formation of the trailing stratiform

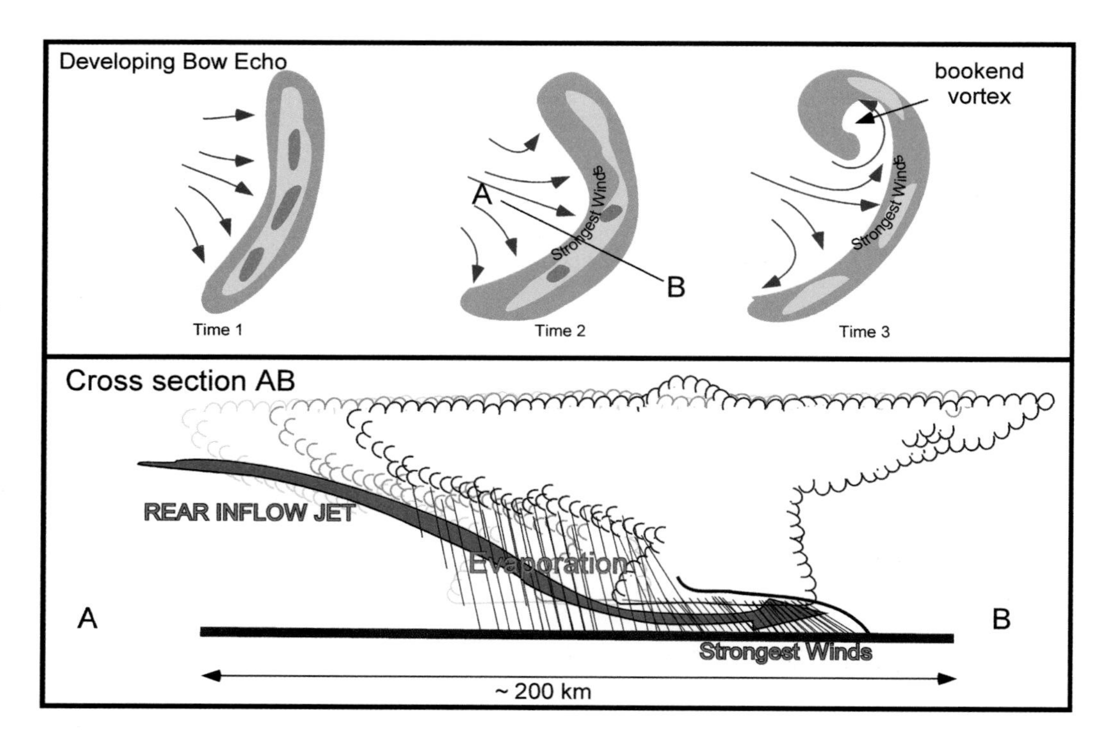

FIGURE 18.8: (Top) Schematic showing the typical evolution of the radar reflectivity field during the evolution of a bow echo. (Bottom) A schematic cross section across a bow echo showing the clouds, rain, and the rear inflow jet descending to the surface from aloft.

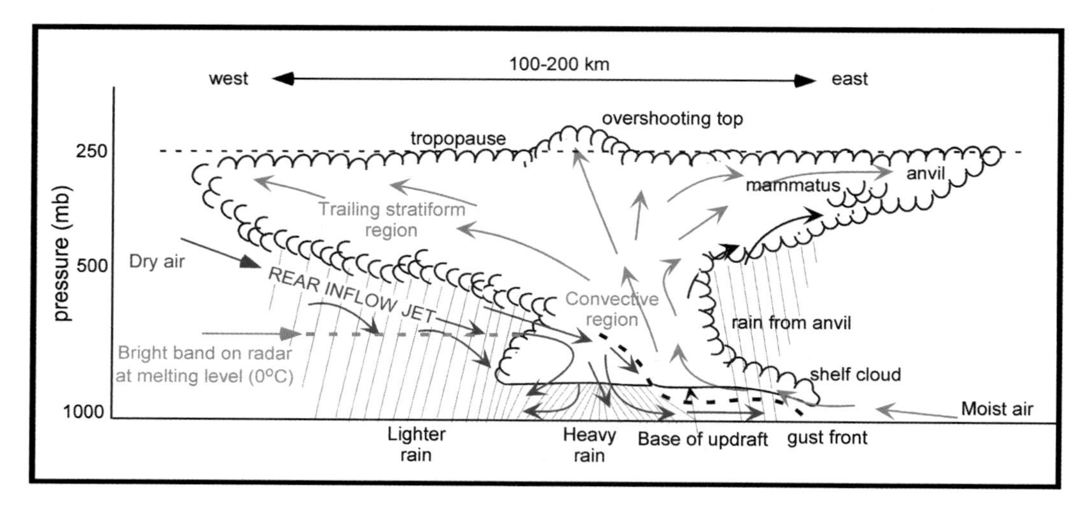

FIGURE 18.9: Schematic cross section through an MCS squall line showing key structural features. The bright band and rear inflow jet can also be seen on Figure 18.10.

region, large quantities of moisture are condensed into rain. Latent heat released during this condensation process often leads to the formation of a weak low-pressure region at middle levels within the trailing stratiform region, causing the clouds in the region to slowly rotate. Recall from Chapter 8 that locally heating air will cause the pressure to lower. As an MCS dissipates, this rotating region of clouds will often remain, slowly spinning as it drifts downwind during the night and following day. These rotating clouds, called a ***mesoscale convective vortex (MCV)*** (see Figure 18.11), are often the focal point for a new thunderstorm outbreak the next day as air heats in the afternoon and primes the atmosphere for another round of convection.

CHECK YOUR UNDERSTANDING 18.2

1. Why are MCSs important to the Central Plains?
2. How large are MCSs and how long do they typically last?
3. What type of severe weather is most common with MCSs?

FOCUS 18.2

BAMEX

Scientists in the spring and early summer of 2003 chased down mesoscale convective systems across the U.S. Midwest using three aircraft and instruments on car-towed platforms in a project called the *Bow Echo and Mesoscale Convective Vortex Experiment*, or *BAMEX*. The BAMEX study was designed to provide the clearest picture to date of how MCSs develop, how the severe winds that accompany bow echoes are generated, how mesoscale convective vortices can trigger new convection long after an MCS has dissipated, and how forecasters can better predict these weather systems. Between January 1995 and July 2000, high winds from U.S. mesoscale convective systems caused over $1.4 billion in damage, 72 deaths, and over 1,000 injuries. BAMEX studied how these damaging winds develop at night, when low-level air usually cools and stabilizes.

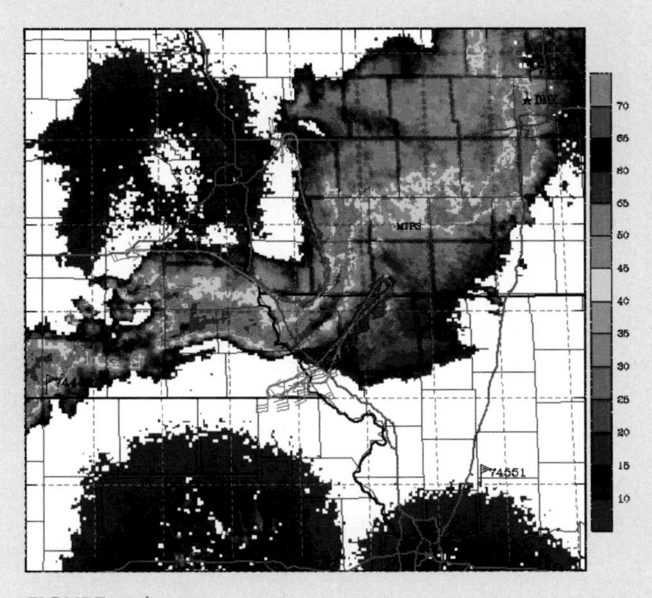

FIGURE 18A: Flight tracks of the NOAA P-3 Aircraft (pink) and the Navy Research Laboratory P-3 aircraft (red) as they studied a developing bow echo on June 10, 2003 over the border between Missouri, Kansas and Nebraska during BAMEX. The radar reflectivity scale is shown at the right. Black lines are state borders, blue lines are interstate highways. The location of the University of Alabama Mobile Integrated Profiling System is indicated by the word MIPS.

The BAMEX study area encompassed most of the Midwest. Field operations included three aircraft, a ground-based mobile team that included the University of Alabama's Mobile Integrated Profiling System and special soundings. When thunderstorms began to form, the fleet of aircraft would fly out from their base near St. Louis to meet them—anywhere from Texas to North Dakota and Kansas to Ohio! Two of the aircraft were equipped with unique Doppler radar systems that allowed them to measure the wind fields at high resolution (about every 500 meters) throughout the depth of the storms. These two large aircraft (NOAA's P-3 Hurricane Hunter and the Navy Research Laboratory's P-3) flew along either side of the line of thunderstorms pointing their radars into the heart of the storms (Figure 18A), while a Learjet flew over the top of the MCS dropping dropwindsondes—devices that are similar to rawinsondes except that they record measurements while the instrument package is descending rather than ascending. The BAMEX datasets are now allowing scientists to obtain unique views of the internal structure of MCSs (Figure 18B) and are leading to a better understanding of the generation of severe winds that accompany these storms.

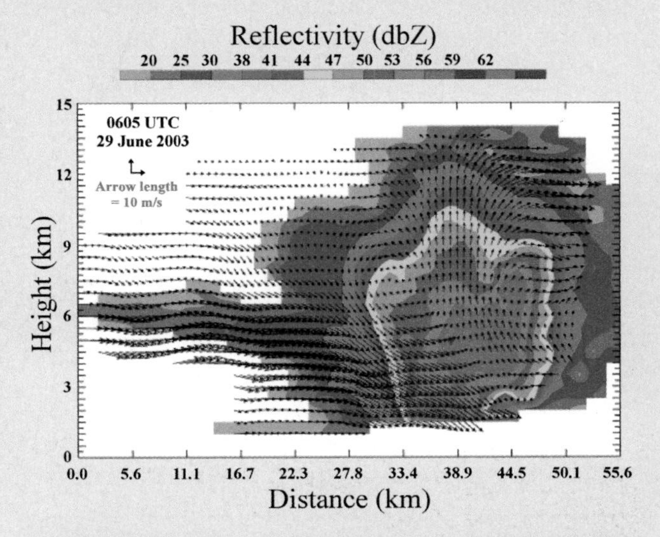

FIGURE 18B: A cross section through a BAMEX MCS that shows the radar reflectivity and the winds, updrafts and downdrafts within the storm. The winds were derived through analysis of data from the airborne Doppler radars.

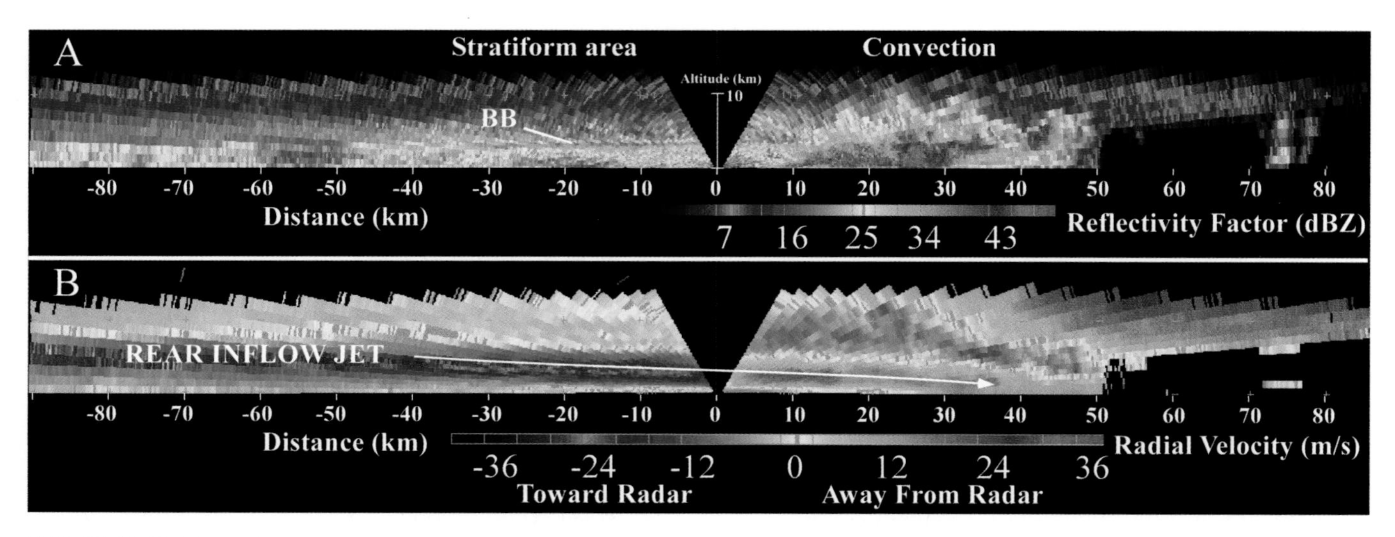

FIGURE 18.10: Vertical cross sections of radar reflectivity and radial velocity through a mature squall line that passed over Oklahoma and Kansas on 11 June 1985. The radar is located at the base of the cone of missing data near the center of the figure. (A) The radar reflectivity shows the convective region (red reflectivity) and the bright band (denoted by the symbol BB) in the trailing stratiform region of the storm on the left side of the figure. (B) The cross section of radial velocity through the squall line shows the rear inflow jet.

ONLINE 18.1

An MCS Squall Line

A mature squall line over central Missouri on 15 April 1994 produced strong straight-line winds over a broad area and a weak tornado near New Haven, Missouri. A radar animation of this squall line, which appears online, shows many of the key features of squall line structure. Note first the southwesterly winds ahead of the line. These winds can be inferred from the rapid northeastward movement of the weak echoes well to the east of the line of thunderstorms. The strong thunderstorms, indicated by the high (red) reflectivity, occur at the leading edge of the line. Behind the thunderstorms a broad area of weaker reflectivity stretches to the west—the trailing stratiform region of the squall line. A gust front, evident as a thin line of radar reflectivity to the east of the convection, moves out just to the east of the line of storms. The weak reflectivity at the leading edge of the gust front is associated with the shelf cloud forming over the rain-cooled outflow. The squall line develops a weak bow echo structure on its north end and a cusp in the center at the south end of the bow. Strong straight-line winds occurred within the bow, and the weak tornado developed near the cusp.

FIGURE 18.11: Visible satellite image of a mesoscale convective vortex (MCV) near the Arkansas-Missouri border on 24 May 2003. Note the swirling pattern of the cloud, indicative of cyclonic flow around the region of lower pressure in the center of the MCV.

FRONTAL SQUALL LINES

Frontal squall lines form in the warm, moist air ahead of surface cold fronts and dry lines, or just ahead of an upper-level front. These squall lines are typically hundreds of kilometers long, as illustrated on Figure 18.12, a radar image of a squall line extending from northwest Missouri south into central Texas on 10 November

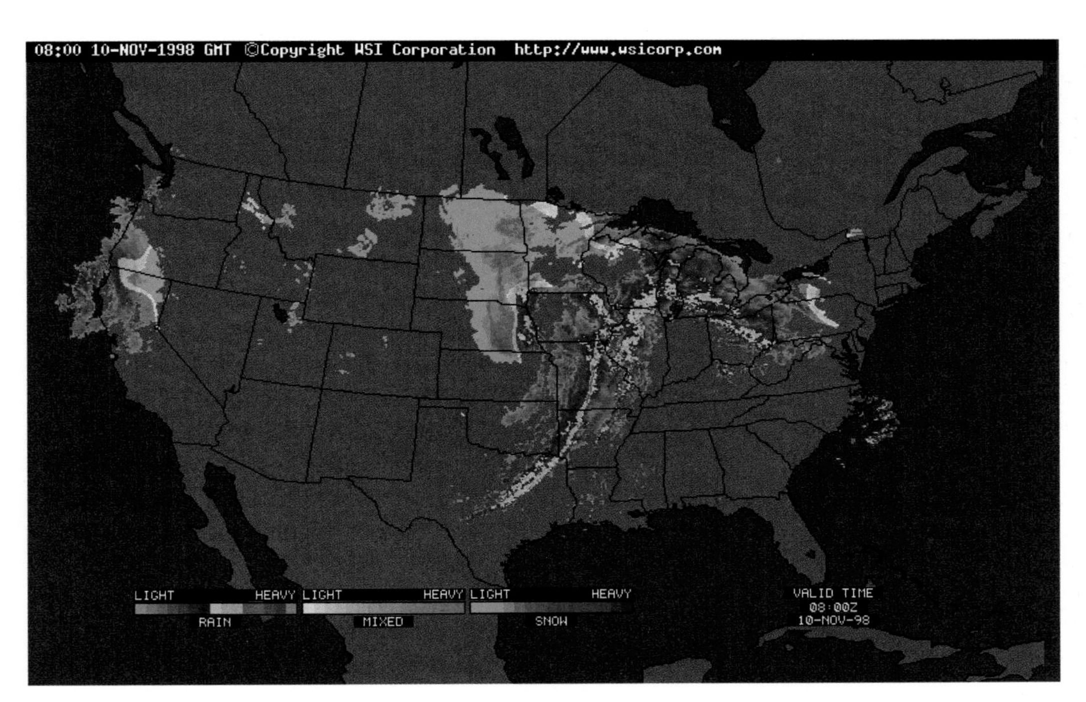

FIGURE 18.12: Radar image of a frontal squall line that developed on 10 November 1998 along a cold front. The squall line crosses Missouri, western Arkansas, and central Texas, and is identified by the high radar reflectivity indicated by yellow and red colors.

1998. Frontal squall lines commonly form the "tail" of the ***comma cloud*** pattern in extratropical cyclones, particularly those that form over land. A typical frontal squall line appears on radar as a line of high reflectivity, characteristic of heavy convective precipitation, with lighter, more stratiform precipitation both to the rear (typically west) and ahead (typically east) of the convection. From a satellite perspective, a frontal squall line appears as a long line of deep clouds as in the infrared image in Figure 18.13. Frontal squall lines can have long lifetimes, lasting many hours to days, since new thunderstorms can be continually triggered along the line as the front advances into the warm moist air ahead of the front.

Figure 18.14 shows the typical large-scale environment in which cool season squall lines develop. The squall line is normally part of the larger circulation of an ***extratropical cyclone***. The comma cloud of the cyclone is shown in gray. The cyclone's surface low-pressure center, indicated in Figure 18.14A, is under the divergent quadrant of the upper-level trough as shown in Figure 18.14B. In the example in Figure 18.14A, a cold front is shown as the ***trigger mechanism*** for the squall line. Low-level winds east of the squall line are southeasterly or southerly. Middle and upper tropospheric winds in the environment east of the squall line are typically southerly to southwesterly and much stronger than the surface winds.

Important structural features of a frontal squall line appear in Figure 18.15. Many of these features

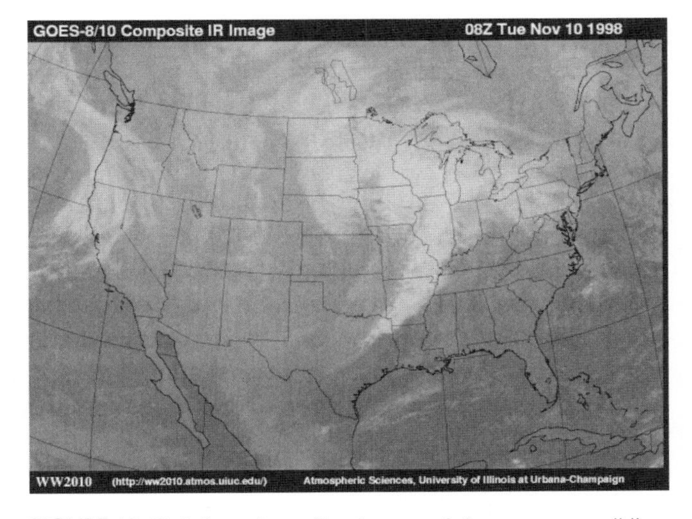

FIGURE 18.13: Infrared satellite image of the same squall line appearing in Figure 18.12. Note that the squall line comprises the tail of the comma cloud. The head of the comma cloud is over Nebraska and Iowa where the center of the low pressure is located.

are similar to squall lines associated with MCSs. As the cold airmass to the west advances forward, warm, moist air ahead of it is forced to rise. Once the moist air is lifted to its ***level of free convection***, the air rises spontaneously, creating deep thunderstorms. The ***stratosphere*** acts as a lid on the storms because environmental temperatures in the stratosphere increase with height (similar to a temperature inversion in the

FIGURE 18.14: A typical environment for frontal squall line development. (A) At the surface, ahead of the front, winds from the south-southeast transport warm, moist air into the line of thunderstorms. The front provides the lift required to trigger the squall line. (B) In the upper atmosphere a trough is located west of the squall line. Winds east of the squall line aloft typically are southwesterly or southerly. West of the squall line the winds are more westerly.

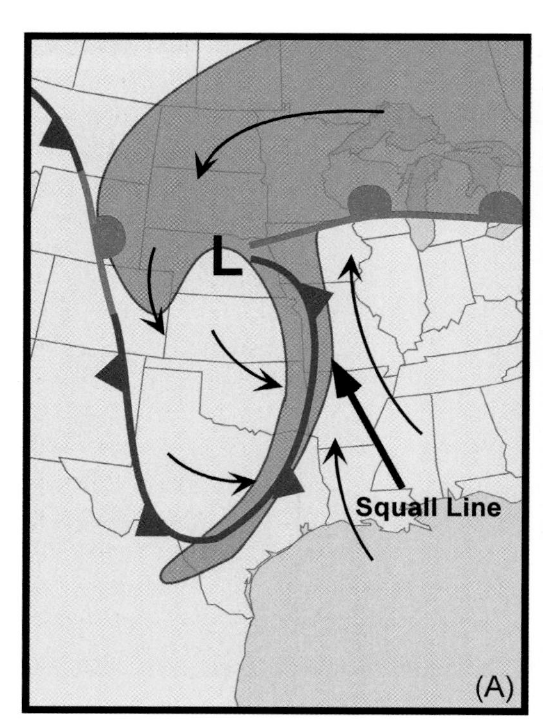

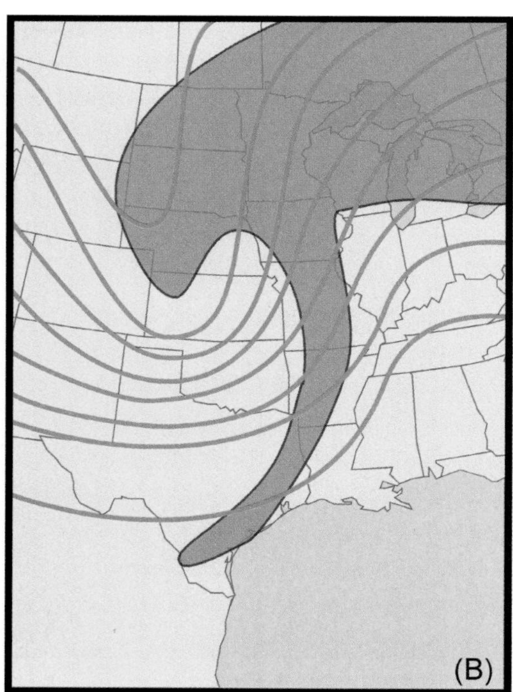

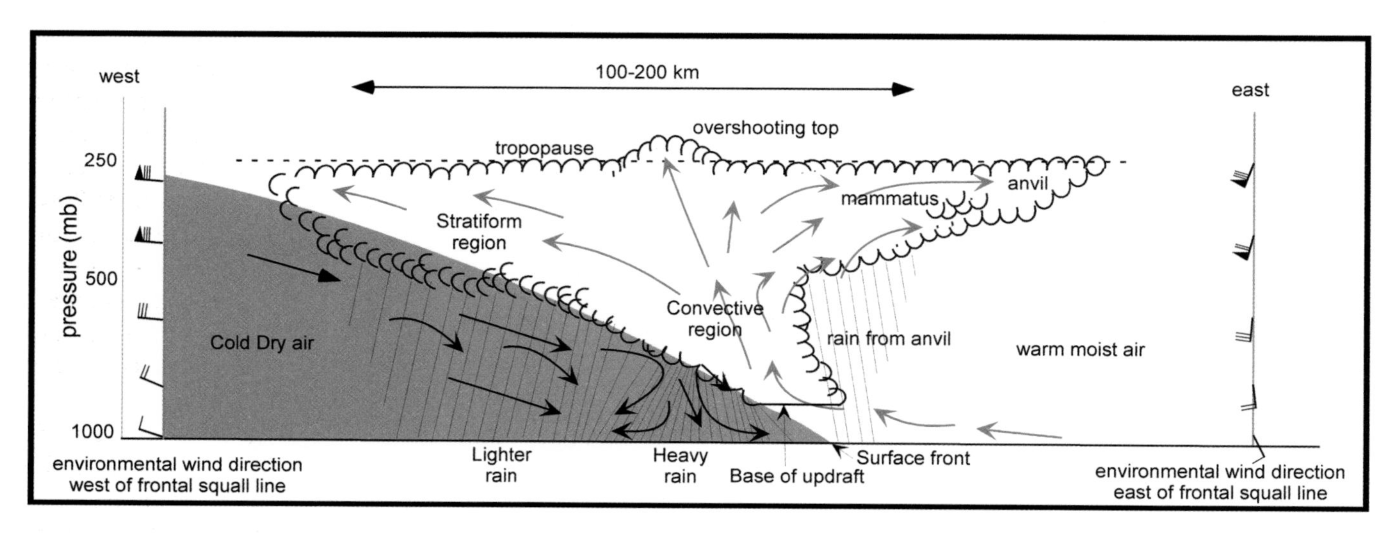

FIGURE 18.15: Schematic cross section through a frontal squall line showing key structural features. The blue shading denotes the cold airmass behind the front. Typical environmental winds west and east of the squall line are shown on the left and right sides of the figure, respectively.

troposphere). Nevertheless, when air rising through the thunderstorm reaches the tropopause, it often is still moving rapidly upward. Because of its momentum, thunderstorm air will intrude into the stratosphere creating the overshooting top. As the rising air approaches the tropopause, it spreads both rearward (west) along the frontal surface, and forward (east) to form the anvil. Colder, dry air typically approaches the line of thunderstorms from the west behind the front. The dry air is normally associated with strong middle and upper tropospheric westerly winds, which descend as they approach the convective region aloft in a manner similar to the rear inflow jet in MCS squall lines.

Although frontal squall lines can produce weak tornadoes and hail, they tend to be associated most often with strong straight-line winds. In some cases, when the front that forces the squall line is moving slowly or is nearly stationary, the individual cells composing the line move parallel to the front rather than advancing eastward. This situation, commonly called ***training*** in the media because cells move along the same track like the boxcars of a train, can lead to flash flooding (see Chapter 25).

ONLINE 18.2

Parallel Squall Lines

Squall lines sometimes evolve into several parallel lines of convection. The way these parallel lines form can be understood by viewing the online animation of the radar reflectivity during the initial triggering of a frontal squall line over the state of Illinois on 25 June 1997. On this day, a line of storms initially develops along a cold front. A gust front, associated with rain-cooled air generated by the storms, rushes eastward ahead of the line. The leading edge of the gust front appears in the radar reflectivity field as a very fine line of weak reflectivity, a signature of the shelf cloud over the gust front. Lifting of moist air by the gust front rapidly triggers a second line of convection along the gust front's leading edge. This line of convection in turn produces a new gust front, which triggers another line of convection. In a short time, several parallel lines of thunderstorms appear.

CHECK YOUR UNDERSTANDING 18.3

1. Along what type of fronts do frontal squall lines typically form?
2. During what time of year are frontal squall lines most common?
3. How large are frontal squall lines and how long can they last?

SUPERCELL THUNDERSTORMS

Supercell thunderstorms are the most intense thunderstorms in Earth's atmosphere. Supercell thunderstorms always rotate. A supercell thunderstorm's circulations so dominate its immediate area that the entire storm behaves as a single entity, rather than as a group of cells. These storms account for most tornadoes, virtually all severe tornadoes, damaging straight-line winds, and most large hail. Hail falling from supercells can grow as large as a grapefruit (see Chapter 20). Updrafts in supercells, determined using multiple Doppler radar measurements, are typically 20 to 40 meters/second (45 to 90 mph), but have been estimated to approach 50 meters/second (> 100 mph)!

Like other severe thunderstorms, supercell thunderstorms require four key ingredients to form: (1) an environment that is conditionally unstable—humid and warm at the surface and dry and very cold aloft; (2) very moist air in the lower troposphere; (3) moderate to strong vertical wind shear through the depth of the unstable layer; and (4) a trigger mechanism—lifting along a boundary. When these four ingredients come together, storms that form will rotate and organize as supercells. We consider first the initial three of these ingredients, and then we will examine different weather patterns in which supercells may be triggered.

Conditional instability develops in the atmosphere when air is heated near the surface while colder air aloft moves over the heated region. A direct measure of instability is the ***Convective Available Potential Energy (CAPE)*** (see Chapter 6), which is a measure of the rising parcel's positive buoyancy accumulated over its trajectory above its level of free convection. Simply put, CAPE is a measure of the maximum upward speed a rising air parcel will attain along its trajectory. CAPE values range from 0 to over 3000 joules/kilogram. Thunderstorms forming in low CAPE environments (< 500 joules/kg) rarely produce tornadoes and hail, while those forming in moderate to high CAPE environments (> 1500 Joules/kg) are more likely to have supercelluar characteristics and produce severe weather.

Figure 18.16 shows a sounding with characteristic high-CAPE. A warm moist layer is present in the lower troposphere. Aloft above the warm moist air, a dry airmass is present. Over the Great Plains, where supercells are most common, the moist airmass typically originates over the Gulf of Mexico, while the dry airmass originates over the mountains to the west. In the supercell environment, these two airmasses are often separated by an inversion, called a ***capping inversion***. The capping inversion exists because the dry air aloft has descended as it moved eastward away from the Rockies.

This inversion acts as a weak lid on updrafts originating in the warm moist layer, and keeps the first thunderstorm from forming too early, before the afternoon sun has a chance to heat the surface layer and provide more energy to the supercells. Once a parcel

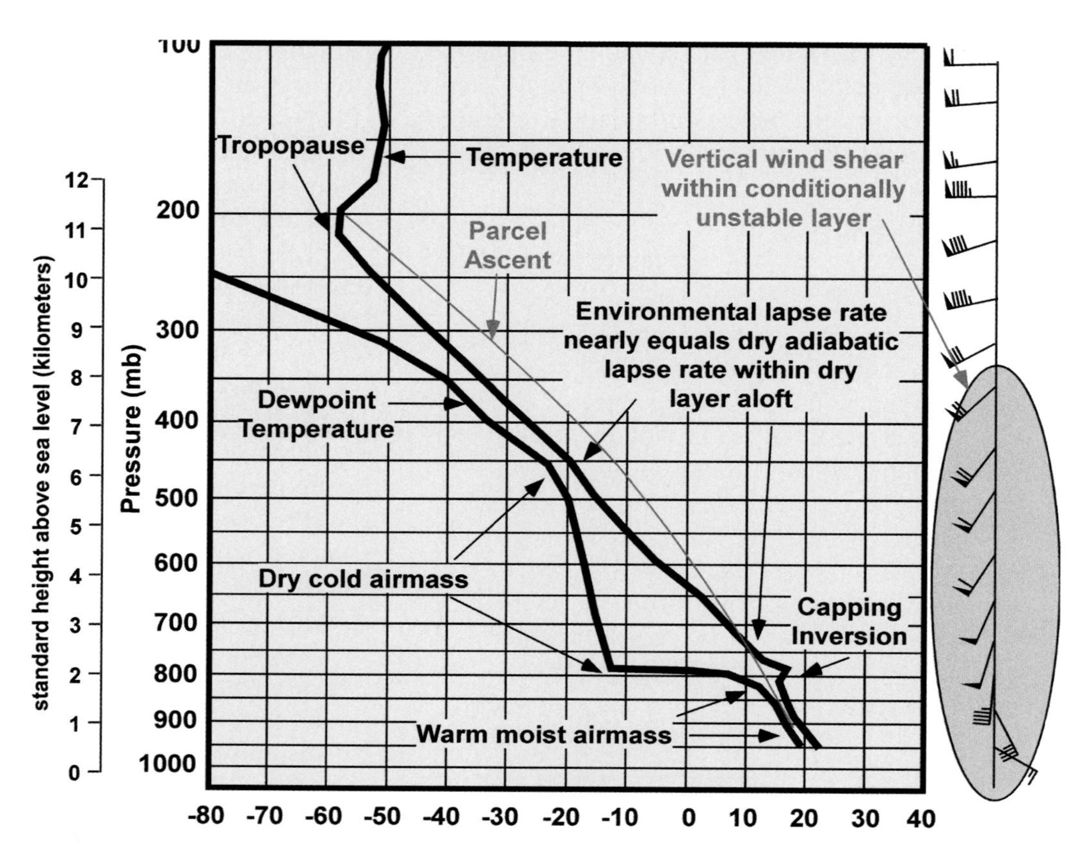

FIGURE 18.16: A typical sounding in the supercell environment showing a conditionally unstable atmosphere (high CAPE) with warm moist air in the lower troposphere, dry cold air in the middle troposphere, a capping inversion, and strong shear. The path of an unstable parcel ascending through this environment is shown in red.

of air breaks through the capping inversion, it will cool at the moist adiabatic lapse rate—but remain much warmer than its environment—and will rush upward as a violent updraft toward the tropopause. The rising air derives its energy from the release of latent heat as enormous quantities of water vapor are converted to liquid water droplets and ice particles within the rising updraft. The moist layer at the surface feeds this updraft, providing the fuel necessary to drive the rapidly growing supercell.

In a supercell environment, is it critical that the environmental winds increase rapidly with height within the layer of maximum instability. A relatively narrow band of very strong winds often develops just above the surface in the warm moist air on the Great Plains. This feature, called the ***low-level jet***, can be forced by differences in nighttime cooling rates across the Plains, or can be forced by strong pressure gradients developing along frontal boundaries. Winds can exceed 50 knots in the core of the low-level jet, which typically extends from just above the ground (friction keeps it from extending to the ground) to just above the capping inversion, typically no higher than about 3 km (about 700 mb). The low-level jet rapidly transports warm, moist air northward in the lower atmosphere and provides the low-level vertical wind shear that supports storm rotation (see Chapter 20). Aloft, winds typically continue to increase with height, and often a jetstream will be present just below the tropopause.

Provided the instability, moisture, and shear are in place, supercells can form in a number of weather patterns. In any of these weather patterns a trigger mechanism, some type of boundary, must be present to lift the warm moist air to its level of free convection. Figure 18.17 shows three examples of weather patterns where supercells may erupt. In Figure 18.17A, the warm moist air is located ahead of a dry line and south of a warm front. Lifting can occur along either or both of these boundaries. Figure 18.17B illustrates a situation where an old outflow boundary, from thunderstorms that occurred the previous day, remains across the Plains. As a cold front advances, lifting may occur along the front and/or along the outflow boundary, triggering supercells. In Figure 18.17C, an upper level front and warm front provide potential lifting mechanisms. Supercells have a tendency to first erupt where ***boundary intersections*** occur, such as the intersection of the dry line and warm front in Figure 18.17A, the outflow and the cold front in Figure 18.17B, or the upper level front and warm front in Figure 18.17C. Once a single supercell develops, gust front outflows from the first storm will often trigger new supercells. The new cells often trigger in rapid succession along the boundaries, creating a line of supercells (Figure 18.18).

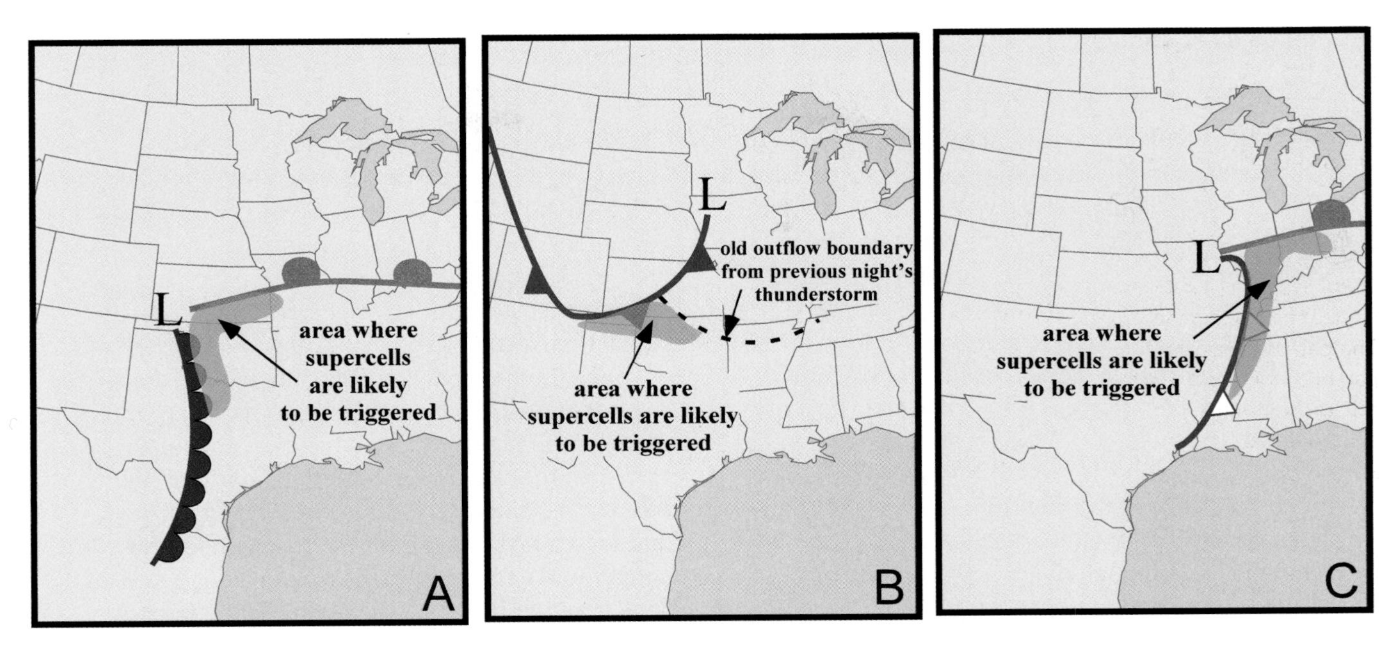

FIGURE 18.17: Examples of weather patterns favorable to supercell development. Supercells develop on warm side of the boundaries, particularly where they intersect one another.

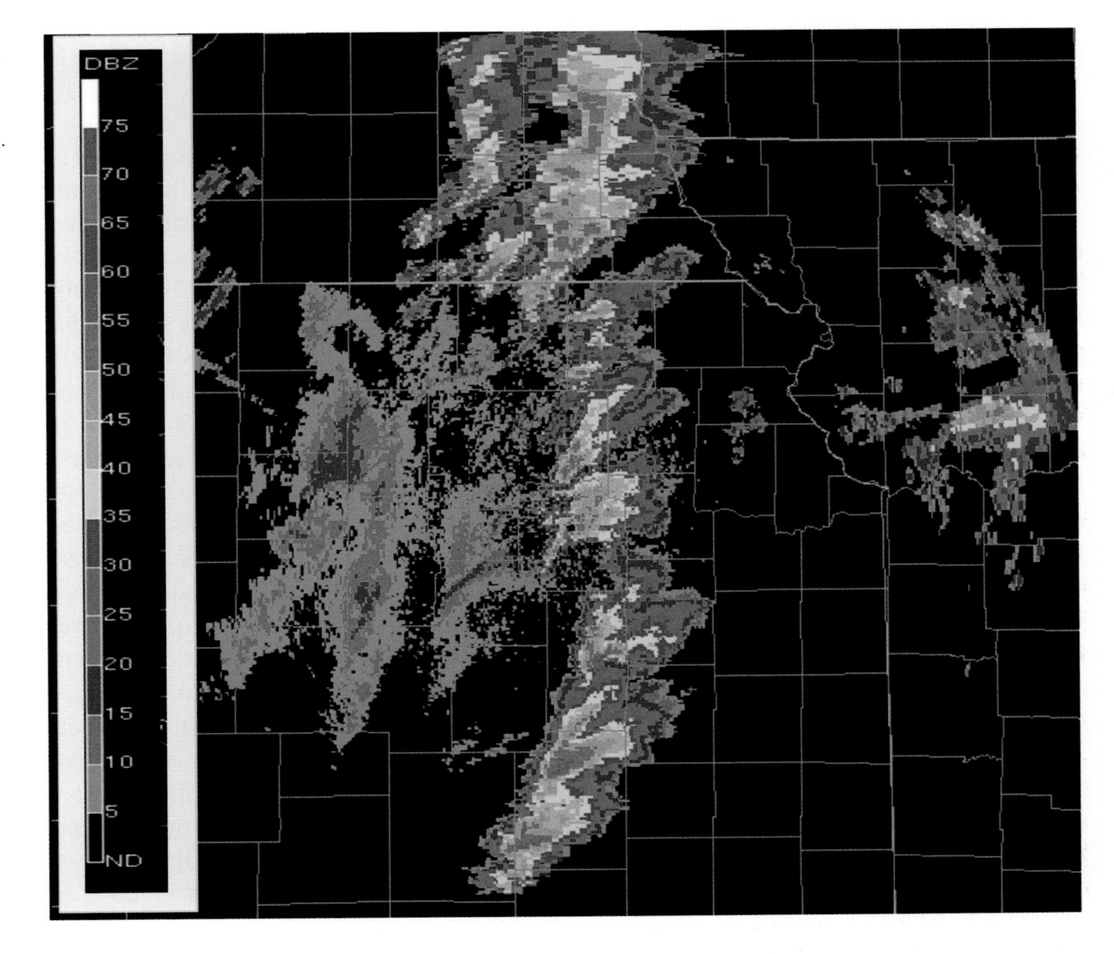

FIGURE 18.18: A line of supercells erupts along a frontal boundary in eastern Kansas and Nebraska.

ONLINE 18.3

Vertical Wind Shear and Storm Structure

Wind shear in the environment of a developing thunderstorm determines whether the storm will become a non-severe, short-lived storm or a supercell capable of producing severe weather. The contrast between thunderstorms growing in sheared and non-sheared environments can be dramatic. Online are two simulations of thunderstorms developed from numerical models similar to the models discussed in Chapter 4. The environmental temperature and moisture in the model atmospheres in the two simulations are identical; however the environmental winds in the two simulations differ substantially. In the first simulation, winds do not change with height, characteristic of the airmass thunderstorm environment. In the second simulation, winds increase and change direction with height as occurs in the supercell environment. The "cloud" in the simulations is actually the rendering of the concentration of cloud water calculated by the model. The 'surface' consists of wind vectors, with purple vectors indicating weak winds and red vectors indicating strong winds. In the "no shear" simulation, the cloud grows vertically, downdrafts develop, the precipitation falls out, and the storm dissipates—an airmass thunderstorm. A gust front can be seen emerging from the precipitation area as the storm dies. In the "strong shear" simulation, the storm grows and intensifies. Rotation is obvious. A rain-free cloud base develops under the updraft and a flanking line of clouds along the rear-flank downdraft. An anvil appears. The model creates a supercell thunderstorm.

Figure 18.19 shows a classic supercell thunderstorm as it would appear when viewed from the southeast. Individual storms will vary somewhat depending on environmental wind patterns. Figure 18.19 is typical of most supercells that occur on the Central Plains of the United States. The rotating updraft, also called the ***mesocyclone***, is located on the southwest side of the storm and typically ranges from about 5 to 10 km (3 to 6 miles) in diameter, and tilts northeastward with altitude. The air in a supercell thunderstorm updraft rises from the surface to the tropopause. The bulge of clouds at the top of the storm, called the overshooting top, is where the rapidly rising updraft air briefly penetrates into the stratosphere before descending back to the tropopause. A ***rain-free base*** is found beneath the updraft region. The rain-free base exists because cloud particles forming there are very small and are carried aloft by the strong updraft. In the middle and upper troposphere, the winds are strong and typically from the southwest, so precipitation particles are carried northeastward. As a result, precipitation rarely appears in the core of the updraft region, leaving a rain-free base. Just above the rain-free base, cloud particles are so small that they produce only weak or no radar echoes. The volume of the cloud in which weak radar echoes appear structurally resembles a gothic cathedral vault, hence the name ***echo-free vault***. This area is also called the ***Bounded Weak Echo Region (BWER)***.

A region of lower cloud, called the ***wall cloud***, will often appear hanging from the rain-free cloud base. The wall cloud can take on many different sizes and shapes—it may appear ragged, sometimes shallow, other times very large (Figure 18.20), and often will be visibly rotating. The wall cloud was once thought to be associated with the low pressure that develops at the base of the rotating updraft. Recent evidence suggests that the lowering of pressure can only account for a fraction of the wall cloud's extension below cloud base. Doppler radar studies suggest that the lowering of cloud base at the wall cloud actually develops as low-level moist air originating near cloud base altitude is wrapped around the rotating updraft from the region of the forward flank downdraft (see below). If a tornado develops, it will most often emerge from the wall cloud. Tornadoes from classic supercells nearly always occur outside the area where it is raining because they occur in the vicinity of the intense updraft circulation.

Supercells typically occur where strong middle and upper tropospheric jetstreams are present. These strong upper-level winds blow the ice particles in the upper part of the storm downstream (northeastward) into the storm's anvil. Anvils can extend hundreds of kilometers downstream of the parent storm. Far from the storm, the anvil is near the tropopause, but closer to the storm, the anvil can extend over a large depth, and, in some cases, can produce a large amount of rain at the ground. Often precipitation falling out of the leading edge of the anvil evaporates into drier air aloft, creating visible streaks called ***virga***. Mammatus clouds frequently form

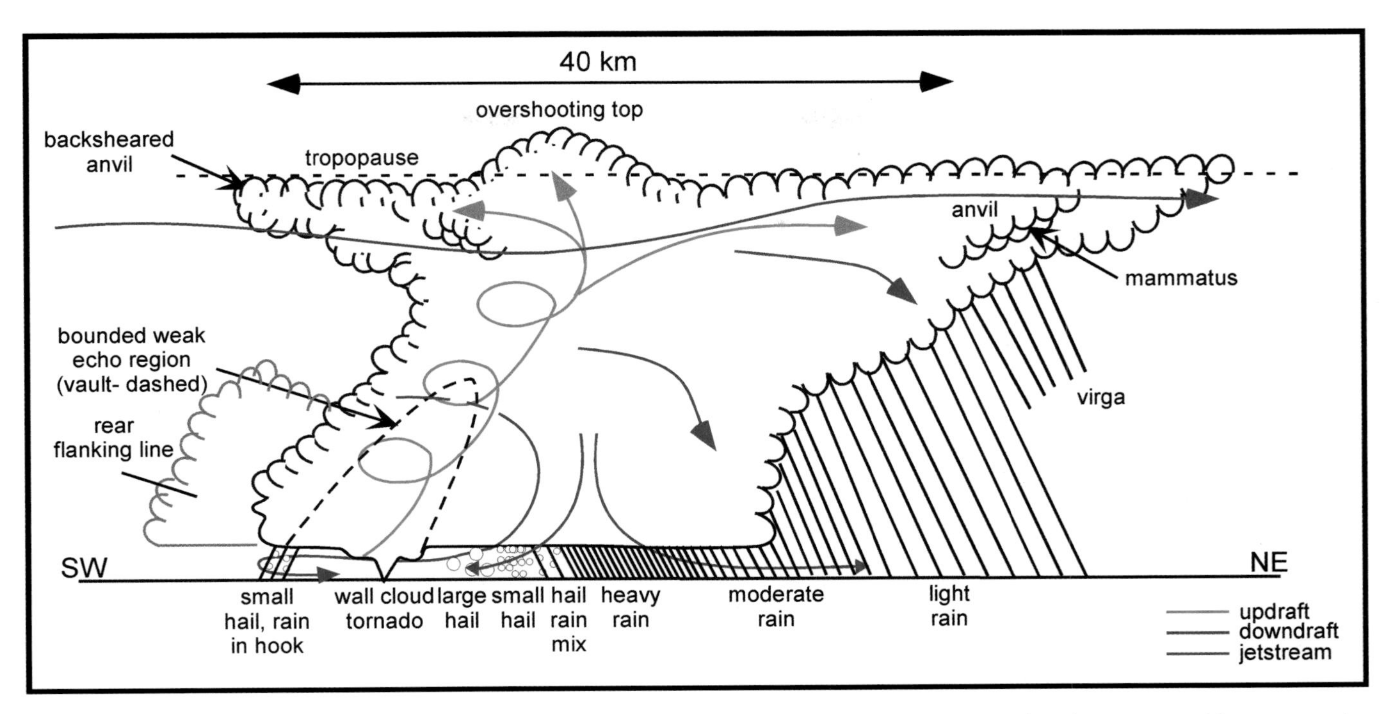

FIGURE 18.19: Cross section of a classic supercell thunderstorm from southwest to northeast showing structural features and the typical precipitation pattern. Red arrows indicate the rotating updraft (mesocyclone) and blue arrows show the downdraft. The purple arrow near the top denotes the jetstream flow diverting around the storm.

FIGURE 18.20: A wall cloud.

at the base of supercell anvils. To the southwest of a supercell, a line of convective cells, called the ***rear flanking line***, will often form. These cells align with the storm's rear flank gust front (see below).

Supercells can produce heavy precipitation, including large hail. Figure 18.21, a horizontal cut through a supercell, shows the location and type of precipitation within the storm. Compare this figure to the radar image of the supercell that produced a devastating tornado in Oklahoma City on 3 May 1999 shown in Figure 18.22. Note that the region of rainfall occurs to the north, northeast, and east of the updraft

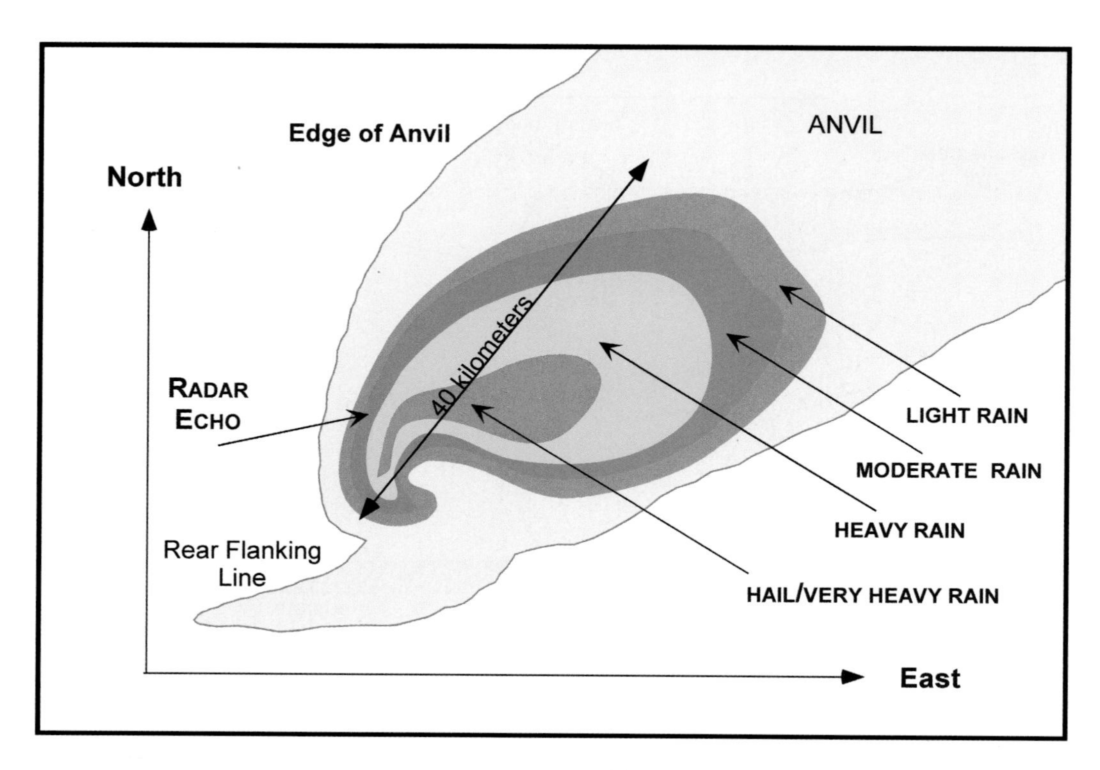

FIGURE 18.21: Plan view of a classic supercell thunderstorm showing the distribution of precipitation (colors) and clouds (gray). Within the colored region precipitation particles are large enough to be easily detected with radar. The cusp of the hook denotes the updraft region.

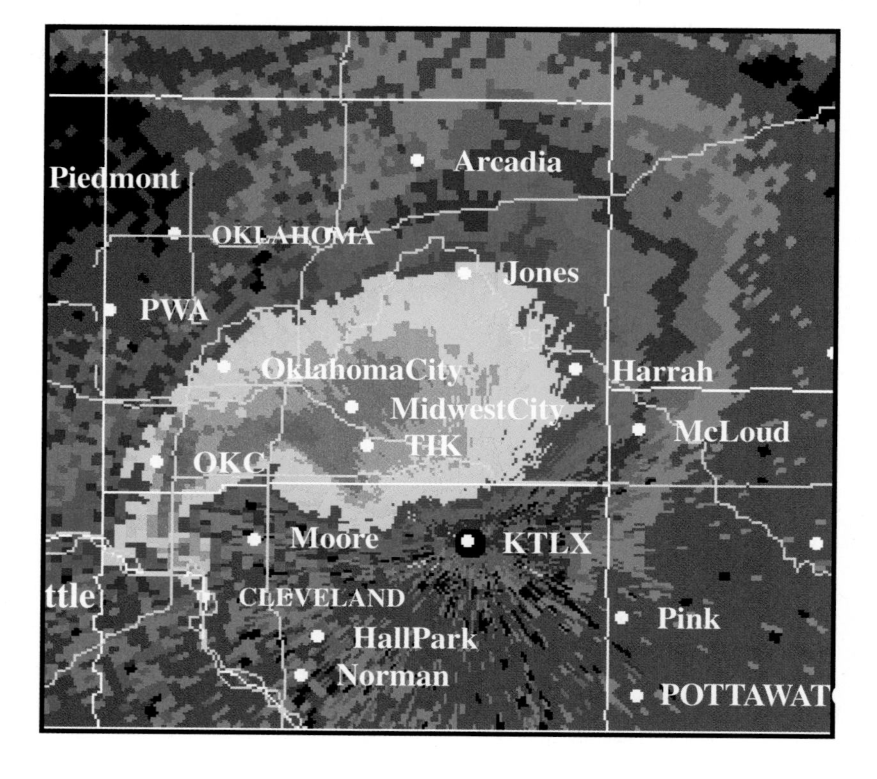

FIGURE 18.22: Radar reflectivity image of the supercell thunderstorm that devastated Oklahoma City, Oklahoma, on 3 May 1999. Note that the hook-shaped echo is similar to the schematic in Figure 18.21.

ONLINE 18.4

Views of Supercells

Views of all the key structural features of supercell thunderstorms appear online. Examine the photos of the rain-free base, the wall cloud, the hail and rain shafts, the back sheared anvil, the primary anvil, the updraft, the gust front, arcus and roll clouds, mammatus clouds, and other features of these magnificent storms. Compare these features with the storm model presented in the text.

region, often over an area extending about 30 to 50 km (19 to 30 miles) northeast of the updraft, which in Figure 18.22 is located at the center of the hook-shaped reflectivity region on the southwest side of the storm. Hail, if it occurs, will fall on the perimeter of the updraft, right next to the rain free base. The largest hail falls closest to the rain free base (see Chapter 20). Moving northeast through the thunderstorm, away from the rain free base one would encounter smaller hail, heavy rain, moderate rainfall, and then light rain.

Airflow patterns within a supercell thunderstorm are closely tied to cloud formation and precipitation processes. Latent heat release in areas of condensation and freezing contributes to the strength of updrafts. Cooling in areas of evaporation and the downward drag force produced by the falling precipitation contribute to the strength of downdrafts. The environmental winds strongly influence the structure of supercells. Supercells move in the direction of the mid-tropospheric winds, which are generally southwesterly. As a supercell moves over a town, the people in the town would progressively experience the conditions on the northeast side of the storm, then the center, then the southwest side. Meteorologists call the northeast side the ***forward flank*** of the storm and the southwest side the ***rear flank*** because the forward flank is the first, and the rear flank the last part of the storm an observer normally experiences. Meteorologists also use this terminology to describe the downdrafts and gust fronts in these sectors of a supercell.

Figure 18.23 shows the flow in a supercell at various levels, including updrafts and downdrafts, and how these flows evolve in time. Figure 18.24 shows a

FOCUS 18.3

Low Precipitation (LP) and High Precipitation (HP) Supercells

Classic supercell thunderstorms produce heavy precipitation and have widespread radar echoes similar to Figure 18.20. There are some supercells, particularly those forming in the vicinity of the dry line in the Southern and High plains of the United States, that do not produce heavy precipitation. These storms take on strange appearance, the clouds within the relatively narrow storm updraft rotating upward from cloud base to the tropopause, an anvil extending downstream aloft, and relatively little precipitation emerging from the storm. The acronym LP (for "low precipitation") is used to describe these storms. Since *LP supercells* produce little precipitation, they lack the strong precipitation-produced cold downdrafts and associated gust fronts that characterize classic supercells. A different class of supercells produce extremely heavy precipitation. Spotting tornadoes in these high precipitation, or *HP supercells,* is much more difficult that in classic supercells, since rain will often be wound into the mesocyclone, obscuring the view from the southwest. HP supercells tend to form in more moist environments and are often found along warm frontal boundaries.

FIGURE 18C: A low precipitation supercell thunderstorm.

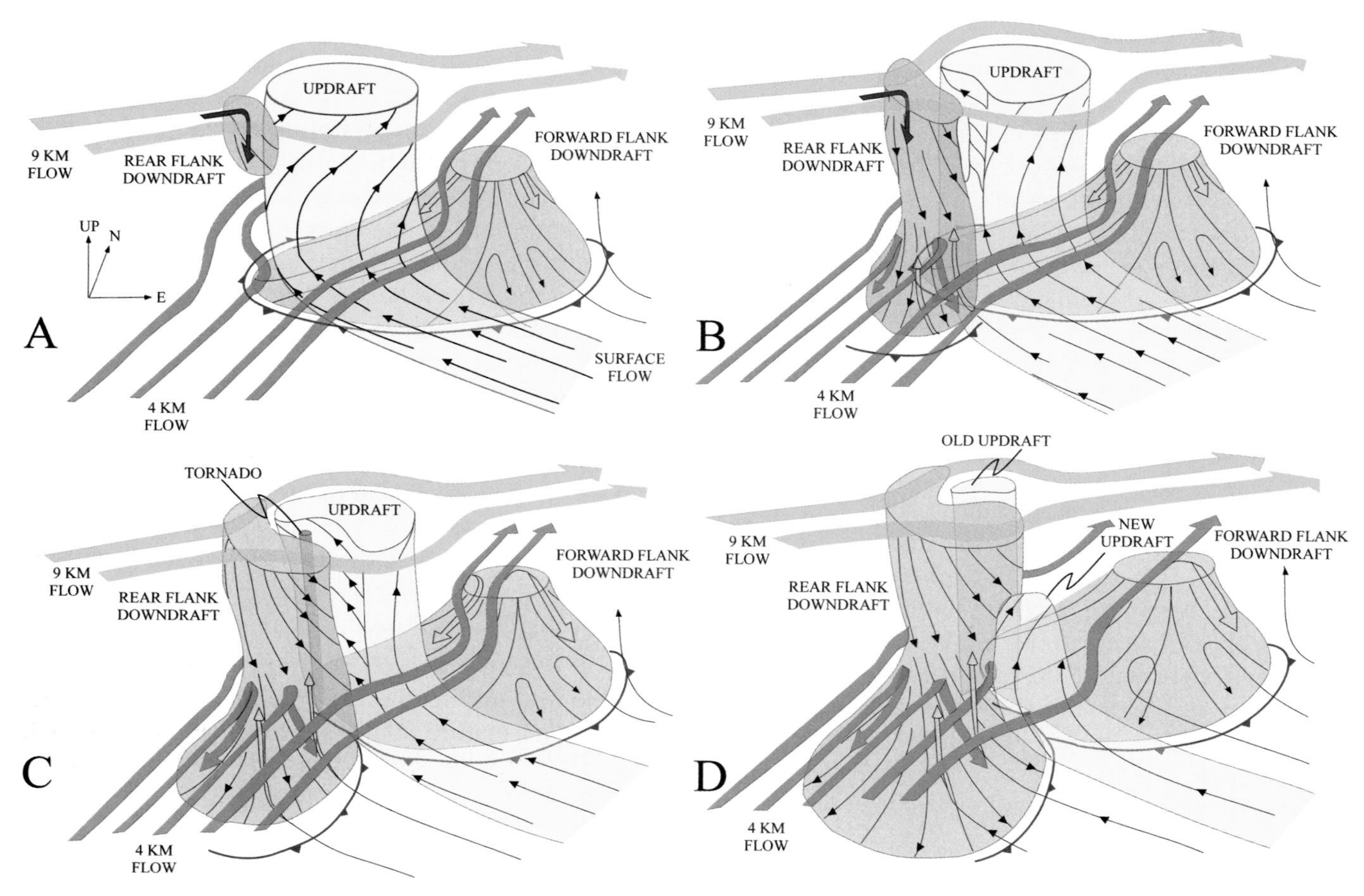

FIGURE 18.23: Schematic three-dimensional depiction of evolution of a supercell thunderstorm. Updrafts/downdrafts and the tornado location are identified. Conventional frontal symbols are used to denote outflow boundaries (gust fronts) at the surface. Lines with arrows denote the flow relative to the storm. Blue regions denote downdrafts and yellow regions updraft. The orange and green arrows denote the flow at 9 km and 4 km respectively. (A) shows the initial formation stage of the supercell with the forward flank downdraft formed and the rear flank downdraft developing. (B) shows the full formation of both forward and rear flank downdrafts. (C) models the mature supercell with a strong rotating updraft and tornado (red) located at the coupling of the updraft and rear flank downdraft. (D) depicts the decaying supercell storm—the rear flank downdraft wrapping around the updraft and cutting off the supply of warm, moist air. A new updraft is forming to the southeast of the previous updraft.

slice through the supercell at low levels identifying regions of ascending (updraft) and descending (downdraft) air.

The inflow to the updraft initially approaches the core of the updraft from all directions. Precipitation quickly begins to fall to the north, northeast, and east, as precipitation particles within the storm are carried downwind of the updraft core by the middle- and upper-level winds. Figure 18.23A shows this early stage of a supercell. The middle- and upper-level flow, shown by the long arrows at 4 and 9 km altitude, is from the southwest and west, respectively. Note that the upper-level flow must divert around the updraft, which blocks the flow within the rapidly moving jetstream aloft. Where precipitation falls, evaporative cooling and drag lead to the formation of downdrafts.

The first downdraft to form is the ***forward flank downdraft (FFD)***. Most of the air in the downdraft has its origin in the lower to middle levels of the storm from altitudes of about 2 to 4 km (~ 800 to 650 mb). The downdraft air reaches the surface and spreads rapidly outward in all directions, creating at its leading edge the ***forward flank gust front*** (see Figure 18.24). Gust fronts are shown in Figures 18.23 and 18.24 using cold front symbols. As mid-level air approaching the storm from the southwest encounters the updraft, cloud and precipitation particles on the rear flank of the storm mix with the dry air and evaporate. This air cools in this region, and

ONLINE 18.5

Record-Breaking Supercell on March 12, 2006

Numerous supercell thunderstorms occurred across the central United States during the period from 9–13 March 2006. One of these supercells, which occurred on March 12, set the record for both longevity and track length. Early that day, several supercells originated in northern Oklahoma and southeastern Kansas, including the one that would set the record. The storms moved northeastward as a group, tracking across Missouri during the afternoon and early evening. The two southernmost supercells of this group retained their supercellular structure while traveling in tandem into Illinois. At this point, the northern storm of the pair merged into the southern supercell. This supercell continued across Illinois, passing into Indiana during the later evening. The storm finally lost supercellular characteristics over Indiana, but continued on into Michigan where it finally dissipated. The track of this remarkable supercell was nearly 1300 km long (~800 miles). The storm lasted over seventeen hours, making it the longest path and duration supercell on record. Radar animations of the supercell, and the larger scale outbreak, from formation to dissipation, appear online, along with severe weather reports along the storm's track.

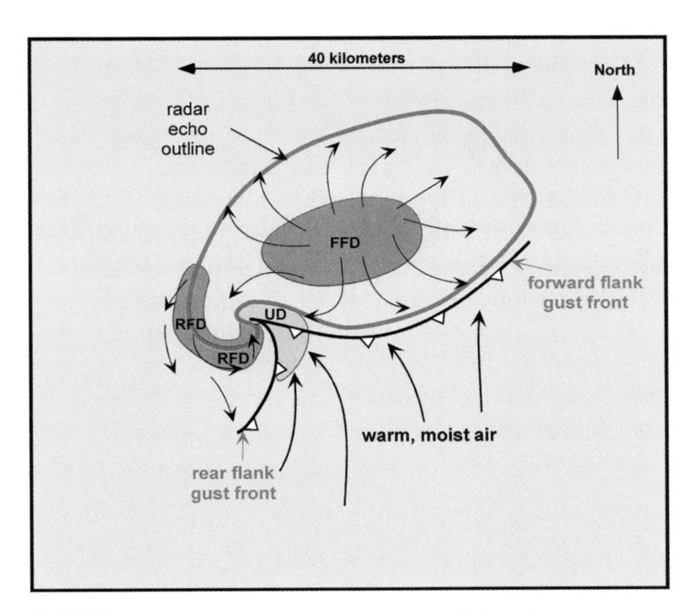

FIGURE 18.24: A slice through a supercell thunderstorm parallel to the Earth's surface showing low-level airflow patterns. FFD and RFD denote the forward flank downdraft and rear flank downdraft, respectively. UD indicates the location of the updraft. The *dashed lines* indicate the location of the forward- and rear-flank gust fronts. The solid outline indicates where radar detects precipitation or large cloud particles.

descends to the surface creating the ***rear flank downdraft (RFD)*** (see Figures 18.23B, 18.23C, and 18.24). As air from the rear flank downdraft reaches the ground and spreads out, it creates the ***rear flank gust front*** (Figure 18.24). New cells are often triggered along the rear flank gust front as it advances, creating a flanking line of convection that extends southwest of the updraft region (Figure 18.19). The forward and rear flank gust fronts move outward and rotate around the supercell. As the supercell reaches its mature stage, shown in Figure 18.23D, the rear flank downdraft wraps around the updraft, cutting off its supply of warm moist air. It is at this time that a tornado may form, as discussed in the following chapter.

CHECK YOUR UNDERSTANDING 18.4

1. What is a supercell thunderstorm? What characteristic distinguishes supercells from other types of severe thunderstorms?
2. What type of severe weather do supercells produce?
3. What are boundary intersections and what role do they have in supercells?

THUNDERSTORMS AND GLOBAL WARMING

From the global perspective, thunderstorms are small features in Earth's atmosphere. While cloud-resolving climate models are now being developed, the current generation of global climate models used to simulate future climates of the Earth under conditions of increased greenhouse gases simply do not have the computational capability to simulate individual thunderstorms. For this reason, few studies exist that examine changes in the frequency, intensity or organization of thunderstorms in the warmer world predicted for the future. We can, however, provide some speculation based on predictions of future climate and the information we know about thunderstorms, as presented in this chapter.

Global climate models predict that the length of the winter season will be reduced in the middle latitudes as a result of global warming. Since thunderstorms are rare in the winter and common in summer, one could speculate that the earlier onset of spring in the middle latitudes should lead to an early onset of the thunderstorm season, and an increase in the total number of thunderstorms in the middle latitudes. A shift in the primary severe weather season (April–June) should also occur as spring arrives earlier in the middle latitudes. However, this assumption is only true if the other conditions required for thunderstorms and severe thunderstorms, including moisture, a lifting mechanism, and vertical wind shear, are not altered in a way that might reduce thunderstorm frequency.

As we learned in Chapter 7, the temperature contrast between the polar and tropical regions is directly related to the intensity of the jetstream. Global climate models predict a significant reduction in the pole to equator temperature contrast as a result of global warming, suggesting that the jetstream will, on average, have a reduced intensity in the future. The reduction in the temperature contrast is expected to be greatest over continents, suggesting weakened airmass contrasts, and associated fronts. As we learned in Chapter 10, a reduced jetstream intensity is likely to cause a reduction in the frequency and intensity of continental extratropical cyclones. These larger-scale changes in cyclones have direct implications for the occurrence of both ordinary and severe thunderstorms.

Thunderstorms, particularly severe supercell thunderstorms, develop most often within the circulations of extratropical cyclones. Cyclones provide the ingredients, moisture, and shear. The predicted reduction in cyclone and jetstream intensity suggests that, climatologically, vertical wind shear will be reduced, leading to a general reduction in the frequency of thunderstorms with supercell characteristics, and that thunderstorms occurrence itself will be reduced.

The behavior of thunderstorms in the summer season is even harder to predict. The longer summer season should lead to an increase in the annual number of mesoscale convective systems. Again, this depends on the presence of the other thunderstorm ingredients, moisture and lift. In the current climate, these vary substantially from season to season, leading, for example, to the extremes experienced two decades ago in 1988 (a severe drought with virtual no MCSs) to 1993 (a severe flood, with several MCSs occurring almost every day). There is no reason to expect that this variability will change in a warmer climate, so the frequency and distribution of these thunderstorms is very difficult to assess. Finally, there may be local changes in summer thunderstorm frequency associated with specific geographic features. For example, along the east coast of North America in summer, thunderstorms often erupt during the day as cool air moves inland over the coastline and lifts air that has been heated over the adjacent land, a phenomena known as the sea breeze. Greater land-sea temperature contrasts will enhance sea breeze circulations, and may lead to an increase in thunderstorm frequency along the coastline.

In short, our understanding of the behavior of thunderstorms in warmer climates of the future is speculative at best. Many factors influence thunderstorm frequency and intensity, and these factors are linked together in complex ways that make it difficult to sort out their individual effects. It is for this reason that the 2007 Intergovernmental Panel on Climate Change reports are vague about thunderstorms.

If our understanding of the behavior of thunderstorms in warmer climates is speculative, our understanding of their smaller-scale features, tornadoes, hail, lightning, and strong straight-line winds, is even worse. For this reason, we will not discuss the effects of global climate change on these phenomena in the upcoming four chapters. Scientists understand that more research is required before we can make definitive statements about the future behavior of these magnificent and illusive phenomena associated with thunderstorms in Earth's atmosphere.

CHECK YOUR UNDERSTANDING 18.5

1. Do global climate models simulate thunderstorms? Why or why not?
2. How is the transition from winter to summer likely to change the frequency of thunderstorms in a global warming scenario?
3. Will global warming increase the frequency of thunderstorms?

TEST YOUR UNDERSTANDING

1. Summarize the average annual impact of thunderstorms on the United States.

2. List the four necessary elements for severe thunderstorm formation. What types of weather maps or images would you examine to see if each element was present?

3. What are the three ways in which severe thunderstorms typically organize?

4. Draw a vertical cross section of an airmass thunderstorm in all three stages. Label all the important features. Briefly describe each stage.

5. What two mechanisms lead to the formation of downdrafts in thunderstorms?

6. What time of year are mesoscale convective systems most common?

7. Describe how a bow echo develops within an MCS.

8. Describe how the cold pool of an MCS develops. What is the role of the cold pool in the generation of new updrafts?

9. Define and distinguish between the following: bow echo, mini-bow, rear inflow jet, derecho, and bookend vortex.

10. Sketch a vertical cross section of a thunderstorm within an MCS. Identify all associated features including: anvil, updraft, downdraft, location of the convective region, stratiform region, melting level, cold pool, gust front, rain shaft, overshooting top, shelf cloud, rear inflow jet, front-to-rear flow, and tropopause.

11. Explain why there is a bright band in a vertical cross section of the reflectivity field of a radar observing an MCS.

12. If you were asked to determine where severe winds were occurring in an MCS, what feature would you look for in the reflectivity field from a radar?

13. What causes the development of a mesoscale convective vortex in the last stages of an MCS?

14. How would you identify a frontal squall line on an infrared satellite image? How about a radar reflectivity image?

15. Sketch a vertical cross section of a frontal squall line thunderstorm and identify the anvil, overshooting top, convective region, location of the surface front, regions of heavy and light rain.

16. Describe how the speed and direction of the environmental winds change throughout the troposphere (a) ahead of a frontal squall line, and (b) behind a frontal squall line.

17. What types of severe weather are typically associated with frontal squall lines?

18. What does the term "training" mean in relation to frontal squall lines?

19. Which direction does a frontal squall line usually move? Why?

20. Explain why frontal squall lines often form ahead of advancing fronts but seldom behind them.

21. Why would a "high CAPE environment" be favorable for supercell formation?

22. What is the low-level jet and what role does it have in supercell development?

23. What is a mesocyclone and where is it found in a supercell thunderstorm?

24. What sequence of precipitation types would you experience if the core of a supercell were approaching you?

25. What are the horizontal dimensions of a supercell?

26. Draw a vertical cross section through a classic supercell thunderstorm. Identify all associated features including: anvil, backsheared anvil, wall

cloud, updraft, downdraft, rain shaft, rain-free base, overshooting top, approximate height of tropopause, location of tornado formation, location of hail, and mammatus clouds.

27. Draw a diagram of a classic mature supercell thunderstorm viewed from above. Identify: the forward flank downdraft, the rear flank downdraft, the forward flank gust front, the rear flank gust front, the updraft, and the region of heaviest precipitation.

28. What physical processes lead to the formation of the forward-flank and the rear-flank downdrafts?

29. Rank the four types of thunderstorms by size.

30. How does the anvil of a severe thunderstorm differ from the shape of an airmass thunderstorm anvil? Explain.

31. During a particularly active spring with many severe thunderstorms reported, your friend comments to you that he is not surprised—after all, global warming is occurring. How do you respond?

TEST YOUR PROBLEM-SOLVING SKILLS

1. Suppose you are in the path of an idealized thunderstorm in which heavy precipitation is falling over a circular area, 10 km in diameter, at a rate of 5 cm per hour. If there is no precipitation outside of this area and the storm's intensity does not change during the next few hours, how much precipitation will you receive in each of the following cases?
   (a) The storm is moving at 40 km per hour and its center passes directly overhead.
   (b) The storm is moving at 40 km per hour, and its center passes 3 km away from your location.
   (c) The storm is moving at 10 km per hour and its center passes directly overhead.
   (d) The storm is moving at 10 km per hour and its center passes 3 km away from your location.

2. The weight of a typical thunderstorm's water (liquid and ice) is about 3.4 trillion grams (Focus 18.1). Using the following information about the density of air, estimate the weight of the *air* in a thunderstorm that is 10 km tall. What percentage of the total weight (air plus water) does the thunderstorm's water represent?

Elevation (km)	Density of Air (kg per cubic meter)
0 (sea level)	1.20
2	1.01
4	0.82
6	0.66
8	0.53
10	0.41

3. At 1 PM local time, a squall line develops along the Hudson River from Albany, New York, through New York City and southward to Atlantic City, New Jersey. The winds at the tropopause level are from the west (270°) at 100 mph. The individual thunderstorm cells containing the precipitation and lightning, steered by the mid-tropospheric winds, move from the southwest (225°) at 40 mph. Use a road atlas to estimate distances and determine the times when the following will arrive at Boston, Massachusetts:
   (a) the leading edge of the anvil clouds
   (b) the precipitation and lightning

4. The following wind profile was obtained from a location at which thunderstorms were about to develop:

Pressure (mb)	Wind Direction	Wind Speed (mph)
1000 (surface)	160°	20
900	180°	40
800	200°	50
700	220°	60
600	230°	70
500	240°	80
400	250°	95
300	260°	110
200 (tropopause)	270°	125

(a) Construct a hodograph (arrows pointing in direction of airflow, lengths proportional to speed) of the winds as a function of altitude.
(b) Suppose the thunderstorms move with the mean wind (averaged over pressure) in the troposphere. What will be the speed and direction of the thunderstorm motion?
(c) What is the inflow speed (relative to the moving thunderstorm) of the air entering the thunderstorm at the surface? At 900 mb?

USE THE SEVERE AND HAZARDOUS WEATHER WEBSITE

http://severewx.atmos.uiuc.edu

1. Radar is generally the most useful tool for determining the geographical distribution of thunderstorm activity at a particular time and for identifying the type of thunderstorms that are present. Airmass thunderstorms often have a "popcorn" appearance on radar (and satellite) images with the thunderstorms distributed randomly over a large area. Airmass thunderstorm activity varies diurnally, developing during the afternoon and dissipating after sunset. Navigate the *Severe and Hazardous Weather Website* to the National Weather Service radar imagery.
   (a) Monitor the radar images and animations over a period of two weeks during spring or summer. Record the locations and periods of airmass thunderstorm activity.
   (b) Do the outbreaks of airmass thunderstorms occur during the times expected, based on the diurnal cycle of heating and cooling? If not, why not?
   (c) Where are boundaries located relative to where the thunderstorms developed?
   (d) What was (were) the trigger(s) responsible for the airmass thunderstorms? How did you determine this?
   (e) Examine the radar closely for any outflows from the thunderstorms (they will appear as radar fine lines). Where do they develop relative to the original thunderstorm?

2. Monitor the weather and on a day when thunderstorms are expected to develop. Navigate the *Severe and Hazardous Weather Website* to the "Current Weather" page and examine radar and satellite imagery of the region expecting the storms. Visible satellite imagery allows the identification of thunderstorms by the development of their anvils, which can expand rapidly over one or two hours. Radar reflectivity can help identify the intensity and type of thunderstorms.
   (a) Use the visible satellite imagery to track the development of thunderstorms by their anvil clouds. Record the location and time of the storms. Estimate the size of the anvil clouds.
   (b) Does the satellite image provide any evidence of an overshooting top?

(c) Examine a radar reflectivity image for the same time.
   (i) What is the approximate size of the precipitation region?
   (ii) What is the most intense reflectivity appearing on the image?

(d) What is the ratio of anvil to precipitation area?

(e) Animate both the satellite and radar imagery. Compare and contrast the direction of movement of the storms on both sets of imagery.

3. Watch the weather patterns during spring or fall and find at least one example of (a) a frontal squall line, (b) an MCS, and (c) a supercell. Record the date, time, and location of each storm. For each storm, explain how you identified the storm type.

4. The Storm Prediction Center (SPC) issues convective outlooks several times per day for the contiguous United States. The outlook is a categorical forecast that specifies the perceived level of threat of severe thunderstorms (slight, moderate, and high risk). Use the *Severe and Hazardous Weather Website* to navigate to the SPC website.
   (a) On a day when there is at least a slight risk of severe thunderstorms, record the region under risk and the time the forecast is valid.
   (b) Examine at least two forecast soundings from within the risk area for the period when thunderstorms are expected. (You can find forecast sounding links on the "Forecast Links" page.) Record each sounding location and forecast time.
   (c) Examine the soundings for environmental conditions conducive to severe thunderstorms: low level moisture, instability and wind shear. Record whether each is present. Explain how you arrived at your conclusion.
   (d) Record the forecast values of CAPE, CINH, SWEAT, Total Totals, Lifted Index, and Showalter Index. Compare the values to those in Table 6.2 in Chapter 6.
   (e) Examine the forecast hodograph for the same time and location. Does the hodograph suggest strong wind shear will develop? Explain.
   (f) Based on the information provided in the forecast soundings and hodographs, do you agree with the SPC forecast? Why or why not?
   (g) Monitor the weather, and after the forecast period compare the actual event with the SPC forecast and your forecast. Did the forecasts verify? Why or why not?

5. On a day when there is at least a slight risk of severe thunderstorms, read the Storm Prediction Center's discussion of the convective outlook. The discussions are typically written in a type of shorthand using abbreviations to save space. "Translate" the discussion. Note all elements of the discussion that match the conceptual models of thunderstorm development provided in this chapter.

CHAPTER NINETEEN

TORNADOES

An EF3 tornado approaches Lyndon, Kansas on May 8, 2003.

KEY TERMS

cold air funnel
conservation of angular momentum
Convective Available Potential Energy (CAPE)
Doppler on Wheels (DOW)
decay stage
dust devil
dust swirl stage
dynamic pipe effect
Energy Helicity Index (EHI)
enhanced Fujita scale
fire whirl
Fujita scale
funnel cloud
gustnado
hook echo
horizontal wind shear
landspout tornado
mature stage
mesocyclone
mesocyclone signature
mesovortices
non-supercell tornado
occlusion downdraft
organizing stage
severe thunderstorm watch
shrinking stage
Storm-Relative Helicity (SRH)
storm spotter
stretching
suction vortices
tilting
tornado
tornadogenesis
Tornado Alley
tornado family
tornado vortex signature
tornado watch
vertical wind shear

vortex breakdown
vortex stretching
warning
watch
waterspout

Tornadoes are violently rotating columns of air that extend from a thunderstorm cloud to the ground. Exceptionally strong tornadoes can destroy steel-reinforced structures, throw automobiles over 30 meters (100 feet), and sweep trains off their tracks. On average, over 1,000 tornadoes are reported in the United States each year. Over the 12 year period ending in 2006, an average of 56 people have been killed and another 975 injured in the United States annually by tornadoes. In this same period, tornadoes were responsible for $855 million in property losses.

Tornadoes primarily develop within supercell thunderstorms, but also form in thunderstorms along squall lines, near the ends of thunderstorm bow echoes, and within landfalling hurricanes. Tornadoes typically range in width from about 50 to 800 meters (150 ft to 0.5 mile) and have wind speeds that range from 65 mph to over 200 mph (57 knots to over 174 knots). A few rare tornadoes have had exceptionally wide damage paths, approaching 1600 m (1 mile). About 75 percent of all tornadoes occur in the United States. Most tornadoes are short lived, although some can remain on the ground for over an hour and produce damage swaths that extend over 50 km (30 miles).

Long-lived tornadoes typically undergo a five-stage life cycle. When a tornado is forming, the first indication might be a ***funnel cloud*** emerging from the wall cloud. The tornado itself begins when the rotation makes contact with the ground ***dust swirl stage*** (Figures 19.1A and 19.1B). During the ***organizing stage*** the funnel cloud descends to the ground and increases intensity (Figure 19.1C). The intensity of the vortex peaks during the ***mature stage.*** It is during this time that the tornado is at its largest size and is often nearly vertically erect (Figure 19.1D). Although most tornadoes remain small, some expand into giants during this stage, reaching diameters approaching or even exceeding 0.8 km (0.5 miles). During the ***shrinking stage*** the vortex tilts over more (Figure 19.1E), often beginning to take on a rope-like appearance. Eventually, tornadoes stretch into rope-like formations in their ***decay stage*** until they finally dissipate (Figure 19.1F).

ONLINE 19.1

A Tornado Gallery

Photographs of tornadoes at various stages of their lifetime appear online. The photographs were selected to illustrate all the phases of a tornado life cycle, from the initial appearance of the condensation or debris funnel, through the multiple vortex stage to the final rope stage.

TORNADO FORMATION IN SUPERCELLS

Supercell thunderstorms rotate about a vertical axis as a result of a process called ***tilting,*** which is illustrated in Figure 19.2. Recall that supercells typically form in an environment with strong vertical wind shear where winds rapidly increase with height above the ground. Imagine for a moment a ball placed in a flow where the air passing over the top of the ball is moving faster than the air passing under the bottom of the ball. The ball would begin to rotate, its axis of rotation parallel to the ground. Just like with the ball, regions of ***vertical wind shear*** (wind speed increasing with height) induce rotation within air. The axis of rotation is parallel to the ground, similar to the axis of rotation of the ball. In a supercell thunderstorm, warm air within the shear layer is drawn into the thunderstorm updraft. As this happens, the axis of rotation is tilted into the vertical, as shown in Figure 19.2. Tilting of the air within the shear layer is the ultimate source of rotation of the updraft.

Figure 19.3 illustrates the location of the rotating updraft within supercell thunderstorms. The rotating updraft is part of a storm circulation that scientists describe as a ***mesocyclone*** (a small-scale cyclone) because it has some structural characteristics analogous to an extratropical cyclone. Recall from Chapter 18 the location of the rear flank downdraft, the forward flank downdraft, the updraft, gust fronts, surface flow fields, and the rain as observed by radar. An area of lower pressure is co-located with the center of rotation. The gust front associated with the forward flank downdraft occupies the position that a warm front would occupy in an extratropical cyclone (although its structural characteristics have no similarity to a warm front). The rear flank downdraft gust front circulates around the low-pressure center, analogously to an

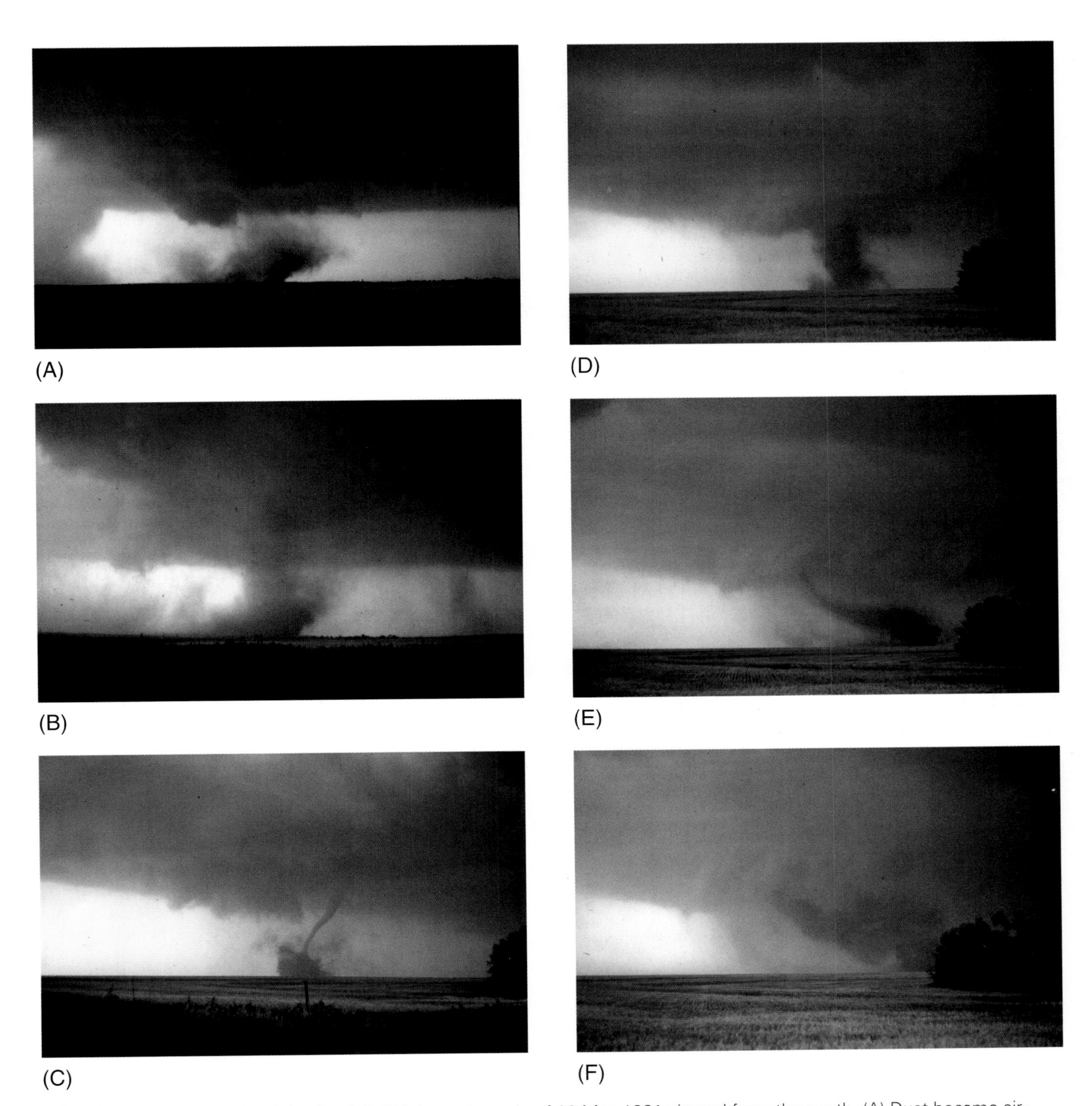

FIGURE 19.1: The life cycle of the Cordell, Oklahoma tornado of 22 May 1981 viewed from the south. (A) Dust became airborne underneath the wall cloud at 5:20 pm; (B) a rotating dust column formed under the wall cloud and obscured any existent condensation funnel at 5:22 pm; (C) at 5:26 pm a narrow condensation funnel became well developed and visible as the rotating dust column disappeared, perhaps because the tornado had moved away from a recently plowed field; (D) the tornado picked up a dust sheath again, surrounding the condensation funnel at 5:26 pm; (E) at 5:28 pm the tornado roped out; (F) the tornado dissipated at 5:28 pm.

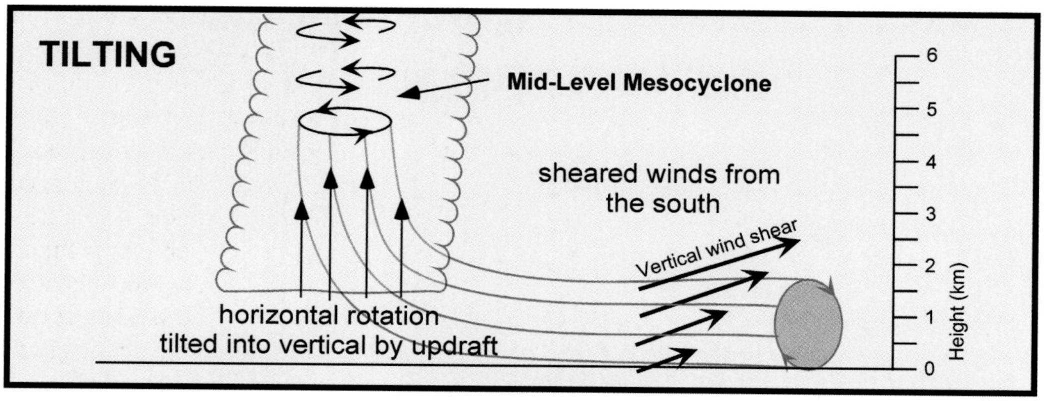

FIGURE 19.2: Supercell rotation requires the presence of wind shear in the layer where air is buoyant, particularly in the lower atmosphere. Vertical wind shear creates rotation with the axis of rotation parallel to the ground. The rotation is tilted into the vertical by the developing updraft.

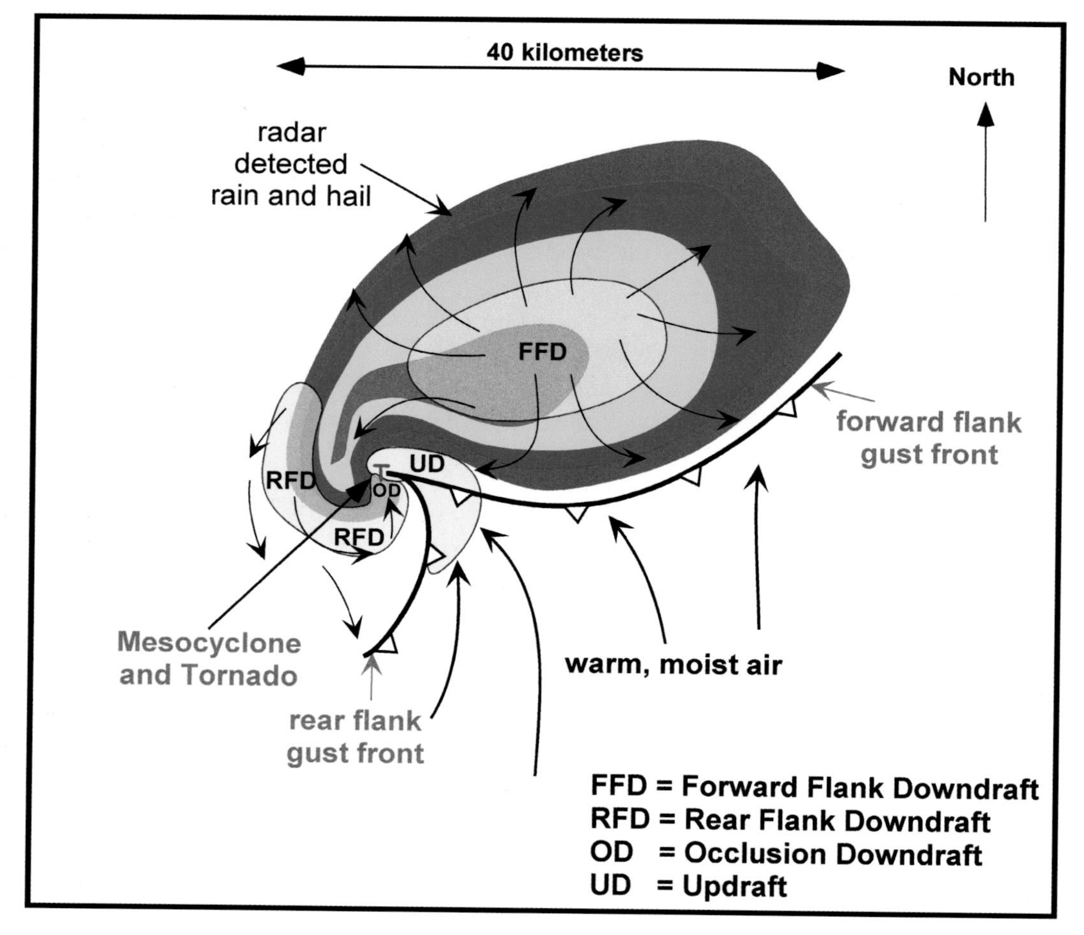

FIGURE 19.3: A plan view of a supercell thunderstorm. Color-coded areas denote precipitation (blue, light rain; green, moderate rain; yellow, heavy rain; red, very heavy rain and hail). The positions of the updraft, downdrafts, gust fronts, mesocyclone, and tornado are noted.

extratropical cyclone's cold front, eventually "catching up" to the forward flank downdraft gust front, creating an "occlusion." The updraft air is rapidly rising aloft above the occlusion. Keep in mind that the width of a mesocyclone is typically about 5 km (3 miles), unlike the extratropical cyclone, which may extend thousands of kilometers.

The source of rotation of the supercell is well understood and the structural features of the mesocyclone have been documented in many supercells. However, scant information exists about how tornadoes form within a mesocyclone. Scientists recognize that a key process, called ***stretching*** (Figure 19.4) is required to concentrate the mesocyclone rotation. Just as skaters spin up by drawing in their arms and legs toward the axis of rotation, air stretched in a narrower and narrower column will rotate faster and faster. Physicists call this principle the ***conservation of angular momentum***. Conservation of angular momentum requires that an object rotating about an

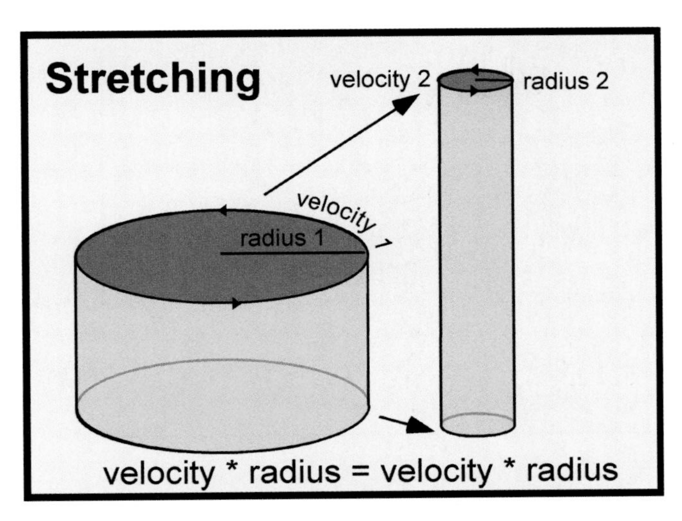

FIGURE 19.4: The rotating column in the supercell is stretched into a narrow column. As the rotating air shrinks into a narrower column, wind speeds increase following the law of conservation of angular momentum.

axis maintain a speed such that the product of its rotational speed and its distance from the axis of rotation remains constant.

Tornadogenesis, the formation of a tornado, typically occurs near the time the supercell's forward and rear flank downdrafts meet under the mesocyclone. At this time, the rear flank downdraft air surrounds and isolates the rising updraft at the mesocyclone center. The low-level part of the updraft circulation, cut off from the warm air source, now rises more slowly than the updraft aloft, so the entire air column stretches. Just like the ice skater pulling in her arms, the low-level rotation is increased dramatically in this process, which is called ***vortex stretching***.

Exactly how stretching proceeds and leads to the narrow tornado vortex is still unknown. One reason is the relative infrequency of tornadoes in supercells. Although vortex stretching undoubtedly occurs in almost all mesocyclones, available statistics suggest that less than 30 percent of all supercells that produce distinct mesocyclone signatures on radar actually produce a tornado. Another reason is the scarcity of measurements in the lowest levels of the atmosphere where tornadoes develop. Radar normally does not observe the region close to the ground where tornado genesis is believed to occur. Very few supercells have ever been observed in sufficient detail to study tornadogenesis. Much of what we have learned comes from model simulations, theoretical studies and a few well-observed storms.

Tornadogenesis appears to occur in one of three ways. The first of these may be thought of as a top down process, where the tornado descends from midlevels within the storm and then emerges from the base of the wall cloud. Scientists have described this process as a ***dynamic pipe effect***. To understand the dynamic pipe concept, think of a narrowly constricted flow in the middle atmosphere that might develop when the mid-level mesocyclone is stretched. Imagine this as a pipe. Air entering this narrowly constricted pipe region from below must itself constrict as it approaches the entry point. That constriction in effect lowers the pipe—the constriction grows downward. This process can continue to the ground as long as air below the pipe is rotating as illustrated in Figure 19.5. When it reaches the ground, tornado touchdown occurs.

A second theory can be considered a bottom-up approach. This process is believed to occur as a result of tilting of the horizontal circulation along the forward flank gust front as it moves under the ascending updraft. To understand how this works, examine Figure 19.6. Air behind the gust front is cool, negatively buoyant, and descending. In the updraft air adjacent to the gust front, air is warm, positively buoyant, and ascending. This leads to a sense of rotation along the interface between the gust front and warm air. If this region advances under the strong updraft of the mid-level mesoscyclone, it can be tilted into the vertical, leading to rapid rotation very close to the earth's surface. With further vortex stretching, this rotation can spin up to become a "bottom up" tornado.

Another explanation for the process of tornadogenesis has emerged from field studies. In the spring tornado seasons of 1994 and 1995, scientists carried out the Verification of the Origins of Rotation in Tornadoes Experiment (VORTEX), a highly instrumented field campaign designed to capture the tornadogenesis process within a mesocyclone. To accomplish this difficult task, the project deployed mobile ground-based radars, meteorological instruments on chaser cars, and a special dual Doppler radar called ELDORA on a research aircraft operated by the National Center for Atmospheric Research. Eight tornadoes were examined at close range. Unprecedented data, such as the remarkable tornado column that extends from cloud base to top in Figure 19.7, were obtained with the ELDORA radar. One of the best data sets collected occurred on 16 May 1995, when the aircraft sampled a supercell throughout the development of both the mesocyclone and a tornado near Garden City, Kansas. The analysis, by tornado specialist Roger Wakimoto and colleagues, held surprises for the tornado research community. To understand what was learned, it is

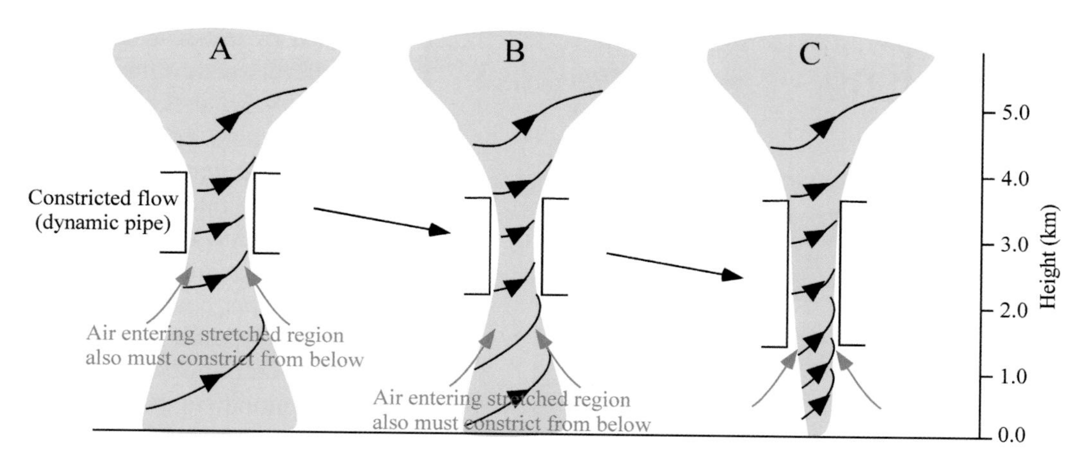

FIGURE 19.5: Illustration of a dynamic pipe effect. The stretched vortex in A causes air below it to constrict its circulation, leading to a tighter vortex in B. This process proceeds earthward until the vortex is narrow at the ground in C, creating a tornado.

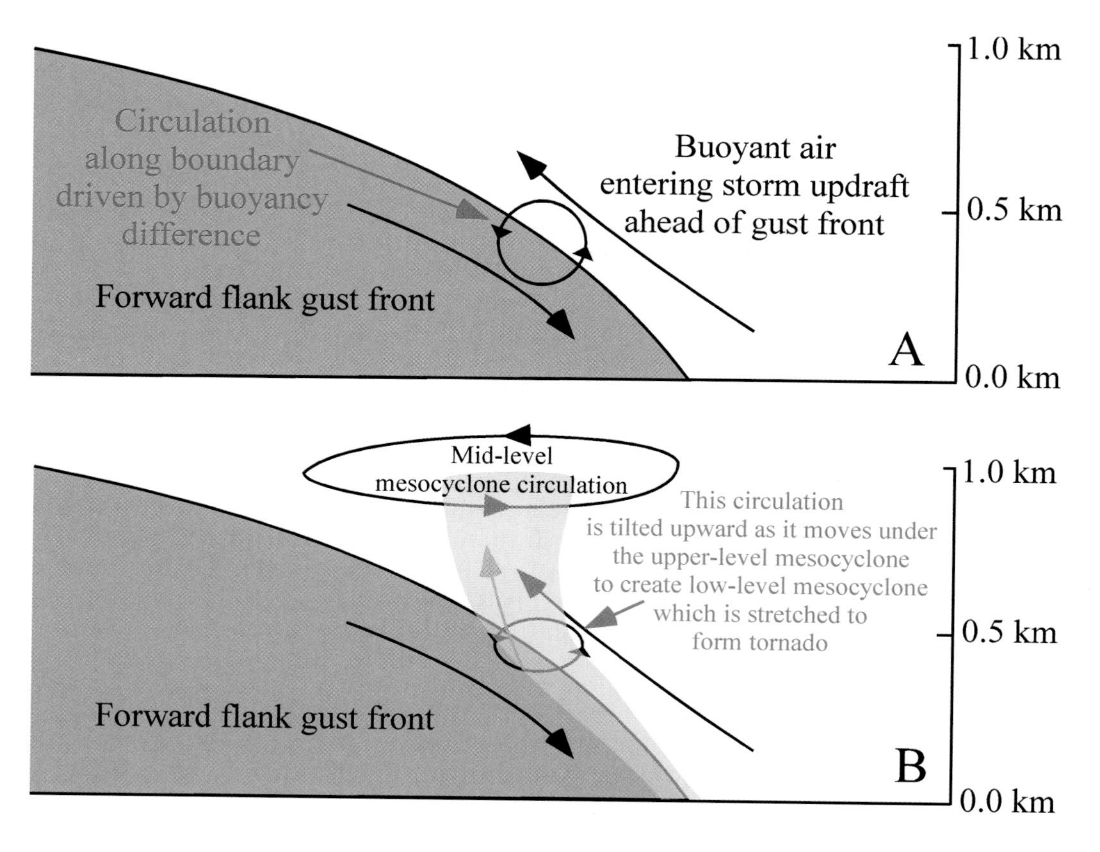

FIGURE 19.6: The "bottom-up" process of tornadogenesis. (A) Descending air behind the gust front and ascending air ahead of the gust front create a region of rotation about a horizontal axis along the gust front. (B) As the gust front moves under the mesocyclone updraft, the horizontal axis of rotation is tilted into the vertical, resulting in an intense low-level mesocyclone circulation that is stretched to form the tornado.

worthwhile first to examine the process by which a small tornado grows to become a giant tornado.

Most tornadoes remain as narrow columns of rising rotating air. In some tornadoes, a process called ***vortex breakdown*** occurs (see Figure 19.8A–C), which can cause the tornado to expand to a very large size. Laboratory studies of tornadoes have shown that during vortex breakdown, a downdraft develops at the center of the tornado vortex. The downdraft results when low pressure at the center of the tornado becomes so extreme that the air in the center of the tornado is no longer in hydrostatic balance (the low pressure causes air in the center to descend, forming a core downdraft, see Figure 19.8B). Recent measurements by storm chaser and engineer, Tim Samaras, made by deploying a probe called the "Tinman" directly in the path of a tornado near Manchester, South Dakota, showed that this pressure deficit can reach 100 mb! As the downdraft progresses down the vortex to the surface, the tornado vortex expands around the central downdraft. When the downdraft reaches the ground (Figure 19.8C), the tornado can become 400 to 800 meters (0.25 to 0.5 miles) wide. These large tornadoes often develop smaller vortices,

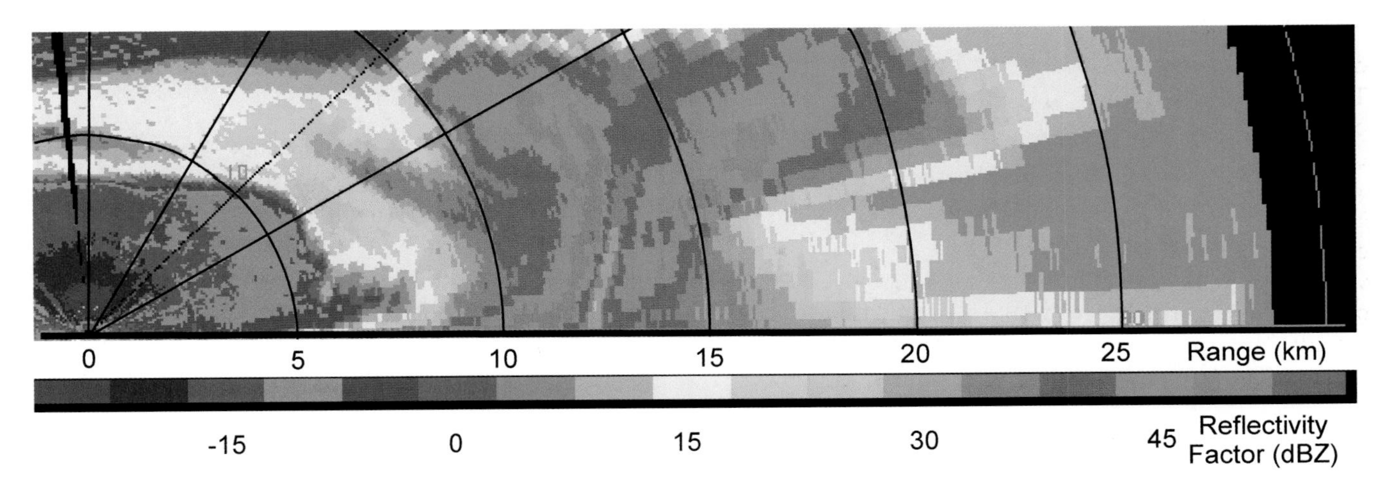

FIGURE 19.7: Vertical cross section of the radar reflectivity through a supercell thunderstorm and tornado near Friona, Texas, on 2 June 1995 measured by the ELDORA radar onboard the National Center for Atmospheric Research Electra aircraft. The data were collected during VORTEX, the Verification of the Origins of Rotation in Tornadoes Experiment.

called ***suction vortices***, within their circulations (Figure 19.8D). The suction vortices form as the central downdraft outflow merges with the rotating air outside the downdraft in the manner shown on Figure 19.8D. The most violent parts of a tornado are the suction vortices, where winds can approach or even exceed 250 knots (290 mph). Tornadoes having as many as five suction vortices have been captured on video.

Wakimoto's analysis of the Garden City tornado showed that a similar process occurred in the mesocyclone during tornadogenesis. In the Garden City case, a central downdraft occurred within the mesocyclone circulation. The tornado developed as the central downdraft outflow within the mesocyclone merged with rotating air in the outer part of the surface mesocyclone. In a similar manner to the suction vortex formation mechanism in Figure 19.8D, one of the resulting vortices spun up to form the Garden City tornado. The downdraft occurring in the vicinity of the mesocyclone, which had been observed in other cases but not associated directly with tornadogenesis, is called the ***occlusion downdraft*** by scientists studying tornadoes.

At present, the Garden City tornado is the only tornado where the genesis process has been observed in such detail. Whether this process is common or rare in unknown, and how its frequency compares with tornadoes forming through the dynamic pipe effect or through tilting of the buoyancy gradient along the gust front remains speculative at best. It is possible that there are other ways that tornadoes form in supercells. Scientists are still studying the VORTEX data, other data sets, and numerical modeling simulations to gather additional evidence about tornado formation.

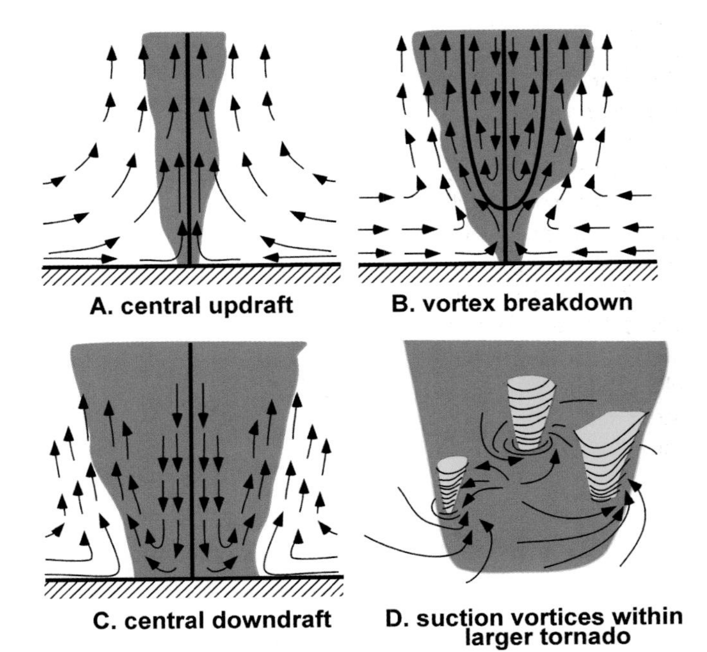

FIGURE 19.8: Some large tornadoes are believed to develop through a process known as vortex breakdown. In this process, the tornado vortex (shaded) transforms from a rotating updraft (A) to a structure with a downdraft at its core and updraft displaced to the outside of the central downdraft (B, C). When the central downdraft reaches the ground, strong shear between the downdraft and updraft areas leads to the formation of smaller vortices within the tornado called suction vortices (D). The strongest winds in tornadoes occur in suction vortices.

FOCUS 19.1

Tornado Winds

The violent winds in a tornado result from the contraction of the mesocyclone vortex during the time when the updraft undergoes stretching. This can be understood by examining Figure 19A. Using the law of *conservation of angular momentum,* this figure shows how the rotational velocity of a rotating column changes as it contracts to the scale of a tornado. The law of conservation of angular momentum states that the product of the rotational velocity (v) and the distance from the center of rotation (r) is a constant. A typical rotational wind speed at a radius (r) of 4,000 meters (2.4 miles) from the center of a developing mesocyclone might be 2.5 meters/second (~ 5 knots). In this case, v × r = 10,000 meters²/second. If the radius of the circulation contracts to 100 meters, v × r still equals 10,000 meters²/second so now the rotational speed must now equal 100 meters/second (200 knots!).

The wind speed within a tornado actually varies across the tornado itself, as demonstrated in Figure 19B. The total wind speed measured at the ground is the combination of the rotational velocity of the tornado and its forward (translational) velocity. If one looks in the direction the tornado is moving, the right side of the tornado has a rotational velocity in the same direction as the translational motion. The left side of the tornado has a rotational velocity in a direction opposite to the translation motion. The actual wind speed is the sum of the translational part and the rotational part of the motion. On the right side, these components add, while on the left side they subtract. For example, if a tornado was rotating at 160 knots, and moving at

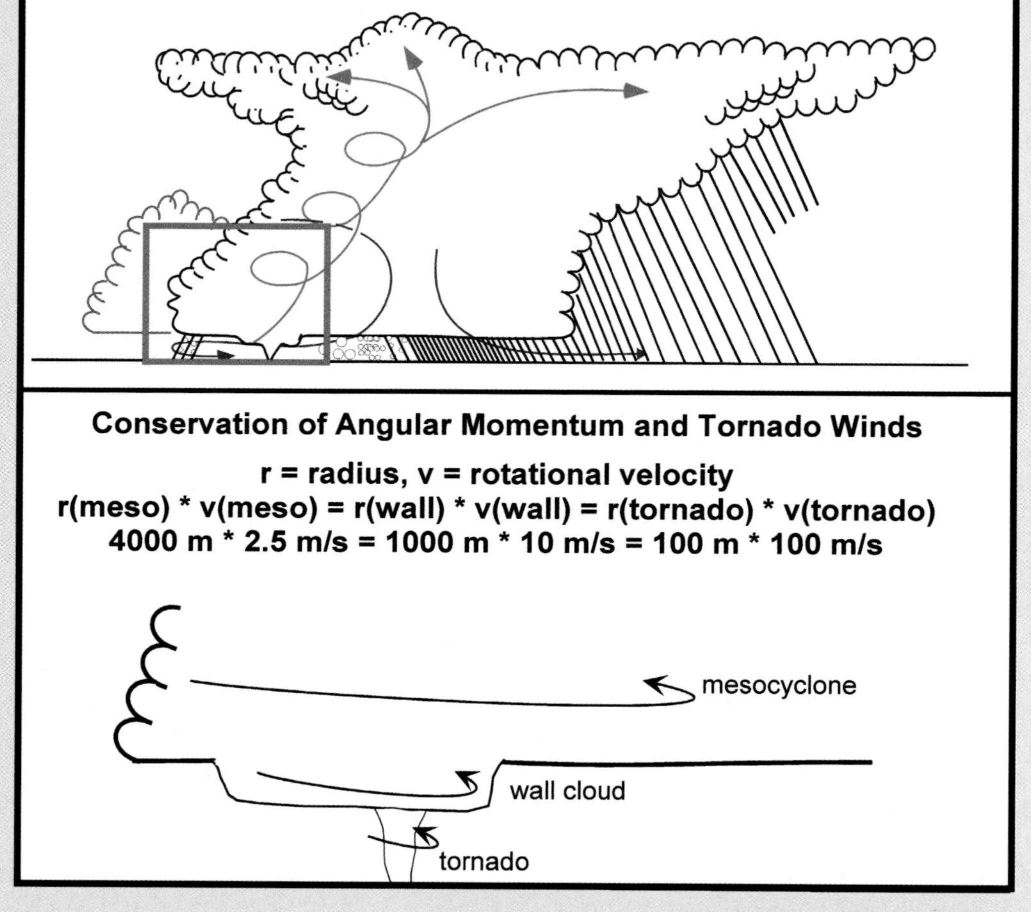

FIGURE 19A: The strong winds of a tornado arise from conservation of angular momentum as the rotation within the mesocyclone contracts to the scale of the tornado.

(continued)

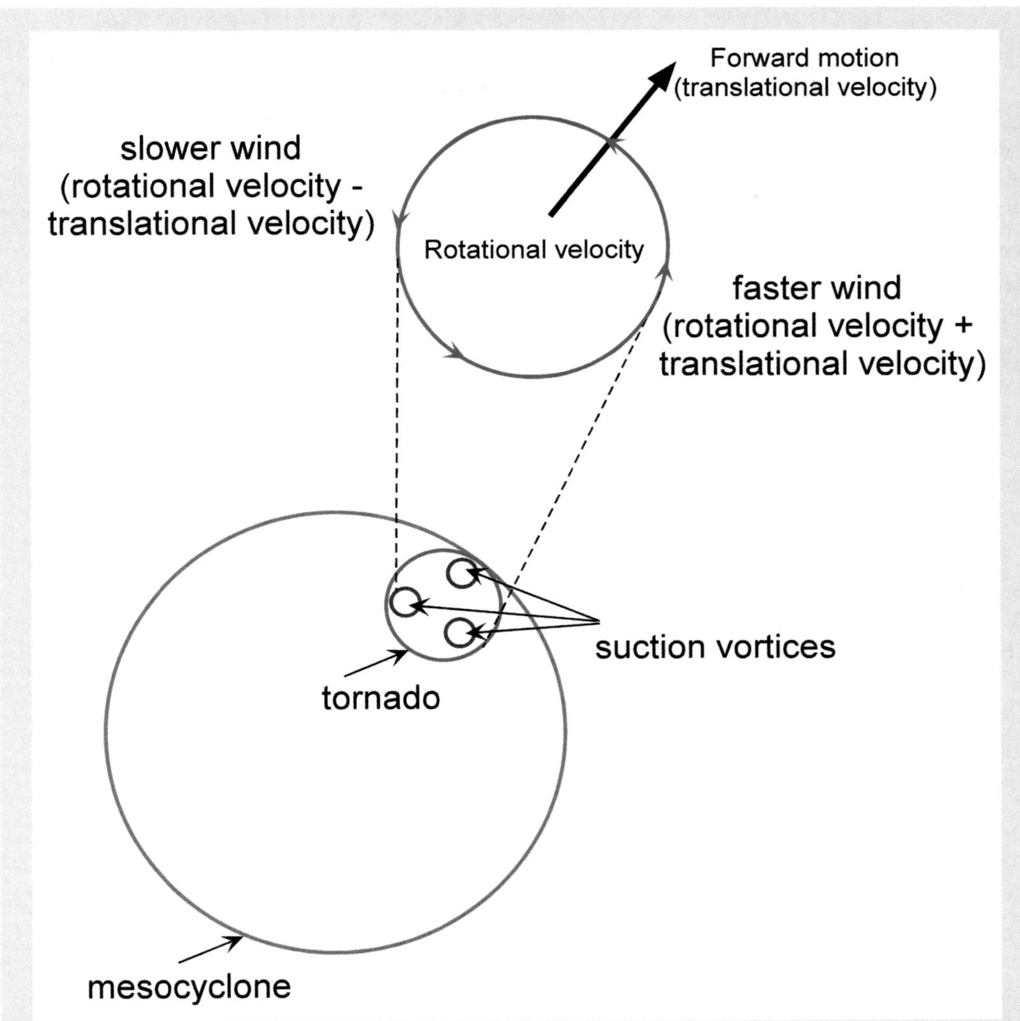

FIGURE 19B: The violent winds in a large tornado vary substantially across the tornado itself. Large tornadoes often have forward speeds of 25 to 50 knots (29 to 58 mph). This forward translation combines with rotation to make the winds stronger on the right side of the tornado vortex. The strongest winds are found in the suction vortices of large tornadoes. Suction vortices rotate at speeds much faster than other parts of the tornado.

a forward speed of 40 knots, the total wind on the right side would be 160 + 40 knots = 200 knots, while on the left side it would be 160 – 40 knots = 120 knots. As this example illustrates, winds in a tornado are considerably stronger on its right side when the tornado is moving rapidly across the ground. In a large tornado, smaller vortices, called suction vortices, sometimes develop. The most violent winds near the Earth's surface, approaching and sometimes exceeding 175 knots, occur in these vortices.

A tornado may be on the ground for a few minutes to as long as an hour. The typical tornado life cycle concludes as the rear flank downdraft wraps completely around the tornadic circulation. In this process, cool, dense air encircles the tornado, eventually weakening and finally eliminating the tornado's updraft. As the supercell moves, typically northeastward, the upper part of the updraft containing the tornado's circulation is tilted downstream by the mid-level winds relative to the lower part of the tornado. As this occurs, the tornado is stretched into a narrow vortex with a thin rope-like shape (Figure 19.1E). Often the tornado vortex can stretch across the sky well over a kilometer during this stage. Eventually, without an updraft to sustain the tornado, the tornado spins down, leaving a weak low-level circulation behind.

A tornadic supercell may pass through this sequence several times in its lifetime. Each cycle is associated with a new updraft that forms just outside

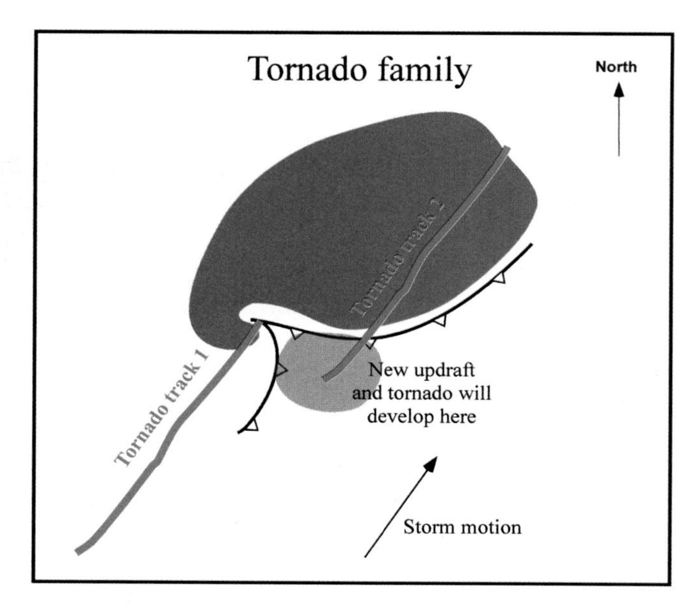

FIGURE 19.9: Tornado track 1 denotes the path of a tornado generated by a supercell. Supercells may undergo regeneration when a new updraft cell develops in the warm air just east of the point where the rear flank downdraft and forward flank downdraft intersect. This updraft cell will undergo a life cycle similar to the first cell, with a new tornado developing and following Tornado track 2.

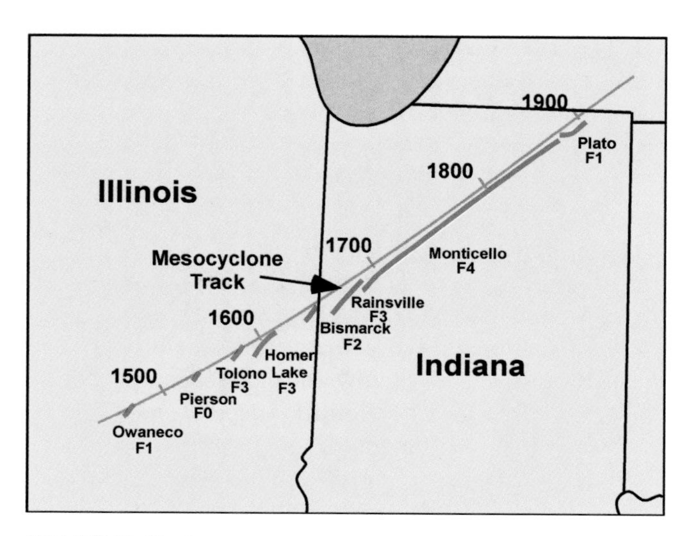

FIGURE 19.10: Example of a tornado family generated by a supercell thunderstorm that moved over Illinois and Indiana on 3 April 1974. The supercell produced eight tornadoes within five hours. The light gray line shows the track of the mesocyclone. The local time that the mesocyclone passed an area is noted along the track.

and southeast of the previous cell that produced the earlier tornado (Figure 19.9). The tornadoes emerging from the supercell over its lifetime constitute a ***tornado family***, as shown on Figure 19.10. Often, more than one member of the family can be on the ground simultaneously. Videos of tornado families have even documented cases where an older rope-shape tornado in the family will merge into a newer tornado.

CHECK YOUR UNDERSTANDING 19.1

1. Describe typical tornado characteristics such as width, range of wind speeds, and time on the ground.
2. What environmental condition is required for supercell thunderstorms to acquire rotation?
3. What is a mesocyclone? What is a typical width of a mesocyclone?
4. How does a "tornado family" develop?

TORNADO FORMATION WITHIN NON-SUPERCELL THUNDERSTORMS

Tornadoes sometimes develop within squall line thunderstorms aligned along fronts or along outflows from mesoscale convective systems (MCSs) (Chapter 18). To differentiate these tornadoes from the generally more violent supercell tornadoes, tornado experts sometimes refer to them as ***non-supercell tornadoes, landspout tornadoes, mesovortices,*** or ***gustnadoes***. The term landspout tornado is used because some of these tornados have visual (and probably dynamic) similarity to waterspouts, another tornado-like vortex that is sometimes observed off shorelines. For convenience, we will use the term "landspout tornado" to refer to this class of tornado vortices. Landspout tornadoes are generally short-lived and not as intense as their supercell tornado counterparts. The thunderstorms that produce landspouts typically do not display the strong pre-tornadic mid-tropospheric rotation commonly observed in supercells. The thunderstorms are triggered by lifting associated with the advancing front or gust front and are located along the frontal boundary. The mechanism of formation of a landspout tornado is illustrated in Figure 19.11. Some fronts are

FOCUS 19.2

The Tri-State Tornado of 1925

The Tri-State Tornado was the worst tornado disaster in U.S. history. The tornado, which occurred on 18 March 1925, killed 695 people, injured 2,027, and caused $16.5 million dollars in damage in 1925 dollars. Figure 19C shows its path. The tornado originated near Ellington, Missouri, and terminated near Princeton, Indiana, maintaining a northeast heading and traveling 294 km (179 miles). In Missouri, its forward speed was about 60 knots (69 mph). The forward speed in Illinois was about 50 knots (58 mph). In Missouri, damage patterns suggested that the funnel was 0.4 km (0.25 mile) wide, while in Illinois, the damage suggested a width of 0.8 to 1.6 km (0.5 to 1.0 mile) wide. Surveys indicated that the tornado damage path was unbroken. Pictures of the towns in its path, such as Figure 19D, were reminiscent of pictures of towns after a saturation bombing during World War II. The death toll of this tornado is unprecedented, a record unlikely to be broken given current advanced warning systems. Today there is controversy about this tornado. We know from our current understanding that tornadoes normally do not exist for such a long time. However, many supercells will produce tornado families—a series of tornadoes that have paths that are almost continuous, the second beginning near the end of the first, and the third after the second, such as the paths of tornado families in the April 1974 outbreak (Figure 19E). Was the Tri-State Tornado one tornado, or was it a family of tornadoes with nearly continuous damage paths? We will never be sure, but the Tri-State Tornado is in the record books as the worst killer tornado in history.

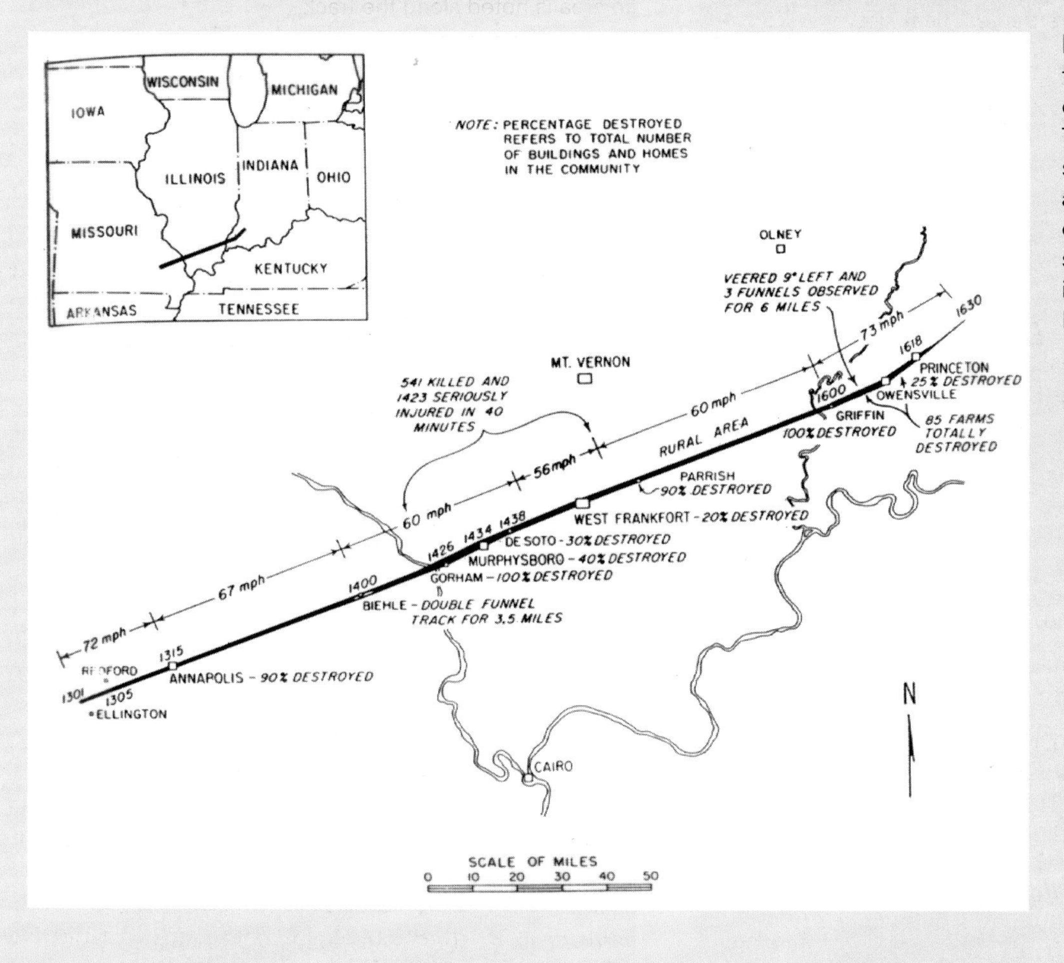

FIGURE 19C: Path of the Tri-State Tornado on 18 March 1925. The tornado began in southeastern Missouri and traveled a distance of 219 miles across southern Illinois and into southwestern Indiana.

(continued)

(continued)

FIGURE 19D: Damage in Murphysboro, Illinois, caused by the 18 March 1925 Tri-State Tornado.

FIGURE 19E: Map showing the location of the 148 tornadoes that occurred during the superoutbreak of 3 to 4 April 1974.

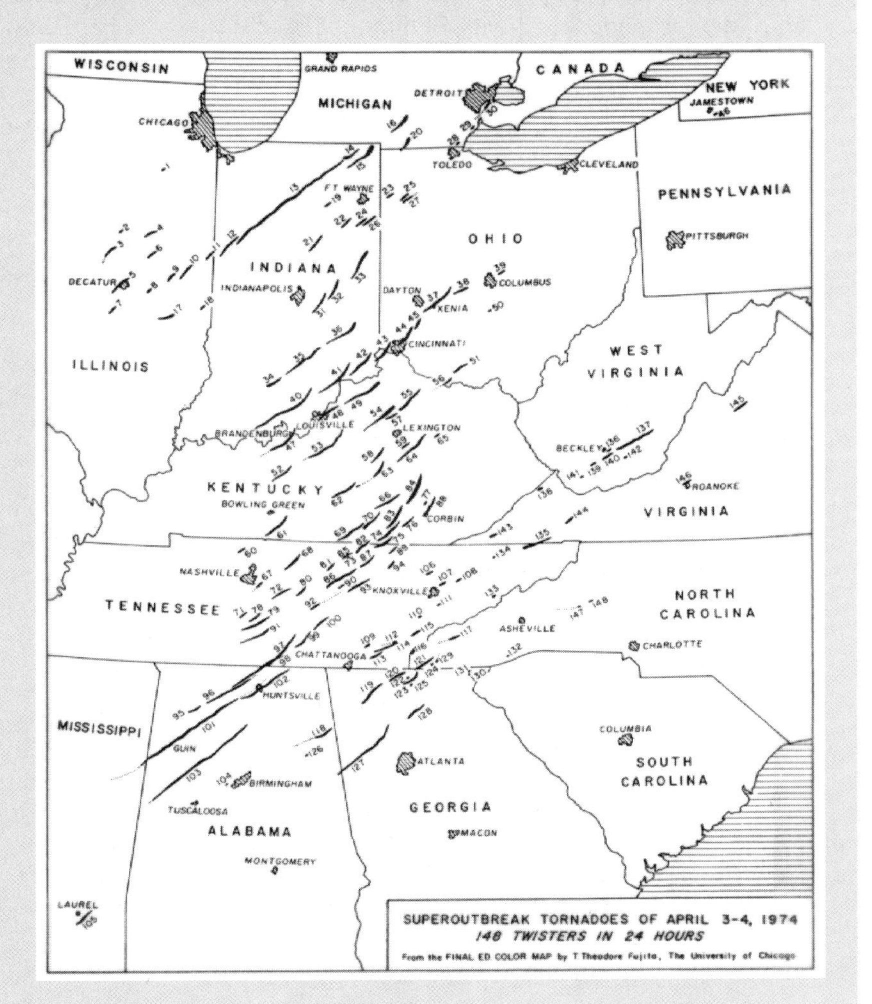

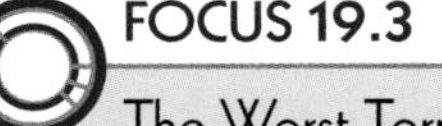

FOCUS 19.3

The Worst Tornado Outbreak in Recorded History: 3 to 4 April 1974

The superoutbreak of 3 to 4 April 1974 was the largest tornado outbreak on record in the United States. The outbreak was the most severe ever recorded in terms of number of tornadoes, total length of tornado tracks, area affected, and damage. More than 5,500 people were injured, 335 people were killed; property losses were estimated at more than $600 million and more than 27,000 families were affected. Aerial and ground surveys identified 148 tornadoes, and at one time at least fifteen different tornadoes were on the ground simultaneously (see Figure 19E). Tornadoes developed continuously for sixteen hours. Most of the 148 tornadoes of 3 to 4 April caused relatively little damage. Over half of the deaths were caused by less than 5 percent of the tornadoes; the worst in Xenia, Ohio, where thirty-four people died and 1,150 were injured. The six most destructive tornadoes, all with EF5 ratings, each had paths longer than 50 km, and two exceeded 160 km. Seventy-six percent of the individual tornadoes occurred in families, and more than 95 percent of the deaths were from tornadoes in families.

The weather conditions necessary for supercell formation—a strong jetstream aloft, a strong low-level jet, very warm moist surface air, and dry cold air aloft—all came together over a large region of the Midwest. Several lines of supercell thunderstorms developed and moved ahead of an advancing cold front. In the superoutbreak, hook echoes were associated with 81 percent of the ninety-three tornadoes studied. All EF4- and EF5-intensity tornadoes were associated with hook echoes. This outbreak occurred before the installation of the WSR-88D Doppler radar network. The only data forecasters had to determine the presence of tornadic thunderstorms was the hook echo on radar. If this same outbreak were to have occurred today, Doppler radar images of reflectivity and radial velocity would likely allow for warnings to be issued several minutes earlier than in 1974. While warning times were short by today's standards, the threat of the April tornado outbreak was perceived early. The first severe thunderstorm watch was issued in the early morning of 3 April and by noon, portions of 11 states were included in tornado or severe thunderstorm watch areas. During this outbreak 172 tornado warnings, 133 severe-thunderstorm warnings, and fifteen tornado watches were issued in 18 states.

characterized by a very sharp change in wind speed and direction between the cool and warm air on either side of the front (Figure 19.11A). Periodically at the frontal boundary, this ***horizontal wind shear*** across the front will be strong enough that the flow will break down and a series of small vortices will develop along the front as shown in Figure 19.11B.

When these vortices develop under an updraft, the updraft stretches the vortex into a tighter and tighter circulation until a landspout tornado forms (Figure 19.11C). Nearly all tornadoes in California in wintertime develop along cold fronts by this process. Landspout tornadoes also form within thunderstorms along squall lines in the central and eastern United States in this manner. Vortices sometimes develop every few kilometers along a front (Figure 19.11C).

In one famous video, six landspout tornadoes were simultaneously present within the field of view of the photographer! Although the characteristic hook echo of the supercell is not present for these tornadoes, some hints do exist in the reflectivity field. Breaks in the line of higher reflectivity along the front typically appear at the locations of the landspout tornadoes. These breaks are associated with the reduction in rain under the stronger updrafts. The best means of radar detection of landspout tornadoes is in the radial velocity field. As with supercells, rotation appears as a tight couplet of inbound and outbound velocities on a display of the radar radial velocity. It is now believed that a significant number of the small tornadoes occurring across North America every year may be landspout tornadoes.

Waterspouts are a class of weak tornadoes that are commonly observed off coastlines, particularly in tropical regions such as the Florida Coast and the Gulf of Mexico (Figure 19.12). Although many photographs of waterspouts exist, their mechanism of formation is not well understood. Scientists suspect that waterspouts are associated with spin up of circulations created by breakdown of the flow in regions

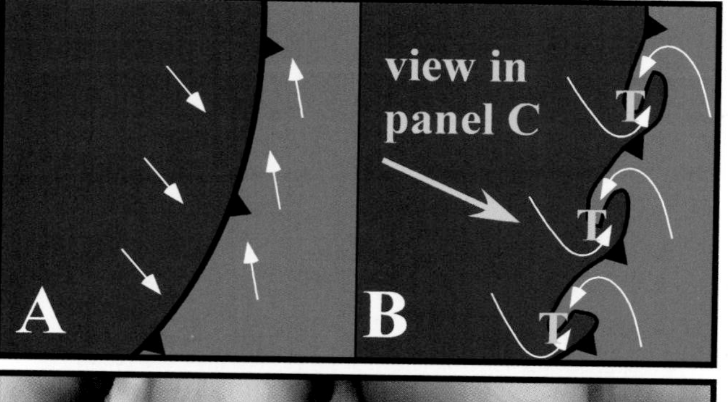

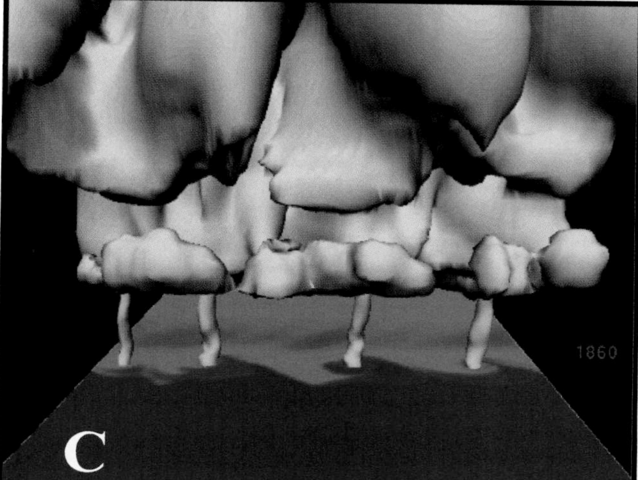

FIGURE 19.11: Landspout tornadoes develop along fronts characterized by strong horizontal wind shear. From the top view: (A) shear along the front leads to (B) vorticies forming along the shear zone. From a side view (C), as the vortices move under an updraft associated with thunderstorms along the front, they are stretched and weak tornadoes can develop.

FIGURE 19.12: Waterspout off the Florida coast.

ONLINE 19.2

Non-Supercell Tornadoes

Online, you can view two animations of high-resolution, three-dimensional modeling simulations, developed at the University of Illinois, that show how landspouts form. The first simulation shows a perspective view of the developing landspout tornadoes and deep convection along the leading edge of an outflow boundary. The yellow tubes, marking the position of the landspouts, are regions of strong rotation. The clouds are rendered by shading cloud water calculated by the numerical model. The second simulation shows a close-up view of the landspout family visualized by 6,500 weightless tracer particles released into the model flow field at low levels. In this simulation, green particles are in the updraft and red particles are in the downdraft. The ribbon of red particles in the center of the three major circulations is a downdraft that developed as the landspouts matured (simulations courtesy of Bruce Lee).

of low-level horizontal wind shear, much like landspout tornadoes. The shear is probably generated along outflow boundaries created by previous thunderstorms. The circulation is concentrated into a waterspout in the same manner that landspouts form—through vertical stretching of the rotation by an updraft within a developing thunderstorm or deep cumulus cloud.

Tornadoes can also develop in the thunderstorms generated by landfalling hurricanes. These tornadoes are often embedded in very heavy rain and are difficult to see approaching. Most of these tornadoes occur in the northeast quadrant of hurricanes. Detailed measurements in the vicinity of these tornadoes are rare, and the mechanism by which they form is not fully understood and is an area of active research.

TORNADO STATISTICS

Figure 19.13 shows the annual number of tornadoes observed per 10,000 square miles in each of the fifty states over a fifty year period ending in 2004. The states with an annual average exceeding five tornadoes per 10,000 square miles, highlighted in red, are

FOCUS 19.4

Other Tornado-Like Vortices

Cold air funnels are vortices that emerge from the base of elevated convective clouds that develop over cool surface air (Figure 19F). These funnels emerge from cumulus clouds that are found in the circulation of large-scale, upper-level cutoff low pressure centers (see Chapter 10). These cutoff lows are associated with unusually cold air in the upper troposphere. The only case study of these funnels suggests that they develop in a manner similar to landspout tornadoes, but the vortices are much weaker and very short-lived. They rarely touch down and cause very little damage when they do. *Dust devils* are vortices associated with dry convection (Figure 19G). Dust devils develop over hot surfaces, and are common in desert regions of western North America. Dust devils can sometimes extend over a kilometer in depth and can reach wind speeds similar to a weak tornado. *Fire whirls* (Figure 19H) are tornado-like vortices that develop in association with fires. These vortices develop in the heated air that is rising above a raging fire, and are often observed during forest fires.

FIGURE 19F: Two cold air funnels observed near Champaign, Illinois, on 23 May 1988.

FIGURE 19G: Two dust devils over an Arizona field.

(continued)

(continued)

FIGURE 19H: A fire whirl in a forest fire.

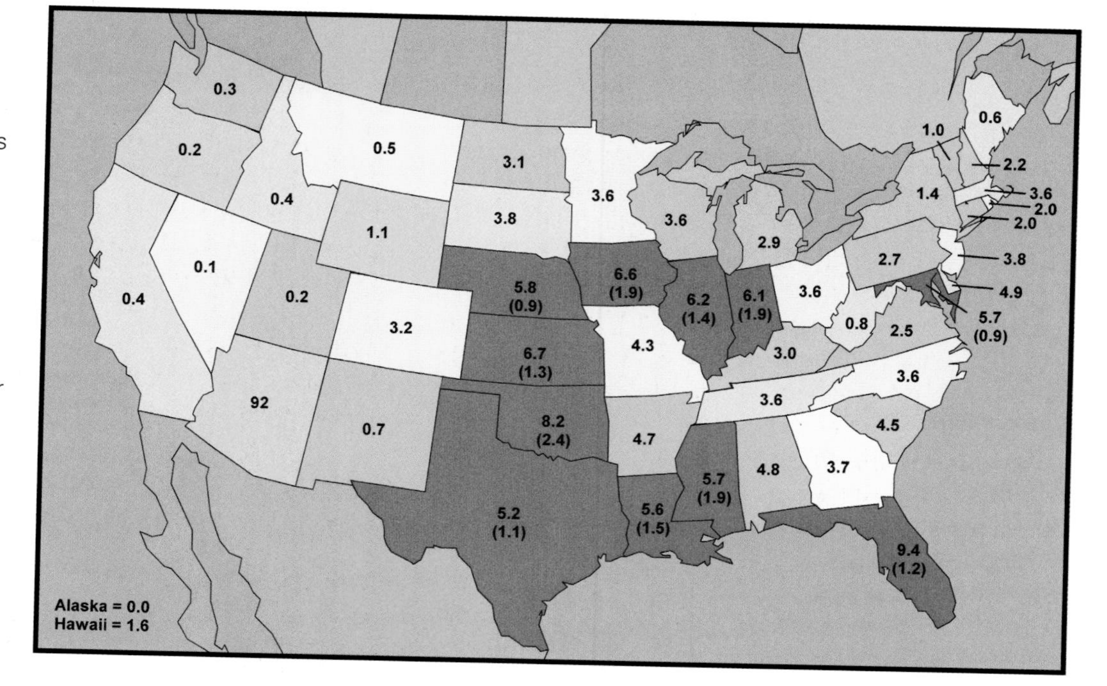

FIGURE 19.13: Average annual number of tornadoes per 10,000 square miles occurring in each state in the United States over the fifty-year period from 1953 to 2004. The red states are those that average greater than five per 10,000 square miles. The average number of strong to violent tornadoes (EF2–EF5 rating) per 10,000 square miles is given in parentheses.

Texas, Oklahoma, Kansas, Nebraska, Iowa, Illinois, Indiana, Maryland, Louisiana, Mississippi, and Florida. For these states, the average number of strong to violent tornadoes (EF2–EF5 rating, see below) per 10,000 square miles is given in parentheses. Somewhat surprisingly, the state with the largest number of tornadoes per 10,000 square mile area is not Oklahoma or Texas, but Florida. Florida, and other Gulf States on the list, experience so many tornadoes primarily because winter is short in these states and the tornado season extends over more months of the year. Florida has the most because the sea breeze, an

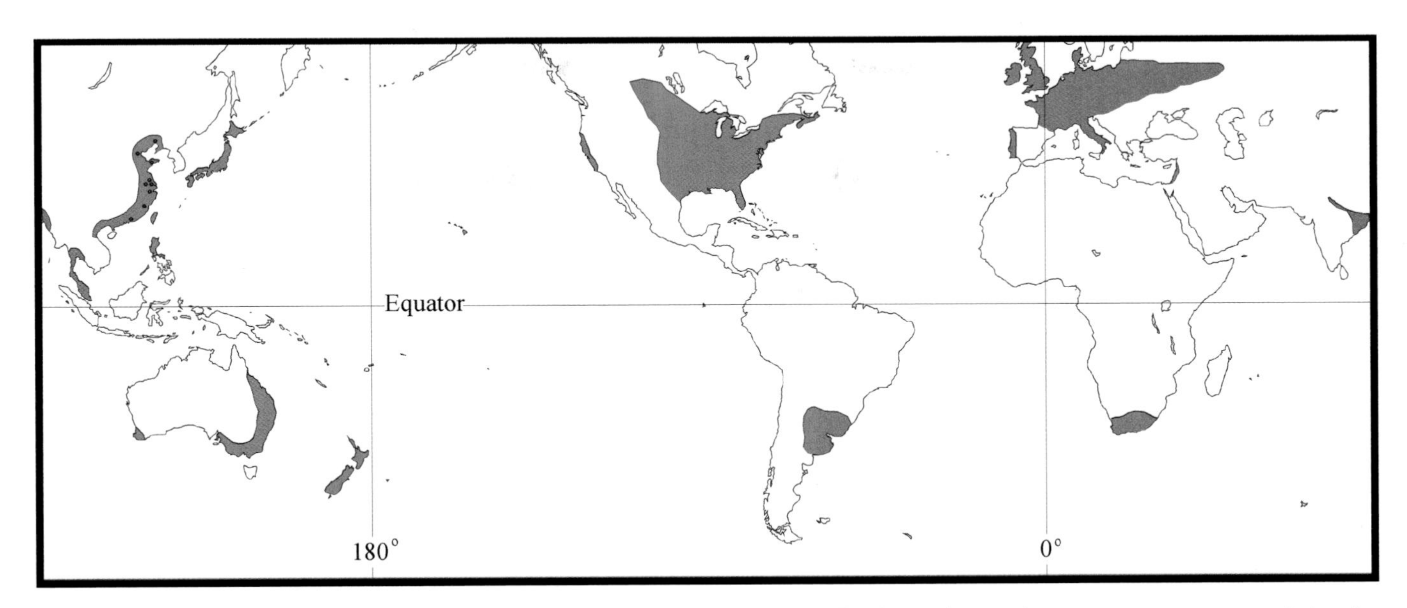

FIGURE 19.14: Locations worldwide where tornadoes have been observed. Regions of tornado occurrence correspond closely to the world's primary agricultural areas.

onshore flow of cool air that occurs most afternoons, triggers thunderstorms nearly every day in summer. Non-supercell tornadoes form along shear zones generated by storm outflows. Florida and the other Gulf States also experience tornadoes associated with landfalling hurricanes.

Aside from the Gulf States, tornadoes occur most frequently over the Great Plains and Midwestern states. These states are oriented along a southwest-northeast line called ***Tornado Alley***. This line corresponds to the typical orientation of strong fronts as they move across the central United States in the spring and early summer, and to the typical orientation of the upper tropospheric jetstream that accompanies these fronts. The largest number of violent tornadoes occurs in Oklahoma, where very sharp dry lines and cold fronts trigger supercell thunderstorms in the spring and early summer.

Only about 25 percent of all tornadoes occur outside the United States. Figure 19.14 shows regions where tornadoes occur worldwide. These regions are coincident with agriculturally productive areas of the world. This correspondence is no accident—the same thunderstorms that produce tornadoes provide the rain necessary to grow crops.

Until recently, tornadoes were classified according to a scale developed in 1971 by the late Dr. Theodore Fujita, formerly of the University of Chicago. The ***Fujita scale***, or F-scale, was based on the damage caused by a tornado, and served as a measure of tornado intensity from its introduction through February 2007. The F-scale was recognized over time to have many weaknesses, the primary problem being that it significantly overestimated the winds in the more violent tornadoes. Also, the guidelines suggested by Fujita for damage assessment were not detailed and did not account for differences in construction techniques that are common in many structures. These and other problems with the F-scale led wind engineers, university scientists, private sector meteorologists and NOAA meteorologists from across the country to partner with scientists at the Texas Tech University Wind Science and Engineering Research Center to develop a new ***enhanced Fujita scale*** (EF-scale) that can be used to more accurately rate the damage and winds associated with a tornado.

Now, meteorologists trained to conduct damage surveys have clear descriptions of the nature of damage that winds of various speeds will cause in a wide range of structures when they assign a tornado a value from the new scale. The EF-scale was designed so that tornadoes ranked in the past should have the same numerical rating, thus avoiding the problem of having to reassign all past tornadoes. Because of this association, in the current edition of this book, we will exclusively refer to the EF-scale to represent data generated either using the current EF-scale or the former F-scale.

The EF-scale, and the older F-scale, are shown in Table 19.1 for comparison. The EF-scale went operational in February 2007. The first EF5 tornado occurred three months later in Greensburg, Kansas on May 4, 2007. Table 19.1, 19.2 and 19.3 together show

TABLE 19.1 **EF-Scale Wind Speed Ranges and Corresponding Wind Speeds from the Fujita Scale**

Fujita Scale	3-Second Gust Speed (mph)	Operational Enhanced Fujita Scale	3 Second Gust Speed (mph)
F0	45–78	EF0	65–85
F1	79–117	EF1	86–110
F2	118–161	EF2	111–135
F3	162–209	EF3	136–165
F4	210–261	EF4	166–200
F5	262–317	EF5	>200

TABLE 19.2 **Damage Indicators for Establishing EF-Scale Ratings**

For the structures listed below, damage assessors use detailed tables that describe the degree of damage, together with example photographs from damaged structures to establish the likely wind speed and EF-scale rating for a tornado

No.	Damage Indicator
1	Small barns or farm outbuildings
2	One- or two-family residences
3	Manufactured home—single wide
4	Manufactured home—double wide
5	Apartments, condos, townhouses
6	Motel
7	Masonry apartment or motel
8	Small retail building
9	Small professional building
10	Strip mall
11	Large shopping mall
12	Large isolated retail building
13	Automobile showroom
14	Automobile service building
15	Elementary school
16	Junior or senior high school
17	Low-rise building (1–4 stories)
18	Mid-rise building (5–20 stories)
19	High-rise building (> 20 stories)
20	Institutional building
21	Metal building system
22	Service station canopy
23	Warehouse building
24	Electrical transmission lines
25	Free standing towers
26	Free standing light poles, luminary poles, flag poles
27	Trees (hardwood)
28	Trees (softwood)

how an EF rating is determined. Table 19.2 shows twenty-eight types of structures (damage indicators) that storm surveyors typically encounter in a tornado damage path. Each of these structures have a separate "damage table" similar to Table 19.3 that is used to assess the probable wind speed in the tornado. For example, Table 19.3 shows the damage table for a one or two family residence (the second damage indicator on Table 19.2). Based on this table, if a damage assessor found that a house had its exterior walls collapsed, but interior walls were standing, the winds were likely to be between 113 and 153 mph, with the average (expected value) of 132 mph. The expected value is used to establish the EF rating, and corresponds to EF2, based on Table 19.1.

Figure 19.15 shows the distribution of past tornadoes using the EF-scale. The majority were EF1, with slightly under 90 percent of all tornadoes in the categories EF0 to EF2. Only 0.3 percent achieved EF5 wind speeds. Figure 19.16 shows where U.S. tornadoes rated F5 (and now EF5) occurred during the last fifty-seven years. All occurred on the Plains between the Rocky and Appalachian Mountains, with many along a line extending from central Texas northward to Iowa. Over 45 percent occurred east of the Mississippi River. Tornadoes classified as EF5 typically exhibit EF5 winds only briefly in their lifetime, and only over a small area relative to the entire damage path of the tornado. For example, Figure 19.17 shows a damage analysis for the Plainfield, Illinois, tornado. This destructive tornado, which occurred on 28 August 1990, killed 30 people and caused at least 350 injuries as it moved across Kendall and Will Counties in northeastern Illinois. The track length of this tornado was 16.4 miles. As Figure 19.17 shows, the tornado produced EF5 damage only over a narrow area as it approached western Plainfield.

TABLE 19.3 **Estimating the EF-Scale Rating from Damage to a One- or Two-Family Residence**

(Indicator 2 in Table 19.2)

Degree of Damage	Damage Description	Expected Wind Speed (mph)	Lowest Wind Speed (mph)	Highest Wind Speed (mph)
1	Threshold of visible damage	65	53	80
2	Loss of roof covering material (<20%), gutters and/or awning; loss of vinyl or metal siding	79	63	97
3	Broken glass in doors and windows	96	79	114
4	Uplift of roof deck and loss of significant roof covering material (> 20%); collapse of chimney, garage doors collapse inward, failure of porch or carport	97	81	116
5	Entire house shifts off foundation	121	103	141
6	Large sections of roof structure removed; most walls remain standing	122	104	142
7	Exterior walls collapsed	132	113	153
8	Most walls collapsed, except small interior rooms	152	127	178
9	All walls collapsed	170	142	198
10	Destruction of engineered and/or well-constructed residence, slab swept clean	200	165	220

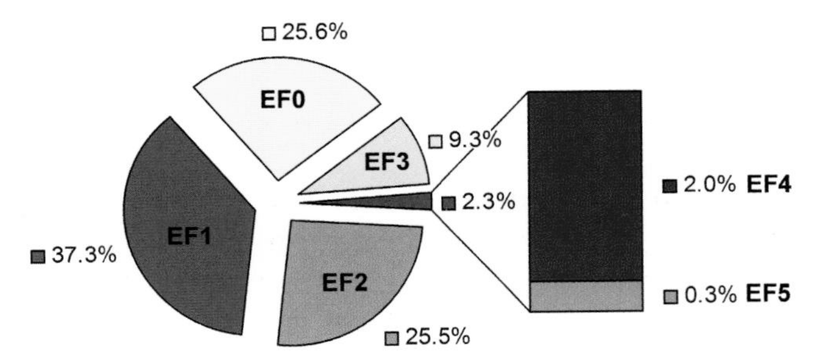

FIGURE 19.15: Percentage of all tornadoes that occurred in each EF-scale category.

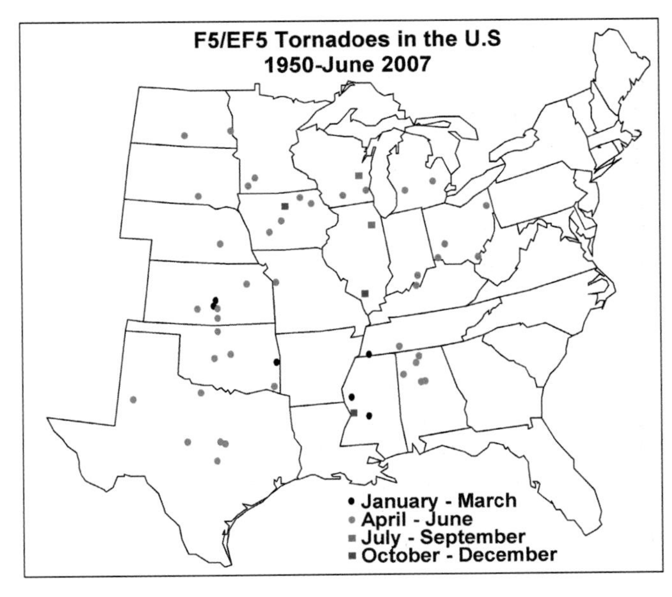

FIGURE 19.16: Location of all tornadoes classified EF5 between 1950 and 2007.

Figure 19.18 shows the number of tornadoes ranked EF0 to EF5 observed for each month. Tornadoes occur in all months of the year, but are most frequent in April, May, and June, the time when the thermal contrast between tropical and polar airmasses is most extreme. The strongest tornadoes occur most often from April through June (see also Figure 19.16). Figure 19.19 shows how the number of tornadoes varies with the time of the day. Tornadoes occur most often in the afternoon and evening, with a peak around 5 P.M., slightly after the warmest hours of the day. A minimum in tornado occurrence occurs around 5 A.M. The path length covered by tornadoes of each category appears in Figure 19.20. The most severe tornadoes tend to stay on the ground longest. For example, the mean path length for an EF5 tornado is 57.2 km (34.9 miles), while weaker (EF0 and EF2) tornadoes have mean path lengths less than 9 km (5.5 miles).

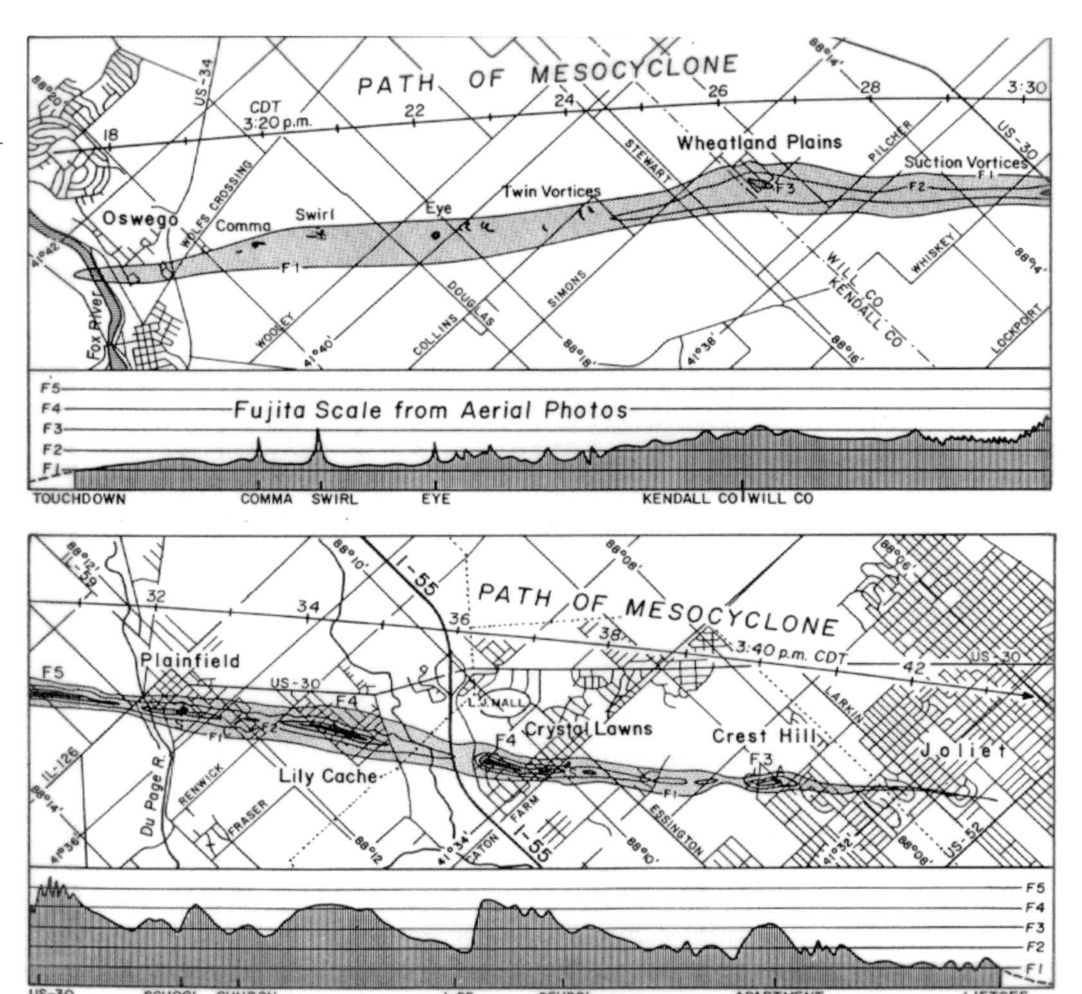

FIGURE 19.17: Detailed map along the path of the Plainfield/Crest Hill tornado showing the distribution of damage based on F-scale rating along the path.

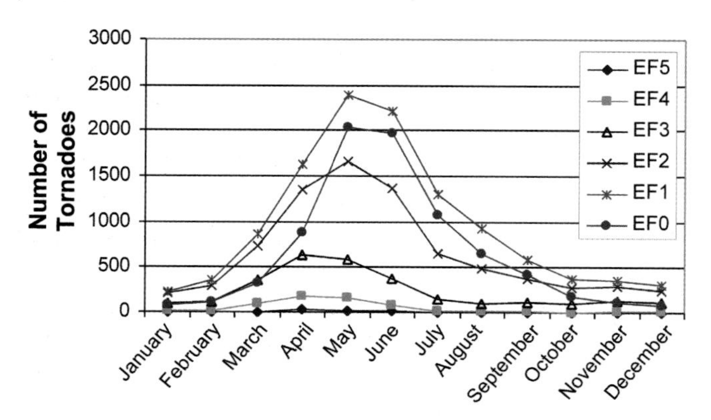

FIGURE 19.18: Number of tornadoes by month and EF-scale rating.

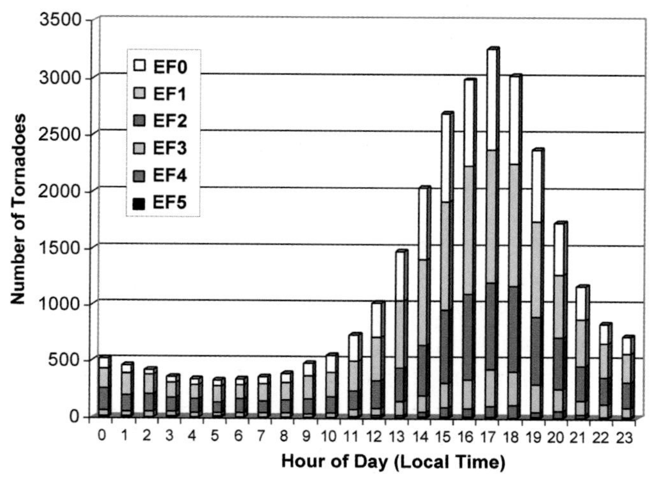

FIGURE 19.19: Number of tornadoes by time of day and EF-scale rating.

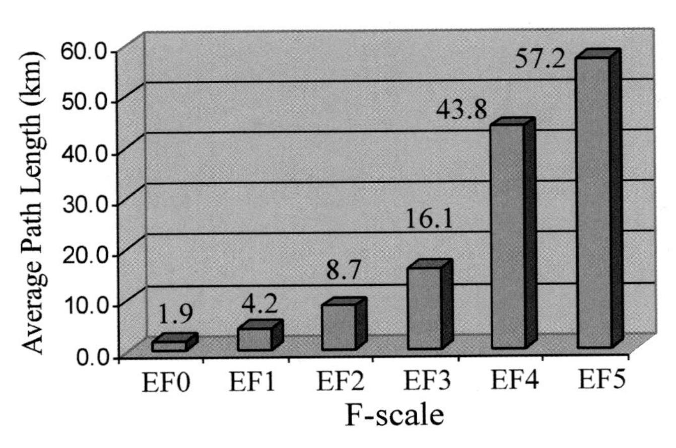

FIGURE 19.20: Average tornado path length (km) as a function of EF-scale rating.

ONLINE 19.3

Seasonal Variation of Tornado Frequency in the United States

Where do tornadoes occur in different seasons? Online you can view an animation of monthly maps showing the location and frequency of tornado occurrence and graphs showing the number of tornadoes in each EF-scale category for each month. The fewest tornadoes occur in the mid-winter months of December, January, and February. Winter tornadoes most often occur in the southeast United States, in storms forming along surface cold fronts or upper level fronts associated with winter cyclones. Winter tornadoes also occur occasionally in California as cold fronts move onshore from the Pacific. As spring arrives, tornadoes become more numerous, with the highest frequency in Tornado Alley through Texas, Oklahoma, Kansas, Nebraska, and Iowa. Tornadoes in summertime become more frequent in the Northern Plains states. As fall approaches and passes, the frequency of tornadoes decreases and the area most affected shifts southward. A maximum along the coast of the Gulf of Mexico appears in September associated with tornadoes in landfalling hurricanes. Tornadoes also occur in central Florida through much of the year.

CHECK YOUR UNDERSTANDING 19.2

1. How does a landspout form?
2. What is the EF-scale? How is an EF-scale rating assigned to a tornado?
3. What two states experience the most tornadoes per 10,000 square miles?

TORNADO DETECTION

Tornado detection has traditionally been anchored by trained networks of ***storm spotters***, volunteers who are deployed at key locations around threatened cities during severe storm outbreaks. These spotters, in cooperation with police, emergency management personnel, and the National Weather Service, report dangerous weather conditions and tornado locations, providing an important line of defense during severe weather.

The installation of the national WSR-88D Doppler network in the 1990s has provided meteorologists with new information that allows advanced warnings to be issued, often minutes before an actual tornado touchdown occurs. The rear flank downdraft often contains precipitation particles visible on radar. As the rear flank downdraft wraps around the echo-free base of the updraft, the echo on radar appears like a "hook," as shown on Figure 19.21A. The tornado is always near the center of the hook. Debris generated by the tornado sometimes leads to very strong radar echoes in the hook itself. Prior to the widespread use of Doppler radar, the ***hook echo*** was the only way to identify a tornado with radar. Unfortunately, hook echoes do not appear in all supercells, and do not exist with non-supercell tornadoes. Furthermore, not every hook echo will contain a tornado. The reason that some supercells do not exhibit hook echoes is that the air composing the leading edge of the rear flank downdraft is sometimes precipitation free. In such cases, ordinary (non-Doppler) radar cannot determine a tornado's location.

Doppler radar provides clear evidence of rotation in thunderstorms long before the collapse of the rotation

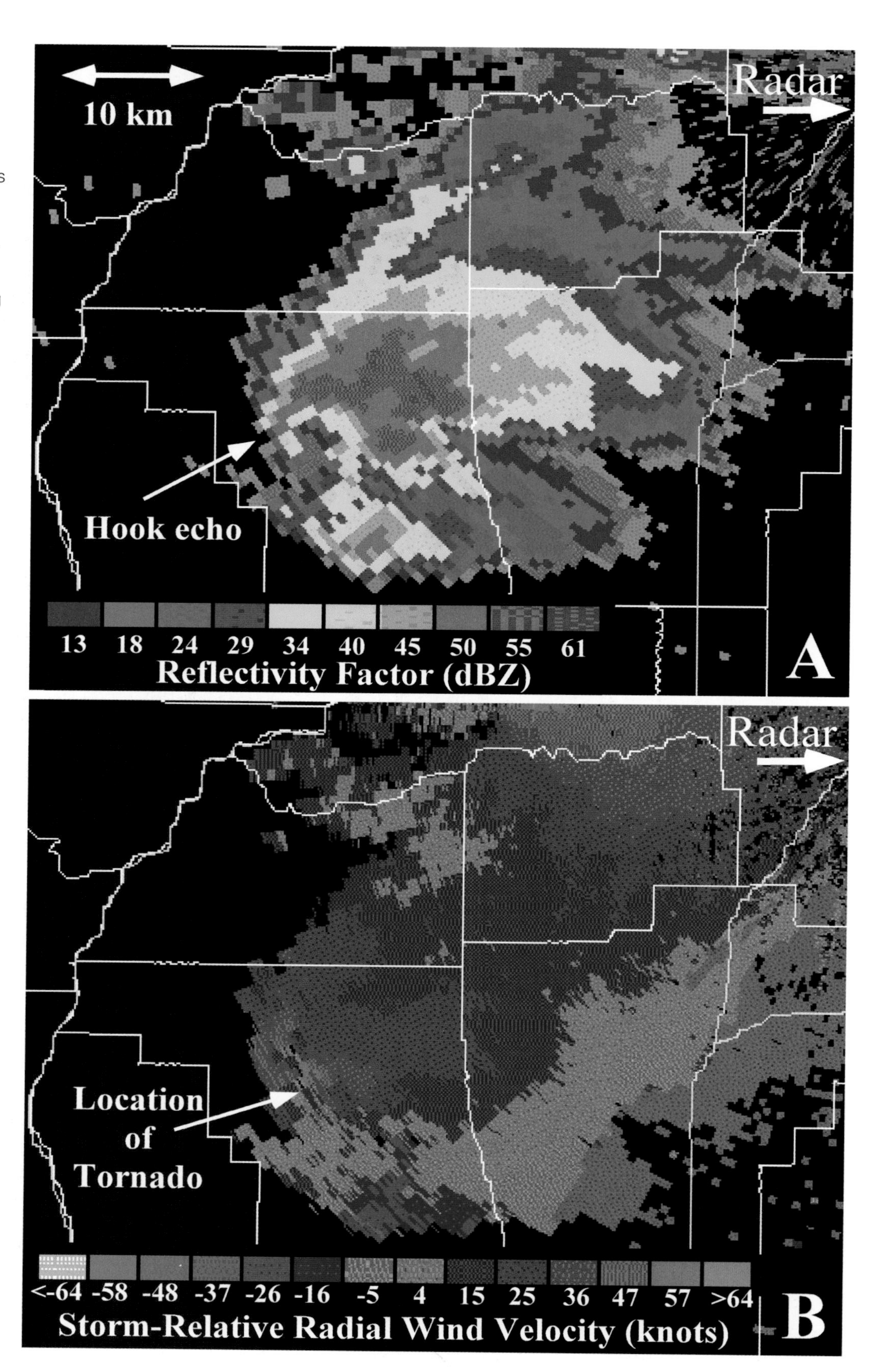

FIGURE 19.21: Radar images of a supercell thunderstorm that produced a tornado near Jacksonville, Illinois on 19 April 1996. White lines are county boundaries. The radar is located just off the upper right corner of the panel. (A) Radar reflectivity (dBZ) showing the hook echo associated with the supercell. (B) Storm relative radial wind velocity (knots) showing the mesocyclone signature. Red (green) colors represent air motion away from (toward) the radar.

FOCUS 19.5

Mobile Doppler Radars

In the past, researchers wishing to conduct detailed studies of tornadoes with Doppler radar had to site the radars and then hope that a tornado would pass nearby during their field campaigns—an event that rarely happened. In the late 1990s two Doppler radars, operated by the University of Oklahoma and the National Center for Atmospheric Research, were placed on trucks. These radars, called the *Doppler on Wheels* or *DOW* (Figure 19I), are driven into storms and parked strategically to view tornado circulations from vantage points that allow scientists to measure the winds within the tornado circulation itself. Figure 19J shows the radial velocity and reflectivity measured by a DOW in a tornado that occurred in Scottsbluff, Nebraska, on 21 May, 1998. The radar reflectivity is associated with debris and dust orbiting within the tornado. Note the "eye" in the reflectivity pattern where the reflectivity is lower. This "eye" is unlike the eye of a hurricane—it exists because the debris within the tornado is centrifuged outward by the rotating column. The DOWs obtain unprecedented data that lead to a better understanding of tornadoes and many other atmospheric phenomena that are hard to observe with fixed radars. Recently joining these mobile radars are a second pair of mobile Doppler radars called the Shared Mobile Atmospheric Research and Teaching (SMART) radars. The SMART radars, operated by the National Severe Storms Laboratory, Texas A&M University, Texas Tech University, and the University of Oklahoma, have joined the DOWs in the search for severe storms.

Airborne meteorological radars also provide a means to make measurements of the structure and dynamics of difficult-to-observe weather phenomena such as tornadoes. Currently, three research aircraft flown by the meteorological community have scanning Doppler radars: two P-3 aircraft operated by the National Oceanic and Atmospheric Administration and a P-3 operated cooperatively by the National Science Foundation and the Naval Research Laboratory through the National Center for Atmospheric Research (NCAR). With the DOWs, SMART, and airborne radars, tornado chasing has taken on a whole new dimension in the meteorological community.

FIGURE 19I: A Doppler on Wheels radar near a large tornado.

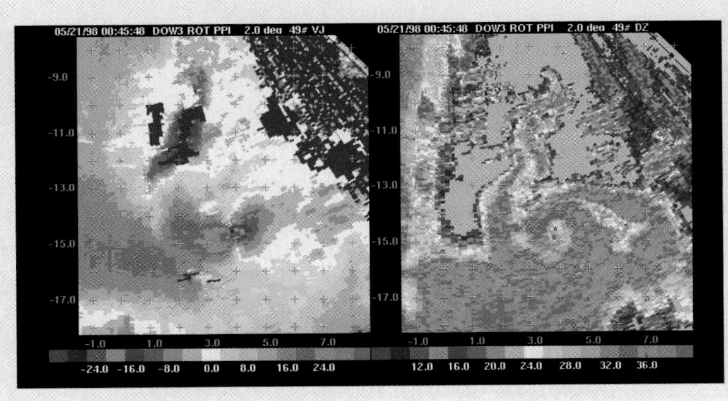

FIGURE 19J: Image of radial velocity (meters/second, left) and reflectivity (dBZ, right) taken by a Doppler on Wheels radar located near a tornado in Scottsbluff, Nebraska, on 21 May 1998. The radar was located just above the top-center edge of the panel. Green and blue colors denote inbound velocities on the radial velocity image.

into a violent tornadic core. Recall that a Doppler radar can measure the component of the wind that is moving toward or away from the radar. Rotation appears on the Doppler radar screen as a relatively small (~5 km in diameter) area where radial winds are moving toward the radar on one side of the area (as viewed from the radar) and away from the radar on the opposite side (Figure 19.21B). The winds on either side of the center of the circulation both appear relatively strong (typically about 20 to 40 knots). Such an area, called a ***mesocyclone signature*** in the radial velocity data, is often a precursor to tornado formation. When the mesocyclone appears on a Doppler radar, forecasters track its progress and determine its path. From this, they extrapolate forward to estimate the future location of the potential tornado. This now can all be done in seconds using computers that control the radar display. Once the path is determined, warnings are issued for the towns and parts of counties in the path of the potential tornado.

Doppler radar normally cannot resolve the details of the tornado itself. The tornado's size is smaller than the resolution (width) of the radar beam. However, the beam that contains the tornado, or the two adjacent beams that contain halves of the tornado, show abruptly different winds from the nearby beams viewing the mesocyclone rotation. A tiny area, representing one pulse volume (the smallest volume of space the radar can "see"), will show up on the screen with unusually large velocity next to a pulse with a large velocity in the opposite direction. This area is called a ***tornado vortex signature*** and marks the location of the tornado. The radar software also tracks the tornado vortex signature and warnings are issued along its projected future path. Doppler radar now provides National Weather Service forecasters with an opportunity to warn the public of an impending tornado an average of twelve minutes before the tornado actually forms, allowing people more time to find shelter. Prior to the installation of WSR-88D Doppler radar systems, the warning time averaged only six minutes.

TORNADO FORECASTING

Forecasting the precise location where a tornado will occur is impossible with current technology. However, identifying regions where tornadic storms are likely to form is done routinely using numerical forecast models. Forecasters begin by examining several indices that are calculated from vertical profiles of the atmosphere predicted by numerical forecast models. A vertical profile in a model atmosphere is analogous to a sounding in a real atmosphere. In fact, the same indices that are calculated from soundings are also routinely calculated from model data. The advantage of model data is that information can be analyzed for future conditions, while soundings can only be used to analyze conditions after they exist and only at 1200 or 0000 UTC. In addition, models provide "soundings" for locations between the relatively sparse network of rawinsondes.

One of the most important indices is the ***Convective Available Potential Energy***, or ***CAPE***. CAPE measures how unstable the atmosphere is and how strong a thunderstorm's updraft will be. It is a measure of the kinetic energy (the energy of motion) that buoyant air parcels will obtain as they rise through the atmosphere. The higher the CAPE value, the greater the potential for severe thunderstorms. A second index is the ***Storm-Relative Helicity (SRH)***. SRH measures the vertical wind shear in the lower atmosphere relative to the motion of a thunderstorm. High values of SRH mean that there is strong shear in the lower atmosphere, which favors storm rotation and supercell formation. The ***Energy-Helicity Index (EHI)*** combines CAPE and SRH. The EHI is simply the CAPE × SRH / 160,000, and represents a combination of shear and instability. Table 19.4 relates values of CAPE, SRH, and EHI to storm intensity and tornado potential.

TABLE 19.4 **Indices Used to Forecast Tornado Potential**

CAPE: Convective Available Potential Energy (Joules/Kilogram)	
0 – 500	Very weak instability
500 – 1500	Weak instability
1500 – 2500	Moderate instability
2500 – 4000	Strong instability
> 4000	Extreme instability
SRH: Storm Relative Helicity (meters^2 / seconds^2)	
0 – 150	Weak shear
150 – 300	Moderate shear
300 – 500	Strong shear
> 500	Extreme shear
EHI: Energy Helicity Index	
0 – 2	Tornadoes unlikely
2 – 4	Tornadoes possible
>4	Significant tornado threat

ONLINE 19.4

The 3 May 1999 Oklahoma City Tornado

On 3 May 1999, a series of supercell thunderstorms that developed over southwestern Oklahoma produced over seventy tornadoes in Oklahoma and Kansas. The tornadoes killed forty-four people and caused over 800 injuries. Over 11,000 structures, including homes, mobile homes, apartments, schools, businesses, and churches, were destroyed or damaged in the outbreak. Over 8,000 vehicles were damaged or destroyed and 116,000 customers were without electricity. The worst of the tornadoes was an EF5 tornado that passed through the town of Moore, Oklahoma and across the southern part of Oklahoma City. Figure 19K shows damage from this tornado. Online you can examine an animation of the WSR-88D radar reflectivity and radial velocity measurements as they appeared to National Weather Service forecasters on the day the storm occurred. The parent storm of the tornado was a powerful supercell thunderstorm. Note the distinct hook echo on the southwest side of the storm. A tornado vortex signature in the radial velocity data can be easily tracked as the storm moves across the radar display. A second supercell with similar signatures in the reflectivity and velocity field appears northwest of the Oklahoma City storm. The Oklahoma City tornado was probably the most well-observed EF5 tornado in history. Occurring only kilometers away from the National Severe Storm Laboratory and the University of Oklahoma, this tornado was surrounded by meteorological instrumentation, such as the Doppler on Wheels radars, and video cameras, and was filmed by a traffic helicopter. Nearly the entire evolution of the tornado was broadcast live on television. Timely tornado warnings, responsible media coverage, and tornado preparedness saved many lives in this storm.

FIGURE 19K: Section of the damage swath of the EF5 tornado that passed through Moore and southern Oklahoma City, Oklahoma on 3 May 1999.

Forecasters at the National Weather Service's Storm Prediction Center in Norman, Oklahoma, monitor the distribution of CAPE, SRH, EHI, and other indices such as the lifted index (see Chapter 6) predicted by numerical models, and focus their attention on regions where the values of these parameters suggest that severe storms are possible. Within these regions, forecasters look for fronts and other boundaries as focal points for the triggering of thunderstorms, since lifting of warm, moist air to its level of free convection can occur along boundaries (see Chapter 6, 18). Forecasters also monitor the distribution of cloud cover, since surface air heats more rapidly in clear regions. Forecast soundings and maps are also examined carefully in these regions to determine whether mid-level inversions are likely to delay or retard thunderstorm development. Forecasters also assess the availability of moisture and evaluate the potential for triggering of thunderstorm updrafts by rising motion associated with jetstream processes (see Chapter 8).

From their analyses, the Storm Prediction Center identifies regions of slight, moderate, and high risk of tornadoes. These areas are forecast twelve hours to three days in advance of the potential severe storm event. The Storm Prediction Center also issues a probability forecast for the 0- to twenty-four-hour period that shows the probability of a tornado occurring within twenty-five miles of a point (Figure 19.22). As the time of greatest risk approaches, the National Weather Service issues ***watches*** for the public. Two types of watches are issued by the Storm Prediction Center. A ***severe thunderstorm watch*** means that conditions are right for the development of thunderstorms containing strong winds, hail, frequent lightning, heavy rain, and possible tornadoes. A ***tornado watch*** indicates that conditions are favorable for tornado formation. Finally, when the storms are in progress, local National Weather Service offices issue ***warnings***. A severe thunderstorm warning means that a severe thunderstorm is occurring in or near the warning area. A severe thunderstorm warning can be based on radar, reports from local officials such as state police and spotters, or information from the public. A tornado warning means that a tornado is believed to be present in the warning area. The warning can be based on radar (a hook echo or tornado vortex signature), or on sightings by the police, storm spotters, or the public. Because tornadoes develop quickly, spotter networks are often the first line of defense. In many cases, such as the 3 May 1999 Oklahoma City tornado, spotters tracked the tornado from its first appearance through its dissipation.

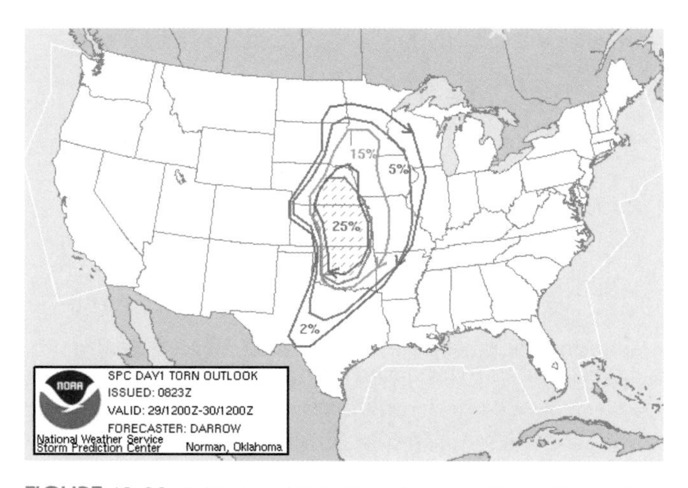

FIGURE 19.22: A National Weather Service Storm Prediction Center forecast for the probability of a tornado occurring within 25 miles of a point on 29 May 2004.

TORNADO SAFETY

Tornado safety is the responsibility of both individuals and organizations. When threatening weather is imminent, attention should be paid to radio and television for warnings in the threatened area. The National Weather Service broadcasts warnings directly on NOAA Weather Radio and on local media. NOAA Weather Radios should be the first line of defense for the public. These radios are relatively inexpensive and can be programmed directly for your county warning area. When the National Weather Service issues a watch or warning a NOAA weather radio immediately sounds an alarm, alerting you that there is a potentially dangerous situation. The National Weather Service encourages families, schools, and businesses to develop safety plans for tornadoes and conduct frequent drills. In a home or building, people should move to a predesignated shelter, such as a basement. If an underground shelter is not available, the safest place is an interior room or hallway on the lowest floor, particularly interior bathrooms where plumbing provides some support to walls. Windows are dangerous and should be avoided. A simple rule is to put as many walls between you and the tornado as possible. Mobile homes, even if tied down, offer little protection from tornadoes and should be abandoned. If caught outdoors, move as far away from potential airborne objects as possible and lie in the lowest spot available. It is better to not try to outrun a tornado in a car; instead, abandon the car and move as far away from it as possible. A video that showed people surviving a

ONLINE 19.5

Record May 2003 Tornadoes

May 2003 went into the record books for tornadoes by a number of measures. The largest number of tornadoes ever recorded in a single month, 516, eclipsed the previous record of 399 set in June 1992, and the previous May record of 391. An amazing 354 tornadoes occurred in the first twelve days of May 2003. Illinois had seventy-four tornadoes, and Missouri seventy-one tornadoes, exceeding their May records of fifty-four and twenty-nine respectively. One Weather Service Forecast Office, Lincoln, IL had forty in its warning area alone during May. Despite the huge number of tornadoes in May 2003, only thirty-eight fatalities were recorded, a remarkable achievement that can be attributed in part to the WSR-88D Doppler radar network and to the increased awareness of the public to the dangers of tornadoes. Online, radar reflectivity and storm-relative radial velocity animations from many of the May 3–10, 2003 events can be viewed. Try to find the mesocyclone signatures by looking in the hook echo region of the supercells.

tornado under a highway overpass has led many people to abandon houses and seek shelter under overpasses. Overpasses in general do not provide safety—exceptionally strong winds generated in the channel below the overpass can lead to severe injuries and death. In all cases, early action when severe weather threatens is the key to safety and survival.

CHECK YOUR UNDERSTANDING 19.3

1. Before the advent of the WSR-88D Doppler radar network, what was the primary means of tornado detection?
2. What is the characteristic signature of a tornadic thunderstorm in a radar reflectivity image?
3. What is the "tornado vortex signature" in a radial velocity field obtained from Doppler radar?
4. How does a tornado warning differ from a tornado watch?

TEST YOUR UNDERSTANDING

1. List and describe the different stages of the life cycle of a long-lived tornado.
2. What is the difference between a funnel cloud and a tornado?
3. What process concentrates the rotation in a supercell thunderstorm?
4. What is the role of low-level shear in the formation of tornadic supercells?
5. What is "vortex stretching"? What physical principle is used to explain vortex stretching?
6. Describe conceptually the three different mechanisms that scientists believe lead to tornadogenesis.
7. Where does a tornado form relative to the primary features of a supercell thunderstorm?
8. How do suction vortices form in a tornado?
9. What is a typical ratio of the diameters of a mesocyclone and a tornado?
10. In what type of thunderstorms do landspout tornadoes typically develop? How do they develop?
11. In which months of the year are tornadoes most common? Why do these months have the most tornadoes?
12. At what time of day are tornadoes most common?

13. What are typical path lengths of EF0 and EF1 tornadoes? Of EF4 and EF5?

14. How would you expect the frequency and geographical distribution of tornadoes to change seasonally? Explain.

15. Why are the primary regions of tornado occurrence generally in productive agricultural areas?

16. Why can it be difficult to assign an EF-scale rating to a tornado?

17. The annual average number of fatalities due to tornadoes has decreased since the 1950s. Explain why.

18. Do all tornadic thunderstorms produce hook echoes in the radar reflectivity field? Why or why not?

19. Can a Doppler radar detect a tornado funnel? Explain.

20. What is the "eye" of a tornado observed in high-resolution radar reflectivity images from DOW radars? How does it develop?

21. What tools are available for meteorologists to detect tornadoes?

22. Describe the process that meteorologists use to forecast tornadoes. In your discussion, include a description of model sounding indices that are typically used.

23. If caught outside in the path of a tornado, what should a person do?

24. Why is it not a good idea to seek shelter under a highway overpass when a tornado is approaching?

TEST YOUR PROBLEM-SOLVING SKILLS

1. Figure 19.13 shows the average annual number of tornadoes per 10,000 square miles occurring in each state over a fifty-year period. By using the following information on the areas of the states, rank these states in Tornado Alley according to the total number of tornadoes per state per year.

State	Area (sq. miles)
Texas	261, 914
Kansas	81,823
Nebraska	76, 878
Missouri	68,898
Oklahoma	68,679
Iowa	55,875
Illinois	55,593
Arkansas	52,075
Indiana	35,870

2. In each of the following examples, consider the rotation of a unit mass:

Rotating object	Distance from axis of rotation	Time for one complete rotation
Air parcel in a cyclone	500 km	2 days
Air parcel in a mesocyclone	1.6 km	10 minutes
Air parcel in a tornado	200 m	12 seconds
Figure skater's elbow	0.3 m	0.2 seconds

Calculate the angular momentum of each rotating object in order of decreasing angular momentum. (Remember to convert to similar units.)

3. The center of a large tornado comes down the middle of your street, which is oriented southwest-to-northeast. On your side of the street, some roof structures are removed and outdoor storage sheds are flattened, but your brick house has lost no walls and your car is still right-side-up in your driveway. On the opposite side of the street, houses have lost most walls except small interior rooms.
   (a) Is your house on the northwest or the southeast side of the street?
   (b) Estimate the wind speeds experienced on each side of the street.
   (c) What is your best estimate of the tornado's forward speed as it moved down your street? Explain your answer.
   (d) What EF-scale rating would you give this tornado? Why?

4. Consider a random location in Oklahoma, and estimate the average time (in years) between tornado occurrences at that precise location by the following procedure. (Use Question 1 in this section and Figures from this chapter to help with your calculation.)
   (a) Determine the average number of tornadoes per year in Oklahoma.
   (b) Determine the average numbers of tornadoes per year in each of the six Enhanced-Fujita categories.
   (c) Estimate the area covered by the average year's tornadoes by using the path length information in this chapter, together with assumed "average" widths of 100 m for EF0 and EF1 tornadoes, 200 m for EF2 and EF3 tornadoes, and 400 m for EF4 and EF5 tornadoes.
   (d) Divide Oklahoma's total area (Problem #1) by your answer to (c) to determine how many years would be required for the cumulative tornado path area to become equal to the state's area. This provides you with an estimate of the expected number of years between tornadoes at a particular point, assuming that tornado occurrences are distributed evenly across the state. (In reality, the latter assumption is not strictly correct, i.e., tornadoes are less common in the panhandle of Oklahoma than in the south-central part of the state.)

USE THE SEVERE AND HAZARDOUS WEATHER WEBSITE

http://severewx.atmos.uiuc.edu

1. A detailed summary of the Oklahoma City tornado event of 3 May 1999 has been prepared by the Norman, Oklahoma Weather Forecast Office. A link to this site is provided on the "Tornadoes" page of the *Severe and Hazardous Weather Website.*
   (a) Examine the information provided on the map that shows the storm survey. Note the location and track of tornadoes along with their EF-scale rating.
   (b) Approximate the track length of each tornado. Are your findings comparable to the typical path lengths given in Figure 19.20?
   (c) Examine the radar reflectivity loop. Locate the region where the tornado would be expected based on the radar. Find the tornado track in the storm survey map that corresponds to the hook echo.
   (d) Examine the radial velocity loop. Identify the mesocyclone signature and compare its track with the track of the tornado.
   (e) Compare damage photos available from the web site for the tornado associated with the hook echo identified in (c). How well do the photos match the damage descriptions in the Enhanced Fujita-scale?

2. The Storm Prediction Center (SPC) produces a daily summary of severe storm reports. The National Severe Storms Laboratory (NSSL) has a Severe Thunderstorm Climatology available online. You can access both of these sites from the *Severe and Hazardous Weather Website* under "Thunderstorms."
   (a) Examine the climatological data for tornadoes for last month and record your findings.

(b) Go through the storm reports for each day of the last month. Record the locations of all reported tornadoes.
(c) How well does the tornado activity from last month match climatology? Explain why the actual occurrences did not necessarily exactly match the climatology.

3. The National Weather Service and Storm Prediction Center issues watches, warnings and advisories for hazardous weather across the United States. Monitor the convective outlook from the Storm Prediction Center, paying special attention to the tornado probability from the Day 1 Outlook.

   (a) Each outlook is accompanied by a several-paragraph discussion. The discussion explains the rationale for the forecast and summarizes the most significant hazards. Read the discussion carefully.
   (b) Discuss how forecasters used some of the concepts discussed in this chapter, including boundaries to focus lifting, CAPE, storm relative helicity, capping inversions and the low-level jet.
   (c) Examine surface maps, upper air maps and soundings and identify some of the features discussed in the outlook discussion. Provide a brief discussion of your findings.

4. The University of Illinois Department of Atmospheric Science's WW2010 Online Guides to Meteorology provides a module describing tornado characteristics and the processes associated with their evolution. You can access this site from the "Tornadoes" page of the *Severe and Hazardous Weather Website.* Use this website to enhance your understanding of tornadoes. Summarize the additional knowledge about tornadoes you gained from this site.

CHAPTER **TWENTY**

HAILSTORMS

Large hail collects on streets and grass during a severe thunderstorm.

KEY TERMS

- beneficial competition
- bounded weak echo region (BWER)
- cloud droplets
- differential reflectivity
- dry growth regime
- echo-free vault
- embryo curtain
- graupel
- hail
- Hail Alley
- hail cascade
- hail embryo
- hailshaft
- hailstone
- hailstorm
- hailstreak
- hailswath
- polarization
- polarization diversity radar
- supercooled cloud droplets
- wet growth regime

Hail is one of the more spectacular phenomena associated with strong thunderstorms. Unlike lightning, which is present in every thunderstorm, hail reaches the ground in only a small fraction of thunderstorms. People experiencing large hail for the first time are awestruck by the size of the stones falling from the sky. The impacts of hail can be dramatic. Figure 20.1, for example, shows the main street of Brush, Colorado, after a severe hailstorm. The photograph, taken three hours after a supercell thunderstorm passed over the town, shows the trees stripped of their late-May foliage. Most of the town's streets were impassable until snowplows removed the accumulation of hailstones.

Hail causes more than $1 billion in crop damage annually in the United States. Hail also results in significant property damage, primarily to automobiles, aircraft, and the roofs, siding, and windows of structures. In recent years, average annual property damage from hail has risen to between $1 and $1.5 billion in the United States alone, and to considerably larger totals worldwide. The Denver, Colorado, area suffered more than $600 million in property losses during a single hailstorm in 1990, and a severe hailstorm in 2000 caused approximately $2 billion in property losses in Fort Worth, Texas (Figure 20.2). Several hail events also cause livestock deaths and human injuries each year, although human fatalities are rare. The 28 March 2000 Fort Worth hailstorm, for example, resulted in approximately 200 injuries and one death. In 1990, 29 people died when a hailstorm in Nepal struck an outdoor stadium during a soccer match, causing fans to trample each other in a stampede toward the exit gates. Hailstorms in China in 1932 and in India in 1888 are reported to have each caused 200 or more deaths.

The potential for damage from a falling hailstone increases with a stone's size and its corresponding fall speed. In still air, a 2-cm (0.8-inch) diameter hailstone falls at approximately 20 meters per second (39 mph), and a 5-cm (2-inch) diameter hailstone falls at approximately 46 meters per second (89 mph). Most hailstones are solid and hard. However, some hailstones may have liquid water trapped within them and are more soft and spongy. Individual hailstones can be spherical, conical, or simply irregular in shape. Figure 20.3 shows typical sizes of hailstones collected under ***hailshafts*** in eastern Colorado. Most of the collected hailstones had sizes of about 1 cm. No hailstones greater than 2.5 cm (~ 1 inch), the diameter of a quarter, were encountered during the collection period. Hailstones with diameters greater than about 2.5 centimeters are relatively rare, although diameters as large as 17.8 cm (7 inches) have been recorded. The hailstone depicted in Figure 20.4, which fell in Aurora, Nebraska on the evening of 22 June 2003, is the largest hailstone ever recovered in the United States. It had a record 7-inch diameter and a circumference of 18.75 inches. An animation of the radar reflectivity from the storm that produced this stone can be found online.

Because of the extreme danger of flying through a hailstorm, the interior structure of hailstorms and

FIGURE 20.1: Main Street of Brush, Colorado three hours after a hailstorm moved through the town.

FIGURE 20.2: Window damage to office buildings during the Fort Worth, Texas, hailstorm of 28 March 2000. Broken windows were temporarily replaced with plywood. Hail from this storm was reported to have sizes up to three inches in diameter.

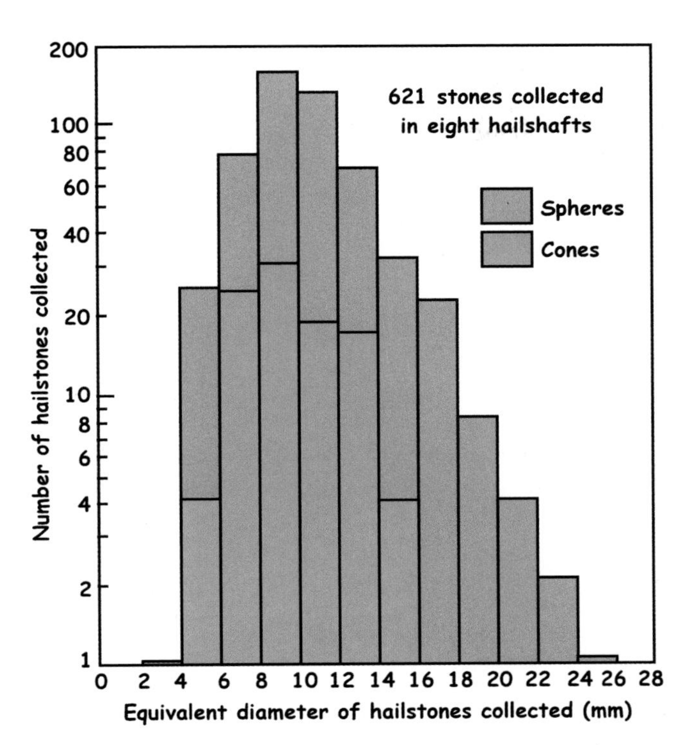

FIGURE 20.3: Distribution of hailstone sizes measured under eight hailshafts in storms in eastern Colorado.

growth processes of hail have been difficult to observe directly. Limited studies have been conducted by flying an armored T-28 aircraft through thunderstorms and using aircraft-mounted radar to obtain information about hail within storms. Scientists have been able to deduce the airflow within hailstorms using multiple Doppler radar analysis techniques (Chapter 2) and to understand the manner in which hail grows by simulating hail growth in laboratory wind tunnels. The results of these laboratory studies have been used to develop numerical models of hail growth. Our current understanding of hail formation and growth in thunderstorms has been derived from studies that combine these models with wind fields derived from multiple Doppler radar.

Hail develops in thunderstorms with very strong updrafts that extend well above the freezing level. Hail growth can be thought of as occurring in two steps, the formation of a ***hail embryo***, and the formation of the ***hailstone***. Hail embryos are the ice particles that occupy the center of hailstones and serve as cores for their initial growth. Hailstones are the final large stones composed of hard or spongy ice. Each step, hail embryo formation and hailstone formation, requires one up-down cycle through the storm clouds. It is easiest to understand these growth cycles in the context of the structure of a supercell thunderstorm, as illustrated in the vertical and horizontal cross sections shown in Figures 20.5 and 20.6.

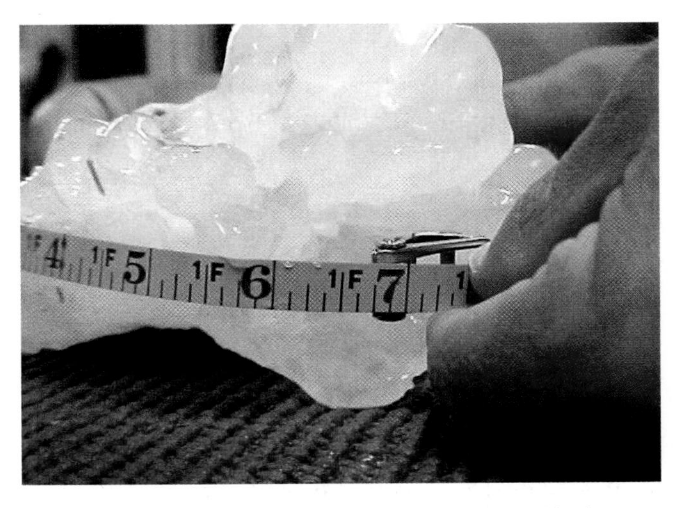

FIGURE 20.4: The largest hailstone ever recorded in the United States. The stone fell at Aurora, Nebraska on 22 June 2003. It measured 17.8 cm (7 inches) in diameter and was 47.6 cm (18.75 inches) around its largest circumference.

HAIL EMBRYO FORMATION

As air rises in a developing thunderstorm updraft (Figure 20.5), ***cloud droplets*** composing the cloud are very small, on the order of 0.02 millimeters in diameter. These tiny water droplets are present in large numbers within the strong updraft region. They are swept from the cloud base to high altitudes and cold temperatures by the rising updraft. Once cloud droplets are carried above the 0° C level, they supercool and can remain liquid even at temperatures colder than –20° C (–4° F). In late spring and summer, when ***hailstorms*** are common, the freezing level typically can be found about 3 to 4 km (2 to 2.5 miles) above the ground (Figure 20.5). Most supercooled water is found between altitudes where the temperature ranges from 0° C to about –15° C (5° F). Cloud droplets survive as liquid at these cold temperatures because of the absence of ice nuclei and because they are transported across these altitudes so rapidly. For example, air in a violent thunderstorm updraft can rise 30 meters/second (67 mph), a velocity that would transport a small cloud droplet upward 3 km in less than two minutes. If air containing the droplet started at the 0° C level, it can cool at the moist adiabatic lapse rate to –20° C in this short time. When illuminated by radar, the region of the updraft between

FOCUS 20.1

How Big Is a Hailstone?

Meteorologists rarely quote the size of hail using standard units like centimeters or inches. Instead most adopt more colorful and graphic comparisons to round objects. Thus, we have pea-sized hail (~0.5 cm diameter), marble-sized hail (~1 cm), golf-ball–sized hail (~ 4 cm), baseball-sized hail (~7 cm), grapefruit-sized hail (~10 cm), and even softball-sized hail (12 cm)! The gap between marbles and golf balls is most commonly filled with coins—hail the size of dimes (~1.5 cm), pennies (~1.75 cm), nickels (~2 cm), quarters (~2.5 cm), half dollars (~3 cm), and "silver dollars" (~3.5 cm). The gap between golf balls and baseballs most often is filled with billiard balls (~5.7 cm) and tennis balls (~6 cm).

FIGURE 20A: Hail the size of tennis balls.

Fruits are quoted less often, probably because most fruits, like apples, range in size. Aside from grapefruits, the most common hail comparisons are with grapes (1.5 cm), lemons (ovals ~4.5 to 6 cm), and oranges (~6 to 7 cm). Some hailstones, like the stone in Figure 20B are hard to size because of their irregular shape. Even here a comparison is in order—in this case, with a $20 bill!

FIGURE 20B: A spiked hailstone compared to a $20 bill.

ONLINE 20.1

The Aurora Supercell That Produced World-Record Hailstones

A new official record for the size of a hailstone was set on 22 June 2003 thanks to alert residents of Aurora, Nebraska who contacted the National Weather Service and preserved the giant hailstone depicted in Figure 20.4. Hailstones like the one in the picture produced large craters as they slammed into the ground around Aurora (Figure 20C). The WSR-88D Doppler radar near Hastings, Nebraska monitored the intense supercells that produced these large stones. Online you can view these thunderstorms. Note the small area where the radar reflectivity reaches values exceeding 65 dBZ, a clear signature of the giant hail. On the animation, the large storm near the top of the cluster of storms produced the world record stone.

FIGURE 20C: Impact craters left after the Aurora hailstorm of 22 June 2003.

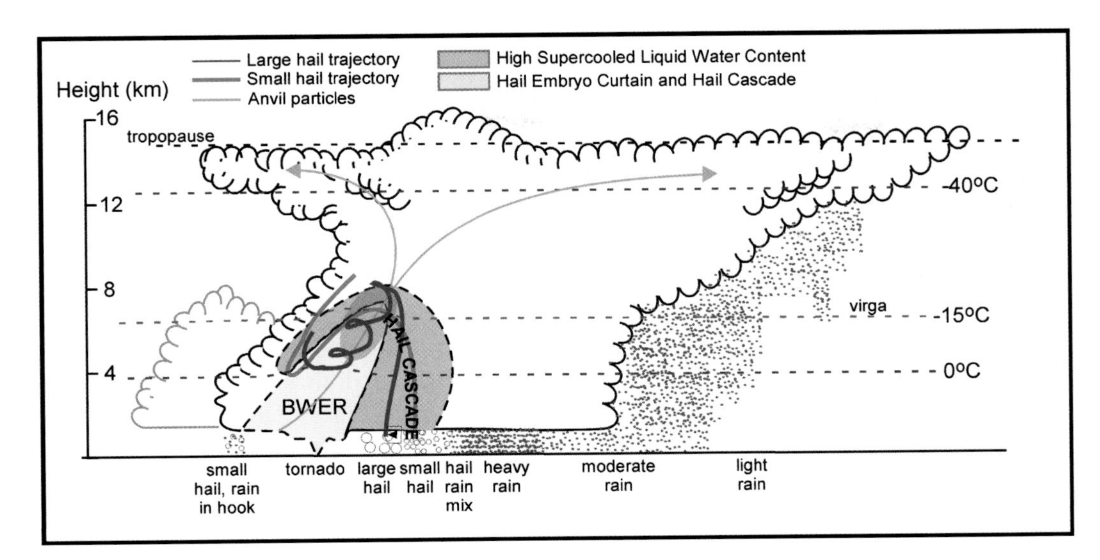

FIGURE 20.5: A southwest to northeast vertical cross section through a supercell showing the bounded weak echo region, the hail embryo curtain, precipitation types at the ground, and selected trajectories for hailstones.

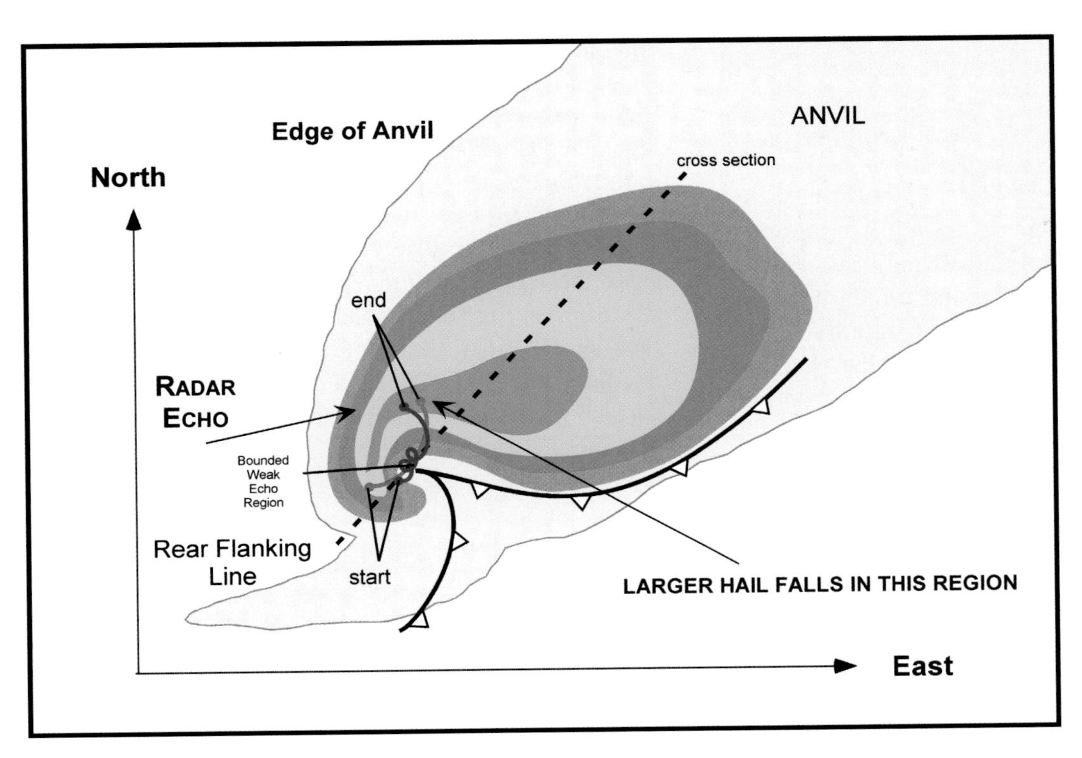

FIGURE 20.6: Horizontal cross section through the supercell shown in Figure 20.5. The distribution of precipitation types (colored regions) and hail trajectories (red and blue lines) corresponding to Figure 20.5 are shown. The dashed line denotes the cross section in Figure 20.5.

cloud base and about the –15° C level produces very weak radar returns because the only objects intercepted by the beam are tiny cloud droplets. Meteorologists call this weak echo region of the supercell updraft the ***bounded weak echo region (BWER)*** or the ***echo-free vault*** (Figures 20.5 and 20.6). The term "vault" is analogous to the tall narrow vaults of gothic cathedrals. This region appears as the cusp of the hook echo in supercells.

When a developing thunderstorm rises to levels where the temperatures fall below about –15° C, ice crystals begin to form in the storm. Ice particles that form in the stronger parts of the updraft are carried upward and ejected into the anvil of the storm, as shown on the thin orange trajectory in Figure 20.5. However, some ice particles near the periphery of the updraft, and in weaker updraft cells forming in the flanking line of clouds along the rear flank downdraft, begin to fall and grow by collecting ***supercooled cloud droplets*** (red trajectory in Figures 20.5 and 20.6). These ice particles, as they fall through the supercooled water, typically grow to sizes of about 1 to 5 mm and form ***graupel*** particles. Graupel are soft ice particles with diameters of one to a few millimeters that have the consistency of a snowball. The largest concentration of graupel particles immediately surrounds the BWER, shown on Figure 20.5. The graupel particles surrounding the BWER form an ***embryo curtain*** representing the initial source of particles that grow to become hailstones.

CHECK YOUR UNDERSTANDING 20.1

1. What types of damage and injuries typically occur due to hail?
2. Why is supercooled water important to the formation of hail in a thunderstorm?
3. At what altitudes in a thunderstorm does hail growth normally occur? What parameter determines these altitudes?

HAILSTONE GROWTH

As the storm continues to grow, some of the graupel particles in the embryo curtain are swept back into the updraft circulation by the horizontal winds. Once in the updraft, the graupel particles grow rapidly to hail size by collecting supercooled water droplets. Graupel particles located in the updraft but close to its periphery have ideal trajectories for hail growth. Those embryos that are located in a position where their fall velocity is marginally less than the updraft velocity will rise slowly through the updraft while collecting large numbers of supercooled cloud droplets. A favorable trajectory is shown in red in Figures 20.5 and 20.6. Along this trajectory, the growing hailstone slowly ascends as it crosses the core of the updraft along the base of the embryo curtain. It reaches the core of the updraft and experiences the strongest vertical motions just at the time when its size is large enough to "float" in the updraft. Located in the region of the high liquid water content and cold temperatures, the "floating" stone grows rapidly to large size. Winds aloft eventually carry the stone northeast. As the hailstone moves toward the periphery of the updraft on the northeast side of the storm, the stone begins to fall rapidly. This region, the so-called ***hail cascade*** that flanks the updraft on the east and northeast side of typical supercells, appears visually from the ground as a white curtain of hail (Figure 20.7). Most hail is believed to grow along trajectories such as the red trajectory in Figures 20.5 and 20.6. Somewhat surprisingly, the ideal trajectory for a hailstone embryo to grow into a hailstone is typically not directly up through the core of the updraft. Graupel particles carried into the core of the updraft early in their growth cycle would be swept into the upper part of the cloud much too rapidly for significant growth to occur.

Multiple Doppler radar studies suggest that very large hailstones may follow a somewhat modified trajectory, shown in blue, in Figures 20.5 and 20.6. Graupel that serve as hail embryos for very large hailstones are believed to enter the updraft along the periphery of the supercell's hook echo. These particles grow within the updraft, spiraling upward within the storm's rotating mesocyclone. Exposed to exceptional liquid water contents while "floating" in the storm's rotating updraft, these stones can grow to sizes 5 cm (2 inches) or larger and fall close to where the hailstone embryos first entered the updraft. Stones that float and spiral within an updraft core of the mesoscyclone apparently have the greatest potential to grow to exceptional size—the legendary grapefruit- and softball-sized stones reported in the most violent of hailstorms.

Hailstones have two modes of growth that depend on the surface temperature of the hailstone. To understand these different modes of growth, consider a hailstone collecting supercooled water droplets on its surface in a thunderstorm updraft where the temperature is colder than 0° C. As each cloud drop freezes on the surface of the hailstone, latent heat is released (recall that latent heat is required to melt ice, and therefore must reappear as heat when water freezes). This heat warms the surface of the hailstone, raising its surface temperature. If the surface temperature of the hailstone remains colder than 0° C, then supercooled

FIGURE 20.7: Photograph of a supercell thunderstorm with a prominent hail curtain east-northeast (to the right) of the rain-free base.

droplets will freeze on contact and the stone's surface will remain dry. This mode of growth is called the ***dry growth regime***.

During some segments of a hailstone's trajectory, so much supercooled water freezes on the stone's surface that the released latent heat will raise the surface temperature to 0° C before conduction can carry that heat away. Once the surface temperature of the stone reaches 0° C, the freezing of the water impacting the hailstone's surface will be delayed. This water can then spread across the surface of the stone and drain into porous regions, or, if the accumulation of liquid is sufficient, even be shed from the stone's surface. This mode of growth is called the ***wet growth regime***. Many of the drops shed from the stone during wet growth later freeze and serve as new hailstone embryos. In this way, as a storm approaches a mature, steady state, a storm becomes populated with greater concentrations of hail.

As hailstones pass through thunderstorms, they may grow by one mode and then the other. Large stones typically experience both modes of growth over their lifetime. The appearance of ice is different in the two growth regimes, clearer in wet growth and opaque or milky in dry growth. The milky appearance is associated with tiny bubbles of air trapped in the ice during rapid freezing. The bubbles coalesce and escape when water freezes slowly in the wet growth regime, so the ice appears clear. Hailstones, when sliced in half, often display layers indicative of different rates of freezing (e.g., Figure 20.8).

It is noteworthy that, aside from the initial recycling necessary to create the graupel particles that serve as hail embryos, no additional recycling of hailstones is required to produce large hail. This is counter to early explanations of hail growth that were heavily based on this notion of recycling, even to the point of including multiple vertical "loops" in the hail trajectory. Part of the reason why recycling was favored as a mechanism for large hail growth was the internal layering of many hailstones (Figure 20.8). It is now known that the turbulent motion within an updraft causes hail embryos to accrete supercooled liquid at varying rates over an extended period of time. The varying rate of accretion accounts for the layered structure of hailstones, including the layers of different opacity indicative of changes between the wet and dry growth modes.

CHECK YOUR UNDERSTANDING 20.2

1. What is a hail embryo? Where is the "hail embryo curtain" in a supercell thunderstorm?
2. What trajectories do hailstones typically follow between the hail embryo curtain and the ground?
3. Where is hail generally found on the ground relative to a thunderstorm's strongest updraft?
4. Explain physically the difference between the dry and wet growth regimes of hail.

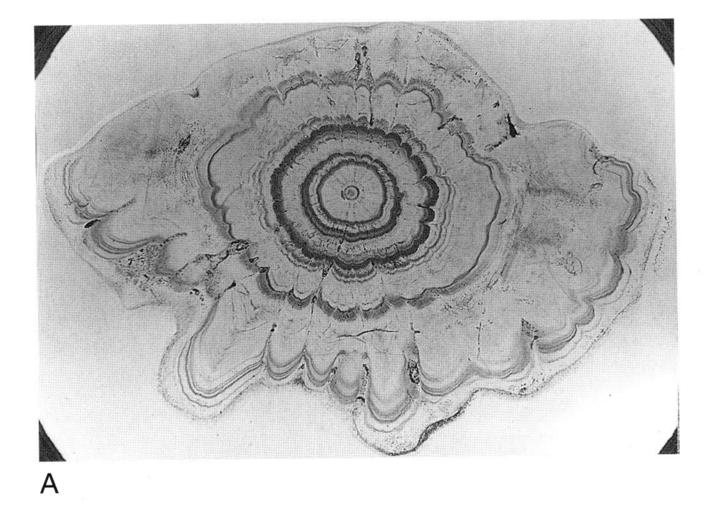

A

B

FIGURE 20.8: View of a thin slice through a large hailstone shown (A) in natural light and (B) in cross-polarized light. The long dimension of the stone is 14.3 cm (5.6 inches). The photograph in Panel (A) illustrates the layers of alternate air bubble density indicative of wet and dry growth modes. The photograph in (B) shows the crystal structures within the hailstone.

FOCUS 20.2

Hail and "Green" Thunderstorms

The ominous green tint sometimes seen in the clouds of a thunderstorm has been interpreted by many people as a sign that the thunderstorm contains hail. Indeed, greenish-tinged thunderstorms often are severe and generally are accompanied by hail. But does the green tint result from the presence of hail?

This question has long puzzled meteorologists, and has led to some systematic studies of thunderstorm color. Frank Gallagher III and Kimbra Kutlip, in an article entitled "The Green Menace,"[1] describe an experiment where light from thunderstorms was measured with a spectrophotometer. This instrument measures the intensity of light in different wavelength or "color" bands, and indicates which color dominates the spectrum of light reaching the observer. The results of these measurements, and also a separate study based on mathematical calculations, led to the conclusion that hail does not cause the green color. Hail occurred in some thunderstorms that the spectrophotometer showed were not "green" while some "green" thunderstorms did not produce hail. The spectrophotometer measurements also disproved the theory that the green tint results from light reflecting off green vegetation.

Then why do some thunderstorms appear green? The answer appears to lie in the large liquid water content of deep thunderstorms. Water acts as a filter, absorbing longer wavelengths of light (i.e., red, yellow), leaving blue and green. In deep thunderstorms, insufficient amounts of blue light survive the long path through the atmosphere and cloud, so the dominant "surviving" wavelength that emerges from the cloud is green. Filtering of colors also occurs in large water bodies, giving even clear water a greenish blue color when illuminated by the sun. In order to produce the green appearance, the path length of light through liquid water drops must be at least several miles, implying that the cloud must be deep and the sun angle must be favorable.

[1] *Weatherwise*, May–June 2001.

FORECASTING AND DETECTION OF HAIL

Hail formation requires strong updrafts. We know from Chapters 6 and 19 that an environment that supports strong convective updrafts is characterized by large values of Convective Available Potential Energy, or CAPE. Forecasters at the National Weather Service's Storm Prediction Center in Norman, Oklahoma, monitor the distribution of CAPE and other stability indices predicted by numerical models to determine where hailstorms are likely to occur. The forecasters also monitor the position of frontal boundaries, the jetstream and jetstreaks, and other atmospheric features that may be associated with thunderstorm development. From these analyses, the SPC issues forecast maps such as Figure 20.9, which shows the probability that hail 3/4-inch diameter or larger will occur within 25 miles of a location. These maps are valid for a 24-hour period from the time of issue. Forecasts for subsequent days (Day 2 and 3) are not hail-specific, showing instead the probability of severe weather within 25 miles of a point. The National Weather Service also does not currently issue hail-specific watches; severe thunderstorm watches are issued for storms that may include hail and/or damaging winds.

Radar is presently the best tool available to meteorologists for real-time hail detection. Estimations of hail location and hail intensity using the current U.S. National Weather Service Doppler radar network (WSR-88D) rely on interpretation of the radar reflectivity. Radar reflectivity is very sensitive to the size of the particles intercepted by the radar beam. The reflectivity is also sensitive to the composition of particles, in general less sensitive to ice and more to water. In rain, where drops rarely exceed 3 to 4 mm in diameter, radar reflectivity values rarely exceed about 50 dBZ. Rain mixed with small hail will generally produce reflectivity values between 50 and 60 dBZ. Above 60 dBZ, the radar reflectivity can normally be attributed to hail, with values approaching 70 dBZ indicating very large hail. Unfortunately, there is a good deal of uncertainty in these values. The uncertainty arises because hail can grow in both wet and dry mode, the wet hail being much more reflective than dry hail. The concentration of rain and hail will also vary from storm to storm, creating further uncer-

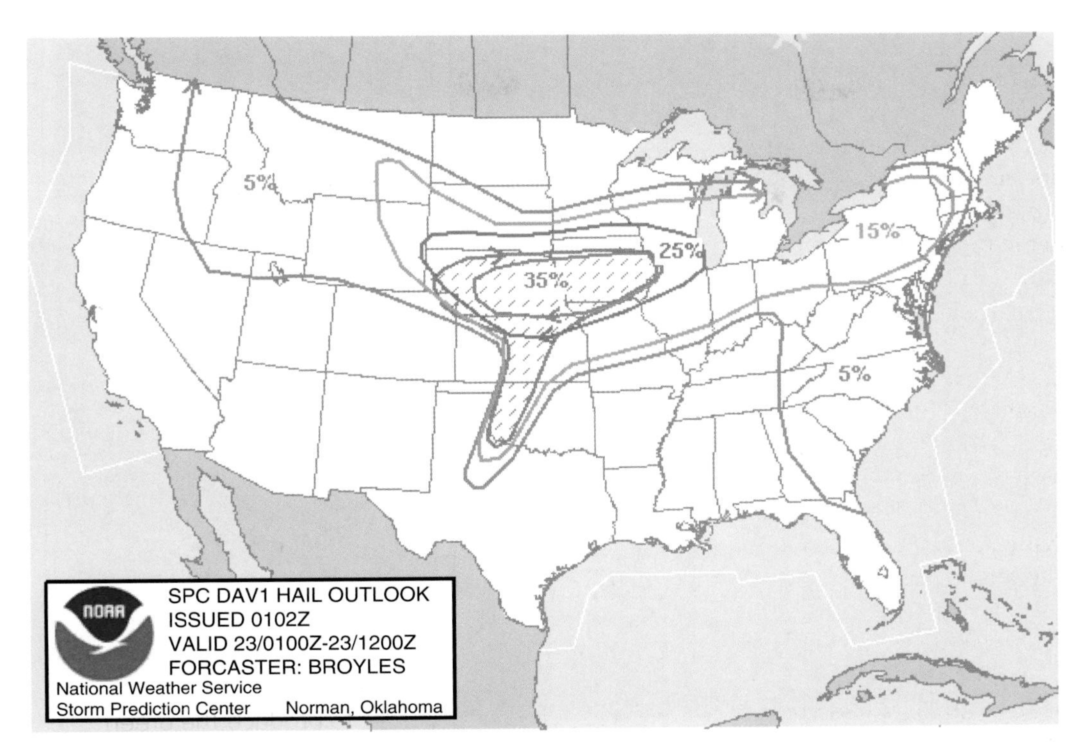

FIGURE 20.9: A National Weather Service Storm Prediction Center forecast for the probability of hail with diameter ¾ inches or greater within 25 miles of a point on 23 May 2004.

tainty concerning the interpretation of the radar reflectivity measurement. For these reasons, determining the location of hail with current operational radars is an inexact science. Consequently, as in the case of severe thunderstorm watches, severe thunderstorm warnings currently issued by the U.S. National Weather Service are not hail-specific. Rather, warnings incorporate all potential types of severe weather, including strong straight-line winds, hail, lightning, and possible tornadoes.

Advanced radar technology should soon improve warnings for hail. A special type of radar, called a ***polarization diversity radar,*** is capable of distinguishing regions of hail from regions of heavy rain (see Focus 20.3). Polarization diversity radars measure many parameters related to the polarization state of the transmitted and received radar energy. The set of measurable quantities that can be derived from polarization diversity radars together can be used to discriminate the types of particles (hail, rain, snow, small ice crystals, etc.) within a storm.

A dramatic example of the potential of polarization diversity radars to discriminate hail and other particles in clouds appears in the cross-sections in Figure 20.10. This figure shows a cross-section of (A) the radar reflectivity for horizontally polarized radiation, (B) the differential reflectivity (Focus 20.3), a measure of the shape of the particles within the beam, and (C) a particle classification based on all of the polarization variables. Note in (C) that the hailshaft (in yellow) has high reflectivity and a low differential reflectivity, while the rain to the right of the hailshaft has lower reflectivity and higher differential reflectivity. The reflectivity and differential reflectivity, combined with other polarization measurements, allow meteorologists to identify the locations of various types of particles within a storm. For example, the upper part of the storm in Figure 20.10 contained hail and graupel, while the anvil top of the storm extending to the right of the diagram is composed of dry snow and irregular ice crystals.

A forecaster confronted with the information in Figure 20.10 would immediately recognize that the storm contained a large region of hail. Most of the hail is above 4 km, which is the approximate altitude of the 0° C isotherm (horizontal line on Figure 20.10C). However, the downward-pointing finger of hail to the left of the storm's center indicates the location at which hail is most likely to reach the ground. Real-time warnings would be keyed to this portion of the storm. As of this writing, polarization diversity radars are only used in meteorological research. However, the U.S. National Weather Service is planning to upgrade its Doppler radar network to include polarization capability in the near future. When this occurs, we can expect to hear more accurate and precise warnings related to the onset, location, and intensity of hail.

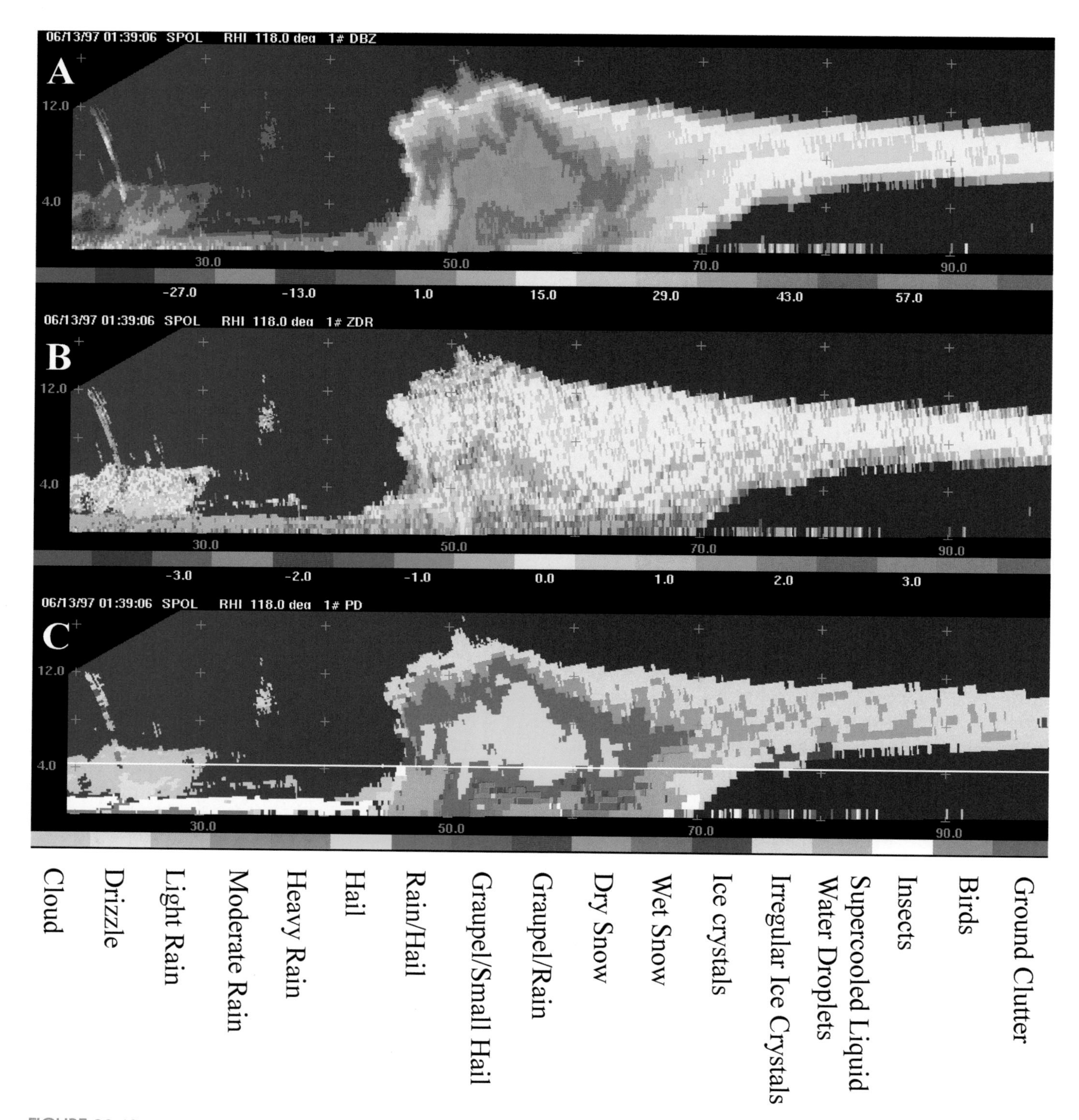

FIGURE 20.10: Vertical cross sections through a hailstorm showing (A) the radar reflectivity; (B) the differential reflectivity; and (C) a particle classification based on analysis of all polarimetric parameters.

FOCUS 20.3

Radar Polarization

The microwaves that radars transmit, like all forms of electromagnetic energy, consist of oscillating electric and magnetic fields. The electric and magnetic fields are oriented perpendicular to each other and to the direction of the radar beam (see Figure 20D). Most conventional radars transmit microwave energy such that the electric field oscillates in a plane oriented horizontally (parallel to the earth's surface) as shown in Figure 20D(A). However, radars can be designed so that the electric field is oriented vertically, as in Figure 20D(B). Polarization diversity radars are designed so that the orientation of the electric field, or *polarization,* is switched back and forth between horizontal and vertical orientation with each successive pulse of energy transmitted by the radar.

Why is this important for hail detection? When large raindrops fall, aerodynamic forces acting on the drops cause them to become squeezed into a shape approaching a hamburger bun (Figure 20E). The distortion begins when a drop's diameter approaches about 1 mm, and becomes quite large as a drop grows to sizes exceeding 3 mm. The electric field of a radar beam that is transmitted at horizontal polarization, when passing through a distorted drop, will sense a drop that appears to have a large diameter (the long, horizontal axis on Figure 20E). As a result more energy will be scattered back to the radar and the radar reflectivity will be larger. On the other hand, the electric field of a radar beam transmitted at vertical polarization, when passing through the same drop, will sense a drop

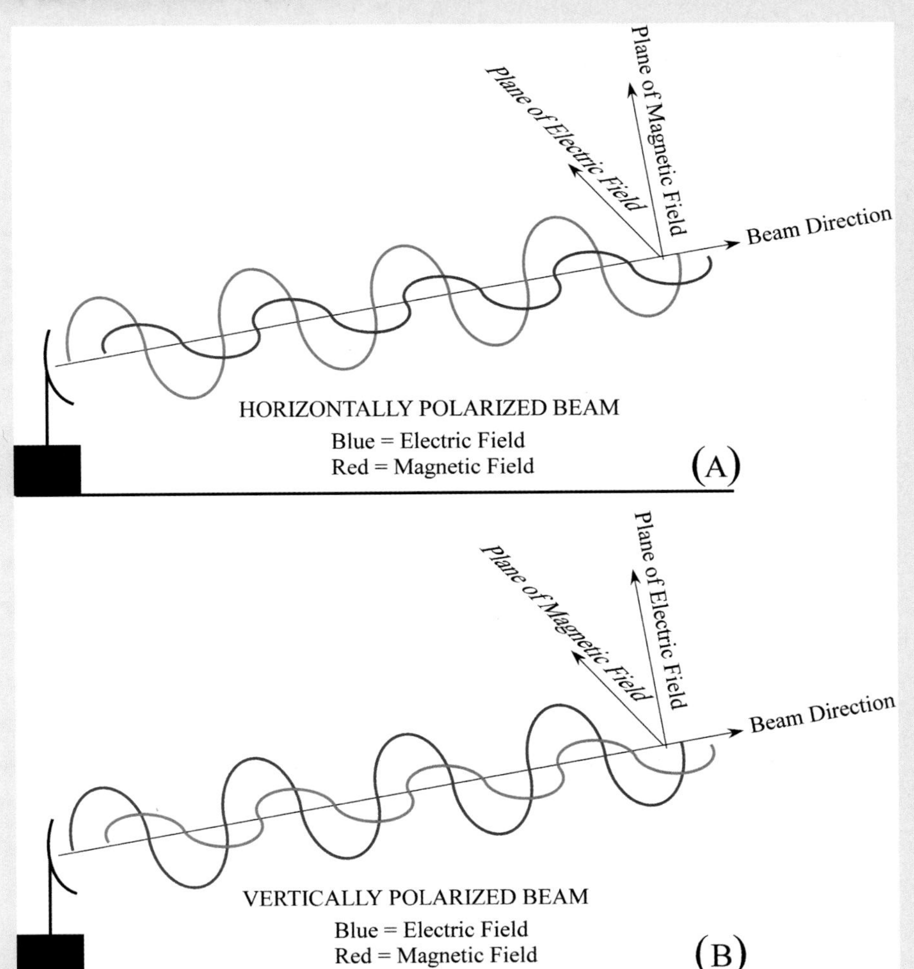

FIGURE 20D: A propagating radar beam: (A) for horizontally polarized radiation with the electric field oscillating in a plane parallel to the earth's surface (the configuration for the current WSR-88D radars); and (B) for a vertically polarized radar beam, with the electric field oscillating in the vertical plane.

(continued)

(continued)

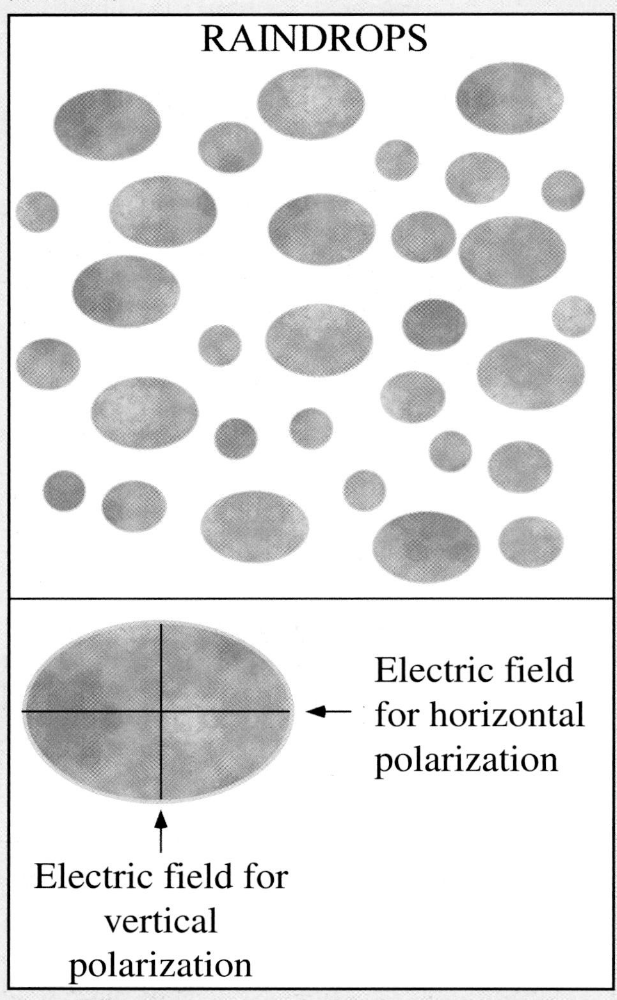

FIGURE 20E: A field of large falling raindrops. Note that all the raindrops are oriented the same way, in contrast to hailstones, which are randomly oriented. The two axes encountered by the electric field of a horizontally and vertically polarized radar beam are shown on the expanded drop in the lower window.

with a smaller diameter (the short, vertical axis on Figure 20E). In this case, less energy will be scattered back to the radar and the radar reflectivity will be smaller. Radar meteorologists define a quantity called the *differential reflectivity*, as the ratio of the reflectivity measured at horizontal polarization to that at vertical polarization. Differential reflectivity, which is measured in logarithmic units called decibels (dB), has values of about 1 to 2 dB in moderate rain and 3 to 4 dB in heavy rain.

What happens to the differential reflectivity when a radar observes hail? When a hailstone grows, it typically tumbles. Sme hailstones are spherical, others have different shapes, but overall, hailstones fall randomly with no preferred orientation. When radar views a large number of hailstones, the "average" shape of the stones appears nearly spherical. As a result, the radar reflectivities at horizontal and vertical polarizations are nearly equal and the differential reflectivity is close to one or, in logarithmic units, near 0 dB. The difference in differential reflectivity between rain and hail allows the differential reflectivity to be a good discriminator between these two precipitation types.

CHECK YOUR UNDERSTANDING 20.3

1. What type of forecast product does the Storm Prediction Center routinely issue for hail?
2. Why does the National Weather Service not issue hail-specific warnings?
3. What advantage will polarization radar give to forecasters in the future?

DISTRIBUTION AND IMPACTS OF HAIL

Hail occurs in specific sectors of a thunderstorm, and hailfall on the ground occurs in relatively narrow regions, a few hundred meters to a few kilometers in width, and several kilometers to as long as 60 km in length. A commonality between hail and tornadoes is the intermittency of their paths on the ground. Intermittent hail is often produced by long-lived supercell thunderstorms and less often by MCS thunderstorms. It is generally difficult to map the hail that falls from a storm, since it may be spread over a broad area—much

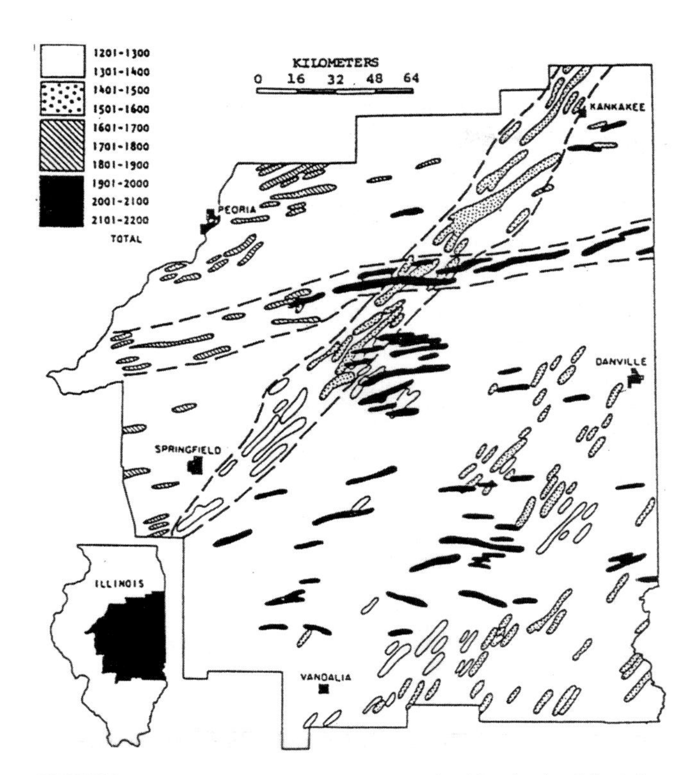

FIGURE 20.11: The hailswaths (areas inside dashed lines) and hailstreaks associated with two major hail systems lasting for ten hours in a series of May storms across central Illinois. Shading of hailstreaks denotes the local time of hail fall.

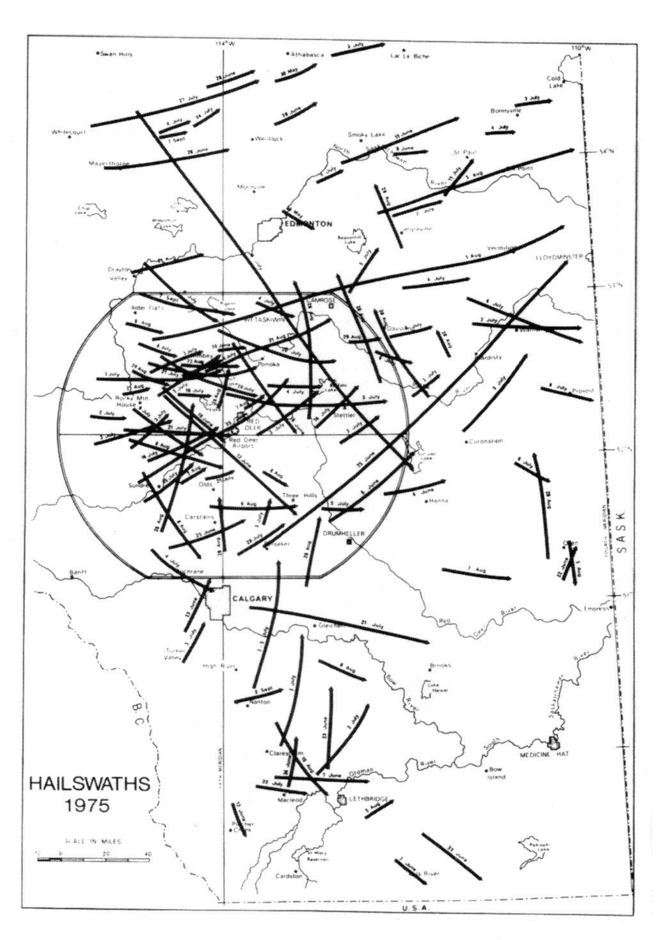

FIGURE 20.12: Map showing the location of all hailswaths in Alberta, Canada, during the 1975 summer hail season.

of which is rural. A few studies give perspective on the distribution of hail from single storms. Figure 20.11 shows a map of the hail across eastern Illinois during a ten-hour period when several hailstorms passed through the region. The smaller continuous regions of hail coverage, referred to as ***hailstreaks,*** are typically 1 to 2 km in width and vary in length from a few to over 60 km (~35 mi). These continuous areas of hailfall are often part of a somewhat wider, longer area of hailfall called a ***hailswath,*** two of which are located within the dashed lines on Figure 20.11. The hailswath is a reflection of the track of an individual supercell, while hailstreaks are associated with a specific updraft. Figure 20.12 shows all the hailswaths that occurred in 1975 in Alberta, Canada. This figure gives some perspective on the overall coverage of hail that can be expected during a single season in locations in the High Plains east of the Rockies.

Impacts of hail in the United States are best understood in the context of the larger geographical distribution of hail. Figure 20.13 shows the distribution of the annual mean number of days with hail of (left) 0.75 inches diameter or greater, and (right) 2.0 inches diameter or greater. In both cases, the maps denote the frequencies of hail occurrence within 25 miles of a point. A circle with a 25-mile radius has an area comparable to a typical county in most states. Hail occurrences in which the diameters did not exceed 0.75 inches are not included in Figure 20.13. These occurrences of small hail, which are common in the lee of the Rockies during summer and even along the West Coast during winter, are often included in other hail climatologies and would increase the numerical values in Figure 20.13.

Figure 20.13 (left) shows that most points east of the Rocky Mountains average at least one 0.75-inch or greater hail event annually within 25 miles of their location. Large hail is most common in the Great Plains states, with maximum frequencies in Oklahoma and Texas. Hail also is common north of the U.S. border through Alberta and Saskatchewan in Canada. The distribution of hail with diameters of two inches or more (Figure 20.13 right) outlines "Hail Alley," a region roughly defined as extending northward from Texas to the Dakotas in the United States.

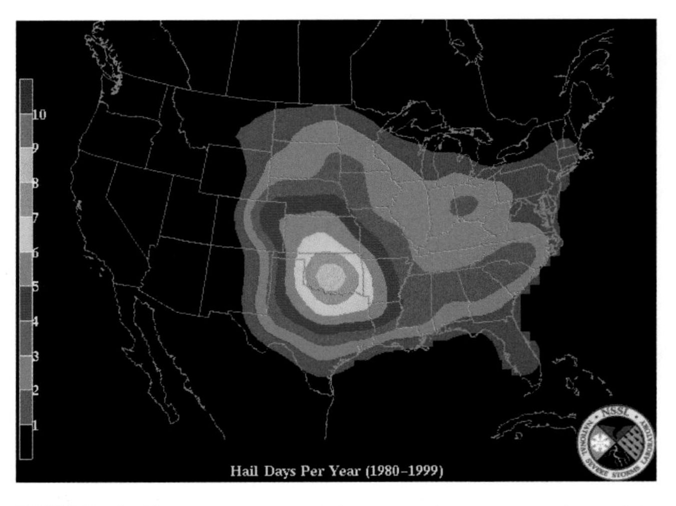

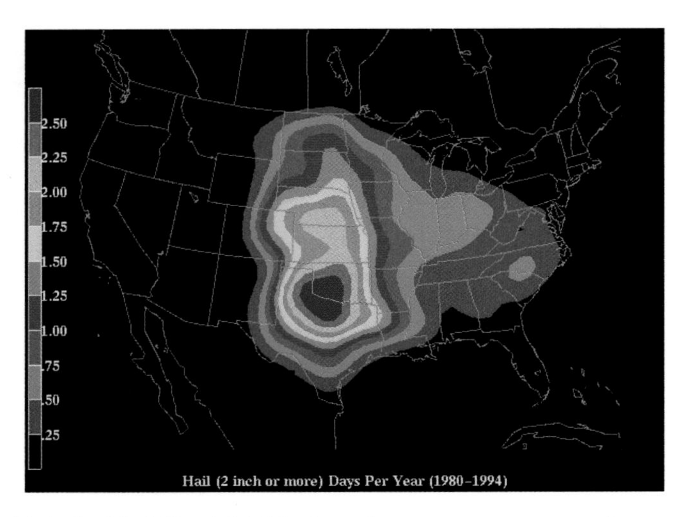

FIGURE 20.13: Distribution of the annual mean number of days with hail of (left/right) 0.75 inches diameter or greater, and (right) 2.0 inches diameter or greater in the United States. Note that the two panels have different color scales.

FOCUS 20.4

Hail Suppression: Can Hail Damage Be Reduced?

Given the catastrophic losses to crops and property that can result from hail, it is not surprising that hail suppression has been an appealing notion for many decades. The approaches taken with hail suppression range from the rather primitive attempts to shatter hailstones by firing cannons into thunderstorms, to more scientific attempts based on hypotheses that the microphysical processes important to the growth of hail might be modified by introducing cloud-altering substances into the cloud.

One hypothesis draws upon the notion of *beneficial competition.* To suppress hail, microscopic particles of silver iodide are injected into the base of the cloud's updraft. Silver iodide is a very effective ice nucleant. The proposed idea was that the silver iodide would create large numbers of ice crystals, which all might act as hail embryos. These embryos together would rapidly reduce the supply of supercooled liquid, creating many small ice particles instead of fewer, larger hailstones. The smaller ice particles would melt when they fall, the result being that hail emerging from the cloud would be reduced in both size and number.

The extent to which this and other methods of hail suppression is effective has been studied in hail suppression experiments such as the National Hail Research Experiment (NHRE) in Colorado (1972–76), and the Grossversuch IV experiment in Switzerland (1976–83). These studies were all inconclusive.

Scientific proof of the effectiveness of hail suppression is still lacking, primarily because the natural variability among thunderstorms is tremendous and the microphysical and thermodynamic processes at work in severe thunderstorms are exceedingly complex, especially with regard to the formation and growth of hail embryos.

Although the effectiveness of hail suppression lacks any scientific proof, hail suppression programs are still being carried out today by private and public organizations in regions of the world where economic losses to agriculture due to hail are significant.

The damage caused by hail falls into two broad categories, agricultural (crop) and structural. The latter includes damage to buildings, automobiles, aircraft, and other vehicles. The geographical patterns in Figure 20.13 indicate that occurrences of the most damaging (largest) hail are most common in the agricultural areas of the high Plains and Midwest, and least common along the heavily populated East and West Coasts. As a result, agricultural damage from hail is more frequent than structural damage. Nevertheless, ***Hail Alley*** includes some large cities, including Dallas, Fort Worth, Oklahoma City, Denver, and Kansas City. When a severe hailstorm strikes an urban area, the damage totals can quickly exceed several

ONLINE 20.2

Seasonal Variation of Hail in the United States

The occurrence of tornadic thunderstorms varies strongly on a regional and seasonal basis (Chapter 19). Since hail is also associated with severe thunderstorms, similar geographical and seasonal variations can be expected. Online, you can view an animation that captures these variations in terms of the probability (in percent) that 3/4-inch or larger hail will occur within 25 miles of any location in the United States on a given day. The probabilities are color-coded, ranging from less than 1 percent (black) to about 7 percent (yellow-orange). Notice how the probabilities remain low throughout the country during the winter months and increase sharply during the spring. A northward migration from spring to summer is also apparent, in much the same way that the frequencies of severe thunderstorms and tornadoes migrate northward. Nevertheless, the maximum hail probabilities remain in the Oklahoma-Kansas region throughout the year. Note how the probability of 3/4-inch hail never reaches even 1 percent at any time of the year over northern New York/New England, over southern Florida, and over nearly the entire region west of the Rockies. Can you think of reasons why large hail is so rare in these areas?

hundred million dollars, while the agricultural damage from individual hailstorms is much smaller. However, the agricultural area impacted by hailstorms far exceeds the urban area, so agricultural and structural losses are about equal, about $1.5 billion annually, which is about 1 to 2 percent of the total crop value in the United States.

Agricultural losses from a hailstorm depend on several factors, including the type of crop and its stage of development. Crops such as corn and soybeans are generally more vulnerable to hail damage in the middle of the growing season, when plants are still developing, than in the early or late phases of the season. A key factor in the severity of crop damage is the wind speed during hailfall. Strong horizontal winds give hail a lateral component of motion that greatly enhances the likelihood of direct contact with (and damage to) a vertically oriented plant. Mitigation strategies for minimizing damage in these situations include windbreaks adjacent to fields and the planting of crops in rows parallel to the most common wind direction. Such strategies can only marginally reduce crop damage from a severe hail

FIGURE 20.14: Damage to an automobile from a supercell near Ogallala, Nebraska.

event. Because hail damage to crops is largely random and unpreventable, many farmers in the central United States purchase hail insurance on their crops. At present, about 25 percent of crop value in the United States is covered by hail insurance, although the percentage is closer to 50 percent in the Plains and Midwest.

Unlike crop losses, property losses attributable to hail have risen sharply in recent decades. The increase in property damage has been fueled by the rapid expansion of urban areas and the increase in the number of automobiles and other vehicles. A specific factor in the increase of property damage has been the occurrence during recent years of severe hailstorms in several major urban areas including Denver and Fort Worth, each of which suffered on the order of a billion dollars of damage from single events.

Property damage by hail is a strong function of the horizontal wind speed during hailfall. Siding and windows, for example, may escape damage when hail falls vertically in the absence of wind, but the damage can be considerable if large hailstones have a horizontal component of motion. Millions of dollars of damage in urban hailstorms results from damage to automobiles. Hailstones of two-inch diameter will generally crack (or pass through) an automobile windshield (Figure 20.14). Vehicular damage represents an example of the potential value of timely warnings of hail, since only a few minutes of advance warning will enable many automobile owners to move their vehicles to the shelter of a garage. Even the relocation of a car to the shelter of a tree can reduce damage. When the effects of increased population, number of vehicles, areal coverage of structures, and changes of structural materials are taken into account, the occurrence of hail does not appear to be increasing systematically. Thus the increase of hail-induced property damage is primarily an indication of changes in societal vulnerability rather than any changes in meteorology.

CHECK YOUR UNDERSTANDING 20.4

1. Where in the United States does large hail occur most frequently?
2. What are the two broad types of damage caused by hail? How do the annual damage totals in these categories typically compare in the United States?
3. Approximately what percentage of crop value in the United States is lost to hail damage in a typical year?
4. How is hail damage related to wind speed? Explain your answer.

TEST YOUR UNDERSTANDING

1. How do the numbers of fatalities from hail compare to the numbers from lightning and tornadoes?
2. How large can hailstones grow?
3. Why are direct measurements of hail difficult to obtain?
4. Differentiate between a hail embryo and a hailstone.
5. What is graupel and how is it different from hail?
6. How does the location of graupel in an updraft influence the size that a hailstone will become?
7. Describe the journey that a large hailstone takes from the time it first forms until the time it hits the ground in a supercell thunderstorm.
8. Why are hailstones layered.
9. Describe where hail would most likely occur relative to other types of precipitation in a thunderstorm. Explain why the precipitation types have this spatial distribution.
10. How does the amount of growth experienced by a hail embryo depend on its trajectory in a thunderstorm?
11. Why is the trajectory of a hail embryo directly through the center of an updraft core less likely to produce a large hailstone than a trajectory along the base of the embryo curtain?
12. Are multiple vertical loops of a trajectory necessary for the formation of large hail in a supercell thunderstorm? Explain why or why not.
13. If you experience large hail during a thunderstorm, has the threat of a tornado passed? Explain your answer.
14. Describe the difference between the wet and the dry growth regime. Be sure to include a thorough discussion of latent heat.
15. What causes ambiguity when trying to identify hail with radar reflectivity alone?
16. What is a polarization diversity radar capable of measuring that the current Doppler radars cannot measure?
17. Explain, in simple terms, why a polarization diversity radar can discriminate between hail and rain.
18. Does the incidence of hail in the United States appear to have changed in recent decades?
19. Explain the difference between the recent trends of hail-related crop losses and hail-related property losses.
20. Why is an "overshooting top" often a good sign that a thunderstorm contains hail?
21. Why is large hail relatively rare west of the Rocky Mountains?

TEST YOUR PROBLEM-SOLVING SKILLS

1. You are driving on the western Great Plains and are caught in a hailstorm that produces golf ball–sized hailstones. The surface temperature before the hail started falling was 77° F. You look southwest as the hail begins to fall and note that the supercell thunderstorm producing the hail has a rain-free base. You estimate the rain-free base of the cloud to be 1.5 km above ground. Estimate the range of altitudes where the hailstones underwent most of their growth. (Hint, consider the temperature of a rising air parcel in the updraft.)

2. Diameters of cloud droplets collected by hailstones are about 0.02 mm. Assume a hailstone grows from a frozen raindrop that is 2 mm in diameter. How many supercooled cloud droplets must be collected by a spherical hailstone in order for the hailstone's diameter to reach 2.0 cm (0.8 in.)? Assume that the hailstone has no air bubbles within it so that its density is the density of pure ice, 0.9 grams per cubic centimeter.

3. A first-order approximation for the fall speed, V, of a spherical hailstone of diameter, d, in still air is

$$V = k\sqrt{d}$$

where k = 20 if d is given in centimeters and V is in meters per second.

(a) Use this formula to estimate the fall speed of a 5-cm (2-inch) hailstone in still air.

(b) Suppose the horizontal wind speed is 20 meters per second (about 39 mph) in the region in which the hailstone is falling. How fast could the hailstone be moving?

(c) According to the laws of physics, velocity is the product of acceleration and the time period over which the acceleration occurs. If the 5-cm hailstone starts to fall and is affected only by the acceleration due to gravity (9.8 meters per sec^2), how long will it take to acquire the fall speed you obtained in (a)?

(d) Why doesn't the gravitational acceleration cause the hailstone to acquire a fall speed larger than your answer to (a)?

4. The notion of *beneficial competition* (i.e., "beneficial" for hail suppression) is based on the premise that precipitation particles will be smaller when larger numbers of them must share the available moisture. Suppose that a cloud contains 10 hail embryos per cubic meter, and enough supercooled liquid to enable the hail embryos to grow to hailstones with diameters of 2.5 cm (1 inch). If seeding of the hail growth region increases the number of embryos to 80 per cubic meter with the same moisture availability, how large would the hailstones grow if they collected the same amount of supercooled water as the original 10 embryos collected?

5. It requires about 2,100 joules of energy to raise the temperature of 1 kg of ice one degree Celsius. When water freezes, about 334,000 joules of heat are released for every kilogram frozen. Suppose a hail embryo of diameter 2 mm and temperature –15° C accretes enough supercooled liquid (also at –15° C) to increase its diameter to 2 cm. If 10 percent of the latent heat of fusion is retained by the growing embryo and 90 percent is released immediately to the surrounding air, by how much would the temperature of the hailstone increase due to the freezing of the accreted water?

6. The damage caused by hailstones is proportional to their kinetic energy (KE), which is the product of half the stone's mass (M) and the square of its speed (V), or

$$KE = \tfrac{1}{2} MV^2.$$

Using the approximation for fall speed in Problem 3, determine the factor by which a vertically falling hailstone's kinetic energy increases when the stone's diameter increases from 2 cm to 4 cm. Assume that the density of the stone is 0.8 grams per cubic centimeter.

USE THE SEVERE AND HAZARDOUS WEATHER WEBSITE

http://severewx.atmos.uiuc.edu

1. The Storm Prediction Center issues convective outlooks several times per day. These outlooks contain the probabilities of severe weather over the contiguous United States for each of the next three days. The Day 1 outlook includes separate sets of probabilities within 25 miles of a point for tornadoes, large hail (diameters > 0.75 inch) and damaging winds. Use the *Severe and Hazardous Weather Website* to navigate to the Storm Prediction Center website.
   (a) Monitor these outlooks during the convective season comparing the Day 1 outlooks for the three severe weather threats. Compare and contrast the location and spatial extent of each of the forecasts.
   (b) Examine the same forecast later in the same day after it has been updated. What changes were made to the forecast? What justifications for changes are provided in the accompanying forecast discussion?
   (c) The Day 2 and Day 3 outlooks only provide categorical and probabilistic forecasts of the threat of severe weather. Compare the Day 1, 2 and 3 forecasts and summarize any patterns in the location, intensity and probability of severe convection.

2. Reports of severe weather are complied by the Storm Prediction Center and made available on their web site. You can access the SPC web site via the *Severe and Hazardous Weather Website.* Monitor the site during the convective season for reports of large or significant hail.
   (a) When severe hail is reported, record the location and approximate time of the events.
   (b) Use the SPC current radar and satellite data to examine the severe thunderstorms. Compare the location of the thunderstorms to the hail reports.
   (c) Examine the radar reflectivity of the thunderstorms associated with the hail reports. Based on your knowledge of thunderstorm structure, what type of thunderstorm produced the hail? How did you determine this?
   (d) What are the highest radar reflectivity readings (in dBZ) associated with the hail events? How does this compare to the information provided in the text of this chapter?

3. Use the Storm Prediction Center's archived Storm Reports to identify a recent hail event.
   (a) Record the date, time and location of the hail event(s).
   (b) Navigate the *Severe and Hazardous Weather Website* to the "Archived Data" page to access at least two soundings in the vicinity of the hail event. Save or print the soundings. Answer the following:
      (i) At what altitude was the freezing level? How might this have affected hailstone growth?
      (ii) Compare and contrast the lapse rates of the two soundings in the layer of air where hail growth was most likely.
      (iii) Examine the stability parameters (as discussed in Chapter 6). What was the stability of the atmosphere in the layer of air where hail growth was most likely?
      (iv) Comment on the similarities and differences of the data for the two locations, keeping in mind the location of the sounding site relative to the storm.

4. Navigate the *Severe and Hazardous Weather Website* to the "Hailstorm" page and examine the animations of "probability of severe hail" and "probability of significant hail."
   (a) Determine the calendar day of the year when severe hail is most likely at your location.
   (b) Determine the calendar day of the year when significant hail is most likely at your location.
   (c) Provide an explanation for any differences you found in your answers to (a) and (b).

5. Use the *Severe and Hazardous Weather Website* "Hailstorm" page to navigate to the Hastings, Nebraska National Weather Service Office web site of the "22 June 2003 Record Hail Event."
   (a) Examine the radar base and composite reflectivities provided and discuss the storm structure.
   (b) Examine the mesoscale analysis. The analysis contains locations of the dry line and outflow boundaries as well as surface observations. Compare the mesoscale analysis to the satellite and radar images provided at the site. Discuss how the various mesoscale features can be identified on the imagery.

CHAPTER TWENTY-ONE

LIGHTNING

Time lapse photograph of several lightning strikes.

KEY TERMS

ball lightning
bead lightning
bolt from the blue
charge
cloud-to-cloud lightning
cloud-to-ground lightning
conductor
current
dart leader
electric field
electron
elves
fair weather electric field
feedback mechanism
heat lightning
image charge
in-cloud lightning
induction charging
insulator
interface charging
ion
ionized
ionosphere
blue jets
lightning
National Lightning Detection Network
negative charge
neutral charge
neutron
non-inductive charging
positive charge
positive polarity stroke
proton
red sprites
return stroke
screening layer
sheet lightning
supercooled water
St. Elmo's Fire
stepped leader
thunder
traveling spark
vapor deposition
voltage
volts
volts/meter

About 100 lightning strikes occur each second on our planet. Data on the worldwide distribution of lightning flashes, derived from satellite measurements, shows a remarkable fact about lightning: most flashes occur over land, with the largest number of strikes in central Africa, central South America, southeast Asia, northern Australia, and the southeast United States (Figure 21.1). Over the contiguous United States, an average of 20 million cloud-to-ground flashes occur annually, based on statistics compiled from the ***National Lightning Detection Network*** (see Chapter 2). Nearly half of all flashes strike the ground at more than one location simultaneously, so, on average, about 30 million locations are struck each year. Figure 21.2 shows the distribution of these lightning strikes. The greatest lightning flash density occurs across central Florida from Tampa, through Orlando to Cape Canaveral, as well as near Ft. Myers and along the Ft. Lauderdale-Miami corridor. Lightning also occurs frequently along the Gulf and south

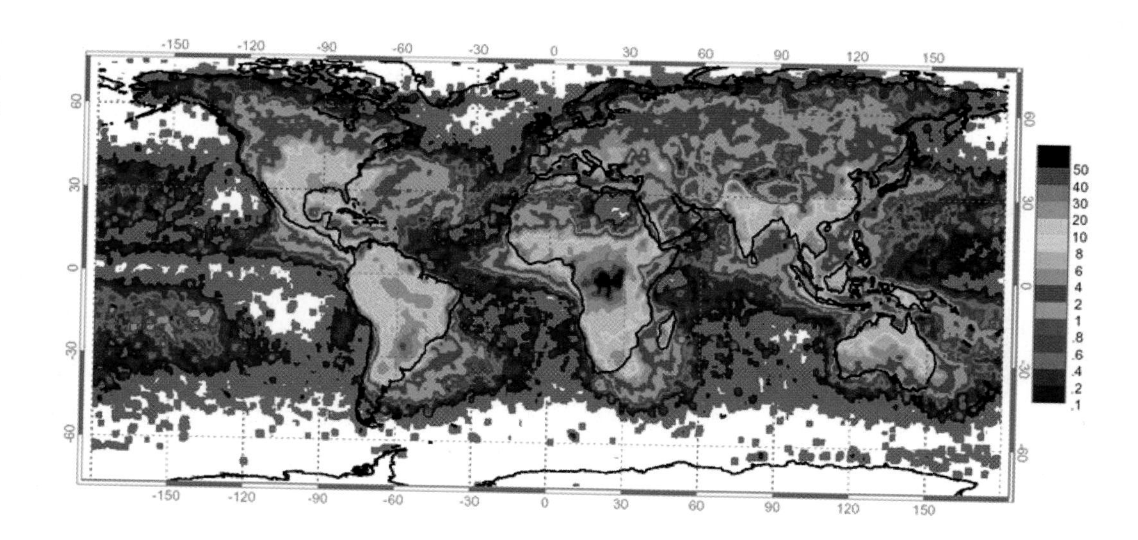

FIGURE 21.1: Average annual number of lightning flashes per square kilometer worldwide.

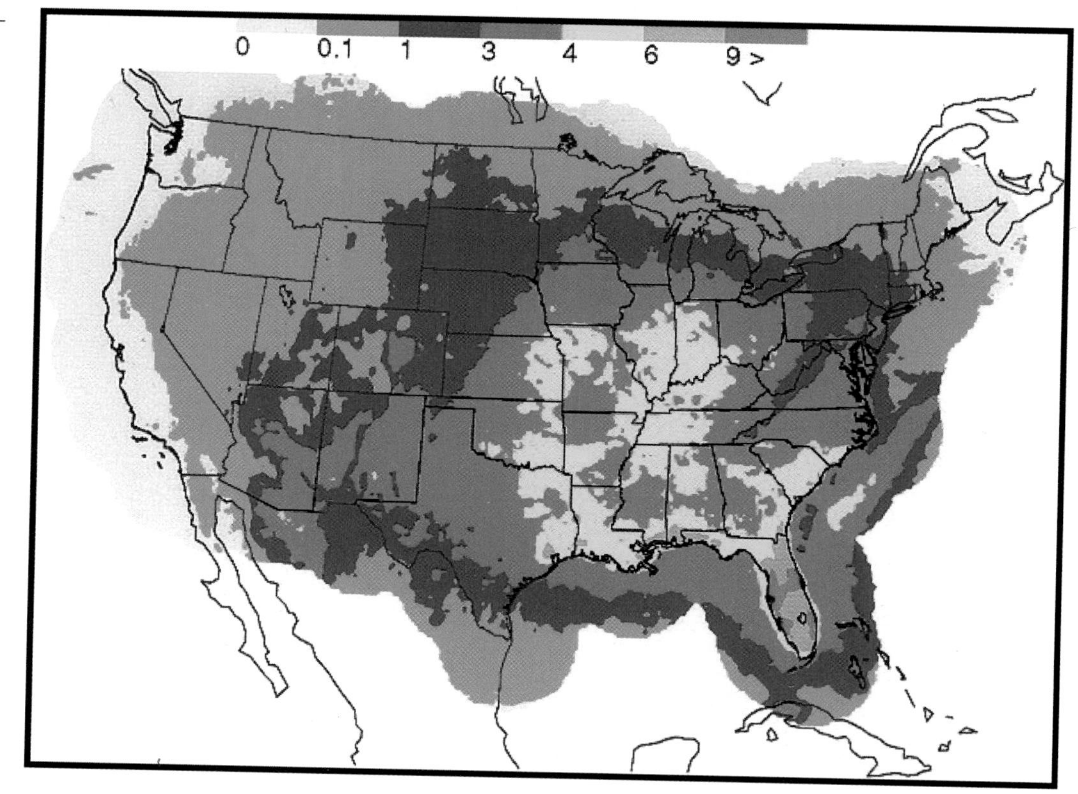

FIGURE 21.2: The distribution of lightning strikes per square kilometer per year in the ten-year period from 1989 to 1998.

Atlantic Coasts, and over a broad area of the Midwestern United States. Lightning frequency decreases westward, with very few lightning strikes occurring along the West Coast. Lightning frequency also decreases northward, as might be expected since thunderstorms require warm, moist air in the lower atmosphere to form. The peak months for lightning strikes in the United States are June, July, and August, coincident with the peak in thunderstorm occurrence.

According to the National Climatic Data Center, in the five year period ending with 2006 nearly fifty people were electrocuted each year by lightning in the United States, while 300 people were injured. Curiously, about 84 percent of all victims are male. Of the 211 deaths in the United States due to lightning in the period 2002–2006, about half occurred in open areas including ballparks, 22 percent were under trees, 7 percent near water, 4 percent on golf courses, 3 percent on or near heavy equipment, 4 percent in campgrounds, 1 percent while talking on corded telephones, and the remainder in other circumstances. Survivors often suffer serious long-term effects and permanent disabilities. The most fatalities occur in Florida, where summer activities draw people outdoors and thunderstorms quickly develop in the afternoon on most summer days. Accurate statistics concerning lightning losses are difficult to obtain, but available estimates range as high as $4–5 billion damage annually, mostly due to lightning induced fires.

FOCUS 21.1

Lightning and Forest Fires

Lightning is a major cause of forest fires. Millions of dollars in timber and property losses are recorded annually in the United States, the result of about 12,000 lightning-related forest fires. These losses have sparked the interest of the U.S Forest Service, which has been a major sponsor of lightning research. According to the National Interagency Fire Center in Boise, Idaho, lightning started 12,363 wildfires annually in the six-year period from 2001 through 2006. In contrast, fires related to human activity averaged 62,609 in the same period. Although human activity was related to 84 percent of all fires during these six years, the average annual acreage burned in lightning-induced fires, 4.6 million acres, is over two million acres more than that due to fires that humans caused. Forest fires started by lightning often occur in remote areas (Figure 21A), and are not always detected as quickly. As a result they are often harder to reach and harder to control. Remote fires are especially common in Alaska, where the lack of road access means most fires burn freely. In 2004, a dry spring followed by a few days with 5,000–10,000 lightning strikes over the boreal forest led to a severe fire season. More than 6.5 million acres (an area the size of Massachusetts)

FIGURE 21A: Fire in the Bitterroot National Forest.

(continued)

(continued)

burned between late June and September. Smoke blanketed much of the state, resulting in hazardous air quality and health advisories for many communities in the interior of the state. Figure 21B shows the shroud of smoke and the active fires in early July, 2004.

The National Lightning Detection Network now provides key information to U.S. fire-fighting agencies such as the U.S. Forest Service to determine where lightning-induced fires may develop. For example, Figure 21C shows the location of lightning strokes over several western states on a single day. The circles denote the location where fires were triggered by the lightning strokes. With this information, forest rangers can decide whether a lightning-induced fire should be extinguished to protect property and lives or allowed to burn as a natural component of forest clearing and regeneration.

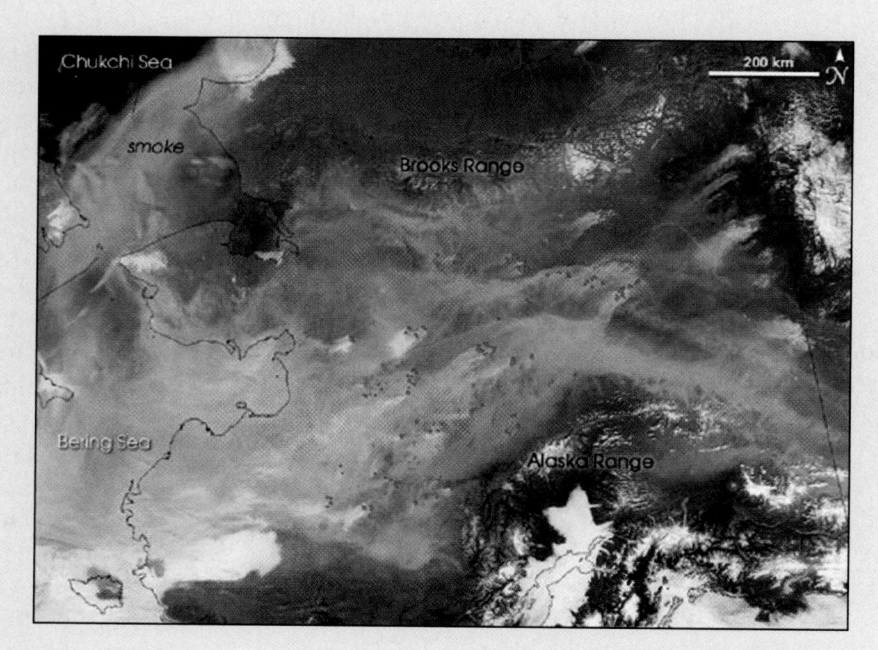

FIGURE 21B: Visible satellite image of Alaska in early July, 2004. Active wildfires are shown by red dots. Fires were started by lightning. Horizontal extent of smoke shield is more than 1000 km (600 miles). Bright white areas at bottom of image are glaciers and clouds.

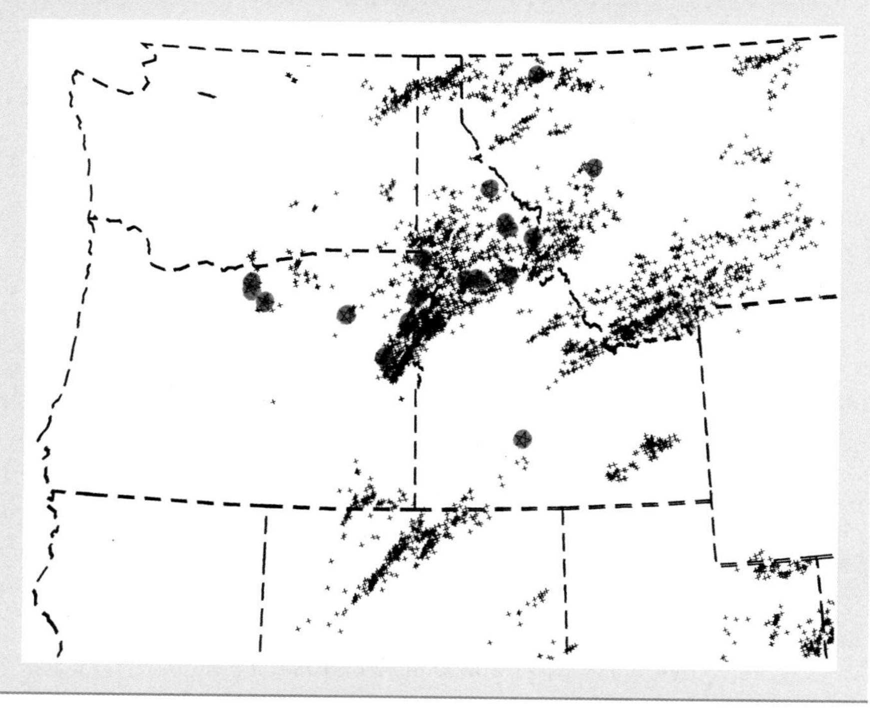

FIGURE 21C: Lightning strikes (+) and associated fires (red dots) in the northwest United States on a single day in summer.

THE LIGHTNING STROKE

Lightning is an electrical discharge in the atmosphere, a form of static electricity similar to the spark created by rubbing your shoes on a carpet and then touching a metal object. The spark created between your finger and the metal object is very short, on the order of a millimeter. By contrast, a lightning stroke typically extends about 5 km (3 miles) and is about 2 to 3 cm (~ 1 inch) in diameter. The electric field of the atmosphere just before a lightning stroke rises locally to values between 1 million and 3 million volts/meter, and the current in a lightning stroke ranges between 15,000 and 30,000 amperes. The temperature in the lightning stroke channel reaches 30,000° C (54,000° F)—five times as hot as the surface of the sun. Energy radiates from the lightning channel in a brilliant flash, which is 10 to 20 times wider than the actual 2 to 3 cm lightning channel. This energy heats the air and generates the sound waves we hear as booming thunder.

To understand the energy output of lightning, compare the power output of lightning to the power requirements of a typical household. The energy output of a typical 5 km lightning channel has been estimated to be 1 billion to 10 billion joules. However, not all the energy lightning produces is electrical. Electrical energy is a small fraction of the total energy, estimated at about 10 million joules. A kilowatt-hour is 3.6 million joules. A typical household might use 500 kilowatt hours of electricity per month, or 1.8 billion joules of energy. If *all* the energy associated with a bolt of lightning were used to power a house, it would supply between half a month and five and a half months of power. If only the electrical energy were used, it would power the house for about six hours.

ELECTRICITY AND CHARGE

To understand lightning and how it forms, we need to examine the cloud at a molecular level. Recall that there are two types of charged particles, protons and electrons. ***Protons*** each carry a ***positive charge*** and are contained in the nucleus of atoms. ***Electrons*** carry ***negative charge*** and orbit the nucleus of an atom. When an atom is ***neutral*** it contains the same number of protons and electrons. An atom is positively charged if it has fewer electrons than protons, and negatively charged if it has more electrons than protons. When the numbers of protons and electrons are not the same, an atom (or molecule) is called an ***ion***. Electrons are mobile, jumping easily from one atom or molecule to another, while protons are immobile—the whole ion has to come along for the ride for positive charges to move. For this reason, electrical currents normally consist of moving electrons. Historically, scientists and others who deal with electricity have defined electrical ***current*** as the direction that the positive charge moves. Therefore, *the current, by definition, is always in the opposite direction of the flow of electrons.*

An ***electric field*** is present in any region exposed to charges. A charge, such as a positive or negative ion, when placed in an electric field, experiences an attractive or repulsive force. Charges of the same sign repel each other and opposite charges attract, so electrons will migrate away from other electrons and toward positive ions. The magnitude of the force of attraction is measured in ***volts***, and the strength of an electric field is measured in ***volts/meter***. Current flows easily through ***conductors*** such as metal and water, and poorly if at all through ***insulators***, such as plastic and air. Because air is an excellent insulator, a very strong electric field must exist before charges can move freely through the atmosphere.

The Earth has a ***fair weather electric field*** (Figure 21.3), an electric field that exists in the absence of clouds. The fair weather electric field exists because there are an excess number of positively charged ions in the atmosphere and an excess number of negatively charged ions on the Earth's surface.

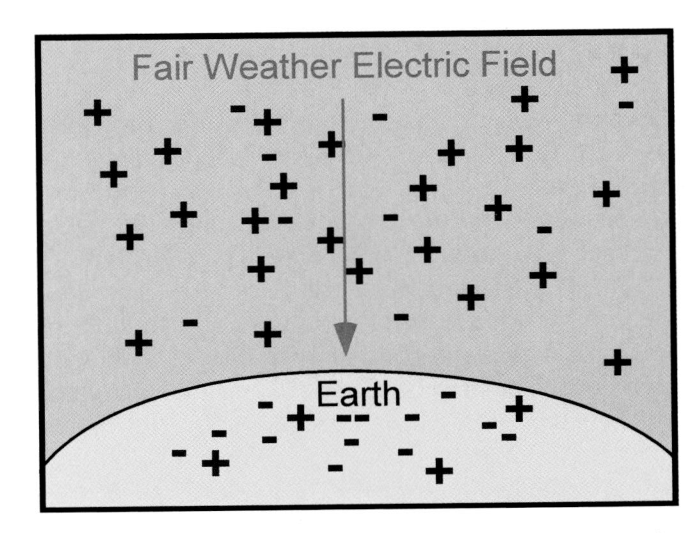

FIGURE 21.3: In clear air, the atmosphere carries a net positive charge and the Earth a net negative charge. This charge distribution creates the fair weather electric field of the atmosphere.

This distribution of ions is a consequence of the action of thunderstorms, which deposit electrons on the Earth's surface and remove electrons from the atmosphere. The fair weather electric field is about 100 volts per meter, which, compared to the field just before a lightning strike (1,000,000 volts per meter), is quite small. Air is an excellent insulator, so its conductivity, the ability to conduct an electrical current, is close to zero.

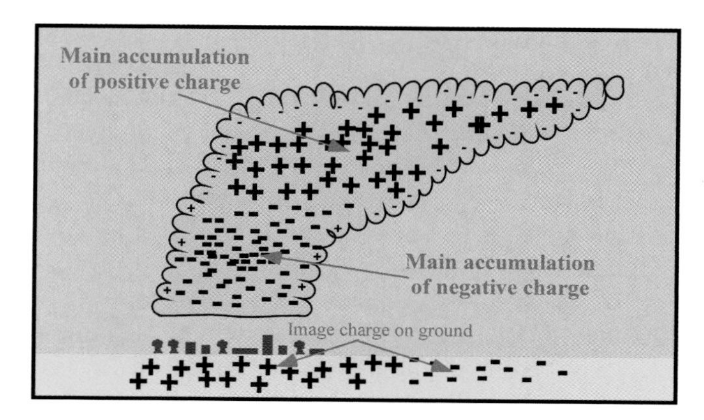

FIGURE 21.4: The distribution of charge in a thunderstorm and on the ground prior to a lightning stroke.

CHARGE DISTRIBUTION AND CHARGING MECHANISMS IN THUNDERSTORMS

Scientists have conducted extensive research and undeveloped theories to the charge distribution in thunderstorms and the charging mechanisms that power them. Before we examine these mechanisms, it is helpful to understand how charges are distributed in a thunderstorm and on the ground just before a lightning stroke. Figure 21.4 shows the distribution of positive and negative ions in a typical thunderstorm prior to a lightning discharge. The upper part of the storm, including the anvil, has an excess of positive ions and is positively charged, while the lower part of the storm has an excess of negative ions, and is negatively charged. The ground beneath the main part of the storm is positively charged, while the ground beneath the anvil is negatively charged. Let us now try to understand how this charge distribution comes about.

Interface Charging

In recent years, evidence from laboratory investigations and from modeling studies of thunderstorms has pointed toward a mechanism called ***interface charging*** or sometimes ***non-inductive charging*** as the likely mechanism for the ***initial charging*** in thunderstorms.

Every conductor, whether it is copper, aluminum, ice, or water, has a specific average arrangement of electrons within the atoms on its surface. Although each conductor may be electrically neutral (have the same number of protons as electrons), the distribution of electrons on the surface of the conductor will differ for each conductor. As a result, an electric field will exist whenever two conductors consisting of different materials are brought into close proximity, and a brief current will flow as electrons transfer from one conductor to the other should the conductors be brought into contact. Since the charge transfer occurs at the interface between the two conductors, the charging mechanism is called interface charging.

How does interface charging occur in a cloud? Recall from Chapter 20 that graupel and hail grow by the collection and freezing of supercooled water droplets. On the other hand, tiny ice crystals in clouds form by ***vapor deposition***, the direct transfer of water vapor molecules from air to the ice crystal. Ice that forms by these two growth processes has surface properties that are quite different, different enough that the two types of ice particles act like two conductors composed of different substances. The electrons on the surface of these different types of ice are distributed differently within and across the surface of the ice (Figure 21.5A). As a result, a charge potential, or ***voltage***, exists between the two types of particles. As a small ice crystal approaches a falling graupel particle or hailstone, an electric field will develop between their surfaces, and, in a brief moment of collision (Figure 21.5B), electrons will jump from the small crystal to the larger particle. In this manner, the larger ice particles, which have large fall speeds, become negatively charged, while the small ice crystals, which have small fall speeds, become positively charged (Figure 21.5C). The charge transferred between individual particles is very small. However, when one considers the countless numbers of ice particles growing within a developing thunderstorm, the total charge transferred is very large indeed.

In the strong updrafts of a thunderstorm, the small, positively charged ice particles, which are relatively light, are swept upward into the anvil, while the heavier negatively charged hail and graupel remain suspended in the updraft or fall toward the Earth's surface. Since ***supercooled water*** concentrations are highest between the freezing level and about –15° C

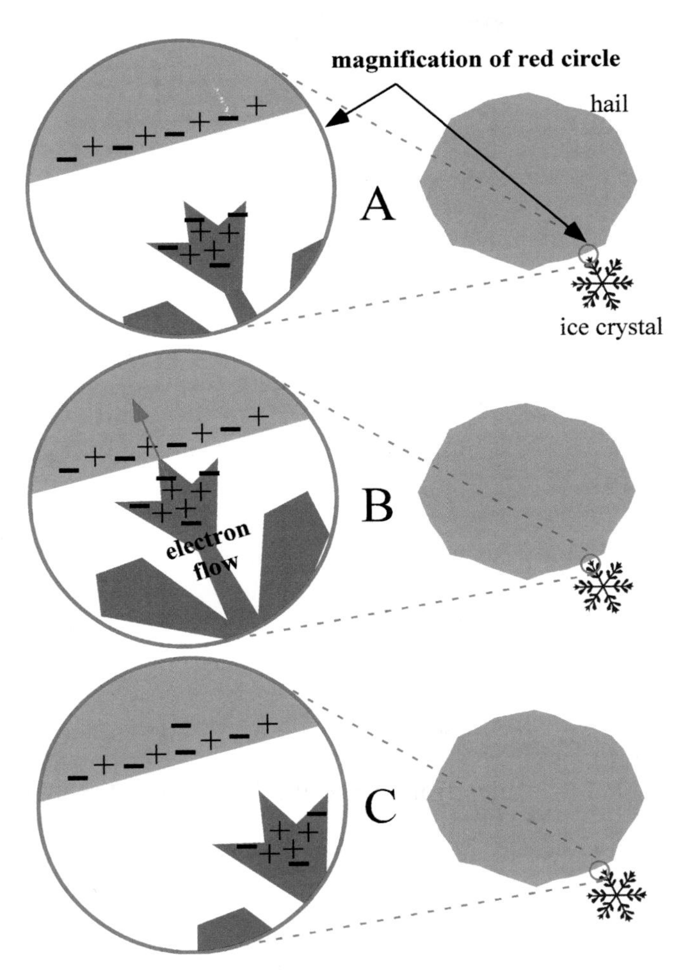

FIGURE 21.5: Interface charging occurs when ice crystals (dark blue snowflake in figure) grown by vapor diffusion collide with hail or graupel particles (light blue large particle) that formed by collection of supercooled water. The surface distribution of electrons is different in the two ice particles, leading to a transfer of electrons from the ice crystal to the graupel particle during the collision.

(+5° F), it is between these levels that growth by collection of supercooled water droplets predominates. It is also between these levels that most negative charge accumulates. The positive charges, residing on the small crystals, are swept into the highest, coldest parts of the cloud by the updraft, and most are carried into the storm's anvil.

Induction Charging

As the thunderstorm charges, a second mechanism, called ***induction charging***, can act to accelerate the charging process. Induction charging is a ***feedback mechanism*** that involves the charge distribution on particles and the electric field present within the developing thunderstorm (recall that a feedback mechanism is one in which one process causes a second process, which in turn reinforces (or opposes) the initial process). As the small, positively charged particles rise to the top of the storm and the larger negatively charged particles fall, strong electric fields develop within a storm and each particle in the storm comes under the influence of this field. Because of the electric field, the electrons *within each particle* will migrate toward the positively charged upper part of the cloud, which means they will, on average, reside more toward the top of each particle than the bottom. Even if each particle was electrically neutral, the top of each particle would be negative and the bottom positive (Figure 21.6A). Now consider what happens when a hailstone and ice crystal collide as they pass each other. In the brief time of contact, the electrons on the top of the ice crystal come in direct contact with the positive base of the hailstone. The electrons on the top of the ice crystal will be attracted to the positively charged side of the hailstone and "jump" to the big particle at the instant of contact (Figure 21.6B). The hailstone and crystal separate and continue on their way, the hailstone now negatively charged and the ice crystal positively charged (Figure 21.6C). Because the ice crystal carries positive charge upward and the hailstone carries negative charge downward, the electric field within the storm is reinforced. The next collision between another hailstone and an ice crystal is even more effective at charge transfer, since the electric field is stronger. On a grand scale, billions of ice particles are colliding with larger graupel and hailstones, transferring more and more charge with each collision. Eventually the storm charges to the point where lightning can occur.

The theories to explain charging described here are not comprehensive—there are other proposed mechanisms and no doubt other ways that charge may be transferred that are not fully understood. By one or more of these mechanisms a storm eventually develops the charge distribution appearing in Figure 21.4, with positive charges accumulating in the upper parts of the cloud while negative charges accumulate in the lower part of the cloud.

Note in Figure 21.4 the very thin layer of positive charge that appears along the sides of the clouds adjacent to the large negative charge region, and the thin layer of negative charge that appears along the upper cloud boundaries adjacent to the positive charge region. The atmosphere outside of clouds always contains both positive and negative ions that are created

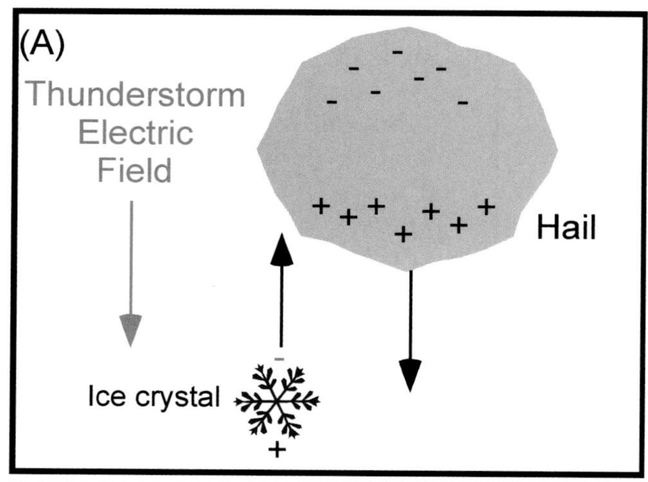

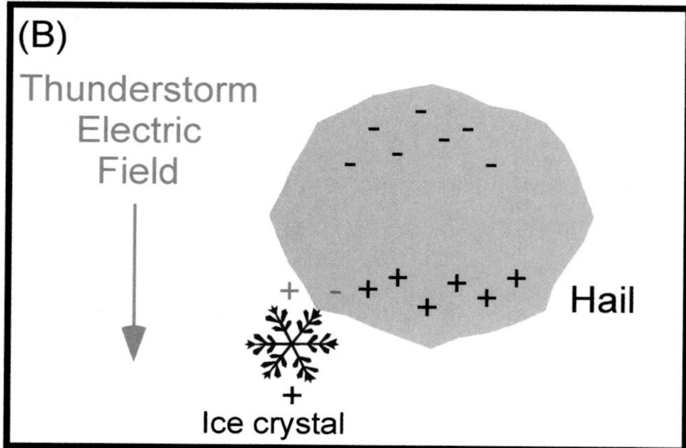

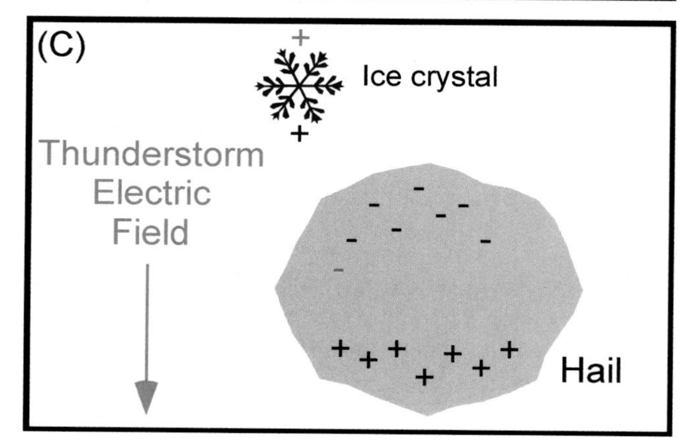

FIGURE 21.6: Induction charging occurs as charge is transferred between large and small ice particles during collisions in the presence of a strong electric field. The smaller ice particles obtain positive charge and the larger particles, such as hail, obtain negative charge. The small particles are swept upward by the updraft, while the larger particles descend toward the cloud base. In this way, the lower part of the cloud becomes negatively charged and the top of the cloud becomes positively charged.

by continuous bombardment of the Earth by cosmic rays. As the negative charge builds up in the lower part of the cloud, some of the background positive ions become attracted to the cloud boundary and are captured by cloud droplets. They create the thin layer of positive charge. As the positive charge builds up in the upper part of the cloud, background negative ions are attracted to the cloud boundary in this region. The thin layer of charge is called the ***screening layer***. The screening layer is not critical to lightning formation, but is a common feature of the charge distribution in a thunderstorm.

Let us now consider what happens on the Earth's surface as the strong negative charge is accumulating in the lower part of the storm. Recall that the Earth's surface is normally negatively charged in the fair weather field. However, as the strong negative charge develops at the base of the storm, the negative charges on the Earth's surface are repelled away from the storm's base, leaving a positively charged region called an ***image charge*** on the earth underneath the storm. Locations of the Earth closest to the thunderstorm base, such as rooftops, treetops, water towers, and golfer's heads, become the most devoid of electrons, and therefore the most positively charged. The same process happens under the thunderstorm anvil, except that the image charge in this region is negative.

CHECK YOUR UNDERSTANDING 21.1

1. What is lightning?
2. What types of charged particles are found in thunderstorms?
3. Approximately how long and how wide is a typical lightning stroke?
4. Where are the regions of positive and negative charge typically found in a thunderstorm just prior to a lightning stroke?
5. How can a charge distribution develop on a precipitation particle that has no net charge?

FOCUS 21.2

Lightning in Hurricanes

Hurricanes are the most destructive storms on the planet, but surprisingly they produce very little lightning. For example, when Hurricane Hugo made landfall on the South Carolina coast in 1989, the National Lightning Detection Network detected only sixteen lightning strikes in an eight-hour period. Why are these violent storms such poor producers of lightning? The answer lies in the strength of the updrafts in the eyewall. Hurricanes are tropical storms, and the freezing level typically is found above about 5 km altitude. The charging mechanisms in thunderstorms all require ice particles, so charging must occur above this level. Charging mechanisms are most efficient in clouds after small hailstones begin to form. Hailstones grow by collecting supercooled water drops (see Chapter 20). Air ascending through the eyewall of a hurricane rises vertically at speeds that rarely exceed 10 meters/sec (23 mph) and are generally much weaker. In contrast, air in thunderstorms of the middle latitudes rarely rises at speeds less than 10 meters/sec and vertical speeds are often much greater. Above the freezing level in a hurricane, the relatively weak upward velocities are generally insufficient to produce much supercooled water. Without supercooled water, and graupel or hail, charging mechanisms become inefficient. Although hurricane clouds charge, they rarely charge sufficiently to permit a lightning discharge. This is particularly true in well-developed hurricanes with a distinct eyewall. Storm clouds along the outer bands of hurricanes (see Chapter 24) have been observed to produce more lightning than the eyewall. Nevertheless, even in these regions, the number of lightning strikes is far less than that observed in other thunderstorms.

FIGURE 21D: Hurricane Hugo at landfall on the South Carolina coast. Hugo produced only sixteen lightning strikes in an eight-hour period as the storm devastated South Carolina's coast.

THE LIGHTNING STROKE

Lightning can occur within clouds ***(in-cloud lightning)***, between clouds ***(cloud-to-cloud lightning)***, or between a cloud and the ground ***(cloud-to-ground lightning)***. Over 80 percent of all flashes are the in-cloud or the cloud-to-cloud variety. These can occur between the positive and negative cloud regions, such as from the main body of the cloud into the anvil.

A cloud-to-ground stroke begins when the electric field in a local area of the cloud reaches about 3 million volts/meter. At this field strength, air no longer acts as an insulator, and electrons begin to move freely. The electrons surge toward the cloud base and to the ground in a series of steps called the ***stepped leader*** (see Figure 21.7A). Each step is about 50 to 100 meters long, with a pause of a few millionths of a second between steps. The charges take the "path of least resistance" downward, jumping, for example, through raindrops which are good conductors of electricity. The charges can take several paths simultaneously (this leads to the "forked" look of lightning strokes as in Figure 21.8). As one of the branches of the stepped leader approaches the ground, the electric field between the ground (or surface object) and the stepped leader becomes so great that positive charges (charged molecules from the object itself) jump upward off the object to meet the descending stepped leader. When this ***traveling spark*** connects with the stepped leader, the channel for electron flow opens and the powerful ***return stroke*** occurs (Figure 21.7B and 21.8). At this time the full current is reached and the flash occurs. The light flash is associated with ionization of the molecules in the path of the lighting

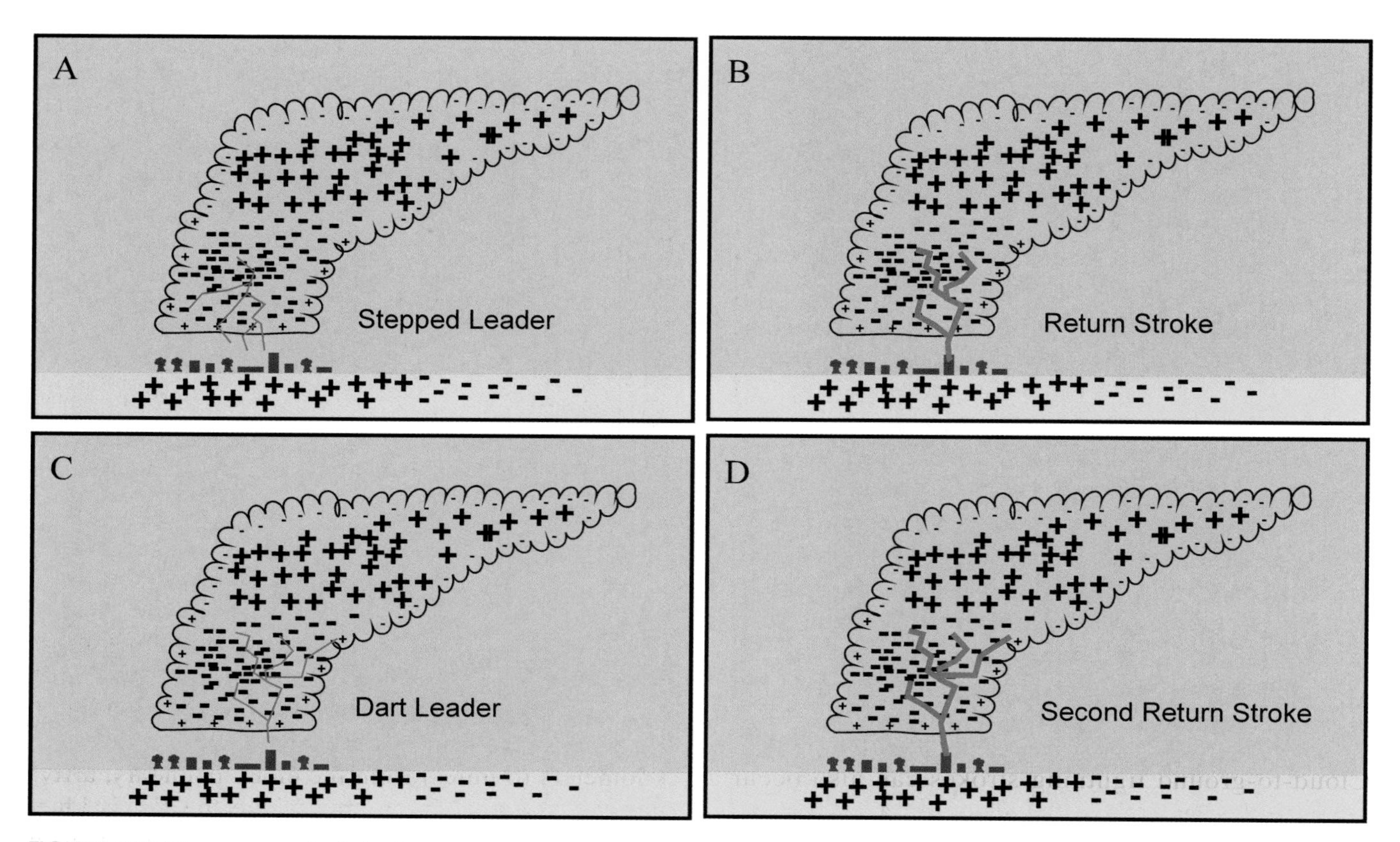

FIGURE 21.7: The stages of a lightning stroke. (A) Negative charges search for the path of least resistance to the ground creating the stepped leader; (B) A powerful return stroke develops when the traveling spark of positive charge surges upward to meet the stepped leader; (C) a second surge of negative charge descends along the ionized path of the previous stroke creating the dart leader; and (D) a second return stroke follows the first. This process repeats until the cloud is discharged.

FIGURE 21.8: Time-exposed lightning stroke that shows the forked path characteristic of the descending stepped leader and the bright return stroke near the ground.

stroke. Ionization begins within the traveling spark and proceeds upward along the lightning channel. The flash of light accompanying the ionization therefore proceeds upward from the ground toward the cloud base. The current is upward as well, since the flow of electrons is downward from the cloud to the ground. Your eyes can't see any of this detail, since it all happens in about 10 microseconds (1/100,000 of a second). What you can see are the branches of the lightning associated with the discharged paths of all the stepped leaders. These all appear to point downward toward the Earth, giving the impression that the lightning bolt descended from the clouds.

It is common that the same process, leader and return stroke, will occur in the same channel one or more times until that portion of the cloud is discharged. The subsequent leader, called a ***dart leader***, moves more rapidly because the channel through which it flows is ***ionized***. In the ionized channel, the atoms and molecules of air have been stripped of some of their electrons, so that the atmosphere within the channel contains many charged particles. As before, when the leader approaches the Earth, a traveling spark, typically less energetic than the first, will jump up to meet it, initiating a return stroke (Figures 21.7C and 21.7D). These repeat strokes can occur rapidly, with the time between them only a few hundredths of a

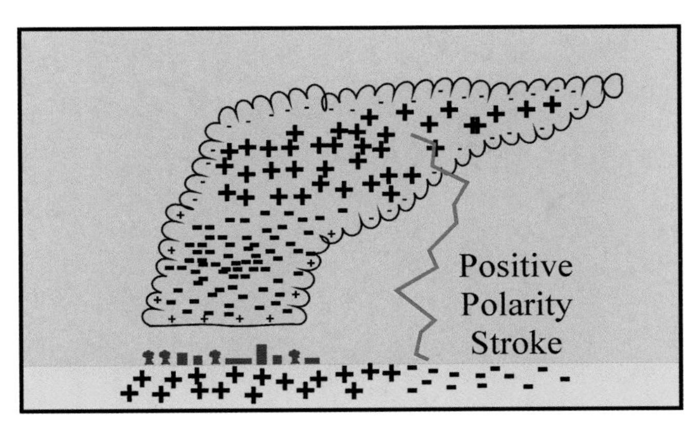

FIGURE 21.9: Cloud-to-ground lightning between the anvil of the thunderstorm and the Earth is called positive polarity lightning because the charge on the Earth's surface is negative, the charge in the cloud is positive, and positive charge is lowered from cloud to ground.

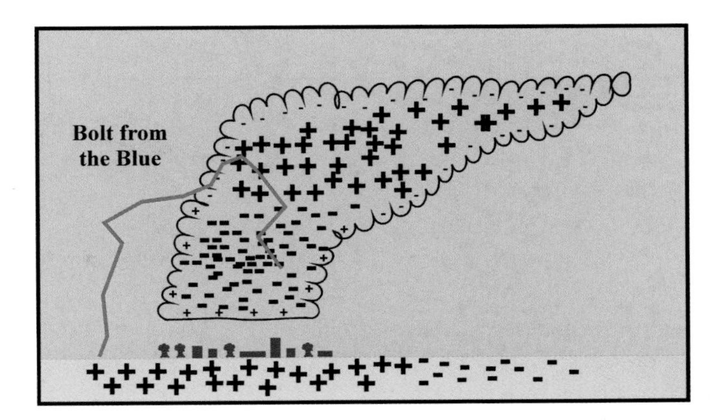

FIGURE 21.10: A bolt from the blue, a lightning strike that occurs far from the cloud boundary, striking the earth while the sky directly overhead is blue.

second. In rare cases, over 20 strokes have occurred within a single discharge event.

Most cloud-to-ground lightning strokes occur between the main body of the cloud and the ground. Cloud-to-ground lightning strokes can also occur between the base of the thunderstorm anvil and the Earth's surface (see Figure 21.9). In this case, the charge on the Earth's surface is negative and the charge in the cloud is positive. These strokes are called ***positive polarity strokes*** because positive charges travel downward (the current downward) to the Earth's surface during these strokes. About 8 percent of all cloud-to-ground lightning strokes are positive polarity. Although positive strokes are less common, they are more dangerous because they require stronger electric fields and discharge more current. The reason for this is that the anvil is much higher above the surface, and the discharge must occur over a longer path.

A ***bolt from the blue*** is cloud-to-ground lightning that exits from the side of a thundercloud and comes to the ground away from the thundercloud (Figures 21.10). These flashes are normal (negative) polarity, and begin as in-cloud discharges, with the initial lightning discharge between the main negative charge and the upper positive charge within the cloud. Negative charges propagate upward into the upper positive charge region, and spread horizontally through the upper positive charge. After neutralizing much of the upper positive charge, an excess of negative charge remains in the cloud, and the discharge continues toward the positive charge at the ground. The discharge progresses to the edge of the cloud, and often follows a screening layer along the cloud boundary for a while. The discharge then develops downward away from the cloud as a stepped leader. The ground strike can be over ten kilometers (6 miles) from the cloud boundary, arriving at the ground in an area where the sky is blue. These strokes are particularly dangerous because people generally have yet to seek shelter, since the storm appears some distance away.

ONLINE 21.1

Lightning from Space

Astronauts on the space shuttle regularly observe lightning as the shuttle passes over thunderstorms. An example online, provided by the Engineering Photo Analysis Group of the Marshall Space Flight Center Propulsion Laboratory, shows a view of lightning discharges observed by the shuttle *Discovery* as it passed over the coast of France. In this movie, low illumination moonlight provides minimal background. Clouds cannot be seen under this illumination, but the lightning flashes are easily seen. The coast of France is on the left side of the image, and the city of Algiers appears later in the imagery. The average diameter of the areas illuminated by lightning is about 35 km and the estimated flash rate within one area is about 130 flashes per minute.

FOCUS 21.3

Ball Lightning

Ball lightning is a luminous, small, glowing ball of gas, typically 10 to 40 cm (4 to 16 in) in diameter. Ball lightning appears very rarely, and always follows a lightning stroke. There have been thousands of observations of ball lightning, dating back to the time of ancient Greece. According to one statistical study of ball lightning, ball lightning has been seen by about 5 percent of the people on Earth, about the same percentage of the population that has witnessed the direct point of a lightning strike. Observers report that the ball of light typically becomes bright and then disappears with an explosion. Ball lightning behaves strangely. It floats in the air, ranges in color from red, orange, and yellow to blue, and has an "electrical" odor, indicating that ozone may be present. It sometimes floats from cloud to cloud, or cloud to ground or ground to cloud, or just horizontally. It appears to be attracted to open windows and tends to be attracted to grounded objects. Its lifetime varies, ranging from a few seconds to several minutes; with the average duration about 25 seconds. Ball lightning has caused property damage and most certainly has frightened many who witness it.

After Benjamin Franklin conducted his famous experiment to determine if lightning was an electrical current by attaching a metal key to a kite during a thunderstorm, he proposed an additional experiment where an elevated rod or wire would be used to "draw down the electric fire" from a cloud. Before Franklin conducted the experiment, Frenchman Thomas Francois D'Alibard used a fifty-foot long vertical rod to draw down the "electric fluid" of lightning in Paris on 10 May 1752. Later, a physicist named Georg Wilhelm Reichmann attempted to reproduce the experiment, according to Franklin's instructions, but standing inside a room. A glowing ball of charge traveled down the string, jumped to Reichmann's forehead and killed him instantly—history's apparent first scientific documentation of, and death from, ball lightning.

In recent times, two New Zealand scientists, John Abrahamson and James Dinniss of the University of Canterbury, have proposed an intriguing theory about how ball-lightning might form. They suggested that when lightning strikes soil, a silica-rich vapor can be created that may condense into particles and combine with oxygen in the air to slowly burn. Two Brazilian Scientists from the Federal University of Pernambuco, Antonio Pavão and Gerson Paiva, have now tested this idea using electrodes to shock silicon wafers with enough electricity to create a silicon vapor. In their laboratory experiments, they created glowing orbs that lasted two to eight seconds—approximating natural ball lightning. Scientists are now focusing on other materials besides silicon that may also form ball lightning when struck by a lightning stroke.

OTHER PHENOMENA ASSOCIATED WITH ELECTRICAL DISCHARGES IN THE ATMOSPHERE

The charging of thunderstorms and discharge of lightning produces other effects at the ground, at cloud level, and even above the clouds. Some phenomena are frightening, while others are visual treats for those who witness them. The most interesting of these phenomena are described next.

Thunder

When lightning occurs, the air in the channel of the stroke is heated to 30,000° C (54,000° F). At this high temperature, the air expands explosively, creating a shock wave that evolves rapidly into crashing sound waves. Sound travels approximately 330 meters/sec (1100 ft/sec). This means that it takes sound about three seconds to travel a kilometer and five seconds to travel a mile. Light travels so fast that it essentially arrives at our eyes in an instant. Therefore, you can roughly estimate how far a stroke is from your location by counting the number of seconds between the flash and the first sound of ***thunder***.

Close to the stroke, these sound waves create a sharp explosive bang. Farther from the stroke thunder sounds like a bang followed by a rumble. Farther still, one only hears a deep rumble, which may go on for several seconds (Figure 21.11). The differences in the characteristic sounds of thunder occur because the speed at which sound waves travel in the atmosphere

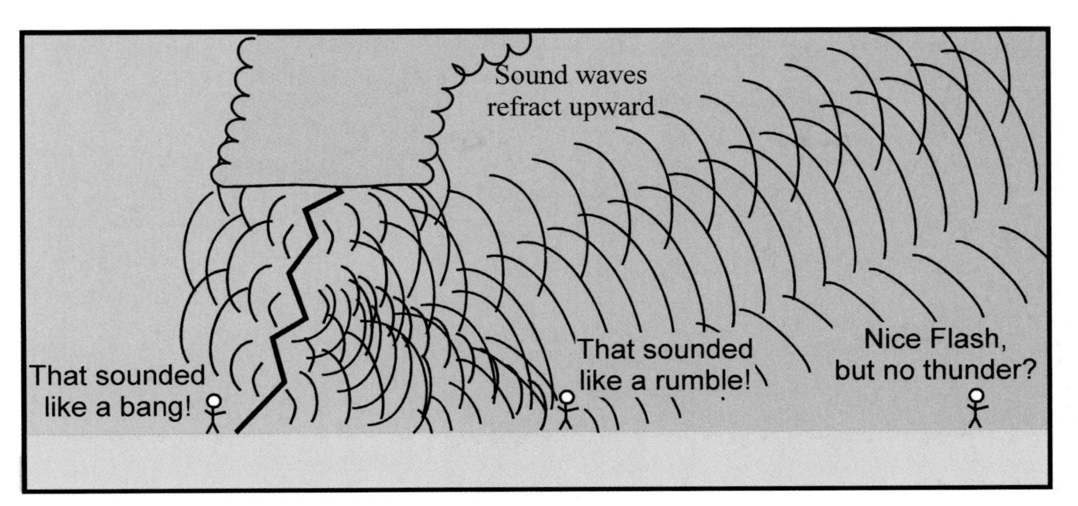

FIGURE 21.11: Thunder begins as a shock wave created by the intense heating that evolves into sound waves. The waves emerge from the lightning channel, travel outward, are reflected off surfaces, and refracted upward.

depends on temperature; the lower the temperature, the slower the speed of sound waves. Since temperature decreases with height in the troposphere, the atmosphere acts like a lens, bending sound waves upward (toward the slower speed) as they move forward. The higher frequencies bend more rapidly upward away from the earth's surface, leaving the lower deeper sounds. Sounds also emanate from different parts of the lightning stroke, echo from hills and buildings, are scattered by small-scale turbulent areas of the atmosphere, and dissipate as they move through the air. As a result, what sounds like a bang becomes a rumble and then no sound at all. More than about 5 km (3 miles) from a lightning stroke, one generally will only see the flash, but not hear the thunder.

Heat lightning

Heat lightning is a misnomer. On sultry summer nights, a thunderstorm may be far off in the distance. The light from a distant flash of lightning will be scattered by air molecules, dust particles or reflected from clouds and appear as a light flash in the sky overhead. Many people thought that the light was generated by the summer heat, since the distant storm could be neither seen nor heard, hence the name *heat lightning*.

Bead lightning

Following a lightning stroke, lightning channels sometimes break up very briefly into a series of luminous "beads." The beads are virtually impossible to observe with the eye because they occur immediately following the bright flash. They have been observed with high speed cameras and reproduced in lightning flashes created in the laboratory. The cause of these ***lightning beads*** is likely to be associated with the deionization of the lightning channel.

Sheet lightning

Sometimes lightning occurs within or behind a cloud, illuminating the exterior of the cloud uniformly, giving the appearance of a sheet of light. This phenomenon is called ***sheet lightning***.

St. Elmo's Fire

As the ground charges in the vicinity of a thunderstorm, objects extending above the surface develop strong positive charge on their tips. Positive charge is concentrated on antennas, ice picks of mountain climbers, and other sharp metal objects. If too much charge accumulates, a discharge of small sparks will often occur. Sometimes, a bluish green halo appears as continuous sparking occurs. Sailors on ships with tall masts were very familiar with this phenomena and gave it the name ***St. Elmo's Fire*** after their patron saint. St. Elmo's Fire is often an early sign of an impending lightning stroke.

Red Sprites, Blue Jets, and Elves

Red sprites, blue jets, and elves are optical phenomena that occur between the tops of thunderstorms and the mesosphere (Figure 21.12). ***Sprites*** are red, large, weak luminous flashes that occur simultaneously with lightning strokes, and typically occur over anvil portions of thunderstorms. The brightest regions are typically 65 to 75 km (about 40 to 45 miles) above the earth's surface. Above that, there is often a faint red

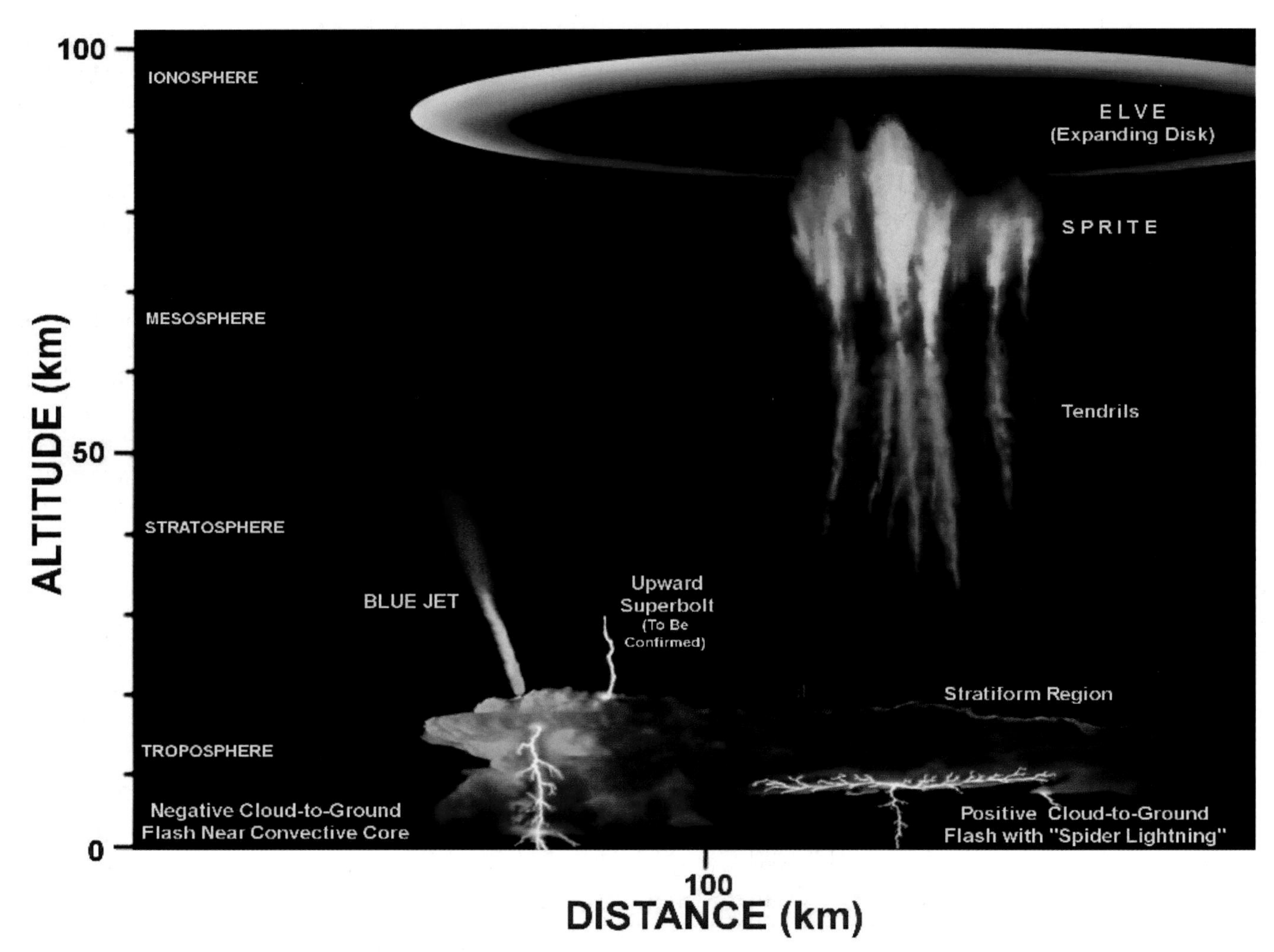

FIGURE 21.12: Sprites and elves are triggered by cloud-to-ground positive polarity strokes that occur in the anvil region of thunderstorms. Blue jets develop over the region of active convection where most lightning strikes occur.

glow or wispy structure that extends to about 90 km (55 miles). Below the bright red region, blue tendril-like filamentary structures often extend downward to as low as 30 km (20 miles). Because of their low surface brightness, they have only been imaged at night, mostly with very sensitive cameras. One can occasionally detect them at night without any visual aid provided there is no background light. Sprites nearly always occur with positive polarity strokes, which are more powerful. For this reason, sprites are most commonly found above a storm's anvil. Amazingly, sprites were predicted by the Scottish scientist C. T. R. Wilson in the 1920s, but were only first observed in the atmosphere in the late 1980s. Sprites develop as charged particles in the mesosphere and stratosphere move in response to rapid changes in the electric field triggered by a lightning discharge in the troposphere. ***Elves*** are disk-shaped regions of light that last less than a thousandth of a second and occur high above energetic cloud-to-ground lightning, primarily of positive polarity. They occur high in the atmosphere in a charged region called the ***ionosphere,*** and are centered on the lightning stroke below. ***Jets*** are blue, and cannot be detected by the eye. They have been observed with low-light television systems. They extend upward from the cloud top in narrow cones and can have upward speeds of 100 km/sec (60 mi/sec). They are coincident with the active portions of thunderstorms.

LIGHTNING SAFETY

Lightning is always dangerous. When a thunderstorm approaches, you should go inside if possible, stay away from electrical appliances, particularly corded phones, and avoid taking a shower or otherwise com-

4. The map in Figure 21.2 shows the geographical pattern of the number of lightning strikes per square kilometer per year in the contiguous United States. In parts of central Florida, the average reaches 10 strikes per square kilometer per year.
   (a) If the land is perfectly flat and free of structures (buildings, poles, etc.), and the lightning strikes are randomly distributed, determine the average number of years between lightning strikes on a soccer-field-sized area (80 meters by 125 meters) in central Florida.
   (b) What would be the answer to (a) if the frequency of lightning strikes was for your location rather than for central Florida?

5. Lightning produces a tremendous amount of electricity in a short burst. Is it reasonable that this energy might someday be harnessed as a domestic energy source? Why or why not?

USE THE SEVERE AND HAZARDOUS WEATHER WEBSITE

http://severewx.atmos.uiuc.edu

1. Use the *Severe and Hazardous Weather Website* link on the "Lightning" page to access the most recent lightning reports from the National Lightning Detection Network. Examine the current image displaying lighting strikes over the past two hours.
   (a) How many strikes have occurred in the past two hours? How does this number compare to the value given in the text? (Note: the text provides the annual number of cloud-to-ground flashes.)
   (b) The flashes are color-coded. Explain what the color coding means.
   (c) Describe the movement of thunderstorms that produced the lightning strokes over the past two hours.
   (d) Locate radar and satellite imagery from the "Current Weather" page of the *Severe and Hazardous Weather Website* that corresponds to the same time as the lightning strike map. What type of organization is apparent in the thunderstorms that produced the lighting strikes? Explain.

2. NOAA's "Hazards Statistics" web page can be accessed from the "Lightning" page of the *Severe and Hazardous Weather Website.*
   (a) Use the past five years of data to construct a graph of (i) the annual total number of lightning deaths in the United States and (ii) the percentage of victims who were male.
   (b) Does the percentage of male victims vary substantially from year to year?
   (c) What is your hypothesis for why this percentage is so much higher than 50 percent?
   (d) Devise a practical plan describing how your hypothesis could be tested.

3. You have recently been hired as a consultant to your local school district to update their safety plans for severe weather. Use the resources available from the "Lightning" page of the *Severe and Hazardous Weather Website* to develop a pamphlet of lighting safety for distribution to school children and their parents.

4. Use the *Severe and Hazardous Weather Website* "Lightning" page to access the University of Alaska website on sprites and jets.
   (a) Use the information provided to identify the common characteristics of sprites and jets, and the differences between the two phenomena.
   (b) The site contains information on strategies for maximizing your chances of witnessing features associated with sprites. How do your chances of visually detecting a sprite compare with your chances of seeing a shooting star or comet? Discuss.

DOWNBURSTS

Leading edge of a downburst near Wichita Falls, Texas, during Project VORTEX, 24 May 1994.

KEY TERMS

airspeed
curl
downburst
downdraft
dry microburst
glide slope
ground clutter
heatburst
inverted-V sounding
lift (aircraft)
Low-Level Wind-Shear Alert System (LLWAS)
microburst
rainshaft
runaway vortex roll
stagnation cone
Terminal Doppler Weather Radar (TDWR)
virga
vortex ring
wet microburst
Wind Index (WI)
wind shear

A ***downburst*** is a strong downdraft that originates within the lower part of a cumulus cloud or thunderstorm and descends to the ground. Downbursts do not require strong thunderstorms to develop. When air within a downburst reaches the surface, it spreads rapidly outward, creating strong straight-line winds. The outspreading of the air at the surface is analogous to the "bursting" of water that falls vertically onto a horizontal surface. Winds from downbursts can exceed 85 knots (~ 100 mph), and can cause damage equivalent to weak EF0 and EF1 tornadoes. Small (< 4 km horizontal dimension) intense downbursts are sometimes called ***microbursts***. Downbursts are extremely hazardous to aircraft, particularly during takeoff and landing. Because a cumulus cloud can evolve from the updraft stage to a downburst in a matter of minutes, special monitoring and warning systems must be deployed in order to alert pilots to the presence of downbursts.

There are important distinctions between downbursts and more typical thunderstorm downdrafts. First, downbursts are much more intense and concentrated over small horizontal areas. Their horizontal compactness is the reason why the term "microburst" is often used (in this chapter, we will use the terms "downburst" and "microburst" interchangeably). Second, downbursts generally develop or intensify in the lower portions of a cloud. It is not necessary that a cloud be deep in order to produce a downburst, since ordinary showers can lead to downbursts. Downbursts often develop even when a cloud's precipitation does not reach the ground. A final distinction, related to the low altitudes at which downbursts form, is the importance of evaporational cooling in downburst intensification.

The prevention of aircraft accidents associated with downbursts represents a "success story" in meteorology. As recently as the 1970s, downbursts were not recognized as phenomena distinct from thunderstorm downdrafts, although it had been long recognized that thunderstorm outflow could produce strong and damaging straight-line winds. Studies of a series of commercial airplane crashes in the 1970s, led by Dr. T. Fujita of the University of Chicago, resulted in the identification and conceptual modeling of downbursts. Several field studies in the 1980s, together with case studies and numerical models, then led to the design of microburst warning systems that capitalized on emerging technology such as Doppler radar. The implementation of these warning systems at major U.S. airports during the 1990s, together with the aviation community's awareness of the phenomenon, has significantly reduced accidents and saved many lives over the past decades.

DOWNBURST FORMATION

Downbursts form by two mechanisms. The first, and most important, is evaporation. Whenever rain falls from a cloud into the unsaturated air below the cloud, the rain will start to evaporate. If the air below the cloud has a low relative humidity, the drops may completely evaporate. When viewing a cloud where this is occurring, darker "streaks" appear to hang down from the cloud base but not reach the ground (Figure 22.1). These streaks, which are actually rain falling from the cloud but evaporating before reaching the ground, are called ***virga***. Recall that evaporation requires energy to convert the liquid drops to vapor. The latent heat required for evaporation must be supplied by the air through which the evaporating drops are falling. Energy is transferred from the air to the drops, so the air containing the drops cools as evaporation continues. Figure 22.2 shows this process in schematic form. The rain-cooled air is now colder (and more dense) than its environment (the air outside of the ***rainshaft***), and it begins to sink. The rate at which it sinks will depend on its temperature relative to the air around it. Often, when a large amount of rain is evaporated, the air in the evaporating downdraft can become several degrees colder and will descend very rapidly, achieving downward speeds of 35 to 50 knots (~ 40 to 60 mph).

FIGURE 22.1: Virga underneath a cumulonimbus cloud. The situation pictured here is often associated with a microburst and can therefore be hazardous to an aircraft.

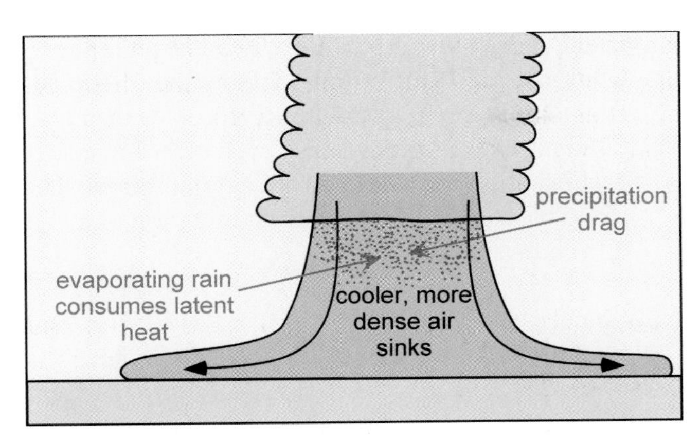

FIGURE 22.2: Summary of the downburst formation process. Downward acceleration of air occurs when evaporating raindrops consume latent heat, cooling the air. The cooler air is denser, so it begins to sink. In addition, falling precipitation drags the air downward, enhancing the downward acceleration.

The second mechanism driving air downward is the drag force of the falling precipitation. Each raindrop pushes air ahead of it as it falls under the influence of gravity. Individually, one raindrop will not do much, but when we consider the millions of raindrops falling in the precipitation shaft, the force can be substantial. The drag of all the drops drives air downward. This same mechanism is at work in all showers and thunderstorms, contributing to their ***downdrafts***. However, in downburst formation, the heaviest rain and its evaporation below the cloud base are concentrated in a relatively small area, from several hundred meters to a few kilometers across.

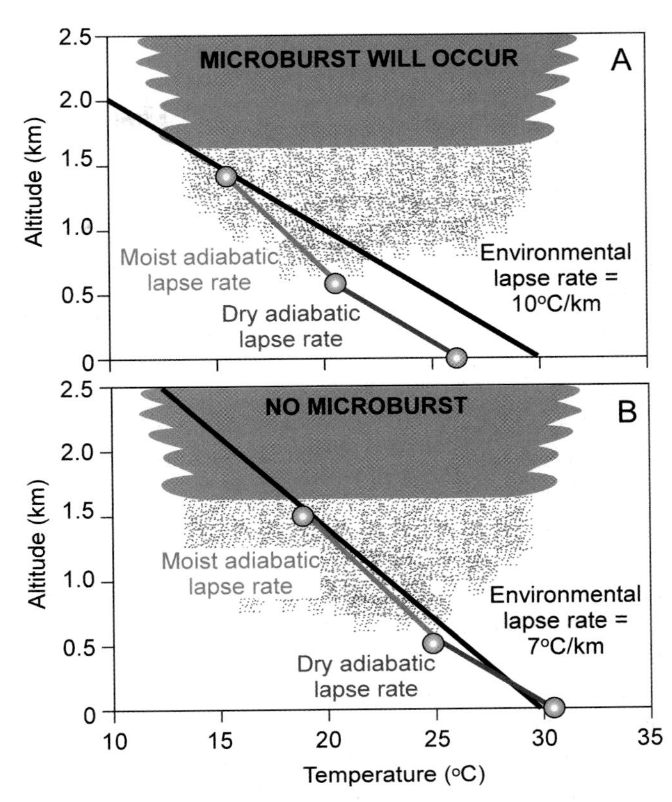

FIGURE 22.3: (A) Lapse rate that favors microburst development. A large environmental lapse rate below cloud base will enhance the descending motion of air because the air will be cooler than the environment and therefore denser. Descending motions will be enhanced in (A) with an environmental lapse rate of 10° C/km. In (B), with an environmental lapse rate of 7° C/km, no microburst will occur because descending air will become positively buoyant before completing its descent. Blue and red line segments show air parcel temperatures during dry and moist adiabatic descent, respectively. Black segments show environmental temperatures at various altitudes (km).

ENVIRONMENTAL CONDITIONS ASSOCIATED WITH MICROBURST DEVELOPMENT

Several characteristics of an atmospheric environment conducive to microbursts have been identified. These characteristics, which are generally consistent with the mechanisms described above, include the following:

1. *A large environmental lapse rate below the cloud.* We can understand why microbursts are favored in regions where the environmental lapse rate is large by considering the two stability diagrams shown in Figure 22.3.

In Figure 22.3A, the environmental lapse rate, shown as a black line, equals 10° C/km, the value of the dry adiabatic lapse rate. Consider the behavior of an unsaturated air parcel below cloud base in a region where rain is falling. As rain evaporates within the air parcel, the air parcel cools because latent heat is required for evaporation. As the parcel cools, the temperature of the parcel falls below that of the environment. Now denser than its environment, the parcel begins to descend. From Chapter 6 we know that descending air compresses and warms. The air parcel, cooling due to evaporation but warming due to compression, experiences a net warming of 6° C/km, the numerical value of the moist adiabatic lapse rate. In Figure 22.3A, the upper part of the air parcel path, labeled with a red line, represents a parcel's temperature as it descends and warms at 6° C/km.

At some point, all raindrops may completely evaporate within the air parcel. At that point the air parcel will only be affected by compression, and will warm at the dry adiabatic lapse rate, 10° C/km, as it descends. This part of the parcel's descent is represented by the blue line in Figure 22.3A. Note in Figure 22.3A that *at every altitude, the parcel of air is colder than its environment.* The air parcel is unstable—it will accelerate *downward* until it approaches the ground, at which point it must diverge outward as a microburst.

Consider now Figure 22.3B, where the environmental lapse rate is 7° C/km. An air parcel originating at the same point as in Figure 22.3A would again warm at the moist adiabatic lapse rate, 6° C/km, until all the rain evaporated. The parcel would then warm at the dry adiabatic lapse rate. However, before completing its descent to the ground, the parcel's temperature in Figure 22.3B would equal the environmental temperature. Indeed, if the parcel should descend further, it will become positively buoyant, its temperature exceeding the temperature of the environment. The downdraft will not reach the ground and no microburst will occur.

Figure 22.3 illustrates the importance of a large environmental lapse rate for microburst formation—air has the greatest potential to accelerate to the ground when the environmental lapse rate below the cloud base approaches the dry adiabatic lapse rate.

2. *Dry air below the cloud base.* Recall from Chapter 1 that when relative humidity decreases well below 100 percent the air is no longer saturated and water droplets will begin to evaporate. If air below the cloud base has a low relative humidity, falling precipitation will easily evaporate. The consumption of latent heat from the environment during evaporation will cool the air, enhancing the downward motions. Therefore, the lower the relative humidity of the air into which precipitation falls, the greater the evaporation rate and the stronger the downburst potential.

3. *An increase of the air's actual moisture content near the surface.* A shallow layer of moist air near the surface will enhance a microburst. Since moist air is less dense than dry air, a moist layer near the surface will make the descending evaporationally-cooled air even more negatively buoyant, accelerating the downward motion.

4. *Below-freezing temperatures in much of the cloud.* Because sublimating ice particles require the latent heat of fusion (melting) in addition to the latent heat of vaporization, air below the cloud is cooled even more by falling ice particles than by falling raindrops. There have even been some suggestions that the melting of ice crystals above the cloud base can contribute to the initiation of a downburst.

CHECK YOUR UNDERSTANDING 22.1

1. What is a downburst?
2. What two processes cause air to descend in a downburst?
3. List four characteristics of an atmospheric environment conducive to a downburst.

DOWNBURST STRUCTURE

The structure of a downburst and its associated outflow depends on the wind profile in the air through which the downburst descends. If the wind is weak, the downburst outflow at the ground is nearly symmetric, so that a toroidal ring of extreme winds will occur around the center of the downburst (Figure 22.4A through C). If the lower atmosphere's background winds are strong, the descending air will be tilted with respect to the vertical, and the most extreme winds will tend to occur on one side, as shown in Figure 22.4D through F. Analogies may be made to two orientations of a hose from which water is streaming onto a horizontal surface of concrete or asphalt. When the hose is pointed straight down, the water spreads symmetrically as in Figures 22.4A through C; when the hose is pointed at an angle 45° relative to the flat surface, the water striking the surface flows primarily in the forward hemisphere, as in Figure 22.4D through F. In the case of downbursts, the fluid is air rather than water, and the downward motion is initiated by mechanisms very different from the water pressure in the hose.

Figures 22.4C and 22.4F show vertical cross sections through mature downbursts. Two oppositely directed outbursts appear in each cross section because the flow "splits" above the ***stagnation cone*** near the central impact point of the downburst. Doppler radar analyses indicate that the outflow speed is several times faster than the downflow speed at 100 m (328 ft) above the ground. A particularly strong area

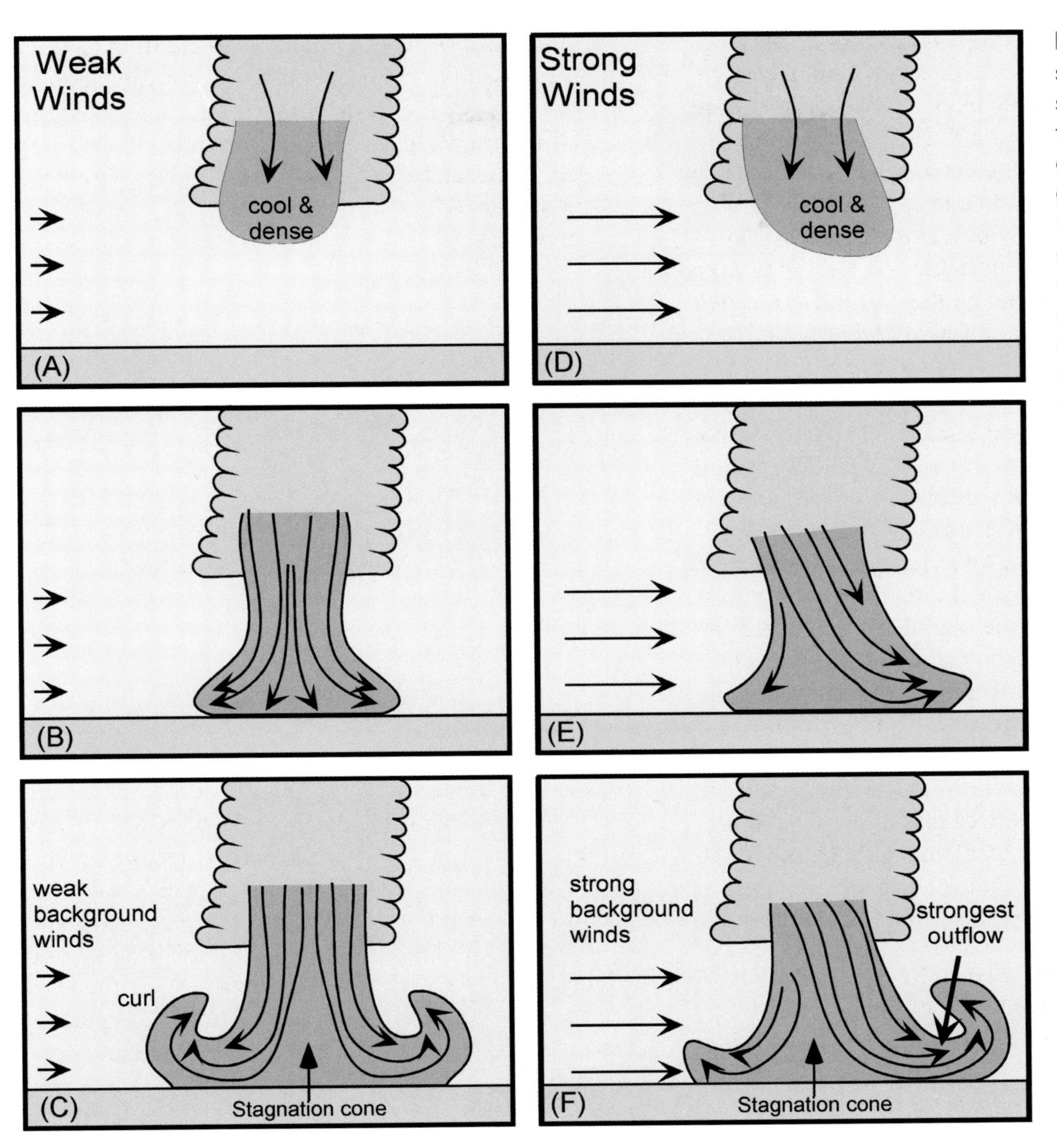

FIGURE 22.4: Downburst structure at various stages of evolution (from top to bottom of each column). When the background wind is weak, as in (A) through (C), the downburst is symmetrical. When the background wind is strong, as in (D) through (F), the downburst is asymmetrical and strongest winds occur on the forward side of the advancing downburst. Note the stagnation cone of weak winds at the surface beneath the center of the downburst in (C) and (F).

of winds typically occurs at the base of the *curl* in the flow, where the strongest winds are generally within 30 to 50 meters (about 100 to 160 ft) of the surface. This area occurs as a ***vortex ring*** that propagates outward away from the downburst (see Figure 22.5). The ring has some resemblance to a large hoop, about which the air circulates in the vertical plane. The most damaging winds occur at the base of the vortex ring. In some cases, there may be several pulses of strong downward motion and outflow, resulting in several rings and several pulses of strong winds at locations affected by the downburst. Portions of a vortex ring sometimes break away from the remainder of the ring, resulting in ***runaway vortex rolls*** that can produce very narrow burst swaths of damage equivalent to that caused by EF0 to EF1 tornadoes (Figure 22.6). Downburst damage can be differentiated from tornado damage because in downbursts, debris is blown in one direction while in tornadoes the debris is typically distributed in swirl patterns.

A photograph of an actual downburst is shown in Figure 22.7. The downburst in this case can be seen by examining the shape of the precipitation (dark area at the right) in the distance. The low-level winds from a downburst similar to the one in Figure 22.7, deduced from multiple Doppler radar, are mapped in Figure 22.8. The outburst of winds from the center of the microburst is obvious in the figure, with enhanced winds characteristic of the curl regions of the microburst indicated by the larger arrows. An example of the curl, containing a horizontal vortex ring marked by dust, is shown in the series of photographs in Figure 22.9. This vortex ring is

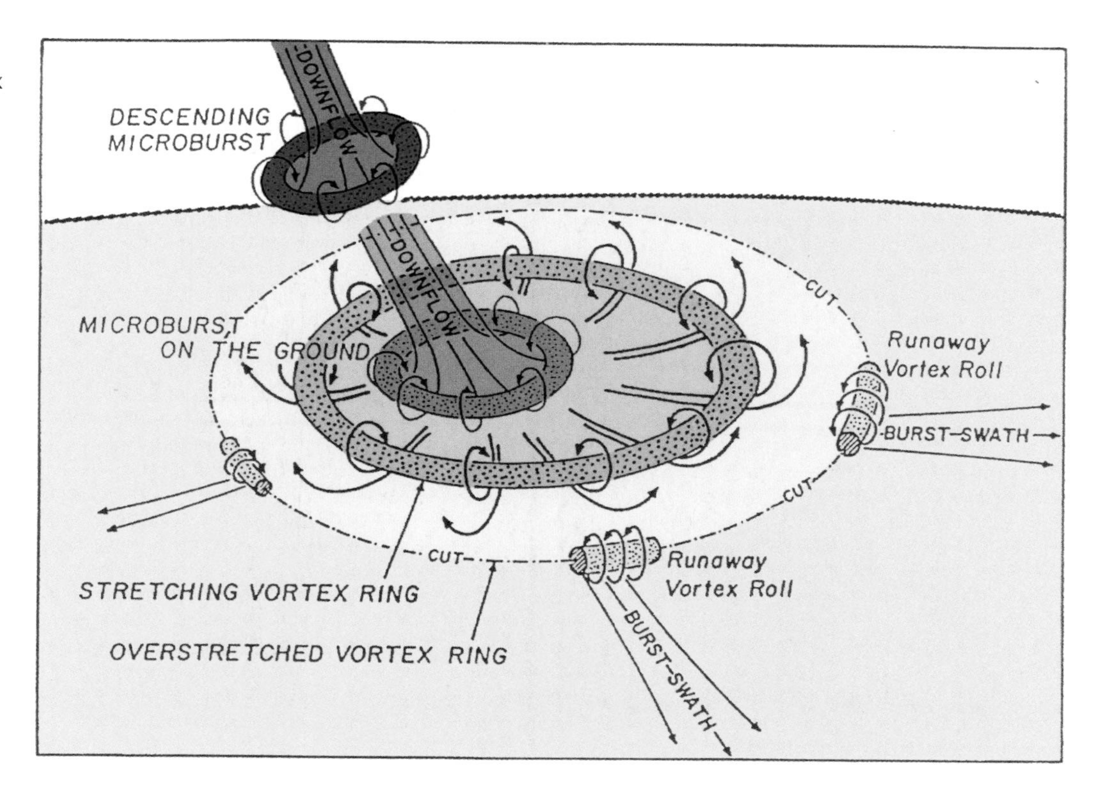

FIGURE 22.5: Schematic representation of a vortex ring at various stages of evolution of a microburst. The vortex ring migrates outward from the center of the microburst. Portions of a ring may break away from the remainder of the ring, creating *runaway vortex rolls* that produce localized wind bursts and damage swaths.

FIGURE 22.6: Photograph of damage associated with runaway vortex rolls in the Andrews Air Force Base microburst (see Focus 22.4) on 1 August 1983.

moving from right to left over the period of 15 seconds spanned by the photographs. The examples shown in Figures 22.7, 22.8 and 22.9 closely match the conceptual diagrams shown in Figure 22.4.

The wind speed and direction change rapidly when the leading edge of the microburst outflow passes. This rapid change in wind is an example of strong horizontal ***wind shear***. Strong values of wind shear are found where there are large changes of wind over small distances—precisely the situation in a microburst. Strong wind shear is a significant problem for aircraft, as discussed later in the chapter.

FIGURE 22.7: Photograph of a downburst near Denver, Colorado. The downburst can be detected by examining the shape of the precipitation (dark area) in the distance.

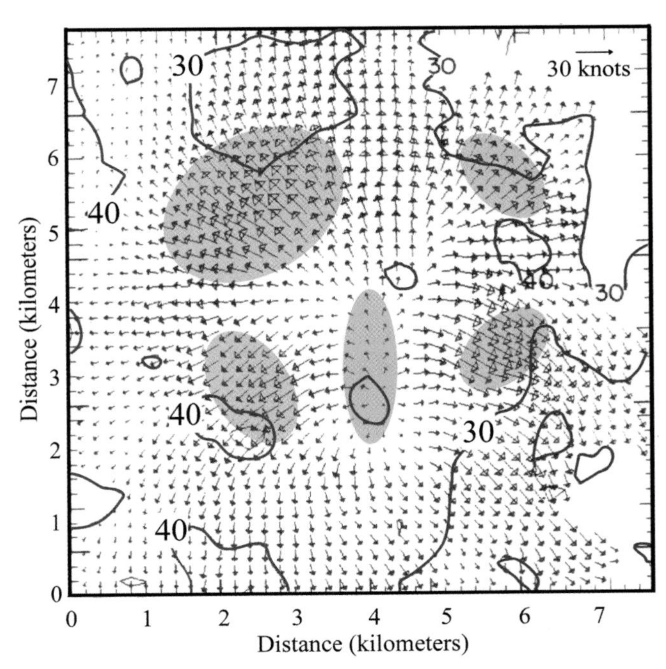

FIGURE 22.8: Low-level wind field of a downburst in Colorado derived from multiple Doppler radar measurements. Scale of the winds is shown by the vector in the upper-right corner. The contours show the radar reflectivity. The radar reflectivity in the core of the downburst, between 30 and 40 dBZ, is characteristic of moderate rain and indicates that this was a wet downburst. The green oval denotes the center of the microburst and the pink ovals denote burst swaths.

FIGURE 22.9: Sequence of pictures showing the curling motion of the dust cloud behind the leading edge of a downburst advancing from right to left over a 15-second period. Support towers for power lines provide spatial scale.

FOCUS 22.1

How Common Are Microbursts?

The small size (diameters ~ 1 to 4 km) and short duration (several minutes) of microbursts make them very difficult to document. The fact that microbursts were not even recognized as distinct phenomena until the 1970s is one indication of their elusiveness. However, statistics compiled from three major field programs designed to document microbursts indicate that microbursts occur surprisingly often, at least in the central and southern United States. The names, locations, and years of the three field programs are the Northern Illinois Meteorological Research on Downbursts (NIMROD, 1978, northeastern Illinois), Joint Airport Weather Studies (JAWS, 1982, Denver area) and the Microburst and Severe Thunderstorm project (MIST, 1986, northern Alabama). These field programs had study areas on the order of forty to sixty miles on a side, dictated primarily by the coverage of multiple Doppler radar arrays. Their durations, several months during the late spring and early summer, were intended to capture a representative sample of weather regimes and microburst occurrences. In each field program, the number of detected microbursts exceeded one per day: 50 microbursts in 42 days of NIMROD, 186 in 86 days of JAWS, and 62 in 61 days of MIST. Since the area covered by each field program was comparable to the size of one to two counties, one may conclude that the frequency of microbursts is surprisingly high in the central and southern states: 50 to 100 microbursts per year in a county-sized area over a typical three- to five-month "convective season."

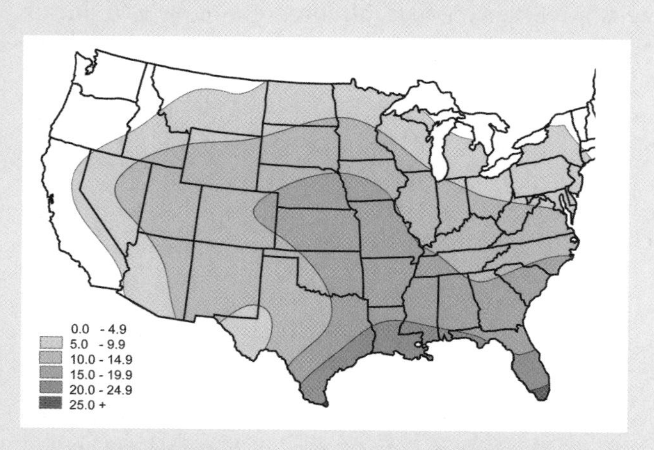

FIGURE 22A: Average number of potential microburst days for the months of July and August.

Another way to determine microburst potential is to determine the frequency of environmental conditions that support microbursts. Figure 22A shows the average number of potential microburst days for the months of July and August compiled from an analysis of thirty years of 0000 UTC rawinsonde data. This analysis shows that the potential for microbursts in the middle of the summer convective season is greatest along the Gulf Coast and in the Great Plains states. The most severe downbursts, when they do occur, are generally found in the Great Plains states and in states in the interior mountains of the western United States.

ONLINE 22.1

Animation of a Downburst Simulated by a Numerical Model

Simulations of mesoscale phenomena by numerical models were described in Chapter 4. The essence of a downburst can be captured in a model simulation by specifying an initial state containing a cool (hence dense) "bubble" or parcel of air that is negatively buoyant relative to its environment. The online simulation shows such a bubble sinking as a downburst, reaching the ground, and spreading out horizontally. (The lapse rate of the environment surrounding the parcel in this simulation was prescribed to be equal to the dry adiabatic lapse rate, 10° C/km.) The colors in the animation denote the temperature relative to the environment (red); yellow, green, blue, and purple denote progressively more negative (colder) temperatures relative to the environment. Note that the downburst does not spread out symmetrically—a background wind, blowing from left to right, was specified in this simulation. Thus the simulation illustrates an asymmetric microburst (see Figure 22.4F) in which the strongest low-level winds in the downburst occur on the viewer's right side.

TYPES OF MICROBURSTS

Two types of microbursts are distinguished according to whether measurable precipitation (0.01 in. or more) reaches the ground during the microburst. ***Wet microbursts,*** accompanied by measurable rain, are more easily visible because their rainshafts appear as curtains reaching the ground (see Figure 22.7). ***Dry microbursts*** have no measurable precipitation, although virga often can be seen descending from the cloud base. Because their precipitation does not reach the ground, dry microbursts are often impossible to detect visually and hence are especially dangerous for aircraft. If the ground is dry, blowing dust can be the only sign of a dry microburst (Figure 22.10). If the ground is wet from previous rains, there may be no visual clues at the surface that a microburst is occurring. Dry microbursts are more common in the western United States and the Great Plains, while wet microbursts are more common in the South, Midwest, and East. For example, during an 86-day field program to observe microbursts near Denver in 1985, 83 percent of the identifiable microbursts were classified as dry and only 17 percent as wet. By contrast, during a forty-two-day program to detect microbursts near Chicago in 1978, 36 percent were classified as dry and 64 percent as wet (see Focus 22.1).

The atmospheric environments associated with dry and wet microbursts are quite different. Examples of actual soundings from microburst environments are presented in Figure 22.11 to illustrate these differences. Figure 22.11A, a sounding associated with the development of a dry microburst near Salt Lake City, Utah shows the ***inverted-V sounding*** feature that is characteristic of dry microburst environments. The downward increase of the difference between the temperature and dewpoint gives the inverted-V appearance to the temperature and dewpoint profiles, and the rapid downward increase of dryness is consistent with the failure of the rain to reach the surface. The high relative humidity aloft is consistent with the formation of clouds that will be sufficiently deep to produce precipitation. Notice also that the environmental lapse rate in Figure 22.11A is equal to the dry adiabatic lapse rate from just above the surface to about 550 mb, and moist adiabatic for most of

FIGURE 22.10: In a dry microburst, the only visible sign might be dust at the leading edge of the strong winds.

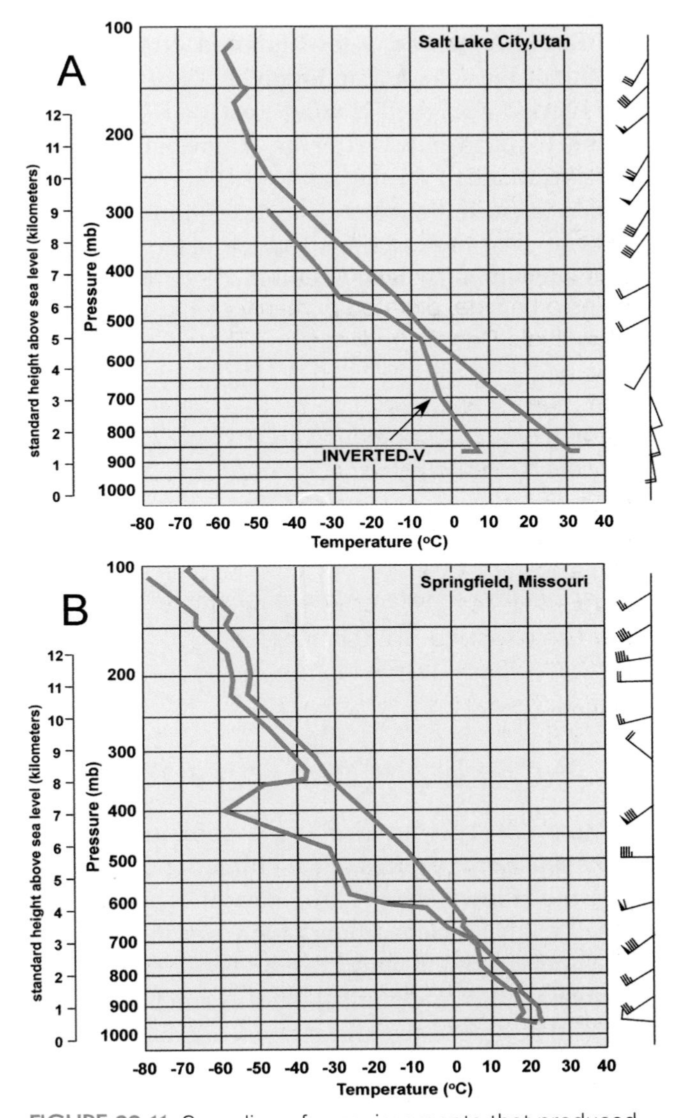

FIGURE 22.11: Soundings for environments that produced (A) a dry microburst in northern Utah (Sounding from Salt Lake City, Utah, 0000 UTC 20 August 1993) and (B) a wet microburst near Little Rock, Arkansas (nearest available sounding from Springfield, Missouri, 0000 UTC 02 June 1999).

the troposphere above 550 mb. In addition, the air is below-freezing from about 600 mb upward, favoring the survival of falling ice crystals into the dry air below cloud base.

The sounding in Figure 22.11B for Springfield, Missouri is characteristic of a wet microburst. In contrast to the dry microburst sounding in Figure 22.11A, the air in this case is quite moist (although not quite saturated) from the surface up to about 700 mb. The level of free convection is 795 mb, and the sounding has a Lifted Index of –4.7, indicating the potential for thunderstorm activity. Later that evening, a wet microburst indeed occurred in the same air mass near Little Rock, Arkansas, a few hundred kilometers to the south of the sounding in Figure 22.11B. The Little Rock Airport recorded a wind gust of 87 mph (39 m/s) only six minutes after an American Airlines plane crash-landed on the runway. While the role of the microburst in the crash is unclear, the crash could have been even more serious had the plane attempted to land a few minutes later than it did.

CHECK YOUR UNDERSTANDING 22.2

1. Where are the fastest winds typically found in a microburst?
2. In the context of a microburst, what is "wind shear"?
3. What is the primary difference between a wet and a dry microburst?

FOCUS 22.2

Heatbursts: Warm Blasts from Downdrafts

Shortly after 9 P.M. local time on 22 May 1996, winds at Chickasha, Oklahoma, gusted to more than 60 mph, the dewpoint dropped from 65° F to the upper 30s, and the temperature rose from the upper 80s to 102° F. The abrupt warming and drying occurred after sunset. How can this odd event be explained? The answer lies in the dynamics of a phenomenon known, appropriately, as a *heatburst*.

Heatbursts share some characteristics with downbursts. Both originate as downrushes of air in showers or thunderstorms, and both produce strong horizontal winds when the downrushing air reaches the ground. However, the outrushing air in most downbursts is relatively cool as a result of the evaporation of substantial amounts of falling precipitation. The high temperatures of a heatburst are an indication that the downdraft contained little liquid water for a substantial portion of its downward trajectory, enabling adiabatic (compressional) warming to dominate the evaporational cooling. For this reason, heatbursts are thought to originate in dying or "collapsing" thunderstorms, in which the incoming moisture is significantly reduced. The air that is warmed may also contain relatively dry environmental air that has been drawn into middle levels of the thunderstorm. It is likely that the air originates at higher levels of a thunderstorm than in most downbursts, so that the available liquid is entirely evaporated at altitudes well above the surface. Nevertheless, the downward motion of the air must be initiated by falling rain (or ice crystals), either through evaporational cooling or frictional drag, because the air must acquire sufficient downward momentum to overcome its positive buoyancy as it approaches the surface. Heatbursts occur occasionally in the Plains during spring and summer, and they often occur in the evening or at night when thunderstorm activity is dying down.

THE PROBLEM FOR AIRCRAFT

The force that lifts an airplane is sometimes erroneously explained in terms of the difference of pressure across the plane's wings. An aircraft's wings are designed so that air must flow faster over the top than over the bottom of the wing. The Bernoulli principle of aerodynamics then implies that the pressure beneath the wings will be greater than the pressure above the wings whenever air flows across the wings, creating an upward pressure gradient that helps to ***lift*** the aircraft. However, the Bernoulli effect does not create sufficient lift to overcome the force of gravity acting on the plane.[1] In fact, aeronautical engineers know that air moving across the lower part of an aircraft's wing is diverted downward by the wing. This creates an upward force (lift) on the plane through a simple application of Newton's third law of motion, which states

[1]For a comprehensive explanation of airplane lift, see *Understanding Flight,* by D. Anderson and S. Eberhardt. New York:McGraw-Hill, 2001, ISBN: 0-07-136377-7.

that "every action has an equal and opposite reaction." When this upward pressure force is exactly equal and opposite to the force of gravity, the plane stays at the same altitude. When planes speed up, the downwash of air and the corresponding upward pressure force on the plane increase, lifting the plane to a higher altitude. When planes slow down, pilots can maintain lift by lowering the wing flaps, which drives air downward faster so that the plane can maintain the desired altitude.

The lift force is typically reduced by about 1.3 percent per knot of tailwind, which is why airplanes almost always take off and land in the direction that provides them with a headwind rather than a tailwind. When an aircraft is in the air, it needs to maintain a certain ***airspeed*** across the wings in order to remain in the air. If the airspeed across the wings is less than this critical value, the aircraft will stall, lose its lift, and go out of control, rapidly descending to the ground. Going below stall speed is a particular problem when an aircraft is near the ground, since there is no room for recovery.

When an aircraft takes off or is landing, it is typically flying closer to its stall speed than at any other time during flight. The aircraft, of course, is also closest to the ground during takeoff and landing. This is when the aircraft is most vulnerable to an encounter with a downburst.

Figure 22.12A shows an aircraft taking off into a downburst that is just beyond the runway. Let's suppose that the aircraft in this example will lift off the ground when its airspeed[2] is 140 mph. In the example, the downburst is producing outflow winds of 40 mph (with respect to the ground). The aircraft, *sitting still on the departure end of the runway,* has a headwind of 40 mph, and therefore, an airspeed of 40 mph. As the aircraft rolls down the runway, it will begin to lift off the ground when its ground speed reaches 100 mph, since its airspeed will then be 100 + 40 = 140 mph. The aircraft takes off. Moments later, it enters the core of the microburst. The horizontal wind in the core is 0, since all the air is descending. This means that the airspeed equals the ground speed, 100 mph. As the aircraft passes to the other side of the downburst, *the wind is now moving in the same direction as the aircraft.* This means that the airspeed across the wings is now 100 – 40 = 60 mph. In a matter of maybe twenty seconds or less, the airspeed dropped from 140 to 60 mph, well below stall speed. In addition, while passing through the core of the microburst, the air's downward momentum or "downwash" drove the aircraft to a lower altitude. The net effect: an out-of-control aircraft close to the ground. Microbursts have been responsible for many aircraft accidents involving large passenger aircraft.

Landing aircraft encounter the same problem. In this case, the aircraft is slowed and descending. As it crosses the downburst, the rapid loss in airspeed and the downwash will cause the aircraft to descend too rapidly and possibly go out of control if the plane's speed falls below stall speed. The tricky aspect for an approach is that the pilot is trying to keep the plane on a path that will put it on the approach end of the runway. This path, called the ***glide slope*** (see Figure 22.12B), is the normal

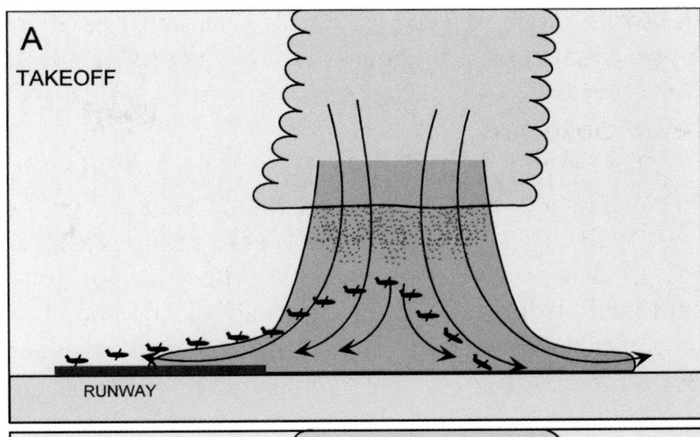

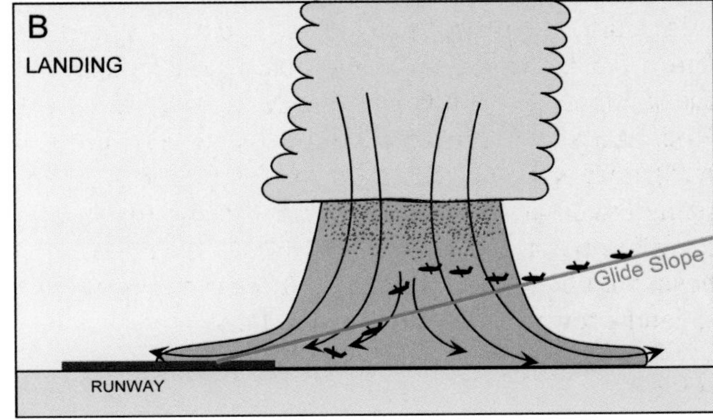

FIGURE 22.12: (A) Schematic depiction of problems encountered by an aircraft during take-off into a microburst. Initial gain of lift resulting from strong headwinds is followed by rapid loss of lift when the plane penetrates the core of downburst. (B) Schematic depiction of problems encountered by an aircraft passing through a microburst on its final approach for landing. If the plane flies through the center of the microburst, it will initially experience a headwind and lift, followed by rapid descent as downward motion of air and the loss of airspeed force the plane below the glide slope.

[2]Note that the plane's airspeed (the speed at which the air flows past the aircraft) is the sum of the plane's ground speed (the plane's speed relative to the ground) and the component of the actual wind blowing toward the plane's nose. The latter is positive for a headwind and negative for a tailwind.

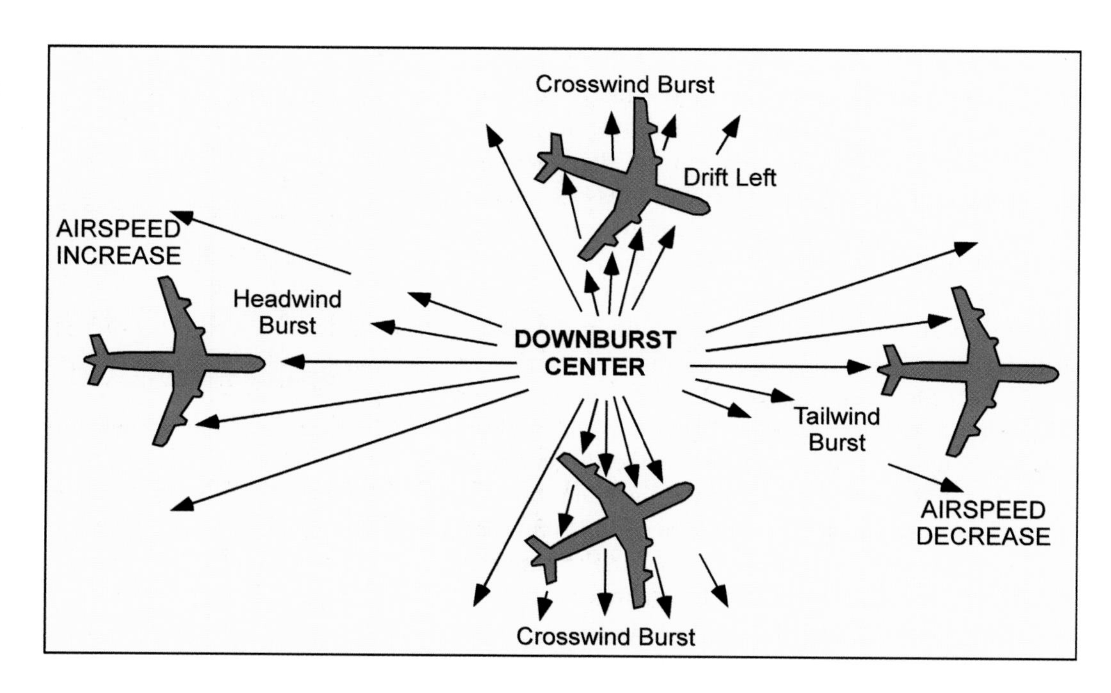

FIGURE 22.13: If a plane passes through the right or left side of a microburst, outflow winds cause lateral drift from the intended path.

FOCUS 22.3

The JFK International Airport Crash of Eastern Airlines Flight 66

Eastern Airlines Flight 66 from New Orleans crashed on 24 June 1975 at 8:05 P.M. local time while on its final approach to New York's John F. Kennedy (JFK) International Airport, killing 112 of the 124 people onboard. The existence of downbursts was established by the late Dr. Theodore Fujita as a result of his analysis of this crash. An intense effort to understand the downburst began in the late 1970s following Dr. Fujita's analysis.[3]

On the evening of 24 June, two thunderstorms were approaching the New York area. The easternmost of these storms moved over the JFK Airport at the time of the crash. Because the airport is near the coast, a sea breeze flow was present prior to the formation of the downburst. The sea breeze is an onshore wind that occurs frequently in this area during the spring and summer. The downburst was over land, and the outflow encountered the sea breeze, resulting in strong horizontal wind shear in addition to the strong downward motion of the microburst—an invitation to disaster near an airport.

[3]The detailed analysis of this event is contained in Dr. T. Fujita's report, "Spearhead echo and downburst near the approach end of a John F. Kennedy Airport runaway, New York City," Satellite and Mesometeorology Research Project Paper #137, Department of Geophysical Sciences, University of Chicago, 1976.

Several aircraft that approached the same runway before and after Eastern Flight 66 were affected adversely in ways that we now know are indicative of downburst encounters. Dr. Fujita deduced that there were actually three downbursts, perhaps resulting from the same downdraft in the parent thunderstorm. The second downburst almost blew one aircraft off the runway, and caused another to abandon approach. The pilot of the aircraft that abandoned its approach had the plane approaching the runway along the glide slope. Upon entering the outer edge of the downburst, the aircraft encountered a headwind, causing it to be lifted above the glide slope. It next encountered heavy rain and started to be pushed downward. As it moved through the core of the downburst, the airspeed rapidly dropped because the headwind diminished, causing the aircraft to fall to within 50 feet of the ground well upstream of the runway. All this happened in fifteen seconds! Fortunately, the pilot applied power in sufficient time so that the aircraft was able to pull up before hitting the ground—just in time to avoid a crash. The plane diverted to Newark, where it landed safely.

Unfortunately, the same luck did not befall Eastern 66 during its approach several minutes

(continued)

(continued)

later. This aircraft encountered similar conditions, first rising above the glide slope and then experiencing a rapid descent, being driven down by the strong downdraft and the sudden loss of airspeed. Power was not supplied sufficiently early to avoid the crash. Twenty seconds passed from the time the aircraft first went above the glide slope to the time of the crash. This short time points to the need for early detection and response by flight crews.

approach. As a pilot approaches the runway, headwinds associated with the microburst will lift the plane above the glide slope. A pilot who is unaware of what is happening might slow the plane in an effort to return to the glide slope. That type of move can be dangerous or tragic, since moments later as the plane crosses the downburst core, the airspeed will drop and the downwash will begin. Without power, the plane is at a loss to recover. When a pilot encounters rapidly increasing headwinds on approach, the safest bet is to increase airspeed, and anticipate the possibility of a missed approach. This is especially true if the pilot observes virga, or precipitation reaching the ground, ahead of the plane.

As a final note, an aircraft passing through a downburst circulation may not always encounter a loss in airspeed and a strong downdraft. Instead, if it passes to one side or the other, it may be blown to the left or right depending on which side of the downburst it traverses. Figure 22.13 shows various possibilities for downburst effects on an airplane.

MODERN TECHNOLOGY TO DETECT MICROBURSTS

To protect aircraft from the hazards of downburst winds, the Federal Aviation Administration, in cooperation with commercial airlines, has installed several types of detection systems at airports around the country to provide warning of microbursts. The primary system is called ***Terminal Doppler Weather Radar (TDWR)***. These radars are like WSR-88D radars, except that the single purpose of each radar is to protect a specific airport by detecting microbursts and wind shear, and alerting tower personnel of severe wind conditions that departing and arriving aircraft will encounter. Microbursts are sufficiently small in size (typically 2–4 kilometers in diameter) that they are near the resolution limits of conventional Doppler radars. Microbursts that are more than twenty to thirty miles (32 to 48 km) from a radar are especially difficult to detect. (Recall that a radar beam's width increases with distance from the transmitting antenna.) In addition, the radar beam can go over the top of a microburst that is not close to the radar. You can see the challenge posed by microburst resolution in Online 22.2, an animation of Doppler radar-derived winds over an area of Utah in which a dry microburst occurred on 8 June 1996. This microburst produced a 62 mph (28 m/s) wind gust in the area just to the east of Magna, Utah. TDWR radars are typically located only 15 to 20 miles (24 to 32 km) from the airport, depending on obstructions to the radar beam and ***ground clutter,*** which is the radar return from buildings and other ground objects. This close distance allows the radar to detect most microbursts. The radar scans around the airport area. Automatic algorithms analyze the radial velocity data looking for sharp wind changes.

Figure 22.14 shows an excellent example of a radial wind "couplet" (adjacent areas of winds toward and away from the radar), indicative of a downburst. The center of the downburst is located in the middle of the couplet. When wind shear and/or microbursts are observed, the radar software tracks the location of the shear zone, determines when it will affect aircraft operations and which runways will be impacted, calculates the expected loss or gain of airspeed an aircraft taking off or landing will encounter, sends this information in a concise compact message to the tower in the "language" pilots use, and sets off a wind shear alarm in the tower so that the controller will pay attention to the message. The controller then contacts the aircraft and says something like "Trans Global 246, wind shear alert, expect an 80 knot decrease in airspeed on landing, clear to land, what is your intention?" The pilot always makes the final decision, which in the case described, would best be to go around and try again later!

Other types of systems have been tested and some have been installed. Chief among these is the ***Low-Level Wind-Shear Alert System (LLWAS),*** a grid of anemometers laid out around the airport to detect wind shear. These are particularly useful in cases where evaporating precipitation does not reach the ground so that microbursts are "invisible" to radar.

FOCUS 22.4

The Andrews Air Force Base Microburst of 1983

A spectacular microburst near Washington, D.C., on 1 August 1983 came within several minutes of leaving a lasting mark on the presidency of the United States. Early that afternoon, President Ronald Reagan was returning to Washington on Air Force One. The plane landed at 2:04 P.M. EDT at Andrews Air Force Base in suburban Maryland, approximately 12 km east of the White House. The runway was dry, and the anemometer at the north end of the runway was recording steady winds of about 17 knots (20 mph) when the plane landed. At 2:10 P.M. EDT, the wind rapidly increased to a peak gust of 130 knots (150 mph) from the northwest (Figure 22B). In the next two minutes, the wind dropped to less than 10 mph, only to increase again to a peak of 84 knots (97 mph) at 2:14 P.M.—this time from the southeast. By 2:15 P.M., the wind speed had dropped back to 17 to 26 knots (20 to 30 mph), and it remained low for the rest of the afternoon.

What caused this incredible wind event lasting only a few minutes? The core of a thunderstorm had moved over the northern portion of Andrews Air Force Base, and a microburst passed directly over the anemometer. The wind trace in Figure 22B contains the classic signature of a microburst. Extreme wind gusts from opposite directions occurred within minutes of each other; separated by the momentary calm as the center or stagnation cone of the microburst passed overhead. A pressure reconstruction based on barometric data from the area indicates that the pressure at the anemometer site at 2:10 P.M. was 17 mb lower than the pressure less than one mile to the northwest, implying an intense pressure gradient consistent with strong northwesterly wind gusts.

This event represents the highest microburst wind speed reported at an airport. Statistics of microbursts suggest that there are probably only a few microburst events approaching this magnitude in the United States in a typical year. However, the small scale of microbursts makes it highly unlikely that an airport anemometer will detect their full intensity. In the case of the Andrews Air Force Base event, for example, the peak wind speeds at an anemometer located only 4 km to the north were 5 to 6 knots. Residents living this short distance to the north of the airport had little reason to suspect that a nearby microburst had come within six minutes of creating a presidential disaster.

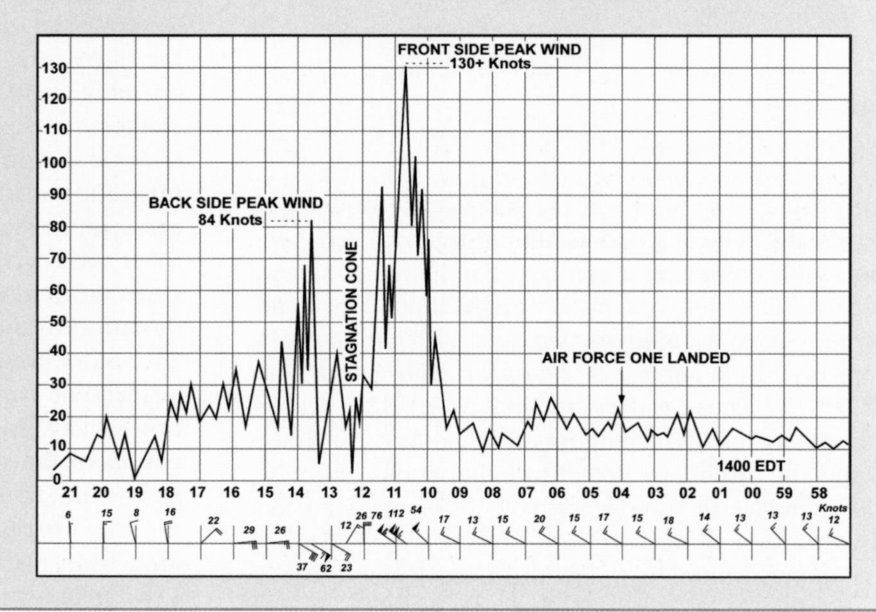

FIGURE 22B: Anemometer recording from northern end of runway during the Andrews Air Force Base microburst event of 1 August 1983. Time advances leftward from 1357 EDT to 1422 EDT. Air Force One landed at 1404 EDT. A peak microburst wind of 130 knots (150 mph) occurred between 1410 and 1411 EDT. Instantaneous winds are plotted every 60 seconds at the bottom of the diagram. All wind speeds are in knots.

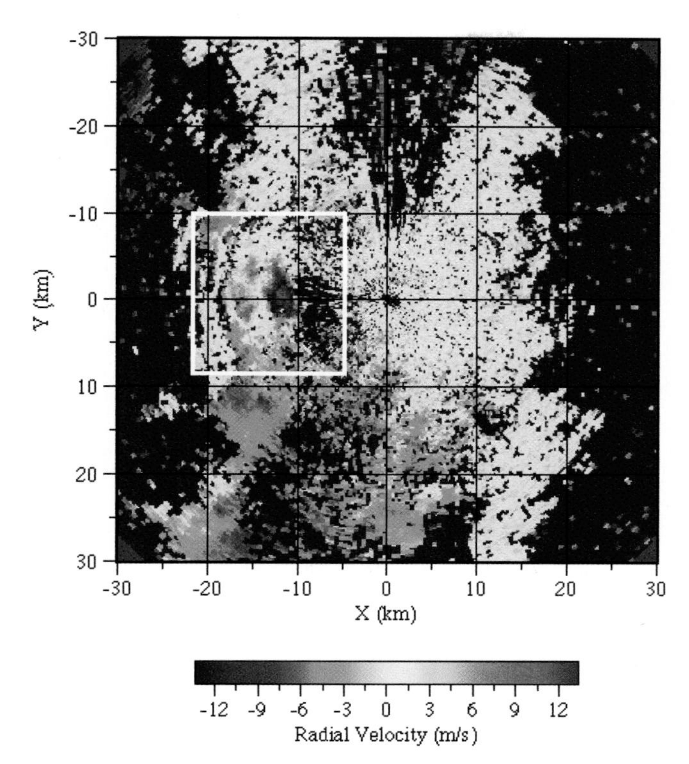

FIGURE 22.14: Radial wind field obtained from Doppler radar near Albuquerque, New Mexico. Negative values (blue, purple) indicate winds blowing toward radar; positive values (yellow, orange) indicate winds blowing away from radar. Wind speeds are in meters per second. Distances from radar are in km. Note the signature of a downburst in the area 10 to 20 km directly to the left (west) of the radar, which is located at the point (0,0).

Downburst-related aircraft accidents have decreased dramatically since the 1970s and 1980s. Warning systems such as those summarized above, together with the aviation community's increased awareness of the hazards of downbursts, have undoubtedly prevented airline disasters in the past 20 years.

PREDICTION OF CONDITIONS FAVORABLE FOR MICROBURSTS

Technologies such as Terminal Doppler Weather Radar and the Low-Level Wind-Shear Alert System serve as warning systems in the sense that they reveal the presence of significant wind shear. In such cases, the wind shear is likely to be associated with downbursts that are already present. While such information is critical for pilots, the focus on present conditions places these warnings in the realm of "nowcasting" rather than "forecasting." A separate challenge is the anticipation of these events prior to their occurrence. As is the case with tornadoes, the science and technology of meteorology do not yet permit advance forecasts (e.g., several hours ahead of time) of the precise locations and times of microburst occurrences. However, approaches have been developed for identifying conditions that are favorable for the development of microbursts, thereby encouraging a higher state of microburst awareness in particular situations. For example, indications that microburst-producing thunderstorms may occur on a particular afternoon can be deduced from the morning (1200 UTC) upper air and surface observations.

One such strategy for anticipating the subsequent development of microburst-producing thunderstorms is based on the so-called ***Wind Index (WI)*** for identifying air masses favorable for microbursts. This index, based on studies of the dynamics of microburst production, was developed at the National Severe Storms Forecast Center (now called the Storm Prediction Center) by D. W. McCann. The WI formula actually provides an estimate of the maximum potential wind gust that can be expected if microburst-producing thunderstorms develop on a particular day. Unlike the Lifted Index (Chapter 6), which is a measure of updraft potential, the WI provides a means to forecast downburst potential. The WI formula includes the environmental lapse rate below the melting level, the height of the melting level above the ground, information about the moisture content in the lowest kilometer above the surface, and the air's moisture content at the melting level. The relevance of these parameters for microbursts should be apparent from our earlier discussion of downburst formation. Note that the parameters require the availability of a sounding for their evaluation, so the index is computed from soundings for conditions at 1200 UTC or 0000 UTC. If a forecaster can anticipate the change of the lapse rate during the day by using the predicted (or observed) surface temperature at a particular location, the WI can be modified for afternoon conditions. The WI can also be calculated from forecast soundings generated from numerical models. Environmental lapse rates smaller than 5.5° C/km are assumed to correspond to "zero likelihood" of microbursts.

Forecasters have found that the WI is particularly useful when a boundary (a front or a thunderstorm outflow boundary) moves into an area in which the WI field contains its maximum values. In such situations, the WI values at the locations of actual microbursts

ONLINE 22.2

Doppler Radar Detects a Microburst over Utah

Because Doppler radars can detect winds from the motion of precipitation particles, the strong wind shear associated with downbursts offers the potential for the detection of downbursts by radar. Specifically, the strong low-level divergence at the core of a downburst can be expected to produce a "couplet" of winds moving toward and away from the radar, as shown schematically in Figure 22C. If radial velocities in opposite directions are color-coded distinctively, the signature of a downburst should appear as in Figure 22C(C). In this radar animation, positive (red) radial winds are blowing away from the radar, while negative (green) radial winds are blowing toward the radar. Hence divergent winds will appear as several green pixels adjacent to (but closer to the radar than) several red pixels, as shown in Figure 22C(C).

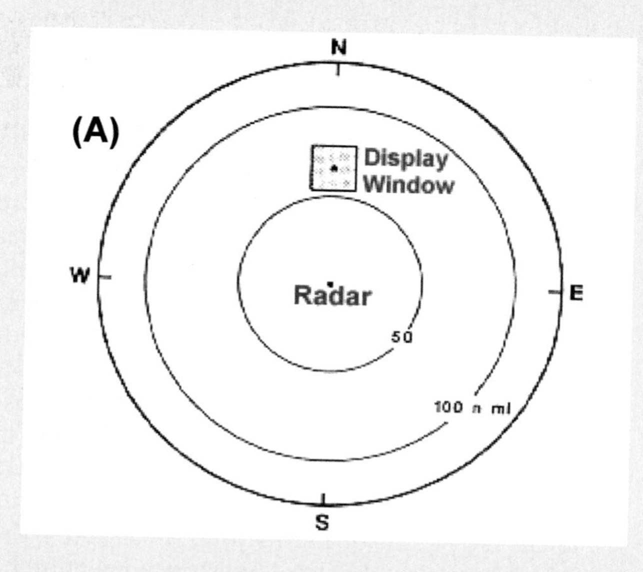

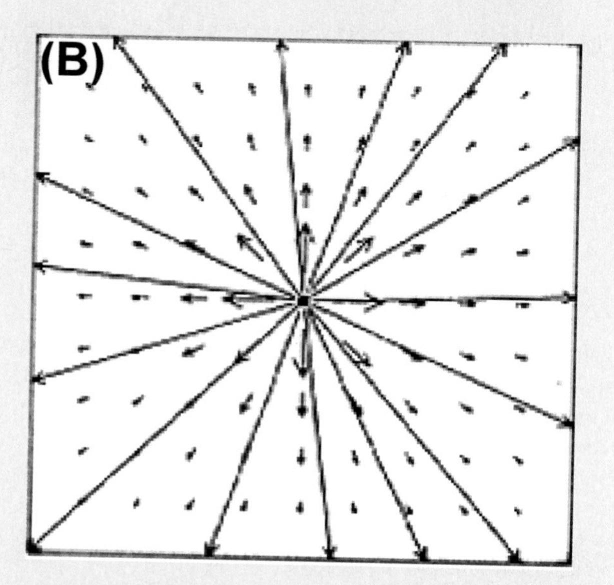

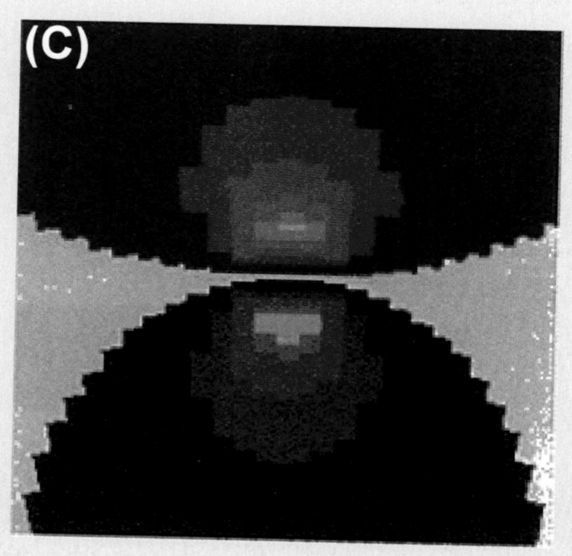

FIGURE 22C: (A) Schematic depiction of the signature of a downburst in the radial velocity field measured by a Doppler radar. Downburst is located in the box shown in the upper portion of (A); divergent velocity field caused by a downburst's low-level outflow in the same box is shown in (B); Doppler-measured radial velocities within the box are shown in (C), where green denotes motion toward radar and red denotes motion away from radar.

correlate highly with the maximum observed wind gusts at those locations. Hence, if a forecaster can monitor the WI field and the movement of frontal or outflow boundaries, the areas and times in which damaging microbursts are most likely can be anticipated up to several hours in advance. This anticipation is analogous to the use of "watch boxes" for areas in which severe thunderstorms and tornadoes are most likely to develop.

CHECK YOUR UNDERSTANDING 22.3

1. Why is wind shear a concern for aircraft?
2. When is an aircraft most vulnerable to wind shear? Why?
3. What new technology exists to detect downbursts and warn pilots of their existence?

TEST YOUR UNDERSTANDING

1. What is the difference between a downburst and a microburst?
2. What is virga? Why is it often associated with downbursts?
3. Explain how each of the four environmental conditions associated with microburst development contributes to microburst formation and intensity.
4. What are typical sizes and lifetimes of downbursts?
5. Compare the damage caused by downbursts and tornadoes.
6. Create a schematic of a downburst and identify the vortex rings, stagnation cone, curl, and runaway vortex rolls.
7. What are the visual signs of a wet downburst? A dry downburst?
8. On a blank Stuve diagram from Appendix B, sketch soundings favorable for the development of (a) wet microbursts and (b) dry microbursts.
9. What importance did the crash of Eastern Airlines Flight 66 at JFK International Airport have with respect to downburst research?
10. About how far in advance can the occurrence of a downdraft at a specific location be predicted? Explain.
11. What factors should forecasters take into consideration in determining whether downbursts are likely in a particular area?
12. What is the Wind Index?
13. Can radar detect microbursts? If so, explain the limitations. If not, explain why not.
14. Consider a square area fifty-miles on a side in the central United States. On the basis of the data from field programs that have studied downbursts, how many downbursts would you expect to occur in that area in an average year?
15. Why are falling ice crystals likely to produce stronger downbursts than falling raindrops?

TEST YOUR PROBLEM-SOLVING SKILLS

1. Consider an idealized downburst shaped like a perfect cylinder with a diameter of 2 km, and suppose that all air within this cylinder is sinking at 20 meters/second. Air at the bottom of the downburst flows outward at a speed "v." Assume v is the same in all directions and at all altitudes from the surface to 100 meters, and that v is zero at all altitudes above 100 meters.

(a) If the downburst is in a "steady state" so that air is neither piling up nor thinning out anywhere, what is v at the cylinder's edge?
(b) Assuming that the outflow remains 100 meters deep, what will be the outflow's speed at a distance of 2 km from the center of the downburst?

2. Figure 22.9 shows the advancing edge of the outflow of a microburst. Assuming that support towers for the high-voltage power lines are 200 feet tall, estimate the speed of propagation of the leading edge of the outflow. Show how you arrived at your answer.

3. Suppose you are a pilot whose small plane is heading directly into a dry (and undetected) microburst. The microburst is stationary and symmetric, with outflow speeds of 60 mph at your flight level. Your plane's stall speed is 90 mph.
   (a) What visual clue(s) might be present to alert you that a microburst is occurring?
   (b) Describe qualitatively the sequence of events that you will experience as you penetrate the microburst.
   (c) What ground speed will you need to maintain in order to keep your plane above its stall speed?
   (d) Suppose that the microburst has the same intensity as above, but is traveling at 20 mph in the same direction as your plane. How will this change your answer to (c)?

4. When air sinks below a cloud base, it warms adiabatically. If the sinking air is unsaturated, the air will warm at a rate of 10° C/km, the numerical value of the dry adiabatic lapse rate. If the sinking air contains rain that is rapidly evaporating, its temperature change will be approximately the numerical value of the moist adiabatic lapse rate. (Latent heat is being consumed by evaporation rather than released by condensation.) Suppose that a cloud base is 2 km above the ground, the *environmental* lapse rate below the cloud base is 10° C/km, and the downdraft air falling through the cloud base is loaded with rain and has the same temperature as the surrounding environmental air at 2 km.
   (a) How much warmer or colder than the surrounding environment will the downdraft air be when it reaches the surface? (Assume evaporation keeps the air saturated as it falls.)
   (b) How would your answer to (a) change if the precipitation falling through the cloud base were ice crystals instead of raindrops? (Hint: The latent heat required to sublimate ice particles is about 1.13 times larger than the latent heat required to evaporate water droplets.)

5. Wind shear is evaluated quantitatively as the change in wind divided by the distance over which the change of wind occurs.
   (a) Evaluate the wind shear in the each of the examples below. Each can be assumed to have no wind at the exact center of the circulation.
   (b) In which example would the wind shear pose the greatest threat to aircraft?

Tornado	wind speed = 100 m/sec at 500 meters from center
Downburst	wind speed = 50 m/sec at 250 meters from center
Hurricane	wind speed = 75 m/sec at 15 kilometers from center of eye
Blizzard	wind speed = 25 m/sec at 100 kilometers from low's center

USE THE SEVERE AND HAZARDOUS WEATHER WEBSITE

http://severewx.atmos.uiuc.edu

1. Access the Microburst Handbook from the "Downburst" page of the *Severe and Hazardous Weather Website.* After reading through the handbook, summarize the visual differences between a gust front, a wet microburst and a dry microburst. What information provided in this handbook would be especially helpful for pilots?

2. Navigate to the Storm Prediction Center website via the "Current Weather" page of the *Severe and Hazardous Weather Website.* Examine three days of storm reports when there were tornadoes and wind events reported.
   (a) Record the dates you examined.
   (b) Is wind damage caused more frequently by tornadoes or by non-tornadic winds?
   (c) Identify at least two straight-line wind events; note their dates, times, and locations.
   (d) Assess the scale of damage from the two severe straight-line wind events. Identify several reports that could be attributable to microbursts and several reports that are most likely not attributable to microbursts. Explain your choices.

3. Denver, Colorado is in the center of microburst activity in the summer. Use the *Severe and Hazardous Weather Website* to examine current or recent soundings from Denver and assess the microburst potential.
   (a) Do the soundings show the "inverted-V"? If so, is the inverted-V more typically found in morning or evening soundings? Explain.
   (b) Microbursts are favorable when CAPE values are high. Do the soundings display this feature? How did you determine this?
   (c) Unlike tornadic thunderstorms, capping inversions are detrimental to microbursts. Do the soundings display a capping inversion?
   (d) What is the environmental lapse below and above cloud base? Is this conducive to microbursts? Why or why not?

4. NOAA provides products to pilots and others involved in aviation that contain information on the potential for microbursts. Access the NOAA Downburst Potential page via the *Severe and Hazardous Weather Website.*
   (a) Briefly define the Wind Index, Dry Microburst Index, and the Dry Microburst Potential Index.
   (b) Examine the indices for your part of the United States. Summarize the current potential for a microburst based on the products available.

5. Clear Air Turbulence (CAT) is a serious hazard for aircraft. Current CAT products and images along with climatology are available from the "Downburst" page of the *Severe and Hazardous Weather Website.*
   (a) Examine the CAT climatology. What region of the United States is most vulnerable to CAT?
   (b) What time of year is CAT most likely?
   (c) What is the current potential for CAT across the United States? Explain how you determined this.

EL NIÑO, LA NIÑA, AND THE SOUTHERN OSCILLATION

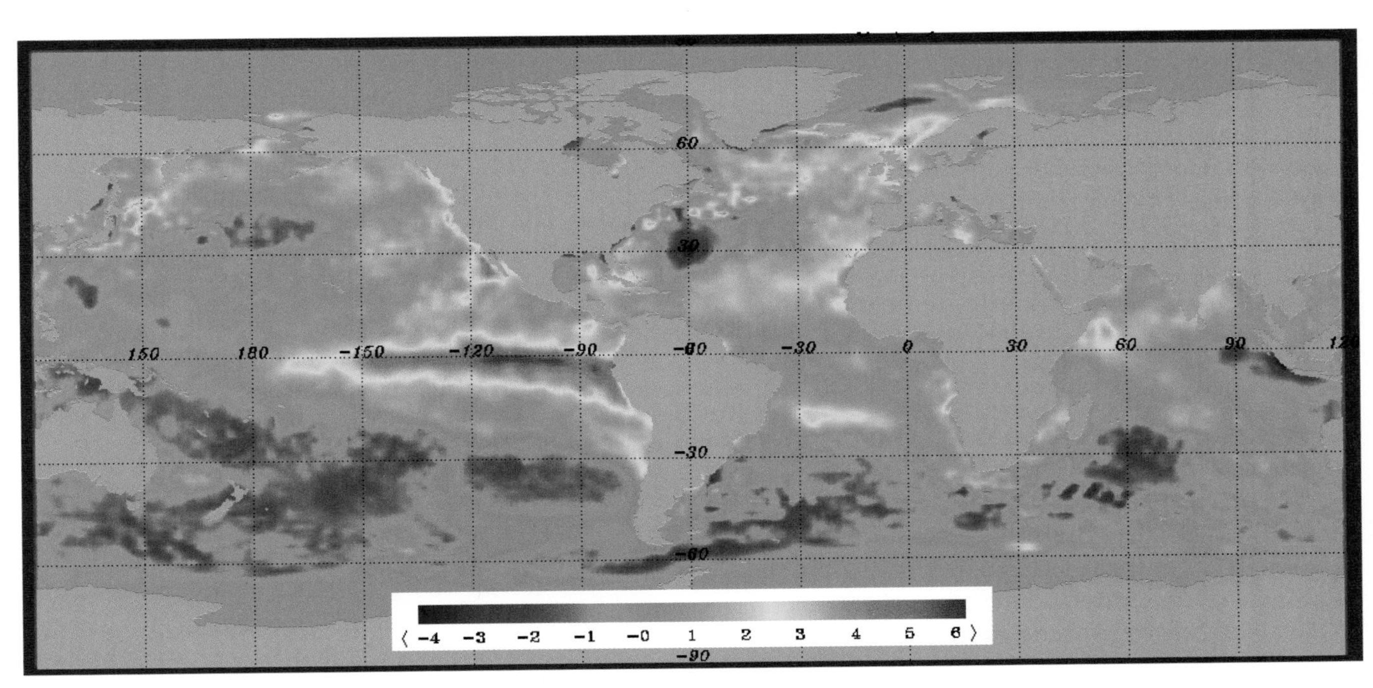

Global ocean surface temperature anomalies in November 1997 during an El Niño.

KEY TERMS

anomaly
El Niño
ENSO
La Niña
Southern Oscillation
Southern Oscillation Index (SOI)
TOGA/TAO array
trade wind
upwelling
Walker Cell

One of the most widely publicized phenomena affecting weather and climate is ***El Niño***. Largely because of the news media's extensive coverage of recent El Niño events, El Niño has almost become a household word. Even the opposite ***La Niña*** phase of the cycle has received considerable media coverage in recent years.

Why do we include El Niño in a book on severe and hazardous weather? First, El Niño and its large-scale counterpart, the ***Southern Oscillation***, are associated with unusual weather patterns that affect much of the world, including North America, and persist for a season or longer, even up to a year, often resulting in a succession of unusual weather events. Second, recent research indicates that the ocean surface temperatures associated with El Niño can significantly affect the intensity of individual extreme weather events, from ice storms in eastern North America to flooding rains along the West Coast. Third, there are indications that El Niños are becoming more frequent; the two strongest El Niños on record have occurred since 1980. Finally, the broader manifestation of El Niño is an excellent illustration of the interconnectedness of the ocean and the atmosphere, and also of the ability of atmospheric heating in the "remote" tropical Pacific to affect the weather and lives of people thousands of miles away. The latter effect not only illustrates the workings of the atmosphere as a dynamic fluid, but also shows how weather and climate forecasters must consider the entire global system if they are to be successful at forecasting severe and hazardous weather.

The term "El Niño," which is Spanish for "the Christ Child," was first used by fishermen to refer to warm ocean waters along the coasts of Ecuador and Peru (Figure 23.1). This warm water is intermittent, generally appearing around the Christmas season (hence the name) and persisting for several months. Fish become relatively scarce when warm water is present, so the local fishing industry experiences a seasonal lull. In some years, however, the water is unusually warm and the interruption of the fishing season persists into May or June. The term El Niño is now used to denote these exceptionally warm intervals. Because the region is so heavily dependent on fishing, the local economic impacts are negative and substantial. Moreover, seabirds that feed on fish also suffer, as do other parts of the coastal ecosystem. The El Niño events are generally accompanied by heavy rains over coastal Peru and Ecuador, suggesting that the phenomenon involves the atmosphere as well as the ocean.

During the past century, more than two dozen El Niño events have affected the western coast of South America (see Table 23.1). Most raised water temperatures not only along the coast but also in a belt stretching 5,000 miles (8,050 km) westward, halfway across the equatorial Pacific (see chapter cover figure). The weaker events, such as the most recent El Niño of

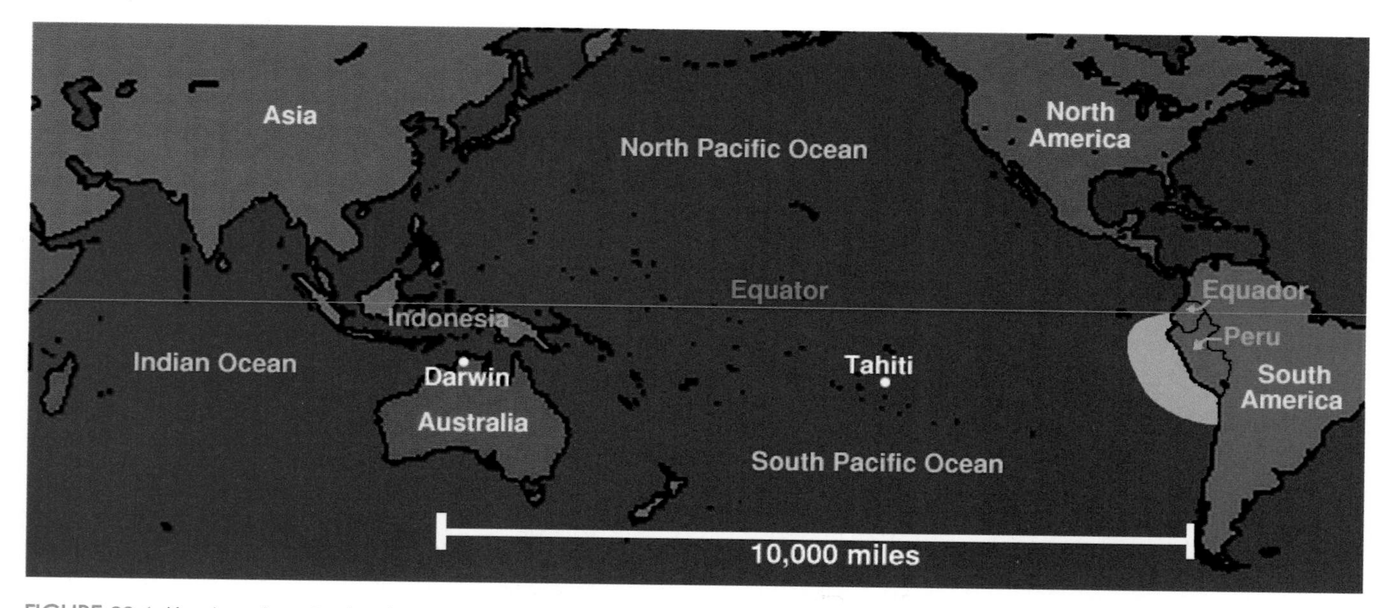

FIGURE 23.1: Key locations in the description and monitoring of El Niño and the Southern Oscillation. The region historically associated with the localized warming of water offshore of Peru during El Niño events is shaded in pink.

TABLE 23.1 El Niño Occurrences since 1875

1877–1878	1932	1977–1978
1888–1889	1939–1941	1979–1980
1896–1897	1946–1947	1982–1983
1899	1951	1986–1988
1902–1903	1953	1990–1993
1905–1906	1957–1959	1994–1995
1911–1912	1963	1997–1998
1913–1914	1965–1966	2002–2003
1918–1919	1968	2004–2005
1923	1972–1973	2006–2007
1925–1926	1976–1977	

Source: Climate Prediction Center/National Centers for Environmental Prediction/NOAA.

2006–2007, warmed the ocean surface by only 1.8 to 3.6° F (1 to 2° C) and had only minor impacts on the South American coastal region. But the stronger events, like the El Niños of 1982–1983 and 1997–1998, increased the sea surface temperature by more than 4° C (7.2° F) in some areas, impacting weather and climate patterns on a much larger scale.

THE SOUTHERN OSCILLATION

Only in the last few decades have scientists come to realize that El Niño is part of a much larger atmospheric phenomenon known as the Southern Oscillation. The Southern Oscillation encompasses most of the equatorial Pacific atmosphere. In order to appreciate the size of the area involved, examine the Pacific sector of a globe. From the South American coast to the Australia–Indonesia region, the equatorial Pacific Ocean extends more than 10,000 miles (16,100 km), or three to four times the width of the United States.

The Southern Oscillation was discovered at the same time scientists in South America began to document the local effects of El Niño in the early 1900s. A British scientist named Sir Gilbert Walker was in India, attempting to develop procedures for predicting the monsoon. While examining world weather data, Walker found an intriguing correlation between the surface pressures on the eastern and western sides of the tropical Pacific Ocean. When the pressure rises on one side of the tropical Pacific, it usually falls on the other side. Walker introduced the term Southern Oscillation to refer to this east-west "seesaw" in the pressures of the equatorial Pacific. The use of the term "Southern" stems from the fact that the seesaw effect is strongest in the tropical Pacific south of the equator. In addition, the tropical Pacific was south of most of the major population centers of Walker's time.

How does the "seesaw" of barometric pressure across the tropical Pacific relate to El Niño? The explanation begins with a look at the major pressure features during winter. Figure 23.2 shows the average sea-level

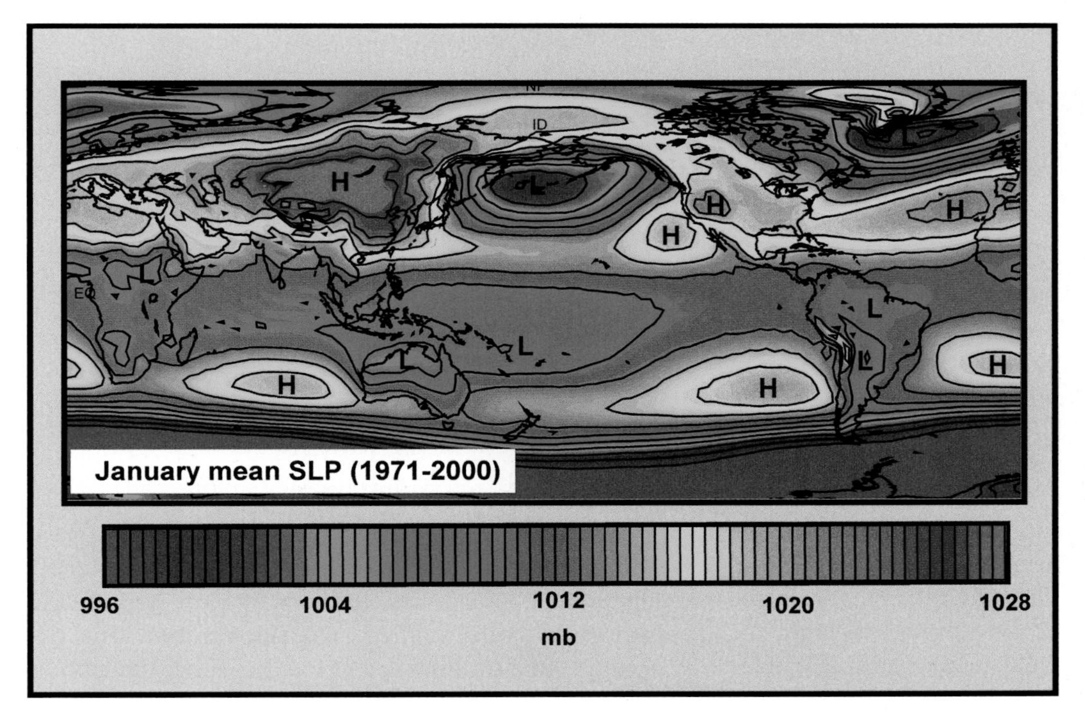

FIGURE 23.2: Average pattern of global sea level pressure (SLP) (mb) in January.

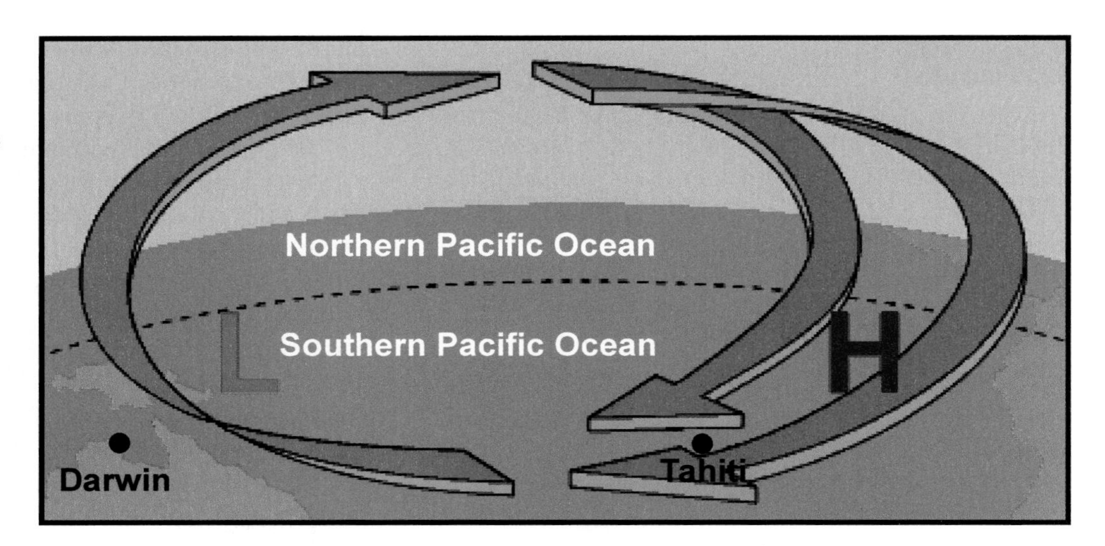

FIGURE 23.3: The Walker Cell over the tropical Pacific Ocean. Axes of the diagram are longitude and altitude. The downward arrows in the figure are "split" because the location of the strongest downward motion varies with the location of the subtropical high-pressure center at the surface.

pressure pattern at the surface during the Northern Hemisphere winter (January). The key features to note are the low-pressure center near Australia and Indonesia, and the subtropical high-pressure center in the eastern tropical Pacific near South America. The general pattern of wind in the tropical Pacific is from east to west, as air flows counterclockwise around the subtropical high (recall that winds around Southern Hemisphere highs and lows are opposite to the Northern Hemisphere directions) and toward the low pressure near Australia. Because the Coriolis force is weak near the equator, the wind pattern is largely a manifestation of the pressure gradient force, which points from east to west in the tropical Pacific. The east-to-west winds are known as the ***trade winds,*** because sailing ships engaged in trade used these winds to cross the oceans from east to west. The vertical motions associated with the two major pressure centers in the tropical Pacific are downward above the high in the east and upward above the low in the west. (The reasoning is exactly the same as in Chapter 8.)

As shown in Figure 23.3, these two branches of vertical motion and the westward surface winds force a "return flow" of air aloft from west to east, forming a gigantic closed circulation cell. This circulation in the east-west and vertical plane is known as the ***Walker Cell,*** in honor of Sir Gilbert. A key feature of the Walker Cell is that it strengthens and weakens over time, with a typical cycle of its intensity lasting several years. This oscillation in the intensity of the Walker Cell is known as the Southern Oscillation. Note that, as the Walker Cell strengthens, so do the intensities of the South Pacific subtropical high, the Australian low, and the trade winds; as the Walker Cell weakens, so do the surface pressure centers and the trade winds. The strengthening and weakening of the Walker Cell explain the "seesaw" of pressure between the eastern and western parts of the tropical Pacific.

The Southern Oscillation is an atmospheric phenomenon. However, it is linked to the more localized El Niño offshore of Peru. This linkage has led to the use of the term El Niño/Southern Oscillation, *ENSO,* in describing the complete cycle of events in the ocean and the atmosphere. ENSO may be regarded as the most appropriate term for describing the large-scale manifestation of El Niño.

CHECK YOUR UNDERSTANDING 23.1

1. What is the importance of El Niño from a severe weather perspective?
2. What is the origin of the term "El Niño"?
3. Which coastal region is most directly affected by El Niño?
4. What are the primary large-scale circulation features that vary in association with ENSO?

EVOLUTION OF AN ENSO EVENT

The circulation features in Figures 23.2 and 23.3 represent the "normal" pattern, which has several important consequences for the weather and the oceanic state of the tropical Pacific. First, the eastern subtropical Pacific normally has little precipitation because of the downward air motion in the vicinity of the high, while the western tropical Pacific is normally much wetter because of the upward motion near the low. Second, the westward surface trade winds drag the ocean water westward, resulting in a sea level that is 10 to 20 cm (4 to 8 in) higher in the western tropical Pacific than in the eastern tropical Pacific. The surface waters in the west are also warmer because they have been exposed to solar heating for a longer time. The waters near South America, on the other hand, are relatively cool because they have just risen from the cold deep abyss to replace the surface waters that were dragged westward by the easterly winds. This rising of colder water from deeper levels to the surface is known as ***upwelling***.

Once every few years, the normal pattern of tropical Pacific pressure tends to weaken or break down. During this breakdown phase, which is what Walker observed during one stage of his Southern Oscillation, several important things happen:

- *Surface pressure systems weaken.* Surface pressure decreases in the eastern high-pressure area (the high weakens), while surface pressure increases in the western low (the low also weakens). This results in a reduction of the cross-Pacific pressure gradient.
- *Trade winds weaken or stop, or even reverse.* The pressure gradient responsible for the trade winds weakens, decreasing wind speeds.
- *Water in the eastern Pacific warms.* Because the trade winds are too weak to maintain the tilt of the sea level, the warm water of the western Pacific flows slowly eastward, returning warm surface water to the eastern Pacific and stopping the upwelling along the South American coast (starting the El Niño). Figure 23.4 shows an example of the change in the distribution of ocean temperatures at the start of the 1997–1998 El Niño.
- *Precipitation shifts eastward.* The area of upward motion in the atmosphere, and its associated clouds and precipitation, move eastward from the Australia/Indonesia region to the central and even the eastern Pacific, in response to the reduction of surface pressure and downward motion aloft in the vicinity of the normal subtropical high pressure.

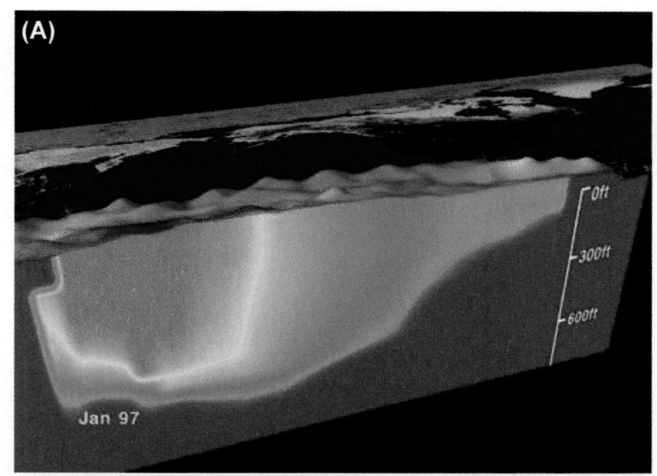

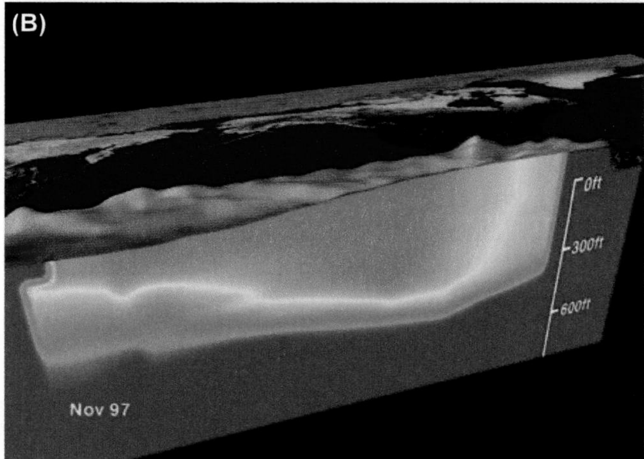

FIGURE 23.4: The oceanic temperature changes that occurred in the upper 600 ft (200 m) of the tropical Pacific Ocean as an El Niño developed during 1997. Colors represent water temperatures, ranging from 8° C (46° F) (dark blue) to 30° C (86° F) (deep red). Note eastward movement of the warm pool of water, and the accompanying change of sea level between (A) January 1997 and (B) November 1997. The figure was produced from a combination of buoy measurements of subsurface water temperatures, and satellite measurements of ocean surface elevation and temperature.

When these conditions coincide for a sufficient time to allow the average sea surface temperature ***anomaly*** (departure from normal) to be at least 0.5° C (0.9° F) for three consecutive months, El Niño is declared to be occurring. Figure 23.5 summarizes schematically the normal and "breakdown" phases of the Southern Oscillation. The figure illustrates that El Niño conditions west of South America are essentially a consequence of the breakdown phase of the Southern Oscillation.

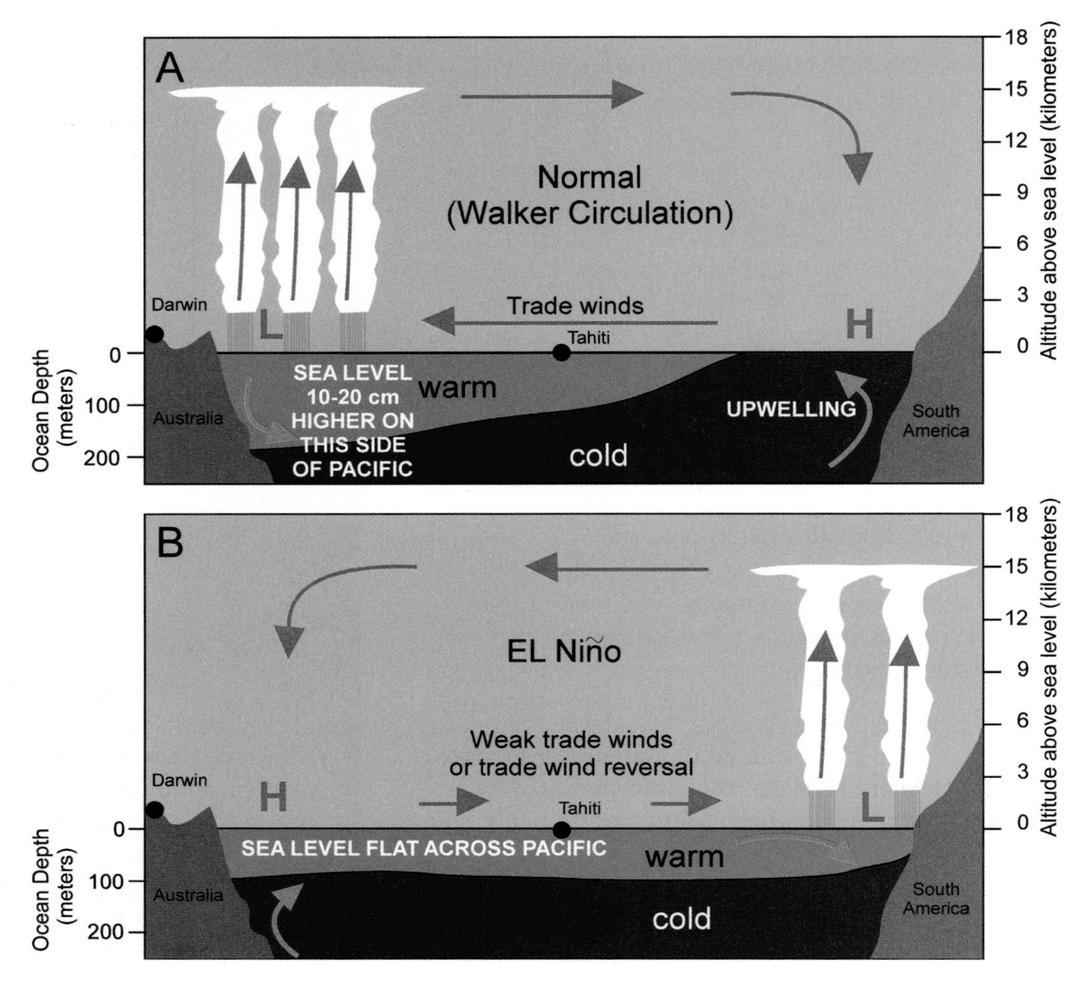

FIGURE 23.5: Schematic representation of ENSO. The figure shows the atmospheric and oceanic states over the equatorial Pacific Ocean during (A) "normal" conditions and (B) "El Niño" conditions in the tropical Pacific Ocean. Note the differences between the two panels in pressure pattern, vertical motions, wind direction, sea level, location of clouds and precipitation, and location and depth of the warm water layer (note different vertical scale for the ocean and atmosphere).

ONLINE 23.1

Animation of the Sea Surface Temperatures during an El Niño Event

The evolution of an El Niño event can be seen in maps of ocean surface temperature plotted as "anomalies," or departures from the climatological monthly averages. Online, you can view an animation of the sea surface temperature anomalies during the strong El Niño event of 1982–1983. The anomalies reached about 7° F (4° C) in early 1983, and were especially large in the region offshore of Peru. A smaller area of positive anomalies of about 4° F (2° C) appeared around Baja California; this secondary warming extended northward to the waters offshore of the West Coast of the United States. While a warming of 4° F (2° C) is considerably smaller than the warming along the coast of Peru, it is large enough to be readily apparent to a swimmer. The 1982–1983 El Niño was the stimulus for an intensive scientific program known as TOGA (Tropical Ocean and Global Atmosphere), which led to major advances in monitoring, understanding, and predicting El Niño.

This particular El Niño was one of the strongest of the past century, comparable only to the 1997–1998 El Niño. Figure 23A shows the departures from normal sea surface temperature in mid-September 1997, prior to the peak of the 1997–1998 El Niño event.

The predictability of El Niño events at lead times of several months stems from the fact that El Niños often show similar patterns in their evolution over the course of several seasons. For example, the spatial patterns of the sea surface temperature anomalies were quite similar at the peaks of

(continued)

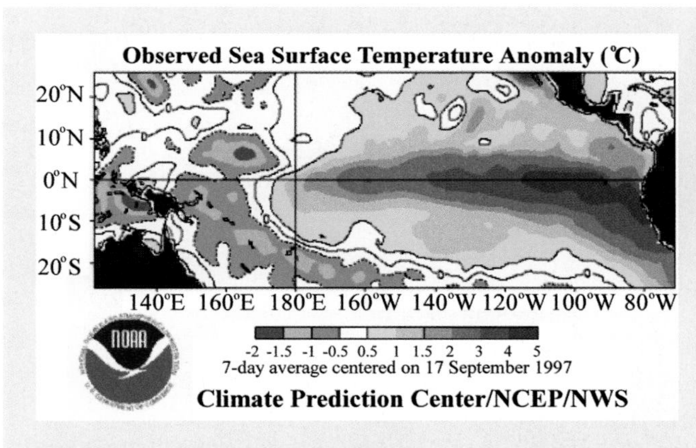

FIGURE 23A: Departures from normal sea surface temperatures (°C) on 17 September 1997, prior to the peak of the 1997–1998 El Niño. Yellow and red denote above-normal temperatures, blue denotes cooler-than-normal temperatures.

(continued)

the El Niños of 1982–1983 (online) and 1997–1998 (Figure 23A). However, exceptions to the pattern have occurred in the past and will almost certainly challenge forecasters in the future.

The last of the points listed above is particularly important for the large-scale circulation of the atmosphere, because clouds and precipitation release large amounts of latent heat into the atmosphere. The eastward shift of clouds and precipitation causes a major change in the pattern of atmospheric heating, which in turn affects the upper-air pressures and the large-scale atmospheric circulation. Ultimately, global weather patterns are affected, and the locations of severe and hazardous weather shift.

CONSEQUENCES OF EL NIÑO

The consequences of the El Niño or breakdown phase of the Southern Oscillation include droughts in Australia and Indonesia, where the upward motion is largely eliminated, and warm rainy weather in the eastern Pacific, where upward motion is enhanced. Along the South American coast, where upward motion is also increased, thunderstorms are often triggered and flooding can occur. Warm water in the eastern Pacific Ocean often extends as far north as the West Coast of the United States, raising the temperatures of coastal waters several degrees above their normal values.

ENSO also affects weather over North America because the altered heating pattern over the Pacific favors a shift in the jetstream and the associated storm tracks. Figure 23.6 shows how this shift typically occurs during the El Niño phase of the cycle. Storms that normally move along the jetstream into the Pacific Northwest (Washington and Oregon) are diverted northward into the Gulf of Alaska. Meanwhile, a secondary storm track, following a stronger-than-normal subtropical branch of the jetstream, extends into California and across the southern United States.

The strengthening of the subtropical jetstream can be viewed as a response to the enhanced heating on the southern (tropical) side of the jetstream since enhanced heating raises the upper-air heights (pressures). This creates an upper-air ridge and strengthens the north-south pressure gradient and the winds aloft (Chapter 8). At the same time, the heating of air over the eastern Pacific triggers thunderstorms that rise into the upper troposphere. The moisture that is carried aloft by these thunderstorms is transported northward into the subtropical jetstream, which in turn carries the moisture northeastward toward North America.

Typical global weather consequences of these shifts in winter appear in Figure 23.7A. There are more storms than usual along the California coast, wetter-than-normal conditions in the southern states, particularly along the Gulf Coast, and warmer than normal conditions in the Pacific Northwest. In addition, stronger west-to-east airflow across the southern United States and Mexico often limits southward penetration of polar airmasses, resulting in milder-than-normal winters across the northwestern and northcentral United States, southern Canada, and southern Alaska. The stronger than normal subtropical jetstream also favors the occurrence of East

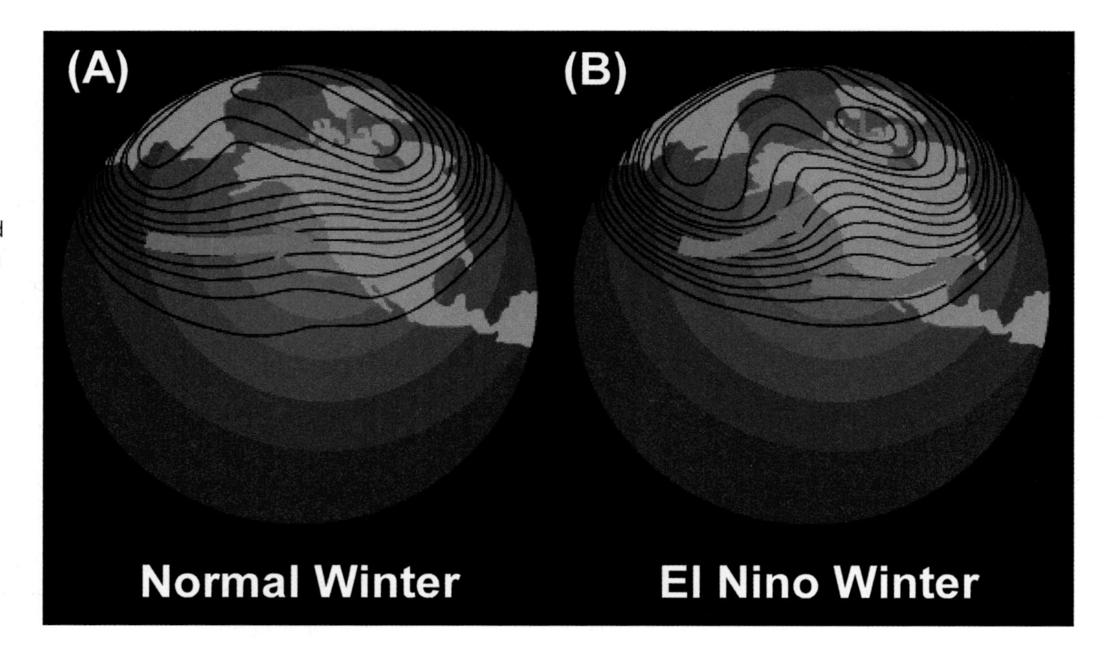

FIGURE 23.6: Northern hemisphere upper-air height patterns and jetstreams during (A) a normal winter and (B) an El Niño winter. The right panel shows an amplified ridge over North America in the northern branch of the jetstream during an El Niño winter.

Coast extratropical cyclones (Chapter 11). Worldwide, the effects of El Niño are quite distinct. Warm and dry conditions prevail over Indonesia and Southern Africa, while dry conditions persist over northern Australia and the Amazon Basin. Wetter weather is experienced in the Eastern Equatorial Pacific and northern Argentina. Warmer conditions are experienced in many parts of the globe. The summertime impacts of El Niño in North America are not well defined (Figure 23.7B). Globally, Australia, Indonesia, India, and parts of South America are the most affected regions, generally experiencing drier conditions in summer as well as winter.

CHECK YOUR UNDERSTANDING 23.2

1. What is the prevailing direction of the wind in the tropics during normal conditions? What about during El Niño?
2. How do sea surface temperature and sea level elevation *normally* vary from east to west across the tropical Pacific Ocean?
3. Why does northern Australia normally have more precipitation than Ecuador and Peru?
4. What is upwelling? What happens to upwelling along the South American coast during an El Niño?

THE LA NIÑA PHASE OF THE SOUTHERN OSCILLATION

The phase of the Southern Oscillation opposite to El Niño is known as La Niña, a term scientists coined in the 1980s. The La Niña phase is characterized by higher-than-normal pressure in the eastern Pacific and lower-than-normal pressure over Australia/Indonesia, creating a stronger-than-normal Walker Cell (see Figure 23.3). This, in turn, results in stronger-than-normal trade winds. Consequences in the Northern Hemisphere winter include colder-than-normal ocean temperatures off the South American coast, an increased slope of sea level from east to west in the Pacific, and above-normal rainfall (even floods) in Australia and Indonesia (Figure 23.8). In middle latitudes, the strongest winds of the North Pacific jetstream are shifted westward, consistent with the westward shift of the tropical heating. La Niña is defined to be underway when the average sea surface temperature is at least 0.5° C (0.9° F) colder than normal for three or more months.

Impacts on North American weather are more variable than in the case of El Niño, although there is a tendency for La Niña winters to be warm and dry in the southern tier of the United States from Arizona to Florida. There is also an increased likelihood of colder-than-normal winters from Montana northwestward to Alaska, and wetter-than-normal winters in the Pacific Northwest (Figure 23.8A).

FOCUS 23.1

ENSO and Hurricane Activity

Because El Niño and the Southern Oscillation affect ocean surface temperatures over large areas, especially in the tropics, one might expect an effect on hurricane activity. Hurricanes do not form unless the ocean surface temperature exceeds a threshold of about 80° F (26.5° C), and the intensity of hurricanes generally increases with the sea surface temperature (see Chapter 24). Statistical studies indeed show that the number of hurricanes in the eastern Pacific Ocean is slightly higher during El Niño years than in other years. The number of intense hurricanes (wind speeds at least 110 mph) averages 5.5 during El Niño years, while the average for all years is 4.8. Hawaii has experienced two of its most damaging hurricanes during El Niño years, Hurricane Iwa in 1982 and Hurricane Iniki in 1992.

However, statistics also show that El Niño is associated with a significant *decrease* of hurricane activity in the Atlantic, the Caribbean Sea, and the Gulf of Mexico. The 2006 hurricane season provides a recent example: Despite forecasts of an active hurricane season in the Atlantic, there were few hurricanes during 2006 as an El Niño developed rather unexpectedly during the spring and early summer months. While the long-term average number of Atlantic hurricanes is 5.8 per year, the average for El Niño years is only 4.0. (The corresponding averages of intense hurricanes are 2.5 and 1.5, respectively.)

Why are Atlantic hurricanes less common during El Niño? The answer appears to lie in vertical wind shear, which adversely affects hurricanes by tilting the thunderstorm updrafts and spreading the released latent heat energy over a larger area (Figure 23B). The anomalous heating pattern of an El Niño results in an increase of upper-level winds over the tropical Atlantic, thereby increasing the vertical wind shear. The heating does not increase the vertical shear over the tropical Pacific, so the effect of the warmer waters "wins out" and hurricane frequency increases in the eastern Pacific during El Niño.

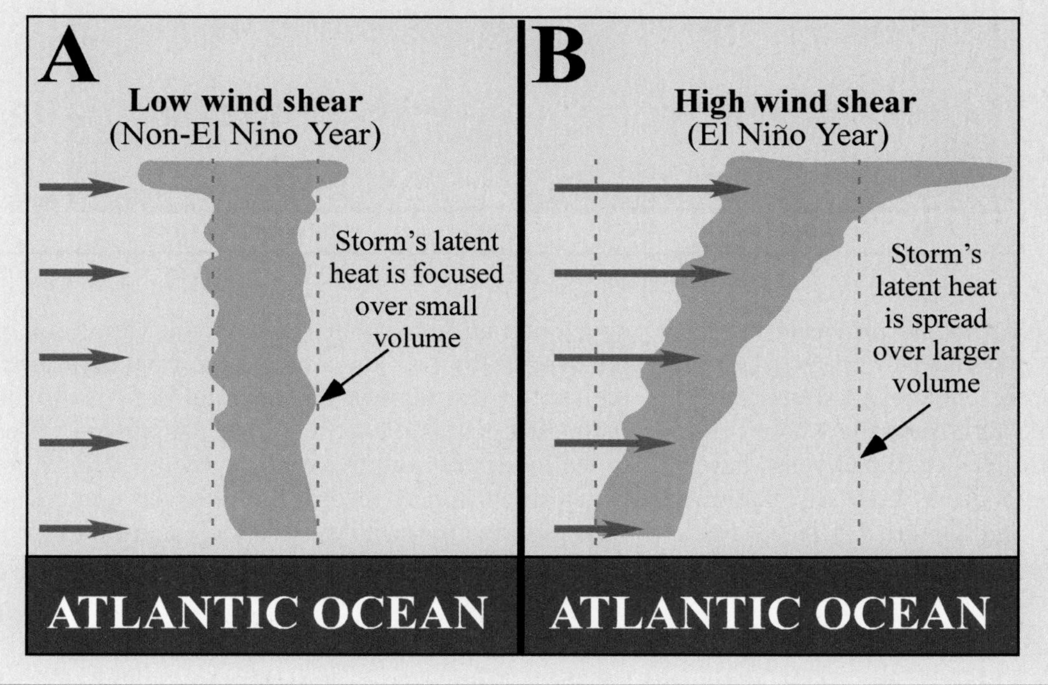

FIGURE 23B: The effect of wind shear on tropical storm development. (A) When wind shear is weak (left panel), the release of latent heat is concentrated over a small area, leading to more intense surface pressure gradients and winds. (B) When wind shear is strong, the release of latent heat is spread over larger areas, resulting in weaker gradients of heating and pressure.

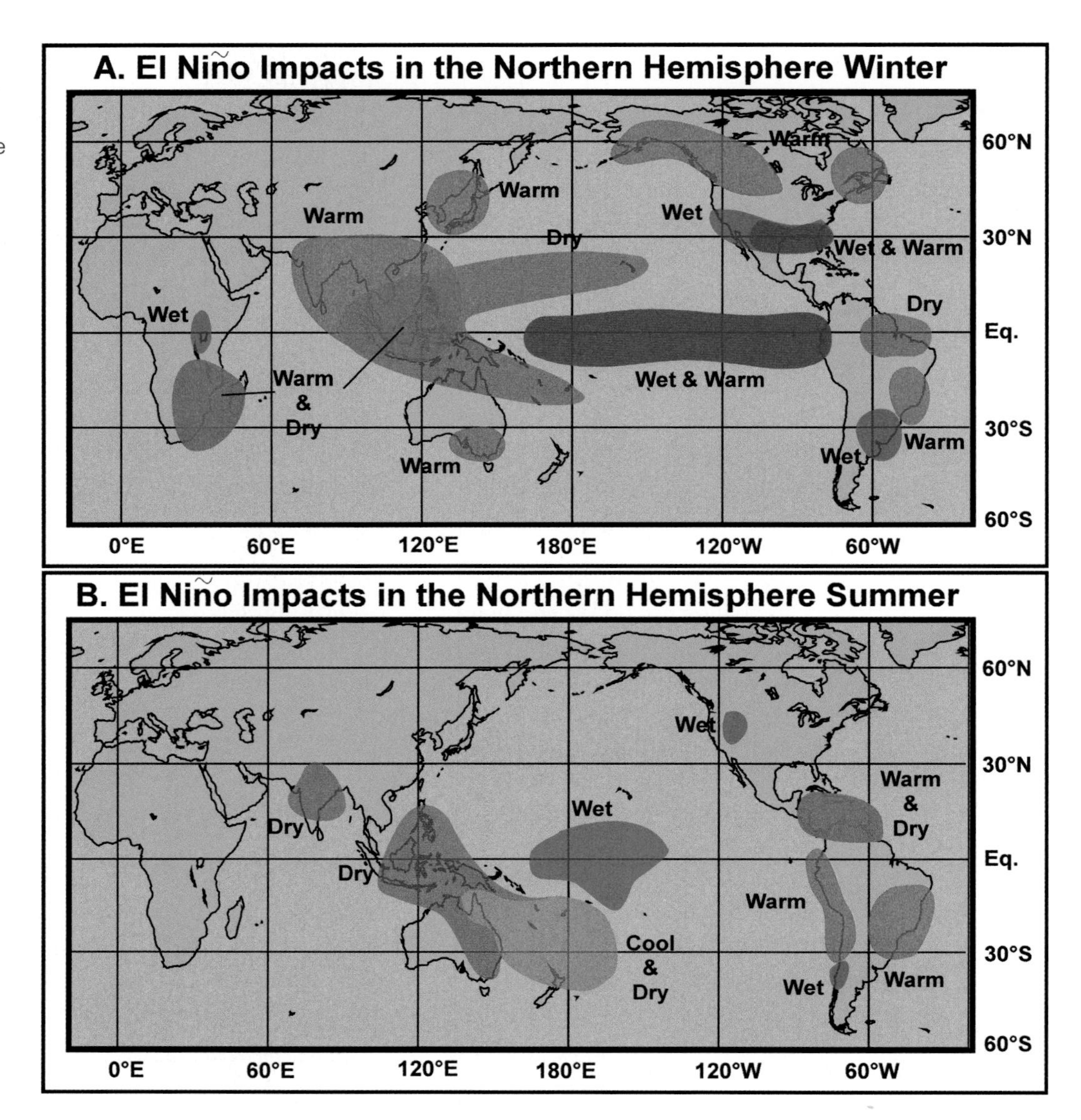

FIGURE 23.7: Global consequences of ENSO during the El Niño phase in the Northern Hemisphere (A) winter and (B) summer. Note that impacts are more pronounced during the Northern Hemisphere winter.

Summertime effects of La Niña are primarily in the Southern Hemisphere, particularly in Southeast Asia, where conditions are cooler and wetter, and the western Pacific and eastern Australia where conditions are warmer and drier. La Niña events have coincided on a few occasions with significant droughts in the central and eastern United States (1988 and 1998–1999, respectively). These droughts are discussed in Chapter 26.

MONITORING THE SOUTHERN OSCILLATION

In order to monitor the Southern Oscillation and its associated El Niño/La Niña events, scientists have developed an index that captures the variations of pressure in the two key areas of the tropical Pacific: the subtropical high in the east and the Australian/Indonesian low in the west. The actual pressures used in the index are from two tropical Pacific stations for which the historical records are relatively long: Darwin, Australia (near the western low), and Tahiti (near the eastern high). The ***Southern Oscillation Index (SOI)*** is defined as

$$\text{SOI} = \text{Pressure at Tahiti} - \text{Pressure at Darwin}$$

where each pressure is a surface pressure expressed as a departure (in standard deviations) from its monthly normal value. When this index is negative, the high- and low-pressure centers are weaker than normal, indicative of the El Niño phase (weak trade

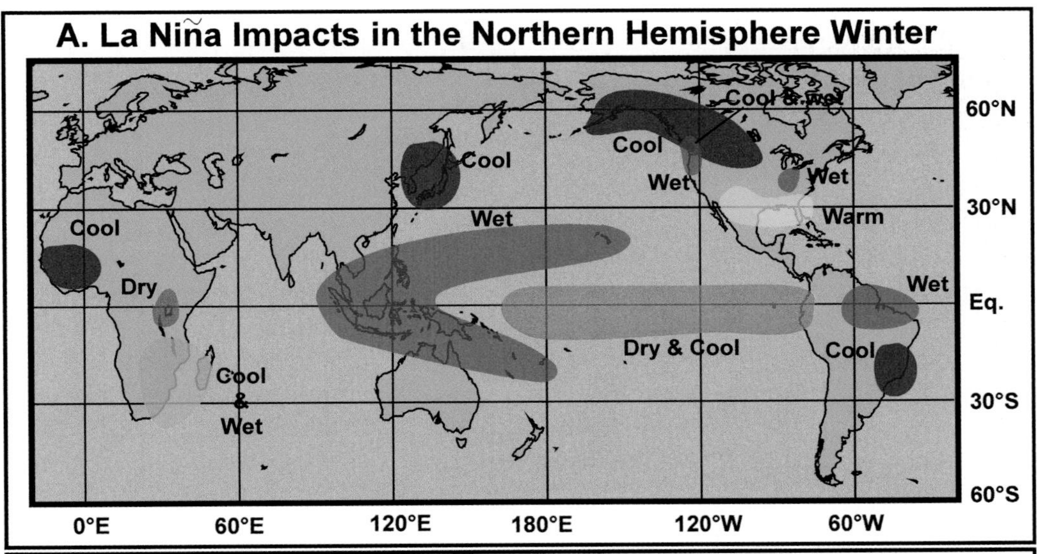

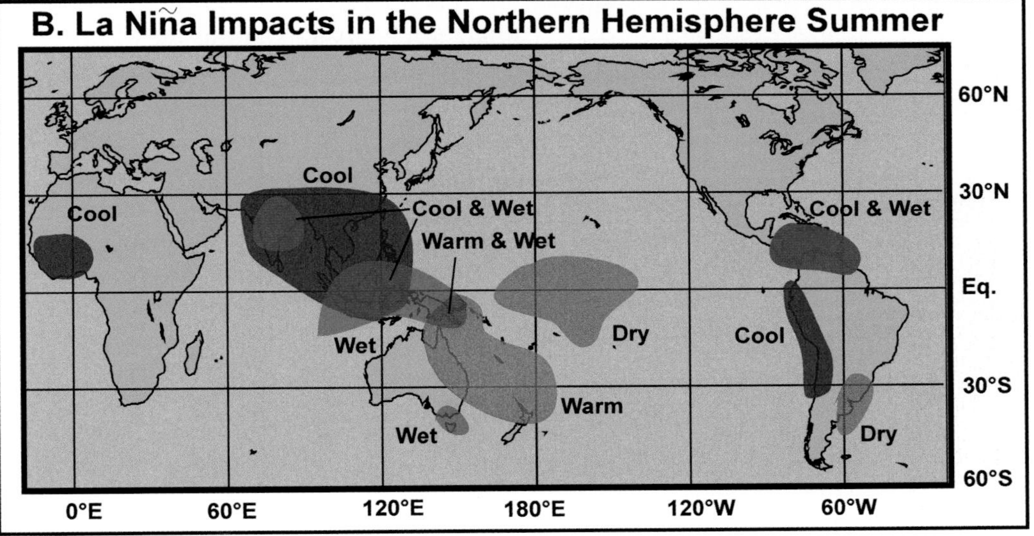

FIGURE 23.8: Global consequences of ENSO during the La Niña phase in the Northern Hemisphere (A) winter and (B) summer. Impacts during both the winter and summer are relatively widespread, especially in the Southern Hemisphere.

winds). When the index is positive, the high- and low-pressure centers are stronger than normal, indicative of the La Niña phase (strong trade winds).

Figure 23.9 shows the departure from normal pressure at Darwin (red) and Tahiti (blue) over the period 1980 to 2007. The "seesaw" effect is clearly evident, as the pressure increases at Darwin when it decreases at Tahiti, and vice versa. Events that stand out are the 1982–1983 and 1997–1998 El Niños. Figure 23.10 provides a longer perspective by showing a similar index (one based on ocean temperature anomalies as well as atmospheric pressures) from 1953 to 2007—although the index in Figure 23.10 is positive during the El Niño phase and negative during the La Niña phase, opposite to the SOI.

EL NIÑO AND GLOBAL WARMING

Figure 23.10 shows an apparent increase in occurrences of the positive phase (warm eastern Pacific) of El Niño over the past several decades. Table 23.1 also shows an increase of El Niño events, although to a lesser degree, in the decades since 1970. Scientists are unsure whether this trend toward warmer water in the eastern Pacific is related to the enhanced greenhouse effect. They do generally agree, however, that a portion of the global warming of the past thirty years is due to the generally warmer conditions in the tropical Pacific Ocean. There are also some indications that the more frequent El Niño-like conditions are affecting the large-scale atmospheric circulation in the

FOCUS 23.2

ENSO and Severe Winter Storms over North America

Is the severity of individual weather events influenced by the El Niño/Southern Oscillation phenomenon? The attention given to El Niño by the news media would lead one to believe so. In order to provide a more rigorous answer to this question, scientists at the National Oceanic and Atmospheric Administration have recently performed a series of numerical weather prediction (NWP) experiments for several severe weather events that occurred during the strong 1997–1998 El Niño.[1] In these experiments, a numerical weather prediction model (Chapter 4) was run for several sixteen-day forecast periods using two sets of prescribed ocean temperatures: (1) actual El Niño tropical Pacific Ocean temperatures, as they were observed at the time, and (2) the climatological normal tropical Pacific Ocean temperatures for the same period, together with the actual observed ocean temperatures in areas outside the tropical Pacific. The differences between the two forecasts represent the effect of the El Niño ocean temperatures in the tropical Pacific.

The investigators found that the use of the actual El Niño ocean temperatures led to a much better forecast of the severe ice storm of January 1998 in the northeastern United States and eastern Canada (see Chapter 12, Online 12.2). The El Niño ocean temperatures also produced a substantial improvement in the forecasts of the strong Pacific storms that led to floods in California in February 1998. However, forecasts of the Colorado blizzard of October 1997 were not improved by the inclusion of the El Niño ocean temperatures. The conclusion is that the severity of some, but not all, storm systems is influenced by El Niño's water temperatures, at least during the winter half of the year. Thus, it appears that El Niño can actually affect the severity of particular weather systems. This influence on individual weather events is in addition to the alteration of the broad weather patterns that make weather conditions generally warmer, colder, wetter, or drier than normal over a period of several months or seasons while an El Niño is occurring.

[1] Barsugli, J. J., J. S. Whitaker, A. F. Loughe, P. D. Sardeshmerkh, and Z. Toth. "The Effect of the 1997/98 El Niño on Individual Large-Scale Weather Events." *Bull. Amer. Meteor. Soc.* 80(1999):1399–1411.

ONLINE 23.2

A Three-Dimensional View of the El Niño/La Niña Cycle in the Ocean

Online you can view a three-dimensional rendition of the sea level elevation and sea surface temperature anomalies during a complete cycle of an El Niño and La Niña. This animation from the NOAA Climate Diagnostics Center is based on a numerical model simulation that captures the essence of the circulations occurring in the upper ocean during an ENSO cycle. Note that sea level rises when the warmer water "sloshes" to either side of the equatorial Pacific Ocean, much as water sloshes in a bathtub when pushed to one end and then left free to move on its own. Unlike the bathtub analogy, the strengthening and relaxation of the westward wind stress on the ocean surface are the keys to the forcing of the cycle of sea level in the tropical Pacific Ocean. In reality, the entire cycle over the Pacific Ocean occurs over a period of several years rather than several seconds as in the animation.

The vertical scale of the rise and fall of sea level is greatly exaggerated in the animation. Sea level in the eastern and western Pacific typically changes by only 10 to 20 cm (4 to 8 inches) during the cycle. While this change may not seem large, the huge expanse of ocean involved in the oscillation makes the volume of water comparable to, or larger than, many of the great flows of water on the Earth's surface.

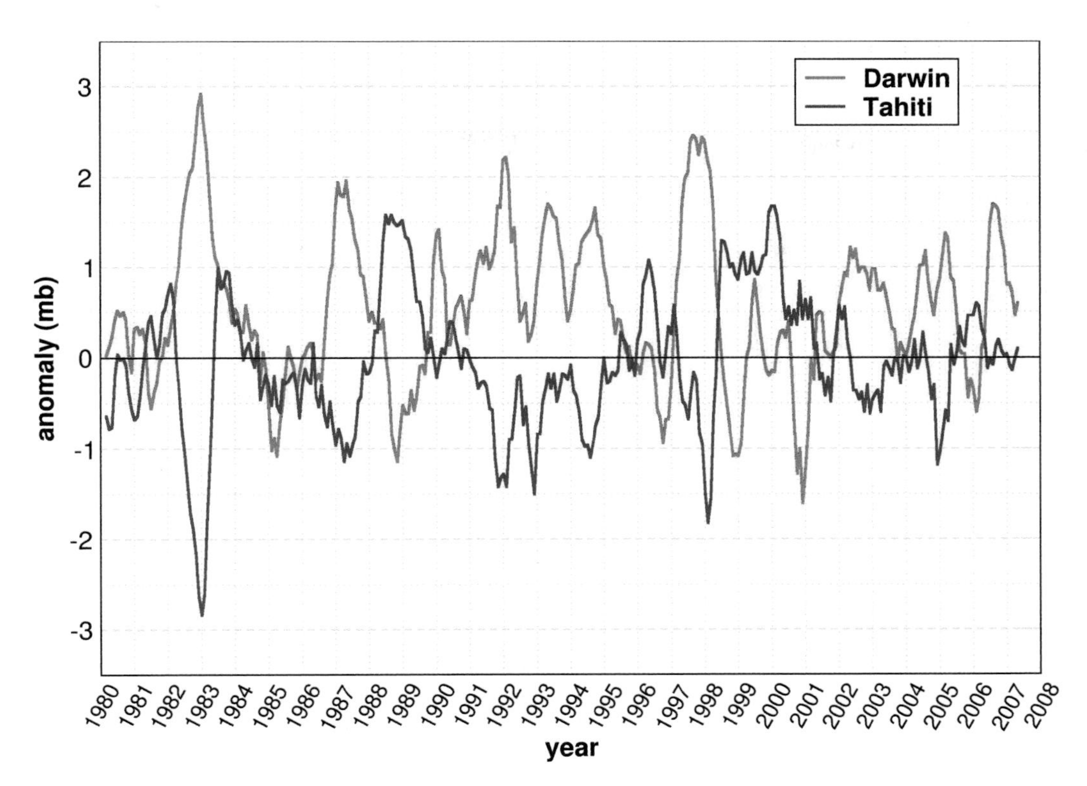

FIGURE 23.9: The pressure anomalies (departures from monthly normal) at Tahiti (blue line) and Darwin (red line). Strong ENSO events occur when pressure departures are large, for example in 1997–98 and 1982–83.

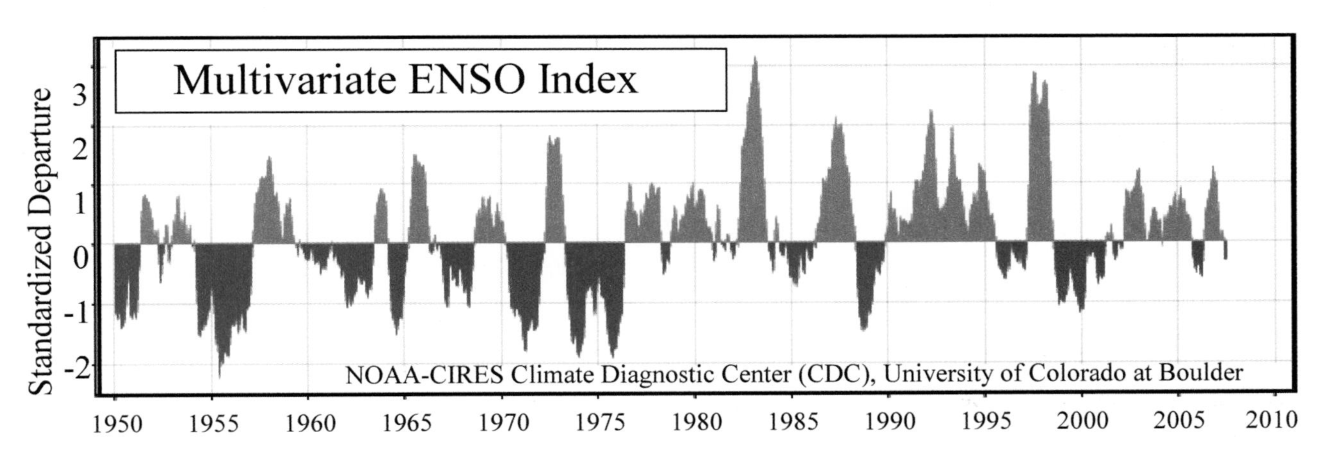

FIGURE 23.10: A history back to 1950 of the Southern Oscillation reconstructed from a combination of oceanic and atmospheric variables. This "multivariate" index has the opposite sign from the SOI based on Tahiti and Darwin pressures, so large positive (red) values denote El Niño events and large negative (blue) values denote La Niña events.

North Atlantic Ocean, thereby affecting eastern North America and western Europe.

Is the El Niño cycle likely to change in the future? The Intergovernmental Panel on Climate Change (IPCC) addressed this question in its latest (2007) assessment. The IPCC stated that future warming of the oceans may leave the tropical Pacific in an average state that more closely resembles El Niño conditions. However, the climate models used by the IPCC for 21st-century projections of climate change gave no consistent signal about changes in the amplitude or the frequency of the El Niño cycle. The frequency and amplitude of El Niño events in the present climate are generally not well simulated by today's global climate models, so it is not surprising that the models' twenty-first-century simulations do not provide a reliable indication of future changes of El Niño. Given El Niño's role in shaping seasonal variations of climate over much of the world, El Niño must be regarded as one of the "wild cards" of 21st-century climate change.

FOCUS 23.3

Monitoring the Tropical Pacific Ocean

While the Southern Oscillation Index is a measure of atmospheric pressure variations relevant to the Southern Oscillation, monitoring of the key oceanic variables is achieved by a very different strategy. A network of several dozen buoys (Figure 23C) provides measurements of the surface winds, air temperatures, and the temperature and currents in the upper several hundred feet of the ocean. This network, installed in the late 1980s and early 1990s, is known as the *TOGA-TAO array* (Tropical Ocean/Global Atmosphere—Tropical Atmosphere-Ocean). The buoys relay information via satellite to climate monitoring centers that process the information into maps and animations of various types. The same data are available to the public via the Internet at various websites.

The ability to monitor the tropical ocean in near-real-time was unheard of prior to the El Niño events of the 1980s, when scientists had to wait several months for compilations of ship reports before they knew that the equatorial Pacific had assumed its El Niño state. Although tremendous distances must be traveled to deploy and maintain these buoys, their contribution to the understanding and prediction of the ENSO phenomenon has made them a technological "success story."

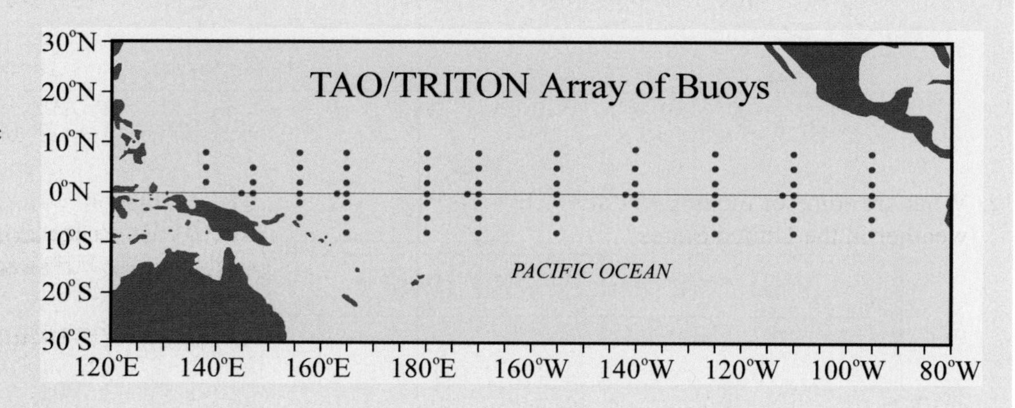

FIGURE 23C: The array of buoys used to monitor the atmosphere and upper ocean in the tropical Pacific Ocean.

CHECK YOUR UNDERSTANDING 23.3

1. What is "La Niña"? How is it related to the Southern Oscillation?
2. Sea level pressures are used from which two locations in monitoring the Southern Oscillation?
3. How does the Southern Oscillation Index change during the transition from El Niño to La Niña?

TEST YOUR UNDERSTANDING

1. What happens to the ocean water temperature along the South American coastline during El Niño?
2. Why does an El Niño generally have an adverse effect on the fishing industry of Peru?
3. How frequently, on the average, do El Niño events occur?
4. Why is the "Southern Oscillation" so named?

5. What is the Walker Cell? What is its direction of circulation over the equatorial Pacific?

6. What is the relationship between the Southern Oscillation and the Walker Cell?

7. How are the trade winds related to the Walker Cell? To the Southern Oscillation?

8. How does the Southern Oscillation affect vertical air motion in the atmosphere?

9. Why is the ocean surface normally warmer in the western tropical Pacific than in the eastern tropical Pacific?

10. What is the mechanism by which El Niño affects the subtropical jetstream?

11. Describe the changes in the jetstream affecting North America during an El Niño.

12. What are some of the impacts of El Niño on weather in the United States?

13. What happens to the water temperatures in the tropical Pacific during La Niña?

14. During a La Niña, how does precipitation in the tropical Pacific differ from normal during December, January, and February? Explain your answer in terms of the primary atmospheric circulation features.

15. How would you expect upwelling to change offshore of Peru as a La Niña develops? Explain.

16. Why are Darwin and Tahiti used to monitor the status of El Niño?

17. By about how many degrees (°C) do ocean temperatures in the eastern tropical Pacific typically change when El Niño conditions are replaced by La Niña conditions?

18. How has the frequency of El Niño events changed during the last few decades?

19. Does hurricane activity in the Atlantic and eastern Pacific Oceans generally increase or decrease during El Niño events? Identify how El Niño can affect hurricanes in each area.

20. If a severe winter storm occurs during an El Niño, is there any validity to the claim that El Niño is responsible for the severity of the storm? Explain.

21. Why are buoys useful in the prediction of El Niño?

TEST YOUR PROBLEM-SOLVING SKILLS

1. Consider the tropical Pacific Ocean as an idealized basin with an east-west extent of 15,000 km and a north-south extent of 4,000 km. Suppose that the strong trade winds of a La Niña raise the sea level by 10 cm in the western half of this basin and lower sea level by 10 cm in the eastern half. Suppose that the trade winds then weaken as an El Niño develops and that the excess water in the western half flows eastward, completely eliminating the sea level difference over a one-year period.
   (a) How many cubic kilometers of water flowed across the east-west midpoint of the tropical Pacific during the year?
   (b) How many "Mississippi River equivalents" of flow are contained in your answer to (a)? (The average discharge rate at the mouth of the Mississippi River is about 535 cubic kilometers per year.)

2. The table below lists the actual values of the Southern Oscillation Index for August, November, February, and May of each year from 1973 to 2007. You are an El Niño forecaster whose job is to make long-range predictions of the wintertime

(February) value of the Southern Oscillation Index.

	SOI Value			
Year	February	May	August	November
1973	−3.2	0.4	1.7	4.7
1974	3.2	1.4	0.8	−0.5
1975	1.0	0.8	3.1	2.1
1976	2.6	0.3	−2.2	1.1
1977	1.7	−1.4	−2.2	−2.5
1978	−5.7	2.1	0.7	−0.2
1979	1.3	0.5	−1.0	−1.0
1980	0.0	−0.5	0.0	−0.8
1981	−1.0	1.1	0.6	0.1
1982	−0.1	−1.1	−4.0	−5.1
1983	−7.6	0.7	−0.3	−0.3
1984	0.9	−0.1	0.1	0.4
1985	1.7	0.3	1.1	−0.5
1986	−2.7	−0.9	−1.6	−2.5
1987	−3.1	−2.8	−2.5	−0.2
1988	−1.4	1.3	2.2	3.0
1989	1.8	1.9	−1.3	−0.6
1990	−3.9	1.9	−1.0	−1.1
1991	−0.1	−2.4	−1.4	−1.4
1992	−2.3	0.1	0.0	−1.4
1993	−2.1	−1.0	−2.4	−0.3
1994	−0.1	−1.7	−3.0	−1.2
1995	−0.8	−1.2	−0.1	−0.1
1996	−0.2	0.2	0.7	−0.3
1997	2.6	−3.0	−3.4	−2.3
1998	−4.4	0.1	1.6	1.7
1999	1.2	0.1	0.1	1.8
2000	2.6	0.3	0.6	3.3
2001	2.4	−1.4	−1.6	1.2
2002	1.5	−2.0	−2.6	−1.0
2003	−2.0	−1.0	−0.5	−0.7
2004	1.8	1.5	−1.2	−1.4
2005	−6.7	−1.9	−1.3	−0.5
2006	−0.3	−1.4	−2.7	0.1
2007	−0.7	−0.6	0.1	1.4

(a) If you attempted to predict the sign (+ or –) of the February SOI, what would be your percentage of successful forecasts if you simply used as your predictor the sign of the previous November's SOI?

(b) How would your answer to (a) change if your predictor was the sign of the previous August's SOI? the previous May's SOI? the previous February's SOI?

(c) The approach used in (a) and (b) is based on "persistence." Is this approach better than simply tossing a coin to predict the sign of the upcoming February's SOI?
(Note: If your answer is yes, then a good measure of a forecaster's skill is the forecaster's improvement over persistence.)

3. Suppose you are a travel consultant, and many customers are relying on you for guidance about the best destination for a winter vacation. Your customers must make their choices in August from the following destinations: Orlando, Florida; Los Angeles, California; and Honolulu, Hawaii. The different destinations are priced in such a way that the customers wish to base their choices on the weather expected at the various locations during the vacation period. State which destination you would recommend to your customers if:

   (a) an El Niño is forecast for the upcoming winter. (Provide your reasoning.)
   (b) a La Niña is forecast for the upcoming winter. (Provide your reasoning.)

4. Suppose that climate changes in such a way that the tropical Pacific locks into a permanent El Niño state in which the entire eastern half of the tropical Pacific Ocean and its overlying atmosphere are warmer than their present "normal" by 5° C, while the temperatures in the western tropical Pacific and its overlying atmosphere remain at their present mean values.

   (a) By assuming the same ocean dimensions as in Problem 1, estimate how much this change would contribute to global warming, i.e., by how much would the global mean surface temperature change as a result of the warming in the tropical Pacific?
   (b) The global warming projected by today's climate models ranges from about 1.5° C to 4.5° C. What portion of these global warming estimates does your answer to (1) represent?

USE THE SEVERE AND HAZARDOUS WEATHER WEBSITE

http://severewx.atmos.uiuc.edu

1. Use the *Severe and Hazardous Weather Website* to examine the tropical Pacific Ocean sea surface temperatures and anomalies every other day over a period of four weeks.
   (a) Record the dates of the four-week period you investigated. Choose three locations (latitude and longitude) that provide a representative sample of the ocean.
   (b) Record the observed sea surface temperature and current sea surface temperature anomaly for each location.
   (c) After you collect all your data, create a chart that plots the temperatures and anomalies over the four-week period. You can put all data on one chart by using different colors or symbols to plot each location.
   (d) What trends in temperature do you observe?
   (e) How do the trends in sea surface temperature anomalies compare to the actual temperature: Is the warming or cooling occurring where the water is normally warmest? How about where the water is normally coolest?
   (f) Do your data show any indications of El Niño, La Niña or neutral conditions? Why or why not?

2. Access the ENSO Diagnostic Discussion from the *Severe and Hazardous Weather Website* "El Niño, La Niña, and Southern Oscillation" page. Read the discussion and examine the accompanying graphs. In your own words, summarize the current status of the tropical Pacific Ocean.

3. The University of Illinois' WW2010 online guides provide a module on El Niño (accessible through the *Severe and Hazardous Weather Website*). Study the discussion of upwelling.
   (a) Discuss how surface winds enhance or hinder upwelling in coastal regions.
   (b) Which wind direction(s) are most favorable for upwelling along the coast of Peru?
   (c) Which wind direction(s) favor upwelling along the California coast?
   (d) Why is upwelling a key determinant of fish populations in coastal regions?

4. Navigate to the NOAA Climate Diagnostic Center's page on ENSO Response using the *Severe and Hazardous Weather Website.* This site allows you to examine historical data to create maps of impacts of El Niño and La Niña. Create maps to answer the following questions. For each question, state what variables you used to create each map.
   (a) How has El Niño impacted your region in the past?
   (b) How do impacts in your region compare to impacts at other locations found at the same latitude?
   (c) How would you expect the temperatures to depart from normal in your region if an El Niño occurs during the next six months?
   (d) How would your answer to (c) change if a La Niña were to occur in the next six months?

TROPICAL CYCLONES

Hurricane Katrina (2005).

KEY TERMS

barometric effect
coastal flood watch/warning
cold core system
conservation of angular momentum
convergence
double eyewall
dropwindsonde
easterly trades
easterly wave
eye

- eyewall
- eyewall replacement cycle
- Hadley Cell
- heavy surf advisory
- hurricane
- Hurricane Rainband and Intensity Change Experiment (RAINEX)
- hurricane strength index
- hurricane watch/warning
- inland flooding
- Intertropical Convergence Zone (ITCZ)
- inversion
- latent heat
- Saffir-Simpson intensity scale
- scud clouds
- secondary circulations
- sensible heat
- spiral rainband
- storm surge
- tornado
- trade winds
- tropical cyclone
- tropical depression
- tropical disturbance
- tropical storm
- tropical storm watch/warning
- typhoon
- upwelling
- vortex
- warm core system
- Wind Induced Surface Heat Exchange (WISHE)

On November 12, 1970, a tropical cyclone moved northward across the Bay of Bengal toward the coast of what is now the country of Bangladesh. The storm's winds drove ocean water toward the shore, creating a surge that caused the sea to rise 15 to 20 feet (5 to 6 meters), flooding a densely populated low-lying region without warning in the early hours of the morning. The flood drowned over a half-million people, with 100,000 additional people never found. Twenty-one years later, a similar storm killed over 140,000 people in the same region. In October of 1998, Hurricane Mitch moved into the mountains of Honduras and Nicaragua in Central America causing floods and mudslides that left over 11,000 people dead and 8,000 missing. And most recently, in 2005, Hurricane Katrina devastated the Gulf Coast, destroyed much of the city of New Orleans, Louisiana, and left 1,836 people dead and over $80 billion in damage (2005 U.S. dollars) in its wake. In this chapter, we will learn why and how these storms form, why they cause so much devastation, and how forecasters use modern technology to warn those in harm's way as these great storms approach the coastlines.

Tropical cyclones are the most destructive storms on the planet. They always originate over tropical oceans, but their paths can take them into the middle latitudes and over land. The name ***hurricane*** is used to describe strong tropical cyclones that occur over the Atlantic and Eastern Pacific Oceans. The storms are called ***typhoons*** when they occur over the western Pacific and ***cyclones*** when they occur over the Indian Ocean. Since this chapter emphasizes tropical cyclones that affect North America, we will refer to tropical cyclones generically as hurricanes.

Meteorologists classify storms in the tropics based on their maximum sustained wind speeds. The maximum sustained wind in the United States is based on one-minute averages. Outside the United States, the maximum sustained wind is based on ten-minute averages. The technical definitions of tropical storm systems are given in Table 24.1. A ***tropical disturbance*** is a tropical weather system of apparently organized thunderstorms, generally about 250 to 600 km (~150–350 miles) in diameter, originating in the tropics or subtropics that maintains its identity for twenty-four hours or more. It may or may not be associated with a detectable perturbation of the wind field. When a cluster of thunderstorms has an identifiable surface pressure drop and a closed wind circulation, but its winds do not exceed 34 knots, it is classified as a ***tropical depression*** and is assigned a number, the first of the year called "tropical depression #1," the second "tropical depression #2" and so on. A storm is given a name when wind speeds increase to ***tropical storm*** strength (>34 knots). It keeps this same name through the rest of its life cycle.

There are name lists for all regions where tropical cyclones develop, including the Atlantic, eastern North Pacific, central North Pacific, Australia's oceanic areas, and the Indian Ocean. Hurricanes that affect North America form in the tropical Atlantic Ocean and adjacent water bodies such as the Gulf of Mexico and Caribbean Sea, and in the eastern Pacific off the coast of Central America. Eastern Pacific hurricanes sometimes make landfall in Mexico, with the hurricane remnants moving over the deserts of the southwest United States and Mexico. Others of these hurricanes drift westward and strike Hawaii. The

TABLE 24.1 **Classification of Tropical Weather Systems**

Organized thunderstorm cluster, no closed circulation	Tropical disturbance
Identifiable pressure drop and closed circulation	
Wind < 34 kts (39 MPH)	Tropical depression
34 kts ≤ wind < 64 kts (74 MPH)	Tropical storm
64 kts ≤ wind	Hurricane

Atlantic name list for hurricanes from 2008 through 2013 is provided in Table 24.2. The list is alphabetical, skips letters that have few names (Q, U, X, Y, Z), alternates male and female names, and uses names from the English, Spanish, and French languages. Only once, in 2005, was the assigned list of names exhausted. When this happens, additional storms are given names from the Greek alphabet: Alpha, Beta, Gamma, Delta, and so on. Names of particularly devastating hurricanes have been "retired" from the list. Recent retired names include Katrina (2005), Wilma (2005), Rita (2005), Jeanne (2004), and Ivan (2004). With the exception of retired names, the list repeats itself after six years, so that the 2008 list will appear again in 2014.

Tropical cyclones reaching hurricane intensity (> 64 knots) are classified according to the ***Saffir-Simpson intensity scale***. The Saffir-Simpson scale, shown in Table 24.3, rates hurricanes from 1 to 5, with 5 being the strongest. The scale is based on sustained wind speeds and the minimum central pressure of the storms. Through early September of the 2007 season, only thirty-one Atlantic Basin hurricanes have been rated category 5, and only a fraction of those have made landfall with that intensity. Until 2007, no more than one hurricane has made landfall as a category 5 in a single year. That record was broken when Dean (2007) struck the Yucatan Peninsula of Mexico and Felix (2007) made landfall in central America within two weeks of each other in late August and early September of 2007. The last category 5 to make landfall in the United States (southern Florida) was Andrew (1992). In 2005, a record four hurricanes, Emily, Katrina, Rita, and Wilma, reached category 5 intensity, with Wilma setting new records for the Atlantic Basin. Wilma reached its peak sustained wind speed of 160 knots on October 19. During the strengthening episode, Air Force reconnaissance observations indicated that the eye of the hurricane contracted to a diameter of 4 km; the smallest eye yet recorded, and the estimated minimum central pressure fell to 882 mb, also a new record.

Worldwide, tropical cyclones cause extreme disaster and loss of life. The worst fatalities occur in Southern Asia, particularly at the North End of the Bay of Bengal in Bangladesh and India, as discussed in the introduction to this chapter. In the United States, loss of life has been greatly reduced due to

TABLE 24.2 Atlantic Hurricane Name List

2008	2009	2010	2011	2012	2013*
Arthur	Ana	Alex	Arlene	Alberto	Andrea
Bertha	Bill	Bonnie	Bret	Beryl	Barry
Cristobal	Claudette	Colin	Cindy	Chris	Chantal
Dolly	Danny	Danielle	Don	Debby	Dean
Edouard	Erika	Earl	Emily	Ernesto	Erin
Fay	Fred	Fiona	Franklin	Florence	Felix
Gustav	Grace	Gaston	Gert	Gordon	Gabrielle
Hanna	Henri	Hermine	Harvey	Helene	Humberto
Ike	Ida	Igor	Irene	Isaac	Ingrid
Josephine	Jaoquin	Julia	Jose	Joyce	Jerry
Kyle	Kate	Karl	Katia	Kirk	Karen
Laura	Larry	Lisa	Lee	Leslie	Lorenzo
Marco	Mindy	Matthew	Maria	Michael	Melissa
Nana	Nicholas	Nicole	Nate	Nadine	Noel
Omar	Odette	Otto	Ophelia	Oscar	Olga
Paloma	Peter	Paula	Philippe	Patty	Pablo
Rene	Rose	Richard	Rina	Rafael	Rebekah
Sally	Sam	Shary	Sean	Sandy	Sebastien
Teddy	Teresa	Tomas	Tammy	Tony	Tanya
Vicky	Victor	Virginie	Vince	Valerie	Van
Wilfred	Wanda	Walter	Whitney	William	Wendy

*Hurricane names retired in 2007 will be replaced on the 2013 list.

TABLE 24.3 **Saffir-Simpson Scale**

Rating	Central Pressure (mb)	Winds mph (knots)	Storm surge ft. (m)	Typical Damage
1	≥ 980	74–95 (64–82)	4–5 (~1.5)	No real damage to building structures. Damage primarily to unanchored mobile homes, shrubbery, and trees. Some damage to poorly constructed signs. Also, some coastal road flooding and minor pier damage.
2	965–979	96–110 (83–95)	6–8 (~2.0–2.5)	Some roof, door, and window damage to buildings. Damage to shrubs and trees with some trees blown down. Considerable damage to mobile homes, poorly-constructed signs, and piers. Coastal and low-lying escape routes flood two to four hours before arrival of the hurricane center. Small craft in unprotected anchorages break moorings.
3	945–964	111–130 (96–113)	9–12 (~2.5–4.0)	Some structural damage to small residences and utility buildings. Large trees blown down. Mobile homes and poorly-constructed signs are destroyed. Low-lying escape routes are cut by rising water three to five hours before arrival of the hurricane center. Flooding near the coast destroys smaller structures, with larger structures damaged by battering from floating debris. Terrain continuously lower than 5 ft (1.5 m) above mean sea level may be flooded inland 8 miles (13 km) or more. Evacuation of low-lying residences within several blocks of the shoreline required.
4	920–944	131–155 (114–135)	13–18 (~4.0–5.5)	Complete roof structure failures on small residences. Shrubs, trees, and all signs are blown down. Complete destruction of mobile homes. Extensive damage to doors and windows. Low-lying escape routes may be cut by rising water three to five hours before arrival of the hurricane center. Major damage to lower floors of structures near the shore. Terrain lower than 10 ft (3 m) above sea level may be flooded requiring massive evacuation of residential areas as far inland as 6 miles (10 km).
5	< 920	> 155 (> 135)	> 18 (> 5.5)	Roof failure on many residences and industrial buildings. Some complete building failures with small utility buildings blown over or away. All shrubs, trees, and signs down. Complete destruction of mobile homes. Severe window and door damage. Low-lying escape routes cut by rising water three to five hours before arrival of hurricane center. Major damage to lower floors of all structures located less than 15 ft (4.5 m) above sea level and within 500 yards of the shoreline. Massive evacuation of residential areas on low ground within 5 to 10 miles (8 to 16 km) of the shoreline required.

TABLE 24.4 Ten Deadliest Atlantic Hurricanes through 2006

Hurricane	Year	Deaths[1]
1. Martinique, St. Eustatius, Barbadous, offshore	1780	>20,000
2. Mitch (Honduras, Nicaragua, Guatemala)	1998	>18,000[2]
3. Galveston, Texas	1900	>8000
4. Fifi (Honduras)	1974	>8000
5. Dominican Republic	1930	>8000
6. Flora (Haiti, Cuba)	1963	>7000
7. Guadeloupe	1776	>6000
8. Newfoundland	1775	>4000
9. Florida, Guadeloupe, Puerto Rico, Turk Islands, Martinique	1928	>4000
10. Puerto Rico, North and South Carolina	1899	>3000

[1]The number of deaths differ substantially in various data sources.
[2]Includes both confirmed deaths and missing persons.

TABLE 24.5 Ten Costliest Hurricanes in the United States (1900–2006) (Adjusted to 2005 dollars)

Hurricane	Year	Category (maximum intensity)	Category (landfall in U.S.)	Damage Costs (U.S. only)
1. Katrina (Louisiana, Gulf Coast, Florida)	2005	5	3	$81,200,000,000
2. Andrew (Florida and Louisiana)	1992	5	5	$44,900,000,000
3. Wilma (Florida)	2005	5	3	$20,600,000,000
4. Charley (Florida, S. Carolina)	2004	4	4	$15,400,000,000
5. Ivan (Alabama, Florida)	2004	5	3	$14,200,000,000
6. Hugo (South Carolina)	1989	5	4	$12,600,000,000
7. Agnes (Northeast United States)	1972	1	1	$11,600,000,000
8. Betsy (Florida and Louisiana)	1965	4	3	$11,100,000,000
9. Rita (Texas, Louisiana)	2005	5	3	$10,000,000,000
10. Frances (Florida)	2004	4	2	$ 9,100,000,000

new technological advances, such as satellites and computer forecast models, as well as excellent warnings, evacuation procedures, and public education. Yet in extremely vulnerable urban areas, these can all fail, as was clearly demonstrated when Hurricane Katrina devastated New Orleans and the Gulf Coast in 2005. In the early part of the previous century, when hurricane detection was still a problem and little was known about these storms, over 8,000 people died in Galveston, Texas—the worst number of fatalities in a hurricane in the United States. Loss of life continues to be a serious problem in poorer Caribbean nations despite the excellent warning systems. For example, Mitch in 1998 was the second most deadly hurricane in the Atlantic basin in history, leaving over 18,000 people dead or missing. The ten deadliest hurricanes in the Atlantic Basin through 2006 are listed in Table 24.4. Hurricane Katrina (2005), with 1836 deaths confirmed, currently ranks twenty-eighth on this list.

The costs inflicted by hurricanes have been increasing. This can be attributed almost entirely to the explosion in coastal population and real estate development in areas vulnerable to hurricanes. The ten costliest tropical cyclones in the United States through 2006 appear in Table 24.5. The costs in Table 24.5 are limited to damage in the United States. The total in many of these storms is higher due to damage in other countries, but information is not consistently available for these regions. Note that Agnes (1972) barely reached hurricane intensity. This storm did enormous damage in the Northeast United States due to inland flooding. Six of the ten storms on this list occurred in the 2004 and 2005 seasons, a fact that emphasizes both the large number of Atlantic tropical cyclones in these years and the increased property values as coastal populations have grown in this century and the latter part of the last century.

FOCUS 24.1

Hurricane Katrina (2005)—A Warning Bell for Our Vulnerable Coastal Cities

Hurricane Katrina exposed fundamental flaws in hurricane preparedness in the United States. Studies by the Federal Emergency Management Agency and the Army Corps of Engineers, completed long before Katrina, clearly showed that a direct hurricane strike on New Orleans would lead to destructive flooding. *Scientific American* in 2001 published an article titled *Drowning New Orleans,* which detailed the destruction that occurred four years later! The *Houston Chronicle* published a story which predicted that a severe hurricane striking New Orleans "would strand 250,000 people or more, and probably kill one of ten left behind as the city drowned under twenty feet of water. Thousands of refugees could land in Houston." In fact, the previous edition of *Severe and Hazardous Weather,* published prior to Katrina, stated: *"New Orleans, protected by levees, lies below sea level. Should hurricane-driven floods top or break the protecting levees, the city would be inundated with seawater."* In short, the risk of catastrophic flooding in New Orleans was very well known. Despite many warnings and clear government and societal understanding of the threat, nothing was done to protect the city from the inevitable hurricane strike. Will this happen again?

The threat of future destruction on the magnitude of Katrina is very real. All urban areas along the Atlantic and Caribbean coastlines and the entire Gulf of Mexico are threatened by storm surge and flooding when a major hurricane approaches, but several large urban areas, in addition to New Orleans, are particularly vulnerable. Tampa, Florida, for example, has no seawall or other storm protection and lies only feet above sea level. A major Gulf of Mexico hurricane moving eastward and making landfall just north of Tampa Bay could create a surge in the bay exceeding 5 m (16 ft) or more, putting much of the city under churning seawater. This nightmare scenario was almost realized in 2004, when Hurricane Charley approached Tampa from the south along the west coast of Florida. Charley, category 4 at landfall, made an unexpected turn inland south of Tampa, instead striking Punta Gorda and Ft. Myers, causing over $10 Billion in damage. The urban corridor on the Atlantic Coast between Miami and West Palm Beach, Florida is especially vulnerable to a major hurricane. Hurricane Andrew (1992), which struck south of this urban corridor, was the second most expensive hurricane on record. A direct hit by a hurricane of Andrew's strength on Miami or Ft. Lauderdale will dwarf the human and financial costs incurred in that storm.

FIGURE 24A: Flooded neighborhood in New Orleans following Hurricane Katrina.

Long Island, the large island east of New York City, currently has a population of 7.5 million. If the eye of a major hurricane would make landfall just west of the island, Long Island would experience major coastal flooding and destructive winds. The shoreline is very smooth and many parts of the island are only a few feet above sea level. Evacuation is a serious problem. Few highways cross to the mainland and traffic jams, which are common when there is not a hurricane, are a certainty. One of the greatest fears of emergency managers is that thousands of people trying to evacuate the island will be trapped in their cars in traffic jams as

(continued)

(continued)

a hurricane strikes the coast. Manhattan Island in New York City is also vulnerable to coastal flooding associated with hurricanes. The financial and human costs of a catastrophic hurricane striking New York City are hard to imagine. Despite our understanding of the threats associated with hurricanes, preparation for the inevitable strike on an urban area continues to be a low priority for government and society.

FOCUS 24.2

The Record Setting Hurricane Season of 2005

The 2005 hurricane season was the most active season in recorded history and was remarkable for the number of records that were broken. Twenty-seven named storms formed during 2005, the most named storms in a single season, breaking the record 21 set in 1933. Fourteen hurricanes formed during 2005, the most hurricanes in a single season, breaking the old record of 12 set in 1969. Seven major (category 3 or higher) hurricanes formed during 2005, tying the record set in 1950. Four category 5 hurricanes formed during 2005 (Emily, Katrina, Rita, and Wilma), the most ever recorded in a single season, breaking the old record of two set in 1960 and 1961. Seven named storms made United States landfall during 2005 (Arlene, Cindy, Dennis, Katrina, Rita, Tammy, and Wilma), a tie for second place behind the eight that occurred in 1916 and 2004.

The 2005 season was the most destructive for United States landfalling storms, largely due to Hurricane Katrina. Damage estimates for all hurricanes for 2005 exceeded $100 billion dollars. Five named storms (Cindy, Dennis, Emily, Franklin, and Gert), and two major hurricanes (Dennis and Emily) formed in July, the most on record in both categories for that month. Five named storms (Harvey, Irene, Jose, Katrina, and Lee) formed in August, surpassing the number of August storms in all years but 1990, 1995 and 2004. Five hurricanes (Maria, Nate, Ophelia, Philippe, and Rita) formed in September, tying 1955, 1969, 1981, 1998, and 2000 for the most hurricanes to form during that month. Six named storms (Stan, Tammy, Vince, Wilma, Alpha, and Beta) formed in October, tying 1950 for the most named storms forming during the month of October. Four hurricanes (Stan, Vince, Wilma, and Beta) formed in October, second only to 1950. Two intense hurricanes (Wilma and Beta) formed in October, tying 1950, 1961, 1964, and 1995 for the most intense hurricanes to form during that month. Finally, Epsilon was only the sixth hurricane to ever form in the month of December.

Individual hurricanes also set records or near records. Dennis became the most intense hurricane on record before August with a central pressure of 930 mb. Emily immediately eclipsed the record set by Dennis for lowest pressure in a pre-August hurricane with 929 mb. Emily was also the earliest category 5 storm on record. Katrina's central pressure dropped to 902 mb, at the time, the fourth lowest pressure ever measured in the Atlantic basin. Katrina became the most destructive storm on record with an estimated $81.2 billion dollars in damage, breaking the old record of approximately $44.9 billion dollars (normalized to 2005 dollars) set by Hurricane Andrew (1992). Rita's central pressure dropped to 897 mb, at the time, the third lowest pressure measured in the Atlantic basin.

Vince was the furthest north and east that a storm has ever developed in the Atlantic basin, and was the first tropical cyclone in recorded history to strike the Iberian Peninsula in Europe. Wilma's central pressure dropped to 882 mb, the lowest pressure ever measured in the Atlantic basin, eclipsing the record of 888 mb set by Hurricane Gilbert (1988). Zeta became the twenty-seventh named storm of the 2005 season, breaking the old record of twenty-one named storms set in 1933 and Epsilon became the fourteenth hurricane of the 2005 season, breaking the old record of twelve

(continued)

(continued)

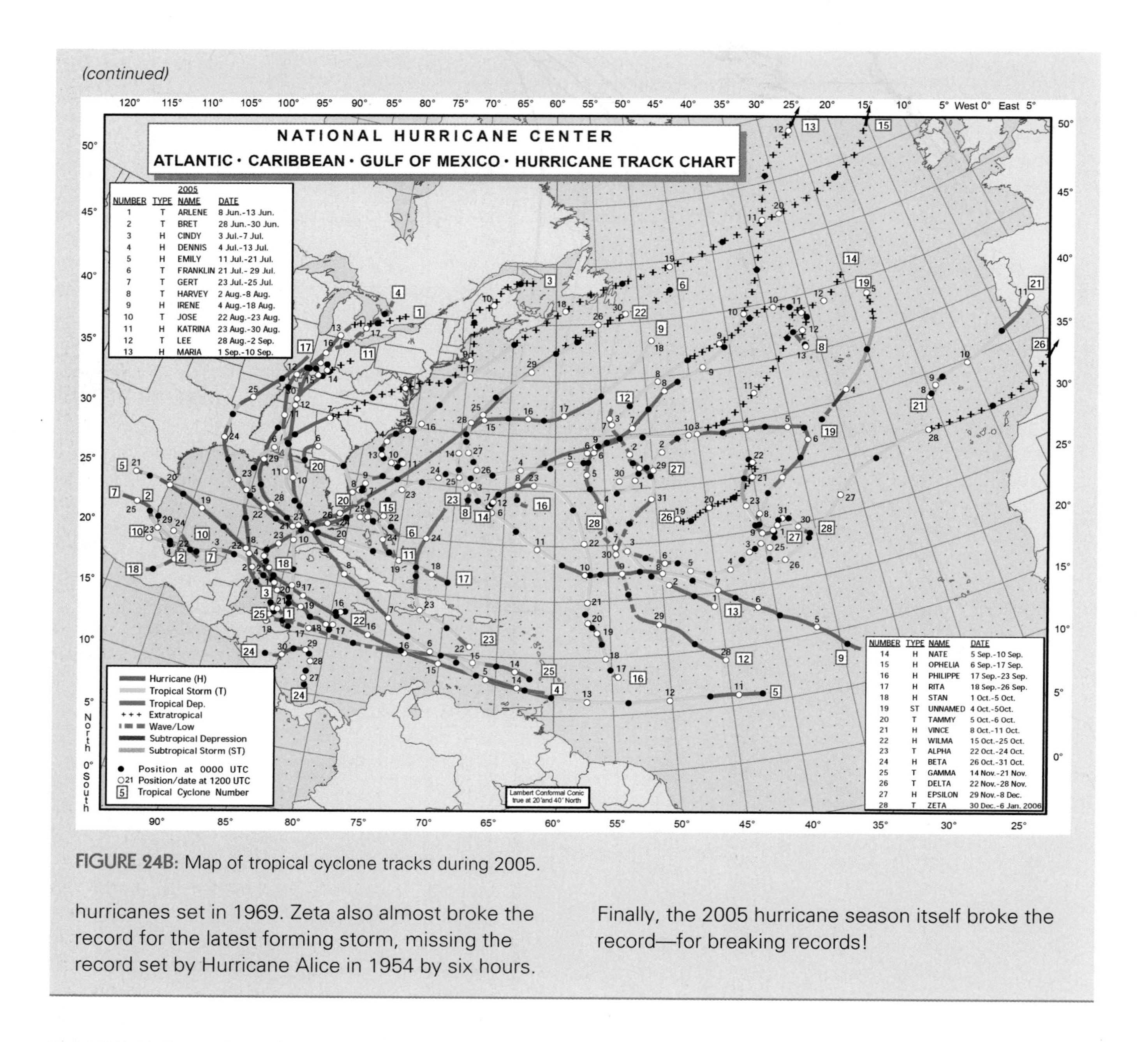

FIGURE 24B: Map of tropical cyclone tracks during 2005.

hurricanes set in 1969. Zeta also almost broke the record for the latest forming storm, missing the record set by Hurricane Alice in 1954 by six hours. Finally, the 2005 hurricane season itself broke the record—for breaking records!

WORLDWIDE HURRICANE OCCURRENCES

Figure 24.1 shows a map of tropical cyclone origins and tracks for a twenty-year period from 1985 through 2005. There are six important points related to this figure. (1) About 66 percent of all tropical cyclones occur in the Northern Hemisphere. All Atlantic and eastern Pacific, and most western Pacific and Indian Ocean tropical cyclones occur in the Northern Hemisphere summer and fall. Southern Hemisphere tropical cyclones develop in the Southern Hemisphere warm season. This prevalence of tropical cyclones in the Northern Hemisphere is because ocean temperatures in the Northern Hemisphere are generally warm over broader areas than in the Southern Hemisphere. (2) Hurricanes never originate within about 5 degrees of the equator, or cross the equator. The direction of air circulation about the low-pressure center of the hurricane is a result of both the pressure gradient and Coriolis forces. At the equator, the Coriolis force is zero, so the circulation of a hurricane could not persist. In fact, a hurricane would have to reverse direction to cross hemispheres. (3) Hurricanes

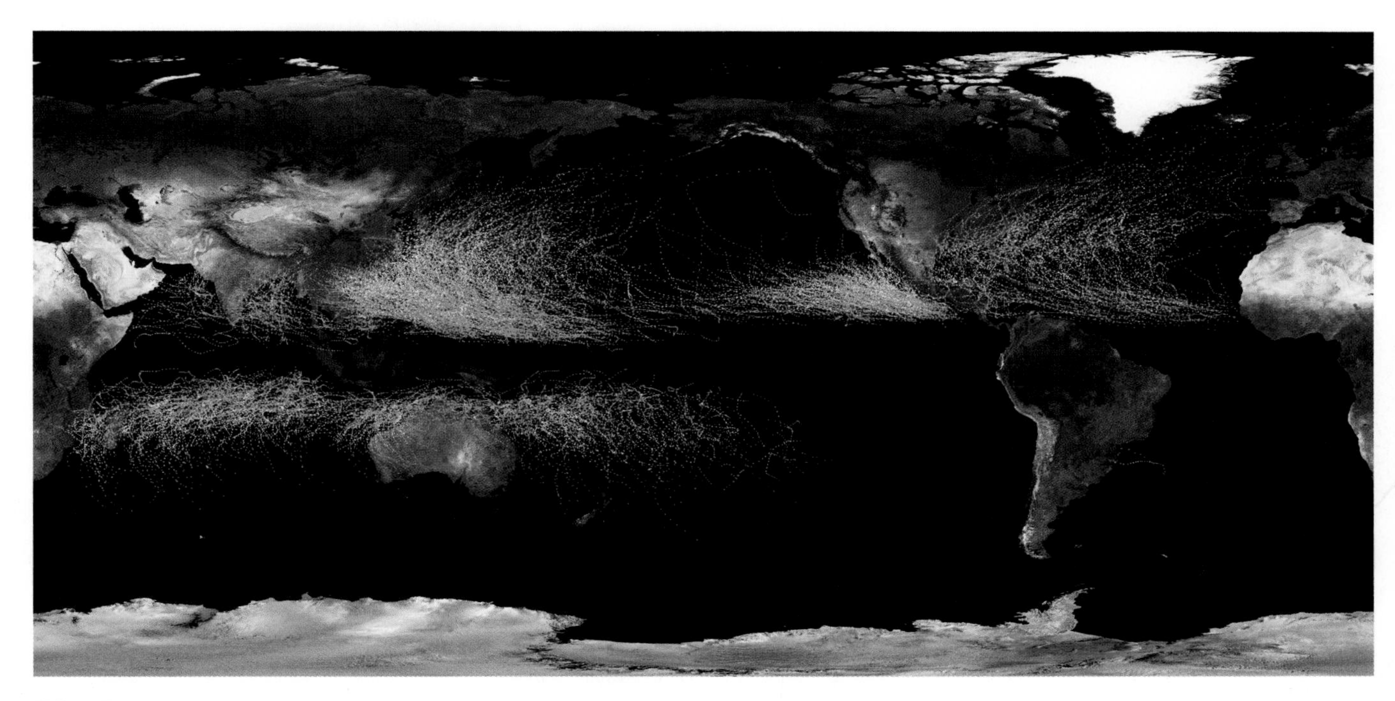

FIGURE 24.1: Tracks of all tropical cyclones that formed worldwide from 1985 to 2005.

rarely originate north (or south, in the Southern Hemisphere) of about 25 degrees latitude. The ocean water is generally too cold to provide the energy required for hurricane development. However, once formed, tropical cyclones often move to much higher latitudes before losing their tropical structure. (4) About 80 tropical storms develop annually around the globe. Between 50 to 70 percent of these develop into hurricanes. (5) The western Pacific in the Northern Hemisphere produces the largest number of tropical cyclones. The high frequency of tropical cyclones over the western Pacific is related to the high sea-surface temperatures in this region. 6) Hurricanes form over all tropical oceans except the south Atlantic and southeast Pacific. The surface water temperature is either too cold, or other atmospheric conditions are unfavorable to support hurricane development in the southeast Pacific and south Atlantic, with one very notable exception.

During the week of 22 to 28 March 2004, an extremely rare meteorological event took place in the southern Atlantic Ocean. A tropical cyclone developed from a cutoff extratropical system off the coast of Brazil. The system grew into a hurricane, with a clear eye and eyewall, as it moved toward the coast of Brazil (Figure 24.2). An unknown number of fishermen were lost at sea during the storm. Based on news reports, 35,000 homes were damaged or destroyed upon landfall, with damage to Santa Catarina estimated at $335 million. This storm, now referred to as Cyclone Catarina, became the first southern Atlantic hurricane in recorded history.

HURRICANE TRACKS

In summer in both hemispheres there are strong high-pressure centers located over the Atlantic and Pacific Oceans north of 30 degrees. Surface winds are normally easterly over the tropical oceans on the equator side of these high-pressure centers. This easterly wind flow is so persistent that early sailing ships depended on these winds to carry them westbound across the Pacific and Atlantic Oceans. Because of their early association with trade routes, the winds are called the ***trade winds*** or the ***easterly trades***. Hurricanes almost always form within the belt of trade winds, and normally move westward with these winds during their early lifetime. The Coriolis force also influences a hurricane's movement. A hurricane vortex is sufficiently large that the strength of the Coriolis force varies across the vortex from its equatorward to poleward side. This effect would cause a hurricane to drift westward and poleward, even if there was no background flow such as the trade winds. Eventually, hurricanes cross from the belt of easterly winds in the tropics into the belt of westerly winds in the mid-latitudes.

FIGURE 24.2: Cyclone Catarina off the coast of Brazil in the South Atlantic on 26 March 2004.

Westerly winds steer a hurricane back toward the east around the north side of the high-pressure center.

Although many hurricane tracks will conform to the "normal" pattern described above, many take highly variable tracks that are anything but predictable. Panels A and B of Figure 24.3 show the paths of six different hurricanes. The three on Panel A had ordinary tracks, east to west, while drifting northward in tropical latitudes, and then west to east in middle latitudes. In contrast, the hurricane tracks shown on panel B are quite unusual.

Hurricane Kate (2003) began very far south in the trades and moved north-northwest, then turned northeastward toward Europe. It then abruptly turned westward after reaching the latitude of Georgia, and then finally turned northeast again after drifting westward. Olga (2001) formed at an unusually high latitude

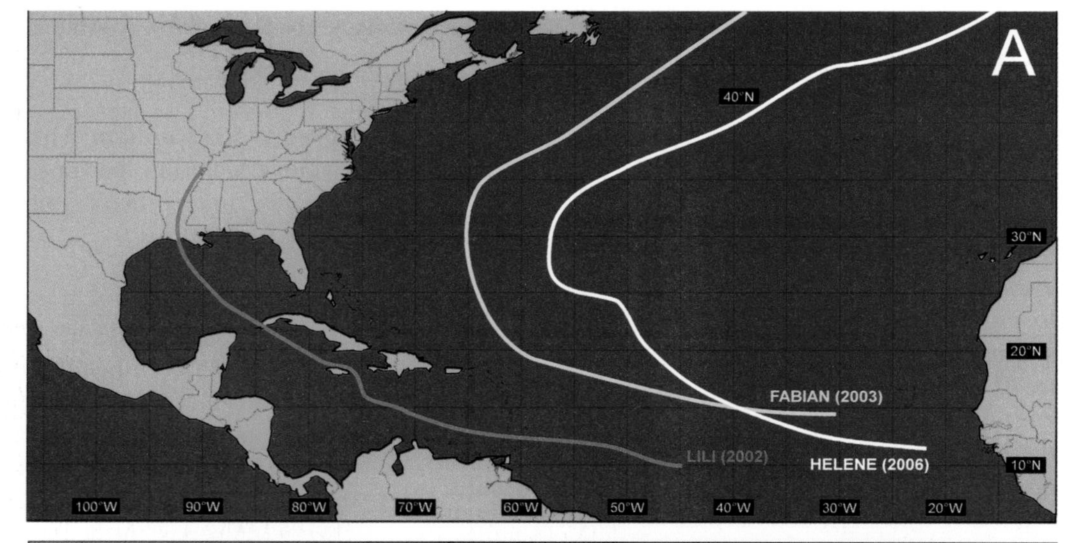

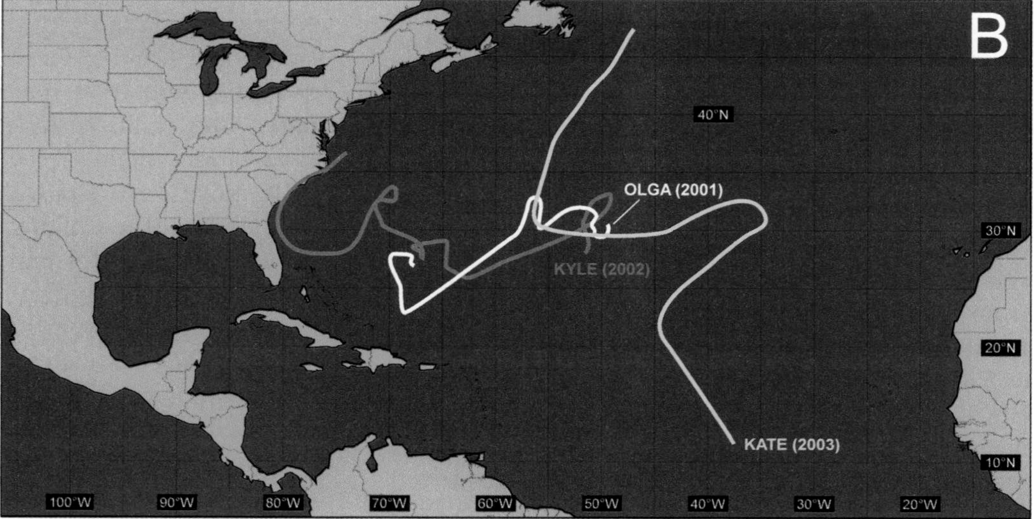

FIGURE 24.3: Typical (A) and unusual (B) tracks of some Atlantic hurricanes. The name and year of each hurricane are placed at the start of the track.

(30° N), and meandered over the Atlantic while drifting westward. The storm then took an abrupt turn to the southwest, approached the Bahamas, and then meandered again east of the Bahamas before dissipating. Perhaps the strangest track was Kyle (2002), which formed in the same area as Olga, drifted westward for weeks in the Atlantic while looping three times, then southwest to near the Florida coast, and finally northeast along the coastline to North Carolina. In general, hurricanes respond to the large-scale flows in which they are embedded, "drifting" within the background flows while interacting with them. Erratic tracks are most common when the background flow is weak.

CHECK YOUR UNDERSTANDING 24.1

1. By what weather variables are hurricanes and tropical storms ranked or categorized?
2. What is the Saffir-Simpson scale?
3. Describe the general trends in cost and fatalities associated with hurricanes.
4. Where do tropical cyclones form? Where do they not form?

HURRICANE STRUCTURE

In this section, we will examine the physical structure of a hurricane—its wind fields, air trajectories, pressure distribution, temperature distribution, cloud structure, rainband structure, the eyewall, and the eye. Figure 24.4 shows a satellite view of Hurricane Katrina (2005) while it was over the Gulf of Mexico. A hurricane typically appears as a large circular storm with spiral cloud bands extending away from its center. At the center of the hurricane is a prominent, nearly cloud-free area called the ***eye***.

The eye of a hurricane is one of the most spectacular locations in the Earth's atmosphere. The eye may be clear all the way to the ocean surface, but often has low-level clouds (referred to as ***scud clouds***) near its base. In strong hurricanes, such as Hurricane Isabel (2003) when it was category 5, these clouds are sometimes organized within small-scale vortices that rotate within the eye itself. Figure 24.5, for example, shows the "pinwheel" structure of the eye of Isabel (2003) at 1812 UTC on 13 September 2003.

FIGURE 24.4: Visible satellite image of Hurricane Katrina (2005). Important features of the hurricane are noted.

FIGURE 24.5: Visible satellite image showing four vortices within the eye of Hurricane Isabel at 1812 UTC on 13 September 2003.

ONLINE 24.1

The Eye of Isabel (2003)

The eye of Hurricane Isabel during the time the storm achieved category 5 intensity had four clearly identifiable small vortices rotating with it, giving the cloud pattern a pinwheel structure. The reason why these vortices form is still not well understood. Numerical models sometimes show transient features similar to these vortices. During Isabel, a NOAA research aircraft actually flew though these vortices and measured their structure, so we should be learning more about the vortices as research based on observations and modeling progresses. You can view these vortices in visible satellite imagery in the spectacular sequence of images of the eye of Isabel online.

ONLINE 24.2

Hurricane Katrina (2005)

Hurricane Katrina was the costliest and one of the deadliest hurricanes in the United States. The storm's evolution was captured by an array of satellites and radars as it made landfall in Florida, moved across the Gulf of Mexico, and made a second landfall in Louisiana. Online animations of radar reflectivity show Katrina from the perspective of the Miami and Key West, Florida radars as it moved across the state as a category 1 hurricane. These animations show the classic radar features of a hurricane—the eye, eyewall, and spiral rainbands. Animations of satellite data show the evolution of the hurricane from its formation east of Florida to its dissipation over the central United States, and its massive size as it became a category 5 storm over the Gulf of Mexico. Katrina's final landfall near New Orleans was documented by the New Orleans radar, until the radar itself succumbed to the storm's winds. Katrina's history and impacts are detailed online with the animations.

When viewed with radar, the hurricane's precipitation structure becomes more apparent. Surrounding the eye is a ring of deep convective clouds called the ***eyewall*** (see Figure 24.6), which extends from near

FIGURE 24.6: The eyewall of Hurricane Katrina at the time of maximum intensity. The clear eye, funnel shape, and sharp rotation are all evident from the cloud forms in the photograph that was taken from inside the eye.

the ocean surface to an altitude of about 15 km (9 miles). A ring of very heavy precipitation occurs within the eyewall, as can be seen from Figure 24.7, a radar image of the eyewall of Hurricane Andrew (1992) just before landfall in Florida. Air in the convective clouds that form the eyewall rises upward and radially outward, so that the convective clouds making up the eyewall form a shape like a funnel, with the storms violently rotating around the eye. Outside the eyewall, precipitation can be quite light, although the winds are still moderately strong.

Heavier rain also occurs in ***spiral rainbands*** that extend outward from the eyewall. These bands are evident in the satellite image in Figure 24.4 and in the radar data from Hurricane Georges (1998) as it crossed Key West (Figure 24.8). Hurricanes are not necessarily as symmetric as implied by the eyewall of Andrew in Figure 24.7. In Georges, for example, almost all the rain was on the east side of the storm.

An oversimplified description of the airflow in a hurricane would be that the air spirals in at the surface,

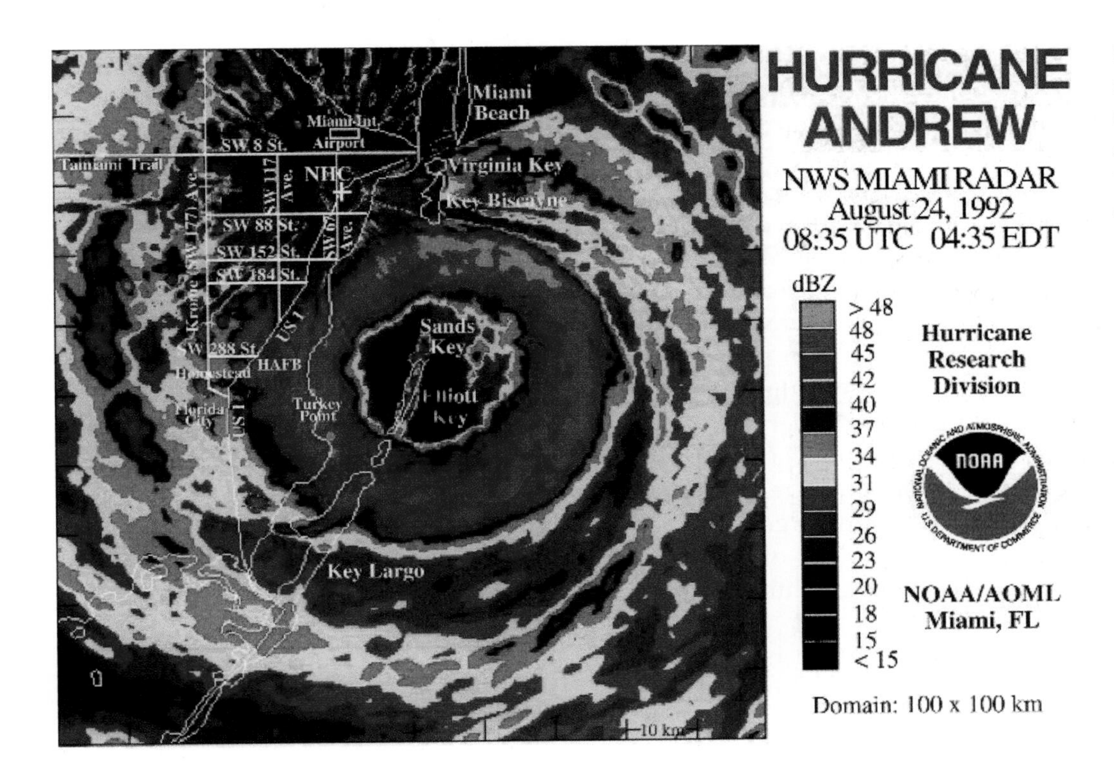

FIGURE 24.7: Radar reflectivity image of the eyewall of Hurricane Andrew as it made landfall on the coast of Florida as a category 5 storm.

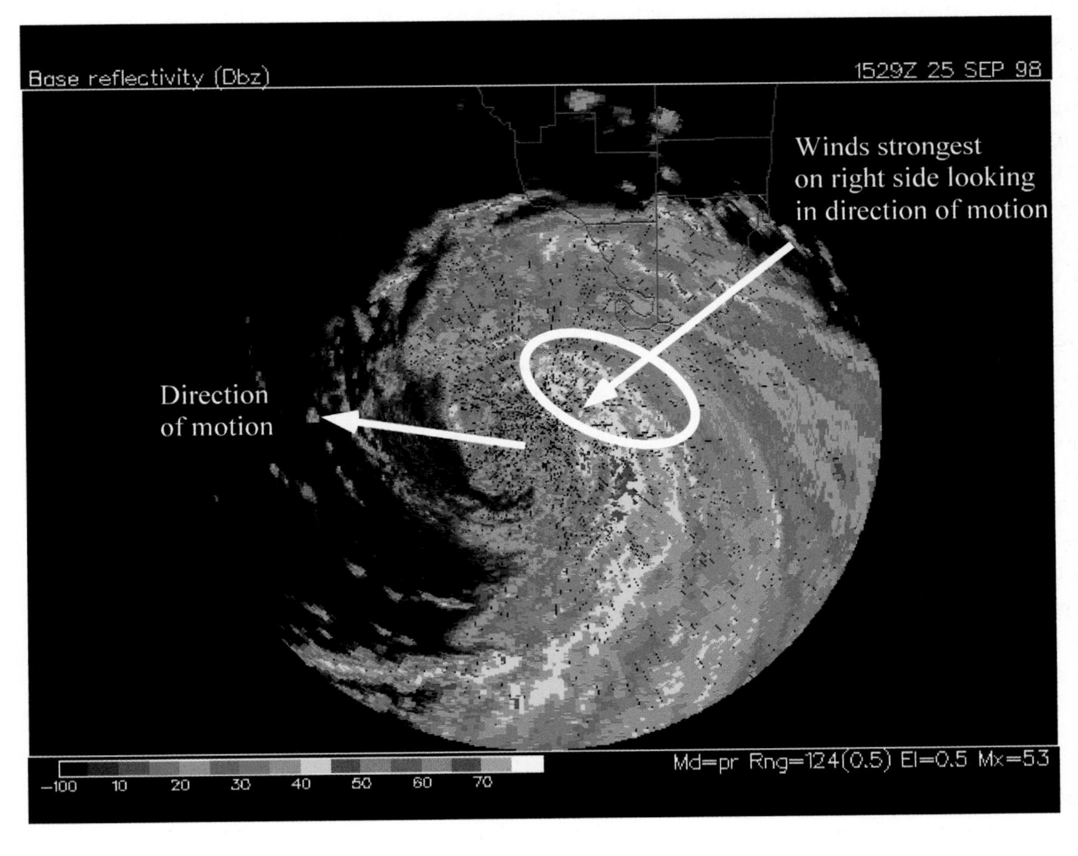

FIGURE 24.8: Hurricane Georges (1998) as viewed by Key West, Florida, radar. Note the asymmetry of the storm, and spiral rainbands extending outward from the center of the hurricane. The oval denotes the location of the strongest winds in Georges.

rises from the surface to the tropopause in the eyewall, and then spirals outward from the eyewall at the tropopause level. This description actually captures much of the process—as is demonstrated by the three-dimensional trajectories in Figure 24.9. Hurricanes also have ***secondary circulations*** associated with the eye, the spiral rainbands, and the regions between the spiral rainbands and the eyewall. These circulations, and other important features of hurricane structure, are shown in Figure 24.10, a cross section through a hurricane. The top portion of the figure shows the location of the eye, eyewall, spiral rainbands, and vertical air motion within a hurricane. Below this diagram are five diagrams showing how the pressure, wind speed, rainfall, temperature, and storm surge vary across a hurricane. The diagram to the right of the top panel shows how the wind speed varies with height within the eyewall.

Air in the upper and middle troposphere is descending within the eye of a hurricane. Air descends into the eye from the stratosphere at the top of the eye (note the descent of the tropopause on Figure 24.10). Most of the air in the eye descends very slowly, taking days to traverse the eye's depth. However, air within the eye that is immediately adjacent to the eyewall descends more rapidly. In this region, mixing of dry air in the eye with cloudy air in the eyewall leads to evaporative cooling, which makes the air denser and prone to sinking. Air sinking in the eye warms at the dry adiabatic lapse rate, and is much warmer than in any other region of the storm. A reasonable question to ask is 'Why, if the air is so warm in the eye, does it sink? Shouldn't this air be buoyant?' Indeed the air *is* buoyant and would rise immediately except that another force is acting to prevent the air from rising. The extreme low pressure at the base of the eye of a hurricane significantly reduces the magnitude of the upward directed pressure gradient force. This causes the downward force of gravity to exceed the upward pressure gradient force, so much so that the downward acceleration of air by gravity slightly exceeds the upward acceleration of air due to the air's buoyancy. The net effect: air gradually descends in the eye, despite its warmth.

Air within the upper part of the eye does not descend completely to the ocean surface, but rather to about the 1.5 to 2.0 km (0.8 to 1.2 mile) level. The base of the descending air is marked by an ***inversion***. Within the layer between the inversion and the ocean surface, weak convection typically occurs, producing the scud clouds visible at the base of the eye in some hurricanes.

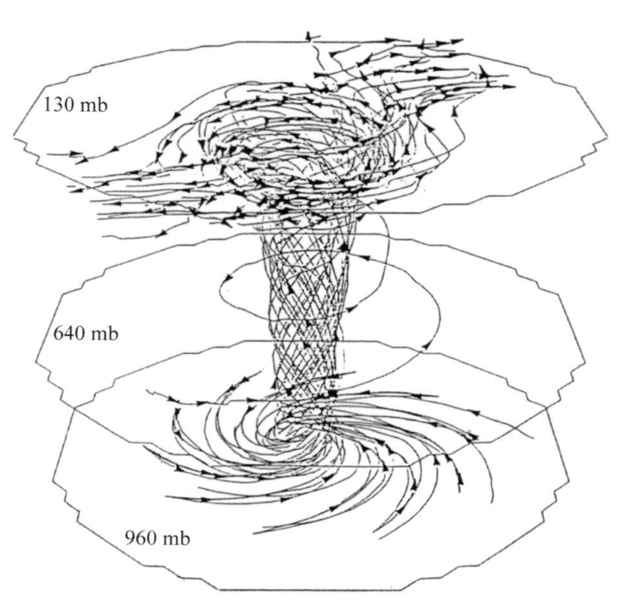

FIGURE 24.9: Trajectories of air parcels as they move through a numerically modeled hurricane. A full trajectory covers a period of eight days and each arrow head along a trajectory denotes a nine-hour interval.

In the eyewall air ascends from the surface to the tropopause. Most air enters the base of the eyewall after spiraling inward over the ocean surface toward the eye. A small amount of air also enters the eyewall from the eye side (Figure 24.10). When air rises in the eyewall, it eventually encounters the tropopause and can no longer rise farther. Most of the air exhausts outward away from the center, creating the large shield of clouds visible on satellite images (e.g. Chapter cover figure and Figures 24.2, 24.4 and 24.5). Some air exhausts inward, cools as the cloud evaporates, and then descends along edge of the eyewall. The strongest winds in a hurricane are found in the eyewall about 0.5 to 1.0 km (0.3 to 0.6 miles) above the ocean surface (Figure 24.10 top right diagram). Friction with the ocean surface keeps the strongest winds from extending to the surface.

Other secondary circulations in hurricanes involve the spiral rainbands. Some of the air converging toward the center of the hurricane rises along preferred regions of ***convergence*** in the flow. These convergence zones spiral out from the center and are the locations of the spiral rainbands. Except for the spiral rainbands, the air outside of the eyewall slowly descends to compensate for the ascent in the eyewall and spiral rainbands.

Moving toward the center of a hurricane from the clear region around its perimeter, the pressure drops

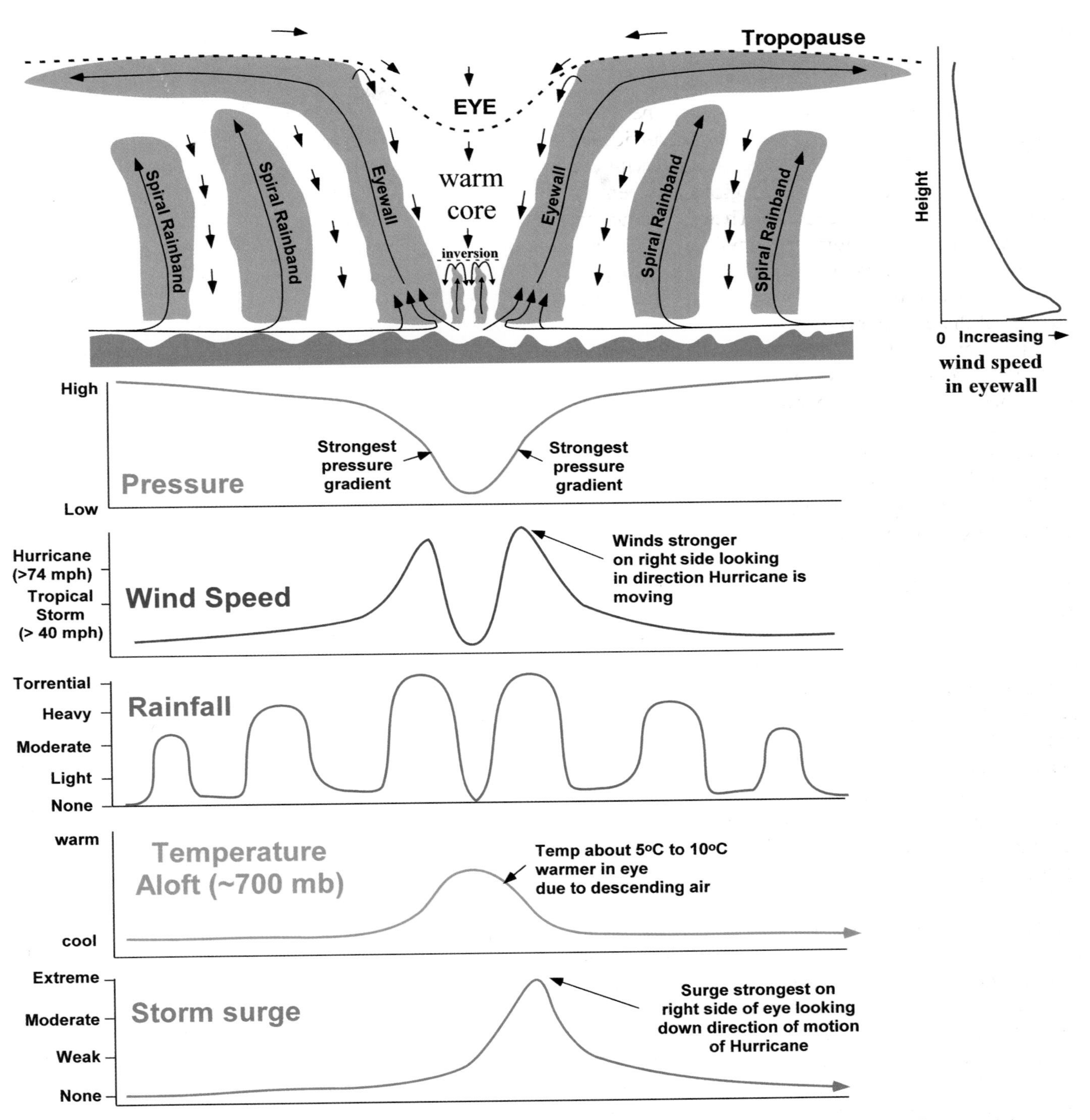

FIGURE 24.10: Top: A cross section through a hurricane showing the clouds (gray) and circulation (arrows). The remaining panels, from top to bottom, show the distribution of surface pressure, wind speed, rainfall, temperature at 700 mb, and the sea-level height (storm surge) across a hurricane. The cross section is drawn such that the hurricane is moving into the paper. The winds are strongest on the right side looking in the direction of hurricane motion. The rightmost top panel shows the variation of wind speed with altitude at the radius of maximum surface winds in the eyewall.

slowly at first, but then drops very rapidly from the outer radius of the eyewall to the inner radius. This strong pressure gradient within the eyewall creates the powerful winds of the hurricane. For example, compare Andrew's extremely strong pressure gradient (see Figure 7.3) with its eyewall during landfall (Figure 24.7). Although the most violent winds are confined to the eyewall, strong winds can exist well over 100 km

FOCUS 24.3

RAINEX, Eyewall Replacement Cycles, and the Double Eyewall

Intense hurricanes such as Katrina (2005) and Rita (2005) often develop a *double eyewall.* The outer eyewall forms as outer rainbands contract into a closed ring about the inner eyewall. A wind maximum appears within each eyewall. Both the outer and inner eyewalls contract with time. Hurricanes typically reach peak intensity during the inner eyewall contraction period. The inner eyewall eventually weakens and dissipates, as the outer eyewall continues to contract and intensify. Once the inner eyewall disappears, the intensity of a hurricane will often decrease. This *eyewall replacement cycle* may occur more than once during a storm's lifetime.

The dynamics of hurricane intensification, particularly the eyewall replacement cycle, remain poorly understood. In 2005, hurricane researchers undertook an ambitious project to close this gap in our knowledge about hurricanes. The project, called the *Hurricane Rainband and Intensity Change Experiment, (RAINEX),* was carried out in the heart of the record 2005 hurricane season, and involved three large aircraft (two NOAA P-3 hurricane hunter aircraft and the Naval Research Laboratory P-3). Scientists flew these aircraft, each equipped with dropwindsondes and dual-Doppler radar capability, across the eye and eyewalls of hurricanes to measure winds and reflectivity, as well as the storms' thermal, wind and moisture structure. One of their ambitions was to document the eyewall replacement cycle as it happens. On the afternoon of 22 September, the scientists hit the meteorological jackpot over the Gulf of Mexico—they flew the aircraft through the eye and double eyewall of Hurricane Rita while category 5 and while undergoing an eyewall replacement cycle (Figure 24C). They documented a clear double peak in the wind profile across the hurricane (Figure 24D), and obtained detailed data about the wind structure throughout the eyewalls. With the help of complementary hurricane numerical modeling studies they are pursuing, these unprecedented data will shed new light on how hurricanes reach category 5 intensities.

Over the course of RAINEX, scientists flew thirteen multi-aircraft missions through three hurricanes, Katrina, Ophelia, and Rita, sampling both Katrina and Rita while category 5. In addition to their many records, these storms are now the best documented in the history of hurricane research. Teams of scientists are now analyzing the RAINEX datasets to develop a better understanding of the eyewall replacement cycle and the process of hurricane intensification.

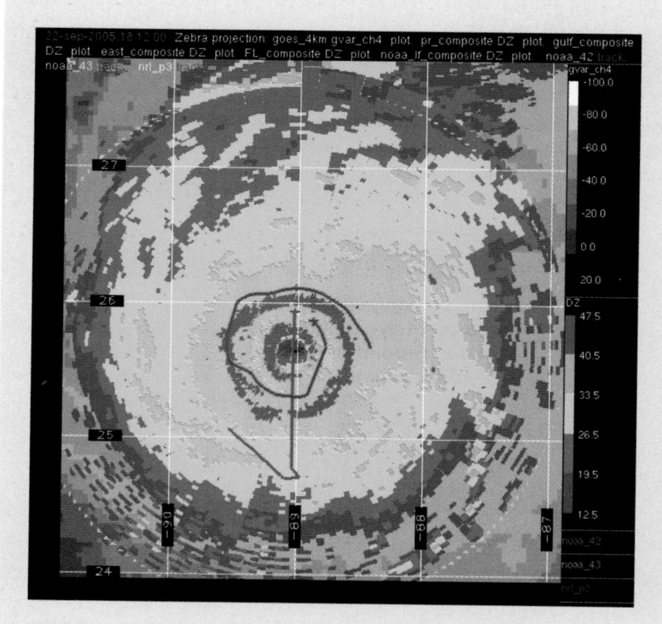

FIGURE 24C: Overlay of radar (color) and satellite (gray) data collected during RAINEX during aircraft penetrations of the double eyewall of Hurricane Rita (2005) in its category 5 stage. The track of the Navy Research Laboratory aircraft is blue, and the NOAA aircraft is red.

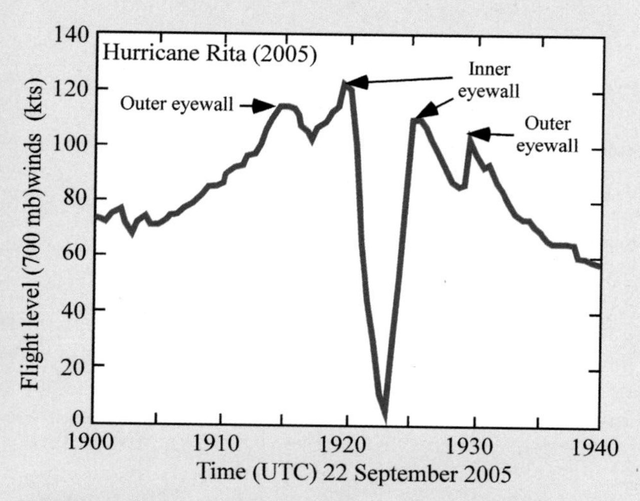

FIGURE 24D: Flight-level winds measured aboard a NOAA aircraft on 22 September 2005 while crossing concentric eyewalls in Hurricane Rita.

(~60 miles) from the center of the storm. Atmospheric pressure is lowest in the eye—the center of the hurricane. Inside the eye, the winds are nearly calm.

Note from Figure 24.10 that the wind speeds are higher on the right side of the eye, when looking in the direction the storm is moving. For example, if a storm were moving northward, the "right side" would be the east side. To understand why this is so, examine Figure 24.11. Winds in a hurricane circulate counterclockwise in the Northern Hemisphere. We can think of a hurricane as having "rotational velocity," the speed it is rotating, and a "translational velocity," the forward speed of the hurricane itself.

Let us assume that a hurricane is moving due north at 20 knots and rotating at 100 knots. Taking into account these two velocity components, the wind to the right of the eye at point B will have a translational component directed north and a rotational component also directed north. We therefore add 100 + 20 to obtain 120 knots. However, on the left side of the eye at point A, winds will have a northward translational velocity but a southward rotational velocity. We must subtract 100 – 20 = 80 knots to determine the wind speed on the left side. *Winds are always strongest to the right of the storm's direction of motion in the Northern Hemisphere.* Figure 24.8 outlines where the strongest winds were in Hurricane Georges as it crossed the Florida Keys.

Returning to Figure 24.10, we see that rainfall intensity varies substantially across a hurricane. No rain falls in the eye. The most violent rainfall occurs under the eyewall. Rain is light, if it is raining at all, about 80 to 120 km (50 to 75 miles) outside the eyewall, except in the spiral rainbands. Moderate to heavy rainfall can fall within these rainbands, with heavier rain typically occurring in the innermost spiral bands and lighter rain in the outer bands.

Descending air within the eye contributes to sharp warming of the core of a hurricane. Temperatures are significantly higher in the eye, with aircraft measurements of temperature showing increases as large as 11° C (20° F). Hurricanes are often referred to as ***warm core systems***, in contrast to extratropical cyclones, which are ***cold core systems***, having cold air aloft above the surface low during their most intense stages.

Storm surge is an abnormal rise in sea level associated with the movement of a hurricane over a coastal region. Notice in Figure 24.10 that storm surge is greatest to the right of the eye of the hurricane, again looking in the direction the storm is moving.

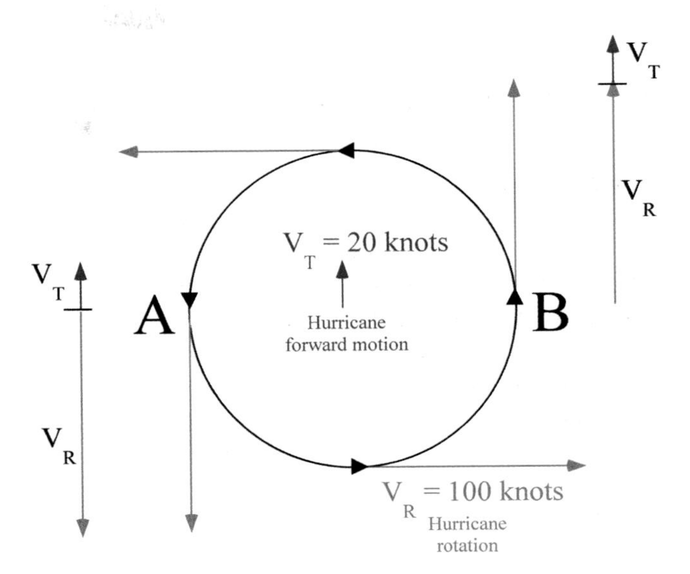

FIGURE 24.11: Illustration showing why winds are strongest on the right side of a hurricane looking in the direction of storm motion. In this figure, the hurricane has a rotational velocity (V_r) of 100 knots, while moving with a translational velocity (V_t) of 20 knots. At Point A, the wind = rotation – translation = 100 knots – 20 knots = 80 knots. At Point B, the wind = rotation + translation = 100 knots + 20 knots = 120 knots.

The strong surge on the right side is directly related to the strength and direction of the wind. When a hurricane nears land, winds to the right of the eye are blowing onshore causing water to "pile up" along the coastline and the sea level to rise, while winds to the left of the eye are blowing offshore. Low pressure at the center of the eye also causes sea level to rise, enhancing the surge to the right of the eye, and compensating for decreases in sea level due to offshore winds to the left of the eye. Storm surge will be discussed in more detail later in the chapter.

CHECK YOUR UNDERSTANDING 24.2

1. What path would an air parcel have if it were injected into a hurricane at the surface?
2. Why are wind speeds across a forward-moving hurricane not symmetric?
3. Where would you expect to find the heaviest rainfall rates in a hurricane?
4. Does the center of a hurricane have a warm or cold core? Why?

TROPICAL CYCLONE DEVELOPMENT

A cluster of thunderstorms must exist for a tropical cyclone to develop. In the tropics, thunderstorms seldom form unless convergence is occurring in the low-level flow to force air upward far enough to become unstable. Even when these thunderstorms form, the environmental conditions must be right for the thunderstorms to organize themselves into a tropical cyclone. In this section, we consider how tropical thunderstorms form, the environmental conditions that support the organization of the thunderstorms into a tropical cyclone, and the mechanism by which that organization occurs.

Trigger Mechanisms for Thunderstorms in the Tropics

Low-level convergence is required in the tropics to lift air sufficiently to trigger tropical thunderstorms. These local areas of convergence develop in three primary ways. To understand the first of these mechanisms, we must step back and examine the basic general circulation in the tropics. The top diagram of Figure 24.12 shows average horizontal and vertical

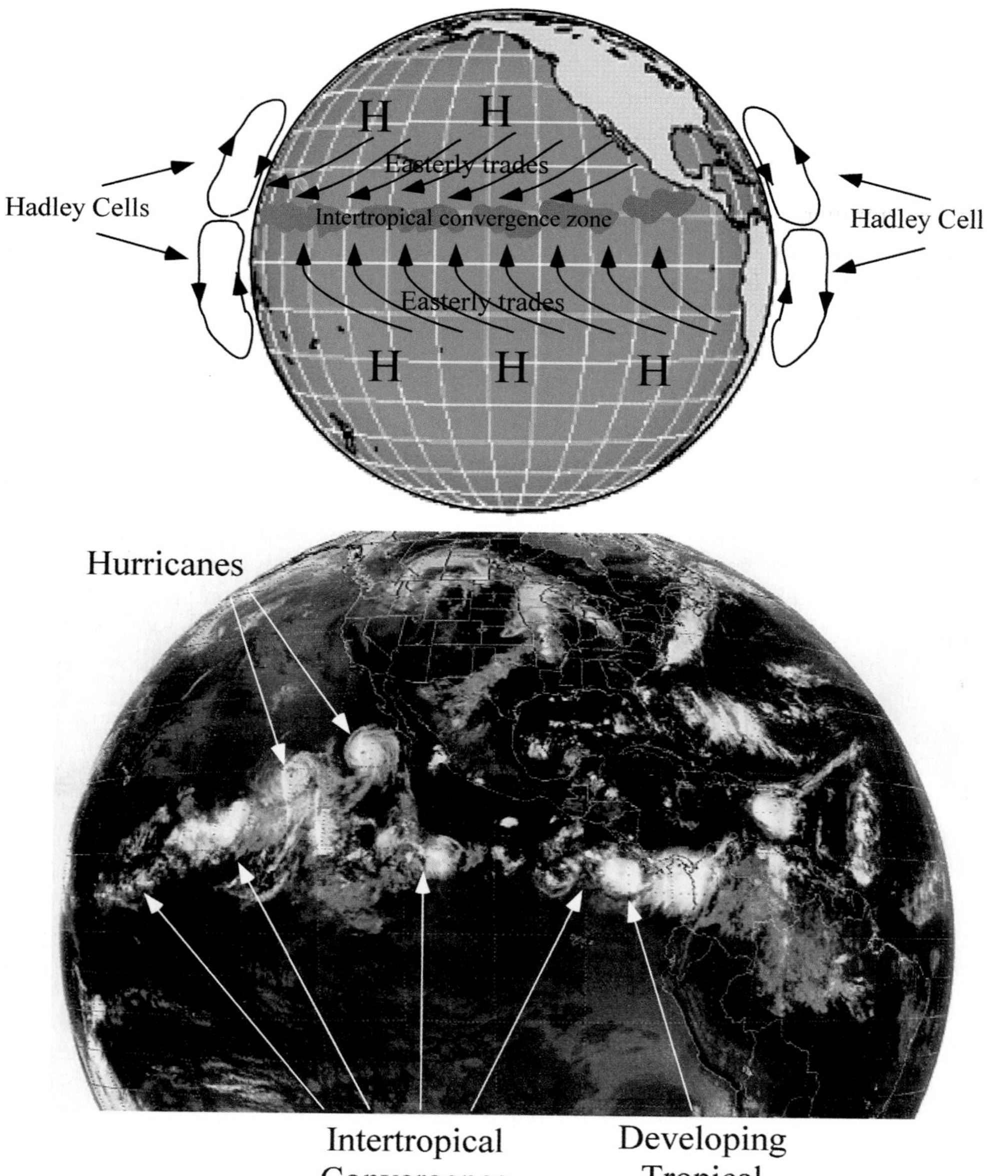

FIGURE 24.12: Top: Schematic showing the Hadley Cells, easterly trade winds, and the intertropical convergence zone. Bottom: An infrared satellite image from 26 July 1990 showing the ITCZ over the Pacific and several tropical storms in different stages of development.

circulations in the tropics. The most prominent features of the general circulation of the atmosphere in the tropics are the ***Hadley Cells***. These cells, one in each hemisphere, consist of rising motion along a belt near the equator, a poleward flow of air in the upper troposphere, descending motion centered around 25° north and south latitudes, and return flow near the surface toward the equator. This flow occurs in direct response to solar heating, which is maximum near the equator, and is a persistent feature of the tropics. Because of the Coriolis force, the poleward upper tropospheric branch of the flow turns eastward (westerly winds), and the lower equatorward branch of the circulation flows westward (easterly winds). The easterly trade winds discussed earlier are one branch of the Hadley circulation. The two Hadley Cells shift north and south with the seasons so that the point where they meet approximately follows the sun on its annual journey north and south.

Figure 24.12 shows the location of the Hadley Cells in late July, the middle of the Northern Hemisphere summer. At the surface, the flows in the north and south Hadley Cells collide in a zone of convergence called the ***Intertropical Convergence Zone (ITCZ)***. The ITCZ is a preferred location for thunderstorm formation (see bottom of Figure 24.12). When the ITCZ is far enough north (or south) of the equator so that the Coriolis force is sufficiently strong, a cluster of thunderstorms forming along it may sometimes organize into a hurricane, as illustrated in the lower panel of Figure 24.12. Worldwide, the ITCZ is the greatest source of hurricanes.

The second mechanism forcing convergence in the tropics is provided by "waves" in the easterly flow. These ***easterly waves*** are similar to the waves in mid-latitude flow, except that they are in the low-level trade wind air. Figure 24.13 shows a schematic diagram illustrating several easterly waves moving across the Atlantic Ocean, and the position of cloud clusters relative to the waves. Examples of these cloud clusters, including one that intensified to become Hurricane Erika (1997), are shown in the satellite image of the Atlantic Basin in Figure 24.14. To understand how easterly waves create low-level convergence, we must return to the principles of Chapter 8 concerning changes in flow curvature and their effect on air acceleration.

Figure 24.15A shows a schematic of an easterly wave. Easterly waves are most apparent near the surface in the low-level airflow. Within the Hadley Cell in the Northern Hemisphere tropics, the highest pressure lies to the north at the oceanic subtropical high and the lowest pressure to the south at the ITCZ. In an easterly wave, *high-pressure ridges bulge southward, while low-pressure troughs extend northward,* the exact opposite of what appears in upper atmosphere mid-latitude flow.

Recall from Chapter 8 that air flows clockwise around ridges in the Northern Hemisphere at a velocity that exceeds the geostrophic value. Flow around a trough will be counterclockwise in the Northern Hemisphere

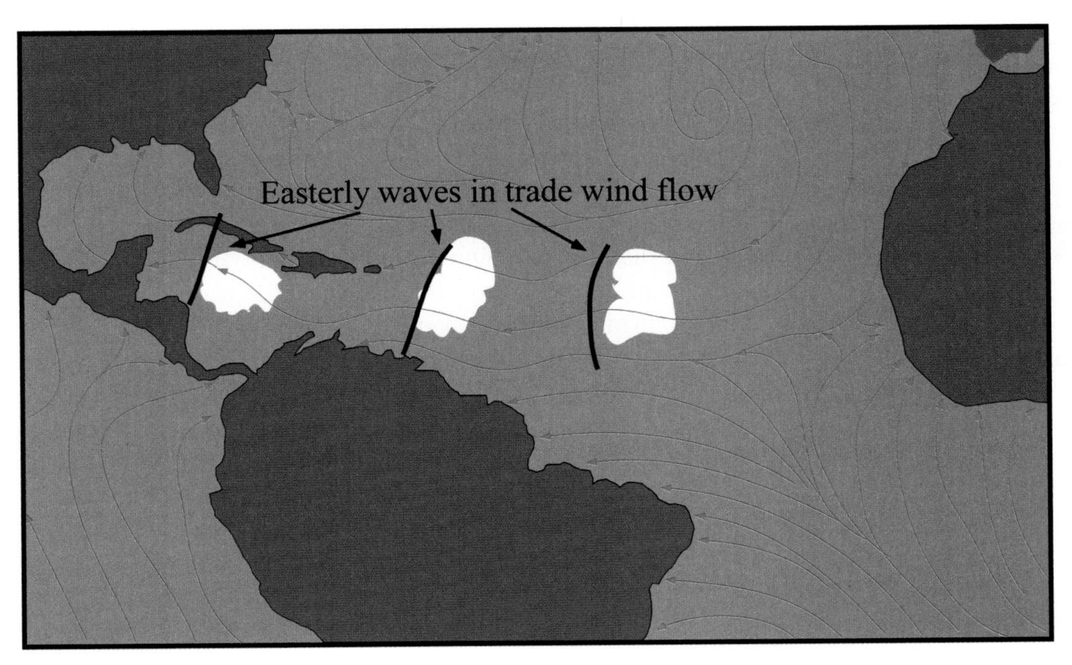

FIGURE 24.13: Schematic of the typical summertime wind pattern over the Atlantic Ocean. Note the three easterly waves and associated cloud patterns. Thunderstorms forming on the convergent (east) side of these waves have the potential to organize into hurricanes.

FIGURE 24.14: Satellite image of the Atlantic Ocean showing several easterly waves and Hurricane Erika (1997), a tropical storm that developed from an easterly wave.

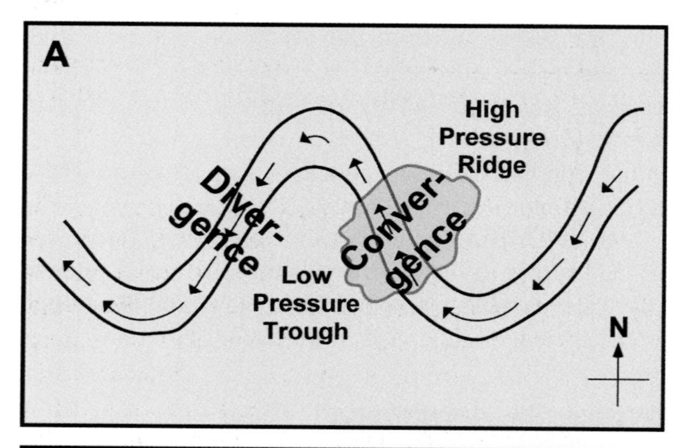

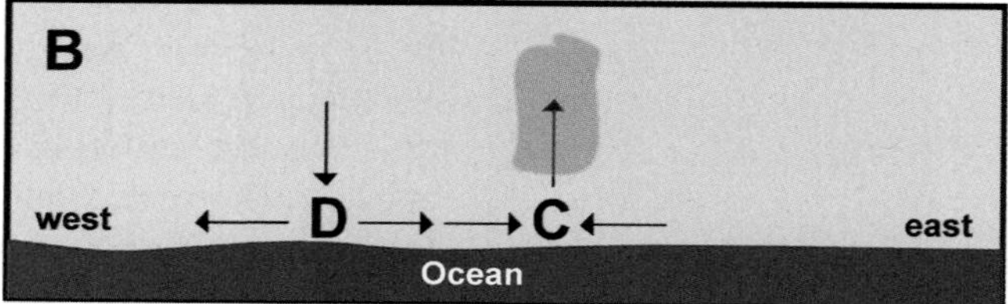

FIGURE 24.15: (A) Schematic of a lower tropospheric easterly wave embedded in the trade winds. Convergence, clouds, and precipitation occur on the east side of the trough. (B) Vertical cross section through the easterly wave in panel A showing convergence (C), divergence (D), circulations relative to the wave, and clouds associated with these circulations.

and have a velocity that is less than its geostrophic value. As shown in Figure 24.15A, air flowing from the crest of the ridge to the base of the trough will decelerate and undergo convergence. Air in this region must rise to compensate for the convergence (Figure 24.15B), since the convergence occurs in the low-level flow. It is this rising motion that can trigger thunderstorm formation. In contrast, air accelerates on the west side of the trough, leading to low-level divergence, downward air motion, and clearing skies.

Thunderstorms will erupt in the convergent region if the convergence is sufficiently strong. Easterly waves generally originate over continents as air moves across mountains and/or deserts, particularly the Sahara of Africa. Storms develop as the waves propagate over the ocean. Clusters of thunderstorms within these waves will sometimes organize into hurricanes. Easterly waves moving off of the continent of Africa are the main source of tropical thunderstorm clusters over the North Atlantic Ocean.

The final source of tropical thunderstorms actually has mid-latitude origins. Occasionally cold fronts from the mid-latitudes will progress far enough south that they will intrude into tropical latitudes. Thunderstorms will frequently occur on the warm side of these fronts, triggered by lifting over the cooler air to the north. This process of tropical thunderstorm formation is most common in the fall, when cold fronts are stronger than in summer, and have the energy to make it far enough south into the Gulf of Mexico or over the Atlantic near the Bahamas. Thunderstorms along a cold front will occasionally organize into a hurricane as illustrated in Figure 24.16, the weather pattern near the time of development of Hurricane Kyle in 2002. Cold fronts are particularly important for triggering thunderstorms that develop into late-season Gulf of Mexico hurricanes.

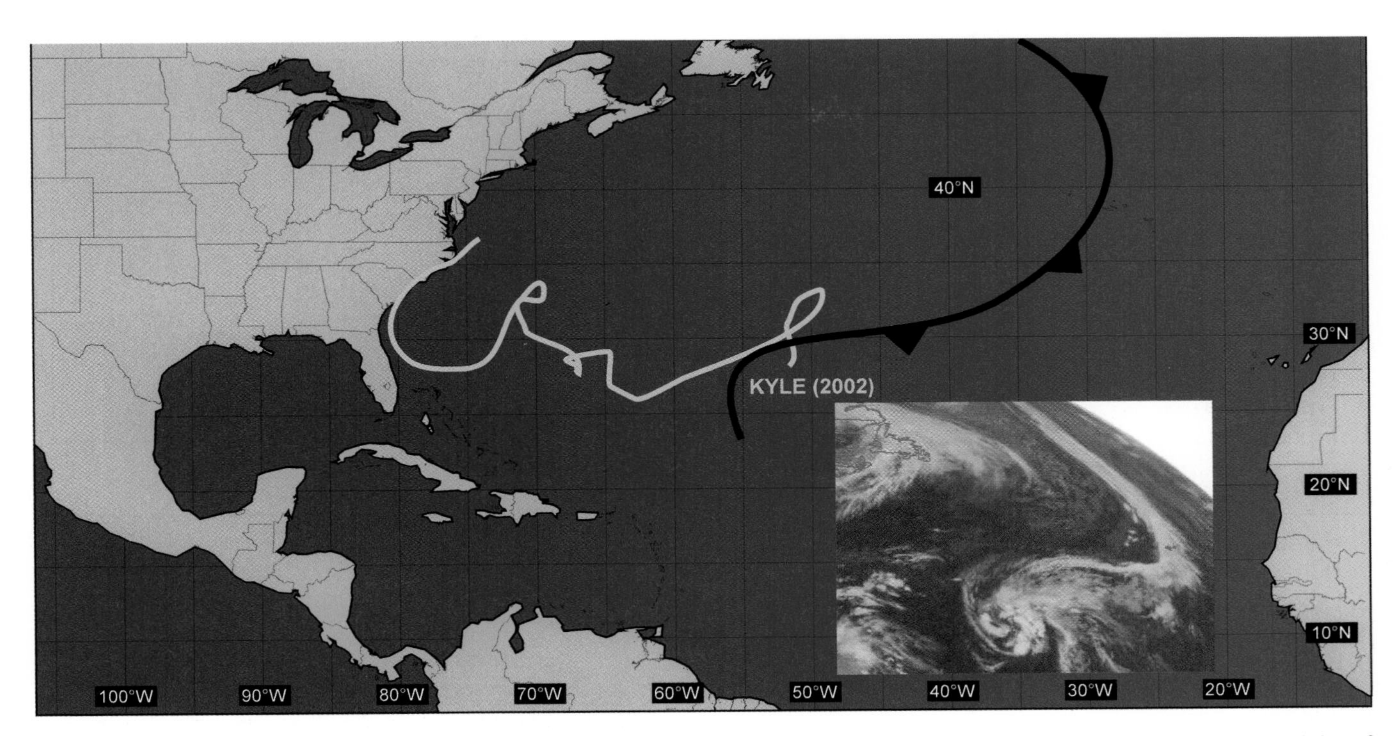

FIGURE 24.16: Hurricane Kyle (2002) is an example of a hurricane that formed from thunderstorms developing in the vicinity of a weak cold front that intruded into tropical latitudes. The satellite inset shows the cloud pattern associated with the cold front extending to the position where Kyle was forming on 21 September 2002.

The Environment Required for Tropical Cyclones to Form from Thunderstorm Clusters

Tropical cyclones form from thunderstorm clusters in environments that have four specific characteristics. If these environmental conditions are not met, tropical thunderstorm clusters, regardless of their trigger mechanism, cannot develop into hurricanes. The required conditions are:

1. the sea-surface temperature must exceed about 80° F (26.5° C);
2. the surface layer of warm water in the ocean must be sufficiently deep, typically about 60 meters (~ 200 ft.) or more;
3. the winds in the atmosphere must not change substantially with height (weak vertical wind shear); and
4. the location must be at least 5 degrees north or south of the equator.

Sea-Surface Temperature

The energy for a hurricane comes from heat and moisture supplied by the ocean that is released into the atmosphere as latent heat during condensation. The higher the temperature of the uppermost layers of the ocean, the more energy can be supplied to the atmosphere. There is a direct relationship between the central pressure of the most intense hurricanes and the temperature of the sea surface over which the storms are moving. When the sea-surface temperature falls below about 80° F (26.5° C), the ocean cannot supply sufficient moisture and heat to support hurricane formation. The North Atlantic and northeast Pacific Oceans generally only achieve sufficiently warm temperatures in late July to early November. In the Western Pacific, the ocean is above 80° F (26.5° C) for much of the year, so typhoons can occur in both the Northern Hemisphere summer and winter. This oceanic region is unique in this regard.

Pressure readings from exceptionally strong typhoons in the Western Pacific Basin and hurricanes in the Atlantic Basin are listed in Table 24.6. Western Pacific storms have not been routinely sampled using aircraft since 1987 when the Air Force disbanded its Pacific Typhoon Chasers squadrons. Maximum winds are now estimated from satellite data, and the central pressure determined from an equation relating central surface pressure and wind speed. The five storms listed in Table 24.6 occurred prior to 1987.

TABLE 24.6 **Low-Pressure Readings in Exceptionally Strong Tropical Cyclones**

Tropical Cyclone	Season	Central Pressure (mb)
Western Pacific Basin		
1. Typhoon Tip	1979	870
2. Typhoon June	1975	876
3. Typhoon Nora	1973	877
4. Typhoon Ida	1958	877
5. Typhoon Rita	1978	878
Atlantic Basin		
1. Hurricane Wilma	2005	882
2. Hurricane Gilbert	1988	888
3. Labor Day Hurricane	1935	892
4. Hurricane Rita	2005	895
5. Hurricane Allen	1980	899

Comparing Atlantic and Pacific storms, note that the lowest pressures recorded all occurred in the western Pacific where the ocean temperatures are warmest. A theory that predicts the lowest possible central pressure of a hurricane based on sea-surface temperature is remarkably accurate (Figure 24.17). The colors on this figure show the predicted extreme minimum pressures possible in hurricanes. The values closely correspond to observations of the minimum pressure in strong hurricanes (Table 24.6).

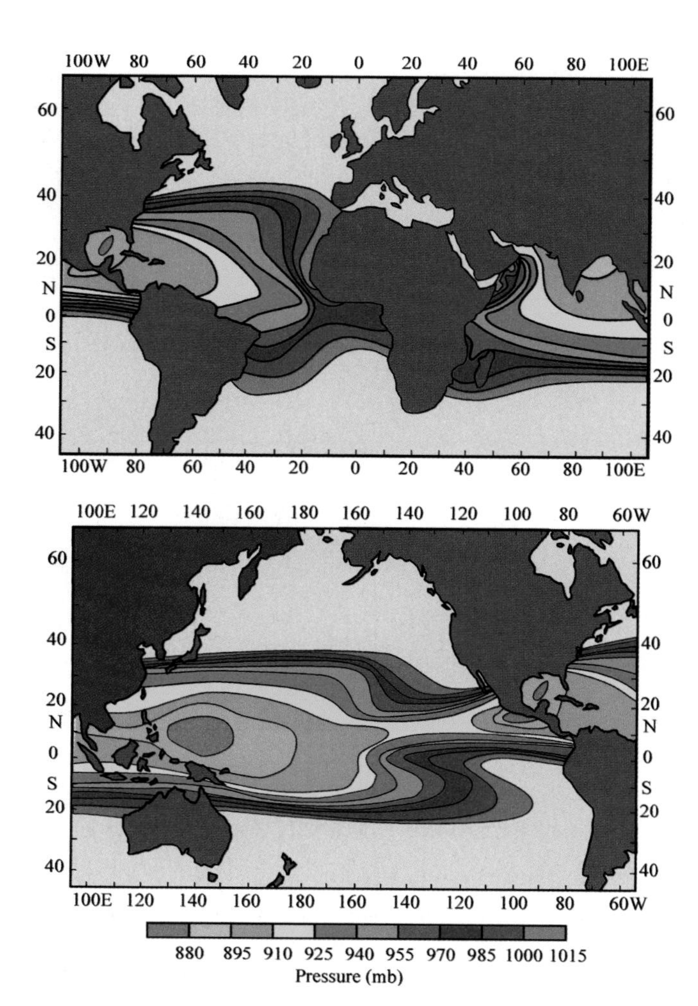

FIGURE 24.17: Predicted maximum intensity of a hurricane based on average ocean surface temperatures in the Atlantic (upper) and Pacific (lower) regions. The agreement between the predicted values shown here and observed values from Table 24.6 is excellent, showing the importance of a warm ocean for hurricane formation.

Depth of Warm Water

A deep layer of warm water is required for hurricane development because a hurricane "stirs" the ocean

ONLINE 24.3
Global Sea-Surface Temperature

The sun modulates sea-surface temperature changes during the year. As the Northern Hemisphere spring progresses to summer, and summer to fall, the most direct rays of the sun progressively march north from the equator to the Tropic of Cancer, and back again toward the equator. During this time, the Northern Hemisphere ocean temperatures slowly rise, reaching their peak around late August to early September. This is the peak of hurricane season in the Northern Hemisphere.

Online you can watch an animation of the annual evolution of global sea-surface temperature. Watch the sea-surface temperature in the Gulf of Mexico and Caribbean Sea and observe how ocean temperatures reach their peak in late summer. Notice where and when the ocean temperatures exceed the threshold (80° F, 26.5° C) for hurricane formation. Notice also that the tropical western Pacific has the warmest water in the world.

water much like water in a teacup. This leads to ***upwelling*** of water in the vicinity of the eye and eyewall. If the warm layer is shallow, colder water will rise to the surface, and the developing storm will no longer have its warm water energy source. This is one reason why hurricane season typically does not start in the Atlantic until late summer, peaking in September. The summer's heat is required to warm the water to the depth necessary to support hurricane formation.

Weak Wind Shear

Hurricanes derive their energy from the release of latent heat in the eyewall. For a hurricane to develop, this energy must be locally concentrated as explained in Focus 23.1. The clouds cannot be carried downstream by high winds in the upper atmosphere. It is also difficult for a vortex of wind to form in the presence of strong flow in the upper atmosphere. The vortex is simply torn apart by the strong winds aloft, which tilt the circulation and carry it downstream. Hurricanes form best in an atmosphere where the background winds are very weak aloft.

ONLINE 24.4

Wind Shear

How do meteorologists determine whether the wind shear profile is favorable over the oceans for tropical storm development when there are no stations to collect upper atmosphere data? Animation of satellite data permits meteorologists to observe the movement of cloud features, and features in the water vapor field. The approximate altitudes of these features are estimated from the intensity of the radiation received by the satellite, and the speed of the features is measured from changes between successive frames in the animation. Pattern recognition programs are used to track features and convert the data to wind measurements. Online you can see an animation of infrared satellite data over the Atlantic taken during Hurricane Isaac (2000). The category 4 hurricane is drifting northward toward a jetstream that originated at higher latitudes. The jetstream, which is characterized by strong wind shear, quickly tears at the hurricane, disrupting its circulation and causing its central pressure to rise.

Away from the Equator

The Coriolis force is important for storm rotation. At the equator the Coriolis force is zero. About 5 degrees north and south of the equator, the Coriolis force becomes significant enough that rotation can occur within developing thunderstorm clusters.

How Thunderstorms Organize into a Hurricane

Most clusters of thunderstorms in the tropics do not organize into a tropical cyclone. If the environment (the four conditions described above) is right, however, the individual thunderstorms can quickly organize into a ***vortex***, and rapidly develop into the eyewall surrounding a newly-formed eye. How does this happen? Scientists are still studying this process. The most promising theory about how this happens, called ***Wind Induced Surface Heat Exchange (WISHE)***, considers the feedback that occurs between the heat and moisture transfer from the ocean surface and the development of the vortex (Figure 24.18).

We know that there is a close association between the sea-surface temperature and hurricane strength. The sea provides the energy for the hurricane. The transfer of energy from the ocean to the lower atmosphere occurs in two ways. The first, less important way is through direct transfer of ***sensible heat*** when the ocean is warmer than the atmosphere. The more important way is through evaporation and the transfer of ***latent heat***. Evaporation of water from the ocean surface requires energy, which comes directly from the ocean. As water evaporates, heat is extracted from the water, cooling the ocean surface. This heat, called latent heat, reappears when condensation occurs in tropical thunderstorms. Enormous quantities of latent heat are released as condensation occurs and precipitation falls out of the clouds.

The rate at which water evaporates from the ocean surface is a strong function of wind speed. At high wind speeds the spray generated by the rough ocean surface significantly enhances the evaporation rate so that the moisture (energy) transfer from the ocean to the atmosphere can be 100 to 1,000 times greater than when the winds are weak.

Figure 24.18A shows a cloud cluster over a tropical ocean in the absence of wind shear. As air rises in a thunderstorm cluster, most of the air exhausts outward away from the cluster, but a small amount

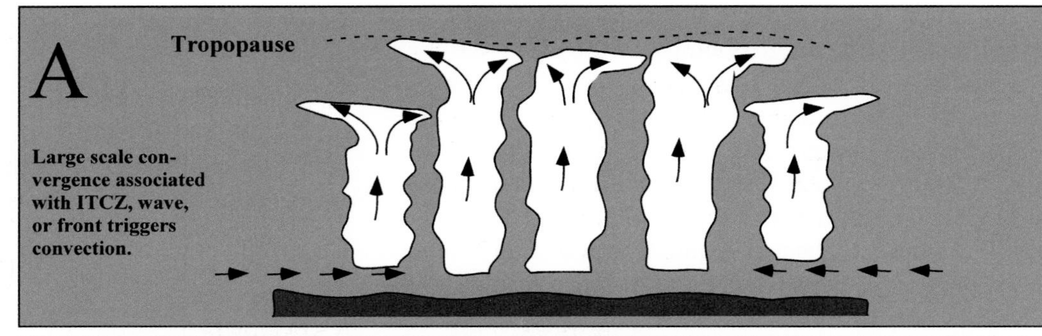

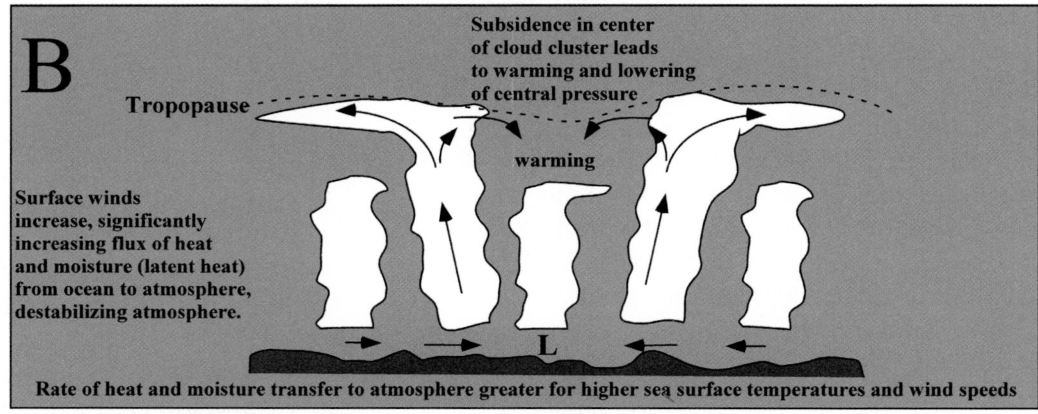

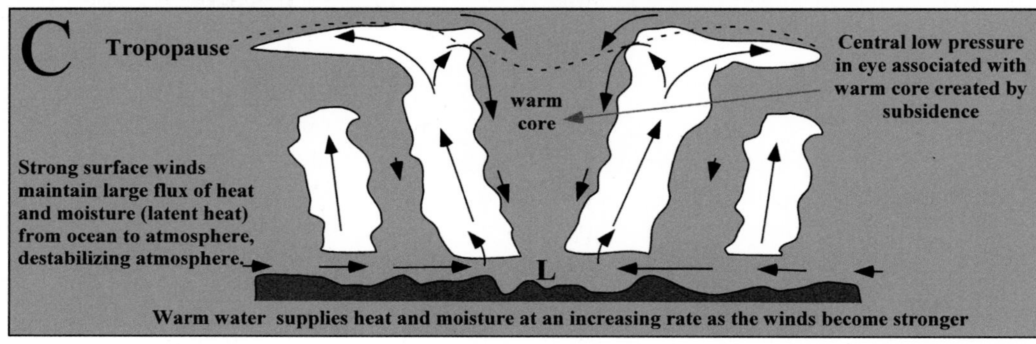

FIGURE 24.18: Illustration of Wind Induced Surface Heat Exchange (WISHE), a mechanism for tropical cyclone intensification. Tropical cyclone intensification occurs as sensible and latent heat and moisture are extracted from the ocean surface by the action of the wind, and carried into the core of the tropical depression. Subsidence in the center of the cloud cluster leads to adiabatic warming and the lowering of surface pressure, intensifying the surface winds, and significantly increasing the rate of transfer of heat from the ocean to the atmosphere. The heat is transferred upward to the tropopause within the developing eyewall.

exhausts inward (Figure 24.18B). The inward branch of the flow causes weak descent in the center of the cluster. Air in this weakly descending branch of the flow warms adiabatically, raising the temperature at the center of the cluster. Recall that any time heating occurs in the atmosphere, divergence will occur in the heated column (see Figure 8.12). This divergence induces the formation of a weak surface low-pressure center within the cluster.

As the newly-formed low increases the low-level pressure gradient, the winds increase at the surface, agitating the ocean, and increasing the evaporation rate. More moisture and heat can now be supplied to the clouds in the cluster, causing the updrafts to intensify. The system also begins to rotate under the combined influence of the pressure gradient and Coriolis forces as the central low pressure develops. Rotation has a centrifugal effect, which acts with heating to reduce the central pressure further.

Outflow from the convection near the tropopause now becomes more focused in two directions, most outward away from the center, but some inward toward the center of rotation (Figure 24.18B). The inward branch descends within the developing eye as the low-pressure center develops near the surface, and the descending air further increases the core temperature through adiabatic warming. The pressure at the surface lowers in response to this warming, centrifuging of the rotating air is enhanced, and the cycle of pressure lowering continues. The lower central pressure also increases the downward-directed pressure gradient force within the eye, further enhancing the descending motion within the eye and the associated adiabatic warming. Stronger wind, more heat transfer into the convection, stronger updrafts, and stronger descent of air in the developing eye all rapidly transform a cluster of storms into a hurricane.

As the whole system of thunderstorms rotates, the updrafts organize to form the eyewall. The descending air, which initially covered much of the eye region, becomes preferentially concentrated just inside the eyewall with slower descent across the remainder of the eye (Figure 24.18C). The system becomes a very efficient simple heat engine, the fuel being the moisture (latent heat) supplied to the storm from the ocean through evaporation, the "combustion" being the release of latent heat in the eyewall, the exhaust being the outflow at the tropopause (which cools as it radiates heat to space), and the overall maintenance of the circulation enhanced by the adiabatic compression and warm core in the descending air in the eye. The storm will roar along as long as the environment—*warm ocean, weak or no shear in the environment*—supports it. Tropical cyclones have maintained hurricane force winds for over twenty days while moving thousands of kilometers over tropical oceans.

The primary limiting factor in the process is how much energy can be transferred to the atmosphere at the sea-surface, and that is related to the-sea-surface temperature. The other limiting factor is the vertical wind shear. With strong winds aloft, the vortex circulation, particularly the warm core of air in the eye, cannot be protected and will be dispersed downstream, weakening the surface low-pressure center and weakening the surface winds. Cold water, land with its increased friction and lack of moisture, or strong environmental wind shear shut down the engine and quickly destroy the circulation. Ultimately, one or more of these factors destroy a hurricane.

The Source of Strong Rotating Winds

The strong rotating winds in the core of a hurricane can be understood by examining a simple physical principle called the ***conservation of angular momentum***. As we learned in previous chapters, the angular momentum for an object rotating around a point is defined as:

$$\textit{Angular momentum} = \textit{mass} \times \textit{radius} \times \textit{rotational velocity}$$

where the mass is the mass of the object, the rotational velocity is its velocity along the path of rotation, and the radius is its distance from the axis of rotation. Consider a simple experiment. Let's suppose we have a ball on a long string and we are whirling the ball slowly around in a circle. While we are whirling the ball, let's rapidly pull the string into the center of rotation so that the string is now one fourth of its original size. What happens to the rotation rate of the ball? The common experience we have as kids playing this game is that the ball rotates faster. Skaters apply the same principle by drawing their arms and legs into the center of rotation, speeding up their rotational velocity. A simple law, called the conservation of angular momentum, governs this process. The law states that in the absence of a torque (an external force that would act to change the rotation rate of an object), the angular momentum of an object will remain constant. In Figure 24.19, we examine this law for a hurricane.

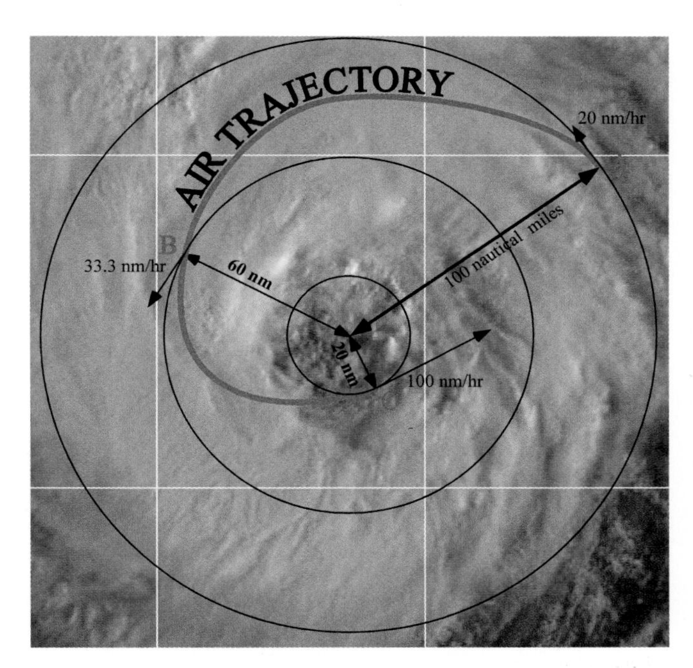

FIGURE 24.19: Illustration of the principle of conservation of angular momentum: the rotational speed × distance = constant. At Point A 20 nm/hr × 100 nm = 2000 nm^2/hr. At Point B: 33.3 nm/hr × 60 nm = 2000 nm^2/hr. At Point C: 100 nm/hr × 20 nm = 2000 nm^2/hr.

Our "object" will be a parcel of air spiraling into a hurricane. Let's assume for the moment that our parcel is not subject to torque forces. Since its mass does not change, conservation of angular momentum depends only on its radius and rotational velocity. Let's suppose, when the air is 100 nautical miles from a hurricane, it is moving 20 knots. Our "constant" is rotational velocity × distance from center = 20 × 100 = 2000 nautical miles^2/hour. When air spirals inward to a radius of 60 nautical miles, its velocity will be 2000/60 = 33.3 knots. When air spirals in to 20 nautical miles from the center of rotation, its speed will be 2000/20 = 100 knots. Like the ball on the string, the rotational velocity of the air increases as it approaches the axis of rotation. In real hurricanes, friction with the ocean surface acts as a torque force that reduces the rotation rate of air. The magnitude of the force of friction increases with wind speed because of ocean roughness. Although conservation of angular momentum cannot be strictly applied in this situation, it illustrates the basic principle explaining why air speeds up considerably as it approaches the center of rotation in a hurricane.

Figure 24.20 summarizes all the conditions and processes associated with hurricane formation. Tropical thunderstorms must be present. These thunderstorms form in three preferred locations: along the ITCZ, within easterly waves, and along cold fronts. The tropical environment, both atmosphere and ocean, must meet four critical conditions for these thunderstorms to organize into a hurricane. These include a very warm and sufficiently deep layer of water in the tropical ocean; little vertical wind shear in the atmosphere; and a location far enough from the equator for the Coriolis force to influence storm rotation. If all environmental conditions are met and thunderstorms are present, latent heat release in the storms, combined with the transfer of energy from the ocean surface to the atmosphere through the action of the wind, will lead to a lowering of the central pressure within the thunderstorm cluster. As air is drawn into the circulation, conservation of angular momentum will then lead to rapid spin-up of the winds.

CHECK YOUR UNDERSTANDING 24.3

1. What are three trigger mechanisms for tropical thunderstorms?
2. What are the four characteristics of the hurricane environment?
3. How does wind shear work against hurricane development?
4. Why don't hurricanes form on the equator?

DESTRUCTIVE FORCES IN A HURRICANE

Storm Surge

Storm surge is an abnormal rise in sea level associated with the movement of a hurricane over a coastal region. Storm surge is one of the greatest concerns to coastal communities as it is responsible for a large percentage of structural damage and coastal flooding.

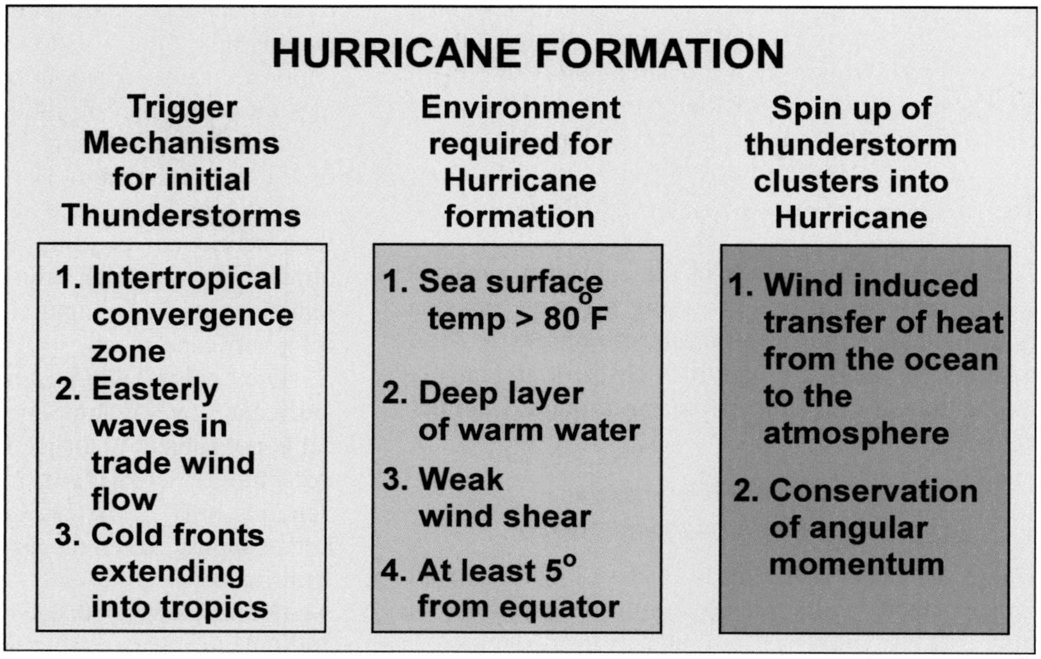

FIGURE 24.20: Summary of trigger mechanisms for tropical thunderstorms, the environmental conditions required for these storms to organize into a hurricane, and the mechanisms by which the thunderstorms organize into hurricanes.

HURRICANE FORMATION

Trigger Mechanisms for initial Thunderstorms
1. Intertropical convergence zone
2. Easterly waves in trade wind flow
3. Cold fronts extending into tropics

Environment required for Hurricane formation
1. Sea surface temp > 80°F
2. Deep layer of warm water
3. Weak wind shear
4. At least 5° from equator

Spin up of thunderstorm clusters into Hurricane
1. Wind induced transfer of heat from the ocean to the atmosphere
2. Conservation of angular momentum

Figure 24.21 illustrates the factors that contribute to the height of storm surge. There are two primary causes of storm surge, and three factors that enhance its destructiveness.

As a hurricane makes landfall, winds to the right of the eye are blowing onshore. The first cause of storm surge is the wind, which pushes ocean water toward shore, raising sea level (Figure 24.21A). The stronger the winds, the deeper the ocean water piles up along the shoreline. The second cause of storm surge is the air pressure difference between the eye and areas surrounding the hurricane. Low pressure in the eye of the hurricane causes the ocean to rise, much like water in a straw rises when air is removed from the straw. This ***barometric effect*** causes the sea level along the coast to rise as this raised region of seawater approaches land (Figure 24.21B).

The factors that can enhance storm surge are natural tides, waves, and shoreline shape. If the hurricane makes landfall during high tide, the problem of storm surge is exacerbated because sea level is higher already. Particularly high tides occur during a new or full moon (Figure 24.21C). Strong winds within the eyewall of the hurricane cause surface waves, which are normally small, to become extraordinarily large, sometimes as high as 10 m (32 ft) or more. These large waves are moving on top of the already raised sea (Figure 24.21C). Finally, the shape of the coastline influences the intensity of the storm surge. Inland bays, especially with relatively flat shorelines, enhance the sea-level rise, much like waves grow in height as they move toward shore into shallower water (Figure 24.21E).

The Galveston hurricane of 1900 came ashore during high tide, was enhanced by the shallowness of Galveston Bay, and submerged much of the city under several feet of water. Tropical cyclones in the Bay of Bengal, particularly near Bangladesh, inundate entire islands populated by poor people who have no chance to evacuate. These disasters often claim over 100,000 lives. Figure 24.22 shows the storm surge during Hurricane Hugo in September 1989 as recorded by the gage in the Charleston, South Carolina, harbor. The tide in the harbor was 3 m (~ 10 ft) above normal and 4.5 m (~ 15 ft) above low tide level, with the peak coincident with the passage of the east part of the eyewall as the eye passed directly over the city. Many cities along the East and

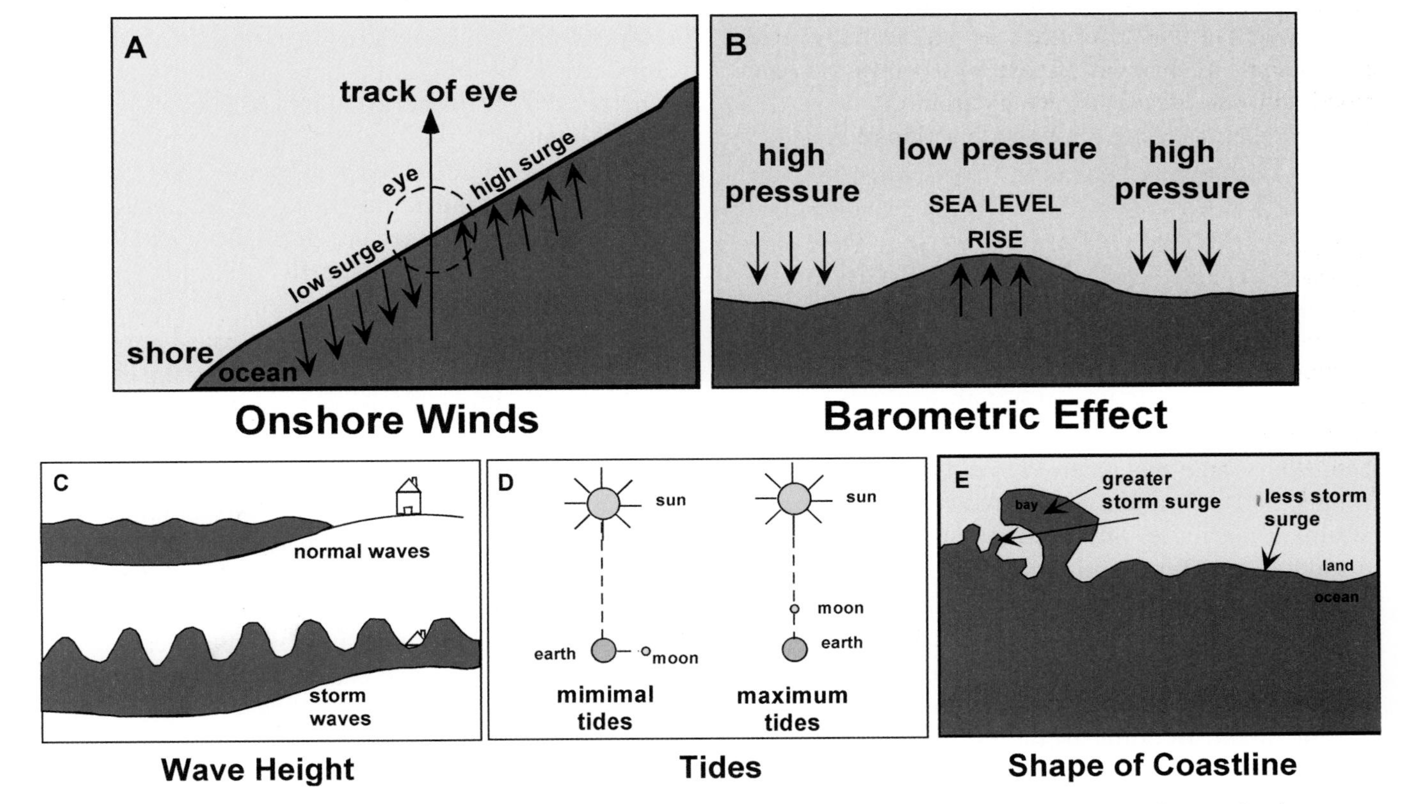

FIGURE 24.21: The two primary causes of storm surge (A and B) are onshore winds and the barometric effect, a rise in sea level associated with low-pressure. Factors that enhance storm surge (C, D, and E) include wave heights, tides, and the shape of the coastline.

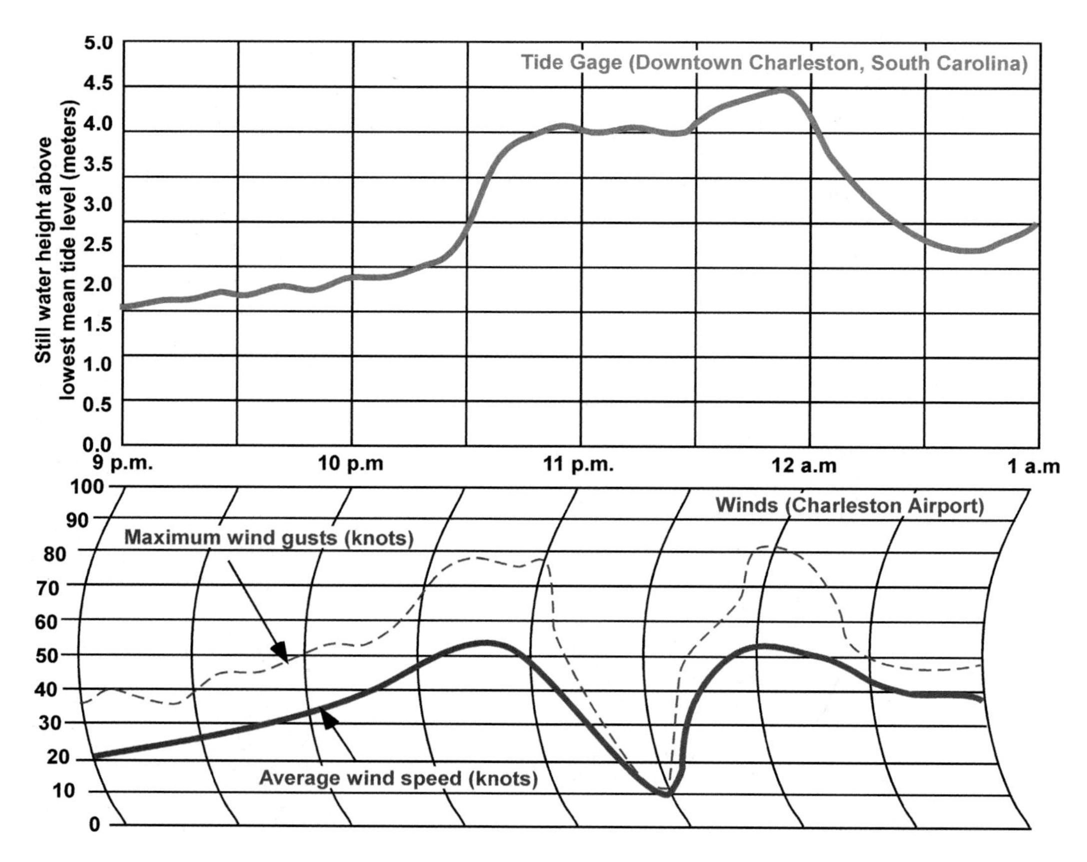

FIGURE 24.22: Top: Variation in the tide level in the Charleston, South Carolina, harbor during the landfall of Hurricane Hugo on 21 to 22 September 1989. Bottom: Winds measured at the Charleston airport.

Gulf Coasts of North America lie on the flat coastal plain, so a 3 to 4 m (10 to 13 foot) rise in sea level can flood areas far inland (see Focus 24.1).

Other Disastrous Forces in Hurricanes

Hurricanes produce heavy rains as they move over land (see Chapter 25). In some cases, where the movement is slow, 30 to 40 inches of rain have been recorded in two days. For example, Figure 24.23 shows the radar-estimated rainfall from Hurricane Georges during landfall on the Gulf of Mexico Coast. Local values as high as 24 inches (61 cm) were recorded. A large area received over 12 inches (30 cm). In July 2001, Tropical Storm Allison, which never became a hurricane, produced over 30 inches of rain in southeastern Texas and Louisiana (see Chapter 25). Heavy rainfall such as this produces incredible floods along rivers and streams. In many hurricanes, it is the ***inland flooding***, rather than the surge-caused coastal flooding, that leads to most of the fatalities. The worst case of inland flooding in the Western Hemisphere occurred with Hurricane Mitch in 1998. Flooding killed over 18,000 people as the hurricane winds slammed into the steep topography of Nicaragua and Honduras. The rain actually filled a volcano cone, causing it to collapse and bury several villages in mud. Although storm surge caused extensive damage, the deaths were attributed mostly to inland flooding and the associated mudslides. Hurricane Katrina (2005) was the most recent, grim illustration of the devastation caused by inland flooding during hurricanes (see Focus 24.1).

High winds have the potential to cause great damage, but generally not as significant as the damage caused by flooding. There are exceptions where high winds cause most of the destruction, as was the case of Hurricane Andrew (1992) as it moved over Florida. Hurricanes often also have ***tornadoes*** embedded within the storms in the spiral bands and occasionally within the eyewall. These tornadoes are generally in the EF0 to EF2 range, but are hard to see because of the heavy rain. The manner in which they form is still under study. The damage caused by Gilbert (1988) in the United States was mostly attributed to tornadoes spawned by the storm. Hurricane Opal (1995) also produced a large number of tornadoes as it made landfall along Florida's Gulf Coast.

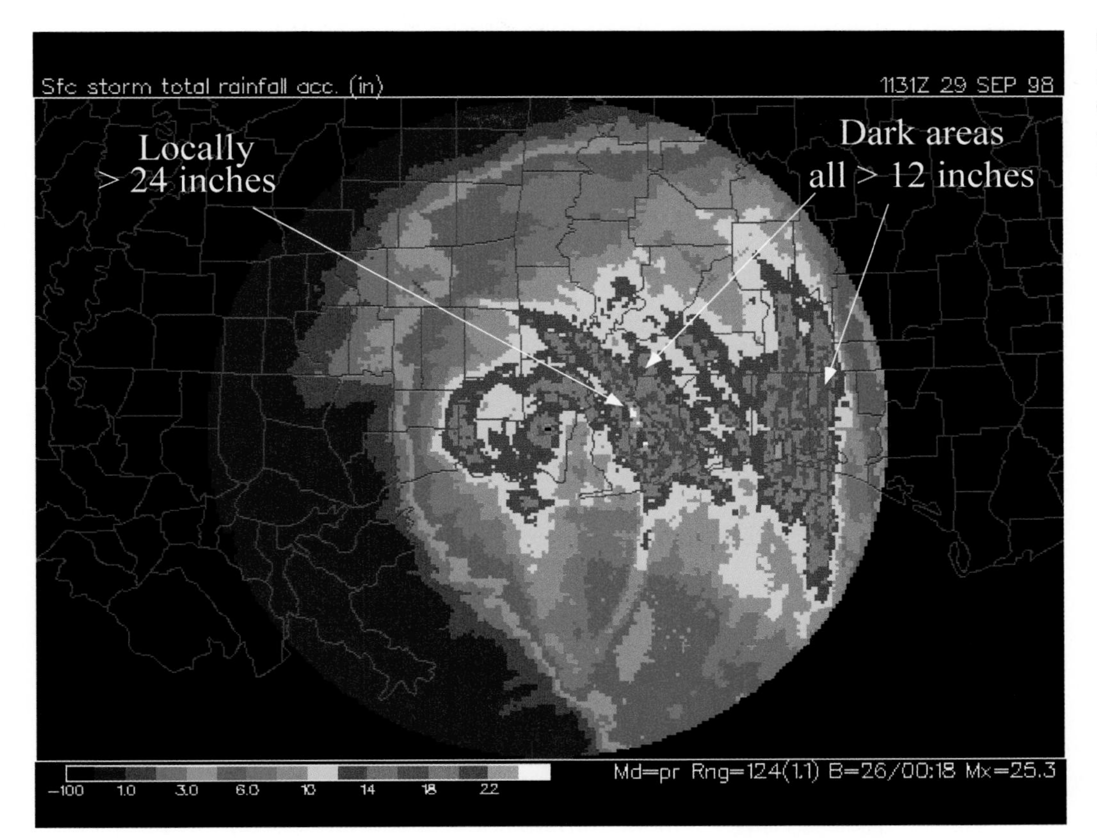

FIGURE 24.23: Radar estimated rainfall from Hurricane Georges (1998) as it moved on the Gulf Coast near Mobile, Alabama.

FORECASTING HURRICANE TRACKS AND INTENSITY

The National Hurricane Center (NHC), one of the National Centers for Environmental Prediction of the U.S. National Weather Service (see Focus 2.2), is responsible for issuing watches, warnings, forecasts, and analyses of hazardous weather conditions in the tropics. The National Hurricane Center maintains a continuous watch on tropical cyclones over the Atlantic, Caribbean, Gulf of Mexico, and the Eastern Pacific from 15 May through 30 November or longer. Forecasting hurricane movement and intensity begins with data collection. Over the oceans, this is not an easy task—it must be accomplished with reconnaissance aircraft and satellites. The three types of aircraft currently used in reconnaissance are a fleet of 10 U.S. Air Force C-130 Hercules aircraft, two National Oceanic and Atmospheric Administration (NOAA) Lockheed P-3 aircraft, and a NOAA Gulfstream-4 jet aircraft.

The C-130 and P-3 aircraft fly directly into the eye of hurricanes, measuring the storm's location and intensity by dropping ***dropwindsondes*** into the storm. Dropwindsondes are rawinsondes that fall from an aircraft rather than rise with a balloon (Figure 24.24). The P-3 aircraft also carry a suite of advanced instruments, such as scanning Doppler radar, that collect data used for scientific investigations of hurricanes. The G-4 has a long range and can fly at high altitude. This aircraft flies in the environment around the hurricane, also dropping dropwindsondes.

The data from these aircraft and from satellites are used at the NHC to initialize numerical models (see Chapter 4) that calculate the future evolution of a storm. Several models are used. Because of uncertainty in the initial state of the atmosphere, differences in the ways models incorporate physical processes, such as evaporation of moisture from the ocean surface, and different ways the models are constructed, a range of predictions for the future behavior of the hurricane is obtained. These predictions are assessed from a statistical point of view, and strike probability maps are developed. These maps estimate the probability that the storm will strike a particular region within a specific time.

Nevertheless, the paths hurricanes take are always somewhat uncertain and sometimes defy predictions. For example, Figure 24.25 shows the predicted and actual tracks of Hurricanes Georges and Mitch in 1998. As shown in the figure, the location of landfall for Hurricane Georges, both in Florida and Missis-

FIGURE 24.24: Photograph of a NOAA P-3 aircraft (foreground) and an Air Force C-130 (background). The inset photograph shows the dropwindsonde station on the NOAA P-3.

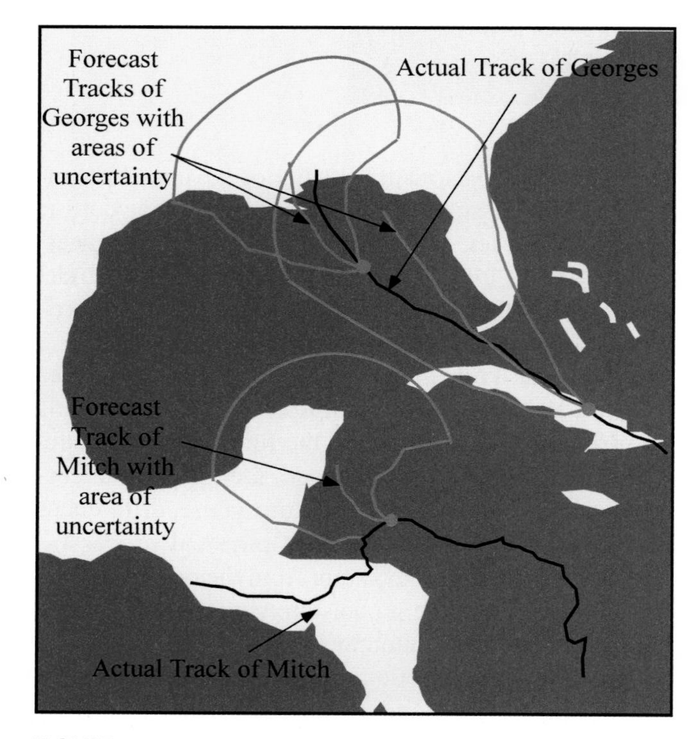

FIGURE 24.25: The tracks (black lines) of Hurricanes Georges (1998) and Mitch (1998). The National Hurricane Center forecast tracks and areas of uncertainty are shown in red and green at select positions along the tracks.

sippi, was very close to model predictions. Hurricane Mitch, on the other hand, took an unexpected southwestward turn, moving not only off track, but completely outside the area of uncertainty in its predicted path. The primary reason for unpredicted behavior is poor initialization of the models due to the lack of observations available throughout the tropics.

The NHC issues watches and warnings based on probabilities from prediction models, but always warns the public outside the area of low-strike probability to stay alert for the unexpected changes hurricanes often display. ***Hurricane watches*** are issued for tropical storms or hurricanes that pose a potential threat within thirty-six hours. ***Hurricane warnings*** are issued for these storms when hurricane conditions or tropical storm conditions will be present in an area within twenty-four hours. ***Coastal flood watches*** and ***warnings*** are also issued if significant flooding is a possibility with landfalling storms. ***Heavy surf advisories*** are issued for coastal areas not expected to be impacted directly from the landfalling storm but that can expect high surf that may threaten life or property. (Formal definitions can be found in Appendix D.)

In the past, the NHC issued forecasts out to three days, but since 2003, they have extended their forecasts to five days. This extension is both a reflection of improvements in forecast models and the need to plan evacuations farther in advance due to increased coastal populations. Public attention and response to these storms remain key to saving lives and protecting property. This is particularly true as the population along coastlines increases, and the length of time and expense of full coastal evacuations grow. It is extremely important that residents on the coastlines heed evacuation orders, move to inland shelters, and remain indoors during hurricanes. Timely evacuations today remain the best way to prevent loss of life in hurricanes, the most dangerous storms on Earth.

GLOBAL WARMING: POTENTIAL EFFECTS ON HURRICANES

Concerns about the catastrophic effects tropical cyclones have on society have motivated scientists to undertake many investigations in the last two decades to determine if global warming is affecting tropical cyclone frequency, duration, and intensity. The record 2005 season in the Atlantic Basin, and the large number of intense landfalling hurricanes in the United States in 2004 and 2005, has raised public awareness of the hurricane threat and led to heightened interest into questions concerning global warming and its influence on hurricanes. This increased threat is particularly important in light of the observation that global ice sheets and glaciers are rapidly melting and that sea level is expected to continue to rise through the 21st century (see Chapter 5).

We know from very accurate measurements that the average sea-surface temperature of the tropical oceans has been increasing over the last several decades. For example, the red curves in Figure 24.26 shows the September sea-surface temperature change since 1930 for the prime hurricane genesis region over the Atlantic Basin (Panel A), the July–November sea-surface temperature change since 1930 in the genesis region of the west Pacific Basin (Panel B), and annual mean sea-surface temperature change since 1930 averaged over all tropical oceans between 30 degrees north and south of the equator (Panel C). In each of these panels, an upward trend in sea-surface temperature is obvious. There are also smaller time-scale cycles associated with phenomena such as El Niño and other oceanic circulations. Theory, modeling studies, and worldwide observations of hurricanes all show a close relationship between maximum hurricane intensity and sea-surface temperature. It is therefore relevant and important that we ask the question "Does global warming cause more hurricanes, stronger hurricanes, and/or longer duration hurricanes?" Scientific evidence collected to date indicates that the answer to the first question is no, and to the second and third questions is yes.

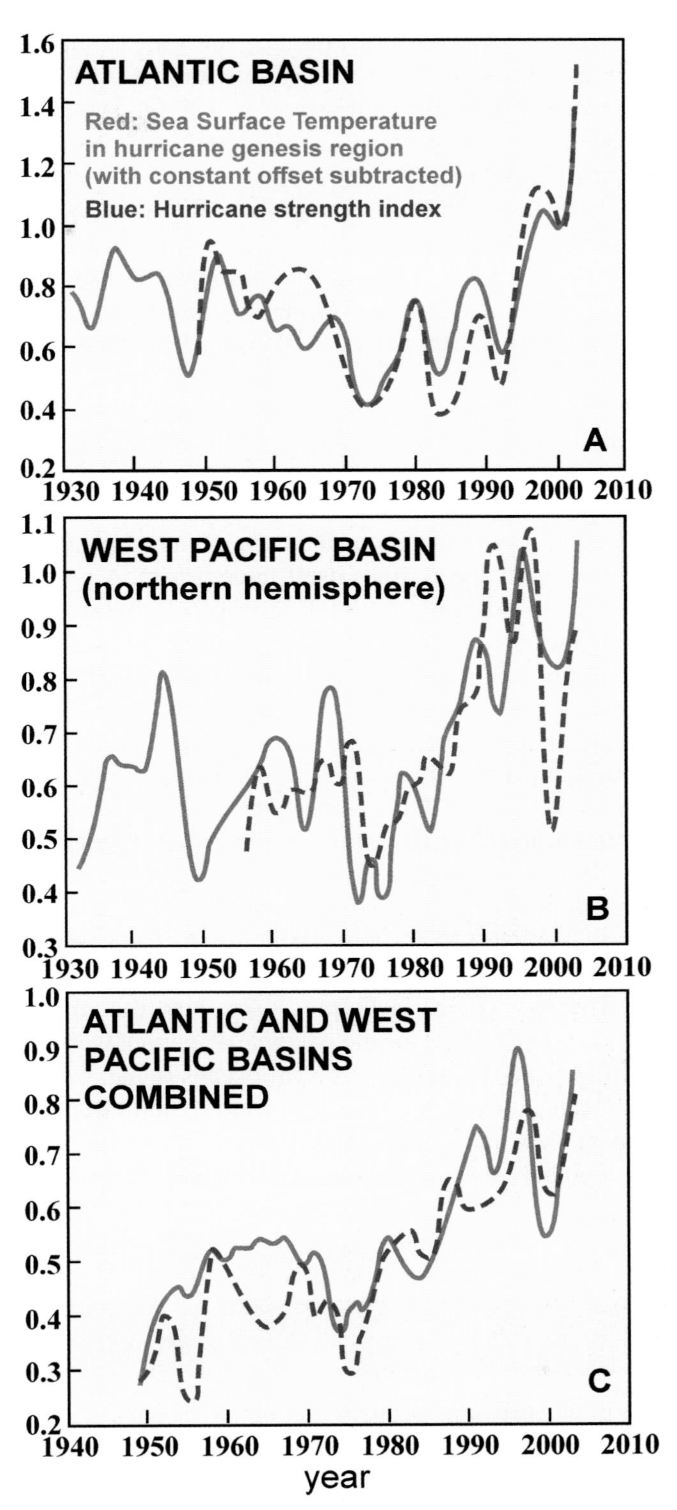

FIGURE 24.26: The sea-surface temperature data and hurricane strength index of hurricane regions of the (A) Atlantic Basin, (B) west Pacific Basin, and (C) Atlantic and West Pacific Basins combined. Red lines show departures from the average of the measured sea-surface temperature. Blue dashed lines show the hurricane strength index.

According to several studies, the global annual frequency of tropical cyclones has shown no trend over the period of reliable record (1949–present). There *has* been a distinct increase in North Atlantic tropical cyclone frequency that corresponds with an increase in sea-surface temperature in that basin. However, North Atlantic tropical cyclones account for only 11 percent of all tropical cyclones. Worldwide, the number of tropical cyclones continues to range between 80 and 100 per year, with no clear trend over the period of record.

In contrast to cyclone frequency, there have been clear and discernable upward trends in both the maximum wind speed and duration of hurricanes. We know from past studies that damage due to wind increases approximately as the cube of the wind speed (V^3, the wind speed times itself three times). A potential way to estimate the total destructive effects of a hurricane is to simply add together the cube of the maximum wind speed at equal time intervals for all tropical cyclones in a given year. Fortunately, this type of calculation is possible because positions and maximum sustained surface winds for all tropical cyclones since 1949 are available at six hour intervals from the National Hurricane Center for the Atlantic, eastern and central Pacific basins, and by the United States Navy Joint Typhoon Warning Center for the western Pacific, Indian Ocean, and the Southern Hemisphere basins.

The blue curves in Figure 24.26 show this ***hurricane strength index*** for the Atlantic (Panel A), western Pacific (panel B), and all tropical oceans (panel C). In each panel, a clear discernable upward trend in tropical cyclone destructive power is evident. Furthermore, this trend closely corresponds to the increase observed in sea-surface temperature. These data are clear evidence that the destructive power of hurricanes worldwide is increasing as the tropical sea-surface temperatures increase. The trends in Figure 24.26 are due to both increases in hurricane intensity and duration.

Sea-surface temperatures in the tropics over the last several thousand years have been estimated from geologic evidence derived from cores taken from the sea floor. Studies of these cores suggest that the increases in sea-surface temperature observed in the recent decades are unprecedented in the historical record. Such rapid change is so unusual that it is compelling evidence that the increase in sea-surface temperature is a direct consequence of global warming. The close connection between global warming and human production of greenhouse gases suggest that human activities may indeed be behind the trend toward stronger and longer duration hurricanes over our tropical oceans.

CHECK YOUR UNDERSTANDING 24.4

1. What are the four destructive forces of a hurricane?
2. What unique information do aircraft provide about hurricanes as they fly through the eyes of these storms?
3. What is the primary reason that hurricane track forecasting is difficult?
4. Does global warming impact hurricane frequency, duration, and intensity?

TEST YOUR UNDERSTANDING

1. In what regions of the world do hurricanes, typhoons, and cyclones develop?
2. Summarize in your own words the six key features that characterize global hurricane occurrences.
3. If you wanted to take a vacation on a tropical oceanic cruise in September, what bodies of water would be safest in terms of avoiding hurricanes?
4. Why do hurricanes typically travel east to west in the tropics?
5. Why do hurricane tracks typically curve to the right in the Northern Hemisphere?
6. Is the state of Florida more likely to be struck by a category 2 hurricane or a category 5 hurricane? Why?

7. Of all the states on the East Coast of the United States, the one that is struck the least by landfalling hurricanes is Georgia. Why? (Hint: consider hurricane tracks.)

8. What are scud clouds and where are they found in some hurricanes?

9. Draw a vertical cross section of the clouds within a hurricane. Draw a diagram to show how each of the following variables changes as you pass through the hurricane: pressure, winds, precipitation, temperature aloft, rainfall, and storm surge.

10. If a hurricane is moving to the west, which side of the storm would have the strongest winds and highest storm surge?

11. Where does the air in the eyewall of a hurricane stop rising? Why?

12. Explain why air in the eye of a hurricane sinks even though it is surrounded by cooler air.

13. What is storm surge?

14. How does air motion in the Hadley Cells lead to the formation of tropical thunderstorms?

15. How do easterly waves favor the formation of thunderstorms?

16. Is water in a public swimming pool typically warm enough to support a hurricane?

17. Why does the Northern Hemisphere hurricane season peak in September rather than in July, when air temperatures are warmest, or June, when the sun is the highest?

18. How are stronger surface wind speeds related to latent heat release in the eye wall of a hurricane?

19. How does the law of conservation of angular momentum relate to the winds in a hurricane?

20. You visit the Texas Gulf Coast and discuss hurricanes with a beachfront hotel owner. The hotel owner tells you that hurricanes are not a problem. After all, when Hurricane Rita (category 3) came through, the hotel only suffered minor damage from wind, and there was no flooding to speak of at the hotel. You ask the owner which way the winds were blowing during the hurricane passage and she says, "From the north."
    (a) Explain briefly to the owner how storm surge occurs and why her hotel was not flooded.
    (b) The owner tells you that she owns a second hotel on a bay nearby and is sure that a hurricane won't cause damage there because it is away from the ocean shoreline. Explain to the owner *all* the destructive factors associated with hurricanes and why the second hotel is no safer from flooding than the hotel you are visiting.

21. What data do tropical forecasters have available to aid in predicting the path and intensity of tropical cyclones?

22. Differentiate between watches and warnings for tropical storms and hurricanes. Differentiate between coastal flood watches and warnings, and heavy surf advisories.

23. According to studies, the frequency of hurricanes has not increased despite the increase of sea-surface temperature over the last several decades. How might global warming suppress the formation of hurricanes? (Hint: what other environmental conditions are required for tropical storm genesis and how might they be affected by global warming?)

24. Scientific analyses suggest that global warming has led to an increase in hurricane intensity. Briefly explain the rationale behind this.

TEST YOUR PROBLEM-SOLVING SKILLS

1. Suppose a ship is designed to withstand seas during 120 knot winds. As the weather officer on the ship, you note that the winds outside are 30 knots. From a satellite image and Global Positioning System data, you pinpoint your position at 60 nautical miles from the center of a hurricane. You note that the eye of the hurricane hasn't changed position in hours, and that the diameter of the eye is 20 nautical miles. The National Hurricane Center has (correctly) forecast that the hurricane will not intensify. Your captain comes in and asks "Can we take'er through the eye without bustin' her to pieces?" What is your reply? Explain your reasoning.

2. You are the director of Emergency Management Services for the Florida Keys. A category 4 hurricane is approaching from the southwest at 10 miles/hour. The eye of the storm is predicted to first pass about 20 miles west of Key West, and then about 20 miles west of Key Largo 12 hours later, causing the right side of the eyewall to track along the islands and effectively destroy the entire island chain. There is one two-lane road connecting the Keys with the mainland. The southbound lane must remain open for emergency vehicles, so only one lane is available for evacuation. In summer, 100,000 residents live on the Keys. Past data taken on the bridge that connects the Keys to the mainland suggests that cars, during emergencies, typically hold three people, move no faster than 5 miles/hour, and maintain a spacing of about one car length. You want to have the Keys clear of people when the hurricane strikes Key West on the south end of the chain of islands. How far must the hurricane be from Key West when you order an evacuation? If the hurricane followed a straight path, where would it have been located geographically at the time you issued the evacuation order? (Assume the average car is 15 feet in length.)

3. Imagine that you are Captain Ahab's deckhand on his new ship *Typhoon*. While chasing a white whale across the Gulf of Mexico, Captain Ahab forgot to look at the weather forecast and has no idea that there is a category 3 hurricane in the Gulf of Mexico. The hurricane is slowly drifting northward. Captain Ahab fears nothing and exclaims that nothing is going to prevent him from capturing the whale and orders you to keep a westbound course through the eye of the hurricane. The mad captain takes the ship right toward the eye! Document your voyage through the hurricane by doing the following:
   (a) Plot a graph showing the qualitative pressure profile and wind speed profile that Ahab's onboard instruments measured during passage across the hurricane.
   (b) Describe the precipitation intensity and wind direction you experienced as you passed from east to west across the storm.
   (c) Draw a sketch of the hurricane as it might have looked from the Mobile, Alabama, WSR-88D Doppler radar site.

4. A cluster of thunderstorms moves off of Africa and over the Atlantic Ocean in mid-August. As a forecaster for the National Hurricane Center, you are responsible for forecasting whether or not the storm will organize into a hurricane. What weather products would you examine and what features would you look for on each product?

5. Suppose you are the emergency management services coordinator for Tampa, Florida. A category 5 hurricane is in the middle of the Gulf of Mexico. The probability that Tampa will have a direct strike within 36 hours is 10 percent. There are approximately 335,000 residents in Tampa and 2.7 million residents in the Tampa Bay Area (including Tampa, St. Petersburg and Clearwater). Residents can expect the evacuation to take 10 to 12 hours, provided there are no major problems. A typical evacuation will cost the city slightly more than $1 million per mile of coastline. Would you order an evacuation? On what basis would you make your decision?

USE THE SEVERE AND HAZARDOUS WEATHER WEBSITE

http://severewx.atmos.uiuc.edu

1. Track an Atlantic hurricane. Make a copy of the hurricane tracking chart from Appendix C. Navigate to the National Hurricane Center (NHC) website via the *Severe and Hazardous Weather Website.* At the NHC site you can view information on all current tropical cyclones.
   (a) Record the latitude, longitude, maximum sustained wind speed, and central pressure at twelve-hour intervals during the lifetime of the storm.
   (b) Examine the sea-surface temperature (SST) in the vicinity of the storm. (You can get this information from the "Current Weather" page of the *Severe and Hazardous Weather Website.*) How do you think the SST will influence the storm's development and intensification?
   (c) Access satellite imagery with derived upper-air wind fields, also available from the "Current Weather" page. Are the winds suitable for further development of the storm?
   (d) If the storm approaches the coast of the United States, examine the local NWS Doppler radar site near it and identify as many features as possible (eye, eyewall, spiral rainbands, regions of weak descent, etc.).

2. Examine satellite imagery via the *Severe and Hazardous Weather Website.* Focus your attention on the tropics in either the Atlantic or Pacific.
   (a) Print or save the image.
   (b) Locate the Intertropical Convergence Zone.
   (c) Locate any tropical disturbances, tropical depressions, tropical storms, or hurricanes.
   (d) Animate several images and comment on the direction of propagation of storms. At what latitudes are storms organizing most rapidly?

3. Access the Naval Research Laboratory (NRL) Tropical Cyclone Page from the "Current Weather" page of the *Severe and Hazardous Weather Website.* Find a typhoon in the western Pacific or a cyclone in the Indian Ocean. What countries are most threatened by these storms? Research the region and outline the vulnerabilities of these countries along with their warning systems.

4. Examine a satellite water vapor loop in the vicinity of a tropical cyclone. Summarize what information the images provide and discuss how these images can be used to identify regions of strong upper-level wind shear.

5. The next time a hurricane makes landfall in the United States, examine the radar-derived precipitation totals and local newspaper reports from the vicinity of the storm.
   (a) Over what region does the intense rainfall accumulate?
   (b) How does this region match with the area of the eye and eyewall?
   (c) Using the local newspaper reports, record where the worst damage occurred during the storm.
   (d) Do the worst damage reports coincide with the areas you would have expected to have the highest storm surge? Explain your findings.

CHAPTER TWENTY-FIVE

FLOODS

Flooding of New Orleans in the aftermath of Hurricane Katrina, August 2005.

KEY TERMS

arroyo
coastal flood
coastal flood warning
flash flood
flash flood warning
flash flood watch
flood stage
floodplain
frontal overrunning
Hawaiian Fire Hose
levee
monsoon
North American Monsoon
Pineapple Express
river flood warning
small stream flood advisory
stage
storm total precipitation
training
tropical storm
watershed
widespread flood

Flooding is the number one weather-related cause of property damage in the United States and the number two weather-related cause of deaths worldwide (drought is number one). Personal property and agricultural losses in the United States exceed $1 billion in most years, although catastrophic flood events can cause far more damage. The Great Flood of 1993 in the Mississippi/Missouri basins cost over $15 billion, and most of Hurricane Katrina's $80 billion of damage in 2005 was the result of flooding. According to the Federal Emergency Management Agency, more than 20,100 communities in the United States currently enforce local floodplain management ordinances pertaining to new and existing development. These floodplain management ordinances reduce flood damage by an estimated $1 billion per year. Nevertheless, the impact of flooding continues to increase as the population increases and more people move closer to waterways and coastlines.

The duration and intensity of floods depend on many factors, only some of which are related to the weather. The intensity, duration, and number of rain events that occur in a region, and the size of the rainfall area and its orientation with respect to a river drainage basin, are directly related to weather systems. Snowmelt from previous storms, ice jams on previously frozen rivers, and the saturation level of soils are indirect weather effects that can enhance floods. Non-weather-related factors include land use, levees and dams and their effect on drainages, and the topography along a drainage, which can vary from wide plains to narrow canyons. The failure of levees and the below-sea-level elevation of New Orleans were largely responsible for the magnitude of the Katrina disaster in 2005.

Hydrologists often use terms like "100-year flood" or "500-year flood" to convey the idea of a flood's magnitude. These numbers are developed by extrapolating historical data to longer time periods. The term "100-year flood" means that, in any given year, there is a one in 100 chance of a flood of that particular magnitude. In other words, the probability of a flood of that magnitude in any given year is 1/100 or 1 percent (see Table 25.1). The actual amount of water that causes a particular flood (e.g., a 100-year flood) varies from river to river and even along a particular river. A danger in the use of such statistics is that an unprecedented flood can occur, especially in areas where land use has changed substantially, redefining the criteria for "100-year flood" and "500-year flood" at a particular location.

TABLE 25.1 Flood Probabilities for Any Single Year

"X"-Year Flood	Probability of Occurrence in a Year
5	20%
10	10%
25	4%
100	1%
500	0.2%

FOCUS 25.1
Worldwide Floods

Floods occur on all continents except Antarctica. Figure 25A shows the areas that experienced flooding in 2006. This map is typical of the longer-term pattern: the largest number of flooding events in the world occurs in Southeast Asia in a region stretching along the south side of the Himalayan Mountains from India to China, and then eastward from the Himalayas along the Yangtze River Basin in China. Southeast Asia receives heavy rainfall during the Asian summer monsoon.

(continued)

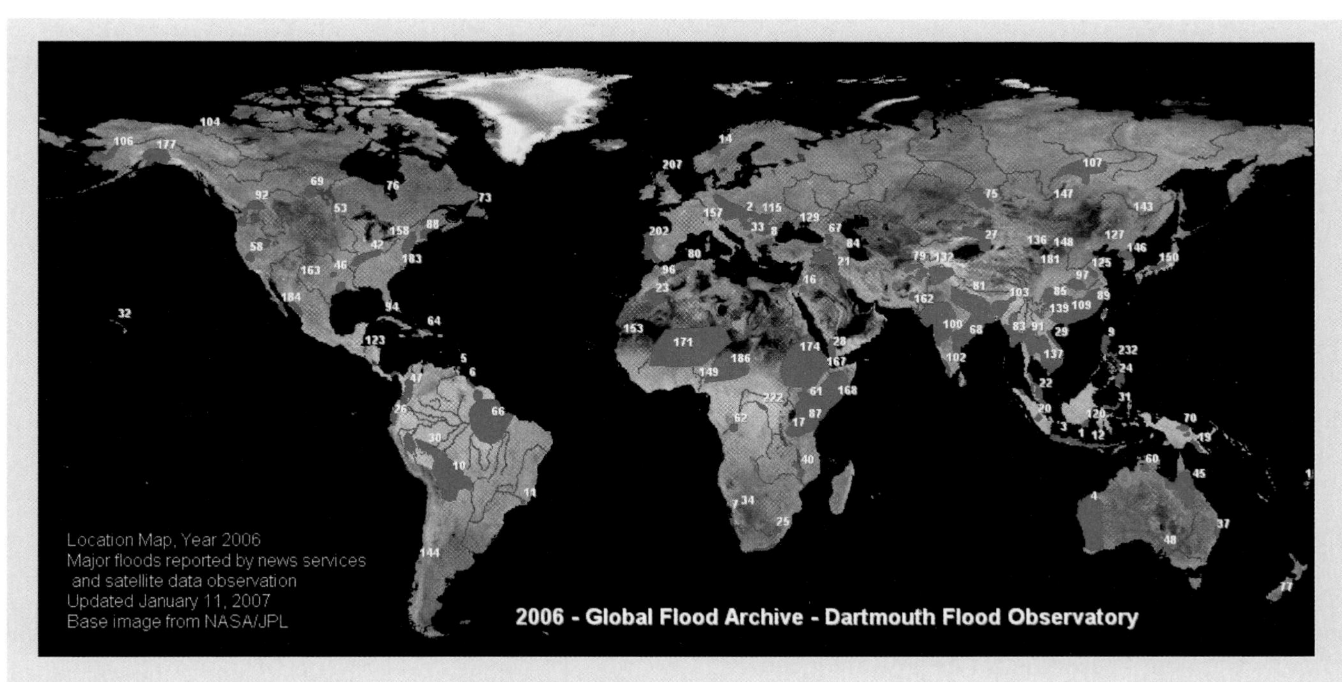

FIGURE 25A: Worldwide locations of significant floods during 2006.

The *monsoon* is a seasonal circulation driven by the large contrast in heating between the Asian landmass and Indian Ocean to its south. During summer, the Asian landmass is much warmer than air over the Indian Ocean. In response to this temperature difference, hot humid air flows northward from the tropical Indian Ocean into Southeast Asia. As the air flows toward the Himalayas, it rises first along the terrain, and then buoyantly as deep thunderstorms are triggered. Thunderstorms form day after day, creating extreme rainfall. For example, the heaviest monthly rainfall ever recorded on Earth occurred in Cherrapungi, India, when 336 inches of rain fell in July 1861. Areas of India and Bangladesh are often deluged, creating disastrous floods on major rivers such as the Ganges, Brahmaputra, and their tributaries. Flooding due to the monsoon in 2004 resulted in over 1500 deaths and 20 million people stranded on high ground.

The Yangtze River and its tributaries have been the site of enormous floods. During the twentieth century, disastrous floods in 1931 and 1959 together produced nearly 6 million fatalities. Floods in 1908, 1911, 1935, 1938, and 1939 each led to between 100,000 and 500,000 deaths. These floods occur east of the Himalayan landmass when air moving northward from the Indian Ocean in the summer monsoon circulation meets cool, dry air moving eastward across Northern China from central Asia. The boundary between the two air–masses, called the Mei-Yu (plum-rain) front, is nearly stationary in July and is often oriented directly over the east-west running Yangtze River. Large thunderstorm systems regularly erupt along the Mei-Yu front, especially in July, creating near-continuous heavy rainfall. As in most years, flooding in China caused hundreds of deaths and affected hundreds of thousands of people in the summers of 2004, 2005, and 2006. The impacts of flooding in China are worsened by decades of deforestation and intensive farming, and by the fact that millions of people live on reclaimed floodplains. Summer flooding also affects other countries in eastern Asia, including North Korea, where thousands of people reportedly died from flooding in July, 2006.

Other continents also experience many floods. In Europe, extreme flooding commonly occurs along the Alps and Carpathian Mountains. Italy is particularly prone to disastrous flooding when deep extratropical cyclones develop over the Mediterranean Sea south of Genoa. East of these low-pressure centers, very warm moist air is transported directly into the Alpine Massif from the Mediterranean, creating large thunderstorms and violent flash floods in the Alpine Valleys.

(continued)

(continued)

In northern Europe, floods occur as large slow-moving extratropical cyclones move into the continent off the North Atlantic Ocean and produce heavy rain. African floods are concentrated on the eastern side of the continent where summer season thunderstorms along the Intertropical Convergence Zone (see Chapter 24) develop over the mountains and are fed by moisture from the west Indian Ocean. Flooding in Ethiopia caused hundreds of deaths in the summers of 2005 and 2006. Farther south, tropical cyclones make landfall on the island of Madagascar and on the mainland in Tanzania and Mozambique, creating flooding in these regions. Many of Australia's floods are also associated with landfalling tropical cyclones, mostly along the northern and eastern sides of that continent. South American floods occur within the Amazon River Basin, and south of the Amazon Basin over Bolivia and Argentina (e.g., Figure 25A). Floods periodically occur along the west side of the Andes, particularly during El Niño years.

TYPES OF FLOODS

Floods are differentiated based on their predictability and their location. ***Flash floods*** occur rapidly with little or no warning. These floods typically are localized, short-duration floods, mostly on smaller rivers and streams. Slow-moving thunderstorms are the primary precursors of flash floods; high rainfall rates over short time periods trigger most flash flooding events. The importance of thunderstorms is evident from Figure 25.1, which shows the monthly distribution of flash floods that occurred during the 1970s. The vast majority occurred during the summer months, when thunderstorms are common. Occasionally, nonmeteorological events, such as a dam break during an earthquake, can also lead to a flash flood. Because flash floods are unanticipated, they often result in many injuries and fatalities.

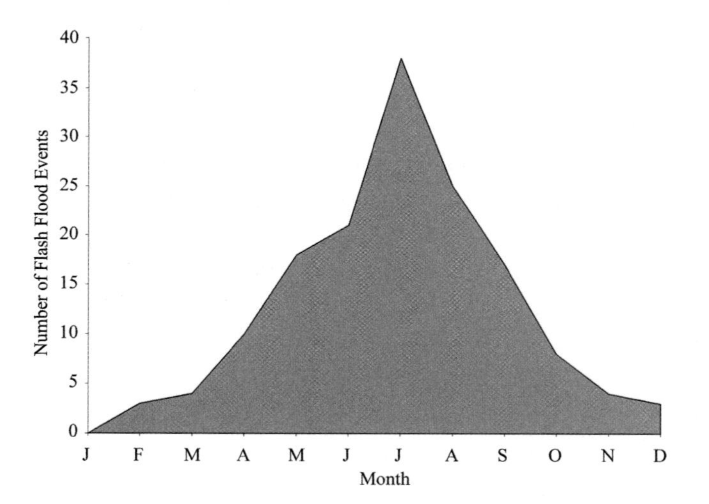

FIGURE 25.1: The monthly distribution of 151 flash floods occurring over a five-year period in the United States. The peak in flash flood activity coincides with the time of year when thunderstorms are most common.

Widespread floods occur when a large amount of rain falls over a ***watershed*** for many days so that significant portions of river basins draining the watershed are inundated for long periods of time. Water levels along rivers rise slowly, ultimately resulting in water overflowing the natural or artificial confines of the river. Widespread flooding events develop more slowly than flash floods and tend to last a week or more. They typically cover a large area, often along a major river. Because widespread floods are slow to develop, forecasters are better able to predict their extent and warn residents to move to safer ground. Widespread floods are sometimes described as "leisurely disasters," slowly and predictably destroying homes and property. The size and duration of widespread floods lead to much more property damage and economic losses than flash floods. Fatalities are generally minimal because warnings are issued well in advance of the disaster.

Coastal floods occur when a rise in the ocean surface due to storm surge develops during hurricanes and strong extratropical cyclones. The inland extent of coastal flooding depends both on storm intensity and the geographic and topographic features of the region. In general, coastal flooding is more extreme along the East and Gulf Coasts of North America than along the West Coast because of the greater number of tropical and extratropical cyclones and the generally flat coastal topography. This problem is compounded in areas such as New Orleans, where land that is below sea level has been reclaimed by engineering projects such as the construction of the city's system of levees. Like flash floods, coastal floods can develop as a result of non-meteorological phenomena, such as earthquake-generated tsunamis.

While it is often possible to identify a flood as falling under one of the three types defined above: flash, widespread, and coastal, not every flood event can be categorized so clearly. It is possible, and not uncommon, for a particular flood event to be comprised of a combination of these flood types. For example, landfalling tropical cyclones cause coastal flooding, but most of the deaths from flooding in tropical systems occur inland (e.g., 150 deaths in the Virginia mountains from Hurricane Camille's remnants in 1969; 80 deaths in southern Ontario from Hurricane Hazel in 1954). Another example of a flood that cannot be placed into a single category is the Great Flood of 1993 (see Focus 25.4), which was a widespread flood but had many local flash floods within it. The combination of different flood types into a single event can increase property damage, injuries, and loss of life.

FOCUS 25.2

Watersheds, River Stages, and Floodplains

The geographic area that drains into a river or stream is called a *watershed*. The Mississippi River watershed covers much of the central United States, while a small stream may have a watershed that consists of a single valley. Rivers draining watersheds fluctuate in height depending on rainfall and the rate of drainage of water stored in the soils, lakes, and tributaries of the watershed. The height of the water surface of a river is called its *stage*. A river's stage naturally fluctuates as rain falls and drainage occurs within a watershed. When a river rises to a level where it begins to flood agricultural lands or has potential to damage property, the river is said to be at *flood stage*. Flood stage varies from river to river and along each individual river. Hydrologists often use flood stage as a reference level to express the intensity of a flood, since flood stage is the level at which significant economic losses begin to be incurred by residents of the riverbanks.

The *floodplain* of a river is the land that, from historical and geologic records, has been repeatedly inundated by the river's floodwaters. Floodplains are attractive because of their natural beauty and because land on the plain is fertile. Each time a river rises onto its floodplain, it takes with it nutrient-rich sediments from the riverbed and deposits them onto the soil. Residents of the floodplain live with the threat that the river will reclaim their land during a flood. Over history, residents of floodplains have made many attempts to modify the natural floodplains by channeling, damming, and otherwise controlling the natural flow of rivers.

A method commonly used to control a river is to construct *levees*. A *levee* is an embankment created to prohibit the flow of water onto a floodplain (Figure 25B). A levee protects a floodplain, but also increases the flow rate and height of the river because it confines the water to the river channel. Some levees are concrete, but most are made of earthen materials such as soil, clay, and rock. Vegetation quickly covers earthen levees and many are hardly discernable as man-made structures. Levees are constructed to hold back a particular level of the river. In many agricultural areas, this level typically corresponds to the estimated height of a 50- or 100-year flood. Problems arise when the drainage exceeds the capacity of the levee. The water will then spill over the levee, often destroying the levee at the point of the break (Figure 25C). When this

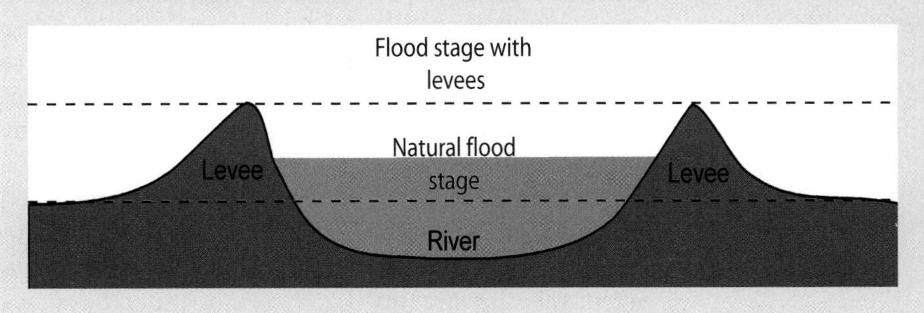

FIGURE 25B: A cross section through a river being channeled by levees.

(continued)

(continued)

FIGURE 25C: A levee break during the Great Flood of 1993. The size of the levee break can be better understood by comparing it to the farm buildings near the break.

occurs, the floodplain in the vicinity of the break may be extended beyond its normal area in the absence of the levee system. This is because the river is artificially much higher, trapped between the levees. Another potential problem for a levee arises when river levels remain high for an extended period of time, but do not exceed the levees. The water can produce very high pressure at the base of a levee, which can create a leak and ultimately a break. Depending on the size of the river and the height of the floodwaters, a levee failure may lead to a localized flash flood.

CHECK YOUR UNDERSTANDING 25.1

1. Explain what is meant by the term "50-year flood."
2. List differences between flash floods, widespread floods and coastal floods.
3. How do the terms watershed, flood plain, river stage, and levee relate to flooding?

NORTH AMERICAN FLOOD WEATHER PATTERNS

A number of weather patterns that affect North America are associated with flooding. These weather patterns vary geographically and seasonally, and are among the many factors that ultimately affect the intensity and duration of a flood. A common feature of many of these weather patterns is their slow movement, a characteristic that leads to rainfall over a watershed for a long duration.

Flooding Following Landfall of Tropical Cyclones

Between June and November in North America, tropical cyclones (hurricanes and tropical storms) affect the Gulf Coast of the United States, the east coasts of the United States, Canada, Mexico, and Central America, as well as the Caribbean Islands. These storms are well-known for the devastation they cause due to storm surge and associated coastal flooding (Chapter 24). However, tropical cyclones that are slow-moving during and after landfall can also produce extreme amounts of rainfall inland and inundate streams and rivers. Flooding from Atlantic tropical cyclones particularly affects the Eastern Coastal Plain of the United States and interior regions into the Appalachian Mountains (Figure 25.2). Communities hundreds of miles from the coast can experience extreme flooding, especially when the remains of the cyclone are slow-moving or stall.

Tropical cyclones produce exceptional rainfall in short time periods. For example, Hurricane Floyd (1999), which caused record flooding, produced 15 to 20 inches of rain in 24 hours along the coastal plains of both North Carolina and Virginia. Hurricane Georges (1998) produced nearly 20 inches of rain inland of the Gulf Coast near Mobile, Alabama (see Figure 2E). Disastrous floods from tropical cyclones moving onshore along the East Coast are enhanced by the Appalachian Mountain chain, which lifts the moist tropical air as it flows into the ridges. The enhanced vertical air motion provided by the flow up the mountain slopes creates additional rain, which is then channeled into the streams and rivers in the valleys. An example of rainfall enhancement in the Appalachians occurred during Hurricane Fran (1996). Fran produced record flooding on the coastal plain of North

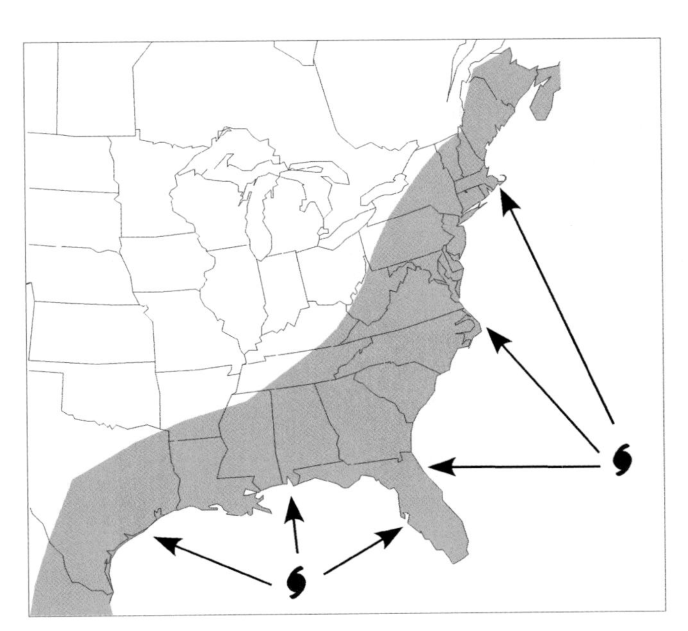

FIGURE 25.2: The region of the United States affected by flooding from tropical cyclones.

Carolina, and even more rain as the storm moved inland and air was lifted by the mountains in Virginia.

Occasionally, the moisture from decaying tropical cyclones can become incorporated into the circulations of extratropical cyclones as the tropical systems move northward into the mid-latitudes. In June 1972, Hurricane Agnes made landfall in Florida as a category 1 hurricane. As it moved up the East Coast to New England it combined with an extratropical cyclone and eventually caused heavy rainfall and flooding from Virginia to New York. In all, 122 deaths were attributed to flooding associated with Agnes. Note that Agnes caused more fatalities than Hurricanes Andrew (1992; category 5) and Hugo (1989; category 4) combined. This flood was the deadliest and most expensive on record for the northeast United States.

A tropical cyclone does not need to achieve hurricane intensity to cause substantial flooding. For example, in tropical storm Allison (2001) Houston experienced the most severe flooding in the city's history with rainfall rates as high as 26.5 inches in 24 hours (see Online 25.1). During the five days of the storm, precipitation totals ranged across the area from two inches to almost three feet. The storm moved slowly and tracked over the Houston area twice (Figure 25D) while continuously drawing moisture from the Gulf of Mexico to sustain its high rainfall intensities. The Houston metro area is highly urbanized, so the absorptive capacity of the surface is significantly diminished. Virtually all precipitation became runoff; and the streams, rivers, and bayous were not able to drain the area quickly enough to prevent massive flooding.

ONLINE 25.1

Tropical Storm Allison

Early in the hurricane season of 2001, Tropical Storm Allison, which never reached hurricane intensity, produced record rainfall over portions of southeast Texas and southern Louisiana, resulting in catastrophic flooding. Allison is the costliest *tropical storm* in United States history. Allison was named a tropical storm on 5 June and later that same day made landfall south of Houston, Texas. The storm moved inland very slowly and then drifted out to the Gulf of Mexico, making landfall again on 10 June in Southern Louisiana before moving across the southeastern United States. Allison is the only tropical storm in recorded history to move back over the Gulf of Mexico after making landfall in Texas (Figure 25D). This unusual track enabled Allison to reinforce its moisture supply, creating a deluge in Houston, and it undoubtedly was a factor in a "double maximum" in rainfall produced by Allison in Texas and Louisiana. In these two states alone, Allison's flooding caused $5 billion in damage and 22 fatalities.

Tropical Storm Allison was a persistent feature of the surface weather maps for two weeks. Online, you can observe the evolution of the accumulated precipitation as compiled by the Southeast Regional Climate Center as Allison slowly made its way eastward and northward from Texas to the Carolina coast during 4 to 17 June 2001. A second animation shows the evolution of the radar reflectivity during Allison. By the end of the period, the total precipitation exceeds 27 inches in two regions, one in the Houston area and other in southern Louisiana. Eastern North Carolina also

(continued)

(continued)

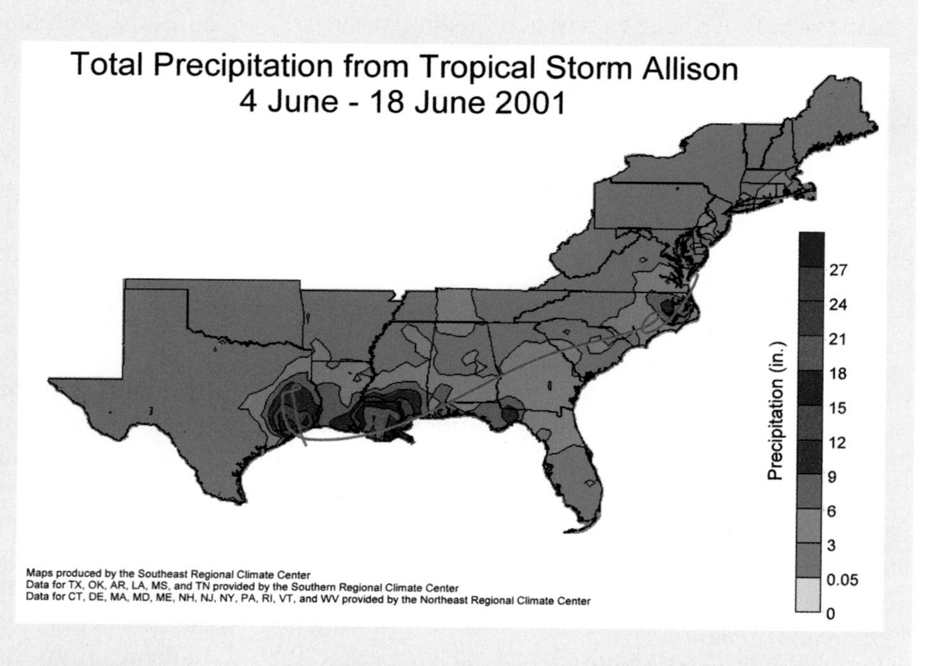

FIGURE 25D: Total precipitation between 4 and 18 June 2001 as Tropical Storm Allison moved across the southern United States. The track of Allison is also shown in red.

accumulated rainfall in excess of a foot during this remarkable storm's final few days on land. An intriguing feature of the animation is the localized nature of the very heavy precipitation, which has a "bull's-eye" appearance in the most severely affected areas.

Flooding from Other Tropical Weather Systems

Most flooding in Central America, the Caribbean Islands, and Hawaii occurs during the passage of tropical cyclones. However, flooding can also occur in these tropical regions when clouds with embedded thunderstorms anchor over higher topography during periods when the atmosphere is moist and conditionally unstable. These conditions arise most often during the passage of easterly waves in the late spring, summer, and early fall, or when cold fronts move southward into tropical regions, weaken, and stall. The problem of flooding in many regions of the tropics is exacerbated by deforestation and poor infrastructure.

In the country of Haiti on the island of Hispaniola for example, more than 90 percent of the country is deforested, in large part because most of its 8 million people use charcoal to cook. Deforestation destabilizes hillsides, which can turn into mudslides during heavy rain. The slides can be large enough to bury villages. For example, extreme rainfall on the Island of Hispaniola in late May of 2004 left more than 3,300 people dead or missing in Haiti and the neighboring Dominican Republic. The magnitude of disasters such as this is compounded by poor infrastructure.

Flooding from Mesoscale Convective Systems

Thunderstorms organized as either frontal squall lines or mesoscale convective systems (MCSs, Chapter 18) most commonly generate floods between the Rocky Mountains and the Appalachian Mountains. More rarely, this type of flooding also occurs over the Appalachians or along the East Coast. The thunderstorms associated with these floods typically form in the conditionally unstable air on the warm side of the boundaries that trigger the storms. The thermal and moisture contrasts across fronts or other boundaries that produce floods often are very weak, and the boundaries are near-stationary. These conditions are common in late spring and summer, and it is during these seasons that floods associated with frontal squall lines and MCSs have their greatest frequency.

A frontal squall line consists of a long line of thunderstorms that develop as conditionally unstable air is lifted along a frontal boundary (Chapter 18). Flash floods associated with frontal squall lines occur when the frontal boundary is nearly stationary and the winds, both at the surface and aloft, flow essentially parallel to the front. Under these conditions, individual thunderstorm cells located along the frontal boundary, and any new thunderstorm cells that are triggered along the front, move essentially parallel to the front. Since the front does not advance, new thunderstorm cells continually move over the same region. The media have popularized the term ***training*** to describe this process, since, as boxcars of a train pass over the same location along a track, individual thunderstorms pass over the same location along a front (see Figure 25.3).

In extreme cases, frontal squall line flooding events may last several days. During this time the front may move very slowly, so the location of the flooding will shift eastward and southward, ultimately spreading the flooding over a wider area. In the short term, the thunderstorms produce flash flooding, but if the storms persist for several days they often lead to widespread flooding in the larger rivers that drain the watershed. Often the intensity of the thunderstorms and rainfall will fluctuate with the diurnal cycle of solar heating, the storms intensifying in the afternoon and into the night and then weakening in the morning, only to regenerate the next day. In this manner, the saturation of the soil and the increased levels of streams and rivers can precondition the area to serious flooding from each of the subsequent rain events. This type of preconditioning was one key factor in the "Great Flood" of 1993 in the Mississippi River Basin (see Focus 25.4) and in the flooding that affected the southern Great Plains during 2007 in June and July.

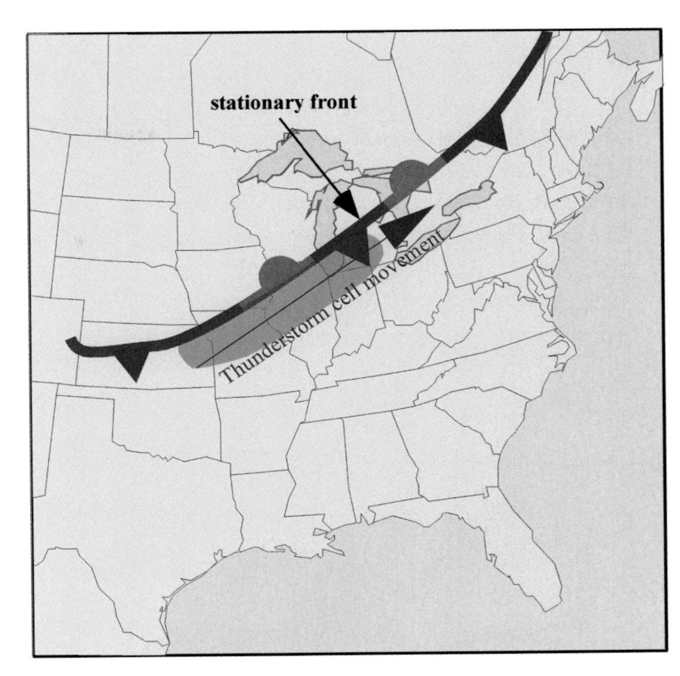

FIGURE 25.3: An example of a weather pattern during a flood associated with a squall line along a stationary front. Individual thunderstorm cells move along the front producing rain over the same location in a process called *training*. The weather pattern shown here is based on an actual flood that occurred on 13 to 14 August 1987. The storms produced over 9 inches of rain in 18 hours in Chicago, Illinois. The green area denotes the region experiencing flooding.

Mesoscale convective systems also cause flash floods. The motion and intensity of these storm complexes are harder to predict because they are not

FOCUS 25.3

Taming the Mighty Mississippi?

In the early part of the 20th century, the United States underwent significant development and expansion, and discovered that the nation's waterways were an efficient, low-cost way to transport goods and property. The nation's waterways also provided an inexpensive source of water for industries, which located facilities along the riverbanks. The potential of floods to disrupt the nation's economy grew with the industries along the rivers. Nowhere was this development more important than along the Mississippi River and its major tributaries, the Ohio, Tennessee, and Missouri Rivers. A tragic flood on the lower Mississippi River in 1927 moved the country to action. During its peak, the flood inundated over 26,000 square miles of land in seven states. Over 600,000 people were evacuated and in some places the river ran 80 miles wide.

As a result of this flood, Congress passed the Flood Control Act in 1928. Water control projects were initiated along the river and its tributaries. The U.S. Army Corps of Engineers was placed in charge of controlling the Mississippi River and other major rivers and waterways east of the Rockies. This is when President Hoover gave the famous charge to "tame the mighty Mississippi."

(continued)

(continued)

The result was an intricate system of levees and dams that today control a significant portion of the Mississippi watershed's waterways. For 65 years, the Mississippi and its tributaries survived seasons of heavy rains without a major catastrophe like the disaster of 1927. The dams, levees, and flow controls placed on the rivers saved many communities from disaster time and time again. However, in 1993, rains of extraordinary magnitude and duration occurred throughout much of the upper Mississippi watershed for over a month. Estimated as a once-in-500-year occurrence, the rains raised the river to extreme heights, causing levees to break along the tributaries of the Mississippi and the Mississippi itself. The "Mighty Mississippi" again laid claim to its floodplain (Figures 25C, 25F, 25G).

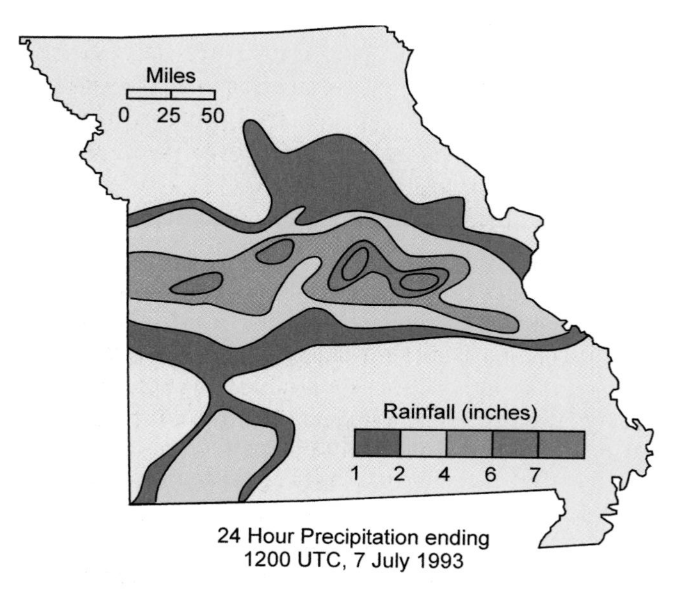

FIGURE 25.4: Rainfall in Missouri during the 24-hour period ending at 1200 UTC, 7 July 1993. An MCS that moved across the state produced the rain. The flooding was one event of many during the great 1993 Mississippi Flood.

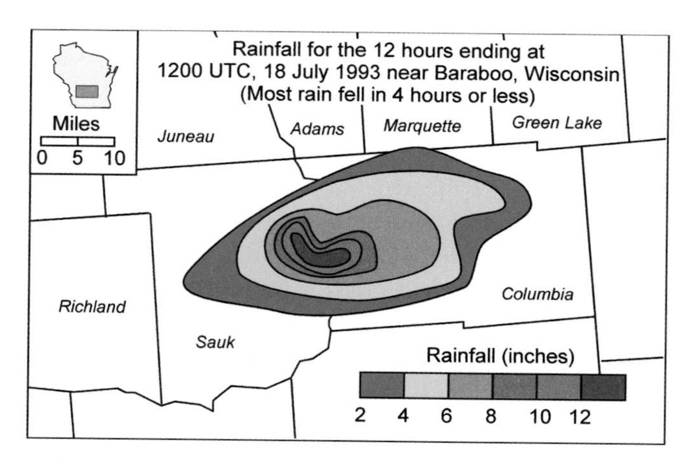

FIGURE 25.5: Rainfall for the 12 hours ending at 1200 UTC, 18 July 1993 in south central Wisconsin. A nearly stationary MCS produced the rain, and was part of the Great 1993 Mississippi Flood.

always associated with distinct frontal boundaries. Mesoscale convective systems can move very slowly, particularly in summer when the upper tropospheric flow is weak. New thunderstorm cells in the storm complexes are triggered by the gust front outflow boundaries of older thunderstorms. In this way, thunderstorms keep regenerating. The result is a heavy rain accumulation over a localized area. Mesoscale convective systems were an important component of the "Great Flood" of 1993 in the Mississippi River Basin. Figures 25.4 and 25.5 show examples of rainfall patterns during individual one-day rainfall events that were part of the two-month-long deluge in the Mississippi Basin. In both these cases, MCSs produced the rainfall. Figure 25.4 shows the rainfall that fell in Missouri during a twenty-four-hour period on 6 to 7 July 1993. During this period, an MCS moved across the state from west to east, locally producing 5 to 7 inches of rainfall. In Wisconsin on 18 July 1993, an MCS produced over 12 inches of rain in the 12-hour period ending at 1200 UTC (Figure 25.5). Most of the rain fell in 4 hours. During the early summer of 1993, events such as these occurred nearly every day somewhere in the Mississippi River Basin, creating local flash floods and one of the greatest widespread floods in U.S. history. A more recent example of frequent convective events, including MCSs as well as many single-cell thunderstorms over a two-month period, occurred in the southern Great Plains, especially Texas, during the late spring and early summer of 2007. Online 25.2 contains many examples of convective signatures in daily precipitation totals during June and July, 2007.

FOCUS 25.4

The Great Flood of 1993

One of the greatest floods in the history of the United States occurred in 1993 along the Mississippi River and its tributaries. The flood, identified as a 500-year event, inundated more than 20 million acres of land in nine states (Figure 25E). Economic losses were estimated between $15 and $20 billion. The effects of the flood were far-reaching. The flood eliminated barge and rail transportation and decimated agricultural production, effectively halting the export of grain and other commodities along the Mississippi River for months. More than 75 towns were completely inundated, including some that were never rebuilt and one, Valmeyer, Illinois, that was completely moved to a safer location on a nearby bluff. Forty-eight people lost their lives during the flood, and another 54,000 were forced to evacuate their homes.

By the time the flood waters receded, over 50,000 homes were completely destroyed or substantially damaged. A large portion of the area was declared a federal disaster area, including the entire state of Iowa and portions of the eight other states affected by the flood. A spectacular LANDSAT image of the 1993 flood appears on the bottom of Figure 25F. The image shows the junction of the Missouri, Mississippi, and Illinois Rivers just north of St. Louis, Missouri. Contrast this image with the top image, which shows a LANDSAT image of the same location during the drought year of 1988 (see Chapter 26). The magnitude of the flood along these major rivers in 1993 is immediately apparent. The disaster can be further appreciated by considering that the larger rivers visible on this image are fed by many smaller streams, virtually all of which experienced record flooding. A view of the Mississippi River west of Illinois is shown in Figure 25G, in which a levee (breached upstream) extends diagonally across the photograph.

Although the Great Flood occurred during the summer of 1993, the stage was set for the flood months earlier. Between 1991 and 1993, a persistent El Niño enhanced the moisture influx from the Pacific Ocean. The summer of 1992 saw record

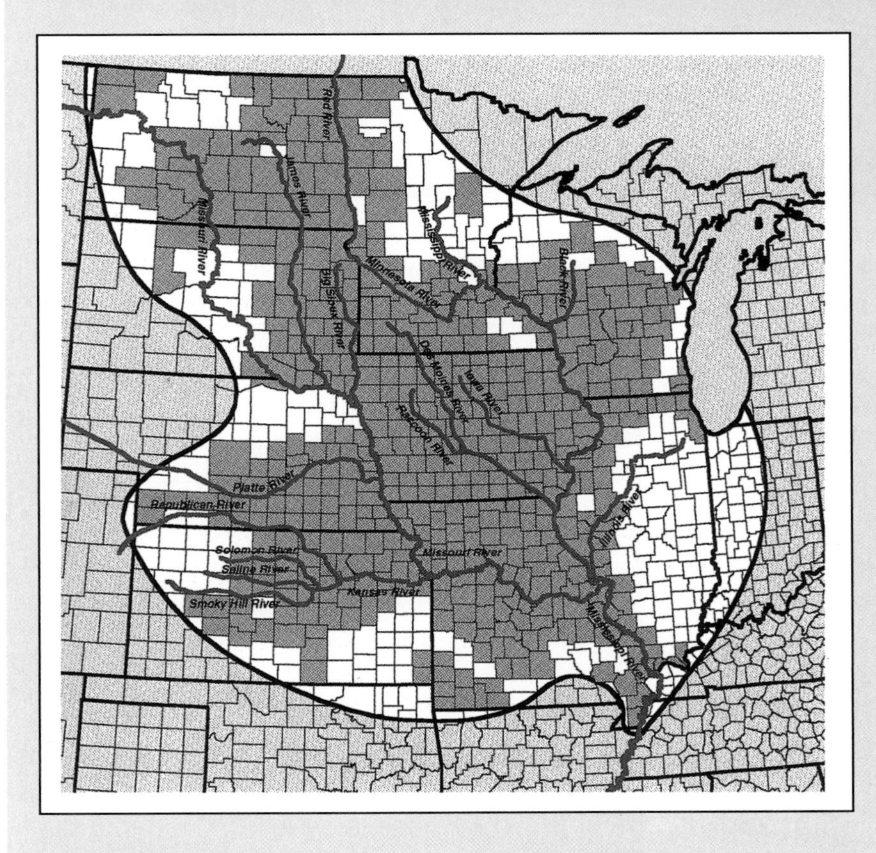

FIGURE 25E: The area within the heavy line is the upper Mississippi drainage area affected by the 1993 flood. Flood-affected counties that received federal disaster assistance are highlighted in orange.

(continued)

(continued)

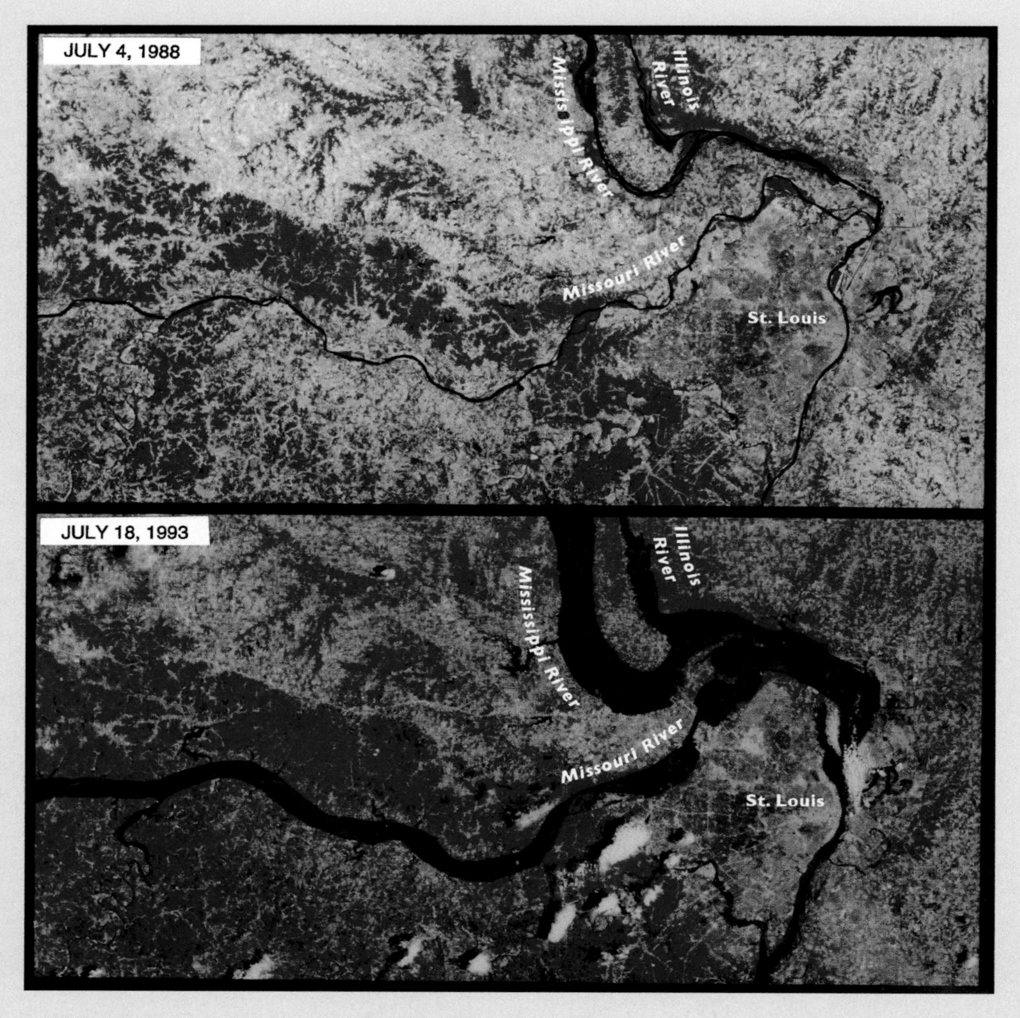

FIGURE 25F: LANDSAT images of the junction of the Mississippi, Missouri, and Illinois Rivers in Missouri and Illinois on (top) 4 July 1988, during the severe drought of 1988, and (bottom) 18 July 1993, during the Great Flood of 1993.

rainfall for many stations in the Midwestern United States. In addition, the spring months of 1993 were the wettest on record for the Upper Midwest. In the late spring of 1993, the Bermuda High, a semitropical high-pressure center normally located well off the southeast coast of the United States near Bermuda, strengthened and moved westward over the southeastern coast. The strong high pressure off the coast provided for an influx of warm, moist Gulf air into the Midwest. The abnormal strength of this high-pressure system is evident from Figure 25H, which shows (A) the average 850 mb flow between 5 June and 19 July 1993, and (B) the departure of the 1993 850 mb flow from the long-term climatological flow. Panel (B) shows a strong, persistent, southerly flow from

FIGURE 25G: The Mississippi River during the Great Flood of 1993. A levee is visible as a linear feature extending diagonally across the photo.

(continued)

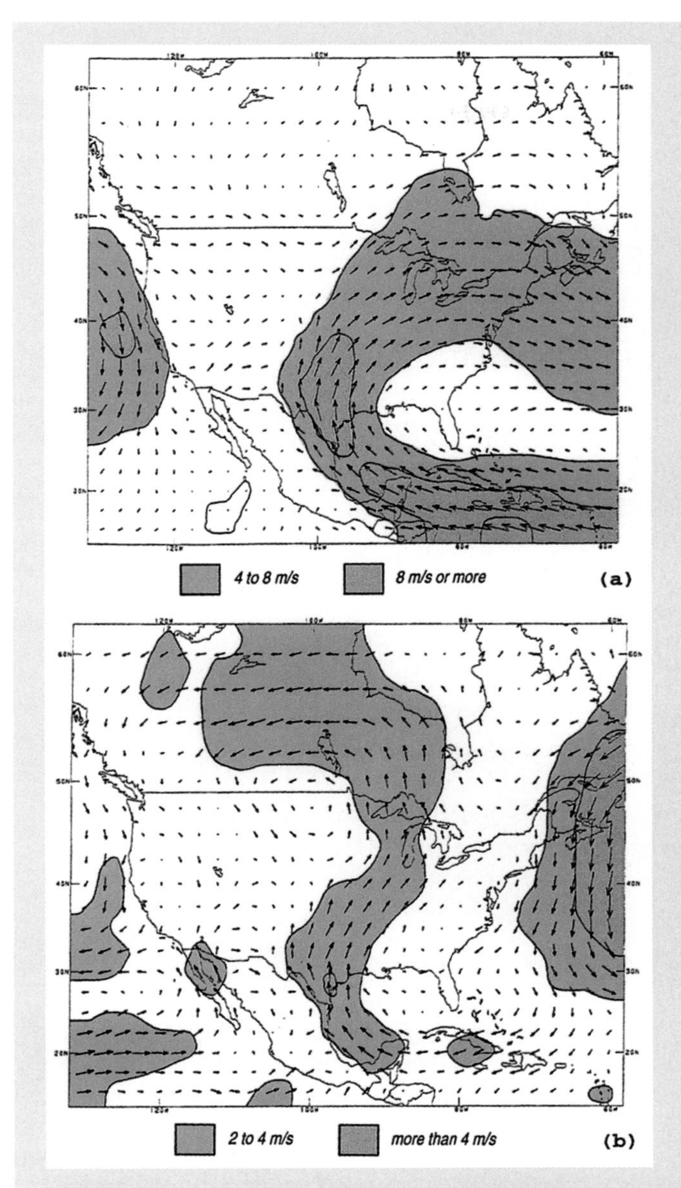

FIGURE 25H: Mean 850 mb flow for the period 5 June – 19 July 1993 for (A) the actual wind field, and (B) the departure from normal (base period 1979–1983). The arrows represent the direction and relative strength of the (A) wind or (B) wind anomaly.

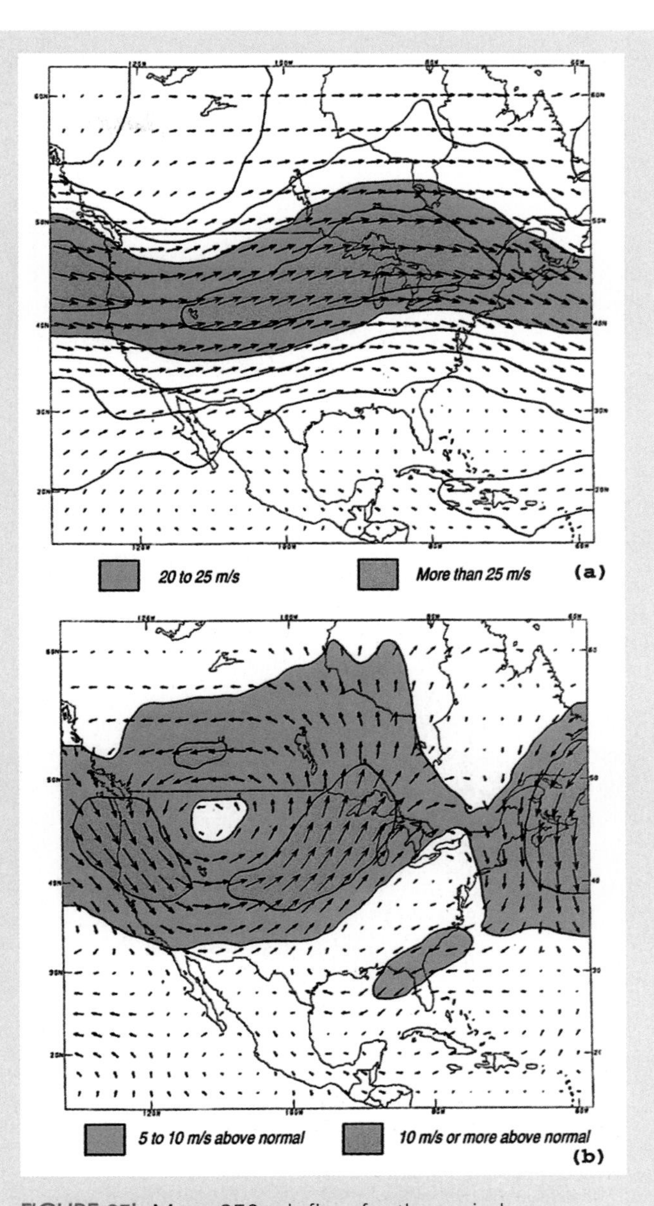

FIGURE 25I: Mean 250 mb flow for the period 5 June – 19 July 1993 for (A) the actual wind field, and (B) the departure from normal (base period 1979–1988). The arrows represent the direction and relative strength of the (A) wind or (B) wind anomaly.

the Gulf of Mexico that illustrates the unusual strength of the Bermuda high-pressure center in 1993 and its effect on moisture transport into the Mississippi Valley.

At the same time the jetstream, which normally is located in Canada during the early summer, formed a trough over the Rockies with the eastern branch over the Upper Mississippi River watershed. The unusual nature of this jetstream pattern can be seen by examining Figure 25I and comparing (A) the average 250 mb flow between 5 June and 19 July 1993 with (B) the departure of the 1993 250 mb flow from the long-term climatological flow. The trough is evident in panel (B). Beneath the jetstream, a nearly stationary front over the Missouri-Mississippi Valley separated two airmasses, a cool airmass over the western High Plains and the warm, moist, conditionally unstable

(continued)

(continued)

air arriving from the Gulf of Mexico. The front provided the lifting to trigger an almost daily series of frontal squall lines and/or MCSs over the basin. The resulting flash floods and the ensuing widespread flood were unprecedented.

The Great Flood was the first major flood event to test the WSR-88D Doppler radar system of the National Weather Service. The Radar System did an exceptional job of providing river forecasters with early warning capabilities. The radar data were available immediately to forecasters, while gauge data were only available after floods were already in progress. Gauge-measured rainfall and radar estimates of rainfall were generally in good agreement. The radar data gave forecasters the lead time necessary to save many lives.

ONLINE 25.2

Frequent Convective Rain Events in 2007 Produce Flooding in Texas

The spring and early summer of 2007 brought flooding rains to the southern Great Plains of the United States. Particularly hard hit were Texas, Oklahoma, and Kansas, including the Dallas–Ft. Worth, Oklahoma City, and Kansas City metropolitan areas. During the month of June, monthly rainfall totals exceeded a foot in many locations. The heavy rain was the result of repeated convective events, including mesoscale convective systems and single-cell thunderstorms that formed in a moist airmass that remained over the area for much of the period from early May through mid-July. The online animation shows the daily radar-derived precipitation totals for the period from mid-June to mid-July, 2007. The broader swaths resulted from the mesoscale convective systems, including several squall lines, while the narrower (~10 km) swaths were produced by single-cell storms, including supercells. An interesting feature of this period is that the saturated ground led to much more evaporation from the surface than is usual in this area, adding to the moisture for rain on subsequent days. This recycling of moisture represents a positive feedback effect that enhances the severity of widespread flooding.

CHECK YOUR UNDERSTANDING 25.2

1. How can a tropical storm cause more damage and fatalities than a strong hurricane?
2. How does topography enhance rainfall from tropical cyclones that make landfall in the eastern United States?
3. Where in the United States does flooding associated with squall lines and MCSs typically occur?

Flooding from Frontal Overrunning

During the cold season, east-west oriented warm fronts often develop across the southeastern United States. Sometimes these fronts stall, or move so slowly that they are essentially stationary. When such a front is present, warm, moist air originating over the Gulf of Mexico flows northward over the cool air–mass north of the front. This process, called ***frontal overrunning***, produces clouds and rain north of the frontal boundary. If the air rising over the front is conditionally unstable, heavy rain and thunderstorms can develop north of the surface frontal position. Storm cells typically form over the front and move northeastward with the upper level flow. If the front is present over a region for several days, the rain accumulation can lead to local flash flooding or even widespread flooding on some rivers. These problems are exacerbated when rain falls over the Appalachian Mountains, since rain is quickly channeled into the valleys. Additional problems occur because of cold weather, since the flooded region is north of the front where the air can be quite cool. People trapped in flooded areas may suffer from hypothermia, and rescuers have to work in cold conditions.

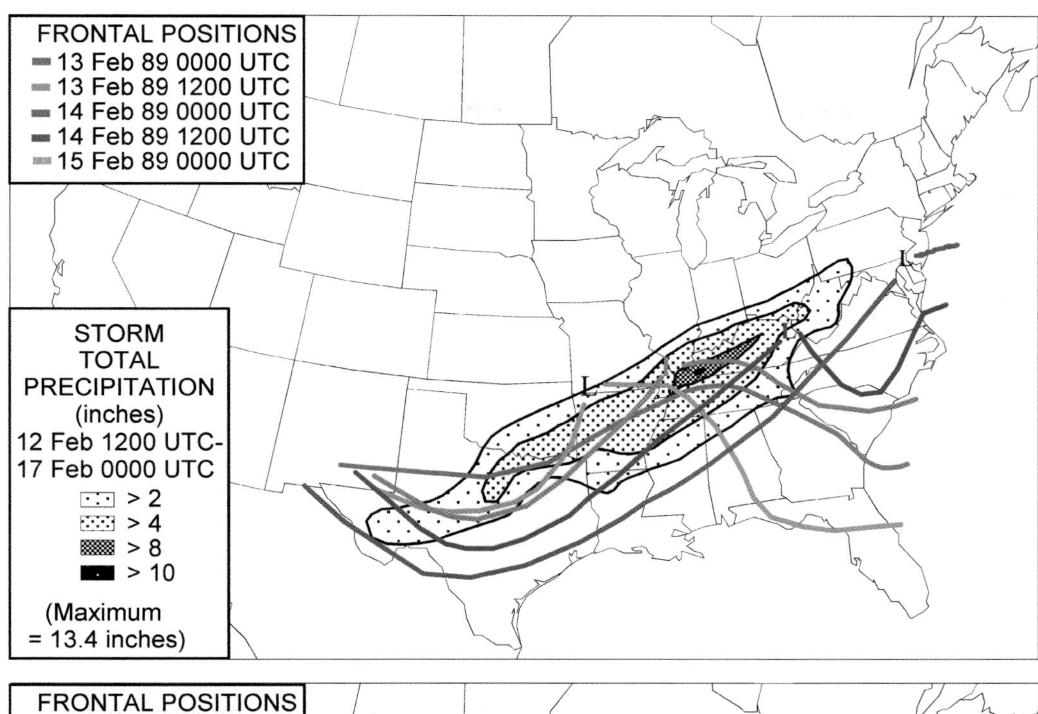

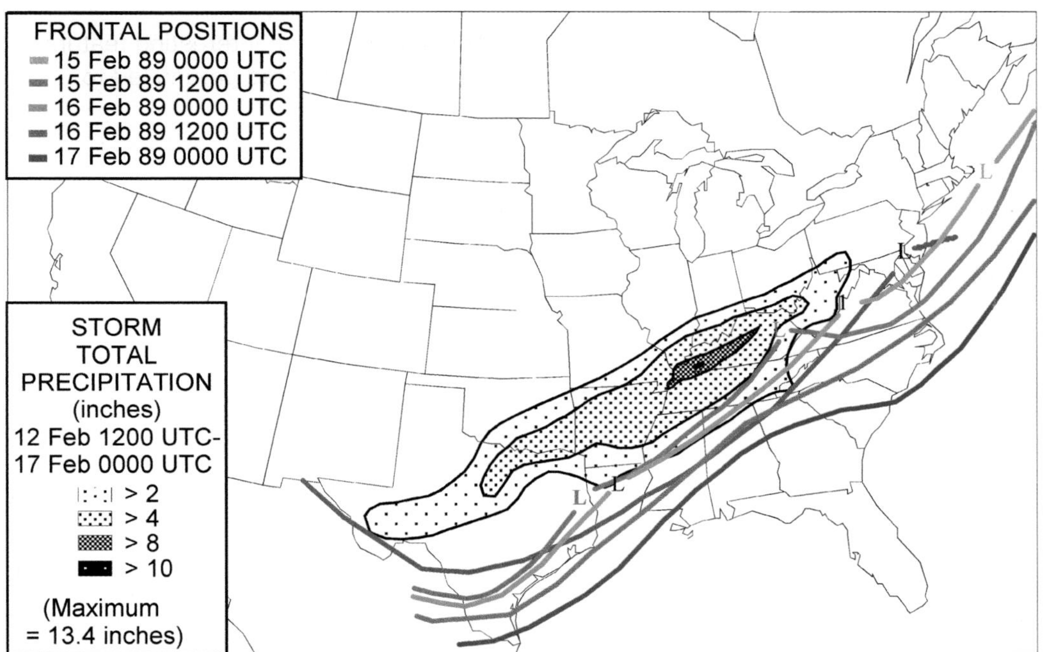

FIGURE 25.6: Frontal positions every 12 hours for four days beginning at 0000 UTC, 13 February 1989. The total precipitation over the four days is also shown on both panels. Over 13 inches of rain in Kentucky led to over $50 million in losses and to three fatalities.

An example of flooding from frontal overrunning occurred during the period 13 to 17 February 1989. During this period, much of Kentucky experienced continuous rainfall and flooding as warm, moist air overran a slowly evolving warm front. Figure 25.6 shows the position of the front over a five-day period, and the distribution of rainfall during the same time. Note that the front is located south of the region that experienced flooding. During the course of the event, small low-pressure centers propagated along the frontal boundaries and heavy rain fell north of the front in an elongated band parallel to the front. Because the front was stationary, the same narrow area experienced a succession of heavy rain events. The greatest amount of rain and worst flooding occurred in the Appalachian region of Kentucky, where the water was channeled into the valleys and flooding was enhanced.

Flooding Compounded by Snowmelt

Many damaging widespread floods in the northern United States occur when the rapid melting of a winter snowpack enhances runoff from heavy spring rains. Heavy rain and rapid snowmelt often occur together when a subtropical moist airmass moves over a snow-covered region during the late winter or early spring. Not only is there a "double supply" of water, from rainfall and snowmelt, but the ground is often frozen beneath the snow, preventing infiltration and enhancing runoff. The situation is often made worse by ice jams on rivers that have accumulated significant ice cover over the preceding weeks or months.

Flooding often occurs when a series of extratropical cyclones follows a similar track during the late winter. Regions northwest of the primary track experience a series of snowstorms (Chapter 15), followed by cold air that preserves the snow cover between storms. With the transition to spring, the jetstream and storm track shift northward, increasing the likelihood that warm air will penetrate northward over the snowpack. In such cases, frontal overrunning and/or convective precipitation in the warm sector of a cyclone can produce heavy rain over the snowpack. Heat transfer from the warm air moving over the snowpack, and somewhat from the rain itself, can quickly reduce the snow depth and increase runoff. Rivers in the north-central and northeastern United States are particularly susceptible to this type of flood because extratropical cyclones commonly pass through these regions in spring, and the air in the warm sector of the cyclones is moist because it often originates over the Gulf of Mexico. In some cases, extreme flooding from snowmelt can occur even without heavy rain.

Such was the case in 1997 along the Red River, which separates Minnesota and the Dakotas. An unprecedented flood in April 1997 followed the winter of blizzards (Chapter 15). During the 1996–1997 winter, blizzards produced a record-breaking 117 inches of snow in Bismarck, North Dakota. As early as February, river forecasters predicted that spring floods would reach unprecedented levels. The forecasts through March called for a crest of about 49 feet, far above the river's natural flood stage of 28 feet. The levees around Grand Forks were raised to 52 feet. However, the final onslaught of snow from the blizzard in early April disrupted the normal springtime melt, worsened the ice jams (Figure 25.7), and added more water to the eventual runoff pulse. The Red River crested at slightly over 54 feet, spilling over levees into Grand Forks on 18 April and forcing over 60,000 residents to flee their homes. Even though there was little precipitation during the previous week, the forecast height of the river crest was revised several feet upward in the two days before the crest. The sudden worsening of the Red River flood highlights the difficulties of forecasting floods when ice and snow dynamics are confounding factors.

FIGURE 25.7: Ice jam on the Red Lake River, a tributary that flows into the Red River at Grand Forks, North Dakota, during the flood of 1997.

CHECK YOUR UNDERSTANDING 25.3

1. Where does heavy precipitation fall relative to a front during floods associated with frontal overrunning?
2. Why are floods associated with frontal overrunning more likely to occur in the east-central United States than in other regions?
3. How does snowfall in February and March contribute to flooding in April?

Flash Floods of the Desert Southwest

Most of the year, the southwest deserts of the United States are hot, dry, and sunny and receive very little rainfall. However, dried riverbeds called ***arroyos***, common throughout the region, testify to the fact that floods occasionally invade the desert landscape. In the summer months, the thunderstorms that produce these floods develop over the southwestern United States following the onset of a seasonal circulation called the ***North American Monsoon***. The North American Monsoon develops at the time that the high plateaus of the desert Southwest undergo their maximum summer heating. Heating by the sun creates a semipermanent low-pressure center over the desert in summer (recall the effect of heating on low pressure formation from Chapter 8). The low-pressure center, in turn, draws moist air from the Pacific Ocean west of Mexico into the plateau region. The moisture within this flow and the daily heating destabilize the atmosphere, creating an environment where thunderstorms can develop. Some of these thunderstorms produce short-duration, very high intensity rainfall. In much of the desert, rain runs off quickly and channels rapidly into very narrow canyons. Within the canyons, the height of the water often builds rapidly, forming a wall of water that roars down the canyon with no warning. The wall of water can arrive suddenly in the lower part of a canyon miles from the thunderstorm. For example, on 13 August 1997 a group of tourists were hiking with a guide through a narrow portion of the Antelope Canyon in northern Arizona. A thunderstorm developed over a plateau 2,000 feet higher than the canyon and five miles upstream. The National Weather Service issued a severe thunderstorm warning. At the time the warning was issued, only a trace of rain fell in the area near the hikers. Forty-five minutes later an eleven-foot wall of water rushed through the canyon, sweeping away everyone except the tour guide, who barely survived by wedging himself against the canyon wall. Because the desert is sparsely populated, it is likely that many flash floods in this region go unreported.

Flash Floods along the East Slope of the Rocky Mountains

Water from the Rocky Mountains drains eastward through a number of deep canyons and on to the plains east of the mountain massif. In summertime, either when weak low pressure develops on the plains southeast of the mountains, or high pressure develops to the northeast, the flow on the plains east of the Rockies becomes easterly. If the air on the plains is moist and conditionally unstable, afternoon heating of the mountain slopes can combine with the moist easterly flow toward the mountains to produce towering thunderstorms that anchor over the foothills just east of the mountain peaks. When these storms remain near-stationary over the foothills, flash floods produced by their heavy rainfall can be disastrous. Killer floods of this type occurred on July 28, 1997 west of Fort Collins, Colorado; on 31 July 1976 in the Big Thompson Canyon west of Longmont, Colorado; and on 9–10 June 1972 on Rapid Creek in the Black Hills of South Dakota. The Big Thompson Flood today remains one of the most studied storms and an excellent example of the dangers posed by these floods.

The Big Thompson River begins on the east side of the continental divide in Rocky Mountain National Park. The river flows from the park eastward to the Great Plains. During its transit out of the mountains, the river passes through a very steep canyon that is a favorite summer area for tourists who camp, fish, and hike along the river banks. The Big Thompson River is more like a stream on normal days. On 31 July 1976 a very large, extremely slow-moving thunderstorm complex developed over the upper portion of the Big Thompson River watershed (Figure 25.8). The thunderstorm formed in an environment with virtually no vertical wind shear, allowing the thunderstorm to remain stationary for several hours. The storm was fed by warm, moist, easterly upslope flow that was conditionally unstable. Surface dewpoint temperatures in excess of 60° F (15.5° C), a strong indicator of afternoon thunderstorms in this normally dry region, were present along a weak frontal boundary in eastern Colorado. A high-pressure system located near the North Dakota/Canada border reinforced the easterly flow into the Colorado mountains. By afternoon, the upslope motion of the air was enhanced by solar heating, maximizing the buoyancy of the air. The rain and downdraft of the thunderstorm were located over the higher elevations and the updraft over the lower elevations. This storm circulation remained in place, forcing the precipitation to fall over the same location for many hours. In addition, the middle troposphere was quite moist; the mid-level moisture originated from the Pacific Ocean west of Mexico as part of the North American Monsoonal circulation. This mid-level moisture limited evaporation aloft, preventing cloud dissipation and enhancing rainfall.

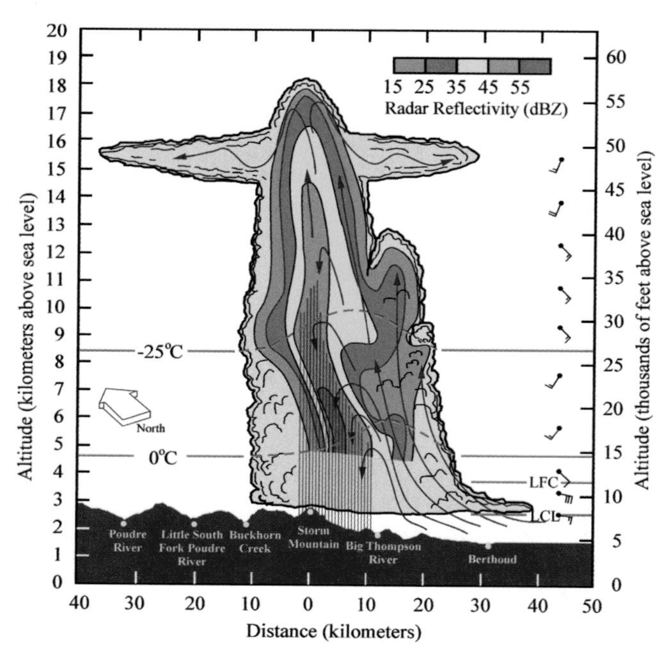

FIGURE 25.8: A physical model of one of the initial cells of the Big Thompson storm complex. The Lifting Condensation Level (LCL) and Level of Free Convection (LFC), winds, and level of the 0° C and –25° C isotherms are shown as red lines. Radar reflectivity measured by a radar on the plains east of the Rockies at Grover, Colorado, is also shown with contours intervals of 10 dBZ beginning at the 15 dBZ level.

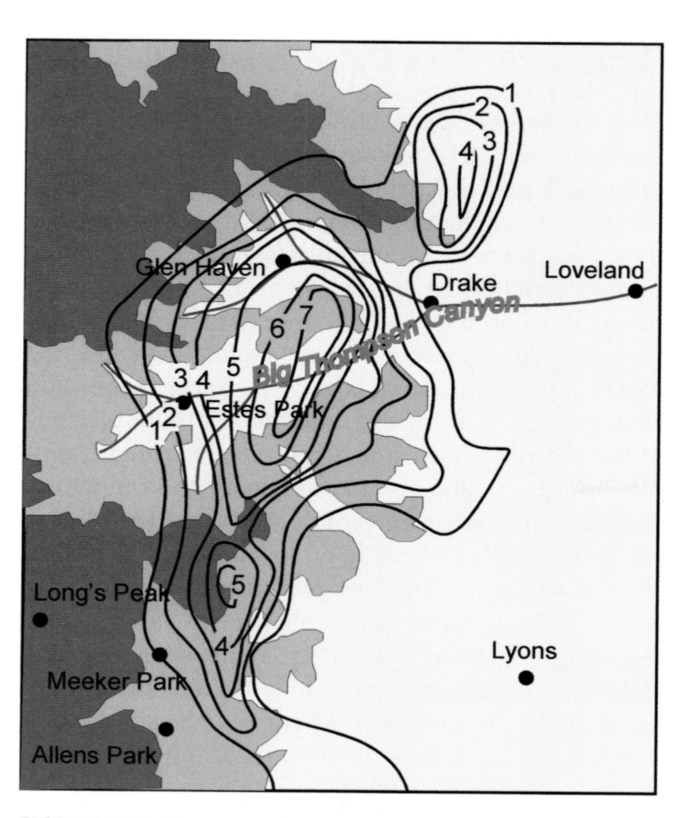

FIGURE 25.9: Heavy rainfall accumulated throughout the lifetime of the Big Thompson Storm between 5 and 10 P.M. local time on 31 July 1976. The yellow regions denote locations with elevations below 8,000 feet and the dark orange denotes regions above 9,000 feet. Contours are total rainfall in inches.

Normally, upper-level winds are sufficiently strong that thunderstorms forming over the eastern Rockies move eastward over the Plains during the evening. In this case, however, the winds were weak up to 50,000 ft, as shown in Figure 25.8. The result was a storm circulation that was essentially stationary. In just under 5 hours, a large portion of the watershed received over 6 inches of rain (Figure 25.9). Precipitation rates exceeded 3 inches per hour at Glen Haven, Colorado, and virtually all this rain became runoff, causing the river to rise rapidly. As a river flows, the water near the bottom is slowed by friction with the ground, while layers of water above move faster. Layers near the top move the fastest. This has the effect of causing a large volume of water flowing down the canyon to build up its leading edge into a turbulent wall of water. This happened on the Big Thompson. People in the canyon turned to find a 10 to 15 foot wall of water rushing downhill toward them. Those who thought quickly climbed the side of the canyon and got above the approaching wall of water. Unfortunately, many people who could have escaped this way instead got into their cars and tried to outrun the water to the bottom of the canyon. The final death toll was at least 139. The river carved the canyon, rolling boulders as large as 275 tons down the channel. Hydrologists estimate that in one day the river eroded as much of the canyon as it normally does in at least a century. Geologists now believe that the many canyons in the Rockies were formed primarily by these events. Now, as visitors enter any of the many canyons leading westward to the high peaks of the Colorado Rockies, they are faced with sign after sign reading "Climb to safety in case of a flash flood." The National Weather Service estimates that a flood of this magnitude probably occurs in one of the canyons of the Front Range at least every 200 years.

West Coast Floods

While summer flooding is typical of the eastern Rockies and intermountain west, wintertime flooding is the greatest danger in the mountains along the West Coast. Flash flooding in the river valleys of the Sierra Nevada, Cascade, and Coastal ranges often leads to more widespread flooding in the larger valleys at the

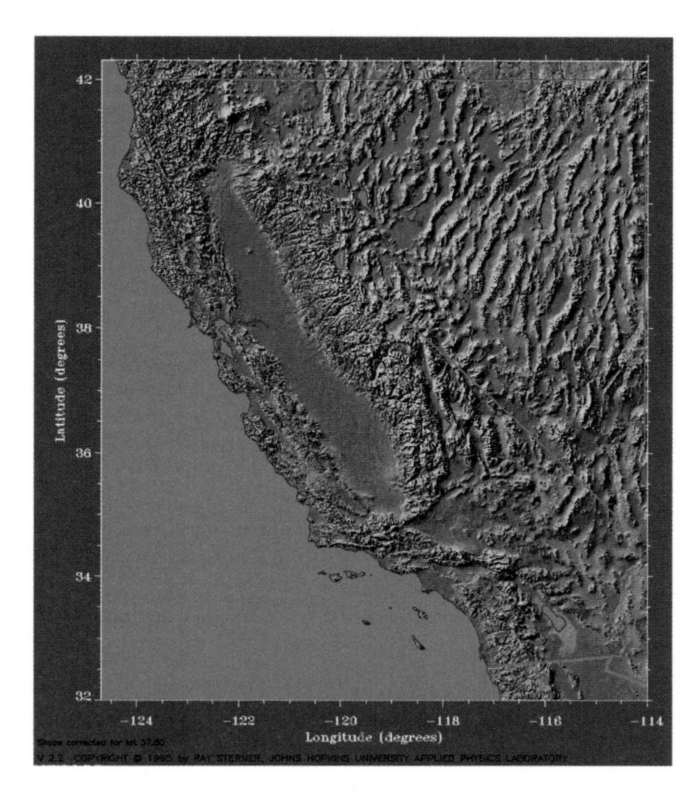

FIGURE 25.10: The topography of California. The Central Valley appears as the green area in the center of the state. Yellow, red, and gray regions denote progressively higher elevations. The Sierra Nevada are to the east of the Central Valley and the Coast Range is to the west.

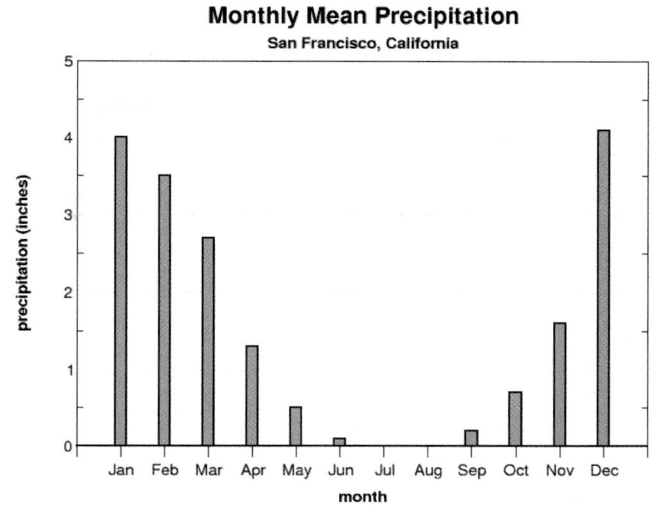

FIGURE 25.11: Average monthly precipitation (inches) at San Francisco, California.

base of these ranges. The Sierra Nevada and the Central Valley of California experience wintertime flooding events every few years. The Central Valley is a low-lying interior region of California surrounded by mountain ranges: the Sierra Nevada to the east, the Cascade Mountains to the northeast and north, the Coastal Range to the west, and the Tehachapi to the south (Figure 25.10). The only natural outlet for water from the Central Valley is just north of San Francisco, where the Sacramento River cuts through the Coast Range into San Francisco Bay. The water finally reaches the Pacific through the Golden Gate. Prior to habitation by European settlers, the entire Central Valley was a wetland. Precipitation falling anywhere on the west slopes of the Sierra Nevada, the east slopes of the Coastal Range, or in the Central Valley itself drained through the Central Valley.

Today the Central Valley contains an intricate network of levees and dams. Because of its mild climate, inhabitants recognized early on that California's Central Valley had the potential to be one of the prime agricultural areas of the world—provided the wetlands could be drained and the flow of water from the mountains controlled. Dams were placed at the base of most streams feeding out of the Sierra Nevada, serving both as flood control systems and as reservoirs to supply a steady stream of water during the dry California summers. Levees were built to channel the water across the Central Valley and distribute it to the farms that occupy the valley, and also to provide large metropolitan areas of California with much-needed water.

The climate of California is such that much of the state receives nearly all of its precipitation during the winter months (e.g., Figure 25.11). The heaviest precipitation occurs when a particular weather pattern brings repeated deluges to the state (Figure 25.12). This pattern, dubbed by local meteorologists as the ***Pineapple Express*** or the ***Hawaiian Fire Hose***, requires the middle- and upper-level flow to take on a relatively uncommon configuration over the Pacific, one in which the jet–stream splits into two branches.

Before this pattern sets up, storms must drive cool air sufficiently southward over the western Pacific so that the jetstream approaches tropical latitudes. At the same time, high pressure must be present over the northeastern Pacific. As the faster jetstream winds approaching from the west side of the Pacific encounter the high-pressure system in the eastern Pacific, part of the jetstream deviates northward around the high. This branch flows up into Alaska and returns southward over the West Coast or Rocky Mountains. The other branch flows east-northeastward around the south side of the high, directly into the West Coast (Figure 25.12).

The winds in the southern branch of the flow pick up moisture from the tropics, and bring it directly into the mountains along the coast. Storms develop within the southern branch of the jet and can move into the coast every 36 to 48 hours. On satellite imagery this flow appears as a band of clouds stretching from Hawaii to the West Coast (Figure 25.13)—hence the names Pineapple Express and the Hawaiian Fire Hose. The mountains force air to rise, leading to heavy precipitation. Because the airstream is tropical in origin, the freezing level is high and most precipitation falls as rain in the mountains, often on top of the snowpack laid down by earlier storms. Rain, falling for days or even weeks, can swamp reservoirs at the base of the mountains. Excess water must be discharged into the river systems, which may fill over capacity or cause levees to break at weak points. When levees break, water engulfs large areas of the essentially flat land of the Central Valley, allowing the water to spread rapidly. Significant flooding events of this type occurred most recently in 1997, 1995, and 1986. This weather pattern also leads to other disasters such as flash flooding and mudslides in hilly regions.

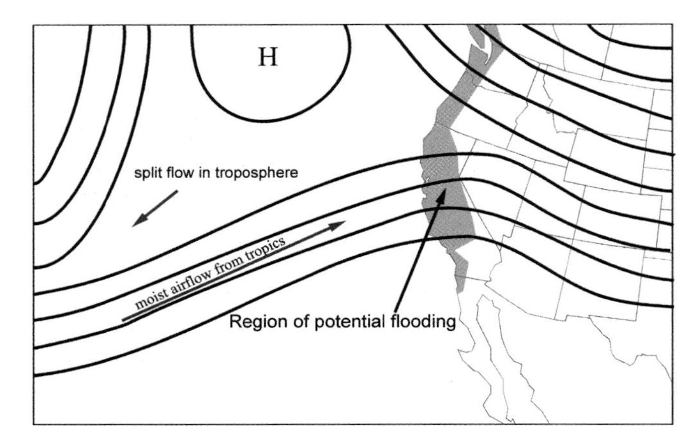

FIGURE 25.12: Split flow often present in the middle troposphere during flooding events in California. The lower branch of the flow brings tropical moisture into the Coastal Mountains and the Sierra Nevada.

CHECK YOUR UNDERSTANDING 25.4

1. How does the North American monsoon develop?
2. Identify at least two ways mountains enhance flooding.
3. What is the *Pineapple Express?*
4. Why is the Central Valley of California susceptible to floods?

FIGURE 25.13: Visible satellite image of clouds over the Pacific Ocean illustrating the "Pineapple Express" plume of moisture flowing into the West Coast from the tropics. The image is from 2100 UTC, 1 January 1997.

ONLINE 25.3

Flooding in the Central Valley of California

Between 29 December 1996 and 4 January 1997, a classic "Pineapple Express" flow pattern established itself in the Pacific Ocean and caused heavy, persistent, warm precipitation throughout northern California (Figure 25.13). Prior to this time, precipitation for much of the lower elevations of northern California was near normal. However, earlier that winter the Sierra Nevada and Cascade Mountains had experienced above normal snowfall. The warm temperatures and rainfall caused widespread melting of the snowpack throughout the mountains. The snowmelt, combined with the persistent heavy precipitation, caused several minor flash floods and a record-breaking widespread flood in the Central Valley.

The animation available online shows the evolution of the relative humidity field at 850 mb through this period. With the continuous flow of warm, moist air from the tropical Pacific, storms were able to develop within the moist airstream and move onshore. The rugged topography helped to lift the warm moist air, enhancing precipitation not only in the high elevations of the Sierra Nevada but also in the Central Valley as the air was lifted over the Coastal Range. During the New Year holiday period, precipitation totals of up to 24 inches were recorded across northern California. This resulted in some of the worst widespread flooding in the region since the time it became heavily agricultural.

Most rivers, especially those in the Central Valley, were flooded. The regions that were hit particularly hard were in the foothills of the Sierra Nevada and areas downstream. The large flood-control reservoirs limited the flooding on the larger rivers, but levee failures on the Cosumnes, Mokelumne, Tuolumne, and Feather Rivers resulted in widespread flooding and damage. Along the coast, small streams and rivers also caused flooding in the smaller valleys of the Coastal Range.

FLOOD FORECASTING AND SAFETY

The National Weather Service Hydrologic Prediction Center provides river and flood forecasts and warnings, conducts research to implement and improve forecasts and warnings, and provides basic hydrologic forecast information. Thirteen River Forecast Centers monitor all aspects of river flooding. River and flood forecasts are made after data about weather conditions, soil saturation, evaporation rates, and river stages are collected and analyzed. River conditions are monitored using streamflow gauges and water stage (depth) monitoring systems. For example, Figure 25.14 shows river conditions across the United States in July 2007. The high river stages in Oklahoma and Texas are consequences of the preceding weeks of heavy rain (Online 25.2). Such information allows meteorologists and hydrologists to forecast more accurately the times and locations of flooding along rivers.

Despite this monitoring, the rapid development and local nature of flash floods make them a difficult forecast problem. However, the skill in identifying potential flash flood events increased considerably in the 1990s with the deployment of the WSR-88D Doppler radar network. These radars routinely use reflectivity data to estimate precipitation rates and then multiply these rates by the time interval over which they occur to estimate the total precipitation. Forecasters can obtain estimates of precipitation over any time period, including the ***storm total precipitation*** (see Chapter 2) and from this information determine which river basins are likely to flood. The first test of the new Doppler radars and their role in predicting flooding was during the Great Flood of 1993 in the Mississippi and Missouri River Basins. The radar's capability to pinpoint potential flash flood regions was immediately apparent.

If a storm is slow-moving or is expected to produce exceptionally heavy precipitation, flash flood watches may be issued in advance of the actual flooding event. Local National Weather Service forecast offices issue a ***flash flood watch*** when there is the potential for flash flooding in a specified area, but the occurrence of a flood is neither certain nor imminent. Flash flood watches are typically issued anywhere from 12 to 36 hours in advance of possible flooding. When a flash flood is occurring or imminent, the local forecast office issues a ***flash flood warning***. Watches and warnings are issued on a county-by-county basis (Figure 25.15).

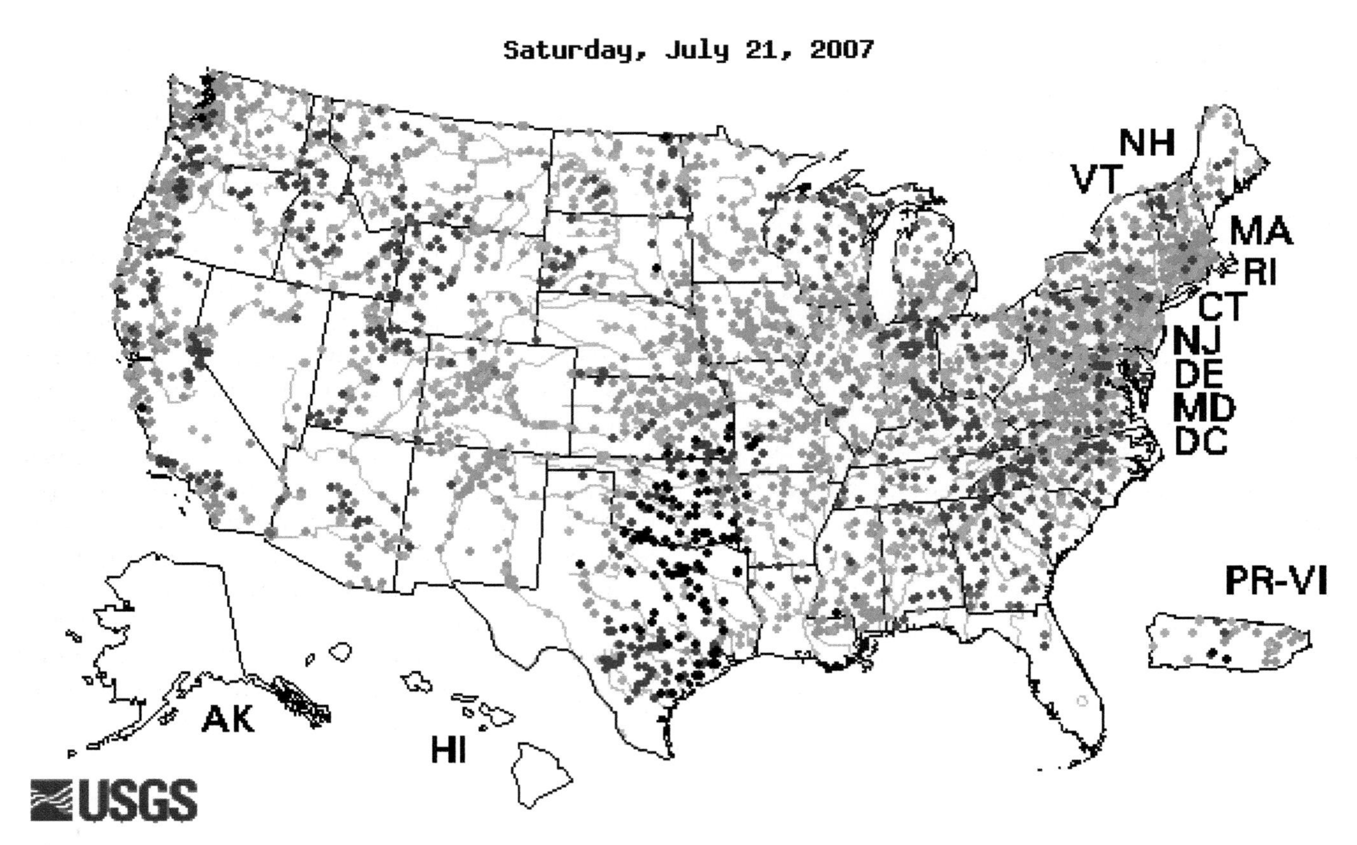

FIGURE 25.14: Streamflow in July 2007 relative to the historical average streamflow for the same time of year. Flows are color-coded as percentiles, ranging from the lowest 10 percent (red, maroon) to the highest 10 percent (blue, black). All percentiles are based on the historical values for mid-July at the corresponding locations.

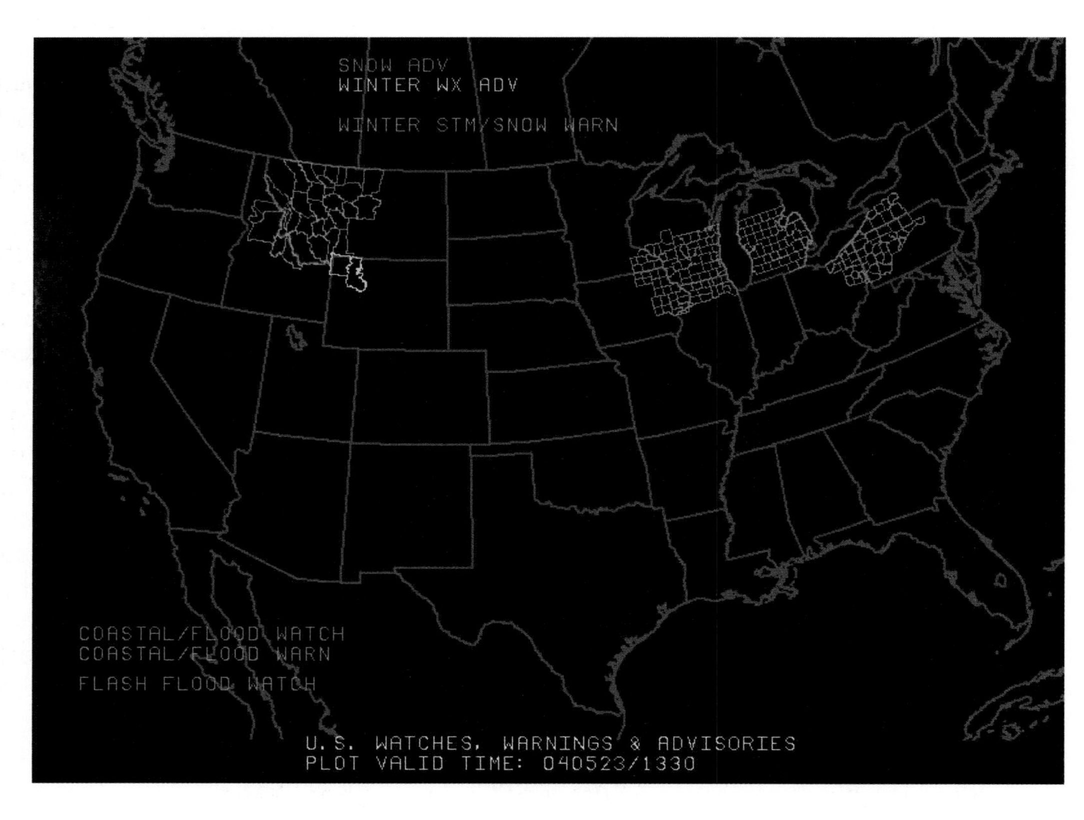

FIGURE 25.15: An example of flash flood watches the National Weather Service issued on 23 May 2004. Watches and warnings are issued on a county-by-county basis. In the example, the counties highlighted in green across Wisconsin, Michigan, Pennsylvania, and in Michigan and New York are under a flash flood watch.

With widespread floods, there is generally ample time to issue warnings as river flooding develops relatively slowly. Along rivers and streams, ***river flood warnings*** are issued when main stem rivers are expected to reach a level above flood stage. When smaller streams and creeks are expected to rise above flood stage, a ***small stream flood advisory*** is issued. Large-scale flooding can occur along coastal areas as well. Coastal flooding occurs when sea water rises above normal tidal levels. Coastal flooding is normally caused by prolonged strong onshore winds and/or high tides. These conditions may occur in association with tropical or extratropical cyclones. When widespread coastal flooding is expected within twelve hours, a ***coastal flood warning*** is issued.

Flooding, whether it is minor street flooding or major flooding of a large river, should be treated with respect and caution; the force of water is almost always underestimated. As little as 6 inches of rapidly moving water can sweep a person off his or her feet. A mere foot of water can wash a car away. Never drive through a flooded area or try to outrun a flash flood in a vehicle; in a hilly area leave your vehicle and climb to higher ground. If the vehicle stalls in water, abandon it immediately—many people who are killed in flash floods are trapped in their cars. If you live in an area that is prone to flash flooding, your best line of defense is to be conscious of the weather and listen for the latest storm information. Be especially alert at night, when it is harder to recognize the signs of flooding. During widespread flooding, local officials may evacuate an area at high risk for flooding. When residents are told to evacuate, most do; however, some try to return to their homes to save personal belongings. This not only puts the returning residents in danger, but endangers the lives of rescue crews who attempt to save their lives. Residents should evacuate quickly, calmly, and immediately following an evacuation order. Many of the fatalities in Hurricane Katrina in 2005 occurred because some residents were either unwilling or unable to leave New Orleans when evacuation orders were issued.

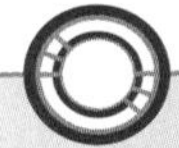

FOCUS 25.5

The Cleanup—Who's Responsible?

Widespread floods are spectacular events that receive significant media attention. Due to the slow nature of a widespread flood, people from surrounding communities often travel many miles to help reinforce levees and protect areas with sandbag walls. However, after the flood is over, the residents of the flooded areas are often left to deal with the aftermath. Low-lying water and sewer systems often fail during floods and wastewaters often are deposited into the flood waters. As the water recedes, wastewaters and silt from the river are deposited in homes, in yards, and in the streets. Where sandbag levees fail, thousands of cubic feet of sand are left behind. Even if the sandbag levees do not fail, the dismantling of the levees can be hazardous. Sandbags are waterlogged with a combination of river water and wastewater and must be disposed of in a safe manner. Currently there are no federal regulations on the removal and disposal of temporary structures (such as sandbag levees) designed to hold back the floodwaters.

Each time flooding occurs on a particular river or waterway, questions arise concerning who should pay for the damages incurred. Typically, if the flooding is severe enough, state or federal aid is offered to the affected residents. Since widespread flooding often occurs repeatedly in the same areas, the same residents can receive financial aid multiple times. The federal and many state governments have recognized this problem and have developed property buyout programs. For example, the state of Missouri implemented a state buyout program in which acquired property is used as wetlands, open space, or recreational facilities. The property is permanently taken off the tax rolls and is not eligible for federal assistance in the event of future flooding. The Federal Emergency Management Agency (FEMA), state emergency management agencies, and county and local governments have moved more than 17,000 homes and businesses out of floodplains since the late 1980s. These bailouts have cost more than $500 million in local, state, and federal tax money. However, in the long term the amount of money saved by not having to provide disaster relief to these areas will be substantially more than the costs of the buyout.

(continued)

(continued)

Nowhere was this problem more dramatically illustrated than in New Orleans, where most of the city was flooded by Hurricane Katrina in 2005 (Chapter cover photo). Despite the infusion of billions of dollars of relief funds, the city remains a long way from recovery even three years after the event. Many residents simply have not returned to their flooded property, leaving city, state, and federal agencies to handle the city's recovery. The loss of a large portion of the population base as well as the unprecedented infrastructural damage is compounded by the fact that the levees are still unreinforced, leaving the city vulnerable to more flooding should a hurricane or tropical storm strike in the next few years.

GLOBAL WARMING: POTENTIAL EFFECTS ON FLOODS

One of the expected consequences of global warming is an increase in the intensity of the heaviest rain events. This expectation arises from the fact that the saturation vapor pressure increases with temperature (Chapter 1), increasing the amount of moisture available for precipitation when lifting and condensation occur. Such reasoning implies heavier rainfall amounts from all of the different types of flood-producing weather systems: coastal storms, frontal cyclones, and convective systems. While scientists agree that the available moisture will increase, there is greater uncertainty about future trends in the frequency of occurrence of tropical storms (Chapter 24), frontal cyclones (Chapters 10, 11), and thunderstorms (Chapter 18).

The anticipated trend toward heavier precipitation events has already been borne out by several studies of precipitation over the past several decades. In Chapter 5 (Figure 5.16), we saw that there has been a general increase of precipitation over most of the world's land areas, especially North America, from 1900 to 2000. Other studies have shown that heavy-precipitation events have undergone even greater increases than has the average precipitation over the contiguous United States, especially in the East. This increase of heavy precipitation has been greatest in the past several decades. In other parts of the world, intense precipitation events have also increased disproportionately beyond what we would expect from corresponding changes in the average precipitation.

The linkage between future changes in flooding and the projected increase of precipitation is complicated by the relative proportions of rain and snow during winter, which is the season for which climate models project the largest percentage increases of precipitation. As the climate warms, most areas in middle latitudes will receive more rain and less snow during winter. Shallower snowpacks will reduce the likelihood of snowmelt-driven floods. However, the increased frequency or wintertime rain events, some of which will fall on snow or frozen ground, will increase the likelihood of winter flood events.

While heavy-precipitation events are increasing in frequency and are likely to continue to increase, flooding and its consequences also depend strongly on land use and other human activities. The paving of urban areas increases runoff, while conversion of forests and wetlands to agricultural lands also increases runoff and downstream flooding, as was the case in the 1993 Mississippi River Flood. China's high flood frequency has its roots in similar trends of land use. In addition, population migration to coastal regions and river floodplains is leaving larger fractions of the population in flood-prone areas. These trends make it virtually certain that economic losses from flooding will increase during the coming decades, even if there were no changes in the meteorological factors responsible for flooding.

CHECK YOUR UNDERSTANDING 25.5

1. What technology has enhanced forecasters' ability to pinpoint potential flash flood regions?
2. What depth of rapidly moving water can sweep a person off his or her feet? What depth can sweep a car away?
3. What are the recent trends of precipitation over the world's land areas?
4. Identify two factors that are complicating the linkage between projected changes of precipitation and future changes in flooding.

TEST YOUR UNDERSTANDING

1. What meteorological and non-meteorological factors influence the intensity of floods?
2. If a 50-year flood occurs in your town, could another flood of that magnitude occur the following year? Explain.
3. How do flash floods and widespread floods compare in terms of (a) economic losses and (b) fatalities?
4. Flood insurance rates are often based on 25-year or 100-year flood statistics for a particular area. If you are an insurance executive setting rates for flood insurance in a new suburban subdivision, why should you be cautious in your use of the statistics?
5. What are advantages and disadvantages of putting levees along a river?
6. Tropical cyclones can produce two very different types of floods. What are they and where do they occur?
7. Discuss what typically causes floods in tropical regions, other than tropical cyclones.
8. How do individual thunderstorms within a squall line move relative to the entire squall line and front during flooding events?
9. How did the large-scale weather patterns favor flooding of the Mississippi River Basin in 1993? What other factors contributed to the flood?
10. How was the central United States "preconditioned" for the summer flood of 1993?
11. What role did levees play in the Mississippi River flood of 1993?
12. What emerging technology aided river forecasters during the Great Flood of 1993?
13. Why are floods associated with overrunning in the winter season particularly dangerous for trapped flood victims and rescuers?
14. What was unusual about the precipitation that preceded the 1997 flood of the Red River Valley?
15. Can frontal overrunning and snowmelt influence the intensity of the same flood? Explain.
16. Why do flash floods often surprise hikers and other recreational users of canyons in the West and Southwest?
17. Characterize the meteorological weather pattern that resulted in the Big Thompson Canyon flood of 1976.
18. It is said that there is a feedback relationship between floods and the canyons of the Colorado Rockies. Explain.
19. Compare and contrast West Coast flooding with flooding over the interior mountains.
20. How does the elevation of the rain/snow line affect flooding in California?
21. What is the "Pineapple Express" weather pattern? Why is it conducive to flooding in California?
22. If you are driving an automobile in a mountain canyon, how should you respond to a flash flood?
23. Why is there some controversy surrounding payments for damages incurred by flood victims?
24. How is Doppler radar used as an aid in flood forecasting?
25. What types of information do river forecasters use?
26. What is the difference between a flash flood watch and a flash flood warning?
27. What are some of the problems that accompany the aftermath of a flooding event?

28. You are driving your car and suddenly find yourself in a severe thunderstorm with frequent intense lightning and rising water in the street. Explain your dilemma from the perspective of a personal safety expert.

29. What are the most likely trends in flood damage during the twenty-first century? Provide at least one meteorological reason and at least one non-meteorological reason for your answer.

TEST YOUR PROBLEM-SOLVING SKILLS

1. Suppose a major flood occurs in each of the following cities during the months listed below.

(i)	Minneapolis, Minnesota	March
(ii)	Fort Collins, Colorado	June
(iii)	New Orleans, Louisiana	September
(iv)	Atlanta, Georgia	January
(v)	Detroit, Michigan	August
(vi)	Sacramento, California	February

   (a) For each city, identify the most likely weather pattern to have caused the flood based on the season and location of the city. Provide a brief justification for your choice.
   (b) Identify any non-meteorological factors that could have contributed to each flood.

2. Suppose a 100-year flood occurs at your location this year.
   (a) What are the odds that your location will experience another flood of the same magnitude in the next twelve months?
   (b) What are the odds that your location will experience two 25-year floods in the year 2012?
   (c) State any assumptions you make in your answers to (a) and (b).

3. Consider a squall line of strong thunderstorms 40 miles wide. Suppose for simplicity that the rainfall rate everywhere within this band of thunderstorms is 2 inches (5 cm) per hour. Assume that the squall line is oriented in a north-south direction and that it moves due east. How much rain will fall at a location over which the storms pass if the line of storms moves at the speeds below?
   (a) 50 miles/hour
   (b) 25 miles/hour
   (c) 5 miles/hour

4. The Mississippi River's annual average discharge rate is about 0.5 million cubic feet per second. The area drained by the river and its tributaries is about 1.24 million square miles.
   (a) Estimate the average annual runoff (equivalent depth of water) from the land comprising the drainage area.
   (b) Over the long term, your answer to (a) represents the difference between average precipitation and average evaporation from the surface. Assume that 70 percent of the precipitation that falls on the drainage basin evaporates or is used by plants. By how much would the Mississippi's discharge rate change if precipitation increased by 10 inches (25 cm) per year over the entire drainage basin?
   (c) Suppose that precipitation exceeds its present average by 5 inches (12.5 cm) *per month* over 20 percent of the drainage area (a rough approximation to the 1993 summer situation). If none of the additional precipitation is evaporated from the surface, by how much would the river's discharge rate increase?

5. Suppose an urban drainage culvert is 20 feet deep and 40 feet wide, and the culvert drains an area that is 10,000 feet by 10,000 feet (about 2 miles on a side). If all the rainwater that falls on this area drains into the culvert, and the water's flow speed in the culvert is 10 feet per second, how large can the rainfall rate become before the culvert overflows? (Express your answer in inches per hour, and compare your answer to Problem 3.)

USE THE SEVERE AND HAZARDOUS WEATHER WEBSITE

http://severewx.atmos.uiuc.edu

1. Examine the current river flood potential:
   (a) Navigate to the River Flood Outlook issued by the Hydrometeorological Prediction Center (HPC). From the *Severe and Hazardous Weather Website* choose the HPC link under "Additional Resources" for "Floods." Identify all regions where significant flooding is expected in the next five days.
   (b) From the same site, review the "HPC Discussions" for precipitation. What supporting evidence do the forecasters provide for the flood outlook?
   (c) During the outlook period, monitor the actual precipitation and flooding that occurs. Were the forecasters accurate for each of the regions where they predicted flooding? What regions experienced flooding that was not predicted? If areas of flooding were not well predicted, what reasons can you think of for the poor prediction?

2. Current watches and warnings from the National Weather Service can be viewed in graphical format from the "Current Weather" page of the *Severe and Hazardous Weather Website.* During a time when flood watches and warnings are issued, examine Doppler radar reflectivity and storm total precipitation for the areas to be affected.
   (a) For what region of the country did the watches or warnings apply?
   (b) Based on the reflectivity pattern and animations, what type of flooding event are you observing?
   (c) What was the total storm precipitation in the watch or warning areas?
   (d) Would you expect that this amount of precipitation could cause flooding? If not, what other factors were likely responsible?

3. The website of the National Weather Service's Natural Hazards Statistics can be linked from the *Severe and Hazardous Weather Website,* "Floods" page. Examine the statistics with a focus on floods.
   (a) What period of time do the records cover?
   (b) How has the total number of deaths from floods varied?
   (c) How has the flash-flood portion varied as a percentage of the total?
   (d) What activity were the largest numbers of victims engaged in?
   (e) Compare the number of flood victims with the numbers of victims of other severe weather types. Which other type of weather takes more lives?

4. Extreme flood events can be examined using the Dartmouth Flood Observatory website. The *Severe and Hazardous Weather Website* "Flood" page links to this site. Use this site and the tools available to examine recent flood events in the United States from a satellite perspective.
   (a) What is the role of the Observatory in studying extreme flood events?
   (b) In meteorology, satellites are primarily thought of as a way to view cloud structures and water vapor in the atmosphere. How are satellites used to study floods?
   (c) Where have recent flooding events occurred in the United States?
   (d) Compare and contrast the recent U.S. floods with other major worldwide floods.

5. The Federal Emergency Management Agency (FEMA) provides information on preparedness and safety tips for severe weather. Examine the FEMA website for flooding and create a safety pamphlet that could be distributed to people in your community. The pamphlet should outline the basic information about floods in your area as well as the hazards associated with flooding and information on safety measures to take before, during, and after a flood. Highlight specific areas within your community that are especially vulnerable to flooding.

The aftermath of the Dust Bowl in South Dakota.

KEY TERMS

agricultural drought
Bermuda High
climatological drought
Crop Moisture Index
drought
Dust Bowl
feedback
hydrological drought
jetstream
megadrought
meteorological drought
Palmer Drought Severity Index (PDSI)
socio-economic drought
subtropical high-pressure system
upper-air wave

orldwide, drought leads to more fatalities than any other weather phenomenon. It affects larger areas, over longer timescales, than all other types of hazardous weather. Every region of the United States is adversely impacted by drought at one time or another, as are many portions of Eurasia, Africa, South America, and Australia. Drought produces complex effects that accumulate slowly and interact with the demand humans and other forms of life place on the water supply.

Yet the beginning of a drought often goes unnoticed by those who will be profoundly affected by it—what appears to be just another heavy rain may turn out to be the last significant rain for weeks or even months to come. Even the end of a drought can be hard to pinpoint, since an apparent "droughtbuster" rain is sometimes followed by more abundant rains but other times by a return to dry weather. Meteorologists have identified factors that maintain and intensify a drought. However, the dynamical or physical "triggers" of drought and its termination have remained elusive.

While drought is associated with water shortages and a lack of precipitation, there is no universal defini-

FOCUS 26.1

Drought Disasters Occur Worldwide

This chapter focuses on drought in the United States. However, nearly all the world's populated areas are subject to drought. Within the lifetimes of most people now alive, significant droughts will have occurred on every continent except the barren Antarctic ice sheet (which receives little precipitation even under normal conditions). Droughts in the "Virgin Lands" of north-central Russia during the 1950s and 1960s had major economic and social consequences during a time when food production and the expansion of agriculture were priorities of the Soviet regime. Like the Great Plains of the United States and the Prairies of Canada, the Virgin Lands of the former Soviet Union receive marginal precipitation for agriculture in normal years, so any deficit results in agricultural drought.

Drought in the sub-Saharan Sahel from the 1960s to the 1980s caused widespread human tragedy and generated debate about the role of land use in the exacerbation of the drought. Other parts of Africa, including Morocco, Ethiopia, Somalia, and Kenya, suffered severe drought in the 1970s, 1980s, and 1990s. Since the early 2000s, a long-term drought has affected parts of Ethiopia, Somalia, Kenya, Uganda, and Tanzania. This drought was particularly severe in 2006, when food shortages attributable to the drought affected more than 10 million people in the region. In early 2007, South Africa also experienced its worst drought in 15 years.

Nearly 30 percent of India was affected by severe drought that began in 2000 and continued for several years. The drought, which was the worst in 100 years in some areas, affected 50 million people in four Indian states.

England was affected by serious droughts in the summers of 1976 and 1996–1997; in both cases, the British government imposed restrictions on water usage. Eastern Europe suffered extensively from droughts in 2000.

The "Nordeste" region of Brazil is periodically affected by drought, with significant agricultural and economic consequences, including the migration of thousands of residents. A recent drought in Brazil during 2005 was the worst in 60 years, causing water levels in the Amazon River to fall to their lowest levels in more than three decades.

Drought, and blowing sand and dust, plagued China and Australia in 2001–2002. Drought in China during 2006 affected more than 12 percent of the nation's agricultural land and impacted the drinking water supplies of more than 10 million people, while a severe drought in Australia during 2002–2007 has led to a ban on irrigation in Australia's major agricultural areas. In both China and Australia, changes in land use are thought to have worsened the environmental and societal impacts of the drought. The tendency for drought to occur as human settlement and agriculture expand into areas of marginal precipitation has led to the notion that "drought follows the plow." As human populations and agricultural pressures continue to increase, drought will almost certainly continue to be the meteorological phenomenon with the most serious social, economic, political, and human consequences.

tion of ***drought***. The use of a variety of definitions can lead to some ambiguity concerning whether a particular area is experiencing a drought. A further complication is that drought is "relative" to normal climatological conditions: the weather that accompanies a normal summer in Phoenix or Tucson, Arizona, is taken in stride by residents of those cities, but the same conditions would produce a severe drought and major economic disruptions in New York City or Chicago, Illinois. A ***meteorological*** or ***climatological drought*** refers to an unusually long period during which precipitation is below normal for a particular area. ***Hydrological drought*** describes an unusual deficiency of groundwater and/or streamflow, i.e., water levels are below normal for the area. ***Agricultural drought*** refers to a period of deficient moisture in the soil layers from which crops and other plants normally draw their water. Agricultural drought often precedes hydrological drought, since water in the near-surface soil layers is generally most important for plants. Finally, ***socioeconomic drought*** occurs when the moisture shortage is sufficiently large that it affects people. Effects on people can occur through the availability of, or the demand for, some economic goods. Socioeconomic drought generally lags the other types of drought because the impacts are consequences of the moisture shortage rather than a direct measure of the moisture shortage. An additional complication is that the impacts of drought may strongly depend on human actions, such as the land use practices that exacerbated the impacts of the drought of the 1930s.

An insidious feature of drought is its ability to feed upon itself. Indeed, one of the characteristics of a drought is the failure of forecasted or approaching weather systems to produce the rain that one would normally expect. The expression, "all signs fail in times of drought," is an admission of uncertainty by meteorologists who must decide whether or not to forecast rain during a drought.

INDICES OF DROUGHT

Because the normal values of precipitation and temperature vary regionally and seasonally, it is difficult to quantify drought in terms of actual precipitation and temperature. Standardized measures have come into use in order to develop maps of drought severity over a complex climatic domain such as the United States. One of the most widely used indices is the ***Palmer Drought Severity Index (PDSI)***, developed in the mid-1960s by Wayne Palmer of the U.S. Weather Bureau (now the National Weather Service). This index (often called simply the *Palmer Index*) is a measure of moisture deficiency standardized to local climate conditions. It is based on a groundwater balance that includes the supply of water by precipitation and stored water, and the depletion of water by temperature-dependent evaporation, recharge of subsurface water, and runoff.

Two key characteristics of the Palmer Index are that (1) it is based on departures from the normal conditions for a location, thereby providing a "relative" measure that is comparable across different regions, and (2) it is cumulative, in the sense that each period's precipitation and temperature provide an incremental change to the prior value of the index. Since it is a hydrologic or persistent drought index, it does not change substantially from week to week. The various categories of the index, after standardization for a particular location and time, appear in Table 26.1.

During extreme hydrologic events, the PDSI can reach values of +/– 6.0. During rare events, the magnitude can be even larger. Because the PDSI changes relatively slowly, it is sometimes called the *Long-Term Palmer Index*. It is most useful for tracking extended droughts and less useful for short-term changes, especially those affecting the upper soil layers. Figure 26.1 is an example of the PDSI map produced on a weekly basis by NOAA's Climate Prediction Center. The figure shows that much of the United States was in the grip of a drought during August 2007, with large areas of "severe" drought (PDSI < –3) and "extreme" drought (PDSI < –4) in the West, where the preceding winter was abnormally dry. Dry conditions were also indicated over much of the Southeast and also in northern Minnesota and Wisconsin. At the same time, a large portion of the Southern Plains, especially Texas, ranged from "unusually" moist (PDSI > +2) to "extremely" moist (PDSI > +4).

TABLE 26.1 Categories of Palmer Drought Severity Index

PDSI Range	Drought Severity
–4.0 or less	extreme drought
–3.9 to –3.0	severe drought
–2.9 to –2.0	moderate drought
–1.9 to +1.9	near normal
+2.0 to +2.9	unusually moist
+3.0 to +3.9	very moist
+4.0 and above	extremely moist

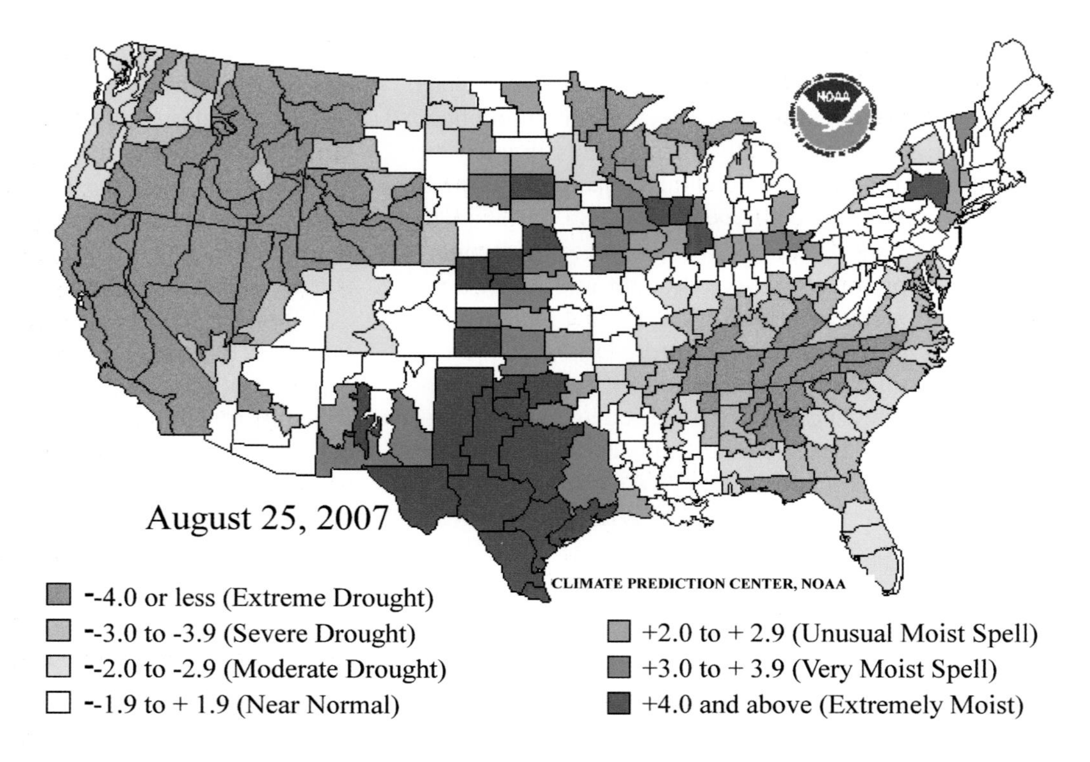

FIGURE 26.1: Palmer Drought Severity Index for the week ending August 25, 2007.

Palmer also developed a shorter-term index, the ***Crop Moisture Index,*** which is a measure of the moisture in the crop rooting zone. This more-rapidly varying index is designed for agricultural uses and is updated weekly. Other indices are also used, such as the amount of rainfall needed to bring the soil to saturation, the Standardized Precipitation Index (based on the probability of precipitation), and localized indices such as the Surface Water Supply Index (Colorado) and the Reclamation Drought Index (Oklahoma), among others.

CAUSES OF DROUGHT

Droughts are associated with persistent departures of the large-scale weather pattern from its normal pattern. The large-scale ***upper-air waves,*** the ***jetstream,*** and ***subtropical high-pressure systems*** often provide the key atmospheric signatures of a drought episode. The season during which these drought signatures may develop varies regionally within the United States. While spring and summer are the key seasons for drought to develop from precipitation deficits in the central and eastern regions, winter precipitation holds the key to drought in the West. The normal features of the atmospheric circulation vary by season, so the meteorology of drought must be considered on a seasonal and regional basis. Nevertheless, some commonalities will be apparent in the circulation characteristics most relevant to drought in the various regions of the United States.

What causes the drought pattern to develop in the first place? The initiation mechanism is difficult to determine. No two droughts are the same, and a single cause of drought has not been identified despite many statistical and model-based studies over the past century. Since our understanding of drought initiation mechanisms is so deficient, the prediction of the onset of a drought is as challenging as the prediction of the end of a drought.

Nevertheless, research has identified at least one factor, ocean surface temperature, that may play a role in some drought-producing shifts of the atmospheric circulation. For example, abnormal water temperatures in the Pacific Ocean (or even the Atlantic Ocean) appear to influence precipitation over the United States. Recent studies have related the Dust Bowl drought of the 1930s to a pattern of abnormally cool ocean surface temperatures in the tropical Pacific Ocean and abnormally warm ocean temperatures in the tropical Atlantic Ocean. However, the associations between drought and sea surface temperatures are far from consistent over time. Abnormal circulation patterns also can develop via the atmosphere's natural variability, which is independent of any forcing by the surface and which introduces a random component into shifts of the circulation. This natural variability may trigger a short dry spell that may then be exacerbated by a feedback that results from a drying of the soil, as discussed below.

We begin with a look at drought in the central United States, where droughts have had major impacts on the land use and settlement patterns of the Plains. We then discuss drought on the East Coast and West Coast, where large populations have been significantly impacted by drought in recent years.

CHECK YOUR UNDERSTANDING 26.1

1. What are the four types of drought?
2. What are some of the indices used to identify drought?
3. How do drought signatures vary seasonally across the United States?

DROUGHT IN THE CENTRAL UNITED STATES

The normal summertime pattern of sea-level pressure contains two dominant features in the middle latitudes of North America: the subtropical high-pressure centers of the North Atlantic and the North Pacific. These features are apparent in Figure 26.2A, which shows the sea-level pressure averaged over the summer months (June–July–August) of the thirty-year period 1971–2000. As altitude increases from the surface up through the troposphere, the subtropical high centers generally weaken (their pressures decrease) and yield to the west-to-east circulation pattern of the jetstream. The jetstream during summer, while considerably weaker than during winter, is typically located to the north of the subtropical surface highs.

We illustrate the average summer upper-air flow by showing the June–July–August mean 700 mb

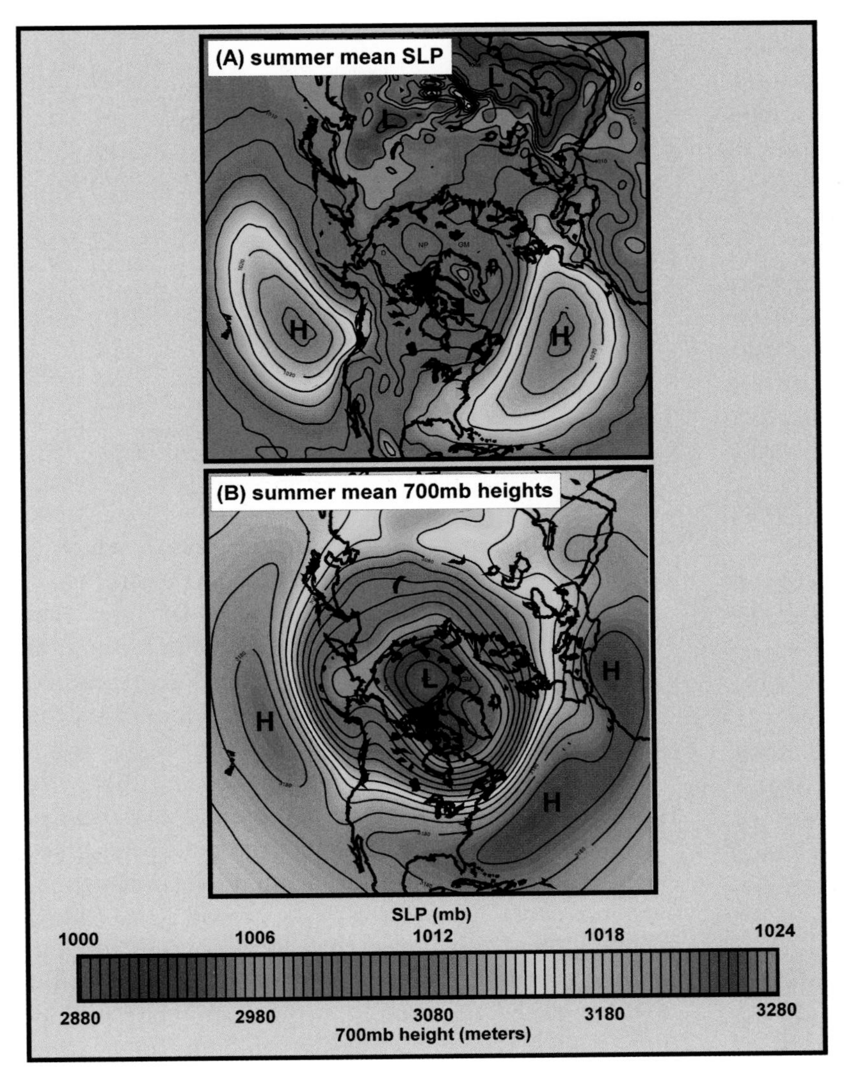

FIGURE 26.2: Northern Hemisphere view of sea-level pressure (SLP) and 700 mb heights. (A) Normal summer sea-level pressure (mb). Two strong high-pressure systems are located off each coast of North America; the central United States is not dominated by high pressure. (B) Normal summer 700 mb height (meters). Both diagrams show higher values as yellow, orange, and red; lower values as green, blue, and purple.

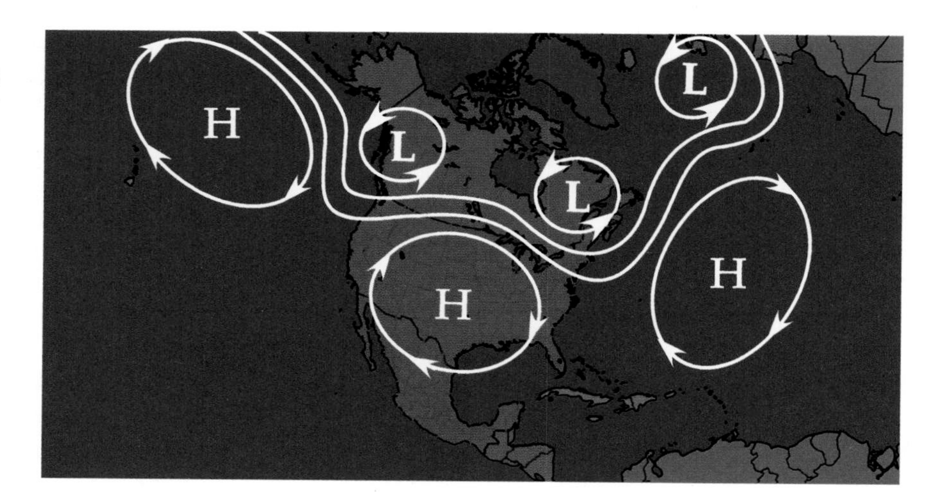

FIGURE 26.3: Summer pattern of three upper-air (700 mb), high-pressure centers during drought. The high-pressure center located over the central United States is absent in a normal summer.

height in Figure 26.2B. (We will use 700 mb maps rather than the 500 mb or 300 mb maps to depict upper-air flow patterns because strong heating of the air near the surface often results in a closed high-pressure center that extends upward to the 700 mb level over land areas experiencing drought.) As shown in Figure 26.2B, the summer jetstream normally follows a west-to-east trajectory across southern Canada and the northern United States. In a typical summer, the jetstream occasionally migrates southward into the central United States, bringing periods of precipitation in the vicinity of the associated surface fronts.

During a summer drought, a third high-pressure center or ridge develops aloft over the interior of the United States, as shown in Figure 26.3. The jetstream, a focus for summer thunderstorm activity, moves northward well into Canada when this third high-pressure area dominates the flow over the central or eastern United States. High pressure is associated with generally clear skies, so the stage is set for dry weather when the pattern in Figure 26.3 develops. Moreover, fronts and extratropical cyclones are generally confined to the region north of the continental high-pressure center, so there is little chance for precipitation to migrate into the drought-stricken area from regions to the north and west.

Once high pressure develops aloft over the continent during spring or summer, the dry weather pattern leads to ***feedbacks*** that reinforce the drought-producing circulation pattern. First, dry air is denser than moist air (Chapter 1). This higher density offsets any tendency for the formation of low surface pressure beneath the warm air aloft. The pressure at the surface is sufficiently high such that it blocks the influx of low-level moisture from the Gulf of Mexico and the southeast Atlantic. Additional feedbacks have their

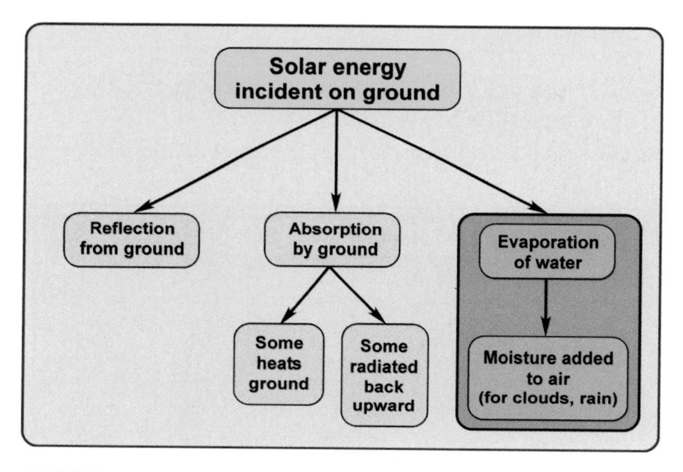

FIGURE 26.4: Schematic diagram showing distribution of solar radiation reaching the Earth's surface, with partitioning into reflection, absorption, and evaporation of water. During a drought, evaporation of water (orange box) is drastically reduced, leading to greater absorption of radiation by the ground and often to warmer temperatures.

origins in the energy budget of the ground, which is driven by incoming radiant energy from the sun.

The high-pressure system has clear skies that allow warming of the lower atmosphere in the core of the high pressure. This warming causes an expansion of the lower atmosphere and hence increases the pressure aloft, strengthening the upper-air ridge (Chapter 8). The absence of clouds, and therefore precipitation, also leads to dry ground and reduces the evaporation rate. Evaporation normally consumes a substantial portion of the incoming solar radiation, which is either (1) reflected by the ground, (2) absorbed by the ground, or (3) used to evaporate water from the ground (Figure 26.4). The reduction of evaporation and latent heat consumption (the orange box in Figure 26.4) leaves more energy available for heating

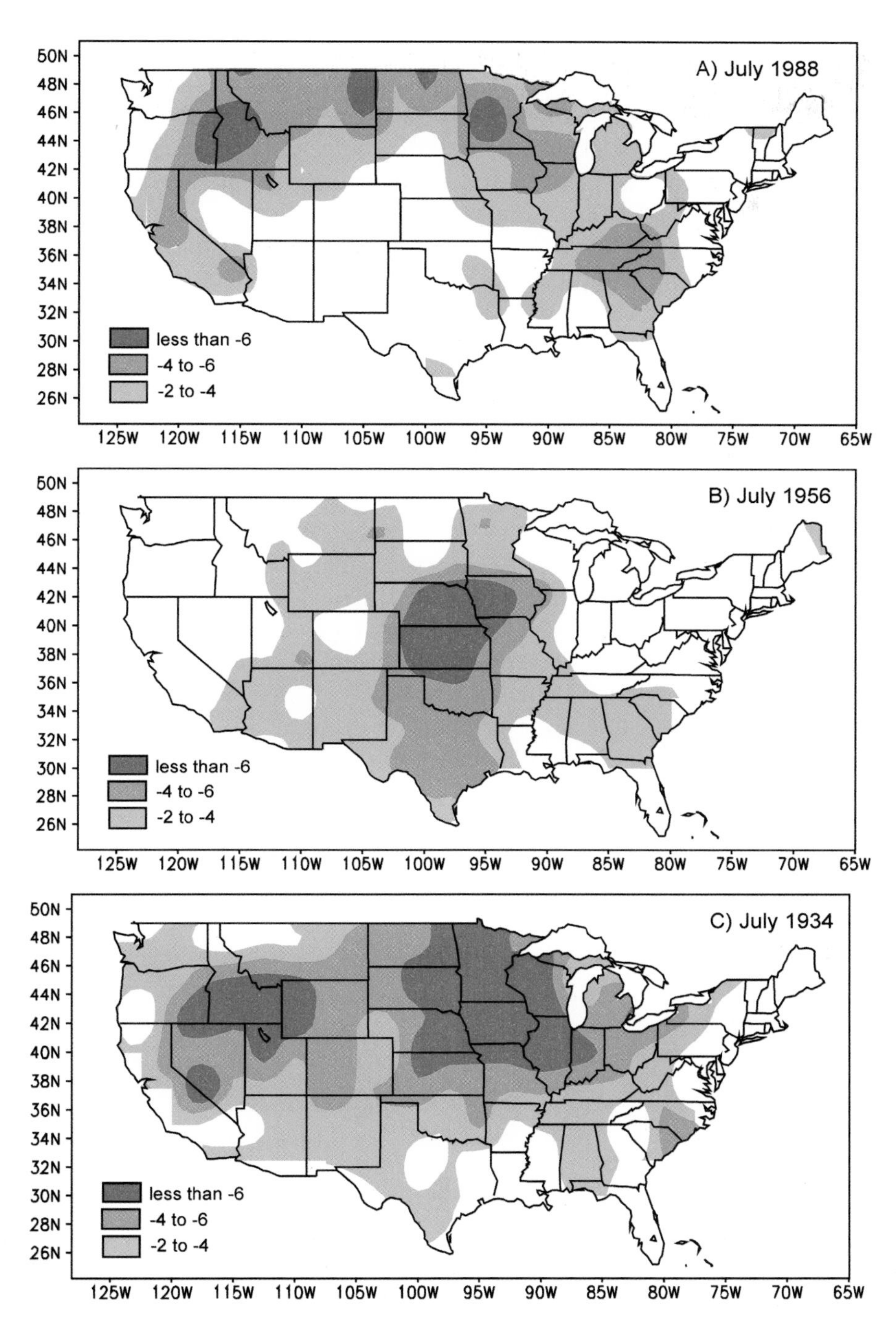

FIGURE 26.5: Palmer Drought Severity Index (PDSI) showing areas affected by droughts in July of (A) 1988, (B) 1956, and (C) 1934. PDSI values were between –2 and –4 in yellow areas, between –4 and –6 in orange areas, and less than –6 in red areas. (See text accompanying Figure 26.1 for interpretation of the PDSI.)

of the ground by absorption. This results in additional warming of the near-surface air and strengthening of the ridge aloft. The additional heating caused by the reduction of evaporation is one the main reasons why droughts are often accompanied by abnormally high temperatures—heat waves (Chapter 27). Since these feedback mechanisms can reinforce or strengthen an existing drought, there is indeed a physical basis for the tendency of a drought to "feed upon itself."

Drought has been a fact of life in the central United States for thousands of years. The ***Dust Bowl*** drought of the 1930s has been immortalized in history and in American literature. While droughts continue to occur, the region's vulnerability has changed: agriculture has become more intensive in the central United States, yet the development of efficient transportation systems has reduced residents' reliance on local growing conditions. Irrigation has become widespread in marginal growing areas such as the Great

Plains, yet long-term droughts pose serious threats to the water table in these areas.

Against this backdrop of changing vulnerabilities, we examine three significant droughts that affected the United States during the twentieth century: the droughts of the late 1980s, the 1950s, and the 1930s (the Dust Bowl period). As shown in Figure 26.5, the most severely affected areas varied considerably among these three droughts, although a significant portion of the Great Plains region was affected by each. Note that surface, upper-air, and satellite data were available for the 1980s, but that the droughts of the 1930s and 1950s preceded the satellite era. Hence the measurement of vegetative parameters by remote sensing was possible only during the drought of the 1980s. The drought of the 1930s preceded the implementation of the rawinsonde network over much of the world, so the tracking of the Dust Bowl drought must rely primarily on surface data. If one attempts to place the twentieth-century droughts into a longer climatological context extending back several hundred years, even surface weather maps are unavailable for much of the period.

The Drought of 1988

The drought of 1988 affected the United States in an arc extending from the southern Appalachians through the Ohio Valley, the Midwest, and the northern Great Plains to the western United States (Figure 26.5A). While the winter of 1987–1988 was relatively moist in the central United States, the weather turned dry with a vengeance in the spring of 1988. Precipitation for the agriculturally critical April through June period was the lowest since 1895 for more than 10 percent of the United States. Less than half of the normal precipitation for April through June fell over much of Illinois and Iowa, the leading corn-growing states. Only three other growing-season months (June 1933, May 1934, and June 1936—all during the Dust Bowl period) were drier than June 1988 in the Great Plains and Midwest. By July, approximately 40 percent of the United States was experiencing severe or extreme drought, according to the Palmer Drought Severity Index. Ironically, while the central United States was gripped by drought, the Southwest was much wetter than normal, mainly because enhanced

FOCUS 26.2

Hotter Days and Cooler Nights during a Drought

An interesting feature of the weather pattern of dry summer months is that many locations experience both warmer-than-normal daytime temperatures and cooler-than-normal nighttime temperatures. Figure 26A shows an example from the summer of 1988. The average daily high temperature in this case was 3.3° F (1.8° C) warmer than normal, while the average daily low temperature was 2.0° F (1.1° C) cooler than normal. Why did this occur? The air was so dry that the daytime solar radiation penetrated easily to the dry ground, which heated rapidly because there was so little evaporation (Figures 5.1 and 26.4). The dry air and clear skies enabled infrared energy emitted by the Earth to escape to space at night, resulting in rapid cooling of the ground and the low-level air. Wider-than-normal daily ranges of temperature are one of the characteristics of the weather regime of a drought.

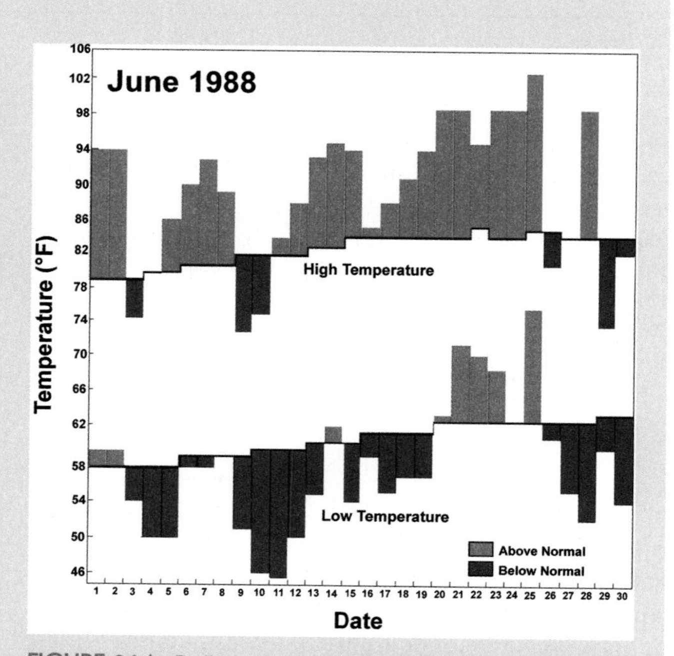

FIGURE 26A: Daily high and low temperatures (°F) at Champaign, Illinois, during the June 1988 drought. Above-normal values are shown in red; below-normal values are in blue. The data show the large daily range of temperature that occurs during drought.

moisture inflow from the Pacific was favored by the circulation pattern in Figure 26.3.

The impacts on vegetation, including crops, were dramatic in the major grain-producing regions of the central United States. Figure 26.6 shows images of a satellite-derived index of vegetation in late June 1987 and 1988. In contrast to the more typical greenness of 1987, broad areas of stressed vegetation are indicated by the brown shades in the 1988 image. United States corn production in 1988 was only 55 percent of its average for the preceding several years and corn prices nearly doubled from their prices a year earlier. Total grain production in the United States was down 31 percent, and the production of fruits and vegetables declined by 10 percent. Had it not been for grain surpluses built up during the previous years, the impacts on prices and supplies would have been far greater.

Streamflow rates in major rivers of the central United States declined by more than 50 percent during 1988, resulting in a severe curtailment (and at times a cessation) of barge traffic along the Mississippi River, which reached record-breaking low levels during the spring and summer of 1988. Ironically, record high levels would occur only five years later during the 1993 floods (see Figure 25F). Groundwater levels also dropped significantly, severely limiting water supplies and requiring the drilling of new wells in some areas. The total cost of the drought in the United States has been estimated at $39 billion in 1988 dollars (approximately $70 billion in 2007 dollars).

What might have triggered the dry weather pattern that developed in the spring of 1988? Climatologists are not in agreement about whether the ocean temperatures played a critical role, but the equatorial eastern Pacific Ocean was definitely in its cold or "La Niña" phase at this time (Chapter 23). Some climate models react to this sea surface temperature anomaly pattern by building a ridge over central North America, as actually happened. However, when a similar pattern of equatorial Pacific water temperatures was in place during 1999 and 2000, the focus of the dry weather was in the East and Southeast, respectively. While these shifts of the possible response to ocean temperatures are subtle, the consequences for a particular region are tremendous. The geographical shifts of the core of the drought highlight the difficulties inherent in seasonal forecasting of drought. Drought prediction continues to be one of the major challenges in atmospheric science.

As severe as the drought of 1988 was, it was not as severe by most measures as the droughts of the 1950s and the 1930s. For the United States as a whole, the precipitation deficits of the 1930s and 1950s were multiyear events, unlike 1988 when the precipitation deficit was essentially a one-year event (Figure 26.7). Consequently, the drought of the 1950s and the more protracted drought of the 1930s dominate twentieth-century drought history.

FIGURE 26.6: Satellite image of NDVI (Normal Difference Vegetation Index) over Illinois in (A) late June 1987 and (B) late June 1988. NDVI is a parameter that indicates the health of vegetation. June 1987 had normal growing conditions, while June 1988 was a drought month. Data are based on measurements from the NOAA-10 satellite.

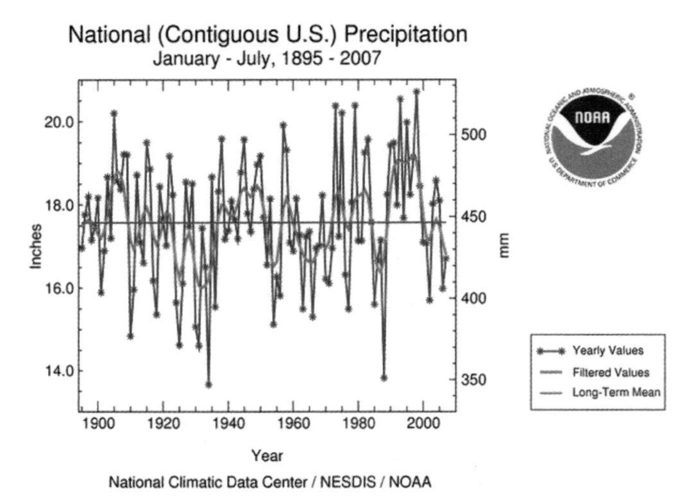

FIGURE 26.7: Yearly precipitation during the period 1895–2007, averaged over the contiguous United States. Red line is a smoothed version representing averages over several previous and subsequent years. Significantly drier than normal years occurred in the 1930s and mid-1950s, as well as in 1988.

The Drought of the 1950s

The United States as a whole experienced some of its driest summers of the past century during the multiyear drought of the 1950s (Figure 26.7). The core of the 1950s droughts was more toward the Southern Plains and the Southwest than in 1988 or the 1930s (Figure 26.5), although the period ended with a severe drought in the northeastern and mid-Atlantic states in 1957.

The drought period of 1952–1957 provides useful illustrations of several characteristics of prolonged droughts. First, a multiyear drought does not worsen relentlessly to its peak, and then steadily improve. Rather, periods of abnormally low precipitation are punctuated by returns to normal or even above-normal precipitation. However, soil moisture and groundwater supplies often do not recover during the breaks in the meteorological drought, so the hydrological or socioeconomic drought persists. Figure 26.7 shows that several years were indeed very dry, while other years were less dry. The three-cell pattern of high pressure at 700 mb during July 1954 and 1957, shown in Figure 26.8, illustrates the classic drought pattern. The third high-pressure cell, located over the central United States, was accompanied by a northward shift of the jetstream to the vicinity of the United States-Canada border and southern Canada. This shift of the jetstream displaced cyclones and their associated precipitation northward as well.

FIGURE 26.8: 700 mb heights (meters) averaged over (A) July 1954 and (B) July 1957. Orange and red denote high pressure.

While the meteorological drought of the 1950s was comparable by some measures to the drought of the 1930s, it was far less severe as a socioeconomic drought. One reason is that the precipitation deficit was more protracted in the 1930s than in the 1950s. More important, changes in agricultural practices and the nation's economic situation acted to mitigate the consequences of drought in the 1950s. Farming practices such as deep plowing, the use of cover crops and shelterbelts, and irrigation prevented much of the wind erosion that characterized the 1930s drought.

The absence of a depression in the 1950s meant that many farmers were able to turn to off-farm jobs rather than abandon their holdings altogether. Nevertheless, losses to the farmers in the Southern Plains totaled billions of dollars, and tens of thousands of farm families were forced to move off their land.

A significant consequence of the drought of the 1950s was an increased reliance on irrigation in the Southern and Central Plains, as is readily apparent from numerous circular patches of cultivated ground that can be seen from airplanes. Figure 26.9 shows clearly that the drilling of irrigation wells in the Texas High Plains peaked in the years during and immediately after the drought of the 1950s. Since the 1950s, irrigation has resulted in groundwater-level drops in some areas of more than 15 m (42 ft) and more than 30 m (98 ft) at a few locations, calling into question the sustainability of present-day agriculture in this drought-prone region.

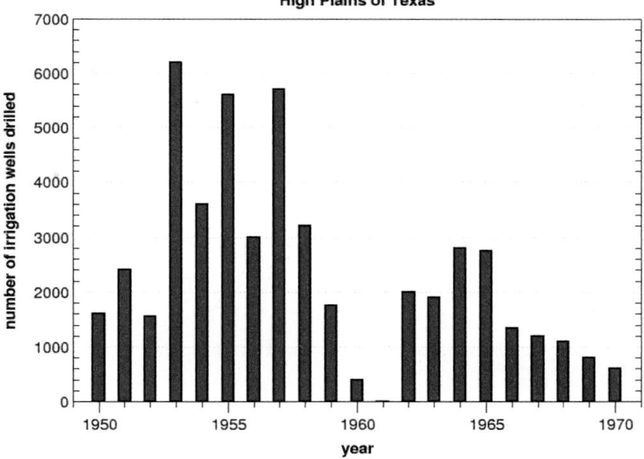

FIGURE 26.9: Yearly number of irrigation wells drilled in the High Plains of Texas from 1950 to 1970.

The Drought of the 1930s: The Dust Bowl

The consensus of meteorological historians is that the drought of the 1930s was the "greatest disaster in American history attributable to meteorological factors."[1] Indeed, in a recent ranking by *Weatherwise* magazine of twentieth-century weather events according to their human, social, and economic impacts in the United States, the drought of the 1930s ranked first, ahead of such events as the Galveston hurricane of 1900 (over 6,000 deaths), the super-outbreak of tornadoes in 1974, and the East Coast's "Storm of the Century" in 1993. While the Dust Bowl disaster was exacerbated by agricultural, social, and economic factors, including its coincidence with the Great Depression, weather was the ultimate driver in the eyes and minds of millions of people whose lives were changed by the events of the 1930s.

The drought of the 1930s was unique among twentieth-century droughts in its extent as well as its duration. The drought actually began in the late 1920s and continued through the late 1930s, making it the longest drought of the twentieth century in the United States (Figure 26.7). The drought's geographical extent, for which Figure 26.5C shows a typical example, was significantly greater than during the droughts of the 1950s and 1988. Like the drought of the 1950s, the drought of the 1930s was punctuated by months and seasons with above-normal precipitation in most areas, although these wet interludes were not sufficient to alleviate the hydrologic drought that became progressively more severe through the early and middle 1930s. The driest calendar years in the central United States were 1934 and 1936, although 1930 and 1931 were also dry, and perhaps more significantly dry, since they initiated the soil moisture depletion that was worsened by the agricultural practices of the 1930s, setting the stage for the blowing-dust events of the subsequent years.

A striking analogy has been noted between 1931 and 1988 in terms of the precipitation during the winter and spring months. In both years, a relatively mild and moist winter was followed by a sudden shift to dry conditions in late March or early April. A key difference between 1988 and 1931, aside from the agricultural and economic background factors, is that 1988 was followed by relatively abundant precipitation during 1989 and 1990 in the drought-affected regions. There was no such relief after the dry year of 1931. The fact that the dry summer of 1931 was not even the worst of the dry summers of the 1930s attests to the uniqueness of the 1930s in terms of twentieth-century precipitation.

The magnitude of the 1930s drought is illustrated by the Palmer Drought Severity Index (PDSI) map for July 1934 (Figure 26.5C). A PDSI value of –4 is the accepted threshold for "extreme drought," the most severe category of the PDSI values. It is apparent from Figure 26.5C that the hydrometeorological conditions were "off the scale"—large portions of the north-central and western states had PDSI values of –5, –6,

[1]Ludlam, D. M. *The American Weather Book* (Boston: Houghton-Mifflin, 1982) p. 182.

FOCUS 26.3

Drought and Population Migration in the United States

There is evidence that droughts worse than those of the 1930s and 1950s have affected the United States in previous centuries. One need only look back as far as the 1890s to find droughts that impacted the population of the Northern Plains more significantly than any of the more recent droughts. Since the westward advances of settlements from 1870 to the mid-1880s occurred during a period of relatively abundant precipitation over the Northern Plains, the stage was set for severe human impacts when the weather turned dry in the late 1880s and remained so through much of the 1890s. Figure 26B shows that large regions from the Dakotas and Montana to Oklahoma and Texas lost over 75 percent of their population in the 1890s. These percentage losses dwarf those of the Dust Bowl period of the 1930s, reinforcing the notion that areas in which precipitation is marginally adequate for agriculture are extraordinarily vulnerable to decadal-scale swings of weather patterns.

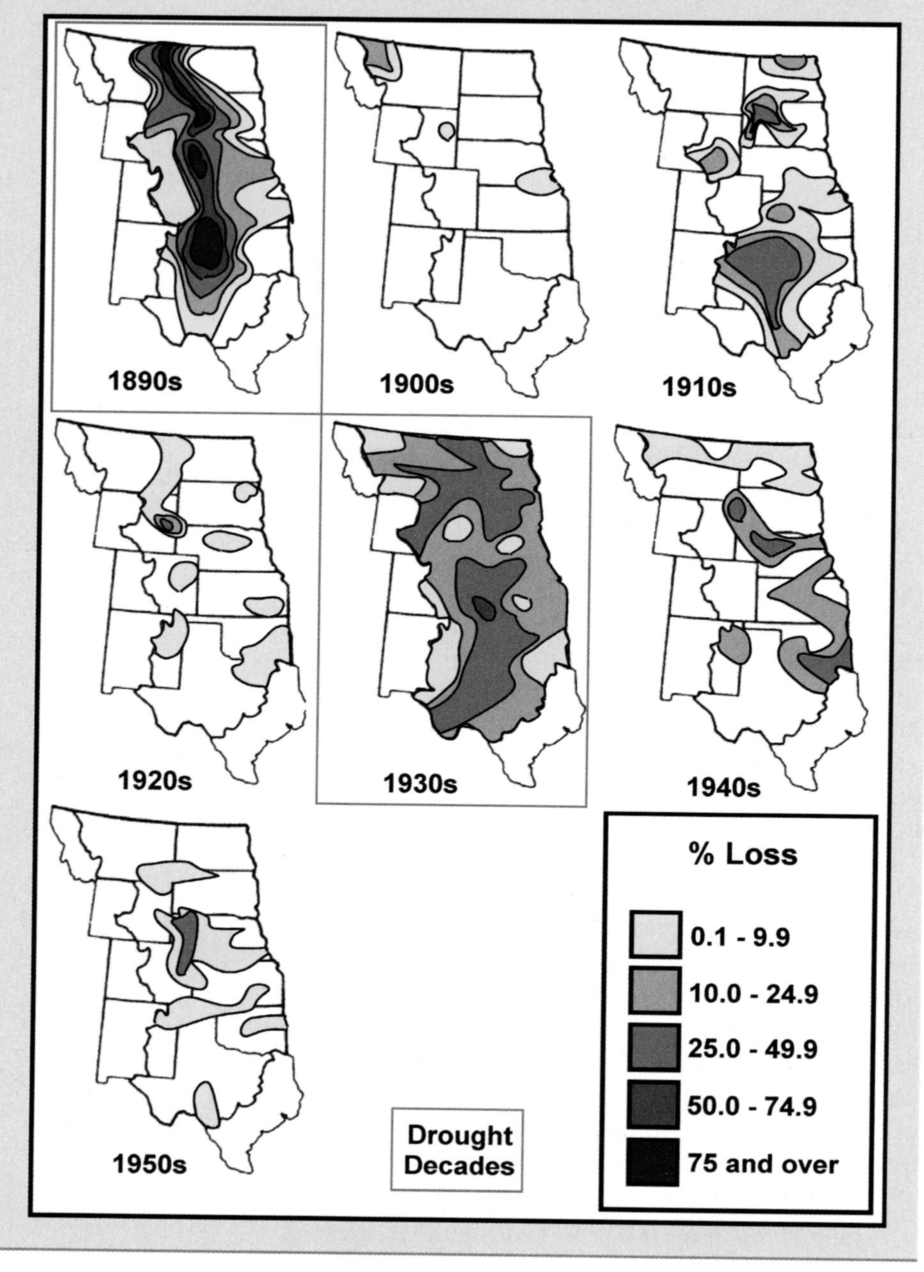

FIGURE 26B: Population loss by decade in the Great Plains from the 1890s through the 1950s.

–7, and even –8. The dry soil conditions implied by Figure 26.5C were associated with relatively hot summer weather (see Chapter 27: Heat Waves of the 1930s), consistent with the physical arguments presented earlier in this chapter. The 1930s were indeed the warmest decade of the century in the United States, and many daily high-temperature records, especially ones set in 1934 and 1936, still stand today.

CHECK YOUR UNDERSTANDING 26.2

1. What weather pattern predominates during drought in the central United States?
2. Where was the 1988 drought most severe? In which season did it intensify?
3. What was a key difference between the droughts that affected the central United States in the 1930s and in the 1950s?

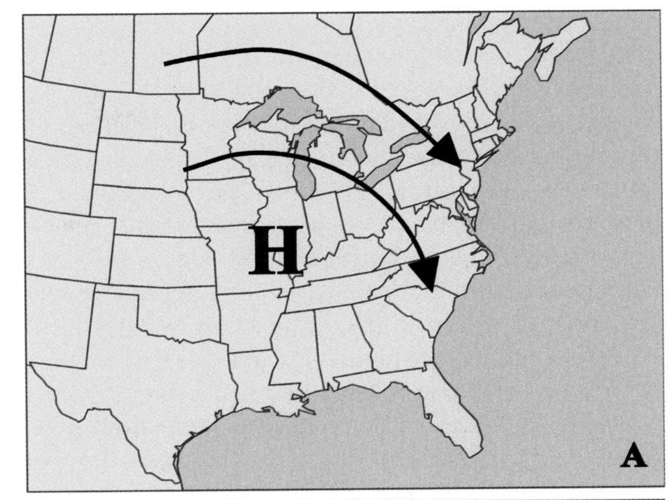

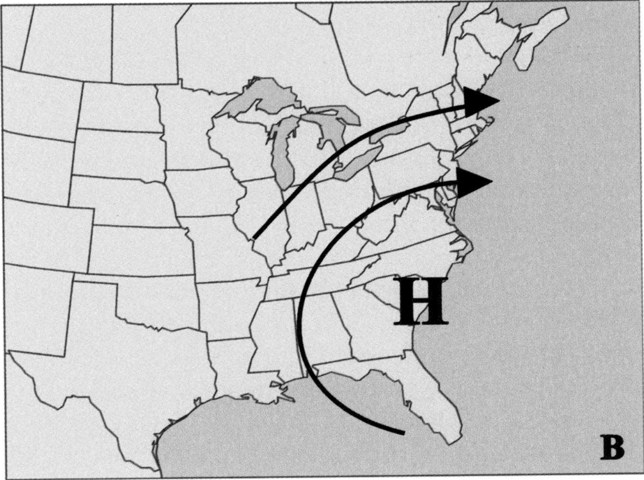

FIGURE 26.10: Schematic representations of summer circulation patterns favoring drought in the East: (A) high pressure over central United States, (B) high pressure along the southeastern coast.

DROUGHT IN THE EASTERN UNITED STATES

Drought in the eastern United States shares some characteristics with drought in the central United States. Specifically, a surface high-pressure system and a ridge aloft tend to block the influx of moisture for precipitation. Air also descends in the vicinity of high-pressure systems, which suppresses the formation of clouds and precipitation. There are two ways in which a surface high-pressure center can inhibit moisture fluxes to the East Coast. In both cases, the airflow has a west-to-east component, giving the air a downslope, and therefore warming, component on the east side of the Appalachians.

First, a large continental high-pressure system, corresponding to the third high-pressure cell (over the central United States) in Figure 26.3, will generally bring air to the East Coast from the west or the northwest, as shown in Figure 26.10A. This air will generally be dry because of its continental origin. Second, the semipermanent high-pressure cell over the western Atlantic in Figure 26.2 can migrate westward to the vicinity of the southeast coast. This feature, often referred to as the ***Bermuda High*** because of its normal position near Bermuda, can result in a broad southwesterly flow over the entire eastern portion of the United States (Figure 26.10B). While this airmass is a maritime tropical airmass, and hence warm and humid, its downslope component of motion as it descends the Appalachians tends to inhibit the development of precipitation over the East Coast, especially from the Carolinas northward. In both patterns shown in Figure 26.10, the downslope motion of air and the general absence of fronts are usually associated with below-normal precipitation along the East Coast.

As in the case of droughts in the central United States, there is an intermittency to East Coast droughts. While the patterns in Figure 26.10 may prevail or simply be more frequent than usual, there are almost always interludes of precipitation associated with fronts, airmass thunderstorms, and occasionally, tropical cyclones. Nevertheless, there are periods of several seasons to several years when the patterns in Figure 26.10 are sufficiently frequent that precipitation is less than normal, and precipitation deficits over these several seasons or years can become substantial. Two such periods were 1962–1965 and 1998–1999. During the former period, the low reservoir and streamflow levels in many locations in the mid-Atlantic region, including the New York City area, approached or broke records that had been set in the early 1930s. Many of these records were threatened again in the summer of 1999, when the recent drought peaked. Pockets of the 1999 drought persisted into 2002.

The 1999 drought extended from Virginia to southern New England, although drought conditions of lesser severity affected much of the Southeast as far west as Texas. Figure 26.11 shows that the non-forested areas of the mid-Atlantic region appeared strikingly brown in a satellite image from August 1999. This drought's origin can be traced back to the dry conditions in the late summer and autumn of 1998, when a general deficit of soil moisture developed. The deficit persisted into spring, when the dry conditions were worsened by frequent high-pressure systems in the Midwest as in Figure 26.10A, producing the westerly airflow that effectively cut off the Northeast from its maritime moisture sources.

The April through July period of 1999 was the driest on record for several states (Rhode Island, New Jersey, Delaware, Maryland), and the second driest for Massachusetts, Connecticut, and New York. In many areas, precipitation deficits of 12 to 18 in (30 to 45 cm) developed over the twelve-month period ending in July 1999. In the region extending from the mid-Atlantic to the Southeast and Gulf States, 30 to 50 percent of river and stream gauge readings were in the lowest 10 percent of their historical readings for the date. Figure 26.12 illustrates this by showing a trace of the streamflow of the Potomac River near Washington, D.C., together with the normal range. The precipitous decline in the spring and summer of 1999 in the flow of the Potomac and other rivers led to serious concern about the up-river penetration of salt water from the Atlantic Ocean, threatening water supplies of large cities.

The southeastern United States was affected by a serious drought from late 2006 through much of 2007. Water levels in many streams and lakes, including Lake Okeechobee, dropped to unprecedented levels, and numerous wildfires threatened structures and inhibited highway traffic in Georgia and Florida (Figure 26.13). Pastureland in much of the Southeast deteriorated, and corn and cotton crops were severely impacted by the absence of precipitation and the depletion of soil moisture (Figure 26.1). As often occurs during summer droughts, the dry soil led to excessively high temperatures (Chapter 27). Many cities in Alabama and Arkansas experienced their hottest August on record, with high temperatures above 100° F (38° C) for ten or more consecutive days.

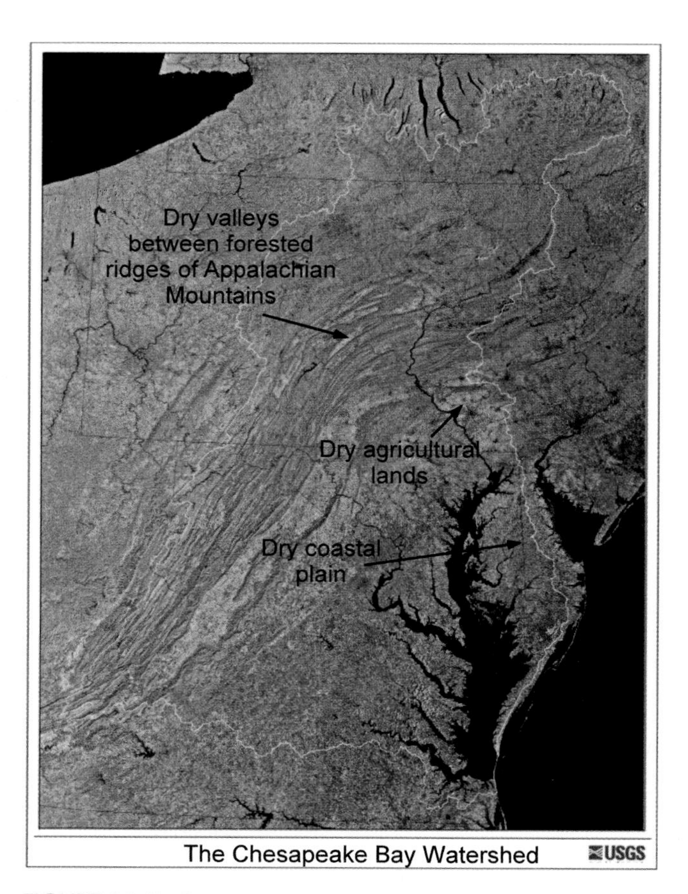

FIGURE 26.11: Satellite image of the mid-Atlantic region in August 1999, showing drought-stressed vegetation in non-forested areas. A thin, white line encloses the Chesapeake Bay watershed, which suffered ecologically from the drought.

DROUGHT IN THE WESTERN UNITED STATES

The western United States experiences a vastly different precipitation regime from the rest of the United

ONLINE 26.1

Evolution of the 1999 Drought in the Eastern United States

Drought conditions are monitored throughout the United States using measures such as the Palmer Drought Severity Index, which is recomputed on a weekly basis for climatological divisions that are typically about 100 km on a side. Online, you can view an animation of the Palmer Index during the 1999 drought. Red colors denote drought conditions (PDSI values of –2 or lower), and green colors denote moist soil conditions (PDSI values of +2 or greater). The animation begins in May, when the Northeast was already slightly dry because of the precipitation deficit of the previous year. The drought then worsens through July and early August, when the maps of the Palmer Index indicate extensive drought conditions. Subsequent spotty rains, followed by heavy rainfall from Tropical Storms Dennis and Floyd in September, alleviated conditions in coastal regions, where moist conditions are indicated at the end of September. Farther to the west and south, however, the drought persisted well into the autumn months, and it remained especially severe in the Ohio Valley. Note also that the Northern Plains experienced generally wet conditions throughout the summer of 1999.

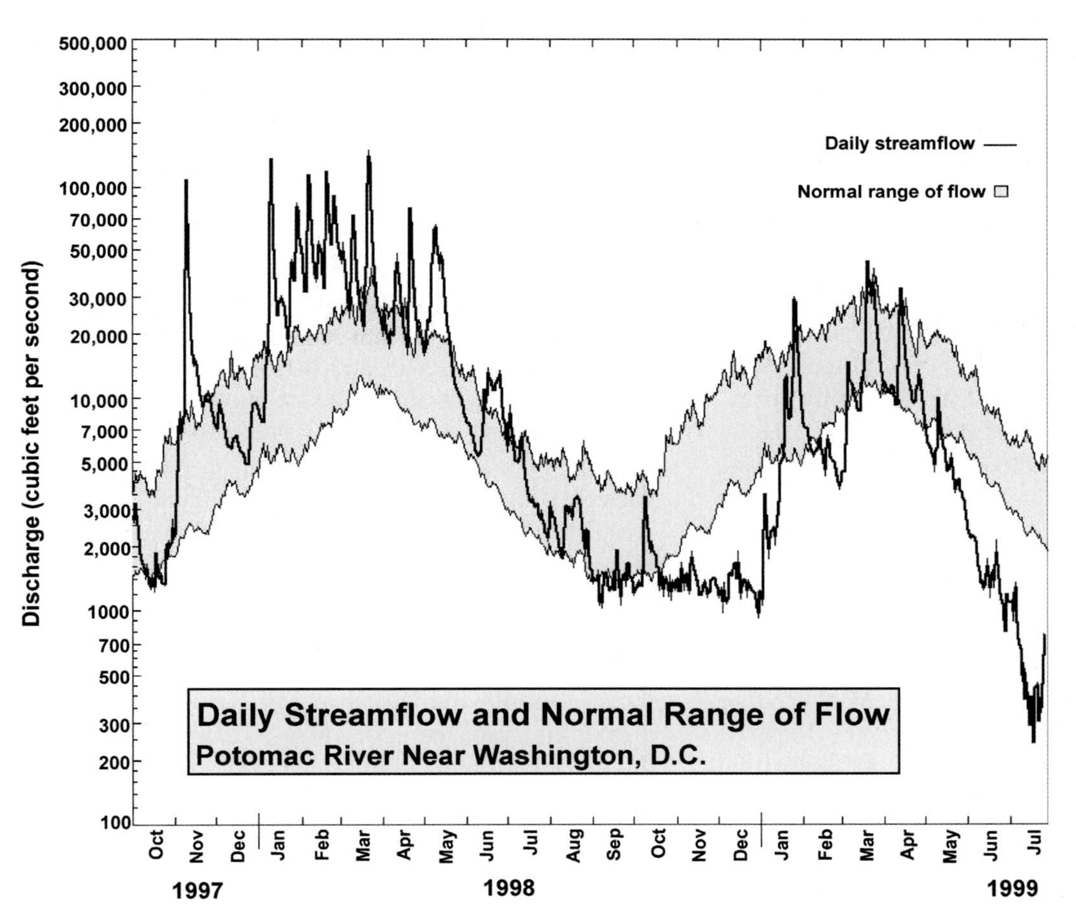

FIGURE 26.12: Daily streamflow (cubic feet per second) of the Potomac River near Washington, D.C., from October 1997 through July 1999. The normal range of streamflow values is shaded gray. Note that the discharge is plotted on a logarithmic scale.

States. Nearly all precipitation in California, and most of the precipitation in all the mountain states, falls during the period from October to April (see Figure 25.11). The heaviest amounts typically occur with Pacific storms that come ashore during the winter months, December through March. This precipitation, much of which falls as snow on the Sierra Nevada, the Cascades, and interior mountains including the Rockies, represents much of the water supply for the western states. The onshore flow of moisture-laden air in

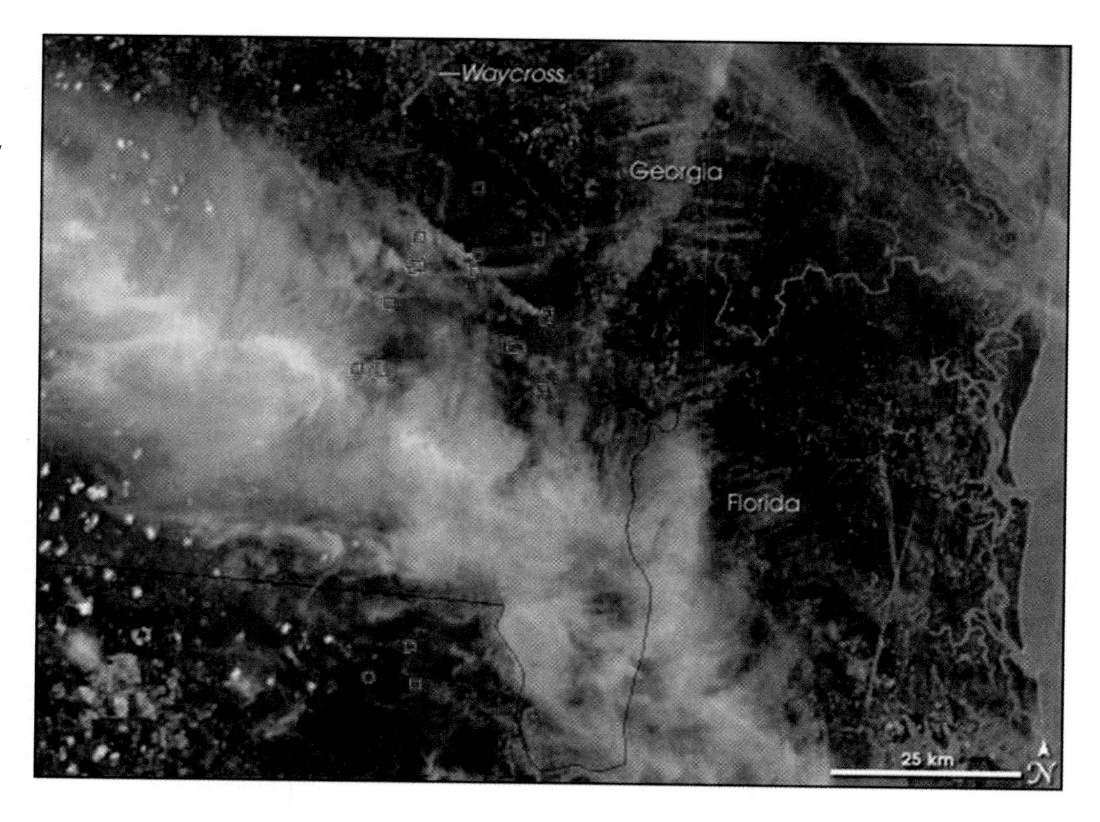

FIGURE 26.13: Wildfires in Georgia and Florida as captured by NASA's Aqua Satellite on May 30, 2007. Actively burning fires are outlined in red. The Georgia-Florida coastline is at far right.

Pacific storms, together with the orographic uplift provided by the mountains of the West, can result in heavy precipitation and flooding (Chapter 25). Snow depths measured in 10's of feet result in subsequent melt, runoff, and storage in reservoirs that support the growing demand for water in the West (Chapter 16). However, when winter storms fail to come onshore with sufficient frequency, deficient mountain snowpacks lead to water shortages throughout the western states.

The characteristic feature of the weather pattern conducive to drought in the West is a persistent ridge in the jetstream over the West Coast, shown schematically in Figure 26.14. When such a ridge is present, the storm track on the western side of the ridge extends from the eastern Pacific into the Gulf of Alaska. Storms move onshore well to the north of the West Coast of the United States. In addition, an upper-air ridge over the West Coast is generally associated with a surface high-pressure center to the east of the ridge axis (Figure 26.14). The corresponding southeasterly surface winds reaching the West Coast have a downslope component that enhances the dryness along the coast. If the axis of the ridge is shifted westward, the West Coast is more directly influenced by the airflow around the surface high under the east side of the ridge.

The upper-air ridge is often accompanied by a downstream trough over the central and eastern United States. When the pattern is amplified and persistent during the winter, a closed (cutoff) low aloft is often found in the vicinity of Hudson Bay. The overall pattern of the jetstream is similar to the configuration that accompanies a major cold outbreak in the central and eastern United States (Chapter 14, Figure 14.3). Dry weather in the West indeed accompanies nearly all periods of exceptional cold in the central United States.

What causes a persistent ridge along the West Coast? As with droughts in the central United States, studies have suggested links to ocean surface temperatures in the Pacific Ocean. An El Niño was occurring during the dry winters of 1976–1977 and 1986–1987, but a La Niña prevailed during 1988, during the heart of the extended western drought of 1986–1991. Since correlations with mid-latitude ocean temperatures are similarly mixed, the prediction of seasonal precipitation in the western United States remains a challenge to forecasters.

An extended drought that resulted from a series of dry winters from 1986–1987 through 1990–1991 seriously impacted western water supplies. Precipitation during all five of these winters was significantly below normal over California. For some areas along the cen-

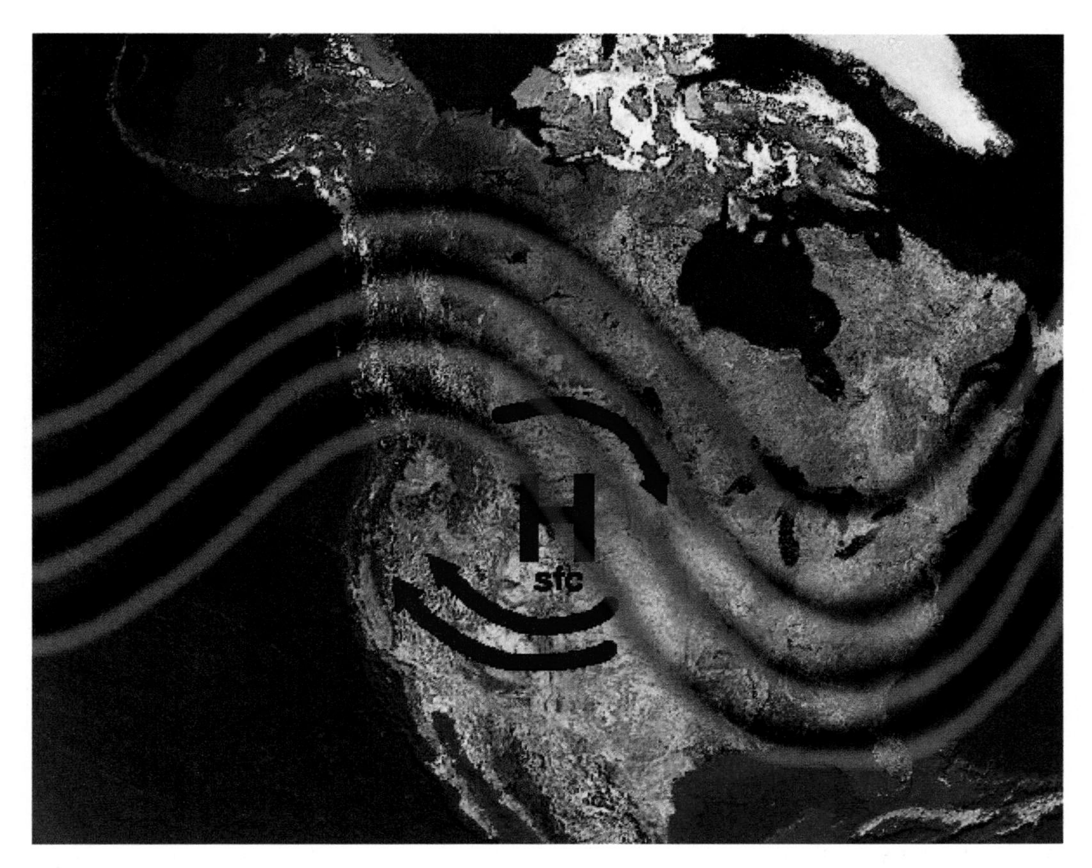

FIGURE 26.14: Schematic diagram showing a West Coast ridge and jetstream (orange), conducive to drought in the West. The surface high-pressure center and associated surface wind pattern are shown in blue.

ONLINE 26.2

The West Coast Ridge Prevails, Then Yields, during the 1991 Drought

The climax of the five-year dry spell in California occurred in early 1991, when the upper-air ridge over the West Coast was a recurring feature of the upper-air circulation. Precipitation was well below normal over most of the West during January and February. Finally, in March 1991, the ridge gave way to a succession of troughs that brought much-needed rain to California. Online, you can compare the evolution of the circulation at 500 mb during February and March. The ridge is prominent during February, especially in the early and middle parts of the month. A key feature in the transition is the appearance of a subtropical branch of the jetstream, which impinges on California at the beginning of March. The frequent large changes of the ridge-trough configuration during March are characteristic of a month with near- or above-normal precipitation in California. Notice also that many of the upper-air troughs can be traced upstream to the eastern Pacific. Troughs that deepen just before reaching the coast are often associated with vigorous surface systems and heavy precipitation.

tral California coast and the central Sierra Nevada, 1987–1990 was the driest four-year period on record. Reservoirs throughout the state fell below 50 percent of normal by early 1991, and groundwater levels dropped by amounts ranging from several feet to as much as 30 feet (9 m). October through September runoff in the Southern California hydrologic region was 32 percent, 40 percent, 28 percent, and 16 percent of normal in 1987, 1988, 1989, and 1990, respectively.

Water rationing was imposed in all major California cities, with cutbacks ranging from 10 to 45 percent. Agricultural water deliveries for irrigation were reduced by 50 percent in early 1991, threatening the economy of the Central Valley, as well as the fruit and vegetable supplies for much of the nation. Fortunately, the persistent ridge finally broke down in March 1991, as shown online, bringing relief to the agricultural industry and boosting reservoir levels to about 60 percent of normal.

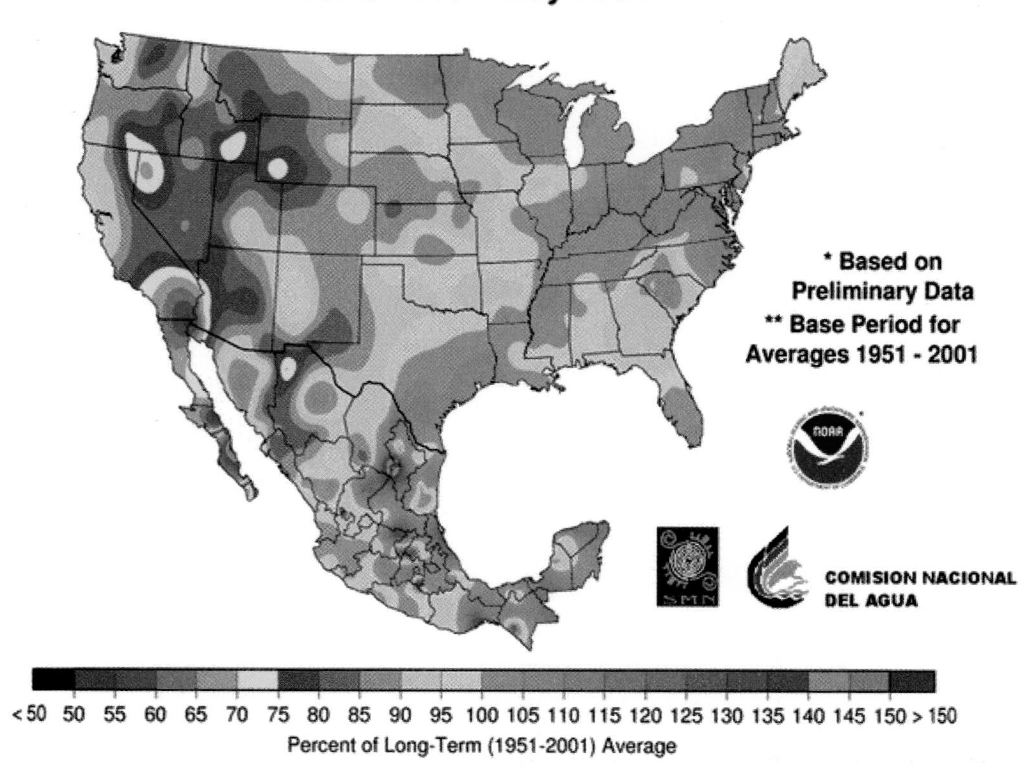

FIGURE 26.15: Precipitation for the 5-year period, June 1999–May 2004, shown as a percentage of normal. Purple, red, yellow and brown represent precipitation deficiencies, green and blue precipitation surpluses (see color scale at bottom).

Even though the subsequent winters were relatively moist, the respite proved to be temporary.

Drought returned to the West with a vengeance in the first years of the twenty-first century. A prolonged period of below-normal precipitation began in 1999 and continued through the next five years, affecting nearly the entire region from the Rocky Mountains to the Pacific Coast (Figure 26.15). The precipitation deficit was especially severe in the Southwest, which had seen tremendous growth in its population and agricultural water usage over the preceding decades. As is characteristic of drought in the West, winter snowpacks were far below their normals during the winters and springs of 2002–2004, especially over the southern Rockies in 2002 and 2003.

The flow rate of the Colorado River was lower during 2001–2003 than during the Dust Bowl drought of the 1930s, and perhaps even the lowest of the past 500 years, according to reconstructions of the flow rates based on tree rings. Outbreaks of wildfires increased in severity during this period, especially in Colorado, New Mexico, Arizona, and southern California, where fires in the autumn of 2003 caused twenty-four fatalities and several billion dollars in property damage.

The western drought has persisted into 2007 (Figure 26.1), although with interruptions such as the wet winter of 2004–2005. That wet winter was a fortunate occurrence, because it replenished reservoirs, at least temporarily, prior to the generally dry subsequent years. To what can such a persistent drought be attributed? As noted earlier, the persistence may be associated with ocean temperature anomalies in the Pacific Ocean. However, the 1999–2004 drought spans the period of both a La Niña (1999–2000) and two El Niño events (2003–2004 and 2006–2007). Some scientists have sought links to decadal-scale oscillations of Pacific Ocean temperatures, while others have looked for clues in the North Atlantic Ocean. The ocean's role is not clearly understood. The perpetuation mechanism most likely involves not only the ocean, but also interactions between the atmosphere and the desiccated land surface, as well as human factors such as land use and water diversions (see Focus 26.4). Moreover, as described in the following section, there are indications that the climate system varies naturally over multiyear timeframes, providing an ominous reminder that prolonged droughts are a fact of life in the West.

FOCUS 26.4

River Flows and Water Rights in the West

As demands for water grow and river flow rates decline in the water-limited western United States, competition for the available water supplies will only increase. Water management, which includes a host of dams and reservoirs constructed during the past century, is becoming an increasingly important economic and legal issue. The problems are particularly acute in the Colorado River Drainage Basin, the largest in the Southwest. Areas from southern California to Colorado rely on the Colorado River for their water supplies. The rights of the various states to this vital water supply are governed by the Colorado River Compact, an agreement reached in 1922. Ironically, the river's highest flow rates of the 20th century occurred from 1905 to 1922. The flow rates during those years were used to determine how much water the various western states would receive. Water rights in various areas of the Southwest are bought and sold at "market prices," which have (not surprisingly) increased steeply in recent years. For example, the price of water rights in the Middle Rio Grande River Basin of New Mexico increased more than four-fold between 1994 and 2001.

Efforts to renegotiate the water distribution rights have been frustrated by the enormous demand for water and to date have not progressed very far. The redistribution of water in the west will be one of the most complex environmental and legal issues that the United States (and Mexico, which shares the Colorado River) will face in this century.

ONLINE 26.3

A Satellite Captures the Drying of Lake Mead

Lake Mead, created by the construction of Hoover Dam on the Colorado River between Arizona and Nevada, is 110 miles (177 km) long and up to 8 miles (13 km) wide. It is an important water source for irrigation and homes in Arizona, California, and Nevada, and its flow over Hoover Dam is a major hydroelectric power source for the region. Because snowmelt provides most of its water, Lake Mead is adversely impacted by deficient snowpacks such as those that occurred in the winters of 2000–01 through 2002–03. Online, you can view an animation of Landsat Satellite images (from the NASA Goddard Space Flight Center Visualization Laboratory and the U.S. Geological Survey) showing the shrinkage of Lake Mead between May 2000 and May 2003. The lake level fell by approximately 60 feet between 2000 and 2003, exposing large areas that had previously been submerged.

CHECK YOUR UNDERSTANDING 26.3

1. What is the "Bermuda high"? Where is it positioned during drought along the East Coast?
2. Where are the two locations a surface high-pressure center might be located during the development of a drought along the East Coast?
3. When is precipitation most common in the western United States?
4. What is a typical pattern of airflow aloft during a West Coast drought?

HISTORICAL PERSPECTIVE ON DROUGHT

Most of the discussion in this chapter has concerned droughts in the United States during the twentieth century, when droughts illustrated the vulnerability to water shortages in every region of the country. However, even though agricultural practices may have exacerbated drought's impacts of the 1930s, drought is a naturally occurring phenomenon, and the twentieth-century droughts are by no means unique.

Evidence from tree rings, lake sediments, and other sources indicates that multiyear and even much longer droughts, spanning several decades, have occurred repeatedly in the central and western United States during the past 2,000 years. Proxy data of these types suggest that droughts comparable to those of the 1930s and the 1890s (Focus 26.3) have occurred one or two times per century back to at least 1600, and that droughts lasting a decade or longer have occurred once every 500 years. None of the droughts of the past century has lasted longer than a decade, raising the ominous prospect that the United States could someday be affected by a drought that is unprecedented in the period since Europeans settled the region. There are even indications of multidecadal droughts, termed ***megadroughts***, in the late 1200s and in the late 1500s. The former occurred at about the same time as the abandonment of the famed Anasazi settlements in the Southwest.

GLOBAL WARMING: POTENTIAL EFFECTS ON DROUGHT

While the indications of multidecadal droughts in earlier centuries the central and western states is serious enough, an additional concern is the impact of the climate changes that are expected to occur with global warming. There is general agreement among climate models, when driven by plausible scenarios of increasing greenhouse gases, that the southwestern United States will dry significantly in the twenty-first century. The recent dryness in the region, highlighted earlier in this chapter, may be an indication that the drying trend is already underway. Water shortages already exist in parts of this region, in which the population is growing more rapidly than in any other area of the United States. The climate models suggest that the levels of aridity of the Dust Bowl and the 1950s drought could become the new norm for the southwestern United States. However, unlike the historical droughts that appear to be related to ocean surface temperatures, the increasing frequency of future droughts is the result of a shift of the atmospheric circulation, including a greenhouse-driven poleward expansion of the Hadley Cells and the subtropical dry zones (Chapter 5). The drying of the Southwest is part of a broader pattern of summer drying that extends into the major agricultural belt of the central United States, including the Midwest as well as the central Southern Plains. If the projections are correct, there would be a long-term lowering of the water table in areas of the Plains now reliant on irrigation, calling into question the sustainability of agriculture in this region.

Climate models also project summer drying over western and central Europe. Ironically, the northward shift of the wetter climate zones, together with a lengthening of the growing season in a warmer climate, could expand the potential for agriculture in Canada and Russia. Nevertheless, the Intergovernmental Panel on Climate Change has stated with high confidence in its 2007 assessment that the negative impacts of climate change on freshwater systems will outweigh the benefits. Drought leads to more fatalities than any other weather phenomenon. Since drought already affects larger areas, over longer timescales, than all other types of hazardous weather, it is not surprising that the increased occurrence of drought looms as the most serious threat of greenhouse-driven climate change.

CHECK YOUR UNDERSTANDING 26.4

1. List at least two types of evidence used to determine whether drought occurred in an area prior to the time when instrumental records began.
2. According to climate models, how will increasing greenhouse gas concentrations affect the occurrence of drought in the United States?
3. To what does the term *megadrought* refer?

TEST YOUR UNDERSTANDING

1. Why is there no universal criterion for a drought?
2. How does "meteorological drought" differ from "hydrological drought" and "agricultural drought"?
3. Is there any validity to the expression "all signs fail in times of drought"? Explain.
4. What is the Palmer Drought Severity Index? On what is it based?
5. What is the Crop Moisture Index? What does it indicate? How does it differ from the Palmer Drought Severity Index?
6. What causes drought to develop?
7. What type of vertical air motion would you expect in an area experiencing drought? Why?
8. How does the drying of the ground alter the surface energy budget?
9. How does the loss of soil moisture enable a drought to reinforce itself?
10. How does the absence of clouds lead to the strengthening of an upper-air ridge in an area of drought?
11. How does the daily range of temperature typically compare with normal in a region experiencing drought?
12. Where is the jetstream located during a summer drought in the Central Plains? How does this differ from normal?
13. Why were the socioeconomic impacts of drought less severe in the 1950s than in the 1930s?
14. What was the driest decade of the twentieth century in the United States?
15. Does a multiyear drought, such as those that occurred in the 1930s and 1950s, increase steadily in severity? Explain.
16. How is the vulnerability of the central United States to drought different from 100 years ago?
17. What conditions preceded the 1999 drought in the eastern United States?
18. What ended the 1999 drought in the mid-Atlantic coastal areas?
19. What role do the Appalachian Mountains play in the development of a drought along the East Coast?
20. In which season(s) is a precipitation deficit most conducive to drought in the eastern United States? In the western United States?
21. The water levels in California's reservoirs generally increase most rapidly several months after the heaviest precipitation. Explain.
22. When California and Oregon are unusually dry, where would you expect Pacific storms to make landfall? Why?

24. During which months of a typical year would you expect fires to pose the greatest danger in the western United States? Explain.
25. Can ocean surface temperatures be used to predict droughts in advance? Explain.
26. Compare and contrast the data available for meteorological analyses of droughts in the 1930s and the 2000s.
27. In greenhouse simulations of future climate change, how do changes in the large-scale atmospheric circulation affect the occurrence of drought in the southwestern United States?

TEST YOUR PROBLEM-SOLVING SKILLS

1. Suppose your county is in a weather regime in which the only source of precipitation is a daily batch of airmass thunderstorms that produce 0.5 inches of rain over 20 percent of the county's area each day. The remainder of the county receives no precipitation on that day. Suppose this pattern continues for a week, and that the area receiving the rain is randomly distributed each day.
   (a) What will be the county's average precipitation for the week?
   (b) What are the chances that there will be no precipitation during the week at your specific location?

2. Assume that abnormal heating of the atmosphere by warmer-than-normal ocean surface temperatures favors ridging in the jetstream above the warm ocean area, and that abnormally cold ocean surface water favors a trough in the jetstream. Copy the blank map of the United States from Appendix C and indicate where abnormally cold or warm water would be located in order to favor drought during the winter season along:
   (a) the West Coast
   (b) the East Coast

   In each case, sketch the jetstream pattern associated with the ocean surface temperatures, and explain why the pattern is conducive to drought in the respective regions.

3. Consider an agricultural area in which 60 percent of the land is used to grow corn and the remainder is pasture. Suppose that a healthy corn crop requires approximately 12 inches of precipitation over the growing season (May through August), and that any shortage is offset by irrigation water pumped from an underground aquifer. If the normal precipitation during September is insufficient to provide any recharge of the aquifer, how far will the water table drop because of irrigation if there is a drought for ten consecutive years with each of the following precipitation deficits during May through August? (Assume the aquifer underlies the entire agricultural area at the same depth and no water runs off the area.)
   (a) 25 percent
   (b) 50 percent
   (c) 75 percent

4. Consider a hypothetical Southern California suburban area, Lemon County, containing three million residents. The average resident uses 100 gallons of water per day. Suppose that all the county's water is supplied by the storage and diversion of runoff (including snowmelt) from a 30-mile by 30-mile drainage basin of the Sierra Nevada, where the average annual precipitation is 30 inches (75 cm). Assume that 50 percent of the precipitation falling on this drainage basin is captured in reservoirs for diversion, and the remainder is lost to evaporation and soil infiltration.
   (a) How does the water supply compare to the usage requirements of Lemon County?
   (b) How much of a precipitation deficit in the Sierra Nevada would it take to create a water shortage (i.e., reservoir levels would drop)?
   (c) Suppose agriculture consumes an amount of water equivalent to the residential usage in Lemon County. How would your answers to (a) and (b) change?

   (Useful conversion: 1 cubic meter contains about 250 gallons of water.)

5. Consider the following simple model that illustrates the working of a positive feedback of drought:
   - 35 percent of days have some rain, 65 percent of the days are dry.
   - If a day has rain, the enhanced ground wetness and evaporation increase the likelihood that the following day will have rain to 45 percent.
   - If a day has no rain, the reduction of soil moisture and evaporation decreases the likelihood that the next day will have rain by 5 percent (e.g., after two consecutive dry days, there is a 75 percent chance that the third day will be dry).

   (a) According to this idealized model, how many consecutive dry days are required to reduce the likelihood of rain from its average value to zero, thereby "locking in" a drought?
   (b) If today is rainy, what is the probability that the following day is the start of a drought that "locks in"?

USE THE SEVERE AND HAZARDOUS WEATHER WEBSITE

http://severewx.atmos.uiuc.edu

1. Navigate to the "Drought" page of the *Severe and Hazardous Weather Website* and access the Hydrologic Information Center site on drought statements.
   (a) Read the most recent statement for the location closest to you. Summarize it in your own words, noting the date and primary concerns.
   (b) This site also contains current maps of the Palmer Drought Severity Index and the Crop Moisture Index. Examine these maps for evidence of the conditions described in the drought statements. Do these indices reflect the statements made in the drought statement you examined? Why or why not?
   (c) Identify regions of the country on the Palmer Index and Crop Moisture Index that are experiencing drought but do not have recent drought statements issued.
   (d) What reasons might there be for why no drought statements have been issued for these areas?

2. Examine archived maps of the Palmer Drought Severity Index (PDSI) using the Climate Prediction Center "Drought Monitoring" website via the *Severe and Hazardous Weather Website.* Examine the recent changes in drought indices in your region of the country.
   (a) Has dryness increased in the past several weeks? Support your answer with specific data, including index value and dates.
   (b) Are the most recent changes consistent with the recent distribution of precipitation? Explain.
   (c) Look back over the past year to determine when the PDSI was at its highest and lowest values at your location. Record the dates and the index values when these occurred.

3. Use the *Severe and Hazardous Weather Website* to navigate to the "Water Watch" drought site of the U.S. Geological Survey. This site provides up-to-date information on river flow rates throughout the country.
   (a) Examine streamflow rates in your area. How do they compare to normal?
   (b) Are the recent variations of streamflow in your area consistent with the occurrence (or absence) of significant rain events? Support your answer.
   (c) Identify an area in which the Palmer Index indicates drought conditions. Are the dry conditions reflected in the flow rates of rivers and streams in that area? Explain.

4. The amount of precipitation required to end a drought is a function of various parameters: drought intensity, location, time of year, and the duration and amount of precipitation. Access the National Climate Data Center's "Drought Termination and Amelioration" website via the *Severe and Hazardous Weather Website.*
   (a) Experiment with the simulator to determine the quantity and duration of precipitation required to terminate droughts of various intensities (as measured by the Palmer Index) in your area or a nearby area experiencing drought. Summarize your findings in a table or graph.
   (b) Discuss how the required amount of precipitation varies with season.
   (c) Explain the seasonal dependence of the required amount of precipitation.

HEAT WAVES

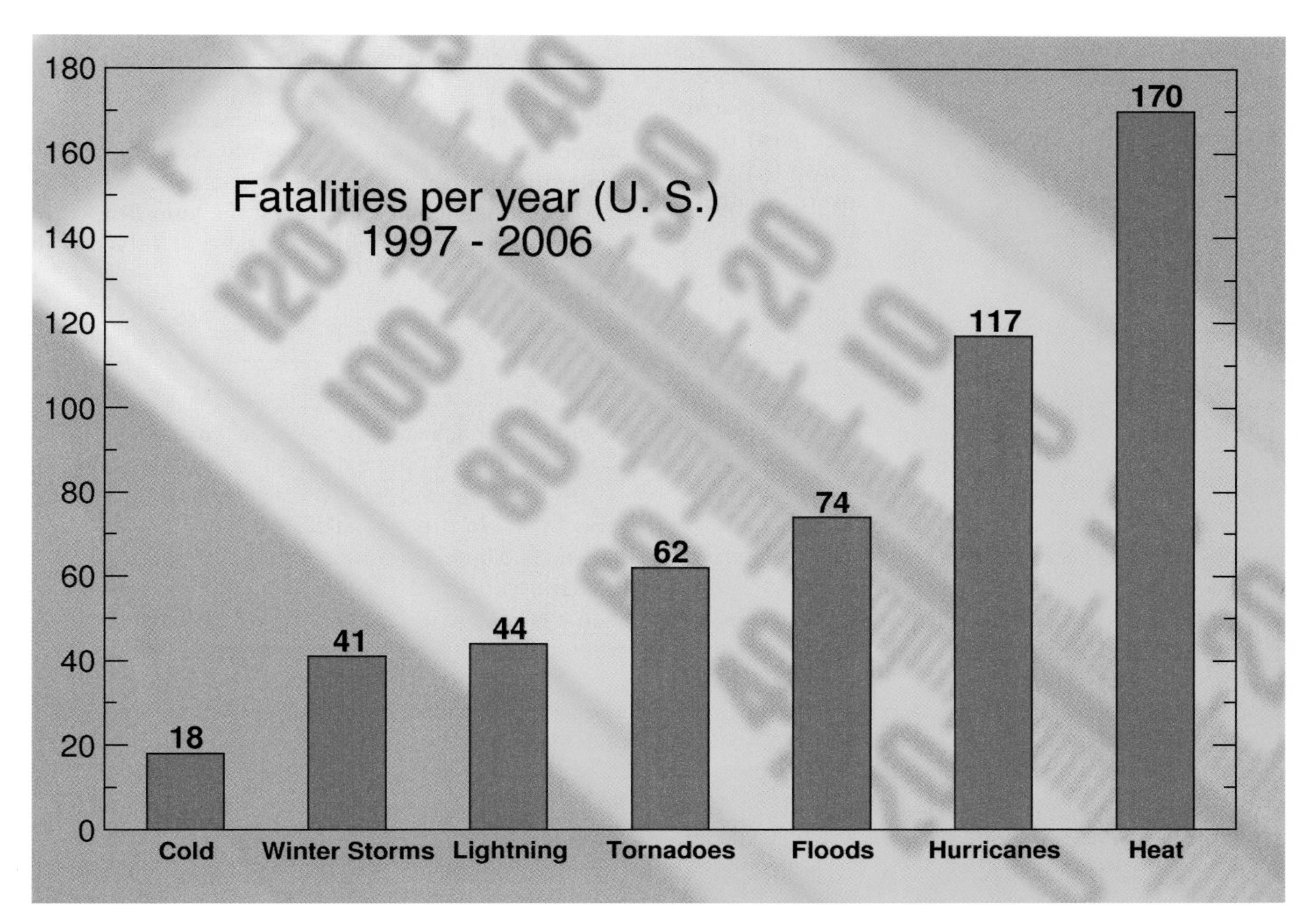

Yearly fatalities (averaged over 1997–2006) in the United States from different types of severe weather.

KEY TERMS

apparent temperature
Dust Bowl
evaporational cooling
excessive heat warning
feedback
global warming
heat advisory
heat cramps
heat exhaustion
Heat Index
heat stress
heat stroke
heat wave
subsidence
sunburn
sunstroke
ridge
upper-air ridges
urban heat island effect
warm-air dome

In recent years, the leading cause of weather-related deaths in the United States has been heat, which since 1995 has resulted in nearly 2,800 fatalities, for more than each of the next two leading killers, hurricanes and floods. The summer of 2006 was no exception, as record summer heat caused several hundred deaths and widespread power outages from coast to coast (see Focus 27.1). Heat-related death tolls have been even higher in the past: over 4,700 in the United States in 1936 and over 9,500 in 1901. Unlike many other forms of severe weather, heat is not a visually striking phenomenon. The impacts of heat are severe but deceptive in the sense that heat's toll in deaths and injuries can be very high in proportion to the damage to property and infrastructure. Extreme heat is often associated with drought, primarily because heat and drought can both result from similar atmospheric weather patterns. However, serious heat waves can occur without drought, especially if the heat does not persist for more than a few days. Similarly, droughts can occur without heat waves, although drought is often exacerbated by heat. For these reasons, separate chapters are devoted to drought and heat waves, despite some similarities in their atmospheric signatures.

Like drought and cold waves, the term ***heat wave*** has a geographic relativism. A normal July day in

FOCUS 27.1

The Extreme Summer Heat of 2006

While 2006 was the warmest year on record for the United States, it will be most remembered for its summer heat waves. The worst of the heat waves developed in mid July, when high temperatures engulfed the Northern Plains. Pierre, South Dakota reached 117° F on July 15, within 3 degrees of South Dakota's record high temperature set in 1936. The heat then shifted to the West Coast and the desert Southwest. Los Angeles had its highest temperature ever recorded, 119° F, in Woodland Hills on July 23. Four days later, Needles, California had an overnight low temperature of 100° F. Temperatures reached 110 to 115° F for nearly two weeks in California's Central Valley, threatening California's multibillion dollar agriculture industry—which, ironically, would suffer severe damage from a hard freeze six months later. More than 25,000 cattle and 700,000 fowl perished; livestock carcasses accumulated, and milk production declined by 15 to 20 percent. The number of human fatalities directly attributed to the heat in California exceeded 150.

As the end of July approached, the heat spread eastward across the country. St. Louis recorded a heat index of 118° F, while many residents were left without power by severe thunderstorms. Through the first week of August, heat indices exceeding 110° F and actual air temperatures exceeding 100° F were common in the central and eastern states. Horse racing and many other outdoor activities were cancelled in early August because of the extreme heat and humidity.

The extreme heat, together with large areas of drought, led to a record wildfire season for the nation, as 9.5 million acres burned. The nation's residential energy demand was 13 percent higher than would have been expected with normal temperatures, straining the nation's electrical utility system. Record power consumption also occurred in Canada, which also experienced the extreme heat. Winnipeg had its highest average maximum temperature of any July, and Toronto set a new record for its highest overnight low temperature (81° F) on August 1.

At the same time that much of North America was sweltering, parts of Europe were experiencing unprecedented heat. July 2006 was the hottest month since records began in Germany, Belgium, the Netherlands, England, and Ireland. September was the warmest on record in the U.K. and the Netherlands. Many European locations also set records for the amount of sunshine; the absence of precipitation led to drought conditions over much of central and northern Europe by the end of the summer. The European heat wave came only three years after more than 15,000 fatalities were reported in the disastrous heat wave of 2003 (discussed later in this chapter), raising the possibility that climate change may be altering the frequency of heat waves.

Phoenix, Arizona, has a maximum temperature above 100° F (38° C). Several days with such temperatures would qualify as a heat wave in many cities in the northern and eastern United States. However, even normally hot locations have episodes of extreme heat by their own standards. The maximum recorded temperatures in the United States and the world, respectively, are 134° F (56.6° C) (Death Valley, California, in 1913) and 136° F (57.7° C) (El Azizia, Libya, in 1922). One may safely assume that anyone present during those events would have regarded them as extremely hot by any standard. Even at a particular location, however, there are no rigid criteria or precise thresholds (°F or °C above normal, duration, etc.) that distinguish heat waves from conditions that are simply warmer than normal.

MEASURES OF HEAT STRESS

The impacts of heat are determined by several factors in addition to the actual air temperature. ***Heat stress*** on humans and animals is a strong function of the air's moisture content, measured in terms of the relative humidity or the dewpoint temperature. The importance of humidity arises from the fact that ***evaporational cooling*** (via perspiration) is the body's primary mechanism for preventing an excessive buildup of heat. The rate of evaporation of sweat varies inversely with the air's relative humidity—when the relative humidity is 100 percent, the net evaporation is effectively zero because the air is already saturated with water vapor. On the other hand, sweat evaporates rapidly when the relative humidity is low, resulting in effective cooling of the body because the body supplies much of the latent heat for evaporation.

While temperature and humidity are the most important determinants of heat stress and human discomfort, other atmospheric factors that affect the level of heat stress include radiant energy, wind speed, and atmospheric pressure. Exposure to direct sunlight can increase skin's "perceived" temperature (later defined as the "apparent temperature") by as much as 15° F (8° C) since the skin's temperature must increase in order to lose the additional heat gained via solar radiation. Hence there is the popular expression often heard on a summer day, "It's cooler in the shade."

Wind is an effective cooling agent, as long as the air temperature is less than the body temperature of about 98° F (37° C). Wind effectively replaces body-heated air near the skin by cooler air blowing over the body, thereby enhancing heat loss from the body by conduction and evaporation. However, when the air temperature is hotter than the body's skin temperature (95 to 98° F, 35 to 37° C), wind actually heats the body by replacing air that has cooled slightly through heat conduction to the skin. While not usually a major factor in heat stress, very low atmospheric pressure can increase heat stress by reducing the supply of air available to the body. Other determinants of heat stress are physiological factors (that vary among individuals), physical activity, and clothing. Heat stress is particularly severe among the elderly and the very young.

Various indices have been developed to measure the effects of heat on the human body. A widely used measure today is the ***apparent temperature,*** which takes into account the air temperature, relative humidity, radiant energy, and wind speed. When average radiative energy inputs and wind speeds are assumed, the apparent temperature is equivalent to the commonly reported ***Heat Index***. Table 27.1 lists the Heat Index for various combinations of the air temperature and relative humidity. For example, the table shows that an air temperature of 96° F and a relative humidity of 65 percent will produce a Heat Index of 121° F, which is the apparent temperature if the wind speed is typical (5 to 6 mph), the pressure is average (1013 mb), and there is no direct solar radiation, i.e., the person is in the shade.

This same combination of air temperature and relative humidity will result in a much higher apparent temperature (as much as 15° F higher) if a person is in direct sunlight on a summer day. Figure 27.1, for example, shows the distribution of the average noontime apparent temperature for July over the contiguous United States, based on average values of the relative humidity, wind, and solar radiation. These apparent temperatures represent what a body would "feel" if it were outside (*not* in the shade) at noontime on an average July day. Note that, in the middle latitudes of the United States (35° N to 45° N), the highest apparent temperatures are generally found in the central United States, especially the Mississippi Valley and the Great Plains. The average noontime apparent temperatures exceed 100° F (38° C) in large portions of the Gulf Coast states.

As the summer season progresses, people within a region can acclimate to the heat, at least to some extent. Early season heat waves, especially in urban areas, are associated with higher mortality because the city population has not yet acclimated to the heat and humidity. More generally, weather variability can be more important than heat intensity in determining

TABLE 27.1 Heat Index as a Function of Temperature and Relative Humidity

Heat Index (Apparent Temperature)

Relative Humidity (%)

Air Temperature (°F)	40	45	50	55	60	65	70	75	80	85	90	95	100
110	136												
108	130	137											
106	124	130	137										
104	119	124	131	137									
102	114	119	124	130	137								
100	109	114	118	124	129	136							
98	105	109	113	117	123	128	134						
96	101	104	108	112	116	121	126	132					
94	97	100	102	106	110	114	119	124	129	136			
92	94	96	99	101	105	108	112	116	121	126	131		
90	91	93	95	97	100	103	106	109	113	117	122	127	132
88	88	89	91	93	95	98	100	103	106	110	113	117	121
86	85	87	88	89	91	93	95	97	100	102	105	108	112
84	83	84	85	86	88	89	90	92	94	96	98	100	103
82	81	82	83	84	84	85	86	88	89	90	91	93	95
80	80	80	81	81	82	82	83	84	84	85	86	86	87

With Prolonged Exposure and/or Physical Activity

Extreme Danger	Heat stroke or sunstroke highly likely
Danger	Sunstroke, muscle cramps, and/or heat exhaustion likely
Extreme Caution	Sunstroke, muscle cramps, and/or heat exhaustion possible
Caution	Fatigue possible

Courtesy of the National Weather Service.

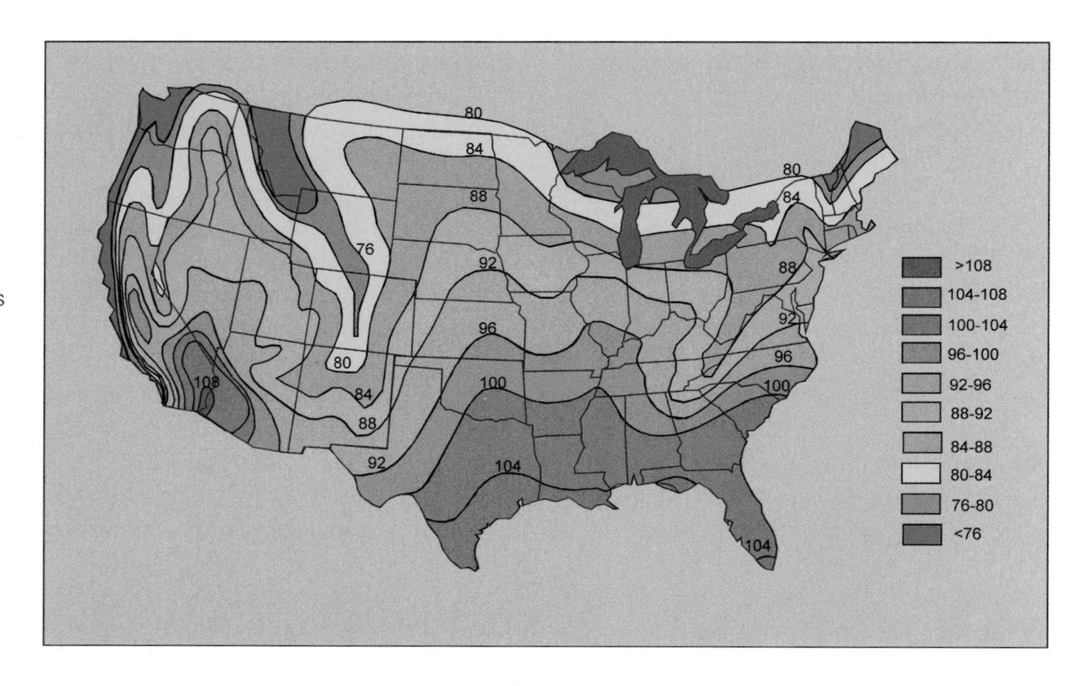

FIGURE 27.1: Map of average noontime apparent temperature (°F) during July. The highest apparent temperatures are found in the central United States and extreme desert southwest; cooler temperatures occur over mountain ranges and along the West Coast.

human sensitivity to heat. For this reason, cities in the northeastern and Midwestern United States, where variable summer climates do not enable residents to adapt to extreme heat, generally have higher heat-related mortality rates than cities such as Miami, Dallas, and Phoenix, where hot weather is the norm in the summer months. Because heat and its impacts vary regionally, there are no absolute criteria for a heat wave. The National Weather Service generally issues a ***heat advisory*** when a threshold value of the heat index is expected to be reached or has been present for two or more consecutive days. The specific thresh-

TABLE 27.2 Examples of Criteria for Issuance of Heat Advisories and Excessive Heat Warnings by the National Weather Service

Western United States (Seattle region)

Heat advisory

May 1 through June 30: Maximum heat index 98 to 105° F for one day or longer
July 1 through Sept. 30: Maximum heat index 100 to 105° F for one day or longer

Excessive heat warning

One of two conditions must be met:
(1) Maximum heat index 106° F or higher for 3 hours or more for two consecutive days, and a minimum heat index of 90° F or higher at night
(2) Maximum heat index of 115° F or higher regardless of duration or nighttime minimum

Central United States (Chicago region)

Heat advisory

Maximum heat index of 105° to 110° F with a minimum of 75° F or higher for two or more consecutive days

Excessive heat warning

One of three conditions must be met:
(1) Three consecutive days with maximum heat index 100 to 105° F with (a) at least 85 percent sunshine on two of the days, or (b) minimum heat index of 75° F or higher each day
(2) Consecutive days with maximum heat index of 105° to 110° F
(3) One day with the maximum heat index greater than 110° F

old value varies by region. An ***excessive heat warning*** is issued when prolonged periods of high heat indices are expected or when extremely high heat indices are expected for a single day. Again, the threshold values and duration vary geographically and seasonally, although an expected heat index of 105° F generally results in a heat advisory or warning. Table 27.2 shows the specific criteria and thresholds used by the National Weather Service for heat advisories and excessive heat warnings in the western United States (Seattle region) and the central United States (Chicago region).

For larger cities, where most heat-related fatalities occur, the National Weather Service is now implementing a "Heat/Health Watch/Warning System". This system is tailored to each major urban area, based on specific meteorology of each locale, as well as urban structure and demographics. First implemented in Philadelphia in 1997, where it estimated to have saved hundreds of lives already, it is being expanded from its current seventeen cities (as of 2007) to include all municipalities with populations exceeding 500,000.

CHECK YOUR UNDERSTANDING 27.1

1. Is there a threshold used to define the occurrence of a heat wave? Explain.
2. List five factors that affect the level of heat stress experienced by the human body.
3. How does the *apparent temperature* differ from the *Heat Index?*

METEOROLOGICAL CONDITIONS ASSOCIATED WITH HEAT WAVES

By definition, heat waves require high temperatures. The term heat wave, as used here, refers to an occurrence of unusually high temperatures for a particular region (i.e., temperatures well above the climatological normals) during the warm season, when above-normal temperatures are likely to be associated with human discomfort and heat stress.

FOCUS 27.2

Heat Disorders and Safety Tips

Excessive exposure to heat can cause a variety of adverse physiological responses, ranging from fatigue and *sunburn* to increasingly serious disorders: *heat cramps, heat exhaustion,* and *heat stroke* (also known as *sunstroke*). The progression from yellow to red in the Heat Index chart (Table 27.1) corresponds to increasing likelihood of the more serious disorders such as heat exhaustion and heat stroke. The danger posed to an individual depends not only on the environmental conditions, e.g., the Heat Index, but also on the individual's level of activity, overall health and physical condition, clothing, and other factors. Table 27A summarizes the symptoms and recommended treatments for heat disorders.

TABLE 27A Heat-Related Medical Problems

Heat disorder	Symptoms	First aid
Sunburn	Redness and pain; in severe cases, swelling of skin.	Ointment for mild cases; dry sterile dressing if blisters break; extensive cases should be seen by physician.
Heat Cramps	Painful spasms, usually in muscles of legs, abdomen; heavy sweating.	Firm pressure on cramping muscles, or gentle massage to relieve spasms; sips of water (unless nauseous).
Heat Exhaustion	Heavy sweating, weakness, cold, clammy skin; thready pulse; fainting or vomiting.	Get victim out of sun; lay down, loosen clothing, apply wet cloths; fan or move to air-conditioned area; sips of water; if nausea persists, seek immediate medical attention.
Heat Stroke	High body temperature (106° F (42° C) or higher); hot dry skin; rapid and strong pulse; possible unconsciousness.	Move victim to cool area; cold bath sponging; remove clothing; use fan or air-conditioning; do not give fluids; seek immediate medical emergency aid.

Source: National Weather Service

Four factors favor the occurrence of unusually high summer temperatures in mid-latitude areas such as the United States. While all four need not be present simultaneously for the development of a heat wave, the likelihood and the severity of a heat wave increase with the number of the factors that are present.

The first is an absence of polar airmasses that might move into the region from the north or west. Since the jetstream is generally found above the boundary between cooler polar airmasses and warmer tropical airmasses, the jetstream must generally be well to the north of the area affected by the heat wave. Hence ***upper-air ridges*** are characteristic features of regions in which heat waves occur. Figure 27.2 shows an upper-air ridge in a location favorable for a heat wave in the central United States. (The ***ridge*** may be regarded as the "wave" in the "heat wave.")

A second factor contributing to a heat wave is strong heating of the surface, which typically occurs when the sky is relatively cloud-free during the summer. Since ***subsidence*** (sinking motion) occurs above surface high-pressure centers, a nearly stationary surface high-pressure center is generally present near a region experiencing the sustained solar heating that results in unusually high temperatures during the summer. During the heat wave, the surface high-pressure center is generally found to the east of the hottest air, i.e., beneath the eastern portion of the ridge, because southerly winds west of high-pressure centers transport hot, moist air northward in the Northern Hemisphere. The upper-air flow supports the surface high through convergence aloft on the eastern side of the ridge, while the surface high reinforces the ridge by the inflow of warm air at low levels. The combination

FIGURE 27.2: Schematic representation of the atmospheric circulation at the surface (blue, yellow) and aloft (orange) during a heat wave over the central United States. A strong upper-air ridge is present over the central United States.

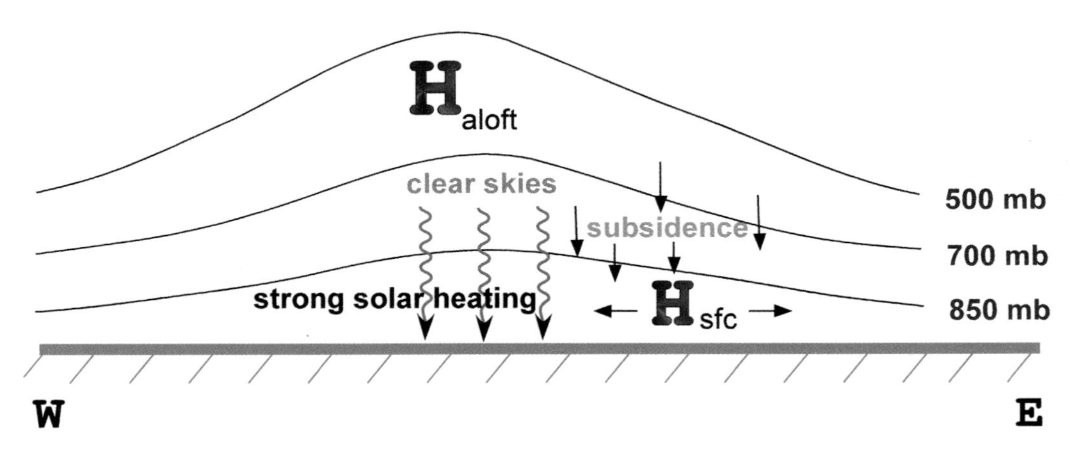

FIGURE 27.3: Cross-sectional view of the factors contributing to a heat wave in the central United States. Air to the west of the surface high flows northward, transporting warm air into the region below the upper-air ridge (see Figure 27.2).

of a strong ridge aloft and a surface high to its east often leads to a ***warm-air dome***, which tends to persist over time (Figure 27.3). Examples in the following sections will illustrate this atmospheric structure.

A third factor that favors a heat wave is dry ground, which provides a linkage or ***feedback*** between drought and extreme heat. If the ground is dry, the absence of significant evaporation leaves more energy available for direct, sensible heating of the air through conduction of the energy obtained by the surface from solar radiation (see Figure 5.1). In addition, an absence of evaporation reduces the likelihood of clouds that might otherwise reflect solar energy before it reaches the ground. The feedback enters the process through the tendency for hot air to desiccate the ground rapidly, especially during the summer, as residents of Oklahoma and Texas are well aware. A shortage of soil moisture (i.e., dry ground) leads to rapid heating of the low-level air, which favors further drying of the ground. Hence it is not a coincidence that the decade of the 1930s was the driest and also the warmest decade of the past century in the central United States.

Ironically, the presence of moisture (high relative humidity or dewpoint temperature) in the near-surface air increases human discomfort and heat stress even though it tends to lower the daily maximum temperatures. Thus, among the heat waves examined in the following section, the one that produced the sharpest spike in deaths (the 1995 Chicago heat wave) had the highest dewpoints but lower air temperatures than did the famous heat waves of the 1930s, which had the highest temperatures and the driest surface conditions.

A fourth factor that affects the intensity of the near-surface heat, especially the heat stress on humans, is the amount of vertical mixing of air. When vertical mixing near the surface is weak or is confined to a shallow depth, there is little chance for drier air aloft to reach the surface and mix with the warm, moist air near the ground. A layer with strong stability, favored by warm air and subsidence aloft, limits the vertical mixing of the near-surface air. Some of the most severe heat waves are characterized by strong stability in the form of a persistent inversion, only several hundred meters deep, at altitudes of 1 to 3 km (0.5 to 2 miles). If the air below the inversion is hot and moist, the trapping of the air near the surface will lead to a perpetuation of high heat indices at the surface. Figure 27.4 illustrates these various characteristics of a sounding conducive to oppressive surface conditions: (1) hot air with a small dewpoint depression in a shallow (typically 1 to 3 km) layer above the surface, (2) a sharp inversion marking the boundary between the shallow surface layer and the overlying air that has been warmed by subsidence, and (3) the large dewpoint-depressions aloft, characteristic of subsiding air above a surface high-pressure center.

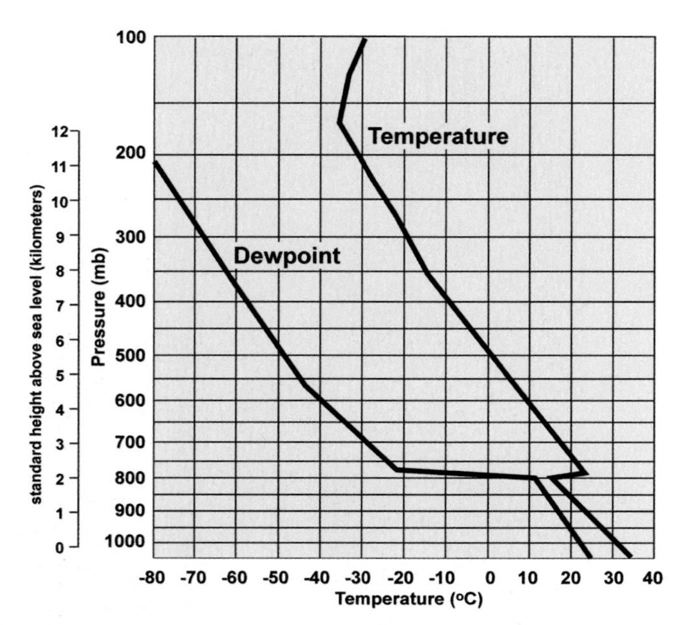

FIGURE 27.4: Sounding characteristic of a heat wave. Heavy solid lines represent temperature (right) and dewpoint (left).

URBAN HEAT ISLAND EFFECT

During intense and protracted heat waves, the heat is often worse in cities, especially the downtown regions. This phenomenon is referred to as the ***urban heat island effect.*** Differences between the comfort levels of rural and urban areas are especially noticeable on hot summer nights, when sleeping conditions can be difficult without air-conditioning. The urban-rural temperature differences in summer are often 5 to 10° F (3 to 5.5° C) in cities such as New York City, New York; Philadelphia, Pennsylvania; and Washington, D.C. (Figure 27.5). Why are urban areas hotter than the surrounding rural areas on summer nights? Several factors are at work.

First, locations that have more vegetation, such as rural areas, experience more cooling than urban areas. Not only is soil moisture more readily available for evaporation when the surface is unpaved, but the transpiration of water by trees, shrubs, and grasses augments the evaporation of moisture directly from the soil. This process is especially effective at night, when heat from solar radiation cannot replace the energy lost in the evaporation process.

Second, the asphalt and concrete of urban areas store greater amounts of heat during the day than do vegetated areas. Heat penetrates more readily into asphalt, concrete, and bricks during the daytime when solar radiation is absorbed. This heat is then released to the air in the city at night, preventing large drops of temperature in the city.

Third, heat sources such as air-conditioners and vehicular combustion contribute to the warming of cities relative to the countryside.

Finally, tall buildings create urban "canyons," where multiple reflection of incoming solar radiation enhances the likelihood that solar radiation will be absorbed in one of its encounters with either a horizontal or a vertical surface. The aggregate warming effect due to these factors increases as the population density increases and as a city's vegetative cover decreases (Figure 27.6).

There are also indications that urban areas have significant effects on precipitation. Studies show higher rainfall amounts downwind of cities such as Dallas, San Antonio, and Nashville, confirming results of earlier studies for cities such as St. Louis. The likely mechanisms include the tendency of urban heat islands to destabilize marginally unstable airmasses

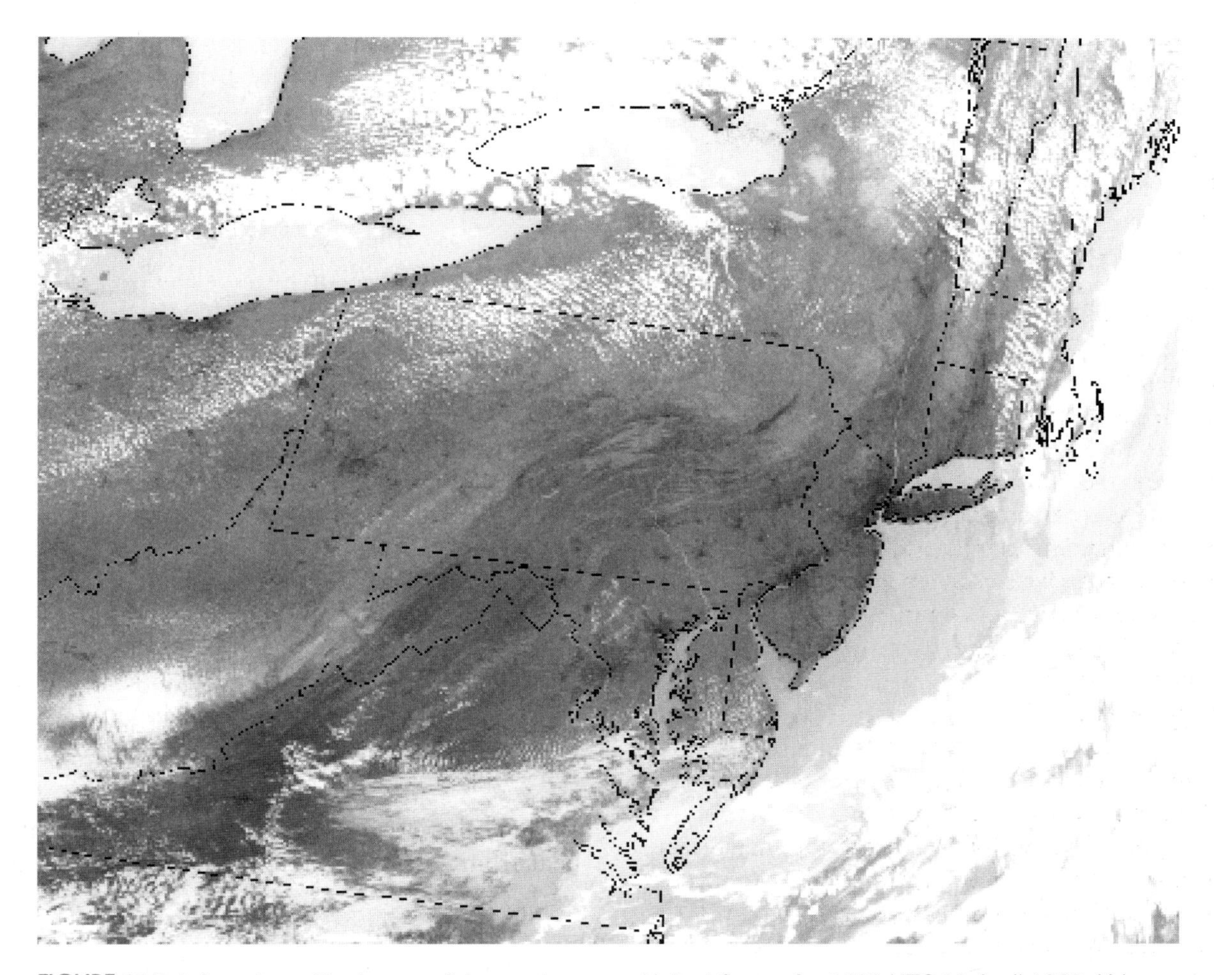

FIGURE 27.5: Infrared satellite image of the northeastern United States for 1200 UTC 13 April 1999. Urban heat islands are apparent as darker gray areas, e.g., New York City; Philadelphia, Pennsylvania; Baltimore, Maryland; Washington D.C.; Pittsburgh, Pennsylvania; and smaller cities in southeastern Pennsylvania.

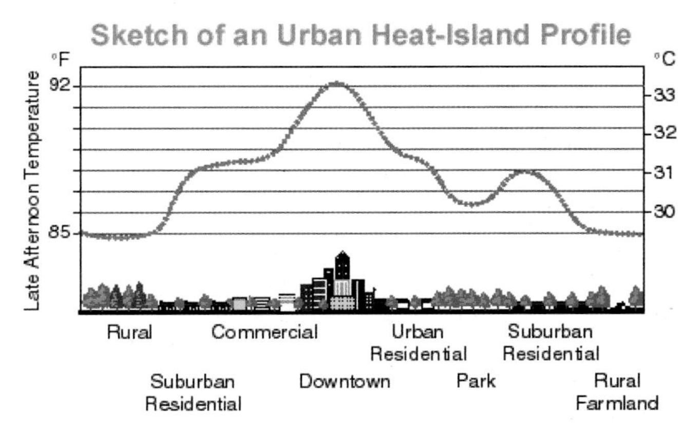

FIGURE 27.6: Schematic representation of late afternoon temperature variations across urban and rural areas on a summer day.

and to affect the convergence/divergence patterns of low-level winds. The precipitation enhancement appears to be largest on the downwind side of cities, within 30 to 50 miles (50 to 80 km) of the city centers. Such enhancements of summer rains can cool the surrounding countryside by providing additional moisture for evaporation.

CHECK YOUR UNDERSTANDING 27.2

1. What four factors favor the development of a heat wave?
2. Where is the jetstream typically located during a heat wave over the United States?
3. What is the urban heat island effect?

FOCUS 27.3

The Heat Strikes in 2003: First in India, Then in Europe

Heat waves gained widespread attention during 2003. First, eastern India suffered temperatures of 113 to 117° F (45 to 49° C), claiming more than 1000 lives during the pre-monsoon month of May. The heat wave exacerbated a serious drought in India's eastern states. A similar pre-monsoon heat wave in May 2002 also took more than 1000 lives in India.

The focus of attention then shifted to Europe, where the summer of 2003 brought abnormal heat to the entire region from the U.K., France, and Spain to western Russia (Figure 27A). For the entire summer, temperatures exceeded their averages by 6° F (3 to 4° C) over large portions of France, Italy, Switzerland, and Germany. The abnormal conditions culminated with an unprecedented heat wave during the first two weeks of August. Temperatures in France reached 104° F (40° C), and, on 10 August, England recorded a temperature above 100° F (38° C) for the first time in the several hundred years of its instrumental observations. The unprecedented nature of the heat is apparent in Figure 27B, which shows that the summer (June–August) temperature averaged over all of France was higher by approximately 3° F (1.7° C) than any other summer temperature of the 1901–2003 period.

The number of lives lost to the heat was also unprecedented. In France alone, the estimated death toll was as high as 15,000. Hospitals and mortuaries were overwhelmed, and it became apparent that France and its neighboring countries were not prepared for a disaster of this magnitude. For Europe as a whole, the estimated number of fatalities directly or indirectly related to the heat ranged up to 35,000. Forest fires raged in Italy, Spain, Portugal, and France. Authorities in France were forced to spray cold water on the walls of nuclear reactors, which came close to the temperature requiring emergency shutdowns. Britain's trains operated under severe speed restrictions due to the risk of buckled tracks.

What weather pattern was responsible for this heat wave? Consistent with the schematic depictions in Figures 27.2 and 27.3, a large upper-air ridge dominated the western European region. As shown in Figure 27C, the axis of this ridge extended from Spain northward to the North Sea on 5 August as the heat wave built to its climax. The maximum 500 mb height, 5950 m over France, is an exceptional occurrence. At the same time, a 1030 mb surface high-pressure center was located over southwestern Norway—far from the normal location of maximum pressure in the central North Atlantic Ocean. The clockwise airflow around this high pumped hot air northward over western Europe, helping to build and maintain the upper-air ridge and providing a European analog to the pattern in Figure 27.2.

(continued)

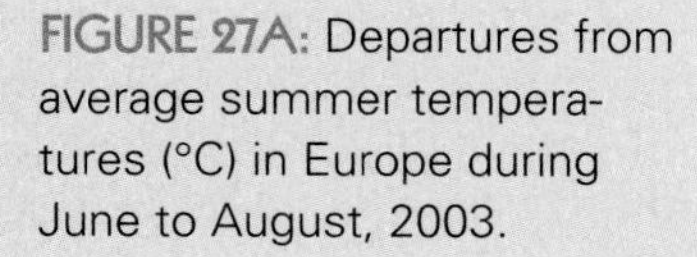

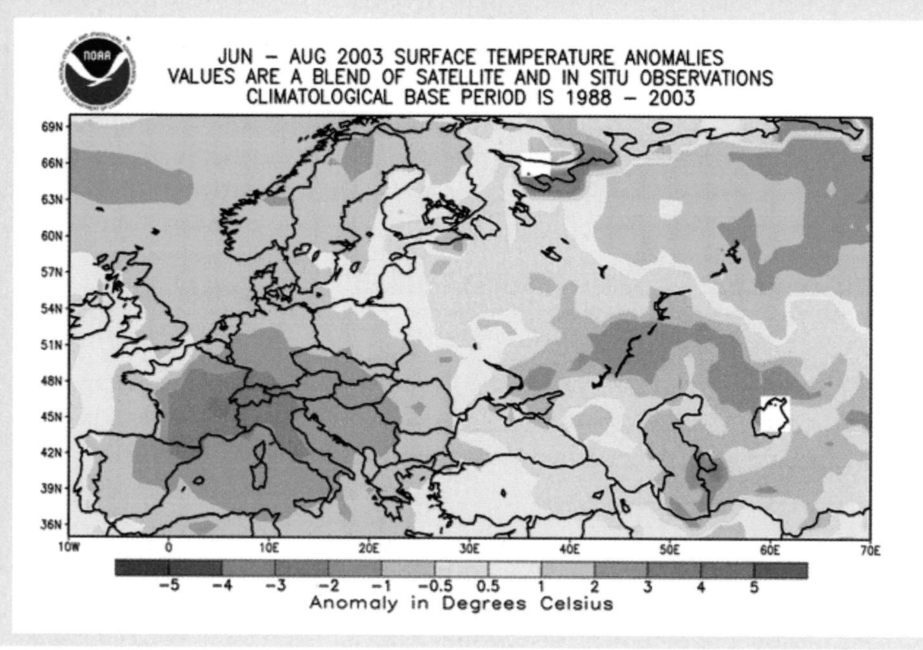

FIGURE 27A: Departures from average summer temperatures (°C) in Europe during June to August, 2003.

(continued)

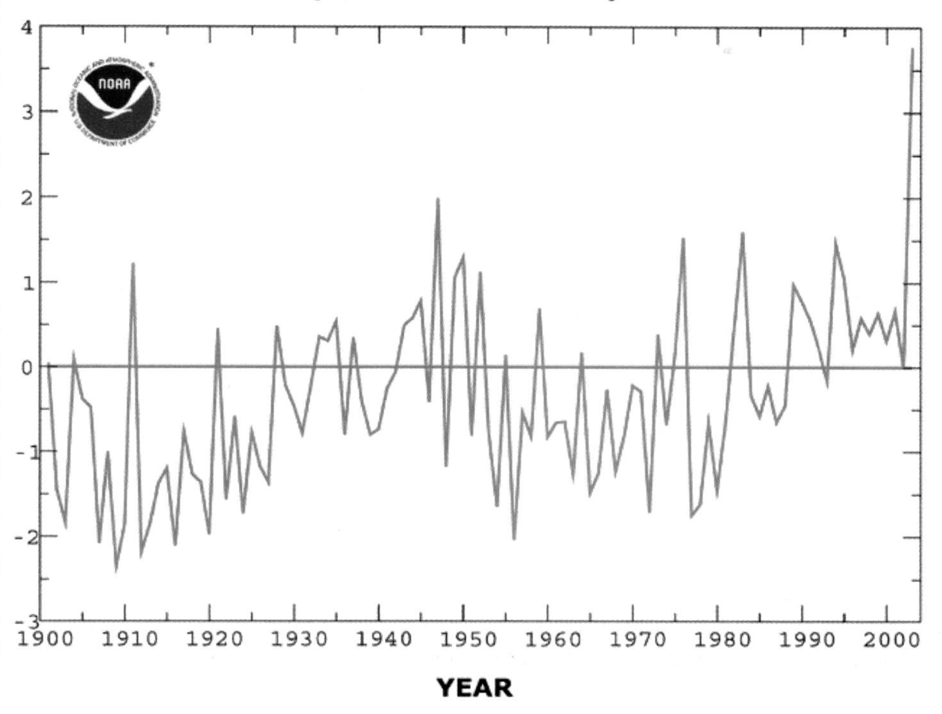

FIGURE 27B: Departures from average summer temperatures (°C) in France, 1901 to 2003.

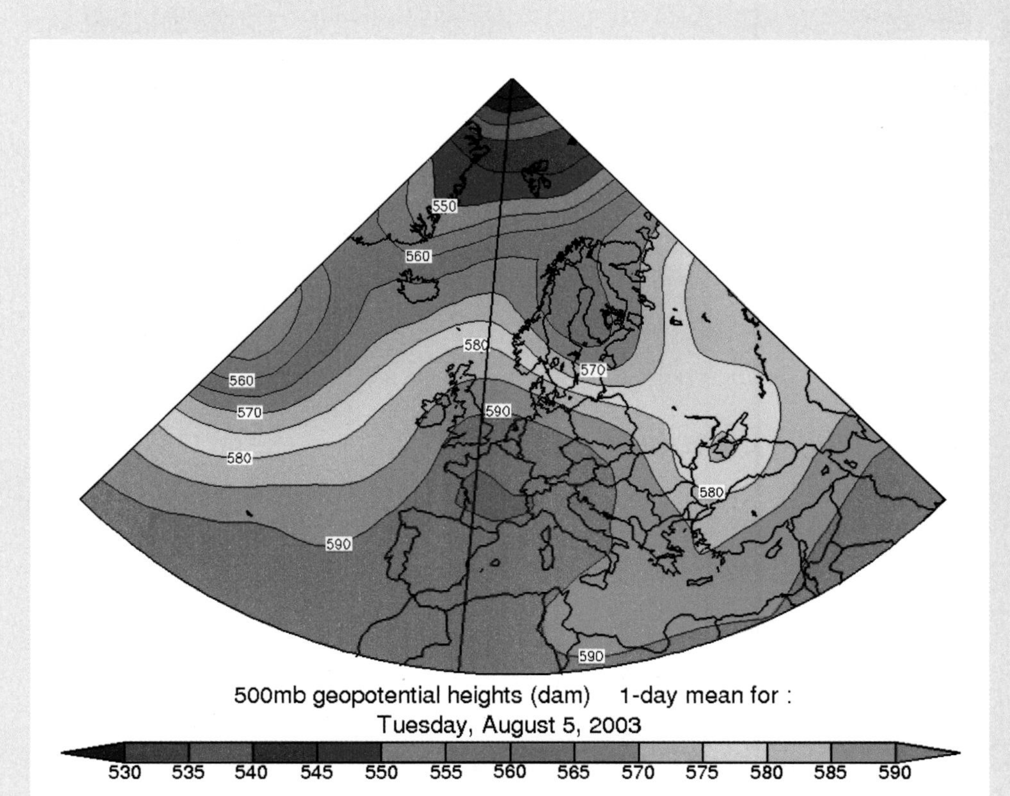

FIGURE 27C: 500 mb heights (m) over Europe at 1200 UTC 5 August 2003. Highest heights are shown in red.

MEMORABLE HEAT WAVES IN THE UNITED STATES

The July 1995 Heat Wave: Chicago and the Midwest

A short, intense, and deadly heat wave struck the midwestern and eastern United States during July 1995. Despite the brief duration (12 to 15 July) of this event, estimates of its death toll range from 500 to 1,000. The range of these estimates is large because heat can aggravate pre-existing illnesses, making it difficult to attribute many deaths to a single cause. Because most of the fatalities occurred in the Chicago, Illinois, area, this event has come to be known as the "Chicago Heat Wave of 1995." However, the affected region extended from the Great Plains to the New England States. This event was a stimulus for improved warning and response systems in major urban areas of the midwestern and northeastern United States.

The outstanding—and atypical— meteorological feature of this heat wave was the air's high humidity, manifested in the record-high dewpoint temperatures at many locations. The combination of these high dewpoints and unusually high temperatures pushed the apparent temperature (or Heat Index) to deadly levels, even at night when there is ordinarily some relief from dangerous heat levels.

The fact that Chicago was located in the core of the zone of most extreme meteorological conditions set the stage for the human disaster. Exacerbating factors in the Chicago area were (1) the nighttime urban heat island effect and (2) the large number of elderly residents of Chicago, which ranks behind only New York and Los Angeles in its elderly population.

The central United States had experienced relatively abundant precipitation during June and early July 1995. The availability of surface moisture was high. Then, during the second week of July, a mid-continent ridge developed in the large-scale circulation, providing the signature of a summer heat wave. Figure 27.7 shows the 500 mb maps for 1200 UTC, 12 July and

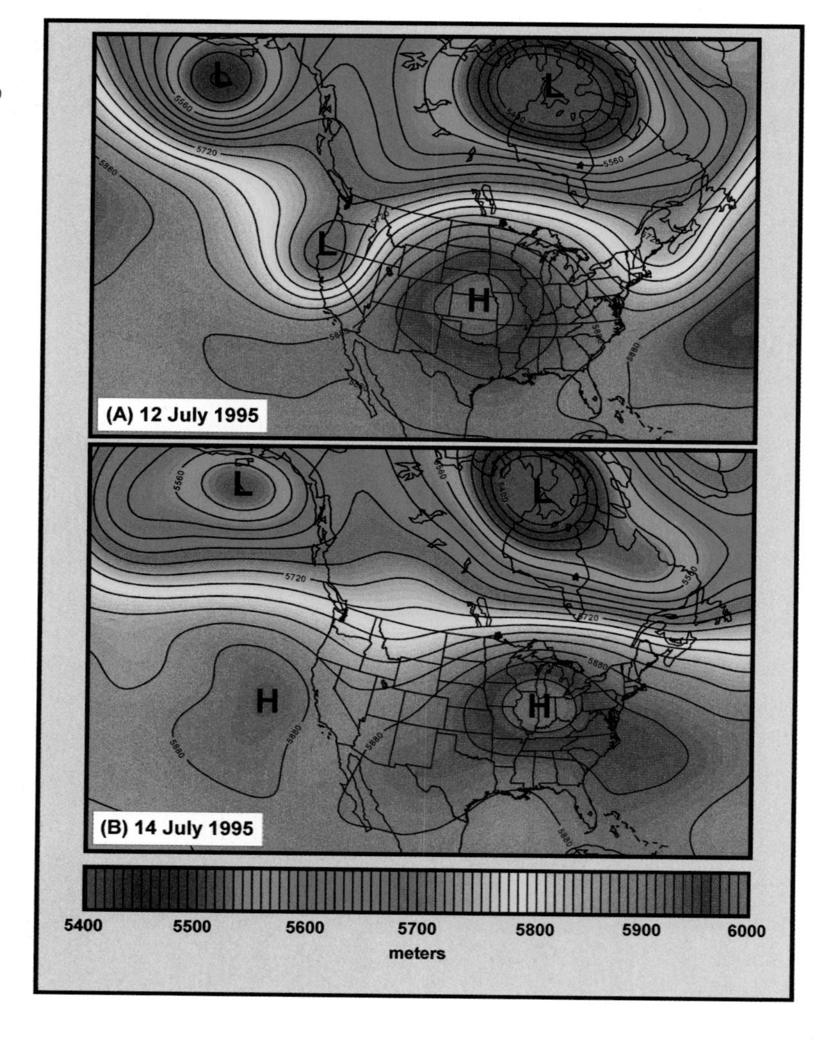

FIGURE 27.7: 500 mb maps showing height contours (meters) for 1200 UTC on (A) 12 July 1995 and (B) 14 July 1995. The largest values of 500 mb height, shown in deep red, shifted eastward during 12 to 14 July.

1200 UTC 14 July when the heat wave reached its greatest intensity. The 500 mb heights in the core of the upper-level ridge over Missouri-Iowa-Illinois reached 5,980 meters, which is close to their highest values in this area for the past sixty years. The ridge slowly migrated eastward during the 10 to 15 July period, taking with it the core of the highest temperatures.

Figure 27.8 shows the surface map for 1200 UTC 14 July. Particularly noteworthy is the very weak pressure gradient (and hence surface winds) over the Midwest on the back side of the surface high-pressure center, which by this time had moved to the Tennessee/North Carolina border. Also apparent in Figure 27.8 is a stationary front beneath the jetstream to the north of the heat-dominated Midwest and East. The lower temperatures and dewpoints poleward of this front moved southward by 16 to 17 July, bringing relief to the areas in which the death toll had been climbing.

The dewpoints reached by many Midwestern cities were unprecedented in the periods of their available data. Table 27.3 shows that the dewpoints reached or exceeded 80° F (26.6° C) at locations from Missouri to Ohio. The dewpoints of 82 to 83° F (27.8 to 28.3° C) in Missouri, Illinois, and Ohio were the highest on record at some locations (Table 27.3). For perspective, air having a dewpoint of 83° F (28.3° C) contains approximately twice as much moisture as air with a dewpoint of 65° F (18.3° C), a value which itself is higher than the average July dewpoints of 62 to 64° F (17 to 18° C) for the Midwestern United States.

The apparent temperatures also reached unprecedented levels at some locations. In the core of the oppressive heat and humidity over Illinois and Wisconsin, the apparent temperatures were near 118° F (48° C) by day (and that was in the shade!), and did not fall below 86° F (31° C) at night on 13 to 14 July in Chicago. On the basis of the apparent temperatures, this heat wave was the most intense—although certainly not the longest—on record for the Chicago area. Studies have suggested that there is a threshold value of the apparent temperature above which the mortality rate accelerates rapidly. Chicago's threshold was evidently exceeded during the 1995 heat wave.

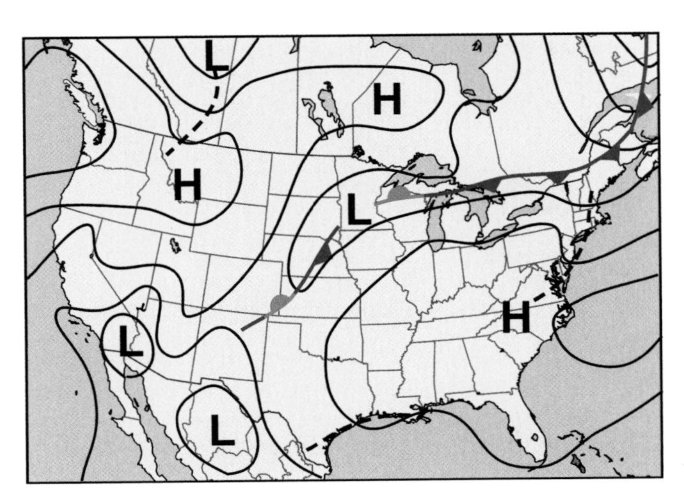

FIGURE 27.8: Map of sea-level pressure and surface fronts, 1200 UTC 14 July 1995.

TABLE 27.3 **Dewpoint Records Set during July 1995 Heat Wave. (Periods of record extend back to the middle or late 1940s in all cases)**

City	1995 Dewpoint Temperature (°F)	Previous Record for Dewpoint Temperature (°F)
Cleveland, Ohio	82	79
Columbia, Missouri	82	80
Eau Claire, Wisconsin	80	79
Flint, Michigan	80	79
Green Bay, Wisconsin	81	80
Moline, Illinois	82	80
Madison, Wisconsin	81	80
Rockford, Illinois	83	80
St. Louis, Missouri	83	82
Toledo, Ohio	80	79

Courtesy of Kenneth Kunkel and the American Meteorological Society.

ONLINE 27.1

The Heat Index Soars in July 1995

The outstanding feature of the heat wave of mid-July 1995 was the combination of heat and humidity, which manifested itself in record or near-record apparent temperatures. Online, you can see the evolution of this event in an animation of maps of the Heat Index over the contiguous United States. Notice how the highest values of the Heat Index develop over the Missouri-Iowa-Illinois-Wisconsin area on 12 to 13 July, then reach a peak on 14 July. The "bull's-eye" of purple coincides with the extremely high readings from Chicago, Illinois, and Milwaukee, Wisconsin. Notice also how the high values spread eastward briefly before the end of the event on 15 July, when a frontal system brought somewhat drier air to the Midwest and the East.

Although the daytime temperature was higher in Chicago than in rural areas, the apparent temperature was actually higher in rural areas during the day because the relative humidity was higher, a direct consequence of evapotranspiration of water from plants. At night, the apparent temperatures were 6 to 8° F (3 to 4° C) lower in rural areas. The fact that the mortality rates were lower in the rural areas points to the importance of the nighttime conditions for the effect of heat stress on the human body. The absence of even a brief respite from the oppressive conditions adds significantly to the stress on the body.

Finally, Figure 27.9 shows the soundings from the surface to 700 mb over Davenport, Iowa, on the evenings of 12 to 14 July (0000 UTC 13 to 15 July). These soundings illustrate the key features of the vertical profiles of temperature and dewpoint that enhance the oppressiveness of the air near the surface (compare to the sounding in Figure 27.4). On all three evenings, an inversion is apparent between 850 and 950 mb (0.7 to 1.5 km above the surface). This inversion represents the boundary between the slowly subsiding air aloft and the low-level air circulating around the west side of the surface high. The inversion prevented the upward penetration of air parcels from near the surface. Thus the hot, humid air near the ground was unable to mix with the much drier air aloft. (Note that the dewpoint decreases across the inversion by an amazing 40° F [22° C] on the 0000 UTC 13 July sounding, and by about 23° F [13° C] on the 0000 UTC 14 July sounding). If this dry air aloft had been able to reach the surface in response to the rising of air parcels from near the surface, the surface dewpoints would have decreased due to the vertical mixing, and the apparent temperature would have been much less dangerous. Instead, the depth of the trapped air actually decreased over time from about 1.0 km on the 13 July sounding to about 0.7 km on the 14 July sounding. Beneath the inversions, the environmental lapse rates were equal to the dry adiabatic lapse rate of 10° C/km, indicating neutral stability and the occurrence of mixing of the air up to the base of the inversion. However, the "capping" of these shallow mixing layers by the subsidence-induced inversion effectively trapped the humid air near the surface, thereby sustaining the high dewpoints and contributing to the severity of the event until the inversion was finally broken on 15 July.

In summary, the 1995 heat wave was an intense, albeit relatively short, event that was exacerbated by unprecedented humidity. Hot, humid air was trapped near the surface by a subsidence inversion in the low-

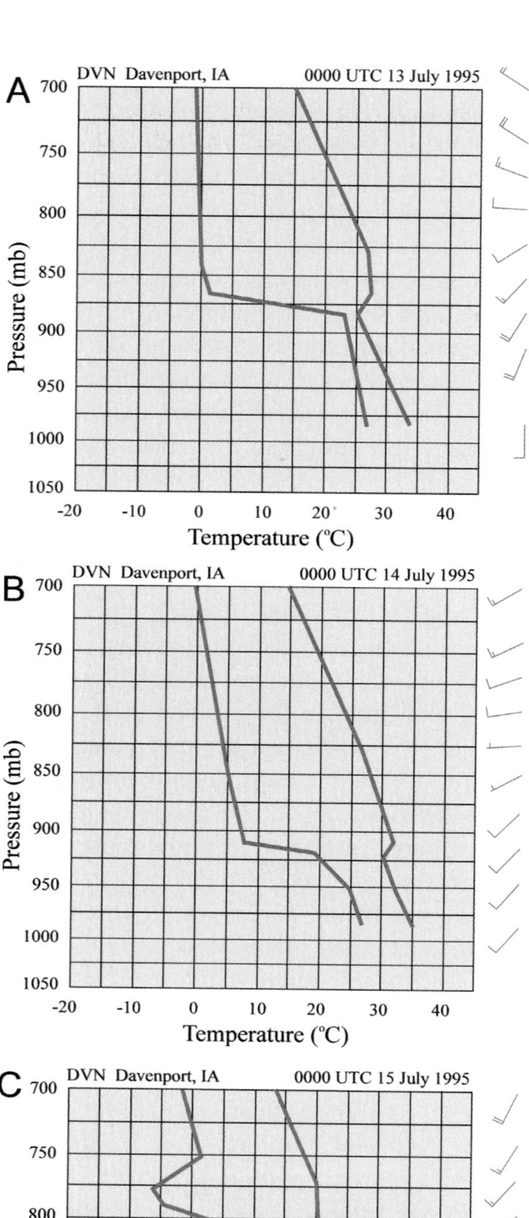

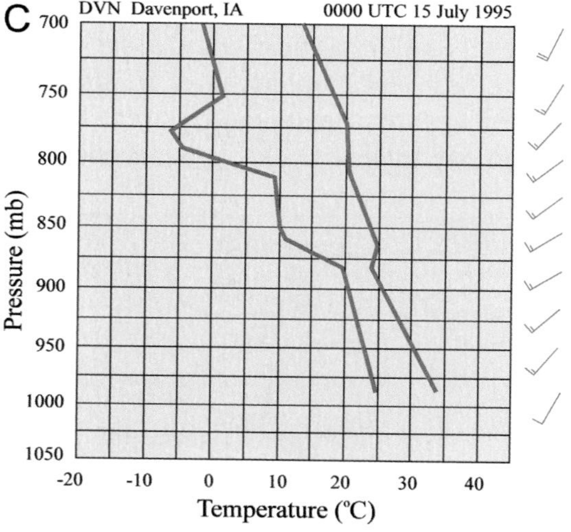

FIGURE 27.9: Soundings of temperature (right red line) and dewpoint temperature (left red line) between the surface and 700 mb over Davenport, Iowa, at 0000 UTC (7 P.M. CDT) on (A) 13 July 1995, (B) 14 July 1995, and (C) 15 July 1995. The 900 mb pressure is approximately 1 km above the surface. Note that the temperature increases and the dewpoint decreases upward through the inversion.

est 1.5 km (5,000 ft). In terms of the apparent temperature, and in terms of the number of fatalities over a short period, this heat wave was the worst on record for Chicago and the surrounding area. Slightly less intense heat waves, including some with higher temperatures (e.g., in the 1930s), have lasted longer. Since mortality rates are known to increase with the duration of a heat wave, it is indeed fortunate that this 1995 event was relatively short-lived.

The Summer of 1980: Texas and the Southern Plains

The summer of 1980 stands out in the meteorological records as the hottest summer since the 1930s and 1950s. Because it was the first widespread and prolonged heat event in more than two decades, the intense heat in the southern and eastern United States generated considerable media attention. Much of this attention centered on the Dallas–Forth Worth area of Texas, which experienced an amazing stretch of forty-two consecutive days (23 June to 3 August) when the temperature reached 100° F (38° C or higher (Figure 27.10A), with a peak of 113° F (45° C) occurring twice in late June. The temperature surpassed 100° F (38° C) every single day during July in this area. Parts of Missouri, Kansas, Oklahoma, and Texas had their highest summer (June through August) average temperatures on record. Dallas' average temperature for the three months, including nighttime as well as daytime

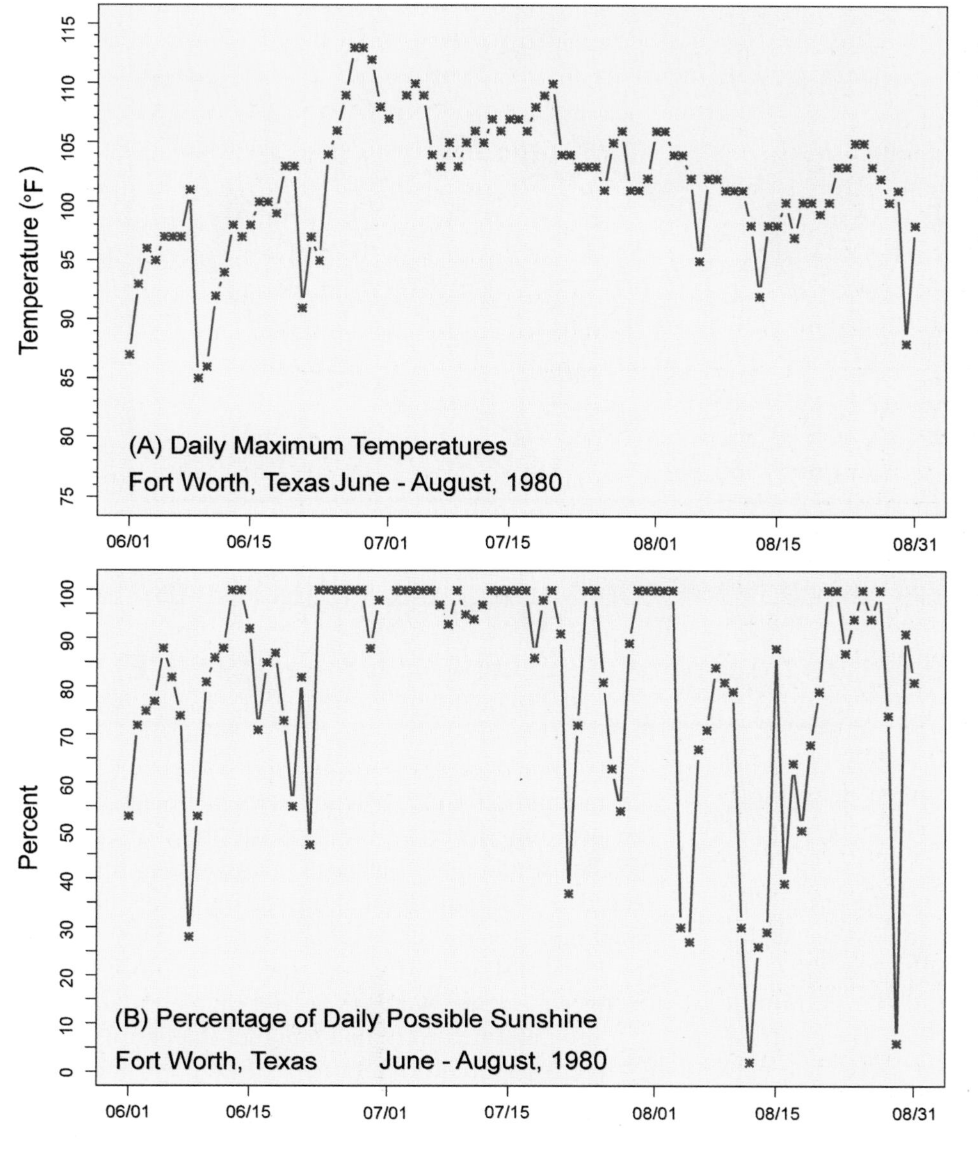

FIGURE 27.10: (A) Daily maximum temperatures (°F) from June through August 1980 for Fort Worth, Texas. (B) The amount of sunshine expressed as a percent of the daily possible amount for the same time period. High percentages indicate clear skies.

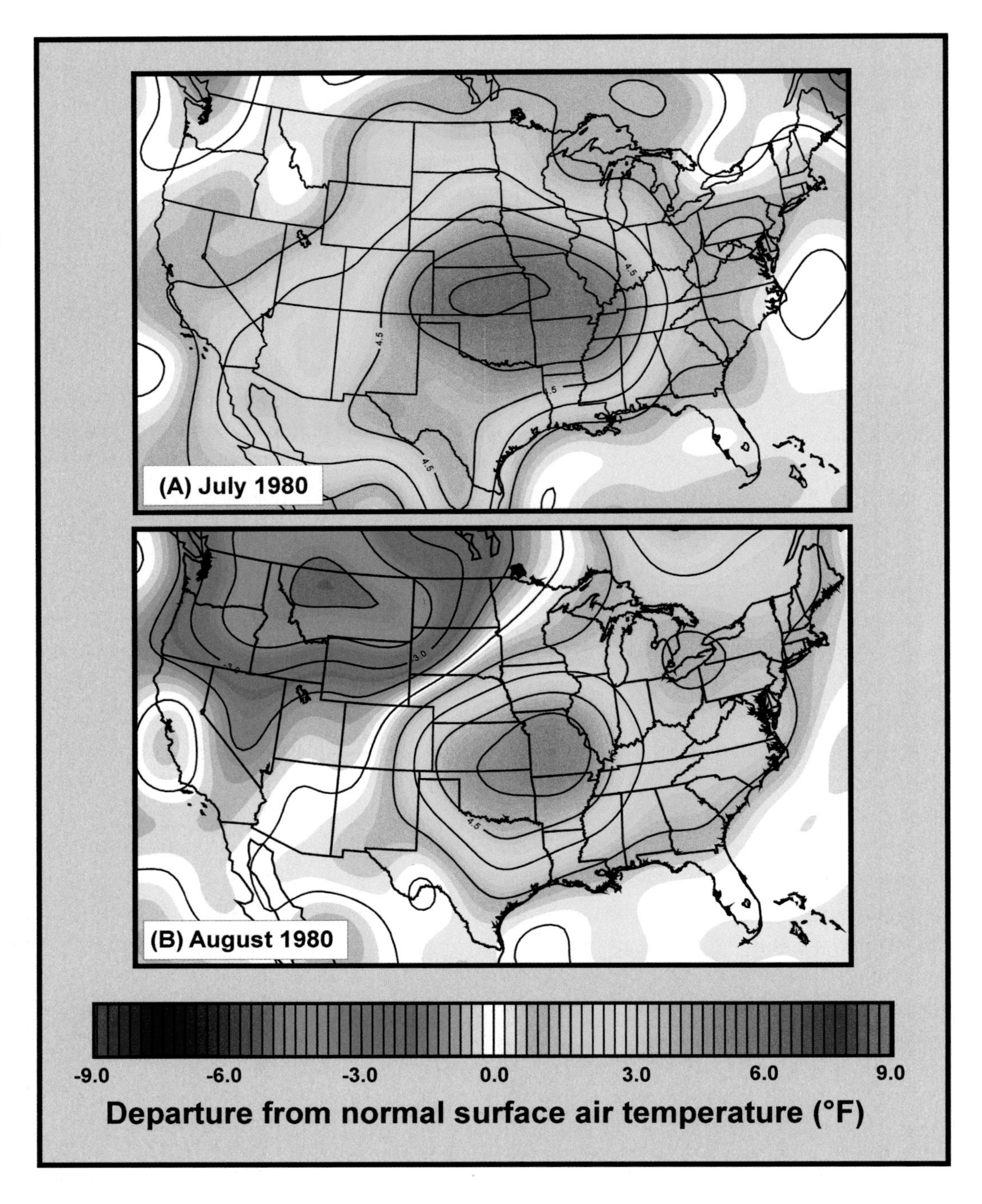

FIGURE 27.11: Departures from normal temperatures (°F) over the United States during (A) July 1980 and (B) August 1980. Yellow, orange, and red denote increasingly warmer than normal temperatures, while green and blue denote colder than normal temperatures.

temperatures, was more than 89° F (32° C), breaking the old summer record of 87.7° F (31° C) set in 1954.

Altogether, the 1980 heat wave and associated drought caused over 1,200 deaths and cost the nation more than $20 billion (about $55 billion in 2007 dollars), a total that is comparable to the economic impact of Hurricane Andrew in 1992. By contrast, Andrew's death toll was only about sixty. The economic cost of the 1980 heat and drought was high in the agricultural sector, where losses were enhanced by the prolonged period (several months) of extreme weather conditions. Electrical energy usage set a new record, and hundreds of miles of highways buckled. Road damage in Illinois alone exceeded $100 million (in 1980 dollars). Heat-related automobile breakdowns and repairs were far greater than normal.

Temperatures were much higher than normal over a large area in July (Figure 27.11A), the month in which the 1980 heat wave was most intense. The

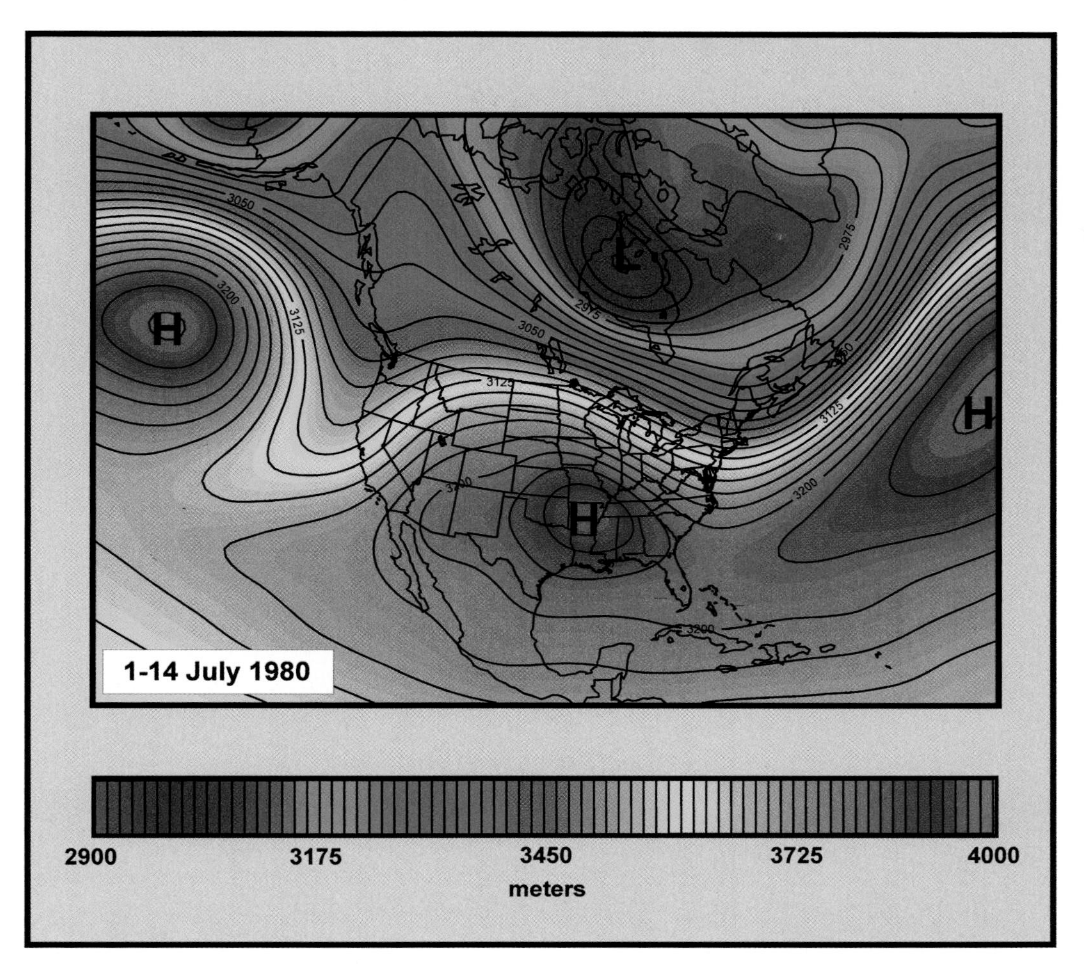

FIGURE 27.12: Map of 700 mb heights (meters) averaged over the period 1 to 14 July 1980, illustrating the pattern of three upper-air high-pressure centers, two over the oceans and one over the interior of the United States. Largest heights are shown in red.

monthly average temperatures exceeded their normal July values by 6 to 10° F (3 to 6° C), which is highly unusual for July, over much of the southern Great Plains. Figure 27.12 shows that the Northern Hemisphere's upper-air pattern for July contained a three-cell configuration of closed high-pressure centers: one in the North Atlantic, one in the North Pacific, and one over the United States. The latter is normally not present as a closed center aloft, but it is the most important feature of this three-cell configuration, which is the classic signature of a prolonged heat wave or drought over the United States (Chapter 26). In this case, the center of the high pressure aloft was located near the Texas-Arkansas border, just east of the Dallas–Fort Worth area. The largest temperature anomalies shifted slightly eastward in August (Figure 27.11B), consistent with the migration of the core of the upper-air ridge (Figure 27.13). An unusual feature of Figure 27.13 is the persistence of the upper-air circulation from July through September. This persistence is likely due at least in part to the feedback from the intense surface heat, which raises upper-air pressures and hence provides support to a ridge aloft. The tendency of the upper-level ridge to suppress clouds and precipitation, thereby exacerbating the buildup of heat, is apparent in Figure 27.10B, which shows that the percent of possible sunshine was near 100 percent during much of the protracted heat wave, especially in late June and early July when the temperatures were highest.

How did the summer of 1980 compare with other hot summers? The duration of the heat in 1980 was far greater than in the 1995 Midwestern heat wave, and the areal coverage was comparable to the heat waves that accompanied the droughts of the 1950s (Chapter 26). However, the drought severity was far less than in the 1950s. Despite negative anomalies of precipitation during the summer of 1980, Texas did not find itself in the extreme or even severe category of the Palmer Drought Severity Index. In contrast to the 1950s and the Dust Bowl years of the 1930s, spring rains over much of the southern Great Plains,

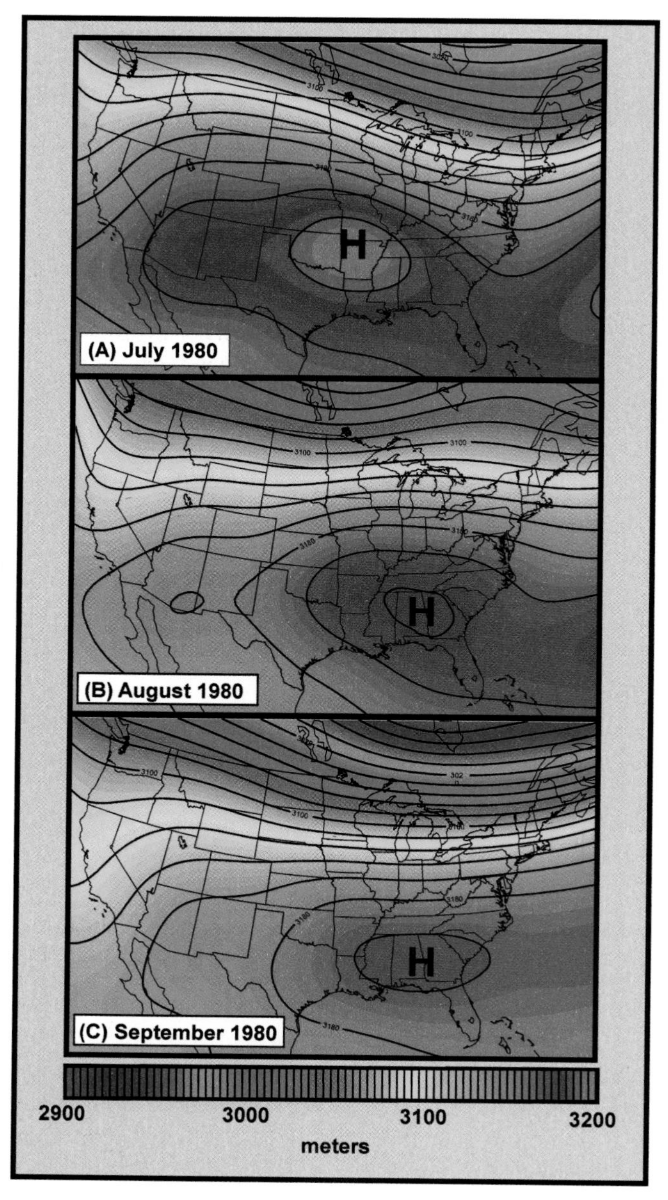

FIGURE 27.13: Maps of 700 mb height (meters) over the United States during July (top panel), August (middle panel), and September (bottom panel) 1980. The high-pressure center (deep red) migrates slowly eastward but remains over the United States during the period.

especially Texas, had been normal to above-normal. While the drought levels in 1980 reached severe levels by September in a broad belt from the Dakotas to Louisiana, the drought situation would have been far worse had the previous spring not been so wet. In this respect, the insidious feedbacks associated with soil desiccation were probably not able to realize their full potential in 1980. We turn next to the 1930s, a time period when this potential was realized.

ONLINE 27.2

The Heat Returns in August 2000

As a reminder of the susceptibility of Texas and the south-central United States to oppressive heat, a vintage heat wave dominated the weather headlines in the late summer of 2000. Online, you can view the afternoon (4 P.M. CDT) temperatures across the United States for each day of the month of August 2000. (Note that the 4 P.M. temperatures are often not the highest temperatures of each day.) The most striking feature of the sequence is the persistence of high temperatures in the south-central United States. Nearly every day had temperatures in the 95° to 105° F (35° to 40.5° C) range in the Texas–Oklahoma region, and the extremely high temperatures occasionally spread eastward into Arkansas and northward into the Great Plains.

On most days, the Texas–Oklahoma area had even higher temperatures than the normally hot desert Southwest. The highest temperatures of all (in excess of 105° F [40.5° C]) tended to occur toward the end of the month, and the area of extreme heat also expanded during the last week of August. Since incoming solar energy decreases throughout August, the higher temperatures in late August are an indication that the heat wave tended to "feed on itself." Indeed, there was essentially no rain in the affected area, so the dry ground and overlying air heated up sufficiently to reinforce the upper-air ridge of high pressure. The warm surface air was effectively anchored over the south-central states, since there was no forcing to change the upper-level winds.

Heat Waves of the 1930s

When nationally averaged temperatures are examined for the summer season (Figure 27.14), one decade stands out as the hottest: the 1930s. This warmth is the result of a series of extremely hot, dry summers that occurred during the period known as the ***Dust Bowl***—a time when air-conditioning was virtually unknown. Dry surface conditions were intertwined with the hot summers. Indeed, the hottest summers of the 1930s were generally the driest summers.

The heat waves of the 1930s are outstanding in terms of both their actual air temperatures and their apparent temperatures. Table 27.4 shows the twenti-

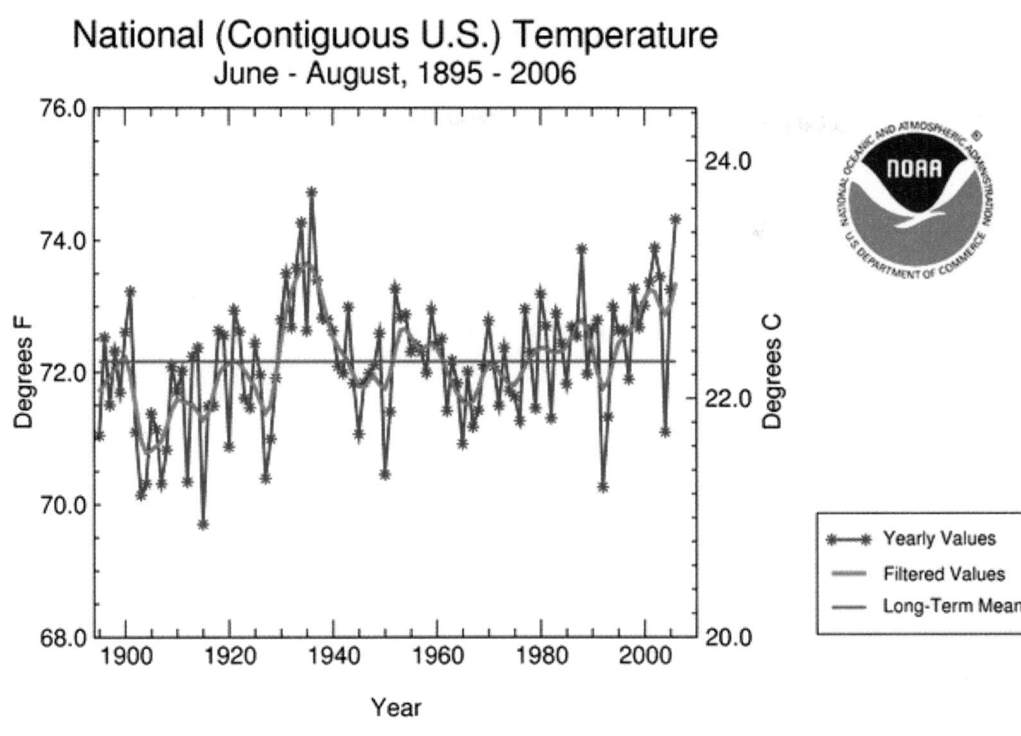

FIGURE 27.14: Summer (June through August) temperatures (°F) averaged over the United States for the period 1895 to 2006. The blue line shows temperatures for individual years, the red line is a smoothed version representing averages over several preceding and subsequent years.

TABLE 27.4 Hottest Four-Day Periods on Record in Chicago, Illinois

Dates	Four-day Mean Temperature (°F)	Four-day Mean Dewpoint Temperature (°F)	Four-day Mean Apparent Temperature (°F)	Maximum Apparent Temperature (°F)
11–14 July 1936	89.4	64.6	91.6	109.0
21–24 July 1934	88.9	66.2	94.5	114.1
12–15 July 1995	88.5	73.8	96.7	118.6
1–4 August 1988	87.4	71.6	93.4	111.0
28 June–1 July 1931	87.3	72.0	93.6	112.5
7–10 July 1936	86.7	63.5	91.6	108.7

Courtesy of Kenneth Kunkel and the American Meteorological Society.

eth-century's six most intense four-day heat waves according to actual air temperatures in Chicago, Illinois, where the 1995 event was centered. Four of the top six events, including the top two, occurred during the 1930s. The 1995 event ranks third. The events of the 1930s are less dominant if Table 27.4's ranking is based on dewpoint or apparent temperature, supporting the contention that the most severe heat waves of the 1930s were examples of "dry heat," albeit "dry heat" in the extreme.

The heat waves of the 1930s, particularly those of 1934 and 1936, were also much more widespread geographically than the 1995 event, and slightly more so than the 1980 event. Figure 27.15 shows that, in terms of the Palmer Drought Severity Index (PDSI; see Chapter 26), the middle-to-late 1930s were characterized by severe drought over the western and central United States. The moister pre-summer conditions that mitigated the 1980 event, as noted in the previous section, are apparent from a comparison of the 1980 PDSI with the PDSI distributions of the 1930s (Figure 27.15).

The heat wave of July 1936 was the worst heat wave in the United States during the twentieth century. July 1936 remains the hottest month on record in terms of nationally averaged temperature (excluding Alaska and Hawaii). Many individual station records, of which Table 27.4 provides but one example, still

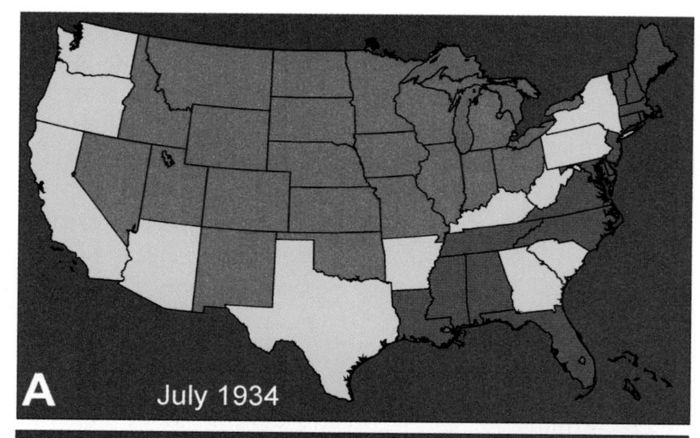

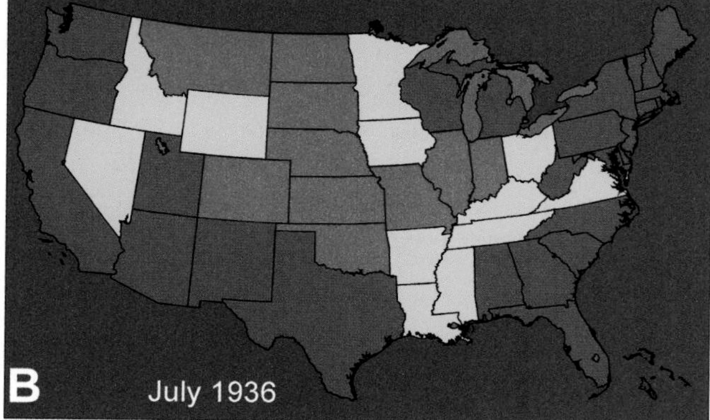

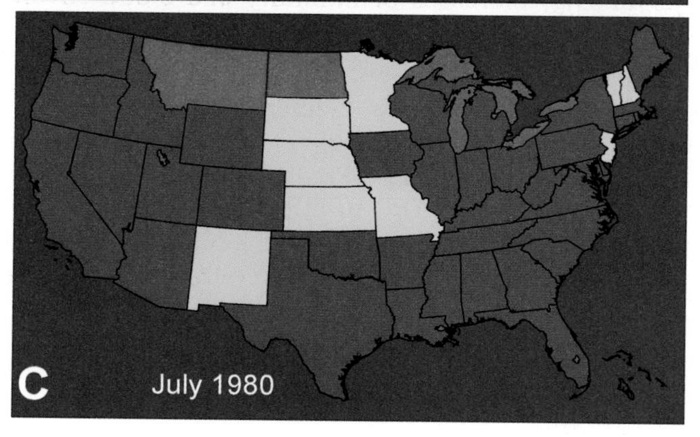

FIGURE 27.15: Areas in which the Palmer Drought Severity Index was in the severe (yellow) or extreme (red) categories for July of (A) 1934, (B) 1936, and (C) 1980. Criteria for severe and extreme drought are PDSI values below –3.0 and –4.0, respectively.

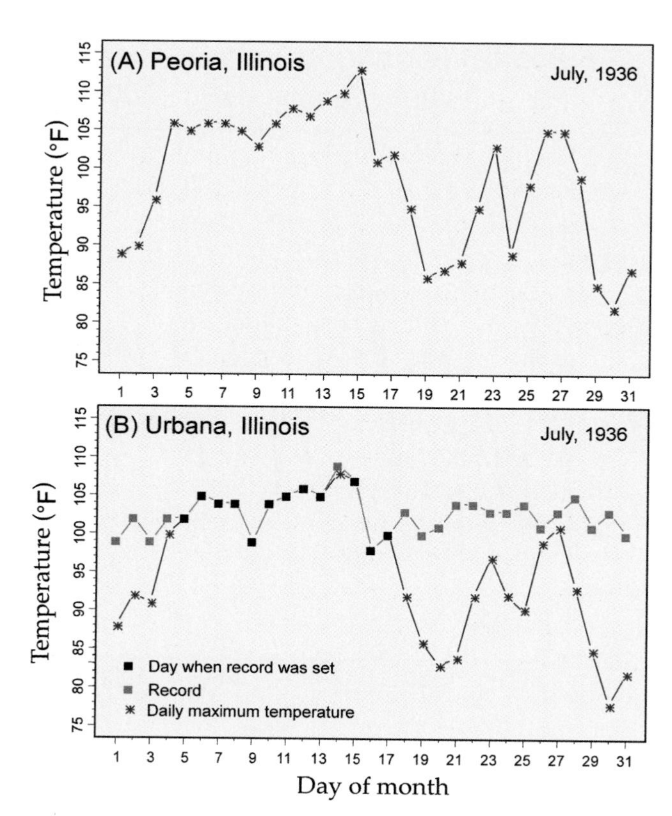

FIGURE 27.16: Daily maximum temperatures (°F) during July 1936 at (A) Peoria, Illinois, and (B) Urbana, Illinois. Daily records for the 1888 to 2007 period are indicated in (B) by red squares; records established and still held by 1936 are indicated by black squares.

stand today. A striking example of the uniqueness of the 1936 event is presented in Figure 27.16. Figure 27.16A shows the daily high temperatures during July 1936 at Peoria, Illinois. The temperature of 113° F (45° C) on 11 July 1936 was the highest ever recorded at this station, and other days in the same month were nearly as hot. Perhaps even more impressive is Figure 27.16B, the corresponding time series for July 1936 at Urbana, Illinois. While the highest temperature reached at Urbana was only 107° F (42° C), the string of daily high temperature records that were set during the 5 to 17 July period of 1936 is unique. No other year of the past century can claim more than five to six daily records (of maximum or minimum temperature) for a calendar month, yet 1936 has a near-monopoly on the high-temperature records during the thirteen-day period when daily temperatures are normally at their highest. This concentration of records belonging to the year 1936 may be compared to one baseball player being the major leagues' home-run champion twelve times during a thirteen-year period.

GLOBAL WARMING: POTENTIAL EFFECTS ON HEAT WAVES

Will extreme heat waves increase in frequency during the coming decades? A simplistic answer, drawing upon the likelihood of greenhouse-driven ***global warming***, is that heat waves are likely to become more severe and/or more frequent. In its 2007 assess-

FOCUS 27.4

The 1930s and the Advent of Air-Conditioning

The heat waves of the 1930s are vivid memories of those who experienced them, especially because air-conditioning had yet to come into residential use. In the early 1930s, movie theaters and department stores provided the only air-conditioned respites from the heat. On the hottest of the nights, many homeowners and apartment dwellers actually slept outside in grass fields in order to escape the oppressive heat that had built up in their homes in response to the intense solar radiation of the daylight hours.

Coincidentally, the 1930s saw the first systematic experiments to address the feasibility and economics of central air-conditioning in houses. The early experiments included the use of tons of ice stored in coal bins to provide cooling for coils over which forced-air ventilation systems circulated air. Although this method of home cooling was found to be too costly to be practical, more economical systems using compressors and refrigerants were developed in the 1940s. By the 1950s, central air-conditioning for homes began to come into general use, and the stage was set for the rapid growth of cities such as Phoenix, Arizona, and Dallas, Texas. The southward migration of the population continues to this day, largely in response to the pervasiveness of air-conditioning in homes.

The air-conditioning of automobiles was unheard of in the 1930s. Not until the 1950s and 1960s did automobile air-conditioning become common. Today, it is unusual to find an automobile that is *not* air-conditioned.

ment report, the Intergovernmental Panel on Climate Change has stated that "severe heat waves, including consecutive nights with high minimum temperatures, are likely to intensify in magnitude and duration over portions of the United States and Canada" during the twenty-first century. The increase of heat waves is consistent with the projections from climate models indicating summer warming and drying over much of the United States (Figures 5.17 and 5.18). A recent model study that focused specifically on heat waves highlighted a trend toward more intense, more frequent, and longer heat waves in Chicago and Paris, two cities that have suffered major fatalities in recent heat waves covered in this chapter.

For Chicago, the worst predicted three-day heat waves of the late twenty-first century show a projected rise of nighttime temperatures by more than 5° F (3°) relative to recent decades. The projected number of heat waves in Chicago increases by about 25 percent in the model simulations. While most models project drying of the continental interior and an increased likelihood of the type of heat wave that accompanied the drought of the 1930s, some models project an increase of precipitation that could counter the occurrence of "dry" heat waves. In this alternative scenario, the high-humidity type of heat such as that which occurred in Chicago in 1995 could become more common.

While the warmest decade on record for the United States remains the 1930s (Figure 27.14), there are indications that nighttime temperatures are becoming warmer in the United States, largely in response to increases of humidity and cloud cover. As a result, apparent temperatures appear to show a more detectable increase than actual air temperatures. Unfortunately, the unambiguous detection of such changes will continue to be confounded by urbanization and by changes in the location of "official" temperature measurement sites. Perhaps the most likely trend for the future, given the models' projections of a general warming during summer, is that nighttime temperatures in urban areas will become even warmer.

CHECK YOUR UNDERSTANDING 27.3

1. What distinguished the heat wave of July 1995 from previous heat waves in the Midwest?
2. What were the unusual features of the atmospheric pressure pattern during the 1980 heat wave?
3. Which decade of the twentieth century had the highest average summer temperatures in the United States?
4. Based on computer model projections, how are summer heat waves in the United States likely to change by the end of the present century?

TEST YOUR UNDERSTANDING

1. On average, how many fatalities in the United States each year are attributed to extreme heat?
2. Where in the United States is the apparent temperature generally highest on a summer afternoon?
3. Explain why the phrase "It's cooler in the shade" is more appropriate for Colorado than Louisiana.
4. How does wind affect heat stress?
5. How are a heat advisory and an excessive heat warning similar? How do they differ?
6. What is the difference between heat cramps and heat exhaustion?
7. How does heat exhaustion differ from heat stroke?
8. You are running a road race on a July day. As you round a turn on a deserted portion of the course, you encounter a collapsed competitor whose skin is hot and dry, and whose pulse is rapid. What should you do?
9. Why is an extreme heat event generally referred to as a heat "wave"?
10. Where is a surface high-pressure center generally located relative to a region that is experiencing a heat wave?
11. What role does subsidence aloft play in a heat wave?
12. Explain the "feedback" between drought and extreme heat.
13. How does the raininess of the preceding spring affect the likelihood of a summer heat wave in that area?
14. What role does vertical mixing of air play in a heat wave?
15. What might a sounding conducive to heat wave conditions look like?
16. Describe the four factors that lead to the development of urban heat islands.
17. How do urban heat islands affect precipitation?
18. Concrete surfaces in urban areas generally reflect more incoming sunlight than rural areas. Why does this not result in an "urban cool island"?
19. Would you expect the urban heat island effect to be stronger during a drought or a wet period? Explain.
20. Why was the death toll so high in Chicago during the heat wave of July 1995?
21. What was the role of cloudiness in the 1980 heat wave over the southern Plains?
22. What was a key difference between the 1980 heat wave and the heat waves of the 1930s?
23. How have urban-rural temperature differences changed over the past century?
24. What types of heat waves could become more common in the United States as greenhouse gases increase? Explain in terms of future changes of precipitation.
25. Has there been a significant trend toward hotter summers in the United States?

TEST YOUR PROBLEM-SOLVING SKILLS

1. Typical daily high temperatures and afternoon relative humidities for July at several cities in the United States are listed below.

City	Maximum Temperature (°F)	Relative Humidity (%)
Washington, D.C.	88	58
Dallas, Texas	95	49
Miami, Florida	89	64
Chicago, Illinois	82	60
Denver, Colorado	85	31
New Orleans, Louisiana	90	66

   (a) Rank the cities in order of decreasing apparent temperature.
   (b) In which city would July afternoons be most comfortable? In which city would they be least comfortable?
   (c) What factors affecting human comfort are neglected in your answer to (b)?

2. The oppressive conditions of the 1995 heat wave were worsened by the trapping of extremely humid air near the surface. The soundings in Figure 27.9 show the "trapping" inversion at Davenport, Iowa, over the period 0000 UTC 13 July to 0000 UTC 15 July 1995. For the sounding on 0000 UTC 13 July (Figure 27.9A):

   (a) Estimate the average dewpoint in the lowest 50 mb (about 500 meters) above the surface. (Express your answer in °C and °F.)
   (b) If convection (thermals) had been able to mix the air up to 700 mb, the dewpoint of the layer from the surface to 700 mb would have been approximately equal to the average dewpoint for this layer before the convection began. Calculate the average dewpoint of the air below 700 mb in Figure 27.9A.
   (c) If the surface temperature remained at 34° C (95° F), how, qualitatively, would the hypothetical mixing in (b) have changed the relative humidity at the surface?
   (d) In order to produce the mixing to 700 mb, what surface air temperature would have been required for the sounding in Figure 27.9A? (Assume dry adiabatic ascent for all rising air parcels.)

3. One of the human body's most valuable defenses against heat is perspiration, since the latent heat for evaporating sweat comes partially from the body. Consider a 155 lb (70 kg) person working outside in Phoenix, Arizona, on a July day when the air temperature is 110° F (73° C) and the humidity is low enough that all sweat evaporates from his skin, enabling him to maintain an (average) body temperature of 100° F (38° C). The man sweats at a rate of 1 liter (1,000 g) per hour and 50 percent of the latent heat of evaporation is supplied by his body (the other 50 percent is supplied by the air). If the man suddenly dehydrates and stops sweating, how long will it take for his body temperature to rise to a soon-to-be-fatal 110° F (43° C)? (Assume that it takes 4,200 Joules of energy to warm each kg of the body by 1° C [1.8° F], and that the latent heat of vaporization of the water in sweat is 2,400,000 Joules/kg.)

4. Figure 27.14 shows the yearly summer temperature averaged over the contiguous United States for the period 1895 through 2006.

   (a) Identify the years of the four hottest summers in the period of record.
   (b) If each summer's temperature was independent of the previous summer's temperature and there were no non-random factors at work, what is the probability that the two hottest summers of the 20th century would occur within a three-year period? (Hint: What fraction of the years is within two years of any randomly selected year?)
   (c) Under the same assumption as in (b), what is the probability that three of the four hottest summers of the twentieth century would occur in the same decade? (Hint: If a randomly selected year is in a particular decade, what are the chances that two of the next three randomly selected years will be in that same decade?)

USE THE SEVERE AND HAZARDOUS WEATHER WEBSITE

http://severewx.atmos.uiuc.edu

1. Use the "Heat Waves" page of the *Severe and Hazardous Weather Website* to navigate to the "Heat Wave Awareness Project". Using the links and information provided on this site, create a safety pamphlet regarding extreme heat. The pamphlet should be accurate and complete with regard to safety information but also aesthetic so that people would read it.

2. Urban heat islands can sometimes be identified in infrared satellite imagery. Examine satellite images from the "Current Weather" of the *Severe and Hazardous Weather Website* to detect urban heat islands in large metropolitan areas.
   (a) What area of the country did you choose to investigate? Explain your choice.
   (b) Examine the infrared satellite image for the urban area you selected. Record the date and time of the image and print or save the image.
   (c) What evidence exists of the urban heat island effect?
   (d) Examine surface station temperatures for the same time. What was the temperature difference between the urban area and surrounding rural areas? Is this what you would expect?
   (e) Locate the dewpoint temperature for the urban and surrounding rural area for the time you are investigating. Use this information to compare the heat index for the urban area and the rural area. Summarize your findings.

3. During a period of above average summer temperatures, examine the surface and upper-air data for evidence of features that are characteristic of a heat wave.
   (a) List the meteorological weather pattern and environmental conditions conducive to a heat wave.
   (b) Which weather products would you examine for each of the variables you identified in (a)?
   (c) Examine maps and imagery that show each of the variables you identified in (a). Summarize what evidence is present to support the existence of extreme heat.
   (d) What characteristics of a heat wave are not present?

4. The National Weather Service provides a "hazards statistics" website containing annual summaries of fatalities from various types of severe weather in the United States. Examine the summaries for the past five years.
   (a) What percentage of severe weather fatalities were heat related?
   (b) How did the number of heat-related fatalities compare with other severe weather events in each year?
   (c) Are the distributions of fatalities and injuries by age group and gender consistent with distributions of the general population and with other severe weather events? Explain.

UNITS, CONVERSIONS, AND CONSTANTS

UNITS

Units	Metric	U.S. / English
Temperature	Celsius (°C) Kelvin (K)	Fahrenheit (°F)
Length	meter (m) kilometer (km)	foot (ft) mile (mi) nautical mile (nm)
Speed	meters per second (m/s) kilometers per hour (km/hr)	feet per second (ft/s) miles per hour (mph) knots (kt)
Mass	gram (g) kilogram (kg)	ounce (oz) pound (lb)
Time	second (s)	second (s)
Density	kilogram per cubic meter (kg/m^3)	pound per cubic foot (lb/ft^3)
Pressure	millibar (mb) Pascal (Pa)	pounds per square inch (lb/in^2) inches of mercury (in Hg)
Force	Newton (N)	pound (lb)
Energy	Joule (J)	calorie (cal)
Power	Watt (W)	horsepower (hp)
Electric current	ampere (A)	ampere (A)
Electric potential	volt (V)	volt (V)

CONVERSIONS

Temperature

°F = °C 9/5 + 32
°C = 5/9 (°F – 32)
K = °C + 273.15

Length

1 kilometer (km)	=	1000 m
	=	0.62 mi
	=	3281 ft
	=	0.539 nm
1 mile (mi)	=	5280 ft
	=	1.61 km
	=	1609 m
	=	0.869 nm
1 nautical mile (nm)	=	6072 ft
	=	1.15 mi
	=	1.855 km
	=	1855 m
1 meter (m)	=	100 cm
	=	3.28 ft
	=	39.37 in
1 foot (ft)	=	12 in
	=	30.48 cm
	=	0.305 m
1 centimeter (cm)	=	0.394 in
	=	0.01 m
	=	0.033 ft
1 degree latitude	=	111 km
	=	69 mi
	=	60 nm

Area

1 square meter (m^2)	=	10.76 ft^2
1 square foot (ft^2)	=	0.09 m^2
1 square inch (in^2)	=	6.45 cm^2
1 square centimeter (cm^2)	=	0.15 in^2
1 acre (ac)	=	43,560 ft^2
	=	3920 m^2

Volume

1 cubic centimeter (cm^3)	=	0.06 in^3
1 cubic inch (in^3)	=	16.39 cm^3
1 cubic meter (m^3)	=	35.31 ft^3
1 cubic foot (ft^3)	=	0.03 m^3
1 liter (l)	=	1000 cm^3
	=	0.264 gal

Speed

1 mile per hour (mph)	=	0.45 m/s
	=	0.87 kt
	=	1.61 km/hr
1 meter per second (m/s)	=	2.24 mph
	=	1.94 kt
	=	3.60 km/hr
1 knot (kt)	=	1.15 mph
	=	0.51 m/s
		1.85 km/hr

Mass

1 gram (g)	=	0.035 oz
1 ounce (oz)	=	28.57 g
1 kilogram (kg)	=	1000 g
	=	2.2 lb
1 pound (lb)	=	0.45 kg
	=	454 g
	=	16 oz

Density

1 kilogram/meter^3 (kg/m^3)	=	0.06 lb/ft^3

Pressure

1 millibar (mb)	=	1000 $dynes/cm^2$
	=	0.0295 in Hg
	=	0.75 mm Hg
	=	0.0145 lb/in^2
	=	100 Pa
1 inch of mercury (in Hg)	=	33.86 mb
1 mm of mercury (mm Hg)	=	1.33 mb
1 pascal (Pa)	=	0.01 mb
	=	1 N/m^2
1 hectopascal (hPa)	=	1 mb

Energy

1 joule (J)	=	1 Nm
	=	0.239 cal
1 calorie (cal)	=	4.186 J

Power

1 Watt (W)	=	1 J/s
	=	14.34 cal/min
1 cal/min	=	0.07 W

CONSTANTS

Gravitational force	=	9.8 m/s^2
	=	32.2 ft/s^2
Mean sea-level pressure	=	1013.25 mb
Dry adiabatic lapse rate	=	10 °C/km
Diameter of Earth at equator	=	12,756 km
Latent heat of fusion (liquid water-ice transition)	=	333,700 J/kg
Latent heat of vaporization (liquid-vapor transition)	=	2,501,000 J/kg
Latent heat of sublimation (ice-vapor transition)	=	2,834,700 J/kg

BLANK MAPS AND CHARTS

THERMODYNAMIC DIAGRAMS

Stuve Diagram

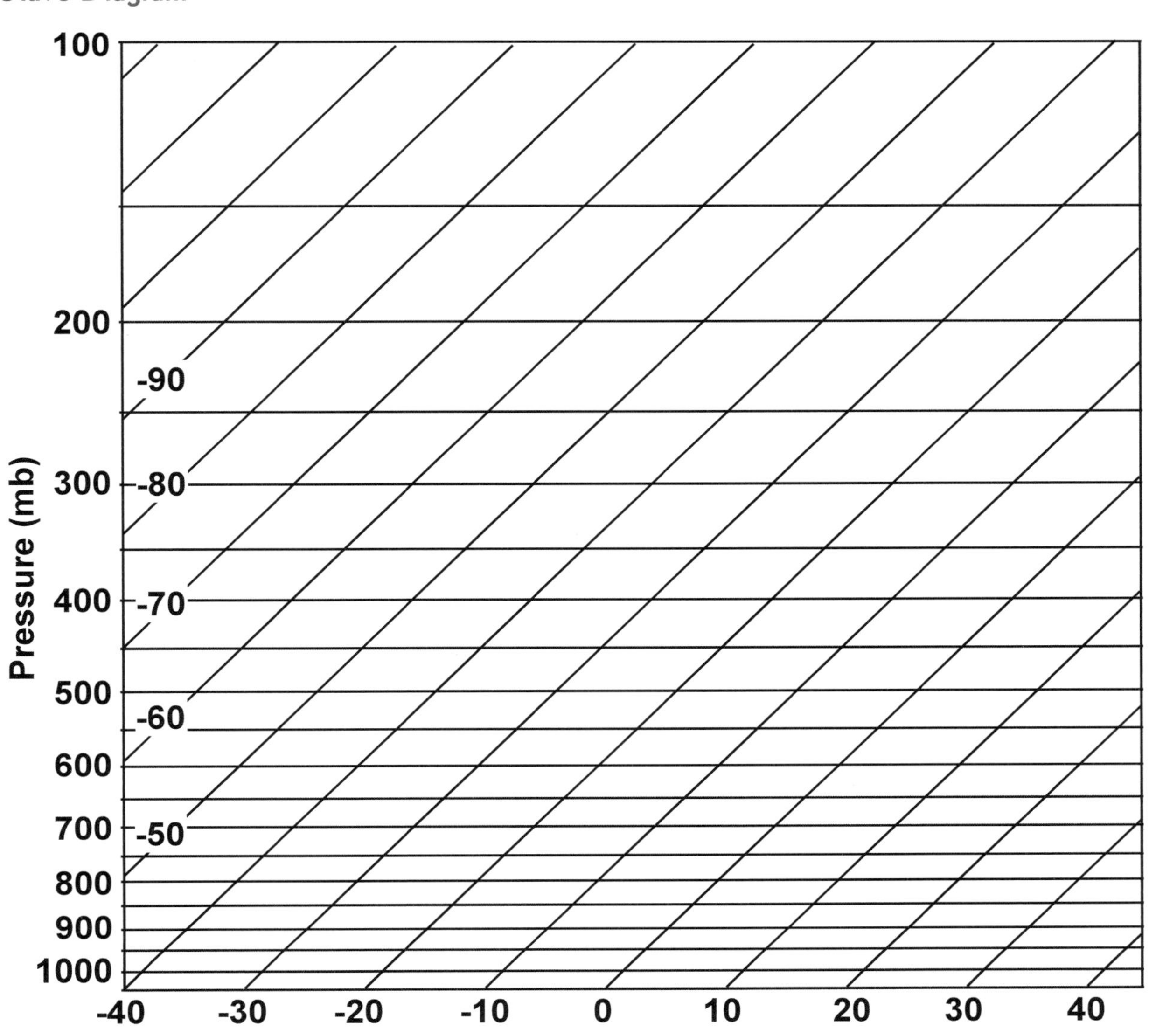

Skew-T Log P Diagram

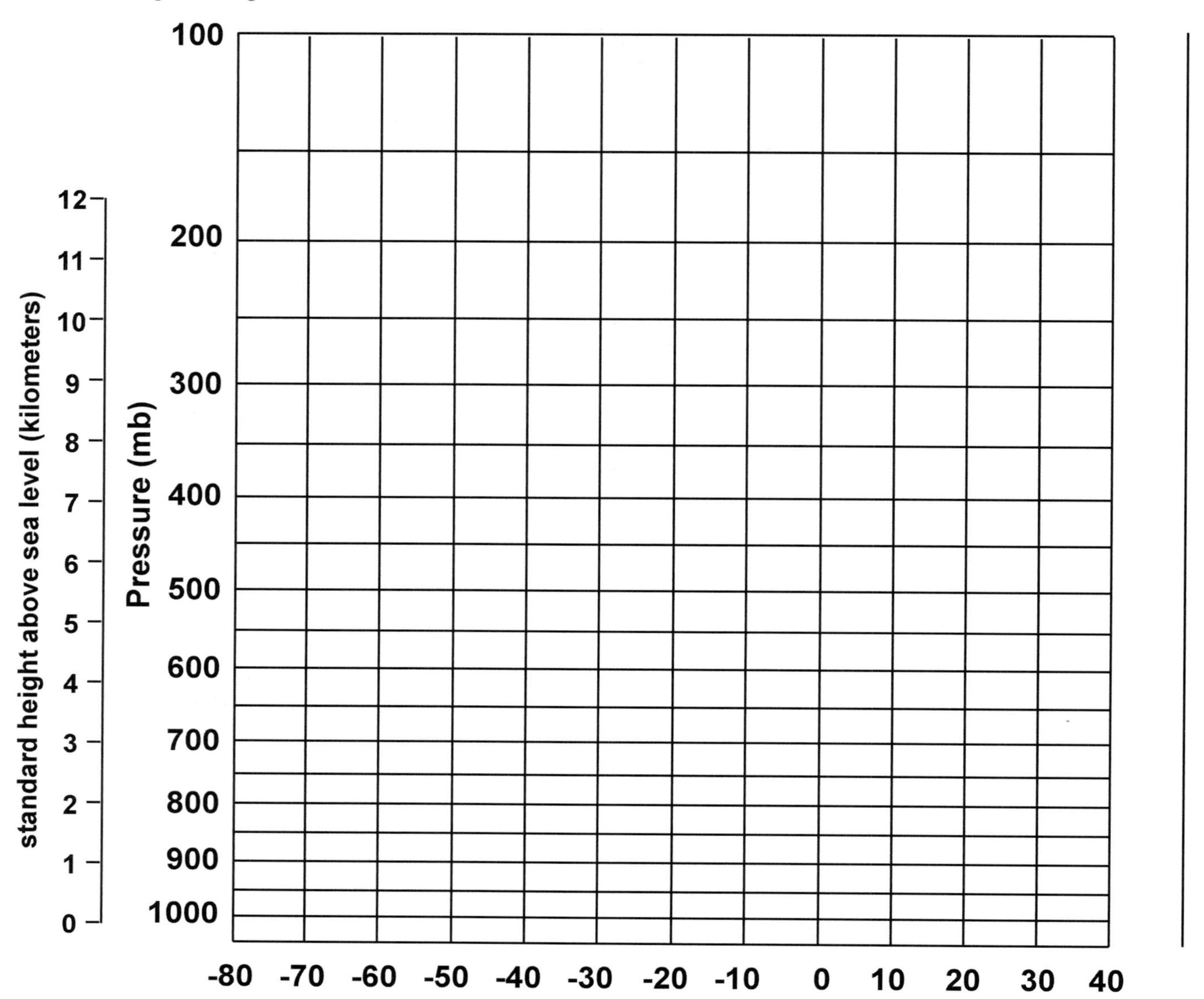

Hodograph

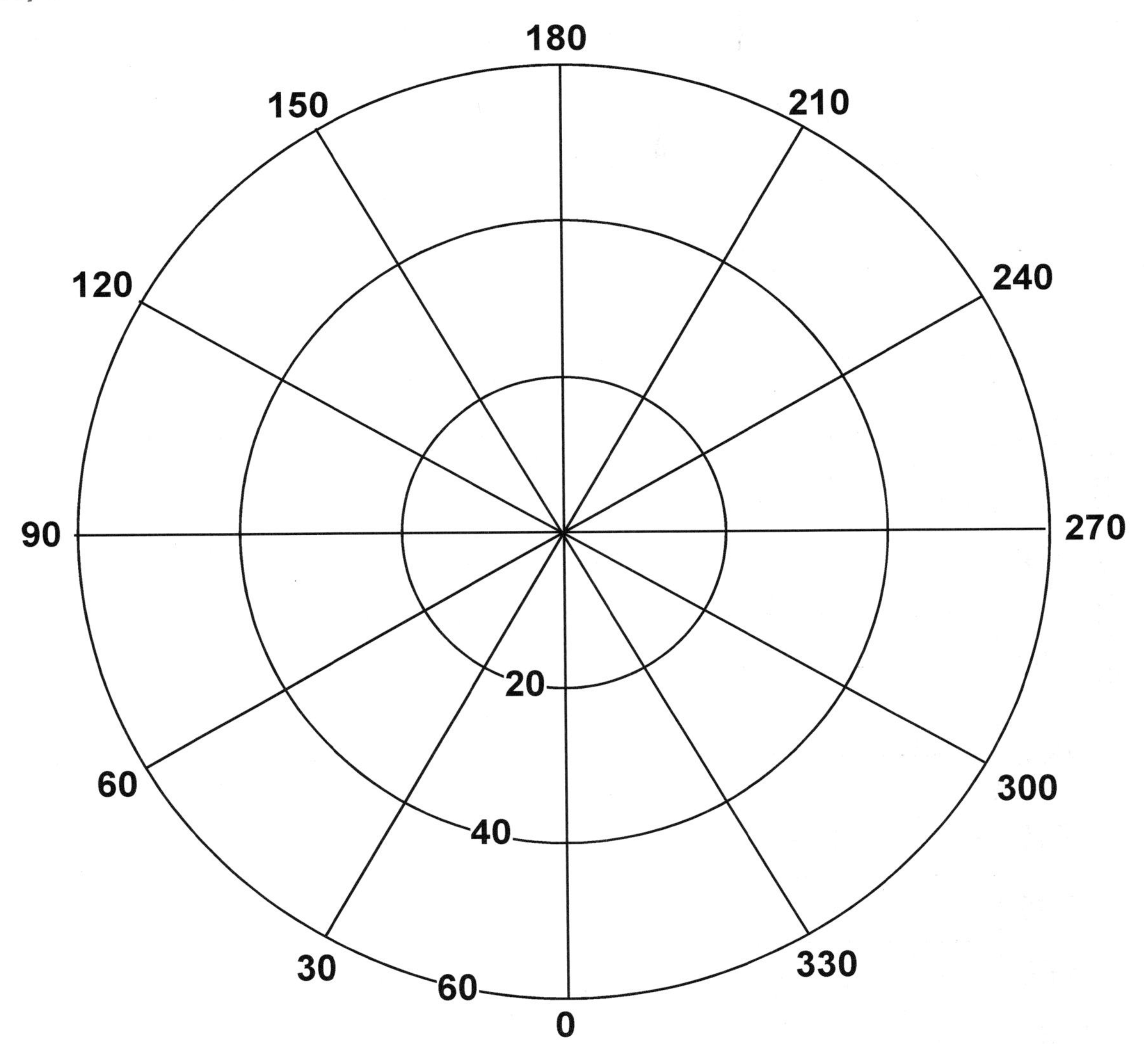

MAPS

United States

North America

GEOGRAPHY OVERVIEW

Physical Geography of North America

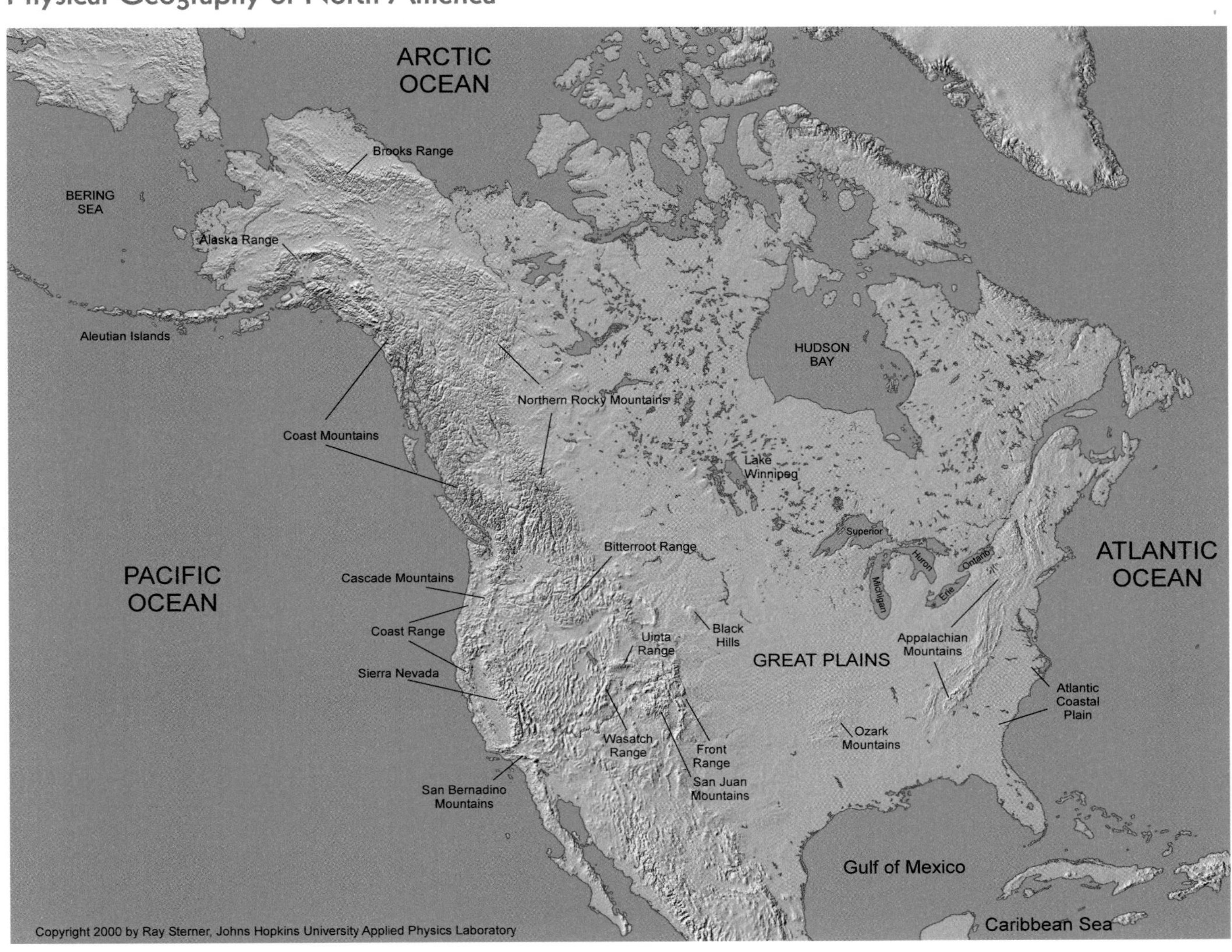

British Columbia
Alberta
Saskatchewan
Manitoba
Ontario
Quebec
Nova Scotia
New Bruns-wick
Maine
Vermont
New Hampshire
Rhode Island
Massachusetts
Connecticut
New Jersey
Delaware
Maryland
Washington
Oregon
California
Nevada
Idaho
Montana
Wyoming
Utah
Arizona
Colorado
New Mexico
North Dakota
South Dakota
Nebraska
Kansas
Oklahoma
Texas
Minnesota
Iowa
Missouri
Arkansas
Louisiana
Wisconsin
Illinois
Michigan
Indiana
Ohio
Kentucky
Tennessee
Miss-issippi
Alabama
Georgia
Florida
New York
Pennsylvania
West Virg-ina
Virginia
North Carolina
South Carolina
Mexico

Tropical Atlantic Hurricane Tracking Chart

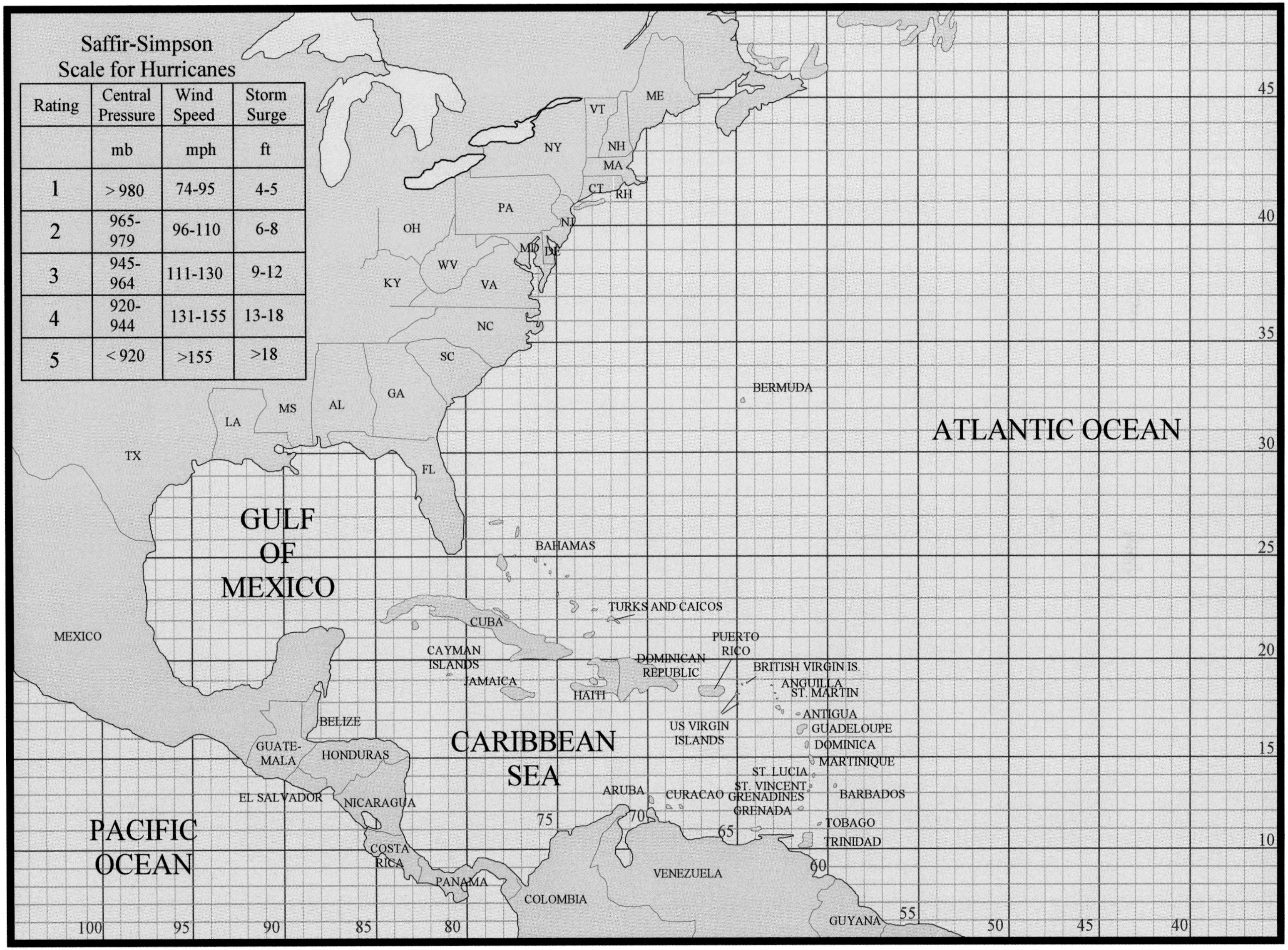

Saffir-Simpson
Scale for Hurricanes

Rating	Central Pressure	Wind Speed	Storm Surge
	mb	mph	ft
1	> 980	74-95	4-5
2	965-979	96-110	6-8
3	945-964	111-130	9-12
4	920-944	131-155	13-18
5	< 920	>155	>18

ME
VT
NH
MA
NY
CT
RH
PA
NJ
OH
MD
DE
WV
KY
VA
NC
SC
GA
AL
MS
LA
TX
FL
BERMUDA
ATLANTIC OCEAN
GULF OF MEXICO
BAHAMAS
TURKS AND CAICOS
CUBA
MEXICO
PUERTO RICO
CAYMAN ISLANDS
DOMINICAN REPUBLIC
BRITISH VIRGIN IS.
JAMAICA
ANGUILLA
ST. MARTIN
HAITI
ANTIGUA
BELIZE
US VIRGIN ISLANDS
GUADELOUPE
CARIBBEAN SEA
GUATE-MALA
HONDURAS
DOMINICA
MARTINIQUE
ST. LUCIA
ST. VINCENT
ARUBA
CURACAO
GRENADINES
BARBADOS
EL SALVADOR
NICARAGUA
GRENADA
TOBAGO
TRINIDAD
PACIFIC OCEAN
COSTA RICA
VENEZUELA
PANAMA
COLOMBIA
GUYANA

45
40
35
30
25
20
15
10

100
95
90
85
80
75
70
65
60
55
50
45
40

WATCHES, WARNINGS, AND ADVISORIES FOR SEVERE AND HAZARDOUS WEATHER

GENERAL

Advisory Issued for weather situations that cause significant inconveniences but do not meet warning criteria and, if caution is not exercised, could lead to life-threatening situations.

Watch Forecast issued in advance to alert the public of the possibility of a particular hazard (e.g., tornado watch, flash flood watch).

Warning Issued when a particular hazard is imminent or already occurring (e.g., tornado warning, flash flood warning).

Local Storm Report Used to distribute severe weather reports to the media, emergency managers, and other NWS offices. It is issued as reports are received, and may also be issued as a collection of all reports received after an event is over. Delayed reports are disseminated after an event is over as well.

Special Weather Statement Issued to convey updated information about severe weather watches and warnings.

THUNDERSTORMS AND TORNADOES

Severe Thunderstorm Watch Conditions are favorable for the development of severe thunderstorms in and close to the watch area.

Severe Thunderstorm Warning Issued when a thunderstorm exists and has the potential to produce hail with diameters three-fourths of an inch or larger and/or winds which equal or exceed 58 mph. Information in this warning includes the storm location, towns that will be affected, and the primary threat associated with the storm.

Tornado Watch Conditions are favorable for the development of tornadoes in and close to the watch area.

Tornado Warning Tornado is indicated by radar or sighted by storm spotters. The warning will include the location of the tornado and towns and counties in its path.

RIVER FLOODING

Flood Watch Indicates that flooding is a possibility in and close to the watch area. Those in the affected area are urged to be ready to take quick action if a flood warning is issued or flooding is observed.

Flood Warning Issued when flooding is occurring or imminent. Can be issued for rural or urban areas as well as for areas along small streams and creeks.

Flood Statement Used as a follow-up to Flood Warnings and Watches. The statement will contain the latest information on the event.

(Source: National Weather Service)

Flash Flood Watch Indicates that flash flooding is a possibility in and close to the watch area. Those in the affected area are urged to be ready to take quick action if a flash flood warning is issued or flooding is observed. These watches are issued for rapid flooding that is expected to occur during heavy rainfall or within 6 hours after heavy rain has ended.

Flash Flood Warning Issued when rapid floods that threaten life or property are occurring or expected within 6 hours after heavy rain has ended. Warning can be issued for rural or urban areas as well as for small streams and creeks. The degree of flash flooding depends on local terrain, ground cover, degree of urbanization, any man-made changes to the natural river banks, and initial ground or river conditions. Dam breaks or ice jams can also create flash flooding.

Flash Flood Statement Used as a follow-up to Flash Flood Warnings and Watches. The statement will contain the latest information on the event.

River Flood Warning Used to inform the public of long-term (more than 12 hours) flooding along major streams and rivers that is a threat to life and/or property. Usually contains river stage forecast and crest information, and the history and impact of the flood.

River Flood Statement Used to update and expand information in a River Flood Warning. May also be used in lieu of a Flood Warning if flooding is expected or imminent, but does not pose a threat to life and/or property.

Urban and Small Stream Flood Advisory Alerts the public to flooding that is generally only an inconvenience and does not pose a threat to life and/or property. Issued when heavy rain will cause flooding of streets and low-lying places in urban areas, or if small rural or urban streams are expected to reach or exceed their normal banks.

HURRICANES AND COASTAL FLOODING

Hurricane Watch Issued when a hurricane or hurricane-related hazard poses a possible threat for a specified coastal area generally within 36 hours.

Hurricane Warning Issued when a hurricane is expected in specified coastal areas within 24 hours or less. A hurricane warning can remain in effect when dangerously high water and/or exceptionally high waves continue even though winds may be less than hurricane force.

Tropical Storm Watch An announcement that a tropical storm or tropical storm conditions pose a threat to coastal areas generally within 36 hours. A tropical storm watch is not usually issued if a tropical cyclone is forecast to attain hurricane strength.

Tropical Storm Warning A warning that sustained winds between 34 to 63 knots (39 to 73 mph) associated with a tropical cyclone are expected in specified coastal areas within 24 hours.

Coastal Flood Watch A coastal flood watch is issued 12 to 36 hours in advance of the expected potential for coastal flooding. Flooding is not imminent or expected but is possible based on current trends.

Coastal Flood Warning Issued when coastal flooding is occurring or expected within 12 hours.

Coastal Flood Statement Keeps the public and cooperating agencies informed of the status of existing coastal flood watches and warnings, as well as provides an update on local conditions. Also used to cancel a watch or warning.

Heavy Surf Advisory Advises people that heavy (high) surf may pose a threat to life or property. Such advisories may be issued alone or in conjunction with coastal flood watches or warnings.

WINTER STORMS

Winter Storm Watch Issued when conditions are favorable for the development of hazardous weather elements, such as heavy snow, sleet, blizzard conditions, significant accumulations of freezing rain or drizzle, or any combination thereof. Watches are usually issued 24 to 48 hours in advance of an event.

Winter Storm Warning Issued when hazardous winter weather conditions are imminent or very likely to develop in the next 12 to 24 hours. Conditions include any occurrence or combination of heavy snow, wind-driven snow, sleet, and/or freezing rain or drizzle. Winter Storm Warnings are usually issued for up to a 12-hour duration, but can be extended out to 24 hours if the situation warrants. The term "near-blizzard" may be incorporated into the "winter storm warning" for serious situations that fall just short of official blizzard conditions.

Winter Weather Advisory Used when a mixture of precipitation is expected such as snow, sleet, and freezing rain or freezing drizzle, but will not reach warning criteria. Typically issued 12 to 24 hours in advance of the event.

Winter Storm Outlook Issued when there is sufficient confidence that a major winter storm may cause a significant hazard to public safety. This product is generally issued from 3 to 5 days in advance of an event.

Heavy Snow Warning Issued for expected snowfall amounts of 4 inches or more in 12 hours or 6 inches or more in 24 hours. Snow is the only precipitation type expected.

Snow Advisory Used when snowfall amounts are below warning criteria, but nonetheless pose a hazard to the public. Issued for average snowfall amounts of less than 4 inches in a 12-hour period.

Blowing/Drifting Snow Advisory Used when wind-driven snow intermittently reduces visibility to a quarter mile or less. Travel may be hampered. Strong winds create blowing snow by picking up old or new snow.

Freezing Rain or Freezing Drizzle Advisory Generally used only during times when the intensity of freezing rain or drizzle is light and ice accumulations are less than a quarter inch.

Ice Storm Warning Issued when damaging ice accumulations are expected during freezing rain situations so that walking or driving becomes extremely dangerous. Ice accumulations are usually a quarter inch or greater.

Sleet Warning Issued when accumulations of sleet (ice pellets) covering the ground to a depth of a half inch or more are expected. This is a relatively rare event.

Sleet Advisory Issued for expected sleet (ice pellet) accumulations of less than a half inch.

Blizzard Warning Issued for winter storms with sustained winds or frequent gusts of 35 miles per hour or greater and considerable falling and/or blowing snow reducing visibility to less than a quarter mile. These conditions are expected to last at least 3 hours.

Wind Chill Warning Issued when wind chill temperatures are life threatening. Actual threshold values for warnings are geographically dependent.

Wind Chill Advisory Issued when wind chill temperatures are potentially hazardous. Actual threshold values for advisories are geographically dependent.

HEAT

Heat Advisory Issued when a threshold value of heat index is expected or has been present for two or more consecutive days. Actual threshold values are geographically dependent.

Excessive Heat Warning Issued when prolonged periods of high heat indices are expected or when extremely high heat indices are expected for a single day. Actual threshold heat index values are geographically dependent.

WIND

Wind Advisory (also Lake Wind Advisory) Wind Advisory issued when sustained winds of 30 mph or greater are expected to last for 1 hour or more, or for gusts of 45 to 57 mph for any duration. Lake Wind Advisory issued for area lakes when sustained winds of 30 mph or more are expected.

High Wind Warning Used when sustained winds of 40 mph or greater are expected to last for 1 hour or longer, or for non-thunderstorm winds of 58 mph or greater for any duration.

Gale Warning Warns the public that sustained winds exceeding 34 knots over oceanic waters are expected or occurring, and are not directly associated with tropical cyclones.

GLOSSARY

Absolute vorticity Vorticity associated with both local rotation within a fluid and fluid rotation induced by rotation of the Earth.

Absolute zero The zero point on the Kelvin temperature scale. Corresponds to –273.15° C. The temperature at which molecules slow to a point where they are essentially not moving and no more energy can be extracted.

Absorbed In reference to solar radiation, the transformation of radiant energy to heat energy by a substance, resulting in a rise in temperature.

Adiabatic process A process in which an air parcel does not mix, or exchange heat, with its environment.

Advisory Issued for weather situations that cause significant inconveniences but do not meet warning criteria and, if caution is not exercised, could lead to life-threatening situations.

Aerosol Microscopic particles in the Earth's atmosphere. Aerosol consist of windblown particles, sulfates, organics, sea salt, pollutants, and other particles that may be natural or man-made.

Agricultural drought A period of deficient moisture in the soil layers from which crops and other plants normally draw their water. Often precedes hydrological drought.

Aircraft icing The formation of ice on aircraft, typically on the wings or nose.

Aircraft Meteorological Data Report (AMDAR) Automated weather report from a commercial aircraft.

Airmass A large body of air with relatively uniform thermal and moisture characteristics. Typically several hundred thousand square kilometers in area, as shallow as 1 km to as deep as the troposphere. Forms over relatively flat regions with homogeneous surface characteristics.

Airmass thunderstorm Short-lived thunderstorm that forms far from frontal boundaries and typically does not contain severe weather.

Air parcel A small volume of air that, for conceptual purposes, is assumed to retain its identity as it moves through the atmosphere.

Airspeed The speed of air moving relative to an aircraft. The sum of the speed of the headwind and the aircraft's speed relative to the ground.

Alberta Clipper Extratropical cyclone that forms in wintertime east of the Canadian Rockies and rapidly moves across southern Canada and the northern United States to the East Coast.

Aleutian Low A semi-permanent surface low-pressure center located over the Pacific Ocean south of Alaska.

Altimeter Instrument that uses pressure to estimate altitude. Normally used on aircraft.

Altocumulus Middle level clouds (~2 to 6 km) with a puffy shape.

Altocumulus lenticularis Lens-shaped clouds generated by airflow over mountains.

Altostratus Middle level clouds (~2 to 6 km) with layered shape.

AMDAR See "Aircraft Meteorological Data Report."

Analysis A depiction of the current weather on a weather map. The weather patterns may be depicted with isolines, color bands, shading or other visual aids.

Anemometer Instrument used to measure wind speed.

Anemometer height The official height of wind measurements at surface stations, 10 meters (~33 feet).

Angular momentum The product of an object's mass, rotational velocity, and radius from the axis about which it is rotating.

Annual average energy budget Global yearly average amount of energy received and absorbed by the Earth and atmosphere, considering all forms of energy exchange.

Anomaly A deviation from normal for a particular variable such as temperature, pressure, or precipitation.

Antenna The component of a radar that broadcasts microwaves into the atmosphere and collects microwaves that have been scattered toward the radar by precipitation particles.

Anthropogenic Alterations to the environment that are human induced or result from human activities.

Anticyclone High-pressure system with surface airflow clockwise around the center in the Northern Hemisphere.

Anticyclonic curvature A curved pattern either in the pressure or height field in which air flows clockwise in the Northern Hemisphere (counterclockwise in the Southern Hemisphere).

Anvil Cloud formation found at high elevations of a thunderstorm; most extensive on the forward portion of a moving thunderstorm.

Apparent temperature The perceived temperature felt by humans when taking into account air temperature, relative humidity, radiant energy, and wind speed. Often used as a measure of heat induced discomfort.

Arctic airmass An airmass characterized by bitter cold temperatures. In some cases, for differentiation from a polar airmass, an airmass is considered "arctic" if it forms over the Arctic Ocean.

Arctic front A cold front that marks the boundary of an extremely cold airmass in wintertime.

Arcus cloud A cloud that forms over a gust front as the cool, dense air lifts warm moist air ahead of the gust front. Sometimes called a "shelf cloud" or "roll cloud."

Arroyo In mountainous regions, a river bed that is typically dry the majority of the time as a result of infrequent rainfall.

ASOS See "Automated Surface Observing System."

Atmospheric river A relatively narrow flow of moist air in the lower and middle troposphere, often appearing on satellite pictures as either a long stream of clouds (visible and infrared image) or high water vapor concentration (water vapor image).

Automated Surface Observing System (ASOS) Automated instrumentation of the U.S. National Weather Service for continuous collection of information about surface weather conditions. Instruments include rain gauge, temperature and dewpoint temperature sensor, device to determine precipitation type, wind vane, anemometer, ceiliometer, freezing rain sensor, and visibility sensor.

Automated Weather Observing System (AWOS) Automated instrumentation of the Federal Aviation Administration and Department of Defense that continually collects information about surface weather conditions. Instruments are similar to ASOS.

Avalanche In mountainous regions, the catastrophic collapse of a large amount of snow down a mountain slope either as loose snow (powder avalanche) or as a large slab (slab avalanche).

Aviation Weather Center One of the National Centers for Atmospheric Prediction. The Aviation Weather Center is responsible for forecasts supporting aviation activities.

AWOS See "Automated Weather Observing Systems."

Ball lightning A luminous, small, glowing ball of gas, typically 10 to 40 cm in diameter. Appears very rarely, and always follows a lightning stroke.

BAMEX See "Bow Echo and MCV Experiment."

Barometer An instrument that measures atmospheric pressure.

Barometric effect A local rise in the height of sea level due to lower atmospheric pressure above that region. Most important in the vicinity of the eye of hurricanes where extremely low pressure occurs.

Beach erosion The deterioration or destruction of beaches due to wave action induced by severe storms.

Bead lightning Brief, luminous dots of light following a lightning stroke. Created when the lightning channel breaks up.

Beneficial competition In hail suppression, the premise that precipitation particles will be smaller when larger numbers of them must share the available moisture.

Bermuda High The semi-permanent surface high-pressure center located over the North Atlantic Ocean. Positioned in subtropical latitudes during winter and middle latitudes during summer.

Black ice Ice that is difficult to see on a road because of either dirt within the ice or because it is not visible on the road.

Blizzard A severe weather condition characterized by high winds and reduced visibilities due to falling or blowing snow.

Blizzard zone The region of North and South Dakota and western Minnesota where blizzards are most common in the United States.

Blue Jet Electrical discharges that extend upward from the cloud tops of thunderstorms in narrow cones and can have upward speeds of 100 km/sec.

Bolt from the blue A lightning stroke that emerges from the side of a thunderstorm and strikes the ground several kilometers from the thunderstorm edge, in a region where the sky overhead is blue.

Bomb cyclone Extratropical cyclone in which the central pressure falls at least 24 millibars in 24 hours.

Bookend vortex A vortex that develops on either end of a bow echo, often marked by a circular curl of higher reflectivity on a radar image.

Bora A cold, downslope wind that develops along the Adriatic Sea just southeast of the Dinaric Alps of Yugoslavia. More generally, any cold downslope windstorm.

Boundary A generic word used by meteorologists to describe fronts, the leading edge of thunderstorm outflow, the leading edge of a sea breeze, or any other line marking the junction of two airmasses.

Boundary intersections Geographic locations where outflows or fronts from thunderstorms and weather systems meet.

Boundary layer Layer of the atmosphere adjacent to the Earth's surface where friction is important. Depth can vary from a few hundred to a few thousand meters. Also called "friction layer."

Bounded Weak-Echo Region (BWER) Radar feature of severe thunderstorms containing only small cloud droplets above the rain-free base of the storm's updraft.

Bow echo Radar reflectivity feature of severe thunderstorms having an arc or bow shape. Often associated with strong straight-line winds and derechos.

Bow Echo and MCV Experiment Field experiment where data was collected within and surrounding mesoscale convective systems including bow echoes and mesoscale convective vortices. Conducted in the Central United States during summer, 2003.

Breaking waves In airflow over mountains, waves in the flow that become turbulent and break, similar to what water waves do along a shoreline.

Bright band Radar feature of high reflectivity that identifies the level where snowflakes falling from aloft melt into raindrops.

Buoy An automated weather station floating on a lake or ocean that reports meteorological data and/or data about the state of the ocean.

BWER See "Bounded Weak-Echo Region."

California Norther' Hot, dry downslope windstorm that forms along the west slope of the Sierra Nevada.

Canadian Lightning Detection Network (CLDN) Network of remote lightning sensors in Canada that monitor cloud-to-ground lightning activity.

CAPE See "Convective Available Potential Energy."

Capping inversion A temperature inversion present in the lower troposphere that prevents convection from developing early in the day.

Celsius scale A temperature scale in which the melting point of ice is set at zero and the boiling point of water at sea-level pressure is set at 100.

Centigrade scale See "Celsius scale."

Centripetal acceleration The acceleration toward the center of rotation experienced by a parcel of air in a rotating fluid.

Chain law The law in western states that requires motorists driving over mountain passes to use chains or snow tires on vehicles during snowstorms.

Channeling effect The channeling of airflow by topography.

Charge A net accumulation of either positive or negative ions on a substance.

Chinook Warm, dry downslope wind that forms along the east slope of the Rocky Mountains.

Chinook wall Cloud that forms on the windward side of the Rocky Mountains during a Chinook wind.

CINH See "Convective inhibition."

Cirrocumulus High clouds (> 6 km) with a puffy shape, often occurring in clusters.

Cirrostratus High clouds (> 6 km) with a layered shape.

Cirrus High, wispy clouds (> 6 km) composed entirely of ice crystals.

CLDN See Canadian Lightning Detection Network."

Clear air mode Mode of operation of Doppler radar used when there is no precipitation in the region. In this situation, the radar primarily receives energy scattered back from insects, birds and turbulence. Used to monitor winds in the atmosphere.

Clear air turbulence Bumpy, sometimes severe, motions experienced by aircraft flying in clear air.

Climate Average weather conditions, typically temperature and precipitation, for a particular location, averaged over a long time period.

Climate Change Long-term change in climate conditions, typically occurring over a period of several decades, centuries, or longer.

Climate model A numerical model used to simulate climate conditions.

Climatological drought See "meteorological drought."

Cloud droplets Tiny liquid water droplets in clouds.

Cloud seeding Process by which ice nuclei are activated or injected into a cloud to enhance precipitation (or suppress hail growth). Dry ice pellets and silver iodide are the most common seeding materials.

Cloud to cloud lightning Lightning occurring between cloud towers that does not extend to the ground.

Cloud to ground lightning Any lightning stroke that extends from a cloud to the ground.

Coastal cyclone Extratropical cyclone that forms along the East or Gulf Coasts of North America. Most common during late fall, winter, and early spring.

Coastal flood Flooding along coastal regions resulting from storm surge created by tropical and strong extratropical cyclones, or non-meteorological phenomena such as tsunamis. Flooding can extend well inland depending on storm intensity and the geographic and topographic features of the region.

Coastal front A stationary boundary between cold air dammed on the east side of the Appalachian Mountains and warm air over the Atlantic Ocean. This boundary is the focal point for the development of East Coast cyclones.

Cold air damming Process by which cold, dry air is forced against the east side of the Appalachian Mountains when strong onshore flow is present from the Atlantic Ocean. Often associated with East Coast cyclones and freezing rain events along the east coast of the United States.

Cold air funnel Vortex of condensed water that emerges from the base of elevated convective clouds. Develops within cool low-level air, often in association with large-scale, upper-level cutoff low-pressure centers.

Cold air trapping The process by which cold air is trapped in the valleys of the Appalachian Mountains while warm air advances northward on either side of the Appalachians.

Cold core system Low-pressure system in which cold air is located aloft, above a surface low-pressure center.

Cold front The boundary between a cold airmass and warm airmass when the cold air is advancing forward, lifting the warmer airmass. The frontal boundary is typically shaped like a dome.

Cold front aloft See "upper-level front."

Cold occlusion An occluded front in which air behind the cold front is more dense (colder) than air behind the warm front.

Cold pool The cold surface air that develops in association with thunderstorm downdrafts due to evaporation of rain.

Cold wave An influx of unusually cold air into middle or lower latitudes.

Colorado cyclone Extratropical cyclone originating east of the Front Range of the Rocky Mountains, within or near Colorado.

Comma cloud Cloud structure associated with extratropical cyclones, often evident on satellite imagery as a comma shape. Clouds at the head of the comma are associated with the low-pressure center and the region north of the warm front; the tail of the comma contains clouds along the easternmost front south of the low.

Computer model Systems of mathematical equations that describe the behavior of the atmosphere and are used to forecast the weather or study storms. Run on computers.

Condensation The process by which water changes from vapor to liquid, as in cloud formation. During this process latent heat is released into the atmosphere, warming the air.

Condensation level Height in the atmosphere where condensation first occurs as a result of rising air. Denotes the base of a cloud.

Conditionally unstable Atmosphere in which environmental lapse rate is between about 6° C/km and 10° C/km. An air parcel will be stable if it is unsaturated and unstable if it is saturated.

Conduction The transfer of heat via molecular contact. In the atmosphere, conduction is important in the several millimeters closest to the surface.

Conductor A substance in which electricity can easily flow.

Consensus forecast The average of various forecasts from the same numerical model, different models, or other sources. Generally more accurate than individual forecasts.

Conservation of angular momentum Physical principle by which angular momentum is neither created nor destroyed over time.

Constant pressure map Weather map on which pressure is a constant value.

Continental airmass Airmass that forms over land and is characterized by low moisture content (low dewpoint temperature).

Continental climate The type of climate typically observed over the continents of the world.

Continental drying For airmasses, the reduction of moisture that airmasses experience over a continent as rain removes water from the atmosphere, but evaporation does not replenish it.

Contour A line depicting a constant value of a variable on a weather map.

Convection Vertical motions within the atmosphere caused by buoyancy. Convection results in mixing of air and transport of warm air aloft.

Convective Available Potential Energy (CAPE) A measure of the positive buoyancy of an air parcel accumulated over its trajectory above its level of free convection on a sounding.

Convective complex Thunderstorms that form in unorganized clusters; sometimes called multicell thunderstorms.

Convective inhibition (CINH) A measure of the near-surface negative buoyancy of an air parcel that must be overcome by lifting before the parcel can rise spontaneously from its level of free convection.

Convective region In reference to thunderstorms, the region of the cumulonimbus cloud or thunderstorm complex with a strong updraft.

Convergence Net inflow of air molecules into a region of the atmosphere with the result of increasing surface pressure. Convergence can be caused by changing wind speeds and/or changing direction of air flow.

Cooling degree-day The average of a day's high and low Fahrenheit temperature minus 65. Over a summer season, cooling degree-days are used by power companies to estimate energy usage from air conditioning.

Coriolis force An apparent force associated with the rotation of the Earth. Causes air to deflect to the right of motion in the Northern Hemisphere.

Crop Moisture Index A short-term drought index that is a measure of the moisture in the crop rooting zone. Designed for agricultural uses.

Cross section A diagram of the atmosphere oriented in a vertical plane, e.g., from the surface to the tropopause.

Crustal plates Pieces of the outer layers of Earth's surface that move and interact with each other over long time scales.

Cumulonimbus Vertically developing cloud in which precipitation has developed. Often associated with heavy rain and sometimes severe weather. A thunderstorm cloud.

Cumulus Vertically developing cloud with puffy lobes and clear skies surrounding the cloud.

Cumulus stage Early stage of development of a thunderstorm (cumulonimbus) cloud.

Curl The part of the flow at the leading edge of a microburst (downburst) where air curls upward as it encounters less dense air ahead of it.

Current A relatively narrow fast moving flow of water within the ocean.

Curvature effect The creation of divergence within the jetstream due to the change of flow curvature from cyclonic flow (around a trough) to anticyclonic flow (around a ridge). Also applies to the creation of convergence when flow changes from anticyclonic to cyclonic.

Cutoff low A cyclone in its decaying stage, when the upper-level low becomes a deep vortex with a cold center. The low at the surface is often directly underneath the low aloft, which is cut off from the main jetstream.

Cyclone Large low-pressure system with surface airflow counterclockwise around the center in the Northern Hemisphere. Large storms forming in the middle latitudes are called "extratropical cyclones," while storms forming in the tropics are "tropical cyclones." Hurricanes that form over the Indian Ocean are called cyclones.

Cyclonic curvature A curved pattern either in the pressure or height field in which air flows counterclockwise in the Northern Hemisphere (clockwise in the Southern Hemisphere).

Dart leader The downward phase of a lightning stroke following the stepped leader and initial return stroke.

Decay stage The last stage of a large tornado in which the vortex weakens and dissipates.

Density Mass per unit volume of a substance. Density decreases with height in the atmosphere.

Deposition Phase change from vapor directly to solid. In the atmosphere, deposition occurs during ice cloud formation and frost formation.

Derecho Widespread windstorm created by severe thunderstorms. Also called "straight-line winds."

Destabilization The process of making the atmosphere more unstable, typically through heating or moistening.

Dewpoint depression The difference between the temperature and the dewpoint temperature. As the dewpoint depression increases, relative humidity decreases.

Dewpoint temperature The temperature at which air will become saturated if it is cooled at constant pressure with no change in its moisture content.

Diabatic process A process in which an air parcel mixes with its environment and/or exchanges heat with its environment.

Differential reflectivity Quantity used in radar meteorology that describes the ratio of the reflectivity measured at horizontal polarization to that at vertical polarization. Measured in decibels (dB). Typical values are near zero for hail, about 1 to 2 dB in moderate rain, and 3 to 4 dB in heavy rain.

Divergence Net outflow of air molecules from a region of the atmosphere with the result of decreasing surface pressure. Divergence can be caused by changes in wind speeds and/or direction of airflow.

Doppler on Wheels Doppler radars mounted on trucks that are transported to the location of severe thunderstorms and hurricanes to collect data for research purposes.

Doppler radar Instrument that detects precipitation and wind motions along the direction of the radar beam. Uses the Doppler effect. U.S. National Weather Service Radars are called WSR-88D or NEXRAD.

Doppler shift The shift in frequency of radar energy caused by the movement of precipitation or other objects in the radar beam toward or away from the radar. Used by Doppler radars to estimate winds.

Double eyewall Concentric rings of organized cumulonimbus clouds within a hurricane.

Downburst A strong downdraft that originates within the lower part of a cumulus cloud or thunderstorm and descends to the ground, often resulting in strong straight-line winds. A significant hazard to aircraft.

Downdraft Region of air with marked sinking vertical motion.

Downslope flow The descent of air on the leeward side of a mountain. Air warms adiabatically as it descends, often resulting in warm flow on the lee side of mountains.

Downslope windstorm General term given to windstorm in which air flows down the side of the mountain. Common North American examples include the Chinook and Santa Ana.

Drizzle Liquid water droplets with diameters between 0.2 and 0.5 mm that usually fall from stratus or stratocumulus clouds.

Dropwindsonde An instrument package deployed from aircraft. Collects data on air temperature, pressure, moisture, and wind.

Drought Condition developing due to extreme deficit of precipitation, often associated with water shortages. See also "agricultural drought," "meteorological drought," and "socioeconomic drought."

Dry adiabatic lapse rate The rate (10° C/km) at which an unsaturated parcel of air will change temperature if it is displaced vertically in the atmosphere.

Dry growth regime Process in hail formation in which the surface temperature of the hailstone remains colder than 0° C. Supercooled droplets freeze on contact and the stone's surface remains dry.

Dry line A front characterized by a sharp moisture contrast, but little temperature change. Often found in the Southern Plains.

Dry microburst A microburst (or downburst) that has no measurable precipitation at the ground. Virga is often seen descending from the cloud base. Blowing dust may be visible on the ground if the surface is dry. Very difficult to identify and are especially dangerous for aircraft. Most common in the western United States and the Great Plains.

Dry slot A region of exceptionally clear air in the southern quadrant of an extratropical cyclone, just west of the tail of the comma cloud. Often evident on satellite imagery as a clear wedge.

Dust Bowl Severe drought of the 1930s affecting much of the central United States.

Dust devil Vortex associated with dry convection that develops over hot surfaces. Can extend over a kilometer in depth and can reach wind speeds similar to a very weak tornado. Common in desert regions of western North America.

Dust whirl stage The stage in a tornado lifecycle where the rotation makes contact with the ground and stirs up dust.

Dynamic pipe effect A mechanism for the formation of a tornado in which the tornado vortex descends from aloft.

Dynamic process A process that involves forces.

East Coast cyclone An extratropical cyclone that develops over the central Atlantic Coast of the United States, often near Cape Hatteras, North Carolina.

Easterly trades See "trade winds."

Easterly wave A wave in the low-level airflow within the trade winds in the tropics. Typically originates over continents as air moves across mountains and/or deserts. Convergence in the wave may create clusters of thunderstorms that sometimes organize into hurricanes.

Eccentricity In reference to Earth's orbit around the Sun, the deviation of the orbital path from circularity.

Echo-free vault Region above the rain-free base of a supercell thunderstorm where small cloud particles produce weak, or no, radar echoes. Structurally resembles a gothic cathedral vault.

Eddy A small disturbance in a flow field that derives its energy from a flow of much larger scale.

Electric field A force field that exists in any region exposed to charges. Any charged body will experience an attractive or repulsive force in an electric field.

Electromagnetic energy Radiative energy that includes visible light, x-rays, ultraviolet energy, microwaves, radio waves, and infrared energy.

Electron Negatively charged subatomic particle.

El Niño A significant increase in sea-surface temperatures that occurs at irregular intervals in the eastern equatorial Pacific Ocean.

Elves Disk-shaped regions of light that last less than a thousandth of a second and occur high above energetic cloud-to-ground lightning of thunderstorms.

Embryo curtain The region of a hailstorm in which the initial ice particles that will grow into hailstones are first produced.

Embryos See "hail embryo."

Energy helicity index An index used by forecasters to anticipate locations where tornadoes are possible, numerically equal to the product of the convective available potential energy and the storm-relative helicity divided by 160,000.

Enhanced Fujita scale Scale for rating the destruction caused by tornadoes, a revised version of the original Fujita scale. Takes into account structural integrity of structures using degree of damage Indicators.

Ensemble forecasting Forecasting technique in which a numerical weather prediction model is run repeatedly for the same case, but with slight changes to the initial state and/or slight changes to the model's formulation.

Ensemble member One of the numerical model runs used in constructing an ensemble forecast.

ENSO See "El Niño Southern Oscillation."

Entrainment The process by which dry air outside of a cloud is mixed into a cloud.

Entrance region The side of a jetstreak in which the height gradient is increasing along the flow direction and the air is accelerating.

Environment For stability determinations, all the air outside of an air parcel.

Environmental lapse rate The rate at which the environmental temperature changes with height in the atmosphere.

Environmental Modeling Center The National Center for Environmental Prediction responsible for the development and testing of numerical forecast models.

Evaporation The process by which water changes from liquid to vapor, as in cloud dissipation. During this process latent heat is taken from the atmosphere, cooling the air.

Evaporational cooling Cooling of air temperature due to the consumption of latent heat during evaporation of water.

Evapotranspiration The process of transfer of water from plants, open water surfaces, ice surfaces, and bare soil to the atmosphere.

Exit region The side of a jetstreak in which the height gradient is decreasing along the flow direction and the air is decelerating.

Explosive cyclogenesis The formation or deepening of an extratropical cyclone where the cyclone intensity increases very rapidly.

Extratropical cyclone A large swirling low-pressure system that forms along the jetstream between about 30° and 70° latitude. Also called "cyclone."

Eye Area in the center of a hurricane, coinciding with the low-pressure center, that is either cloud free or contains low-level scud clouds.

Eyewall A ring of strong convective storms surrounding the eye of a hurricane. Contains the hurricane's most violent winds, heaviest rain, and strongest updrafts.

F-scale See "Fujita scale."

Fahrenheit scale A temperature scale in which the melting point of ice is set at 32 and the boiling point of water at sea-level pressure is set at 212.

Fair weather electric field The atmosphere's electric field in the absence of thunderstorms.

Feedback A sequence of processes in the atmosphere in which the first process initiates the second process, and the second process in turn reinforces the first process (positive feedback), or opposes the first process (negative feedback).

Feedback mechanism A mechanism by which one process results in the enhancement or destruction of a second process that subsequently reinforces the initial process.

Fine line On radar, a very narrow long line of weak echoes typically associated with the leading edge of an outflow boundary.

Fire whirl Tornado-like circulation that forms during wildfires.

Flash flood Localized, short-duration floods that occur rapidly with little or no warning.

Floodplain Land adjacent to a river that is repeatedly inundated by the river's floodwaters.

Flood stage The level at which a rising river begins to flood agricultural lands or has potential to damage property.

Foehn Warm, dry downslope windstorm that forms along the Alps. More generally, a strong warm downslope wind.

Fold See "tropopause fold."

Forward flank The side of a supercell thunderstorm that typically first approaches an observer located to the storm's northeast. The forward flank is typically associated with heavy rain, and in most supercells is on the northeast side of the storm.

Forward-flank downdraft Region of descending air in a supercell. Normally found on the northeast side of the thunderstorm and co-located with heavy precipitation.

Forward-flank gust front The leading edge of the cold air outflow emerging from the heavy rain region associated with the forward flank downdraft of a supercell thunderstorm.

Four Corners low An extratropical cyclone that develops over the region where Arizona, Utah, Colorado, and New Mexico meet.

Freezing Phase change of water whereby liquid changes to solid.

Freezing drizzle Light, misty precipitation droplets with diameters between 0.2 and 0.5 mm that freeze on contact with the surface.

Freezing rain Liquid precipitation that freezes on contact with the surface, often producing an ice glaze.

Friction layer See "boundary layer."

Frictional force Drag force acting in a direction opposite the motion of air; always acts to reduce the speed of the flow. Most important in the friction layer.

Front Boundary between airmasses of differing density (due to different temperatures and/or moisture). Classified on the basis of the thermal and moisture characteristics of the airmasses, the direction of movement of the airmasses, and whether the boundary between the airmasses is in contact with the ground or can only be found aloft.

Frontal overrunning The flow of warm moist air northward over the cool airmass north of a warm or stationary front, producing clouds and precipitation north of the front.

Frontal squall line A line of thunderstorms forming along a frontal boundary such as the cold front, dry line, or upper-level front of an extratropical cyclones.

Front to rear flow In a mesoscale convective system, the upper part of the storm system flow that moves away from the convective region toward the trailing stratiform region.

Frostbite A medical condition in which the affected part of the body is frozen.

Fujita scale Scale use to classify tornadoes based on damage. Damage assessments are used to estimate a rank between F0 (weakest) and F5 (strongest). Wind speeds estimated from the scale ranking. Also called "F-scale."

Funnel cloud A funnel-shaped cloud of condensation, usually extending from a deep convective cloud and associated with a violently rotating column of air that is not in contact with the ground. Not a tornado, although it may later descend and become a tornado.

General circulation The global scale pattern of winds, characterized by semi-permanent features such as the subtropical highs, the subpolar lows, the trade winds, and the jetstream.

Geostationary Operational Environmental Satellite (GOES) An operational satellite in geostationary orbit above longitudes that provide views of the United States. Numbered sequentially as GOES-10, GOES-11, etc. Also referred to as Geosynchronous Operational Environmental Satellite.

Geostationary orbit An orbit 35,800 km (22,300 miles) above the Earth's surface in which a satellite remains over a fixed point on the Earth's equator.

Geostrophic balance The balance between the pressure gradient force and the Coriolis force.

Geostrophic wind The wind that would exist if the pressure gradient and Coriolis forces were in balance.

Geosynchronous satellite A satellite in a circular orbit lying in the Earth's equatorial plane in which the satellite has the same rotational velocity as Earth. The satellite remains essentially motionless relative to a point on Earth's equator.

Glacier A mass of ice originating as snow that survives the summer melt season. Successive years of accumulation enable large masses of snow to be compressed into ice that can survive for many centuries.

Glaze A coating of ice resulting from freezing rain or freezing drizzle.

Glide slope The ideal path a landing aircraft follows during approach toward the runway.

Global warming A persistent increase of the worldwide average temperature. The term is generally used in reference to the effect of increasing concentrations of greenhouse gases.

GMT See "Greenwich Mean Time."

GOES See "Geostationary Operational Environmental Satellite."

Gradient The ratio of the change of a quantity and the distance over which the change occurs. Units are °C per km, mb per 100 km, etc.

Gradient wind balance The state of curved flow in which the residual between the pressure gradient and Coriolis forces exactly balances the centrifugal acceleration experienced by an air parcel.

Graupel A small (< 3–4 mm), soft ball of ice that results when an ice crystal collects supercooled water droplets, which freeze on the surface of the growing particle.

Gravitational force Attractive force between any two objects. Earth's attractive force acting on air parcels has essentially a constant magnitude throughout the atmosphere and is directed toward the center of Earth.

Greenhouse effect The absorption of upward infrared radiation, and the subsequent re-emission of infrared radiation, by gases such as water vapor, carbon dioxide, methane, and ozone. The downward re-emission of the radiation warms the surface of the Earth.

Greenhouse warming The result of an enhancement of downward-directed infrared radiation in the lower atmosphere as a consequence of increasing concentrations of carbon dioxide and other gases such as methane and ozone.

Greenwich Mean Time (GMT) A universal time referenced to the meridian of Greenwich, England (0° longitude). Also called Universal Coordinated Time (UTC) or Zulu time (Z).

Ground blizzard A blizzard caused solely by blowing snow. Can occur under otherwise clear conditions.

Ground clutter Radar return from the Earth's surface and objects on it.

Gulf Coast cyclone An extratropical low-pressure system that develops along the coast of the Gulf of Mexico, often near the Texas-Louisiana border. Most common during the winter months.

Gust front The leading edge of the outflow of a thunderstorm's rain-cooled downdraft air. Passage is often marked by a sudden increase of wind speed.

Gustnado Short-lived and weak tornado-like vortex that can develop along gust fronts.

Hadley Cell Dominant circulation in Earth's tropical atmosphere, consisting of rising air near the equator, poleward flow at high altitudes between the equator and 30° N and S, descending motion near 30° N and S, and low-level equatorward flow (trade winds) from 30° to the equator in each hemisphere.

Hail Frozen precipitation particles, with diameters generally exceeding 3 to 4 mm, resulting from the accretion of supercooled liquid by ice crystals and graupel. Formation requires the strong upward motion characteristic of thunderstorms.

Hail Alley The area in the central United States extending from Texas to the Dakotas where damaging hailstorms commonly occur.

Hail cascade The region of large hailstones on the flanks of the core updraft on the east and northeast side of a typical supercell thunderstorm.

Hail embryo An ice particle that serves as the nucleus of a hailstone.

Hailshaft The relatively narrow region of hail in a vertical cross-section of a thunderstorm. Detectable by radar as a region of high reflectivity and low differential reflectivity.

Hailstone A spherical or irregularly shaped lump of ice that forms in a thunderstorm containing a strong updraft and abundant amounts of supercooled water droplets. The supercooled liquid accumulates on small hail embryos, forming hailstones.

Hailstorm A thunderstorm producing hail that reaches the ground.

Hailstreak A small continuous region of hail coverage, typically 1 to 2 km in width and varying in length from a few to over a hundred kilometers.

Hailswath A wide, relatively long area of hailfall, generally consisting of more than one hailstreak.

Hawaiian Firehose The atmospheric flow pattern that brings heavy rainfall to California. Characterized by a branch of the jetstream flowing northeastward from the vicinity of Hawaii to California, bringing moisture from the tropics directly into the West Coast. Also called the "Pineapple Express."

Haze droplets Tiny particulates onto which sufficient water has condensed that the atmospheric visibility is reduced.

Heatburst An outrush of warm air near the surface from a thunderstorm downdraft containing little liquid water for most of its downward ascent, allowing adiabatic (compressional) warming to dominate the evaporational cooling.

Heat capacity The amount of heat required to increase the temperature of an object by a certain temperature interval. If the heat capacity is for a warming of one degree Celsius and for a unit mass, it is referred to as the specific heat.

Heat cramps Muscle cramps and spasms caused by exercise and excessive sweating in hot conditions. The mildest form of heat injury.

Heat exhaustion A mild form of heat stroke characterized by faintness, dizziness, and heavy sweating, often precipitated by prolonged exertion in hot, humid conditions.

Heat Index A measure of the effect on the human body of a combination of high air temperature and relative humidity. Same as "apparent temperature" if average radiative inputs and wind speeds are assumed.

Heating degree-day Calculated as 65 minus the average of a day's high and low Fahrenheit temperature; heating degree-days are used by power companies to estimate energy usage over a winter season.

Heat lightning Light from a distant flash of lightning that is scattered by air molecules, dust particles, or reflected from clouds and appears as a light flash in the sky overhead.

Heat stress A group of disorders due to overexposure to, or overexertion in, hot conditions; consequences include heat cramps, heat exhaustion, and heat stroke.

Heat stroke A medical condition in which the body temperature is dangerously elevated, leading to an impairment of the body's temperature-regulating abilities and causing the body's internal organs to begin shutting down; characterized by cessation of sweating, severe headache, high fever, hot dry skin, and, in some cases, collapse and coma.

Heat wave A period during the warm season in which temperatures are substantially above normal for a particular area.

Helicity A measure of the vertical wind shear in the atmosphere. Large values favor thunderstorm rotation and supercell formation.

High precipitation supercell A supercell thunderstorm with a heavy precipitation core that wraps around the mesocyclone; the large rainshaft reaching the ground is visually prominent.

High-pressure center A location at which the pressure exceeds the pressure at all surrounding points; generally associated with subsiding air and clear skies.

Hodograph A circular diagram used to show the vertical variation of wind speed and direction with altitude.

Hook echo The characteristic curved-shape signature of a tornadic supercell on a radar reflectivity image. A tornado is often located near the center of the hook.

Horizontal wind shear A change in wind speed in the horizontal, divided by the distance over which the change in wind occurs.

Hurricane A strong tropical cyclone over the Atlantic or Eastern Pacific Oceans in which sustained wind speeds reach 64 knots (74 mph) or higher.

Hydraulic jump An unusually turbulent and abrupt change in the velocity of air downwind of a mountain range caused by atmospheric conditions upstream of the flow.

Hydrological drought An unusual deficiency of groundwater and/or streamflow for a particular area. The below-normal water levels result from a deficit of precipitation over a period of several weeks to several seasons, or even longer.

Hydrometeorological Prediction Center The National Center for Environmental Prediction responsible for issuing precipitation forecasts focused on heavy rain, heavy snow, and areas of potential flash flooding.

Hydrostatic balance The balance between the upward pressure gradient force and the downward force of gravity.

Hypothermia A medical condition in which a person's body temperature is lowered to the point where it can be life threatening.

Ice age A period of expanded continental ice sheets, polar ice sheets, and mountain glaciers; ice ages alternate with intervening periods when the Earth has no permanent ice sheets. Durations of past ice ages have been several tens to hundreds of million years.

Ice nuclei Particles that promote the formation of ice crystals in air or in liquid water. Particles generally have a crystalline structure similar to ice.

Ice pellets Liquid drops that have frozen or refrozen. Often referred to as "sleet."

Ice storm A winter storm in which there is a substantial accumulation of freezing rain or freezing drizzle at the surface.

Icelandic low A semi-permanent surface low-pressure center located over the subpolar North Atlantic Ocean, generally centered near Iceland.

Image charge During a thunderstorm, the positive (negative) charge region on the ground that lies beneath the negative (positive) region of the thunderstorm.

In-cloud lightning An electrical discharge within a cloud that does not extend to the Earth's surface.

Induction charging A feedback mechanism in which the atmosphere's background electric field causes positive and negative charge to migrate to opposite ends of solid or liquid particles, enabling the transfer of charge during collisions of particles.

Infrared channel A specific band of frequency of infrared radiation observed by weather satellites. The intensity of radiation in this frequency band is related to the temperature of the underlying surfaces and cloud tops.

Infrared energy Radiation emitted by objects at temperatures typically found in the atmosphere and at the Earth's surface; infrared radiation has longer wavelengths than visible light and is not detectable by the human eye.

Infrared radiation Electromagnetic radiation emitted by the Earth and the atmosphere. Wavelengths of this radiation are generally longer than those that are visible to the human eye. As objects warm, they emit more infrared radiation.

Initialization The process by which a numerical weather prediction model incorporates the most recent observations and balances the fields of different variables, providing a smooth transition to the computation of the future state of the atmosphere.

Inland flooding Flooding that occurs during hurricanes and other coastal storms, but at a distance of tens to hundreds of miles from the coastline.

Insulator A material that does not conduct electricity.

Interface charging The transfer of electrical charge from one conducting object to another when the conductors are brought into contact. Electrons flow across the interface between the two conductors.

Intertropical Convergence Zone (ITCZ) A zone of convergence of winds near the surface in the vicinity of the equator. The converging winds are the low-level branches of the Hadley Cells of the Northern and Southern Hemispheres.

Inversion An increase of temperature with altitude. An inversion represents the opposite of the more common tropospheric situation in which temperature decreases with altitude.

Inversion cap An inversion at the top of a layer in which vertical mixing of air is occurring. The inversion prevents air parcels from rising farther, thereby limiting the vertical extent of the mixing of the air.

Inverted-V sounding A sounding characterized by a downward increase of the difference between the temperature and the dewpoint, representing an environment that is increasingly dry toward the surface.

Ion A charged atom or molecule, or an electron.

Ionized Containing ions.

Ionosphere The region of the upper atmosphere (above 60 to 100 km) that has a high concentration of ions.

Isobar A contour connecting points of equal pressure.

Isodrosotherm A contour connecting points of equal dewpoint temperature.

Isotach A contour connecting points of equal wind speed.

Isotherm A contour connecting points of equal temperature.

ITCZ See "Intertropical Convergence Zone."

Jet (electrical) See "Blue Jet."

Jet (wind) A region of exceptionally strong winds in the atmosphere.

Jetstreak A region of exceptionally strong winds within the jetstream. Convergence and divergence associated with jetstreaks directly influence the development of cyclones.

Jetstreak effect The creation of divergence and convergence in the upper troposphere due to the imbalance between the pressure gradient and Coriolis forces as air moves through the entrance or exit region of a jetstreak.

Jetstream A band of strong winds encircling the Earth in middle latitudes. Generally strongest in the upper troposphere.

Katabatic wind Downslope wind caused by gravitational drainage of very cold, dense air. Most common at the edges of ice sheets (Antarctica, Greenland) and large glaciers.

K Index A measure of atmospheric stability used to assess the likelihood of non-severe convective precipitation. Based on temperature and dewpoint data from 850, 700, and 500 mb.

Kelvin scale A temperature scale in which the melting point of ice is set at 273.15, the boiling point of water at sea-level pressure is set at 373.15, and zero represents the lowest possible energy state of a substance.

Knot A unit of wind speed, numerically equivalent to 1.15 mph.

La Niña Condition characterized by colder-than-normal surface waters in the eastern equatorial Pacific and by a strengthening of the Walker Cell and the trade winds. The phase of the Southern Oscillation opposite to the El Niño phase.

Lake-effect snow Snow falling over and immediately downstream of water bodies such as the Great Lakes. Triggered by the flow of very cold air over relatively warm lake water, which supplies moisture and destabilizing heat to the atmosphere. Most common during late autumn and early winter.

Lake-enhanced snow The augmentation of an extratropical cyclone's snowfall by the gain of heat and moisture during the air's trajectory over a large lake; the total snowfall is greater downwind than upwind of the lake.

Landspouts Short-lived and relatively weak non-supercell tornadoes that sometimes develop within squall line thunderstorms aligned along fronts.

Lapse rate The rate at which temperature decreases with altitude. (Note: If the temperature increases with altitude, the lapse rate is negative.)

Latent heat The energy required for, or released by, a transition between two phases (solid, liquid, gaseous) of a substance.

Lead A crack in ice cover on a large water body where the water beneath the ice is exposed.

Lee trough An area of low pressure east of the Rocky Mountains, formed in part by the warming of air as westerly airflow descends the east side of the mountains. Can also form downstream of other mountain ranges.

Lee wave Waves in airflow created as air approaches a mountain range, rises on the windward side, descends on the leeward (downslope) side, and continues to oscillate up and down as the air moves downstream of the mountain. Waves form downstream of the mountains, especially if the air is stable, and are often identifiable by lens-shaped clouds at the crests of the waves.

Leeward The downwind side of a mountain or lake. In the case of a mountain range, the leeward side is characterized by downslope airflow.

Lenticular Lens-shaped cloud forming at the crests of lee waves.

Levee An embankment created near a river to prevent the flow of water onto a floodplain.

Level of free convection The altitude at which a vertically displaced air parcel first becomes buoyant, i.e., the elevation at which the parcel's temperature first exceeds the surrounding environment's temperature.

Lift (aircraft) An upward force exerted on an aircraft, arising from the aircraft-induced downwash of air and the corresponding upward pressure force on the plane.

Lifted Index A measure of atmospheric stability, defined as the difference between the temperature of the environment at 500 mb and an air parcel lifted from near the surface to 500 mb. Negative values indicate the potential for strong ascent, showers, and thunderstorms.

Lifting condensation level The altitude at which a lifted air parcel's relative humidity first reaches 100 percent. Corresponds to the altitude of the cloud base.

Lifting mechanism A process that causes air parcels to rise in the atmosphere. Examples include the ascent of air at a frontal boundary, airflow over mountains, convergent winds near the surface, and the heating of low-level air to create buoyancy.

Lightning An electrical discharge in the atmosphere, representing a rapid flow of electrical charge between a cloud and the ground, between two clouds, or between two portions of the same cloud.

Long range outlook A thirty day or longer forecast issued by the National Weather Service for above or below normal conditions such as temperature or precipitation.

Long wave A wave appearing in the flow on a map such as a 500 or 300 mb map that has a large wavelength (distance from ridge crest to ridge crest) and generally moves slowly if at all through the flow.

Low earth orbit A satellite orbit that is several hundred to several thousand kilometers above the Earth.

Low-level jet A band of strong winds in the lower troposphere, typically strongest within the 700 to 900 mb layer. Most common during spring and summer in the Great Plains, extending from the Gulf Coast to the north-central United States.

Low level wind shear alert system A grid of anemometers deployed at airports to detect wind shear events.

Low-precipitation (LP) supercell A supercell thunderstorm that produces little precipitation. Most common in the vicinity of dry lines in the Southern and High Plains of the United States.

Low-pressure center A location at which the pressure is less than the pressure at all surrounding points; generally associated with rising air, clouds, and precipitation.

Mammatus Downward protrusions of cloud at the base of the anvil of a thunderstorm, formed as evaporation of particles in the anvil results in pockets of cool air that descend.

Marine Forecast Center The National Center for Environmental Prediction that issues forecasts and warnings for marine interests.

Maritime airmass An airmass that has formed over the oceans and hence is relatively moist.

Maritime climate A climate strongly influenced by the ocean; characterized by relatively small seasonal changes and by relatively high humidity.

Mature stage The phase in a tornado lifetime when it is at its greatest size and intensity.

MCS See "Mesoscale Convective System."

MCV See "Mesoscale Convective Vortex."

Mean sea level Average elevation of the surface of the world's oceans. Used as a standard level to convert station pressure to a common altitude.

Mean sea-level pressure The atmospheric pressure after conversion from the value measured at a station to a corresponding value for mean sea level.

Mechanical turbulence Rapid variations of air motion that develop when moving air encounters obstructions associated with ground roughness.

Medieval warm period A period of unusually warm climate lasting from about the 10th to the 14th century A.D. The warmth is most clearly documented in the North Atlantic and European regions.

Medium range model A numerical forecast model that predicts weather in the range of 7 to 15 days.

Megadrought A prolonged dry spell lasting several decades.

Melting The process by which a substance changes from its solid phase to its liquid phase.

Melting level The altitude at which falling precipitation changes from the ice phase to liquid drops.

Melting process The formation of freezing precipitation when snowflakes fall into an atmospheric layer in which the temperature exceeds 0° C, and continue their fall back into a sub-freezing (< 0° C) layer of air near the ground.

Mesocyclone Cyclonic circulation within the updraft region of a supercell thunderstorm. Normally coincides with the region where tornadoes form. Typically several kilometers in diameter.

Mesocyclone signature The Doppler radar radial velocity pattern of a mesocyclone. Appears as a small couplet of pixels indicating strong inbound velocities adjacent to pixels indicating strong outbound velocities.

Mesoscale Reference size of weather systems with horizontal dimensions ranging from several kilometers to several hundred kilometers.

Mesoscale Convective System (MCS) A weather system the size of a large state that occurs in connection with an ensemble of thunderstorms. Commonly consists of a leading squall line, a trailing stratiform region, a cold pool beneath the precipitation, a rear inflow jet, and front-to-rear flow aloft.

Mesoscale Convective Vortex (MCV) A mid-level vortex that forms in a mesoscale convective system (MCS) and often remains after the convection of the MCS dissipates.

Mesosphere The layer of the atmosphere above the stratosphere between approximately 50 km and about 85 km. Characterized by a decrease of temperature with increasing altitude.

Mesovortex In a mesoscale convective system, a rotational circulation in the middle troposphere that develops in the trailing stratiform region as the system matures. The vortices are often several hundred kilometers in diameter.

METAR See "Meteorological Aviation Report."

Meteogram A diagram that shows how several atmospheric properties change with time. Typically used to depict variations over periods of several hours to several days.

Meteorological Aviation Report (METAR) A coded report of weather conditions measured at an airport. The data are normally from automated surface or weather observing systems.

Meteorological bomb A cyclone in which the central pressure decreases by at least 24 mb in a 24-hour period.

Meteorological drought An unusually long period during which precipitation is below normal for a particular area.

Microburst An intense downdraft, or downburst, occurring in a shower or thunderstorm. The term "microburst" is sometimes reserved for downbursts having diameters smaller than about 4 km.

Middle latitudes Latitudes between 30 and 60 degrees.

Milankovich cycles Periodic variations in the Earth's orbital parameters, including the eccentricity of the Earth's orbit, the tilt of the Earth's axis, and the precession of the Earth's axis (which determines the season of the shortest Earth-sun distance). The periods of these cycles are tens of thousands to hundreds of thousands of years. The cycles appear to be associated with climate variations, including the alternation of glacial and interglacial intervals.

Millibar A unit of atmospheric pressure.

Mini Bow A small bow echo associated with squall line thunderstorms, typically of length less than 20 to 30 km.

Model domain The geographical area covered by a numerical model simulation.

Model grid The array of points or locations for which a numerical model computes the variables by its time-stepping procedure. In many models, the grid points are spaced at regular intervals of distance or latitude and longitude.

Model Output Statistics (MOS) A forecast technique in which forecast variables at cities, airports, and other locations are determined through statistical evaluation of the output from numerical forecast models.

Model resolution The distance between the grid points used by a numerical model. Weather features having dimensions smaller than the model resolution cannot be represented explicitly by model.

Model timestep The time interval over which rates of change of weather variables are assumed to remain constant in the "marching" procedure used by numerical models. The timestep is the interval between model's successive estimates of the variables.

Moist adiabatic lapse rate The rate (°C per km) at which a saturated air parcel will cool as a result of expansion during its ascent if there is no exchange of heat with the surrounding environment. The rate is variable, typically about 6° C per km at the surface and >6° C per km higher in the atmosphere.

Monsoon A seasonally-changing wind circulation. The large-scale circulation pattern in summer is one in which the airflow is onshore at low levels and upward over land, producing extensive cloudiness and rainfall. In winter, the air descends over the land and flows offshore.

MOS See "Model Output Statistics."

Mountain wave A wave in the airflow generated by flow over a mountain.

Multiple Doppler radar A technique of using the radial velocity measurements from two or more radars to deduce the three-dimensional wind field at high resolution in storms.

National Centers for Environmental Prediction (NCEP) A group of nine specialized centers of the National Weather Service that analyze and forecast the global atmosphere and maintain the overall national warning system for severe and hazardous weather.

National Lightning Detection Network (NLDN) A network of over 100 remote, ground-based sensing stations that monitor cloud-to-ground lightning activity across the United States.

Natural variability Variations of climate that cannot be associated directly with external driving factors such as changes of solar radiation, greenhouse gases, volcanoes, and other aerosols. Natural variations are a manifestation of the chaotic behavior of the atmosphere.

NCEP See "National Centers for Environmental Prediction."

NDVI See "Normalized Difference Vegetation Index."

Negative charge An excess of electrons relative to the positively charged protons in a gas, liquid, or solid.

Neutral charge A state of balance between negative and positive charge; in such a state, the number of electrons is equal to the number of protons.

Neutral stability The condition in which a parcel of air, when displaced vertically, will remain in its new position.

Neutron A particle in the nucleus of an atom with a mass of one atomic unit and no electrical charge.

Nimbostratus A widespread layered cloud from which rain is falling.

Nimbus The Latin word for "rain."

NLDN See "National Lightning Detection Network."

Non-inductive charging See "interface charging."

Non-supercell tornado A tornado that forms along a squall line or shear zone rather than within a supercell thunderstorm.

Nor'easter An intense extratropical cyclone offshore of the northeastern United States in a location such that it produces strong northeasterly winds along the coast.

Normalized Difference Vegetation Index (NDVI) A satellite-measured parameter that is an indicator of the level of photosynthetic activity (greenness) in the vegetation from which upwelling radiation originates. The parameter is proportional to the difference between the radiation in two different wavelengths measured by the satellite.

North American monsoon The summertime semi-permanent low-pressure center, moisture inflow, and associated precipitation over the southwestern United States, occurring in response to the strong solar heating of the Desert Southwest during the summer months.

Nor'wester An extratropical cyclone northeast of New England in a location such that it produces strong northwesterly winds over New England. Similar to a Nor'easter, except the storm's more northerly position gives the winds a westerly component over New England.

Numerical model A system of mathematical equations that describe the behavior of the atmosphere by representing the major physical and dynamical processes. The model can be used for weather prediction or for research experiments aimed at improved understanding of atmospheric phenomena.

Numerical weather prediction Predicting future weather by the use of atmospheric numerical models run on a computer.

Occluded front The airmass boundary that occurs when the air behind a cold front comes in direct contact with the cool air north or east of a warm front. Develops in the latter stages of an extratropical cyclone's life cycle when the cold front catches up to the warm front.

Occlusion downdraft A downdraft that forms near the core of the mesocyclone in a supercell thunderstorm.

Ordinary thunderstorm A thunderstorm that does not become severe or become part of an organized thunderstorm complex; often referred to as an airmass thunderstorm.

Organizing stage The phase of development of a tornado where the funnel cloud descends to the ground and increases intensity.

Orographic clouds Clouds that result from processes occurring over mountains, where large-scale upslope flow and solar heating of mountain slopes can enhance upward motion of air.

Orographic lifting The forced ascent of air that occurs on the windward (upslope) side of mountains.

Orography Mountains or other hilly features of a landscape.

Overrunning The gradual upward glide of warm moist air when it encounters a denser cold airmass at a frontal boundary. The upward motion of the warm air above the cold airmass leads to clouds and precipitation.

Overshooting top Intrusion of a thunderstorm's updraft and its condensed (ice) particles into the stratosphere as a result of the air's upward momentum.

Ozone A gas in which each molecule consists of three oxygen atoms. The largest natural concentrations are found in the stratosphere.

Pacific high The semi-permanent surface high-pressure center located over the North Pacific Ocean. Positioned in subtropical latitudes during winter and middle latitudes during summer.

Palmer Drought Severity Index A measure of moisture deficiency standardized to local conditions. Based on a groundwater balance that includes precipitation, evaporation, recharge by subsurface water, and runoff.

Parameterize In numerical modeling, the mathematical representation of a complex physical process by simple equations that can be solved using the model's predicted variables.

Phase change The change of state of a substance, such as water to ice, or water to vapor.

Phased array antenna A radar antenna that consists of an array of cables and is steered by slightly delaying the transmission from one side of the antenna to the other. Typical antenna used with profilers.

Pineapple Express The atmospheric flow pattern that brings heavy rainfall to California. Characterized by a branch of the jetstream flowing northeastward from the vicinity of Hawaii to California, bringing moisture from the tropics directly into the West Coast. Also called the "Hawaiian Firehose."

Pleistocene A period of the Earth's geological history beginning about two to three million years ago, characterized by cycles of extensive glaciation and interglacial ice retreats.

Polar airmass A large mass of cold, dry air that develops over the northern part of the North American continent. The Eurasian counterpart forms over Siberia.

Polar amplification The tendency for climatic warming and cooling to be greater in the polar regions than in middle or lower latitudes.

Polar front jetstream The band of strong winds found above the polar front and encircling the middle or high latitudes of each hemisphere in a wavelike pattern; most prominent during the winter months, although it is found in all seasons.

Polar jetstream A band of strong winds found in the upper troposphere above the boundary between polar air and warmer air of tropical or maritime origin. When more than one jetstream is present, the polar jetstream is generally farthest north.

Polarization The orientation of the electric field in electromagnetic radiation.

Polarization diversity radar A special type of radar capable of distinguishing regions of hail from regions of heavy rain. Uses radar beam in which the orientation of the electric field, or polarization, alternates between the horizontal and vertical planes.

Positive charge A deficit of electrons relative to the positively charged protons in a gas, liquid, or solid.

Positive polarity lightning Lightning in which positive charge is transferred from the cloud to the ground, typically between the anvil region of a thunderstorm and the ground.

Powder snow avalanche An avalanche consisting of loose snow that blows high in the air as the snow cascades down the slope.

Precession The variation of the tilt of the Earth's axis with a cycle of about 22,000 years; determines the season in which the Earth is closest to the sun.

Precipitation mode Mode of operation of Doppler radar used when there is significant precipitation in the region.

Precipitation rate The amount of precipitation (inches, centimeters) falling in a unit of time (minute, hour, day).

Pressure The force applied by air on a unit area of surface. Equivalent to the weight of a column of air above a unit area.

Pressure gradient A horizontal or vertical change in pressure divided by the distance over which the change occurs.

Pressure gradient force The force applied to a small parcel of air due to the variation of pressure over a small distance around the parcel.

Pressure surface An imaginary surface in the atmosphere consisting of all points that have the same atmospheric pressure.

Prog A map showing a forecast of one or more weather variables.

Projection A prediction that is contingent on the correctness of an underlying assumption; a projection of climate is a prediction contingent on the assumed scenario of future increases of greenhouse gas concentrations.

Proton A particle in the nucleus of an atom with a mass of one atomic unit and a positive electrical charge of one electrostatic unit.

Proxy data Data that paleoclimatologists gather from natural recorders of climate variability, e.g., tree rings, ice cores, fossil pollen, ocean sediments, coral, and historical data; proxy data are distinguished from instrumental measurements, which provide much of the climate information from the past 100 to 200 years.

Radar A device that transmits pulses of microwave energy. The amount and distribution of the energy returned to the radar provide information about precipitation and wind in the surrounding area.

Radar echo A region of enhanced signal strength in a radar display, indicating enhanced reflection of the radar beam by precipitation particles.

Radar reflectivity (also called Radar reflectivity factor) The intensity of the energy scattered back to a radar by raindrops, ice particles, and other objects. Measured on a logarithmic scale.

Radial velocity The component of wind along a line directed radially outward from a central point, e.g., the location of a radar. By convention, winds directed away from (toward) the central point are assigned positive (negative) signs.

Radiation Energy in the form of electromagnetic waves. Solar radiation is energy from the sun, much of it in the form of visible light, while terrestrial radiation is energy emitted by the Earth and atmosphere, nearly all as infrared energy.

Radiational cooling The emission of thermal energy by a substance, resulting in a decrease of the temperature of the substance. The primary mechanism by which the ground cools at night when incoming solar radiation is absent.

Radio Acoustic Sounding System (RASS) An instrument that provides information about the atmosphere's vertical temperature profile by transmitting sound waves vertically and measuring the rate at which they recede from the transmitter. This rate, the speed of sound, is related to temperature.

Rain Droplets that fall from a cloud to the ground. Light rain consisting of very small droplets is called drizzle.

Rain free base In a supercell thunderstorm, a non-precipitating cloud base beneath the strong updraft.

Rainshaft The region of heavy rain in a thunderstorm. Detected on radar as a narrow region of strong reflectivity.

Rain-snow line The boundary in a winter storm between precipitation reaching the ground as rain and precipitation reaching the ground as snow.

RASS See "Radio Acoustic Sounding System."

Rawinsonde A balloon-borne instrumentation system that measures pressure, temperature, dewpoint temperature, wind direction, and wind speed at altitudes between the surface and the lower stratosphere.

Rear flank The side of a supercell thunderstorm that is farthest from an observer as a supercell approaches. The rear flank is typically associated with lighter rain, and in most supercells is on the southwest side of the storm.

Rear-flank downdraft Downdraft that develops on the rear (typically southwest) side of a supercell thunderstorm.

Rear flank gust front The leading edge of the cold air outflow emerging from the region of evaporating cloud and rain associated with the rear-flank downdraft of a supercell thunderstorm.

Rear flanking line A line of cumulus clouds that often develops along the rear flank gust front in a supercell thunderstorm.

Rear inflow jet In a mesoscale convective system, a descending airstream in the middle troposphere that flows toward the convective line.

Receiver The component of a radar that converts microwaves collected by the antenna into an electrical signal.

Relative humidity The ratio of the amount of water vapor in the atmosphere to the atmosphere's capacity for moisture at a given temperature (expressed as the vapor pressure divided by the saturation vapor pressure).

Relative vorticity Vorticity associated with local rotation within a fluid.

Research model A computer code that incorporates the equations governing atmosphere processes and is used to simulate atmospheric phenomena for research purposes.

Residence time The time that particles or gases remain in the atmosphere after they are first introduced by natural or human processes.

Return stroke The powerful current flow associated with lightning that immediately follows the joining of a descending stepped leader with an upward traveling spark of opposite charge from the ground.

Ridge An elongated area of high atmospheric pressure, normally accompanied by anticyclonic flow in the Northern Hemisphere.

Ridge axis The line along which the pressure locally reaches its maximum values within a ridge.

River stage The height of the water surface of a river above a standard measurement point on the bottom of the river channel.

Rocky Mountain cyclone An extratropical cyclone that develops on the plains just east of the Rocky Mountains.

Roll cloud A low-level, horizontal, tube-shaped arcus cloud associated with air rising over a gust front.

Rotational velocity That part of an air parcel's total velocity that is associated with rotation about a central point in the flow.

Rotor A turbulent vortex that forms downwind of a mountain range during downslope windstorms.

Runaway vortex rolls In a downburst, regions of strong outflow and associated curl that maintain strength over a longer distance than adjacent regions of outflow.

Saffir-Simpson scale A scale that relates a hurricane's minimum sea-level pressure and maximum wind speed to the coastal flooding and damage that a hurricane can inflict as it makes landfall. Ranges from 1 (weakest) to 5 (strongest).

Santa Ana wind A warm, dry, downslope wind that occurs when air flows westward along the slopes of the mountains east of the Los Angeles basin in southern California.

Satellite An object in orbit around a planet. Satellites launched into orbit around Earth (and other planets) carry instruments to study the properties of the planet's atmosphere and surface.

Saturated Air that contains the maximum amount of water vapor for a given temperature and pressure. Has a relative humidity of 100 percent.

Saturation A state of the atmosphere where the amount of water vapor it contains is equal to its capacity for water vapor. When air is saturated, the relative humidity is 100 percent.

Saturation vapor pressure The vapor pressure at which air is saturated at a given temperature.

Scattered The process by which radiation is absorbed by a molecule and then reradiated in a random direction.

Screening layer Thin layer along the boundaries of thunderstorm clouds that has the opposite charge of the interior of the cloud.

Scud clouds Generic name for clusters of residual small cumulus clouds in the vicinity of a larger storm. Used to describe clouds at the base of the eye of a hurricane.

Sea breeze A circulation that develops along shore lines during daytime in which warm air rises over land and moves shoreward aloft, while cool air offshore descends and moves onshore to replace the warm air.

Sea breeze front The leading edge of local coastal wind that blows from the ocean over the shore, primarily during daytime.

Sea-level pressure The estimated pressure that would occur at the elevation of mean sea level. Sea-level pressure is estimated using surface pressure and temperature data at stations located above or below sea level and a standard lapse rate.

Secondary circulations Air circulations induced by force imbalances in the primary circulation of a weather system. For example, in a hurricane, the air rotating about the center of the storm is the primary circulation, while the secondary circulation consists of air converging toward the center at low levels, rising in the eyewall, and diverging outward at the tropopause.

Semi-permanent high A high-pressure system that is persistent over an entire season, such as the Bermuda high over the North Atlantic, and the Pacific High over the North Pacific, both which occur in summer.

Semi-permanent low A low-pressure system that is persistent over an entire season, such as the Icelandic low over the North Atlantic, and the Aleutian low over the North Pacific, both which occur in winter.

Sensible heat Heat energy that can be transferred directly from one substance to another without involving a change of phase of water (i.e., without evaporation or melting).

Shear-induced turbulence Turbulence created in the atmosphere under conditions where winds increase rapidly with altitude.

Sheet lightning The visual phenomena observed when lightning occurs within or behind a cloud and illuminates the exterior of the cloud uniformly, giving the appearance of a sheet of light.

Shelf cloud A low-level, wedge-shaped arcus cloud associated with air rising over a gust front.

Shooting flow The accelerated flow along a mountain slope during a downslope windstorm.

Shore-parallel band A single precipitation band that aligns approximately parallel to the shoreline along the long axis of a lake in lake-effect storms over the Great Lakes.

Short range model A numerical forecast model that is designed to produce forecasts in the 0- to 3-day range.

Short wave A wave appearing in the flow on a map such as a 500 or 300 mb map that has a small wavelength (distance from ridge crest to ridge crest) and generally moves quickly through the flow.

Showalter index A measure of thunderstorm potential and severity. Especially useful when a shallow, cool layer of air below 850 mb conceals greater convective potential above.

Showers Intermittent rain from convective clouds that may range in intensity from very light to heavy.

Shrinking stage The stage in a tornado's lifetime when the tornado stretches into a narrow rope-like formation.

Skew-T/Log P diagram A standard diagram used to plot atmospheric soundings and determine atmospheric stability. Pressure on the vertical scale of the diagram is logarithmic and temperature is skewed upward to the right side of the diagram.

Slab avalanche An avalanche consisting of dense packed snow that slides down a mountain slope in large slabs.

Sleet Precipitation consisting of frozen raindrops.

Snow Precipitation consisting of ice crystals or aggregates of ice crystals.

Snow roller A rolled up large snowball that develops as wind blows across a field of snow.

Snow squall Heavy snow accompanied by wind. Usually short-lived or intermittent.

Snowbelt A region that receives exceptionally heavy snow each year, such as the regions on the downwind shores of the Great Lakes.

Snowflake A loosely packed collection of individual ice crystals. Also called an aggregate.

Socioeconomic drought A drought in which the moisture shortage is sufficiently large that it affects the supply of, or the demand for, economic goods.

Solar energy Energy in the form of radiation that comes from the sun.

Sounding The measurement of a vertical profile of atmospheric properties, typically made with a rawinsonde.

Southern Oscillation The rise and fall of sea-level pressure that occurs on opposite sides of the equatorial Pacific Ocean over a period of several years. Associated with El Niño and La Niña events.

Southern Oscillation Index An index used to determine the phase of the Southern Oscillation. Found by determining the difference between the deviations from normal of mean sea-level pressure at Tahiti and Darwin, Australia.

Specific heat The amount of energy required to raise a unit mass (e.g., one gram) of a substance one degree Celsius.

Spiral rainband An outer rainband within a tropical cyclone that normally assumes the shape of a spiral emanating out from the center of the cyclone.

Sprite A weak emission of light that appears directly above an active thunderstorm and is normally coincident with a cloud-to-ground lightning flash. Predominant color is red.

Squall line A long line of thunderstorms in which adjacent thunderstorm cells are so close together that the heavy precipitation from the cells falls in a nearly continuous line.

St. Elmo's fire An electrical discharge from objects subjected to the strong electrical fields present in the vicinity of thunderstorms.

Stable The condition in which a parcel of air, displaced vertically, will return to its original position.

Stagnation cone At the center of a microburst, the region near the Earth's surface characterized by near-zero wind speed.

Station model The standard depiction of data from a weather station on a weather map.

Stationary front A boundary between airmasses where the colder airmass is neither advancing nor retreating at the surface.

Steering flow The flow in the middle to upper troposphere that has the strongest influence on thunderstorm motion.

Stepped leader The first stage of a lightning stroke in which an initial flow of electrons proceeds from the base of a thunderstorm toward the ground in a series of rapid steps.

Storm Prediction Center The National Center for Atmospheric Prediction responsible for monitoring and forecasting severe and non-severe thunderstorms, tornadoes, and other hazardous convective weather phenomena across the Unites States.

Storm relative helicity An index that measures the vertical wind shear in the lower atmosphere relative to the motion of a thunderstorm. High values mean that there is strong shear in the lower atmosphere, which favors storm rotation and supercell formation.

Storm relative radial velocity On a Doppler radar display, the radial velocity adjusted to remove the component of the storm motion toward or away from the radar site.

Storm spotter A trained person who watches for tornadoes or other dangerous phenomena in thunderstorms and then reports observations back to the National Weather Service or Emergency Management offices.

Storm surge The abnormal rise in sea level and subsequent coastal flooding that occurs as a hurricane or intense extratropical cyclone approaches and crosses a coastline.

Straight-line winds In thunderstorms and downbursts, the strong to severe winds that flow outward in a straight line from the storm's downdraft region.

Stratiform region A cloudy region in an organized weather system, such as a mesoscale convective system, where updrafts are very weak or non-existent.

Stratosphere The layer of air between approximately 10 and 50 km (above the troposphere and below the mesosphere) in which temperature generally increases with altitude.

Stratus Low, generally widespread, layered clouds with a relatively uniform base.

Streamline A line that is parallel to the airflow at an instant in time.

Stretching The process where a column of air becomes narrower and taller. Stretching leads to an intensification of rotation if the column of air is initially rotating before the stretching occurs.

Stuve Diagram A standard diagram used to plot atmospheric soundings and determine atmospheric stability. Temperature is on the horizontal axis of the diagram and pressure, scaled logarithmically, is on the vertical scale.

Subgeostrophic Wind speed that is less than its geostrophic value as determined from the pressure gradient and latitude.

Sublimation The process by which ice is converted directly to vapor without passing through the water phase.

Subpolar low A high latitude low-pressure system that is south of the primary polar region.

Subsidence The slow sinking of air.

Subsidence inversion A temperature inversion created at the base of an airmass aloft caused by the airmass sinking and undergoing adiabatic compression and warming.

Subtropical front A temperature contrast in the upper troposphere that develops aloft where the Hadley Cell circulation meets cooler air in the middle latitudes. The subtropical front is associated with the subtropical jetstream.

Subtropical high-pressure system Large semi-permanent high-pressure systems that develop over the subtropical oceans and are most prominent in the summer season.

Subtropical jetstream A band of strong winds typically found between 20° and 35° latitude. The maximum speeds are typically found at altitudes between 12 and 14 km.

Suction vortices Small, extremely violent, rotating air columns embedded within the rotating circulation of large tornadoes.

Sunburn Blistering and reddening of the skin associated with overexposure to the sun's radiation.

Sunstroke A loss of consciousness associated with heat exhaustion due to long exposure to the sun.

Sun synchronous A low Earth satellite orbit designed so that the satellite passes over the same latitude at the same local time (e.g., local noon).

Supercell thunderstorm A large rotating thunderstorm in which the updraft and downdraft circulations maintain the storm's structure for a long time period. Supercells often produce hail, strong straight-line winds, and tornadoes.

Supercool The process by which water is cooled to a temperature lower than 0° C but remains in the liquid state.

Supercooled droplets Cloud droplets that have a temperature cooler than 0° C but remain in a liquid state.

Supercooled warm rain process The process by which tiny cloud droplets in a subfreezing (< 0° C) cloud grow to precipitation size by colliding and coalescing with each other.

Supercooled water Water that has a temperature below 0° C but is in a liquid state.

Supergeostrophic Wind speed that is greater than its geostrophic value as determined from the pressure gradient and latitude.

Surface drifter A buoy that is free to drift with ocean currents or on an ice floe.

SWEAT index The Severe Weather Threat Index, a stability index that is a weighted sum of the 850 mb temperature, the "Total Totals" index, the wind speeds at 850 mb and 500 mb, and the change in wind direction between 850 mb and 500 mb.

Swell A general rise in the sea surface, normally associated with large waves, but also under the center of low pressure in a hurricane.

Synoptic Simultaneous. In meteorology, refers to standard measurements such as rawinsondes that are launched simultaneously worldwide to document and provide an overview of weather occurring on scales comparable to extratropical cyclones.

Synoptic meteorology The study and analysis of weather occurring on scales comparable to extratropical cyclones and waves in the jetstream.

Temperature A measure of the average speed of molecules in a fluid or vibration of molecules in a solid. Commonly, the hotness or coldness of a substance as measured with a thermometer.

Temperature gradient A change in temperature over a horizontal or vertical distance, divided by the distance over which the temperature change occurs.

Temperature inversion A region of the atmosphere where temperature increases with increasing altitude.

Terminal Doppler Weather Radar Doppler radar deployed near an airport specifically to observe wind shear and microburst events that threaten aircraft takeoffs and landings.

Thermal turbulence Turbulence created in the atmosphere by the ascent of buoyant air parcels and subsequent mixing of the parcels with environmental air.

Thermistor An electronic device used to measure temperature.

Thermodynamic process A process involving heating or cooling of the atmosphere.

Thermometer An instrument used to measure temperature, often constructed by partially filling a sealed glass tube with a liquid such as alcohol.

Thermosphere The atmospheric layer above the mesosphere where temperature increases rapidly with height.

Thunder The booming sound created by the rapid expansion of air within a lightning channel.

Thunderstorm A cumulonimbus cloud that produces lightning and thunder.

Timestep In numerical forecast or research models, the time interval between two successive calculated states of the modeled atmosphere.

Tilting Process by which an updraft converts vertical wind shear into horizontal wind shear, inducing rotation in thunderstorms.

TOGA-TAO array An array of buoys in the tropical Pacific Ocean used to monitor El Niño and La Niña.

Tornado A violently rotating column of air, in contact with the ground. Tornadoes are pendant from or directly underneath cumulonimbus clouds and often, but not always, visible as funnel clouds.

Tornado Alley A region of high frequency of tornado occurrence on the Great Plains of the United States that extends from Texas and Oklahoma northward to Nebraska and northeastward to Illinois and Indiana.

Tornado family A group of tornadoes that originate from the same supercell.

Tornadogenesis The process by which a tornado forms.

Tornado vortex signature The Doppler radar velocity signature of a tornado, marked by a sharp (pixel to pixel) change in radial velocity on a radar screen within a broader region where inbound radial velocities border outbound radial velocities.

Total totals index A stability index defined as the sum of the temperature difference between 850 mb and 500 mb, plus the 850 mb dewpoint, minus the 500 mb temperature.

Trace gas A gas that exists in very low concentrations in the atmosphere.

Trade winds The low-level winds found throughout the tropics that flow from east-to-west with a component toward the equator.

Trailing stratiform region In a mesoscale convective system, the stratiform cloud region to the rear of the convective line. Most evident during the system's mature and dissipating stage.

Training The situation in which individual thunderstorms organized in a squall line move parallel to the line so that storm after storm moves over a fixed location on the ground.

Trajectory The path that an air parcel takes as it moves through the atmosphere.

Translational velocity The part of an air parcel's total velocity that is associated with the forward movement of the storm system in which the air parcel is embedded.

Transmitter The component of a radar that generates and transmits microwaves to the antenna.

Traveling spark In a developing lightning strike, the initial discharge from a surface object that rises to meet the descending stepped leader.

Trigger mechanism A front, airmass boundary or other phenomena that lifts air sufficiently to release convective instability and cause a thunderstorm to form.

Tropical airmass An airmass that originates in tropical regions and is characterized by very warm to hot temperatures.

Tropical cyclone A storm complex that originates in the tropics, and has organized rotation and low pressure at its center.

Tropical depression A cluster of thunderstorms of tropical origin that has a pressure drop and a weak organized circulation and sustained winds between 20 and 34 knots.

Tropical disturbance An unorganized cluster of thunderstorms over the tropical oceans that has wind speeds between 0 and 20 knots.

Tropical Prediction Center The National Center for Environmental Prediction that is responsible for forecasting the behavior of tropical storms and hurricanes and issuing watches and warnings for these storms.

Tropical storm A cluster of thunderstorms of tropical origin that has an organized circulation and sustained winds between 35 and 64 knots.

Tropics Latitudes of the Earth between 30° N and 30° S where cold weather is rare. The higher latitudes between 20° and 30° are sometimes referred to as the subtropics since cold air occasionally reaches these latitudes in winter.

Tropopause The boundary between the troposphere and the stratosphere.

Tropopause fold A location where the tropopause is distorted downward toward the Earth's surface due to the secondary circulations about a jetstream. Folds are associated with intrusions of stratospheric air into the troposphere.

Troposphere The layer of Earth's atmosphere that extends from the Earth's surface to the tropopause and contains all of Earth's weather.

Trough An elongated area of low atmospheric pressure, normally accompanied by cyclonic (counterclockwise) flow in the Northern Hemisphere.

Trough axis The line along which the pressure locally reaches its minimum values within a trough.

Trowal Short for "trough of warm air aloft," the warm region aloft that coincides with the wraparound band of clouds and precipitation north and northwest of the low-pressure center of an extratropical cyclone.

Turbulence Irregular motions occurring in atmospheric flow.

Turbulent eddies Small-scale disturbances in a flow field. See "eddy."

Typhoon A hurricane that forms over the western Pacific Ocean.

Ultra high frequency (UHF) A band of electromagnetic wave frequencies in the radio range that are commonly used in wind profiler systems.

Unit area An area that has a value of one in a given set of units. For example, one square centimeter, one square meter, one square foot, and one square mile are all unit areas.

Universal Coordinated Time (UTC) The official time standard used worldwide for meteorological measurements. Also called Greenwich Meridian Time (GMT) and Zulu time (Z).

Unstable The condition in which a parcel of air, displaced vertically, will accelerate away (upward or downward) from its original position.

Updraft Region of air with marked rising vertical motion.

Upper air wave A wave that appears in the flow on an upper-level map, normally at pressure altitudes of 500 mb and above.

Upper-level front A boundary between two airmasses that is present in the middle- or upper-troposphere but does not extend to the Earth's surface (also called a "cold front aloft").

Upslope flow Wind directed up a slope, such as a hill or mountain.

Upslope storm A snowstorm along the east slope of the Rocky Mountains caused by air flowing westward across the plains and up the slope of the Rockies.

Upwelling The rising of deep, cold ocean water to the ocean surface.

Urban heat island The locally warmer temperatures that often occur in urban areas caused by the retention of solar heat by urban structures and by the generation of heat by human activities, such as fuel combustion.

UTC See "Universal Coordinated Time."

Vapor deposition The process by which water vapor becomes incorporated into a water droplet or ice crystal, causing the drop or ice crystal to grow.

Vapor pressure That part of the total atmospheric pressure due to water vapor molecules. The force exerted by water vapor molecules on a unit area.

Vertical wind shear The rate of change of horizontal wind speed or direction with altitude.

Very high frequency A band of electromagnetic wave frequencies in the radio range that are commonly used in profiler systems.

Virga Wisps or streaks of water or ice particles falling out of a cloud but evaporating before reaching the Earth's surface as precipitation.

Visibility The farthest distance that an observer can see and identify large objects.

Visible channel A specific frequency band of visible radiation observed by weather satellites. The intensity of the radiation in this frequency band is related to the amount of sunlight reflected from cloud and ground surfaces below the satellite.

Volt The unit of electrical potential; volts are a measure of the charge difference between two points.

Voltage A measure of the electrical potential energy, which is related to the charge difference between two points.

Volts/meter The unit used to measure an electric field.

Vortex An atmospheric circulation, such as a tornado or hurricane, in which air flows in a generally circular pattern around a center of low pressure.

Vortex breakdown The process by which a downdraft develops aloft in the center of a tornado and descends to the surface. Following vortex breakdown, a tornado widens considerably and often develops suction vortices.

Vortex ring A curl in the flow at the leading edge of a downburst that is often characterized by strong surface winds.

Vortex stretching The process by which a vortex narrows as the upper part of the vortex rises more rapidly than the lower part, leading to an increase in rotation.

Vorticity A measure of the rotation rate of a small fluid element in a flow field.

Walker cell An atmospheric circulation in the equatorial regions of the Pacific Ocean characterized by rising motion over the western Pacific, eastward flow near the tropopause, sinking motion over the eastern Pacific, and westward flow near the surface.

Wall cloud A region of rotating cloud that extends below the rain-free base of a supercell thunderstorm. The formation of the wall cloud often precedes tornado formation.

Warm air dome A large region of exceptionally warm air, typically associated with heat waves.

Warm-core system A low-pressure center in which air aloft above the low is warmer than air at the same level at the circulation's periphery.

Warm front A boundary between airmasses where the cold airmass is retreating and the warm air is advancing at the surface.

Warm occlusion An occluded front in which air behind the cold front is less dense (warmer) than air behind the warm front.

Warm rain Rain that forms in the absence of ice particles.

Warning An alert issued by the National Weather Service when specific severe or hazardous weather phenomena are occurring and threaten the region under the warning.

Watch An alert issued by the National Weather Service when conditions are right for the possible occurrence of a severe or hazardous weather phenomenon.

Water equivalent The depth of water that would be obtained if snow is melted (typically expressed as a ratio of snow depth to water depth, e.g., ten-to-one).

Watershed The drainage area of a river basin.

Waterspout A non-supercell tornado that forms over a body of water.

Water vapor Water in its gaseous form.

Water vapor channel A specific frequency band of infrared radiation observed by weather satellites. The intensity of the radiation in this frequency band is related to the total amount of water vapor in an atmospheric column below the satellite.

Wave clouds Clouds that form at the crest of waves in atmospheric flow and are typically found over and just downwind of mountains.

Weather modification The practice of attempting to increase rainfall, reduce hail, or otherwise alter natural atmospheric processes through cloud seeding with silver iodide or other chemical agents.

Wet growth regime The mode of hail growth where so much supercooled water freezes on a hailstone's surface that the released latent heat raises the stone's surface temperature to 0° C. In the wet growth regime, water can spread across the surface of the stone, drain into porous regions, and shed from the stone's surface while the freezing process continues.

Wet microburst A microburst that occurs simultaneously with significant rainfall reaching the ground.

Widespread flood A flood that occurs when a large amount of rain falls over a watershed for many days so that significant portions of the watershed are inundated for long periods of time.

Wind The movement of air.

Wind barb A symbol used on weather maps to indicate wind speed. A wind report is plotted as a staff oriented to indicate wind direction and a combination of pennants and barbs to indicate wind speed.

Wind chill factor A parameter reported by the National Weather Service in winter to account for the effect of both temperature and wind on the rate at which exposed flesh will cool. Reported numerically as the wind chill temperature.

Wind chill index An index that accounts for the effect of both temperature and wind on the rate at which exposed flesh will cool.

Wind chill temperature The temperature that human skin feels due to heat loss caused by the combined effects of both cold and wind.

Wind index Index used by the National Weather Service to forecast the likelihood of microbursts (downbursts) occurring in a region.

Wind Induced Surface Heat Exchange (WISHE) A theory explaining the mechanism by which a hurricane forms from a cluster of thunderstorms over tropical oceans.

Wind-parallel rolls Parallel bands of precipitation that form over the Great Lakes in winter as airflow develops alternating rising and sinking motions perpendicular to the direction of the wind.

Wind profile A collection of wind measurements at altitudes above a point on the Earth at a specific time.

Wind profiler A Doppler radar that operates in clear as well as cloudy air. The radar measures winds at altitudes above the radar site by determining the Doppler shift in frequency caused by the scattering of radar energy by turbulent air moving with the wind.

Wind shear The rate of change of wind speed and/or direction with distance (horizontal wind shear) or altitude (vertical wind shear).

Windward The side of a mountain toward which the wind is blowing.

Windward slopes The slope of a mountain on the side of a mountain toward which the wind is blowing.

WISHE See "Wind Induced Surface Heat Exchange."

Wrap-around band The major band of precipitation in an extratropical cyclone that wraps around the north and northwest side of the cyclone's low-pressure system.

WSR-88D Abbreviation for "Weather Service Radar 1988 Doppler," one of a network of Doppler radars currently in use by the National Weather Service (see also "NEXRAD").

Zonal flow A wind flow that has a predominant west-to-east component (parallel to a latitude circle).

Zulu time (Z) A universal time used worldwide for meteorological measurements. Also called Greenwich Mean Time (GMT) and Universal Coordinated Time (UTC).

CREDITS

Chapter 1

CHAPTER 1 TITLE PAGE: (Courtesy of NOAA.)
FIGURE 1.7: (A: Apollo 17, Courtesy of NASA; B: Courtesy of Glen Romine; C: Courtesy of Snowcrystals.com.)

Chapter 2

FIGURE 2.2: Courtesy of Larry Oolman, Department of Atmospheric Science, University of Wyoming.
FIGURE 2.4: Courtesy of the National Weather Service, Lincoln, IL.
FIGURE 2.5: Courtesy of Larry Oolman, Department of Atmospheric Science, University of Wyoming.
FIGURE 2.12: Courtesy of NOAA.
FIGURE 2.15 A-C: Courtesy of the Dept. of Atmospheric Sciences, University of Illinois at Urbana.
FIGURE 2.16: Courtesy of NOAA.
FIGURE 2.17: Courtesy of NOAA/FSL.
FIGURE 2.18: Courtesy of Richard E. Orville, Texas A&M University.
FIGURE 2.19: © American Meteorological Society. Used with permission. American Meteorological Society, used with permission.
FIGURE 2A: Courtesy of NCEP.
FIGURE 2F: 2007 Copyright, © University Corporation for Atmospheric Research.

Chapter 3

CHAPTER 3 TITLE PAGE: Courtesy of NOAA.
FIGURE 3.2: Courtesy of the Department of Atmospheric Science, University of Illinois at Urbana–Champaign.
FIGURE 3B: Courtesy University of Wisconsin–Madison.

Chapter 4

CHAPTER 4 TITLE PAGE: Courtesy of the Dept. of Atmospheric & Oceanic Sciences, University of Wisconsin–Madison.
FIGURE 4.1: Courtesy of Brian Jewett and the National Center for Supercomputer Applications, University of Illinois.
FIGURE 4.3: Courtesy of NOAA/NCEP.
FIGURE 4.4: Courtesy of NOAA/NCEP.
FIGURE 4.5: Courtesy of NOAA/NCEP.
FIGURE 4.6: Courtesy of Dept. of Atmospheric & Oceanic Sciences, University of Wisconsin–Madison.
FIGURE 4.7: Courtesy of NOAA/NCEP.
FIGURE 4.8: Courtesy of Dept. of Atmospheric & Oceanic Sciences, University of Wisconsin–Madison.
FIGURE 4.9: Courtesy of Dept. of Atmospheric & Oceanic Sciences, University of Wisconsin–Madison.
FIGURE 4.10: Courtesy of NOAA.
FIGURE 4.11: Courtesy of NOAA/NCEP.
FIGURE 4.12: Courtesy of Glen Romine and Robert Wilhelmson, University of Illinois at Urbana-Cahmpaign and National Center for Supercomputing Applications(NCSA).
FIGURE 4.13: (Courtesy of Robert Wilhelmson, Glen Romine, and the National Center for Supercomputing Applications at the University of Illinois at Urbana Champaign.
FIGURE 4B: Courtesy of NOAA Climate Diagnostics Center.

Chapter 5

CHAPTER 5 TITLE PAGE: Courtesy of NASA Science Mission Directorate.
FIGURE 5.8: (Courtesy of Global Precipitation Climatology Project (GPCP), World Meteorological Organization.)
FIGURE 5.9: Adapted from Petit et al, 1999: Climate and Atmospheric History of the Past 420,000 Years from the Vostok Ice Core, Antarctica, Nature, 399: 429–436.
FIGURE 5.10: (Courtesy of Oak Ridge National Laboratory.)
FIGURE 5.12: Fig. 1 of FAQ 2.1, p. 135 of Intergovernmental Panel on Climate Change, 2007: Climate Change 2001, The Physical Science Basis, Cambridge University Press, Cambridge, U.K., page 135.
FIGURE 5.13: Figure 2.2 from p. 135 of Intergovernmental Panel on Climate Change, 2001: Climate Change 2001, The Scientific Basis, Cambridge University Press, Cambridge, U.K.
FIGURE 5.14: Courtesy of NASA Goddard Institute for Space Studies.
FIGURE 5.15: Courtesy of NASA Goddard Institute for Space Studies.
FIGURE 5.16: Fig. 2.25(ii) from p. 144 of Intergovernmental Panel on Climate Change, 2001: Climate Change 2001, The Scientific Basis, Cambridge University Press, Cambridge, U.K.

Chapter 6

CHAPTER 6 TITLE PAGE: Courtesy of Glen Romaine.
FIGURE 6A: Copyright Adam Houston. Used with permission.

Chapter 7

CHAPTER 7 TITLE PAGE: (Courtesy of Dept. of Atmospheric Sciences, University of Illinois at Urbana.
FIGURE 7.3: Adapted from Mayfield et al. Atlantic hurricane season of 1992. *Monthly Weather Review*, 122 (1994): 517–538. Courtesy of the American Meteorological Society.
FIGURE 7.10: Courtesy of the Department of Atmospheric Sciences at the University of Illinois at Urbana–Champaign.

FIGURE 7.11: Courtesy of Dept. of Atmospheric & Oceanic Sciences, University of Wisconsin–Madison.

Chapter 8

CHAPTER 8 TITLE PAGE: Courtesy of Dept. of Atmospheric Sciences, University of Illinois at Urbana.
FIGURE 8.11: Courtesy of Dept. of Atmospheric Sciences, University of Illinois at Urbana.

Chapter 9

FIGURE 9.10: Courtesy of the Department of Atmospheric Sciences at the University of Illinois at Urbana–Champaign.

Chapter 10

CHAPTER 10 TITLE PAGE: Courtesy of the Department of Atmospheric Sciences, University of Illinois at Urbana–Champaign.
FIGURE 10.1: Courtesy of the Department of Atmospheric Sciences, University of Illinois at Urbana–Champaign.

Chapter 11

CHAPTER 11 TITLE PAGE: Courtesy of NOAA.
FIGURE 11.1: Data from Northeast Snowstorms of the Twentieth Century: Volume 1 Overview, Paul J. Kocin and Louis W. Uccellini, American Meteorological Society Monograph, 2004.
FIGURE 11.5: Courtesy of NOAA.
FIGURE 11.8: Courtesy of the Department of Atmospheric and Oceanic Sciences, University of Wisconsin–Madison.
FIGURE 11B: Collection of the New York Historical Society.
FIGURE 11C: Adapted from Kocin, Paul J., and Louis W. Uccellini, 1990: Snowstorms along the northeastern coast of the United States: 1955 to 1985. Meteorological Monographs, Vol. 22, No. 44. Used with permission.
FIGURE 11D: Adapted from Kocin, Paul J., and Louis W. Uccellini, 1990: Snowstorms along the northeastern coast of the United States: 1955 to 1985. Meteorological Monographs, Vol. 22, No. 44. Used with permission.
FIGURE 11Q2: Adapted from Kocin, Paul J., and Louis W. Uccellini, 1990: Snowstorms along the northeastern coast of the United States: 1955 to 1985. Meteorological Monographs, Vol. 22, No. 44. Used with permission.

Chapter 12

CHAPTER 12 TITLE PAGE: Photo: Eric Snodgrass.
FIGURE 12.3: Courtesy of NASA.
FIGURE 12.4: Rauber R. et al (2001), A Synoptic weather pattern and sounding-based climatology of freezing precipitation east of the Rocky Mountains, Journal of Applied Meteorology, 40, 1724–1747. Courtesy of the American Meteorological Society, used with permission.)
FIGURE 12.5: (Rauber R. et al (2001), A Synoptic Weather Pattern and Sounding-Based Climatology of Freezing Precipitation East of the Rocky Mountains, *J App Met*, 40, 1724–1747. Courtesy of the American Meteorological Society. Reprinted with permission.
FIGURE 12.6: Adapted from Cortinas J. et al., 2004: An analysis of freezing rain, freezing drizzle, and ice pellets across the United States and Canada: 1976–90. *Weather and Forecasting*, 19, 377–390: American Meteorological Society, used with permission.
FIGURE 12.7: Adapted from Cortinas J. et al., 2004: An analysis of freezing rain, freezing drizzle, and ice pellets across the United States and Canada: 1976–90. *Weather and Forecasting*, 19, 377–390: American Meteorological Society, used with permission.
FIGURE 12C: Copyright © The Canadian Press/Jacques Boissinot.

Chapter 13

CHAPTER 13 TITLE PAGE: Courtesy of NASA SeaWIFS Project.
FIGURE 13.1: Adapted from Scott and Huff. *J Great Lakes Res* 22(1996): 845–863.
FIGURE 13.4: Courtesy of Cooperative Institute for Meteorological Satellite Studies, University of Wisconsin.
FIGURE 13.5: Courtesy of David Kristovich.
FIGURE 13.6: (Adapted from Eichenlaub, *V. Weather and Climate of the Great Lakes Region*. University of Notre Dame Press; 1979; Figure 49.
FIGURE 13.8: Image courtesy of Ray Sterner, Johns Hopkins University.
FIGURE 13.9: Adapted from Eichenlaub, *V. Weather and Climate of the Great Lakes Region*. University of Notre Dame Press; 1979; Figures 44 and 45.
FIGURE 13.11: (Courtesy of the GOES project, NOAA/NASA.
FIGURE 13.12: Courtesy of Neil Laird.
FIGURE 13.14: Courtesy of Neil Laird.
FIGURE 13.15: From Laird, N. F. (1999): Observation of coexisting mesoscale lake-effect vortices over the western Great Lakes. *Monthly Weather Review*, 127, 1137–1141, Courtesy American Meteorological Society. Reprinted with permission.
FIGURE 13B: Photo: Tom Niziol, National Weather Service.
FIGURE 13F: Courtesy of Cooperative Institute for Meteorological Satellite Studies, University of Wisconsin.

Chapter 14

CHAPTER 14 TITLE PAGE: © Sam Harrel/Fairbanks News-Miner.
FIGURE 14.1: Data Courtesy National Climatic Data Center, NOAA.
FIGURE 14A: Data courtesy of National Centers for Environmental Prediction/National Center for Atmospheric Research Reanalysis.
FIGURE 14C: Courtesy of NOAA.
FIGURE 14D: From Walsh et al. (2001), Extreme cold outbreaks in the United States and Europe 1948–99, *Journal of*

Climate, 14, 2642–2658, Courtesy of the American Meteorological Society. Reprinted with permission.)
FIGURE 14E: Adapted from Coastal Weather Research Center.
FIGURE 14F: Adapted from Kocin et al. The Great Arctic Outbreak and East Coast Blizzard of February 1899. *Weather and Forecasting* 3(1988), 305–318; American Meteorological Society.

Chapter 15

CHAPTER 15 TITLE PAGE: Photo by Bill Koch, Courtesy of the NOAA photo library collection of Dr. Herbert Kroehl.
FIGURE 15.1: From Schwartz, R. M., and T. W. Schmidlin (2002) Climatology of blizzards in the coterminous United States, 1959–2000. *Journal of Climate* 15:1765–1772; Used with permission.
FIGURE 15.2: From Schwartz, R. M., and T. W. Schmidlin (2002) Climatology of blizzards in the coterminous United States, 1959–2000. *Journal of Climate* 15:1765–1772; Used with permission.
FIGURE 15.6: Courtesy of the Department of Atmospheric Sciences, University of Illinois at Urbana–Champaign.
FIGURE 15.9: Courtesy of Mike Hardiman, National Weather Service, Lincoln, IL.
FIGURE 15A: Courtesy of Jeff Stith.
FIGURE 15C: Courtesy of Chris Geelhart, National Weather Service, Lincoln IL.
FIGURE 15D: Courtesy of Paul White, Petersburg, Illinois.

Chapter 16

CHAPTER 16 TITLE PAGE: Courtesy of the California Department of Transportation.
FIGURE 16.1: Source: Department of Energy.
FIGURE 16.2: Copyright, Colorado Climate Center, Colorado State University, Courtesy of Nolan Doesken.
FIGURE 16.4: Courtesy of the Department of Atmospheric Sciences, University of Illinois at Urbana–Champaign.
FIGURE 16.6: Courtesy of George Taylor, Oregon Climate Service.
FIGURE 16.8: Topographic map courtesy of Ray Steiner, Johns Hopkins University.
FIGURE 16.11: Background map courtesy of Ray Steiner, Johns Hopkins University.
FIGURE 16A: Courtesy of Northwest Weather Avalanche Center, US Forest Service.
FIGURE 16B: Courtesy of the Desert Research Institute, University of Nevada.
FIGURE 16C: Photo: Mary J. Brodzik.
FIGURE 16D: Courtesy of Paul Wolyn, National Weather Service, Pueblo, Colorado.
FIGURE 16E: Copyright, University Corporation for Atmospheric Research. Used with permission.
FIGURE 16F: Courtesy of NOAA.
FIGURE 16G: Courtesy of NOAA.

Chapter 17

FIGURE 17.2: Courtesy of the National Oceanic and Atmospheric Administration and the National Center for Atmospheric Research.
FIGURE 17.7: Photo by Glen Romine.
FIGURE 17A: Courtesy of Bruce Lee.
FIGURE 17B: 2007 Copyright ,© University Corporation for Atmospheric Research.
FIGURE 17C: Courtesy of V. Grubisic, DRI Mesoscale Dynamics and Modeling Lab, and AMS.

Chapter 18

CHAPTER 18 TITLE PAGE: Courtesy of Bruce Lee.
FIGURE 18.1: Courtesy of the NOAA Storm Prediction Center.
FIGURE 18.2: Photo: Alan Moller, National Weather Service.
FIGURE 18.7: Courtesy of NOAA Photo Library and the National Severe Storms Laboratory.
FIGURE 18.10: Courtesy of M. Biggerstaff, University of Oklahoma, with changes.
FIGURE 18.12: Courtesy of Weather Services International Corp.
FIGURE 18.13: Courtesy of the Department of Atmospheric Sciences, University of Illinois at Urbana–Champaign.
FIGURE 18.20: Courtesy of Joseph Grim.)
FIGURE 18.23: Adapted from Lemon and Doswell (1979) Severe Thunderstorm Evolution and Mesocyclone Structure as related to tornadogenesis. *Monthly Weather Review* 107(1979): 1184–1197. Courtesy of the American Meteorological Society. Meteorological Society, used with permission.
FIGURE 18.24: Adapted from Lemon and Doswell (1979) Severe Thunderstorm Evolution and Mesocyclone Structure as related to tornadogenesis. *Monthly Weather Review* 107(1979): 1184–1197. Courtesy of the American Meteorological Society.
FIGURE 18C: Courtesy of Bruce Lee.

Chapter 19

CHAPTER 19 TITLE PAGE: Photo: Glen Romine.
FIGURE 19.1: All photos copyright Howard B. Bluestein.
FIGURE 19.7: From *Bulletin of the American Meteorological Society*, 77(7) 1465–1481 by Wakimoto et al. Copyright © by the American Meteorological Society. Preprinted with permission.
FIGURE 19.8: Adapted from Davies-Jones, 1986. Tornado Dynamics. *Thunderstorms: A Social and Technological Documentary—Volume 2*, 2nd ed., ed. E. Kessler, Norman: University of Oklahoma Press, 1986. Courtesy of the American Meteorological Society.

FIGURE 19.10: Adapted from Abbey and Fujita, 1986. Tornadoes: the tornado outbreak of 3–4 April 1974. In *Thunderstorms: A Social Scientific, and Technological Documentary*, Volume 1, 2nd ed., ed E. Kessler, Norman: University of Oklahoma Press.
FIGURE 19.11: (Panel C from Lee, B and R. Wilhelmson (1997). The numerical simulation of nonsupercell tornadogenesis. Part II: Evolution of a family of tornadoes along a weak outflow boundary. Journal of the Atmospheric Sciences, 54, 2387–2415 Courtesy of the American Meteorological Society, used with permission.)
FIGURE 19.12: Photo by Joe Golden, courtesy of NOAA.
FIGURE 19.17: Analysis by T. T. Fujita.
FIGURE 19.22: Courtesy of NOAA Storm Prediction Center.
FIGURE 19C: Courtesy of the Illinois State Water Survey.
FIGURE 19D: Courtesy of the Jackson County Historical Society, Murphysboro, Illinois.
FIGURE 19E: Courtesy of the Theodore Fujita family.
FIGURE 19F: From R. M Rauber and R. W. Scott. Central Illinois cold air funnel outbreak. *Monthly Weather Review*, 129, 2815–2821. Reprinted with permission.
FIGURE 19G: Courtesy of Nilton O. Rennó.
FIGURE 19H: Courtesy U.S. Forestry Service.
FIGURE 19I: Courtesy of J. Wurman.
FIGURE 19J: Courtesy of J. Wurman.
FIGURE 19K: Photo by John Jarboe, National Weather Service.

Chapter 20

CHAPTER 20 TITLE PAGE: Courtesy of NOAA/DAR/ERL/NSSL.
FIGURE 20.1: Courtesy of Bruce Lee.
FIGURE 20.3: Adapted from Matson and Huggins (1980) The direct measurement of the sizes, shapes and kinematics of falling hailstones. *Journal of the Atmospheric Sciences* 37:1107–1125. Courtesy of the American Meteorological Society.
FIGURE 20.4: Courtesy of NOAA.
FIGURE 20.7: Courtesy of Glen Romine.
FIGURE 20.8: National Center for Atmospheric Research/University Corporation for Atmospheric Research/National Science Foundation.
FIGURE 20.9: Courtesy of NOAA/SPC.
FIGURE 20.10: Courtesy of J. Vivekanandan and the American Meteorological Society.
FIGURE 20.11: From S.A. Changnon. "The scales of hail," *Journal of Applied Meteorology* V16, June 1997, p626–648. Copyright © by American Meteorological Society. Reprinted with permission.
FIGURE 20.12: From Kessler, E. 1986. *Thunderstorm Morphology and Dynamics*, 2nd ed. University of Oklahoma Press. Reprinted by Permission.
FIGURE 20.13: Courtesy of the National Oceanic and Atmospheric Administration.
FIGURE 20.14: Courtesy of Bruce Lee.
FIGURE 20A: Courtesy of Bruce Lee.
FIGURE 20B: Courtesy of NOAA.
FIGURE 20C: Courtesy of NOAA.

Chapter 21

CHAPTER 21 TITLE PAGE: Courtesy of NOAA Photo Library, NOAA Central Library; OAR/ERL/National Severe Storms Laboratory.
FIGURE 21.1: Courtesy of NASA.
FIGURE 21.2: From *Monthly Weather Review, Vol. 129, Issue 5, May 2001* by Richard E. Orville and Gary R. Huffines. Copyright © 2001 by American Meteorological Society. Reprinted by permission.
FIGURE 21.8: Courtesy NOAA.
FIGURE 21.12: Originally published in *EOS Newsletter*, used by permission of Carlos Miralles/Walter Lyons.
FIGURE 21.13: Courtesy of NOAA.
FIGURE 21A: Courtesy of Alaskan Type I Incident Management Team photographer: John McColgan, Bureau of Land Management, Alaska Fire Service.
FIGURE 21B: Courtesy of NASA Goddard Spaceflight Center and Geographic Information Network of Alaska.
FIGURE 21C: Courtesy of Miriam Rorig, USDA Forest Service, Pacific Northwest Research Station.
FIGURE 21D: Courtesy of NOAA.

Chapter 22

CHAPTER 22 TITLE PAGE: Courtesy National Severe Storms Laboratory.
FIGURE 22.5: Courtesy of the Theodore Fujita family, with changes
FIGURE 22.6: United States Air Force photo.
FIGURE 22.7: National Center for Atmospheric Research/University Corporation for Atmospheric Research/National Science Foundation.
FIGURE 22.8: Courtesy of Jim Wilson and the American Meteorological Society, with changes.
FIGURE 22.9: From *Andrews AFB Microburst* by Tetsuya Theodore Fujita.
FIGURE 22.10: National Center for Atmospheric Research/University Corporation for Atmospheric Research/National Science Foundation.
FIGURE 22.13: Courtesy of the Theodore Fujita family, with changes.
FIGURE 22.14: Earle Williams and Dennis Boccippio/MIT Radar Laboratory.
FIGURE 22A: Courtesy of James Walter.
FIGURE 22B: Courtesy of the Theodore Fujita family, with changes.
FIGURE 22C: Courtesy of the National Oceanic and Atmospheric Administration.

Chapter 23

CHAPTER 23 TITLE PAGE: Courtesy of the United States Navy Fleet Numerical Meteorology and Oceanography Center.
FIGURE 23.4: Courtesy of NOAA Pacific Marine Environmental Laboratory.
FIGURE 23.6: Adapted from "El Niño and Climate Prediction" Our Changing Planet: Reports to the Nation, NOAA Office of Global Programs/UCAR.
FIGURE 23.7: Adapted from Climate Prediction Centers/ NCEP/NWS/NOAA.
FIGURE 23.8: Adapted from Climate Prediction Centers/ NCEP/NWS/NOAA.
FIGURE 23.10: Courtesy of NOAA Climate Diagnostics Center.
FIGURE 23A: (Courtesy Climate Prediction Centers/NCEP/ NWS/NOAA.
FIGURE 23B: Courtesy of Department of Atmospheric Sciences, University of Illinois.
FIGURE 23C: Courtesy of NOAA Pacific Marine Environmental Laboratory.

Chapter 24

CHAPTER 24 TITLE PAGE: Courtesy of NASA.
FIGURE 24.2: MODIS image courtesy of NASA.
FIGURE 24.4: Courtesy of the GOES project, NASA Goddard Space Flight Center and NOAA.
FIGURE 24.5: Courtesy of the GOES project, NASA Goddard Space Flight Center and NOAA.
FIGURE 24.6: Courtesy of NOAA.
FIGURE 24.7: Courtesy of the National Climatic Data Center, NOAA.
FIGURE 24.9: From Anthes, R. A., J. W. Trout and S. S. Ostlund. 1971. Three-dimensional particle trajectories in a model hurricane." Weatherwise 24:176. Reprinted with permission of the Helen Dwight Reed Educational Foundation and Heldref Publications, 1319 18th Street, NW, Washington, DC 20036-1802.
FIGURE 24.17: Adapted from Emanuel, K. (1986), An Air Sea Interaction Theory for Tropical Cyclones. Part I: Steady State Maintenance. Journal of the Atmospheric Sciences 43, 585–605, Courtesy of the American Meteorological Society, used with permission.
FIGURE 24.24: Courtesy of NOAA.
FIGURE 24.26: Adapted from Emanuel, K., 2005: Increasing destructiveness of tropical cyclones over the past 30 years. Nature, Vol. 436, 4 August 2005, doi:10.1038/nature03906. Used with permission.
FIGURE 24A: Courtesy of NOAA.
FIGURE 24B: Courtesy of NOAA.
FIGURE 24C: From *Bulletin of the American Meteorological Society, Vol. 87, Issue 11, November 2006* by Robert A. Houze, Jr. et al. Copyright © 2006 by American Meteorological Society. Reprinted by permission.
FIGURE 24D: From *Bulletin of the American Meteorological Society, Vol. 87, Issue 11, November 2006* by Robert A. Houze, Jr. et al. Copyright © 2006 by American Meteorological Society. Reprinted by permission.

Chapter 25

CHAPTER 25 TITLE PAGE: Courtesy of the United States Coast Guard.
FIGURE 25.4: Adapted from "The Great Flood of 1993," Natural Disaster Survey Report, courtesy of NOAA.
FIGURE 25.5: Adapted from "The Great Flood of 1993," Natural Disaster Survey Report, courtesy of NOAA.
FIGURE 25.7: Courtesy of the U.S. Army Corp of Engineers.
FIGURE 25.8: Adapted from Caracena et al. (1979) Mesoanalysis of the Big Thompson Flood. *Monthly Weather Review* 107, 1–17, Courtesy of the American Meteorological Society. Used with permission.
FIGURE 25.9: (Adapted from Caracena et al. (1979) Mesoanalysis of the Big Thompson Flood. *Monthly Weather Review* 107, 1–17, Courtesy of the American Meteorological Society. Used with permission.)
FIGURE 25.10: Courtesy of Ray Sterner, Johns Hopkins University, Applied Physics Laboratory.
FIGURE 25.15: Courtesy of NOAA/SPC.
FIGURE 25A: Courtesy Dartmouth Flood Observatory.
FIGURE 25C: Courtesy of the U.S. Army Corps of Engineers.
FIGURE 25D: Courtesy of the Southeast Regional Climate Center, with changes.
FIGURE 25E: From "The Great Flood of 1993," Natural Disaster Survey Report, Courtesy of NOAA.
FIGURE 25F: Courtesy of Space Imaging.
FIGURE 25G: Courtesy of the Illinois State Water Survey.
FIGURE 25H: From "The Great Flood of 1993," Natural Disaster Survey Report. Courtesy of NOAA.
FIGURE 25I: From "The Great Flood of 1993," Natural Disaster Survey Report. Courtesy of NOAA.

Chapter 26

CHAPTER 26 TITLE PAGE: Courtesy of NOAA Photo Library, Russell Lord, U. S. Department of Agriculture, 1938.
FIGURE 26.1: Courtesy of NOAA Climate Prediction Center.
FIGURE 26.6: Courtesy of ITT Industries, Aerospace Engineering Division.
FIGURE 26.7: Courtesy of National Oceanographic Data Center/NOAA.
FIGURE 26.9: Adapted from Glantz, Drought Follows the Plow.
FIGURE 26.11: Courtesy of United States Geological Survey.
FIGURE 26.12: Courtesy of United States Geological Survey.
FIGURE 26.13: Courtesy of NASA Earth Observatory.
FIGURE 26.15: Courtesy of NOAA/National Climate Data Center.

FIGURE 26B: Adapted from Bowden, M. et al. "The effect of climate fluctuations on human populations: two hypotheses." In: *Climate and History: Studies in Past Climates and Their Impact on Man*, Eds. Wigley, Ingram, and Farmer (From Climate and History by Bowden et al. Copyright © 1981 by Cambridge University and Press. Reprinted with the Permission of Cambridge University Press.)

Chapter 27

CHAPTER 27 TITLE PAGE: Source: NOAA/National Weather Service, Office of Climate, Water and Weather Services.
FIGURE 27.1: Adapted from Quayle, R. and F. Doehring. "Heat stress: a comparison of indices. *Weatherwise*, 34(1981)120–124.
FIGURE 27.5: Courtesy of University of Wisconsin, Space Science and Engineering Center/CIMSS.
FIGURE 27.6: Courtesy of Heat Island Group, Lawrence Berkeley National Laboratory.
FIGURE 27.14: Courtesy NOAA, National Climate Data Center.
FIGURE 27A: Courtesy of NOAA/National Climate Data Center.
FIGURE 27B: Courtesy of NOAA/National Climate Data Center.

SOURCES AND ADDITIONAL READING

GENERAL METEOROLOGY REFERENCES

Ackerman, Steven A., and John A. Knox. *Meteorology: Understanding the Atmosphere.* 2nd Ed., Pacific Grove, CA: Brooks/Cole, 2006.

Aguado, E. and J. E. Burt. *Understanding Weather and Climate.* 4th ed. Upper Saddle River, NJ: Prentice Hall, 2007.

Ahrens, C. Donald. *Essentials of Meteorology, An Invitation to the Atmosphere.* 4th ed. Pacific Grove, CA: Brooks/Cole, 2005.

Ahrens, C. Donald. *Meteorology Today: An Introduction to Weather, Climate, and the Environment,* 8th ed. Pacific Grove, CA: Brooks/Cole, 2006.

Carbone, Greg. *Exercises for Weather and Climate.* 6th ed. Upper Saddle River, NJ: Prentice Hall, 2006.

Danielson, Eric W., James Levin, and Elliot Abrams. *Meteorology.* 2nd ed. New York: McGraw-Hill, 2003.

Grenci, Lee M. and Jon M. Nese. *A World of Weather: Fundamentals of Meteorology.* 4th ed. Kendall/Hunt Publishing, 2006.

Lutgens, Fredrick. K. and Edward J. Tarbuck. *The Atmosphere: An Introduction to Meteorology.* 10th ed. Upper Saddle River, NJ: Prentice Hall, 2007.

Moran, Joseph M., and Michael D. Morgan. *Meteorology, the Atmosphere and the Science of Weather.* 5th ed. Upper Saddle River, NJ: Prentice Hall, 1997.

Stull, Roland B. *Meteorology for Scientists and Engineers.* Pacific Grove, CA: Brooks/Cole, 2000.

POPULAR BOOKS ON SEVERE AND HAZARDOUS WEATHER

Abley, M. *The Ice Storm, An Historic Record in Photographs of January 1998.* Toronto: The Gazette (Montreal), McClelland & Stewart Inc., 1998.

Allen, O. E. *Atmosphere.* Alexandria, Virginia: Time-Life Books, 1983.

Bedard, R. *In the Shadow of the Tornado, Stories and Adventures from the Heart of Storm Country.* Norman, Oklahoma: Gilco Publishing, 1996.

Bluestein, H. *Tornado Alley, Monster Storms of the Great Plains.* New York: Oxford University Press, 1999.

Caplovich, J. *Blizzard! The Great Storm of '88.* Vernon, CT: VeRo Publishing Company, 1987.

Caviedes, C. N. *El Niño in History: Storming Through the Ages.* Gainsville, Fla.: University Press of Florida, 2001.

Clark, C. *Flood.* Alexandria: Time-Life Books, 1982.

Drye W. *Storm of the Century: The Labor Day Hurricane of 1935.* National Geographic, 2002.

Egan, T. *The Worst Hard Time.* Mariner Books/Houghton Mifflin Company, 2006.

Farrand, J., Jr. *Weather.* New York: Stewart, Tabori & Chang Inc., 1990.

Felknor, P. S. *The Tri-State Tornado, The Story of America's Greatest Tornado Disaster.* Ames: Iowa State University Press, 1992.

Frater, A. *Chasing the Monsoon.* New York: Knopf Inc, 1991.

Henson, R. *The Rough Guide to Weather.* London, U.K., Rough Guides Limited, 2002.

Inglish, H., Ed. *Year of the Storms, The Destructive Kansas Weather of 1990.* Hillsboro, KS: Hearth Publishing, 1990.

Keen, R. A. *Skywatch, The Western Weather Guide.* Golden, CO: Fulcrum Inc, 1987.

Klineberg, E. *Heat Wave: A Social Autopsy of Disaster in Chicago.* University of Chicago Press, 2002.

Laskin, D., *The Children's Blizzard.* HarperCollins Publishers, 2004.

Ludlum, D. M. *Early American Hurricanes 1492–1870.* Boston: American Meteorological Society, 1963.

Ludlum, D. M. *Early American Tornadoes 1586–1870.* Boston: American Meteorological Society, 1970.

Ludlum, D. M. *Early American Winters 1604–1820.* Boston: American Meteorological Society, 1966.

Ludlum, D. M. *Early American Winters 1821–1870.* Boston: American Meteorological Society, 1968.

Macdougal, J. D. Frozen Earth: *The Once and Future Story of Ice Ages.* University of California Press, 2004.

National Geographic Society. *Nature on the Rampage: Our Violent Earth.* Washington: National Geographic Society, 1986.

National Geographic Society. *Restless Earth, Disasters of Nature.* Washington: National Geographic Society, 1997.

Pitt, S. *Rain Tonight: The Story of Hurricane Hazel.* Tundra Books, 2004.

Toomey, D. Stormchasers: The Hurricane Hunters and Their Fateful Flight into *Hurricane Janet.* Norton, Va., 2002.

Weems, J.E. *A Weekend in September, The Galveston Hurricane of 1900—The Nation's Deadliest Natural Disaster.* College Station: Texas A&M University Press, 1999.

Whipple, A. B. C. Storm. Alexandria: Time-Life Books, 1982.

Zebrowski, E., and J. A. Howard, *Category 5: The Story of Camille.* The University of Michigan Press, 2005.

ADVANCED METEOROLOGY TEXTS

Battan, L. J. *Radar Observation of the Atmosphere.* Chicago: University of Chicago Press, 1973.

Bluestein, H. *Synoptic Meteorology in Midlatitudes Volume 1.* New York: Oxford University Press, 1992.

Bluestein, H. *Synoptic Meteorology in Midlatitudes, Volume 2.* New York: Oxford University Press, 1992.

Brock, F. V., and S. J. Richardson. *Meteorological Measurement Systems.* New York: Oxford University Press, 2001.

Curry, J. A., and P. J. Webster. *Thermodynamics of Atmospheres and Oceans.* San Diego: Academic Press, 1998.

DeFelice, T. P. *An Introduction to Meteorological Instrumentation and Measurement.* Upper Saddle River, NJ: Prentice Hall, 1998.

Doviak, R. J., and D. S. Zrnic. *Doppler Radar and Weather Observations.* 2nd ed. San Diego: Academic Press, 1993.

Holton, J. R. *An Introduction to Dynamic Meteorology.* 3rd ed. San Diego: Academic Press, 1992.

Kidder, S. Q., and T. H. Vonder Haar. *Satellite Meteorology.* San Diego: Academic Press, 1995.

Krishnamurty, T. N., and L. Bounoua. *An Introduction to Numerical Weather Prediction Techniques.* Boca Raton, FL: CRC Press, 1996.

MacGorman, D. R., and W. David Rust, *The Electrical Nature of Storms.* New York: Oxford University Press, 1998.

Martin, J. E., *Mid-Latitude Atmospheric Dynamics.* John Wiley and Sons, 2006.

Rinehart, R. E., *Radar for Meteorologists.* 3rd ed. Grand Forks, ND: Reinhart Publications, 1997.

Rogers, R. R., and Yau, M. K. *A Short Course in Cloud Physics.* 3rd ed. Oxford: Pergamon Press, 1991.

Ruddiman, W. F. *Earth's Climate: Past and Future.* W. H. Freeman and Company (2001).

Tsonis, A. A. *Introduction to Thermodynamics.* Cambridge University Press, 2002.

Wallace, J. M., and P. V. Hobbs. *Atmospheric Science, and Introductory Survey.* San Diego: Academic Press, 1977.

Washington, W. M., and C. L. Parkinson. *Introduction to Three Dimensional Climate Modeling* (Second Edition). University Science Books, 2005.

Vasquez, T. Weather Forecasting Handbook. 4th ed., Garland, TX: Weather Graphics Technologies, 2001.

TECHNICAL MONOGRAPHS, BOOKS, AND JOURNAL ARTICLES

New References

Abbey, R. F. and T. T. Fujita. "Tornadoes: The Tornado Outbreak of 3–4 April 1974." *Thunderstorms: A Social Scientific, and Technological Documentary—Volume 1,* 2nd ed., Ed. E. Kessler. Norman: University of Oklahoma Press, 1986. (Chapter 18, 19)

American Meteorological Society. "AMS Policy Statement on Meteorological Drought." *Bulletin of the American Meteorological Society* 78(1997): 847–849. (Chapter 26)

American Meteorological Society. *Glossary of Meteorology.* 2nd ed. Boston: American Meteorological Society, 2000. (All chapters)

Anderson, D., and S. Eberhardt. *Understanding Flight.* New York: McGraw-Hill, 2001. (Chapter 22)

Andrews, J.F. "The Weather and Circulation of July 1955: A Prolonged Heat Wave Effected by a Sharp Reversal in Circulation." *Monthly Weather Review* 83(1955): 147–153. (Chapters 26, 27)

Andronova, N. G., E. V. Rozanov, F. Yang, M. E. Schlesinger and G. L. Stenchikov (1994): Radiative forcing by Volcanic Aerosols from 1850 through 1994. J. Geophys. Res., 104: D14, 16,807–16,826. (Chapter 5)

Anthes, R. A., J. W. Trout and S. S. Ostlund. "Three Dimensional Particle Trajectories in a Model Hurricane." *Weatherwise* 24(1971): 174–178. (Chapter 24)

Atlas, D., ed. *Radar in Meteorology.* Boston: American Meteorological Society, 1990.

Baker, M. B., and J. G. Dash. "Mechanism of Charge Transfer Between Colliding Ice Particles in Thunderstorms." *Journal of Geophysical Research* 99(1994):10,621–10,626. (Chapter 21)

Barsugli, J. J., J. S. Whitaker, A. F. Loughe., P. D. Sardeshmerkh and Z. Toth. "The Effect of The 1997/98 El Niño on Individual Large-Scale Weather Events." *Bulletin of the American Meteorological Society* 80(1999):1399–1411. (Chapter 23)

Batty, K. "National Weather Service Marks Centennial of Benchmark Cold Wave." NOAA Press Release 99-R219 (1999). (Chapter 14)

Beard, K. V., and H. T. Ochs. "Charging Mechanisms in Clouds and Thunderstorms." *The Earth's Electrical Environment.* Studies on Geophysics. Washington: National Academy Press, 1986. (Chapter 21)

Beniston, M. The 2003 heat wave in Europe: A shape of things to come? An analysis based on Swiss climatological data and model simulations. Geophys Res Lett 31 (2004): 2022–2026 (Chapter 27)

Bentlet, M. "Monstrous Mitch." *Weatherwise* 52(1999):14–18. (Chapter 23)

Bentley, W. A., and W. J. Humphreys. *Snow Crystals.* New York: Dover Pub, 1962. (Chapter 1)

Bernstein, B. C., T. A. Omeron, F. McDonough, and M. K. Politovich. "The Relationship Between Aircraft Icing and Synoptic-Scale Weather Conditions." *Weather and Forecasting* 12(1997):742–762. (Chapter 12)

Bhowmik, N. G. ed. *The 1993 Flood on the Mississippi River in Illinois.* Miscellaneous Publication 151. Champaign, IL: Illinois State Water Survey 1995. (Chapter 25)

Bluestein, H. B., and C. R Parks. "A Synoptic and Photographic Climatology of Low-Precipitation Severe Thunderstorms in the Southern Plains." *Monthly Weather Review* 111(1983):2034–2046. (Chapter 18)

Bowden, M. J., R. W. Kates, P. A. Kay, W. E. Riebsame, R. A. Warrick, D. L. Johnson, H. A. Gould, and D. Weiner. "The Effect of Climate Fluctuations on Human Populations: Two Hypotheses." *Climate and History.* Ed. T. M. Wigley, M. J. Ingram and G. Farmer. Cambridge: Cambridge University Press, 1981. (Chapter 26)

Braham, R. R., Jr., and M. J. Dungey. "Quantitative Estimates of the Effect of Lake Michigan on Snowfall." *Journal of Climate and Applied Meteorology* 23(1984):940–949. (Chapter 13)

Branick, M. L. "A Climatology of Significant Winter-Type Weather Events in the Contiguous United States, 1982–94." *Weather and Forecasting* 12(1997):193–207. (Chapter 12, 14, 15)

Brinkmann, W. A. R., "What is a Foehn?" *Weather* 26(1971):230–239. (Chapter 17)

Brinkmann, W. A. R., "Strong Downslope Winds at Boulder, Colorado." *Monthly Weather Review* 102(1974):592–602. (Chapter 17)

Bromwich, D. H., E. R. Toracinta, R. J. Oglesby, J. L. Fastook and T. J. Hughes, 2005: LGM summer climate on the southern margin of the Laurentide ice sheet: Wet or dry? J. Climate, 18, 3317–3328. (Chapter 5)

Bromwich, D. H., E. R. Toracinta, H. Wei, R. J. Oglesby, J. L. Fastook and T. J. Hughes, 2004: Polar MM5 simulations of the winter climate of the Laurentide ice sheet at the LGM. J. Climate, 17, 3415–3433. (Chapter 5)

Brooks, H. E., A. Witt and M. D. Ellis. "Verification of Public Weather Forecasts Available by the Media." *Bull. Amer. Meteor. Soc.* 79(1997):2167–2177. (Chapter 5)

Brown, Rodger A., and V. T. Wood. *A Guide for Interpreting Doppler Velocity Patterns.* Norman, OK: National Severe Storms Laboratory, 1987. (Chapters 3, 19, 22)

Caplan, P. M. "The 12–14 March 1993 Superstorm: Performance of the NMC Global Medium-Range Model." *Bulletin of the American Meteorological Society,* 76, (1995):201–212. (Chapter 5, 11)

Carcena, F., R. A. Maddox, L. R. Hoxit, and C. F. Chappell. "Mesoanalysis of the Big Thompson Storm." *Monthly Weather Review* 107(1979):1–17. (Chapter 25)

Changnon, S. A. "Impacts of 1997–98 El Niño Generated Weather in the United States." *Bulletin of the American Meteorological Society* 80(1999):1819–1828. (Chapter 23)

Changnon, S. A. "Impacts of Hail in the United States." *Storms – Volume II.* Eds. Pielke, R., Jr. and Pielke, R. Sr. London: Routledge, 1999. (Chapter 20)

Changnon, S. A. "Data and Approaches for Determining Hail Risk in the Contiguous United States." *Journal of Applied Meteorology* 38(1999):1730–1739. (Chapter 20)

Changnon, S. A., *The Great Flood of 1993, Causes, Impacts, and Responses.* Boulder, CO: Westview Press, 1996. (Chapter 25)

Changnon, S. A. *Thunderstorms Across the Nation.* Mahomet, IL: Changnon Climatologist, 2001. (Chapters 18, 20)

Changnon, S. A., K. E. Kunkel and B. Reinke. "Impacts and Responses to the 1995 Heat Wave: A Call to Action." *Bulletin of the American Meteorological Society* 77(1996):1497–1506. (Chapter 27)

Changnon, S. A. "The 1988 Drought, Barges and Diversion." *Bulletin of the American Meteorological Society* 70(1989):1092–1104. (Chapter 26)

Charney, J. G. "Dynamics of Deserts and Drought in the Sahel." *Quarterly Journal of the Royal Meteorological Society* 101(1975):193–202. (Chapter 26)

Clark, T. L., W. D. Hall, and R. M. Banta. "Two- and Three-Dimensional Simulations of the 9 January 1989 Severe Boulder Windstorm: Comparison with Observations." *Journal of the Atmospheric Sciences* 51(1994):2317–2343. (Chapter 17)

Davies-Jones, R. P. "Tornado Dynamics." *Thunderstorms: A Social and Technological Documentary—Volume 2,* 2nd ed., Ed. E. Kessler. Norman: University of Oklahoma Press, 1986. (Chapter 18, 19)

Dean, J. S. "The Medieval Warm Period in the Southern Colorado Plateau." *Climate Change* 26(1994):225–241. (Chapter 26)

Doesken, N. J. and A. Judson. *The Snow Booklet—A Guide to the Science, Climatology and Measurement of Snow in the United States.* 2nd ed. Colorado Climate Center. Colorado State University, Fort Collins, CO, 1997. (Chapters 15, 16)

Doswell, C. A. III, ed. *Severe Convective Storms.* Meteorological Monograph Series, 27. Boston: American Meteorological Society, 2001. (Chapters 19, 20)

Doswell, C. A. III, and D. W. Burgess. "Tornadoes and Tornadic Storms: A Review of Conceptual Models," *The Tornado: Its Structure, Dynamics, Prediction and Hazards.* Eds. C. Church, D. Burgess, C. Doswell, R. Davies-Jones. *Geophysical Monographs* 79(1993):161–172. (Chapter 17, 18)

Doswell, C. A. III. *The Operational Meteorology of Convective Weather (Vol. 1): Operational Mesoanalysis.* NOAA Tech. Memo. NWS NSSFC-5, U.S. Dept. of Commerce, NOAA, NWS, Kansas City, MO (1985). (Chapter 18, 19)

Easterling, D. R., J. L. Evans, P. Y. Groisman, T. R. Karl, K. E. Kunkel and P. Ambenje. "Observed Variability and Trends in Extreme Climate Events: A brief Review." *Bulletin of the American Meteorological Society* 81(2000):417–425. (Chapter 14, 27)

Eichenlaub, V. *Weather and Climate of the Great Lakes Region.* Notre Dame, IN: University of Notre Dame Press, 1979. (Chapter 13)

Elmore, K. L., D. J. Stensrud, and K. C. Crawford. "Ensemble Cloud Model Applications to Forecasting Thunderstorms." *Journal of Applied Meteorology* 41(2002):363–383. (Chapter 4)

Elsner, J. B., and A. B. Kara, *Hurricanes of the North Atlantic.* New York: Oxford University Press, 1999. (Chapter 24)

Emanuel, K. "An Air Sea Interaction Theory for Tropical Cyclones. Part I: Steady State Maintenance." *Journal of the Atmospheric Sciences* 43(1986):585–604. (Chapter 24)

Emanuel, K. A., 2005: Increasing destructiveness of tropical cyclones over the past 30 years. *Nature,* 436, 686–688.

Fischetti, M. Drowning New Orleans. *Scientific American,* 79 (2001): 77–85 (Chapters 24, 25)

Fitzjarrald, D. R. "Katabatic Wind in Opposing Flow." *Journal of the Atmospheric Sciences* 41(1984):1143–1158. (Chapter 17)

Foote, G. Brante, and C. A. Knight, eds. *Hail: A review of Hail Science and Hail Suppression.* Boston: American Meteorological Society, 1977. (Chapter 20)

Forbes, G. S., R. A. Anthes and D. W. Thomson. "Synoptic and Mesoscale Aspects of an Appalachian Ice Storm Associated With Cold-Air Damming." *Monthly Weather Review* 115(1987):564–591. (Chapter 12)

Fujita, T. *Spearhead Echo and Downburst Near the Approach End of a John F. Kennedy Airport Runway, New York City.* Satellite and Mesometeorology Research Project Paper #37, Dept. of Geophysical Sciences. University of Chicago, 1976. (Chapter 22)

Fujita, T. *The Downburst.* Chicago: University of Chicago Press, 1985. (Chapter 22).

Fujita, T. T. *Memoirs of an Effort to Unlock Mystery of Severe Storms during the 50 Years, 1942–1992.* Department of Geophysical Sciences. Chicago: University of Chicago, 1992. (Chapter 19)

Fujita, T. T. *U.S. Tornadoes. Part 1, 70 Year Statistics.* Chicago: University of Chicago, 1987. (Chapter 19)

Gaffen, D. J., and R. J. Ross. "Increased Summertime Heat Stress in the United States." *Nature* 296(1998):529–530. (Chapter 27)

Gallagher, F. W., III, and K. Cutlip. "The Green Menace." *Weatherwise* 54(2001): 24–31. (Chapter 20)

Glantz, M. H., "Drought, Desertification and Food Production." *Drought Follows the Plow: Cultivating Marginal Areas.* Ed. M. H. Glantz. Cambridge: Cambridge University Press, 1994. (Chapter 26)

Gray, W. M., C. W. Landsea, P. Mielke and K. Berry. "Predicting Atlantic Basin Seasonal Tropical Cyclone Activity by 1 June." *Weather and Forecasting* 9(1994):103–115. (Chapter 24)

Grim, J., R. M. Rauber, M. K. Ramamurthy, B. F. Jewett, and M. Han. "High-Resolution Observations of the Trowal–Warm-Frontal Region of Two Continental Winter Cyclones." *Monthly Weather Review* 135(2007): 1629–1646 (Chapter 9, 10)

Gyakum, J. R., and P. J. Roebber. "The 1998 ice storm—analysis of a planetary-scale event." *Monthly Weather Review* 129(2001):2983–2997. (Chapter 11, 12)

Hawkins, H. F., Jr. "The Weather and Circulation of July 1954: One of the Hottest Months on Record in the Central United States." *Monthly Weather Review* 82(1954):209–217. (Chapters 26, 27)

Heggli, M. R., and R. M. Rauber. "The Characteristics and Evolution of Supercooled Water in Wintertime Storms Over the Sierra Nevada: A Summary of Microwave Radiometric Measurements Taken During the Sierra Cooperative Pilot Project." *Journal of Applied Meteorology* 27(1988):989–1015. (Chapter 16)

Heim, R. R., Jr. "About That Drought." *Weatherwise* 41(1988):266–271. (Chapter 26)

Heldorn, K. C., 2004: Eighteen hundred and froze to death: The year there was no summer. http://www.islandnet.com/~see/weather/history/1816.htm (Chapter 5)

Hickox, D. H. "How Hot Can it Get?" *Weatherwise* 41(1988): 157–158. (Chapter 27)

Hjelmfelt, M. R. "Numerical Study of the Influence of Environmental Conditions on Lake-Effect Snowstorms Over Lake Michigan." *Monthly Weather Review* 118(1990):138–150. (Chapter 13)

Hobbs, P. V., J. D. Locatelli, and J. E. Martin. "Cold Fronts Aloft and the Forecasting of Precipitation and Severe Weather East of the Rocky Mountains." *Weather and Forecasting* 5(1990):613–626. (Chapters 9, 10)

Hobbs, P. V., J. D. Locatelli, and J. E. Martin. "A New Conceptual Model for Cyclones Generated in the Lee of the Rocky Mountains." *Bulletin of the American Meteorological Society* 77(1996):1169–1178. (Chapters 9, 10)

Hoerling, M. P., J. W. Hurrell and X. Tu. "Tropical Origins for Recent North Atlantic Climate Change." *Science* 292(2001):90–92. (Chapter 23, 26)

Hoskins, B. J., 2003: Climate change at cruising altitude? Science, 301, 469–470. (Chapter 5)

Huffman, G. J., and G. A. Norman Jr. "The Supercooled Warm Rain Process and the Specification of Freezing Precipitation." *Monthly Weather Review* 116(1988):2172–2182. (Chapter 12)

Hughes, P. "The Worst Droughts of the 20th Century." *Weatherwise* 41(1988):268. (Chapter 26)

Hughes, P., and D. Le Comte. "Tragedy in Chicago." *Weatherwise* 49(1996):18–27. (Chapter 27)

Huo, Z, D.-L. Zhang, J. Gyakum, A. Staniforth. "A Diagnostic Analysis of the Superstorm of March 1993," *Monthly Weather Review* 123(1995):1740–1761. (Chapter 11)

Imbrie, J. A theoretical framework for the ice ages. *Journal of the Geological Society* 142(1985): 417–432 (Chapter 5)

Intergovernmental Panel on Climate Change, Climate Change 2001: The Scientific Basis. Contribution of Working Group I to the Third Assessment Report of the Intergovernmental Panel on Climate Change [Houghton, J. T., Y. Ding, D. J. Griggs, M. Noguer, P. J. van der Linden, X. Dai, K. Maskell, and C. A. Johnson (eds.)]. Cambridge University Press, Cambridge, United Kingdom and New York, NY, USA, 881 pp. (Chapter 5, 27)

Intergovernmental Panel on Climate Change, Climate Change, 2007: The Physical Science Basis, Cambridge University Press, Cambridge, U.K, 1056 pp., ISBN-13: 9780521705967 (Chapter 5)

Kageyama M., P. J. Valdes, G. Ramstein, C. Hewitt, and U. Wyputta, 1999: Northern Hemisphere storm tracks in present day and Last Glacial Maximum climate simulations: A comparison of the European PMIP models. *J. Climate,* **12,** 742–760. (Chapter 5)

Kalkstein, L. S., and R. E. Davis. "Weather and Human Mortality: An Evaluation of Demographic and Inter-Regional Responses in the U. S." *Annals of the Association of American Geographers* 79(1989):44–64. (Chapter 27)

Kalkstein, L. S., P. S. Dunne and R. S. Vosse. "Detection of Climatic Change in the Western North American Arctic Using a Synoptic Climatological Approach." *Journal of Climate* 3(1990):1153–1167. (Chapter 14)

Kalkstein, L. S., P. F. Jamason, J. S. Greene, J. Libby and L. Robinson. "The Philadelphia Hot Weather—Health Watch/Warning System: Development and Application, Summer 1995." *Bulletin of the American Meteorological Society* 77(1996):1519–1528. (Chapter 27)

Kalnay, E., S. J. Lord and R. D. McPherson. "Maturity of Operational Numerical Weather Prediction: Medium Range." *Bulletin of the American Meteorological Society* 79(1998):2753–2792. (Chapter 4)

Karl, T. R., and R. W. Knight. "The 1995 Chicago Heat Wave: How Likely is a Recurrence?" *Bulletin of the American Meteorological Society* 78(1997):1107–1119. (Chapter 27)

Karl, T. R., and R. G. Quayle. "The 1980 Summer Heat Wave and Drought in Historical Perspective." *Monthly Weather Review* 109(1981):2055–2073. (Chapter 27)

Kaspar, D. T. "Santa Ana Windflow in the Newhall Pass as Determined by an Analysis of Tree Deformation." *Journal of Applied Meteorology* 20(1981):1267–1276. (Chapter 17)

Kessler, E. ed. *Thunderstorm Morphology and Dynamics.* 2nd ed. Norman: University of Oklahoma Press, 1986. (Chapter 18, 19, 20)

Kessler, E. ed. *The Thunderstorm in Human Affairs.* 2nd ed. Norman: University of Oklahoma Press, 1983. (Chapters 18, 19, 20)

Keyser, D., and M. A. Shapiro. "A Review of the Structure and Dynamics of Upper-Level Frontal Zones." *Monthly Weather Review* 114(1986):452–499. (Chapter 8)

Klemp, J. and D. K. Lilly. "The Dynamics of Wave-Induced Downslope Winds." *Journal of the Atmospheric Sciences* 32(1975):320–339. (Chapter 17)

Knight, C. A., and P. Squires. *Hailstorms of the Central High Plains, I: The National Hail Research Experiment.* Boulder, CO: Colorado Associated University Press, 1982. (Chapter 20)

Knight, C. A., and P. Squires. *Hailstorms of the Central High Plains, II: Case Studies of the National Hail Research Experiment.* Boulder, CO: Colorado Associated University Press, 1982. (Chapter 20)

Kocin, P. J., P. N. Schumacher, R. F. Morales Jr., and L. W. Uccellini. "Overview of the 12–14 March 1993 Superstorm," *Bulletin of the American Meteorological Society* 76(1995):165–182. (Chapter 11)

Kocin, P. J., and L. W. Uccellini. *Snowstorms Along the Northeastern Coast of the United States: 1955–1985.* Meteorological Monograph 44. Boston: American Meteorological Society (1990). (Chapter 11)

Kocin, P. J., A. D. Weiss and J. J. Wagner. "The Great Arctic Outbreak and East Coast Blizzard of 1899." *Weather and Forecasting* 3(1988):305–318. (Chapter 14)

Konrad, C. E., and S. J. Colucci. "An Examination of Extreme Cold Air Outbreaks Over Eastern North America." *Monthly Weather Review* 117(1989):2678–2700. (Chapter 14)

Kristovich, D. A. R. and R. A. Steve III. "A Satellite Study of Cloud-Band Frequencies Over the Great Lakes." *Journal of Applied Meteorology* 34(1995):2083–2090. (Chapter 13)

Kunkel, K. E., S. A. Changnon, B. C. Reinke and A. W. Arritt. "The July 1995 Heat Wave in the Midwest: A Climatic Perspective and Critical Weather Factors." *Bulletin of the American Meteorological Society* 77(1996):1507–1518. (Chapter 27)

Laird, N. "Observation of Coexisting Mesoscale Lake-Effect Vortices over the Western Great Lakes." *Monthly Weather Review* 127(1999):1137–1141. (Chapter 13)

Laird, N. F. "The Structure and Intensity of Winter Mesoscale Lake-Effect Circulations Associated with an Isolated Lake." Ph.D. Thesis, University of Illinois, 2001. (Chapter 13)

Lee, B., and R. B. Wilhelmson. "The Numerical Simulation of Nonsupercell Tornadogenesis. Part II: Evolution of a Family of Tornadoes along a Weak Outflow Boundary." *Journal of the Atmospheric Sciences* 54(1997):2387–2415. (Chapter 19)

Lee, B. D., and R. B. Wilhelmson. "The Numerical Simulation of Non-Supercell Tornadogenesis. Part I: Initiation and Evolution of Pretornadic Misocyclone Circulations Along a Dry Outflow Boundary." *Journal of the Atmospheric Sciences* 54(1997):32–60. (Chapter 187)

Lee, T. J., R. A. Pielke, R. C. Kessler, and J. Weaver. "Influence of Cold Pools Downstream on Downslope Winds and Flushing." *Monthly Weather Review* 117(1989):2041–2058. (Chapter 17)

Lemon, L. R., and C. A. Doswell III. "Severe Thunderstorm Evolution and Mesocyclone Structure as Related to Tornadogenesis." *Monthly Weather Review* 107(1979):1184–1197. (Chapter 18, 19)

Lilly, D. K. "A Severe Downslope Windstorm and Aircraft Turbulence Event Induced by a Mountain Wave." *Journal of the Atmospheric Sciences* 35(1978):59–77. (Chapter 17)

Lilly, D. K., and E. J. Zipser. "The Front Range Windstorm of 11 January 1972, A Meteorological Narrative." *Weatherwise* 25(1972):56–63. (Chapter 17)

List, R. "Properties and Growth of Hailstones." *Thunderstorm Morphology and Dynamics.* 2nd ed. Ed Kessler, E. Norman: University of Oklahoma Press, 1986. (Chapter 20)

Locatelli, J. D., R. D. Schwartz, M. T. Stoelinga, and P. V. Hobbs. "Norwegian-Type and Cold Front Aloft–Type Cyclones East of the Rocky Mountains." *Weather and Forecasting* 17(2002):66–82. (Chapters 9, 10)

Ludlam, D. M. "A Century of American Weather: Decade of 1891–1900." *Weatherwise* 23(1970):187–191. (Chapter 14)

Ludlam, D. M. *The American Weather Book.* Boston: Houghton Mifflin Co., 1982. (Chapter 26)

Ludlam, D. M. *Weather Record Book.* Boston: Houghton Mifflin Co., 1982. (All Chapters)

Maddox, R. A., L. R. Hoxit, C. F. Chappell, and F. Carcena. "Comparison of Meteorological Aspects of the Big Thompson and Rapid City Flash Floods." *Monthly Weather Review* 106(1978):375–389. (Chapter 25)

Maddox, R. A., C. F. Chappell, and L. R. Hoxit. "Synoptic and Meso-Alpha Scale Aspects of Flash Flood Events." *Bulletin of the American Meteorological Society* 60(1979):115–123. (Chapter 25)

Mass, C. F., D. Ovens, K. Westrick and B. A. Colle. "Does Increasing Horizontal Resolution Produce More Skillful Forecasts?" *Bulletin of the American Meteorological Society* 83(2002):407–430. (Chapter 4)

Marks, F. D., and R. A. Houze. "Inner Core Structure of Hurricane Alicia (1983) from Airborne Doppler Radar Observations." *Journal of Atmospheric Sciences* 44(1987):1296–1317. (Chapters 3, 24)

Martin, J. E. "The Structure and Evolution of a Continental Winter Cyclone. Part I: Frontal Structure and the Occlusion Process." *Monthly Weather Review* 126(1998):303–328. (Chapters 10, 11, 15)

Martin, J. E. "The Structure and Evolution of a Continental Winter Cyclone. Part II: Frontal Forcing of an Extreme Snow Event." *Monthly Weather Review* 126(1998):329–348. (Chapters 10, 15)

Martin, J. E. "Quasigeostrophic Forcing of Ascent in the Occluded Sector of Cyclones and the Trowal Airstream." *Monthly Weather Review* 127(1999):70–88. (Chapters 10, 11, 15)

Mayfield, M., L. Avila, and E. N. Rappaport. "Atlantic Hurricane Season of 1992." *Monthly Weather Review* 122(1994):517–538. (Chapter 6, 24)

McCann, D. W. "WINDEX—A New Index for Forecasting Microburst Potential." *Weather and Forecasting* 9(1994):532–541. (Chapter 22)

Moller, A. "The Improved NWS Storm Spotters' Training Program at Ft. Worth, Texas." *Bulletin of the American Meteorological Society* 59(1978):1574–1582. (Chapter 18, 19)

Morgan, G. M., Jr., and P. W. Summers. "Hailfall and hailstorm characteristics." *Thunderstorm Morphology and Dynamics.* 2nd ed. Ed. Kessler, E. Norman: University of Oklahoma Press, 1986. (Chapter 20)

Namias, J. "Written in the Winds: The Great Drought of '88." *Weatherwise* 42(1989):85–87. (Chapter 26)

Nebeker, F. *Calculating the Weather: Weather in the 20th Century.* San Diego: Academic Press, 1995. (All chapters)

Neiman, Paul J., P. T. May, and M. A. Shapiro. "Radio Acoustic Sounding System (RASS) and Wind Profiler Observations of Lower- and Midtropospheric Weather Systems." *Monthly Weather Review* 120(1992):2298–2313. (Chapter 2)

Neiman, P. J., F. M. Ralph, M. A. Shapiro, B. F. Smull, D. Johnson, "An Observational Study of Fronts and Frontal Mergers over the Continental United States." *Monthly Weather Review* 126(1998):2521–2554. (Chapter 10)

Niziol, T. A., W. R. Snyder, and J. S. Waldstreicher. "Winter Weather Forecasting Throughout the Eastern United States. Part IV: Lake Effect Snow." *Weather and Forecasting* 10(1995):61–77. (Chapter 13)

Niziol, T. A. "Operational Forecasting of Lake Effect Snowfall in Western and Central New York." *Weather and Forecasting* 2(1987):310–321. (Chapter 13)

NOAA, Office of Global Programs/University Corporation for Atmospheric Research. *El Niño and Climate Prediction, Reports to the Nation on Our Changing Planet.* 1994 (Chapter 23)

NOAA U.S. Department of Commerce/National Weather Service. *The Great Flood of 1993.* Natural Disasters Survey Report, 1994. (Chapter 25)

Orville, R. E., and G. R. Huffines. "Cloud-to-Ground Lightning in the United States: NLDN Results in the First Decade, 1989–98," *Monthly Weather Review* 129(2001):1179–1193. (Chapters 1, 21)

Peppler, R. A., and P. J. Lamb. "Tropospheric Static Stability and Central North American Growing Season Rainfall." *Monthly Weather Review* 117(1989):1156–1180. (Chapter 6)

Pielke, R., Jr., and R. E. Carbone. "Weather Impacts, Forecasts and Policy: An Integrated Perspective." *Bulletin of the American Meteorological Society* 83(2002):393–403. (Chapters 26, 27)

Pobanz, B., J. Marwitz, and M. Politovich. "Conditions Associated with Large-Drop Regions." *Journal of Applied Meteorology* 33(1994):1366–1372. (Chapter 12)

Powell, M. D. "Tropical Storms During and After Landfall." *Storms – Volume 1* Eds. R. Pielke, Jr., and R. Pielke, Sr. New York: Routledge, 2000. (Chapter 243)

Powell, M. D., S. H. Houston, T. A. Reinhold. "Hurricane Andrew's Landfall in South Florida. Part I: Standardizing Measurements for Documentation of Surface Wind Fields." *Weather and Forecasting,* 11(1996):304–328. (Chapter 24)

Pruppacher, H. R., and J. D. Klett. *Microphysics of Clouds and Precipitation.* New York: Kluwer Academic Publishers, 1997. (Chapters 1, 2)

Quayle, R., and F. Doehring. "Heat stress: A Comparison of Indices." *Weatherwise* 34(1981):120–124. (Chapter 27)

Rauber, R. M. "Characteristics of Cloud Ice and Precipitation During Wintertime Storms Over the Mountains of Northern Colorado." *Journal of Climate and Applied Meteorology* 26(1987):488–524. (Chapter 16)

Rauber, R. M., and L. O. Grant. "The Characteristics and Distribution of Cloud Water Over the Mountains of Northern Colorado During Winter-Time Storms. Part II: Spatial Distribution and Microphysical Characteristics." *Journal of Climate and Applied Meteorology* 25(1986):489–504. (Chapter 16)

Rauber, R. M., and R. W. Scott. "Central Illinois Cold Air Funnel Outbreak." *Monthly Weather Review.* 129(2001):2815–2821. (Chapter 19)

Rauber, R. M., M. K. Ramamurthy, and A. Tokay. "Synoptic and Mesoscale Structure of a Severe Freezing Rain Event: The St. Valentine's Day Ice Storm." *Weather and Forecasting* 9(1994):183–208. (Chapter 12)

Rauber, R. M., D. Feng, L. O. Grant, and J. B. Snider. "The Characteristics and Distribution of Cloud Water Over the Mountains of Northern Colorado During Wintertime Storms. Part I: Temporal Variations." *Journal of Climate and Applied Meteorology* 25(1986):468–488. (Chapter 16)

Rauber, R. M., L. S. Olthoff, M. K. Ramamurthy, and K. E. Kunkel. "A Synoptic Weather Pattern and Sounding-Based Climatology of Freezing Precipitation in the United States East of the Rocky Mountains." *Journal of Applied Meteorology* 40(2001):1724–1747. (Chapter 12)

Rauber R. M., L. S. Olthoff, M. K. Ramamurthy, and K. E. Kunkel. "The Relative Importance of Warm Rain and Melting Processes in Freezing Precipitation Events." *Journal of Applied Meteorology* 39(2000):1185–1195. (Chapter 12)

Riebsame, W. E., S. A. Changnon, Jr., and T. R. Karl. *Drought and Natural Resource Management in the United States: Impacts and Implications of the 1987–89 Drought.* Boulder, CO: Westview Press, Inc., 1991 (Chapter 26)

Rogers, J. C., and R. V. Rohli. "Florida Citrus Freezes and Polar Anticyclones in the Great Plains." *Journal of Climate* 4(1991):1103–1113. (Chapter 14)

Rorig, M. L., and S. A. Ferguson. "The 2000 Fire Season: Lightning Caused Fires." *Journal of Applied Meteorology* 41 (2002) 786–791. (Chapter 21)

Rosenberg, N. R. ed. *North American Droughts.* Boulder, CO: Westview Press, Inc., 1978. (Chapter 26)

Samsury, C. E., and R. E. Orville. "Cloud-to-Ground Lightning in Tropical Cyclones: A Study of Hurricanes Hugo (1989) and Jerry (1989)." *Monthly Weather Review* 122(1994):1887–1896. (Chapter 21)

Santer, B. D., M. F. Wehner, T. M. L. Wigley, R. Sausen, G. A. Meehl, K. E. Taylor, C. Ammann,

J. Arblaster, W. M. Washington, J. S. Boyle and W. Bruggermann, 2003: Contributions of anthropogenic and natural forcing to recent tropopause height changes. Science, 301, 479–483. (Chapter 5)

Schiesser, H. H., and A. Waldvogel. "Hailstorms." *Storms—Volume II.* Eds. Pielke, R., Jr. and Pielke, R. Sr. London: Routledge, 1999. (Chapter 20)

Schubert, S. D., M. J. Suarez, P. J. Pegion, R. K. Koster and R. J. Bacmeister. On the cause of the 1930s Dust Bowl. Science 303 (2004): 1855–1959 (Chapter 26)

Schultz, P. "Relationships of Several Stability Indices to Convective Weather Events in Northeast Colorado." *Weather and Forecasting* 4(1989):73–80. (Chapter 6)

Schwartz, R. M., and T. W. Schmidlin. "Climatology of Blizzards in the Coterminous United States, 1959–2000." *Journal of Climate,* 15 (2002) 1765–1772. (Chapter 15)

Scott, R. W. and F. A. Huff. "Impacts of the Great Lakes on Regional Climate Conditions." *Journal of Great Lakes Research* 22(1996):845–863. (Chapter 13)

Shapiro, M. A. and S Grøn??s, Ed. *The Life Cycles of Extratropical Cyclones.* Boston: American Meteorological Society, 1999. (Chapters 10, 11)

Sheets, B., and J. Williams. *Hurricane Watch: Forecasting the Deadliest Storms on Earth.* New York: Vintage Books, 2001. (Chapter 24)

Shuman, F. G. "History of Numerical Weather Prediction at the National Meteorological Center." *Weather and Forecasting* 4(1989):286–296. (Chapter 4)

Sommers, W. T. "LFM Forecast Variables Related to Santa Ana Windstorm Occurrences." *Monthly Weather Review* 106(1978):1307–1316. (Chapter 17)

Steadman, R. G. "The Measurement of Sultriness. Part I: A Temperature-Humidity Index Based on Human Physiology and Clothing Science, and Part II: Effects of Wind, Extra Radiation and Barometric Pressure on Apparent Temperature." *Journal of Applied Meteorology* 18(1979):861–885. (Chapter 27)

Stenhoff, M. *Ball Lightning, An unsolved problem in Atmospheric Physics.* New York: Kluwer Academic/Plenum Publishers, 1999. (Chapter 21)

Ting, M., and H. Wang. "Summertime U.S. Precipitation Variability and its Relation to Pacific Sea Surface Temperature." *Journal of Climate* 10(1997):1853–1873. (Chapter 26)

Uccellini, L. W., P. J. Kocin, R. S. Schneider, P. M. Stokols, R. A. Dorr. "Forecasting the 12–14 March 1993 Superstorm." *Bulletin of the American Meteorological Society* 76(1995):183–200. (Chapter 11)

Van de Kamp, D. W. *Profiler Training Manual,* Vols. 1–4. Washington: National Weather Service Office of Meteorology, National Oceanic and Atmospheric Administration. (Chapter 3)

Wagner, A. J. "Worst Heat Wave in 26 Years Grips South-Central United States." *Weatherwise* 33(1980):168–169. (Chapter 27)

Wakimoto, R., and J. W. Wilson. "Non-Supercell Tornadoes." *Monthly Weather Review* 117(1989):1113–1140. (Chapter 18)

Wakimoto, R. M. "Forecasting Dry Microburst Activity Over the High Plains." *Monthly Weather Review* 113(1985):1131–1143. (Chapter 22)

Wakimoto, R. M. and P. G. Black. "Damage Survey of Hurricane Andrew and its Relationship to the Eyewall." *Bulletin of the American Meteorological Society* 75(1994):189–200. (Chapter 24)

Wakimoto, R., C.-H. Liu, W.-C. Lee, P. H. Hildebrand, and H. B. Bluestein. "ELDORA Observations During VORTEX 95." *Bulletin of the American Meteorological Society* 77(1996):1465–1481. (Chapter 19)

Walsh, J. E., A. S. Phillips, D. H. Portis and W. L. Chapman. "Extreme cold outbreaks in the United States and Europe." *Journal of Climate* 14(2001):2642–2658. (Chapter 14)

Wandishin, M. S., S. L. Mullin, D. J. Stensrud and H. E. Brooks. "Evaluation of a Short-Range Multimodel Ensemble Prediction System." *Monthly Weather Review* 129(2001):729–747. (Chapter 4)

Warner, T. T., R. A. Peterson and R. E. Treadon. "A Tutorial on Lateral Boundary Conditions as a Basic and Potentially Serious Limitation on Regional Numerical Weather Prediction." *Bulletin of the American Meteorological Society* 78(1997):2599–2617. (Chapter 4)

Weatherwise. "The 20th Century's Top Ten Weather and Climate Events." *Weatherwise* 52(1999):14–19. (Chapters 8, 12, 19, 24, 26, 27)

Weatherwise. "The 1980 Heat Wave: Deadly and Costly." *Weatherwise* 34(1981):125. (Chapter 27)

Wild, R. "Historical Review of the Origin and Definition of the Word 'Blizzard'." *Journal of Meteorology* 22(1997):331–340. (Chapter 15)

Williams, E. R. "Sprites, Elves, and Glow Discharge Tubes." *Physics Today* 54(2001):41–47. (Chapter 21)

Willoughby, H. E., and P. G. Black: "Hurricane Andrew in Florida: Dynamics of a Disaster." *Bulletin of the American Meteorological Society* 77(1995):543–549. (Chapter 24)

Wilson, J., and S. A. Changnon, Jr. "The Great Tornado." *Illinois Tornadoes* Circular 103. Champaign, IL: Illinois State Water Survey, 1971. (Chapter 18)

Woodhouse, C. A., and J. T. Overpeck. "2000 Years of Drought Variability in the Central United States." *Bulletin of the American Meteorological Society* 79(1998):2693–2714. (Chapter 26)

Young, K. C. *Microphysical Processes in Clouds.* New York: Oxford University Press, 1993. (Chapter 13)

Zishka, K. M. and P. J. Smith. "The Climatology of Cyclones and Anticyclones over North America and Surrounding Environs for January and July 1950–1977." *Monthly Weather Review* 108(1980):387–401. (Chapters 10, 11)

INDEX

F

I

J

N

O

P

T

U

V

W

Z